ENCYCLOPEDIA OF ENTOMOLOGY

Editorial Board

ENCYCLOPEDIA OF ENTOMOLOGY

Edited by

John L. Capinera
University of Florida

Volume 3

P - Z

KLUWER ACADEMIC PUBLISHERS
DORDRECHT / BOSTON / LONDON

A.C.I.P. Catalogue record for this book is available from the Library of Congress

ISBN 0-7923-8670-1 (HB)
ISBN 0-306-48380-7 (e-book)

Published by Kluwer Academic Publishers,
P.O. Box 17, 3300 AA Dordrecht, The Netherlands.

Sold and distributed in North, Central and South America
by Kluwer Academic Publishers,
101 Philip Drive, Norwell, MA 02061, U.S.A.

In all other countries, sold and distributed
by Kluwer Academic Publishers,
P.O. Box 322, 3300 AH Dordrecht, The Netherlands.

Printed on acid-free paper

Cover credits:

The photographs on volume 1 are *Buprestis lineata* Fabricius (Coleoptera: Buprestidae)
The photographs on volume 2 are *Chrysis serrata* Taylor (Hymenoptera: Chrysididae).
The photographs on volume 3 are *Solenopsis invicta* Buren (Hymenoptera: Formicidae).

The photographer for all the cover illustrations is David T. Almquist, University of
Florida.

Printed in the Netherlands

P

PACHYGRONTHIDAE. A family of bugs (order Hemiptera, suborder Pentamorpha). See also, BUGS.

PACHYNEURID FLIES. Members of the family Pachyneuridae (order Diptera). See also, FLIES.

PACHYNEURIDAE. A family of flies (order Diptera). They commonly are known as pachyneurid flies. See also, FLIES.

PACHYNOMIDAE. A family of bugs (order Hemiptera). See also, BUGS.

PACHYTROCTIDAE. A family of psocids (order Psocoptera). See also, BARK-LICE, BOOK-LICE, OR PSOCIDS.

PACKARD, ALPHEUS SPRING. Alpheus Packard was born on February 19, 1839, at Brunswick, Maine, USA. He graduated from Bowdoin College in 1861, and served from 1864 to 1865 as an assistant surgeon during the American Civil War. After the war he was employed by the Boston Society of Natural History, the Essex Institute, and the Peabody Academy of Science. He was a member of the United States Entomological Commission from 1877 to 1882, and was a professor of zoology and geology at Brown University from 1878 to 1905. Packard was a well-rounded naturalist, and wrote on several topics including crustaceans, myriapods, systematic entomology, and economic entomology. He served as chief editor of the journal American Naturalist for 24 years, from 1881 to 1905. He is remembered for his authorship of books, including introductory entomology books such as "Guide to the study of insects" (1969), "A text book of entomology" (1898); systematics books such as "Monograph of the geometrid moths or Phalaenidae of the United States" (1876), "Monograph of the bombycine moths of America north of Mexico" in three volumes (1895, 1905, 1914); and economic entomology books such as "Injurious insects, new and little known" (1870), and "Insects injurious to forest and shade trees" (1881). He also authored "Lamarck: the founder of evolution; his life and work, with translations of the writings on organic evolution" in 1901. Packard died at Providence, Rhode Island, on February 14, 1905.

References
Anonymous. 1905. Alpheus Spring Packard, M.D., Ph.D. *Entomological News* 16: 97–98.
Essig, E. O. 1931. *A history of entomology.* The Macmillan Company, New York. 1029 pp.
Mallis, A. 1971. *American entomologists.* Rutgers University Press, New Brunswick, New Jersey. 549 pp.

PAECILOMYCES. The genus *Paecilomyces* includes a number of plant, nematode, and insect pathogenic species. Of the fifteen entomopathogenic species, *P. farinosus* and *P. fumosoroseus* are used most often in laboratory studies and are therefore the best described. *Paecilomyces* is characterized by having flask-shaped phialides, or phialides with swollen base structures; the phialides taper into a distinct neck and generate conidia that are dry, hyaline, or slightly pigmented. These conidia adhere end-to-end, forming chains as they emerge. Synnematous growth often occurs on insects.

Paecilomyces farinosus has a broad host range, attacking a number of lepidopterans, coleopterans, hymenopterans and hemipterans. Larval stages of these insects are most readily infected, and a recent report on the identification of fungi infecting gypsy moth larvae shows that *P. farinosus* is actually the most prevalent hyphomycete species in these insects. Eggs of some hosts also can be invaded; for example, *P. farinosus* blastospores were observed to adhere to the sticky surface of sawfly eggs, germinate, and penetrate the egg chorion.

Paecilomyces is cultured easily on agar or under submerged conditions. Media such as Sabouraud dextrose or potato dextrose commonly are used. In one study, it was found that *P. farinosus* conidia produced on Sabouraud dextrose medium containing trehalose accumulated this carbohydrate. Trehalose is known to enhance desiccation tolerance so that propagules with a high content of this sugar may be stored longer, and therefore better suited for biocontrol purposes. It is possible that the trehalose in insect hemolymph provides enhanced survival capabilities to *in vivo*-produced conidia. The pathogenicity of *P. farinosus* towards sawfly larvae is improved by *in vivo* passaging and virulence lost during *in vitro* culture (attenuation) can be restored by passage through living hosts. Additionally, it has been suggested that *in vivo* growth improves infectivity because this process selects for propagules with relatively high levels of cuticle-degrading enzymes.

References

Hajek, A. E., J. S. Elkington, and R. A. Humber. 1997. Entomopathogenic Hyphomycetes associated with gypsy moth larvae. *Mycologia* 89: 825–829.

Prenerová, E., and F. Weyda. 1992. An ultrastructural study of the fungus, *Paecilomyces farinosus* (Deuteromycotina) infecting the eggs of the spruce sawfly *Cephalcia abietis* (Insecta, Hymenoptera). pp. 231–242 in B. Bennettová, I. Gelbic, and T. Soldán (eds.), *Advances in regulation of insect reproduction.* Czech Academy of Science, Ceske Budejovice, Czech Republic.

Samson, R. A. 1974. *Paecilomyces* and some allied Hyphomycetes. *Studies in Mycology* 6: 38–41.

PAEDOGENESIS. An unusual form of reproduction in which reproduction commences in the immature stage, usually by activation of the ovaries of immature parthenogenetic females.

PAENIBACILLUS. Although there are several species of endospore-forming bacteria in the genus *Paenibacillus*, only a few are the causative agents of disease in insects. These are not as well known or studied as the important insecticidal species, *Bacillus thuringiensis*, but they affect economically important insects and therefore have drawn some attention. These bacteria are *Paenibacillus larvae* subsp. *larvae*, *P. larvae* subsp. *pulvifaciens*, *P. popilliae* and *P. lentimorbus*.

Paenibacillus larvae subsp. *larvae* and *P. larvae* subsp. *pulvifaciens* are closely related, having about 90% DNA similarity. However, they are distinguishable at the subspecies level by SDS-PAGE of whole cell proteins, by DNA fingerprinting (AFLP analysis), by a few biochemical tests and by the pathology of the disease produced. *P. larvae* subsp. *larvae* causes American foulbrood and *P. larvae* subsp. *pulvifaciens* causes powdery scale disease in larval honeybees. The bacteria are gram positive, catalase negative and have fastidious nutrition. Larvae are infected by ingesting bacterial spores contaminating their food. Spores germinate in the larval midgut, penetrate the midgut epithelium by phagocytosis and produce a systemic bacteremia resulting in larval death. Vegetative cells sporulate in the dying larvae and provide a source of spores for further infections. Spores can survive for years in larval food, in soil, or in larval cadavers. Adult bees can carry bacterial spores in their digestive tract but are immune to the disease. The bacteria are susceptible to a variety of antibiotics, and oxytetracycline has been used prophylactically or to treat diseased colonies to suppress symptoms. There are reports of the development of bacterial resistance to this antibiotic, and its use is not allowed in some countries.

Paenibacillus popilliae and *P. lentimorbus* are the causative agents of milky disease in various larvae of the family Scarabaeidae. These two species are distinguishable by DNA similarity studies but are quite similar in the usual bacteriological tests. They are facultative, nutritionally fastidious, spore-forming, and present a gram positive cell wall profile although the Gram stain is reported to be negative during vegetative growth. It has been suggested that there should be a single species (*P. popilliae*) and that *P. lentimorbus* should be a subspecies within that species, a situation similar to that of *P. larvae*. The species were originally separated by the production of a parasporal inclusion by *P. popilliae* and the absence of this body in *P. lentimorbus*. However, the DNA similarity study demonstrated that most isolates (although not the type strain) of *P. lentimorbus* also produce paraspores. *P. popilliae*, the subject of most physiological studies, appears to lack a complete tricarboxylic acid cycle and is catalase negative. Although *P. popilliae* grows better in air than without air, growth is better in 10% oxygen than in the 21% oxygen present in air. It is possible that the enzyme deficiencies may be related to the inability of these bacteria to sporulate *in vitro*.

There is no direct evidence that the protein composing the parasporal body plays any role in the course of the disease process. The gene encoding the parasporal protein has been cloned and sequenced, and was shown to have significant similarity to genes encoding the parasporal proteins of *Bacillus thuringiensis*.

The course of infection in scarab larvae is initiated following ingestion of spores. Over the period of 2 to 4 weeks, spores germinate in the larval gut, vegetative cells penetrate the epithelium, proliferate in the hemolymph, and finally sporulate in the hemolymph. The dead larva displays a turbid or milky hemolymph giving the name 'milky disease'. As the larval cadaver disintegrates, spores are released into the soil and may be consumed by larvae feeding on plant roots. Insecticides containing *P. popilliae* spores have been produced by collecting larvae in the field, infecting them in the laboratory, blending the spores (and larval debris) with inert carrier, and dispersing the product into soil. Although the bacteria can be grown in bacteriological media, they sporulate poorly outside larvae, and this has frustrated attempts to produce large volumes of spores by fermentation techniques.

The inability of all these insect pathogenic paenibacilli to produce large quantities of spores outside larvae has been the subject of much research. A variety of empirical methods to achieve *in vitro* sporulation have been attempted without success. Until the underlying relationships between sporulation and the peculiarities of the metabolism of these bacteria are better understood, *in vitro* sporulation is likely to remain elusive.

Allan A. Yousten
Virginia Polytechnic Institute and State University
Blacksburg, Virginia, USA

References
Davidson, E. W. 1973. Ultrastructure of American foulbrood disease pathogenesis in larvae of the worker honeybee, *Apis mellifera*. *Journal of Invertebrate Pathology* 21: 53–61.

Hanseh, H., and C. J. Brodsgaard. 1999. American foulbrood: a review of its biology, diagnosis and control. *Bee World* 80: 5–23.

Stahly, D. P., R. E. Andrews, and A. A. Yousten. 1991. The genus *Bacillus*: insect pathogens. pp. 1697–1745 in A. Balows, H. Truper, M. Dworkin, W. Harder, and K. Schleifer (eds.), *The Procaryotes, a handbook on the biology of bacteria* (2nd ed.). Springer-Verlag, New York, New York.

PAINTER, REGINALD HENRY. Reginald Painter was born on September 12, 1901, at Brownswood, Texas, USA. He received a B.A. (1922) and M.A. (1924) from the University of Texas, and a Ph.D. from Ohio State University (1926). He immediately joined the faculty at Kansas State University, where he remained for his entire career except for brief periods in Honduras and Guatemala. Painter became widely recognized as the leading authority of plant resistance to insects. He worked cooperatively with plant breeders in the production of sorghum, wheat, and alfalfa varieties resistant to insect pests. He also documented the existence of insect biotypes that could overcome host plant resistance. He is remembered for his authorship of ''Insect resistance in crop plants,'' which was the major synthesis and leading work on the subject for decades. Painter also had a strong interest in Bombyliidae, and he and his wife described several new genera and numerous new species from North and Central America, and redescribed many European species. Painter was a fellow of the Entomological Society of America and the American Association for the Advancement of Science, and was awarded the Gamma Sigma Delta International Award for Distinguished Service to Agriculture. He died on December 23, 1968, in Mexico City.

Reference
Anonymous. 1969. Reginald Henry Painter 1901–1969. *Journal of Economic Entomology* 62: 759.

PALAEOSETIDAE. A family of moths (order Lepidoptera). They also are known as minature ghost moths. See also, MINIATURE GHOST MOTHS, BUTTERFLIES AND MOTHS.

PALAEPHATIDAE. A family of moths (order Lepidoptera). They also are known as Gondwanaland moths. See also, GONDWANALAND MOTHS, BUTTERFLIES AND MOTHS.

PALEARCTIC REALM. The Palearctic realm is a zoogeographic region encompassing Europe and Asia except for Southeast Asia. The fauna consists of such animals as vireos, wood warblers, deer, bison and wolves, and is quite similar to the fauna of the Nearctic realm (North America). Thus, the Palearctic and Nearctic realms often are combined into a larger region called the Holarctic realm.
See also, ZOOGEOGRAPHIC REALMS

PALE LICE. Members of the family Linognathidae (order Siphunculata). See also, SUCKING LICE.

PALEOPTEROUS. Lacking the ability to fold the wings backward over the abdomen.

PALE WESTERN CUTWORM. See also, WHEAT PESTS AND THEIR MANAGEMENT.

PALIDIUM. A portion of the raster in scarab larvae.

PALINDROME. A DNA sequence in two strands that reads the same in both directions.

PALLOPTERIDAE. A family of flies (order Diptera). They commonly are known as flutter flies. See also, FLIES.

PALM BEETLES. Members of the family Mycteridae (order Coleoptera). See also, BEETLES.

PALMETTO BEETLES. Members of the family Smicripidae (order Coleoptera). See also, BEETLES.

PALM INSECTS. The Palmae are one of the largest plant families, with about 2,600 species. The family is concentrated in the humid lowland tropics, extending to extra-tropical warm regions of the world, including deserts and regions with a Mediterranean climate. A few cold-hardy species can be grown outdoors as far north as Britain. Members of the palm family are easily recognized by their large leaves, or fronds, which occur in two general forms, pinnate and palmate. Palms range in size from less than a meter tall at maturity to arborescent forms such as *Ceroxylon* spp., which may reach a trunk height of more than 65 m and are among the tallest plants in the plant kingdom. Different species have solitary or multiple stems, and stem diameters range from less than a centimeter (e.g., *Chamaedorea* spp.) to 1.8 m in Jubaea chilensis (Molina) Baillon.

Palms are important components of the plant communities of warm regions. Their broad fronds and deep leaf axils provide shelter for birds, reptiles, mammals and other animals. The fruits of many species of palms are a prime food source for vertebrate animals. The fronds, leaf axils and stems support a diversity of epiphytes.

Many of the species are sources of economic products, including fruits, beverages, and fiber products. Some species have been cultivated since prehistoric times, notably the coconut palm, *Cocos nucifera* L., the African oil palm, *Elaeis guineensis* Jacquin, the date palm, *Phoenix dactylifera* L., and the peach-palm, *Bactris gasipaes* Kunth. Commercial palm cultivation has expanded along with a general expansion in agriculture beginning in the 19th century, and the use of palms as landscape and interior plants has increased dramatically in recent decades. Increasing interest in the culture of palms, along with recognition of the importance of tropical life in the world ecosystem, has stimulated a need for knowledge of the insects associated with these plants.

Palms as hosts of insects
The foliage, stems, roots, flowers, and fruits of palms provide food for insects that are undoubtedly important components of their respective ecosystems. Palms constitute a highly stable resource for insects that are adapted to them. They produce foliage

periodically throughout the year, so that not only is green tissue continually available to leaf-feeders, but also is present both as fresh foliage and in subsequent stages of maturity. Thus, palms may be food sources for both insects that feed on young foliage and those adapted to feed on mature foliage. The broad fronds are relatively easy targets for insects searching for food sources. Other matters being equal, insects can more readily colonize a large palm leaf than the multitude of small leaves of dicotyledons. The stiff foliage of palms provides superior protection from heavy rain and sunlight. On the other hand, the rigidity of palm foliage is largely due to its high fiber content, which poses a challenge to phytophagous insects.

The stems of palms are highly fibrous, and in some species are extremely hard, especially in the peripheral zone ('rind'). As in monocotyledons in general, the stems, once formed, do not undergo radial growth. There is no cambium or bark, thus no habitat for bark beetles. Some important insect pests of palms are species that bore in the fleshy tissue of buds or rachises. Some insects that bore in the petiole or bud may enter the stem tissue. Few insect species bore directly into palm stems.

The roots of palms are poor nutritionally except at the root tips, and few insects are associated with them.

Palm flowers are generally entomophilous. The inflorescences typically support massive numbers of small, shallow flowers of pale color, often with a strong fragrance. Flowers are produced at different intervals in different species, e.g., monthly in coconut, yearly in date palm, and at the end of the life of the plant in *Corypha* spp.

Although many palm fruits serve as important food resources for vertebrate animals, they are probably less important as food sources for insects.

Severe damage by insects to palms in the wild has been reported, but this is probably rare. As in other crops, palms seem to be most susceptible to destruction by insects when grown in dense monocultures.

The insect fauna of palms

There are striking similarities in the insect fauna associated with palms in different regions. Most of the significant palmivorous insect species are in one of six orders: Orthoptera, Phasmatodea, Hemiptera, Thysanoptera, Coleoptera, and Lepidoptera. Within each of these orders, palmivorous insects are concentrated in particular families, most of which are represented by different species in disparate regions.

Insects that are restricted to palms usually attack a range of several different species of this family, although some insects are restricted to a single genus or species e.g., *Xylastodoris luteolus* Barber (Hemiptera: Thamastocoridae), which is known only on Cuban royal palms, *Roystonea regia* [Kunth] O. F. Cook. Probably no species of insect is adapted to feed on all species of palms. Some palmivorous species also feed on bananas, pandans, or other large monocotyledons, but not on dicotyledons. Many of the pests of palms in cultivation are highly polyphagous species that also attack dicotyledons, such as citrus, coffee, cacao, etc.

The diversity of palmivorous insects in different regions tends to reflect the respective diversity of palms: greater numbers of palmivorous species are known in regions rich in palms, including Southeast Asia and South America, and fewer in regions poor in palm diversity such as tropical Africa.

Defoliators

Caterpillars (Lepidoptera) that feed on palm foliage are found in virtually all palm-growing regions except desert regions, such as in North Africa and the Middle East. Psychidae, Gelechioidea, Zygaenidae, Limacodidae, Hesperiidae, and Nymphalidae are represented by species on palms in most regions.

Important species of bagworms (Psychidae) on palms include *Oiketicus kirbyi* Guilding in tropical America, and *Metissa plana* Walker, *Cremastopsyche pendula* Joanna, and *Mahasena corbetti* Tams (the coconut case caterpillar), in Southeast Asia.

Species of Gelechioidea include the coconut blackheaded caterpillar, *Opisina arenosella* Walker (Oecophoridae), of southern Asia, the coconut flat moth, *Agonoxena argaula* Meyrick (Agonoxenidae) of Oceania, and the palm leaf skeletonizer, *Homaledra sabalella* (Chambers)(Coleophoridae) of the southeastern U.S. and western Caribbean.

The coconut leaf moth, *Artona catoxantha* Hampson, of Southeast Asia, and *Homophylotis catori* Jordan, a pest of coconut palms in West Africa, are two of several important species of palmivorous Zygaenidae. The levuana moth, *Levuana iridescens* Bethune-Baker, decimated coconut plantations in Fiji in the early part of the 20th century, but became nearly (or possibly completely) extinct as a result of a famous biological control campaign in which *Bessa remota* Aldrich (Diptera: Tachinidae), a natural enemy of *A. catoxantha* in the Malay Peninsula, was imported and established in Fiji.

Species of the zygaenoid family Limacodidae occur on palms in probably all humid tropical regions, and are richest in Southeast Asia, where more than 60 species have been reported on palms. Certain species of Limacodidae are among the most damaging of palm defoliators.

Larvae of many species of skippers (Hesperiidae) feed on monocotyledons, and species of this family are defoliators on palms in most humid tropical regions.

The brushfooted butterflies (Nymphalidae) are represented on palms in the eastern hemisphere by species of Amathusiinae, and in the western hemisphere by Brassolinae and Satyrinae. The adults are large, showy butterflies, much sought after by collectors, and their large, gregarious larvae are highly destructive to foliage.

Worldwide, beetles (Coleoptera) are second in importance to Lepidoptera as palm defoliators, but in certain localities they may equal or surpass Lepidoptera in this respect. Most are in the subfamily Hispinae of the leaf beetle family Chrysomelidae. Most are leaf miners, but a few are superficial leaf feeders.

Several species each of walking-sticks (Phasmatodea) and longhorned grasshoppers (Orthoptera: Tettigoniidae) are important pests of coconut palms in some islands of Oceania. They are polyphagous, their status as pests of coconut palms reflecting the overwhelming importance of these palms on some Pacific islands.

Sap-feeders

Relatively few true bugs (Hemiptera: Heteroptera) feed on palm foliage, but several species are important pests. *Stephanitis typica* Distant (Tingidae) is widely distributed on coconut palms, bananas, and probably other arborescent monocotyledons in Asia. It is sometimes considered a pest in coconut palm nurseries, and has been shown to be a vector of Kerala coconut decline, also known as coconut root (wilt) disease, of mature coconut palm in southern Asia.

Thaumastocoridae are a small family closely related to Tingidae with 17 known species, six of which are found in the western hemisphere, specifically on palms in tropical America. The 11 eastern hemisphere species are associated with various dicotyledons.

There are only two species of Miridae, the largest family of Heteroptera, of significance on palms: *Carvalhoia arecae* Miller and China, a pest of betel-nut palm, *Areca catechu* L., in India, and *Parasthenaridea arecae* Miller, which has a similar biology in the Malay Peninsula.

Several species of *Lincus* (Pentatomidae) are considered vectors of protozoans that cause marchitez sorpresiva (sudden wilt) of African oil palm, and heartrot of coconut palms, respectively, in northern South America.

The Auchenorrhyncha on palm foliage are much more diverse than the Heteroptera. As in other major taxa, palmivorous species are concentrated in certain families, notably the superfamily Fulgoroidea (planthoppers). The fulgoroid family Derbidae has the highest diversity and widest distribution of Auchenorrhyncha on palms. Only the adults visit palms, typically in sparse numbers. Their nymphs are believed to feed on fungi, and the numbers on palms of one species, *Cedusa inflata* (Ball), were shown to be related to the proximity of decaying plant debris.

Although Cixiidae are less diverse on palms than Derbidae, a few palm-associated species of this family are widely distributed. Attention has been drawn to this family because two species are considered vectors of palm diseases. *Myndus crudus* Van Duzee is a vector of lethal yellowing (LY), which affects almost 40 species of palms in Florida and parts of the Caribbean Region. This insect passes its immature stages on grasses and the adults move to palms. Thus, ground cover management in plantations has been investigated as a method of reducing LY vector populations. *Myndus taffini* Bonfils is a vector of foliar decay of coconut palm in Vanuatu.

Although Cicadellidae are the largest family of Auchenorrhyncha, leafhoppers are not well represented on palm foliage.

The Sternorrhyncha are the best represented suborder on palms, Coccoidea accounting for most of this diversity. Of the few aphid species (Aphididae) reported on palms, two very similar species of *Cerataphis* of an unusual aphid subfamily, Hormaphidinae, have been spread to many tropical countries where they are known only on palms and considered occasional pests. Often their presence is signaled by a thick crust of sootymold over the palm foliage, along with multitudes of honeydew-seeking ants. In their native Southeast Asia, they alternate between palms and certain dicotyledonous trees.

Whiteflies (Aleyrodidae) are more diverse in the tropics than the Aphididae, and nearly 50 species have been reported on palms. Whiteflies probably can be found on palms in most tropical localities, but are not known on date palms in arid regions. Their populations are typically sparse, except in a few recorded cases in which whiteflies recently introduced into new areas have achieved dense enough populations to be considered pests.

Ten of the 20 families of the superfamily Coccoidea have species reported on palms, and it is the superfamily of insects represented on palms by the most species. Palms appear to be particularly favorable hosts for these insects. Coccoidea invade new host plants primarily via the wingless crawler stage (1st instar), thus the large fronds of palms and their 'evergreen' quality (i.e., continual availability) facilitate the establishment of these passively dispersed insects, and their fibrous tissues provide firm surfaces upon which Coccoidea fabricate their scales. The larger coccoid families, viz., mealybugs (Pseudococcidae), soft scale insects (Coccidae), and armored scale insects (Diaspididae), each contain numerous species that infest palms, and species of each group are known only on palms. As discussed below, several taxa of Coccoidea have special relationships with palms.

The armored scale insects comprise the largest family of Coccoidea. Of the more than 100 species of this family reported from palms, about 15 are recognized as important pests of these plants. Most of the latter are widely distributed, polyphagous insects that are pests of various crop trees and ornamental plants in addition to palms. Among the most notorious of these is the coconut scale, *Aspidiotus destructor* (Signoret), which originated in the tropics of the eastern hemisphere and is now pantropical. Classical biological control of this insect has been successful in many countries but not in others.

The white date scale, *Parlatoria blanchardi* (Targioni-Tozzetti), is an example of an armored scale insect reported exclusively on palms. It is especially frequent on palms of the date palm genus, *Phoenix*. Native to North Africa and the Middle East, where it is sometimes a serious pest, it was introduced accidentally into the date-growing region of the southwestern U.S. during the late 1800s, but was eradicated after a long campaign.

Several families of Coccoidea have special relationships with palms and are thus of exceptional interest to coccidologists. The pit-scale family, Asterolecaniidae, has about 400 species, 29 of which are reported only from palms. Many species of this family induce pit-like galls on their hosts, a characteristic not shared with the palm-infesting species. Phoenicococcidae and Halimococcidae are restricted to palms and related monocotyledons, as is *Comstockiella*, which is a genus of uncertain familial status. Beesoniidae are a family of nine species, four of which are found on palms in tropical America. Curiously, the five species native to the tropics of the eastern hemisphere are gall-makers on dicotyledonous trees.

Thysanoptera are usually not particularly common on palms, but certain species, including the greenhouse thrips, *Heliothrips haemorrhoidalis* (Bouché), a ubiquitous and highly polyphagous pest, sometimes causes superficial damage to palm foliage.

Pests of flowers and fruits

The inflorescences of palms may be attacked by various insects, whose feeding causes immature dropping of flowers and fruits. Examples include caterpillars of *Tirathaba* spp. (Galleriidae) and *Batachedra* spp. (Coleophoridae), and a stink bug, *Axiagastus cambelli* Distant (Hemiptera: Pentatomidae). Some species of weevils consume floral parts of palms, in the process serving as pollinators.

Fruits of palms are more important food sources for birds and mammals than for insects. Certain true bugs (Hemiptera: Heteroptera) including *Amblypelta cocophaga* China in the Solomon Islands and *Pseudotheraptus wayi* Brown (Coreidae) in Africa, attack fruits of various plants including young coconut fruits, causing premature fruit-drop. Scale insects that infest palm fruits are generally eurymerous species that also infest leaves, stems, etc.

Nitidulid beetles such as *Carpophilus hemipterus* (L.), *C. dimidiatus* (F.), *C. humeralis* (F.), and *Haptoncus luteolus* (Erichson), are economically important pests of ripening and curing dates in the Coachella Valley of California. Certain varieties of dates are very susceptible to the date stone beetle, *Coccotrypes dactyliperda* (F.) (Scolytinae). Females of this beetle oviposit in unripened fruit and the larvae penetrate the seed and develop inside, while subsequent generations may develop in fruit tissue.

Fruit flies (Diptera: Tephritidae), a major fruit infesting family, do not normally infest palm fruits.

Bud, petioles, stem and root borers

Several species of scarab beetles (Scarabaeidae) in the subfamily Dynastinae are pests of palms, the most notorious of which is the coconut rhinoceros beetle, *Oryctes rhinoceros* L., a pest of coconut palm in Oceania. The adults bore into the bud, so that when the leaves unfold, large portions have been consumed. The grubs live in decaying vegetation. Several additional species of *Oryctes* are pests of palms in Asia and Africa. In tropical America, several species of *Strategus* are known to bore in the stem bases, or in roots of seedling palms.

Jebusea hammerschmidti Reich (Coleoptera: Cerambycidae) is an important pest of date palms. The adult longhorned beetle females oviposit on palm foliage, and larvae bore into petioles and eventually may enter the trunk.

Dinapate wrighti Horn (Coleoptera: Bostrichidae) bores into the crown and then down into trunk of mature palms, including *Washingtonia* spp. and *Phoenix* spp. Known until recently only in southern California, this species has extended its range in the southwestern U.S. This largest bostrichid in the world (30-50 mm long) makes extensive galleries that weaken the trunk so that it may break in high winds. *Dinapate hughleechi* Cooper is a similar species on palms in Mexico.

The grubs of several species of *Rhynchophorus* (Coleoptera: Curculionidae) bore in the meristem (bud) and sometimes from there into the stems of palms. These weevils include *R. palmarum* (L.) in the American tropics, *R. cruentatus* (F.) in the southeastern U.S., *R. phoenicis* (F.) in Africa, and *R. ferrigeneus* Olivier, the latter which is currently the most widely distributed, having been spread from Asia and Oceania to Africa, the Middle East, and more recently to Southern Europe. *Rhynchophorus palmarum* is a significant pest by itself but it can also vector the red ring nematode, *Bursaphelenchus cocophilus* (Cobb), which causes red ring, a lethal disease of coconut and African oil palms in the American tropics, and little leaf, a chronic disease of these palms.

A second group of weevils important on palms is the tribe Sphenophorini, the New and Old World billbugs. *Metamasius* has 15 species that are reported on palms in tropical America. *Metamasius hemipterus* (L.), considered probably the most damaging, bores into petiole bases, often causing enough damage that they break off. Sometimes they penetrate far enough into the stem to cause superficial damage. *Rhabdoscelus obscurus* (Boisduval) causes similar damage to palms in Queensland, Australia.

Ambrosia beetles (Curculionidae: Scolytinae and Platypodinae) are among the few beetles that bore directly into palm stems (not via the bud or petioles). Extremely small insects, they make narrow galleries in which they culture specific fungi on which they feed. They do not feed on the tissue of palms themselves. They usually attack only stressed palms.

Castnia daedalus Cramer (Lepidoptera: Castniidae) is a large moth whose caterpillars bore between petiole bases and the trunk of palms, causing the fronds to buckle and sometimes damaging the stem surface. It is widely distributed in South America.

Opogona sacchari (Bojer)(Lepidoptera: Tineidae) is a small moth whose larvae bore in stems of various monocotyledons, including some palms, mostly under nursery conditions. The larvae are apt to begin the attack by feeding on damaged stem tissue, and then may continue feeding and making a gallery in healthy stem tissue, at times penetrating into the roots.

The larvae of *Sagalassa valida* Walker (Lepidoptera: Glyphipterigidae) bore in roots of African oil palm in Colombia. Females oviposit near the base of the stems, from which the larvae penetrate the roots, hollowing them out and sometimes causing extensive death of primary roots.

Pollinators

Although in date palms and other species, pollen transfer is partially or exclusively anemophilous, i.e., by air currents, the vast majority of palm species have entomophilous flowers, i.e., that are pollinated by insects. Thus, palm flowers are often scenes of intense insect life. Coleoptera, Hymenoptera, and Diptera are the best-represented orders. Weevils (Coleoptera: Curculionidae) and stingless bees (Hymenoptera: Apidae: Meliponinae)are particularly important. Lepidoptera are of far less importance in palm pollination. Other insects, e.g., Thysanoptera, may be significant in pollinating some species in some localities. Some pollinators, e.g., certain weevils, also consume somatic floral tissue, the damage of which may be vastly offset by their benefits as pollinators. *Elaeidobius kamerunicus* (Faust) (Curculionidae), the major pollinator of African oil palm in

West Africa, has been introduced into other tropical countries to increase fruit production in this palm.

F. W. Howard and R. M. Giblin-Davis
University of Florida
Ft. Lauderdale, Florida, USA

References

Broschat, T. K., and A. W. Meerow 2000. *Ornamental palm horticulture*. University Press of Florida, Tallahassee, Florida. 255 pp.

Carpenter, J. B., and H. S. Elmer. 1978. *Pests and diseases of the date palm*. Agriculture Handbook No. 527, U.S. Department of Agriculture, Washington, DC. 42 pp.

Cock, M. J. W., H. C. J. Godfray, and J. D. Holloway (eds.) 1987. *Slug and nettle caterpillars. The biology, taxonomy and control of the limacodidae of economic importance on palms in South-east Asia*. CABI Publications, Wallingford, United Kingdom. 270 pp.

Corner, E. J. H. 1966. *The natural history of palms*. Weidenfeld and Nicholson, London, United Kingdom. 393 pp.

Howard, F. W., D. Moore, R. M. Giblin-Davis, and R. G. Abad. 2001. *Insects on palms*. CABI Publications, Wallingford, United Kingdom. 400 pp.

Jones, D. L. 1995. *Palms throughout the world*. Smithsonian Institution Press, Washington, DC. 410 pp.

Lepesme, P. 1947. *Les insectes des palmiers*. Paul Lechavalier, Paris, France. 903 pp.

Tothill, J. D., T. H. C. Taylor, and R. W. Paine 1930. *The coconut moth in Fiji*. Imperial Bureau of Entomology, London, United Kingdom.

PALM MOTHS (LEPIDOPTERA: AGONOXENIDAE).

Palm moths, family Agonoxenidae, are a small family of 68 known species, in all faunal regions for subfamily Blastodacninae but only South Pacific for four of the species of subfamily Agonoxeninae (plus one from Argentina). The family is part of the superfamily Gelechioidea in the section Tineina, subsection Tineina, of the division Ditrysia. Adults small (six to 15 mm wingspan), with head smooth-scaled; haustellum scaled; labial palpi recurved; maxillary palpi small, 1-segmented. Wings elongated, usually with very narrow hindwings with long fringes. Maculation varied, often yellow or brown shades with some markings, but a few are colorful and with iridescent markings. Adults are diurnal. Larvae are leaf skeletonizers, or borers in leaves, stems, and fruits; rarely gall-makers. Host plants mostly in Rosaceae for Blastodacninae and Palmae for Agonoxeninae. A few are economic.

John B. Heppner
Florida State Collection of Arthropods
Gainesville, Florida USA

References

Bradley, J. D. 1965. A comparative study of the coconut flat moth (*Agonoxena argaula* Meyr.) and its allies, including a new species (Lepidoptera, Agonoxenidae). *Bulletin of Entomological Research* 56: 453–472.

Diakonoff, A. N. 1939. Notes on Microlepidoptera. II. Remarks on some species of the genus Blastodacna Wocke (Fam. Cosmopterygidae). *Tijdschrift voor Entomologie* 82:64–77.

Hodges, R. W. 1997. A new agonoxenine moth damaging Araucaria araucana needles in western Argentina and notes on the Neotropical agonoxenine fauna (Lepidoptera: Gelechioidea: Elachistidae). *Proceedings of the Entomological Society of Washington* 99: 267–278.

Karsholt, O. 1997. The genus *Chrysoclista* Stainton, 1854 in Europe (Lepidoptera, Agonoxenidae). *Entomologiske Meddelelser* 65: 29–33.

Lucchese, E. 1942. Contributi alla conoscenza dei lepidotteri del melo 5. *Blastodacna putripennella* Zell. *Bolletino Rurale del Laboratorio di Entomologia Agraria Portici* 5: 175–195.

PALM SCALES. Members of the families Phoenicococcidae and Halimococcidae, superfamily Coccoidea (order Hemiptera). See also, SCALES AND MEALYBUGS, BUGS.

PALM, THURE. Thure Palm was born near Ystad in southern Sweden on January 30, 1894. He became interested in insects as a child, and in 1918 became a forestry officer. About 1926 he began to study beetles of importance to forestry and in 1951 published a book on insects of importance to wood and bark in northern Sweden, and in 1959 another for middle and southern Sweden. He also published over 200 papers on systematics, ecology, and faunistics, and was one of the most recognized Swedish coleopterists of modern time. He received an honorary doctoral degree from the University of Lund in 1953. He died at Malmö, Sweden, on May 2, 1987.

Reference

Herman, L. H. 2001. Palm, Thure. *Bulletin of the American Museum of Natural History* 265: 121–122.

PALP. (pl., palps) Small, paired, segmented sensory appendages attached to the maxilla or labium; more correctly called palpus.

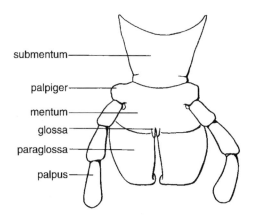

Fig. 759 External aspect of the labium in an adult grasshopper, showing some major elements.

See also, PALPUS, MOUTHPARTS OF HEXAPODS.

PALPIFER. A small sclerite that bears the maxillary palpus, and is connected to the stipes.
See also, MOUTHPARTS OF HEXAPODS.

PALPIGER. A small sclerite that bears the labial palpus, and is connected to the mentum.
See also, MOUTHPARTS OF HEXAPODS.

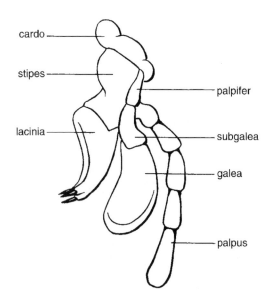

Fig. 760 External lateral aspect of the left maxilla in an adult grasshopper, showing some major elements.

PALPUS. (pl., palpi) Small, paired, segmented sensory appendages attached to the maxilla or labium; also called palp.
See also, MOUTHPARTS OF HEXAPODS.

PAMPHILIIDAE. A family of sawflies (order Hymenoptera, suborder Symphyta). They commonly are known as web-spinning or leaf-rolling sawflies. See also, WASPS, ANTS, BEES, AND SAWFLIES.

PANOISTIC OVARIES. Ovaries in which the oocytes lack nurse cells (contrast with meroistic ovaries).

PANORPIDAE. A family of scorpionflies (order Mecoptera). They are known as common scorpionflies. See also, SCORPIONFLIES.

PANORPODIDAE. A family of scorpionflies (order Mecoptera). They commonly are known as short-faced scorpionflies. See also, SCORPIONFLIES.

PANTHOPHTHALMIDAE. A family of flies (order Diptera). See also, FLIES.

PANZOOTIC. A condition wherein a disease affects all, or a large proportion of the animals of a region; extensively epizootic.

PAPER WASPS. Members of the family Vespidae (order Hymenoptera). See also, WASPS, ANTS, BEES, AND SAWFLIES.

PAPILIONIDAE. A family of butterflies (order Lepidoptera). They commonly are known as swallowtails. See also, SWALLOWTAILS, BUTTERFLIES AND MOTHS.

PAPILLA. (pl., papillae) A small nipple-like elevation.

PARABIOSIS. Use of the same nest and sometimes the same odor trails by colonies of different species, which nevertheless maintain separate broods.

PARAGLOSSA. A paired labial structure, often jointed, found at each side of the ligula.
See also, MOUTHPARTS OF HEXAPODS.

PARAJAPYGIDAE. A family of diplurans (order Diplura). See also, DIPLURANS.

PARALLEL EVOLUTION. The evolution along similar lines by taxa that were separated geographically at an earlier stage in history.

PARAMERES. Structures in the male genitalia of insects; lobes at the base of the aedeagus.
See also, ABDOMEN OF HEXAPODS.

PARAPHRYNOVELIIDAE. A family of bugs (order Hemiptera). See also, BUGS.

PARAPHYLETIC GROUP. Taxa that do not contain all the recent descendents of a single past species. Insect orders are now not clearly polyphyletic, though some artificial classifications in the past or present are polyphyletic (e.g., grouping all wingless insects into a single taxon) (contrast with polyphyletic and monophyletic groups)

PARAPROCT. One of the two lobes bordering the anus and formed from the ventrolateral parts of the epiproct.
See also, ABDOMEN OF HEXAPODS.

PARASITE. An organism that obtains its food by feeding on the body of another organism, its host, without killing the host.

PARASITIC CASTRATION. Any process that interferes with or inhibits the production of mature ova or spermatozoa in the gonads of an organism.

PARASITIC FLAT BARK BEETLES. Members of the family Passandridae (order Coleoptera). See also, BEETLES.

PARASITIC HYMENOPTERA (PARASITICA). The Order Hymenoptera has traditionally been separated into 3 groups: the suborder Symphyta (sawflies), and the suborder Apocrita, which is subdivided into the Aculeata (bees, wasps and ants) and Parasitica (= Terebrantia). Whereas the Symphyta is monophyletic and the Aculeata holophyletic, the Parasitica is a paraphyletic assemblage of taxa that does not have a formal taxonomic status. But the families comprising the Parasitica share a similarity in basal biology and fill important ecological niches that are of great economic importance, which makes the group a useful one to distinguish. When discussing this group, most modern workers use the term 'parasitoid' to denote insects whose larvae feed on and usually kill an arthropod host, to distinguish them from true parasites, which generally do not kill their hosts. The higher classification is still unsettled, but one current scheme recognizes 11 superfamilies and 48 families. Other hymenopteran parasitoids are found in both the Symphyta (family Orussidae) and Aculeata (superfamily Chysidoidea and families Scoliidae, Tiphiidae, Mutillidae, Sapygidae, Pompilidae, Rhopalosomatidae and Bradynobaenidae).

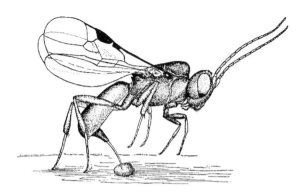

Fig. 761 Adult of *Chelonous shoshoneanorum* Viereck (Hymenoptera: Braconidae) ovipositing within egg of moth (from U.S. Department of Agriculture).

Families of the Parasitica, their mode of parasitism, and hosts.

Superfamily	Family	Mode		Host			
		1°	2°	Hemi	Holo	Arach	Phyto
Ichneumonoidea	Braconidae	+ +	+	+ +	+ +		+
	Ichneumonidae	+ +	+ +	+ +	+ +	+ +	
Evanioidea	Evaniidae	+ +		+ +			
	Gasteruptiidae	+ +			+ +		
	Aulacidae	+ +			+ +		
Stephanoidea	Stephanidae	+ +			+ +		
Megalyroidea	Megalyridae	+ +			+ +		
Trigonalyoidea	Trigonalyidae	+ +	+ +		+ +		
Cynipoidea	Ibaliidae	+ +			+ +		
	Liopteridae	+ +			+ +		
	Figitidae	+ +	+ +		+ +		
	Cynipidae	N/A					+ +
Proctotrupoidea	Vanhorniidae	+ +			+ +		
	Pelecinidae	+ +			+ +		
	Austroniidae						
	Maamingidae						
	Monomachidae	+ +			+ +		
	Peradeniidae						
	Renyxidae						
	Roproniidae				?		
	Heloridae	+ +			+ +		
	Diapriidae	+ +	+		+ +		
	Proctotrupidae	+ +			+ +		
Platygastroidea	Scelionidae	+ +	+	+ +	+ +	+ +	
	Platygastridae	+ +		+ +	+ +		+ +
Ceraphronoidea	Ceraphronidae	+ +	+ +	+ +	+ +		
	Megaspilidae	+ +	+ +	+ +	+ +		
Mymarommatoidea	Mymarommatidae						
Chalcidoidea	Mymaridae	+ +		+ +	+ +		
	Chalcididae	+ +	+ +		+ +		
	Leucospidae	+ +			+ +		
	Eurytomidae	+ +	+	+ +	+ +	+	+ +
	Pteromalidae	+ +	+ +	+ +	+ +	+	+
	Agaonidae	N/A					+ +
	Torymidae	+ +	+	+ +	+ +		+
	Ormyridae	+ +	+ ?		+ +		
	Perilampidae	+ +	+ +	+ +	+ +		
	Eucharitidae	+ +			+ +		
	Eupelmidae	+ +	+ +	+ +	+ +	+ +	
	Tanaostigmatidae	+ +			+ +		+ +
	Encyrtidae	+ +	+ +	+ +	+ +	+ +	
	Aphelinidae	+ +	+ +	+ +	+ +		
	Signiphoridae	+ +	+ +	+ +	+ +		
	Tetracampidae	+ +			+ +		
	Rotoitidae						
	Eulophidae	+ +	+ +	+ +	+ +	+ +	+
	Elasmidae	+ +	+ +		+ +		
	Trichogrammatidae	+ +	+	+ +	+ +		

Mode of parasitism: 1° − Primary parasitoids; 2° − Hyperparasitoids

Host: Hemi − Hemimetabolous insects; Holo − Holometabolous insects; Arach − Arachnida; Phyto − Phytophagous and inquilines

Entries: + + − Common; + − Uncommon or rare; ? − Suspected; None − Unknown

Adult morphology

Given their paraphyletic status, the Parasitica has no unique morphological character, though they are usually smaller and have reduced wing venation compared to the symphyta and Aculeata. Additionally, the Symphyta have no constriction between their first (propodeum) and second (petiole) abdominal segment (although in some very small Parasitica the petiole is reduced in size, and the constriction is not easily evident), while the ovipositor in the Aculeata has been modified to form a sting and no longer acts as an egg-placing device. Adult Parasitica range in body length (excluding antennae and ovipositors) from 0.2 mm to 6 cm, but the vast majority are 5 mm or less. The reduction or absence of wings is found in many families. Sexual dimorphism occurs in several groups as well, but is usually limited to differences in antennae, abdominal morphology and color patterns; the most extreme cases are found in the Agaonidae, which have been highly modified as pollinators of figs.

Immature stages

There are several morphological egg types. The hymenopteriform is several times longer than wide with rounded poles, and is the most common. The acuminate egg is typically long and narrow, and is adapted for extrusion from longer ovipositors. The stalked egg is elongate with a constricted stalk-like projection from one or both poles. The pedicellate egg is a modification of the stalked egg, in which one end is anchored to the host. The encyrtidiform egg resembles a double-bodied dumbbell while in the ovary, but after oviposition one end collapses and it then resembles a stalked egg. Eggs also can be classified as lecithal (which are relatively larger and more yolky), or alecithal (which are smaller and physiologically less expensive to produce).

There are one to five larval instars. The greatest variation in morphology occurs in the first instar, where up to 14 types have been distinguished. The most common is hymenopteriform, which is spindle-shaped and maggotlike, and is generally smooth without any conspicuous structures. Subsequent instars of almost all families take this form as well, and their identification can be very difficult, though families may be distinguished by their scleriterized head structures. Another larval form is the planidium of Eucharitidae and Perilampidae, which is a free-living stage that actively seeks out its host, and has heavier body scleritization. Adult solitary parasitoids optimally deposit only one egg per host. Under some conditions, when more than one egg is deposited, superparasitism may occur, leading to competition between the hatching larvae. Some larvae have well-developed mandibles which can destroy competitors physically, while physiological suppression, selective starvation and suffocation may suppress the supernumerary larvae, although

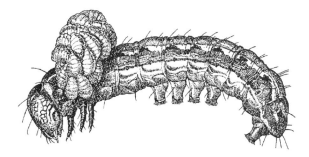

Fig. 762 Larvae of an ectoparasitic wasp, *Euplectrus* sp. (Hymenoptera: Eulophidae) feeding externally on a caterpillar (from USDA).

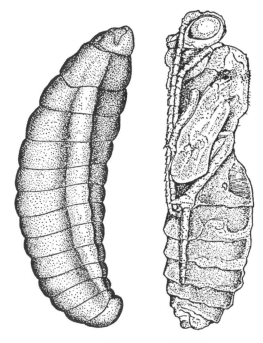

Fig. 763 The larval (left) and pupal (right) stages of an endoparasitic wasp, *Agathis gibbosas* (Say) (Hymenoptera: Braconidae) (from USDA).

the exact mechanisms are not yet fully understood. In some polyembryonic broods, a precocious 'guard' morph is produced that defends its siblings from other larvae, but fails to mature itself. Larvae have a closed midgut, so they cannot expel waste which may foul their environment. Upon completion of feeding, the larvae enter a prepupal stage, when the midgut opens, and the stored wastes are finally expelled and bundled with the last larval skin as a meconial pellet.

The larvae of many species pupate within the host remains, and do not spin cocoons. Upon death, some hosts become mummified, which effectively provides the same protection to the parasitoid pupa as a cocoon does, but without the physiological expenditure of producing silk. However, most of the Ichneumonidea and a few Chalcidoidea do produce silk cocoons, some of the most notable being gregarious microgastrine braconids, which feed internally in their Lepidoptera hosts, but emerge to pupate on the outside of their host by the dozens.

Biology

Like other Hymenoptera, most of the Parasitica have a haplo-diploid reproductive strategy, where a fertilized (diploid) egg produces a female, and an unfertilized (haploid) egg produces a male (arrhenotokous parthenogenesis). Thus, by controlling fertili-zation, ovipositing females can choose what sex egg to allocate after inspecting the prospective host. In some groups, thelytokous parthenogenesis occurs, where males are unknown and unfertilized females produce only females. In several groups of Chalci-doidea, thelytoky can be induced through infection by *Wolbachia* bacteria. Finally, deuterotokous parthenogenesis occurs in a few species, where unfertilized eggs can produce both males and females; in this case, males appear to be sexually non-functional. Two strategies of egg-production have been characterized in the Parasitica: synovi-genesis, where females successively develop a number of eggs (generally lecithal), throughout her lifetime, and proovigenesis, where ovigenesis is completed soon after females emerge from the pupa, and thus have only a fixed number of eggs (generally alecithal) to lay.

In some groups, adult morphology varies based on environmental factors, typically depending upon the size or choice of host. The most dramatic changes occur in the gall-inducing Cynipidae, which undergo an alternation of generations (heterogony) – a sexual generation of females and males, and a non-sexual (agamic) generation of females. Often the adult morphology and the gall structure and site differs considerably between the two generations.

Parasitoids can be classified in several ways. 1) Endoparasitoids develop internally in the host, and ectoparasitoids develop externally (though a few

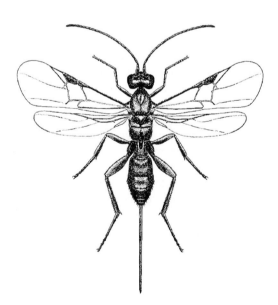

Fig. 764 Adult of *Agathis gibbosas* (Say) (Hymenoptera: Braconidae) (from USDA).

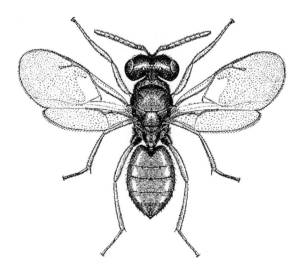

Fig. 765 Adult of *Pteromalus eurymi* Gahan (Hymenoptera: Pteromalidae) (from USDA).

species may begin their development internally and finish externally). 2) They also can be separated based on the variety of hosts they may attack, ranging from monophagy (specialists attacking only one species or a few closely related species) through oligophagy to polyphagy (generalists attacking hosts in a wide variety of taxa). This range of hosts is a continuum with few well-defined parameters, and very few species are properly placed at either extreme. Nevertheless, the concept of a generalist versus a specialist plays an important role in biological control applications. 3) Most parasitoids develop as a single individual per host; they are called solitary parasitoids. When two or more larvae can successfully develop in a single host, they are called gregarious parasitoids. The most extreme example of gregarious parasitism involves polyembryony, the production of several larvae from a single egg. Polyembryony is found in a few species of Platygastridae and Braconidae, and virtually the entire tribe of Copidosomatini (Encyrtidae), where one egg may produce up to 3,000 larvae. 4) Parasitoids also can be classified by the stage of host attacked and the span of host stages during parasitoid development, e.g., egg, egg-larval, egg-pupal, larval, larval-pupal, etc. 5) When a host is killed without undergoing further development, the parasitoid is classified as an idiobiont. Conversely, koinobiont parasitoids allow their hosts to continue developing after the initial parasitism, thus ultimately providing a larger food source for their larvae. 6) Parasitoids which do not attack other parasitoids are primary parasitoids; those which develop at the expense of other parasitoids (exclusive of those which may kill competing parasitoids but do not rely upon them for development) are hyperparasitoids. Hyperparasitism can be obligate or facultative. Most hyperparasitoids attack primary parasitoids and are called secondary parasitoids, but tertiary and even quaternary parasitoids can occur facultatively. Hyperparasitoids which attack their definitive host after it has killed its own host may be referred to as pseudohyperparasitoids. Finally, an unusual form, heteronomous hyperparasitism, occurs in a few Aphelinidae; in these species the females always develop as primary endoparasitoids of scales or whiteflies, but the conspecific males can develop in several different ways: as diphagous parasitoids (primary ectoparasitoids of the same host), as heteronomous hyperparasitoids (obligate or facultative hyperparasitoids on their own [= autoparasitism] on other species), or

remarkably, as heterotrophic parasitoids (primary parasitoids of Lepidoptera eggs).

Hosts have evolved several lines of defense against parasitism, using morphological, behavioral and physiological adaptations. Chief among the latter is the encapsulation of foreign objects mediated by haemocytes. Endoparasitoids in return have developed several strategies to ensure their success, including: (1) avoidance by oviposition into specific sites of the host not accessible to its hemocytes; (2) attack though the injection of viruses, teratocytes (= giant cells, trophic cells or trophoserosa cells), or venom along with the egg; and (3) passive defense through morphological adapations of the egg which inhibit encapsulation.

The primary source of nutrition for parasitoids is the host of the larva. However, many species also feed as adults to maximize their reproductive capacity and life-span, typically feeding on honeydew, nectar and other plant secretions. Additionally, the females of many species will feed on the hosts as well (host-feeding), in some cases leading to the death of the host.

Although most of the Parasitica are true parasitoids, in some groups a single larva will feed on a number of prey found in an enclosed area (typically an egg mass or several small larvae in a gall or cell), and thus are actually predators. In a few groups (Agaonidae, Eurytomidae, Pteromalidae, Torymidae, Tanaostigmatidae, Eulophidae, Braconidae, Cynipidae, and Platygastridae) there have been revisions to phytophagy, in many cases marked by gall formation.

Adults have developed some remarkable abilities to parasitize otherwise protected hosts. Two species (a mymarid and a trichogrammatid) are fully aquatic, using their wings and legs as oars, while an ichneumonid species can stay underwater for 30 minutes searching for prey. The deliberate use of another species for transport is termed phoresy, and several Parasitica species practice it. Scelionid and trichogrammatid females will ride on the female adult of their hosts, so that they may immediately parasitize newly laid eggs. Phoresy has also been recorded in the Torymidae and Pteromalidae.

Biological control

In biological control programs, it is essential to use agents which do not attack non-target organisms, and parasitoids of insect pests in general tend to be more host-specific than predators. Thus the parasitic

Hymenoptera, which comprise more parasitoid species than any other group, provide the greatest pool of potential agents against insect pests. In classical biological control programs, the majority of successes have been through importations of agents from three parasitic families: Aphelinidae, Encyrtidae and Braconidae. In mass-culturing and release programs, the Trichogrammatidae, Braconidae and Pteromalidae have been heavily relied on. The most important families (and some of the principal taxa of pests they are used against) are as follows:

Chalicidoidea
    Aphelinidae (Aphoidea, Diaspidae, Aleyrodidae and Lepidoptera)
    Encyrtidae (Pseudococcidae and Coccidae)
    Trichogramatidae (Lepidoptera)
    Eulophidae (Chrysomelidae and Lepidoptera)
    Mymaridae (Cicadellidae)
    Pteromalidae (Muscidae)
Ichneumonoidea
    Braconidae (Aphoidea, Lepidoptera and Tephritidae)
    Ichneumonidae (Lepidoptera and Curculionidae)
Platygastroidea
    Scelionidae (Pentatomidae and Noctuidae)
    Platygastridae (Aleyrodidae)

Fossil parasitica

The parasitica appears to have arisen in the Jurassic Period, based on the fossil remains from that time of 13 extinct families reputably assignable to the group. By the Cretaceous period, several modern families had appeared, while almost every extant family is known from the oligocene period. Parasitic Hymenoptera fossils are known from both amber and sedimentary deposits, and from Australia, Asia, Europe and North and South America.

Robert L. Zuparko
University of California – Berkeley and California Academy of Sciences – San Francisco
California, USA

References

Gauld, I., and B. Bolton. 1988. *The Hymenoptera*. Oxford University Press and British Museum (Natural History), Oxford, United Kingdom. 332 pp.
Godfray, H. C. J. 1994. *Parasitoids. Behavioral and evolutionary ecology*. Princeton University Press, Princeton, New Jersey. 473 pp.
Goulet, H., and J. T. Huber (eds.). 1993. *Hymenoptera of the world: an identification guide to families*. Agriculture Canada Publications, Ottawa, Canada. 668 pp.
Gupta, V. K. (ed.). 1988. *Advances in parasitic Hymenoptera research: Proceedings of the II Conference on the Taxonomy and Biology of Parasitic Hymenoptera held at the University of Florida, Gainesville, Florida, November 19–21, 1987*. E.J. Brill, Leiden, The Netherlands. 546 pp.
Hanson, P. E., and I. D. Gauld. 1995. *The Hymenoptera of Costa Rica*. Oxford University Press and British Museum (Natural History), Oxford, United Kingdom. 893 pp.
Waage, J., and D. Greathead (eds.). 1986. *Insect parasitoids*. Academic Press, London, United Kingdom. 389 pp.

PARASITIC WOOD WASPS. Members of the family Orussidae (order Hymenoptera: suborder Symphyta). See also, WASPS, ANTS, BEES, AND SAWFLIES.

PARASITISM. Living on or in another organism, and using that host to obtain food. The parasitic relationship is usually debilitating for the host, and sometimes fatal, though fatal conditions are perhaps more correctly termed parasitoidism.

PARASITISM OF LEPIDOPTERA DEFOLIATORS IN SUNFLOWER AND LEGUME CROPS, AND ADJACENT VEGETATION IN THE PAMPAS OF ARGENTINA. Pampean grasslands were dramatically modified after the introduction of exotic plant species for livestock industry and agriculture during the last century. A succession of wheat, maize, linseed, green pastures (mainly alfalfa and clover), sunflower, grain sorghum, and lastly soybean crops have changed the original grassland community structure and function. To date, the landscape in the Pampas comprises an assortment of secondary natural grasslands, and farming and crop fields with differential degrees of disturbance (i.e., use of agrochemical, planting techniques, etc.). Soybean and wheat are the major summer and winter crops, respectively. Extensive cropping and grazing are dominant practices, but allocation of land for each fluctuates according to changes in market demand and technology. Relicts of native plus introduced flora and fauna can be found in small areas of abandoned land, and in corridors alongside roads and railways. Together they cover a significant proportion of the pampean landscape.

Most regions of the world suffer the effects of invasions by introduced crops. In the case of insect communities, crops are colonized by herbivorous species; the recruitment is influenced by the natural reservoir of fauna associated with native (wild) plants chemotaxonomically related to that exotic crop, as well as by the deficiency of a broad spectrum of defensive allomones that make plants vulnerable.

Predators, parasitoids and pathogens attack herbivorous insects. Parasitoid insects are defined as entomophagous insects. Parasitoid larval stages feed internally (endoparasitoid) or externally (ectoparasitoid) upon arthropods, usually other insects, killing their hosts. One or more individuals can be obtained from a single host (solitary, gregarious or embryonic parasitoids). Parasitoids can attack a taxonomically broad range of insect orders and all developmental stages, and they themselves often are subject to attack by parasitoids; thus, they can be distinguished as primary, secondary or hyperparasitoids. Given their importance and ubiquity (probably over 1 million species), and because they can play a critical role in limiting the abundance of their hosts, parasitoids are considered to be very reliable candidates for biological control of pests.

Unfortunately, changes in natural insect communities during the agricultural expansion in the Pampas are poorly documented. However, it is interesting to investigate current insect complexes in a landscape scale, and to examine both historical and ecological processes. Insect community structure, life cycles, and interactions allow reconstruction of part of the past native communities and enable us to understand contemporary dispersion and refuges of insects in the region. It also indicates the potential use of natural enemies in the context of the biological control of pests through conservation tactics.

Larval stage of Lepidoptera naturally supports a large number of parasitoids, and non-crop habitats associated with agricultural fields provide these and other beneficial arthropods with alternate hosts or prey, food and water resources, shelter, favorable microclimates, overwintering sites, mates and refuges from pesticides.

In crops such as soybean, sunflower, alfalfa and mixed clover-grass pastures, and in mono-specific patches of *Melilotus alba* L. and *Galega officinalis* L., two naturally occurring legumes alongside roads and railroads, and in crop borders in the northwest of Buenos Aires province (34°4' S, 58–60°00' W), a total of 28 Lepidoptera defoliator species can be found. These species are polyphagous and multivoltine, some are migratory and some may hibernate or overwinter. They belong to the following families: Noctuidae (21 species), Pyralidae and Geometridae (2 species each), and Tortricidae, Pieridae and Arctiidae (1 species each). Soybean supports 10 species, pastures 9, legume patches 8, and sunflower crop only 2.

Lepidoptera species inhabiting crops and corridor vegetation in northwest Buenos Aires Province, Pampean Region, Argentina.

Habitat plants	Lepidoptera species
Soybean	*Anticarsia gemmatalis* (Hübner), *Heliothis* spp. complex, and *Spodoptera frugiperda* (J.E. Smith), *Rachiplusia nu* (Guenée), and *S. ornithogalli* (Guenée) (Noctuidae), *Elasmopalpus lignosellus* Zeller and *Loxostege biffidalis* (Fabricius) (Pyralidae), *Colias lesbia* (Fabricius) (Pieridae), *Eulia loxonepes* Meyrick (Tortricidae), *Spilosoma virginica* (Fabricius) (Arctiidae)
Sunflower	*R. nu, S. virginica*
Alfalfa	*Paracles vulpina* (Hübner), *Leucania jaliscana* Schaus, and *Pseudoleucania minna* (Butler) (Noctuidae), *R. nu, C. lesbia*, 1 unidentified species (Geometridae), 3 unidentified species (Noctuidae)
Mixed – Clover	*Faronta albilinea* (Hübner), and *Mocis phasianioides* (Guenée) (Noctuidae), *P. minna, R. nu, C. lesbia, P. vulpina*, 3 unidentified species (Noctuidae)
M. alba	*R. nu, P. vulpina, F. albilinea, S. virginica*, 3 unidentified species (Noctuidae), 1 unidentified species (Geometridae)
G. officinalis	*Elaphia repanda* (Schaus), and *Helicoverpa gelotopoeon* (Dyar) (Noctuidae), *P. minna, M. phasianoides, S. virginica, F. albilinea, R. nu, P. vulpina*

The parasitoid complex of defoliator Lepidoptera species on soybean crops in northwest Buenos Aires Province, Pampean Region, Argentina.

Parasitoid complex	Hosts				
	R. nu	A. gemmatalis	S. virginica	C. lesbia	L. biffidalis
Egg prepupal endoparasitoids					
Encyrtidae					
Copidosoma floridanum	x				
Early larval endoparasitoids					
Braconidae					
Apanteles lesbiae				x	
Apanteles or Cotesia sp.					
Cotesia marginiventris	x				
Cotesia a					
Cotesia b			x		
Meteorus rubens		x			x
Microgaster a	x				
Microgaster b					
Rogas nigriceps	x				
Ichneumonidae					
Campoletis grioti	x				
Campoletis a	x				
Campoletis b					
Campoletis c					
Hymenoptera					
Unidentified sp. 1					
Unidentified sp. 2					x
Late larval endoparasitoids					
Tachinidae					
Voria ruralis complex	x				
Chetogena a and Lespezia aletie			x		
Patelloa similis and Lespezia sp.			x		
Sturmia sp.			x		
Unidentified sp. 3	x				
Unidentified sp. 4	x				
Unidentified sp. 5			x		
Chetogena c				x	
Larval-pupal endoparasitoids					
Tachinidae					
Winthemia sp.	x				
Chetogena b			x		

Rachiplusia nu Guenée (Noctuidae) is the dominant species, being recovered from all six habitats; *Paracles vulpina* (Hübner) (Noctuidae) is found in four (both pastures and non-crops); and *Colias lesbia* Fabricius (Pieridae), *Pseudoleucania minna* (Butler), *Faronta albina* (Hübner) (Noctuidae) and *Spilosoma virginica* (Fabricius) (Arctiidae) in three (soybean, and both pastures; alfalfa, clover and *G. officinalis*; and soybean, sunflower and *G. officinalis*, respectively). But in general, most species are observed in one or two out of six plant habitats.

Numerous parasitoid species attack the larval stage of defoliators of soybean. Nine out of ten host species are attacked at least by one primary endoparasitoid species; all categorized as koinobionts (parasitoids associated with exposed hosts and which permit them to continue to move, feed and defend themselves). Hyperparasitoids are not found. The parasitoid assemblage comprises 30 species; they belong to Hymenoptera (4 families, 17 species) and Diptera (1 family, 13 species). *Campoletis grioti, Copidosoma floridanum, Cotesia marginiventris, Rogas nigriceps, Voria*

The parasitoid complex of *R. nu* larvae on crops and corridor vegetation in northwest Buenos Aires Province, Pampean Region, Argentina.

Parasitoid complex	Plant habitats				
	Sunflower	Alfalfa	Clover	*M. alba*	*G. officinalis*
Egg prepupal endoparasitoids					
Encyrtidae					
Copidosoma floridanum	x	x	x	x	x
Early larval endoparasitoids					
Braconidae					
Cotesia marginiventris		x			
Microgaster sp.	x	x	x		x
Rogas nigriceps	x	x	x		x
Rogas sp.		x			
Ichneumonidae					
Campoletis grioti	x	x			
Campoletis d	x				
Campoletis e			x		x
Casinaria sp.	x	x	x	x	x
Hymenoptera					
Unidentified sp. 6 ^	x				
Unidentified sp. 7 ^	x				
Unidentified sp. 8				x	
Unidentified sp. 9*					x
Late larval endoparasitoids					
Tachinidae					
Voria ruralis complex	x		x		x
Unidentified sp. 10 ^	x				
Unidentified sp. 11^		x			
Unidentified sp. 12*				x	
Larval-pupal endoparasitoids					
Tachinidae					
Winthemia sp.	x				
Hyperparasitoids					
Chalcididae					
Unidentified sp. 13 from *Casinaria* sp.)	x				
Unidentified sp. 14 (from Tachinidae)	x				

^ only the pupa was obtained, * only larvae were obtained.

ruralis, Cotesia spp., *Microgaster* spp., and *Chetogena* spp. are the most common species. Most species are solitary, some gregarious, and one polyembrionic. The number of parasitoid species attacking each host is different, highest for *R. nu* (11) and lower for *Eulia loxonepes* and *Elasmopalpus lignosellus* (1). Likewise, differences are found in parasitoid abundance (number of individuals) when comparing host species; in particular for *R. nu*, the high number of parasitoid species and number of individuals reveal the importance of parasitoids as mortality agents and point out for their preservation in protected areas. When parasitoid guild analysis is used to explore the determinants of species richness and structure of parasitoid complexes, it can be observed that the number of parasitoid guilds differs among hosts: four parasitoid guilds were recorded for *R. nu*: egg-prepupal, early larval, late larval, and larval-pupal; two for *S. virginica*: late larval, and larval-pupal, and *C. lesbia*: early larval and larval-pupal; and one for *A. gemmatalis* and *L. biffidalis*: early larval. *Elasmopalpus lignosellus* and *E. loxonepes* are parasitized by one species each (*C. marginiventris* and *Apanteles* or *Cotesia* sp., respectively).

Parasitoid assemblages from each host differ in species composition, levels of parasitism and number of guilds. Besides historical processes, some of these differences could be associated to ecological factors, such as host abundance patterns, including constancy and predictability in time and space.

Rachiplusia nu has the most complex parasitoid assemblage, and this could be explained by its higher abundance and constancy than the other host species, and the development of two generations within the crop cycle. We might expect to find more similarity in parasitoid assemblages between *R. nu* and *A. gemmatalis*, considering they belong to the same family. However, *A. gemmatalis*, a tropical species, is possibly in the limit of its geographical distribution, having one generation a year at the end of the crop cycle and sporadic occurrence in the study area. Its lower predictability, compared to *R. nu*, could account, at least to some extent, for the observed differences. The fact that *S. virginica* does not support parasitoid guilds composed of hymenopteran species could be attributed to the long setae in the host body that, in general, would prevent the attack of wasps lacking a long ovipositor, but not for tachinids, since they can lay eggs in foliage or attach eggs on the host integument. The simplicity of the parasitoid guild structure associated with *C. lesbia* and *L. biffidalis* could be related to their low abundance.

From a regional perspective, *R. nu* population abundance is higher in summer crops, followed by multi annual crops and lower in non-crop habitats, though some low densities are registered for alfalfa and mixed clover-grass pastures. *Rachiplusia nu* larvae support 18 primary endoparasitoid species in plant habitats other than soybean, including two hyperparasitoids species registered in sunflower crops. Comparatively, the number of parasitoid species and the number of parasitized hosts is greater in crop than in non-crop habitats, ranging from 11 in sunflower (as in soybean) to 4 in *G. officinalis* and *M. albus*. Some species are coincidental with those registered for the soybean crop. *Copidosoma floridanum*, *R. nigriceps*, and *V. ruralis* again are the species consistently reared in five plant habitats, whereas *Casinaria* sp. is found for the first time for *R. nu* and in the five sites. The most frequently abundant parasitoid species in crop habitats is *C. floridanum*; the remaining species have lower incidence on *R. nu*. The parasitism in non-cultivated habitats is very low.

Parasitoid guild analysis yields differences for *R. nu* parasitoid assemblages among plant habitats:

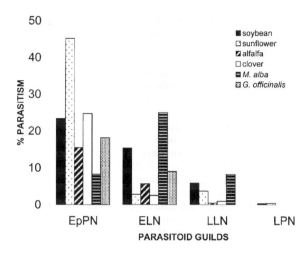

Fig. 766 A comparison of levels of parasitism for the four larval parasitoid guilds associated with *Rachiplusia nu* (Guenée) (Lepidoptera: Noctuidae) in six Pampean plant habitats (Argentina).

four parasitoid guilds are found in sunflower and in soybean (egg-prepupal, early larval, late larval and larval pupal); three in mixed clover-grass, alfalfa and *M. albus* (egg-prepupal, early larval, and late larval), and two in *G. officinalis* (early larval and late larval).

In this region, only four out of seven potential parasitoid guilds defined for Lepidoptera are recorded, suggesting that potential host niches are not totally utilized. Parasitoid species belonging to guilds that attack earlier host stages are responsible for higher levels of parasitism than those attacking later developmental stages. At present, Lepidoptera defoliators occur infrequently at damaging levels and only in limited areas. The action of parasitoids seems to be, in part, responsible for maintaining defoliator populations in crops at low levels.

Norma E. Sánchez
Universidad Nacional de La Plata La Plata, Argentina
and
María G. Luna
University of California at Irvine
Irvine, California, USA

References

Hall, A. J., C. M. Rebella, C. M. Ghersa, and J. P. Culot. 1992. Field-crop systems of the Pampas. pp. 413–450 in C. J. Pearson (ed.), *Ecosystems of the world, 1. Field crop ecosystems*. Elsevier Science Publishers, Amsterdam, The Netherlands.

Hawkins, B. A., and W. Sheehan. 1994. *Parasitoid community ecology*. Oxford University Press, Oxford, United Kingdom. 516 pp.

Luna, M. G., and N. E. Sánchez .1999a. Specific composition and abundance of the soybean defoliator Lepidoptera community in northwest Buenos Aires province, Argentina. *Revista de la Sociedad Entomológica Argentina* 58: 67–75.

Luna, M. G., and N. E. Sánchez. 1999b. Parasitoid assemblages of soybean defoliator Lepidoptera in north-western Buenos Aires province, Argentina. *Agricultural and Forestry Entomology* 1: 255–260.

Mills, N. J. 1994. Parasitoid guilds: defining the structure of the parasitoid communities of endopterygote insect hosts. *Environmental Entomology* 23: 1066–1083.

Quicke, D. L. J. 1997. *Parasitic wasps*. Chapman and Hall, New York, New York. 485 pp.

PARASITIZATION. Parasitism. This is not a widely accepted term.

PARASITOID. A parasite that kills its host at about the time the parasite completes it development.

PARASITOIDISM. Parasitism resulting in death of the host. This is not a widely accepted term.

PARASOCIAL BEHAVIOR. A level of social behavior less developed than eusocial behavior. This type of sociality includes cooperative behaviors within the same generation of insects, namely communal, quasisocial, and semisocial behavior.

See also, SOLITARY, SUBSOCIAL, COMMUNAL, QUASISOCIAL, SEMISOCIAL, EUSOCIAL BEHAVIOR.

PARASPORAL BODY. A particle which lies alongside the spore or is included in the sporangium along with the spore, formed during sporulation of a number of *Bacillus* and *Paenibacillus* species. If the inclusion is a crystalloid, the species is called crystalliferous. It is usually diamond-shaped (bipyramidal). When digested it releases an endotoxin.

PARATYPE. Any specimen in the series from which the species description was completed, other than the holotype specimen.

PARENTAL INVESTMENT. Behaviors displayed by the parent insects to increase the probability that their offspring will survive and reproduce, at

Fig. 767 Example of parnassian moths (Pterothysanidae), *Hibrildes ansorgei* Kirby from Mozambique.

the expense of the parent's ability to produce more offspring. Females are considered to invest more in offspring than males. Social behavior is considered to be a high degree of parental investment because parents invest in brood care.

PARIETALS. The lateral areas of the insect head between the frontal and occipital areas. Each parietal consists of the antenna, compound eye, and lateral ocelli.

PARNASSIAN MOTHS (LEPIDOPTERA: PTEROTHYSANIDAE). Parnassian moths, family Pterothysanidae, include 19 species from southern Africa (seven sp.) and Southeast Asia (12 sp.). There are two subfamilies: Pterothysaninae and Hibrildinae (the latter subfamily also considered in Eupterotidae in some classifications). The family is in the superfamily Calliduloidea, in the section Cossina, subsection Bombycina, of the division Ditrysia. Adults medium size (55 to 75 mm wingspan), with head scaling average; haustellum naked; labial palpi porrect; maxillary palpi vestigial; antennae filiform or bipectinate. Wings rounded. Maculation pale and spotted, sometimes translucent (Hibrildinae); some with long hair-like setae from hindwing margin (Pterothysaninae). Adults diurnal; possibly also crepuscular. Biologies and larvae remain unknown.

John B. Heppner
Florida State Collection of Arthropods
Gainesville, Florida USA

References

Hampson, G. F. 1892. Family Pterothysanidae. In W. T. Blanford (ed.), *Fauna of British India, including Ceylon and Burma*. moths, 1: 430–432. Taylor & Francis, London.

Hering, E. M. 1926. Familie: Pterothysanidae. pp 123–125 In A. Seitz (ed.), *Die Gross- Schmetterlinge der Erde. 10. Die afrikanischen Spinner und Schwärmer*, pl. 19. A. Kernen, Stuttgart.

Minet, J. 1987. Description d'une chrysalide de Ptero-thysaninae (Lep Callidulidae). *Nouvelle Revue d'Entomologique (n.s.)* 4: 312.

Seitz, A. (ed.) 1926. Gattung: Pterothysanus. In *Die Gross-Schmetterlinge der Erde* 10: 277, pl. 26, 30. A. Kernen, Stuttgart.

PARONELLIDAE. A family of springtails in the order Collembola. See also, SPRINGTAILS.

PARTHENOGENESIS. Development from an egg that has not been fertilized. Reproduction without fertilization.

PARTHENOGENETIC REPRODUCTION. A-sexual reproduction. (contrast with sexual reproduction)

See also, PARTHENOGENESIS.

PARVOVIRUS. Parvoviruses, the smallest DNA viruses, are contained within the genera *Parvovirus*, *Erythrovirus*, *Dependovirus*, *Iteravirus*, *Contravirus*, and *Densovirus*. Members of the *Parvovirus*, *Ery-throvirus*, and *Dependovirus* infect vertebrates, whereas the host range of genera *Contravirus*, *Itera-virus*, and *Densovirus* is restricted to arthropods. The dependoviruses require a helper virus for replication; the vertebrate parvoviruses and the Densovirinae are autonomous viruses. Autonomous parvoviruses repli-cate through a variety of double-stranded linear DNA intermediates in mitotically active host cells. The canine and human B-19 parvoviruses are simple ico-sahedral viruses that are constructed from 60 protein subunits or protomers that contain three structural proteins. The insect parvoviruses produce a small, non-enveloped icosahedron (20 to 24 nm diameter) composed of four structural polypeptides (VP1-VP4). These viruses, like other parvoviruses, possess a relatively high DNA/protein ratio (about 37%), which confers a characteristic heavy buoyant density (about 1.40 g/cm^3) in CsCl$_2$ gradients. These viruses are very stable and are resistant to exposure to pH 3 to 9, solvents (CHCl$_3$), and temperature (58°C, 1h). Insect parvoviruses, like their vertebrate counterparts,

replicate only in actively multiplying insect cells. The parvovirus replication takes place in the cell nucleus and is closely affiliated to cellular DNA replication events. Parvoviruses are not capable of stimulating DNA replication in resting cells; replication of these viruses requires host cells to go through the S phase. The insect parvoviruses undergo a non-lytic cycle in cell culture.

The subfamily Densovirinae (DNVs) contains members that have been isolated mainly from dip-teran and lepidopteran hosts. The DNVs in the genus *Densovirus* contain a 6 kb genome which codes for structural and non-structural proteins on separate strands. Many of these viruses are polytropic and replicate in the nuclei of all insect tissues except the midgut. The *Bombyx mori* densovirus, the sole member of the genus *Iteravirus*, contains a smaller genome (about 5.0 kb) that codes for all proteins on one strand and is able to replicate only in midgut cells. Members of the third genus, the *Contravirus*, have been reported to cause persistent, non-lytic infections in mosquito cell lines and have been shown to be vertically transmitted in *Aedes*. These viruses possess a 4.0 kb genome and have open-reading frames (ORFs) on the plus strand and/or negative strands. The mosquito DNVs have been examined for potential vectors for the delivery and expression of foreign genes in mosquito cells. These polytropic DNVs share characteristics of both the *Densovirus* and *Iteravirus* groups. However, unlike the lepidop-teran DNVs that encapsidate plus and negative strands at equal frequency, the *Aedes* DNVs encapsi-date only 15% of the plus polarity strand.

The best-studied insect parvoviruses include the *Bombyx* DNVs, the causal agents of *densonucleosis* in the silkworm *Bombyx mori*. Two forms, DNV-1 and DNV-2, have been detected in silkworm popula-tions. Both DNV-1 and DNV-2 replicate in the nuclei of midgut columnar cells. The DNV-1 (*Ina* isolate) induces infected midgut cells to be discharged into the gut lumen. DNV-2 infected cells are not as read-ily discharged as those infected with DNV-1. In nat-ure, both DNV-1 and DNV-2 are able to cause chronic infections in larvae of the mulberry pyralid, *Glyphodes pyloalis*. More than 50% of field-collected *G. pyloalis* larvae screened with anti-DNV rabbit antisera were infected with DNV and/or the infec-tious flacherie virus. It is believed that this DNV overwinters in diapausing *G. pyloalis* and is trans-mitted to silkworms via contamination of mulberry foliage by prior generations of DNV-infected *G. pyloalis* larvae.

References
Bando, H., T. Hayakawa, S. Asano, K. Sahara, M. Naka-gaki, and T. Iizuka 1995. Analysis of the genetic information of a DNA segment of a new virus from silkworm. *Archives of Virology* 140: 1147–1155.
Berns, K. 1990. Parvovirus replication. *Microbiology Reviews* 54: 316–329.
O'Neill, S. L., P. Kittayapong, H. R. Braid, T. G. Andreadis, J. P. Gonzalez, and R. B. Tesh 1995. Insect denosoviruses may be widespread in mosquito cell lines. *Journal of General Virology* 76: 2067–2074.

PASSALIDAE. A family of beetles (order Coleoptera). They commonly are known as bess beetles. See also, BEETLES.

PASSANDRIDAE. A family of beetles (order Coleoptera). They commonly are known as parasitic flat bark beetles. See also, BEETLES.

PASSIVE DISPERSAL. The redistribution of animals caused by external agents such as wind or movement of seeds. The small size of insects allows passive dispersal frequently (contrast with active dispersal).

PATAGIUM. (pl., patagia) A small flap or lobe at the anterior edge of the forewing of some insects. It is also known as the tegula.
See also, WINGS OF INSECTS.

PATCH DYNAMICS. The concept that communities are not homogeneous, rather consisting of a mosaic of patches, with differing rates of biotic and abiotic interactions and disturbances.

PATCH, EDITH MARION. Edith Patch was born at Worcester, Massachusetts, USA, on July 27, 1876. Her family moved to Minnesota in 1884, where she had ample opportunity to live with, and observe nature. She was particularly fascinated with the monarch butterfly and at an early age came to love entomology. She entered the University of Minnesota in 1897 and received her B.S. in 1901. It was here that she was introduced to aphids, a subject that came to dominate her life. Unable to gain employment because entomology was not yet regarded an appropriate field for women, she taught high school for two years. However, in 1903 she was invited to organize a Department of Entomology at the University of Maine. During leaves of absence from Maine she studied at Cornell University, and was granted a Ph.D. in 1911. She served as head of the entomology department at the University of Maine until her retirement in 1937. Edith Patch wrote about 80 technical publications on insects, mostly concerning aphids, but also including many of Maine's important pests. She was interested both in taxonomy and economic entomology. Among her important works were "Aphididae of Connecticut" (1923) and "Food-plant catalogue of the aphids of the world" (1938). She was especially supportive of other students and scholars, and a significant amount of her time was devoted to assisting others to learn aphidology or in identifying aphids for others. Patch also wrote about 40 popular articles on scientific topics, about 100 nature stories for children, and 17 books on natural history for children. She was elected the first woman president of the Entomological Society of America and received many other honors and recognitions associated with her expertise in aphids. She died at Orono, Maine, on September 27, 1954.

References
Adams, J. B., and G. W. Simpson. 1955. Edith Marion Patch 1876–1954. *Annals of the Entomological Society of America* 48: 313–314.
Mallis, A. 1971. *American entomologists*. Rutgers University Press, New Brunswick, New Jersey. 549 pp.
Stoetzel, M. B. 1990. Edith Marion Patch: her life as an entomologist and as a writer of children's books. *American Entomologist* 36: 114–118.

PATCHY ENVIRONMENT. A habitat within which occurs significant variability in suitability for an organism of interest.

PATELLA. In arachnids, a leg segment between the femur and the tibia.

PATHOGEN. A virus, bacterium, parasitic protozoan or other microorganism that causes disease by invading the body of a host; infection is not always disease because infection does not always lead to injury of the host.

PATHOGENESIS. The ill health or death of an organism caused by a pathogenic microorganism.

PATHOGENICITY. The ability of a pathogen to cause disease.

PATHOGEN TRANSMISSION BY ARTHROPODS. Vector arthropods are those on, or in, which pathogenic organisms can survive and be transferred. Parasites transmitted by arthropods range widely from eukaryotes, including helminths, to prokaryotes and viruses. Some of them, for example, the sporogenic *Anthrax* bacterium (a prokaryote) and myxoplasmosis (a virus that affects rabbits) survive long enough on the mouthparts of bloodsucking insects such as horseflies and fleas that they are capable of contaminating vertebrate hosts. Regurgitation or the feces of their arthropod host disperse other pathogens that survive in the gut of non-bloodsucking insects such as cockroaches and domestic flies. Representatives of these groups are called contaminators. The efficiency of contamination depends upon the ability of the pathogenic agents to survive in, or on, the body of their temporary vector. The food of man or of animals can also be infected by pathogens transferred by these arthropods. In most cases, this type of transmission is occasional. However, for example, diarrhea agents can be transmitted in poorly managed hospitals where flies or cockroaches are abundant. There are some examples of persistent transmission. For example, the myxomatosis virus, which is capable of surviving for a long time on the mouthparts of *Spilopsyllus cuniculi* fleas, is transmitted from rabbit to rabbit by this vector only. In this case, the flea is an obligate, specific vector. However, this is an exception to the rule.

The term specific vector is mainly used to define the large group of vectors, mostly bloodsuckers, whose body provides not only a pathogen's survival, but also its propagation to the level of an invasive stage capable of infecting a vertebrate host.

Specificity of bloodsucking vectors is defined by the following criteria:

1. Any pathogen that consumes the energy sources accumulated by its bloodsucking host. The parasite propagates to enhance its chances of transmission.
2. A specific stage in the pathogen's life cycle is developed inside the arthropod host. This developmental stage can be distinguished morphologically (helminths, Protozoa, agent of plague) or antigenically (bacteria, viruses).
3. Highly efficient mechanisms of transmission through injection or other specific contamination are characteristic.
4. The pathogen transmitted is relatively harmless to the vector either at the individual or at the population level.
5. It possesses both a) an adequate level of pathogen for the vector to be infected on a vertebrate host, and b) the ability for vector infection on an aviremic host. (An aviremic host is one that is tolerant to infection or it develops the infection much later than bloodsuckers are able to exchange the pathogen between infected and naive specimens).

The malaria agent *Plasmodium* is the best example of items 1 and 2. Only sexual forms that are not very abundant in the host blood can develop in the mosquito host. In the sporocysts of *Plasmodium*, the numbers increase a thousand-fold. Individual sporozoites flooding out with mosquito saliva can infect a vertebrate host.

There are many mechanisms of parasite transmission. The most widespread mechanism is pathogen transmission through the bite of a bloodsucker. Pathogens are transmitted in the ectoparasite's saliva. All mites and ticks, mosquitoes, gnats, sand flies and tsetse flies demonstrate this mechanism of transmission. Helminthes infest the host during the bite of a fly, but not with saliva. Using their 'knife and scissors' mouthpart they are able to perforate the bloodsucker's soft labium that remains in tight contact with the wound they make in the host. Passive specific contamination is typical of triatomid bugs, also called 'kissing bugs' because they often sting a sleeping person's mouth where the skin is thin. Their feces infected by the Chagas disease agent are dropped over a small wound made by the bug's 'dagger-like' mouthparts. Lice-borne pathogens contaminate scratches in human skin either with infected feces or with crushed louse bodies whose guts are full of *Rickettsia* (red typhus) agents. Louse saliva is a highly itchy substance and stimulates both the scratching and the subsequent lice crushing. Bugs and lice are, therefore, not only specific hosts, but specific contaminators also.

The mouthparts of ectoparasitic mites, ticks and insects are highly varied in construction, but

Fig. 768 Specific contamination and cannibalistic routes of Chagas disease agent. I — specific contamination: infected feces contaminate the wound made by the bug; II — cannibalistic route: hungry nymph consumes trypanosomes from the fed adult's gut; III — contamination route: hungry nymph consumes agent from the drop of fresh feces with trypanosomes.

invariably well adapted to blood sucking. The stiff 'daggers and scissors' of a horse-flea's mouthparts enable rapid blood-sucking in spite of the rough skin of its host. The fine, delicate vein-opening apparatus of sand flies account for their Latin name, Phlebotominae, 'phlebo-tomeo' meaning 'vein dissection.' The long, sword-like, flexible, often serrate mouthparts of mosquitoes and fleas enable them not only to pierce the host's skin, but also to penetrate the blood vessels. The rough, scissors-like tick chelicerae provide a relatively large hole in the host skin, inside which the hooked hypostome is inserted to ensure tick attachment. Very often, a special kind of saliva, which hardens in contact with air and is thus termed cement, anchors the tick mouthparts in the skin. This is necessary to fix a hard tick whose feeding takes several days. Saliva of bloodsucking arthropods is rightly called a pharmacological laboratory, as it either serves to irritate the host skin (lice) or to anesthetize the place of the bite (triatomid bugs, ticks and mites), or to dilate host's blood vessels and suppress blood clotting. Tick saliva is perhaps the most multifunctional in suppressing not only blood clotting, but also host immune reactions. The more detailed the study of bloodsuckers' saliva, the more diverse are the features disclosed. The presence of specific insect- or tick-borne pathogens not only alters arthropod host behavior (e.g., tick-borne ence-

phalitis virus enhances the vector's locomotor activity) but also its saliva. For example, the malaria agent appears to disable the production of the enzyme apyrase by the mosquito salivary glands. As a result, the duration of bloodsucking is extended, and a considerable increase in sporozoite transmission to the vertebrate host is observed.

Blood consumption from vertebrate hosts is an absolute prerequisite to survival of bloodsucking arthropods. Yet, blood is the main, but not the only, pathway of pathogen transmission. Devouring the gut content of older individuals is relatively common among bloodsuckers. For example, the larvae and nymphs of triatomid bugs very often behave as cannibals, piercing the body of adults and sucking their blood content, which might be infected by *Trypanosoma cruzi*. Sucking the adult feces is also typical of these contaminators. Four pathways of pathogen transmission can be seen in this picture: two types of contamination (both from and to, the insect) and two routes of pathogen acquisition (with blood of a vertebrate or an invertebrate host).

Often enough, subadults of soft ticks suck the gut content of older individuals which have fed, thus engulfing relapsing tick fever pathogens (e.g., *Borrelia duttoni*). Likewise, there are several modes of transmission and circulation of the plague agent *Yersenia pestis* which is usually transmitted through a flea bite (donor rodent — to flea — to recipient rodent). However, it can also be acquired by the direct contact of a healthy person with an ill one in cases of the pulmonary form of the disease, or, carried from a bloodfed, infected, soft tick to a hungry flea. At present, a telluric way of plague agent maintenance is in debate. *Yersenia pestis* pathogens can be preserved and are capable of re-entering the animal/flea circle at a later time. *Anthrax* bacilli in the spore form can remain preserved in the soil for decades and then infect wild animals or livestock. *Anthrax* kills the vertebrate host while horse flies feeding on the dying or freshly dead animals may disseminate this agent as contaminators. Horse flies carry the spores on their mouthparts. Even a typical vector-borne pathogen such as the malaria agent can circulate by transmission from an infected mother to a child or from an infected person to a healthy one by blood transfusion. In these cases, asexual forms of the parasite are transmitted.

The pathways of tick-borne pathogen transmission seem the most complex. The most studied examples

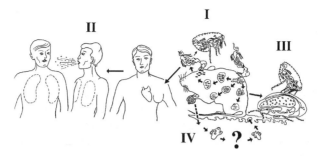

Fig. 769 Plague agent routes of transmission and circulation. I – system: flea, *Yersenia pestis* and susceptible rodent; II –generalized and aerial forms of plague; III – *Ornithodoros* as a 'living can' of *Y. pestis* for fleas; IV – possible phytophase of the *Y. pestis* cycle (telluric hypothesis).

concern the tick-borne encephalitis virus, an agent that is the most dangerous to man.

The first, 'classical' pathway demonstrates infected blood consumption by an adult female tick feeding on a vertebrate showing viremia (high concentrations of the virus in host blood). For a long time, this pathway was believed to be the main one. It includes:

(a) Transovarial transmission, implying infected blood consumption and pathogen transmission to the egg (beginning of the F1 progeny). Transovarial virus transmission has also been proven for Diptera: mosquitoes as vectors of yellow and dengue fevers, and sand flies as vectors of phleboviruses (for example, papattasi fever).

(b) Transphasic transmission must include at least one of the following steps: transmission of a pathogen from fed and infected larvae to nymphs; survival of the pathogen in hungry and fed nymphs and its transmission to adults following the nymph's molt. Adults can transmit the virus to humans and animals. Transphasic transmission is also typical of *Borrelia*, the relapsing tick fever agent transmitted by soft ticks, and of the Lyme disease agent transmitted by hard ticks of the genus *Ixodes*.

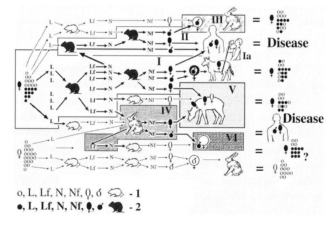

Fig. 770 Tick-borne encephalitis virus routes of transmission. I – 'classical' (transmissive) route from egg to egg by feeding on the susceptible vertebrate animals with threshold quantity of virus in the blood; Ia – human infection by virus contained in goat or cow milk; II – transphasic transmission: naïve larva gets virus on infected susceptible vertebrate animal; III – sexual transmission: infected male transmits virus to naïve female during copulation; IV – virus exchange between infected nymphs and naïve larvae co-feeding not near each other on the vertebrate animal without viremia (distant transmission); V – virus exchange between specimens co-feeding near each other on aviremic (not susceptible to the virus) animal (transsalival transmission); VI cannibalistic route of transmission: infected male consumes hemolymph from naïve female and injects virus-infected saliva into her body; 1 – naïve tick and vertebrate animal: o – ova, L – larva, Lf – fed larva, N – nymph, Nf – fed nymph.

The sexual transmission of tick-borne encephalitis (TBE) virus and Lyme disease agents from males to females has been proved as well.

A cannibalistic type of pathogen exchange cannot be excluded as, for example, *Ixodes* tick males sometimes feed not only on fed females but also on hungry specimens.

At present, the classical pathway of pathogen transmission is no longer considered to be the main one. Pathogen exchanges between co-feeding vector specimens belonging to various age groups, for example, nymphs and larvae of ticks, or different generations, is now theoretically the most efficient way of spreading the viruses and other pathogens from infected individuals to uninfected ones. Ticks issuing saliva near each other may transfer regurgitated pathogens while feeding in the same site of inflammation in the host skin. This type of transmission is called transsalival and may occur both on viremic and aviremic vertebrate hosts.

Similar exchanges of pathogens, not only of viruses but of borreliae as well, occur when ticks feed separately on infected and aviremic animals. This type of transmission is termed infection on an aviremic host, or distant infection. The mechanism by which this exchange occurs is enabled by the host immune cells, which transfer absorbed pathogens from one area of inflammation to another.

The many forms of pathogen transmission ensure the effective transmission of infection by bloodsucking insects, mites and ticks.

Andrey N. Alekseev
Russian Parasitological Society
St. Petersburg, Russia

References

Alekseev, A. N. 1993. *Tick-tick-borne pathogen system and its emergent qualities.* Zoological Institute RAS, St. Petersburg, Russia. 204 pp.

Alekseev, A. N., and S. P. Chunikhin. 1991. Virus exchange in ticks feeding on vertebrate host in the absence of viremia (Distant transmission). *Med. Parazitologiya i parazitamye bolezni* (Moscow) 2: 50–54.

Alekseev, A. N., and Z. N. Kondrshova. 1985. *Organism of arthropods as environment for pathogens.* Academy of Science of the USSR Urals Branch, Sverdlovsk, USSR. 181 pp.

Jones, L. D., C. R. Davies, G. M. Steele, and P. A. Nuttall. 1987. A novel mode of arbovirus transmission involving a nonviremic host. *Science* 237: 775–777.

Kennedy, C. R. 1975. *Ecological animal parasitology.* Blackwell Science Publishers, Oxford, United Kingdom. 163 pp.

Ribeiro, J. M. C. 1995. Blood-feeding arthropods: live syringes or invertebrate pharmacologists? *Infectious Agents and Disease* 4: 143–152.

PATHOGENS OF WHITEFLIES (HEMIPTERA: ALEYRODIDAE).

Whiteflies are tropical and subtropical in origin, and can be serious pests in greenhouses and on house plants. In the United States, whiteflies were considered secondary pests in agricultural field crops until outbreaks of the silverleaf whitefly (*Bemisia argentifolii*) began to occur in the late 1980s. Before the development of the silverleaf whitefly problems in the U.S., whiteflies were mainly greenhouse pests, and occasionally pests on certain other subtropical crops, such as citrus. The silverleaf whitefly has a wide host range, including cotton, melon, soybean, and a variety of vegetable crops.

Whiteflies are very small insects, and like other Hemiptera, they have sucking mouthparts. Insect pathogens (also known as entomopathogens) in the groups bacteria, viruses, rickettsia, and protozoans nearly always infect hosts through the alimentary tract, and so the host must ingest the pathogen to become infected. Because whiteflies suck the phloem from plants, the only means by which they might ingest these pathogen groups is if they occur in the plant phloem. Entomopathogens have been known to occur and even grow within plants, but it is not very common, and for that reason, these pathogens are rare in whiteflies. This situation should not be confused with its reverse, that is, where the insect acts as a vector for plant pathogens, a much more common phenomenon. Whiteflies do act as vectors for plant pathogenic viruses, particularly the Gemini viruses.

Nearly all known whitefly pathogens are fungi, which can infect their hosts directly through the exoskeleton. Fungi in the phylum Deuteromycotina are the asexual fungi, and most all pathogens of whiteflies belong to this group. Conidia, asexually produced spores, are the infective units. These spores attach to the outside of the host, where they germinate, producing a germ tube. The germ tube will grow along the cuticle of the host until it finds a suitable site to penetrate and cause infection. The invasion process varies somewhat between species, but a good general description can be found by Hajak and St. Leger (1994).

Aschersonia spp. are fungal pathogens commonly found in whiteflies in the genus *Trialeuroides*. These fungi belong to the class Coelomycetes and are specific to whiteflies and coccids. Twenty-five species have been isolated from whiteflies. They produce conidia in structures called pycnidia. Different species of *Aschersonia* can be distinguished by the color of the conidia – usually reddish, orange, or yellow. *Aschersonia* conidia are coated in a hydrophilic mucus layer. This mucus is somewhat sticky and prevents them from readily being dispersed by wind, so *Aschersonia* conidia are usually dispersed by rainfall. Young whitefly nymphs are the most susceptible life stage, with the older nymphal stages being progressively less susceptible, and the adults resistant, to infection.

Other fungi that have been found in populations of *Trialeuroides* whitefly nymphs include *Verticillium lecanii*, *Paecilomyces fumosoroseus*, *Aphanocladium album*, and *Beauveria bassiana*. All these belong to the class Hyphomycetes. The nymphs are most commonly infected, but adults have been found infected with *Paecilomyces farinosis*, *Paecilomyces fumosoroseus*, *Verticillium fusisporum*, and *Erynia radicans*. Unlike the other fungi mentioned so far, *Paecilomyces* spp. and *Beauveria bassiana* produce dry conidia without a mucous coating. These may be dispersed by either rain or wind, but the spores are dried and shriveled and require high levels of moisture for germination. This moisture may come from rain, dew, or high levels of relative humidity that sometimes occur at the plant surface.

Aschersonia has rarely been isolated from *Bemisia* whitefly species. Pathogens in the genera *Verticillium* and *Paecilomyces* are much more common. *Beauveria bassiana* has been shown to infect nymphs of *Bemisia* whiteflies if they are sprayed with conidial suspensions, but no epizootics in untreated whitefly populations have been reported.

Pathogens of whiteflies have been used as biological control agents to control pest outbreaks. Using pathogens as biological control agents is also called microbial control, and is done two different ways: by natural epizootics and augmentative use. When whitefly populations reach outbreak levels and climatic conditions are right (usually during times of high rainfall), then fungal pathogens can cause epizootics. Epizootics of fungal pathogens in whiteflies can be quite striking and lead to a rapid decline in host populations. The limitation of natural epizootics as a pest control strategy is that growers have little control over their occurrence, and they often do not occur until host populations are high, which means

that crop damage is likely to have already occurred. However, natural occurrences of fungal pathogens may control whitefly populations more often than is recognized.

Augmentative control is when a natural enemy is mass cultured and then released for biological control. The control agent may or may not be expected to establish in the region. It is the mass release that leads to control. This approach is common for microbial control agents; *Aschersonia*, *Verticillium lecanii*, *Paecilomyces fumosoroseus*, and *Beauveria bassiana* have all been produced for commercial sale as myco-pesticides against whiteflies. Most myco-pesticides used for whiteflies contain fungal spores, either conidia or blastospores, as the active ingredient. Conidia have a longer shelf- and field-life, but can only be produced on a solid substrate. Blastospores are produced during liquid fermentation. Liquid fermentation tends to be faster and less labor intensive than fermentation on a solid substrate.

These myco-pesticides have tended to work better in greenhouses than in field crops. The reasons for this are surely many-fold, but the fact that greenhouses tend to be more controlled environments is probably a major factor. Temperature and relative humidity can have a dramatic impact on infection levels, and these factors are more controlled in greenhouses. The high levels of humidity in greenhouses are conducive to infection. Conidia and blastospores also are very sensitive to ultraviolet radiation, which is generally lower in a greenhouse than the field, just by the nature of a greenhouse (greenhouse materials such as glass, plastic, shade cloths, etc., block some

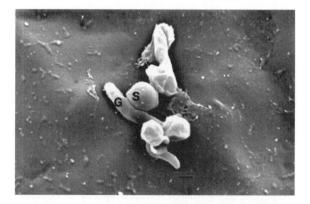

Fig. 771 Four conidia of the fungus *Beauveria bassiana* germinating on the cuticle of a nymph of the silverleaf whitefly. S = conidial spore, G = germ tube, Bar = 1 μm.

light from entering). Lower ultraviolet light levels extend the life of the spores and their activity. In the field, mass immigration of new adult whiteflies into a treated field can lead to whitefly population increases if conditions are not conducive to disease transmission of the fungus from treated insects to new, uninfected hosts.

Rosalind R. James
USDA ARS
Logan, Utah, USA

References

Fransen, J. 1990. Natural enemies of whiteflies: fungi. pp. 211–225 in D. Gerling (ed.), *Whiteflies: their bionomics, pest status and management*. Intercept, Ltd., Andover, United Kingdom.

Gindin, G., N. U. Geschtovt, B. Raccah, and I. Barash. 2000. Pathogenicity of *Verticillium lecanii* to different developmental stages of the silverleaf whitefly, *Bemisia argentifolii*. *Phytoparasitica* 28: 229–239.

Hajek, A. E., and R. J. St. Leger 1994. Interactions between fungal pathogens and insect hosts. *Annual Review of Entomology* 39: 293–322.

Lacey, L. A., J. J. Fransen, and R. Carruthers. 1996. Global distribution of naturally occurring fungi of Bemisia, their biologies and use as biological control agents. pp. 401–433 in D. Gerling (ed.), *Bemisia 1995: taxonomy, biology, damage control and management*. Intercept, Ltd., Andover, United Kingdom.

Wraight, S. P., R. I. Carruthers, S. T. Jaronski, C. A. Bradley, C. J. Garza, and S. Galaini-Wraight. 2000. Evaluation of the entomopathogenic fungi *Beauveria bassiana* and *Paecilomyces fumosoroseus* for microbial control of the silverleaf whitefly, *Bemisia argentifolii*. *Biological Control* 17: 203–217.

PATHOLOGY. The science that deals with all aspects of disease.

PATHOTYPE. An intraspecific variant of a pathogen, distinguished by variation in pathogenicity on a specific host relative to other pathogens of the same type.

PATROLLING. In honey bees, the act of investigating the interior of the nest by workers. Such behavior allows the nest to detect and respond to problems quickly.

PAUROMETABOLOUS DEVELOPMENT. This is a type of incomplete metamorphosis (hemimetabolous development) found in some aquatic insects (Odonata, Ephemeroptera, Plecoptera).

Unlike terrestrial insects displaying the typical form of incomplete metamorphosis, in which the immature and adult stages are substantially the same in body form (differing principally in the presence of fully formed wings among the adults), immature and adult stages of these aquatic insects differ slightly to significantly in appearance as compared to their adults. However, they lack a pupal stage, which is characteristic of insects with complete metamorphosis (holometabolous development). Because these insects depart from the typical pattern of hemimetabolous development, they sometimes are said to have paurometabolous development or gradual metamorphosis. Consistent with this differentiation, the immature are sometimes called naiads rather than nymphs. (contrast with hemimetabolous development, holometabolous development) See also, METAMORPHOSIS.

PAYKULL, GUSTAF. Gustaf Paykull was born at Stockholm, Sweden, on August 21, 1757. Educated initially by tutors, Paykull was likely influenced by Linnaeus, and developed interest in natural history, literature, and poetry. His literary successes were limited at the University of Uppsala, and he began to concentrate more on natural history. Beginning in 1779 he worked in government administration, though he found pleasure in collecting, and he also learned that this activity earned him respect from colleagues. He went on to build an extensive private collection of natural objects. He produced monographic treatments on several groups of beetles in the late 1700s, culminating in his ''Insecta Svecica'' (1798 to 1800). Paykull traveled widely, and collected birds, mammals, and fish in addition to insects. His eagerness to amass a large collection earned him the ire of some colleagues, however, as some borrowed specimens apparently were not returned. He was honored by election to the Swedish Academy of Sciences in 1791. Paykull died at Stockholm on January 28, 1826.

Reference

Herman, L. H. 2001. Paykull, Gustaf. *Bulletin of the American Museum of Natural History* 265: 122–123.

PBAN. A neuropeptide that controls synthesis of pheromone in glands of some female moths. See also, REPRODUCTION.

PCR.   See polymerase chain reaction.

PCR-RFLP.   A technique that combines the PCR and RFLP analyses. Genomic DNA is amplified by traditional PCR. Once the DNA is amplified, it is cut with restriction enzymes, electrophoresed, and visualized by ethidium bromide staining. Because the DNA was amplified by the PCR, the DNA fragments can be visualized without having to blot and probe the labeled probe, making PCR-RFLP more efficient and inexpensive than traditional RFLP analysis.

PEA APHID.   See also, APHIDS.

PEACH SILVER MITE.   See also, FOUR-LEGGED MITES.

PEAIRS, LEONARD MARION.   Leonard Peairs was born at Leonard, Kansas, USA, on June 5, 1886. He obtained his B.S. and M.S. degrees from Kansas State University in 1905 and 1907, respectively. For several years he worked at the University of Illinois, University of Maryland, and Kansas State, but in 1912 he accepted employment at West Virginia University where he remained until retirement in 1952. In 1925 he received a doctorate from the University of Chicago. Peairs is best known as a co-author of the text ''Insect pests of farm, garden, and orchard'' (with E.D. Sanderson, 1921 and subsequent editions). He served as editor of the ''Journal of Economic Entomology'' from 1940 to 1953. Peairs died on January 29, 1956.

Reference

Mallis, A. 1971. *American entomologists*. Rutgers University Press, New Brunswick, New Jersey. 549 pp.

PEA LEAFMINER, *Liriomyza huidobrensis* **(BLANCHARD) (DIPTERA: AGROMYZIDAE).**
What is known today as the pea leafminer was originally described as five different fly species: *Agromyza huidobrensis* Blanchard (1926, Brazil), *Liriomyza huidobrensis* Blanchard (1938, Argentina), *L. cucumifoliae* Blanchard (1938, Argentina), *L. langei* Frick (1951, California), and *L. dianthi* Frick (1958, California). In 1973, Spencer synonymized the California and South American species under the name *Liriomyza huidobrensis*. Recent genetic

evidence (in addition to observations of behavioral differences) strongly suggests that the species occurring in California and Hawaii is different from the South American species and that the name *L. langei* should be resurrected.

The pea leafminer is a small fly, less than 2.5 mm long, with a generally black body, and yellow face, frons and scutellum. Salient characteristics that distinguish *L. huidobrensis* from other members of the genus are as follows: the vertical bristles on the head are on a dark background contiguous with the black hind margin of the eye; the antennal segments are yellowish-brown and the distal third of the third segment is sometimes darkened; and the hind-corners of the mesonotum adjoining the scutellum are black. The posterior spiracles of the puparia have 6 to 9 pores.

Liriomyza huidobrensis was not considered a pest prior to the 1970s, when in South America, large amounts of insecticides were applied against the gelechid potato moth, *Scrobipalpula absoluta* (Meyrick). The leafminer became resistant to these insecticides and emerged as a secondary pest. In the late 1980s it spread from South America to Europe on cut flowers, quickly spread throughout Europe, and arrived in Israel in the early 1990s. The South American species also has spread to Kenya, South Africa and Indonesia.

The pea leafminer is truly polyphagous, feeding on a large number of flowers, vegetables and weeds. Some of the more important economic plants are Cucurbitaceae (gherkin, cucumber, melon), Leguminosae (numerous bean species), Solanaceae (pepper, tomato, potato, eggplant), Caryophyllaceae (*Dianthus* spp., *Gypsophilla* spp.), Chenopodiaceae (spinach, beet), Compositae (thistle, endive, aster, *Chrysanthemum* spp. *Gerbera* spp., lettuce), Cruciferae (Chinese cabbage, radish), Umbelliferae (carrot, celery, parsley), and Violaceae (*Viola* spp.).

Damage to plants occurs in two forms: aesthetic damage from feeding punctures and tunnels, and reduction of photosynthesis. Adult females puncture top and bottom surfaces of leaves with their ovipositor and feed on the plant juices. Adult males also feed at these same holes. The amount of stippling varies with plant species, varying from about 50 to 300 per day. Additionally, females lay eggs on the underside of leaves; larvae hatch and mine the leaves. Larval mines are typically along the midrib and lateral veins, although not exclusively so. Even in leaves with one larval mine, photosynthetic rates are significantly reduced because larval tunnels are in the spongy mesophyll where chloroplasts are located. The feeding punctures also significantly reduce photosynthetic rates and stomatal and mesophyll conductance.

Female flies live less than three weeks and males about a week. During her lifetime, a female lays an average of 8 to 14 eggs per day after a 1 to 2 day pre-ovipositional period. The egg stage lasts about 2 to 3 days depending on temperature and host plant. There are three instars in the leaf, and the mines become progressively larger with each molt. The duration of the larval stage depends on temperature and host plant, but averages about one week. The third instar chews a hole in the leaf surface and emerges completely from the leaf to pupate. There is a prepupal stage of 4 to 5 hours duration between formation of the puparium and actual pupation. Pupae vary in color from a light brown to almost black, and live about one week.

Control measures for greenhouse and open field crops include biological, cultural/mechanical, chemical, and use of resistant plants. Biological control measures in greenhouses with releases of *Dacnusa sibirica* Telenga, *Opius pallipes* Wesmael and *Diglyphus isaea* Walker wasp parasitoids have been very successful on various crops. While natural parasitism in the field sometimes may be quite high, it is insufficient to control the leafminer. Release trials have shown that it is economically unfeasible to release sufficient parasitoids to achieve control in open field situations in celery. Some success has been achieved in applying entomopathogenic nematodes (*Steinernema feltiae* (Filipjev)) for control of the leafminer.

Cultural/mechanical control measures are relatively few in number. It is always important to maintain good agricultural practices – removing harvested plants from fields, maintaining control over weeds, etc. – to prevent the buildup of populations in these reservoirs. However, to date, trials with traps plants have been unsuccessful. Attempts have been made at mass trapping flies by circumscribing greenhouses with yellow sticky sheets or by dragging sheets of oil coated yellow plastic just above crops in open fields; these methods have reduced the numbers flies but have not significantly controlled the situation. Similarly, vacuum removal of flies has had limited success because they quickly reinvade the field. In South America, trials have been proceeding on developing cultivars of potatoes that are resistant to the leafminer by developing plants with a high density of glandular trichomes that interfere with feeding and oviposition.

Chemical control measures have had mixed results. The leafminer was discovered in Israel when conventional insecticides failed to control what were thought to be adult *L. trifolii*. Growers in Indonesia spray 2 to 3 times a week with a variety of 35 conventional insecticides in attempts to control the fly, with unsatisfactory results. On the other hand, efforts to control the larvae with translaminar insecticides (neem, cyromazine, abamectin, and spinosad) have been successful. Timing of insecticide application with these translaminar insecticides is critical and requires educating growers to treat the (unseen) larval population and not wait until large numbers of adults are observed, when insecticides essentially are useless.

Phyllis G. Weintraub
Agricultural Research Organization
Gilat Research Station Negev, Israel

References

Parrella, M. P., V. P. Jones, R. R. Youngman, and L. M. Lebeck. 1985. Effect of leaf mining and leaf stippling of *Liriomyza* spp. on photosynthetic rates of chrysanthemum. *Annals of the Entomological Society of America* 78: 90–93.

Rauf, A., B. M. Shepard, and M. W. Johnson. 2000. Leafminers in vegetables, ornamental plants and weeds in Indonesia: surveys of host crops, species composition and parasitoids. *International Journal of Pest Management* 45: 257–266.

Scheffer, S. J. 2000. Molecular evidence of cryptic species within the ea leafminer *Liriomyza huidobrensis* (Diptera: Agromyzidae). *Journal of Economic Entomology* 93: 1146–1151.

Spencer, K. A. 1973. Agromyzidae (Diptera) of economic importance. *Series Entomology (The Hague)* 9: 1–444.

Weintraub, P. G., and A. R. Horowitz 1995. The newest leafminer pest in Israel, *Liriomyza huidobrensis*. *Phytoparasitica* 23: 177–184.

PEAR DECLINE. See also, TRANSMISSION OF PLANT DISEASES BY INSECTS.

PEAR PSYLLA, *Cacopsylla pyricola* **(FOERSTER) (HEMIPTERA: PSYLLIDAE).**
At least seven species of pear-feeding psyllids in the genus *Cacopsylla* are recognized from Europe and North America, with unknown numbers of additional species occurring in Asia. Several of these

species are pests of commercial pear (*Pyrus*), most notably *Cacopsylla pyricola* (Foerster) in North America and Europe, and *C. pyri* (L.) in Europe. The North American psyllid, *C. pyricola*, is not native to North America, but was introduced into the eastern United States in the early 1800s on infested pear seedlings imported from Europe. The pest rapidly dispersed from the eastern U.S., ultimately reaching the pear-growing regions of western North America in the early 1900s. The species apparently now occurs in all pear-growing regions of North America. With codling moth, *Cydia pomonella* (L.), pear psylla is the major insect pest attacking commercial pears in North America. It is estimated that 50 to 80% of the costs associated with controlling arthropod pests in pear orchards are directed at pear psylla and a few other soft-bodied arthropods.

The adult pear psylla is 2 to 2.5 mm in length, and in appearance resembles a miniature cicada. The insect is greenish to dark brown in color, depending upon time of year, as discussed below. Wings are held roof-like over the abdomen. Eggs are yellow and elongate, having a curled filament extending from one end. The eggs are inserted partially into the host plant, anchored by a small pedicel that appears to function at least partially in uptake of water from the host plant. Nymphs are yellow at hatch, becoming darker with successive molts. Late instar nymphs are flat and oval in shape, having large obvious wing pads. Nymphs of all ages have a pair of conspicuous red eyes at the front of the head.

Pear psylla has a somewhat unusual life cycle. The species is seasonally dimorphic, producing a large, dark overwintering adult, or winterform, that is quite distinct from the smaller and light-colored summerform adult. The dimorphism is controlled by photoperiod; nymphs experiencing short-day condition develop into winterform adults, whereas those experiencing long-day conditions develop into summerform adults. The dimorphism is striking enough that the two morphotypes were once considered to be separate species. Pear psylla overwinters both in the pear orchard and outside of the orchard, often on other tree fruit species. Dispersal from the pear orchard in autumn by Pacific Northwest populations of psylla occurs in October and November, coinciding with leaf fall in pear. Re-entry into the pear orchard by post-diapause winterforms begins in February. Pear psylla reallocates among orchards between fall and spring associated with this dispersal behavior.

Winterform adults overwinter in reproductive diapause, characterized by immature ovaries and a lack of mating. Diapause ends in mid-winter; in the Pacific Northwest, cold temperatures keep the insect quiescent until late winter. Egg-laying by post-diapause winterforms begins in late February or early March in Oregon and Washington growing regions. The first eggs are deposited directly in wood at the base of fruit and leaf buds. Oviposition then shifts to newly emerging foliage as it becomes available. Fecundity of the overwintered female is quite high, exceeding 1,000 eggs per female. There are 5 nymphal instars, requiring 3 to 4 weeks to complete development at moderate (20 to 25°C) temperatures. Offspring of the overwintered adults eventually eclose as summerform adults, in Washington growing regions first appearing in early May. Pear psylla has 2 to 4 summerform generations per year depending upon latitude. Fecundity of the summerform adult is lower than that of the winterform adult.

Pear psylla causes several types of damage in pear orchards: fruit russet, psylla shock, and pear decline. Of these, russet is of most concern to growers, and management protocols are generally aimed at preventing this injury. Fruit russet is caused by the feeding activities of nymphs. Immature pear psylla ingest large quantities of plant juices that are eliminated as honeydew during the digestive process. Nymph-produced honeydew is in the form of a syrupy liquid, and if the product is in contact with fruit for extended periods of time it causes dark blotches or streaks on the fruit surface. If the marking is highly noticeable, the fruit may be downgraded at harvest. The damage

Fig. 772 Adult pear psylla.

is exacerbated by a sooty mold fungus that colonizes honeydew.

Psylla shock is also caused by nymphs. At high densities, pear psylla causes tree stunting, premature leaf drop, reduced fruit size, and premature fruit drop, symptoms collectively termed psylla shock. Substantial losses in yield may accompany this damage, and the effects may actually carry-over from one year to the next, even in the absence of high psylla densities the second year.

Lastly, adult pear psylla vector a mycoplasma-like organism that is responsible for pear decline disease. The pathogen causes sieve-tube necrosis at the graft union, preventing tree-synthesized nutrients from reaching the roots. Symptoms of the disease include a slow to rapid decline in tree vigor, accompanied by reduced yield and often eventual death of the tree. Resistant rootstock has largely eliminated this problem from North American pear production.

Control recommendations for pear psylla generally emphasize destruction of the overwintered generation and offspring of the overwintered generation, as it is very difficult to control the summer generations. Successful control in early spring is necessary to prevent unmanageable problems during the growing season. Also, it is necessary to manage the spring generation because contamination of fruit by honeydew early in the growing season almost ensures that the fruit will be damaged by russet. A typical program for overwintered adults is characterized by the use of horticultural mineral oil, sulfur, and insecticide, the latter generally being a synthetic pyrethroid or chlorinated hydrocarbon. The initial applications occur in late winter when pear is still dormant, with additional applications repeated as necessary as the tree breaks dormancy. The initial spray should be delayed until re-entry by psylla into the orchard is mostly complete, but before significant levels of egg-laying has occurred. The oil applications cause delays in egg-laying by the returning winterform adults, resulting in increased synchronization of egg hatch in early spring. This increased synchrony simplifies subsequent pesticide-based control efforts. Summer control of pear psylla has become increasingly difficult because of resistance to insecticides. Currently, summer chemicals used include Mitac (amitraz), Agri-Mek (avermectin), and a neonicotinyl, Provado. Other products found to be useful against pear psylla include insect growth regulators, applied in spring, and kaolin particle film, used from late-winter through bloom. Kaolin is a wettable powder that dries to a white film on the tree and is an extremely effective feeding and oviposition deterrent to pear psylla. In some pear production areas of Washington State, the kaolin product Surround has substantially replaced the more traditional spring chemical controls.

Growers have certain horticultural practices that they use to simplify management. These practices include primarily strategies that make the tree less suitable or less attractive to psylla. For example, it is recommended that growers use only enough nitrogen fertilizer to produce good fruit set and fruit size, as excessive nitrogen leads to outbreaks of pear psylla. Similarly, pear trees should not be overpruned, as this may prompt extensive growth of new foliage which is highly attractive to adults. Suckers and water sprouts should be removed from limbs, as these are a source of highly nutritious foliage.

The primary means of monitoring pear psylla for management decisions are by use of a beat tray (for sampling adults) and by taking leaf samples (for immature stages). The beat tray method involves holding a white, cloth-covered tray (generally 45×45 cm in size) beneath a horizontal limb, and sharply jarring the limb with a piece of stiff rubber hose. Psylla that are dislodged onto the tray are then counted. Recommended sample sizes are 25 trays (trees) per 10 to 20 acre (about 4 to 8 ha) pear block. Samples should be taken at random locations throughout the block. For the immature stages, samples taken during the first generation should consist of fruit spurs, examined either with a hand lens or beneath a dissecting microscope. Recommended sample sizes are at least 10 spurs per block. Additional samples taken just after bloom may also be necessary. For these samples, one fully expanded leaf should be taken from each sampled spur, with 2 spurs taken per tree and 20 trees sampled per 20 acre pear block. Mid-season leaf samples for psylla eggs and nymphs should be taken from actively growing shoots, near the top of the tree if possible. Extension personnel recommend taking 50 leaves per block, comprising a mix of young and old leaves.

Treatment thresholds tend to be fairly inexact and assume that the type of damage to be prevented is fruit russet (having a lower threshold than that for psylla shock, the latter requiring high densities of nymphs). Thresholds for adult psylla are 1 or 2 per 10 trays during late winter before the dormant spray, and 1 or 2 per tray during the spring and summer growing seasons. For the immature stages, treatment thresholds are 1 to 2 eggs or nymphs per fruit cluster in spring, and 0.25 to 0.5 nymphs per leaf during the summer sampling periods. It should be noted that susceptibility of fruit to russet damage depends upon a number of factors, including pear variety (naturally russetted varieties such as 'Bosc' are less susceptible than clear-skinned varieties such as 'Anjou'), fruit destination (russet is of little importance in fruit destined to be canned), and age of the fruit when marked. Thus, threshold decisions for psylla control depend upon factors other than just psylla density.

Lastly, natural enemies have the potential to play a large role in controlling pear psylla. Many scientists consider pear psylla to be an induced pest, in that insecticide use destroys natural enemies that would otherwise keep psylla in check. Pear psylla is attacked by a variety of specialist and generalist natural enemies. One of its most important enemies is a specialist parasitoid, *Trechnites insidiosus* (Crawford) (Hymenoptera: Encyrtidae), that may cause very high levels of parasitism in reduced-insecticide orchards. Predatory true bugs, including species of Anthocoridae (*Anthocoris* spp.) and Miridae (*Deraeocoris brevis* (Uhler), *Campylomma verbasci* (Meyer)) are very efficient predators of pear psylla and other soft-bodied pests in pear orchards. Ladybug beetles (Coleoptera: Coccinellidae), lacewing larvae (Neuroptera), and European earwig (*Forficula auricularia* L.) also prey on eggs and nymphs of pear psylla. The relative importance of the various taxa in controlling pear psylla remains unexplored, and merits study.

David Horton
U.S. Department of Agriculture,
Agricultural Research Service
Yakima, Washington, USA

References

Beers, E. H., J. F. Brunner, M. J. Willet, and G. M. Warner. 1993. *Orchard pest management: a resource book for the Pacific Northwest*. Good Fruit Grower, Yakima, Washington.

Horton, D. R. 1999. Monitoring of pear psylla for pest management decisions and research. *Integrated Pest Management Reviews* 4: 1–20.

Unruh, T. R., P. H. Westigard, and K. S. Hagen. 1994. Pear psylla *Cacopsylla pyricola* (Forster) Homoptera: Psyllidae. pp. 95–100 in L. Andres, R. D. Goeden, G. Jackson, and J. Beardsley (eds.), *Biological control in the western region*. DANR Publications, University of California, Berkeley, California.

VanBuskirk, P. D., R. J. Hilton, N. Simone, and T. Alway. 1999. *Orchard pest monitoring guide for pears: a resource book for the Pacific Northwest*. Good Fruit Grower, Yakima, Washington.

Westigard, P. H., and R. W. Zwick. 1972. The pear psylla in Oregon. Oregon State University, Agricultural Experiment Station Technical Bulletin 122, Corvallis, Oregon.

PEBRINE. A disease of the silkworm caused by the microsporidian *Nosema bombycis*.

PECAROECIDAE. A family of sucking lice (order Siphunculata). This family is also known as peccary lice. See also, SUCKING LICE.

PECCARY LICE. Members of the family Pecaroecidae (order Siphunculata). See also, SUCKING LICE.

PECK, WILLIAM DANDRIDGE. William Peck was born at Boston, Massachusetts, USA, on May 8, 1763. He is sometimes known as America's first native entomologist, as those preceding him were immigrants from Europe. Peck graduated from Harvard University in 1782 and though he aspired to be a physician, he entered business, an occupation he found to be unsatisfactory. He and his father (a naval architect who also was discontented) moved to a small farm at Kittery, Maine, were he spent the next 20 years isolated but studying nature. He became an authority on plants, birds, fishes, and insects. In 1794 he published the first American paper on systematics, a description of four fishes. He began writing on insects in 1795 and won a prize from the Massachusetts Society for Promoting Agriculture for his publication on cankerworm in 1796. In 1805 he became a professor of Natural History at Harvard University and he went to Europe to study botanic gardens in preparation for establishment of a botanic garden at Harvard. There he became acquainted with important European naturalists including Linnaeus. He described a few species of insects, but unlike most naturalists and entomologists of the period, was more interested in the economic aspects of entomology. Peck taught a course on entomology while at Harvard, undoubtedly the first course of its kind in the United States. He published such articles as "The description and history of the cankerworm" (1795),

"Natural history of the slugworm" (1799), and "Some notice of the insect which destroys the locust tree" (1819). He died at Cambridge, Massachusetts, on October 8, 1822.

References

Essig, E. O. 1931. *A history of entomology*. The Macmillan Company, New York. 1029 pp.
Mallis, A. 1971. *American entomologists*. Rutgers University Press, New Brunswick New Jersey. 549 pp.

PECTEN. Any series of bristles arranged like a comb. This term (or pecten teeth) is sometimes applied to rows of spines on the legs of pollen-gathering bees.

PECTINATE. Having a series of slender projections from a slender shaft. Comb-like.
See also, ANTENNAE OF HEXAPODS.

PEDICEL. The constricted region of the abdomen in Hymenoptera. In ants the pedicel bears one or more upright lobes; the first segment is the petiole, the second the post-petiole. This term also is applied to the second segment of the antenna.
See also, ABDOMEN OF HEXAPODS, ANTENNAE OF HEXAPODS.

PEDICINIDAE. A family of sucking lice (order Siphunculata). See also, SUCKING LICE.

PEDICULIDAE. A family of sucking lice (order Siphunculata). This amily is also known as body lice. See also, SUCKING LICE.

PEDIPALPS. The second pair of appendages of an arachnid.

PEDUNCULATE. This term refers to a structure on a slender stalk.

PELECINIDAE. A family of wasps (order Hymenoptera). See also, WASPS, ANTS, BEES, AND SAWFLIES.

PELECORHYNCHID FLIES. Members of the family Pelecorhynchidae (order Diptera). See also, FLIES.

PELECORHYNCHIDAE. A family of flies (order Diptera). They commonly are known as pelecorhynchid flies. See also, FLIES.

PELORIDIIDAE. A family of bugs (order Hemiptera, suborder Coleorrhyncha). They also are known as moss bugs. See also, BUGS.

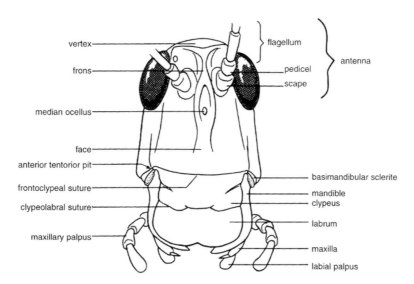

Fig. 773 Front view of the head of an adult grasshopper, showing some major elements.

PELTOPERLIDAE. A family of stoneflies (order Plecoptera). They sometimes are called roach stoneflies. See also, STONEFLIES.

PEMPHREDONIDAE. A family of wasps (order Hymenoptera). See also, WASPS, ANTS, BEES, AND SAWFLIES.

PENIS. (pl., penes) The male intromittent organ. The aedeagus.

See also, ABDOMEN OF HEXAPODS.

PENTATOMIDAE. A family of bugs (order Hemiptera). They sometimes are called stink bugs or shield bugs. See also, BUGS

PENULTIMATE. The next to the last. For example, the penultimate instar for a lepidopteran with six larval instars is the fifth instar.

PEPPER WEEVIL, *Anthonomus eugenii* CANO (COLEOPTERA: CURCULIONIDAE). The pepper weevil, *Anthonomus eugenii* Cano (Coleoptera: Curculionidae), is a serious pest of peppers, *Capsicum* spp., causing millions of dollars damage annually in the southern U.S.A., Mexico, Central America, Hawaii and several Caribbean islands. Pepper weevil damages the pepper crop by directly feeding and ovipositing in the fruit, causing premature abscision of all stages of peppers from flower buds to maturing pods. This insect is generally more prevalent in warm climates and, based on the earliest records of this insect and the corresponding first records of a principal host plant group, the domesticated *Capsicum* spp. around 7000 BC in Mexico, it is probably native to Mexico or surrounding regions. The reason pepper weevil is such an important pest relates to the difficulty of controlling this insect once it has become established in the field. Larvae feed within pepper pods, and are protected from insecticide sprays. Also, since the level of weevils required to cause economic loss is so low, detection of this pest when it reaches an action threshold is time consuming and relatively difficult. The economic impact of pepper weevil as a pest of peppers in the U.S.A. is

conservatively estimated at $5 to $20 million annually and the impact of pepper weevil on Mexican pepper production is probably much greater. Current factors that keep potential losses due to pepper weevil from reaching 50% to 100% in more areas in the U.S. are the availability of effective insecticides and local eradications from harsh winters or the absence of host plants causing local extinctions.

The introduction of pepper weevil from Mexico into Texas is thought to have occurred through fresh market transport. In the USA, it occurs in Hawaii, California, Arizona, Texas, Louisiana, Florida, Georgia, North Carolina, and New Jersey. Outside the U.S. it occurs in Mexico, Guatemala, and El Salvador. Additional areas infested include Honduras, Puerto Rico, and Costa Rica. There is a report of pepper weevils in greenhouse production of pepper in Canada, apparently introduced on pepper transplants from southern California. This illustrates that pepper weevil can be introduced into any area that will sustain host plants for the pepper weevil and that re-infestations can even occur in areas where pepper weevil is periodically eradicated or does not normally exist because of climatic conditions.

The pepper weevil shares the same biological characteristics of many of the other 330 species of the genus *Anthonomus*. Adults are oligophagous, and females lay eggs in flower buds or fruit. Larvae complete development within immature buds or fruit, causing premature bud and fruit abscision (fruit drop). The pepper weevil has three larval instars and multiple generations per year. Reported generation times for the pepper weevil varies widely among different investigators from 10 to 40 days, with perhaps the largest differences occurring between summer and fall observations. Generation time, the number of generations per year (5 to 8), and longevity of adults (79 to 316 days) are determined primarily by host availability and temperature. The oviposition period ranges from 16 to 129 days and averages 30 days. Feeding and development of pepper weevil is limited to two host plant genera, *Capsicum* and *Solanum*, and limited in its northern distribution by temperature with the exception of introduction of infested plant material into northern greenhouses. The development time in pepper can be as short as 12 days, and in nightshade as short as 11 days.

Temperature data associated with development time were reported by several investigators.

Emergence on artificial diet was observed in 17 to 18 days at 25.7 to 27.7°C and 70% relative humidity (RH). An average generation time of 17.5 days at 23.9 to 26.7°C and 60% to 85% RH. A mean generation time on bell pepper of 14.2 days at 25.7 to 27.7°C and 40% to 100% RH. Pepper weevil development has been related to growth stage and temperature by a regression of 1/days development (Y) to °C (X) for the period of adults to new emergence as

$$Y = -1.075 + 0.059X, \ R^2 = 0.95$$

This data has been used to develop predictive models for pepper weevil population development in the field.

Several efforts have been made to define the relationship of crop damage to weevil infestation level. A pepper weevil population density exceeding 1 pepper weevil/100 terminal pepper buds can be economically damaging. Infestation levels of 5% damaged terminals, 1 pepper weevil/10 whole pepper plants, and 1 pepper weevil/400 terminal buds also were shown to result in significant economic loss in high-yielding bell peppers. The frequency of pepper weevil oviposition per fruit is a factor of time, pepper weevil population density, pepper fruit size and the plant variety. Varietal differences in peppers may affect host preference and these effects are thought to be negligible under heavy feeding and oviposition pressure. However, studies indicate that plant structure, number of fruiting buds per plant, etc., may affect the plant's ability to maintain greater numbers of unaffected mature pods.

Early attempts at scouting pepper weevil populations were simply presence determinations and, in many cases, the presence of fallen fruit was the first indication of a pepper weevil infestation. Examination of fallen buds for presence of pepper weevil immatures early in the season is a means of determining the presence of pepper weevil in the field. Counting fallen fruit or using a beat cloth are unacceptable sampling methods for timing pepper weevil control. However, inspecting terminal buds or bud clusters for pepper weevil adults was effective in predicting subsequent infestations. Yellow sticky trap captures correlate well with direct adult pepper weevil counts, using whole plant inspections, and the use of traps is an alternative to the more intensive and costly direct count method. It is also possible to monitor feeding or ovipositional damage to terminal bud clusters as

a method for determining the presence of pepper weevil in commercial pepper fields.

The dispersion of pepper weevil has been investigated in several locations. The dispersion index developed for pepper weevil adults in Puerto Rico indicated that adult populations exhibited a moderately clumped dispersion in the field with a negative binomial k = 2.5. In Florida, weevils concentrate along field margins. Clumping patterns of pepper weevil have been confirmed in non-bell peppers. Early in the season, pepper weevil distribution is moderately to highly clumped and that there is generally a concentration of weevils along pepper field margins.

The majority of research literature on pepper weevil control is associated with chemical control. Natural enemies of the pepper weevil include various predators (e.g., *Solenopsis geminata*, *Strunella magna*, *Tetramorium guineese*) and parasites (e.g., *Pyometes venticosis*, *Catolaccus incertus*, *Pediculoides ventricosus*, *Bracon mellitor*, *Habrocytus piercei* and *Zatropis incertus*, *Catolaccus hunterii* and a new species of Braconidae, *Triaspis eugenii*). Classical biological control for pepper weevil was attempted with *Eupelmus cushmani* (Crawford) and *Catolaccus hunterii* Crawford in Hawaii, and with *Bracon vestiticida* (Vierick) in southern California with some establishment, but no documented success.

Movement to and from secondary host material and cull sites has been implicated as one of the primary causes for periodic re-infestation in commercial pepper fields. Also, because pepper weevil has been reported to reproduce on nightshade, *Solanum* spp., movement from nightshade to pepper is an important management consideration. Crop residue destruction, transplant sanitation, and nightshade management have been recommended as effective cultural controls of pepper weevil. Even so, caution should be taken with the timing of destruction of host material since the lack of oviposition sites may trigger the migration of weevils, as has been suggested with the boll weevil.

Considerable research needs to be conducted on pepper weevil. Pepper weevil pheromonal sex attraction and the use of yellow sticky traps are important in pepper weevil sampling and management. The males produce an airborne chemical that is attractant to the female, and males are not as attracted. More definitive information is needed on alternative management tactics, such as biological control, mating disruption with pheromones, natural mortality factors, and general ecology of pepper weevil, which ultimately will prove critical for managing this pest.

David G. Riley
University of Georgia
Tifton, Georgia, USA

References

Burke, H. R., and R. E. Woodruff. 1980. *The pepper weevil Anthonomus eugenii Cano (Coleoptera: Curculionidae) in Florida*. Florida Department of Agriculture and Consumer Services, Division of Plant Industry, Circular 219, Gainesville, Florida. 4 pp.

Riley, D. G., and E. King 1994. Biology and management of the pepper weevil: a review. *Trends in Agricultural Science* 2: 109–121.

Riley, D. G. 1992. *The pepper weevil and its management*. Texas A&M Agricultural Extension Leaflet-5069. 6 pp.

Wilson, R. J. 1986. *Observations on the behavior and host relations of the pepper weevil Anthonomus eugenii Cano (Coleoptera: Curculionidae) in Florida*. M.S. Thesis, University of Florida, Gainesville, Florida. 94 pp.

PERADENIIDAE. A family of wasps (order Hymenoptera). See also, WASPS, ANTS, BEES, AND SAWFLIES.

PERENNIAL. A plant that lives at least three years and reproduces at least twice.

PERGIDAE. A family of sawflies (order Hymenoptera, suborder Symphyta). See also, WASPS, ANTS, BEES, AND SAWFLIES.

PERICARDIAL CELLS. Loose clusters of cells that are found on or near the external surface of the heart. They are believed to have a phagocytic function, removing particulate matter from the hemolymph.

PERICARDIAL SINUS. A space around the heart, bordered below by the dorsal diaphragm.

PERIKARYON. (pl., perikarya) The cell body in a nerve cell. Also known as the soma. See also, NERVOUS SYSTEM.

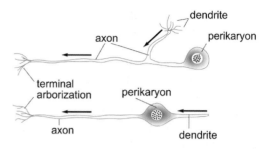

Fig. 774 Diagrams of insect nerve cells showing direction of nervous impulse (adapted from Chapman, The insects: structure and function).

PERILAMPIDAE. A family of wasps (order Hymenoptera). See also, WASPS, ANTS, BEES, AND SAWFLIES.

PERILESTIDAE. A family of damselflies (order Odonata). See also, DRAGONFLIES AND DAMSELFLIES.

PERIPSOCIDAE. A family of psocids (order Psocoptera). See also, BARK-LICE, BOOK-LICE, OR PSOCIDS.

PERISCELIDID FLIES. Members of the family Periscelididae (order Diptera). See also, FLIES.

PERISCELIDIDAE. A family of flies (order Diptera). They commonly are known as periscelidid flies. See also, FLIES.

PERISSOMMATIDAE. A family of flies (order Diptera). See also, FLIES.

PERISTOME. In Diptera, the membranous area surrounding the base of the mouth.

PERITREME. Any sclerite bearing a body opening, particularly a spiracle. A sclerotic ring around a spiracle.

PERITROPHIC MEMBRANE. A porous sheath lining the midgut and protecting the midgut cells from abrasion by the food, penetration by microbial pathogens, and the degradatory properties of digestive enzymes. It is secreted continuously along the length of the midgut (type I peritrophic membrane) or from the anterior portion of the midgut only (type II peritrophic membrane). See also, ALIMENTARY CANAL AND DIGESTION.

PERLIDAE. A family of stoneflies (order Plecoptera). They sometimes are called common stoneflies. See also, STONEFLIES.

PERLODIDAE. A family of stoneflies (order Plecoptera). See also, STONEFLIES.

PERORAL. By way of or through the mouth; *per os.*

PER OS. By way of the mouth; peroral.

PERSISTENT VIRUS. A virus that passes through the body of the vector, and that usually persists for the remainder of the vector's life.

PESTICIDE. A material that kills pests. This term is often used to describe insecticides, which are pesticides that kill insects.

PESTICIDE HORMOLIGOSIS. Hormoligosis (from the Greek: *hormo* = excite and *oligo* = small quantities) is a term applied to the phenomenon in which sublethal quantities of stress agents such as chemicals, antibiotics, hormones, temperature, radiation, and minor wounds are stimulatory to an organism by providing increased efficiency to develop new or better systems to cope in a suboptimum environment.

The occurrence of pesticide hormoligosis in agriculture is well documented and may be a common phenomenon, but it is rarely monitored so we are uncertain of its importance in fostering outbreaks of certain pests or in accounting for failures in pest control programs. The purpose of this section is, therefore, to review the state-of-the-art of pesticide hormoligosis and to discuss controversial hypotheses on the reasons for the resurgence of certain arthropod pests and the drastic increase in their populations. Medical studies show that (1) stimulation of tissue repair after a sublethal dose of CCl4 (i.e., chemically induced hepatotoxicity) appears to be the central mechanism in protection against death from a subsequent large dose; (2) sublethal environmental stresses by exposure to ethanol (5%, v/v), acid (HCl, pH 4.5 to 5.0), H2O2 (500 ppm) or NaCl (7%, w/v), protects *Listeria monocytogenes* against lethal preservation factors by increasing its resistance to lethal doses of the corresponding factors; and (3) pre-exposure of *Vibrio cholerae* cells to sublethal dose of 1.7 Gy X-rays makes these cells 3- to 38-fold more resistant to the subsequent challenge by X-rays.

In spite of a good initial kill at the time of treatment, different pesticides and even carrier materials and diluents used on orchard or field crops are reported to cause a tremendous increase of the pest against which they have been applied. Furthermore, spray applications are sometimes followed by serious outbreaks, not of the pest against which they were applied, but of other phytophagous insects and mites which, prior to the treatment, were in very small numbers too low to be of economic importance. Both

Oviposition of *Panonychus ulmi* Koch on leaf discs treated with different concentrations of DDT dust during a 8- to 10-day observation (1 = field application rate amounting to 2.5 g/m^2 of dust containing 5% DDT; see Hueck *et al.*, 1950, for more details).

Treatment	Total eggs produced by ten females[1]	Egg production (index)
Control	37	100.0
1/8	78	210.8
1/4	24	64.9
1/2	31	83.8
1	23	62.2

[1]Average of five replicates.

types of increases, usually called pest resurgence, have been recorded in different climatic conditions (namely, temperate, subtropical and tropical regions) and for many species of the following families of mites or insects: Tetranychidae, Eriophyidae, Tarsonemidae, Coccidae, Aphidae, Aleyrodidae, Cicadellidae, Noctuidae, Tortricidae (= Olethreutidae), Yponomeutidae, Trypetidae, Agromyzidae, and the order Collembola. Spread of this phenomenon over a number of very different groups of phytophagous arthropods, and its occurrence after application of pesticides using different modes of action and under different climatic conditions, indicates that chemical control in these cases upsets the population dynamics of the pests in question.

For an explanation of this phenomenon, three hypotheses usually are suggested: (a) The reduction of natural enemies by the pesticides; (b) pesticide-induced reproductive stimulation of phytophagous arthropods; and (c) the removal of competitive species. Two additional factors also have been proposed: (1) altered host plant quality caused by agrochemicals, and (2) the pesticide-induced irritation causing pest dispersal. Although sporadic literature confirms the validity of the second hypothesis (i.e., pesticide-induced reproductive stimulation of phytophagous arthropods), and its importance in pest resurgences, most information on the subject refers chiefly to the destructive effects of pesticides on the natural enemies of the phytophagous species and, to a lesser extent, to the removal of competitive fauna. Critical information needed to substantiate the importance of natural enemies in regulating pest populations includes: (a) a negative correlation between the number of natural enemies and pests, (b) proof that the natural enemies were the limiting factor to the density of the population, and (c) an exclusion of stimulating influences of the pesticide on the pest directly or via the plant. The less well-known phenomenon, pesticide-induced reproductive stimulation of phytophagous arthropods, comprises the backbone of pesticide hormoligosis and is the focus here.

There is considerable evidence of a favorable effect of some pesticides on the biotic potential of arthropods. For example, a drastic increase in the population of the fruit tree red spider mite, *Metatetranychus ulmi* (Koch) [*Panonychus ulmi* Koch], following the application of DDT has been demonstrated in The Netherlands since the late 1940s. Studies on the sublethal effect of DDT on oviposition of this mite revealed stimulation of egg production at a DDT concentration much lower than the recommended rates.

In California, the results of a 3-year field and greenhouse study with DDT on natural populations and fecundity tests in small cages on the European red mite, *Paratetranychus pilosus* C. & F., failed to support the theory of destruction of the natural enemies as the primary cause of the mite outbreaks. Some of the results were, however, consistent with the idea of a physiological stimulus to reproduction under DDT influence.

Although some scientists at first thought that the mite resurgence was solely attributable to an elimination of the natural enemies of the mites, others posed the question of the possible involvement of a change in the physiology of the mites, or in their host, or both. It was found that DDT treatments produced an increase of mite populations above that occurring on check trees which was not explainable by a reduction of the natural enemies.

In order to find the reason for the mite's resurgence, apple trees with a moderate infestation of red spider mites were sprayed with parathion which killed all mites that were not in the egg stage, and all predators. Following the application of parathion, a number of trees were sprayed with DDT. Afterward, the numbers of all stages of red spider mites were counted every second day. An examination of the resulting data showed that egg production had been substantially higher on DDT-sprayed trees than on the check trees. Laboratory experiments also showed that the egg production of the mites feeding on leaf discs dusted with 5 percent DDT was significantly greater than those feeding on untreated discs. Both experiments, however, were not designed to explore whether the stimulating factor was caused by a direct influence of the pesticide, or via the plant. Coding moth adults, exposed to various sublethal doses of several insecticides, demonstrated stimulation of oviposition at concentrations where the females are under stress. Wherever the stress factors are not remarkable, oviposition is not stimulated.

More recent studies on population changes of spider mites (Acari: Tetranychidae) following insecticide applications in corn have shown that the exposure of twospotted spider mite (*Tetranychus urticae* Koch) to sublethal doses of permethrin leads to an increase in dispersal behavior (under laboratory conditions) and to an increase in population densities (under field conditions).

Stimulation of some vegetable plants by DDT with an action resembling that of some plant hormones suggests direct or indirect effect of sublethal doses of pesticides on phytophagous pests, but no field experiments or laboratory/greenhouse tests have been conducted either on intact or on excised plants (i.e., leaf discs). Nevertheless, numerous scientists have suggested the host-plant physiology hypothesis for the observed data on the reproductive stimulation of mites/insects whenever the hypothesis on the reduction in natural enemies is refuted. This hypothesis warrants additional research.

Cyrus Abivardi
Swiss Federal Institute of Technology
Zurich, Switzerland

References

Abivardi, C., D. C. Weber, and S. Dorn. 1998. Effect of azinphos-methyl and pyrifenox on reproductive performance of *Cydia pomonella* L. (Lepidoptera: Tortricidae) at recommended rates and lower concentrations. *Annals of Applied Biology* 132: 19–33.

Abivardi, C., D. C. Weber, and S. Dorn. 1999. Effects of carbaryl and cyhexatin on survival and reproductive behaviour of *Cydia pomonella* (Lepidoptera: Tortricidae). *Annals of Applied Biology* 134: 143–151.

Chapman, R. K., and T. C. Allen. 1948. Stimulation and suppression of some vegetable plants by DDT. *Journal of Economic Entomology* 41: 616–623.

Hueck, D. J., D. J. Huenen, P. J. Den Boer, and E. Jaeger-Draafsel. 1950. The increase of egg production of the fruit tree red spider mite (*Metatetranychus ulmi* Koch) under influence DDT. *Physiologia comparata et oecologia* 2: 371–375.

Luckey, T. J. 1968. Insecticide hormoligosis. *Journal of Economic Entomology* 61: 7–12.

PESTICIDE RESISTANCE MANAGEMENT.

"Integrated Pest Management" (IPM) and "Management of Pesticide Resistance in pest arthropods" (MPR) are usually perceived to be distinct topics, but have equivalent goals and methods. Effective management of resistance to pesticides and effective IPM programs require that we employ a holistic and multitactic management strategy. A key component of a holistic and multitactic strategy will include enhancing the compatibility of pesticides and biological control agents.

Resistance to pesticides is an extremely important problem. At least 440 arthropod species have become resistant to insecticides and acaricides, with many species having become resistant to all the major classes of such products. Resistance to pesticides in weeds, plant pathogens, and nematodes also is increasing, although somewhat more slowly.

Developing and registering new pesticides is an elaborate, and increasingly expensive, business in most parts of the world. Thus, pesticide producers should be increasingly interested in extending the economic life of their products in order to maximize a return on their investment. Likewise, pest management specialists want to preserve registered pesticides. This is especially true for products that are effective against arthropod pests in minor crops such as vegetables, which may be ignored by pesticide companies because they are a relatively small market compared to field crops. Registration of new pesticides is likely to be more difficult and expensive in the future, which could leave some pest management specialists with extremely limited options for managing certain recalcitrant pests.

A few environmentalists have argued that we do not need pesticides, that they will soon be outlawed, or that they will become unimportant because transgenic crops resistant to pests and diseases will dominate the market, and that pesticide resistance will no longer be an important issue. However, it is unrealistic to eliminate all pesticides from agriculture; there are significant arthropod pests for which we have no other effective control tactic. Pesticides are effective tools for fighting outbreaks and emergency pest problems. Pesticides are often required to control plant pathogens, weeds, or nematodes because they cannot be controlled by alternative methods at this time.

Research on resistance in arthropods

Scientists have approached the problem of pesticide resistance in a variety of ways. Fundamental research over the past 40 years has produced insights into resistance mechanisms and the mode of inheritance of resistance in arthropods. Simulation models have been developed to evaluate different options for managing resistance, but these models may not be applicable to a specific situation.

The debate over whether to recommend alternation of different pesticides or to recommend mixtures of different pesticides as appropriate methods for slowing the development of resistance remains controversial. Each management approach is dependent upon specific genetic and operational assumptions associated with the resistance mechanisms and mode of inheritance of the resistance in the specific pest population, as well as application methods and timing, and details of the biology of the pest. If one or more of these assumptions are not valid, then the management recommendations may not result in the intended preservation of the pesticide. Furthermore, few field data are available to support adopting either alternation or mixtures, or other resistance management concepts.

The hypothesis that reduced fitness, which is often associated with pesticide resistance alleles, could be used in resistance management programs continues to be controversial. The concept of reduced fitness may have limited application because not all resistance alleles confer lowered fitness and selection for modifying genes that restore fitness to individuals carrying resistance alleles can occur in the field.

Various monitoring techniques have been developed to identify resistant insects or mites and detect their establishment and spread. These monitoring methods are particularly useful for documenting that resistance has developed, but monitoring methods that would allow us to detect rare resistant individuals in populations in sufficient time that operational pest management programs could be altered remains difficult and expensive to execute.

Resistance management research programs and IPM research programs have had fairly distinct identities to date. The current scenario usually goes something like this: Once a pesticide has been registered and used, and resistant individuals have been detected in populations, people begin to discuss developing and implementing a resistance management program. This approach is short sighted because an effective program is exceedingly difficult to execute in sufficient time to have the desired results.

Developing a resistance management program may take several years; studies typically are conducted to develop an appropriate monitoring method, estimate the frequency of resistant individuals in populations, detect cross resistances, and evaluate mode of inheritance and stability of the resistance. Meanwhile, unless pesticide applications are discontinued, selection for resistance continues. Because the initial detection of resistance usually requires that resistant individuals comprise 5% of the population, by the time resistant individuals are detected additional pesticide applications are likely to increase significantly their frequency in the population. This scenario is particularly familiar with pests such as aphids, spider mites, whiteflies, and leafminers with a high reproductive rate and multiple generations a year.

The experiences of scientists studying pesticide resistance in ubiquitous pests in other geographic regions can alert pest managers to a potential problem, but this seems to be an inefficient method for managing resistance in arthropods and it is possible for two different populations of the same species to respond to selection for resistance in different ways because different resistance alleles may be present in the different populations. Mechanisms may vary, the mode of inheritance may vary, and the degree of reduction in fitness associated with the resistance may vary because resistance alleles at each site are different. Because it is difficult to sample rare individuals in natural populations, monitoring programs, if employed other than as a method to document a problem once it has developed, may not be cost effective. Waiting until the pest becomes resistant before instituting a resistance management program is ineffective.

Resistance management and IPM

A better paradigm for managing resistance in arthropods involves altering pesticide use patterns. Nearly everyone will agree that reducing pesticide use is an effective resistance management tactic. What is more controversial is whether resistance management programs should include altering the way pesticides are developed and registered. Also, many pest managers have been slow to recognize that resistance management must be a broad-based, multi-tactic endeavor.

It seems reasonable to assume that nearly all major insect and mite pests will develop resistances to all classes of pesticides given sufficient selection pressure over sufficient time. There may be some exceptions, but this generalization appears reasonable given the documented record of resistance development in arthropod pests during the past 40 years. Resistance to stress is a fundamental and natural response by living organisms. On an evolutionary time scale we should expect insects to have developed multiple and diverse mechanisms to survive extreme temperatures, allelochemicals, and other environmental stresses. We should expect most insects and mites to develop resistance to most pesticides if subjected to appropriate and sustained selection. While new pesticide classes have been proclaimed to be a potential 'silver bullet' and not amenable to resistance development, these hopes have been misplaced repeatedly. It seems appropriate to assume that the development of resistance is nearly inevitable and the issue is not whether resistance will develop, but when. With this assumption, resistance management programs have the goal of delaying, rather than preventing, resistance.

Growers and pest management experts cannot afford to rely on pesticides as their primary management tool. There are increasing social, economic, and ecological pressures to reduce pesticide use and to increase the use of nonchemical control tactics such as host plant resistance, cultural controls, biological controls and biorational control methods such as mass trapping, sterile insect release programs and mating disruption. There is an increasing emphasis to use pesticides that are nontoxic to biological control agents, and that have minimal impacts on the environment and on human health. The compatibility of pesticides with natural enemies and other nonchemical management tactics is critical for improving pest management programs and environmental quality, as well as for managing resistance to pesticides in pest arthropods. Enhancing the compatibility of pesticides and biological control agents is complex and sometimes difficult, but can play handsome dividends in improved pest control and pesticide resistance management.

Changing the way pesticides are registered could become part of an effective resistance management strategy and would have the benefit of achieving improved integrated pest management. For example, some pesticides are relatively nontoxic to important natural enemies in cropping systems when applied at low rates, but the recommended application rates are too high. Use at the high rates thus disrupts effective natural enemies, leading to additional pesticide applications, which exert unnecessary selection for resistance in the pest. Under these circumstances, it may be appropriate for the label to contain two different directions for use; one rate could be recommended for the strategy of relying on pesticides to provide control. A lower rate could be recommended for use in an IPM program when it is known that effective natural enemies are present but need help to suppress the pest population(s). This approach to labeling could reduce the number of pesticide applications and reduce rates, which would reduce selection for resistance in both target and nontarget pests in the cropping system.

Another innovation in pesticide registration would require that the toxicity of the pesticide to a selected list of biological control agents be determined for each cropping system. This information should be provided, either on the label or in readily available computerized data bases, perhaps via the internet. Without such information, pesticides may be used that disrupt effective natural enemies. This often results in unnecessary use of pesticides, leading to more rapid evolution of pesticide resistance in pests. Enhancing biological control not only can lead to improved pest management, but is an essential tool in managing pesticide resistance.

How could information about the toxicity of pesticides to biological control agents best be made available to the end user? How should bioassays be conducted to evaluate pesticide selectivity? There are no simple answers. Data on pesticide toxicity to

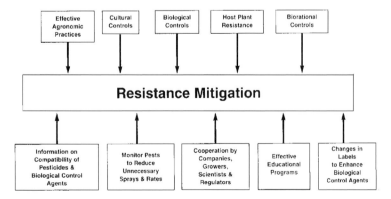

Fig. 775 Effective resistance management programs incorporate multiple tactics as part of a fully integrated pest management program. Pesticides should be used only when needed, at the lowest rates possible, and in a manner that reduces negative effects on arthropod natural enemies. Ideally, pesticide labels would include information about their effects on natural enemies.

natural enemies often is buried in publications or reports that pest managers and farmers cannot find easily. Unfortunately, even if the data can be found, it is not easy to interpret bioassay data obtained by different scientists using different methods. Different bioassay methods can produce different conclusions about the toxicity of pesticides to natural enemies, and it is often difficult to predict the effects of pesticides under field conditions based on laboratory assays. Thus, the recommendation that labels or data bases be developed with information on the impact of pesticides on natural enemies requires considerable discussion and additional research. Should pesticide companies conduct the research using standard bioassay methods? Should a consortium of pest management scientists conduct the assays? Who should pay for the research? What species of natural enemies should be tested? This concept is not new; in Europe standardized bioassays already are conducted on selected natural enemy species by a scientific working group concerned about managing pests in glasshouses. Whether this model can be used elsewhere should be explored. Increased international consultation and cooperation between scientists, regulatory agencies, and pesticide companies could resolve many of the questions raised above.

Evolution of pesticide resistance in pests has been shown by computer simulations of predator and prey interactions to be slowed by reduced pesticide use. There is general agreement that reduced pesticide use is one of the essential elements of any resistance management program. Thus, the compatibility of pesticides and biological control agents is a crucial issue in pesticide resistance management.

Attempts to manage pesticide resistance in pest arthropods generally has involved making relatively minor tactical shifts in use patterns. A major shift in how we think about pesticide development and use could provide more effective resistance management tactics. The strategy should be to manage use of the pesticide, even before it is fully developed and registered. If this strategy is adopted, decisions on application rates and the numbers of applications per growing season would be made with the understanding that they affect the speed with which resistance will develop. In some cases, new products may not be registered for specific crops because they are toxic to biological control agents in that crop and thus could disrupt effective IPM programs in place.

Pesticide-resistant natural enemies

Pesticide-resistant natural enemies are a special category of pesticide selectivity. Relatively few natural enemies have developed resistance to pesticides through natural selection, but several have been important in effective IPM programs. Artificial selection of phytoseiid predators for pesticide resistance can be a practical and cost effective tactic for the biological control of spider mites. Field tests have been conducted with several laboratory-selected phytoseiid species and some are being used in IPM programs.

Selection of pesticide-resistant strains of parasitoids and predatory insects have been conducted and some strains have been evaluated for incorporation

Important assumptions of resistance management models; violation of the assumptions could make management decisions based on these models ineffective (modified from Hoy 1998).

Model type	Assumptions
Mixtures	Resistance to each product is monogenic
	No cross resistance occurs between products in the mixture
	Resistant individuals are rare in the population
	Products have equal persistence
	Some of the population remain untreated (refuge)
	Resistance is functionally recessive (only homozygotes survive exposure)
Mosaics	Susceptible individuals are maintained and able to move into surrounding patches
	Negative cross resistance or fitness costs associated with the resistance may be required
Rotations	The frequency of individuals resistant to one product will decline during application of the alternative product, which is true if there is negative cross resistance (rare), a substantial fitness cost associated with resistance, or immigration of susceptible individuals occurs
High-dose strategy	Assumes complete coverage, effective kill of all individuals, ignores negative effects on secondary pests and their natural enemies or the environment

into IPM programs. The use of mutagenesis and recombinant DNA techniques could improve the efficiency of such genetic improvement projects. However, development of pesticide-resistant natural enemies is time consuming and expensive and should not be considered before exploring other, less expensive, options for IPM and pesticide resistance management.

Pesticides are powerful and effective pest management tools. Some can be highly selective, rapid in their impact, adaptable to many situations, and relatively economical. Thus, preserving pesticides is essential. Effective paradigms for resistance management are not yet deployed because resistance management and IPM have been considered separate issues. Effective programs will acknowledge the role of biological control, host plant resistance, cultural controls, and biorational controls such as mating disruption, insect growth regulators, and mass trapping. A key issue should always be whether pesticides can be used in a precise and selective manner without disrupting natural enemies.

Marjorie A. Hoy
University of Florida
Gainesville, Florida, USA

References
Croft, B. A. 1990. *Arthropod biological control agents and pesticides.* Wiley-Interscience Publishers, New York, New York.

Georghiou, G. P., and T. Saito (eds.). 1983. *Pest resistance to pesticides.* Plenum Press, New York, New York.

Hoy, M. A. 1998. Myths, models and mitigation of resistance to pesticides. *Philosophical Transactions of the Royal Society of London B* 252: 1787–1795.

Hull, L. A., and E. H. Beers. 1985. Ecological selectivity: modifying chemical control practices to preserve natural enemies. pp. 103–122 in M. A. Hoy and D. C. Herzog (eds.), *Biological control in agricultural IPM systems.* Academic Publishers, Orlando, Florida.

National Academy of Sciences. 1986. *Pesticide resistance: strategies and tactics for management.* National Academy Press, Washington, DC.

Roush, R. T., and B. E. Tabashnik (eds.). 1990. *Pesticide resistance in arthropods.* Chapman and Hall, New York, New York.

Tabashnik, B. E., and B. A. Croft. 1985. Evolution of pesticide resistance in apple pests and their natural enemies. *Entomophaga* 30: 37–49.

PEST RESURGENCE.
Increase in abundance of pests, rapidly and sometimes to higher than previous levels, following actions made to suppress them.

PESTS AND THEIR NATURAL ENEMIES (PARASITOIDS AND/OR PREDATORS) IN THE MIDDLE EAST.
The Middle East is a geographic region comprising a territory in which Asia, Africa, and Europe converge and which includes the Mediterranean Sea, the Red Sea, and the Persian Gulf. To the south, the Sahara Desert divides it from Tropical Africa; to the north, its outer limits lie in the latitude of the Black and Caspian Seas. On the east it extends as far as the Indian subcontinent, while its western limits lie at about the longitude of the Aegean Sea.

Agriculture is the most important activity in the Middle East in terms of the number of people it employs; in some countries it provides a substantial part of the gross domestic economy. Rainfed farming is possible only in these limited areas that receive more than 8 inches (800 mm) of rain annually, and then only for winter crops, because the summer is dry everywhere. A perennial supply of irrigation water is available from the Middle East's two great river systems, the Nile in Egypt and the Tigris-Euphrates system in Iraq and Syria. The agriculture these rivers support is highly productive. Irrigation water is also obtainable from underground water resources. These include renewable ground water resources, continually resupplied from precipitation and subterranean flow, and nonrenewable resources such as the fossil water deposits of the Sahara and Arabian peninsula. Another important agricultural resource of the Middle East is the region's grazing lands.

The largest vegetable production area in the world is in Asia, with 61% of total production and an annual growth rate of 5.1% per year. Over 100 types of vegetable crops are grown and consumed. Developing countries account for 61% of world vegetable production. China has the highest proportion of vegetable production, with South Asia is second with 19.4% of developing countries' vegetable production, followed by the Near East (8.5%) and East and South East Asia (9.1%). In terms of income generation, vegetable production has an important economic significance in developing countries. In many developing countries, the return per hectare for vegetable crops is twice the return for rice crops. Small-holder producers with limited mechanization carry out most production. This also increases the opportunities for on-farm employment. In 1996, developing countries accounted for 72% of world fresh fruit production.

According to FAO (1998), citrus is the most important fruit crop with world production at 87 million tons and increasing at an annual rate of 3.4% per year. Bananas are the second most important fruit commodity with 82 million tons and an annual growth increase of 2.7%. Most of the world banana production occurs in the developing countries of Latin America, Asia and Africa. Tomatoes are the third most important crop with 80 million tons of annual production and an annual production increase of 2.9% per year. Grapes are the fourth most important fruit crop with 55 million tons of annual production. Apples are the fifth most important fruit crop with 49 million tons of annual production.

Following is a list of key pests, sorted by crop known to be important in the Middle East. Indented beneath each pest is a list of natural enemies of each pest found in this region, and where they are known to occur.

Key insect and mite pests and their determined natural enemies in the Middle East

---

[a] Information for each crop is arranged as follows: name of the crop; scientific name of pest(s); scientific name of natural enemy(ies) (parasitoid or predator), if available; country where species were encountered and studied.
[b] Reference in Arab Journal of Plant Protection.
[c] Seventh Arab Congress of Plant Protection. 22–26 October 2000. Amman, Jordan. Abstract Book.
[d] Gerling, D. 1990. Whiteflies: their Bionomics, Pest Status and Management.

Alliaceae
Garlic

Thrips tabaci Lindeman (Thysanoptera: Thripidae); Egypt[c], Lebanon

Onion

Thrips tabaci Lindeman (Thysanoptera: Thripidae); Jordan[c], Saudi Arabia

Anacardiaceae
Pistachio tree

Agonoscena targionii (Licht.) (Hemiptera: Psyllidae)
  Adonia variegata Goetze (Coleoptera: Coccinellidae); Lebanon, Syria
  Anthocorus minki Dohrm (Hemiptera: Anthocoridae); Syria
  Camphylomma sp. (Hemiptera: Miridae); Syria
  Chilocorus bipustulatus (L.) (Coleoptera:Coccinellidae); Jordan, Lebanon, Syria
  Chrysoperla carnea (Stephens) (Neuroptera: Chrysopidae); Jordan, Syria

  Coccinella novemnotata Herbst (Coleoptera:Coccinellidae); Syria
  Coccinella septempunctata (L.) (Coleoptera:Coccinellidae); Iraq, Lebanon, Syria
  Coccinella tredecimpunctata (L.) (Coleoptera: Coccinellidae); Syria
  Coccinella undecimpunctata (L.) (Coleoptera: Coccinellidae); Iraq, Lebanon, Syria
  Geocoris sp. (Hemiptera: Lygaeidae); Syria
  Hyperaspis quadrimaculatus Redt. (Coleoptera: Coccinellidae); Syria, Turkey
  Nabis ferrus L. (Hemiptera: Nabidae); Syria
  Orius horvathi (Reuter) (Hemiptera: Anthocoridae); Syria
  Psyllaephagus pistaciae (Hymenoptera: Encyrtidae); Syria
  Rhizophagus sp. (Coleoptera: Rizophagidae); Syria
  Scymnus sp. (Coleoptera: Coccinellidae); Syria
  Stethorus sp. (Coleoptera: Coccinellidae); Syria
  Syrphus sp. (Diptera: Syrphidae); Lebanon, Syria

Capnodis cariosa Pall. (Coleoptera: Buprestidae); Lebanon, Syria

Hylesinus vestitus M. & R. (Coleoptera: Curculionidae: Scolytinae); Syria[c]
  Cheiropachus quadrum F. (Hymenoptera: Pteromalidae); Syria
  Clerus mutillarius Fisch. (Coleoptera: Cleridae); Syria
  Dendrosoter protuberans (Nees) (Hymenoptera: Braconidae); Syria
  Denops albofasciata K. (Coleoptera: Cleridae); Syria
  Iponemus sp. (Acari: Tarsonemidae); Syria, Lebanon, Turkey
  Tarsonemus sp. (Acari: Tarsonemidae); Syria
  Thansimus formicarius (L.) (Coleoptera:Cleridae); Syria

Thaumetopoea solitaria (Fr.) (Lepidoptera:Thaumetopoeidae); Iraq, Iran, Lebanon, Syria, Turkey

Arecaceae/Palmae
Palm

Apate monachus F. (Coleoptera: Bostrychidae); Algeria[c]
Arenipses sabella Hampson (Lepidoptera: Pyralidae); Algeria, Iran, Iraq, Libya, Saudi Arabia
Batrachedra amydraula Meyrick (Lepidoptera: Cosmopterygidae); Egypt, Iraq, Libya, Saudi Arabia
Ecotomyelois ceratoniae (carob moth)

Habrobracon hebetor Say (Hymenoptera: Braconidae); Tunisia[c]

Ephestia cautella (Walker) (Lepidoptera:Pyralididae); Iraq and Libya[c]

  Bracon hebetor Say. (Hymenoptera: Braconidae); Iraq[c]

Icerya purchasi Mask. (Hemiptera: Margarodidae); Libya

Jebusea hammerschmidti Reiche (Coleoptera: Cerambycidae); Iraq, Saudi Arabia

Ommatissus binotatus Berg. (Hemiptera: Tropiduchidae); Algeria, Egypt, Libya

  Chilocorus bipustulatus (L.) (Coleoptera: Coccinellidae); Iraq

  Chrysopa carnea Steph. (Neuroptera: Chrysopidae); Iraq

  Coccinella septempunctata L. (Coleoptera: Coccinellidae); Iraq

  Coccinella undecimpunctata L. (Coleoptera: Coccinellidae); Iraq

Oryctes elegans Prell (Coleoptera: Scarabaeidae); Iraq[c] & Qatar[c], Saudi Arabia

Paratetranychus afrasiaticus (McG.) (Acari: Tetranychidae); Saudi Arabia

Parlatoria blanchardi Targioni-Tozzetti (Hemiptera: Diaspididae); Egypt, Iran, Iraq[b], Jordan, Libya, Palestine, Saudi Arabia

Phoenococcus marlatti Cock (Hemiptera:Phoenicococcidae); Libya

Phonapate frontalis Farhr. (Coleoptera:Bostrychidae); Iraq, Libya, Saudi Arabia

Planococcus citri (Risso) (Hemiptera: Pseudococcidae); Libya

Pseudophilus testaceus Gahan (Coleoptera:Curculionidae); Iraq & Qatar[c]

Rhynchophorus ferrugineus Fabricius (Coleoptera: Curculionidae); Egypt[c], Qatar[c], & Saudi Arabia

Schistocerca gregaria Forsk. (Orthoptera: Acrididae); Iraq, Jordan, Lebanon, Libya, Palestine, Saudi Arabia, Syria, Turkey

Brassicaceae (Cruciferae)
Cabbage

Aporia crataegi (L.) (Lepidoptera: Pieridae)

  Apanteles (*Cotesia*) *glomeratus* L. (Hymenoptera: Braconidae); Lebanon, Syria

Bemisia tabaci Genn. (Hemiptera/Hemiptera: Aleyrodidae)

  Clitostethus arcuatus Rossi (Coleoptera: Coccinellidae); Syria[c]

Brevicoryne brassicae L. (Hemiptera: Aphididae)

  Aphidius brassicae March. (Hymenoptera: Braconidae); Lebanon

  Diaeretiella rapae Mc'Intosh (Hymenoptera: Braconidae); Jordan, Lebanon, Syria

Pieris brassicae L. (Lepidoptera: Pieridae)

  Cotesia glomeratus L. (Hymenoptera: Braconidae); Lebanon, Syria

  Brachymeria intermedia Nees (Hymenoptera: Chalcididae); Lebanon

  Compsilura sp. (Diptera: Tachinidae); Syria

  Hyposoter sp. (Hymenoptera: Ichneumonidae); Syria

Spodoptera littoralis (Boisd.) (Lepidoptera: Noctuidae); Egypt, Lebanon, Iraq[b]

Cauliflower

Bemisia tabaci Genn. (Hemiptera: Aleyrodidae)

  Clitostethus arcuatus Rossi (Coleoptera: Coccinellidae); Syria[b]

Spodoptera littoralis Boisduval (Lepidoptera: Noctuidae); Iraq[b]

Kohlrabi (*Brassica oleraceae* var. *caularapa*)

Atherigona orientalis (Schinner) (Diptera: Muscidae); Iraq[b]

Baris opiparis Jacquilin DuVal (Coleoptera: Curculionidae); Iraq[b]

Hellula undalis (Fabricius) (Lepidoptera: Pyralidae); Iraq[b]

Chenopodiaceae
Sugarbeet

Agrotis segetum (Schiff) (Lepidoptera: Noctuidae); Lebanon, Syria

Spodoptera exigua (Hübner) (Lepidoptera: Noctuidae); Lebanon, Iraq, Syria

Cucurbitaceae
Cucumber

Aphis gossypii Glover (Hemiptera: Aphididae)

  Aphidius colemani Vierck (Hymenoptera: Aphidiidae); Algeria[c]

  Egypt[c], Lebanon, Saudi Arabia, Syria

Bemisia tabaci Genn. (Hemiptera: Aleyrodidae); Egypt[c], Lebanon, Syria[b]

Empoasca decipiens Paoli (Hemiptera: Cicadellidae); Egypt[c], Jordan, Lebanon, Syria

    Henosepilachna elaterii Rossi (Coleoptera: Coccinellidae); Sudan[b]

Liriomyza cicerina Rond. (Diptera: Agromyzidae)

    Diglyphus isaea (Walker) (Hymenoptera: Eulophidae); Syria

Liriomyza huidobrensis (Blanchard) (Diptera: Agromyzidae)

    Diglyphus isaea (Walker) (Hymenoptera: Eulophidae); Syria

    Hemiptarsenus sp. (Hymenoptera: Eulophidae); Syria

    Neochrysocharis formosa (Westwood) (Hymenoptera: Eulophidae); Jordan, Lebanon, Syria

    Halticoptera sp. (Hymenoptera: Pteromalidae); Syria

Liriomyza trifolii (Burgess) (Diptera: Agromyzidae)

    Diglyphus isaea (Walker) (Hymenoptera: Eulophidae); Syria

Tetranychus urticae Koch (Acari: Tetranychidae)

    Phytoseiulus persimilis (Acarina: Phytoseiidae); Lebanon & Syria[b]

Thrips tabaci Lindeman (Thysanoptera: Thripidae); Egypt[c], Syria[b]

Trialeurodes vaporariorum Westwood (Hemiptera: Aleyrodidae); Syria[b]

Melon

Acytopeus (Baris) *curvirostris* (Tourn.) (Coleoptera: Curculionidae)

    Sarcophaga sp. (Diptera: Sarcophagidae); Saudi Arabia

Aphis gossypii Glover (Hemiptera: Aphididae)

    Aphelinus gossypii (Timberlake) (Hymenoptera: Aphelinidae); Jordan[c], Lebanon, Syria

Dacus ciliatus Lw. (Diptera: Tephritidae); Egypt Lebanon, Saudi Arabia, Syria, Sudan

Eudioptes (*Margaronia*) *indica* Sn. (Lepidoptera: Pyralidae); Saudi Arabia, Syria

Myopardalis pardalina Bigot (Diptera: Tephritidae); Lebanon, Syria

Squash

Aphis gossypii Glover (Hemiptera: Aphididae); Egypt[c], Lebanon, Syria

Bemisia tabaci Genn. Genn. (Hemiptera: Aleyrodidae); Egypt[c], Lebanon, Syria

Empoasca decipiens Paoli (Hemiptera: Cicadellidae); Egypt[c], Lebanon, Syria

Watermelon

Acytopeus (Baris) *curvirostris* (Tourn.) (Coleoptera: Curculionidae)

    Sarcophaga sp. (Diptera: Sarcophagidae); Saudi Arabia

Aphis gossypii Glover (Hemiptera: Aphididae)

    Aphelinus gossypii (Timberlake) (Hymenoptera: Aphelinidae); Jordan[c], Lebanon, Syria

Trichoplusia ni (Hübner) (Lepidoptera: Noctuidae); Lebanon

Fagaceae
Cork oak

Cerambyx cerdo L. (Coleoptera: Cerambycidae); Algeria[c]

Platypus cylindrus F. (Coleoptera: Platypodidae); Algeria[c]

Oak

Malacosoma neustria L. (Lepidoptera: Lasiocampidae); Libya[c]

Gramineae/Poaceae
Barley

Cephus pygmaeus L. (Hymenoptera: Cephidae)

    Collyria coxator Villers (Hymenoptera: Ichneumonidae); Syria[b]

    Collyria orientator Aubert (Hymenoptera: Ichneumonidae); Syria[b]

    Bracon terebrella Wesmael (Hymenoptera: Braconidae); Syria[b]

Cnephasia pumicana Zeller (Lepidoptera: Tortricidae); Syria[c]

Diuraphis noxia Kurdjumov (Hemiptera: Aphididae)

    Syrphus sp. (Diptera: Syrphidae); Syria[b]

    Leucopis sp. (Diptera: Chamaemyiidae); Syria[b]

    Coccinella septempunctata L. (Coleoptera: Coccinellidae); Syria[b]

    Diaretiella rapae (Hymenoptera: Aphidiidae); Syria[b]

    Praon sp. (Hymenoptera: Aphidiidae); Syria[b]

Aphidius colemani (Hymenoptera: Aphidiidae); Syria[b]

Eurygaster integriceps Puton (Hemiptera: Scutelleridae); Iran[b], Iraq, Syria[c], Turkey

  Asolcus sp. (Hymenoptera, Scelionidae); Iran, Syria

  Microphanurus vasilievi (Mayr) (Hymenoptera); Iran, Syria

Haplothrips tritici Kurd (Thysanoptera: Phlaeothripidae); Syria[c]

Oria musculosa Hübner (Lepidoptera: Noctuidae); Syria[c]

Porphyrophora tritici Bodenheimer (Hemiptera: Margarodidae); Syria[b]

Rhopalosiphum maidis Fitch (Hemiptera:Aphididae); Iraq, Lebanon, Syria[c]

Rhopalosiphum padi L. (Hemiptera: Aphididae); Iraq[b]

Schizaphis graminum (Rondani) (Hemiptera:Aphididae); Iraq

Sitotroga cerealella (Olivier) (Lepidoptera:Gelechiidae); Iraq

Tracheus judaicus Konow (Hymenoptera: Cephidae)

  Collyria coxator Villers (Hymenoptera:Ichneumonidae); Syria[b]

  C. orientator Aubert (Hymenoptera: Ichneumonidae); Syria[b]

  Bracon terebrella Wesmael (Hymenoptera: Braconidae); Syria[b]

Trachelus libanensis Andre (Hymenoptera:Cephidae)

  Collyria coxator Villers (Hymenoptera:Ichneumonidae); Syria[b]

  C. orientator Aubert (Hymenoptera: Ichneumonidae); Syria[b]

  Bracon terebrella Wesmael (Hymenoptera: Braconidae); Syria[b]

Corn

Cicadulina chinai Ghauri (Hemiptera: Cicadellidae); Egypt[c]

Cicadulina bipuncellazea China (Hemiptera:Cicadellidae); Egypt[c]

Empoasca descipiens Paoli (Hemiptera: Cicadellidae); Egypt[c], Jordan, Lebanon, Syria

Empoasca decedens Paoli (Hemiptera: Cicadellidae); Egypt[c]

Ostrinia (*Pyrausta*) *nubilalis* (Hübner) (Lepidoptera: Pyrlidae); Lebanon

Sogatella vibix (Haupt) (Hemiptera: Delphacidae); Egypt[c]

Sogatella furicifera Horv (Hemiptera: Delphacidae); Egypt[c]

Rhopalosiphum maidis Fitch (Hemiptera: Aphididae)

  Coccinella septempuncata L. (Coleoptera: Coccinellidae); Iraq, Lebanon, Saudi Arabia, Syria

Sesamia cretica Led. (Lepidoptera: Agrotidae); Jordan, Saudi Arabia

  Bracon spp. (Hymenoptera: Braconidae); Iraq, Syria

  Coccinella septempuncata L. (Coleoptera: Coccinellidae); Iraq, Lebanon, Syria

  Coccinella undecimpuncata L. (Coleoptera: Coccinellidae); Iraq, Lebanon, Syria

  Orius albidepennis (Reut.) (Hemiptera: Anthocoridae); Iraq

  Platytelenomus hylas Nixon (Hymenoptera: Scelionidae); Egypt[c]

  Telenomus sp. (Hymenoptera: Scelionidae); Iraq

Tribolium castaneum (Herbst) (Coleoptera: Tenebrionidae)

  Platytelenomus hylas Nixon (Hymenoptera: Scelionidae); Iraq[b]

Rice

Sitophilus oryzae L. (Coleoptera: Curculionidae); Iraq[b]

Trogoderma granarium Everts (Coleoptera: Dermestidae); Iraq[b]

Sorghum

Atherigona yorki Deeming (Diptera: Muscidae); Yemen[b]

Chilo partellus Swinhoe (Lepidoptera: Pyralidae); Yemen[b]

Sesamia cretica Lederer (Lepidoptera: Noctuidae); Yemen[b]

Sugarcane

Chilo agamemnon Bleszynski (Lepidoptera: Pyralidae)

  Trichogramma evanescens West. (Hymenoptera: Tricogrammatidae); Egypt[b]

Wheat

Cephus pygmaeus L. (Hymenoptera: Cephidae)

  Collyria coxator Villers (Hymenoptera: Ichneumonidae); Syria[b]

  Collyria orientator Aubert (Hymenoptera: Ichneumonidae); Syria[b]

Bracon terebrella Wesmael (Hymenoptera: Braconidae); Syria[b]

Diuraphis noxia (Mordvilko) (Hemiptera: Aphididae)

  Coccinella septempuncata L. (Coleoptera: Coccinellidae); Lebanon and Syria[b]

  Leucopis sp. (Chamaemiiydae: Diptera); Lebanon, Syria

  Syrphus sp. (Diptera: Syrphidae); Lebanon, Syria

  Aphidius colemani Viereck (Hymenoptera: Braconidae); Lebanon, Syria

  Diaeretiella rapae Mc'Intosh (Hymenoptera: Braconidae); Jordan, Lebanon, Syria

  Praon sp. (Hymenoptera: Braconidae); Lebanon, Syria

Eurygaster integriceps Puton (Hemiptera: Scutelleridae); Iraq, Lebanon, Turkey

  Asolcus sp. (Hymenoptera: Scelionidae); Iran, Syria

  Microphanurus vasilievi (Mayr) (Hymenoptera); Iran, Syria

  Trissolcus sp. (Hymenoptera: Scelionidae); Syria[c]

Exaeretopus tritici Williams (Hemiptera: Coccidae); Iraq[b]

Lytta vesicatoria (L.) (Coleoptera: Meloidae); Algeria[c]

Mayetiola destructor (Say) (Diptera: Cecidomyiidae); Morocco, Tunisia[c]

Metopolophium dirhodum (Walker) (Hemiptera: Aphididae)

  Coccinella septempuncata L. (Coleoptera: Coccienllidae); Lebanon, Syria

Oria musculosa Hübner (Lepidoptera: Noctuidae); Syria

Oulema melanopus (L.) (Coleoptera: Chrysomelidae); Syria[b]

Rhopalosiphum padi (L.) (Hemiptera: Aphididae)

  Coccinella septempuctata L. (Coleoptera: Coccinellidae); Syria[c]

  Harmonia axyridis Pallas (Coleoptera: Coccinellidae); Syria[c]

Sitobion avenae (Fabricius) (Hemiptera: Aphididae)

  Coccinella septempuncata L. (Coleoptera: Coccinellidae); Lebanon, Syria

Trachelus judaicus Konow (Hymenoptera: Cephidae)

  Collyria coxator Villers (Hymenoptera: Ichneumonidae); Syria[b]

  Collyria orientator Aubert (Hymenoptera: Ichneumonidae); Syria[b]

  Bracon terebrella Wesmael (Hymenoptera: Braconidae); Syria[b]

Trachelus libanensis Andre (Hymenoptera: Cephidae)

  Collyria coxator Villers (Hymenoptera: Ichneumonidae); Syria[b]

  Collyria orientator Aubert (Hymenoptera: Ichneumonidae); Syria[b]

  Bracon terebrella Wesmael (Hymenoptera: Braconidae); Syria[b]

Leguminoseae
Alfalfa

Sitona discoideus Gyllenhal (Coleoptera: Curculionidae)

  Anaphes diana (Girault) (Hymenoptera: Mymaridae); Syria

Sitona lineatus (L.) (Coleoptera: Curculionidae); Lebanon, Syria

Spodoptera exigua (Hübner) (Lepidoptera: Noctuidae)

  Chrysopa vulgaris L. (Hymenoptera: Chrysopidae); Libya[b]

  Coccinella novemnotata Herbst (Coleoptera: Coccinellidae); Libya[b]

  Coccinella septumpunctata L. (Coleoptera: Coccinellidae); Libya[b]

  Coccinella undecimpunctata L. (Coleoptera: Coccinellidae); Libya[b]

  Hippodamia tredecimpunctata tibialis (Say) (Coleoptera: Coccinellidae); Libya[b]

  Nabis ferrus L. (Hemiptera: Nabidae); Libya[b]

  Syrphus corollae Fabricius (Diptera: Syrphidae); Libya[b]

  Tripolitanus sp. (Diptera: Asilidae); Libya[b]

Spodoptera littoralis (Boisd.) (Lepidoptera: Noctuidae); Egypt, Iraq, Lebanon, Saudi Arabia, Syria

Therioaphis maculata (Buckton) (Hemiptera: Aphididae)

  Chrysopa vulgaris L. (Hymenoptera: Chrysopidae); Libya[b]

  Coccinella novemnotata Herbst (Coleoptera: Coccinellidae); Libya[b]

  Coccinella septumpunctata L. (Coleoptera: Coccinellidae); Libya[b]

  Hippodamia tredecimpunctata (L.) (Coleoptera: Coccinellidae); Libya[b]

  Coccinella undecimpunctata L. (Coleoptera: Coccinellidae); Libya[b]

Nabis ferrus L. (Hemiptera: Nabidae); Libya[b]

Syrphus corollae Fabricius (Diptera: Syrphidae); Libya[b]

Tripolitanus sp. (Diptera: Asilidae); Libya[b]

Therioaphis trifolii Monell (Hemiptera: Aphididae)

  Aphelinus asychis Walker (Hymenoptera: Aphelinidae); Syria

  Aphelinus semiflavus Howard (Hymenoptera: Aphelinidae); Syria

  Coccinella septumpunctata L. (Coleoptera: Coccinellidae); Iraq, Lebanon, Syria

  Malacocoris sp. (Hemiptera: Miridae); Syria

  Praon exoletum (Needs) (Hymenoptera: Aphidiidae); Lebanon, Syria

  Praon palitans Muesebeck (Hymenoptera: Aphidiidae); Lebanon, Palestine, Saudi Arabia

  Trioxys complanatus Quilis Pérez (Hymenoptera: Aphidiidae); Lebanon

  Trioxys utilis Muesebeck (Hymenoptera: Aphidiidae); Lebanon, Palestine, Syria

Beans

Aphis craccivora Koch (Hemiptera: Aphididae); Egypt[b]

Aphis fabae Scopoli (Hemiptera: Aphididae)

  Coccinella septempunctata L. (Coleoptera: Coccinellidae); Syria[c]

  Harmonia axyridis Pallas (Coleoptera: Coccinellidae); Syria[c]

  Chrysoperla carnea (Stephens) (Neuroptera: Chrysopidae); Jordan, Syria

  Didea sp. (Diptera: Syrphidae); Syria

  Epistrophella sp. (Diptera: Syrphidae); Syria

  Episyrphus balteatus De Geer (Diptera: Syrphidae); Lebanon, Syria

  Eristalis sp.; Syria

  Lysiphlebus confusus Tremblay & Eady (Hymenoptera: Aphidiidae); Lebanon

  Melanostoma sp. (Diptera: Syrphidae); Syria

  Scaeva sp. (Diptera: Syrphidae); Syria

  Scymnus apetzi Mulsant (Coleoptera: Coccinellidae); Jordan, Lebanon, Syria

  Trioxys angelicae (Haliday) (Hymenoptera: Aphidiidae); Lebanon

Bemisia tabaci (Gennadius) (Hemiptera: Aleyrodidae); Iraq, Jordan, Lebanon

  Clitostethus arcuatus Rossi (Coleoptera: Coccinellidae); Syria[c]

Liriomyza cicerina Rondani (Diptera: Agromyzidae)

Diglyphus isaea (Walker) (Hymenoptera: Eulophidae); Syria

Liriomyza huidobrensis (Blanchard) (Diptera: Agromyzidae)

  Chrysocharis orbicularis (Nees) (Hymenoptera: Eulophidae); Lebanon

  Crossopalpus sp. (Diptera: Empididae); Syria

  Diglyphus isaea (Walker) (Hymenoptera: Eulophidae); Syria

  Halticoptera sp. (Hymenoptera: Pteromalidae); Syria

  Hemiptarsenus sp. (Hymenoptera: Eulophidae); Syria

  Neochrysocharis formosa (Westwood) (Hymenoptera: Eulophidae); Jordan, Lebanon, Syria

  Nordlanderia sp. (Hymenoptera: Eucoilidae); Lebanon

  Opius sp. (Hymenoptera: Braconidae); Syria

  Pediobius acantha (Walker) (Hymenoptera: Eulophidae); Syria

  Platypalpus sp. (Diptera: Empididae); Syria

Liriomyza trifolii (Burgess) (Diptera: Agromyzidae)

  Diglyphus isaea (Walker) (Hymenoptera: Eulophidae); Syria

Lixus algirus L. (Coleoptera: Curculionidae); Syria[b]

Melanagromyza (*Agromyza*) *phaseoli* Coq. (Diptera: Agromyzidae); Lebanon

Rhopalosiphum padi L. (Hemiptera: Aphididae)

  Coccinella septempuctata L. (Coleoptera: Coccinellidae); Syria[c]

  Harmonia axyridis Pallas (Coleoptera: Coccinellidae); Syria[c]

Sitona crinitus H. (Coleoptera: Curculionidae); Syria[c]

Tetranychus urticae Koch (Acari: Tetranychidae)

  Phytoseiulus persimilis (Acarina: Phytoseiidae); Lebanon & Syria[b]

Trialeurodes ricini Misra (Hemiptera: Aleyrodidae)

  Clitostethus arcuatus Rossi (Coleoptera: Coccinellidae); Jordan[c]

Chickpeas

Heliocoverpa (*Heliothis*) *armigera* Hb. (Lepidoptera: Noctuidae); Jordan, Lebanon

  Cynopterus sp. (Hymenoptera: Braconidae); Syria

  Bracon hebetor (Say.) (Hymenoptera: Braconidae); Iraq, Syria

  Campoletis sp. (Hymenoptera: Ichneumonidae); Syria

Liriomyza cicerina Rondani (Diptera: Agromyzidae)

Opius sp. (Hymenoptera: Braconidae); Syria

Dygliphus isaea (Walker) (Hymenoptera: Eulophidae); Syria

Lixus algirus L. (Coleoptera: Curculionidae); Syria[b]

Sitona lineatus (L.) (Coleoptera: Curculionidae); Lebanon, Syria[b]

Lentils

Heliothis spp. (Lepidoptera: Noctuidae); Syria[b]

Liriomyza cicerina Rondani (Diptera: Agromyzidae); Syria[b]

Lixus algirus L. (Coleoptera: Curculionidae); Syria[b]

Sitona lineatus (L.) (Coleoptera: Curculionidae); Lebanon, Syria[b]

Smynthurodes betae Westwood (Hemiptera: Aphididae)

    Chrysotoxum intermedium Meigen (Diptera: Syrphidae); Syria, Turkey

    Hyperaspis quadrimaculatus Redt. (Coleoptera: Coccinellidae); Syria, Turkey

Soybeans

Aphis gossypii (Glover) (Hemiptera:f Aphididae); Egypt[b]

Bemisia tabaci Gennadius (Hemiptera: Aleyrodidae); Egypt[b]

Earias insulana Boisduval (Lepidoptera: Noctuidae); Egypt[b]

Empoasca lybica De Bergevin (Hemiptera: Cicadellidae); Egypt[b]

Platyedra (*Pectinophora*) *gossypiella* (Saunders) (Lepidoptera: Gelechiidae); Egypt[b]

Spodoptera littoralis Boisduval (Lepidoptera: Noctuidae); Egypt[b]

Tetranychus cinnabarinus (Boisduval) (Acari: Tetranychidae); Egypt[b]

Lythracaea
Pomegranate

Aphis punicae Passerini (Hemiptera: Aphididae)

    Aphidius sp. (Hemiptera: Aphidiidae); Libya[c]

    Coccinella novemnotata Herbst (Coleoptera: Coccinellidae); Libya[c]

    Scymnus nubilus Mulsant (Coleoptera: Coccinellidae); Libya[c]

    Scymnus syriacus Mars. (Coleoptera: Coccinellidae); Libya[c]

Syrphus corollae Fab. (Diptera: Syrphidae); Libya[c]

Ectomyelois ceratoniae Zeller (Lepidoptera: Pyralidae)

    Apanteles angaleti Muesebeck (Hymenoptera: Braconidae); Iraq[c]

Siphoninus phillyreae (Haliday) (Hemiptera: Aleyrodidae)

    Eretmocerus sp. (Hymenoptera: Aphelinidae); Jordan, Syria

    Encarsia formosa Gahan (Hymenoptera: Aphelinidae); Jordan, Syria

Tenuipalpus punicae P. & B. (Acari: Tenuipalpidae); Iraq[b]

Virachola (*Deudorix*) *livia* Klug (Lepidoptera: Lycaenidae)

    Brachymeria sp. (Hymenoptera: Chalcididae); Jordan

Malvaceae
Cotton

Aphis gossypii Glov. (Hemiptera: Aphididae)

    Aphelinus gossypii (Temberlack) (Hymenoptera: Aphelinidae); Jordan

    Geocoris megacephalus (Rossi) (Hemiptera: Lygaeidae); Syria

    Hyperaspis variegata Goeze (Coleoptera: Coccinellidae); Lebanon

    Orius laevigatus (Fieber) (Hemiptera: Anthocoridae); Syria

    Scymnus levaillanti Mulsant (Coleoptera: Coccinellidae); Jordan, Syria

    Scymnus quadriguttatus Capra (Coleoptera: Coccinellidae); Syria

    Scymnus subvillosus (Goeze) (Coleoptera: Coccinellidae); Jordan, Lebanon, Syria

    Trioxys angelicae (Haliday) (Hymenoptera: Aphidiidae); Lebanon

Bemisia tabaci Gennadius; Egypt[c]

    Campylomma diversicornis Reuter (Hemiptera: Miridae); Syria

    Deraeocoris punctulatus (Fallen) (Hemiptera: Miridae); Syria

    Encarsia lutea (Masi) (Hymenoptera: Aphelinidae); Jordan, Syria

    Eretmocerus mundus Mercet (Hymenoptera: Aphelinidae); Jordan, Syria

    Macrolophus sp. (Hemiptera: Miridae); Syria

Earias insulana Boisduval (Lepidoptera: Noctuidae); Egypt[b], Iraq, Iran, Turkey

Trichogramma brassicae Bezdenko (Hymenoptera: Trichogrammatidae); Syria

Helicoverpa armigera Hübner (Lepidoptera: Noctuidae)

Bracon (*Haprobracon*) *brevicornis* Wesmael (Hymenoptera: Braconidae); Syria

Haprobracon hebetor Say (Hymenoptera: Braconidae); Syria

Trichogramma brassicae Bezdenko (Hymenoptera: Trichogrammatidae); Syria

Trichogramma chilonis Ishii (Hymenoptera: Trichogrammatidae); Syria

Trichogramma principium Sugonyaev-Sorokina (Hymenoptera: Trichogrammatidae); Syria

Platyedra (*Pectinophora*) *gossypiella* (Saunders) (Lepidoptera: Gelechiidae); Egypt[b], Sudan, Syria

Spodoptera litoralis Boisduval (Lepidoptera: Noctuidae); Eygpt[b]

Tetranychus cinnabarinus (Boisduval) (Acari: Tetranychidae)

Chrysoperla carnea (Stephens) (Neuroptera: Chrysopidae); Syria

Exochomus pubscens Kuster (Coleoptera: Coccinellidae); Jordan

Okra

Bemisia tabaci Gennadius. (Hemiptera: Aleyrodidae)

Clitostethus arcuatus Rossi (Coleoptera: Coccinellidae); Syria[c]

Earias insulana Boisduval (Lepidoptera: Noctuidae); Lebanon, Saudi Arabia

Oleaceae
Olives

Acaudaleyrodes olivinus Silvestri (Hemiptera: Aleyrodidae)

Encarsia spp.; Jordan

Bactrocera (*Dacus*) *oleae* Gmelin (Diptera: Tephritidae); Libya[c]

Carpophilus mutilatus Er. (Coleoptera: Nitidulidae); Lebanon

Cyrtoptyx dacicida Masi (Hymenoptera: Pteromalidae); Lebanon

Cyrtoptyx latipes (Rondani) (Hymenoptera: Pteromalidae); Egypt[c]

Eupelmus sp. (Hymenoptera: Eupelmidae); Egypt[c]

Eupelmus urozonus Dal. (Hymenoptera: Eupelmidae); Lebanon

Eurytoma rosae Nees (Hymenoptera: Eurytomidae); Lebanon

Eurytoma martelli M. (Hymenoptera: Eurytomidae); Egypt[c], Syria

Macroneura sp. (Hymenoptera: Eupelmidae); Egypt[c]

Opius concolor Szépligeti (Hymenoptera: Braconidae); Egypt[c], Lebanon, Palestine, Syria

Platygaster sp. (Hymenoptera: Scelionidae); Lebanon

Pnigalio agraules W. (Hymenoptera: Eulophidae); Egypt[c]

Pnigalio mediterraneus Ferriere & Delucchi (Hymenoptera: Eulophidae); Lebanon, Syria

Prolasioptera berlesiana Paoli (Diptera: Cecidomyiidae); Lebanon

Synopeas sp. (Hymenoptera: Scelionidae); Lebanon

Clinodiplosis oleisuga Targ. (Diptera: Cecidomyiidae)

Inostema sp. (Hymenoptera: Platygasteridae); Syria, Jordan, Lebanon

Leptacis sp. (Hymenoptera: Platygasteridae); Syria

Dasyneura (*Perrisia*) *oleae* (Loew) (Diptera: Cecidomyiidae)

Aprostocetus sp. (Hymenoptera: Eulophidae); Jordan, Lebanon

Platygaster apicalis Thomas; Syria

Platygaster oleae Szeleny; Jordan, Syria

Euphyllura straminea Loginova (Hemiptera: Psyllidae)

Anthocoris nemoralis (Fabricius) (Hemiptera: Anthocoridae); Syria

Anthocoris nomorum L. (Hemiptera: Anthocoridae); Lebanon, Syria

Crysoperla carnea (Stephens) (Neuroptera: Chrysopidae); Jordan, Lebanon, Syria

Euphyllura olivina Costa (Hemiptera: Psyllidae)

Anthocoris nemoralis (Fabricius) (Hemiptera: Anthocoridae); Syria

Anthocoris nomorum L. (Hemiptera: Anthocoridae); Lebanon, Syria

Parlatoria oleae (Clovée) (Hemiptera: Diaspididae)

Chilocorus bipustulatus (L.) (Coleoptera: Coccinellidae; Jordan, Lebanon, Syria

Phloeotribus scarabeoides (Bern.) (Coleoptera: Curculionidae: Scolytinae); Lebanon, Morocco[c], Syria

Prays oleellus F. (Lepidoptera: Hyponomeutidae)

Ageniaspis fuscicollis (Dalman) (Hymenoptera: Encyrtidae); Lebanon

Apanteles dilectus Haliday (Hymenoptera: Braconidae); Lebanon

Chelonus elaeaphilus Silv. (Hymenoptera: Braconidae); Lebanon

Chrysocharis sp. (Hymenoptera: Eulophidae); Lebanon

Cirrospilus elongatus Boucek (Hymenoptera: Eulophidae); Lebanon

Elasmus flabellatus Fonscolombe (Hymenoptera: Elasmidae); Lebanon

Himertosoma sp. Schmiedeknecht (Hymenoptera: Ichneumonidae); Syria

Phytomyptera sp. Rond. (Diptera: Tachinidae); Lebanon

Trichogramma oleae Voegelé-Pointel (Hymenoptera: Trichogrammatidae); Syria

Saissetia oleae (Bern.) (Hemiptera: Coccidae)

Chilocorus bipustulatus (L.) (Coleoptera: Coccinellidae); Jordan, Lebanon, Syria

Scutellista cyanea Motschulsky (Hymenoptera: Ormyridae); Lebanon, Syria

Pedaliaceae
Sesame

Asphondylia sesami Felt (Diptera: Cecidomyiidae); Yemen[b]

Antigastra catalaunalis Duponchel (Lepidoptera: Pyralidae); Yemen[b]

Pinaceae
Cedars

Cephalcia tannourinensis Chevin (Hymenoptera: Pamphiliidae); Lebanon

Dasineura cedri Coutin (Diptera: Cecidomyiidae); Lebanon

Thaumetopoea pityocampa Schiff. (Lepidoptera: Thaumetopoeidae)

Anastatus bifasciatus Fonsc. (Hymenoptera: Eupelmidae); Lebanon, Morocco[b]

Baryscapus (*Tetrastichus*) *servadeii* (Dom.) (Hymenoptera: Eulophidae); Lebanon, Morocco[b]

Ooencyrtus pityocampae Mercet (Hymenoptera: Encyrtidae); Lebanon, Morocco[b]

Trichogramma embryophagum Hartig (Hymenoptera: Trichogrammatidae); Morocco[b]

Pine

Aspidiotes hederae (Vallot) (Hemiptera: Diaspididae)

Chilocorus bipustulatus (L.) (Coleoptera: Coccinellidae); Jordan, Lebanon, Syria

Eulachnus rileyi (Williams) (Hemiptera: Lachnidae); Iraq[b]

Thaumetopoea pityocampa (Den. and Schiff.) (Lepidoptera: Thaumetopoeidae)

Anastatus bifasciatus Fonsc. (Hymenoptera: Eupelmidae); Morocco[b]

Baryscapus (*Tetrastichus*) *servadeii* (Domenichini) (Hymenoptera: Eulophidae); Syria[c]

Calosoma sycophanta L. (Coleoptera: Carabidae); Lebanon, Syria

Ooencyrtus pityocampae (Mercet) (Hymenoptera: Encyrtidae); Morocco[b], Syria[c]

Phryxe caudata (Rondani) (Diptera: Tachinidae); Syria

Trichogramma embryophagum Hartig. (Hymenoptera: Trichogrammatidae); Morocco[b], Syria[c]

Thaumetopoea wilkinsoni Tams. (Lepidoptera:Thaumetopoeidae)

Anastatus bifasciatus Fonsc. (Hymenoptera: Eupelmidae); Lebanon[c]

Baryscapus (*Tetrastichus*) *servadeii* (Domenichini) (Hymenoptera: Eulophidae); Lebanon

Compsilura concinnata (Meigen) (Diptera: Tachinidae); Lebanon

Ooencyrtus pityocampae (Mercet) (Hymenoptera: Encyrtidae); Lebanon

Phryxe caudata (Rondani) (Diptera: Tachinidae); Lebanon

Rosaceae
Almond

Aporia crataegi L. (Lepidoptera: Pieridae)

Apanteles glomeratus (L.) (Hymenoptera: Braconidae); Lebanon, Syria

Microbracon kikpatricki Wilk. (Hymenoptera: Braconidae); Syria

Trichogramma semblidus (Aurivilius) (Hymenoptera: Trichogrammatidae); Syria

Brachycaudus amygdalinus (Schout.) (Hemiptera: Aphididae)

Aphidius matricariae Hal. (Hymenoptera: Aphidiidae); Lebanon, Syria

Aphidoletes aphidimyza (Rondani) (Diptera: Cecidomyiidae); Lebanon, Syria

Deraeocoris pallens (Reuter) (Hemiptera: Miridae); Lebanon

Exochomus quadripustulatus (L.) (Coleoptera: Coccinellidae); Lebanon, Syria

Leucopis sp. (Diptera: Chamaemyiidae); Lebanon, Syria

Scymnus subvillosus (Goeze) (Coleoptera: Coccinellidae); Jordan, Lebanon, Syria

Synharmonia conglobata (L.) (Coleoptera: Coccinellidae); Lebanon, Syria

Cimbex quadrimaculatus Muell. (Hymenoptera: Tenthredinidae); Cyprus, Lebanon, Syria, Turkey

Spilocryptis cimbicis Tschek. (Hymenoptera: Ichneumonidae); Palestine

Eriogaster amygdali talhouki Wilts. (Lepidoptera: Lasiocampidae); Lebanon

Eurytoma amygdali End. (Hymenoptera: Eurytomidae); Palestine

Ascogaster sp. (Hymenoptera: Braconidae); Syria

Mesochorus nigripes Ratz. (Hymenoptera: Ichneumonidae); Syria

Plastotorymus amygdali n.sp. (Hymenoptera: Torymidae); Jordan, Syria

Saltis sp. (Araneae); Lebanon

Syntomaspis sp. (Hymenoptera: Torymidae); Syria

Neurotoma nemoralis L. (Hymenoptera: Pamphiliidae)

Sinophorus (*Limnerium*) *crassifemur* (Thomson) (Hymenoptera: Ichneumonidae); Syria

Pterochloroides persicae (Cholod) (Hemiptera: Lachnidae); Lebanon, Syria

Pauesia sp. (Hymenoptera: Braconidae); Yemen[b]

Apple

Aphis pomi DeGeer (Hemiptera: Aphididae)

Adalia bipunctata (L.) (Coleoptera: Coccinellidae); Lebanon, Jordan, Syria

Adalia decempunctata (L.) (Coleoptera: Coccinellidae); Lebanon, Syria

Harmonia quadripunctata (Melsheimer) (Coleoptera: Coccinellidae); Jordan

Bryobia sp. (Acari: Tetranychidae)

Typhlodromus kettanehi Dosse (Acari: Phytoseiidae); Lebanon

Capnodis tenebrionis L. (Coleoptera: Buprestidae); Lebanon

Cenopalpus pulcher (Canestrini and Fanzago) (Acari: Tenuipalpidae)

Amblyseius finlandicus Ondemans (Acari: Phytoseiidae); Lebanon[b]

Phytoseius ocellatus Bayan (Acari: Phytoseiidae); Lebanon[b]

Typhlodromus invectus Chant (Acari:Phytoseiidae); Lebanon[b]

Typhlodromus kettanehi Dosse (Acari:Phytoseiidae); Lebanon[b]

Typhlodromus pyri Scheuten (Acari:Phytoseiidae); Lebanon[b]

Ceratitis capitata Weid. (Diptera: Tephritidae); Jordan, Lebanon, Syria[c]

Cydia pomonella L. (Lepidoptera: Olethreutidae)

Ascogaster quadridentata Wesml. (Hymenoptera: Braconidae); Syria[c]

Dibrachys cavus (Walker) (Hymenoptera:Pteromalidae); Lebanon

Itoplectis maculator F. (Hymenoptera:Ichneumonidae); Syria[c]

Liotryphon caudatus Ratz. (Hymenoptera:Ichneumonidae); Syria[c]

Microdus rufipes Nees (Hymenoptera:Braconidae); Syria[c]

Perilampus sp. (Hymenoptera: Perilampidae); Lebanon

Perilampus tritis Mayr (Hymenoptera: Perilampidae); Syria[c]

Pristomerus vulnerator Gravenhorst (Hymenoptera: Ichneumonidae); Syria[c]

Trichogramma sp. (Hymenoptera: Trichogrammatidae); Syria[c]

Trichomma enecator Rossi (Hymenoptera:Ichneumonidae); Syria[c]

Dysaphis plantaginea (Passerini) (Hemiptera:Aphididae); Lebanon

Dysaphis pyri Boy. (Hemiptera: Aphididae); Iraq[b]

Eriosoma lanigerum Hausm (Hemiptera:Aphididae)

Adalia bipunctata L. (Coleoptera: Coccinellidae); Jordan, Lebanon, Syria

Aphelinus mali (Haldeman) (Hymenoptera:Aphelinidae); Lebanon

Coccinella septempunctata L. (Coleoptera:Coccinellidae); Lebanon, Jordan, Syria

Oenopia conglobata (L.) (Coleoptera:Coccinellidae); Syria, Jordan

Leucoptera scitella Costa (Lepidoptera: Lyonetidae); Algeria[c]

Malacosoma neustria (L.) (Lepidoptera: Lasiocampidae); Libya[c]

Panonychus ulmi (Koch) (Acari: Tetranychidae)

Amblyseius andersoni Chant (Acari: Phytoseiidae); Algeria[c]

Phytoseiulus persimilis Athias-Henriot (Acari: Phytoseiidae); Algeria[c]

Phytoseius plumifer (C. & F.) (Acari: Phytoseiidae); Lebanon

Stethorus gilvifrons (Muslant) (Coleoptera: Coccinellidae); Iraq, Jordan, Lebanon, Syria

Zetzellia mali Oud. (Acari: Stigmaeidae); Lebanon, Syria

Zetzellia talhouki Dosse (Acari: Stigmaiedae); Lebanon

Pterochloroides persicae (Cholod) (Hemiptera: Lachnidae); Lebanon, Syria

 Pauesia sp. (Hymenoptera: Aphidiidae); Yemen[c]

Tetranychus urticae Koch (Acari: Tetranychidae)

 Amblyseius andersoni (Acari: Phytoseiidae); Algeria[c]

 Phytoseiilus persimilis Athias-Henriot (Acari: Phytoseiidae); Algeria[c]

 Stethorus gilvifrons (Muslant) (Coccinellidae); Iraq, Jordan, Lebanon, Syria

 Zetzellia spp. (Acari: Stigmaeidae); Lebanon, Syria

Tetranychus turkestani Ugarov & Nikolski (Acari: Tetranychidae)

 Stethorus gilvifrons (Muslant) (Coccinellidae); Iraq, Jordan, Lebanon, Syria

 Tydeus californicus Banks (Acari: Tydeidae); Lebanon[b]

Zeuzera pyrina (L.) (Lepidoptera: Cossidae) Algeria[c], Lebanon, Libya[c], Syria

Peach

Anarsia lineatella Zell. (Lepidoptera: Gelechidae); Palestine, Syria

 Aetecerus discolor Wesm. (Hymenoptera: Ichneumonidae); Lebanon

Aulacorthum solani (Kaltenbach) (Hemiptera: Aphididae)

 Aphidius matricariae Hal. (Hymenoptera: Aphidiidae); Lebanon, Syria

Bactrocera zonata (Saunders) (Diptera: Tephritidae); Egypt, Middle East[c]

Brachycaudus amygdalinus (Schout.) (Hemiptera: Aphididae)

 Episyrphus balteatus (De Geer) (Diptera: Syrphidae); Lebanon, Syria

Brachycaudus helichrysi (Kalt.) (Hemiptera: Aphididae); Iraq, Jordan, Saudi Arabia

Episyrphus balteatus (De Geer) (Diptera: Syrphidae); Lebanon, Syria

Ceratitis capitata Weid. (Diptera: Tephritidae); Egypt, Jordan, Lebanon, Syria[c]

Myzus persicae (Hemiptera: Aphididae)

 Praon sp. (Hymenoptera: Aphidiidae); Syria

 Syrphus ribessii (L.) (Diptera: Syrphidae); Lebanon, Syria

 Trioxys angelicae (Haliday) (Hymenoptera: Aphidiidae); Lebanon, Tunisia[c]

Neurotoma nemoralis L. (Hymenoptera: Pamphiliidae)

 Sinophorus (*Limnerium*) *crassifemur* (Thomson) (Hymenoptera: Ichneumonidae); Syria

Pterochloroides persicae (Cholod) (Hemiptera: Lachnidae)

 Syrphus sp. (Diptera: Syrphidae); Lebanon, Syria

 Pauesia antennata (Mukerji) (Hymenoptera: Braconidae); Yemen

Pear

Apiomyia bergenstammi (Wachtl.) (Diptera: Cecidomyiidae)

 Oxyglypta rugosa Rushka (Chalcididae); Lebanon, Syria

Cacopsylla bidens (Sulc.) (Hemiptera: Psyllidae); Lebanon, Syria

Capnodis tenebrionis L. (Coleoptera: Buprestidae); Lebanon

Ceratitis capitata Weid. (Diptera: Tephritidae); Egypt, Jordan[b], Lebanon, Syria[c]

Dysaphis pyri Boy. (Hemiptera: Aphididae); Iraq[b]

Zeuzera pyrina (L.) (Lepidoptera: Cossidae); Algeria[c], Lebanon, Libya[c], Syria

Plum

Acalitus phloeocoptes Nal. (Acari: Eriophyidae)

 Typhlodromous invectus Chant (Acari: Phytoseiidae); Lebanon[b]

 Zetzellia talhouki Dosse (Acari: Stigmaeidae); Lebanon[b]

Aculus fockeui (Nalepa and Trouessart) (Acari: Eriophyidae)

 Typhlodromous invectus Chant (Acari: Phytoseiidae); Lebanon[b]

 Zetzellia talhouki Dosse (Acari: Stigmaeidae); Lebanon[b]

Bryobia rubrioculus (Scheuten) (Acari: Tetranychidae)

Typhlodromous invectus Chant (Acari: Phytoseiidae); Lebanon[b]

Zetzellia talhouki Dosse (Acari: Stigmaeidae); Lebanon[b]

Cenopalpus lanceolatisetae Attiah (Acari: Tenuipalpidae)

   Typhlodromous invectus Chant (Acari: Phytoseiidae); Lebanon[b]

   Zetzellia talhouki Dosse (Acari: Stigmaeidae); Lebanon[b]

Diptacus gigantorhynchus (Nalepa) (Acari: Diptilomiopidae)

   Typhlodromous invectus Chant (Acari: Phytoseiidae); Lebanon[b]

   Zetzellia talhouki Dosse (Acari: Stigmaeidae); Lebanon[b]

Eotetranychus carpini (Oudemams) (Acari: Tetranychidae)

   Typhlodromous invectus Chant (Acari: Phytoseiidae); Lebanon[b]

   Zetzellia talhouki Dosse (Acari: Stigmaeidae); Lebanon[b]

Hyalopterus pruni (Geoffroy) (Hemiptera: Aphididae)

   Adonia variegata (Goeze) (Coleoptera: Coccinellidae); Lebanon, Syria

   Ephedrus plagiator (Nees) (Hymenoptera: Braconidae); Lebanon, Iraq[b]

   Syrphus ribesii L. (Diptera: Syrphidae); Lebanon, Syria

Phyllocoptes abaenus Keifer (Acari: Eriophyidae)

   Typhlodromous invectus Chant (Acari: Phytoseiidae); Lebanon[b]

   Zetzellia talhouki Dosse (Acari: Stigmaeidae); Lebanon[b]

Tetranychus cinnabarinus Boisd. (Acari: Tetranychidae); Saudi Arabia, Syria

   Typhlodromous invectus Chant (Acari: Phytoseiidae); Lebanon[b]

   Zetzellia talhouki Dosse (Acari: Stigmaeidae); Lebanon[b]

Tetranychus urticae Koch (Acari: Tetranychidae); Saudi Arabia, Syria

   Typhlodromous invectus Chant (Acari: Phytoseiidae); Lebanon[b]

   Zetzellia talhouki Dosse (Acari: Stigmaeidae); Lebanon[b]

Quince

Capnodis tenebrionis L. (Coleoptera: Buprestidae); Algeria

Ceratitis capitata Weid. (Diptera: Tephritidae); Jordan[b]

Recurvaria nanella Hb. (Lepidoptera: Gelechidae); Lebanon

   Macrocentrus abdominalis F. (Hymenoptera: Braconidae); Syria

   Orgilus obscurator Nees (Hymenoptera: Braconidae); Syria

Strawberry

Phytonemus (*Steneotarsonemus*) *pallidus* (Banks) (Acarina: Tarsonemidae); Lebanon

Tetranychus cinnabarinus (Boisduval) (Acari: Tetranychidae); Lebanon

Tetranychus urticae Koch

   Phytoseiulus persimilis Athias-Henriot (Acari: Phytoseiidae); Egypt[c], Lebanon

Rutaceae
Citrus

Acaudaleyrodes citri (Priesner & Hosni) (Hemiptera: Aleyrodidae)

   Cales noaki Howard (Hymenoptera: Aphelinidae); Syria[c]

   Encarsia lahorensis (Howard) (Hymenoptera: Aphelinidae); Syria[c]

   Encarsia lutea (Masi) (Hymenoptera: Aphelinidae); Middle East[d]

Acrythosiphon lactucae (Tlja.) (Hemiptera: Aphididae)

   Scymnus apetzi Mulsant (Coleoptera: Coccinellidae); Jordan, Lebanon, Syria

Aleurothrixus floccosus (Maskell) (Hemiptera: Aleyrodidae)

   Cales noaki Howard (Hymenoptera: Aphelinidae); Syria[c]

   Encarsia lahorensis (Howard) (Hymenoptera: Aphelinidae); Syria[c]

Aonidiella aurantii (Maskell) (Hemiptera: Diaspididae)

   Aphytis chrysomphali (Mercet) (Hymenoptera: Aphelinidae); Lebanon, Syria

Aphis citricola (Van der Goot) (Hemiptera: Aphididae); Syria[c]

Aphis gossypii (Glover) (Hemiptera: Aphididae); Iraq, Jordan, Lebanon, Saudi Arabia, Syria[c]

Aspidiotus hederae Vallot (Hemiptera: Diaspididae)

Rhizobius lophanthae (Blaisdell) (Coleoptera: Coccinellidae); Jordan, Syria

Bemisia tabaci Genn. (Hemiptera: Aleyrodidae)

Encarsia lutea (Masi) (Hymenoptera: Aphelinidae); Middle East[d]

Euseius scutalis (Athias-Henriot) (Acari: Phytoseiidae); Jordan[b]

Brevipalpus lewisi McGregor (Acari: Tetranychidae)

Euseius scutalis (Athias-Henriot) (Acari: Phytoseiidae); Jordan[b]

Ceratitis capitata Wied. (Diptera: Tephritidae); Jordan, Lebanon, Syria[c]

Ceroplastes floridensis Comostock (Hemiptera: Coccidae)

Scutellista cyanea Motsch (Hymenoptera: Pteromalidae); Lebanon, Syria[c]

Chrysomphalus ficus Ashm (Hemiptera: Coccidae)

Aphytis chrysomphali (Mercer) (Hymenoptera: Aphelinidae); Lebanon, Syria

Dialeurodes citri Ashmead (Hemiptera: Aleyrodidae)

Cales noaki Howard (Hymenoptera: Aphelinidae); Syria[c]

Encarsia lahorensis (Howard) (Hymenoptera: Aphelinidae); Syria[c]

Encarsia sp. (Hymenoptera: Aphelinidae); Syria

Diaphorina citri Kuw. (Hemiptera: Psyllidae)

Chilomenes vicina (Muls.) (Coleoptera: Coccinellidae); Saudi Arabia

Diaphorencyrtus aligarhensis (Shafee, Alam & Agaral) (Hymenoptera: Encyrtidae); Saudi Arabia

Duraphis noxia (Kurdjumov) (Hemiptera: Aphididae)

Aphidius colemani Viereck (Hymenoptera: Braconidae); Lebanon, Syria

Aphidius matricariae (Hymenoptera: Braconidae); Lebanon, Syria

Ephedrus persicae Frogatt (Hymenoptera: Braconidae); Lebanon

Lysiphlebus fabarum Marshal (Hymenoptera: Braconidae); Lebanon[b]

Praon volucre (Hymenoptera: Braconidae); Lebanon[b]

Trioxys angelicae Haliday (Hymenoptera: Braconidae); Lebanon[b]

Eriophyes (*Aceria*) *sheldoni* (Ewing) (Acari: Eriophyidae)

Phytoseiulus (*Phytoseides*) sp. (Acari: Phytoseiidae); Lebanon, Syria

Eutetranychus orientalis (Klein) (Acari: Tetranychidae)

Euseius scutalis (Athias-Henriot) (Acari: Phytoseiidae); Jordan[b]

Exochomus nigromaculatus (Goeze.) (Coleoptera: Coccinellidae); Jordan, Lebanon, Syria

Heliothrips haemorrhoidalis Bouche (Thysanoptera: Thripidae)

Franklinothrips megalops (*myrmecaeformis*) (Thysanoptera: Aeolothripidae); Lebanon, Palestine

Icerya purchasi Mask. (Hemiptera: Margarodidae)

Rodalia cardinalis (Mulsant) (Coleoptera: Coccinellidae)

Macrosiphum euphorbiae (Thomas) (Hemiptera: Aphididae); Syria[c]

Microtermes najdensis Harris (Isoptera: Termitidae); Saudi Arabia

Nipaecoccus viridis (Newstead) (Hemiptera: Pseudococcidae); Iraq, Jordan, Lebanon, Saudi Arabia

Anagyrus (*Agraensis*) *indicus* Shafee, Alam, & Agarwal (Hymenoptera: Encyrtidae); Jordan, Syria

Anagyrus kamali Moursi (Hymenoptera: Encyrtidae); Jordan

Anagyrus pseudococci Girault (Hymenoptera: Encyrtidae); Saudi Arabia

Chrysopa nr. *gobiensis* (Tjeder) (Neuroptera: Chrysopidae); Saudi Arabia

Exochomus marginipennis Leconte (Coleoptera: Coccinellidae); Saudi Arabia

Hyperaspis sp. (Coleoptera: Coccinellidae); Saudi Arabia

Nephus bipunctatus (*includens*) Kirsch (Coleoptera: Coccinellidae); Jordan, Syria

Scymnus sp. (Coleoptera: Coccinellidae); Saudi Arabia

Panonychus citri (McGregor) (Acari: Tetranychidae)

Amblyseius (*Iphiseius*) *degenrans* Berlese (Acari: Phytoseiidae); Syria; Lebanon, Saudi Arabia

Neoceiulus (*Amblyseius*) *californicus* McGregor (Acari: Phytoseiidae); Syria

Phytoseiulus (*Phytoseides*) sp. (Acari: Phytoseiidae); Syria

Papilio demodocus (Esper) (Lepidoptera: Papillionidae); Saudi Arabia

Papilio demolus L. (Lepidoptera: Papillionidae); Saudi Arabia

Parabemisia myricae Kowana (Hemiptera: Aleyrodidae)

Cales noaki Howard (Hymenoptera: Aphelinidae); Syria[c]

Encarsia lahorensis (Hymenoptera: Aphelinidae); Syria[c]

Eretmocerus sp. (Hymenoptera: Aphelinidae): Israel[d]

Paraleyrodes minei Iaccarino (Hemiptera: Aleyrodidae)

    Cales noaki Howard (Hymenoptera: Aphelinidae); Syria[c]

    Encarsia lahorensis (Hymenoptera: Aphelinidae); Syria[c]

Phyllocnistis citrella (Stainton) (Lepidoptera: Gracillariidae); Saudi Arabia

    Ageniaspis citricola Logvinovskaya (Hymenoptera: Encyrtidae); Syria

    Chrysoperla carnea (Stephens) (Neuroptera: Chrysopidae); Jordan, Syria

    Cirrospilus ingenus (Gahan) (Hymenoptera: Eulophidae); Syria

    Cirrospilus luteus (Hymenoptera: Eulophidae); Lebanon

    Cirrospilus lyncus (Walker) (Hymenoptera: Eulophidae); Lebanon

    Cirrospilus nr. *lyncus* (Walker) (Hymenoptera: Eulophidae); Syria

    Cirrospilus pictus (Nees) (Hymenoptera: Eulophidae); Jordan

    Cirrospilus quadristriatus (Subba Rao & Ramamani) (Hymenoptera: Eulophidae); Jordan, Syria

    Cirrospilus sp. (Hymenoptera: Eulophidae); Algeria[c]

    Citrostichus phyllocnistoides (Hymenoptera: Eulophidae); Syria

    Neochrysocharis formosa (Westwood) (Hymenoptera: Eulophidae); Jordan, Syria

    Pnigalio spp. (Boucek) (Hymenoptera: Eulophidae); Algeria[c], Jordan, Lebanon, Syria

    Ratzeburgiola incompleta (Boucek) (Hymenoptera: Eulophidae); Jordan, Syria, Saudi Arabia

    Semielacher petiolatus (Girault) (Hymenoptera: Eulophidae); Algeria[c], Syria

    Stenomesius japonicus (Ashmead) (Hymenoptera: Eulophidae); Syria

    Sympiesis spp. (Hymenoptera: Eulophidae); Syria

Phyllocoptruta oleivora (Ashmead) (Acarina: Eriophyidae); Saudi Arabia

    Phytoseiulus (*Phytoseides*) sp. (Acari: Phytoseiidae); Lebanon, Syria

Planococcus (*Psuedococcus*) *citri* (Risso) (Hemiptera: Pseudococcidae)

    Cryptolaemus montruzieri (Mulsant) (Coleoptera: Coccinellidae), Syria[c]

Planococcus spp. (Hemiptera: Pseudococcidae); Saudi Arabia

    Anagyrus (*agraensis*) *indicus* Shafee, Alam, & Agarwal (Hymenoptera: Encyrtidae); Jordan, Syria

Pseudococcus adonidum (L.) (Hemiptera: Pseudococcidae)

    Cryptolaemus montruzieri (Mulsant) (Coleoptera: Coccinellidae), Syria[c]

    Leptomastix dactylopii Howard (Hymenoptera: Encyrtidae); Syria

    Nephus bipunctatus (*includens*) Kirsch (Coleoptera: Coccinellidae); Jordan, Syria

Prays citri Miller (Lepidoptera, Hyponomentidae); Egypt[b]

    Bracon hebetor Say. (Hymenoptera: Braconidae); Iraq, Syria

    Elasmus steffani (Hymenoptera: Elasmidae); Syria

Scirtothrips citri (Moulton) (Thysanoptera: Thripidae)

    Euseius scutalis (Athias-Henriot) (Acari: Phytoseiidae); Jordan[b]

Tetranychus cinnabarinus (Boisd.) (Acari: Tetranychidae)

    Exochomus nigromaculatus (Goeze.) (Coleoptera: Coccinellidae); Jordan, Lebanon, Syria

    Stethorus gilvifrons Muls. (Coleoptera: Coccinellidae); Lebanon

Tetranychus turkestani Ugarov & Nikolski (Acari: Tetranychidae)

    Stethorus gilvifrons (Muslant) (Coleoptera: Coccinellidae); Iraq, Jordan, Lebanon, Syria

    Chrysoperla carnea Steph. (Neuroptera: Chrysopidae); Iraq

Toxoptera aurantii (Boyer de Fonscolombe) (Hemiptera: Aphididae)

    Aphidius colemani Viereck (Hymenoptera: Braconidae); Lebanon, Syria

    Aphidius matricariae (Hymenoptera: Braconidae); Lebanon, Syria

    Ephedrus persicae Frogatt (Hymenoptera: Braconidae); Lebanon

    Lysiphlebus fabarum Marshal (Hymenoptera: Braconidae); Lebanon[b]

    Praon volucre; Lebanon[b]

    Trioxys angelicae Haliday (Hymenoptera: Braconidae); Lebanon[b]

Solanaceae
Eggplant

Aphis gossypii (Glover) (Hemiptera: Aphididae); Iraq, Jordan, Lebanon, Saudi Arabia

Bemisia tabaci Gen. (Hemiptera: Aleyrodidae); Algeria[c]

   Clitostethus arcuatus Rossi (Coleoptera: Coccinellidae); Iraq, Syria[c]

   Orius albidipennis Reut. (Hemiptera: Anthocoridae); Iraq[b]

Leucinodes orbonalis Guen. (Lepidoptera: Pyralidae); Egypt, Lebanon, Saudi Arabia

Tetranychus urticae Koch (Acari: Tetranychidae)

   Phytoseiulus persimilis (Acari: Phytoseiidae); Lebanon & Syria[b]

Potato

Gnorimoschema (*Phthorimaea*) *operculella* Zeller (Lepidoptera: Gelechiidae); Cyprus, Egypt, Iraq[b] & Jordan, Lebanon, Saudi Arabia, Syria

   Chelonus phthoremiaeae Gahan (Hymenoptera: Braconidae); Tunisia

Macrosiphum euphorbiae (Thomas) (Hemiptera: Aphididae); Algeria[c]

Myzus persicae (Sulzer) (Hemiptera: Aphididae); Algeria[c], Lebanon, Saudi Arabia, Syria

Tomato

Aculops (= *Vasates*) *lycopersici* (Massee) (Acari: Eriophyidae); Lebanon, Saudi Arabia

Bemisia tabaci Gennadius (Hemiptera: Aleyrodidae); Egypt[c], Lebanon, Saudi Arabia, Syria, Yemen

   Encarsia spp. (Hymenoptera: Aphelinidae); Jordan

Helicoverpa (*Heliothes*) *armigera* Hb. (Lepidoptera: Noctuidae); Egypt, Lebanon, Saudi Arabia, Syria, Yemen[b]

Liriomyza bryoniae Kalt. (Diptera: Agromyzidae); Iraq[c]

Urticaceae
Figs

Ceratitis capitata Weid. (Diptera: Tephritidae); Jordan, Syria[c]

Ceroplastes rusci L. (Hemiptera: Coccidae)

   Ancytus sp.; Syria; Libya[c]

Coccidophaga scitula Ramb. (Lepidoptera); Lebanon

Scutellista cyanea (Motschulsky) (Hymenoptera: Pteromalidae); Jordan, Lebanon, Syria

Tetrastichus sp. (Hymenoptera: Eulophidae); Jordan, Lebanon, Syria

Ephestia cautella (Walker) (Lepidoptera: Pyralididae)

   Bracon hebetor Say. (Hymenoptera: Braconidae); Iraq, Syria

Hypoborus ficus Er. (Coleoptera: Scolytidae)

   Sycoster lavagnei (Hymenoptera: Braconidae); Lebanon

Rhycaphytoptus ficifoliae K. (Acari: Rhyncaphytoptidae); Iraq[b]

Silba adipata McAlpine (Diptera: Lonchaeidae); Iraq, Lebanon, Turkey

Mulberry

Icerya aegyptiaca Douglas (Hemiptera: Margarodidae)

   Amblyseius swirskii Athias-Henriot (Acari: Phytoseiidae); Egypt[c]

   Euseius scutalis (Acari: Phytoseiidae) (Athias-Henriot); Egypt[c]

Pseudococcus citri Risso (Hemiptera: Pseudococcidae)

   Amblyseius swirskii Athias-Henriot (Acari: Phytoseiidae); Egypt[c]

   Euseius scutalis (Acari: Phytoseiidae) (Athias-Henriot); Egypt[c]

Tetranychus urticae Kock. (Acari: Tetranychidae)

   Amblyseius swirskii Athias-Henriot (Acari: Phytoseiidae); Egypt[c]

   Euseius scutalis (Acari: Phytoseiidae) (Athias-Henriot); Egypt[c]

Vitaceae
Grapes

Colomerus (*Eriophyes*) *vitis* (Pangest) (Acari: Eriophyiidae); Lebanon

   Euseius scutalis Athias-Henriot (Acari: Phytoseiidae); Jordan

Lobesia botrana Denis & Schiffermüller (Lepidoptera: Tortricidae); Iran[c], Jordan[c]

   Ascogaster quadridentata (Wesmael) (Hymenoptera: Braconidae); Jordan, Syria

   Bassus (*Microdus*) *dimidiator* (Nees) (Hymenoptera: Braconidae); Syria

   Bracon brevicornis (Wesmael) (Hymenoptera: Braconidae); Syria

Coccinella septempunctata L. (Coleoptera: Coccinellidae); Iraq, Lebanon, Syria

Dibrachys boarmiae (Walker) (Hymenoptera: Pteromalidae); Syria

Pristomerus sp. (Hymenoptera: Ichneumonidae); Syria

Planococcus (*Pseudococcus*) *citri* (Risso) (Hemiptera: Psuedococcidae); Lebanon[b], Syria

Theresimima (*Procris*) *ampelophaga* B.-B. (Lepidoptera: Zygaenidae); Jordan, Lebanon, Syria

Viteus vitifolii (Fitch.) (Hemiptera: Phylloxeridae); Jordan, Lebanon, Syria

Miscellaneous
Stone fruit trees

Capnodis tenebrionis L. (Coleoptera: Buprestidae); Jordan[c], Lebanon, Syria

Capnodis carbonaria Klug. (Coleoptera: Buprestidae); Jordan[c], Lebanon, Syria

Storage pests
Beans (Leguminosaea)

Bruchus dentipes Baudi (Coleoptera: Bruchidae)
   Triaspis thoracicus (Curtis) (Hymenoptera: Braconidae); Syria

Callosobruchus chinensis L. (Coleoptera: Bruchidae); Syria[b]

Chickpeas (Leguminosaea)

Bruchus dentipes Baudi (Coleoptera: Bruchidae); Syria[b]

Callosobruchus chinensis L. (Coleoptera: Bruchidae); Syria[b]

Trogoderma granarium Everts (Coleoptera: Dermestidae); Iraq[b]

Cotton (Malvaceae)

Tribolium castaneum (Herbst) (Coleoptera: Tenebrionidae)
   Trichogramma principium (Hymenoptera: Trichogrammatidae); Iraq[b]

Cowpeas (Leguminosaea)

Collosobruchus maculatus F. (Coleoptera: Bruchidae); Iraq[b], Syria

Lentils (Leguminosaea)

Bruchus dentipes Baudi (Coleoptera: Bruchidae); Syria[b]

Callosobruchus chinensis L. (Coleoptera: Bruchidae); Syria[b]

Sesame (Pedaliaceae)

Trogoderma granarium Everts (Coleoptera: Dermestidae); Iraq[b]

Sunflower (Asteraceae/Compositae)

Trogoderma granarium Everts (Coleoptera: Dermestidae); Iraq[b]

Wheat (Gramineae/Poaceae)

Sitotroga cereallella (Olivier) (Lepidoptera: Gelechiidae); Iraq

Trogoderma granarium Everts (Coleoptera: Dermestidae); Iraq[b]

Vegetables

Agrotis ipsilon (Rottemburg) (Lepidoptera: Noctuidae); Lebanon, Saudi Arabia, Syria
   Apanteles ruficrus Hal. (Hymenoptera: Braconidae); Egypt

Earias biplaga Walker (Lepidoptera: Noctuidae); Egypt, Lebanon, Saudi Arabia, Syria

Empoasca lybica De Bergevin (Hemiptera: Cicadellidae); Egypt[b], Lebanon, Saudi Arabia, Sudan, Syria

Tetranychus cinnabarinus (Boisduval) (Acari: Tetranychidae); Egypt[b]

Spodoptera exigua (Hb.) (Lepidoptera: Noctuidae); Egypt, Lebanon, Saudi Arabia, Syria

Trichoplusia ni (Hübner) (Lepidoptera: Noctuidae); Egypt, Lebanon, Saudi Arabia

Efat M. Abou-Fakhr Hammad
American University of Beirut
Beirut, Lebanon

References
Al-Matni, W., and H. Samara. 2001. *Natural enemies of insects recorded in Syria and neighboring countries.* Version 2. Dar An-Nokhba, Damascus, Syria.

Arab Journal of Plant Protection. A scientific journal published by the Arab Society for Plant Protection, Volumes 1 (1983) - 14 (1996), 19 (2001). Beirut, Lebanon.

Gerling, D. 1990. *Whiteflies: their bionomics, pest status and management*. Intercept Ltd., Andover, United Kingdom.

Kawar, N. S., A. M. Al-Ajlan, and M. Yassin. 1995. The most important insect and mite pests on major crops in the west region of Saudi Arabia. Ministry of Agriculture and Water Resources, Agricultural Research Center in Makka Al-Mukarrama and Jaddah, Saudi Arabia, and Food and Agriculture Organization (FAO).

Talhouk, A. S. 2002. *Insects and Mites Injurious to Crops in Middle Eastern Countries*. (2nd ed.) American University of Beirut Press. Beirut, Lebanon.

PEST SPECIES. Any species that humans consider to be undesirable. More often, a pest is considered to be a species that competes with humans for food, fiber, or shelter, or transmits diseases to humans or livestock, or affects the comfort of humans.

PETALURIDAE. A family of dragonflies (order Odonata). They commonly are known as graybacks. See also, DRAGONFLIES AND DAMSELFLIES.

PETIOLATE. Attached by a narrow stem or stalk (a petiole).

PETIOLE. Stalk that connects the leaf to a stem. In Hymenoptera, it is sometimes used to describe the first section of the narrow stalk-like abdominal segment, or 'waist.'
See also, ABDOMEN OF HEXAPODS.

PETRUNKEVITCH, ALEXANDER. Alexander Petrunkevitch was born at Pliski, Russia, in December 1875. He was born into a noble family, and at an early age displayed interest in both zoology and the literary arts. His liberal politics forced him to leave Russia, and he moved to Germany, where he came under the influence of August Weismann. He completed a Ph.D. disertation in 1900 on the cytology and embryonic development of the honey bee, and married an American. They moved to the United States in 1903, and he became a lecturer at Harvard University in Boston, and then acting professor of Zoology at Indiana University, though he commuted regularly to Massachusetts to be with his wife, who had become ill. He moved to Yale University when his wife contracted tuberculosis, becoming an assistant professor in 1911 and a full professor in 1917. Petrunkevitch made important contributions to the study of arachnids. Among his important publications were "A synomic index-catalogue of spiders of North, Central and South America" (1911),"On families of spiders" (1923), "Catalog of American spiders-part I" (1939), and "An inquiry into the natural classification of spiders, based on a study of their internal anatomy" (1933). He died in New Haven, Connecticut, on March 9, 1964.

Reference
Mallis, A. 1971. *American entomologists*. Rutgers University Press, New Brunswick, New Jersey. 549 pp.

PHACOPTERONIDAE. A family of bugs (order Hemiptera, superfamily Psylloidea). See also, BUGS.

PHAGE (BACTERIOPHAGE). A virus that attacks bacteria. Frequently used as vectors for carrying foreign DNA into cells by genetic engineers.

PHALACRIDAE. A family of beetles (order Coleoptera). They commonly are known as shining flower beetles. See also, BEETLES.

PHAGOCYTES. Cells that are capable of moving in the insect body and engulfing or destroying small foreign bodies such as microorganisms.

PHAGOSTIMULANT. A substance that induces feeding.

PHAGOCYTOSIS. The process of ingestion and digestion by cells, especially the ingestion or engulfing of microorganisms and other small particles by blood cells.

PHALLUS. The intromittent (copulatory) organ of insects; the aedeagus and any processes found at its base.

PHANTOM CRANE FLIES. Members of the family Ptychopteridae (order Diptera). See also, FLIES.

PHANTOM MIDGES. Members of the family Chaoboridae (order Diptera). See also, FLIES.

PHARATE. The stage at which molting has started but the insect has not yet cast off the old cuticle.

PHARYNX. The anterior portion of the foregut immediately behind the buccal cavity.
See also, ALIMENTARY SYSTEM, ALIMENTARY CANAL AND DIGESTION.

PHASIC RECEPTORS. Sensory neurons that adapt rapidly to continuing steady stimuli, with the receptor potential falling, and the neuron becoming relatively insensitive. (contrast with tonic receptors)

PHASMATIDAE. A family of walkingsticks (order Phasmatodea). They commonly are known as winged walkingsticks. See also, WALKING-STICKS AND LEAF INSECTS.

PHASMATODEA. An order of insects. They commonly are known as walkingsticks and leaf insects. See also, WALKINGSTICKS AND LEAF INSECTS.

PHENACOLEACHIIDAE. A family of insects in the superfamily Coccoidae (order Hemiptera). See also, BUGS.

PHENGODIDAE. A family of beetles (order Coleoptera). They commonly are known as glow-worms. See also, BEETLES.

PHENOGRAM. A branching diagram that links different taxa by estimating overall similarity based on data from characters. Characters are not evaluated as to whether they are primitive or derived.

PHENOLOGICAL ASYNCHRONY. Lack of simultaneous occurrence between an insect and its host, or lack of correspondence between traits of organisms.(contrast with phenological synchrony)

PHENOLOGICAL SYNCHRONY. Seasonal correspondence between an insect and its host, or seasonal correspondence between traits of organisms. (contrast with phenological asynchrony)

PHENOLOGY. The seasonal life history of a plant or animal, especially in relation to weather and climate.

PHENOLOGY MODELS FOR PEST MANAGEMENT. All living organisms use various 'substances' for their energy, body growth, and tissue maintenance. Enzymes control the use of those

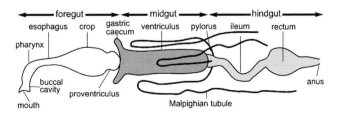

Fig. 776 Generalized insect alimentary system (adapted from Chapman, The insects: structure and function).

substances. Enzymes are very large, complex protein molecules that are formed within cells.

Enzymes have very important roles to play within the life of an organism. In the most general terms, they function to increase the rate of various chemical reactions (catalysts). In more specific terms, they are involved in the removal of electrons from some chemical compounds, the transfer of chemical groups from some compounds to others, the 'breaking apart' of larger molecules (digestion), and the linking together of some molecules. Their function within an organism is affected by several factors, one of which is temperature.

For those reactions in which enzymes are involved, there is a range of temperatures within which they will take place. For temperatures that are less than the least value in that range, no reactions will occur. For temperatures that are greater than the greatest value in that range, the enzyme is denatured and coagulates, so no reactions will occur. For temperatures within that range, the rate of reactions increases with increasing temperatures.

The most general measure of enzymatic activity within an organism is, of course, 'growth.' Growth can be measured in many ways, two of which are the length or height of an organism, and its weight. The growth of an insect to certain discrete stages, such as the larval, pupal, or adult stages, is often referred to as 'development.' When development is linked with temperature, it is referred to as 'phenology.'

Growth and development of some organisms, such as humans, may be independent of the temperature that is outside their bodies because their body temperature is fairly constant and always warm. Such organisms are called 'homoiotherms.' Due to their relatively constant cellular temperatures, the growth of homoiotherms proceeds at a constant rate which can be predicted by 'calendar time' (i.e., days, weeks, months, years).

In contrast, development of some other organisms, such as insects, cannot be predicted using calendar time because their body temperature is not maintained at a constant, warm level. Rather, their body temperature largely fluctuates with environmental temperatures. Such organisms are called 'poikilotherms.' As the cellular temperatures of poikilotherms are not constant, and the level of enzymatic activity is linked with environmental temperatures, their growth and development can usually be predicted through 'thermal time.'

Thermal time is calculated by first determining the amount of heat that is experienced during a 24 hr period. That heat can be expressed as units of heat, or 'heat units,' also known as degree-days, day-degrees, and thermal units (more on that later). Thermal units are daily amounts of the total heat needed for a poikilotherm to develop to a given growth stage. The total heat needed before a poikilotherm will grow to that stage is sometimes called the effective cumulative temperature, required degree-days, or required thermal summation.

Thermal time can be used to predict the growth of an individual insect by having pest managers calculate and sum thermal units, from one 24-hr period to another, until that sum equals the required thermal summation; the date on which that occurs is, theoretically, the date on which the growth of that insect has reached a specific stage. Pest managers often wish to predict insect developmental events, such as egg hatch, which can signal the onset of crop damage.

Growth and development of populations

Knowledge of how temperatures influence the growth and development of just one insect can also be used to predict the growth and development of many insects of the same species (population). Measures used to predict the development of populations, however, necessarily differ from those that are used to predict the development of individuals. This is so because pest managers who want to predict the date on which just one larva will develop to, for example, its third instar would calculate and sum thermal units, beginning at oviposition, and declare that individual as being in the third instar on the date that summed thermal units equals the required thermal summation. For a population, however, numerous eggs will be laid over several days or weeks. Consequently, individuals that hatch from the first eggs laid will develop to the third instar sooner than will individuals that hatch from the last eggs laid. Thus, when predicting developmental events for populations, pest managers may want to determine, in the context of this example, the date on which those first larvae develop to the third instar. Other population events, such as the peak occurrence of third instar larvae, may also be predicted. This process of calculating and summing thermal units, until they reach the required thermal summation, for the purpose of predicting a population event, is called phenological modeling, and the models used for making such predictions are called phenology models. Phenology models can be constructed through the conduct of laboratory experiments or field studies.

Constructing phenology models through the conduct of laboratory experiments

When constructing phenology models through the conduct of laboratory experiments, several environmental chambers are used at the same time, with each being set at a different temperature. Temperatures should vary widely so as to encompass the extremes within which enzymatic activity occurs. Next, a laboratory colony of the insect is accessed, with many individuals being used in the study (experimental subjects). Then, experimental subjects are divided into as many groups as there are chambers, with each group being placed into its respective chamber. Next, all individuals are provided with sustenance (either artificial diet or a host), and monitored daily. The number of days required for each individual to grow to a given stage is then recorded. If the average or

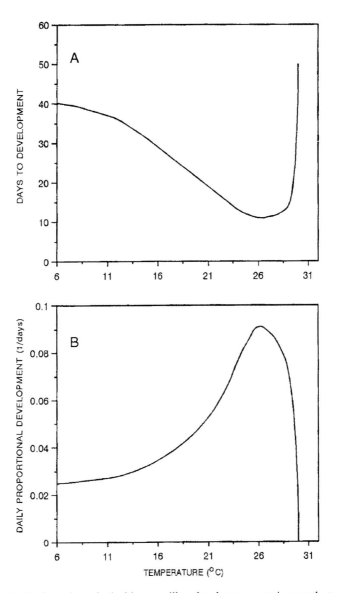

Fig. 777 Average days needed before a hypothetical insect will to develop to a certain growth stage, at various temperatures (A), and the inverse of those days (B), which expresses development on a daily basis.

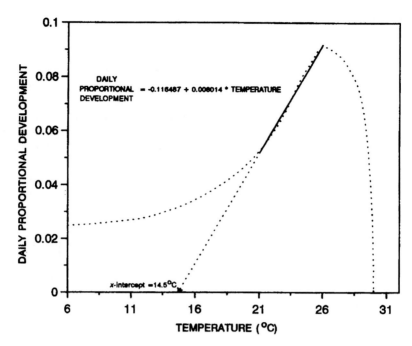

Fig. 778 One way the lowest temperature at which insect development will first occur (lower developmental threshold) can be estimated is to trim the nonlinear portions from the daily developmental curve, then mathematically fit a straight line to the rest; the lower developmental threshold is the temperature where the straight line crosses the x-axis (14.5°C), and the total amount of heat needed for the insect to develop is the inverse of the slope (i.e., 1/0.008014 = 124.8°C).

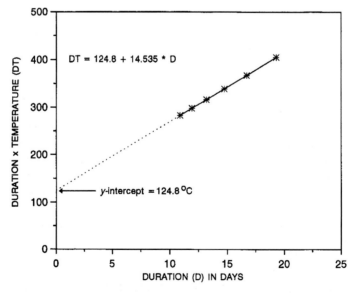

Fig. 779 Another way the lowest temperature at which insect development will first occur (lower developmental threshold) can be estimated is to trim the nonlinear portions from the relationship, multiply the average (or median) days to development (D) by their respective temperatures (T), and mathematically draw a straight line between that product and D; the lower developmental threshold is the slope of that straight line (14.535°C), and the total amount of heat needed for the insect to develop is the value for DT where the straight line crosses the y-axis (124.8°C).

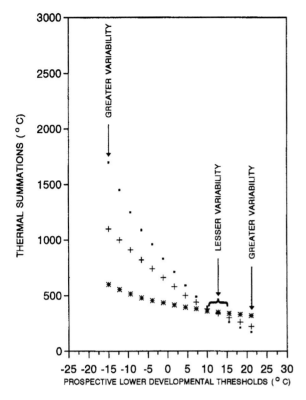

Fig. 780 Typical curve-like relations between prospective thermal summations and prospective lower developmental thresholds that are seen in the temperate climes; the area of lesser variability corresponds with estimates of values for the required thermal summation (y-axis).

median of those days is calculated and graphed on temperatures, the resulting chart should look like that shown in the following figure (responses of experimental subjects that did not develop to the desired growth stage, at the coldest and warmest temperatures, were excluded from this graph). The upper portion of the figure shows that the number of days needed for insect growth to a given stage decreased with increasing temperatures, up to a point, after which the number of days increased; that increase was caused by temperatures being so warm that basic physiological processes were disrupted, causing retarded growth and death.

Such information is not directly used to predict insect development. However, the information can be used to predict insect development if pest managers invert those values to express growth on a daily basis (the lower portion of the figure). Then, either a linear thermal unit function or a nonlinear developmental function is fitted to part or all of the data; that

function is then used to predict insect phenological development.

Predicting insect development with a linear thermal unit function

A linear thermal unit function is calculated by trimming the nonlinear portion from the relationship, and mathematically fitting a straight line to what is left. The resulting 'straight line' is then extrapolated to where it crosses the x-axis (x-intercept); the temperature corresponding with that point is the developmental zero temperature or lower developmental threshold. Hypothetically, that value is the lowest temperature at which insect growth will occur. Next, the required thermal summation is calculated. One way to do that is to invert the value of the slope from the straight line (i.e., $1/0.008014 = 124.8°C$). Another way to do that is to multiply the effective temperature of each growth chamber (effective temperature = temperature of growth chamber − lower developmental threshold) by the average or median number of days that were required for development. A third way to do that is to multiply each average or median number of days that were required for development by their respective temperature, then calculate a straight line between that product and the average or median number of days that were required for development. When that is done, the point where that line crosses the y-axis (y-intercept) ($124.8°C$) is the required thermal summation and the slope of that line ($14.5°C$) is the lower developmental threshold. Regardless of how the required thermal summation and lower developmental threshold are calculated, those values are then used to predict phenological events in the field.

That is done by first obtaining average daily temperatures from locations where the predictions are to be made. One way to do that is to add the daily maximum and minimum temperatures together, then divide that sum by two ('rectangle' method). There are other ways that can be done, one of which is to use a 'sine-wave' method. Next, daily thermal units are calculated by subtracting the lower developmental threshold from the average daily temperature; if that difference is less than zero, daily thermal units are set equal to zero. Thermal units are calculated for each 24-hour period and summed until they equal the required thermal summation. Theoretically, the date on which that happens is the date on which the developmental event occurs.

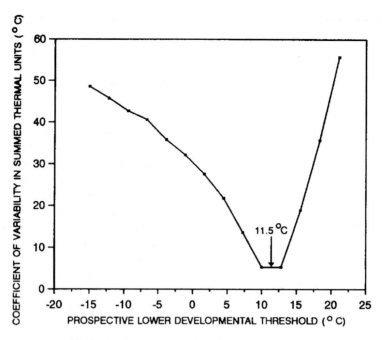

Fig. 781 A typical plot of the variability in the data, on values for prospective lower developmental threshold (x-axis); values for prospective lower developmental threshold at which the variability was least are used as estimates of lower developmental threshold (in this case, their mid-point value, 11.5°C, was selected).

Predicting insect development with a nonlinear developmental function

Nonlinear developmental functions can be constructed by fitting a mathematical equation to both the linear and nonlinear portions of the relationship curve. Several such equations could be used to do this, some of which are based on theory (theoretical equations), and others are based on fitting techniques (empirical equations). In addition, some of these equations will provide estimates of the lower developmental threshold. Whichever type of equation is chosen, it is then used to calculate proportional insect development, given some measure of daily heat. Such calculations are made for each 24 hr period, with those proportions being added, from one day to the next, until that sum equals 1.0. Theoretically, the date on which that sum equals 1.0 is the date on which the phenological event occurs.

Constructing phenology models through the use of field studies

If the insect cannot be reared on artificial diet, or if its host cannot be grown in the greenhouse and placed in environmental chambers, then phenology models must be constructed through the use of field studies. This can be done in several ways, one of which is to find a location at which the insect is known to habit, and visit that location every day until the phenological event occurs; that date is then recorded. This process is repeated for many years. Next, daily maximum and minimum temperatures are procured from the nearest weather station, for each year the observations were made, beginning on a selected starting date and ending on the dates the phenological event occurred. These temperatures are saved in separate computer files, with each year of temperatures comprising a single file (temperature data sets).

Initially, as pest managers do not know what value to use for the lower developmental threshold, several prospective values (prospective lower developmental thresholds) are chosen to calculate and sum prospective thermal units (prospective thermal summations). This is done with each temperature data set. For example, if there were just three years of field observations, and their corresponding temperature data sets were used to calculate prospective thermal summations with each of fourteen prospective lower

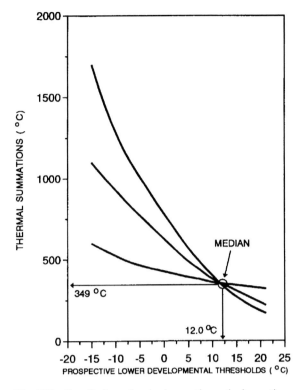

Fig. 782 Results from developing mathematical equations (curves) showing their generalized area of convergence (encircled area); corresponding values for prospective required thermal summation (y-axis) (349°C) and prospective lower developmental threshold (x-axis) (12.0°C) are estimates of the required thermal summation and the lower developmental threshold, respectively.

developmental thresholds, a plot of prospective thermal summations on prospective lower developmental thresholds may look like that shown in the accompanying figure labeled "Typical curve-like relations." Note that the spread between prospective thermal summations is greater at both ends of the plot.

Using this data, there are several methods by which a pest manager can actually calculate the lower developmental threshold. Two methods discussed here are the least variability and the modified regression methods. The least variability method involves calculating a simple measure of variability, or spread, in the required thermal summations, then determines the prospective lower developmental threshold(s) for which that variability is least. This approach clearly shows that prospective lower developmental thresh-

olds which ranged from 10 to 13°C provided the least variability in the required thermal summation and, therefore, would work equally well as the lower developmental threshold. When all temperatures within a range are equally acceptable, their mid-point value is calculated (11.5°C) and used as the lower developmental threshold. Next, the chosen lower developmental threshold is used to calculate and sum thermal units, separately for each temperature data set. An average or median of those values is then calculated and used as the required thermal summation in the phenology model.

The modified regression method is used by first calculating mathematical equations that realistically express the curve-type relations; the use of such equations produces a set of curves. These equations are then used to determine mathematically where those curves cross, or have a point of closest convergence. Each such crossing point, or point of closest convergence, corresponds with an individual estimate of a required thermal summation (y-axis) and lower developmental threshold (x-axis). Then, the average or median of those estimates is calculated and used in the phenology model as the required thermal summation and lower developmental threshold. The modified regression method is extremely difficult to use without the aid of a computer. Fortunately, a computer program (CALFUN) is available and can be used to quickly and easily perform these calculations. CALFUN can be obtained from the World Wide Web at the following URL: http://w3.uwyo.edu/~dlegg/calfun.html.

Once values for the lower developmental threshold and required thermal summation are calculated, the value for the threshold is used to calculate daily thermal units, which are summed until the date on which that sum equals the required thermal summation. Theoretically, that is the date on which the phenological event occurs.

Concluding remarks

There are advantages and disadvantages to developing phenology models in the laboratory and in the field. One advantage to developing phenology models in the laboratory is that values for the lower developmental threshold are biologically meaningful. That is, they represent temperatures at which growth and development will first occur. One potential disadvantage to developing phenology models in the

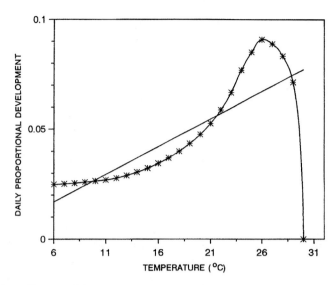

Fig. 783 Results from using a linear model (straight line) and nonlinear model (curve) to describe the observed proportions of daily development (asterisks); note that the nonlinear model best describes observed daily proportional development over a wide range of temperatures.

laboratory, however, is that they must make use of temperatures that actually occur where the insect is found. Those places include soil, dung, fruit, and plant tissues (substrate). As phenology models often make use of air temperatures to calculate either daily thermal units or proportional daily development, it may be necessary to convert those temperatures into substrate temperatures before laboratory-based phenology models will be effective.

One advantage to developing phenology models from field studies may be that converting air temperatures into substrate temperatures will not be necessary. This is so because field-developed phenology models appear to have the air temperature-substrate temperature relationship already factored into the values that are calculated for required thermal summation and lower developmental threshold. This, however, suggests one disadvantage to calculating lower developmental thresholds and required thermal summations from field studies; their values may not be biologically meaningful.

Phenology models that are based on nonlinear thermal unit functions are sometimes thought of as being superior to those that are based on linear thermal unit functions. This is so because nonlinear thermal unit functions best describe the relation between insect growth and a wide range of temperatures. However, many insects have behaviors that, when

exhibited, help regulate their body temperatures to some extent (thermal regulation). For example, many grasshoppers will climb plants to escape uncomfortably warm temperatures near the ground, or will climb plants to better impact sunshine, on cool mornings, so they will warm rapidly. Also, many grasshopper species will turn their bodies in such a way as to 'bask' in sunshine, thereby keeping their cellular temperatures warmer, for longer periods of time, throughout a chilly day. Such behaviors may largely serve to keep insect growth within the linear portion of their developmental curves. If that is true, then phenology models that are based on linear thermal unit functions will be superior to those that are based on nonlinear developmental functions when predicting phenological events in the field. However, if such behaviors do not keep insect development within the linear portion of their developmental curves, then phenology models that are based on nonlinear developmental functions will be superior to those that are based on linear developmental functions when predicting phenological events in the field.

Finally, the performance of many phenology models may be acceptable when they are used to predict insect phenological development in most fields, orchards, or pastures (target populations), but unacceptable in some others. Unacceptable performances may occur when the temperatures recorded at the location of a weather station do not reflect the temperatures that occur at some target populations. Such discrepancies can occur for several reasons, but most involve localized site-specific features such as soil types, plant litter, or slope of terrain relative to the angle of the sun. For some combinations of these features, the area of the target population may be warmed much more than the area of the weather station, resulting in accelerated insect development relative to the output of a phenology model that is using temperatures recorded at the weather station. For other combinations, however, the area of the target population may be much cooler than the area where the weather station is located, thus resulting in retarded insect development relative to the output of a phenology model that is using temperatures recorded at the weather station. For those target populations where such site-specific features can cause phenology models to perform poorly, it may be necessary to make use of on-site temperature recorders. The use of temperatures from these recorders should result in acceptable predictions of insect phenological events.

David Legg
University of Wyoming,
Laramie, Wyoming, USA

References

Higley, L. G., L. P. Pedigo, and K. R. Ostlie 1986. DEGDAY: a program for calculating degree-days, and assumptions behind the degree-day approach. *Environmental Entomology* 15: 999–1005.

Ikemoto, T., and K. Takai. 2000. A new linearized formula for the law of total effective temperature and the evaluation of line-fitting methods with both variables subject to error. *Environmental Entomology* 29: 671–682.

Legg, D. E., S. M. Van Vleet, J. E. Lloyd, and K. M. Zimmerman. 1998. Calculating lower developmental thresholds of insects from field studies. pp. 163–172 in S. G. Pandalai (ed.), *Recent research developments in entomology*, Vol. 2. Research Signpost Press, Trivandrum, India.

Pruess, K. P. 1983. Day-degree methods for pest management. *Environmental Entomology* 12: 613–619.

Wagner, T. L., R. L. Olson, and J. L. Willers. 1991. Modeling arthropod development time. *Journal of Agricultural Entomology* 8: 251–270.

PHENOTYPE. The physical appearance of an individual that are determined by both genotype and environment. (contrast with genotype)

PHENOTYPIC PLASTICITY. The phenomenon of individuals displaying a wide range of phenotypes from the same genotype.

PHEROMONE. Chemical substance secreted externally into the environment that affects the behavior or physiology of other members of the same species. Pheromones are a type of semiochemical. See also, CHEMICAL ECOLOGY, ALARM PHEROMONES, SEX PHEROMONES.

PHEROMONES. When we hear the term 'pheromone,' our thoughts automatically turn to love! Pheromones certainly are involved in getting the sexes together (not only insects, but most animals, and possibly even humans, use sex pheromones to facilitate reproduction) but they have myriad other functions in the lives of insects as well. Males and females, sometimes both sexes, within a species may produce pheromones to coordinate their life histories. Simply defined, pheromones are one type of semiochemical (chemical compound that mediates interactions between organisms) that carry information between organisms of the same species. The term is derived from the Greek verbs 'pherein' (to carry) and 'horman' (to excite or to stimulate). Thus, pheromones are chemical substances that, when secreted to the outside of an organism, cause a specific reaction in a receiving organism of the same species.

Types of pheromones

Pheromones can be classified by the effect that they have on the receiving organism. They may be 'releaser' pheromones, where the receiving organism immediately performs a stereotypical behavior or sequence of behaviors upon perceiving the pheromone. For example, when flying male moths perceive the volatile, sex attractant pheromone of a female moth, they immediately turn upwind, and begin casting, a flight behavior characterized by large-amplitude horizontal excursions with no forward movement. This behavior, and the subsequent upwind flight, help them locate the source of the odor, which is the virgin female. The second type of pheromone is a 'primer' pheromone. These pheromones set in motion complex physiological changes, such as development or sexual maturation, in the receiving organism. For example, mandibular gland pheromone produced by queen honeybees inhibits initiation of queen rearing and regulates division of labor in the worker caste through effects on the worker bees' juvenile hormone levels.

Pheromones may be categorized based on the function they appear to have in the insect. For example, chemicals released by an insect under attack by a predator may be termed an alarm pheromone if other individuals of the same species react to it by dispersing, running around frantically, jumping off the plant, or displaying some other type of alarmed behavior. Many types of pheromones have been described based on their supposed function in a species: sex attractant pheromone, courtship pheromone, marking pheromone, oviposition deterrent pheromone, aggregation pheromone, and the many pheromones used in the lives of social insects for the purposes of recruitment, nest-mate recognition, and trail laying.

The early days of pheromone research

The history of pheromone research is relatively recent. It has been observed since the 1700s that male moths are attracted to female moths of the same species. Jean-Henri Fabre, a French naturalist active in the late 1800s, was the first to perform experiments to test the attraction of male moths to females and to suggest a mechanism to explain his observations. His experiments began after a chance observation, as great scientific discoveries often do. He was keeping a pupa of the Great Peacock moth, *Saturnia pyri*, in his office. One day the moth eclosed and it happened to be female. That night he and his young son went into his office and, because the window had been left open, found twenty male *S. pyri* moths fluttering about the room. He originally thought that the virgin female moth was producing some sort of X-rays or some form of "etheric" waves to signal males. However, after several years of experiments with this species, and several others, he concluded, as astounding as it seemed to him, that female moths were producing an odor that was attractive to males from a great distance.

It was not until the epic research of German chemists, Adolf Butenandt and his colleagues, that the first pheromone was identified. This group was interested in the potential use of insect sex pheromones in management of agricultural pests. They began isolating the sex attractant pheromone of the silkworm moth, *Bombyx mori*, in 1939, choosing this beneficial lepidopteran as a model because of the ease of obtaining large numbers of insects for chemical isolation. Twenty years, and 500,000 female moths later, they had succeeded in isolating and identifying one component of the silkworm's female-produced sex attractant pheromone, which they termed 'bombykol.' Their success was remarkable because of the meager tools that these analytical chemists had at their disposal. They were dealing with much smaller amounts of chemicals than were analyzed routinely at the time. Today, insect pheromones often can be isolated, identified and synthesized within six months, depending on their complexity. Amazingly, with today's analytical instrumentation (gas chromatographs and mass spectrometers), we can measure easily the amount of pheromone a single female insect releases over only an hour's period of time.

Pheromone biosynthesis

Many insects biosynthesize their pheromones de novo, meaning that they break down their food into small molecules which they then use as the building blocks for their pheromones. For example, the female-produced sex attractant pheromones of most moth species analyzed so far are straight-chain hydrocarbon molecules with oxygen-containing functional groups (alcohols, aldehydes and acetates), varying in length from 8 to 18 carbon atoms. The molecules usually are made by the end-to-end addition of acetate molecules, two-carbon-atom building blocks which are end-products of fatty acid metabolism.

Other insects however, biosynthesize their pheromones from specific plant-derived molecules, processing them only minimally before use as a pheromone. This occurs commonly in several species of ithomiine and danaiid butterflies (like the monarch butterfly) and tiger moths (family Arctiidae). In these species, the male butterfly sequesters plant compounds, either from its larval food or, as an adult, from plant species that are visited solely for the purpose of collecting these molecules, and modifies them slightly for use as an aphrodisiac or courtship pheromone. These pheromones are produced in specialized glands at the tip of the male's abdomen and are either sprinkled on the female or wafted in front of her from modified scales called hair pencils or inflatable, eversible sacs called coremata. Perception of the courtship pheromone may allow the female to judge the fitness of the courting male and will result in mating if she deems him 'fit.'

Pheromone components may be made, not by the insect itself, but by symbiotic microorganisms that live within the insect's body. For example, the spruce bark beetle, *Ips typographus* (family Scolytidae), a devastating pest of forest trees in Europe, has a symbiotic relationship with yeast fungi. When pioneer beetles first attack healthy trees, they release a multi-component aggregation pheromone that attracts more beetles to help overwhelm the tree's sticky and toxic resin defense. One of the components of this aggregation pheromone (*cis*-verbenol) is produced as a detoxification product of a terpene defensive compound in the tree (α-pinene) by yeast found in the hindgut of the colonizing beetles. After there is adequate colonization of the tree and its defense has been breached, the yeast transforms *cis*-verbenol to verbenone, which acts as an anti-aggregation pheromone stopping further colonization and reducing the negative effects of beetle overcrowding and competition.

Finally, some insects may use plant compounds directly as pheromones without further modification. The euglossine orchid bees (family Apidae) are pollinators of orchids in the New World tropics, with each species of bee usually visiting and pollinating only one species of orchid. In this group of bees, the males, never the females, pollinate the orchids. The bees gain no nectar or pollen reward but instead collect large quantities of fragrances that the orchids produce. These fragrances are stored in specialized organs in the hind femora of the bees and are hypothesized to be used to facilitate mating. The stored orchid volatiles attract other male bees to a mating display, or a lek, to which female bees are subsequently attracted. Females at these leks can possibly choose specific males to mate with based on the quality or quantity of the orchid fragrance they have collected.

Pheromones are made in exocrine glands, specialized glands that are found on various portions of the body, usually in the epidermis. The glandular secretions are released to the outside of the body. Pheromones usually are released only at certain times of the day or night based on an internal biological rhythm. Most pheromones are volatile, small molecules borne in the air and are perceived by olfactory chemoreceptors on the antennae. Some pheromones are larger, less volatile, or even nonvolatile, molecules that are perceived by contact chemoreceptors found on the ovipositor, antennae, and elsewhere.

Use of pheromones by insects and humans

Chemical communication via pheromones has tremendous benefit to insect species that may not have extensive adaptations for visual or acoustic communication. Thus, pheromone production is extremely common and widespread among the class Insecta. However, there are costs associated with the production of pheromones. One of the major costs is the fact that many predatory and parasitic insects have evolved the ability to perceive these pheromones and can eavesdrop on the communication going on between members of their prey species. For example, several species of predaceous clerid beetles are attracted to the aggregation pheromone of bark beetles. Species of *Trichogramma*, minute hymenopterans that often parasitize the eggs of Lepidoptera, are attracted to the sex attractant pheromone of female moths. Female parasitoids of several species of tephritid fruit flies are stimulated to search a fruit for hidden host larvae when they perceive the nonvolatile host-marking pheromone laid down on the fruit by the ovipositing female fruit fly. Some species of predator have even evolved the ability to produce the pheromone of their prey and thus lure their prey to them. For example, the females of several species of bolas spiders, *Mastophora* spp., produce the sex attractant pheromone of the females of the moth species on which they prey. When the male moth approaches the bolas spider, expecting to find a virgin female moth, he is captured by a sticky bolas that the spider launches at him.

The list of such examples of illicit communication is endless and illustrates the fascinating odorous world in which insects live, and of which we humans largely are unaware due to our poor sense of smell. Despite being unable to directly experience this hubbub of chemical communication, we have taken advantage of it to help control pestiferous insects that harm our food supply. We have used insect pheromones to detect the presence of pest insects as an early warning of their immigration into an area, to time insecticide applications or other control measures, and to document arrival of quarantine pests in pest-free areas. Pheromone monitoring also can be used to quantify insect populations or, at least, to indicate whether populations are declining or increasing. We have used pheromones to disrupt insect's mating behavior. For example, the pink bollworm, *Pectinophora gossypiella*, an economically important pyralid moth pest of cotton in both the

Old and New World, is managed almost exclusively now by the area-wide application of Gossyplure, a synthetic version of the female's sex attractant pheromone. The pheromone is formulated for slow release and is dispersed over large acreages of cotton. The multiple point sources of pheromone in the field make it very difficult for the male bollworm moth to find virgin females and mate with them, thus slowing population increase and the associated yield loss.

Pheromones also are used in mass trapping and attract and kill strategies. In mass trapping of bark beetles for example, large vertical funnel traps are baited with aggregation pheromones. Bark beetles are attracted to the tall, cylindrical funnel traps (which visually mimic trees) and are trapped in the funnels, thus taking them out of the population. In the attract and kill strategy, insects are attracted to a pheromone-impregnated substrate that also has been treated with an insecticide or sterilant so that the insects are excluded from the breeding population. The addition of the house fly sex pheromone, (Z)-9-tricosene, to granular food baits impregnated with an insecticide greatly improved control of house flies in commercial animal production facilities over use of an insecticide-treated food bait alone.

To date, the majority of pheromones that have been used in pest management are female-produced sex attractant pheromones and bark beetle aggregation pheromones. Other pheromones that have potential are the anti-aggregation pheromones of bark beetles that may have potential to prevent infestation of trees, oviposition-deterring pheromones of fruit flies and other insects to prevent oviposition in crops, and alarm pheromones of aphids that increase aphid movement in the area of the alarm pheromone, thus making the aphids more likely to contact a lethal dose of insecticide or to be noticed by visually orienting predators or parasitoids.

See also SOCIAL INSECT PHEROMONES, ALARM PHEROMONES, SEX ATTRACTANT PHEROMONES, MARKING PHEROMONES.

Heather J. McAuslane
University of Florida,
Gainesville, Florida, USA

References

Cardé, R. T., and A. K. Minks. 1995. Control of moth pests by mating disruption: successes and constraints. *Annual Review of Entomology* 40: 559–585.

Hardie, J., and A. K. Minks (eds.). 1999. *Pheromones of non-lepidopteran insects associated with agricultural plants.* CABI Publishing, New York, New York. 466 pp.

Haynes, K. F., and K. V. Yeargan. 1999. Exploitation of intraspecific communication systems: illicit signalers and receiver. *Annals of the Entomological Society of America* 92: 960–970.

Howse, P., I. Stevens, and O. Jones. 1998. *Insect pheromones and their use in pest management.* Chapman & Hall, London, United Kingdom. 369 pp.

Wyatt, T. D. 2002. *Pheromones and animal behaviour: communication by smell and taste.* Cambridge University Press, Cambridge, United Kingdom.

PHEROMONE PARSIMONY. A phenomenon wherein the same pheromone can serve multiple functions. For example, a marking pheromone may also function as an aggregation pheromone, an anti-microbial agent, and a kairomone.

PHIALIDES. Specialized cells with one or more open ends from which conidia are produced in basipetal succession. Thus, the apical conidia are the oldest.

PHILANTHIDAE. A family of wasps (order Hymenoptera). See also, WASPS, ANTS, BEES, AND SAWFLIES.

PHILOPOTAMIDAE. A family of caddisflies (order Trichoptera). They commonly are known as finger-net caddisflies or silken-tube spinners. See also, CADDISFLIES.

PHILOPTERIDAE. A family of chewing lice (order Mallophaga). They sometimes are called bird lice. See also, CHEWING LICE.

PHILOTARSIDAE. A family of psocids (order Psocoptera). See also, BARK-LICE, BOOK-LICE, OR PSOCIDS.

PHLAEOTHRIPIDAE. A family of thrips (order Thysanoptera). See also, THRIPS.

PHLOEIDAE. A family of bugs (order Hemiptera, suborder Pentamorpha). See also, BUGS.

PHOENICIAN BILLBUG. See also, TURF-GRASS INSECTS AND THEIR MANAGEMENT.

PHOENICOCOCCIDAE. A family of insects in the superfamily Coccoidae (order Hemiptera). They sometimes are called palm scales. See also, SCALES AND MEALYBUGS, BUGS.

PHONOTAXIS. Taxis response with respect to sound.

***PHORACANTHA* LONGICORN BEETLES (COLEOPTERA: CERAMBYCIDAE).** The genus *Phoracantha* Newman belongs to the family Cerambycidae of the order Coleoptera.

Order: Coleoptera
  Family: Cerambycidae
    Subfamily: Cerambycinae
      Tribe: Phoracanthini
        Genus: *Phoracantha*

External morphology

Phoracantha adults range from 14 to 48 mm in length. Body is pale, yellowish, reddish to blackish brown with elytra having colored fasciae in most species, usually zigzag, arranged transversely. At least antennal segments 3 to 6 are unispined or bispined at apex. The prothorax is wider than long, with depressed hairs, and the pronotum has a spine or strong prominent conical process at each side. The elytral apex is usually bispined, at least spined at marginal angle. The femora are lineate or gradually thickened. Larvae have an elongate, cylindrical body shape.

Biodiversity and distribution

Phoracantha has 40 known species and is the most speciose genus of the tribe Phoracanthini. This genus is mainly distributed in southern Australia, and some species occur in northern Australia and New Guinea. Only two species are endemic to Papua New Guinea. In Australia, these beetles are predominantly along the coast, with only one species, *P. tuberalis*, restricted to the center of the Australian continent.

About eight widely distributed species occur in both coastal and central areas. Several species are known to be introduced to other parts of the world, for example, *P. semipunctata* and *P. recurva* now occur in all zoogeographic regions except the Oriental Region.

Habitat, host and life cycle

Many *Phoracantha* species are associated with the tree genus *Eucalyptus* and a few with *Acacia*. Adults are active during the night, hiding in daytime under loose bark or in crevices, and can be attracted by light. Larvae bore into trunks or branches of trees. Biologically, the borers can be divided into two groups: (1) dead/dying tree consumers, including *P. semipunctata*, *P. recurva*, *P. tricuspis* and *P. punctata*, and (2) living tree consumers, including *P. mastersi*, *P. acanthocera*, *P. frenchi*, *P. impavida*, *P. synonyma*, *P. solida* and *P. odewahni*. These are clearly defined functional groups, and the species can be allocated clearly to one or other of these.

The members of the dead/dying tree consumers have 1 to 2 generations a year, attacking newly felled and dying trees, of all ages, of *Eucalyptus* species.

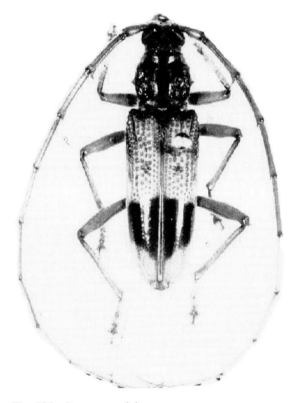

Fig. 784 *P. recurva* adult.

More than 50 eucalypt species are host plants of this group. The group is distributed in both northern and southern parts of Australia. Adults lay eggs in large batches under loose bark, or in crevices, with each egg mass consisting of 23 to 340 eggs. The larvae feed in and under the bark for two to six months, making regular tunnels up to 1.5 meters long, either in a straight line or twisted, radiating from the egg mass in all directions. When mature, the larvae bore into hardwood to pupate at a depth of up to 20 cm from surface. In general, this group of consumers damages trees because of the large number of larvae produced.

The living tree consumers need from two to three years to complete their life cycle. They attack living trees of all ages, but particularly young trees aged from 6 to 20 years. They have a much narrower range of host plants than the dead/dying tree consumers, only about 20 eucalypt species have been recorded as their hosts, with each longicorn species having its own eucalypt hosts. In addition, they are distributed in either southern or northern parts of Australia. Eggs are laid singly or in only small batches, in bark cracks or the sites of injuries, with each egg mass consisting of 1 to 18 eggs. The larvae do not make the radial or parallel tunnels as dead/dying consumers do, but their activities make 1 to 6 large damaged areas under the bark and they may bore into the hardwood several times before pupating there. In general, one tree supports only one species of living tree consumers, and one or a few larvae. In many cases, even a couple of larvae can kill a tree.

Insect natural enemies

Insect natural enemies cause substantial mortality of *Phoracantha*. Parasitoids are mainly recorded from Hymenoptera, including 3 families and 13 species. The genera *Callibracon*, *Syngaster* and *Jarra* of Braconidae are major larval parasitoids and *Avetianella* from Encyrtidae is the only known egg parasitoid of *Phoracantha*. Some of these parasites have been introduced to North America and South Africa for biological control of *P. semipunctata* and *P. recurva*. Known predators are mainly from coleopteran families Cleridae and Colydiidae. The clerid genera *Stigmatium*, *Trogodendron* and *Tenerus* attack *Phoracantha* larvae, and the colydiid genera *Bothrideres*, *Deretaphrus*, *Phormesa* and *Teredalaemus* prey on either adults or larvae. *Stigmatium gilberti* was observed to feed on *Phoracantha* eggs as well.

Economic and ecological importance

Most phoracanthine species of economic importance are placed in this genus and some species are serious pests of eucalypt forests in various states of Australia and many countries around the world. The widely distributed species are more likely to become pests in Australia and elsewhere. For example, *P. semipunctata* and *P. recurva* have now been established in many countries in Africa, South and North America, and the Mediterranean region, and have become serious pests of *Eucalyptus* plantations in these regions. They feed mainly on dead or stressed trees, and can survive lengthy shipment in logs between continents. Other species such as *P. obscura*, *P. punctata* and *P. tricupis*, have similar necessities and life cycles and can cause serious damage to *Eucalyptus* outside Australia if introduced. The dead/dying tree consumers are important in recycling dead and sickly trees in nature and promoting regrowth of forest.

The species *P. solida*, *P. synonyma*, and *P. acanthocera* have wider geographic adaptation than most other *Phoracantha* species, and have become pests of Australian *Eucalyptus* forests. Of these species, *P. acanthocera* has become a serious pest in all southern states of Australia. Because these species mainly live on healthy trees, they may not be able to survive lengthy shipment in logs. So far, none of them has been reported to have established outside Australia. However, if these species are accidentally introduced into other regions, they may be more harmful than dead/dying tree consumers to *Eucalyptus* plantations. Species other than those mentioned above appear not to be able to adapt to

Fig. 785 Tunnels in a eucalyptus tree made by *P. semipunctata*.

Fig. 786 Tunnel in a eucalyptus tree made by *P. acanthocera*.

a wide range of environments and are less likely to become important pests in Australia or other countries, except one species, *P. mastersi*, which is a fairly serious pest in southeastern Australia, particularly in Tasmania.

Qiao Wang
Massey University,
Palmerston North, New Zealand

References

Austin, A. D., D. L. J. Quicke, and P. M. Marsh. 1994. The hymenopterous parasitoids of eucalypt longicorn beetles, *Phoracantha* spp. (Coleoptera: Cerambycidae) in Australia. *Bulletin of Entomological Research* 84: 145–174.

Moore, K. M. 1963. Observations on some Australian forest insects, 15. Some mortality factors of *Phoracantha semipunctata* (F.) (Coleoptera: Cerambycidae). *Proceedings of the Linnean Society of New South Wales* 88: 221–229.

Wang, Q. 1995. A taxonomic revision of the Australian genus *Phoracantha* (Coleoptera: Cerambycidae). *Invertebrate Taxonomy* 9: 865–958.

Wang, Q., I. W. B. Thornton, and T. R. New. 1999. A cladistic analysis of the phoracanthine genus *Phoracantha* Newman (Coleoptera: Cerambycidae: Cerambycinae), with discussion of biogeographic distribution and pest status. *Annals of the Entomological Society of America* 92: 631–638.

PHORESY. A symbiotic relationship with transport of one organism by another. Phoretic relationships display little or no pathology to the host. Many insects, nematodes, and microbial organisms have phoretic associations with insects.

PHORIDAE. A family of flies (order Diptera). They commonly are known humpbacked flies. See also, FLIES (DIPTERA), MYIASIS.

PHRYGANEIDAE. A family of caddisflies (order Trichoptera). They commonly are known as large caddisflies. See also, CADDISFLIES.

PHOSPHORYLATION. The combination of phosphoric acid with a compound. Many proteins in eukaryotes are phosphorylated.

PHOTODYNAMIC ACTION IN PEST CONTROL AND MEDICINE. In most cases, life depends on light. For instance, photosynthesis (the process by which plants derive the energy needed for growth) provides almost all food for both humans and animals. Light has profound implications for the field of medicine. While it works as a cause of disease (e.g., UV damage of DNA) it also may be applied as a therapeutic agent (cf. photodynamic therapy). Additionally, light also acts as a means for plant defense against herbivorous animals including insects.

In contrast to the normal photobiological processes that are essential to the physiology of living organisms such as photosynthesis in green plants, biological systems also can be damaged and destroyed by non-physiological photochemical reactions after light absorption. These reactions are the result of the interference of certain dyes and many secondary plant products known as photosensitizers that are carried out either in the presence or the absence of oxygen. The latter category, which is realized under anaerobic conditions, is much less common.

The question of potential hazards to humans arises with the wide use of photodynamic sensitizers in foods, drugs and cosmetics. In addition, there is considerable experimental evidence on the photodynamic effects of several popular dyes as highly active insecticides, and some clear evidence for photocarcinogenicity of certain phototoxic agents.

The term photodynamic action or "photodynamische Erscheinung" was suggested in 1904 by Tappeiner and Jodlbauer to differentiate this activity from the photosensitization of photographic plates by dyes. Much later, Spikes and Glad defined photodynamic action as the killing or damaging of an organism, cell, or virus, or the chemical modification of a (bio-) molecule in the presence of a sensitizing dye and molecular oxygen. However, considering recent discovery of the photodynamic action in some important groups of phytochemicals, such as furanocoumarins, polyacetylenes and their thiophene derivatives, cercosporin, hypericin, and numerous photosensitizing drugs, which exclusively work at the near UV range (300–400 nm), it is apparent that photodynamic action is not limited to dyes, but includes many biomolecules.

O_2 metabolism

The oxidative processes that are fundamental to photodynamic action differ greatly from normal cellular metabolism. Differences include:

(a) While the respiratory quotient, i.e., the ratio between CO_2 production and O_2, is near unity for normal aerobic metabolism, the ratio during photodynamic action is up to 20-fold lower than the former (e.g., 0.048 vs 1.00).

(b) Normal O_2 metabolism is virtually abolished if the structure of the cell is destroyed. Whereas, the O_2 uptake during photodynamic action has been found to remain the same for both intact or hemolyzed red blood cells.

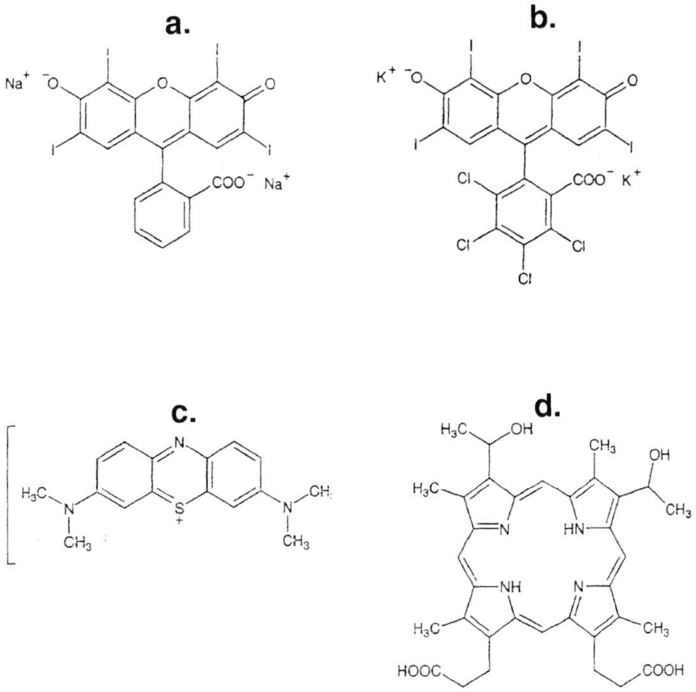

a. **b.**

c. **d.**

Fig. 787 Some examples of several dyes with photodynamic action: i.e., halogenated dyes - e.g., (a) Erythrosin ($C_{20}H_6I_4Na_2O_5$), (b) Rose Bengal ($C_{20}H_2Cl_4I_4K_2O_5$) and (c) Methylene Blue ($C_{16}H_{18}Cl-N_3S$); and non-halogenated dyes – e.g., (d) Hematoporphyrin ($C_{34}H_{38}N_4O_6$).

(c) While cyanide inhibits normal O_2 metabolism, it increases the uptake of O_2 in photodynamic action.

(d) Normal O_2 metabolism is destroyed by heat, whereas boiling does not abolish the photosensitized uptake of O_2.

Light absorption

The first step of photodynamic action is absorption of light by a sensitizer.

An important aspect of the photodynamic effect is that the active wavelengths are longer than 300 nm. Therefore, it is restricted to the regions of the electromagnetic spectrum that penetrates the Earth's atmosphere: i.e., some parts of the UV (300–400 nm), and entire spectrum of the visible light (400–700 nm). The destructive effect of the short-wave lengths (less than 300 or 330 nm) are well documented. However because the ozone layer absorbs almost all energy below 280–290 nm, the destructive region of solar light is confined to a narrow range (i.e., 280–330 nm).

The extent of penetration of ultraviolet (UV) and visible radiation into skin varies considerably. While UV-C (200–280 nm) does not penetrate beyond the epidermis, UV-B (280–320 nm) and UV-A (320–400 nm) enter the dermis; and visible light penetrates further into the subcutaneous tissue. UV penetration into tissue is estimated at less than 0.1 mm, whereas the typical penetration depth in living tissue of the red light used for photodynamic therapy ranges between 1–3 mm.

Mechanisms

When the absorption of a quantum of radiation excites a sensitizer, the energy imparted by this excitation is related to wavelength by the following equation: $E = hv$, where E is the energy of radiation, h is Planck's constant (i.e., 2.8591), and v is the frequency of the radiation. On the other hand, as $v = c/\lambda$, where c is the velocity of light and λ the wavelength, one can substitute to obtain $E = hc/\lambda$ (i.e., $2.8591 \times 10^5/\lambda$). From this relationship, the energy imparted at a given wavelength can be determined. Basically, the shorter the wavelength, the

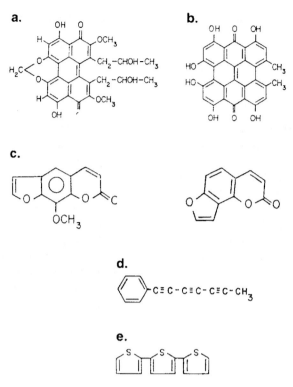

Fig. 788 Chemical structures of several secondary plant products with photodynamic action: (a) cercosporin; (b) hypericin; (c) furanocoumarins (left, linear: e.g., 8-methoxypsoralen, i.e., 8-MOP or xanthotoxin; and right, angular: e.g., angelicin); and finally polyacetylenes (e.g., phyenylheptatriyne) and their thiophenes - e.g., α-terthienyl (d and e, respectively).

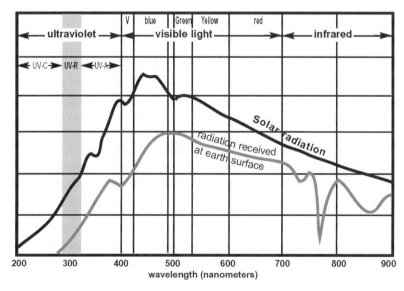

Fig. 789 Solar radiation and its participating spectra received at the Earth's surface: i.e., (1) ultraviolet ($\leq 400\,nm$), (2) visible light including violet (abbreviated by V; ca. 400–420 nm), green (420–490 nm), yellow (490–ca 600 nm) and red (600–700 nm), and finally (3) infrared (700–900 nm). Adapted from The Ozone Layer (UNEP, Nairobi, 1987).

higher the energy. At 300 nm, for example, 95.3 kcal/mole would be imparted to the molecule, which is sufficient to disrupt most covalent bonds. Basic to any energy transfer is that the molecule in question must exhibit an absorption spectrum at the wavelengths of excitation, and this absorption generally follows Beers law ($I = I_0 10^{-Ebc}$).

Phytodynamic action begins when a sensitzer (*Sens*) absorbs light, giving an excited state (*Sens**), often (but not always), the triplet. *Sens** can either react directly with the substrate (Type I reaction; less common) or with oxygen (Type II reaction). Therefore, high substrate concentrations (as well as electron-rich or hydrogen-atom-donating substrates) promote Type I reactions and high oxygen concentrations favor Type II reactions. Furthermore, in biological systems, where binding of sensitizer to substrate is common, Type I reactions are particularly favorable. While a type I reaction results in hydrogen atom or electron transfer, yielding radicals or radical ions, a type II reaction leads mainly to singlet molecular oxygen by energy transfer.

Singlet oxygen can be produced in high yield in the Type II reaction by energy transfer from *Sens**. It is an electronically excited state of oxygen, with a lifetime that varies about 3–4 µsec in water and as long as 0.1 sec in solvents with no hydrogen atoms. In biological lipids and memberanes, it probably has a lifetime considerably shorter than that in most organic solvents because of quenching by proteins and escape from the membrane into cytosol. The mechanism of action of the halogenated xanthene dyes such as rose bengal and erythrosin is considered to occur by a Type II mechanism.

The dye photosensitizer does not enter into the toxic reaction. The photosensitizer is a catalyst, not a participant. A single dye molecule is able to cycle through light absorbance, movement to the excited singlet state, transfer to the excited triplet state, sensitization of the ground state oxygen to the excited singlet state, and eventual return to the ground singlet state of the dye in approximately 10 msec, or less. It is then possible that a single dye molecule could be responsible for the generation of thousands of molecules of singlet oxygen per minute. A single dye molecule is then capable of initiating oxidation reactions which can destroy thousands of different target molecules rather than the single target molecule destroyed by a single organophosphate molecule, for instance.

Photodynamic action (photochemical damage by oxygen) is caused by oxidation of biological target molecules, and can lead to (1) membrane lysis by oxidation of unsaturated fatty acids and cholesterol, (2) enzyme deactivation by oxidation of amino acids (methionine, histidine, tryptophan, tyrosine and cysteine), and (3) oxidative destruction of nucleic acid bases (primarily guanine).

The responses of living organisms to photosensitized modification are varied, ranging from mild irritation to death. The precise nature of the response depends on many factors, but the majority of cases can be explained based on the changes produced in individual cells. Cellular photomodification is in fact a multi-step process which is carried out in six steps.

Step 1: The photosensitizer must get to the target cell. This involves introduction into the organism during feeding, via injection or diffusion. Once in the organism, the photosensitizer may be carried in combination with proteins, lipoproteins, or other molecules. At the target cell, the photosensitizer may need to diffuse through a vascular wall, a cell wall and/or a cell membrane. Ultimately, it must localize at or near the site of photomodification. Variations in any of these processes can significantly alter the final effect.

Step 2: Once the sensitizer has reached its site of action, it may remain free or may be bound to various biomolecules. It can also be metabolized to a form that may be more effective or less effective as a photosensitizer. The use of precursors, such as δ-aminolevulinic acid which is metabolized by the organism to a photosensitizing porphyrin, is an example of this possibility.

Step 3: At the site of action the sensitizer must be able to absorb light. This ability can be altered by the environment and physical state of the sensitizer in the organism. Binding, aggregation, metabolism and altered dielectric constant (as in a lipid membrane) are all variables that may significantly alter the absorption spectrum of the sensitizer.

Step 4: After absorbing light, the sensitizer must make use of the energy in a way that effects a change in the cell. It may transfer energy to oxygen creating an electronically excited state such as singlet oxygen, which can then react with a cellular structure nearby. Alternatively, it can react directly with cellular biomolecules. But, in the cellular environment, it may also be quenched by the impressive array of antioxidants available in cells. Therefore, the excited state properties and reactivities of photosensitizers *in vivo* will not always be the same as they are in simple solution.

Step 5: The next step is the reaction of either the excited state sensitizer or a reactive intermediate, such as singlet oxygen, with cellular biomolecules which critically impair cell function and/or survival. Many biomolecules and cellular structures are photomodifiable, but not all of these affect cells in ways that result in cell death or irreparable damage. For example, it has been known for many years that cell lipids are peroxidized by photosensitized modification, but a definitive link between that peroxidation and cell killing has remained elusive.

Step 6: Finally, if enough critical cells in an organism are affected, then the foregoing steps will result in death or impairment of function. While each of these steps is critical and can significantly affect the degree of cellular modification, an in-depth analysis of Step 5 follows, namely the reactions of photosensitizer and light which lead to the critical impairment of cell function.

Effects of photosensitizers and light on cells

Photosensitizers and light can affect a variety of biomolecules and cellular structures. Proteins, lipids and nucleic acids are all susceptible to photomodification. Carbohydrates are much less sensitive. This means that direct effects on cellular energy stores in the form of glucose or glycogen are unlikely to be significant. Conversely, effects on enzymes,

nuclear DNA and lipid membranes may be important. Several sensitizers are accumulated in lysosomes where they may sensitize swelling and increase permeability of the lysosomal membrane allowing enzymes to be released from the lysosome. Other sensitizers seem to localize in mitochondria where they cause swelling and can affect the function of membrane-bound enzymes involved in energy production for the cell. Finally, a large variety of sensitizers, including many of the porphyrins and xanthene dyes, localize in the plasma membrane (and perhaps other membranes) altering the permeability properties of the surface memberane of the cell when illuminated, either by affecting membrane proteins or lipids. To summarize, photodynamic action involves photooxidation of various substrates which results in inactivation of biological systems, distortion of membranes, inactivation of enzymes, photocarcinogenesis, cell death and losses in other functions.

How do photosensitizers and light kill cells?. The above-mentioned effects of photosensitizers and light on cells offer many possible means by which a cell may be killed. Lysosomes may release degradative enzymes into the cytoplasm, mitochondrial production of ATP may be inhibited, plasma membrane barrier function may be compromised, and DNA damage may prevent cell replication and/or impair transcription required for protein synthesis and normal cell function. But some of these effects are critically

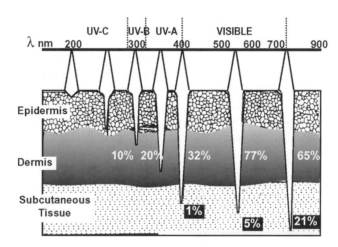

Fig. 790 Extent of penetration of ultraviolet (UV) and visible light into human skin. The drawing is from the late B. E. Johnson, Department of Dermatology, University of Dundee (Scotland, U.K.). Courtesy of Springer Verlag and the editors of Photosensitisation: Molecular, Cellular and Medical Aspects (G. Moreno, R. H. Pottier, T. G. Truscott: 1988).

related to cell death and others are merely consequences of the process of cell death.

Presently two major forms of cell death are recognized. One form, known as necrotic death or necrosis, is a degenerative cell death resulting from severe cell injury. The second form, known as apoptotic death or apoptosis, is a programmed cell death which is a cellularly controlled process. The death can occur under pathological conditions, such as cell injury, but also appears to be the mechanism by which organisms delete certain cell populations when they are no longer needed in the developmental process. While a hallmark of apoptosis is a selective cleavage of DNA, necrosis produces a random degradation of DNA.

An increase in intracellular calcium concentration has been suggested as the trigger of both forms of cell death (i.e., apoptotic and necrotic). Even cells which tolerate transient elevation of intracellular calcium concentration as part of their normal function, e.g., contracting cardiac cells where such elevation of calcium serves as the trigger for contraction, cannot tolerate this increased concentration continuously. It is well known that cardiac cells are susceptible to destruction by calcium overload. It has been documented that cellular photosensitization produces cell death by an apoptotic mechanism. Although apoptosis is a carefully orchestrated series of reactions within the cell leading ultimately to death, calcium is the trigger for this process. Thus, one should look for the mechanism that can increase intracellular calcium concentration as the seminal effect of photosensitization. As it is known that cell membranes are responsible for maintaining the very low calcium concentrations inside normal cells, i.e., 10^{-7} M, an obvious way for such increase is by a change in membrane permeability to calcium.

Multiple mechanisms of xanthene dyes in insects.
Based on the foregoing studies on the toxicity of several xanthene dyes on insects, multiple mechanisms should be considered: e.g., light dependent mechanism, light independent mechanism and developmental toxicity mechanism.

Light dependent mechanism (or photodynamic action) in insects involves the ingestion of the dye by the insect, followed by exposure to a visible light source which results in the death of an insect. It is quite fast and needs comparatively lower concentrations of the photosensitizer and a source of light. In contrast, the light independent mechanism is slow,

needs a higher concentration of sensitizer and operates in the absence of light.

In the developmental toxicity mechanism, the insect is exposed to a sublethal dose of the compound in the earlier stages of development. This results in mortality or some adverse morphological abnormalities during development, such as delayed development, growth retardation, and changes in fecunditiy and fertility.

Photodynamic dyes
With the exception of hematoporphyrin, a substance known to be derivable from hemoglobin or hemin, other dyes studied for their photodynamic action are mainly confined to the synthetic dye molecules with halogenated structures.

The first documented (although qualitative) study in which light was understood to cause an enhancement of a chemically induced toxic effect was that of A. Marcacci in 1888. He reported that alkaloids were more effective against seeds, plants, fermentations, and amphibian eggs in sunlight than in the dark. This appears to be the first formal report that some materials have a light-enhanced toxic effect, even though a few dyes were already used for their anti-bacterial properties without recognition of the involvement of light action.

In 1900, a study on the toxicity of a dye (acridine) toward paramecia reported the results of the first quantitative studies on photodynamic action. Low concentrations of acridine and other dyes (at 50 ppm), which had no effect on microorganisms such as paramecium in the dark, were found to lead to a rapid death on illumination in the presence of oxygen. An increase of 2 to 3 orders of magnitude in the acridine sensitized mortality of paramecia was recorded. Also, those light rays which generated the greatest fluorescence in the dyes induced the largest biological reaction.

Following this discovery, it was soon established that many dyes and pigments could sensitize almost any type of biological system (plants, animals, cells, viruses, biologically important molecules) and result in cell damage, induction of mutations or cancer, and death.

Photodynamic effect of rose bengal and erythrosin against insects
The first reported use of photodynamic action against an insect target was a study of *Anopheles*

and *Culex* mosquito larvae, which were shown to be susceptible to solutions of several classes of dyes in direct sunlight. The most active dyes were the halogenated fluorescein derivatives, erythrosin and rose bengal, alone and in mixture. There were no deaths reported from populations used as control for the effect of dye or light alone. A mixture of rose bengal and erythrosin was found to be the most effective, showing high larvicidal activity at a dilution of one part of dye-mixture in 1.5 million parts of water in direct sunlight. Damage to larval alimentary tract was noted.

Another study found that *Anopheles maculipennis* larvae with a series of dye solutions [10 ppm (W/V)] and exposed to sunlight were sensitive to dyes. Field tests in small ponds demonstrated that rose bengal or acridine red caused 100% mortality, and erythrosin in about 90% mortality. However, eosin and fluorescein were ineffective. The larvae of another species of *Anopheles* (*A. superpictus*) and those of *Aedes aegypti* were also similarly sensitive, whereas there was no effect on the mosquito fish (*Gambusia* sp.) that were present.

The modern era of photodynamic insecticide studies began with the report of T.P. Yoho and associates in 1971. The paucity of literature on the effects of phototoxic agents on insects, until this period, is not surprising. Light often has been a forgotten or underestimated factor in the study of insects. Even until recently little attention has been paid to its role as a parmeter in the mortality of insects.

Yoho's group at West Virginia University published several investigations on the efficacy of photodynamic action against the adult house fly using primarily the halogenated series of dyes. These papers compared toxicological data with the parameters of light source and intensity, dye structure and concentration in the diet, source of light, and length of light exposure. It was observed that all house flies fed on diets containing either rhodamine (625 ppm) or erythrosin B (1,250 ppm) were killled after a 3-hr exposure to light, whereas, when the dye-fed house flies were kept in the dark, no adverse effect was observed. It was also reported that the midgut epithelial cells appeared to be damaged and that the external symptoms associated with toxicity suggested an involvement with the nervous system.

Yoho and associates also studied a series of 14 food, drug and cosmetic dyes for efficacy in photodynamic toxicity to house fly adults. Nine of the

14 dyes tested produced up to 100% mortality in illuminated flies, particularly in those receiving liquid diets with a concentration as low as 0.25% dye (2,500 ppm). Lower concentrations were not tested.

The results of these investigations motivated other scientists to work on insecticidal properties of a series of common dyes. It also led to the registration of a formulation of erythrosin B under the name of "Synerid".

Practical aspects of erythrosin B in insect control

In 1980, G. D. Pimprikar and associates reported on their attempts to control house flies in a commercial caged layer house, under minimal light, using weekly applications of aqueous solutions of erythrosin B directly on the manure (about 650 mg a.i./m^2). They reported decreases of adults and larval house flies up to 90% with respect to pretreatment levels with no change in the beneficial soldier fly larval population. Their additional studies on the effects of several fluorescein derivatives in the diet on each developmental stage of the housefly revealed that the treated adults exhibited lowered fecundity, the eggs had a reduced viability, and mortality was observed in each life stage of the house fly. This experiment which was also conducted under minimal light, required a relatively large amount of dyes for 100% mortality: e.g., 0.5 g a.i./kg diet for rose bengal and 2 g a.i./kg for erythrosin B.

In 1981, T. L. Carpenter and associates reported on the role of fluorescein on enhancement of the phototoxicity of rose bengal, and some other xanthene dyes, towards *Aedes* larvae. Subsequently, a United States patent was issued covering the synergism of a nontoxic dye with a demonstrated toxic dye in both house fly and mosquito systems. Further studies resulted in an over two-fold increase in the toxicity of erythrosin B against house flies.

In 1983, N. C. Respicio and J. R. Heitz reported development of resistance to erythrosin B in the house fly. While a laboratory strain developed only 6-fold resistance after 40 generations, a wild strain experienced a 48-fold resistance after 32 generations of exposure to increasing levels of erythrosin B in the diet. Upon removal of the selection pressure for 20 generations, the resistane remained constant. Furthermore, the erythrosin B-resistant strain was cross-resistant to phloxin B and rose bengal, as it is to be expected since they function by same mechanism.

Whereas, no cross-resistance of the resistant strain to conventional insecticides was observed.

In 1984, Primprikar and Heitz observed an unusually high insecticidal activity in *Aedes* mosquito larvae which had been illuminated after exposure to the insoluble free acid forms of the xanthene dyes. In all previous studies, the larvae had been treated with the water soluble salt forms of the dyes and the larvae consumed the dye as they ingested in the water. With the insoluble dyes they were able to filter feed on dye particles and thereby receive higher levels of these chemicals. Toxicity ratios ranged up to 2 orders of magnitude between the soluble and insoluble forms of the same compound.

A simultaneous report of Carpenter and associates showed that insoluble forms of the xanthene dyes were 10-fold more effective against *Culex* mosquito larvae than the soluble forms, when used at the same dosage. They also reported that when the insoluble forms of the dyes were dispersed with a surfactant, such as sodium lauryl sulfate, the dyes were 50- to 60-fold more effective than the soluble forms. Depending on the concentrations under consideration, the increase in toxicity ranged from 26- to 229-fold for erythrosin B and 13- to 206-fold for rose bengal. Therefore, doses of free acid, dispersed formulations were in the lower ppm to upper ppb range for up to 24 hours of light exposure.

In 1985, Carpenter and associates reported that a series of 8 dispersants for use with the insoluble forms of the dyes were evaluated and none were toxic alone. Erythrosyin B, dispersed with sodium dodecyl sulfate, was the most toxic against *Culex* mosquito larvae. In small-scale field tests, this formulation caused significant reductions in larval and emergent adult populations of *Culex* mosquitoes at concentrations ranging from 0.25 to 8.0 ppm. Nevertheless, possible limitation of the activity in "naturally basic waters," as a result of conversion of Erythrosin B to the soluble ionized form, was suggested.

Following the studies on phototoxic activity of erythrosin B against insects, a formulation of this dye has been registered under the name of "Synerid Fly Control B." Presently, this is the only dye commercially available for insect control.

Mode of action of photosensitizing dyes in insects

Photoactive dyes were found to have numerous effects on insects. Acetylcholinesterase from the black imported ant and the boll weevil was suscepti-

ble to dye-sensitized photooxidation *in vitro*, but levels were not depressed in insects killed by photodynamic action. It was also reported that in the American or Oriental cockroach, photodynamic action induced by rose bengal or erythrosin B caused a significant decrease in hemolymph (over 90% reduction by erythrosin B) and a large increase in the crop volume (over 3-fold increase in erythosin B treatment). Furthermore, photodynamic action induced by erythrosin B resulted in up to 40% reduction of hemocytes relative to controls.

Under fluoroscent light and at rose bengal treatment levels of 1 to 20 ppm, *Culex* larvae were more susceptible than *Aedes* larvae, and early instars were more susceptible than later instars. Physiological and morphological abnormalities were observed in the pupal and adult stages when the mosquito was treated during the larval period, suggesting improper chitin formation in the insect. This sometimes resulted in incomplete release of the pupal stage from the larval cuticle and of the adult stage from the pupal cuticle. Where this was observed, mortality resulted. They also observed larval-pupal intermediates similar to those found after treatment with insect growth regulators.

The abnormalities observed in insects have led to the hypothesis that these photosensitizers may have an effect on the molting hormones. There is experimental evidence that the titers of the two most prominent molting hormones in insects (e.g., α-ecdysone and β-ecdysone) are distinctly different in the erythrosin B-treated insects as compared to the control insects. According to these studies the imbalance of the molting hormone titers during the critical stages of development may contribute to the abortive molting or to the development of morphologically abnormal individuals.

Persistence of erythrosin in biological system. In 1981, T.E. Fairbrother and associates made a comprehensive study of the toxicological effects of erythrosin B and rose bengal on the face fly. These dyes were incorporated with manure, either by hand or by passage of the dye through cattle. When larvae developed on the manure containing either erythrosin B or rose bengal, mortality was observed at each life stage. Wing deformation in some emerging adults and incomplete removal from the puparium were also recorded.

1. Conversion of sensitizer (Sens) to an electronically excited state (Sens*)

a) Sensitizer is either converted to a short-lived singlet state, i.e.

$$Sens \underset{h\nu}{\Rightarrow} {}^1Sens^*, \text{ or}$$

b) It undergoes further to the longer-lived triplet state, i.e.

$${}^1Sens^* \underset{h\nu}{\Rightarrow} {}^3Sens^*$$

2. Production of radicals or singlet oxygen ($^1O_2$)

$$Sens^* ({}^1Sens^* \text{ or } {}^3Sens^*) \underset{\text{Substrate or solvent}}{-\!-\!-\!-\!-\!-\!-\!-\!-\!\Longrightarrow} Radicals \textbf{ (Type I Reaction)}$$

$$Sens^* ({}^1Sens^* \text{ or } {}^3Sens^*) \underset{\text{Molecular oxygen } (O_2)}{-\!-\!-\!-\!-\!-\!-\!-\!-\!\Longrightarrow} {}^1O_2 \textbf{ (Type II Reaction)}$$

3. Oxidation of vital molecules

$$Radicals \underset{\text{Molecular oxygen } (O_2)}{-\!-\!-\!-\!-\!-\!-\!-\!-\!\Longrightarrow} Oxygenated\ Products$$

$$Singlet\ oxygen\ (^1O_2) \underset{\text{Substrate}}{-\!-\!-\!-\!-\!\Longrightarrow} Oxygenated\ Products$$

Fig. 791   A schematized version of reactions involved in the process of photodynamic action. These include (1) conversion of sensitizer to an electronically excited state, (2) production of radicals or singlet oxygen, and (3) oxidation of vital molecules. Energy (E) imparted by this excitation is related to Planck's constant (h) and the frequency of radiation (v) (see the text).

Adults, held from emergence and illuminated with visible light, were observed to have a much higher mortality than controls, thus suggesting that dye sequestered in the insect body during development from larvae to adult was responsible for the toxicity. This is the first report of photodynamic action occurring in a life stage different from the life stage in which the dye was ingested. This was confirmed in 1982 by studies of the inhibition of growth and the photodynamic action caused by rose bengal and erythrosin B in the housefly. House flies which had consumed a nonlethal amount of dye in the larval period exhibited a considerable light-dependent toxicity in the adult stage.

Effect of photosensitizing dyes on other organisms

Although most of the photodynamic studies with the fluoroscein dyes concerned insects, and to some extent other animals, phototoxicity of eosin to chloroplasts and that of erythrosin B to the infectious 3rd stage larvae of gastrointestinal nematodes have been recorded. Furthermore, the effect of photodynamic action on the eggs of sea urchins also has been reported. Simultaneous irradiation of eggs of sea urchins (*Paracentrotus lividus*), incubated in 0.1 mmol/liter solutions of two singlet oxygen sensitizers (i.e hematoporphyrin derivative- Hpd- at

60 ppm or rose bengal- RB- at 102 ppm solutions), led to different phototoxic reactions. For example, inhibition in formation of fertilization membrane (which is required to inhibit polyspermy), shrinkage of the egg surface, appearance of many small holes in the eggs, formation of larger holes, breakage of the eggs, and finally leakage of cytoplasm.

For additional information on properties of photosensitizing dyes refer to the following section.

Photodynamic therapy (PDT)

The use of photoactive dyes as therapeutic agents (i.e., photodynamic therapy) only began in the 1960s. In 1966, Lipson reported on the practical treatment of a patient with metastatic chest wall breast cancer by a hematoporphyrin derivative. Photosensitizing dyes are currently being investigated as potential agents in the photo-chemotherapeutic treatment of tumors and for other medical applications. In Canada, the use of hematoporphyrin derivatives in tumor phototherapy has recently received approval for clinical use. Furthermore, second-generation photosensitizers such as phthalocyanines or benzoporphyrins that absorb in the far-red or near-infrared region are receiving increasing attention because of the ability of light of these wavelengths to penetrate tissue. Such absorption properties would permit deeper penetration of exciting light into the skin and lower doses of photosensitizer.

The laser beam (632 nm) has also been successfully used for endoscopic photodynamic therapy of gastrointestinal neoplasms, following a 60 min infusion of the patient with hematoporphyrin (2.5–5.0 mg/kg of body weight). Furthermore, the controlled application of photosensitization has been exploited for the relief of the symptoms of some skin diseases, and therapy of different types of malignant tumors.

Viral components, including nucleic acids and lipid-rich envelopes, are potential targets for photochemical attack. Indeed, a number of photosensitizers have been able to inactivate viruses at both the DNA and the envelope levels. Nevertheless, it has been shown that enveloped viruses are significantly more sensitive (by several orders of magnitude) to photodynamic destruction than are non-enveloped viruses.

Hematoporphyrin, also known as photofrin or porphyrin, is the only photosensitizer to date to have been used extensively in the clinical treatment of a variety of malignancies. Since it is activated at a wavelengh of 630–635 nm, a wavelength that is seriously decreased by hemoglobin, its antiviral activity in blood (*in vivo*) is expected to be reduced. Notwithstanding this fact, it has been found to inactivate a number of human viruses in tissue culture or blood: i.e., HSV (type 1 & 2), VSV, CMV, measles, SIV, and HIV (all of which are enveloped viruses). In contrast, non-enveloped viruses have been shown to be unaffected: e.g., ECHO 21, poliovirus (P1) and adenovirus (AD12).

Because there is differential photodynamic activity of hematoporphyrin on enveloped and non-enveloped viruses, inactivation of the virus envelope is suggested. Photoactiviation occurs primarily as a result of the oxidative modification of the lipid and protein components of the viral envelope. The inactivated virus is then unable to adsorb to or to penetrate into host cells, hence it cannot initiate an infection or induce the formation of viral antigens. Two other groups of photosensitizers, psoralens and hypericin, also have been found to be virucidal agents.

Extent of penetration

Although UV, visible, or near infrared radiation could be possible energy sources for photodynamic therapy of patients, UV radiation is considered disadvantageous because of low penetration. In the course of studies on photodynamic therapy of tumors in humans, it was soon recognized that by using hematoporphyrin derivative and long wavelength light (e.g., red: above 600 nm), both greater activity and deeper tissue penetration could be achieved. While UV penetration into tissue is estimated at less than 0.1 mm, the typical penetration depth in living tissue of the red light used for photodynamic therapy ranges between 1–3 mm.

Mode of action in photodynamic therapy

Singlet oxygen is the basis for photodynamic killing of cells for several classes of sensitizers, including xanthene dyes such as rose bengal. Although both proteins and lipids are photooxidized, proteins are generally believed to be a critical target leading to the disruption of function in cell membranes. Singlet oxygen is a genotoxic substance which can be produced *in vivo*, inside or outside cells causing severe damage to various biological macromolecules, even to those deeply embedded inside the cells such as DNA.

Aerobic organisms have evolved antioxidant defenses against most reactive oxygen species (e.g., superoxide dismutase, catalase, glutathione peroxidase and DNA repair enzymes, and low molecular weight agents such as α-tocopherol, β-carotene and ascorbic acid), but the antioxidation mechanism does not appear to be always effective. In this case, when active oxygen species are not adequately removed, an oxidative stress situation appears in the cell resulting in major metabolic dysfunctions. These include membrane peroxidation, rise in intracellular free calcium ions, cytoskeletal disruption and DNA damage. Singlet oxygen can mediate the oxidation of major cellular molecules. DNA, proteins and lipids are all at risk. When singlet oxygen is generated inside cells, it is very genotoxic leading to an important mutagenic effect. When singlet oxygen is produced extracellularly, it reacts with the lipids of the cellular membrane as the major target. In addition, DNA also can be attacked. Reactions leading to peroxidation of the membrane and DNA are discussed by Legrand-Poels *et al.* (1993).

Effect on tumor and non-target tissues

Tumor destruction in photodynamic therapy is the result of the combination of direct cellular toxicity and damage to tumor microvasculature. It has long been realized that a significant proportion of photodynamic tissue destruction involves the vasculature. These phenomena appear to be caused by tissue interactions with toxic oxygen compounds which are formed when light interacts with photosensitizing agents.

The targets of photodynamic sensitizers such as hematoporphyrin are predominately the plasma membrane, mitochondria and lysosomes and, to a lesser extent, nuclear sites. Photodynamic treatment has been reported to affect transmembrane transport systems in fibroblasts, to influence lipid bilayer membranes, and to alter membrane potential in renal cells. Although nuclear aberrations, mutagenicity, micronuclei and strand breakage occur in photodynamically treated patients, the induced abnormalities may not be severe enough to cause side effects in these patients. However, there is no question that major changes occur to both tumor and surrounding normal microvasculature during and following photodynamic therapy. Changes in the endothelium, smooth muscle contraction, and increased capillary permeability have also been observed during therapy.

In addition, the oxidative stress which is mediated by intracellular generation of singlet oxygen by rose bengal photosensitization in cell culture, has been found to reactivate HIV-1 from latently infected lymphocytes and monocytes. It is a possibility that HIV-1 reactivation caused by rose bengal leads to AIDS progression by impairing the antigen presentation function or increasing infection of the adjacent T cell population of the skin, *in vivo*.

Cellular porphyrin distribution by fluorescence microscopy shows staining of the nuclear membrane without induction of damage. Recent experiments revealed an initial fast (within minutes) and uniform penetration of the dyes into the nuclear envelope and chromocenters. This may indicate that the matrix between the nuclear membranes is a primary target for the dyes.

Risks and benefits

Analysis of the risks and benefits of photodynamic action is very complex, especially when the photodynamic therapy of patients suffering from malignancies, AIDS or other disorders is taken into consideration and photodynamic effects in humans such as photosensitive porphyrias (i.e., porphyrin metabolism disorders), drug photosensitivity, and photoallergies are included.

With the application of laser beam technology in endoscopic photodynamic therapy, successful treatment of certain internal cancers has been achieved (e.g., gastrointenstinal neoplasms), and experimental results from photoinactivation of viruses by hematoporphorin derivatives in cell-containing blood products also are promising. In fact, with the impressive light-mediated activity of hematoporphyrin, a new chapter in clinical trials for the treatment of HIV-infection and AIDS has been opened. Furthermore, the discovery of the second-generation photosensitizers that absorb in the far-red or near infrared region (e.g., phthalocyanines or benzoporphyrins) is promising because of the ability of these light wavelengths to penetrate tissue. Such absorption properties would permit deeper penetration of exciting light into the skin and lower doses of photosensitizer.

The simultaneous appearance of numerous reports on the success of phototoxins as insecticides also has played a great role in a growing optimism concerning the future of photodynamic agents in agriculture. This is manifested by the recent registration of erythrosin B in the U.S.A. for insect control.

Both the economic and the environmental aspects of using photodynamic pesticides, with special reference to dyes, have been studied. While some literature lists the prospects of using photodynamic dyes, others suggest the potential dangers of food, drug, and cosmetic dyes. Often they are considered to be extremely safe and pose no threat to the health or welfare of the applicator or environment in field usage, with many of them being registered as food additives. Arguments have been presented for speedy registration, requiring few label restrictions, and limited toxicological testing of the photoactive dyes.

An important reason for serious concern about the environmental threat of the photodynamic agents is the method by which their toxicity is evaluated. Since light has not been considered as a parameter, in many toxicological data which are usually presented by LD_{50} for rats or mice, the validity of such data needs to be reexamined before making a general conclusion on safety.

Because the time required for initiation of photoxicity is very low, detoxification does not appear to prevent sublethal toxicity. Therefore, development of new strategies in the marketing of foods, drugs and cosmetics and new techniques in toxicological studies of these compounds (using light as a parameter) is indispensable. More needs to be known about the long-term effects of these materials and their terminal residues, including their use as insecticides.

Cyrus Abivardi
Swiss Federal Institute of Technology,
Zurich, Switzerland

References

Blum, H. F. 1964. *Photodynamic action and diseases caused by light.* Hafner Publishing Co., New York, New York. 309 pp.
Fondu, M., H. van Gindertael-Zegers, G. Bronkers, H. Stein, and P. Carton (eds.). 1988. *Food additives tables,* updated edition. Compiled by Food Law Research Centre, University of Brussels. Elsevier, Amsterdam, The Netherlands. (unpaginated)
Heitz, J. R., and K. R. Downum (eds.). 1987. *Light-activated pesticides.* ACS Symposium Series 339, Washington DC. 339 pp.
Heitz, J. R., and K. R. Downum (eds.). 1995. *Light-activated pest control.* ACS Symposium Series 616, Washington DC. 279 pp.
Legrand-Poels, S., M. Hoebeke, D. Vaira, B. Rentier, and J. Piette. 1993. HIV-1 promoter activation following an oxidative stress mediated by singlet oxygen. *Journal of Photochemistry & Photobiology (B)* 17: 229–237.
North, J., J. Neyndorff, and J. G. Levy. 1993. Photosensitzers as virucidal agents. *Journal of Photochemistry & Photobiology (B)* 17: 99–108.

PHOTOKINESIS. Kinesis response with respect to a gradient of light.

PHOTOPERIOD. Daylength. Length of daylight during a 24-hour day-night cycle. See also, DIAPAUSE.

PHOTOPERIODISM. Response of an organism to periodic changes in day length.

PHOTOSYNTHESIS. The utilization by plants of the energy provided by sunlight to split water and fix (synthesize) carbon dioxide into sugars.

PHOTOTAXIS. The taxis response of a cell or organisms toward or away from light.

PHRAGMA. (pl., phragmata) An internally projecting structure or internal ridge in the insect body, usually an extension of the endoskeleton, but also a membrane that partitions the body.

PHRAGMOSIS. A condition in which the head or tip of the abdomen is used as a plug for the nest entrance. This behavior occurs in ants and termites, usually performed by the soldiers.

PHTHIRAPTERA. A group of ectoparasitic insects affecting mammals and birds. Considered by some to be an order, here it is divided into two orders, Mallophaga (chewing lice) and Siphunculata (sucking lice). See also, CHEWING LICE, SUCKING LICE.

PHYLETIC LINES. Links drawn between present and past taxa that imply their evolutionary relationships.

PHYLETIC SPECIATION. The gradual transformation of one species into another without an increase in species number at any time within the lineage. Also called vertical evolution or speciation.

PHYLOGENETIC TREE. A graphic representation of the evolutionary history of a group of taxa or genes.

PHYLOGENETICS. Classification of insects (or their genes) based on their evolutionary relationships, or reconstruction of their evolutionary history. Increasingly, relationships are being deduced by comparison of DNA sequences.

PHYLOGENY. The history of evolution of a group of taxa or their genes. The ordering of species and their ancestors into higher taxa based on evolutionary relationships.
See also, CLASSIFICATION

PHYLOSPHERE. The microenvironment associated with an individual plant leaf.

PHYLLOXERANS. Members of the family Phylloxeridae (order Hemiptera). See also, BUGS.

PHYLLOXERIDAE. A family of insects in the order Hemiptera. They sometimes are called phylloxerans. See also, BUGS.

PHYLUM. (pl., phyla) A unit of classification, one of the major divisions of the kingdom and containing several classes. Class Insecta generally is placed in the Phylum Arthropoda of the Kingdom Animalia, as follows:

Phylum: Arthropoda
   Subphylum Trilobita – Trilobites (these are extinct)
   Subphylum – Chelicerata
      Class Merostomata – Horseshoe crabs
      Class Arachnida – Arachnids (scorpions, spiders, ticks, mites, etc.)
      Class Pycnogonida – Sea spiders
   Subphylum Crustacea – Crustaceans (amphipods, isopods, shrimp, etc.)
   Subphylum Atelocerata
      Class Diplopoda – Millipedes
      Class Chilopoda – Centipedes

Class Pauropoda – Pauropods
Class Symphyla – Symphylans
Class Entognatha – Collembolans, proturans, diplurans
Class Insecta – Insects

In this classification system, the class entognatha and the class insecta are separate, though they can be placed together into the superclass hexapoda. Formerly, however, entognaths were commonly considered to be in insecta. other variations also exist. See also, CLASSIFICATION, ORDERS.

PHYSICAL CONTROL. Control techniques that are based on physical properties of the environment to kill insects. examples of physical control include cold storage, heating, burning, and modification of the gaseous atmosphere in storage.

PHYSICAL GILL. A bubble of air that adheres to the body of aquatic insects, providing a reservoir of air or plastron.
See also, TRACHEAL SYSTEM AND RESPIRATORY GAS EXCHANGE.

PHYSICAL MANAGEMENT OF INSECT PESTS. Physical control methods aim to prevent or reduce pest invasion into a crop. various physical means function either mechanically or by affecting insects' viability or behavior. For example, insect suction devices, insect glue, and electromagnetic energy suppress insect populations. Insect exclusion screens also reduce insect density, though not through mortality. Also, color and chemicals are used to change insect behavior. Physical control methods may have some shortcomings, which must be weighed against their advantages. Physical control methods generally do not interfere with other control methods. Screens are often crucial for the implementation of Integrated Pest Management (IPM) programs. They enable the use of biological control agents as well as the use of insect pollinators. Generally, most physical control methods are environmentally safe, fit well into IPM strategies, and greatly reduce the use of chemical control.

Insect-proof screens and covers

The physical exclusion of insects from a greenhouse or open field is aimed at preventing direct crop damage by insects in general, and the incidence of insect borne virus diseases in particular. Additional outcomes of the reduction of pest damage are a marked reduction in the use of pesticides, the ability to use biocontrol agents as a complementary control measure, and more effective use of insect pollinators. Pest exclusion can be obtained by applying appropriate fabrics, of mesh aperture smaller than the insects' body width, or by fitting insect repellent fabrics. These fabrics are positioned over the plants in open fields, or attached to ventilators and doorways of greenhouses. Due to the exclusion feature of the technique, the fabrics must be applied before the crop is sown or planted because they do not suppress insect populations. Screens impede ventilation and sunlight; thus, to avoid adverse effects on crops and their susceptibility to diseases, compromises are necessary in the management of air flow, light, temperature and humidity. Various types of screens and plastic covers are known to exclude insects, and the challenge for the grower is to select the type of screening best suited to solve his specific problems. Although screening is a very efficient way to prevent primary pest penetration, some pest individuals manage to enter the greenhouses, despite all efforts. These insects may build up an indoor population and cause damage. When they exceed the economic threshold level it is necessary to apply complementary control measures, such as biocontrol or environmentally safe insecticides, which would be useless without the netting. The success of pest exclusion, especially its success in preventing insect-borne viruses, has led recently to a rapid increase in the area of screened greenhouse, primarily in subtropical climate regions.

Specifications of insect exclusion covers

Insect-proof covers are a general name for many types of plastic sheets and screens that prevent insects from reaching the crop. There are today several companies manufacturing or supplying screens differing from each other in aspects of mesh density and thread gauge. These variations provide differing rates of ventilation but also different rates of insect penetration. The optimal screen will be the one that maintains the indoor population density independent from the outdoor population density, but still allow ventilation. For example, if the accepted economic

threshold level for tomato yellow leaf curl virus infection is 10% of virus-infected plants at the end of the season, the threshold for the whitefly population density in the greenhouse will be 1.4 whitefly/trap/day or 10 whiteflies/trap/week. As long as whitefly catches are below this threshold, growers do not need to apply complementary control means. Screens can solve the problem economically and almost by themselves. To be effective, screens must be installed prior to the pests' appearance; and all openings must be totally covered by screens, including the entrance, which needs to employ air locks. Screen maintenance (repairing rips) is of paramount importance. Furthermore, plants must be quarantined before they go into production areas, to ensure they are pest free before planting. Even with these precautions, which are not without cost, some penetration by pests may occur and require complementary pest control measures.

Microperforated and unwoven sheets. Agronet[R] and FastStart[R] are clear, microperforated polyethylene fabrics. Reemay[R] and Agryl are unwoven, polyester and polypropylene porous sheets. All are light materials ($17 g/m^2$) that can be applied loosely, directly over transplants or seeded soil, without the need for mechanical support. Plants easily support these materials. They are used in the open field, in early spring, as spun-bonded row covers, to enhance plant growth and to increase yield. At the same time, they also proved to protect plants from insects. Reemay[R] is known to efficiently protect squash from *Bemisia*-borne viruses (in California), and pepper from aphid borne viruses (in Spain). FastStart[R] protects plants from flea beetles, root flies, moths and insect transmitted diseases (in the U.S.). Indoors, Agryl[R] is used to protect organically grown tomatoes from virus transmission by *Bemisia* (in Israel). High environmental temperatures may damage plant parts that are in contact with the fabrics.

Woven screens (whitefly exclusion screens). The conventionally-woven "Whitefly Exclusion Screens" – generally known as 50-Mesh[R], Anti Whitefly[R], or AntiVirus[R], and by many other commercial names – are produced from plain, woven, white plastic yarns. These screens are made of yarns 210–230 μm thick; 24–28 yarns/inch in warp (about 10/cm), and 54–56 yarns/inch (about 20/cm) in weft, resulting in slots of rectangular shape. The limiting factor for blocking the whiteflies is the smaller width of the slot, approximately 240 μm. Insects' ability to pass through a barrier cannot be predicted solely from thoracic width and hole size. Screens are three-dimensional fabrics with a specific hole geometry, which is an important element in insect exclusion. The blocking efficacy of any screen must be tested in laboratory and/or field trials. The elongated shape of the slot improves air and light passage. However, elongating the slot more will enable the threads to slide, and whiteflies to penetrate. These screens, though developed primarily to block penetration by *Bemisia* whitefly, also exclude all insects larger than *Bemisia*, e.g., moths, beetles, leafminers, aphids, plant hoppers, and psyllids; thus they also prevent epidemics of aphid-borne viruses.

Knitted shade screens. The mesh of knitted screens is defined by the percentage of shade (e.g., 30, 40 and 50%) that they produce. Because of irregularity in the shape of the holes, whiteflies are not excluded. Reducing the size of the screen holes until they are capable of blocking whiteflies reduces ventilation to an impractical level. However, insects bigger than whiteflies, and birds, might well be excluded.

Knitted-woven screen. A new type of plastic screen (SuperNet®) is under development. It is produced by a combination of knitting and weaving. This technique produces a screening with slots almost 3 times longer than those of the commercial woven screen, while keeping the width of the slot smaller than whitefly body size. This screen possesses high blockage capacity for whiteflies, similar to that of conventional woven screens, but with improved ventilation.

Thrips exclusion. Thrips are the smallest of common insect pests, with a body width of only 245 μm. They move freely through whitefly-proof woven screens. However, thrips are strongly affected by color. As a result, a high proportion of the thrips population (50%) is excluded in the field, due to the optical features of the screen's material. To improve this still-insufficient rate of exclusion, a very fine mesh screen is needed (Bugbed-12®, NoThrips®), but this hampers ventilation too much. Nevertheless, a loose shading screen of aluminum color, through which even whiteflies penetrate freely, reduces thrips penetration by 55%, contrary to a screen, similar in structure but white in color, which attracts many thrips (similarly to a white mulch).

Greenhouse ventilation, insect invasion and screens

The use of insect-proof screens reduces natural ventilation, which may in turn increase temperature and air humidity to harmful levels. This is even more likely if the greenhouse area exceeds about 1,000 m². In this case, natural ventilation may become insufficient, because the area of the side-openings becomes proportionally too small for the greenhouse area. Consequently, natural ventilation becomes ineffective. Forced ventilation may minimize these harmful effects, but the type of ventilation system strongly affects the influx of insects into the greenhouse. The use of vents that exhaust air from the greenhouse causes underpressure, and many insects are sucked into the structure. Alternatively, overpressure, induced by actively pushing air into the greenhouse through an insect-proof filter, reduces the influx of insect pests significantly even in unscreened greenhouses. In large greenhouses (greater than 1000 m²) natural roof ventilation is crucial. Obviously, the roof openings must also be protected by an appropriate insect exclusion screen.

Insect suction devices

Efficient removal of certain flying and non-flying insects (e.g., aphids, Colorado potato beetles, leafminers, and whiteflies) is possible with modern vacuuming machinery. Vacuum treatments, applied either prophylactically or whenever the insect population exceeds a certain threshold level, reduce populations markedly. One of many examples is the Biovac, a tractor-propelled vacuum device that can be used to control tarnished plant bug populations in strawberry crops. Field trials have demonstrated the additional benefits of enhanced pollen dispersal in strawberries coupled with no reduction in beneficial honey bee pollinators (which are able to fly out of the path of the vacuum machine). Greenhouses present unique problems due to their physical structure, but special equipment could be developed, with vacuum machines running automatically on overhead tracks. Vacuum devices are completely compatible with all forms of pest control, and can even be used in biological control by reducing pest levels before releases of natural enemies.

Insect glue (polybutene)

"Insect glues", e.g., polybutenes, are synthetic hydrocarbon polymers. Glue viscosity is proportional to chain length; the longer the chain, the higher the viscosity. Viscosity ranges from 0.3, a liquid with little tackiness, to 600, a very viscous grease-like material with a high degree of tackiness. "Thripstick" is a commercial mixture consisting of polybutene, an emulsifying agent, an insecticide, and water. It is sprayed under greenhouse crops, either on the floor covering or the soil. It does not affect natural enemies. It has been used successfully to control two typical soil-pupating insects: thrips and leafminer flies.

Electromagnetic energy

The effect of electromagnetic energy on organisms is incompletely understood, although in general, longer wavelengths produce heating effects and shorter wavelength radiation produces chemical effects, including ionization of the absorbing media.

Longwaves: radio frequencies (RF) and infrared (IR). The use of RF for pest control has been frequently considered, mainly in grains, foodstuffs, and wood. RF energy was found to be effective against some moth larvae, e.g., the pink bollworm in cotton seeds. Still, no practical large-scale applications have been yet developed. It appears that major improvements in efficiency of RF treatments are necessary for the method to become practical. Potential uses of IR for insect control involve two generally different concepts. One concept is the use of radiation directly applied to the insects or to the infested material. The other concept is based on the insect's suspected ability to sense infrared radiation and man's ingenuity in employing this knowledge in some way to achieve control, e.g.,by attracting specific insects to IR lured traps for monitoring and control purposes.

Ultraviolet (UV) radiation. Insects (e.g., aphids, whiteflies and honey bees) are sensitive to much shorter wavelengths of light than man, and may visualize the part reaching into the UV portion of the spectrum. The responsive region for insects lies, generally, between 700 nm and about 253 nm. Insects respond to visible and UV radiation in many ways, physiologically and behaviorally. The insects may be affected by the quality, intensity, and duration of light, either directly or indirectly through the host-plant. One aspect of major interest for control is that of phototaxis, movement toward or away from light. The effect of UV radiation on insects' photoperiod,

diapause, and sterilization are also of interest. Investigations on insect control with UV radiation have taken two approaches: (i) the use of low-level illumination to manipulate attractive or behavioral responses, and (ii) the use of high intensity irradiation that produces adverse physiological responses. UV (black light) lured traps are used to monitor noctuid insects. UV-absorbing plastic sheets, used as a greenhouse roof cover, reduce to some extent the immigration of whiteflies, probably due to a retarding effect of their UV reflection features. Exposure of newly emerged virgin adults of *Bemisia* for 15 min to the radiation of a "Hanovia model II" ultraviolet lamp, causes complete sterilization of the males and partial sterilization of the females. Sterilization of both male and female adults can also be induced by irradiation of pupae during the last day before emergence, whereas fecundity of females is not significantly affected.

Gamma irradiation – a postharvest treatment. Whiteflies, leafminer flies, and thrips are considered "quarantine pests", even at numbers far below the conventional economic threshold. Gamma irradiation, originating from 60Co, is used with doses of 2.5 to 200 krad per hr (25 to 2,000 GY per hr). For example, 200 krad are lethal to eggs and all immature stages of the leafminer *Liriomyza trifolii* on bean seedlings. At doses of 75 krad or less larvae survive to pupate, but do not give rise to adults. Larval and pupal radiosensitivity decrease with increasing age. Eggs and prepupae are more susceptible to radiation than other stages. Very few flies emerge from eggs or first instar larvae irradiated by 4 to 5 krad, and those that do are all impaired and die within 24 h. Doses of 20 krad kill eggs and all larval stages of *Bemisia* on cotton seedlings. *Spodoptera littoralis* eggs in chrysanthemum flowers are killed by a dose of 50 krad, and neonate larvae by 125 krad. Consequently, the higher dose is suitable to eliminate the pest in flowers for export.

Air ion stimulation. In one study, an air ion generator-emitter complex, installed above tomato plants grown in a commercial soilless culture, caused a marked decrease in greenhouse whitefly infestation and no chemical treatment was needed. When the air ion treatment was halted for 10 days whiteflies appeared. When the air ion generator was reactivated, there was a prompt reduction in whitefly activity. It appears that negative air ions diminish whitefly activity and prevent their population growth.

Visible light

Attractive colors. Color lured traps are an efficient monitoring device. Yellow sticky traps are used for monitoring a insects such as whiteflies, aphids, thrips, and various flies (e.g., leafminer flies, fungus flies, etc.). Blue sticky traps are more specific, attracting mainly flower-infesting insects such as thrips and the flower chafer, *Epicometis squalida*. The highly significant correlation between the number of trapped whiteflies and the number of tomato yellow leaf curl virus infected plants turns the traps into a very helpful monitoring device for control decision making.

Ground colors and whitefly trapping. Ground colors affect the number of whiteflies trapped by yellow sticky traps. The highest numbers are attracted to traps placed on bare soil (100%). Black, white, and glittering transparent plastic mulches, white and aluminum colored insect screens, kaolin sprayed whitened soil, all reduce the insect catches. In the following decreasing order from black to the various white and aluminum colored backgrounds Glittering transparent plastic and Kaolin sprayed soil are the most repellent backgrounds. This indicates the importance of reflective light when mulching is selected as a mean of control, or when traps are exposed for monitoring purpose.

The number of trapped thrips is also strongly affected by background colors. Aluminum colored backgrounds, commercial white-appearing whitefly exclusion screens, or transparent plastic sheets, significantly reduce the numbers of alighting thrips. This may explain the relatively scarce appearance of thrips in structures covered by the commercially used whitefly exclusion screens. Conclusively, thrips are affected not only by the screen mesh but also by its color. Thus, the population can be reduced drastically by using a proper combination of mesh and color screen.

Mass trapping. Attempts to control greenhouse whitefly populations through mass trappings in greenhouses give controversial results. Some claim success while others do not, since many whiteflies are found on the plants despite the fact that many are caught on yellow sticky strips.

Colored screens. The incidence of aphid-borne virus infection in paprika is reduced significantly by using a white screen, even when the holes were larger than aphids and the screens are not a mechanical barrier. Attempts to increase screen holes, by combining the mechanical effect of the screen with a behavioral effect of colors to exclude *Bemisia tabaci*, turn out to be insufficient. Although whiteflies react to colors, it seems that the crucial restricting mechanism of screens is mechanical rather than behavioral. The behavior of the western flower thrips is more affected by colors.

Mulches. The mode of action of mulches is not fully understood. White mulch attracts thrips to the crop, aluminum colored mulch repels them, and yellow mulch delays tomato infection by tomato yellow leaf curl virus vectored by *Bemisia tabaci*. Before using mulches their effects must be carefully investigated for each circumstance.

Whitewashes. Whitewashes are white suspensions which contain 2.5 to 6% Zn. They are used as repellents inert reflective materials like Ca, and various adjuvant like stickers and spreaders. They are sprayed directly onto the crop's leaves to make them white. Whitewash increases potato leaf reflectivity in the visible spectrum by 130 to 250%, and markedly reduces the overall number of aphids landing in whitewash-treated plots. However, different aphid species respond differently; *Myzus persicae* (Sulzer) and *Aphis fabae* Scopoli are repelled, whereas *A. gossypii* Glover is attracted to whitewash-treated pots. A water solution of Kaolin reduces remarkably both the alighting and the development of *Bemisia*, on melons and zucchini plants.

Photoperiod

Many physiological processes depend on the length of the light period. Under long day conditions, whitefly development is quicker, the progenies' body size is larger and the females lay more eggs. Sex ratio is not affected by the length of the day. Finally, the population growth rate is enhanced under long day conditions. Adult emergence time is often synchronized by the diurnal rhythm under which the immature stages had developed. Many insects survive hard living conditions by diapausing (during wintering) or aestivating during summer. In most insects, the length of the day regulates the initiation of diapause or aestivation, and its end. By artificially changing the photoperiod, the process can be disrupted and the insects die in the course of the season.

Chemically induced behavior modifiers

Chemically induced behavior modifications are a well-known phenomenon in nature in general, and in insects in particular. Formulations based on safe active ingredients can act both as repellents and as physical poisons. Nontoxic, or safe active ingredients, such as behavior modifiers, repellents, knockdown agents, and physical poisons, are used to protect people and their animals, and can be used to protect crop plants. However, many of those compounds cannot yet be applied to foliage due to the phytotoxicity of their chemicals, emulsifiers, or other necessary additives. Some soaps and vegetable oils (e.g., sunflower, soybean, pea nuts, and ricinusoil) and many more chemicals are included in this group of insect behavior modifiers.

M. J. Berlinger and Sarah Lebiush-Mordechi
ARO, Gilat Regional Experiment Station,
Beer Sheva, Israel

References
Berlinger, M. J., and S. Lebiush-Mordechi. 1996. Physical methods for the control of Bemisia. pp. 617–634 in D. Gerling and R. T. Mayer (eds.), Bemisia: 1995. *Taxonomy, biology, damage, control and management.* Intercept Ltd., Andover, Hants, United Kingdom. 702 pp.
Berlinger, M. J., W. R. Jarvis, T. J. Jewett, and S. Lebiush-Mordechi. 1999. *Managing the greenhouse, crop and crop environment.* pp. 97–123 in R. Albajes, M. L. Gullino, J. C. van Lenteren, and Y. Elad (eds.), *Integrated pest and disease management in greenhouse crops.* Kluwer Academic Publishers, Dordrecht, The Netherlands. 545 pp.
Cohen, S., and M. J. Berlinger. 1986. Transmission and cultural control of whitefly-borne viruses. *Agriculture, Ecosystems and Environment* 17: 89–97.
Kilgore, W. W., and R. L. Doutt (eds.) 1967. *Pest control: biological, physical, and selected chemical methods.* Academic Press, New York, New York. 477 pp.
Taylor, R. A. J., S. Shalhevet, I. Spharim, M. J. Berlinger, and S. Lebiush-Mordechi. 2001. Economic evaluation of insect-proof screens for preventing tomato yellow leaf curl virus of tomatoes in Israel. *Crop Protection* (in press).
Vincent, C., B. Panneton, and F. Fleurat-Lessard (Coord.). 2000. *La Lutte Physique en Phytoprotection.* INRA – Institut National de la Recherche Agronomique, Paris, France. 347 pp.

PHYTOECDYSONES. Ecdysone-like chemicals found in plants that affect the growth and development of insects, providing a basis for resistance of the plants to insect herbivory.

PHYTOPHAGOUS. Feeding on plants or plant products.

PHYTOSEIID MITES (ACARI: PHYTOSEIIDAE). The family Phytoseiidae consists of approximately 1,700 species of small mites (200–500 microns) mostly predatory, free-living, terrestrial, and known throughout the world, except the Antarctica. They belong to the order Acari, suborder Mesostigmata, Family Phytoseiidae, subfamilies Amblyseiinae, Phytoseiinae, and Typhlodrominae. The life cycle consists of the egg, larva, protonymph, deutonymph and adult. They are found on many plant species, soil, and debris. Published accounts of food sources include pollen, fungi, nematodes, mites, scale insects, whiteflies, and other small arthropods. They have attracted attention due to their role in the biological control of small arthropods and are included in some integrated pest management programs to reduce the need of pesticides. Some of the species that are mass-reared and sold for use in biological control programs are: *Phytoseiulus persimilis* Athias-Henriot is predaceous on spider mites world wide; *Metaseilus occidentalis* (Nesbitt) is predatory on spider mites on apples, grapes, peaches and almonds in California, but is less effective on the genera *Panonychus, Bryobia*, and *Eotetranychus*; *Typhlodromus pyri* Scheuten is predatory on the twospotted spider mite, *Tetranychus urticae* (Koch) world wide and the European red mite, *Panonychus ulmi* (Koch) in Europe and apples in New Zealand; *Neoseiulus alpinus* (Schweizer) is an effective predator of the cyclamen mite on strawberries in Florida; *Euseius hibisci* (Chant) is considered to be an effective predator of the sixspotted mite, *Eotetranychus sexmaculatus* (Riley) in California.

Classification

The evolution in the Phytoseiidae has been marked by the loss of many of the idiosomal setae as shown in the schematic view of a phytoseiid mite. Three subfamilies are recognized by the absence of z3 and s6 in the Amblyseiinae Muma as shown in *Amblyseius mazatlanus* Denmark and Muma, either or both setae z3 and s6 present and setae Z1, S2, S4, and S5 absent in the Phytoseiinae Berlese as shown in *Phytoseius chanti* Denmark, and one of the setae Z1, S2, S4, and S5 present in Typhlodrominae Chant and McMurtry as shown in *Typhlodromus swirskii* Denmark. There have been many attempts to classify

Various approaches to suprageneric classification of the Phytoseiidae.

Berlese[4]	Vitzhum[16]	Baker & Wharton[3]	Karg[9]
Family Laelapidae Tribe Phytoseiini	Family Laelapidae Subfamily Phytoseiinae	Family Phytoseiideae Subfamily Phytoseiinae Subfamily Podocinae[1]	Family Typhlodromidae
Muma[14] Family Phytoseiidae Subfamily Phytoseiinae Subfamily Macroseiinae Subfamily Amblyseiinae Subfamily Aceodrominae[1]	Hirschman[8] Family Gamasidae	Chant[5] Family Phytoseiidae Subfamily Phytoseiinae Subfamily Otopheidomeninae[1]	Karg[10] Family Typhlodromidae Subfamily Phytoseiinae Subfamily Otopheidomeninae[1] Subfamily Blattisociinae[1]
Muma[15] Family Phytoseiidae Subfamily Macroseiinae Subfamily Amblyseiinae Subfamily Phytoseiinae	Wainstein[17] Family Phytoseiidae Subfamily Phytoseiinae Subfamily Macroseiinae Subfamily Treatiinae[1] Subfamily Evansoseiinae Subfamily Gigagnathinae	Chaudhri[7] Family Phytoseiidae Subfamily Gnoriminae	Aruntunjan[2] Family Phytoseiidae Tribe Amblyseiini Tribe Phytoseiini Tribe Typhlodromini Tribe Macroseiini Tribe Iphiseiini
Krantz[13] Superfamily Phytoseioidea Family Phytoseiidae Family Otopheidomenidae[1] Family Ameroseiidae[1] Family Podocinidae[1] Family Epicriidae[1]	Karg[11] Family Phytoseiidae Subfamily Phytoseiinae SubFamily Blattisociinae[1] Subfamily Macroseiinae Subfamily Treatiinae[1]	Chant & McMurtry[6] Family Phytoseiidae Subfamily Amblyseiinae Subfamily Phytoseiinae Subfamily Typhlodrominae	Kolodochka[12] Family Phytoseiidae Subfamily Amblyseiinae Subfamily Phytoseiinae Subfamily Cydnodromellinae Tribe Amblyseiini Tribe Kampimodromini Tribe Phytoseiini Tribe Typhlodromini Tribe Seiulini Tribe Paraseiulini Tribe Anthoseiini

[1]Species in this group are not currently placed in the Phytoseiidae. [2]Aruntunjan, E. S. 1977. *Identification manual of phytoseiid mites of agriculture crops of Armenian S.S.R.* An. Armenian SSR Erevan. 177 pp. [In Russian]. [3]Baker, E. W., and G. W. Wharton. 1952. *An introduction to acarology.* The Macmillan Company, New York, New York. 465 pp. [4]Berlese, A. 1913. Systema Acarorum genera in familiis suis disposita. Acar. *Italica* 1–2: 3–19. [5]Chant, D. A. 1965. Generic concepts in the family Phytoseiidae (Acarina: Mesostigmata). *Canadian Entomologist* 97: 351–374. [6]Chant, D. A., and J. A. McMurtry. 1994. A review of the subfamilies Phytoseiinae and Typhlodrominae (Acari: Phytoseiidae). *International Journal of Acarology* 20: 223–310. [7]Chaudhri, W. M. 1975. New subfamily Gnoriminae (Acarina: Phytoseiidae) with the new genus *Gnorimus*

(continued)

and description of new species *Gnorimus tabella* from Pakistan. *Pak. J. Agric. Sci.* 12: 99–102.[8]Hirschmann, W. 1962. *Gangsystematik der Parasitiformes. Acarologia Schrift.* Vergleichende Milbenkunde, Hirschmann-Verlag, Furth/Bay. 5 (5–6). 80 pp.[9]Karg, W. 1960. Zur Kenntnis der Typhlodromiden (Acarina: Parasitiformes) aus Acker und Grunlandboden. *Z. ang. Entomol.* 47: 440–452.[10]Karg, W. 1976. To the knowledge of the Superfamily Phytoseiodea Karg, 1965. *Zool. Jb. Syst.* 103 (In German).[11]Karg, W. 1983. Systematische Untersuchung der Gattungen und mit der Beschreibung van 8 neuen Arten. *Mitt. Zool. Mus. Berlin.* 59: 293–328.[12]Kolodochka, L. A. 1998. Two new tribes and the main results of a revision of Palearctic phytoseiid mites (Parasitiformes, Phytoseiidae) with the family system concept. 32 (1–2): 51–63.[13]Krantz, G. W. 1978. *A manual of Acarology* (2nd ed.). Oregon State University Book Store, Corvallis, Oregon. 50 pp.[14]Muma, M. H. 1961. Subfamilies, genera, and species of Phytoseiidae (Acarina: Mesostigmata). *Florida State Museum of Biological Science* 5: 267–302.[15]Muma, M. H., H. A. Denmark, and D. De Leon. 1970. Phytoseiidae of Florida. Arthropods of Florida and neighboring land areas, 6. Florida Department of Agriculture and Consumer Services, Division of Plant Industry, Gainesville, Florida. 150 pp.[16]Vitzhum, H. von. 1941. Acarina. Pp. 764–767 in H. Brons (ed.), *Klassen und Ordnungen des Tierreichs* 5, Akad. Verlag. M. B. H. Leipzig: 764-767.[17]Wainstein, B. A. 1962. Revision du genre *Typhlodromus* Scheuten, 1857. Et systematique de la famille des Phytoseiidae (Berlese, 1916) (Acarina: Parasitiformes). Acarologia 4: 5–30.

the phytoseiids as shown in the suprageneric classification of the Phytoseiidae by various taxonomists (see accompanying table).

are routinely collected on fruit and/or vegetable crops. IPM programs can always use additional predators to help reduce pest problems.

Ecology

The important role that the phytoseiids play in the ecology of agriculture crops and ornamental plants is poorly understood because the biology is known only for a few species. Phytophagous mites are involved in losses to agricultural, medicinal, forestry textile, fruit, ornamental and forage crops. While the availability and use of pesticides in agriculture now is essential, alternative control strategies of mites must be developed. The cost of pesticides, the development of pesticide resistance and sustaining competitive food exports to foreign markets dictate the need for new tactics of control. Students with an interest in Phytoseiidae could make a meaningful contribution by researching the biology of some of the species that

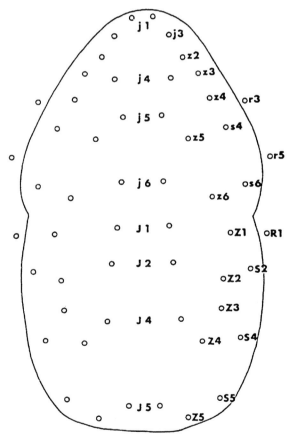

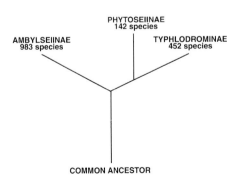

Fig. 792 Probably phylogenetic relationships of the Amblyseiinae, Phytoseiinae, and Typhlodrominae (after Chant and McMurtry).

Fig. 793 Schematic view of a hypothetical adult phytoseiid mite showing the 27 pairs of dorsal setae known to occur in the family (after Chant and Yoshida-Shaul).

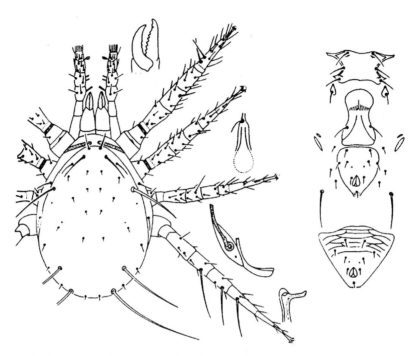

Fig. 794 Diagram of *Amblyseius mazatlanus* Denmark and Muma, an example of the subfamily Amblyseiinae.

Harold A. Denmark
Florida Department of Agriculture and Consumer Services,
Division of Plant Industry,
Gainesville, Florida, USA

References

Baker, E. W., and G. W. Wharton. 1952. *An introduction to acarology*. The Macmillan Company, New York, New York. 465 pp.

Chant, D. A., and J. A. McMurtry. 1994. A review of the subfamilies Phytoseiinae and Trophlodrominae (Acari:

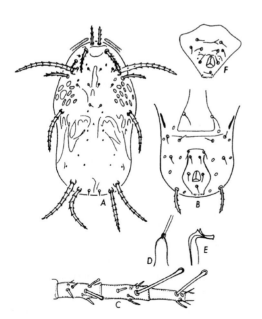

Fig. 795 Diagram of *Phytoseius chanti* Denmark, an example of the subfamily Phytoseiinae.

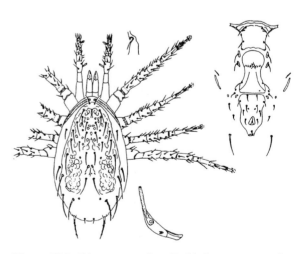

Fig. 796 Diagram of *Typhlodromus swirskii* Denmark, and example of subfamily Typhlodrominae (Cydnodromellinae).

Phytoseiidae). *International Journal of Acarology* 20: 223–310.

Kranz, G. W. 1978. *A manual of acarology* (2nd ed.). Oregon State University Book Stores, Corvallis, Oregon. 509 pp.

Muma, M. H. 1961. Subfamilies, genera, and species of Phytoseiidae (Acarina: Megostigmata). *Florida State Museum Bulletin of Biological Sciences* 5: 267–302.

Muma, M. H., H. A. Denmark, and D. DeLeon. 1970. *Phytoseiidae of Florida*. Arthropods of Florida and neighboring land areas, 6. Florida Department of Agriculture and Consumer Services, Division of Plant Industry, Gainesville, Florida. 150 pp.

PHYTOTELMATA.

PHYTOTELMATA. Phytotelmata are pools of water impounded by terrestrial plants. The structures that impound them are modified leaves, leaf axils, flowers, perforated internodes of plants that have internodes (such as bamboos), rot-holes in tree trunks or branches (henceforth called treeholes), open fruits, and fallen leaves. This expression, derived from the Greek words for plant and pool, was coined by Varga (1928) who wrote in German, with a companion paper in Hungarian. Maguire (1971) popularized its acceptance into English. The plural is phytotelmata (correctly pronounced phyto.TELM.ata, where the capital letters indicate location of the stressed syllable), in keeping with other plural words ending in -ata in Greek such as stemmata and stomata. The singular is phytotelma (compare with soma, stemma and stoma). The singular has been further Anglicized to phytotelm, which serves as a noun and as an adjective.

Phytotelmata formed by leaves, flowers and internodes of living plants are found in members of at least 29 plant families, mostly monocotyledons. They are Agavaceae, Amaryllidaceae, Araceae, Bromeliaceae, Campanulaceae, Cannaceae, Cephalotaceae, Commelinaceae, Compositae, Cyperaceae, Dipsacaceae, Eriocaulaceae, Euphorbiaceae, Gesneriaceae, Gramineae, Heliconiaceae, Hanguaceae, Liliaceae, Marantaceae, Musaceae, Nepenthaceae, Palmae, Pandanaceae, Rafflesiaceae, Sarraceniaceae, Strelitziaceae, Typhaceae, Umbelliferae, and Zingiberaceae. A list of plant families providing fruits or fallen leaves that form phytotelmata has not been compiled, but would be long. Treeholes are formed in just about any plant family containing hardwood trees.

Some of the plants that form phytotelmata are carnivorous. The pitcher plant families (Cephalotaceae, Nepenthaceae, and Sarraceniaceae) are not closely related to one another. Pitcher plants have modified leaves that trap and digest terrestrial arthropods as a source of nutrients for the plant. Despite the digestive fluids produced, specialist aquatic arthropods (mites and insect larvae) dwell in the pitchers of some species of *Sarracenia* (North America) and *Nepenthes* (southeast Asia and Madagascar). Carnivory is also displayed by some members of two genera of bromeliads, *Catopsis* and *Brocchinia*, although because there is no evidence of plant-produced digestive fluids, these plants have been dubbed "protocarnivorous" by purists. In them, prey organisms are broken down by autolysis and then the products became available for uptake by the plant. Specialist aquatic insect larvae dwell in the water-impounding leaf axils of these bromeliads.

Bromeliads are a family (Bromeliaceae) of monocotyledons with about 2,500 described species. Almost all of them are native to the neotropics. They include many species whose growth form impounds water and many that do not. Phytotelmata held by water-impounding species range from small bodies of water in several to many leaf axils (typified by some plants of the genera *Tillandsia* and *Guzmania*), or the formation of a large central tank formed by a few leaves (typified by some plants of the genera *Aechmea* and *Neoregelia*). Very many bromeliad species are epiphytic whereas others grow on the ground. None of the epiphytic species is believed to be parasitic on trees. Instead, the roots serve only as holdfasts, and water and nutrients are absorbed through pores on the leaves from the impounded water. Wet neotropical forests may support dense populations of such epiphytic bromeliads which have been likened to "aquaria in the treetops." Some bromeliads that grow on the ground dwell on rock surfaces, some in rather arid habitats and others in swamps, and in these the roots may absorb nutrients. Some of the ground-dwelling bromeliads provide phytotelmata. One species of one ground-dwelling genus (*Ananas comosus*) is an important crop plant (pineapple). Many genera, species, and hybrids are grown as ornamental plants; although many of these are epiphytic in nature, horticulturists have learned how by use of well-drained media, most may be grown in pots. The food-chains in bromeliad phytotelmata are of two basic kinds. In the phytotelmata provided by epiphytic bromeliads, rainwater falling through tree canopies is there enriched by leachates (to form throughfall), and is impounded by the plants together with debris (leaves, twigs, and seeds) of the trees. The plant obtains its nutrients from the impounded water. Various organisms inhabit the impounded water. In the phytotelmata provided by ground-dwelling bromeliads, debris trapped tends to be sparser (for lack of a tree canopy overhead), algae tend to grow in the water, and these algae form the basis of a food chain. A third kind of food chain is formed in those few bromeliads that grow in sun-exposed bromeliads and have developed carnivorous habits.

Heliconia is a genus of some 200 + species distributed in the neotropics and in some Pacific islands. It is now placed in the family Heliconiaceae, although in the past it has been variously assigned to Musaceae and Strelitziaceae. Floral spikes are pendent in some species, upright in others. Flower bracts in upright flowers may hold from a few drops to about 90 ml of water, depending upon species. Flowers emerge through this water and perhaps are thus protected to some extend from herbivory by it. At all events, the presence of water seems important to some species because they do not rely totally upon rainfall to replenish the water, but are able to pump water into the bracts. After flower fertilization, which is believed to be performed by hummingbirds or bats, the flower petals decompose in this water. The flowers are the basis of a food chain that nourishes various invertebrate animals. The first invertebrates to invade are specialist aquatic insect larvae. As the contents of each bract age, and as bract walls are damaged by vertebrate animals, the water level in the bract declines, and the bract risks invasion by semi-aquatic or even non-aquatic insects. However, the seed walls are extremely hard and thus very resistant to attack by insects. Each seed is enclosed within a fleshy, typically blue, pericarp, which doubtless makes it attractive to vertebrate animals and serves to disperse the seeds. During the flowering season, each floral spike at any time typically presents a succession, from youngest, unopened bracts at the apex, to older bracts with formed seeds.

It is not clear that other plants that impound water in leaf axils or floral parts use the water in any way. Most of these plants are monocotyledons. Fallen leaves may provide ephemeral, nutrient-poor phytotelmata during rainy seasons. Fruit husks may provide nutrient-richer phytotelmata. One example of the latter is the husks of cacao, which are usually left in piles in cacao plantations after harvest.

Treeholes provide phytotelmata in all but one continent. Those in an appropriate position in a tree trunk or large branch may be filled by stemflow. Part of the rainfall striking a tree canopy falls through it (after enrichment) as throughfall. Another part runs down branches and trunks, picking up leached materials and debris, as stemflow. Dead leaves and other debris from the canopy above may be blown into well-placed treeholes. All these materials decomposing inside treeholes provide a nutrient soup, which typically is not wasted by invertebrate animals. Some trees whose stem gives rise aboveground to roots, form bark-lined basal treeholes, which may likewise impound water. Not all treeholes impound water, not just because they are not in an appropriate position to collect stemflow, but because they have more than one opening, and leak. Some treeholes rarely receive stemflow and may be dry or almost so for months of each year. Rainfall fluxes are as important to the inhabitants of treehole phytotelmata as they are to the inhabitants of bromeliad phytotelmata.

Bamboos provide phytotelmata by one of two methods. If the stem is snapped or cut, the standing part, whose lower limit is marked by a node, can accumulate rainwater and debris. In some tropical countries, however, Coleoptera may make a small lateral hole into an internode, and this may accumulate water by stemflow. This small hole may limit access by flying insects, but some insects have adapted to injecting their eggs, giving rise to aquatic larvae, through it. The inner lining of the internode provides some nutrients.

The phytotelmata provide habitats for various aquatic invertebrates, including insects; these aquatic organisms and their habitat are of main concern in this article as "phytotelmatous." However, the plant structures around the phytotelmata provide habitat for some non-aquatic invertebrates that interact as predators with the aquatic fauna, much as do banks of ponds and streams. Other invertebrates (including insects) that dwell around the edges of phytotelmata, but that do not interact with the aquatic organisms, are numerous, but cannot be construed as part of the phytotelm fauna. Thus, lepidopterous larvae that feed on decomposing materials in non-water-holding outer leaf axils of bromeliads are not "phytotelmatous", nor are cockroaches nor ants. A few vertebrate animals, mainly frogs, have evolved to become specialist inhabitants of bromeliad phytotelmata or treeholes, laying their eggs nowhere else. But almost all phytotelm specialists are invertebrates.

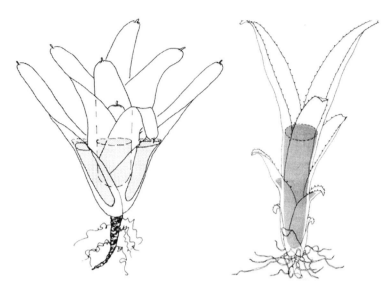

Fig. 797 Figures show how the architecture of two bromeliads affects distribution of phytotelmata: left, *Billbergia pyramidalis*, right, *Aechmea nudicaulis*.

Specialist inhabitants of phytotelmata typically do not harm the plant hosts. Some appear to be mutualists. In bromeliads, invertebrates (and bacteria and fungi) break down debris and may make it more easily taken up as nutrients by the leaves. Algae growing in sun-exposed bromeliad phytotelmata may perhaps compete with the plant host for nutrients, but the action of chironomid larvae is to feed on the algae, thus benefiting the host in two ways. In *Heliconia* bracts, it does not matter that decomposing petals are eaten by syrphid larvae after fertilization of the flowers has occurred. It does not matter that inner walls of bracts are scraped by hispine chrysomelid larvae once the seeds have formed.

Regulation of populations of invertebrates in phytotelmata takes several forms. Rainfall fluxes are, of course, of paramount importance in almost all, for the phytotelmata would not exist without rainfall. Rainfall, in the form of throughfall and stemflow, is also important for the provision of many nutrients in many phytotelmata. With water and nutrients and detritus that form the basis of the food chains, the next level of most food webs consists of saprovores (that feed on detritus) or, in phytotelmata that support algal growth, of herbivores (that feed on algae). Then, if the web is complex enough, come one or two levels of predators. Population regulation of the saprovores is brought about either by competition for resources, or by predation. It is of

course the organisms that are lower in the food web (chain, pyramid) that are the most abundant.

Invertebrate animals that occupy at least some phytotelmata include rotifers, aquatic oligochaete worms, crustaceans (ostracods, copepods, and decapods), aquatic mites, and representatives of some orders of insects. Aquatic insect larvae are best represented in diversity and numbers of individuals by Diptera. For example, over 200 species of Culicidae are known from bromeliad phytotelmata alone. Culicidae are reported from all the other groups of phytotelmata, from pitcher plants through bamboo internodes to *Heliconia* floral bracts and fruit husks and fallen leaves. But Ceratopogonidae, Chironomidae, Psychodidae, Richardiidae, Stratiomyidae, and Syrphidae may be heavily represented in numbers of individuals, and the list of dipterous families does not end there. The abundance of individuals of numerous species of Diptera is because they are low in the food web. Those dipterous larvae that feed as predators are much less abundant. Adults and larvae of aquatic Coleoptera (Hydrophilidae, Dytiscidae, Scirtidae) are known from some bromeliad phytotelmata, as are adults and nymphs of Veliidae (Hemiptera). Fewer aquatic coleopterous families and far fewer species are reported from treeholes and *Heliconia* bracts. Nymphs of Trichoptera are recorded from bromeliad phytotelmata, and nymphs of Odonata from bromeliads and treeholes.

Phytotelmata inhabitants exhibit a full range of specialization. Some species inhabit many kinds of water bodies, whereas others inhabit only small bodies of water. Among the latter, some inhabit various kinds of phytotelmata, whereas the ultimate specialists inhabit only one kind (for example, only bracts of one or more species of *Heliconia*). The means by which most specialist insects select the places in which to oviposit is unknown. However, perception of volatile chemicals plays a role in some, and vision, including color vision, in some. Flightless invertebrate animals may have little control in the matter of dispersing from one phytotelm to another, beyond hitching a ride (phoresy) on some larger, more mobile organism that just may be about to disperse to another phytotelm.

J. Howard Frank
University of Florida
Gainesville, Florida, USA

References

Frank, J. H. 1996. Bromeliad phytotelmata. (available: http://bromeliadbiota.ifas.ufl.edu)

Frank, J. H., and L. P. Lounibos (eds.). 1983.*Phytotelmata: terrestrial plants as hosts for aquatic insect communities*. Plexus, Medford, New Jersey. 293 pp.

Kitching, R. L. 2000. *Food webs and container habitats: the natural history and ecology of phytotelmata*. Cambridge University Press, Cambridge, England. 431 pp.

Kitching, R. L. 2001. Food webs in phytotelmata: "bottom-up" and "top-down" explanations for community structure. *Annual Review of Entomology* 46: 729–760.

Maguire, B. 1971. Phytotelmata: biota and community structure determination in plant-held waters. *Annual Review of Ecology and Systematics* 2: 439–464.

Varga, L. 1928. Ein interessanter Biotop der Biocönose von Wasserorganismen. *Biologisches Zentralblatt* 48: 143–162.

PHYTOTOXEMIA. Symptoms of plant disease that are induced from the effects of insect/mite feeding and introduced salivary compounds (phytotoxins) are called phytotoxemias or plant toxemias. Categories of phytotoxemias include: (1) local lesions at the feeding point; (2) local lesions with development of secondary symptoms; (3) tissue malformations; and (4) translocated phytotoxins. An arthropod that is capable of producing a phytotoxemia is sometimes described as being 'toxiniferous.'

Phytotoxemias are almost entirely produced by arthropods that suck plant fluids, and primarily involve Hemiptera. Some Hemiptera, gall making insects found within other orders, and some mites (Tetranychidae and Eriophyiidae) are also considered capable of producing certain phytotoxemias.

The nature of the toxins involved in a phytotoxemia is little understood, but apparently often includes enzymes adapted to the insect diet that have adverse effects on the host. Various pectinases, proteases, amylases, saccharases, lipases and pheonolases are among the compounds identified that produce phytotoxemias. The effects of such phytotoxic compounds may vary depending on the response of the host plant. For example, the potato/tomato psyllid, *Paratrioza cockerelli*, is a species that produces serious systemic injuries to potatoes and tomatoes, but has little effect when feeding on eggplant and pepper.

Induction of a phtytoxemia is usually related directly to the presence of the insect responsible, and once the insect is removed no further damage occurs. However, if the damage is severe the plant may not recover. To prevent a phytotoxemia, usually the insect must be prevented from feeding, or killed soon after it begins feeding. In some cases, plant varieties are known that are resistant to injury.

Whitney Cranshaw
Colorado State University
Ft. Collins, Colorado, USA

PHYTOTOXICITY. Damage to a plant due to contact with a chemical toxin.

PICEOUS. Very dark, black.

PICTURE-WINGED FLIES. Members of the family Ulidiidae and Platystomatidae (order Diptera). See also, FLIES.

PICTURE-WINGED LEAF MOTHS (LEPIDOPTERA: THYRIDIDAE). Picture-winged leaf moths, family Thyrididae, total 794 species worldwide, nearly all tropical, with nearly half the species Indo-Australian (only a few species are in the Nearctic and Palearctic regions); actual fauna likely exceeds 1,200 species. There are six subfamilies: Simaethistinae, Whalleyaninae, Argyrotypinae, Thyridinae, Siculodinae, and Striglininae. The first three subfamilies are sometimes treated as separate families. The family is in the superfamily Pyraloidea in the section Tineina, subsection Tineina, of the division Ditrysia (sometimes placed in its own superfamily, Thyridoidea). Adults small to large (9 to 90 mm wingspan), with head scaling average; haustellum naked; labial palpi upcurved or slightly porrect; maxillary palpi 1 to 2-segmented. Maculation varied, often dark with light spots, or shades of brown to mimic dead leaves, or very colorful; many with wing margins irregular in shape, or leaf-like. Adults are diurnal or crepuscular. Larvae are leafrollers (one Australian species is gregarious), or borers in stems and flower racemes; a few are gall makers. A number of host plants are used. Only a few species are economic. The subfamily Whalleyaninae and its nominate genus *Whalleyana* are named after the British lepidopterist Paul E. S. Whalley.

John B. Heppner
Florida State Collection of Arthropods
Gainesville, Florida, USA

References

Freina, J. J. de, and T. J. Witt 1990. Familie Thyridae Herrich-Schäffer 1846. Fensterschwärmer (= Thyrididae Hampson 1897). In *Die Bombyces und Sphinges der Westpalaearktis*, 2: 69–71, pl. 10. Forschung & Wissenschaft Verlag, Munich.
Thiele, J. H. R. 1994. Thyrididae. In G. Ebert (ed.), *Die Schmetterlinge Baden-Württembergs. Bd. 3. Nachfalter*, 505–514. Stuttgart: Verlag Eugen Ulmer.
Whalley, P. E. S. 1964. Catalogue of the world genera of the Thyrididae (Lepidoptera) with type selection and synonymy. *Annals and Magazine of Natural History* (13) 7: 115–127.
Whalley, P. E. S. 1971. The Thyrididae (Lepidoptera) of Africa and its islands. a taxonomic and zoogeographic study. *Bulletin of the British Museum (Natural History), Entomology Supplement* 17: 1–198, 68 pl.
Whalley, P. E. S. 1976. *Tropical leaf moths: a monograph of the subfamily Striglininae (Thyrididae)*. British Museum (Natural History), London. 194 pp., 68 pl.

PIERCE'S DISEASE OF GRAPE. See also, TRANSMISSION OF PLANT DISEASES BY INSECTS.

PIERIDAE. A family of butterflies (order Lepidoptera). They commonly are known as yellows-white butterflies, yellows, whites, and sulfurs. See also, YELLOW-WHITE BUTTERFLIES, BUTTERFLIES AND MOTHS.

PIESMATIDAE. A family of bugs (order Hemiptera). They sometimes are called ash-gray leaf bugs. See also, BUGS

PILIFER. A small projection, resembling a small mandible, at each side of the clypeus in Lepidoptera.

PILL BEETLES. Members of the family Byrrhidae (order Coleoptera). See also, BEETLES.

PILLBUGS AND SOWBUGS, OR WOODLICE (ISOPODA). The terrestrial isopods known as pillbugs and sowbugs in North America are collectively known as 'woodlice' in Europe. The term 'woodlice' conveniently depicts their relatedness and preferred habitat, and deserves wider recognition and use in North America. Consistent with this is the fact that many of the pillbugs and sowbugs now found in North America seem to be immigrants from Europe. Indeed, several species have become almost cosmopolitan.

Classification

Woodlice are arthropods in the subphylum Crustacea, whereas insects are in the subphylum Atelocerata. Like many other terrestrial arthropods, they are commonly confused with insects. However, they are more closely related to shrimp and crabs than to insects, as shown below:

Phylum: Arthropoda
    Subphylum: Trilobita – Trilobites (these are extinct)
    Subphylum: Chelicerata
        Class: Merostomata – Horseshoe crabs
        Class: Arachnida – Arachnids (scorpions, spiders, ticks, mites, etc.)
        Class: Pycnogonida – Sea spiders
    Subphylum Atelocerata
        Class: Diplopoda – Millipedes
        Class: Chilopoda – Centipedes
        Class: Pauropoda – Pauropods
        Class: Symphyla – Symphylans
        Class: Entognatha – Collembolans, proturans, diplurans
        Class: Insecta – Insects
    Subphylum: Crustacea – Crustaceans (isopods, shrimp, crabs, etc.)
        Class: Malacostraca
            Superorder: Peracarida – Amphipods, isopods, etc.
            Order: Isopoda

Class Malacostraca contains about two thirds of all crustacean species, including all of the larger forms. The three principal orders of malacostracans are the Isopoda, Amphipoda, and the Decapoda. The latter is best known to most people, as it contains the crayfish, lobsters, crabs and shrip. The woodlice are found in several families of Isopoda.within the suborder Oniscidea. Many Isopods are marine, and a few live in freshwater habitats, but the member of the suborder Oniscidea are terrestrial. The members of the suborder Oniscidea are the most successful (diverse) of the land-dwelling Crustacea. The herbivorous and detritivores of the suborder Oniscidea are considered to be relatively primitive; the more advanced groups of Isopoda are carnivores, predators and parasites.

Life cycle and description

Though superficially similar to insects because they have a rigid exoskeleton and jointed appendages, there are some important differences. As in insects, the body of woodlice is divided into three major regions: the head, which bears the antennae and mouthparts; the thorax or pereion which bears the legs but never wings; and the abdomen or pleon. The head bears two pairs of antennae instead of the one pair found in insects, but one pair of the antennae in woodlice is greatly reduced in size and therefore not often observed. The pereion (thorax) consists of seven segments instead of the three found in insects, with each segment bearing a pair of legs ventrally. The pleon (abdomen) consists of 6 segments, but invariably is much smaller than the pereion. The ventral surface of the pleon bears plate-like structures, and is an important site for gas exchange. A terminal pair of tail-like appendages, called uropods, may be located at the tip of the pleon. Uropods are present in sowbugs but absent in pillbugs. Sowbugs cannot completely roll into a ball, though pillbugs are capable of this behavior. Because they can roll into a ball, pillbugs are sometimes called rolly-pollys.

The female woodlouse carries her eggs and young about with her in a special compartment, called the marsupium, on the underside of her body. Fertilized eggs are inserted into the marsupium where the embryos (and later the young) obtain water, oxygen and nutrients from a nutritive fluid called, appropriately, marsupial fluid. The eggs may be up to 0.7 mm in diameter, and in some species over 100 eggs may be produced. The eggs persist for 3 to 4 weeks, then hatch, but the young remain in the marsupium for another 1 to 2 weeks before crawling out. They are only two mm in length at this stage of development. Woodlice commonly produce offspring 1 to 3 times per year, with spring and autumn broods most common. Woodlice often survive for longer than a year, with longevity of 2 to 5 years not uncommon.

Woodlice often produce 20 to 40 young, and 1 to 3 broods per season. Brood size was positively correlated with female size. The young are highly gregarious, and sometimes cannibalistic. Once they have left the female they molt, usually within 24 hours, acquiring a seventh pereion segment. After an additional 14 days a second molt occurs, and a seventh pair of legs is produced. Thereafter they do not change in morphology, other than to increase in size. The interval between molts is 1 to 2 weeks until the age of about 20 weeks, and molting continues irregularly for the remainder of their life, including adulthood. Molting occurs in two stages, with the posterior portion of the body shedding its old skin first, followed about three days later by the anterior portion.

Woodlice often attain a length of 8.5-18.0 mm as they reach adulthood. The width of the body is about half of its length. Woodlice are somewhat flattened and elongate oval in shape, seven pairs of legs and 13 body segments are apparent, and they have long, jointed antennae. Eyes are evident on the side of the head. They are brownish or grayish in general body color, though often marked with areas of black, red, orange or yellow.

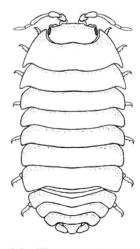

Fig. 798   An adult pillbug, *Armadillidium vulgare*.

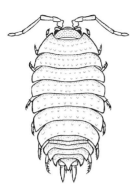

Fig. 799 Adult of dooryard sowbug, *Porcellio scaber.*

Ecology

Woodlice generally feed on dead plant material, though they also accept dead animal remains and dung, and occasionally ingest bacteria, fungi, and living plants. They are best viewed as decomposers, similar to earthworms, breaking down plant material and mixing it with mineral particles to produce soil. However, they are not the first organisms to attack leaf litter, waiting until microorganisms have begun the degradation process. Also, they sometimes have the unfortunate habit of grazing on plants, particularly seedlings. Woodlice occasionally attack seedlings above-ground, feeding especially on stems and young leaves, and below-ground, feeding on roots. Woodlice are most common in soils with neutral or alkaline pH, good crumb structure, high organic matter content, and where soil bacteria and other macrodecomposers such as earthworms and millipedes flourish. They tend to be absent from acid and waterlogged soil. Due either to the disturbance or lack of shelter, woodlice are virtually absent from thoroughly tilled land. On the other hand, straw or other coarse mulch provides good habitat for woodlice and can lead to crop damage. They have adapted well to humans and human habitations, and are often considered to be anthropophilic, but they also survive well in forests and grasslands, particularly if they can find shelter beneath logs and rocks. They are nocturanal.

Woodlice are parasitized by species of tachinids (Diptera: Tachinidae), with most displaying a fairly specific host range. Predation and cannibalism are known to occur, but it is uncertain whether these are important mortality factors in nature. Lizards, salamanders, shrews, spiders, centipedes, and ground beetles (Coleoptera: Carabidae) will eat woodlice. An iridovirus has been found to occur in woodlice populations in California. In addition to causing a slight blue to purple discoloration in infected woodlice, the longevity of infected hosts is greatly reduced when woodlice are infected with iridovirus. Fungus, nematode, and protozoan parasites seem to be of little importance.

References

Paris, O. H. 1963. The ecology of *Armadillidium vulgare* (Isopoda: Oniscoidea) in California grassland: food, enemies, and weather. *Ecological Monographs* 33: 1–22.

Sutton, S. 1972. *Woodlice*. Pergamon Press, Oxford, United Kingdom.

Warburg, M. R. 1993. *Evolutionary biology of land isopods*. Springer-Verlag, Berlin/New York.

PILOSE.   Covered with soft hair.

PINE AND SPRUCE APHIDS.   Members of the family Adelgidae (order Hemiptera). See also, APHIDS, BUGS.

PINE TIP MOTHS, *Rhyacionia* SPP. (LEPIDOPTERA: TORTRICIDAE).   There are approximately 35 known species of pine tip moths distributed throughout the palearctic and nearctic regions of the world. However, the greatest diversity of species within the genus *Rhyacionia* is found in the western United States. As their name implies, pine tip moths feed exclusively on pine trees. A few species are economically important, particularly in the southeastern and western United States.

Tip moth larvae feed within the buds and shoots of their host. Damage caused by tip moths is usually visible by the appearance of small accumulations of pine resin which collect around the entrance wound. The terminal portion of the infested shoot eventually turns brown as it is hollowed out by the burrowing larvae. The tree is normally not killed by this pest unless repeated, severe infestations occur. However, significant losses in growth and wood quality are common. Tip moths are primarily a problem in intensively managed pine plantations where slow growth and poorer wood quality can result in economic losses.

The biologies of tip moths within the genus *Rhyacionia* are similar. Adults normally emerge in spring from infested shoots, duff, or soil. The majority of species studied are univoltine, however, some have multiple generations per year. Adults are often inactive during much of the day, with mating generally occurring at dusk or later during the night. Females release a sex pheromone to attract males for mating. Oviposition occurs on the needles, buds, or shoots of the host tree. Eggs may be deposited singly or in clusters. First instar larvae generally mine the needles while later instars enter the shoot, feeding on pith and cambium tissue. Pupation occurs within the shoot, on or in the ground below the host tree, or in cocoons just below the soil, often attached to the root collar. Univoltine species generally overwinter as pupae on the ground. An important exception is the European pine shoot moth, *Rhyacionia buoliana*, which overwinters as a larvae in the buds, completing feeding the following spring. Multivoltine species usually spend the winter as pupae within the shoots.

The greatest diversity of species occurs in the southern Rocky Mountains in Colorado, New Mexico, and Arizona, most of which are associated with Ponderosa pine. However, the most economically significant tip moth species, the Nantucket pine tip moth, *Rhyacionia frustrana*, occurs in the eastern United States. This species attacks a number of pine species, but is primarily a problem on the vast acreage of intensively managed loblolly pine (*Pinus taeda*) plantations throughout the Southeast. For this reason, an extensive amount of research has been done on this pest compared to most other species of *Rhyacionia*, for which even the most basic information is lacking. Another significant pest is the European pine shoot moth, *Rhyacionia buoliana*, which was accidentally introduced into North America in the early 1900s. It is a sporadic pest of pine plantations in the Northeast, the Lake States, and the Pacific Northwest.

In most cases, prevention of pine tip moth infestations involves planting pines on suitable sites within their native range and avoiding intensive management practices (herbicides, pesticides, fertilizers, etc.). Chemical control is a viable option for preventing damage but has seldom been used historically due to a lack of perceived threat from these non-lethal forest pests. Recent research on Nantucket pine tip moth, however, has demonstrated that substantial growth losses can occur in pine plantations that receive repeated damage, even at seemingly low levels. These pests are likely to become more important in the future as forest management becomes increasingly intensive in order to meet the wood and paper demands of a rapidly growing population.

Chris Asaro
University of Georgia
Athens, Georgia, USA

References

Asaro, C., C. J. Fettig, K. W. McCravy, J. T. Nowak, and C. W. Berisford. 2003. The Nantucket pine tip moth (Lepidoptera: Tortricidae): A literature review with management implications. *Journal of Entomological Science* 38: 1–40.

Berisford, C. W. 1988. The Nantucket pine tip moth. In A. A. Berryman (ed.), *Dynamics of forest insect populations: patterns, causes, implications*. Plenum Publishing Corp., New York, New York.

Miller, W. E. 1967. The European pine shoot moth: ecology and control in the lake states. *Forest Science Monograph* 14: 1–72.

Powell, J. A., and W. E. Miller. 1978. Nearctic pine tip moths of the genus *Rhyacionia*: biosystematic review (Lepidoptera: Tortricidae, Olethreutidae). *USDA Agricultural Handbook* No. 514.

PINE WEEVIL, Hylobius abietis Linnaeus (Coleoptera: Curculionidae).

Hylobius abietis L. feed on the stem bark of conifer seedlings. In the absence of appropriate control measures, the damage may reach a level that rules out planting as a means of conifer regeneration. It is distributed throughout the coniferous forests of northern Eurasia, including the British Isles and Japan. Other *Hylobius* species cause damage of a similar type, both within this area, and in North America. Soon after the introduction of clear felling systems in Central Europe, damage by *H. abietis* became widespread.

Pine weevil adults are blackish brown, robust-looking insects with a body length between 9 and 14 mm. Their antennae are located near the tip of the snout. Small tufts of yellow hair form characteristic patterns on the thorax and the abdomen, but are often worn off in older individuals that have spent a long time burrowing in the soil. The yellowish-white larvae are footless, curved and wrinkled, with a brown head capsule. Fully grown, fifth instar larvae attain a body length of about 15 millimeters.

Migration

Upon logging of coniferous trees, volatile substances such as terpenes emanate from stumps and logging debris, particularly in clear-cuts when temperatures rise at the onset of summer. These volatiles attract weevils that have emerged from their hibernation sites in the forest floor. A main migration flight, usually lasting for a week or two, starts when maximum temperatures approach 25°C. The weevils take off into the wind to drift along. They have been observed flying above the forest canopy and are likely to land when sensing the strong bouquet of conifer volatiles. Migrants may end up in the wrong place, such as lumberyards and newly built wooden houses, and at sawmills, they can be found in great abundance on heaps of fresh sawdust. However, a large part of the migrants reach areas with fresh conifer stumps, where a new generation of weevils can reproduce in the roots. After the spring migration the weevils rarely fly for the rest of the season.

Oviposition

As both sexes are attracted by volatiles from fresh stumps, they will meet in the immigration areas. It is doubtful that sex pheromones play a role, other than maybe at close range. Upon mating, an extensive period of oviposition starts using roots with a diameter of at least 1 cm. Clear-cut areas offer an abundance of suitable breeding substrate. During the growth season, a female may lay several dozen eggs, which are laid singly, either in small cavities that she gnaws in the root phloem, or simply in the soil. When the soil is dry, the latter option seems to be avoided. Egg larvae hatching in the soil apparently use chemical clues to find suitable roots and may cover considerable distances during their underground migration. Throughout the egg-laying period, the weevils feed on local vegetation, including conifer seedlings. In cool areas, the oviposition may come to a halt by the onset of winter, to be resumed the next spring in stumps that remain fresh enough. The adults may survive for two, or even three seasons.

Development

Inside the subterranean stump roots, the offspring is protected against extreme weather and some of their natural enemies. The weevil larvae go through 5 stages during which they dig long, winding galleries in the phloem and outer xylem of the roots. Ambient temperature determines the developmental speed: at a stable 23°C, they are full-grown 40 days after the egg is laid, while at 12°C, the development takes more than 100 days. In their last instar, which in warmer areas may be reached the first autumn, the larvae construct an oblong pupal chamber. In thick-barked roots, this is dug out at the interface between

Fig. 800 *Hylobius abietis*

phloem and xylem; where root bark is thin, the chamber is excavated in the wood and the entrance is closed from the inside with a plug of frass. If temperatures stay below about 21°C during their fifth instar, the larvae enter a diapause, which is broken after hibernation. Thus, a diapause is a regular feature of the weevil life cycle in cold boreal forest soils. After pupation, the adults remain in the pupal chamber while their exoskeleton hardens. They may then make their way to the surface, but if hatching occurs late in the season, they over-winter in the pupal chamber. Pine weevils hibernate at various larval stages or as adults, but not as eggs or pupae. Before migrating to other areas with fresh breeding material, the young adults go through an intensive period of feeding to develop their energy reserves and sexual organs.

The prolonged period of oviposition, the temperature-dependant development, and the facultative diapause lead to a highly variable generation time. In southern Scandinavia and Finland, a 2-year life cycle dominates; in the north, generation time stretches to 4 years or more. A mountainous landscape adds to the variability because the microclimate of steep, sunny and shady slopes are quite different. The survival of the immature stages is related to the generation time; a prolonged development means a higher risk of parasitation and predation, as well as increased competition from root-inhabiting fungi and other insects. A prolonged presence of the parental generation and a highly variable development of their offspring mean that the population remains high for several years. This is reflected in the period of damage.

Damage to the conifer seedlings

The adult pine weevil damages seedlings by eating the phloem of the above-ground stem. Moderate stem feeding often results in small, irregular 'pockmarks' in the bark. A healthy plant may survive this damage and callus tissue soon covers the wounds. Severe attacks can girdle the stem, thereby interrupting photosynthate translocation and impairing water flow in the seedlings. The weevils prefer larger seedlings over smaller, but the former generally survive better. The damage occurs in natural regeneration as well as on planted seedlings, but in the former case, it often goes unnoticed because of an abundance of seedlings. Attack on planted seedlings is more likely to be noticed and generally has more serious consequences because the plants are distributed at an economically set distance. Both endemic and introduced coniferous species are attacked, as well as a number of other trees and weeds. Popular names in Europe often refer to the conifer that is most commonly attacked. As opposed to the stem feeding, larval tunneling in the roots can be considered useful because it accelerates stump decomposition and nutrient recycling.

If planted seedlings are not given appropriate protection, the mortality rate may be unacceptably high (for example, in southern Scandinavia, the mortality frequently exceeds 50%). The damage is often patchy, resulting in gaps in the future forest stand. The severity of damage is influenced by a variety of factors: the local environment and silvicultural practices both play a role. Regionally speaking, the damage is more severe in the southern parts of the pine weevil's range, probably due to more favorable conditions for the immature stages. Ground vegetation may be important, since the weevils can feed on a variety of plants. Upon clear cutting on lush sites, when thick vegetation is cut back, the ground vegetation will often succumb to sun-exposure, leaving new-planted seedlings as an extra welcome food for the weevils. Burned areas are particularly prone to damage. The common practice of extending felling coups from year to year facilitates migration to nearby breeding sites and in such cases, the population may build up to very high levels. Damage may become so severe that planting must be postponed for years. This practice, known in German as "Schlagruhe," may imply serious weed problems on sites with thick vegetation.

Control

Before World War II, a wide variety of control methods were used, including removing stumps, digging trap pits from which the weevils could not escape, using fresh stem sections and bark pieces to lure the insects, and a wide variety of stem treatments, including mechanical as well as chemical. After World War II, stem application of DDT was thought to solve the pine weevil problem for the future. DDT was thought to be both efficient and harmless. At least in more northern regions, the protective effect of DDT lasted for 2 years, protecting the seedlings during the most critical periods. Lindane (γ-BHC) was added to the compounds to accelerate the killing. When these chlorinated hydrocarbons were banned for environmental reasons in the 1970s, pyrethroids such as permethrin became the preferred chemicals. Although pyrethroids do not accumulate in food webs like the chlorinated hydrocarbons, environmental groups oppose their use, even when they are applied in nurseries. The European Commission may not allow the use of permethrin for forest seedling protection in the future.

Currently, the alternatives to insecticides are a mixed bag of silvicultural and technical methods. Substitution of clear felling and planting with natural regeneration methods would reduce the problem. However, this is often difficult in spruce forests because the shallow-rooted seed trees that are left are prone to wind felling. More deep-rooted species, e.g., Scots pine, are indeed often naturally regenerated. Here, a moderately dense shelterwood stand may significantly reduce seedling damage. This is partly due to the weevils' feeding on twig bark in the canopy and on superficial tree roots. Soil scarification also reduces feeding, particularly where seedlings are planted in exposed mineral soil. This effect diminishes gradually as the scarified patches fill up with vegetation and litter. Planting on small mounds of mineral soil is an improvement of the scarification method, though expensive. A variety of stem protecting devices exist, including coats of wax or latex, and various objects made from plastic, paper, etc., that are fitted around the seedling. At present, none of these devices seem to be reliable for a prolonged period of time; some are efficient for one season, but fail the next. Some of the designs may cause damage to the plant roots.

The pine weevil has a variety of natural enemies, such as parasitoids, and invertebrate and vertebrate predators. However, these organisms rarely seem to make significant inroads on the populations. In some cases, this may be because the immature stages are relatively inaccessible in their subterranean hiding places. Fungi and nematodes kill weevils, and nematodes may possibly become an active means to destroy the soil-dwelling stages, provided strains can be found that are active in cool soils. The artificial application of fungi to occupy the phloem of stumps could render this substrate less useful for the larvae, but success depends on how fast the hyphae are able to permeate the substrate. Considerable efforts are allocated to the search for chemicals with repellant and antifeedant properties, potentially to include them in stem protectants. Recently, a repellent substance has been patented, which occurs in excrements that the egg-laying females deposit next to their eggs, probably as a warning to other pine weevils. It seems unlikely that the pine weevil will easily adapt to this natural signal.

A panacea to the pine weevil problem does not seem likely. In areas where clear cutting remains a dominant way of regenerating conifers, the pine weevil is likely to retain its unique position as a headache for foresters and an expensive enemy of commercial forestry.

Erik Christiansen
Norwegian Forest Research Institute
Ås, Norway

References

Bejer-Petersen, B., P. Juutinen, E. Kangas, A. Bakke, V. Butovitsch, H. Eidmann, K. J. Heqvist, and B. Lekander 1962. Studies on *Hylobius abietis* L. I. Development and life cycle in the Nordic countries. *Acta Entomologica Fennica* 17: 1–106.

Leather, S. R., K. R. Day, and A. N. Salisbury. 1999. The biology and ecology of the large pine weevil, *Hylobius abiestis* (Coleoptera: Curculionidae): a problem of dispersal? *Bulletin of Entomological Research* 89: 3–16.

Nordlander, G., H. Nordenhem, and H. Bylund. 1997. Oviposition patterns of the pine weevil *Hylobius abietis*. *Entomologia Experimentalis et Applicata* 85: 1–9.

Ratzeburg, J. T. 1839. Die Forst-Insekten, Berlin.

Örlander, G., G. Nordlander, K. Wallerts, and H. Nordenhem. 2000. Feeding in the crowns of scots pine trees by the pine weevil *Hylobius abietis*. *Scandinavian Journal of Forest Research* 15: 194–201.

PINE WILT. See also, TRANSMISSION OF PLANT DISEASES BY INSECTS.

PINE-FLOWER SNOUT BEETLES. Members of the family Nemonychidae (order Coleoptera). See also, BEETLES.

PIN-HOLE BORERS. Some members of the subfamily Platypodinae (order Coleoptera, family Curculionidae). See also, BEETLES.

PINK HIBISCUS MEALYBUG, *Maconellicoccus hirsutus* GREEN (HEMIPTERA: PSEUDOCOCCIDAE). The pink hibiscus mealybug, *Maconellicoccus hirsutus* (Green), is native within the area encompassing Southeast Asia and nearby Australia. Its host range is very large, exceeding over 200 perennial and annual plant species, many of which are important in agriculture and as ornamentals. During the early part of the twentieth century, its range extended into Central Asia and Egypt. Within Egypt, it was a very significant pest of several common plant species, including cotton. Initially, a predatory beetle species (*Cryptolaemus montrouzieri*) was reared and introduced in large numbers. This provided some short-term relief; however, it was not until several parasite species were introduced that high levels of control were sustained. The pink hibiscus mealybug appeared first in the Western Hemisphere in Hawaii in 1984 and in the Caribbean island of Grenada and neighboring islands in 1994. In subsequent years, it spread to over 27 island countries in the Caribbean and in 1999, it was found in southern California, USA, the adjoining border region of northern Mexico, and the Central American country of Belize. By 2000, it was discovered in the Bahamas and northern South America (i.e., Venezuela/Guyana region), and attained Florida, USA, in 2002.

Female pink hibiscus mealybugs pass through three immature nymphal stages prior to developing into wingless adults. Most commonly, egg masses (i.e., ovisacs) containing several hundred eggs are laid at nearly the same location on a plant where the female developed. They lay one ovisac during their life. Total egg production ranges widely and is dependent on the host plant. Several hundred eggs are produced by females that have developed on

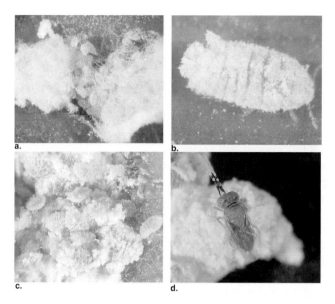

Fig. 801 Pink hibiscus mealybug (upper left) ovisac, (upper right) young adult female, (lower left) cluster of nymphs and adult females, (lower right) parasitoid *Anagyrus kamali* female.

potato sprouts, whereas 600 or more eggs often are laid by females that have developed on pumpkin fruit. At temperatures ranging from 26 to 33°C, the total development time from egg to gravid female takes four to five weeks. This includes approximately five days for egg hatch, seven days for first and second instar female nymphs, eight days for third instars and approximately another week for adult females to fully mature and begin laying eggs. As adults, females undergo a considerable increase in size (approximately 2 to 3-fold) prior to laying eggs. In contrast, although male pink hibiscus mealybugs look essentially identical to female mealybugs as first instars and most of the second life stage, male nymphs take on a considerably different appearance as late second instar nymphs. They become elongate and begin to sport small wing buds that become accentuated in third and fourth instars. As late second instars to the time males emerge as winged adults, they live clustered together as small groups within flocculent mounds of waxy material. During these late stages of development, the males are developing within a cocoon-like structure. Females also produce an abundant white, waxy substance that covers their bodies. Male pink hibiscus mealybugs have four immature nymphal stages. There is a third instar life stage of very short duration (lasting for little more than a day) and a fourth nymphal life stage lasting

for five to six days prior to emerging as a winged adult.

First instar nymphs are referred to as crawlers because they are very actively moving at this stage. The crawlers and all other immature stages range in color from orange to medium pink. Adult females have a very dark pink coloration. Wax is present on all stages; however, it is especially abundant on third instar and adults. Although many mealybug species are field identified by specific patterns of wax filaments extending from the perimeter of the body or very long filaments extending from the posterior, the pink hibiscus mealybug produces no distinct pattern of filaments. As a result of the delicate nature of wax filaments, field identification of mealybug species typically requires viewing the characteristics of numerous individuals in a group as opposed to one isolated mealybug. The life stage of preserved pink hibiscus mealybug specimens can be accurately identified by the number of antennal segments. First and second instar nymphs have six segments, third instar females have seven segments, whereas adult females have nine-segmented antennae.

The pink hibiscus mealybug undergoes numerous generations each year. Population increase is dependent on temperature and suitability of host plants, which is often linked to moisture availability. The pink hibiscus mealybug performs well at moderate to high temperatures (exceeding $38°C$). In the tropical and subtropical regions, this mealybug species reproduces continuously throughout the year, having abundant host plant material and temperatures ranging from $22°C$ and above. In the more temperate areas of its distribution in the northern hemisphere, the pink hibiscus mealybug population is difficult to find from winter to late spring. From June to September, population densities may increase dramatically in the absence of effective natural enemies. From November onward, the population declines as certain host species (especially mulberry) drop their leaves, causing the mealybugs to move from the growing regions of the tree (i.e., branch terminals) to larger branches that often contain crevices in rough bark, well suited for providing protection. Essentially all life stages are present during the winter period, although later instars and adult female mealybugs are more common.

As previously stated, *Cryptolaemus montrouzieri* is cited as an important biological control agent within its presumed native range. It has also been used with some success within new regions where the pink hibiscus mealybug has become established. Parasites also play a very important role in population regulation. *Anagyrus kamali* (family Encyrtidae) is among several agents that have had a long-term impact on the pink hibiscus mealybug in Egypt and has been the most important natural enemy introduced to date to control the pink hibiscus mealybug in the Western Hemisphere. This parasitoid has reduced mealybug population densities by 90% or more in St. Kitts, West Indies; St. Thomas and St. Croix in the U.S. Virgin Islands, Puerto Rico, Belize, Bahamas and California, USA, resulting in a highly successful biological control program.

W. J. Roltsch
California Department of Food and Agriculture
Biological Control Program
Sacramento, California, USA
and
D. E. Meyerdirk
USDA-APHIS, PPQ
National Biological Control Institute
Riverdale, Maryland, USA

References

Ghose, S. K. 1972. Biology of the mealybug *Maconellicoccus hirsutus* Green (Pseudococcidae: Hemiptera). *Indian Agriculture* 16: 323–332.

Hall, W. J. 1926. The hibiscus mealybug (*Phenacoccus hirsutus*, Green) in Egypt in 1925 with notes on the introduction of *Cryptolaemus montrouzieri*, Muls. *Ministry of Agriculture, Cairo, Egypt. Technical and Scientific Service Bulletin No. 70.* 19 pp.

Kairo, M. T. K., G.V. Pollard, D. D. Peterkin, and V. F. Lopez. 2000. Biological control of the hibiscus mealybug, *Maconellicoccus hirsutus* Green (Hemiptera: Pseudococcidae) in the Caribbean. *Integrated Pest Management Reviews* 5: 241–254.

Mani, M. 1989. A review of the pink mealybug – *Maconellicoccus hirsutus* (Green). *Insect Science and Its Application* 10: 157–167.

Persad, A., and A. Khan. 2002. Comparison of life table parameters for *Maconellicoccus hirsutus, Anagyrus kamali, Cryptolaemus montrouzieri* and *Scymnus coccivora*. *BioControl* 47: 137–149.

Sagarra, L. A., C. Vincent, and R. K. Stewart 2001. Body size as an indicator of parasitoid quality in male and female *Anagyrus kamali* (Hymenoptera: Encyrtidae). *Bulletin of Entomological Research* 91: 363–367.

PIOPHILIDAE. A family of flies (order Diptera). They commonly are known as skipper flies. See also, FLIES.

PIPUNCULIDAE. A family of flies (order Diptera). They commonly are known as big-headed flies. See also, FLIES.

PIROPLASMOSIS: Babesia and Theileria. The Piroplasmea, a class of protozoa, are transmitted to vertebrates by ticks. Piroplasms infect erythrocytes (RBCs), and some also attack leukocytes. Species from two genera, *Babesia* and *Theileria*, cause serious diseases in dogs, cattle, sheep, and goats.

Babesia

Some 99 species of *Babesia* are parasites of nine orders of mammals. We shall consider only those species that cause babesiosis in dogs, ruminants and humans.

Canine babesiosis, also known as canine piroplasmosis or malignant jaundice, is caused by *Babesia canis* and *B. gibsoni*. The former is cosmopolitan, especially in warm climates, while the latter is found in Asia and North America. Both are pathogenic in dogs. *B. canis* is transmitted by the brown dog tick *Rhipicephalus sanguineus*. The extent of pathology in the dog is dependent on the virulence of the strain of *B. canis*.

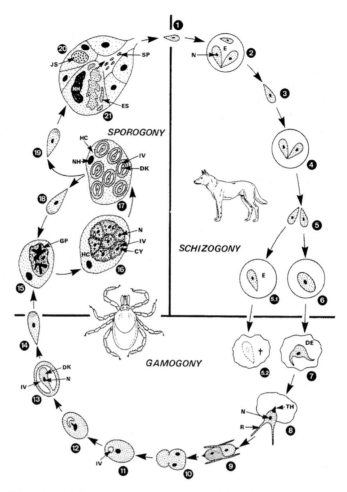

Fig. 802 Diagram of the life cycle of *Babesia canis*. An infectious tick introduces sporozoites (1) into the blood of its host and the sporozoites; they undergo binary fission in the red blood cells (RBCs) (2–4); the merozoites differentiate into gamonts also within RBCs (5, 6). When infectious blood is ingested by a tick, gamogony and fertilization occur in the gut of the tick (7–13), and then the kinete (14) enters cells of the salivary gland where sporogony takes place (15-20). Abbreviations: CY, cytomere; DE, digested RBC; DK, developing kinete; E, RBC; ES, enlarged sporont forming sporozoites; GP, growing parasite or polymorphic stage; HC, host cell; IV, inner vacuole; N, nucleus; NH, nucleus of host cell; R, raylike protrusion; SP, sporozoite; Th, thornlike apical structure; YS, young meront. (From Marquardt et al., 2000, used with permission of Harcourt Academic Press.)

The life cycle of *Babesia canis* is typical of other species of *Babesia*. The sporozoite is introduced into the vertebrate host via the saliva of *R. sanguineus*. Sporozoites invade the host's RBCs where they develop via binary fission into merozoites. Eventually these RBCs rupture, releasing the parasites to invade other RBCs. Merozoites develop into gamonts in some RBCs. Gamonts infect ticks when a blood-meal is taken. The development of sporozoites into gamonts is an asexual form of reproduction known as schizogony. It allows the rapid build-up of the parasite in the host. Within the tick's midgut, sexual reproduction occurs in a process known as gamogony. The fertilized kinete then enters cells of the salivary gland where sporogony takes place. This is a process of asexual reproduction that results in the production of large numbers of sporozoites. The kinete can also infect the ovaries which allows transovarial transmission to new larvae and produces a new generation of infected ticks. *Babesia canis* are also transmitted from one life stage to another in a process named transstadial transmission.

Most of the symptoms of the different babesiosis including canine babesiosis, can be traced to the breakdown of infected RBCs and the release of toxic products. This can result in anemia, enlarged spleen, kidney damage and sometimes impaired circulation of blood. Dogs that recover from canine babesiosis are immune to future infections. As with other babesiosis, diagnosis is made by identifying the parasite within RBCs. Several drugs are effective against canine babesiosis.

Two species of *Babesia* are parasites of sheep: *Babesia motasi* and *Babesia ovis*. Some strains of *B. motasi* will infect goats as well as sheep. *Babesia motasi* produces a severe disease but is even more pathogenic when it is accompanied by infections of *Theileria* and or *Ehrlichia*. *Babesia ovis* is found in equatorial Africa north to the Sahara, in the Mediterranean area throughout Europe, and portions of the former USSR. It is believed to be transmitted by *Rhipicephalus bursa* and *R. evertsi* outside of Asia, and *R. bursa* in the former USSR. The latter tick can transmit *B. ovis* transovarially through at least 44 generations. *Babesia motasi* occurs in equatorial Africa north to the Sahara, the Mediterranean area and Europe. It is believed to be transmitted by species of *Haemaphysalis* in Europe. Sheep infected with *Babesia* can act as carriers for up to 2 years.

Horses and other equids are infected by *Babesia caballi* which causes equine piroplasmosis worldwide. Mortality in domestic horses is less than 50%, but strains with differing virulence exist. Over ten species of ticks can vector *B. caballi*. *Dermacentor nitens* transmits the parasite in the United States, Central and South America.

Bovine babesiosis is caused by seven species of *Babesia* and is transmitted by several species of ticks. Domestic cattle are at high risk from babesiosis worldwide. Two of the seven species that cause bovine babesiosis, *Babesia bovis* and *Babesia bigemina*, are widely distributed in Africa, Asia, Europe, Australia and Central and South America. *Babesia bovis* is transmitted by species of *Boophilus*, especially *B. microplus* and *B. decoloratus* in tropical and subtropical areas, and by *Rhipicephalus bursa* in southern Europe. Transmission by a one-host tick such as *B. microplus* and other species of *Boophilus* is described in the accompanying figure. The female tick becomes infected during feeding and transmits the pathogen transovarially to larvae that infect cattle

as they feed. Many factors are related to the success of this transmission cycle.

Babesia bigemina causes Texas cattle fever, also known as Red water fever or splenetic fever. Feeding ticks transmit the parasite. The African strain of *B. bigemina* is the most pathogenic, the Australian strain less so. The tick vector in the USA is *Boophilus annulatus*. Texas cattle fever is important historically because it is the first disease caused by a protozoan parasite that was shown to be transmitted by a blood feeding arthropod. The disease was first noticed during the cattle drives in the mid 1800s, and was officially described in 1888 as Texas cattle fever. Efforts to eradicate Texas cattle fever started at the turn of the century with the goal of eliminating the disease in the southeast part of the country, especially Texas and Florida. The ability to control cattle movement into Florida resulted in success in eliminating Texas cattle fever in that state by the 1930s. Because of its long border with Mexico, control of cattle movement into Texas was more difficult. Even so, the disease has largely been eliminated as a problem though there is still a sub-symptomatic level of infection in some Texas herds.

Five other species of *Babesia* are not distributed as widely as *B. bovis* and *B. bigemina*. *Babesia major* occurs in Europe and the Middle East where it causes a disease in cattle with limited mortality. *Haemaphysalis punctata* vectors *B. major* which is transmitted transovarially in the tick. *Babesia jakimovi* occurs in Siberia where it causes a disease in cattle that has high mortality. It is transmitted by species of *Ixodes*. The pathogen is transmitted transovarially in its vectors. Adults of the next generation of ticks are the infective stage. *Babesia divergens* occurs in Europe and causes a disease that can produce a parasitemia in cattle with as high as 24% of RBCs infected. The pathogen is transmitted by *Ixodes ricinus*. The adult tick picks up the pathogen during feeding and transmits it transovarially. All feeding stages of the next generation can transmit the pathogen to cattle. *Babesia ocultans* occurs in South Africa, and *Babesia ovata* in Japan.

Clinical symptoms of babesiosis are seen 8 to 16 days following infection. Most of the symptoms result from infected RBCs which eventually break down releasing hemoglobin and metabolic byproducts of the parasite. This results in a fever that can rise to over 40°C and anemia. Hemoglobin creates the characteristic red tinge of the urine. *Babesia bovis* infections can produce a hypertensive shock

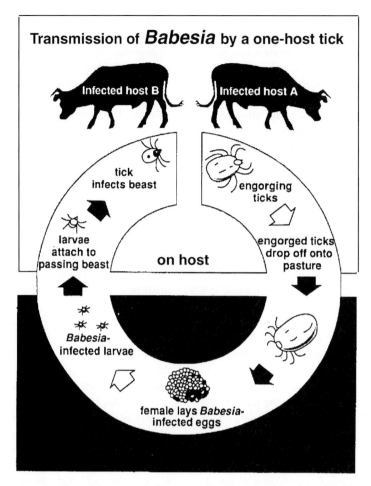

Fig. 803 Diagrammatic representation of the transmission of a *Babesia* species by a one-host tick. The upper half shows the ticks attaching, feeding, and dropping from the host, while the lower half shows the events that occur in the pasture. Factors affecting the success of the transmission cycle include: (1) the concentration of infected erythrocytes; (2) number of infected, replete females dropping from the diseased host; (3) proportion of engorged females that lay eggs; (4) proportion of eggs that hatch; (5) the proportion of the larval population that is infected with the pathogen; (6) the number of infected larvae attaching to susceptible bovine hosts each day; (7) host immunity (natural and acquired) to infection; and (8) host resistance to tick parasitism. (From Sonenshine, 1993, used with permission of Oxford University Press, Inc.)

syndrome where sick animals quit feeding and become listless. Often dry bloodstained feces occur. Parasitized animals may develop sunken eyes and muscle tremors. Symptoms can be severe especially in older animals. Newborn calves are protected by maternal antibodies.

Cattle are able to develop immunity to infection with *Babesia* species. Persistent subclinical infections causes acquired immunity. Calves also get immunity from their mothers that can last several months, and if infected, calves develop an even stronger immunity that can last for up to 4 years. Some breeds of cattle are naturally more resistant to the disease, so they are preferred in areas where there is a high level of infected ticks. Currently the best control of the disease is by dipping and spraying the animals to control tick infestations. A vaccine used in Australia has had some success.

Four species of *Babesia* have been found in humans: *B. bovis* and *B. divergens* whose normal host are oxen, *B. equi* whose normal host are horses, and *B. microti* whose normal host are rodents. Piroplasmosis in humans is rare and those without a spleen are the most susceptible. *Babesia microti* can infect individuals with a normal spleen. *Babesia microti* is transmitted by *Ixodes ricinus* and *I. trianguliceps* in

Europe, and *I. scapularis* in the USA. Symptoms of human babesiosis are similar to those of malaria, but the disease lacks the periodicity found in malaria.

Theileria

Theileria spp. are parasites of domestic and wild ruminants, and domestic and wild felines. Six species infect cattle: *Theileria parva, Theileria annulata, Theileria orientalis, Theileria mutans, Theileria taurotragi* and *Theileria velifera*. The most serious diseases are caused by *T. parva* and *T. annulata*. These two pathogens are considered later. *Theileria orientalis* occurs worldwide and is transmitted by several ixodids. In western Eurasia, through the northwest of Africa, it is transmitted by *Haemaphsalis punctata*, but in the eastern Asia and Australia it is transmitted by *H. longicornis*. Usually *T. orientalis* causes a mild disease, but in susceptible breeds of cattle it can produce a severe anemia. *Theileria orientalis* may be a synonym of *T. buffeli* and *T. sergenti*. *Theileria mutans* occurs throughout tropical Africa. Its principle vector is *Amblyomma variegatum*. *Theileria mutans* is usually not very pathogenic, although a pathogenic strain of the parasite exists in East Africa that can produce severe anemia with up to 45% of the host's RBCs infected. *Theileria taurotragi* infects cattle in eastern, central and southern Africa, and can also infect sheep and goats. The pathogen is transmitted by *Rhipicephalus appendiculatus, R. zambeziensis* and *R. pulchellus*. *Theileria taurotragi* produces a mild or subclinical disease in cattle but can produce a fatal disease in eland. *Theileria velifera* is found throughout Africa south of the Sahara. It is transmitted by *Amblyomma variegatum, A. hebraeum* and *A. lepidum*. The parasite causes a disease of low pathogenicity. *Theileria cervi* is a parasite of cervids in Europe and the United States. In the USA the pathogen is transmitted by *Amblyomma americanum*. The pathogen is common in white-tailed deer (*Odocoileus virginianus*) in the south and southeastern states. No clinical manifestations of these infections occur.

Theileria parva causes East Coast Fever, the most serious of the diseases caused by *Theileria* in cattle. It is trasmitted by *Rhipicephalus appendiculatus*. East Coast Fever was first described in east Africa at the start of the 19th century. By 1901–1902 it had spread to southern Africa. The incubation period for East Coast Fever is 10–25 days. Symptoms include fever, swollen lymph nodes and emaciation. Most of the damage is to the lymphatic system, especially the lymph nodes and spleen, but other organs including the kidney, lungs and liver may also be affected. Older animals are more susceptible to East Coast Fever than younger animals. The most common control of East Coast Fever is to kill the vector ticks. Antigenic diversity exists in *T. parva,* and this can complicate the outcome of the disease. In addition, strains of *T. parva* exist that have different virulence. Virulence is also tied to the tick vector, and some ticks only transmit a 100% fatal form of the parasite. The disease is believed to have originated in African buffalo *Syrlcerus caffer* which acts as a reservoir. East Coast Fever is a benign disease in the buffalo.

An important concept in understanding East Coast Fever is endemic stability, which is defined as the condition in a given cattle herd (or population) where a large majority of individuals become infected and immune by six months of age, with little or no clinical symptoms. Endemic stability occurs when the cattle have a low innate susceptibility to the pathology of East Coast Fever; when *R. appendiculatus* infestation occurs throughout the year; and when most of young calves are exposed to a low *T. parva* challenge. Eradication of the tick vectors in endemic areas can result in the loss of endemic stability. Subsequent relaxation of control measures in these areas has often resulted in a rapid increase in the tick population with the return of severe epidemics of East Coast Fever.

The life cycle of *Theileria parva* is similar to that of *Babesia canis,* but differs in that lymphocytes are invaded as well as RBCs. The infectious stage of the parasite is the sporozoite. These first invade lymphocytes in the lymph nodes, unlike *Babesia canis* sporozoites. In lymphocytes, the sporozoites undergo schizogony and multiply, then invade RBCs. Erythrocytic schizogony is rare or absent. Infected RBCs are transmitted to the tick where they undergo gamogony and fertilization in the midgut, then invade salivary gland cells. Here sporogony occurs, resulting in sporozoites that are transmitted to the vertebrate host with the tick's saliva. In the tick vector, transstadial transmission occurs but not transovarial transmission.

Theileria annulata causes Mediterranean or tropical theileriosis in cattle. The pathogen is found throughout northern Africa, into the Middle East including Saudi Arabia, Turkey, southern Europe through the Caucasian area into India and China. It is transmitted by species of *Hyalomma*. Common vectors include *H. anatolicum anatolicum, H. detritum* and *H. asiaticum*. Tropical theileriosis is a milder disease than ECF and has a much lower mortality in cattle. However, it is an economically important impediment to livestock improvement. In enzootic areas, *Bos taurus* cattle can have a mortality of 40 to 80%, and *Bos indicus* cattle, which are more resistant to *T. annulata*, can have as much as 20% mortality, mostly in calves. Symptoms include fever, swollen lymph glands, weight loss, anorexia, and apathy. A generalized leucocytosis occurs with great reduction in RBCs, resulting in hemolytic anemia. Organs that are damaged are mostly similar to those damaged in East Coast Fever. Erythrocytic schizogony is common in the life cycle of *T. annulata*. Cattle that survive the disease develop a persistent immunity.

Several species of *Theileria* infect small ruminants. *Theileria hirci* occurs in eastern and southern Europe, the Near and Middle East and in northern Africa. It is transmitted by *Hyalomma anatolicum*. It produces a disease that is highly pathogenic in sheep and goats. *Theileria separata*, a non pathogenic parasite of sheep, is found in Tanzania and is transmitted by *Ripicephalus evertsi*. *Theileria ovis* is worldwide in distribution and is transmitted by many different ticks, including species of *Ornithodorus*, *Hyalomma*, *Haemaphysalis*, *Dermacentor* and *Rhipicephalus*. It produces a disease with little or no pathogenicity in sheep and goats.

Wild and domestic cats are infected by *Theileria felis*, which causes feline cytauxzoonosis in the United States. It is transmitted by *Dermacentor variabilis*. The disease can be fatal unless treated. Cougars and bobcats are carriers of the disease, but show no symptoms. See also, TICKS.

Lewis B. Coons and Marjorie Rothschild
The University of Memphis
Memphis, Tennessee, USA

References

Kettle, D. S. 1995. *Medical and veterinary entomology* (2nd ed.). CAB International, Wallingford, United Kingdom. 725 pp.

Kocan, A. A., and K. M. Kocan. 1991. Tick-transmitted protozoan diseases of wildlife in North America. *Bulletin of the Society of Vector Ecology* 16: 94–108.

Norval, R. A. I., B. D. Perry, and A. S. Young. 1992. *The epidemiology of Theileriosis in Africa*. Academic Press Inc., San Diego, California. 481 pp.

Sonenshine, D. E. 1993. *Biology of ticks*, Vol. 2. Oxford University Press, New York, New York. 465 pp.

Sonenshine, D. E., and T. N. Mather (eds.) 1994. *Ecological dynamics of tick-borne zoonoses*. Oxford University Press, New York, New York. 447 pp.

PISTACHIO SEED WASPS. Two species of such wasps are known pests of pistachio because they destroy the fruit. They belong to different families of the Hymenoptera, and their population densities vary with the region and with time. In some orchards the two species co-exist, but one of them usually outnumbers the other. In addition to fruits of the cultivated pistachio, *Pistacia vera*, both these wasps infest also fruits of *P. terebinthus*.

Eurytoma plotnikovi Nikol'skaya (Eurytomidae)

The adult female is 4 to 5 mm long and has a reddish brown head and thorax, yellowish red abdomen and red eyes. The legs and the two basal antennal articles may have a lighter color, but this varies with the region. The antennae have 10 segments, with the basal one, the scape, being three times as long as the next one. The male is usually black and a little shorter than the female. The larva is whitish, tapering in both ends, apodous, curved, and is 6 mm in length when fully grown. It occurs from the Mediterranean region east to central Asia.

This wasp is univoltine. However, some individuals may complete their life cycle in two years because of prolonged diapause. It overwinters as a fully-grown larva inside the fruit, whether the infested fruit remains on the tree throughout winter or has fallen to the ground. Pupation occurs also inside the fruit, sometime in May, and the adult emerges in late May to late June, after boring an exit hole in the fruit with its mandibles. Oviposition occurs in June. Using her ovipositor, the female inserts her egg near the tip of the fruit. The stalked egg is usually placed on the inner wall of the endocarp. The oviposition hole darkens, is easily seen, and may offer entry to plant pathogenic fungi. The larva first feeds on the still tender inner layers of the endocarp, and later, in July to August, on the seed, which it consumes completely or almost so. There it remains in diapause until the following spring. In coastal northern Greece, diapause is terminated between early April and early May, with high temperatures and long days leading to diapause termination. The infested fruits become mummified, and as a rule remain on the tree after the leaves have fallen in autumn. Yet, if the fruits are also infected by a fungus, they mostly drop to the ground, usually in autumn.

This seed wasp is among the most destructive pests of pistachio in many countries. Thus, control measures are generally needed. One such measure is the collection and destruction of all mummified fruits, whether they are on the tree or the ground. This measure to be effective should be applied by all neighboring growers. If fruit collection in a given area cannot be practiced, one or two insecticidal sprays are needed. The insecticide should preferably be systemic, but contact ones of long residual action also proved effective. The proper time is determined by following the exit of adult wasps from infested fruits kept in cages in the orchard. In Cyprus, a spray applied three days after the exit of the first wasps from caged fruits, gave satisfactory protection of the crop.

Megastigmus pistaciae Walker (Torymidae)

The adult female has a golden yellow body, with grey-yellow head, red eyes and reddish reflections on the abdomen. Its length is variable, from 3 to 5.5 or even 6 mm. The front wings have a dark oval spot near the costa. The male is reddish yellow. Males are rare. The larva is greyish white, apodous, curved, narrower in both ends and 6 mm long when fully grown. It occurs in southern Europe, northern Africa, the Black Sea coast, the Middle East, Iran, Central Asia, and the United States.

Oviposition takes place in the fruit and the larva consumes the seed, much like *Eurytoma plotnikovi*, above. In Iran and Tunisia it is reported to be bivoltine. It overwinters as a larva inside the pistachio seeds, pupates in spring, and emerges as an adult in spring or early summer. In Tunisia, the adults of the second generation are seen in April-June, and those of the first in July to August. Overwintering larvae are produced from eggs of both generations. In Iran, adults emerge in mid-April to late May and in mid-June and oviposit shortly before the endocarp is completely hardened.

In California, in addition to pistachios, this wasp often infests the seeds of *Pistacia chinensis*, a common ornamental, as well as fruits of other plants of the same genus. Adult trapping and emergence records indicate two generations per year, with adults being active in June and August to September. Yet,

Fig. 804 Fully-grown larvae of *E. plotnikovi* within infested mummified pistachios (above), and infested mummified fruits remaining on the tree after harvest (below).

adults emerging during the second period are evidently unable to oviposit through the endocarp which is already hard. Therefore, it is possible that only one generation develops in pistachios in California. This is supported by the fact that a high percentage of larvae of the first generation enter diapause, to give adults in the following spring. A possible second generation may develop in fruits of other species of *Pistacia* having an endocarp that hardens later than that of pistachio.

Damage is similar to that by *E. plotnikovi* and, when it reaches economic levels, control should be applied in spring against the adults. Careful collection and destruction of the fruits housing the diapausing larvae are also effective, if applied by all the growers of a given area, and provided that no other species of host trees are in the vicinity.

Minos E. Tzanakakis
Aristotle University of Thessaloniki
Thessaloniki, Greece

References

Davatchi, A. 1956. Sur quelques insectes nuisibles au pistachier en Iran. *Revue de Pathologie Végétale et Entomologie Agricole de France* 35: 17–26.

Haralambidis, C. G., and M. E. Tzanakakis. 2000. Time of diapause termination in the pistachio seed wasp *Eurytoma plotnikovi* (Hymenoptera: Eurytomidae) in northern Greece and under certain photoperiods and temperatures. *Entomologia Hellenica* 13: 43–50.

Jarraya, A., and J. Bernard. 1971. Premières observations bioécologiques sur *Megastigmus pistaciae* en Tunisie. *Annales Institut National de la Recherche Agronomique de Tunisie* 44: 1–28.

Rice, R. E., and T. J. Michailides. 1988. Pistachio seed chalcid, *Megastigmus pistaciae* Walker (Hymenoptera: Torymidae), in California. *Journal of Economic Entomology* 81: 1446–1449.

PIT SCALES. Members of the family Asterolecaniidae, superfamily Coccoidae (order Hemiptera). See also, BUGS.

PLAGUE: BIOLOGY AND EPIDEMIOLOGY.

Plague, also called the black death, is an infectious bacterial disease of humans caused by *Yersinia* (*Pasteurella*) *pestis*. The bacterium is a non-motile, gram negative, plump coccobacillus that is transmitted through the bite of the rat flea, *Xenopsylla cheopis*, from susceptible rat hosts, primarily the domestic rat, *Rattus rattus*. People may sometimes become infected by handling other rodents such as prairie dogs that have sylvatic plague and harbor the infected fleas. The infected rodent populations are often decimated by the disease and the hungry fleas abandon the host carcasses in search of new hosts, often domestic rats, cats or even humans. Direct bacterial infection can occur in humans who handle dead rodents or who care for infected domestic cats. In such cases, the bacterium may enter an open wound. Airborne bacterium in the cough of an infected cat or human can also be the source of infection (pneumonic plague).

In the Middle Ages, plague killed millions of people in Europe and, from 1924–1925, there was an urban plague epidemic in Los Angeles. Since 1974, small scattered foci of the plague were identified in northern regions of China and other Asian countries, Southern Africa and Madagascar, and in Brazil. In the United States, between 1970 and 1998, small foci have been confined to rural areas of Arizona, California, Colorado, and New Mexico with 36 or as many as 191 cases reported from each of these regions.

Plague induces three types of clinical pathologies in human hosts: the bubonic, primary pneumonic, and primary septicemic plague. Of these, the bubonic plague is the most common and is caused by the bite of an infected flea. Within 2–4 days of infection the person experiences chills, high fever, headache, nausea, and vomiting. There is also evidence of rapid pulse and rapid breathing accompanied by anxiety. Physiologically, the neutrophile leukocyte counts are rapidly elevated. Concomitant with these symptoms, single or multiple enlarged hemorrhagic lymph nodes, the "bubo" (hence the name bubonic plague), appear and are painful and tender. If treated early with antibiotics such as streptomycin, gentamicin, tetracycline, or chloramphenicol, the infection is often eliminated. However, if treatment is delayed, death ensues within 24 hours as a result of toxemia even though the bacteria are killed by the antibiotic. About one in seven patients in the United States dies from the disease. Mild variations of the disease (plague minor) may occur and involve minimal toxemia with small "buboes" which contain evidence of the bacterium (thus resulting in the diagnosis). However, these patients frequently survive. Other patients may serve as "carriers" with temporary bacterial infection in the throat (transient pharyngeal carrier) but who may have no manifestations of the disease.

Patients with primary septicemic plague may occur in about 1% of cases. The disease pathogenesis is not as clear-cut as in bubonic plague infections. Sudden fever, meningitis, hemorrhage and pneumonia are common. Presumably, the infection is induced by direct introduction of the bacillin into a blood vessel.

Primary pneumonic plague is contracted by inhalation of the bacterium in aerosol form from coughing patients, and from airborne bacilli released from the remains of infected people or animal carcasses. There is a 48–60 hour incubation period, followed by sudden chills and fever. The sputum and other body exhudates of persons with primary pneumonic plague are replete with *Y. pestis* bacilli but the individual exhibits no symptoms of pneumonia. Death occurs within one or two days after infection, as a result of respiratory distress and toxic shock.

The vector of the bacterium, the (Oriental) rat flea, *Xenopsylla cheopis*, like other fleas, belongs to the insect Order Siphonaptera. Unlike most other insects, fleas have a bilaterally flattened body and possess jumping legs than enable them to jump more than 100 times their body length, and even to heights of more than 30 cm. They have piercing-sucking mouthparts that enable them to pierce the skin of their rodent or human host and siphon the blood. The proteins in the blood are important for flea egg development. Fleas have a holometabolous form of development involving an egg, larval and pupal stages, and the adult male and female.

The eggs (about 0.5 mm long) do not adhere to the rat's body but usually fall off into the nest. After 2–21 days, depending on temperature, one larva hatches/egg and feeds upon the detritus and flea feces. There are three larval instars that require 9–15 days to reach the pupal stage. Larvae have a somewhat cylindrical body, have no legs or other appendages, and are covered with long setae. They have chewing mouthparts. A mature larva may measure up to 4 mm long. The pupal stage lasts 5–7 days but can be prolonged if temperatures are low. The pupa may spin a silken cocoon around itself. Adults may live for 38–100 days, depending on the humidity. Such longevity has epidemiological implications, particularly in relation to plague as they may allow the pathogen to survive long after an outbreak has been though to be eliminated.

Several public health and environmental management strategies have been established in the United States to prevent plague infections. These involve sanitation measures to reduce rodent infestations where people work, securing grain and other food storage areas from rodent entry, use of rodenticides in areas (e.g., ships and docks) where rodents may breed, and applying insecticides to kill fleas, particularly in rodent nests where flea eggs, larvae, or pupae may be sequestered. See also, FLEAS, HISTORY AND INSECTS.

Pauline O. Lawrence
University of Florida
Gainesville, Florida, USA

References

Binford, C. H., and D. H. Connor. 1976. *Pathology of tropical and extraordinary diseases, an atlas*. vol. 1. Armed Forces Institute of Pathology, Washington, DC. 339 pp.

Centers for Disease Control. 2002. *A quick guide to plague*. http://www.cdc.gov/ncidod/dvbid/dvbid.htm. Department of Health and Human Services.

Schmidt, G. D., and L. S. Roberts. 1989. *Foundations of parasitology*. 4th ed. Times Mirror/Mosby College Publishing, St. Louis, Missouri. 750 pp.

Youmans, G. P., P. Y. Paterson, and H. M. Sommers. 1980. *The biologic and clinical basis of infectious diseases*. W.B. Saunders Co., London, United Kingdom. 849 pp.

PLANIDIIFORM LARVA. A flattened body form found in the active first instar of certain parasitoids in the orders Diptera and Hymenoptera. This mobile stage, occurring before the insect penetrates the host, is called a planidium (pl., planidia).

PLANIDIUM. In certain insects that undergo hypermetamorphosis, the first instar larva, which is legless and somewhat flattened. This is found in certain Diptera and Hymenoptera. (contrast with triungulin).

PLANTA. (pl., plantae) The basal joint of the posterior tarsus in pollen-gathering Hymenoptera.

PLANT BUGS (HEMIPTERA: MIRIDAE). The family Miridae, often referred to as plant bugs, is the largest true bug family. Worldwide, more than 10,000 species of Miridae are known, but this number is expected to more than double once the tropical faunas are more thoroughly studied. Plant bugs belong to the superfamily Miroidea, infraorder Cimicomorpha, and suborder Heteroptera, within the order Hemiptera. Also included in Miroidea are the families Joppeicidae, Microphysidae, Thaumastocoridae, and Tingidae.

Eight mirid subfamilies are currently recognized, primarily on the basis of pretarsal structures and male genitalia. Mirinae is by far the largest subfamily, followed in size by Phylinae, Orthotylinae, Bryocorinae, Deraeocorinae, Cylapinae, Isometopinae, and Psallopinae. Mirinae are characterized largely by the apically divergent parempodia on the claws and male genitalia with an inflatable aedeagal membrane; Phylinae, by hairlike or setiform parempodia and straplike male genitalia; Orthotylinae, by apically convergent parempodia, unique secondary gonopore, and frequently complex vesica and parameres; Bryocorinae, by the often stout mouthparts and tarsi, frequently single-celled hemelytral membrane, and fleshy pseudopulvilli on the claws; Deraeocorinae, by the basally toothed claws and generally punctate pronotum and hemelytra; Cylapinae, by the apically notched claws, setiform parempodia, long slender antennae and legs, and often stylate eyes; Isometopinae, by the setiform parempodia, often apically toothed claws, two-segmented tarsi, plesiomorphic ocelli, strongly modified head, and single-celled hemelytral membrane; and Psallopinae, by setiform parempodia, apically toothed claws, two-segmented tarsi, head structure, and single-celled hemelytral membrane.

Morphology

Adults range in size from 1.5 mm or less in some bryocorines, orthotylines, and phylines to more than 15 mm in certain restheniine Mirinae. Plant bugs can be described as delicate and are well known for their tendency to lose appendages when preserved in alcohol. Many Miridae often are brightly colored, ranging from bright, shiny yellow to vivid orange or red and black. Others are more cryptically colored green, gray, brown, or black, closely resembling leaves, stems, flowers, and bark of their hosts. Mirids may be nearly glabrous or may possess diverse kinds of setae, ranging from simple and hairlike to thick and woolly, silky, or even flattened and scalelike.

The family is characterized by a four-segmented antenna, four-segmented labium, primarily three-segmented tarsi (only two segments in Isometopinae, Psallopinae, and some Bryocorinae and Cylapinae); declivent to porrect head, lacking ocelli (except in Isometopinae); trapeziform pronotum, often bearing paired callosities or swellings on the anterior half; paired lateroventral scent glands on the metathorax; and asymmetrical male parameres or clasping structures. Miridae have two pairs of wings, the characteristic hemelytra or forewings, and shorter, membranous hind wings. The hemelytron of fully winged or macropterous Miridae possesses a triangular-shaped cuneus, one or two closed cells on the apical membrane, and a widened costal area frequently referred as the embolium. Miridae can also have varying degrees of hemelytral short wingedness or brachyptery, ranging from only a reduced apical membrane to the absence of even wing stubs. Some taxa exhibit beetlelike (coleopteroid) hemelytral modifications. While brachyptery can be exhibited by either sex, it occurs most often in females. Myrmecomorphy, or a resemblance to ants, is a common adaptation in many groups, generally characterized by a rounded head well separated from a sometimes anteriorly narrowed pronotum, and constricted hemelytra (or a basally constricted abdomen in brachypterous forms) to give the image of three antlike body sections.

Mirid eggs, sometimes described as banana- or cigar-shaped, are elongate and slender, straight to weakly curved, and generally 0.6 to 2.5 mm long. They are typically creamy or white. The egg shell or chorion usually is smooth but may bear faint hexagonal sculpturing. A characteristic operculum or egg cap, bearing two micropyles, is always present. Eggs frequently are inserted singly but can be laid in groups of 2 or 3 to 20 or more on their host and usually are deposited deep within leaf or stem tissue, within structures such as lenticels, or tucked tightly within leaf bundles or sheaths of grasses and other plants. The operculum usually remains visible and flush with the surrounding tissue until eclosion.

Nymphs or larvae (as they often are called outside the U.S.) undergo gradual or paurometabolous metamorphosis. Nymphs resemble adults in general appearance, differing primarily in body proportions (particularly the head, thorax, and appendages), pigmentation, two-segmented tarsi, and the lack of wings and reproductive structures. Each instar possesses a characteristic dorsal abdominal scent gland, the opening of which is visible between abdominal terga III and IV. All but a few species undergo five stages. Instars I and II lack wing pads and the meso- and metanotal segments are of nearly equal length; traces of wing pads appear in instar III, with the metanotum becoming wider than the mesonotum; larger wing pads extending onto the abdomen develop on the metanotum in instar IV; and fully developed pads extending well onto the abdomen are present in instar V. The ovipositor in females or the genital capsule of males sometimes are visible through the cuticle in fifth instars, but are not functional until the adult stage.

Habitats and host plants

As the common name 'plant bugs' suggests, mirids are mainly plant inhabitants. Though often underappreciated for their species richness, they are among the most common insects found on herbs,

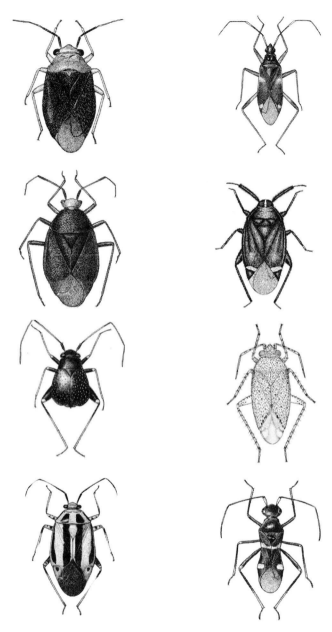

Fig. 805 Some plant bugs: (top row left) the yucca plant bug, *Halticotoma valida* (Bryocorinae). 3.0 to 3.5 mm long. North America (by Elsie Herbold Froeschner, courtesy of the Smithsonian Institution); (top row right) *Fulvius imbecilis* (Cylapinae). 4.00 mm long. North America (by Elsie Herbold Froeschner, courtesy of the Smithsonian Institution); (second row left) *Bothynotus modestus* (Deraeocorinae). 5.0 mm long. North America (by Elsie Herbold Froeschner, courtesy of the Smithsonian Institution); (second row right) *Myiomma cixiiforme* (Isometopinae). 3.00 mm long. North America (by Thomas J. Henry, courtesy of the U.S. Department of Agriculture); (third row left) the garden fleahopper, *Halticus bractatus* (Orthotylinae). 1.5 to 2.0 mm long. North, Central, and South America (by Elsie Herbold Froeschner, courtesy of the Smithsonian Institution); (third row right) the cotton fleahopper, *Pseudatomoscelis seriata* (Phylinae). 3.0 to 3.5 mm long. North and Central America, West Indies (by Linda H. Lawrence, courtesy of the U.S. Department of Agriculture); (bottom row left) the fourlined plant bug, *Poecilocapsus lineatus* (Mirinae). 7.0 to 7.5 mm long. North America (by Elsie Herbold Froeschner, courtesy of the Smithsonian Institution); (bottom row right) *Sericophanes heidemanni* (Orthotylinae). 3.0 to 3.5 mm long. North America (by Elsie Herbold Froeschner, courtesy of the Smithsonian Institution).

shrubs, and trees. Plant bugs are common not only in managed systems and ruderal sites such as vacant lots but also in specialized communities such as granite outcrops, pine barrens, serpentine barrens, and shale barrens. They can be found from below sea level to an altitude of nearly 5,400 meters.

Mirids develop principally on coniferous gymnosperms and dicotyledonous and monocotyledonous angiosperms. Twenty or more mirid species can be found on a single species of oak or pine. In arid regions of the southwestern United States and Mexico, shrubs such as ephedra (*Ephedra*), mesquite (*Prosopis*), rabbit-brush (*Chrysothamnus*), sagebrush (*Artemisia*), and saltbush (*Atriplex*) also are characterized by a diverse plant bug fauna. Some shrub- and tree-inhabiting Miridae are most numerous on isolated plants, in hedgerows, or along forest edges. Other species prefer shaded habitats or are shade tolerant. A few species are associated with fungi, especially those that grow on logs and tree trunks. The first moss-feeding plant bug was recently described from Japan. While no mirids are truly aquatic, those species that live on plants in or on the edge of water, such as water-hyacinth, might be considered semiaquatic. The Miridae also include species that are found in the litter layer, in ant nests, and in webs of subsocial spiders.

The Miridae are mostly a host-restricted group. Diet breadth ranges from strict monophagy, feeding only on one plant species, to polyphagy. Among well-known generalists are the European tarnished plant bug (*Lygus rugulipennis*) in Eurasia and the tarnished plant bug (*L. lineolaris*) in North America. Even large, agriculturally important genera such as *Lygus*, which tend to be dominated by broadly polyphagous species, usually contain specialists of narrow host range. Although predatory mirids typically show broader host associations than do plant feeders, species of mainly predacious genera, such as *Deraeocoris* and *Phytocoris*, often develop only on one plant genus (e.g., pine or oak).

Life history and habits

In temperate regions, most mirids that feed on woody plants overwinter as eggs inserted into various plant parts or in crevices on the host. In spring, the hatching of diapausing eggs often is synchronized with budbreak of host plants and is triggered by water uptake of the host. Plant bugs that feed on trees and shrubs tend to produce only one annual generation

– that is, populations are univoltine. Overwintering in the adult stage is typical of multivoltine species found on herbaceous weeds, including grasses, and field crops. Among other plant bugs overwintering as adults are some species of the predacious genus *Deraeocoris*.

Nymphal development, which generally is inversely proportional to temperature, can be completed in 12 to 35 days, depending on the species, temperature, and diet. Males of many species emerge slightly earlier than females (i.e., exhibit protandry). A sex pheromone released by the female attracts the male in certain species of the subfamilies Bryocorinae, Mirinae, and Phylinae. Olfactory receptors on the antennae allow males to detect the female pheromones. After a short premating period (usually only several days), mating takes place on the host plant. Elaborate courtship behavior may or may not be involved, depending on the species. Multiple matings apparently are common. In many species, mated females deposit about 30 to 100 eggs, but fecundity can be lower or as high as several hundred eggs. The insertion of eggs into host tissue, referred to as endophytic, reduces water loss, minimizes winter mortality, and provides some protection from natural enemies. As noted above, endophytic oviposition also is important in establishing contact with the host's vascular system so that egg hatch of plant-feeding species can be synchronized with host phenology.

Eggs of plant bugs are moved readily in shipments of nursery stock; the usually inconspicuous eggs are unlikely to be detected by plant-regulatory inspectors. Consequently, numerous Old World mirids have been accidentally introduced into North America and have become established in the Nearctic fauna.

Plant-feeding mirids generally use high-nitrogen resources, including buds, meristems, young leaves, pollen, and ovules. Most plant bugs are mesophyll rather than vascular-tissue feeders. Phytophagous members of the family, as feeders on liquefied solids, should not be considered sap feeders. Predatory mirids also suck up liquefied solids. Mirids have been described as lacerate (or macerate) and flush feeders. Their flexible stylets (paired maxillae and mandibles) macerate mesophyll cells, or prey tissues in the case of predators, and digestive enzymes from their saliva liquefy plant or animal material so that it can be imbibed through the narrow food canal. Although mirids do not tap into sieve tubes as aphids and certain other phloem-feeding hemipterans do, some species do feed from vascular tissues of their host plants. Adults of both leaf- and flower-feeding species sometimes disperse to other plant species to feed on nectar and pollen.

Mirids probably exhibit greater trophic plasticity than bugs of any other hemipteran family. Even important plant pests such as lygus bugs can be scavengers and facultative predators. Other plant bugs are mostly or entirely predacious. Predatory species range from multivoltine generalists that during the season track various prey over a wide range of host plants to univoltine specialists that feed only on a certain type of prey on a particular plant genus. Thus, some mirids prey only on oak-infesting scale insects, whereas others specialize on thrips or lace bugs.

Economic importance

The Miridae sometimes are stated to be important vectors of plant pathogens. Although some species help disseminate fungal and bacterial pathogens, plant bugs are not among the principal vectors of plant viruses. The bugs' destruction of plant cells presumably makes it difficult for obligate parasites to infect the host plant, and their salivary secretions might inactivate plant viruses. Moreover, as mesophyll feeders, mirids would not be expected to vector phloem-limited viruses.

Mirids, however, are significant pests of numerous crops, including alfalfa, apple, cocoa, coffee, cotton, guava, peach, sorghum, strawberry, sugar beet, and tea. Symptoms of plant bug injury range from foliar chlorosis, crinkling, and shot holing to lesions, cankers, and abnormal growth patterns such as stunting, bushiness, multiple leadering, and witches'-brooming. Because symptoms of mirid feeding can be so similar to those caused by abiotic factors, as well as by other insects and plant pathogens such as fungi, bacteria, and viruses, their injury often is misdiagnosed.

Millions of dollars are spent each year in attempts to control lygus bugs and other mirids that affect cotton production in the United States. An intensive, and sometimes excessive, use of insecticides against cotton-infesting plant bugs has led not only to outbreaks of secondary pest species but also the development of insecticide tolerance or resistance among mirid populations. Certain species of the subfamily Bryocorinae, which are key pests that limit the pro-

duction of cocoa in West Africa, also have developed resistance to chlorinated hydrocarbon insecticides.

The agricultural importance of plant bugs also extends to predacious species. Some of the most successful examples of long-term biological control involve the use of Miridae that feed on eggs of delphacid planthoppers. For example, the introduction into Hawaii of a mirid (*Tytthus mundulus*) from Australia has been credited with saving the Hawaiian sugar cane industry from an exotic pest, the sugarcane delphacid. Another case of successful classical biological control involves a non-native predatory plant bug that preys on an introduced planthopper that attacks taro in Hawaii. Other predacious mirids help suppress populations of rice-infesting planthoppers. Numerous crops benefit from native species of Miridae that prey on mites, thrips, leafhoppers, psyllids, whiteflies, scale insects, aphids, lace bugs, and the eggs, larvae, and pupae of various beetles, flies, and moths. These naturally occurring predators should be conserved in pest management programs. Since the 1980s, dicyphine Bryocorinae have been studied and used successfully against thrips, whiteflies, and aphids that are pests of greenhouse crops.

In addition, some mirids penetrate mummified aphids, killing the larvae, prepupae, or pupae of the parasitic wasp within the mummies. Plant bugs sometimes also prey on various life stages of insects intentionally released as biological control agents of weeds. Mirids might impair the effectiveness of parasitic wasps that are enemies of crop-infesting aphids and herbivores released for the biological control of weeds, but the potential detrimental effect of such predation on population dynamics of the prey species is unknown.

Natural enemies

In all their life stages, mirids are subject to attack by a diversity of predators, parasitoids, and microbial pathogens. Among vertebrate enemies are lizards, frogs, toads, and birds. An even more diverse group of invertebrate predators helps suppress plant bug numbers. Generalist predators include spiders, bigeyed bugs (geocorids), damsel bugs (nabids), minute pirate or flower bugs (anthocorids), assassin bugs (reduviids), mantids, and ants. Well-known families such as lady beetles' (coccinellids) and flower flies (syrphids) include predators of mirids, but members of these groups are unimportant natural enemies of plant bugs. Despite being partially or completely concealed in crevices or plant tissues, the eggs of mirids are killed by parasitic wasps. The most important egg parasitoids belong to the family Mymaridae. Mirid nymphs are killed by wasps of the family Braconidae, specifically certain specialized parasitoids of the subfamily Euphorinae. At times, euphorine braconids cause population crashes of injurious mirids.

Other biotic agents that can limit plant bug populations are parasitic nematodes (mermithids) and various microbial pathogens. Fungi are more important natural enemies of mirids than are bacteria or protozoa. Laboratory cultures, however, can be decimated by bacteria and microsporidian protozoa. No virus diseases of mirids are known.

A. G. Wheeler, Jr.
Clemson University
Clemson, South Carolina, USA
and
Thomas J. Henry
U.S. Department of Agriculture,
Agricultural Research Service
c/o Smithsonian Institution
Washington, DC, USA

References

Henry, T. J., and A. G. Wheeler, Jr. 1988. Family Miridae Hahn, 1833 (= Capsidae Burmeister, 1835). The plant bugs. pp. 251–507 in T. J. Henry and R. C. Froeschner (eds.), *Catalog of the Heteroptera, or true bugs, of Canada and continental United States*. E. J. Brill, Leiden, The Netherlands. 958 pp.

Schuh, R. T. 1995. *Plant bugs of the world (Insecta: Heteroptera: Miridae): Systematic catalog, distributions, host list, and bibliography*. New York Entomological Society, New York, New York. 1329 pp.

Schuh, R. T., and J. A. Slater. 1995. *True bugs of the world (Hemiptera: Heteroptera): Classification and natural history*. Cornell University Press, Ithaca, New York. 336 pp.

Wheeler, A. G., Jr., 2001. *Biology of the plant bugs (Hemiptera: Miridae): Pests, predators, opportunists*. Cornell University Press, Ithaca, New York. 507 pp.

Wheeler, A. G., Jr., and T. J. Henry. 1992. *A synthesis of the holarctic Miridae (Heteroptera): Distribution, biology, and origin, with emphasis on North America*. Thomas Say Foundation Monograph. Volume 15. Entomological Society of America, Lanham, Maryland. 282 pp.

PLANT COMPENSATION. Feeding by insects does not always result in less productivity of plants, due to compensatory physiological processes in the injured plant. Insect feeding can actually stimulate a plant to grow more than an uninjured plant, but more often results in partial compensation for the removed plant tissue, resulting in less damage than expected. Compensatory processes function best at low levels of damage, and are due to such factors as removal of apical dominance, removal of less productive tissue, increased penetration by light, and reduction in carbohydrate-induced inhibition of photosynthesis.

PLANT EXTRAFLORAL NECTARIES. It is well known that plant flowers produce nectar that is important in encouraging pollination, as well as providing food for bats, birds and insects. However, few people are aware of plant extrafloral nectaries that are nectar-producing glands physically apart from the flower. Extrafloral nectaries have been identified in more than 2,000 plant species in more than 64 families. For example, in nine cerrado areas of Brazil's São Paulo and Mato Grosso states, the plant families Mimosaceae, Bignoniaceae and Vochysiaceae contained the highest frequency of extrafloral nectaries.

Several review articles dealing with extrafloral nectaries or related ecological relationships and at least one book is available. In this article we briefly summarize some aspects of current knowledge about plant extrafloral nectaries and discuss their role and significance to parasitic and predatory arthropods that are important biological control agents.

Extrafloral nectaries were distinguished from floral nectaries in 1762. In 1874 Delpino described extra-nuptial nectary glands. Studies of extrafloral nectaries on plants from various families and habitats provide data on the location, morphology, and abundance of extrafloral nectaries. Approximately 93 angiosperm families with 2,000 species have extrafloral nectaries, but monocotyledons have few taxa with extrafloral nectaries. Plants with extrafloral nectaries tend to be woody and perennial. Extrafloral nectaries are often found on plant species that grow as vines. Aquatic plants appear to be entirely devoid of extrafloral nectaries. The presence or absence of extrafloral nectaries may vary even among closely related species or cultivars. Whereas up to 80% of plants in tropical habitats have extrafloral nectaries, the proportion of plants in temperate climates varies widely. Only 14% of plants sampled in Nebraska had extrafloral nectaries. In plant communities

Fig. 806 Ants frequently visit the nectaries of the vine *Smilax* sp.

similar to those of Nebraska, no plants in northern California were found to have extrafloral nectaries. The percent of plant cover with extrafloral nectaries in a myriad of habitats varies from 0–80%. Plants with extrafloral nectaries are few in number in areas that do not have ants. For example, at higher elevations in Jamaica in comparison to sea level habitats, ants and plants with extrafloral nectaries were significantly lower. Hawaii is an excellent example with few species of native plants with extrafloral nectaries. Extrafloral nectaries often occur on plants noted for toxicity, thorns, or coriaceous leaves. An interesting model predicting the probability of development of mutualistic associations among plants and ants indicates a probability of 1.0 when ants are 'omnipresent' and predicts only low probability when ants are few, and then only under very narrow conditions.

Some plant families with extrafloral nectaries

Asclepiadaceae
Bignoniaceae
Caesalpiniaceae
Caprifoliaceae
Compositae
Convolvulaceae
Curbubitaceae
Euphorbiaceae
Fabaceae
Leguminaceae
Liliaceae
Malvaceae
Mimosaceae
Papilionaceae
Rosaceae
Salicaceae

Location and morphology

Extrafloral nectary glands may be located on leaf laminae, petioles, rachids, bracts, stipules, pedicels, fruit, and most any above ground plant parts. Their size, shape and secretions vary with plant taxa. Extrafloral nectaries on Nebraska plants are most common (73%) on the foliage. Elderberry, *Sambucus nigra*, extrafloral nectaries are stalk-like on the base of leaves with nectar-producing tissue on top with a single central vascular bundle of dense cytoplasm with a well-developed endoplasmic reticulum. Light microscopy shows that very young nonsecreting nectaries are less than 2 mm, still-growing nectaries with nectar droplets are 2 to 4 mm and full-grown nectaries are 5 to 6 mm with small vacuoles fused to large vacuoles.

Apocynaceae, for example *Nerium odorum* L., have extrafloral nectaries on the adaxial surface at the junction of petiole and lamina with short apices, one or two to six to seven nonvascular nectaries develop on each leaf axial. These nectaries are active on young leaves and dry on mature leaves. *Ipomoea*

Fig. 807 The knob-shaped extrafloral nectaries of the passion vine, *Passiflora* sp.

carnea has two types of extrafloral nectaries with one set of two located on the distal end of the petiole. These mature before the lumina. The other set of five in a ring is on the pedicel which develop with flower buds. Nectar is produced continuously with no change in rate. *Rincinus communis* has extrafloral nectaries found on the lower side of petioles and at the base of leaves with normally three to seven per leaf. Nearly all *Prunus* spp. have extrafloral nectaries, notable exceptions include the low-chill peach, *P. persica*, cultivars 'JunePrince' and 'GoldPrince' developed by commercial breeding programs. Peach nectaries are globose or reniform and located on the leaf petioles. Cherry laurel, *P. laurocerasus* L., has histoid extrafloral nectaries on the lower leaf surfaces. *Calotropis gigantea* and *Wattakaka volubilis* (Asclepiadaceae) have multiple nectaries at the junction of the petiole and lamina. *Cipadessa baccifera* (Roth.) (Meliaceae) have 25 to 35 extrafloral nectaries per leaf. The nectaries occur mostly on the abaxial surface of the rachis. Magnified, the extrafloral nectaries appear as ridges and are green on immature leaves and reddish brown on mature leaves. The nectaries of the angiosperm, *Pteraduim aqualinum*, are smooth protrusions located on the stipe and frond and consist of vascular tissue with abundant mitochondria with well-developed vacuoles, a large number of plasmodesmita and less dense cytoplasm. Extrafloral nectaries of cotton, *Gossypium hirsutum*, are closely packed papillae in a pyriform depression on the lower midvein of the leaves. Peak secretion of cotton extrafloral nectaries occurs in July. *Passiflora incarnata* has two sets of extrafloral nectaries, one set is located on the glands of the petiole at the base of each leaf and the other set consists of pairs of glandular bodies on each of three bracts directly underneath the flower or bud. *Dioscorea rotundata* has extrafloral nectaries embedded in the leaf lamina with the pore opening on the lower leaf surface. Each gland is closely associated with leaf veins. Vetches have stipular nectaries present on young plants.

Some species with extrafloral nectaries

Abutilon (Indian mallow)
Ailanthus altissima (silk tree)
Allamanda nerifolia
Aphelandra (tropical herb or shrub)
Callecarpa (beauty berry)
Campsis radicans (trumpet creeper)
Cassia fasciculatus (partridge pea)
Catalpa speciosa (Indian bean)
Cattleya (orchids)
Cissus rhombifolia (ivy)
Clerodendron (tube flower)
Costus (spiral ginger)
Crotolaria striata
Croton
Curcurbits
Dioscorea sp. (air potato)
Fraxinus sp. (ash)
Fritillaria sp. (N. Am. lily)
Gossypium hirsutum (cotton)
Helianthus sp. (sunflower)
Helionthella quinquenervis (W. N. Am. herb)
Hibiscus sp.
Hoya sp.
Impatiens balsamina
Ipomoea pandurata (morning glory)
Osmanthus sp. (devil weed)
Oxypetalum sp. (S. Am. shrub)
Paeonia sp. (peony)
Passiflora incarnata (passion flower)
Pennisetum sp. (tropical grass)
Phaseolus sp. (beans)
Polygonium sp. (knot, smartweed)
Prunus spp.(peach) most of 431 species have
Pteridium aquilinum (bracken)
Ricinus communis (castor bean)
Robinia pseudoacacia (black locust)
Salix sp. (willow)
Sambucus nigra (elderberry)
Smilax macrophylla (green briar)
Thunbergia grandifloria (blue trumpet vine)
Viburnum opalus, V. americanum
Vicia sativa (vetch)
Vigna unguiculata (cowpeas)

The location of some extrafloral nectaries on the plant

Ailanthus: leaf margins
Allamanda: leaf axils
Callecarpa: adaxial surface near veins at leaf base
Cassia: petiole
Cissus: stipule
Costus: outer surface of floral bracts
Crotalaria: flower stalk
Croton: petiole

Curcurbits: lamina, pendunular bracts, abaxial surface of calyx

Fraxinus: glandular trichomes on lower leaf surface

Gossypium: leaf or flower bracts

Helianthus: flower bracts and phyllaries

Hibiscus: sunken, elongate cavity part of midvein adaxial surface

Hoya: upper leaf surface

Impatiens: petiole and leaves

Ipomoea: lower leaf surface, petiole, pedicel just below junction with sepals

Osmanthus: glandular trichomes on lower leaf surface

Passiflora: petiole, bud and flower bracts

Phaseolus: on the cushion-like compressed lateral branches on the inflorescence axis

Prunus: distal part of leaf petiole/leaf blade, adaxial leaf

Pteridium: stipe and fronds

Ricinus: leaf and inflorescence

Robinia: stipules

Salix: leaves

Sambucus: stipules

Smilax: tiny, flattened on lower leaf surface

Thunbergia: sepals

Viburnum: lower leaf surface near petiole

Vicia: stipules

Vigna: stipules and inflorescence stalk

Composition and periodicity

The most prevalent components of floral nectar are a combination of fructose, glucose, and sucrose in various proportions with other sugars such as maltose, trehalose, metazelose, methyl-glucose. The content of extrafloral nectaries differs from floral nectar even on the same plant, varies by taxa, and may or may not flow in a daily pattern. The composition of extrafloral nectaries secretion is about 95% sugar — predominantly sucrose, glucose and fructose — with the other 5% consisting of a wide array of amino acids, proteins and other important nutrients. Extrafloral nectar contains significantly higher numbers of amino acids and non-protein amino acids than does floral nectar. All of the common amino acids in floral nectaries are also equally common in extrafloral nectaries. Other components of extrafloral nectaries may include acids, proteins, lipids, and other organic compounds.

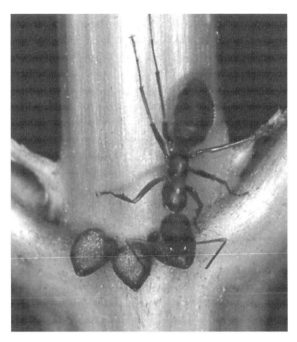

Fig. 808 Ants frequently visit the large extrafloral nectaries of the elderberry, *Sambucus* sp.

Factors which determine the concentration of the nectar solution include proximity of the phloem vessels, proportion of xylem in the vascular trace, photosynthetic rates of associated organs and local climatic conditions. The volume of extrafloral nectar is strongly affected by evaporation. *Ricinus communis*, castor bean, extrafloral nectar is mostly sugars supplied by the phloem, with the xylem supplying some water. The nectar has lower levels of amino acids than the phloem. Cherry laurel has extrafloral nectaries that produce fructose, glucose, saccharose, and 38% dry substances.

Extrafloral nectaries have a longer secretion period than floral nectaries but often do not have a concise diurnal secretion pattern. Studies have shown that the peak secretion of extrafloral nectaries can occur in the morning and the evening or can be relatively constant day and night. Extrafloral nectaries secretion often begins several weeks prior to flowering and may last through the flowering period. The most active extrafloral nectaries usually occur at the same time with and are associated with plant reproductive organs. In temperate zones peak extrafloral nectaries activity occurs during the middle of the growing season, whereas in tropical habitats secretion may occur continuously during the year. Artificial leaf damage to *Macaranga tanarius* L. induced greater flow from extrafloral nectaries for the next three days. Jasmonic acid was found to be the induced chemical responsible for the increase in nectar production rate.

Function and ecological interactions

Two hypotheses have been presented to explain the function of extrafloral nectaries. One theory holds the function of extrafloral nectaries as one of a purely physiological function enabling plant waste elimination from metabolism. This hypothesis suggests that extrafloral nectaries may serve as 'sap valves' to release extra sugars. The other theory suggests that the function of extrafloral nectaries is one of plant defense. Extrafloral nectaries may attract ants to protect plants against herbivores, may attract pollinators, as defense against ant-hemipteran mutualisms or may attract ants away from florets to protect from nectar thieving and interference with pollination. Of the plant species with extrafloral nectaries that have been studied, most of the results, although not all, have supported the plant defense hypothesis. It is well documented that many insects use extrafloral nectaries and it is easy to observe beneficial insects such as parasitic Hymenoptera and ladybird beetles feeding on extrafloral nectaries. Many species of ants are found in association with plants having extrafloral nectaries and are thought to be manipulated by the plant using its extrafloral nectaries. Removal of extrafloral nectaries from cotton decreased field populations of both phytophagous (60% reduction) and predacious insects (17–35% reduction). Interestingly, a great many species of vines have extrafloral nectaries and the evolution and selection for extrafloral nectaries is hypothesized to occur as a direct result of the ants using the vines as natural pathways into the forest canopy. Sparsity of extrafloral nectaries in Hawaii seems consistent with a plant antiherbivore defense system that does not function in the absence of ants.

There are five ways in which plants may benefit from interactions with ants: (1) ants protect plants against herbivores, (2) ants prune neighboring plants which promotes growth and survival of host plants, (3) ants feed plants with essential nutrients, (4) ants disperse seeds and fruit, and (5) ants pollinate plants. Studies of the mutualism between plants and ants show that mutualistic association may vary with time, habitat, aggressiveness of ants, and the ability

Fig. 809 The large extrafloral nectaries of the elderberry, *Sambucus* sp., produce large droplets of nectar.

of herbivores to overcome ant predation. Mutualism is favored when the investment in nectar and nectaries is low with high herbivore damage, effective alternative defenses are not available, and effective defenses are provided by ants. For mutualism using extrafloral nectaries to be favored, ants must be omnipresent. However, direct predation by ants may be unnecessary; the mere presence of the ants may be enough. In order for ants to truly be considered as 'protectors' of plants they must have the characteristics of aggressive behavior against feeding herbivores, daily activity patterns to provide 24 hour protection, actively forage for a plants herbivores, actively forage in large numbers, have nest locations in close proximity to the extrafloral nectaries plant and be able to move (nest mobility) the nest close to extrafloral nectaries plants.

Pasionflora incarnata has two sets of extrafloral nectaries. These extrafloral nectaries are visited by five species of ants and the number of ants was correlated positively with the number of extrafloral nectaries and negatively with herbivores. The number of fruit produced was greater on plants with extrafloral nectaries than without. Studies on *Cassia* sp. showed that ant visitation resulted in decreased herbivore numbers and leaf loss with increased growth but did not result in seed set differences. Continuous production of nectar is necessary to attract ants and ant activity follows the secretion patterns of extrafloral

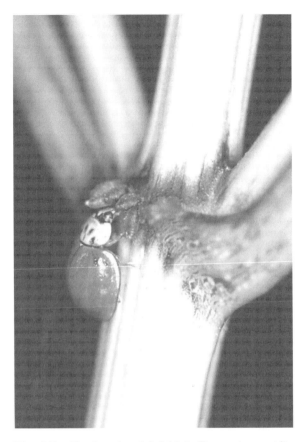

Fig. 810 The introduced ladybird, *Harmonia axyridis*, feeds from the extrafloral nectaries of the elderberry, *Sambucus* sp.

nectaries. This continuous production is evident in extrafloral nectaries, but not floral nectaries.

While ant visitation does not affect the reproductive output of plants, it is advantageous if ants protect plants during growth because pollinators are increased by large floral display, and a surplus of initial fruit and seeds may allow a plant to selectively abort genetically inferior progeny. Another way extrafloral nectaries may function is to protect plants against ant-hemipteran mutualism. By supplying nectar to ants, extrafloral nectaries distract the ants from the honeydew produced by the Hemiptera, and the plant incurs less hemipteran damage. Plant antiherbivore mechanisms are often reduced in commercial crops.

The evolution of nectar rewards may illustrate a correlation between plants and ants that provides a consistent plant defense system. The plant may select for biological control by ant protection which probably reduces the potential of damage by herbivores resistant to chemical or structural barriers provided by the plant. Natural selection may operate to optimize the production of nectar from extrafloral nectaries, balancing the cost for production with benefit in leaf and fruit production. A model for facultative mutualism predicts a positive correlation between extrafloral nectaries and ant abundance. Extrafloral nectaries selection is stronger for species that encounter ants frequently. Extrafloral nectaries are expected to be more abundant in vines and may be favored in seasonal environments. Extrafloral nectaries may result in ant symbiosis, not cause it. In Nebraska the occurrence of ants was not related to the distribution of plants with extrafloral nectaries.

Extrafloral nectaries may also function to attract ants to plants which use the ants to disperse seeds. Plants with ants have significantly fewer predators because ants may interfere with predator oviposition. Major herbivorous insects are more likely to be parasitized on plants with high quality nectar that attract parasites. Ants decrease the amount of time *Eurybia* larvae are on inflorescences. Diptera seem more prevalent on *Aerobe* sp.-infested plants.

Ant exclusion studies have shown that removing ants from plants did not lessen the number of ants visiting extrafloral nectaries; however, sugar would not attract ants to plants. Exclusion of ants increased the number of insects on developing *Capitula* but did not affect pollination. *Mentzelia nuda* (Loasaceae), a short wild perennial, has extrafloral nectaries which secrete after flowering, and attract ants, which significantly enhances seed set. The number of ants increases sharply when extrafloral nectar is available, which indicates that ant foragers can respond to a new food source. On black cherry, extrafloral nectaries are most active when ants are most able to prey on herbivores.

Ants visiting extrafloral nectaries may add to an increase in plant fitness by deterring leaf, seed, and flower herbivores and this in turn may increase the number of seeds produced. The defense of plant parts by ants appears to be a flexible interaction using diverse ant species and repelling various types of herbivores. Extrafloral nectaries on flowers may increase parasitoid survivorship, fecundity, retention and pest suppression. In Costa Rica, studies have shown more buds mature to ripe fruit on plants with extrafloral nectaries and a higher ant visitation rate. Ant activity followed the secretion pattern of extrafloral nectaries. Total number of seeds per cyme was significantly higher on plants with ants.

While ants are the most common visitors to extrafloral nectaries, and thirty species of ants have been observed visiting *Passiflora*, extrafloral nectaries are attractive to many families of arthropods. Over an eight day period, 3,941 insects, including 40 families, 77 genera, and 10 species were observed visiting extrafloral nectaries of *I. carnea* in Costa Rica. The insects included Coleoptera, Hemiptera, Lepidoptera, Neuroptera, Diptera, and Hymenoptera. Extrafloral nectaries on cotton attract parasitoids and increase the retention time and rate of parasitism. Extrafloral nectaries serve as an important supplementary food source of many insects, including predatory mites on cotton during times of stress when usual food sources are scarce, and are an important nutritional source for parasitic Hymenoptera. Even spiders may respond to the presence of extrafloral nectaries as the presence of salticids are known to enhance seed production of *Chamaecrista nictitans* (Caesalpiniaceae). Nineteen genera and 41 species of ladybird beetles, Coccinellidae, have been observed feeding on extrafloral nectaries.

Extrafloral nectaries offer an important supplemental food source for many beneficial insects (and too, to phytophagous insects including some pest species), particularly during extreme weather conditions such as drought and other times of the year when prey are scarce. Extrafloral nectaries may be valuable if not critical components in the ecology of landscapes. Passion flower, *Passiflora* spp., partridge pea, *Cassia* spp., hairy vetch, *Vicia* sp. and elderberry, *Sambucus* spp., are common examples of plants with large extrafloral nectaries on the leaves and/or stems. A great many opportunities exist to further our understanding of extrafloral nectaries as much remains unknown about the important ecological roles extrafloral nectaries may play. Exploiting extrafloral nectaries and other natural ecosystem functions that increase the presence and effect of natural enemies of pests have great potential to help reduce further the need for conventional pest controls in commercial agriculture and urban landscapes.

Russell F. Mizell, III, and Partricia A. Mizell
University of Florida
Quincy, Florida, USA

References

Beatie, A. J. 1985. *The evolutionary ecological ant-plant mutualism*. Cambridge University Press, Cambridge. 182 pp.

Elias, T. S., and B. Bently (eds.). 1979. *The biology of nectaries*. Columbia University Press, New York, New York.

Galetto, L., and L. Bernardello. 1992. Extrafloral nectaries that attract ants in Bromeliacea: structure and nectar composition. *Canadian Journal of Botany* 70: 1101–1106.

Grout, B. W., and A. Williams. 1980. Extrafloral nectaries of *Dioscorea rotnndata* Poir.: their structure and secretions. *Annals of Botany* 46: 255–258.

Keeler, K. H. 1977. The extrafloral nectaries of *Ipomoea carnea* (Convolvulaceae). *American Journal of Botany* 64: 1182–1188.

Oliveira, P. S., and A. Oliveira-filho. 1991. Distribution of extrafloral nectaries in the woody flora of tropical communities in western Brazil. pp. 163–175 in P. Price, T. Lewisohn, G. Fernandes, and W. Benson (eds.), *Plant-animal interactions; evolutionary ecology in tropical and temperate regions*. John Wiley & Sons, New York, New York.

PLANTHOPPER PARASITE MOTHS (LEPIDOPTERA: EPIPYROPIDAE).

Planthopper parasite moths, family Epipyropidae, total 40 described species, with at least another 30 known species awaiting naming. Species are known from all faunal regions, but are most diverse in Australia. Two subfamilies are known: Epipyropinae and Heteropsychinae. The family is in the superfamily Cossoidea (series Limacodiformes) in the section Cossina, subsection Cossina, of the division Ditrysia. Adults are minute to small (4 to 35 mm wingspan), with head scaling mostly roughened; haustellum absent; labial palpi minute; maxillary palpi absent; antennae bipectinate in males and rather conspicuous. Body robust. Wings quadratic and broadly rounded. Maculation mostly black or dark brown; sometimes with some spots and iridescence. Adults are crepucular and nocturnal; females are sedentary. Larvae slug-like with rounded dorsum; parasitic on fulgorids and planthoppers (Hemiptera). Eggs are laid on various plants and upon hatching, larvae search for suitable hemipteran hosts to parasitize.

John B. Heppner
Florida State Collection of Arthropods
Gainesville, Florida USA

Fig. 811 Example of planthopper parasite moths (Epipyropidae), *Fulgoraecia exigua* (H. Edwards) from Florida, USA.

References

Davis, D. R. 2004. *The ectoparasitic moths of the family Epipyropidae of the world (Lepidoptera: Zygaenoidea)*. *Smithsonian Contributions to Zoology* (in press).

Freina, J. J. de, and T. J. Witt. 1990. Familie Epipyropidae Dyar [1903] 1902. In *Die Bombyces und Sphinges der Westpalaearktis*, 2: 72–73, pl. 10. Forschung & Wissenschaft Verlag, Munich.

Kato, M. 1940. A monograph of Epipyropidae (Lepidoptera). *Entomology World* 8(72): 67–94, 4 pl.

Sick, H. 1939. Familie: Epipyropidae. pp. 1313–1315 In A. Seitz (ed.), *Die Gross-Schmetterlinge der Erde*. Teil 6. Die amerikanischen Spinner und Schwärmer, pl. 168. Stuttgart: A. Kernen.

Viette, P. E. L. 1960. Les epipyropides de Madagascar (Lépidoptères parasites). *Lambillionea* 60: 41–46.

PLANTHOPPERS (HEMIPTERA: FULGOROIDEA).

Current cladistic and molecular studies are resolving their placement in the Hemiptera and the placement of odd genera and tribes within families. Perhaps only 20% of the species of the superfamily are described.

Order: Hemiptera
    Suborder: Fulgoromorpha
        Superfamily: Fulgoroidea
            Families: 16 to 21, consensus not yet reached

Morphology

The superfamily is usually identified by the placement of the eyes, antennae, and lateral ocelli on the genae, which are separated from the front of the head by lateral carinae; by the tegula, a small plate covering the base of the wing; a y-shaped claval vein in the forewing; and rows of spines at the apex of the hind tibia and first tarsomere. Specimens measure from 2 mm to 10 cm (4 inches).

Phylogeny

At the moment, but still subject to change, we believe the Sternorrhyncha are the sister group of the rest of the Hemiptera. Next, the Fulgoromorpha (the suborder, which is equal to the superfamily Fulgoroidea) are separated as the sister group of the Heteroptera + Coleorrhyncha + Cicadomorpha (Cicadoidea, Cercopoidea, and Cicadelloidea).

Within the Fulgoroidea there are from 16 to 21 families, three with only one or two genera. Each family will be listed and discussed after some general characters are mentioned. These families are divided into two or three clades, depending upon the author. The Delphacidae and Cixiidae both have a sword-shaped ovipositor. The rest do not. They are divided by the number of spines on the second hind tarsomere into two groups. One has a row of spines and includes the Meenoplidae and its sister group Kinnaridae, Derbidae, Achilidae, Achilixiidae, Dictyopharidae and its sister group Fulgoridae, and also the Cixiidae and Delphacidae. The nymphs of the first five families of this group and of the Cixiidae are thought to live underground or in association with the ground. The last group includes the Acanaloniidae, Caliscelidae, Issidae, Nogodinidae, Ricaniidae, Flatidae, Tropiduchidae, Tettigometridae, Lophopidae, Eurybrachidae, and the two small families Hypochthonellidae and Gengidae.

Habitats

Eight of the families are found on all the continents but Antarctica, and some are found in each major biome, including forests, grasslands, deserts, tropical rainforests, and arctic tundra. Three families are worldwide except for the Palearctic (fulgorids, nogodinids, acanaloniids), one except for the Nearctic (ricaniids), one except for the Holarctic (lophopids). The tettigometrids are not found in the New World or Australia. The Nearctic and Palearctic have 12 families each, Australia 15, the Neotropics 16 or 17, and Africa (lacking kinnarids and achilixiids) and the Orient (lacking gengids and hypochthonellids) each 19.

Most adults feed on plant parts above ground. Nymphs of cixiids are found in cavities underground, feeding on plant roots, often grasses. Nymphs of achilids and derbids have been found in cavities in rotten wood or under bark, where they are thought to feed on fungi. Kinnarid and meenoplid nymphs are supposed to be associated with soil, as are the three smallest families, achilixiids, gengids, and hypochthonellids. The other families, as far as reported, have the nymphs on the leaves and trunks of the same hosts as the adults.

Eggs are placed in several substrates. In delphacids and cixiids, they are inserted into the soil or plant tissue with a sword-shaped ovipositor. Derbids, achilids, kinnarids and meenoplids are thought to have a raking-sweeping type of oviposition, where the substrate is moved and the eggs are attached to small particles such as bark. Fulgoroids, dictyopharids, and eurybrachids are thought to glue eggs to plant surfaces, and the rest have a secondarily derived piercing excavating ovipositor. Most cover their eggs with wax. Two genera of issids (two in Europe, and one in California) carry dry soil in an internal sac and mix it with body fluids to form mud egg cases.

Adults of four families, cixiids, meenoplids, kinnarids, and delphacids, have been found adapted to caves, with eyes reduced or absent, wings often reduced, and loss of pigment, and adults of hypochthonellids burrow through the soil. Cavernicolous species have been found in caves or lava tubes chiefly in Australia and Hawaii, but also in Thailand, Mexico, New Caledonia, the Canary Islands, Galapagos Islands, and Western Samoa.

Fulgoroidea are one of the groups that have been successful reaching and establishing themselves on islands. Delphacids are the most successful and widespread, and obviously traveled as aerial flotsam, as they have been found in samples taken in nets on the prows of ships in the Pacific. Often they have maintained their specific identity, which may indicate repeated gene flow to the islands. In other cases, such as Hawaii, they have speciated successfully. Actually, 13 families have reached oceanic islands, including all of the families with nymphs associated with the soil (except hypochthonellids and gengids with their three species total), suggesting that rafting may be involved in some cases.

Hosts

Fulgoroidea, as far as is known, feed on phloem of plants, through roots, trunks or leaves, except for the Achilidae and Derbidae, which are thought to feed on fungus, at least as nymphs. In a few cases this has been demonstrated in various families of the order through histological studies of plant parts, tracing the stylet sheath secreted by the insect through or between the cells to a cell or to the phloem or xylem tubes. In others it is accepted because of the production of honeydew. Some species are known to be monophagous, but a flatid, *Metcalfa pruinosa* (Say), has been reported from 100 plants in the United States, and now has been accidentally introduced into Italy, where it has been found to be a pest of grapes especially, either the wax from the insect or sooty mold from the honey dew detracting from the appearance of the fruit. Monocots are popular hosts, especially rice, sugarcane, corn, and palms. Wilson et al. (1994) list the hosts for as many species as they could find references, finding 10 families feeding on Pteridophyta, 7 on Gymnospermae, all but 2 of the 16 common families (Eurybrachidae and Nogodinidae) on Monocotyledoneae, and all on Dicotyledoneae. They also give fossil history and differing phylogenies of the group to consider the evolution of host plant use.

Pest species

Fulgoroidea are implicated in rice diseases in the orient, lethal yellowing of palms in the Caribbean and Florida, and Dubas-bug-caused death of date palms in the middle east. An unusual pest is the ricaniid *Scolypopa australis* Walker from Australia, which, although it has many hosts, sometimes feeds on a poisonous plant tutu, *Coriaria arborea* Lindsay, producing a poisonous honeydew. In times of poor nectar supply, honey bees feed on the honeydew and produce a honey poisonous to man. A list of species reported as pests of crops is reported in Wilson and O'Brien (1987).

Communication

Although the cicadas communicate by airborne sounds, the Fulgoromorpha and the rest of the Cicadomorpha communicate by substrate-transmitted sounds scarcely or not audible to the human ear. Males and females in the families studied (cixiids, delphacids, dictyopharids) each have species specific mating sounds, with the males usually initiating the sounds and the females responding. Although the male genitalia are very complex and almost always species specific, the calls are considered significant in mating isolation as well as mate finding. Fulgoroids are not thought to produce pheromones.

Wax production

Wax (actually a 60–80 carbon atom ester of 30 to 40 carbon acids and alcohols) is produced in most nymphs and females and some males in most members of the superfamily (rare in delphacids, only the vision of *Saccharosydne sacchivora* (Westwood) with waxy tufts comes to mind). Chemically it shares compounds found in the wooly alder aphid and the cochineal scale. Most of the wax is produced through hundreds of flower-shaped glands in wax plates on the sixth, seventh, and eighth abdominal tergites. Hollow strands are produced which touch each other to form another hollow tube. Small clumps of gland cells are usually found around the spiracles also. Wax particles as opposed to strands may also simply be excreted through the integument almost anywhere on the body, including the wings in some species, without an external gland visible by means of scanning electron microscopy.

What is the function of wax? Females cover the eggs with it and reports show the wax hides the eggs, repels water running over eggs, and may protect the eggs and nymphs from parasites. It is also thought to keep the nymphs in enclosed places from becoming sticky with their own honeydew, to aid escape from spider webs, to reflect ultraviolet light to aid species recognition (insects see in the ultraviolet), etc. Fulgoroidea feed on sugary phloem which they need to ingest in excess to get N for growth, and they need to get rid of sugars because of their osmotic effect on the hemolymph, but they don't have filter chambers. Might they simply have produced more of these high molecular weight compounds, some of which are similar in wax and in cuticular wax, to get rid of the excess sugar, and retained them for both physical and physiological benefits?

Other behavior

The genera *Megamelus* (Delphacidae) and *Taosa* (Dictyopharidae) are able to live on emergent aquatic plants without all of the jumping nymphs drowning. Some of the fulgorids have favorite trees in the jungle

and may be found on them for weeks at a time and year after year. A tribe of derbids can roll its wings longitudinally, presumably for protection from tropical rain drops. Tettigometrids, in a commensal arrangement with ants, can increase the production of figs. Poison honeydew from a ricaniid can be incorporated into honey, making it poisonous to humans.

Predators and parasites

Four families of insects are parasitoids of planthoppers and leafhoppers, the Pipunculidae (Diptera), Dryinidae (Hymenoptera), Halictophagidae (Strepsiptera), and Epipyropidae (Lepidoptera), the latter inserting their mandibles through the intersegmental membranes and feeding on the haemolymph of several families of Fulgoroidea in the Nearctic, Neotropics, and Asia. Egg parasites include the Hymenoptera Mymaridae, Trichogrammatidae, Eulophidae, and Aphelinidae. Nematodes may also be parasitoids of nymphs and adults.

Birds are important predators. Up to 71% of the insects in stomachs of flycatchers have been found to be fulgoroids. It is assumed that other insectivores also eat them, although perhaps not preferentially. Man, from fear, is a killer of one genus in Brazil (see Fulgoridae, below).

Family synopsis

Each family will be discussed alphabetically (refer to the listing above for phylogenetic sequence). If there is a single character that can separate that family from other Fulgoroidea, it will be mentioned. Otherwise, consult O'Brien and Wilson (1985) for a key, illustrations of morphology, and a general introduction to the superfamily.

Acanaloniidae. This mostly New World family can usually be recognized by the lateral flattening, reticulate venation, and green color in specimens of small to medium size, 4 to 12 mm. long. It was placed in the issids by Fennah (1954) but resurrected by Emelyanov (1999).

Achilidae. This worldwide family can be identified by the tips of the forewings overlapping at rest except in a few genera. A key to genera of the world is provided by Fennah in 1950. One species in Australia is found in termite nests. At present there is a disproportionate number of genera with few species in them in this family.

Achilixiidae. This small family of two genera and 24 species is found in the Neotropics and the Philippines and Borneo. They may be identified by a lateral projection of the abdomen topped by cup-like depressions. They are from 4 to 8 mm long. Bennini in the Cixiidae have a similar projection, but the top surface varies, and the cixiids have a branched spine in each depression. Achilixiids have been placed in the achilids, derbids, cixiids, and kinnarids. The function of the projection is unknown.

Caliscelidae. This worldwide group is found on bunch grasses in South America, grasses and sedges in the U.S., and also palms and bamboo elsewhere. It was only recently raised to family level (removed from the Issidae). They are small (3 to 6 mm long) and brachypterous, and are often taken for weevils or Heteroptera nymphs. Some are associated with ants.

Cixiidae. Female cixiids can be identified by a sword-shaped ovipositor and no moveable tarsal spur (delphacids also have the sword-shaped ovipositor). Males need a composite of characters. They are small or medium sized, from 3 to 13 mm long. Of world-wide distribution, they are probably the second most economically important group of Fulgoroidea because they transmit lethal yellowing of palms and other phytoplasmas.

Some have become of interest recently because they have adapted from epigean adults with nymphs feeding on roots underground to fully cave adapted species, with loss of eyes and pigment. This is particularly interesting in Hawaii, where some *Oliarus* have invaded lava tubes, which we have begun to date. Evidence from their recorded mating calls shows that cave species on Mauna Loa and Mauna Kea are more closely related to the epigean species above ground above them than the troglobitic (cavernicolous) species of the other mountain, even though there are fissures in the lava which would allow them to travel great distances underground.

Delphacidae. The delphacids can be recognized by the movable spur on the hind tibia. When sorting collections by eye, these specimens, 1 to 6 mm long, may usually be picked out of Auchenorrhyncha by their antennae sticking out from the side of their head, more visibly than other families. Economically, they are the most important of the Fulgoroidea. The Delphacidae include at least 55 species that feed on economic plants, including some major pests of agricultural crops, such as rice, sugarcane, maize, taro, and cereals. Plant damage results both from direct feeding, and the transmission of plant diseases, particularly viral diseases (e.g., Fiji disease of sugarcane by *Perkinsiella saccharicida*, rice grassy stunt and rice ragged stunt by *Nilaparvata* spp (*Nilaparvata lugens*, a rice pest, cost China an estimated US $400 million loss in 1990), rice hoja blanca by *Tagosodes* spp., rice yellows by *Sogatella furcifera*, cereal mosaic and oat rosette virus by *Laodelpax striatella*, maize mosaic and maize stripe viruses by *Peregrinus maidus*). Note that the above list of delphacid-vectored plant diseases includes 5 of the top 10 major world food crops (wheat, rice, corn, barley, and sugarcane), plus, at least three delphacid species are known to feed on sorghum.

Delphacids are carried easily by air currents. Rice pests reinvade Japan each year from China, and delphacids have probably reached every island in the world, but sometimes do not form new species on islands, presumably because other introductions of the same species do not allow reproductive isolation. They, and cixiids, seem most tolerant of cold habitats.

Derbidae. The derbids have a number of characteristic shapes, varying in size from 4 to 16 mm. It may be the possibility of tropical raindrops plastering their wings to a substrate with subsequent death that has favored the selection of multiple resting positions. Many have roof-like wings and sit head-up on plants, often grasses. Others, Mysidiini, are moth-like and hold their wings flat against the underside of broad leaves. Specimens of Zoraidini hang upside down, frequently on palm leaves, with the wings held together under the back, forming a T-shape with the body in lateral view. Specimens of Dawnarioidini have the ability to roll each pair of wings into a longitudinal tube, which is held above the body and to the side, forming a V shape in caudal view. Specimens tend to congregate on the same leaf or on the same plant.

Dictyopharidae. The dictyopharids, sister group of the fulgorids, are often green and with a head projection, but do need to be keyed to family. They vary in length from 3 to 33 mm. One genus, *Taosa*, lives on aquatic plants and a species is being tested to see whether it is host specific and may be used for biological control of water hyacinth. The nymphs seem to be able to fall onto the water and get out again, perhaps because of broad spines on the hind tibiae and tarsi. A desert loving subfamily, Orgerinae, is small and brachypterous and lacks the tegulae found in the rest of the Fulgoroidea.

Eurybrachidae. Found in Africa, the Orient, and Australia, these insects usually can be identified by the rectangularly (not triangularly) shaped wings, the frons being 3 x as broad as long, and the 2nd hind tarsomere lacking spines. Usually the forewings are opaque and colored. They measure 8 to 28 mm in length. In Australia they are associated with Eucalyptus and Acacia. In both an Asian and African genus, the forewings are modified with a projection that looks like an antenna, and the hind wings with one that bends downward like a snout, so the insect appears to have two heads, but the false one is the more obvious.

Flatidae. This family has the second most colorful and second largest specimens (4.5 to 32 mm) of the superfamily after Fulgoridae. They can be identified by having pustules in the basal half of the clavus plus the costal area transversed by many parallel veinlets. Many species (46%) are polyphagous, so introduced species may become pests, as *Metcalfa pruinosa* (Say) has in Italy and Southern Europe, introduced from the U.S.; *Siphanta acuta* (Walker) may become a pest in Hawaii and California, introduced from Australia or New Zealand; and *Melormenis basalis* (Walker) may in Florida, introduced from the West Indies.

Fulgoridae. This family, known for large (up to 10 cm) and colorful specimens, is also known for large and colorful legends. It can be identified by cross veins in the hind wings. The largest number of genera and greatest variation in the head shape are in the Neotropics, but the greatest variation in color occurs in Asia. This family does not reach the Palearctic. The nominate genus, *Fulgora*, was called a lantern fly because of its large peanut shaped head that suggested a lantern and was reported to produce light, with the last report in 1951. Now it is accepted

that it does not bioluminesce, but the natives in the South American jungle still fear it, saying it kills anything it touches, perhaps because the head looks like an alligator in lateral view, with a false eye spot, a false nostril on the top of its head and a row of false teeth. Presumably this mimicry works on birds and perhaps monkeys, and perhaps man in the past, when the medicine man carried a dead Fulgora in his amulet bag. But a Brazilian ethnoentomologist, Eraldo Medeiros Costa Neto, says now people get a big stick and try to kill the insects. Male college students, on the other hand, from Mexico to Argentina, have asked if they will die if bitten by a Fulgora if not saved by having sex within 24 hours.

Gengidae. This family of 2 genera and 2 species is found in the Union of South Africa. Their nearest relative is thought to be Eurybrachidae.

Hypochthonellidae. This family is known from adults and nymphs of one species taken underground on the roots of corn, tobacco, and peanuts in Southern Rhodesia. The compound eyes are obsolescent, the ocelli absent. The wings are brachypterous and the adult described as maggot-like. It most seems to resemble Flatidae.

Issidae. This large family of small to medium insects, 2 to 19 mm long, are often brown and a rounded diamond shape from above. There is still a question whether the family is monophyletic. Females of one genus, *Hysteropterum*, which occurs in Europe and the U.S. (and a second genus in Europe), scratch up dirt and store it in a sac near the ovipositor and use it with bodily secretions to make a mud case for the eggs.

Kinnaridae. Kinnaridae and Meenoplidae are sister groups, identified by wax on the females on their chevron-shaped 6th to 8th tergites, or the wax plates, if the wax is lacking. Both are thought to have nymphs that live underground or near the ground. Kinnarids are found primarily in the New World, and are especially common in the West Indies. They are from 2 to 8 mm in length.

Lophopidae. The Lophopidae, found everywhere but in North America and Europe, with one genus (*Carrionia*) in South America, are the first family to have a complete modern cladistic treatment at the generic level. *Pyrilla perpusilla* (Walker) can be a pest of sugarcane, rice, and corn in India. Lophopids vary in shape and color and in length from 6 to 15 mm. A fossil has been found in North America, although no species are present there now.

Meenoplidae. The meenoplids, a small family of small insects, 3 to 7 mm, can be recognized by the combination of chevron shaped wax-bearing 6 to 8 tergites in the females and pustules on one or both claval veins. Kinnarids, their sister group, have the same shape of the wax glands, but lack the pustules on the forewing. Meenoplids are not found in the New World and are associated with the soil in the literature. Cave dwelling species are found in Australia, New Caledonia, the Canary Islands and Western Samoa.

Nogodinidae. Tribes in the nogodinids, lacking in the Palearctic, are still being moved from family to family in the hope of delineating a monophyletic group. Wings are usually broadly oval, usually membranous or with a membranous cell. They vary in size from 4 to 17 mm. Fennah moved some of the issids in California to nogodinids and acanaloniids to issids; Emelyanov (1999) moved both Caliscelinae and Acanaloninae back to families, and moved Bladininae to Issidae, based primarily on the structure of the ovipositor.

Ricaniidae. Sub-triangular fore wings separate most species of this family from the rest of the Fulgoroidea. They are found everywhere but in the Nearctic, range from 6 to 20 mm, and are usually membranous with dark patterns or dark with membranous areas. *Scolypopa australis* can cause honey poisoning.

Tettigometridae. Tettigometrids lack some of the characters usually found in Fulgoroidea. They have been considered the most primitive, but now are considered among the more recent families. They are not found in the New World or Australia, are 3 to 7 mm long, with the wings shaped to the body, and lack a jumping apparatus. They are often associated with ants which remove their honeydew and protect them in return. In tropical Africa *Camponotus* ants which feed on the honeydew of *Hilda undata* (Walker) can reduce the predation on pollinators and figs by other species of ants.

Tropiduchidae. Tropiduchids are found worldwide and can be identified in most species by a transverse groove between the apex and disk of the mesonotum. They are usually green, depressed, and similar to dictyopharids but have only one pair of spines on the hind tarsomere and the venation is different with fewer apical cells. Sizes range from 5 to 13 mm. *Ommatissus lybicus*, the Dubas bug, can kill date trees in the Middle East.

Current status

In no country of the world can one identify all of the species of one family of Fulgoromorpha with a single reference except England, Scandinavia, New Zealand, and perhaps Taiwan. But new species are still being discovered in Taiwan. Europeans say they can identify the species of Middle Europe, but Spain and Portugal, southern Italy, and Greece and Turkey are poorly known. Also, the U.S. and Canadian faunas are known and most species can be identified, with the western flatids and delphacids being the most troublesome. There are many papers on the fauna of national parks in Africa, but nothing I know of that relates the Entomofaunal regions to these papers. European museums have many species from the Orient, so perhaps some sort of monograph might be done, but surely many species have not been discovered. Australia has many species to be described. South America is virtually unknown. Judging from 852 described species in the U.S. and Canada and 750 in Taiwan, their museums are obviously incomplete.

Further information

The Metcalf catalog is superb, with notes on whether each paper cited contains information on keys, description, illustrations, biology, host, etc. Unfortunately, it is out of date, as is the bibliography that accompanies it.

Review papers which provide an introduction to the superfamily (O'Brien and Wilson 1985), to pest species (Wilson and O'Brien), to behavior (O'Brien 2002), to host preferences (Wilson et al.) are listed below. The book ''The Planthoppers and Leafhoppers'' (Nault and Rodriguez eds. 1985) and a volume of Denisia (2002) provide a series of papers of related review articles.

Fennah (1950) keyed the achilid genera of the world. Fennah also has provided keys to Neotropical genera in derbids, dictyopharids, kinnarids, and tropiduchids, and in the "Fulgoroidea of Fiji" a key to Australasian cixiids, delphacids, derbids, and Pacific issids, tropiduchids and ricaniids. Earlier treatments of the world fauna, or a large part, such as Melichar's (1898–1915) can be located in Metcalf's catalogue. A recent paper (1998) keyed the genera of Lophopidae, and subsequent ones, also by Soulier-Perkins (2000, 2001), describe their phylogeny and biogeography. Denno and his students have done many ecological studies to provide a theoretical background for pest management practices, emphasizing studies on the effect of brachyptery in a dimorphic species.

Bourgoin, Campbell, Asche, and Emelyanov are doing phylogenetic studies between and within families. Porion (1994) and Nagai and Porion (1996) have provided photographic atlases and checklists of the Fulgoridae of America, and of Asia and Australia, with many colored plates.

See also, BUGS (HEMIPTERA).

Lois B. O'Brien
Florida A&M University
Tallahassee, Florida, USA

References

Holzinger, W. E. (ed.) 2002. Zikaden – leafhoppers, planthoppers and cicadas (Insecta: Hemiptera: Auchenorrhyncha). *Denisia* 4: 556 pp. Verlag Biologiezentrum, Linz, Austria.

Mason, R. T., H. M. Fales, T. H. Jones, L. B. O'Brien, T. W. Taylor, C. L. Hogue, and M. S. Blum. 1989. Characterization of fulgorid waxes (Homoptera: Fulgoridae: Insecta). *Insect Biochemistry* 19: 737–740.

Metcalf, Z. P. 1942. *A bibliography of the Homoptera (Auchenorrhyncha)*. Vols. 1 & 2. North Carolina State College, Raleigh, North Carolina.

Metcalf, Z. P. 1932–1946. General catalogue of the Hemiptera. Fasc. IV. Fulgoroidea. Parts 1–10. Smith College, Northampton, Massachusetts.

Z. P. 1954–1958. General catalogue of the Hemiptera. Fasc. IV. Fulgoroidea. Parts 11–18. North Carolina State University, Raleigh, North Carolina.

Nault, L. R., and J. G. Rodriquez (eds.). 1985. *The leafhoppers and planthoppers*. John Wiley & Sons, New York, New York.

O'Brien, L. B., and S. W. Wilson. 1985. Planthopper systematics and external morphology. pp. 61–102 in L. R. Nault and J. G. Rodriquez (eds.), *The leafhoppers and planthoppers*. John Wiley & Sons, New York, New York.

Wilson, S. W., R. F. Denno, C. Mitter, and M. R. Wilson. 1994. Evolutionary patterns of host plant use by delphacid planthoppers and their relatives. pp. 7–113 in R. F. Denno and T. J. Perfect (eds.), *Planthoppers: their ecology and management*. Chapman and Hall, Inc., New York, New York.

Wilson, S. W., and L. B. O'Brien. 1987. A survey of planthopper pests of economically important plants (Homoptera: Fulgoroidea). pp. 343–360 in M. R. Wilson and L. R. Nault (eds.), *Proceedings of the Second International Workshop on Leafhoppers and Planthoppers of Economic Importance*, held in Provo, Utah, USA, 28th July–1st August 1986. CIE, London, United Kingdom.

Wilson, M. R. 1988. Ronald G. Fennah, 1910–1987. Obituary and list of publications. *Entomologist's Monthly Magazine* 124: 167–176.

PLANT LICE. Members of the family Aphididae (order Hemiptera). See also, BUGS, APHIDS.

PLANT RESISTANCE TO INSECTS. Plant resistance to insects is a natural phenomenon based on plant self-defense mechanisms. It results from insect-plant co-evolution and is crucial for their co-existence. During plant domestication, some important features of plant resistance may be inadvertently removed by the breeders, increasing their susceptibility. This necessitates additional selection for resistance factors.

When plant breeding was still performed under field conditions, the most susceptible plants were so badly damaged by insects that they were lost from the breeding pool before they produced seeds. Thus, the plant population retained a natural resistance to insects. After World War II, however, massive use of insecticides by plant breeders accelerated the loss of natural resistance in crop plants, because it allowed the conservation of plants that were high yielding but, on the other hand, very sensitive to insects. As a result, these modern, high valuable cultivars must be continuously protected against pests. The solution to this problem lies in an attempt to reincorporate resistance into the modern crop varieties. This task requires a clear definition of the breeding target(s), adequate source(s) of resistance, and development of methods to evaluate resistance that are reliable, inexpensive and rapid. Plants and insects are very dynamic, and highly developed organisms with a good capacity for adaptation to ever-changing environments. Because insects are capable of evolving, and overcoming plant resistance, a suggested strategy is to implement an Integrated Pest Management (IPM) program that combines partial plant resistance with nontoxic (e.g., biological, physical, biorational) control measures.

The nature of insect damage to plants: direct vs. indirect damage

There are two main types of insect damage to plants: (i) Direct damage is caused by insect feeding on the plant resources. This may be accompanied by insect excretion of honeydew, on which a black sooty mold develops. The damage is correlated with pest population size. (ii) Indirect damage is caused by the insect transmission of plant diseases (viruses, mycoplasma, etc.), and can be caused by a rather low vector population. Accordingly, the crop resistance breeding program should include two different breeding concepts: (i) Prevention of direct (quantitative) damage will be achieved by suppressing pest population buildup, to keep it below the "Economic Injury Level" (EIL). (ii) Indirect damage can be prevented by breeding for "vector resistance", minimizing virus transmission by insect vectors.

Definitions of plant resistance to pests

Plant resistance is defined as any reduction in plant acceptance, in pest population growth rate, or in the damage cause by pests, that is due to inherited self-defense mechanisms in the plant. If the EIL is not reached until the end of the crop production season, the plants are considered resistant (R) to that specific pest. If the EIL is exceeded during the crop production season, the plants are considered partially resistant (PR). If the EIL is reached even before the crop production has started, the plants are designated as susceptible (S). Immune plants are plants which are not attacked at all, whereas tolerant plants are plants which possess a high EIL.

Mechanisms of resistance

The mechanisms of resistance can be divided into two major categories: antixenosis and antibiosis, which often occur together. Antixenosis is related to arthropod behavior that leads away from plant damage, whereas antibiosis is related to poor performance or lethal effects on different stages of the target insect. Although complete resistance to insects has been found and used, it is rather exceptional. On the other hand, the more frequently occurring partial resistance tends to be more durable, which is an important advantage for the development of stable agro-ecosystems. Introducing partial resistance requires more sophisticated testing methods.

Plant resistance to insect pests based on recombinant proteinase inhibitors (PIs) could interfere with natural enemies of target pests, as their own proteolytic systems may also be sensitive to broad-spectrum proteinase inhibitors.

Vector resistance

Resistance to vectors is a special case of plant resistance to pests. Vector resistance is the tendency of plants, which are by themselves virus-susceptible, to escape infection by preventing the vector from transmitting the virus. Hypothetically, two mechanisms of plant resistance are recognized: resistance or tolerance to the virus itself, and resistance to the vector which transmits the virus.

Various insects transmit virus diseases to plants. Most insect vectors belong to the Hemiptera (e.g., aphids, whiteflies, mealybugs, psyllids). All are phytophagous, piercing the plant with their mouthparts (stylets) to suck sap from the plant tissues. Viruses are transmitted in a rather short time of inoculation-feeding, usually within minutes when they are "stylet-borne" (non-persistent) viruses, or within few hours if they are "circulative" (semi-persistent or persistent) viruses. Stylet-born viruses are "mechanically transmitted" when the insect probes the plant, whereas circulative virus transmission requires vector feeding upon the plant.

The advantage of breeding resistance to the insect vector is two-fold: it can be integrated with plant resistance, or tolerance, to the virus itself, and plant resistance to insects is expected to be of greater durability than plant resistance to viruses. Several plant features can be responsible for this kind of resistance: hairy leaves, sticky and poisonous secretions, and intrinsic factors in plants which influence the settling, acceptance, and feeding behavior of the insect vector (such as the secretion of an aphid alarm pheromone mimic by the plant).

Some projects have been based on the advantage of breeding vector resistance. A significant reduction of Tungro virus was achieved in rice by the use of cultivars resistant to the leafhopper vector, *Nephotetix impicticeps*. A similar result was obtained by using cultivars resistant to *Nilaoarvata lugens*, the vector of the grassy stunt virus in rice. Resistance to *Aphis gossypii* of *Cucumis melo* prevented the transmission of cucumber mosaic virus (CMV). *Solanum polyadenium*, *S. berthaultii* and *S. tarijense* reduced attacks by aphids, and thereby reduced the viruses they transmit to potatoes. Cassava cultivars, which are partially resistant to *B. tabaci*, significantly reduce the incidence of African cassava virus.

This type of resistance may have also some drawbacks because it is vector specific, but not virus specific. The resistance to CMV transmission in muskmelon, for example, appears to be associated with nonpreference for *Aphis gossypii*. Though this plant was completely resistant to the transmission of several strains of cucumber mosaic virus (CMV) by *A. gossypii*, it was susceptible to inoculation by *Myzus persicae*. Furthermore, once this cultivar became infected, it was a source of CMV for both *A. gossypii*, and *M. persicae*. This cultivar is also resistant to CMV transmission by several clones of *A. gossypii*, and to the transmission of some other viruses as watermelon virus 1 (WMV$_1$), WMV$_2$, and muskmelon yellow stunt virus (MYSV) by this vector. However, it was susceptible to the transmission of these viruses by several other aphid species. Conclusively, this type of resistance to transmission is vector specific, but not virus specific.

Methodology: evaluating plant-resistance to insects

Methods to evaluate levels of plant resistance to both the direct or and indirect types of damage are crucial for any breeding program of plant resistance. The method should be quick, cheap, and reliable.

One of the first steps to a deliberate exploitation of genetic variation in host-plant resistance would be large-scale screening of a wide collection of varieties, breeding materials, or related wild species. The search for resistance should not be limited to free-choice experiments since the differences found are obscured when varieties are grown in monoculture. Thus, at least the most promising materials should be tested by non-choice experiments as well, to make sure that the differences found are based on true resistance and not merely on preference (antixenosis). Results should be carefully interpreted and conclusions limited to the varieties used in the experiments. Laboratory studies must, likewise, be cautiously viewed because laboratory cultures of insects suffer greatly reduced genetic variability. The strain and origin of the insects used should also be specified.

Plant resistance to pests causing direct damage

Direct damage is usually correlated with pest population densities. Hence, host preference, rates of feeding, and rates of the pest population built-up, like the 'innate capacity of population increase' (r_m), provide powerful tools for choosing a suitable source of resistance and for determining the level of resistance among the progenies. A quick and reliable test of resistance, in plants in which a sticky exudate is the mechanism of resistance, is the "sugar-test". Fine-ground crystalized "tea" sugar is spread on the tested leaf, the excess of sugar is shaken off, and a leaf diskette of a determined area is punched out from the treated leaf. The sugar is washed off from the leaf diskette, and the amount of sugar in the aqueous solution is determined with a refractometer by means of a pre-prepared correlation graph.

An additional method to evaluate plant resistance is based on the expression of resistance in in-vitro derived callus tissue rather than in seedlings or complete plants.

Evaluation of vector resistance

Plant resistance to stylet-born viruses may be evaluated by confining viruliferous insects (insects contaminated with plant virus) onto test-plant seedlings. After the virus incubation time has passed, the percentage of plants showing virus symptoms is determined.

A much quicker technique is, once a correlation graph has been produced, to relate virus incidence to the amount of feeding or to the excretion rates. For example, the amount of excreted honeydew often directly reflects the probability of TYLCV transmission by *B. tabaci*. The rate of honeydew excretion can be quantified by counting the number of droplets, or by determining the amount of sugar in the excreted honeydew. The honeydew can be collected by confining the whitefly adults onto the underside of a healthy tomato leaflet by means of a clip-on-cage, or by a modified 'Munger-cell'. If a detached leaf is used, its petiole must be kept moistened to avoid leaf desiccation during the test. The honeydew can then be collected on a piece of filter-paper, and treated with a reagent (Ninhydrin 0.2%) that stains the amino acids of the honeydew blue. The droplets can then be easily counted. To determine the total sugars, the honeydew is collected on a microscope glass cover slide and washed off with 1 ml of distilled water, to which 2 ml of a 0.2% Anthrone solution had been added. The optical density of this solution is then recorded by a spectrophotometer at 620 nm. The absolute amount of sugars in the honeydew is then derived from a pre-prepared calibration curve. Such tests are performed within 4 or 24 hrs.

Sources of resistance

Genetic sources of resistance to insects can be introgressed into modern crop cultivars. Many "old" or "primitive" varieties of crops that have been under cultivation for a long time, such as rice, eggplants, cucumbers, etc., can be investigated as sources for resistance. However, such "new world" crops as tomato, potato, pepper or corn, which have been under cultivation for a relatively short period of time, often lack known resistant cultivars, so resistance must be searched for and acquired from wild plants that can be interbred with these crops. Some examples are:

Soybeans. Leaf pubescence influences oviposition of *Bemisia* on soybean, *Glycine max*. More eggs are laid on hirsute and pubescent than on glabrous isoline. The within-plant distribution of eggs is related to trichome density.

Cotton. Glabrous cotton confers resistance to the *Heliothis* spp., but numbers of tarnished plant bugs and cotton flea-hoppers are greater on pilose lines; damage is reduced depending on the degree of pilosity. *High gossypol*, due to genetic increase of gossypol naturally occurring in cotton, causes the death of larvae through antibiosis and phagodeterrence. It was found to inhibit protease and amylase activity in *Spodoptera littoralis* larvae. High tannin causes antibiosis and feeding deterrence to *Heliothis* spp. *Nectariless* has been shown to be a resistance character for pink bollworm based on antibiosis. (Transgenic cotton, to which the gene responsible to the development of the toxin of *Bacillus thuringiensis* has been transferred by genetic engineering manipulations).

Engineered resistance

There is interest in more rapidly creating plant resistance to insects. An induced method for obtaining resistant mutants might be achieved by irradiation or by transgenic manipulations. Plant genetic engineering offers opportunities for the creation of insect-resistant plants by insertion and expression of entomopathogenic proteins.

Techniques for gene transfer have been developed for most crop plants, but current prospects for engineering resistance to insects are limited by our lack of identified candidate genes to transfer, our elementary stage of understanding of gene regulation in plants, and uncertainty about the acceptance of engineered organisms by society. Most plants genetically engineered for resistance to insects and currently being tested in the field derive their resistance from a protein endotoxin from one of the many strains of *Bt* (*Bacillus thuringiensis*), a bacterium long used as a microbial insecticide. Examples of this are transgenic cotton or potato plants to which the *Bt* toxin gene had been transferred, or transgenic clones of *Bt*-1 that impose resistance to the potato tuber moth. Some tomato varieties, bearing the Mi-1.2 gene, which provides resistance to nematodes (*Meloidogyne* spp.) and to the potato aphid (*Macrosiphum euphorbiae*), are also less preferred by *Bemisia tabaci* than varieties that do not bear this gene.

The potato proteinase inhibitor gene, pin2, was introduced into several rice varieties and inherited. Bioassay for insect resistance with the fifth-generation transgenic rice plants showed that transgenic rice plants had increased resistance to a major rice insect pest, pink stem borer (*Sesamia inferens*).

Environmental factors affecting resistance

Environmental factors such as day-length, light intensity and plant nutrition, affect not only the development and behavior of the insects but also the morphology and physiology of the plant. Temperatures, drought, plant nutrition, plant age and previous virus-infection may also affect the hostplant preference of the insect, which is then expressed as variations in plant resistance by making a susceptible plant to appear "resistant." Furthermore, the conditions under which the resistance-test are performed must also be optimized, standardized and noted.

The effect of light intensity and photoperiod. Some *Lycopersicon pennellii* and *L. hirsutum* f. *glabratum* accessions are always resistant to *B. tabaci*, some are always susceptible, but in some accessions the resistance varies according growth conditions, summer or winter. Both light intensity and day-length (photoperiod) affect their resistance to *B. tabaci*, *Manduca sexta*, and to *Leptinotarsa decemlineata*. Furthermore, 2-tridecanone, a toxin important in the resistance of *L. hirsutum* f. *glabratum* accessions, is significantly more abundant in plants grown under long day than under short-day regime. Accessions of *L. hirsutum* f. *glabratum* varied in their resistance to *B. tabaci* when grown under low light intensity, independent of day-length. In the case of *L. pennellii* accessions, only plants grown under low light intensity and a short-day regime are susceptible. The density of glandular trichomes, which secrete 2-tridecanone, is influenced by an interaction between day length and light intensity. Hence, the transition from resistance to susceptibility and vice versa took about 3–4 weeks and found its expression in the leaves, which had grown under the new conditions. The resistance can be noted clearly by the stickiness of the leaves, as quantified by the "sugar test". Day length also has significant effects on the expression of resistance in *L. hirsutum* leaves to *L. decemlineata*, through the tomatine content of the leaves which affects the rate of feeding.

The effect of plant nutrition. Concerning their effects on insects, plant biochemicals may be divided into nutrients and non-nutrients. The effect of nutrients on hostplant specificity is small. Variations in nutrient value of plants are usually not significant. Furthermore, most species of insects do not differ greatly in their qualitative requirements for nutrients. Thus, although the host plant obviously has to satisfy the nutritional requirements of the insect, it does not seem likely that the insects' nutritional requirements play more than a minor role in host plant specificity. Allelochemics (non-nutritional chemicals produced by an organism) which affect the growth, health, behavior, or population biology of insects, can be extremely important factors in host plant resistance. They may also interact with the nutrients.

In cultivated crops, artificial fertilization may have an important impact on insect-hostplant relations. In many crops, nitrogen fertilizer increases the number of pests because they affect the suitability of the plant. But in some pest-crop systems, the same nitrogen levels may have a negative impact on plant resistance; this can occur in forests and, less often, in grass, where nitrogen decreases pest numbers for reasons which are unclear. Generally, pest populations increase when the host plants are over-fertilized, especially when nitrogen is in excess. High N-levels in the hostplant cause an increase in aphid populations. In whitefly, increasing N-levels in the plant nutrition causes an increase in the intrinsic rate of

natural population growth (r_m) due to an increase in survival, fecundity, respiration rate, net reproductive rate (R_0), followed be an decrease of generation time and in the doubling time of the populations.

Increasing population growth enhances not only direct damage but also the development of resistance to pesticides, which will undoubtedly result in positive feedback by a significant, and completely unwanted, increase in pesticide application.

The effects of P and K, as well as minor and trace elements, are less clear. The idea that K is outstandingly important in conferring 'resistance' to pest attack has found little support. Undoubtedly, the composition of cell sap is affected by the nutrients applied to soil as N, P and K, and may sometimes enhance or reduce the real resistance of the crop to particular pests by modifying non-preference or antibiosis. The effects might well operate in different directions for a variety of pests attacking the same crop.

Glucosinolates and the availability of free amino acids in the phloem affects the feeding behaviour and development of the specialist cabbage aphid, *Brevicoryne brassicae*, and of the generalist peach potato aphid, *Myzus persicae*, on *Brassica* species and cultivated cabbage.

M. J. Berlinger
ARO, Gilat Research Center
Beer Sheva, Israel

References

Berlinger, M. J., M. Tamim, M. Tal, and A. R. Miller. 1997. Resistance mechanisms of *Lycopersicon pennellii* (Corr.) D'Arcy accessions to *Spodoptera littoralis* (Boisduval) (Lepidoptera: Noctuidae). *Journal of Economic Entomology* 90: 1690–1696.

de Ponti, O. M. B., L. R. Romanov, and M. J. Berlinger. 1990. Whitefly-plant relationships. pp. 91–106 in D. Gerling (ed.), *Whiteflies: their bionomics, pest status and management*. Intercept Ltd., Andover, Hants, United Kingdom.

Frutos, R., C. Rang, and F. Royer. 1999. Managing insect resistance to plants producing *Bacillus thuringiensis* toxins [Review]. *Critical Reviews in Biotechnology* 19: 227–276.

Heinrichs, E. A. (ed.) 1988. *Plant stress-insect interaction*. John Wiley & Sons, Inc., New York, New York. 492 pp.

Jouanin, L., M. Bonade-Bottino, C. Girard, G. Morrot, and M. Giband. 1998. Transgenic plants for insect resistance [Review]. *Plant Science* 131: 1–11.

Maxwell, F. G., and P. R. Jennings (eds.) 1980. *Breeding plants resistant to insects*. John Wiley & Sons, Inc., New York, New York. 683 pp.

Smith, C. M. 1989. *Plant resistance to insects: a fundamental approach*. John Wiley & Sons, Inc., New York, New York. 286 pp.

PLANT SECONDARY COMPOUNDS AND PHYTOPHAGOUS INSECTS. Although phytophagy is limited to eight (Coleoptera, Diptera, Hemiptera, Hymenoptera, Lepidoptera, Orthoptera, Phasmatodea and Thysanoptera) of the approximately thirty orders of insects, the diversity of herbivorous insect species is extensive. While the total number of phytophagous insect species is difficult to assess because of the overwhelming diversity of insects as a group and the large number of non-described species, it has been estimated that approximately 46% (or 361,000) of the almost 800,000 species of insects are herbivorous. Because plants and animals differ substantially in their chemical makeup (plants are composed primarily of C-based carbohydrates, while the primary biological macromolecules of animals are N-based proteins), strictly phytophagous animals face two major hurdles during feeding. First, nitrogen accounts for approximately 7–14% of the dry mass of animal cells, while plants rarely exceed 7% and are typically much lower. As a result, plants are nutritionally sub-optimal and N often is limiting for herbivores. Numerous studies have documented the direct positive relationship between N fertilization of host plants and increased size, fecundity and population density of plant-feeding insects.

Second, plants may employ either physical defenses such as thorns or trichomes, chemical defenses, or both for protection from attack by phytophages. Secondary compounds are substances that have no known metabolic function (i.e., they are not constituents of any known primary metabolic pathway). Moreover, secondary compounds are often energetically expensive to produce and, although a defensive role of plant chemicals was proposed more than 100 years ago, it wasn't until Fraenkel in the 1950–60s specifically noted that host-location and feeding behavior of herbivores is regulated by plant secondary compounds. Secondary metabolites can be classified according to their concentration within plant tissues. Generally, plants or plant parts that are low in abundance (i.e., non-apparent to herbivores) produce qualitative defensive compounds that interfere with metabolic pathways of herbivores; whereas, common (i.e., apparent) plants or plant parts

are more likely to be defended by quantitative compounds, which are produced in much higher concentrations and reduce the digestibility of the plant material. The concept of strategic chemical defense of plants, based on the visibility of the plant to searching herbivores, is referred to as the 'plant apparency hypothesis' of chemical defense.

Secondary compounds can be broadly classified into two groups: N-containing and non-N-containing. Among the N-containing compounds, non-protein amino acids, which are commonly found in the seeds of legumes, act as anti-metabolites because they are structurally similar to one of the twenty amino acids required for normal protein synthesis. For example, azetidine 2-carboxylic acid is structurally very similar to the protein amino acid proline and, as a result, it is mistakenly incorporated into the structural and enzymatic proteins of herbivores that consume the seeds. Incorporation of these toxic non-protein amino acids by non-adapted herbivores is likely to be fatal. However, some specialist seed-eating bruchid beetles have evolved the ability to overcome these toxic amino acids. For instance, larvae of *Caryedes brasiliensis* feed exclusively on seeds of the tropical legume *Dioclea megacarpa*, which contains high levels of the non-protein amino acid canavanine. Canavanine is similar structurally to the protein amino acid arginine. However, this bruchid beetle has very specific arginyl t-RNA synthetase, which discriminates between arginine and canavanine so that the latter is not incorporated into the insect's proteins. Glucosinolates, characteristic of the mustard family (Brassicaceae), are another important group of N-containing secondary compounds that also possess sulfur. While the glucosinolates vary in their side chains, they all contain thioglucose and sulfate moieties. Plants such as mustard and horseradish that synthesize glucosinolates also produce thioglucosidase, which enzymatically hydrolyzes glucosinolates by cleaving glucose and HSO_4^- from the parent molecule to produce isothiocyanate, thiocyanate or nitrile. Although glucosinolates are isolated from the thioglucosidase within the intact plant tissues, disruption of the plant structure by chewing or grinding action will cause glucosinolates to mix with the thioglucosidases resulting in hydrolysis and release of mustard oils. These pungent compounds (mustard oils) can cause blistering, irritation of mucous membranes including the alimentary canal, and they result in rejection of plant tissue containing them by non-adapted insects. For instance, a concentration of $\geq 0.1\%$ sinigrin (a glucosinolate) infiltrated into celery leaves was fatal to the non-adapted black swallowtail (*Papilio polyxenes*). Moreover, plant concentrations of as little as 0.01% sinigrin increased the development time of larva to pupa by 21%, reduced pupal mass by approximately 28%, and adult females laid 31% fewer eggs compared to insects reared on control plants containing no sinigrin. In addition, females, which had been reared as larvae on 0.01% sinigrin, only produced half as many viable eggs as females that had been cultured on plants containing no sinigrin. However, specialist herbivores such as the cabbage butterfly (*Pieris brassicae*), which feed exclusively on mustards, may utilize the pungent glucosinolates or their hydrolytic products for host location and to stimulate gustation. The cabbage butterfly may even reject artificial diet unless it contains glucosinolates or mustard oils.

The non-N-containing secondary compounds include some economically important substances such as the rotenoids and pyrethrins, which are used commercially to control insect pests of crops and turf grasses. The rotenoids (such as rotenone), primarily isolated from the roots of tropical legumes, prevent cellular oxygen uptake in insects by inhibiting NADH-dependent dehydrogenase activity associated with the electron transport system of mitochondria. Rotenoids are highly toxic to insects and as little as .003 g per g body mass was lethal to the silkworm (*Bombyx mori*). Pyrethrins, which may disrupt the permeability of nerve cell membranes to sodium ions, causes paralysis and death of insects.

Among the most widespread of the quantitative defensive compounds are the non-N-containing tannins. These C-based defensive compounds can be divided into two categories: hydrolyzable, which are derivatives of phenolic acids and condensed (proanthocyanidins), which are often of higher molecular mass. The name 'tannin' refers to the ability of these substances to render animal hides into leather, which are much more resistant to water infiltration. Their ability to 'tan' leather is based on the formation of hydrogen bond cross-links between tannin and protein molecules. The protein binding ability of tannins also provides protection to the plant from herbivores. Defensive tannins typically are stored in vacuoles, which prevents them from interacting with the plant's proteins. Upon disruption of the vacuoles, which

occurs during herbivore feeding activities, tannins are released from the ruptured vacuoles and they quickly combine with available proteins. As a result, tannins greatly decrease the quality of the plant tissue by combining with two important groups of proteins. First, they combine with plant protein, which reduces the ability of proteases such as trypsin to digest them into simpler polypeptides for N metabolism in the herbivore. Second, they bind to the digestive enzymes of the phytophage, thereby reducing their ability to breakdown plant proteins (and other enzymatically digested compounds). This non-selective cross-linkage of proteins (both host's and herbivore's) can reduce the N availability (and hence quality) of the plant. The negative relationship between tannin levels and insect herbivore abundance was most clearly demonstrated by Feeny (1970) for the winter moth, *Operophtera brumata*, which feeds on leaves of the oak *Quercus robur*. Larvae of the winter moth feed on oak leaves during the spring, but quickly cease feeding on the oaks by mid-June when they search for alternative hosts. This sudden decrease in feeding activity is inversely correlated with levels of leaf tannin, which are low during the spring and rapidly increase during the summer. In laboratory studies, larvae of the winter moth reared on oak leaves collected on May 16 (prior to tannin buildup) were approximately 2.5 times heavier (peak mass) than larvae reared on leaves collected on May 29 (when tannin accumulation had reached approximately 1% of the dry leaf mass). The decrease in larval mass is important biologically because numerous studies have documented the correlation between larval or pupal mass and insect fecundity.

In summary, most vascular plants probably contain one or more secondary compounds. Although a defensive role has been hypothesized for secondary compounds, it is unclear whether this description is suitable in all cases. For instance, tannins (along with lignins) also help strengthen plant structures and prevent invasion by pathogens, in addition to their anti-herbivore role. Moreover, secondary compounds, while protecting the plant from attack by non-adapted herbivores, may actually serve as host selection cues and feeding stimulants for adapted insects. Although many of these substances such as tannins have general anti-herbivore properties and offer the plant protection from most herbivores, others such as such as pyrethrins offer protection from insect phytophages in particular.

Select classes of secondary compounds produced by plants and their major effects on non-adapted insect herbivores (see text for details).

Classification	Class	Example	Common Source	Effects on non-adapted insects
N-containing:				
	non-protein amino acids	canavanine	legume seeds	antimetabolite, malfunctioning proteins
	glucosinolates	sinigrin	mustard family	anti-feedant, blistering, irritation of mucus membranes
non-N-containing:	rotenoids	rotenone	legume roots	inhibition of cellular respiration
	pyrethrins	pyrethrin I	flower heads of *Chrysanthemum cinearifolium*	paralysis
	proanthocyanidins	procyanidin	most classes of woody plants, such as oaks	precipitation of plant and insect proteins, reduction in insect size, performance and fecundity

Anthony Rossi
University of North Florida
Jacksonville, Florida, USA

References
Bernays, E. A., and R. F. Chapman. 1994. *Host-plant selection by phytophagous insects.* Chapman and Hall, New York, New York. 312 pp.
Erickson, J. M., and P. Feeny. 1974. Sinigrin: a chemical barrier to the black swallowtail butterfly, *Papilio polyxenes. Ecology* 55: 103–111.
Feeny, P. 1970. Seasonal changes in oak leaf tannins and nutrients as a cause of spring feeding by winter moth caterpillars. *Ecology* 51: 565–581.
Harbourne, J. B. 1994. *Introduction to ecological biochemistry* (4th ed.). Academic Press, London, United Kingdom. 356 pp.
Slansky, F. Jr., , and J. G. Rodriguez (eds.). 1987. *Nutritional ecology of insects, mites, spiders and related invertebrates.* Wiley Interscience, New York, New York. 1016 pp.
Spencer, K. C. 1988. *Chemical mediation of coevolution.* Academic Press, San Diego, California. 609 pp.
Strong, D. R., J. H. Lawton, and R. Southwood. 1984. *Insects on plants: community patterns and mechanisms.* Harvard University Press, Cambridge, Massachusetts. 313 pp.
Rosenthal, G. A., and D. H. Janzen. (eds.) 1979. *Herbivores: their interaction with secondary plant metabolites.* Academic Press, Orlando, Florida. 718 pp.

PLANT VIRUSES AND INSECTS. The principal families of insect vectors which cause the most damage to agricultural crops through the spread of plant diseases are in the order Hemiptera, and include the aphids, leafhoppers, delphacid planthoppers and whiteflies. Another important group of insect vectors found worldwide is the order Thysanoptera, the thrips. Other insects also spread plant diseases; however, aphids alone are responsible for spreading the majority of known plant viral diseases, followed closely by the leafhoppers, whiteflies, and thrips. The known number of plant disease vectors within these taxa is large, including Cicadellidae (leafhoppers, containing 49 known vector species), Aphididae (aphids, with the majority of 192 vector species), Aleyrodidae (whiteflies, with 3 vector species) and Thripidae (thrips, with 8 known vector species). Of course, these numbers change every year with the description and discovery of new viral diseases and new insect vectors. Furthermore, a group with only a few insect vectors still may be able to carry and spread a huge number of viral diseases to many different host plants, as occurs in the whiteflies and thrips.

In 1997, more than 380 viruses were known to be transmitted by these insect vectors; however, in the last five years the number has increased greatly (about 600 in 2001) and is increasing every year. This dramatic increase has been due in part to our ability to detect and characterize viral diseases more accurately, and also due to increased travel and trade between countries, which often result in the accidental introduction of either diseases or insects from one country into another.

Components involved in disease epidemiology

The epidemiology of plant diseases caused by insect-carried plant pathogens involves four main components: the pathogen, the insects, the plant and the environment. Thus, the transmission of a plant pathogen by an insect to a plant appears relatively simple. However, this situation is highly complex when one examines all the possible elements that can influence these interactions. The availability of the pathogen is affected by its quantity, location, and the strain within the plant. The insect's biology is influenced by population size, number of generations, longevity, dispersal patterns, feeding behavior and interactions with the pathogen. The plant's performance can be influenced by its level of susceptibility to the pathogen, multiple infections of different strains, and/or different pathogens, susceptibility to the insect, the location and stage of growth when exposed to the pathogen and insect. Environmental factors add another level of complexity as temperature, moisture, air currents, and cultural practices come into play. The discovery of a new plant pathogen that is carried and spread by an insect is usually the beginning of a long and difficult task toward understanding its epidemiology (all the elements that influence the development and spread of a plant disease).

Insect feeding mechanisms

The traits of morphology that contributes to the ability of these insects to transmit plant diseases so efficiently is their piercing-sucking feeding style. Insects in the order Hemiptera (aphids, leafhoppers, whiteflies), and Thysanoptera (thrips) have similar basic morphologies of the head and body. In the accompanying scanning electron micrograph you can see the compound eyes, and the proboscis of the insect. In Thrips this proboscis is referred to as a mouthcone due to the thick, short nature of the structure. The proboscis helps support the stylets as the insect works its stylets into the plant cells. The stylets are under muscular control so that they can be extended into the plant tissues. The stylets are each curved and are held against each other so that they go straight, one pushing and sliding against the other. However, when one stylet moves in front of the other the curve pushes the stylet in a lateral, side-

ways direction. Thus, the insect controls the direction in which it moves the stylets. Some leafhoppers feed in a manner whereby they will pierce into the plant tissues and then proceed to feed in a clockwise or counter-clockwise procession, emptying cells as they go, thus creating an emptied out 'spot' in the plant leaf, called a stiple. Others feed directly from the vascular tissues of plants, the phloem or xylem.

All these insects have piercing-sucking mouthparts that allow them to feed on plants while causing minimal damage. This is important for virus transmission, as viruses require a living cell to reproduce. The insects use paired maxillary stylets to form a suction tube that is inserted into plant cells, similar to a flexible syringe needle. In the Hemiptera, these stylets form two canals, the food canal, and a smaller salivary canal where the saliva of the insect comes out during feeding. The Thysanoptera are unique in that thrips stylets form a single canal used for both sucking up plant fluids and to secrete saliva. The insect salivary secretions have several functions. There are at least two types of saliva, one is liquid and aids in the digestion of plant cells and cell debris

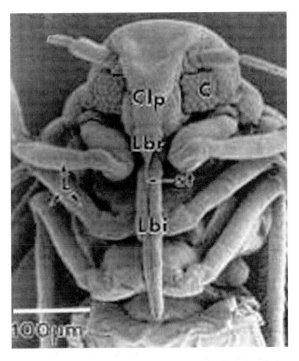

Fig. 812 Scanning electron micrograph (SEM) of the whitefly, *Bemisia tabaci*, showing the ventral surface. Bod parts are as follows: C, compound eye; Clp, clypeus; Lbi, labium; Lbr, labrum; L, legs; St, stylet bundle.

so that they can be ingested, sucked up through the food canal. Another solidifies or hardens during feeding which functions to form a salivary sheath to help prevent leakage around the inserted stylets, and to hold the stylets firmly in place during feeding. The saliva also is thought to prevent or hinder the plant's response to repair its damaged cells so the insect can continue feeding once it finds the desired location inside the plant (i.e., the phloem or xylem).

There are also mandibular stylets. These are paired, thicker stylets on the outside of the maxillary stylets. The aphids, leafhoppers and whiteflies all have symmetrical, paired, matching mandibular stylets which function to pierce the hard epidermis of plants and to assist attaching the insect firmly to the plant surface. Only the thrips have an asymmetrical morphology with one of the mandibular stylets being reduced or absent, and the remaining stylet being closed at the end, forming a needle-like structure, closed at the end. The thrips use the single, mandibular stylet to pierce a hole into the epidermis of the plant surface so that the slender, paired maxillary

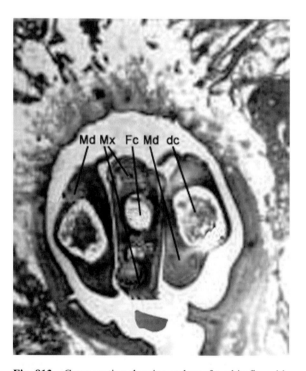

Fig. 813 Cross section showing stylets of a whitefly, with separate food and salivary canals: Dc, dendritic canal; Fc, food canal; Md, mandibular stylet; Mx, maxillary stylet; Sc, salivary canal.

stylets can be inserted to feed in a piercing-sucking manner from plant cells that are deeper.

By being able to feed in such a precise and direct manner, insects that feed in a piercing-sucking manner can avoid many of the plant's natural defenses. These insects also can deposit viruses directly into specific tissues from which they feed, such as into the vascular tissues of a plant. Once a virus has been introduced into the vascular tissues, it can spread rapidly throughout the plant to cause disease. Furthermore, piercing-sucking feeders cause less damage to the plant than a chewing insect, so plant cells that are infected with a virus often survive the feeding and support virus survival within the plant.

Inside the head there are several valves whereby the insects can stop the procession of food into their midgut. The plant sap is drawn up the food canal of the maxillary stylets. The food is then held in place by the precibarial valve where it is tasted by gustatory sensilla. The food is then drawn into the cibarium, the pumping chamber of the mouth. The cibarium also has gustatory sensilla for tasting and evaluating the quality of the food as the insect sucks up the plant sap. The food then passes the esophageal valve and enters the esophagus and passes through to the midgut which is the area of the alimentary tract where most nutrients are absorbed. In most plant sap-feeding insects, there is a region of the midgut where the hindgut coils around and is attached to it. This is the filter-chamber region of the insect's alimentary canal. Plant phloem and xylem, the liquids within plants' vascular tissues, are full of water; insects which feed on these as a primary food source have adapted over time the ability to shunt or direct excess water directly into the hindgut. This allows the insect to concentrate food and nutrients in the midgut for maximum absorption and to release excess water without having to process it through the midgut.

Tools to understand feeding (electronic monitoring of insect feeding)

Scientist have many methods to study insect/virus interactions. One such method is the invention of an electronic feeding monitor system, EMS, that allows someone to measure aspects of feeding as they occur. This is very important in studies where the amount of time an insect spends feeding needs to be measured. The EMS allows the scientist to know how many times an insect inserts its stylets into a plant, how long the insect fed, and if the insect was feeding from

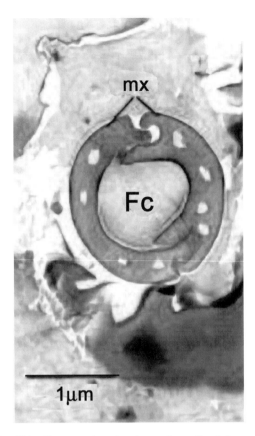

Fig. 814 Cross section showing stylets of a thrips, with single canal used for salivation and food intake: Fc, food canal; Mx, maxillary stylet.

the phloem or xylem tissues within the plant. Being able to examine insect feeding so closely enables the early selection of crop varieties that may have resistance to insect feeding before they are planted in the field. Plant varieties which can disrupt insect feeding may be useful to prevent the transmission of some virus diseases. The EMS works by running a low voltage of electricity through a plant, usually by placing a copper electrode into the moist soil of a potted plant. The insect then has a fine gold wire glued onto its back, using electrical-conductive paint, so that the electricity will pass up through the plant, and then through the insect when it inserts its stylets into the plant. When the insect either salivates out, or ingests plant fluids up the food canal, the electricity passes through the insect, which acts as a variable resistor, and goes back into the EMS, which then amplifies the signal so that it can be recorded.

Insect vector-plant pathogen interactions

There have been two systems of terminology established to describe the association and transmission of plant diseases by insect vectors which feed in a piercing-sucking manner. One is based on how long the virus persists in the insect vector, and the second is based on the route of virus movement through the insect vector. They can be combined as follows: (1) the non-persistently transmitted, stylet-borne viruses; (2) the semi-persistently transmitted, foregut-borne viruses; (3) the persistently transmitted, circulative viruses; and (4) the persistently transmitted, propagative viruses. Using this terminology, virus 'transmission' is referred to as 'non-persistent,' 'semi-persistent,' or 'persistent.'

The way a virus moves through the insect vector then is described by the terms: 'circulative' or 'propagative.' Circulative viruses pass into the insect hemolymph and circulate through the insects before being salivated back out during feeding. This involves the ability of the virus to pass several barriers within the insect, passing through the midgut membranes, and then the salivary gland membranes, to be able to be released back out with the saliva. These types of viruses do not replicate inside their insect vectors but merely pass through the insect. Viruses that reproduce inside the insect are considered propagative. Propagative viruses are able to enter the insect hemolymph but they also replicate once they infect an insect. As one would expect, a virus that is circulative is retained in the insect for a longer period of time than a virus that is non-circulative and merely stuck to the insects' stylets (stylet-borne) or foregut (foregut-borne virus). Viruses that are propagative (replicating in the insect) are retained for the life of the insect.

Non-persistently transmitted, stylet-borne viruses are transmitted into the plant during short durations of feeding. Virus acquisition (the ingestion of a virus that results in the insect's ability to transmit the virus to a plant), is brief, often just a few seconds of feeding. There is no latent period (the time that passes between when the virus is acquired and when it can be transmitted to a plant). Since these types of viruses usually are binding to the insect's stylets for only a brief period of time, the insect does not retain the ability to transmit the virus for long periods. Usually, virus transmissibility is lost after a few minutes of feeding on a non-infected plant. Aphids transmit the majority of non-persistently transmitted viruses. The

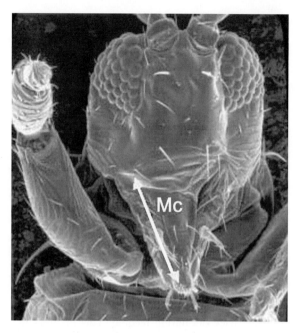

Fig. 815 Scanning electron micrograph of a thrips showing the face and mouthparts: Mc, mouthcone.

ability of viruses to bind to the insect's stylets is aided by a helper component (a virus encoded, non-structural protein produced only in infected plants). During subsequent periods of feeding the virus is released, or washed from the stylets, thus depositing virus into the plant tissues.

Semi-persistently transmitted, foregut-born viruses are transmitted into the plant during longer durations of feeding (minutes). Virus acquisition increases with increased time spent feeding (minutes to hours), and the virus stays in association with the insect for several hours, being able to be transmitted into other plants. The virus is thought to be binding in the anterior areas of the alimentary tract, along the stylets to the foregut, and a few virus particles are released during each act of feeding. There is no latent period, the virus does not replicate, and the insect will lose the ability to transmit the virus during its life.

Persistently transmitted, circulative viruses do not replicate in the insect vector. These types of viruses are acquired and transmitted during long periods of feeding (minutes to hours), and there is a latent period of hours to days before the virus can be transmitted to another plant. This makes sense as the virus must move through the insect body and get into the salivary glands to be salivated back out before

transmission can occur. Virus retention is long, but is dependent upon the amount of virus acquired into the insect body, and may last for the life of the insect, usually around 30 days.

Persistently transmitted, propagative viruses do replicate inside the insect. Virus acquisition time takes hours to days of feeding. The latent period can take weeks before an insect can transmit virus. The virus is retained for the life of the insect and often the virus is passed to the eggs (transovarial transmission).

Some insect-transmitted viral plant diseases
Aphids. Of all known aphids, about 250 are considered serious pests. They are pests because of their feeding, which reduces the vitality of the crops they feed on, but primarily due to the transmission of viral plant diseases. Perhaps the most important aphid pest is *Myzus persicae*, often referred to as the green peach aphid. *M. persicae* is a green or slightly reddish aphid which has peach as its primary host and a wide range of secondary hosts, including many brassicas. *M. persicae* is cosmopolitan in temperate climates occurring in the U.S.A., and a fair portion of Europe including the United Kingdom. Though it seldom occurs in numbers large enough to cause direct damage from feeding pressure, it is capable of transmitting and spreading over 100 viruses including the potato leaf roll, potato virus Y, yellow net and yellows viruses of sugar beet, cauliflower mosaic, plum pox, cucumber mosaic, lettuce mosaic, and turnip mosaic virus.

The pea aphid, *Acyrthosiphum pisum*, is a large green aphid with long antennae and legs. The pea aphid is found on many leguminous plants and transmits Lucerne mosaic virus, pea leaf-roll virus, pea enation mosaic virus and pea mosaic virus in the United Kingdom, and pea enation mosaic virus in the U.S.A. The cabbage aphid, *Brevicoryne brassicae*, is a serious pest of the major cabbage crops, cabbages, cauliflowers and Brussels sprouts. The main cause of its pest status is the transmission of cauliflower mosaic and turnip mosaic virus. The brown citrus aphid, *Toxoptera citricida*, is a dark, black, aphid that is the main vector of citrus tristeza virus in the subtropics and tropics. The melon aphid or cotton aphid, *Aphis gossypii*, also is an important aphid vector of viral diseases in citrus and on many other agricultural crops.

The control of aphid pests currently still involves large amounts of pesticides in most countries, but other more ecologically friendly methods have been used in other places for some time. These generally involve biological control, mostly the use of Hymenopteran parasitoids. These are small wasps that lay their eggs inside the aphids. Other methods include plant improvement, and monitoring aphid dispersal to predict when a pre-emptive spraying in smaller amounts might be effective. The most important element of insect pest control for all of us is education. Farmers as well as the general public need to become better informed as to the alternatives to, and proper uses of, insecticides.

Leafhoppers. A large group of plant viruses, the plant rhabdovirus group, consists of more than 70 members. They are transmitted by aphids, leafhoppers, planthoppers, lacebugs, and mites. These viruses infect and replicate in the insect cells, but each virus is specific to its insect vector. Some of them also can be transmitted mechanically, through artificial means using plant sap from infected plants. Another important leafhopper-transmitted virus is maize chlorotic dwarf virus (MCDV). This virus is a semi-persistently transmitted, foregut-borne virus, and is restricted to the phloem of maize. Transmission of this virus requires a protein that is produced by virus-infected plants. This protein, called the helper component (HC), is suspected to bind to receptor-like structures in the food canal of leafhoppers, thereby forming a matrix to which virus particles attach. Viruses are then slowly released from this matrix during feeding and, consequently, are transmitted to other plants when leafhoppers fly to neighboring plants and then feed. The insect vector of maize chlorotic dwarf virus is the leafhopper, *Graminella nigrifons*.

The beet leafhopper (*Circulifer tenellus*) is one of the most important insect pests of sugarbeets in the western United States because it is the vector of beet curly top virus, BCTV. Curly top virus is a severely devastating plant virus that affects more than 300 broad-leaved plants. Tomato, bean, squash, cucumber, melon, spinach, table beet, pepper, and some flowering plants are the most common cultivated plants affected in the western United States. Leafhopper populations survive on weeds and cultivated plants infected with curly top virus which serve as reservoirs for both the insect and virus. Leafhoppers

are able to acquire the virus during very short feeding times. The leafhopper retains the ability to transmit BCTV for a month or more after acquisition.

Whiteflies. In the past decade, whiteflies as pests and vectors of plant viruses have become one of the most serious crop protection problems in the tropics and subtropics. Yearly losses are estimated in the hundreds of millions of dollars. Several species of whitefly cause crop losses through direct feeding, while others are important in virus transmission. *Bemisia tabaci*, for example, is the vector of African cassava mosaic, bean golden mosaic, bean dwarf mosaic, bean calico mosaic, tomato yellow leaf-curl, tomato mottle, and other *Begomoviruses* in the family *Geminiviridae*, affecting crops worldwide.

With the spread of an especially aggressive biotype of *B. tabaci* into the New World tropics (*B. argentifolii*), crop losses likely will continue to increase, resulting in higher pesticide use on tomatoes, beans, cassava, cotton, cucurbits, potatoes, sweet potatoes and other crops. There is an urgent need to develop integrated pest management systems aimed at reducing insecticide use and which will help re-establish the ecological equilibrium of predators, parasitoids, and microbial controls. Needed are crop varieties with resistance to the whiteflies, and/or to the whitefly-transmitted viral diseases.

This problem is manifested in the fact that whiteflies and the viruses they carry can potentially infect many different host plants, including agricultural crops and weeds. A pest problem on one crop, such as beans, cannot be tackled as a single problem, as neighboring crops or weeds also may be affecting the disease spread. The different viruses and forms of the whitefly also are difficult to identify, and/or separate on the basis of symptoms or morphology. Determining where the problems in food crops are coming from becomes almost impossible. Proper diagnosis of the problem depends on using sophisticated molecular techniques to characterize the viruses and whitefly vectors, followed by epidemiological work, usually based on dynamic modeling, to understand the incidence of disease spread.

Thrips. Thrips species in the genus *Frankliniella* are commonly referred to as flower thrips. The western flower thrips, *F. occidentalis*, has a world-wide distribution and is considered the primary vector of tospoviruses. Thrips feed on over 600 different plants and crops, especially on flowering plants where they also feed on pollen. Many thrips are pests of commercial crops due to their damage to flowers. Also, their feeding causes stunting, deformed and unmarketable fruits and vegetables.

Thrips in the genera *Frankliniella* sp. and *Thrips* sp. also spread plant diseases through the transmission of viruses such as tospoviruses. Tomato spotted wilt virus is the type member of the genus *Tospovirus* in the family *Bunyaviridae*. These enveloped viruses are considered among the most damaging of emerging plant pathogens around the world. Virus members also include the impatiens necrotic spot viruses, which infect many ornamental plants. Tospoviruses can kill plants or reduce yields of marketable fruits and vegetables (i.e., lettuce, tomato, peanut, watermelon and ornamental crops). To transmit tospoviruses, thrips must acquire the virus during the larval stage. Most thrips species over-winter as either adults or as pupae. A typical flower thrips generation time varies from between 7 and 22 days depending on the temperature. The eggs are about 0.2 mm long and reniform (kidney shaped); they take on average three days to hatch. Thrips have two larval stages, then go through a prepupal and a pupal stage. Adults take between one and four days to reach sexual maturity. The females of the suborder Terebrantia are equipped with an ovipositor which they use to cut slits into plant tissue into which they insert their eggs one per slit, while females of the suborder Tubulifera, which lack an ovipositor, lay their eggs on the outside surface of plants, either singly or in small groups.

Wayne B. Hunter
U.S. Department of Agriculture,
Agricultural Research Service
U.S. Horticultural Research Laboratory
Ft. Pierce, Florida, USA

References

Hunter, W. B., and E. A. Backus. 1989. Comparison of feeding behavior of the potato leafhopper *Empoasca fabae* (Homoptera: Cicadellidae) on alfalfa and broad bean leaves. *Environmental Entomology* 18: 473–480.
Hunter, W. B., and D. E. Ullman. 1992. Anatomy and ultrastructure of the piercing-sucking mouthparts and paraglossal sensilla of *Frankliniella occidentalis* (Pergande) (Thysanoptera: Thripidae). *International Journal of Insect Morphology and Embryology* 21: 17–35.
Hunter, W. B., D. E. Ullman, and A. Moore. 1994. Electronic monitoring: characterizing the feeding behavior of western flower thrips (Thysanoptera: Thripidae). pp. 73–85 in M. M. Ellsbury, E. A. Backus, and
D. L. Ullman (eds.), *History, development, and application of AC electronic insect feeding monitors*. Thomas Say Publications in Entomology, Entomological Society of America, Lanham, Maryland.
Hunter, W. B., E. Hiebert, S. E. Webb, and J. E. Polston. 1996. Precibarial and cibarial chemosensilla in the whitefly, *Bemisia tabaci* (Gennadius) (Homoptera: Aleyrodidae). *International Journal of Insect Morphology and Embryology* 25: 295–304.
Nault, L. R. 1997. Arthropod transmission of plant viruses: a new synthesis. *Annals of Entomological Society of America* 90: 521–541.

PLASMATOCYTE. A common type of hemocyte capable of phagocytic, encapsulation and secretory functions.
See also, HEMOCYTES OF INSECTS: THEIR MORPHOLOGY AND FUNCTION.

PLASMID. Circular, dsDNA molecules found in bacteria that are often used in cloning. Plasmids are independent, stable, self-replicating, and often confer resistance to antibiotics. Often used in recombinant DNA work as vectors of foreign DNA.

PLASTERER BEES. Members of the family Colletidae (order Hymenoptera, superfamily Apoidae). See also, BEES, and WASPS, ANTS, BEES, AND SAWFLIES.

PLASTRON. Hairs or a tube-like process in which the cuticle of an insect or egg chorion holds a bubble of air through which gas exchange can occur while the insect is submerged in water. A physical gill.

PLATASPIDAE. A family of bugs (order Hemiptera, suborder Pentamorpha). See also, BUGS.

PLATE-THIGH BEETLES. Members of the family Eucinetidae (order Coleoptera). See also, BEETLES.

PLATYGASTRIDAE. A family of wasps (order Hymenoptera). See also, WASPS, ANTS, BEES, AND SAWFLIES.

PLATYPEZIDAE. A family of flies (order Diptera). They commonly are known as flat-footed flies. See also, FLIES.

PLATYPODIDAE. Considered by some to be a family of beetles (order Coleoptera). They commonly are known as pin-hole borers. Here they are treated as a subfamily (Platypodinae) of Curculionidae. See also, BEETLES.

PLATYSTICTIDAE. A family of damselflies (order Odonata). See also, DRAGONFLIES AND DAMSELFLIES.

PLATYSTOMATIDAE. A family of flies (order Diptera). They (and Otitidae) commonly are known as picture-winged flies, but Otitidae is now considered to be part of Ulidiidae. See also, FLIES.

PLEASING FUNGUS BEETLES. Members of the family Erotylidae (order Coleoptera). See also, BEETLES.

PLEASING LACEWINGS. Members of the family Dilaridae (order Neuroptera). See also, LACEWINGS, ANTLIONS, AND MANTIDFLIES.

PLECOMIDAE. A family of beetles (order Coleoptera). They commonly are known as rain beetles. See also, BEETLES.

PLECOMIDAE. A family of beetles (order Coleoptera). They commonly are known as rain beetles. See also, BEETLES.

PLECOPTERA. An order of insects. They commonly are known as stoneflies. See also, STONEFLIES.

PLEIDAE. A family of bugs (order Hemiptera). They sometimes are called pygmy backswimmers. See also, BUGS.

PLEIOTROPIC. Term used to describe a gene that affects more than one, apparently unrelated, trait.

PLESIOMORPHIC. A character used to reconstruct a phylogeny that is ancestral or primitive.

PLEURITE. A lateral (pleural) sclerite.

PLEURON (PL., PLEURA). The lateral plates of the insect body segments, especially the thoracic segments.
See also, THORAX OF HEXAPODS.

PLICA. A longitudinal fold or wrinkle in the wing of an insect.

PLOKIOPHILIDAE. A family of bugs (order Hemiptera). They sometimes are called web-lovers. See also, BUGS.

PLUMARIDAE. A family of wasps (order Hymenoptera). See also, WASPS, ANTS, BEES, AND SAWFLIES.

PLUM CURCULIO, *CONOTRACHELUS NENUPHAR* HERBST (COLEOPTERA: CURCULIONIDAE). The plum curculio, *Conotrachelus nenuphar* Herbst (Coleoptera: Curculionidae), is an important pest of pome and stone fruit orchards in North America. The insect is distributed from Quebec to Florida, east of the Rocky Mountains and in Utah. Two strains are found: a northern and a southern strain, which have, respectively, one and two generations per year.

Life cycle
C. nenuphar overwinters as adults in plant debris, preferably under maple leaves. The pest is univoltine in the northern part of its range (north of Virginia, USA) and at least partially multivoltine in southern areas including populations present in mid-Atlantic regions of West Virginia and New Jersey. Spring emergence times vary with geographical location. In Quebec (Canada), overwintered adults emerge in

late April when apple trees of cv. McIntosh reach the 'green tip stage'. Emergence reaches a peak from 6 days before full bloom to 10 days after petal fall. In southwestern Quebec emergence may take 3-4 weeks to complete. In Ontario (Canada), emergence begins at end of April and is nearly complete by early June, but continues until late June or early July. In Texas, emergence occurs from late March to early May. Adults begin to emerge and become active when mean daily temperatures reach approximately 13–15°C.

After emergence, adults remain on the surface of the soil for some time before appearing on the trees, where they feed on the new shoots and blossoms until fruit becomes available. In spring, adults invade orchards mainly from the surrounding woodland. In Quebec, the adult population peaks somewhere between the tight cluster stage and 10 days after petal fall on apple. The highest distance covered by the adults, mainly by walking, was recorded from the tight cluster stage until physiological fruit drop in late June. In southern regions in the United States, adult populations on peach and plum peak at least one month earlier.

In Ontario, oviposition begins in late May and continues until early August. The timing of oviposition varies with climate. In New York State, it was estimated that 60% of total fruit damage by oviposition is accomplished when 230 day-degrees (above 10°C) had accumulated after petal fall in apple (mid-June on average in that region).

Eggs are laid in an epidermal cavity in the fruit that the female excavates. The skin is cut into a distinctive crescent-shaped slit, which partially surrounds the egg. The eggs and young larvae are sensitive to internal pressure and other unfavorable effects of fruit growth. Eggs crushed by rapidly growing apple fruit may account for varietal differences in the susceptibility to attack by *C. nenuphar*. The gum exudate from egg-laying scars on half-grown plums can kill the larvae. More than one larva can develop in a single fruit. The abundance of fruit has a significant influence on *C. nenuphar* population dynamics and a poor crop may lead to a marked decrease in population size. The larvae feed in the fruit, which usually drops prematurely unless egg or larval development is interrupted in its early stages. The time spent in the fruit varies from 15 to 18 days. When fully fed, the larvae leave the fruit and pupate in cells in the soil. The time spent in the soil depends on temperature and humidity but varies from 3 weeks to more than 5 weeks, the longer periods generally occurring in the northernmost part of its range.

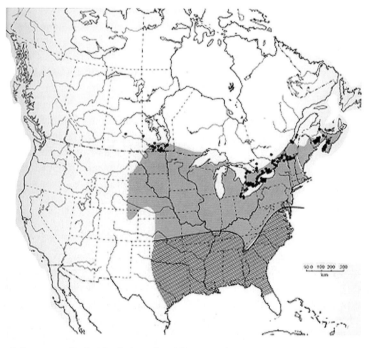

Fig. 816 Distribution of plum curculio in North America. The map shows the area inhabited by both the northern (single generation per year) and southern (more than one per year) strains.

The summer generation of adults emerges over a long period from July to October in Ontario, and Maine, with a peak of emergence in September. In Georgia, approximately 50–75% of females lay eggs in the same year, giving a partial second generation. The adults leave the trees and search for overwintering sites in September and October. In caged experiments in Quebec, 93% of adults overwintered at the soil surface under leaves and other debris, 4% were found in the top inch of the soil and the remainder overwintered deeper in the soil. In Virginia, weevils were found hibernating up to 15 cm deep in loose soil and in clay, at an average depth of 6 cm.

Host plants

Plums, peaches, apricots and nectarines are the preferred hosts of *C. nenuphar*, but apples are also widely affected. Other economically important hosts include sweet cherry, sour cherry and blueberry. Apples are less damaged in areas adjacent to peach orchards than in areas where peaches are not grown. Peaches are often scarred and deformed by the feeding and egg punctures of *C. nenuphar*, with larvae developing successfully and leaving the fruit to pupate. There are varietal differences in the susceptibility of apples, with eggs being destroyed and larval establishment being prevented by fruit growth in some cultivars.

Damage and economic impact

Second only to the codling moth (*Cydia pomonella*), *C. nenuphar* is regarded as the most serious pest of pome and stone fruit in eastern North America. For example, in Quebec up to 85% of harvested apples may be damaged by *C. nenuphar* in unsprayed orchards. Plum curculio damage returns to levels of economic importance 1 to 3 years after cessation of pesticide spraying.

On apple, *C. nenuphar* can cause three types of external damage. In spring, males and females feed on, and females oviposit in, young fruit, leaving behind small round feeding punctures and marking them with characteristic half-moon shaped scars, respectively; and in spring and summer, the adults puncture the fruit causing round (2–3 mm diameter), feeding scars. The appearance of plum curculio damage is highly variable and, of all fruit damage rated by IPM specialists, damage caused by plum curculio had the lowest agreement level, i.e., 71.8%. Internal damage to the fruit is caused by

Fig. 817 Oviposition scar on the surface of an apple.

larval feeding and exit holes. Most infested fruits drop prematurely in June, though cherries rot on the trees. Larvae release pectic enzymes and cellulase while they feed internally aiding in fruit abscission. Adult feeding may also cause marginal damage to leaves and blossoms.

Management of the plum curculio

Insecticide treatments are usually recommended against the adults at petal fall and once or twice thereafter at 10–14 day intervals. No resistance to synthetic insecticides has yet been reported for this pest. Alternatives to insecticides are discussed hereafter.

Fig. 818 Adults and feeding punctures on young fruit.

Behavioral studies

Observations on the behavior of adults, both in the field and in cages, Using Zn^{65} as a marker to track adults has led to the design of better IPM programs for plum curculio. In autumn, most adults labeled with Zn^{65} moved from orchards to surrounding woodlots. After overwintering, the returning plum curculios gradually invade adjacent apple orchards between pink and petal fall, after spending several days on the ground under the perimeter rows of trees, which most of them then climb. From full bloom to 9 days after fruit set, plum curculio adults were found to be active mainly during the night. In field cages, adults labeled with Zn^{65} also showed a similar diel periodicity while foraging on dwarf apple trees. Because adults are most active in the trees at night, it is recommended that insecticide treatments are likely to be most effective if applied during the first hours of darkness.

There have been several attempts to relate adult activity to ambient temperature in order to optimize the timing of insecticide treatments. Two approaches have been investigated: the development of a trap to evaluate adult populations and establish a relationship between population levels to risks and the development of day-degree models to predict the appearance of damage in orchards. Neither of these methods has been used in isolation to manage populations of *C. nenuphar*. A model predicting the nocturnal activity of *C. nenuphar* hourly has recently been developed and is currently under validation.

Treatment of peripheral zones of apple orchards

The strategy to treat 20 m-wide peripheral zones of apple orchards (when needed) in spring is based on the fact that plum curculio damage is frequently more abundant at this time in peripheral zones, and that during the tight cluster stage most plum curculio adults move only 1 to 4 m per day when returning to the orchards from their overwintering sites in adjacent woodlots. During this 5-20 day reinvasion period in southern Quebec, petal fall appears as the most appropriate time for a peripheral zone treatment.

Using this approach in a 1.7 ha experimental orchard (with standard-size trees) in Quebec, fruit damage at harvest was reduced from 57 to 2.4%, while reducing the amount of insecticide used by 70%, and the plum curculio adult population by 83%. These results were consistent over a two-year period in one locality, under high population pressure. The mortality data were based on recaptures of plum curculio adults, radio-labeled with Zn^{65}, that had been released in a woodlot adjacent to an orchard in which the peripheral zones had been treated with insecticides.

Peripheral-zone spraying has been validated in commercial orchards and compared for effectiveness with full-block spraying in four commercial apple orchards in southern Quebec. Plum curculio damage at harvest was less than 0.7% and 0.8% fruit in plots receiving peripheral and full-plot sprays, respectively; and most damaged apples (95%) were found in peripheral zones.

Trapping methods

The best (although tedious) monitoring method available for plum curculio in commercial orchards still remains careful examination of thousands of small fruit to detect fresh egg-laying scars. In Quebec a threshold of 1% damaged fruit, based on careful monitoring of fruit three times a week, was successfully used for managing localized peripheral zone treatments following full-block treatment at petal fall. A lower threshold and daily monitoring would be required, however, in areas where the pest pressure is very high. Limb tapping as a monitoring technique is not popular with growers because beating sticks damage the trees, and accuracy varies with cultivar, tree shape, time of day and scout experience.

There have been several studies aimed at improving the timing of insecticide treatments by monitoring adult plum curculios with traps designed for other curculionids or with novel trapping methods. Inverted polyethylene funnels have been evaluated by hanging them beneath tree trunks to capture falling adults. Unbaited sticky-coated green plastic spheres (3 and 8 cm diameter) and sticky-coated green thinning apples (3 cm diameter) hung in host trees, PVC pitfall traps placed beneath host trees, 5 cm bands coated with a sticky substance encircling tree trunks, and unbaited and baited (baited with boll weevil pheromone, grandlure) boll weevil traps placed on vertical stakes between woods and commercial orchards have also been tested. However, no plum curculios were captured on sticky bands and very few were captured in pitfall and boll weevil traps. Black pyramid traps placed next to apple tree trunks, originally designed to mimic the visual silhouette of the trunk of a pecan tree and used to monitor pecan weevil, *Curculio caryae*, captured significantly more plum curculios than those traps placed between apple trees, between apple trees and an adjacent wood lot, and between apple trees and an adjacent field.

However, occurrence of plum curculios in these pyramid traps did not coincide with temporal occurrence of ovipositional injury nor did number of adults captured coincide with levels of ovipositional injury, possibly because of abiotic factors. In experiments designed to learn how plum curculios move into host fruit trees, adults dislodged from apple trees tended to walk off a small collecting frame when ambient temperatures were below 20°C, but fly to a host tree canopy or inter-tree space when temperatures were 20°C or above. This indicates that plum curculios are likely to bypass any sort of trap designed to intercept crawling individuals such as black pyramid traps placed next to tree trunks when temperatures are above 20°C. Therefore, traps that capture flying plum curculio adults as they enter the orchard or after arrival in the host tree canopy also have been evaluated. Clear Plexiglas panels have been evaluated for their ability to intercept flying adults. The circle trap consists of a wire screen cage capped with an inverted screen funnel and attached to a limb; it is designed to intercept foraging adults walking on tree limbs or on the tree trunk, and not to attract adults based on stimulating visual cues. A third trap type that has been evaluated in host tree canopies is the branch-mimicking cylinder trap, designed to exploit visual cues provided by an upright twig. So far, none of these unbaited trap types have shown sufficient attractiveness or reliability to replace the visual examination of fruitlets as the recommended monitoring technique. However, addition of attractants to trap types is considered to be the next logical step for their improvement.

Attractants

Numerous species of weevils are attracted to host plant volatiles or specific compounds present in host plant volatiles, including the banana weevil, *Cosmopolites sordidus* (Germar), the cabbage seed weevil, *Ceutorhynchus assimilis* (Paykull), the pea weevil, *Sitona lineatus* (L.), and the red weevil, *Rhynchophorus ferrugineus* F.

Attempts to identify potential attractants for adult plum curculios were first published in the 1920s. It was reported that adult plum curculios are attracted to salicyl-aldehyde early in the season and gallic acid

later in the season. More recently, much effort has focused on identification of potential attractants. In the 1980s, fresh apple juice and synthetic apple blossom fragrance were evaluated, and in the 1990s stored apples, fresh apple branchlets, and ammonium carbonate as potential attractants were also evaluated: all proved unsuccessful. Chemically uncharacterized host apple odor from water- and hexane-based extracts was most attractive between bloom and two weeks after bloom. Furthermore, volatiles released from punctured plums were very attractive to plum curculios in laboratory bioassays. Specific compounds identified from plum and/or apple volatiles and found to be attractive to plum curculio as evidenced by laboratory and/or field bioassays include ethyl isovalerate, limonene, benzaldehyde, benzyl alcohol, decanal, E-2-hexenal, geranyl proprionate, and hexyl acetate. Because odors of host fruit were significantly less attractive at 4 and 8 cm than at 2 cm from plum curculios, fruit odor-based traps may not be useful in commercial orchards. However, host plant volatiles often synergize or enhance insect responses to sex and/or aggregation pheromones. In the family Curculionidae, male-produced pheromones have been documented in at least 21 species and enhancement of attraction to them by the presence of host plant volatiles have been documented in at least 18 species.

Some antennal sensory structures on plum curculio are similar to pheromone receptors found on related curculionids. Researchers recently isolated and subsequently synthesized an aggregation pheromone from male plum curculios: $(+)$-(1R,2S)-Methyl-2-(1-methylethenyl) cyclobutaneacetic acid. This pheromone, which they named grandisoic acid, is attractive to both sexes. Attempts to use live adults in pyramid traps were unsuccessful, baited traps being no more attractive than unbaited traps, possibly because of repulsive distress signals emitted by the curculios. However, lures impregnated with a racemic mixture of grandisoic acid have been reported to significantly increase the number of plum curculios trapped in pyramid traps. A two-fold increase in attractiveness when the lure was used in conjunction with small amounts of green leaf volatiles; high amounts showed a repulsive effect. More recently, it has been demonstrated that pyramid and sticky Plexiglas panel traps baited with a combination of grandisoic acid and benzaldehyde and placed close to woods and adjacent to orchards, captured more immigrating plum curculios than grandisoic acid alone or unbaited traps. This combination of attractants has been deployed within apple tree canopies to aggregate plum curculios and to monitor seasonal activity. This monitoring technique is termed a "trap free" approach.

Biological control

Several natural enemies have been recovered from *C. nenuphar* but none are able to provide an effective alternative to chemical insecticides in commercial orchards. Several nematode species have been tested and found to be effective as larvicides against *C. nenuphar*. In the laboratory, 95% larval mortality was reported at 400 *Steinernema carpocapsae*/larva. At 200–400 nematodes/larva, 73.4% larval mortality was achieved in natural sods. Nematode treatments would also affect other pests, such as the larvae of *Hoplocampa testudinea* (apple sawfly). The application of nematode treatments to the soil would not prevent damage to apples, but would lower *C. nenuphar* and *H. testudinea* populations for the subsequent growing season.

Repeated applications of *S. carpocapsae* to the foliage or aerial parts of apple trees were tested to prevent damage to the fruit. In caged environments, localized application of nematodes at the base of tree trunks significantly reduced adult populations (82–100% mortality).

See also, APPLE PESTS AND THEIR MANAGEMENT.

Charles Vincent
Agriculture and Agri-Food Canada Saint-Jean-sur-Richelieu, Quebec, Canada
and
Gérald Chouinard
Institut de recherche et de développement en agroenvironnement Saint-Hyacinthe, Quebec, Canada
and
Tracy Leskey
U.S. Department of Agriculture, Agricultural Research Service Kearneysville, West Virginia, USA

References

Chouinard, G., S. B. Hill, and C. Vincent. 1993. Spring behavior of the plum curculio (Coleoptera: Curculionidae) within caged dwarf apple trees. *Annals of the Entomological Society of America* 86: 333–340.

Lafleur, G., S. B. Hill, and C. Vincent. 1987. Fall migration, hibernation site selection and associated winter mortality of plum curculio (Coleoptera: Curculionidae) in a Quebec apple orchard. *Journal of Economic Entomology* 80: 1152–1172.

Racette, G., G. Chouinard, C. Vincent, and S. B. Hill. 1992. Ecology and management of plum curculio, *Conotrachelus nenuphar* (Coleoptera: Curculionidae), in apple orchards. *Phytoprotection* 73: 85–100.

Vincent, C., G. Chouinard, N. J. Bostanian, and Y. Morin. 1997. Peripheral zone treatments for plum curculio management: validation in commercial apple orchards. *Entomologia Experimentalis et Applicata* 84: 1–8.

Vincent, C., G. Chouinard, and S. B. Hill. 1999. Progress in plum curculio management: a review. *Agriculture, Ecosystems and Environment* 73: 167–175.

PLUME MOTHS (LEPIDOPTERA: PTEROPHORIDAE).

Plume moths, family Pterophoridae, comprise about 1,292 species worldwide, with about a third being Palearctic; actual world fauna probably exceeds 1,500 species. There are five subfamilies: Macropiratinae (placed as a separate family by some specialists), Agdistinae, Ochyroticinae, Deuterocopinae, and Pterophorinae. The family is in the superfamily Pterophoroidea in the section Tineina, subsection Tineina, of the division Ditrysia. Adults small (8 to 38 mm wingspan), with head scaling average; haustellum naked; labial palpi porrect; maxillary palpi 1-segmented. Wings elongated and usually with hindwings split into three fringed plumes; forewings often entire or split into two parts near the termen (a few species have both wings entire; subfamilies Macropiratinae and Agdistinae). Maculation variable but mostly mottled shades of brown or gray, with some spotting, and with a few more colorful species; some entirely white. Adults mostly nocturnal but some crepuscular or in shaded areas during the day. Larvae mostly leaf feeders, or miners and borers of various plant parts; a few are gall makers. Host plants include many families but particularly in Compositae, Scrophulariaceae, Labiatae, and Geraniaceae. Several species are economic.

John B. Heppner
Florida State Collection of Arthropods Gainesville, Florida USA

References

Arenberger, E.. 1995. Pterophoridae. In *Microlepidoptera Palaearctica*. Vol. 9. G. Braun: Karlsruhe. 2 v. (258 pp, 153 pl.)

Barnes, W., and A. W. Lindsey. 1921. The Pterophoridae of America, north of Mexico. In *Contributions to the natural history of Lepidoptera in North America*, 4: 281–452, pl. 41–54. W. Barnes, Decatur.

Gielis, C. 1993. Generic revision of the superfamily Pterophoroidea (Lepidoptera). *Zoologische Verhandlingen* (Leiden), 290: 1–139.

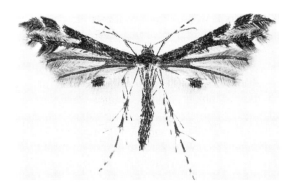

Fig. 819 Example of plume moths (Pterophoridae), *Oxyptilus parvidactyla* (Haworth) from Italy.

Gielis, C. 1996. Pterophoridae. In *Microlepidoptera of Europe*. 1. Apollo Books, Stenstrup. 222 pp (16 pl.).

Landry, B. 1987. *A synopsis of the plume-moths of the subfamily Platyptilinae of eastern Canada*. Montreal. 269 pp, 22 pl.

Razowski, J. 1988. Motyle (Lepidoptera) Polski. 20. Pterophoridae, Carposinidae. In *Monografie Fauny Polski*. 17. Krakow. 140 pp, 6 pl.

PLUMOSE. Feather-like structure with a single thick stem and numerous parallel branches. This term usually is used to describe the antennae of Lepidoptera. See also, ANTENNAE OF HEXAPODS.

PLUTELLIDAE. A family of moths (order Lepidoptera). They commonly are known as diamondback moths. See also, DIAMONDBACK MOTHS, BUTTERFLIES AND MOTHS.

PODURIDAE. A family of springtails (order Collembala). The commonly are known as water springtails. See also, SPRINGTAILS.

POIKILOTHERMIC. Cold-blooded animals; animals (including insects) that lack an internal temperature regulation system, and so tend to have body temperatures that mirror the temperature of their environment. Poikilothermic animals actually use behavior and heat generated by muscles to adjust their temperature to some degree. (contrast with homeothermic)

POLAR FILAMENT. A hollow, thread-like organelle associated with a microsporidian spore that extrudes, and allows the sporoplasm to pass to the exterior and to inoculate a cell.

POLLEN BASKET. A specialized scopa, or pollen holding apparatus, found in bumble bees and honey bees. The pollen basket consists of the broad, concave hind tibia surrounded by a fringe of long hairs. A corbicula.

POLLENOSE. Covered with a loose, dusty yellow material resembling pollen, and which can be rubbed off the surface of an insect.

POLLEN POT. A container made from soft cerumen by stingless bees, and used to store pollen.

POLLEN RAKE. A comb-like row of bristles at the tip of the hind tibia of bees.

POLLINATION AND FLOWER VISITATION. Insects are the most common and abundant pollinators of flowering plants, far surpassing the other winged pollinators, the birds and bats. Other insects (anthophiles) visit flowers but do not necessarily pollinate them. The history of the relationships between insects (both pollinators and anthophiles) and flowers is long and varied. Their modern-day importance in ecosystem functioning and agricultural production has made them the subject of numerous scientific studies.

Evolutionary overview

The relationships of insects and the sexual reproductive organs of plants may be as ancient as the insects and terrestrial plants themselves. Such relationships probably preceded pollination to a time before pollen existed. The first terrestrial plants produced unicellular spores that may have been food for the first terrestrial insects, probably Collembola. Thus, some 400 million years ago in Devonian time, *Rhyniella* (Collembola) and its relatives may have consumed the spores of *Rhynia* and its relatives (primitive vascular plants known only from fossils). Other arthropods have been found in association with the sporangia of plants of that time also. By the time the Carboniferous forests dominated major parts of the globe, insects were well represented in the fauna. Several extinct orders are known to have had elongated mouthparts, reminiscent of those associated today with insects that suck liquids from tubular structures, such as flowers. Moreover, spores (microspores) have been found associated with those insects, particularly on their mouthparts and their wing bases. Reproductive organs of the plants of the Carboniferous, and earlier, had structures (bracts) that formed tubes leading to the micropylar droplet (the germination medium for the microspores) above the megaspore retained on the parent plant. By mid-Mesozoic, some cycadoid gymnosperms appear to have had apparently showy structures embracing the sexual sporangia. Thus, a firm trend towards the insect-plant mutualism of pollination seems to have roots extending into the past well before the advent of the Angiospermae (the flowering plants), a Cretaceous phenomenon of some 160 million years ago.

Insect pollination is generally thought of as being associated with flowering plants, Angiospermae, but it is now realized that most extant Cycadaceae (one family of the non-flowering Gymnospermae) seem to be pollinated by insects also. Both groups of plants produce pollen (i.e., microgametophytes of two or three cells) that must move from the microsporangia to the megasporangia, in which fertilization of ovules takes place, followed by embryogenesis, seed development, and fructification. Which insects were the first pollen vectors (i.e., pollinators) is a matter of debate. Both Diptera and Coleoptera have been suggested (reviewed by Labandiera and Bernhardt, respectively). Perhaps there is no reason to assume that the two suggestions are mutually exclusive.

Within Diptera, short-horned flies are noted having tubular mouthparts that fit with feeding at flowers. Coleoptera as early pollinators are thought to have been associated with heavily constructed flowers (such as those of Magnoliaceae, the magnolias) in which they fed on floral tissues (tepals and ovaries) and pollen, spreading pollen between flowers by 'mess and soil.' Nowadays, Diptera, Coleoptera, Hymenoptera, and Lepidoptera are the most conspicuous and well-known of pollinators and anthophiles (flower visitors that are not necessarily pollinators). These orders and several minor orders associated with anthophily and pollination are discussed below (Taxonomic diversity).

The details of how flowers attract insects are discussed under 'Floral advertising.' The benefits that anthophiles reap from flowers (pollen, nectar, and other foods, oviposition sites, mates, and comfort) are discussed under 'Floral rewards.' Flowers and their symbiont insects comprise evolving and functional ecosystems in a complex interplay of mutualisms and competition (community and co-evolutionary ecology) that are becoming recognized, along with biological diversity, as being crucial to conservation.

Taxonomic diversity

The Hymenoptera are well known as flower visitors and pollinators, primarily because of the bees (Apoidea), which are thought to have co-evolved with reproductive structures of plants, especially flowers. They are adept at handling flowers, gathering pollen (mostly as provisions for their brood) and imbibing nectar. Their specialized, plumose body hairs entrap pollen as they move from flower to flower so that they are effective pollen vectors and pollinators. Honeybees and bumblebees are quick students of floral colors, shapes, and complexity and are efficient foragers. Some bees have special relationships with oil-providing, and gum-providing flowers. The closely related Sphecidae are mostly predatory, but like other families of wasps (Pompilidae and Vespidae), often are seen at flowers feeding mostly on floral or extra-floral nectar. The Masaridae (Vespoidea) are almost all vegetarian and are especially associated with anthophily. The ants (Formicidae) also may frequent flowers. There are only a few examples of ant pollination, but their association with flowers and extra-floral nectaries often protects the flowers and seeds from predation. Most other families in the suborder Apocrita are parasitoids. There are many examples of these kinds of wasps feeding from nectar at flowers. Because they mostly have unspecialized and short mouthparts, they are found mainly on open-bowl shaped flowers with exposed nectar. Indeed, floral nectar may be a crucial resource to fuel their activities, especially as young adults that have not yet found hosts or mates. Among the herbivorous families of Apocrita are the Cynipidae (the gall wasps) and several species of seed-eating and gall-forming Chalcidoidea. Notorious among the chalcidoids are the Agaonidae, the fig wasps, which have tightly co-evolved relationships with *Ficus* species (figs). The wasps oviposit in the enclosed inflorescence (flower). The larvae then mature as the fig inflorescence ripens into a complex infructescence. The newly emerged adult female wasps collect pollen in specialized pouches and seek out new inflorescences in which to lay their eggs. Pollination of the fig inflorescences occurs as a byproduct of the reproductive behavior of the wasp species.

Among the Symphyta (the other suborder of the Hymenoptera which includes the sawflies), anthophily is less well studied. They sometimes are found on flowers, often well dusted with pollen, and feed on nectar. Some species seem to be associated with pollination, especially of orchids in Australia.

The Diptera are highly diverse as flower visitors. Many families of long-horned flies (Nematocera) derive their carbohydrate nutrition from feeding at flowers. This habit is especially well studied in mosquitoes (Culicidae). Males feed on floral nectar for fuel in swarming flight, and females may feed on nectar prior to searching for mates and blood. Some orchids are pollinated by mosquitoes. The idea that blackflies (Simuliidae) are pollinators of blueberries (*Vaccinium* spp.) is not correct, but nectar feeding is known in this group. Nectar feeding and pollination is attributed to most nematocerous families, especially in relation to floral mimicry. Some plant groups, notably Araceae and Aristolochiaceae, attract flies by scents and colors that mimic such fly-attract-

ing substrates as dung, carrion, musk, and fungi. Among the midges (Ceratopoginidae) are a few specialized species that bite open pollen grains to feed. Cacao is pollinated by midges. Among short-horned flies (Brachycera) there are also many flower visiting species. Particularly interesting are the long-tongued Tabanidae (horse flies) and Nemestrinidae (tangle-veined flies) of southern Africa. Some species have highly specialized pollinating relationships with plants in the iris family (Iridaceae). The flower flies, or hover flies (Syrphidae), are probably the best known of the Brachycera for their close associations and pollinating activities with a wide variety of plants. Among the Diptera, the 50 families of acalypterate muscoids are poorly represented as flower visitors, with records associated with only the pomace flies (Drosophilidae) for which yeasts and floral nectars are parts of the symbioses. The calypterate muscoids generally are well known as flower visitors. Their roles as pollinators have been explored in various plants, including the umbels (Apiaceae), and especially the mimetic Araceae, Aristolochiaceae, and Rafflesiaceae.

Almost all Lepidoptera visit flowers and imbibe nectar. The exceptions are those that do not feed as adults (some microlepidoptera) and perhaps some of the specialized moths that feed on animal secretions, and even blood. Some Micropterigidae feed directly on pollen. Also, a few butterflies (for example, *Heliconius*, *Parides*, *Battus*) use pollen as food. They place pollen into floral nectar and make a 'pollen soup.' They then imbibe the nectar enriched with the eluents of the pollen. Most butterflies and moths have elongated, tubular mouthparts and imbibe nectar that is dilute (less than 45% sugars). One of the

Fig. 820 Honeybee, *Apis mellifera*, a common daytime visitor of flowers (photo Andrei Sourakov).

Fig. 821 Moths such as this sphinx moth visit flowers at dusk or during the evening (photo R. J. Barnas).

longest proboscides is that of *Xanthopan morgani* f. *praedicta* (Sphingidae), the existence of which was predicted on the basis of an orchid from Madagascar, *Angraecum sesquipedale*, with its 25 to 30 cm nectariferous spur. Some moths (for example, *Plusia gamma*) secrete saliva and dilute heavy nectar at open-bowl shaped flowers.

Moths, particular Sphingidae, hover while feeding at flowers and have high energy demands. The flowers they visit and pollinate tend to be presented with outward facing tubes (like trumpets) and secrete copious nectar. They are also often heavily scented. The nectar of flowers visited by hawkmoths are weak in amino acids. It has been suggested that too much amino acids ingested with large amounts of nectar could be toxic. Many pestiferous moths feed extensively at flowers, but their activities, being nocturnal, are not well understood. The army cutworm moth (*Euxoa auxilliaris*) is migratory from the U.S. plains to the Rocky Mountain alpine ecosystem. In the summer, the moth does not aestivate (become quiescent), but feeds extensively on floral nectar, converting the sugars to fat before return migration and sexual maturation. The resident moths of the same family (Noctuidae) do not show that pattern of activity and reproductive diapause.

Some special relationships have co-evolved between Lepidoptera and flowering plants. One of the best studied is that between *Yucca* spp. (Liliaceae) and its pollinating moths (*Tegeticula*). These moths gather pollen in their mouthparts and carry it between flowers of different plants, stuff the pollen mass into the stigma of the flower, and then lay their eggs in the floral ovary. As the ovary matures, the larvae eat some of the developing seeds. There are other similar, but less well appreciated examples (for example, *Hadena bicruris* and its oviposition behavior on *Silene alba*).

Butterflies are diurnal and less energy demanding than moths in their flower visiting. Many of the flowers with which they are associated have tubular corollas that flare and provide landing platforms for the settling butterflies. The flowers are often brightly colored, sometimes distinctly reddish, and mostly not strongly scented. Butterflies often forage sporadically, visiting flowers that may be widely spaced and not visiting those along the way. Pollination by butterflies has been rarely proven to be important, mostly because specific studies have not been made. For migratory butterflies, especially the monarch

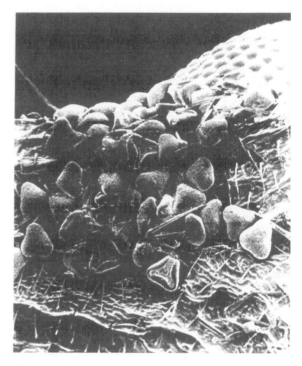

Fig. 822 Scanning electron micrograph of pollen of the oil palm (*Elaeis guineesis*) on the snout and compound eye of a weevil, *Elaeidobius kamerunicus*, the major pollinator of oil palm.

(*Danaus plexippus*), considering the importance of nectar corridors is part of conservation efforts.

The Coleoptera is a hugely diverse order and members of about 30 families have been recorded from flowers. Members of the suborder Adephaga (i.e., Caraboidea) are not flower visitors, although a few records have been made. Among the suborder Polyphaga, flower visiting is widespread among the families, but not in any apparently systematic way. Among the Hydrophiloidea and Staphylinoidea, anthophily is recorded only for a few Hydrophilidae (or Hydraenidae) and Staphylinidae and Ptiliidae, respectively. The Scarabaeoidea has anthophiles in the Scarabaeidae, with pollination relationships ascribed to a few species. There are a few records of anthophiles in the Elateroidea (Elateridae), the Buprestoidea (Buprestidae), the Cantharoidea (especially in the Cantharidae, and a few in the Lycidae) and in Dermestoidea (Dermestidae). In the Cleroidea, flower visiting is well represented in the most diverse families (Cleridae, Melyridae). Among the wide array of families in the Cucujoidea is a smattering of anthophilous families (Nitidulidae, Rhizophagidae, Coccinellidae, Lagriidae especially some Scraptiidae, Mordellidae, Oedemeridae, and few Anthicidae). The Chrysomeloidea has anthophily recorded from a few Cerambycidae, Bruchidae, more Chrysomelidae, and many Curculionidae. Only the Melyridae, Mordellidae, and Oedomeridae seem to be exclusively anthophilous, however, the nature of the relationships of those beetles with flowers is mostly unknown.

A wide variety of other taxa use flowers in various ways. Collembola have been recorded from flowers from all over the world. They seem to be mostly pollenophagous, but some feed on nectar. There are a few records of Plecoptera (stoneflies) on flowers. Among the Orthopteroid orders, flower visiting is known, but rarely studied. Some Dictyoptera visit flowers. For example, cockroaches may be found in the inflorescences of palms and may be involved in some pollination in the canopy of tropical forests. Some mantids use flowers as sites from which to ambush prey and some have floral-mimicking coloration. Dermaptera (earwigs) often are found in flowers, but seem to be mostly destructive, feeding on the stamens and pistils. Among the Grylloptera, some long-horn grasshoppers (Tettigoniidae, especially *Conocephalus*) also feed destructively on flowers. The Australian Zaprochilidae have elongated heads and prognathous mouthparts that suggest floral feeding. The Orthoptera proper (short-horned grasshoppers) are not often found in flowers, but some, as nymphs, may feed destructively there.

Various Hemiptera sometimes are encountered feeding in flowers, notably Nabidae, Miridae, Lygaeidae, Coreidae, Pentatomidae, and Phymatidae. Among the phymatids, both nymphal and adult ambush bugs are well known to take prey at flowers. The way they choose which flowers to use seems to reflect ease of movement, alate adults being less choosy than strictly ambulatory nymphs. The importance of various predatory Hemiptera in biocontrol suggests that much more needs to be known about their nutritional needs outside their prey. Among the hemipterans, flower visiting is not well known, but some mimic flowers or inflorescences in group activity (for example, Flattidae).

The thrips (Thysanoptera) often are found in flowers. Some have specialized asymmetrical pairs of mandibles that allow them to crush pollen grains. Thrips can build up huge populations in flowers,

and may be pestiferous in the horticultural trade. They are invoked as pollinators in some special situations. In the Malaysian forest trees (Dipterocarpaceae), the development of thrips populations has been associated with multi-specific plant flowering phenology, sexual reproductive strategies (breeding systems), and pollination.

The potential for lacewings (Neuroptera) for biocontrol of pest insects has stimulated research into their flower visiting activities and nutritional needs. It seems that nectar or pollen feeding, or both, is important for their longevity and persistence.

Floral advertising

Flowers can be considered as advertisements for plant sex. Primarily, they exploit the senses of vision and smell, but at close range and on flowers, taste and touch come into play. Color vision is general in insects. It is like color vision in human beings because, in most species studied, three primary colors are involved (trichromacy) but, unlike in human beings, the primary colors are ultraviolet (UV), blue (B), and green (G) rather than B, G, and red (R). Some insects seem to have tetrachromatic (UV, B, G, R) color vision, while others are deuteranopes (UV, B + G) Most of our understanding of color vision in insects comes from the western honeybee (*Apis mellifera*). The neural coding is also different from our own, so that brightness is not important and the green receptor is involved in other visual tasks as well as resolving colors. Thus, in studies of colors of objects of interest to insects, the whole insect visual spectrum must be considered, not just UV and not just the B and G parts of the spectrum. Many flowers show UV patterns that are invisible to people. In general, more architecturally complex flowers with hidden rewards and sexual organs tend to have color pattern guides (UV or not) more often than simpler, open flowers.

It has been assumed that floral color and size act as long-distance attraction for floral visitors. However, recent research has shown that discs of about 8 cm diameter become visible to honeybees only at a range of less than a meter. Moreover, the colored discs must have color and green contrast with the background to be even that visible. Without green contrast, the distance for detection declines to a few centimeters. The green receptor is involved in the detection of the edges of shapes, movement of objects, shape recognition, and such tasks as landing after flight. Another component of flower form that has been investigated recently is the importance of the dissectedness of the outline. It was assumed that dissectedness of the outline added attractiveness to flowers, but circular targets are detected over longer distances than those of the same diameter with variously dissected outlines.

Flowers produce a wide variety of scents. Some are mimetic, as noted above under Diptera and Coleoptera as flower visitors and pollinators. 'Heavy' floral scents are associated with flowers that bloom and are pollinated at night, mostly by moths. Those scents are assumed to attract pollinators over long distances, and the highly developed olfactory sense of the moths is well known. Diurnally blooming flowers tend to have more delicate 'floral' scents. These seem to be used by floral visitors at close range. Different parts of flowers emit different smells that are used in orientation. Once the visitor has landed, the definition of scent and taste becomes blurred, as antennation allows the visitors to follow odor/taste guides on the surface of the floral parts.

Coupled to the subtleties of floral size, shape, color, color patterns, scent, and odor/taste guides may be the role of micro-textural features of the floral epidermis. Insects, at least bees, have been shown to use and to learn 'micro-Braille' while foraging. The features on the floral surface match the scale of antennal sensilla in size and spacing.

Floral rewards

Pollen is probably the most important of floral food rewards for anthophiles, but may have been coupled with sugary drinks (nectar) since the association of spores and micropylar droplets in the Carboniferous. Pollen of entomophilous (insect-pollinated) plants is generally highly nutritious. Depending on the plant species of origin, it contains variable amounts of protein (even up to about 60%), many amino acids (including those essential for animal nutrition), lipids (up to about 20%), starch (up to 22%) and most minor, but important, nutrients such as vitamins, sterols, and minerals. Entomophilous pollen is also characterized often by external ornamentation (spines in Asteraceae, Cucurbitaceae, Malvaceae, Cactaceae; tubercules and ridges) and by having an external, oily pollenkitt. Both features enhance adherence to the bodies of pollinating insects for dispersal. Pollen is ingested whole by most insects, and the nutritive protoplast is digested by enzymes that diffuse in so the nutrients diffuse out into the gut lumen. There are some insects that chew or bite pollen grains to extract the nutrition, among them *Atrichopogon* (Diptera: Ceratopogonidae), various thrips (Thysanoptera), some beetles, and Collembola. Pollen ingestion by some Diptera and Coleoptera is associated with maturation of their sexual organs, especially of the ovaries and ova in females. During nesting by bees, pollen is gathered by females and stored as 'bee bread' or larval food. The females in most families have specialized pollen collecting (parts of the legs) and carrying organs (scopae, corbiculae). Most female solitary bees lay their eggs on a loaf of bee bread within the cells of the nests.

Nectar is a sugary liquid secreted by various parts of plants, including flowers. Floral nectar is comparatively well studied and is the primary fuel for flight in many insects. The energy content (5 cal/mg of sugar) of nectar depends on the volume produced and its sugar concentration. The concentration of sugars reflects the capacities of various insects to imbibe liquids of different viscosities. The nectar imbibed by long-tongued insects such as Lepidoptera, some Diptera (e.g., Nemestrinidae, some Tabanidae, Bombyliidae), and some Apoidea (Bombinae, Euglossinae) is usually watery (around 30% sugars), whereas that taken by short-tongued and lapping insects (e.g., Muscidae) may even be crystalline. The main sugars (sucrose (a disaccharide), glucose, and fructose (monosaccharides)) vary with plant family and type of pollinator. Some generalizations can be made about pollinator type and the ratio of sucrose to fructose and glucose in nectar. If this ratio is high, pollination by insects with short proboscides (Diptera, short-tongued bees) is expected; if it is low, pollination by insects with long proboscides (Lepidoptera, long-tongued bees) is expected. Floral nectar contains minor constituents, among which amino acids are the best understood. Floral nectars consumed by insects (e.g., many Diptera, bees) with nitrogen-rich diets tend to have lower titers of amino acids than do those consumed by insects that do not feed on nitrogen-rich diets (e.g., most Lepidoptera, long-tongued Diptera). The role of amino acids in nectar in the nutrition of anthophiles has been investigated rarely, but it is known that some insects have greater longevity when allowed to feed on nectar rather than on syrup of the same sugar concentration. Little is known of the other minor constituents, save their presence. Lipids often are accompanied by anti-oxidants (for example, ascorbic acid) and may be involved in reducing evaporation rates of water from exposed nectars.

Among the Apoidea (especially *Centris* species in the subfamily Anthophorinae) are specialists that collect floral oils from the flowers of certain families of plants, such as Malpighiaceae and Krameriaceae. These bees use the oils for provisioning their nests and collect it on specialized tarsal brushes. Many bees collect resins and gums, but some flowers provide it as a reward for their pollinators. *Dalechampia* species and *Clusia* species are well-known provenders of gums that the bees use in nest construction. Among the Euglossinae (Apidae) are species in which male bees collect floral scents, especially of orchids, that they use as olfactory attractants for mates in lekking behavior.

Other floral rewards: heat, floral tissue, sex

Some blossoms become heated. Among the mimetic Araceae, such as jack-in-the-pulpit and skunk cabbage, endogenous heat production is well known. The pollinators of these plants are various flies and beetles that become temporarily trapped in female phase blossoms that become male as the stigmas senesce and anthers shed pollen, which is when the pollinators are liberated. The heat helps drive off the mimetic scents (dung, carrion, musk), speed floral and insect development, and bring the pollinator sexes together for mating. Other flowers trap the heat of the sun by various means. Parabolic diaheliotropic (sun-following) flowers may warm their insect visitors by six or more degrees Celsius. Such heating improves speed of mobility, and, it seems, optical and other neural processing. Catkins of willows (*Salix* spp.) act as hairy heat traps and offer shelter and warmth to small insects, such as non-biting midges (Chironomidae) in the Arctic. Some insects, especially male bees, sleep in flowers. The males of the hoary squash bee (*Peponapis pruinosa*: Anthophorinae) sleep, often in small bachelor parties, in the spent flowers of squash and pumpkin (*Cucurbita* spp.) that open and close in a single day.

A wide variety of insects feed on floral tissues, apart from pollen. Many, for example, caterpillars, maggots, earwigs, thrips, and some beetles, probably are not pollinators, but some are. Notorious are the highly mutualistic interactions between figs and fig-wasps, yucca and yucca-moths, globeflowers and the anthomyiid flies, oil palms and its pollinating weevils, and some special fly-pollinated flowers (for example, *Siparuna* species (Monimiaceae)). The flowers, mostly the female parts and resulting fruit (except in oil palm and *Siparuna*), serve as oviposition sites and for brood food. The adult insects are the pollinators. Some thrips breed in flowers, becoming serious horticultural pests in some situations. Other sorts of floral tissues seem to be fed on by some anthophiles and pollinators. Special structures, food bodies and staminodes with pollen substitute (for example, in Melastomataceae) are eaten or collected by various insects, including bees.

There are a few flower visiting arthropods that use flowers as sites for prey capture. Well known are some species of crab spider (for example, *Misumena* species (Thomisidae)) and ambush bugs (*Phymata* species (Reduviidae)). These predators choose their ambush sites according to energetic constraints of changing site. For example, adult female crab spiders that walk slowly from place to place are most careful in choice, immature and apterous bugs are also fastidious, but adult alate ambush bugs choose quickly and switch sites readily. Some tropical mantids are cryptically colored to match the flowers they use. Some crab spiders may be able to change color to match their floral host.

Flowers may be mating sites for some insects, as recorded for a few species of bees. Some insects, such as the males of some orchid bees (*Euglossa* spp. (Euglossini)) and some flies collect floral scents that become attractive to mates.

Energetics: time and motion

A great deal has been learned about optimal foraging from insect and flower relations. Bees, especially bumblebees (*Bombus* spp.), are central place foragers (i.e., they forage from a home, the nest), are conspicuous and quite easily followed. Their foraging illustrates the range of behaviors that exemplify maximizing energy gathering while minimizing energy expenditure. Foraging bumblebees tend to follow the same general path from one trip to the next. Thus, they are familiar with their area. They also tend to visit the same patches of flowers on sequential trips, a behavior sometimes called 'trap-lining.' Within a patch, bumblebees usually show a 'forwarding' path, but this may involve sharp turns to left or right and visiting mostly 'next-closest' blooms if resources present are relatively abundant and rich. If resources are sparse, the bees tend to veer back and forth much less between sequentially visited blooms, and often overfly the 'next closest.' Thus, in a rich patch they visit more and stay longer, but in a poor patch, they sample and depart quickly. On the blooming plants, the bees tend to act systematically. For example, on tall inflorescences, they start at the lowermost open flowers and work upwards. The energetic implications are for efficiency, crawling vertically instead of flying, and flying down between adjacent inflorescences. Mostly, such inflorescences have female-phase older flowers and male-phase younger flowers (protandrous flowers), with more dilute nectar below and less, but sweeter, above. Most of the studies that exemplify those movement patterns have been made in meadow-like situations. On trees, with vertically arranged arrays of blooms, the same within-patch pattern has been noted. However bees treat inflorescences close to each other as a unit and continue upward but they treat separated (more than 10 cm apart) inflorescences as individuals and so show a generally descending overall direction. The implications for pollination of trees remains to be explored.

Flowers are variable in their anatomical complexity. Thus, simple, open-bowl-shaped flowers are easily exploited, but complex flowers with intricate mechanisms and hidden rewards require skill on the part of the pollinators to use. Experiments with bumblebees have shown that learning to manipulate complex flowers with skill, accuracy, and speed requires learning and an investment of time. That investment can pay off through the much greater amounts of reward in complex flowers versus simple ones. Of course, if too many bees invest in the learning when there are few flowers, loss in profitability from investing in learning may result.

The models of optimal foraging by bumblebees can be extrapolated to predator and prey interactions, but blossoms do not try to escape (quite the contrary, they advertise their presence), worker bees forage for the colony rather than for themselves, and their demise as individuals is not as drastic a loss for the next generation as in non-eusocial animals. Nevertheless, in such flower-visiting animals as sphinx moths and hummingbirds, the similarities in foraging patterns are remarkable. Butterflies tend to forage in a much more haphazard fashion, sipping nectar at flowers as they need energy.

Floral specialization and constancy in the broad sense include the tight interspecific interactions between specialized pollinators and their partner flowering plants (for example, *Yucca* and *Tegeticula* (Lepidoptera), *Ficus* and Agaonidae (Hymenoptera), oil palm and *Elaeidobius* spp. (Coleoptera), *Siparuna* and Mycetophilidae (Diptera), and the myriad bees and flower specializations. The terms mono-, oligo-, and polylectic (or -tropic) refer to anthophiles that use one, a few, and many floral resources. The equivalent botanical terms for flowers are mono-, oligo-, and polyphilic. Entomologists and naturalists have made observations of apparent, but unexplained, floral specializations in other groups of insects in which generalist anthophily might be otherwise expected (e.g., in Syrphidae (Diptera), other Diptera, and some Coleoptera). Apparent floral 'specialization' (mono- or oligolecty) may be thrust upon anthophiles by lack of a diversity of flowers upon which to forage. Presented with more kinds of flowers, such anthophiles would presumably expand their foraging base. Constancy is a term best reserved for oligo- or polyleges that restrict their visits to the flowers of one species of plant, even when others are in bloom. Such constancy is well explored by studies of social bees (Apidae) in which individuals from a single colony may be constant to the flowers of a particular plant for a short period before switching to another, while other individuals from the same colony may be constant to yet a different species, and so on. The benefits of constancy are clear from the viewpoint of intraspecific pollination, and from the anthophile's viewpoint, in efficiency of learning to forage quickly and accurately. Such foragers, either as colonies as in honeybees, or as individuals, have species of plants that can be called 'majors' and 'minors' so that they sample the floral resources in their areas and can switch as resource availability, type or quality change in the habitat or according to the colonies' needs.

Ecosystems, community and co-evolutionary ecology

Various ecosystems and communities have particular assemblages of plants and pollinators, for example, the Arctic, New Zealand, and alpine communities with pollination predominantly by flies. In general, there appears to be a positive relation between phytosociological progression (by succession, habitat, or geographic area, or combinations thereof) and specialization and diversity in pollination mechanisms. How those various diversities of pollination partnerships came about evolutionarily may be outlined as follows. Starting with the idea that flowers may sometimes compete for pollinator services and sometimes pollinators may compete for floral resources, it can be invoked that some individuals and some species fare better than others. If the direction of competition remains the same over a period of time, one would expect some evolutionary changes to take place in some of the partners. These changes would involve directional selection and character displacement to result in better adaptedness in species of any group, or of groups of species comprising the partnerships. Thus, the assemblage of flowers and pollinators might change in species abundances and diversities as generalists become more specialized through pressures of resource partitioning (increased reliability of pollination for the plants by specialist pollinators, and lesser interspecific competition for floral resources for the pollinators). The direction of competition probably naturally fluctuates over the short term (years to centuries), but with long-term (thousands to millions of years) trends. It is presumably through those processes, and their reversals, that the complexity of pollination, and other ecosystem functions, has arisen. The extreme situation of the evolution of eusocial pollinators (which are found everywhere from tropical forests to high arctic tundra and exhibit constancy and specialization) has depended on a flora of plants with short-lived flowers (from open-bowl shaped ones visited by a wide diversity of pollinators to specialized ones that only the eusocial pollinators can service) blooming in sequence throughout the long life-span of the insect colony. Co-evolution in pollination is a complex process.

Pollination for food and fiber production

Although many agricultural crops are independent of insects for pollination (wind pollination in grains, self-pollination in some seed crops and vegetables), others require insect pollination. Honeybees (*Apis mellifera* and *A. cerana*) are especially important, and stingless bees (Meliponini) are gaining recognition also. The importance of leafcutting bees (especially *Megachile rotundata*) in agriculture came to prominence for production of alfalfa seed and the culture of these bees is now a multi-million dollar industry. In Japan especially, but also in other places, mason or orchard bees (*Osmia* spp. (Megachilidae)) are managed as pollinators for use in fruit production.

Bumblebees (*Bombus* spp. (Apidae)) are now firmly entrenched as part of technology for greenhouse production, especially of tomatoes, around the world. Flies (especially Calliphoridae) are used in pollination of some crops (for example, seed production for onions, carrots, parsnips), and are probably the principal pollinators of mango. Midges (Ceratopogonidae) are the principal pollinators of cacao. Rotting vegetation is used in plantations to provide oviposition sites and food for the maggots to assure adequate pollinator forces for the crop. Some beetles (Nitidulidae) are the principal pollinators of Annonaceae. In these, and other crops for which even less is known, much research is needed to determine how to provide the pollination forces needed from crop production.

The importance of pollinators to forest sustainability has been hardly investigated, but it seems reasonable to suggest, from evidence in Costa Rica, Thailand, and Canada (Ontario, Quebec, New Brunswick) that pollination has been adversely affected by human activities (logging, pesticide applications, fragmentation). Grasslands in South Africa and Canada also seem to be suffering declines in pollinators and pollination through overgrazing.

As mentioned, honeybees (*Apis* spp.) are especially important pollinators in agriculture throughout the world, and where they are native, are part of the natural ecosystem. Beekeeping has come to grips with the issues of pesticide kills, but recently in Europe and North America, has suffered major setbacks with the spread of exotic diseases. *Varroa* mites transmitted from the natural host, the Asiatic hive bee *Apis cerana*, onto the western honeybee, *A. mellifera*, continue to adversely impact natural, feral, and managed colonies. Conversely, diseases of the western honeybee have been transmitted to Asiatic species. In some places now, honeybees are in too short supply to service growers' needs. Moreover, as beekeeping has come to require more intensive management to maintain healthy, strong colonies, the costs of pollination services have risen. In some places, notably in China and parts of Nepal, bees are so scarce that apple pollination is now done by hand.

The economic impact of pollinator shortages are crops failing to reach potential yield levels locally, regionally, nationally, or globally. The effects, depending on scale, may be rising prices or farmers being unable to produce with profit. Whatever the effect, prices to consumers rise. The needs for pollinator conservation are clear.

Insect conservation

There are relatively few insects on lists of rare and endangered life forms by comparison with birds, mammals, and plants. Most insects on such lists are butterflies. Flower relations and pollination are central to conservation because rare and endangered flowering plants must have their mutualists with them in their habitats if the populations are to persist. Thus, pollinators and seed dispersers must be included in conservation and restoration planning. That approach is being embraced in Hawaii for plant conservation, especially for endangered bird-pollinated plants. Also in the Carolinian forest fragments of southern Ontario, sensitivity to the need for pollinators in rare plant conservation is recognized. Several magnificent trees, for example, cucumber magnolia, Kentucky coffee tree, and the native paw paw, fail to set fruit for various reasons such as pollinator limitation and wide spacing of trees.

The vicious cycle involving the demise of pollination mutualisms and the partners of those mutualisms proceeds as follows. The general habitat is stressed (by fragmentation, pollution, etc.) so that one or more components become scarce. If one of the dwindling components is part of a crucial mutualism, such as pollination or seed dispersal, then the partner not directly affected by the stress becomes indirectly affected and dwindles too. As the community become simplified by erosion of diversity and abundance, other components become adversely affected. So the system, as a whole, can be seen to become less and less diverse in terms of species present and the interactions that make for a functioning ecosystem. The complex is simplified, and a different ecosystem becomes established.

There is now global concern for the demise of pollination in agricultural and forest ecosystems. In agriculture, pollination services are variously threatened by adversity, disease, and pests in beekeeping (for example, fluctuating honey prices, mite and other diseases, small hive beetle); pesticide poisonings of pollinators; expansive cultural practices that remove pollinator habitat from proximity to crops (for example, as happened by expanded alfalfa seed production fields in the Canadian prairies in the 1940s and 1950s); and reduction in the amount of natural or semi-natural areas in agricultural regions. In forestry,

large scale clear-cutting, especially now in the tropics (the effects in much of the temperate world are now largely unrecorded history), seems to be adversely affecting populations of various flower visiting insects. Some insects, notably generalist bees, seem to have a remarkable propensity for persistence, but many specialists do not have that capacity.

Pollination deficits in agriculture are documented in various parts of the world and have become common enough to prompt major international concern, especially for sustainable food and fiber production for human life. The International Convention on Biological Diversity acted over 1998–2002 through the International Pollinators Initiative of the Food and Agriculture Organization of the United Nations. Various other national and continental programs also have started (for example, the World Conservation Union, and various regional initiatives such as the North American Pollination Protection Campaign and the African Pollinators Initiative).

See also, POLLINATION BY YUCCA MOTHS, NIGHT BLOOMING PLANTS AND THEIR INSECT POLLINATORS, BEES, HONEY BEE, ALFALFA LEAFCUTTING BEE, APICULTURE.

Peter G. Kevan
University of Guelph
Guelph, Ontario, Canada

References

Bernhardt, P. 2000. Convergent evolution and adaptive radiation of beetle-pollinated angiosperms. *Plant Systematics and Evolution* 222: 293–320.

Dobson, H. E. M., and G. Bergström. 2000. The ecology and evolution of pollen odors. *Plant Systematics and Evolution* 222: 63–87.

Free, J. B. 1993. *Insect pollination of crops* (2nd ed.). Academic Press, London, United Kingdom.

Heinrich, B. 1979. *Bumblebee economics*. Harvard University Press, Cambridge, Massachusetts.

Kevan, P. G., and H. G. Baker. 1983. Insects as flower visitors and pollinators. *Annual Review of Entomology* 28: 407–453.

Kevan, P. G., and H. G. Baker. 1999. Insects on flowers. pp. 553–584 in C. B. Huffaker and A. P. Gutierrez (eds.), *Ecological entomology* (2nd ed.). John Wiley and Sons, New York, New York.

Labandeira, C. 1998. How old is the flower and the fly? *Science* 280: 57–59.

Lloyd, D. G., and S. C. H. Barrett (eds.). 1996. *Floral biology: studies on floral evolution in animal-pollinated plants*. Chapman and Hall, New York, New York.

Proctor, M., P. Yeo, and A. J. Lack. 1996. *The natural history of pollination*. Timber Press, Portland, Oregon.

Roulston, T. H., and J. H. Cane. 2000. Pollen nutritional content and digestibility for animals. *Plant Systematics and Evolution* 222: 187–209.

POLLINATION BY YUCCA MOTHS. Yucca moths provide one of the best understood examples of obligate pollination mutualism, in which the female adult pollinates the flowers of her host plant and her larval progeny consume some of the developing seeds. The yucca moth and its relationship with the yuccas was first described from Missouri in 1872 by Charles V. Riley. At least 25 species of *Tegeticula* and *Parategeticula* (Lepidoptera: Prodoxidae) serve as the exclusive pollinators of an estimated 30 to 40 species of yucca (*Yucca* and *Hesperoyucca*, Agavaceae). Yuccas have white, usually fragrant flowers in a large panicle, but have effectively lost nectar production and produce very little pollen, thus, they only attract a limited range of floral visitors. A modified urn-shaped stigma of the floral ovary makes pollination unlikely by casual brushing, and only the moths have been documented to be pollinators. The plants can reproduce vegetatively, but the moths appear to be critical for sexual propagation. Geographically limited to the Americas, the moths and yuccas are distributed from southwestern Canada southward at least to northern Belize, with the highest species diversity in arid and semiarid portions of western North America. Over the past two centuries, contiguous yucca range expansion through horticulture in the interior of eastern North America led to rapid colonization by the moths. Within decades, they spread northward to southeastern Canada and westward to the edge of the Great Plains.

Fig. 823   Yucca moth pollinating a yucca flower.

Early studies held that four yucca moth species, including one thought to feed on all but two yuccas, served as pollinators of all yuccas. Recent studies based on behavioral, morphological, and DNA sequence data have shown that the polyphagous species, *T. yuccasella*, in fact is a large species complex with species ranging in diet breadth from monophagy to feeding on as many as seven host species. Adult moths have a wingspan of 15 to 35 mm, and range in color from solid white to black, a mix of white and black, or sand-colored. The female has unique tentacular mouthparts at the tip of the first segment of the maxillary palp. With a single exception, all species are nocturnal. Adults, which are estimated to live 2 to 4 days, emerge during the flowering period of their host, and mate during their first night. Flowering host plants are likely found through a combination of visual cues and prominent floral scent cues. The female then visits yucca flowers to gather pollen by dragging her tentacles across the anthers. The highly sticky pollen is compacted and then held under the head. The female next seeks out flowers where she can oviposit. *Tegeticula* species use a cutting ovipositor to insert eggs one at a time at different positions in the pistil, whereas *Parategeticula* females create grooves in petals or pedicels where eggs are laid. After oviposition, the female removes a pollen batch from her load with her tentacles, walks to the stigma, and pollinates the flower by probing 10 to 20 times into the stigmatic cavity. In so doing, she assures that lack of pollination will not prevent fruit formation, which is critical because seeds, or modified seed tissue, is the sole food item of yucca moth larvae. Over-exploitation by the moths is prevented by selective abscission of flowers that have received many eggs, and possibly by other mechanisms as well. Depending on the species, females may lay one or many eggs per flower, but typically only a handful of larvae from any one female will exist in a fruit, in part because of egg mortality. The larva hatches from the egg within a week in most species, and feeds on developing yucca seeds during fruit maturation. The fully fed larva exits the fruit, and diapauses inside a dense cocoon in the soil for one or more years. Pupation generally takes place a few weeks before adult emergence. In one tropical species, there may be no diapause.

DNA-based phylogenetic analyses of the moths suggest an estimated minimum age of 40 million years of the obligate moth-plant mutualism, and also reveal that the pollination behavior arose shortly after the moth lineage had colonized the yuccas from the Nolinaceae, a coexisting family of woody monocots in arid regions. A rapid diversification over the past five million years in the northern part of the range gave rise to about half of the extant species, including two species that have reverted to antagonism by having lost functional tentacles and pollination behavior. Instead, they oviposit directly into yucca fruits, and rely on coexisting pollinator species for the creation of yucca seeds.

Olle Pellmyr
University of Idaho
Moscow, Idaho, USA

References

Riley, C. V. 1892. The yucca moths and *Yucca* pollination. *Annual Reports of the Missouri Botanical Garden* 3: 99–158.
Pellmyr, O. 1999. A systematic revision of the yucca moths in the *Tegeticula yuccasella* complex north of Mexico. *Systematic Entomology* 24: 243–271.
Pellmyr, O., and J. Leebens-Mack. 2000. Reversal of mutualism as a mechanism for adaptive radiation in yucca moths. *American Naturalist* 156: S62–S76.
Pellmyr, O. 2002. Yuccas, yucca moths and coevolution: a review. *Annals of the Missouri Botanical Garden.* 90: 35-55.

POLLINATOR. The agent of pollen transfer in plants, often bees. See also, POLLINATION AND FLOWER VISITATION.

POLLUTION AND TERRESTRIAL ARTHROPODS. Pollution, the unwanted and undesirable presence of a chemical or compound, is unfortunately common. The problem is global, affecting all continents and nations, and frequently crossing natural and political boundaries. Specific occurrences may be quite localized (such as waste-water runoff from mining operations) or cover exceptionally broad geographic areas (like acidic precipitation in northern Europe or the entire northeastern U.S.A. plus adjacent regions of Canada).

Problems with pollution are not new; air and soil contamination have been reported for thousands of years. Over 2,000 years ago, the Roman poet, Horace, complained about soot damaging the walls of temples. However, the problem has become substantially worse since the Industrial Revolution, with the large scale production and transport of many toxic materials. Most countries evolve through a period of intense industrialization, where the primary goal is to raise the standard of living for the population. During this period the environmental effects of pollutants are not considered a primary concern. As countries become more affluent, the desire for improving environmental quality increases, but nationalistic concerns, economic costs of less polluting technologies, and long standing patterns of industrial production work to impede changes that can reduce contaminants. Even in situations where pollution has been largely eliminated, a 'legacy' of contamination may still exist. Thus, the problem of pollution is likely to continue for the foreseeable future.

Solving our existing problems of environmental contamination and mitigating the effects of contaminants on living organisms are difficult because of the incredible variety of sources and forms of pollution. Even an abbreviated list of pollutants would include thousands of industrial by-products, pesticide residues from chemicals that have been banned from use, a variety of toxic metals and chemicals in mining waste, many compounds produced by burning fossil fuels, the by-products of warfare, chemicals used in electrical generation/transport machinery, fuel additives, as well as a host of other materials. Each pollutant has the potential to disrupt ecosystems. Some have minimal effects, others have contaminated soils so that plants or animals from these areas cannot be eaten. A few have created wastelands, where the ground has become too toxic to support even the most basic organisms in an ecosystem.

Terrestrial arthropods are critical to the functioning of ecosystems. Because they are at the base of the food web, changes in population densities of arthropods can have profound effects on higher level organisms that depend on them. Insects and their relatives are used as food by many birds and mammals. Many arthropods are beneficial, serving to keep pest populations under control, thereby preventing damaging outbreaks. Others pollinate plants, disseminate seeds, and produce structures used by countless other organisms. Disruption of any of these activities can have disastrous effects on an ecosystem. Thus, arthropods are often the first animals examined when ecosystems become polluted.

Interestingly, direct contact with most pollutants generally does not harm terrestrial arthropods. Populations of arthropods are more commonly affected when pollutants are ingested, or if the pollutants change the quality or quantity of their food. These effects can be either positive or negative. The following sections summarize the major types of pollutants and their effects on arthropods.

Air pollutants

Air pollution takes many forms. Some of the more common pollutants contain products resulting from fossil fuel combustion include ozone, carbon dioxide (CO_2) and carbon monoxide (CO), acidic precipitation (acidic fogs and acid rain), and many related compounds. In nearly all scientific studies, direct exposure to high concentrations of these contaminants does not physically harm arthropods. However, air pollution can alter plant chemistry, and thereby change the nutritional value of plants or their chemical defenses against arthropods. Examples are provided for ozone, acid fog, and CO_2.

Ozone is a remarkably active compound generated when combustion products from fuel are exposed to sunlight. Even moderate levels of ozone can damage plants and cause modifications in the form and content of plant nutrients. Exposure to ozone often increases availability of a key nutrient, nitrogen, which is critical for arthropod growth. This nutrient is very important, and frequently determines how fast an arthropod can grow, and if it will survive. Thus, insects such as the tomato pinworm, a key agricultural pest throughout the southern United States and Mexico, grow about 10% faster and survive at twice the rate if feeding on plants exposed to ozone. If ozone exposure is very high, then plants can become so damaged that arthropod populations can no longer survive. Like many toxic substances, the concentration of the contaminant, and the duration of exposure will determine if the pollutant is a benefit or detriment to arthropods.

The levels of CO_2 in the earth's atmosphere have been increasing dramatically since the industrial revolution. CO_2 concentration in the atmosphere has increased from 270–280 ppm to the current level of 355–360 ppm. This represents an increase of approximately 27% in a relatively short period of time. Scientists already have shown that increasing levels of atmospheric CO_2 can have substantial effects on plant suitability for arthropods. Because arthropods (like all animals) are mostly made of nitrogen, those feeding on plants have to separate the relatively small amounts of nitrogen from plant material consisting mostly of carbon. Plants grown in elevated concentrations of CO_2 have increased levels of carbon (from the carbon availability in the CO_2), and substantially reduced amounts of nitrogen. Most arthropods respond to this problem by simply eating more plant material. Some eat twice as much. Others cannot cope with the relative lack of nitrogen, and develop more slowly or even die. Changing CO_2 levels will therefore have significant effects on the plants that arthropods can eat, and how much damage is caused by their feeding.

Acid deposition in the form of 'acid rain' or acidic fogs is common in North America. Although terrestrial arthropods are not typically affected by direct exposure, their food plants are often damaged. Typical damage symptoms include lesion development, weathering of leaf surface waxes, foliar leaching, premature leaf fall, changes in plant nitrogen form and content, or even plant death. All of these can impact arthropod populations. Encounters with lesions can change arthropod feeding patterns. Leaf waxes are important cues used by some arthropods to identify a particular plant as a food source, and changes in waxes can make normally acceptable plants unrecognizable. Early leaf loss shortens the time available for leaf-feeding arthropods to develop. Changes in plant nitrogen form and content and plant defensive chemistry generally have profound impacts on arthropods. Some acidic fogs contain high levels of nitrogen, and may act as a fertilizer. In some instances this provides the arthropod with a more nutritious food source, allowing populations to increase. In other cases, the plants use the additional nitrogen to produce defensive chemicals that can suppress insect populations. The death of large areas of trees caused by acid rain and acidic fogs in eastern North America and some parts of Europe have dramatic effects on abundance of many arthropod species that survive on the affected tree species.

Pollutants that transfer from water to soil

Many common water pollutants are readily transferred from water to soil or directly from water to plants. Terrestrial arthropods are then exposed to these materials. Some of the more common waterborne pollutants include hexavalent chromium, MTBE (methyl tertiary butyl ether), and selenium.

Each of these widespread contaminants has different effects on insects.

Hexavalent chromium is one of the most common contact sensitizers in industrialized countries and is associated with numerous materials and processes, including chrome plating baths, chrome colors and dyes, cement, tanning agents, wood preservatives, anticorrosive agents, welding fumes, lubricating oils and greases, cleaning materials, and textile production. Due to the past and present use of chromium in so many industries, it is a widespread pollutant. When ingested, this material has been shown to cause a decrease in growth and fecundity in arthropods.

MTBE is a gasoline additive used to elevate the oxygenate level in gasoline. This helps the gasoline burn more completely, reducing the production of some contaminants associated with automobile exhaust. Unfortunately, this chemical has leaked into the groundwater at over 385,000 sites nationwide due to poorly sealed underground fuel storage tanks. MTBE has now been detected in 21% of 480 wells in regions using MTBE as a gasoline additive. In addition, findings from the National Water Quality Assessment Program indicate that MTBE is the second most frequently detected volatile organic compound in ground water and urban streams. Preliminary data suggest that this material can slow development of some arthropod species.

Selenium is found in contaminated soils throughout western North America. Soil accumulation is associated with agricultural irrigation, geochemical processes, mining, and a variety of other industrial sources and frequently results in significant effects on animal and human health. Although selenium is an essential trace nutrient important to humans and most other animals, toxicity occurs at high concentrations due to replacement of sulfur with Se in amino acids resulting in incorrect folding of the protein and consequently malformed, nonfunctional proteins and enzymes. Remediation strategies include removal of soil selenium by plant accumulation, harvest, and removal. Use of plants in soil remediation programs results in the availability of selenium to plant eating arthropods. Ingestion of selenium in plants by arthropods generally results in slowed growth, reduced egg production, and higher mortality.

Pollutants that transfer from water to the soil are likely to be long term problems for arthropods. Contaminated aquifers will be used for irrigation, and plant feeding organisms will be exposed to these pollutants throughout their lives. The long term effects of such exposures, and the possible interactions between the various water-borne contaminants, is not yet known.

The special case of metals

Eighty-seven of the elements on the periodic table are considered metals or metalloids (elements which act like metals). These metals are toxic at relatively low levels to terrestrial arthropods and many other organisms. Although natural mineral deposits containing metals occur around the world, and the erosion of rocks and volcanoes release metals into the atmosphere, contamination of soils by metals is most often associated with human activity. In our industrial society, metals are one of the most commonly used raw materials. Consequently, waste-water runoff from mining, metal refining, sewage sludge, and other anthropogenic sources contain high levels of metals that pollute water and soil. Additionally, gas exhaust, energy and fuel production, smelters, and foundries emit metals as airborne particulates. Sources of contamination often contain mixtures of several metals or metalloids making analysis of the effects of any one element difficult to determine in a field setting.

Airborne particulates containing metals may land on the surface of food plants of arthropods, damaging plant photosynthetic systems and resulting in altered plant chemistry and nutrition for herbivores. Contamination of soils also allows for plant uptake of many metals making these metals available to herbivores. Additionally, decaying plant materials containing metals are consumed by soil and leaf litter dwelling arthropods. Metals have been shown to accumulate in the tissues of some arthropods making them more available to predatory and parasitic arthropods, as well as higher animals that eat arthropods. This can lead to biomagnification, where pollutants accumulate in the tissues of animals as they consume the contaminated arthropods.

Toxicity of metals to terrestrial arthropods has been demonstrated in the field as well as in the laboratory. Because metals are such a large and diverse group of elements it is not surprising that the mode of action and concentrations resulting in toxicity to terrestrial arthropods are variable. Additionally, arthropods themselves differ in their ability to tolerate environments containing metals. Some arthropods are able to excrete small amounts of metals and thereby avoid toxic effects at low levels of pollution. However, at slightly higher concentrations the presence of metals may result in their incorporation into proteins and enzymes, altering their ability to function properly in the arthropod system. Some metals interfere with metabolic pathways in arthropods resulting in reduced total body protein. Additionally, metals can affect the energy source of insects, the fat body. Collectively, these effects often result in impaired growth and development and the disruption of reproduction. Therefore, not surprisingly, arthropod abundance and species diversity are usually diminished in areas where metal pollution is present.

Conclusion

All of these pollutants, whether airborne, carried by water, or present in contaminated soil, can affect population development and survival of terrestrial arthropods. Because of this, all of these contaminants can influence how communities of arthropods, and the higher animals that feed on them, will function within ecosystems. Scientists are just beginning to understand the long term effects of these pollutants on terrestrial organisms.

John T. Trumble and Danel B. Vickerman
University of California-Riverside
Riverside, California, USA

References

Agrawal, S. B., and M. Agrawal. 2000. *Environmental Pollution and Plant Responses*. Lewis Publishers, Boca Raton, Florida.

Coviella, C., and J. T. Trumble. 1999. Elevated atmospheric CO_2 and insect-plant interactions: implications for insect conservation. *Conservation Biology* 13: 700–712.

Frankenberger, W. T., and R. A. Engberg (eds.). 1998. *Environmental chemistry of selenium*. Marcel Dekker Press, New York, New York.

Heliövaara, K., and R. Väisänen. 1993. *Insects and Pollution*. CRC Press, Boca Raton. Florida.

Legge, A. H., and S. V. Krupa. 1986. *Air Pollutants and their Effects on the Terrestrial Ecosystem*. John Wiley & Sons, New York, New York.

POLYACETYLENES (AND THEIR THIOPHENE DERIVATIVES).

Investigations carried out in the second half of the last century have suggested that certain secondary metabolites from plants, including polyacetylenes and their thiophene derivatives, exert photodynamic action. As discoveries are made, the number of natural photodynamic sensitizers of this group is increasing.

Structure

Polyacetylenes have a conjugated double and triple bond system, or may be biosynthetically cyclized into thiophene compounds such as α-terthienyl. The chemistry and the natural distribution of polyacetylenic compounds have been comprehensively described elsewhere (see ''Naturally occurring acetylenes'' by Bohlmann *et al.* 1973).

During studies on the structure activity relationship in these compounds, more than two dozen polyacetylenes and thiophene derivatives, originating from the plant family Asteraceae, have been extensively tested against various biological systems. In general, aliphatic compounds containing fewer than three conjugated acetylenic bonds do not exhibit phototoxic effects. Furthermore, in a study using the microorganisms *Escherichia coli* and *Saccharomyces cerevisiae*, it was found that thiophene derivatives generally were more phototoxic than polyacetylenes.

Occurrence in plants

Polyacetylenes and thiophene derivatives are a very large group of secondary compounds whose photosensitizing properties in insects were first reported in 1975 by Arnason's team at the University of British Columbia in Canada. These compounds have been considered characteristic of taxonomically advanced plant families such as Asteraceae, Apia-

$$CH_3\text{-}CH=CH\text{-}(C\equiv C)_2\text{-}(CH=CH)_2(CH_2)_4CH=CH_2$$

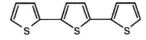

Fig. 824 Chemical structure of polyacetylenes and their thiophenes: above, straight-chain aliphatic acetylenes (e.g., heptadeca tetraene diyne); middle, partly cyclized (e.g., PHT: phenylheptatriyne); and, below, their thiophene derivatives (e.g., α-terthienyl).

ceae, Araliaceae and Campanulaceae, as well as certain groups of Basidomycete fungi. The greatest diversity is found in the Asteraceae where many polyacetylenes occur in roots, and some in the aerial parts of the members of this family.

Activity

Although the first report on the biological activity of α-terthienyl, a thiophene, as a nematocide goes back to 1958, its photoactive properties were not noticed until 1972 when the treated nematodes were exposed to near UV-A light. This was the beginning of a series of studies on the photosensitization of polyacetylenes and thiophene derivatives mediated by UV-A. Since then, numerous articles on different aspects of their activity have been published. With few exceptions, most of the studies concern insects.

Although the photosensitizing properties of polyacetylenes on insects was reported in 1975, the presence of a polyacetylene, 8-cis-decene-4,6-diyonic acid, in insects had been recorded earlier (1968). This compound was isolated from the thoracic and abdominal glands of the cantharid beetle, *Chauliognathus lecontei*.

Twenty-four polyacetylenes (and thiophene derivatives) isolated from species in the Asteraceae were screened by Arnason and associates for their near-UV mediated larvicidal properties to *Aedes aegypti* mosquitoes. One of these, α-terthienyl that was extracted from *Tagetes*, was found to be more toxic than DDT in the presence of UV-A. Since this discovery, studies on the photodynamic action of α-terthienyl have attracted more interest.

Further studies revealed that both phenylheptatriyne (PHT) and α-terthienyl exhibited ovicidal activity against the eggs of the fruit fly, *Drosophila melanogaster*, in the dark. It was reported, however, that irradiation by UV-A respectively enhanced the toxicity by 37- and 4,333-fold.

In 1984, additional investigations on the UV-mediated activity of α-terthienyl against the tobacco hornworm, exhibited delayed and abnormal pupal formation with no subsequent adult emergence. During the same year, the first example of the inactivation of acetylcholinesterase *in vivo* by a photoactive pesticide, α-terthienyl and its two isomers, was documented by Kagan and associates. A high rate of inactivation (ca. 65% to 90%) occurred within a few minutes when fourth instar larvae of *Aedes aegypti*

mosquitoes were treated with this phototoxin or either of its two isomers in the presence of UV-A.

In 1985, a Canadian patent was awarded covering the control of pests (algae, fungi, nematodes, or herbivorous invertebrates) by polyacetylenes. During the same year, the successful synthesis of α-terthienyl was reported. In field trials, effective control of third- and fourth-instar larvae of *Aedes intrudens* mosquitoes was achieved with the application of 0.1 kg/ha to natural breeding pools. This compound (α-terthienyl) was found to have a half-life of about 4 hours in sunlight.

Studies on the photobiological activity of polyacetylenes and thiophene derivatives are relatively new, therefore few reports are available on their phototoxicity to organisms other than insects. Nevertheless, the broad spectrum of activity in different photoactive molecules, including dyes and furanocoumarins, is well documented. Based on a wide range of biological activity reported from several non-phototoxic studies on polyacetylenes, neither photoactive polyacetylenes nor thiophene derivatives are believed to be the exceptions. For instance, α-terthienyl (a thiophene) that has been patented for use against algae, fungi, weeds, nematodes, insects and other herbivorous invertebrates, also has been found as a potent fish poison. It also results in damage to red blood cells and human skin in the presence of UV-A. Furthermore, phenylheptatriyne (PHT), identified as the major polyacetylenic constituent of the leaves and stems of *Bidens pilosa*, is phototoxic to yeasts and bacteria when mediated by the near UV. PHT, which is present in the cuticle of *Bidens pilosa* (up to 600 ppm), strongly inhibits the germination and growth of *Fusarium culmorum* in the presence of UV, but not in the dark.

Mode of action of polyacetylenes and thiophene derivatives

In spite of over a century of research on photodynamic action, little is known about the selectivity of phototoxins. In contrast, most photoactive agents including polyacetylenes and their thiophene derivatives have shown a wide spectrum of activity. For example, many polyacetylenes and thiophene derivatives are reported to be toxic to a wide range of microorganisms and to human skin fibroblasts in the presence of UV-A. A thiophene (α-terthienyl) causes photodermatitis in human skin characterized by immediate severe erythema on exposure to sunlight and long lasting hyperpigmentation. There is evidence that this compound, in the presence of UV-A, damages DNA since it induces unscheduled DNA synthesis.

Unlike the linear furanocoumarins, e.g., 8-MOP, that kill cells by a photoinduced modification of DNA, photoactive polyacetylenes and their thiophenes attack the cell membrane by photodynamic as well as by oxygen-independent mechanisms. However, as in the case of furanocoumarins, photosensitization is mediated by the near-UV region (UV-A: 320–400 nm).

In general, straight chain aliphatic acetylenes, which are well known for their instablility *in vitro*, have a non-oxidative mode of action that probably involves the formation of free radicals upon photoexcitation (i.e., Type I Reaction). Thiophenes, however, are Type II photodynamic photosensitizers that damage membranes via catalytic generation of singlet oxygen.

Generation of superoxide anion radicals by α-terthienyl in the anal gills of *Aedes aegypti* mosquito larvae recently has been observed. On the basis of this observation, the phototoxic action of α-terthienyl is suggested. Other investigators who have examined the metabolism of α-terthienyl in the mosquito *Culex tarsalis* found that pretreatment of the larvae with piperonyl butoxide increased larval susceptibility to α-terthienyl and reduced the rate of elimination of this substance.

Partly cyclized aromatic acetylenes such as phenylheptatriyne, which are intermediate in structure between the aliphatic compounds and the thiophenes, apparently exhibit both photodynamic and non-photodynamic processes. It was subsequently found that phenylheptatriyne and other polyacetylenes are photodynamic toward some organisms, including *E. coli*, but are partially non-photodynamic in other systems, including the microorganism *Saccharomyces*.

Most acetylenes are able to produce singlet oxygen *in vitro* at levels that do not fully account for their phototoxic effects. For example, after removal of oxygen, phenylheptatriyne showed only partial or no decrease in phototoxicity to microorganisms or photohemolysis of erythrocytes.

It should be mentioned that many biologically active acetylenes, such as falcarinone, falcarindiol, oenanthotoxin, capillin, *Matricaria* ester and its derivatives, and cicutotoxin, are not light-activated. Inter-estingly, falcarindiol, recently identifed as a potent phytoalexin in fungal infections, was found to stimulate oviposition by the carrot fly, *Daucus carota*.

All photoactive representatives of dyes, furanocoumarins, polyacetylenes and their thiophene derivatives, including cercosporin and hypericin, have shown a wide spectrum of activity. For example, it has been demonstrated that furanocoumarins, upon activation by light, are powerful antimicrobial agents, nematocides, molluscicides, piscicides and powerful skin photosensitizers against man and animals.

Many polyacetylenes and thiophene derivatives recently have been shown to be toxic to a wide range of microorganisms and to human skin fibroblasts, in the presence of UV-A. A thiophene (α-terthienyl) causes serious photodermatitis in human skin, characterized by immediate severe erythema on exposure to sunlight and long lasting hyperpigmentation. There is evidence that this compound, in the presence of UV-A, damages DNA since it induces unscheduled DNA synthesis. For these reasons, there may be genetic risk associated with polyacetylenes and thiophene derivatives.

Cyrus Abivardi
Swiss Federal Institute of Technology
Zurich, Switzerland

References

Arnason, T., T. Swain, C.-K. Wat, E. A. Graham, S. Partington, G. H. N. Towers, and J. Lam. 1981. Mosquito larvicidal activity of polyacetylenes from species in the Asteraceae. *Biochemical Systematics and Ecology* 9: 63–68.

Arnason, T., G. H. N. Towers, B. J. R. Philogène, and J. D. H. Lambert. 1983. The role of natural photosensitizers in plant resistance to insects. pp. 139–151 in P. A. Hedin (ed.), *Plant Resistance to Insects*. ACS Symposium Series 208, Washington DC. 375 pp.

Bohlmann, F., T. Burkhardt, and C. Zdero. 1973. *Naturally occurring acetylenes*. Academic Press, London, United Kingdom. 547 pp.

Champagne, D. E., J. T. Arnason, B. J. R. Philogène, P. Morand, and J. Lam. 1986. Light-mediated allelochemical effects of naturally occurring polyacetylenes and thiophenes from Asteraceae on herbivorous insects. *Journal of Chemical Ecology* 12: 835–858.

Christensen, L. P. 1998. Biological activities of naturally occurring acetylenes and related compounds from higher plants. *Recent Research Developments in Phytochemistry* 2: 227–257.

Heitz, J. R., and K. R. Downum (eds.). 1987. *Light-activated Pesticides*. ACS Symposium Series 339, Washington DC 339 pp.

POLYACRIDAMIDE GEL ELECTROPHORE-SIS. A process by which molecules are separated based on their size and charge, using a polyacrida-mide gel and electrical current.

POLYANDRY. Mating of a female with several males.

POLYCENTROPODIDAE. A family of caddis-flies (order Trichoptera). They (as well as Psycho-myiidae) commonly are known as trumpet-net and tube-making caddisflies. See also, CADDISFLIES.

POLYCTENIDAE. A family of bugs (order Hemi-ptera). They sometimes are called bat bugs. See also, BUGS.

POLYCULTURE. A mixture of crop plants grown in the same area (field).

POLYDNAVIRUSES (PARASITOID RELATED VIRUSES). Members of the family Polydnaviridae are unique to insects and possess enveloped, quasicy-lindrical (helical symmetry) nucleocapsids which encapsidate multiple (20 to 30) dsDNA molecules having a composite size of 200 to 280 kbp. Histori-cally, these viruses were originally placed in subgroup D of the family Baculoviridae. Polydnaviruses replicate exclusively in the nuclei of the calyx cells located in the female reproductive tract of adult hymenopterans. Two genera of polydna-viruses are recognized: the *Bracovirus* and *Ichnovirus* detected in braconids and ichneumonids, respec-tively. Bracoviruses are rod-shaped, have a width of 35 to 40 nm, and can vary in length from 30 to 200 nm. The nucleocapsids are enveloped by a single unit membrane (one or more nucleocapsids/mem-brane) that is formed de novo in the nuclei of calyx cell. Progeny bracoviruses are released through cyto-lysis of calyx cells. Ichnoviruses have a fusiform morphology (about 100 to 350 nm), possess a lenticu-lar nucleocapsid, and contain a two-unit membrane, the first formed in the nucleus and the second acquired via budding through the cell membrane of infected calyx cells. Both the bracoviruses and ichno-viruses are complex particles and contain 20 to more than 30 structural proteins ranging from 10 to 100 kDa.

The key characteristic of the polydnaviruses is the heterodispersed double-stranded circular DNA gen-ome. In the mid-1970s, numerous enveloped virus particles were detected in the calyx epithelial cells of the accessory glands of female parasitoids. Poly-dnaviruses can be purified easily from dissected calyx tissue using density gradient centrifugation methods. Polydnavirus DNA preparations applied to $CsCl_2$ gradients containing ethidium bromide pro-duce two bands, representing the relaxed circular and superhelical DNA molecules. Kleinschmidt spreads of these preparations revealed a complex of circular DNA molecules ranging from 1.5 to 8.0×10^6 daltons. Agarose gel electrophoresis of polydnavirus DNA preparations produced more than twenty bands ranging from 2 to more than 28 kbp. The bands detected in these gels were present in non-equimolar ratios. Hybridization studies, utilizing DNA extracted from agarose gels and labeled *in vitro* with ^{32}P as probes for Southerns, demonstrated that more than 80% of the electrophoretically separated bands were unique DNA molecules. The relative number of DNA molecules packaged within the poly-dnavirus nucleocapsid differs between the *Ichnovirus* and *Bracovirus*. It has been proposed that the differ-ent *Ichnovirus* DNA molecules are encapsidated together within a single particle. However, certain bracoviruses, characterized by possessing nucleocap-sids of variable length, are believed to encapsidate a single DNA molecule. Kleinschmidt spreads of DNA released from osmotically shocked preparations of the bracovirus of *Chelonus inavitus* revealed that each nucleocapsid released a single DNA molecule, suggesting the presence of a population of virions. The helical symmetry of these viruses allows for dif-ferential packaging of DNA; the longer the helix, the larger the encapsidated dsDNA fragments.

In most cases, polydnaviruses replicate in the calyx cells of female wasps and release progeny virus particles into the calyx lumen. In the calyx cells, polydnavirus morphogenesis is observed at the pupal-adult stage of wasp development and is regu-lated in part by 20-hydroxyecdysone. Within a speci-fic parasitoid population, these vertically transmitted viruses are found in 100% of egg and sperm cells, suggesting a Mendalian transmission mode. In certain ichnoviruses, including the *Campoletis sonorensis* viruses (CsV) and *Hyposoter fugitivus* (HfPV), the polydnavirus DNA is integrated into the chromosomal DNA of both the parasitoid and in selected lepidopteran cell lines. Cloned polydnavirus DNA hybridized to Southern blots detected off-size fragments in REN digests of chromosomal preparations of both parasitoid and selected insect cell lines. The stable integration of complete, unarranged polydnavirus DNA into wasp chromosomes suggests that these viruses may be transmitted as proviruses. Therefore, it has been proposed that the polydnavirus has two replicative pathways. First, it exists as a linear chromosomal provirus responsible for transmission in the wasp and, secondly, as an encapsidated circular DNA produced within the calyx cells during the pupal-adult stages. The circular DNAs, packaged into virus particles, are released during oviposition into the parasitoid's host.

Deposition of polydnaviruses during oviposition into the host plays an important role in the survival and development of the parasitoid egg. Parasitoid eggs explanted from the wasp ovary and implanted into host larvae are readily recognized as non-self, encapsulated by circulating hemocytes, and killed. However, the combination of viable polydnavirus and the egg implants results in the survival and development of the parasitoid. The observed obligatory mutualism observed between polydnaviruses and their respective wasp species is unique. The polydnaviruses delivered with the parasitoid egg during oviposition do not replicate in host lepidopteran cells; nevertheless, polydnaviruses mediate dramatic changes in host physiology. In addition to the polydnavirus, wasps deliver host-modulating substances, including venom, ovarian-secreted proteins, and/or specialized teratocyte cells that may complement the activity of polydnaviruses. The polydnaviruses delivered into the lepidopteran host are able to penetrate various cell types, undergo partial transcription, and produce m-RNA and selected viral proteins within several hours of oviposition. Host granulocytes and plasmatocytes, as well as the hemopoietic tissues (hemocyte stem cells) and prohemocytes, are the primary targets of the polydnaviruses. The presence of polydnavirus causes a marked depletion of immunoresponsive cells, disrupts the actin cytoskeleton of plasmatocytes and granulocytes, and may induce apoptosis (programmed cell death) of targeted hemocytes. The disruption of cellular actin inhibits the ability of these cells to adhere to and spread over

non-self, resulting in the inhibition of the encapsulation response. The polydnavirus-mediated inhibition of the host cellular defense has been shown recently to increase the susceptibility of host larvae to other disease agents such as baculoviruses. Apoptosis of the granulocytes, characterized by cellular blebbing and fragmentation of chromosomal DNA, seen as a ladder of DNA molecules on agarose gels, results in depletion of functional granulocytes. Although the polydnaviruses are able to suppress the host cellular defense, the humoral response which involves the induction and synthesis of the anti-microbial cationic proteins remains functional in the parasitized hosts.

References

Beckage, N. 1998. Modulation of immune responses to parasitoids by polydnaviruses. *Parasitology.* 116 Suppl: S57–64.

Edson, K., S. B. Vinson, D. Stoltz, and M. Summers. 1981. Virus in a parasitoid wasp: suppression of the cellular immune response in the parasitoid's host. *Science* 211: 582–583.

Fleming, J., and M. Summers. 1991. Polydnavirus DNA is integrated in the DNA of its parasitoid wasp host. *Proceedings of the National Academy of Sciences of the USA* 88: 9770–9774.

Lavine, M., and N. Beckage. 1995. Polydnaviruses: potent mediators of host insect immune dysfunction. *Parasitology Today* 111: 368–377.

Whitfield, J. B. 2002. Estimating the age of the polydnavirus/braconid wasp symbiosis. *Proceedings of the National Academy of Sciences of the USA* 99: 7508–13.

POLYDOMOUS. Social insects in which single colonies occupy more than one nest.

POLYEMBRYONY. Production of more than one embryo from a single egg, a condition found in some Hymenoptera in the families Braconidae, Dryinidae, Encyrtidae, and Platygasteridae.

POLYETHISM. Division of labor among members of a colony of social insects. This includes caste-based polyethism, wherein different forms perform different functions, and age-based polyethism, wherein individuals perform different functions as they age.

POLYGENIC RESISTANCE. Resistance of a host to a 'parasite' based on many genes.

POLYHEDRON (PL., POLYHEDRA). Crystal-like inclusion bodies produced in the cells of tissues affected by certain insect viruses; ordinarily the polyhedrosis-virus particles formed in the nuclei of the host cells are rod-shaped, while those formed in the cytoplasm are polyhedral or approximately spherical.

POLYHEDROSIS (PL., POLYHEDROSES). A virus disease of certain insects characterized by the formation of polyhedral inclusions in the tissues of the infected insect; if the inclusion bodies (polyhedra) are formed in the nuclei of the infected cells, the disease is known as a 'nuclear polyhedrosis.' If the inclusions are formed in the cytoplasm, the disease is known as a 'cytoplasmic polyhedrosis.'

See also, BACULOVIRUSES.

POLYGYNY. The existence of more than one egg-producing queen within a single colony. Primary polygyny is a condition wherein queens form a colony simultaneously; secondary polygyny is when supplementary queen are added after the colony is founded.

POLYMER. A chemical compound consisting of a long chain of identical or similar units.

POLYMERASE CHAIN REACTION (PCR). A method for amplifying DNA by means of DNA polymerases such as *Taq* DNA polymerase. PCR fundamentally involves denaturing double-stranded DNA, adding dNTPs, DNA polymerase, and primers. DNA synthesis occurs, resulting in a doubling of the number of DNA molecules defined by the primers. Additional rounds of denaturation and synthesis occur, resulting in a geometric increase in DNA molecules because each newly synthesized molecule can serve as the template for subsequent DNA amplification. Modifications of the PCR reaction have been developed for special purposes. PCR is used to clone genes, produce probes, produce ssDNA for sequencing, and carry out site-directed mutagenesis. DNA sequence differences are used to identify individuals, populations, and species.

POLYMITARCYIDAE. A family of mayflies (order Ephemeroptera). See also, MAYFLIES.

POLYMORPHISM. More than distinct one body form (phenotype) within the same stage of a species. Two or more genetically different classes in the same interbreeding population. (contrast with polyphenism)

POLYPHAGIDAE. A family of cockroaches (order Blattodea). See also, COCKROACHES.

POLYPHAGOUS. Feeding broadly. In herbivores, feeding on more than one family of plants.

POLYPHENISM. More than one body form (phenotype) within the same clone (genotype) of a species. (contrast with polymorphism)

POLYPHYLETIC GROUP. Taxa that do not contain all the recent descendents of a single past species, but those excluded from the group are descended from a species of the group that is younger than the stem-species. Often it is difficult to distinguish paraphyletic from polyphyletic taxa.

POLYPLACIDAE. A family of sucking lice (order Siphunculata). This family is also known as spiny rat lice. See also, SUCKING LICE.

POLYPLOIDY. An increase in the number of copies of the haploid genome. Most individuals are 2n, but species are known that are polyploid (3n, 4n, 5n, 6n), and such species are parthenogenetic because of the difficulty of maintaining normal meiosis. Many insect species have tissues that are polyploid, such as the salivary glands, nurse cells of the ovary, and germ line tissues.

POLYPORE FUNGUS BEETLES. Members of the family Tetratomidae (order Coleoptera). See also, BEETLES.

POLYSTOECHOTIDAE. A family of insects in the order Neuroptera. They commonly are known as giant lacewings. See also LACEWINGS, ANTLIONS, AND MANTIDFLIES.

POLYTENE CHROMOSOMES. Polytene chromosomes, also referred to as giant chromosomes, are huge, transversely banded ribbons of DNA. Compared with typical interphase chromosomes, they are longer by one hundred times or more and have diameters thousands of times greater. Only Collembola and Diptera among the hexapods have polytene chromosomes, which were first described in larval midges in 1881, and in Collembola in 1961. These giant chromosomes are formed in tissues that grow by cellular enlargement rather than by an increase in cell number. In contrast to a typical mitotic cycle, the homologues of polytene chromosomes typically remain paired and do not participate in the mitotic cycle of coiling and uncoiling. The sister chromatids remain paired at the end of each replication cycle, and the nuclear membrane and nucleoli remain intact throughout replication. The end product of the replication cycles is a nucleus with a haploid number of chromosomes, each containing up to 2,000 or more parallel strands. In the Collembola, however, the polytene chromosomes typically remain unpaired and the nuclei contain the diploid number.

Polytene chromosomes of Diptera are found in a wide variety of larval, pupal and adult tissues, but they are best developed in tissues with a high level of secretory activity, such as the salivary glands, midgut, fat body, and Malpighian tubules. Larval salivary glands typically yield the finest preparations, as represented by the photographs of the genus *Droso-*

phila, which often are featured in textbooks. Polytene chromosomes are also present to varying degrees in ovarian nurse cells and trichogen cells. In the Collembola, polytene chromosomes are well developed in the salivary gland cells.

The enormous size of the polytene chromosomes and the constancy of the series of light and dark bands, seen most easily in stained preparations, provide a wealth of taxonomic, phylogenetic, and genetic information. The banding patterns of the chromosomes can be photographed or drawn and then mapped by assigning sectional numbers and letters so that every band or puff can be referenced. The banding sequences tend to be species specific, with at least a segment of the total polytene complement of each species having a unique sequence of bands. These unique segments are created by one or more rearrangements, typically inversions, which reorient the sequence of bands. Nonetheless, examples of homosequential species (i.e., species with identical banding sequences) are known among taxa such as drosophilids and black flies. Inverted sequences often appear in the heterozygous condition, forming knots or loops in the chromosomes, as the sequences of each chromosomal homologue attempt to pair. An absence of heterozygous inversions in the presence of two opposite banding sequences indicates a lack of hybridization, providing evidence of reproductive isolation. Polytene chromosomes, consequently, have been used to reveal morphologically similar species (i.e., sibling species) through the absence of hybrids. The evolutionary relationships of insects also can be reconstructed on the basis of uniquely shared banding sequences. Not all Diptera have polytene chromosomes amenable to band-by-band analysis. In many taxa, the polytene chromosomes are under replicated or sticky and fragmented, making analysis of banding patterns difficult.

Peter H. Adler
Clemson University
Clemson, South Carolina, USA

References

Ashburner, M. 1970. Function and structure of polytene chromosomes during insect development. *Advances in Insect Physiology* 7: 1–95.

Beermann, W. (ed.). 1972. Developmental studies on giant chromosomes. *Results and Problems in Cell Differentiation*, Vol. 4. Springer-Verlag, New York, New York. 227 pp.

Fig. 825 Section of giant, polytene chromosome from salivary gland of larval black fly.

POLYTHORIDAE. A family of damselflies (order Odonata). See also, DRAGONFLIES AND DAMSELFLIES.

POLYUNSATURATED FATTY ACIDS. A dietary source of polyunsaturated fatty acids is required by many insects. Linoleic acid and linolenic acid satisfy their need. Lack of these dietary components often results in slow growth and deformed wings in the adult stage.

POMACE FLIES. Members of the family Drosophilidae (order Diptera). See also, FLIES.

POMPILIDAE. A family of wasps (order Hymenoptera). They commonly are known as spider wasps. See also, WASPS, ANTS, BEES, AND SAWFLIES.

POPULARITY OF INSECTS. In terms of numbers of species, insects are the largest group of animals by far. There are approximately one million named species of insects, and most recent estimates suggest there are three to four million more in the world yet to be discovered. Insects also are among the most abundant in terms of sheer numbers. Dr. E.O. Wilson has estimated that there may be as many as 10,000,000,000,000,000,000 (10 quintillion) individual insects in the world.

Because of the great numbers and spectacular appearance of insects, everyone has some familiarity with them. Throughout history, insects have permeated almost every area of human society, a fact that is well documented on the web site *Cultural Entomology Online*. Most people have some opinion, either positive or negative, about insects. Currently, insects and other arthropods are enormously popular. Everywhere one looks there is evidence of this popularity. It is virtually impossible to visit the jewelry departments of large department stores without seeing insect necklaces, pins, and earrings. Also, insects are commonly featured as prints for fabrics used in women's clothes. Butterflies were proclaimed to be a "fashion statement" in a recent (June 14, 2002) article in the *Life* section of *USA Today*. The article stated, "This season, butterflies are fluttering across ankle straps and purses, up-dos, and jean pockets."

While butterflies are certainly the most popular insects featured in fashion, dragonflies, damselflies, bees, and ladybugs are also extremely popular.

Insects also are prevalent in photography and art. Kjell B. Sandved spent 24 years photographing all the letters of the alphabet and the numerals from one to nine and zero from the wings of butterflies. Many of these photographs are now featured in three magnificent posters and a book available at his butterfly alphabet web site. Even the United States Postal Service has recognized the popularity of insects. In October of 1999, the Postal Service issued a sheet of magnificent 33¢ postage stamps that featured an outstanding selection of sixteen species of insects and four species of spiders. The insects and spiders sheet of stamps was ranked fourth on the Postal Service's list of the "Top 20 Commemorative Stamps of All Time".

An area of growing interest is the use of insects as human food. There are several web sites that have links to insect recipes, nutrition facts and food insect festivals. Also, there are several books devoted to the subject. One, *Man Eating Bugs* by Peter Menzel and Faith D'Aluisio, is a fascinating account (with numerous color pictures) of the use of insects as human food around the world.

Numerous television and magazine commercials and movies have used insects as actors. Recent popular animated movies featuring insects include PDI/Dreamworks' *ANTZ* and Pixar/Disney's *A Bug's Life*. Films with outstanding video of living insects include Galatee Films' *Microcosmos* and the BBC's

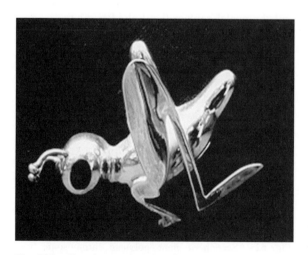

Fig. 826   Grasshopper jewelry pin.

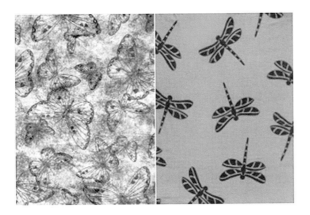

Fig. 827 Fabrics with butterfly and dragonfly designs.

outstanding three part educational video series for television, *Alien Empire*. All of these videos are currently available for purchase from on-line sources. Also, forensic entomology (use of insects as evidence by law enforcement agencies) has become popularized on the *Crime Scene Investigations* television series.

Adding to the popularity of insects is the abundance of excellent insect books for all age groups from young children to adults. For young children there are books that use insects to teach moral values and desired behaviors in addition to those that feature insect identification and behavior. For older children and adults, there are a variety of outstanding field guides to the insects and specialized books that address various specialized topics related to insects. Representative selections of insect books are found in the natural history sections of all of the major bookstores, or more complete lists may be found by using the search engines of on-line booksellers. Also, books as well as many other items (insect videos, posters, t-shirts, puppets, collecting supplies, etc.) of entomological interest are available from entomological supply catalogs, some of which are available on-line.

Fig. 828 Insect and Spider stamps. Stamp Design © 1999. U.S. Postal Service. Reproduced with permission. All rights reserved.

Fig. 829   Front cover of ''Man Eating Bugs.'' Used with permission of Ten Speed Press.

Many societies and organizations have contributed to the popularization of entomology. For children, most state 4-H programs have entomology projects, and the Boy Scouts of America has an ''Insect Study'' merit badge. The Young Entomologists' Society is another excellent choice for children. National and state professional entomological societies have annual meetings, often with special outreach programs to schoolchildren in the localities where meetings are held. Many countries throughout the world have their own entomological societies. In the United States, the Entomological Society of America (ESA) is the leading entomological society. It has a wealth of educational material on its web site and sponsors educational programs for children and teachers at each of its annual meetings. Of special interest is the ESA's web page on frequently asked questions on entomology. The web addresses for ESA, other entomological societies in the United States, and those throughout the world are found in the accompanying table of web addresses.

The success of insect zoos and butterfly pavilions are testimonials to the current popularity of insects. In contrast to most museums that may house millions of dead insects, the zoos and butterfly pavilions display live insects. In the case of the butterfly pavilions, the butterflies are free-flying and often land on the human visitors. Many large cities either have insect zoos or are in the process of building one. At least two new ones are due for completion during 2003. One of these, the Audubon Insectarium in New Orleans bills itself as the largest freestanding museum in the United States devoted to insects and projects an estimated annual visitation of 400,000 and an estimated economic impact for the city of $87,000,000. The larger of the two new facilities is Butterfly Kingdom in Hardeeville, South Carolina. Its 84,000 square foot facility will house the largest butterfly conservatory in North America plus an insect zoo, a nocturnal moth exhibit, a learning center, a research laboratory, and a 3-D giant screen theater.

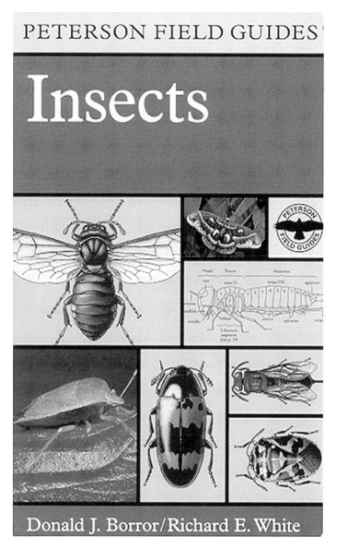

Fig. 830 Front cover of ''Insects'' (Peterson Field Guide Series). Used with permission of Houghton Mifflin Company.

The popularity of butterflies has spawned a lot of interest in creating butterfly gardens in city parks, K-12 schools, and home lawns. Because different butterfly species prefer different species of flowering plants as nectar sources for the adults and host plants for the caterpillars, creating a butterfly garden requires knowledge of the appropriate plants for each geographic area. To meet this need, there are many excellent regional butterfly gardening books. One may find books specific to a region at local bookstores or by searching the web sites of on-line booksellers. Also, there is a wealth of information on the web including a good selection of free downloadable regional butterfly gardening guides on the web site of the North American Butterfly Association (NABA). NABA also sponsors a large number of butterfly counts throughout North America. Dates and localities for the butterfly counts are found on the NABA web site. These butterfly counts are an excellent opportunity to learn the local butterflies. Some excellent national, regional and state butterfly books with color photographs are available to assist with identification of butterflies. On the ''Butterflies of North America'' web site, one may select a state and see a list of links to color photographs and biological accounts of butterflies for that state.

Fig. 831 Roachhill Downs at Purdue University's 'Bug Bowl.'

Releasing butterflies at weddings has become a very popular but somewhat controversial practice. According to the International Butterfly Breeders Association, there are hundreds of commercial butterfly farms in the United States that raise butterflies for schools and for release at special events. Some scientists do not approve of the releases because of the potential harmful effects of the released butterflies (that may not originate from the same area in which they are to be released) mating with local butterflies. There is concern that the progeny resulting from these matings may not be as well adapted to survive. Also, the United States Department of Agriculture is concerned that some of the butterflies may pose a threat as plant pests and has proposed regulations to restrict releases of all but a few species.

Schools and universities have played a major role in educating the public about insects. Elementary school teachers frequently use insects in their teaching. Several major university entomology departments offer entomology courses designed for education majors. Also, entomology courses are offered at many universities as electives or to meet

Fig. 832   Cricket-spitting at Purdue University's 'Bug Bowl.'

Selected entomological web sites

The Entomology Index of Internet Resources	www.iastate.edu/list/
Entomological Society of America	www.entsoc.org
Entomological Societies	www.amnh.org/learn/biodiversity_counts/know_more/ w_entomsoc.htm
World's Entomological Societies	www.sciref.org/links/EntSoc/eslists.htm (case sensitive)
Young Entomologists' Society	members.aol.com/yesbugs/bugclub.html
National 4-H headquarters	www.reeusda.gov/4h
Boy Scouts of America Insect Study Merit Badge	www.usscouts.org/usscouts/mb/mb065.html
Links to University Entomology Departments	www.entsoc.org/education/educ_career/colleges.htm
Frequently Asked Questions on Entomology	www.entsoc.org/education/faqs.htm
Featured Creatures	creatures.ifas.ufl.edu
Best of the Bugs website	pests.ifas.ufl.edu/bestbugs
Insect Zoos, Museums, and Butterfly Gardens in North America	www.entsoc.org/education/links/insect_zoos.htm
Public Butterfly Gardens & Zoos	Butterflywebsite.com/gardens/index.cfm
NABA Butterfly Gardening Guides	www.naba.org/pubs/bgh.html
North American Butterfly Association (NABA)	www.naba.org
Butterflies of North America	www.npwrc.usgs.gov/resource/distr/lepid/bflyusa/ bflyusa.htm
Y.E.S. Minibeast Merchandise Mall	members.aol.com/YESsales/mainmenu.html
Bioquip Products	www.bioquip.com
Carolina Biological Supply Company	www.carolina.com
Butterfly Alphabet Posters & Book	http://www.butterflyalphabet.com
The Cultural Entomology Digest OnLine	www.insects.org/ced/index.html
The Food Insects Newsletter, Inc.	www.hollowtop.com/finl.html/finl.html
The Bay Area Bug Eating Society	www.planetscott.com/babes/frame1.html

general education biology requirements for non-science majors. Some of these courses have enrollments of more than 500 students per semester.

A few university entomology departments conduct insect fairs for the general public. The best of these is Purdue University's "Bug Bowl" organized by Professor Tom Turpin. Over 12,000 people attend Bug Bowl each year. These fairs offer a combination of hands-on educational experiences and entertainment. Two of the most popular events at Bug Bowl are "Roachhill Downs", a miniature race track with preserved cockroaches sitting in the stands, where live cockroaches race and an event that is now sweeping the nation, "cricket-spitting". The current world record for cricket spitting is 32 feet, 1 and $\frac{1}{4}$ inches.

The popularity of insects is certain to continue to grow in the future. As with many other subjects, the world wide web will play a major role. There is already a wealth of information about this fascinating group of animals available online, just a few "clicks" away.

Donald W. Hall
University of Florida
Gainesville, Florida, USA

Selected popular books on insects

Arnett, R. H., and R. L. Jaques, Jr. 1981. *Simon & Schuster's guide to insects.* Simon and Schuster, Inc., New York, New York.

Borror, D. J., and R. E. White. 1970. *Insects.* (Peterson Field Guides). Houghton Mifflin Company, Boston, Massachusetts.

McGavin, G. C. 2000. *Insects spiders and other terrestrial arthropods.* Dorling Kindersley, Inc., New York, New York.

Milne, L., and M. Milne. 1980. *The Audubon Society field guide to North American insects and spiders.* Alfred A. Knopf, New York, New York.

Berenbaum, M. R. 1989. *Ninety-nine gnats, nits, and nibblers.* University of Illinois Press, Urbana, Illinois.

Glassberg, J. 1999. *Butterflies through binoculars: the East.* Oxford University Press, New York, New York.

Glassberg, J. 2001. *Butterflies through binoculars: the West.* Oxford University Press, New York, New York.

Akre, R. D., G. S. Paulson, and E. P. Catts. 1992. *Insects did it 1st.* Ye Galleon Press, Fairfield, Washington.

Wangberg, J. K. 2001. *Six-legged sex − the erotic sex lives of bugs.* Fulcrum Publishing, Golden, Colorado.

Menzel, P., and F. D'Aluisio. 1998. *Man eating bugs − the art and science of eating insects.* Ten Speed Press, Berkeley, California.

POPULATION. A group of individuals of one species that occupy the same area at the same time and generally interbreed.

POPULATION DENSITY. The number of organisms expressed as a unit of space (e.g., insects per plant, per square meter, or per cubic centimeter).

POPULATION DYNAMICS. The changes in population sizes over time, and the factors responsible for these changes.

POPULATION ECOLOGY. The study in time and space of populations, their density and distribution, relative to factors causing changes in the populations.

POPULATION INDEX. A sampling method that attempts to provide an indirect estimate of population density based on an associated product (e.g., frass) or effect (e.g., defoliation).
See also, SAMPLING ARTHROPODS.

POPULATION PYRAMID. A method of illustrating the age structure of a population diagrammatically by placing the youngest members of the population at the base and by stacking successive age classes above it.

POPULATION REGULATION. Maintenance of a relatively consistent population size and density. Some factor causes population density to increase when it is low, and to decrease when it is high.

POPULATION RESILIENCE. The ability of a population to adapt to change in its density or environment.

PORE CANAL. Canals running from the epidermis through the procuticle, and used for transport of waxes, cement, and sometimes other substances to the surface (epicuticle) of the insect. The pore canals do not penetrate the epicuticle, though smaller canals called wax channels do penetrate the epicuticle.
See also, CUTICLE

PORRECT. Labial palps that protrude straight forward (not curved upeward) in a pronounced manner, such as is found in Pyralidae.

POST-EMERGENCE TREATMENT. Treatment of a plant, usually with a pesticide, after the plant has emerged from the soil.

POSTERIOR. The hind region of the body, or referring to the end containing the anus.

POSTOCCIPITAL SUTURE. A suture that marks the presence of an internal ridge, the postoccipital ridge, to which dorsal prothoracic muscles and head muscles are attached.
　　See also, HEAD OF HEXAPODS.

POSTOCULAR AREA. An area of the head behind the eyes.
　　See also, HEAD OF HEXAPODS.

POSTPETIOLE. The second segment of the 'waist' (petiole) of an ant.

POST-PLANT. Reference to treatments applied to a crop after planting.

POSTSCUTELLUM. A small portion of the thoracic notum immediately behind the scutellum.
　　See also, THORAX OF HEXAPODS.

POTAMANTHIDAE. A family of mayflies (order Ephemeroptera). See also, MAYFLIES.

POTAMOCORIDAE. A family of bugs (order Hemiptera). See also, BUGS.

POTATO APHID. See also, POTATO PESTS AND THEIR MANAGEMENT.

POTATO FLEA BEETLE. See also, POTATO PESTS AND THEIR MANAGEMENT.

POTATO LEAFHOPPER. See also, POTATO PESTS AND THEIR MANAGEMENT.

POTATO PESTS AND THEIR MANAGEMENT. Potato is one of the crops that receives the heaviest use of pesticides in the world. This over reliance on insecticides has resulted in the development of pest resistance and the emergence of pests that were previously considered to be of minor economic importance.
　　Potato pest integrated pest management (IPM) programs should incorporate alternatives to chemical control. Pest control decisions using an IPM approach will help to reduce production costs by decreasing pesticide use. The final result of reduced pest control costs will be higher income for growers and reduced health risks for all people involved in crop management.
　　Managing pests successfully, using any method, depends on being able to reliably identify and monitor pest populations at each life stage. In North America, this means principally the Colorado potato beetle, the green peach aphid, and wireworms, which are considered major potato pests in most potato growing areas.

Colorado potato beetle
　　The Colorado potato beetle, *Leptinotarsa decemlineata* (Say) (Coleoptera: Chrysomelidae), is probably the most important pest in all potato-growing areas in the United States and has spread to the rest of the continent, Europe and Asia. Potato yields are reduced and plants sometimes killed by the adult and larval leaf feeding. Larvae cause most of the defoliation, consuming about four times more than the adults. Leaf feeding has the highest effect on yields when it occurs within two weeks of flowering. The few tubers produced by damaged plants are stunted and possibly unmarketable. Potato plants still can tolerate some defoliation without a yield reduction: up to 25% prior to flowering, between 10 and 15% when tubers are beginning to bulk, and up to 20% for the last 3 weeks of the growing season.
　　This beetle was first recognized as a potato pest in 1859. The high reproductive potential and the fact

that this pest has become resistant to almost all the insecticides that have been used against it, make the Colorado potato beetle such an important pest. One possible explanation for the ability to develop resistant to many insecticides is that this beetle evolved on a family of plants (Solanaceae) that contains high concentrations of toxins, and during the evolution process the beetle developed the ability to detoxify ingested toxins.

The adult is about 10 mm long, with yellow rounded and convex wing covers (elytra) marked with 10 black stripes. The eggs are orange-yellow, and found in clusters of about 30 on the underside of leaves. The larvae are 3–12.5 mm long and have slug-like, soft-skinned, brick red to orange, humped bodies with two rows of black spots on each side, six legs and a black head.

Biology. In temperate regions, adults spend the winter buried 10–25 cm deep in the soil. Adults emerge in the spring just as the first volunteer potatoes appear. Potato fields are usually rotated; therefore, beetles emerge and feed in the same field on volunteer potatoes or weed hosts and then fly to find a new host field. Recently emerged beetles either mate close to the overwintering sites or fly to new potato fields. Females are able to retain sperm from the last season's mating, and to produce viable eggs without new copulation in the spring. Beetles are able to fly several kilometers. Colonizing beetles first feed and then oviposit within a week, depending on temperature. Females lay up to 800 eggs over a 4- to 5-week period. The extended egg laying period means that larvae may be present in the field for four to five weeks. In Idaho and the northeastern states, the eggs begin to hatch at the end of May or first week of June and hatch in 4–9 days depending upon temperatures. Young larvae remain close to the egg mass but begin to move throughout the plant as that leaf is consumed. The larvae pass through 4 growth stages in as little as 8–10 days with average temperatures of about 30°C, while it will take longer with lower temperatures. The fourth and final growth stage consumes three times as much as the other three stages together.

The mature larvae drop from the plant and build cells in the soil where they pupate and transform into adults in 1–2 weeks. The new summer generation of adults emerges and lays eggs in that field or an adjacent field, and then migrates to overwinter. At higher elevations, the beetle may have only a single generation or a partial second generation, which feeds only briefly and then migrates to the overwintering sites without reproduction.

Management. There are several effective ways of dealing with this pest, including cultural, biological, physical and chemical practices, several of which work best in combination. Keep in mind that potato beetle control practices in one field may affect nearby fields.

Crop rotations. Crop rotations help in delaying or reducing beetle pressure. Planting cereal grains after potatoes aids in reducing migrations because cereals are poor launch sites for the beetles. Planting new potato fields as far from last year's fields as possible (at least 400 m) will reduce the number of immigrant overwintering beetles in the new field.

Control of volunteers and weeds. Because overwintering adults will need to feed before walking or flying into new fields, controlling volunteer potatoes, and weeds such as nightshades, is important, as they are a preferred and early food source for these emerging adults. This tactic does not provide complete control. Therefore, it is important to scout fields to see whether additional control methods are needed.

Other methods that have proven fairly effective are the use of plastic-lined trenches, propane flamers, and vacuums. Plastic-lined trenches are arranged on the edge of a field and beetles are trapped early in the season when they walk towards the crop and at the end of the season when they move out of the field to overwintering sites.

Chemical control and management of insecticide resistance. We currently have several insecticides that are very effective. However, as mentioned before, this pest has a great ability to develop resistance to almost all different classes of insecticides. In most cases, the choice of an insecticide is based on price, effectiveness, and ease of application. It is well known that the repeated use of the same insecticide or other insecticides from the same chemical class in the same year or in successive years will accelerate the development of resistance. With a little planning and forethought insecticide programs can be developed that will allow usage of older and newly developed insecticides for a prolonged period of time.

Mixtures of insecticides are not generally recommended. Potato beetles can develop tolerance to components of the insecticide mixture and the resultant resistance may be more permanent and more difficult to manage than one developed separately to each of the active ingredients present in the mixture.

Insecticides that are not harmful to beneficial organisms should be used because natural enemies play a major role in the management of resistance. They reduce potato beetle populations regardless of the pest's resistance and act against the selection of resistant populations.

Insecticides should only be used when needed. As mentioned before, potato plants can tolerate some defoliation without a yield reduction. Once populations become damaging, insecticides can be applied either at planting or to the foliage after emergence of the plants. Systemic insecticides applied at planting are effective not only against the beetles, but also in controlling aphids. If the systemic insecticide is no longer protecting the crop when tubers are beginning to bulk (within two weeks of flowering) a foliar insecticide application may be necessary.

It is not always necessary to spray foliar insecticides over the whole field. When crops are rotated, the arriving beetles will concentrates at the edge of the new fields and defoliated plants will be restricted to field edges. Thus, treating only this area may be adequate.

Young larvae are most susceptible to insecticides. Since the older larvae are responsible for 75% of the feeding damage, early treatment will also prevent economic damage. The foliar application will be more effective when 15 to 30% of the beetle eggs have hatched, and will be justified with defoliation from larvae above 10–15%.

Biological control.   Not all the Colorado potato beetle eggs deposited early in the spring will become adults. Several beneficial insects that feed on eggs and young larvae will reduce the number of adults in the first spring generation. Ground beetles, predatory stinkbugs, ladybird beetles, and collops beetles are some of the predators that may reduce potato beetle populations. There also are pathogenic fungi such as *Beauveria bassiana* and *Paecilomyces farinosus* that infect and reduce potato beetle populations. However, the limited knowledge about natural enemies and the reliance on insecticides have limited their usefulness. Natural enemies can be protected by using "biorational" approaches such as microbial insecticides, mineral and vegetable oils, neem, and

fermentation-based products that are toxic only to the Colorado potato beetle. Toxins from strains of the naturally occurring bacterium *Bacillus thuringiensis* (*B.t.*) are commercially available and have good potential for use in combination with natural enemies because of their specificity to potato beetle.

Green peach aphid

The green peach aphid, *Myzus persicae* (Salzer) (Hemiptera: Aphididae), is a European native that occurs throughout the world on a diverse host range of over 875 species of plants. It is the most common and abundant aphids in North America. It is considered one of the most difficult insect species to control due to its high reproductive potential and diverse host range. It seriously damages many crop plants directly by feeding, and also transmits more than 100 viruses to cultivated crops. Aphid numbers may occasionally be high enough to cause damage to potato crops by excessive removal of sap, but main losses occur when it transmits potato leafroll virus (PLRV). The virus causes yield reduction and reduced quality. Green peach aphid is by far the most efficient PLRV vector.

Green peach aphid may be winged or wingless. Wingless forms are yellow-green to pinkish. Winged green peach aphids are pale or bright green with a dark head and thorax. The irregular dark patches on the abdomen are characteristic but not unique. Not all winged aphids in potato fields are green peach aphids. Many species of winged aphids that have developed on other crops or weeds may be present.

Biology.   The life cycle of aphids in general is very unusual and complicated. It includes several body forms, and a sexual and an asexual mode of reproduction. Asexual reproduction occurs during the growing season when females give birth to live females (they do not lay eggs, as it is usually the case with insects) for 10–25 generations. Sexual females produce eggs in the fall.

In the north, green peach aphid overwinters as eggs on the primary host, the peach tree. However, in southern regions and during mild winters aphids can also overwinter on several weeds, such as mustards, nightshade and ground cherries.

When green peach aphid overwinters on peach and nectarine trees, wingless females known as "the stem mothers" hatch from eggs that were laid near the buds in the previous fall, and produce live nymphs without mating. Eggs may hatch in response to warm periods as early as late January or early February. At high elevations egg hatching may not occur until April. The green peach aphid remains on the host tree in the spring until leaves are fully expanded. Then, winged aphids, known as 'spring migrants,' are produced. These spring migrants fly to secondary hosts, which may be weeds, crop plants or ornamentals. Up to 40,000 spring migrants may be produced on one tree.

Although long-distance, wind-aided flights are possible, most winged migrants establish colonies near the winter host if summer host plants are present. They search for suitable hosts by making a series of short random flights. When an acceptable host is located the aphid feeds and deposits as many as 20 nymphs. Then the aphid takes flight and the process is repeated.

Once hosts mature, another winged form is produced, the 'summer migrant.' Summer migrants usually land in greatest numbers on the edges of potato fields. Summer development of populations is strongly influenced by weather. Sudden hot or cold periods, high winds and hard rains or hail can significantly reduce populations. Summer migrants produce nymphs that can complete development in as short as 6 days and they in turn begin producing young 2 or 3 days later. Maximum reproduction occurs at temperatures between 24 and 30°C. Reproduction is sharply reduced at temperatures above 35°C and reproduction slows as temperatures are lowered from the optimum range. Temperatures under the vine canopy, where the aphids are reproducing, are less extreme than in the open. In general, weather conditions that favor high tuber yields also favor aphid reproduction. Each aphid is capable of producing 30–80 nymphs over a period of 10–20 days. They tend to deposit nymphs on a series of plants instead of putting them all on one plant. Where population peaks result in extreme crowding, winged aphids develop and move out of the field, thereby greatly reducing numbers. Declines in populations also may be associated with periods of extremely high temperatures and with a decline in potato vine condition.

The cold weather in the fall and/or the lack of suitable hosts triggers the production of winged aphids, which will be the ones returning to the primary host. These are the 'fall migrants.' One type of winged migrant aphid deposits nymphs on leaves of peach trees. These nymphs develop into wingless sexual females. The other type of winged migrant is the sexual male, which mates with the wingless female. Each female then deposits 5–15 eggs on or near the axillary buds. Total number of eggs per tree is extremely variable but may exceed 10,000 in mild falls when aphid activity continues into late November.

Management. A successful integrated pest management program should include methods aimed to breaking the life cycle of this aphid, such as applying insecticides to control the aphid, and eliminating or treating overwintering and secondary hosts.

Cultural control. The number of green peach aphids present in the spring to infest crops depends upon winter survival. The common means of overwintering is on the winter host in the egg stage. Peach trees are the most common winter hosts, although apricots and other species of *Prunus* are infested on rare occasions. Fields near commercial peach orchards, or urban areas with backyard and abandoned peach trees, usually have higher populations than those in isolated areas. If spring and summer weather are favorable for aphid development a single peach tree can potentially produce enough winged aphids to initiate economic infestations on at least 500 acres of potatoes. Removing and replacing peach and apricot trees and spraying insecticides on commercial peach orchards are valid attempts to prevent aphid buildup.

Seed production areas at high elevations and commercial production areas with severe winters are usually too cold for survival of primary host plants. However, it is not unusual to observe potato fields in these areas close to towns with high aphid populations and high PLRV incidence. This is because many of the bedding plants that homeowners buy in local greenhouses are infested with green peach aphids. In Idaho for example, a 1990 survey of bedding plants that were introduced from surrounding states and were commercially available in all seed production areas of eastern Idaho, revealed that 37% of these plants were infested with green peach aphids and, therefore, home gardens represented a major source of aphid infestations in seed production areas in that state. Aphids moving directly from home gardens to potato plantings often transport viruses because home garden potato plants often have a high rate of disease infection.

Elimination of aphids on bedding plants is a very important part in the success of any integrated pest management program of green peach aphid and

PLRV. Campaigns oriented to educate home gardeners about the importance of buying and planting bedding plants with no aphids are needed.

Many winged aphids from peach trees and other sources appear before crops are available for colonization. The ground cover of overwintering orchards usually includes suitable hosts that may become heavily infested. Early infestations commonly occur on a number of weeds including species of mustards, nightshade and ground cherries. Winged forms produced on these weeds later infest crop plants, including potatoes, and high numbers may appear during a short period when one or more species of weeds dry up or mature.

Where the winters are mild, aphid colonies can survive the winter outside on plants that maintain green growth. In areas where minimum temperatures are too severe for plant growth, plants near canals, springs, or adjacent to heated buildings may be infested because of the higher minimum temperatures at these locations. This kind of overwintering occurs primarily in low elevation potato-growing areas or protected places, and except in unusually mild winters or local situations, is less important than the other two aphid sources discussed.

General insecticide application guidelines based upon aphid numbers are used in some areas. When making counts, sample several areas of a field because aphid numbers are usually highest on field margins, including weeds. Nightshades are one of the preferred weeds.

When scouting, it is important to keep in mind that green peach aphid prefers to infest lower portions of potato plants. After periods of cool cloudy weather tops of plants may be infested. Colonies will also develop on upper portions of plants where crowding occurs. Aphid reproduction is favored by dense vigorous vine growth.

Early aphid migrants not controlled on time will form colonies. Therefore, detection of few aphids and application of insecticides on time will prevent the formation of these colonies. It is important to inspect the fields every 3–4 days and watch for the presence of surviving, wingless aphids.

Biological control. Predators, parasitoids and pathogens affect aphid populations and may keep aphids below economic levels in particular situations. The sudden decline of aphid population late in the season may be associated with several factors of which the action of predators is often dominant. However, applications of insecticides and fungicides against other potato pests reduce or even eliminate populations of natural enemies, allowing aphid populations to increase rapidly. For this reason, high populations of aphids can sometime be observed after the application of an insecticide against Colorado potato beetle and potato leafhopper.

Chemical control and management of insecticide resistance. The green peach aphid is difficult to control because of the high reproduction capacity, and because it has developed resistance to at least 69 insecticides representing all major classes. However, it is critical to control aphid in production areas with PLRV susceptible cultivars such as Russet Burbank. Most of the principles of chemical control of Colorado potato beetle explained above apply also to green peach aphid.

Alternating insecticide use among the major insecticide groups reduces the development of resistance. Some systemics can give adequate aphid control and also reduce Colorado potato beetle, wireworm, or nematode numbers. Cost of the insecticide may be an important consideration.

The effectiveness of any insecticide used intensely in an area will be reduced significantly after a few years. Preventing or reducing speed with which resistance develops depends mainly on reducing insecticide use to the bare minimum necessary for economic crop production. It is important to note that not all insecticides that kill aphids prevent virus transmission.

Timing the application of an insecticide is as critical as selecting it. Applying systemic insecticides to the soil at the time of planting effectively controls aphids. At high elevations where mid- to late-season pressure from winged aphids is light, these applications may provide season-long protection. At lower elevations, one or more foliar applications of insecticide may be necessary after about midseason. Application of foliar insecticides should begin when one to three wingless aphids per 100 leaves are detected. This is a very low threshold for most scouting programs to detect with confidence.

Wireworms

Wireworms are becoming increasingly important in several potato-growing areas in North America. Two of the possible reasons for this are the increased rotation with grasses for the cattle industry, and the removal of insecticides with long residual activity in the soil. The adults, known as click beetles, produce little or no damage and the larval stage causes the damage to seedlings and underground parts of annual crops by feeding on seeds and tunneling potato seeds and tubers. There are 885 species of wireworms (family Elateridae) in the United States. Three species of wireworms commonly damage potatoes in western North America. The sugarbeet wireworm, *Limonius californicus* (Mannerheim), and the Pacific Coast wireworm, *L. canus* LeConte, are found in soils that have been under irrigation for three or more years. The Great Basin wireworm, *Ctenicera pruinina* (Horn), infests soils previously farmed without irrigation, in pasture, or soils recently brought under cultivation. Although crop losses from wireworms are only sporadic, these are substantial (5 to 25%) in some places.

Wireworm larvae are about 2.5 cm long when fully mature, hard-bodied, slender, cylindrical, glossy, small-legged, and yellow-to-light brown. They feed on potato seed pieces and underground stems during the spring. This early feeding opens the seed pieces and stems to rotting organisms, which can result in poor or weak stands of potatoes. Wireworms also burrow into developing tubers. The holes look as if they were made by stabbing the tuber with a nail and usually are lined with potato skin.

Biology. The life cycle of our most common wireworms requires 3-4 years under favorable conditions. Wireworms spend the winter in the soil either as partially grown larvae or as new adults in overwintering cells. Adults work their way to the soil surface in the spring when soil temperatures reach 13°C or above. These adults require little or no food and cause no economic damage. The female mates soon after emerging from the soil, then burrows back into the soil and lays eggs at depths of 2 to 20 cm in several locations. The sugarbeet wireworm prefers vegetated areas for oviposition. Infestations are often spotty because oviposition is not uniform and some localities are more favorable for larval development than others.

Wireworm larvae cause the most severe feeding damage during their 2nd and 3rd years. In the spring, when soil temperatures reach 13°C or above, the larvae move toward the soil surface from depths of 6 to 24 inches where they have spent the winter. When soil surface temperatures reach 27°C or higher, they move downward again. In irrigated fields with complete foliage cover, this higher temperature level may not be reached. During the 3rd or 4th season, mature larvae transform to fragile pupae in earthen cells. In 3 to 4 weeks the pupae change to adults, which remain in the earthen cells until the following spring. Wireworms in all stages may be present in the soil during any growing season.

Management. Because of the similarities in the biology of wireworms, the same management approaches apply to all the species. USDA standards for U.S. No. 1, U.S. Commercial, and U.S. No. 2 potatoes allow only 6% external defects. These includes soil or other foreign matter, sunburn, greening, growth cracks, air cracks, scab, rhizoctonia and mechanical damage, as well as insect damage. If allowance is made for defects other than wireworm damage, only to a small percentage of wireworm injury is allowable.

Detecting wireworm infestations and determining size of wireworm populations is not easy. Baiting gives a very poor estimate of population size but is a quick way to determine whether wireworms are present. Baits have to be buried in the ground one month before planting to determine the need for insecticide treatment. If wireworms are found in baits, soil sampling can be used to estimate the population density. An understanding of wireworm larval movement in soil is needed to design an effective sampling method. There are no reliable economic thresholds for wireworms.

Carrots, corn, wheat, and coarse-ground whole-wheat flour buried about 7 cm in the soil are good baits. Wrap a few grams of flour in a scrap of nylon mesh with the tail end of the mesh protruding from the soil. Randomly place these baits or carrots in a field. Mark the bait locations clearly. The more bait locations used, the greater the chance of discovering an infestation. After 2 to 3 days, dig up the bait and check for wireworms. Baits are not effective in soils that are very dry, wet, or cold, or if excessive organic residue is present. Covering bait stations with dark plastic will allow sampling earlier in the season when soil temperatures are cooler.

Most of the insecticides used to control wireworms are relatively old. Wireworms can be suppressed by broadcast or band treatments, by fumigation, or with seed treatments. Usually, controlling wireworms in one crop of a 2 to 4 year rotation will reduce wireworm damage in the other crops. For broadcast treatments, apply granules or emulsifiable concentrates evenly over the soil and incorporate immediately. Depth of incorporation varies depending upon the insecticide selected. Granular insecticides may be used as band treatments at planting time. These should be applied in narrow bands 7 to 10 cm below the seed piece in the seed piece furrow at planting time. Fumigants may be used to control high wireworm populations but a combination of broadcast and band treatments may be more economical to use, depending on the pest complex. Seed treatment insecticides used to control Colorado potato beetle and green peach aphid have also proven effective at reducing wireworm damage. Keep in mind that even the best insecticides will not kill all the wireworms and a small percent of a large population could still cause economic damage.

Certain cultural control practices can also be effective. One practice is to avoid rotations that include clovers and grasses. Because soil dryness can kill many wireworms in an infested field, fallowing a field will reduce wireworm numbers, but the control achieved must be weighed against the income lost from missing a crop year. It is important to keep in mind that when soils dry out in potato fields, wireworms may seek moisture from tubers, therefore increasing the wireworm damage. Plowing a dry field during the first 10 days of August can break up the pupal cases. In fields where populations have been reduced, potatoes, a susceptible crop, should be planted the first year in rotation followed by less susceptible crops such as sugarbeets, beans or corn in the ensuing years. If wireworm populations are very high in a certain field, perhaps potatoes should not be planted in that field. Avoid planting potatoes in fields that have had several successive years of cereal grains and/or corn.

Known natural enemies of wireworms include birds, carabid and staphylinid beetles, entomopathogenic nematodes, and pathogenic fungi such as *Beauveria* sp. and *Metarhizium* sp. However, there is not much information on the real effect of these natural enemies.

Secondary pests in alphabetical order
Blister beetles. Four species of blister beetles commonly damage potato: the spotted blister beetle, *Epicauta maculata* (Say), the ash-gray blister beetle,

Epicauta fabricii (LeConte), the Nuttall blister beetle, *Lytta nutalli* Say, and the punctured blister beetle, *Epicauta puncticollis* Mannerheim. They are elongate (1.5 to 3 cm) with conspicuous heads and necks. The wing covers are soft and do not completely cover the tip of the abdomen. The beetles cluster on the tips of the plants causing leaf ragging and stunted plants. Severe damage, however, is not common. The adults first appear in the summer and live about 45 days. They are usually abundant only in areas adjacent to rangeland where the larval stages are predatory on grasshopper eggs.

Check field edges in years of heavy grasshopper infestations. The beetles are strong flyers and often fly from an area before damage is detected and controls can be applied. If beetles remain in the field and continue to defoliate field edges, border sprays will eventually alleviate the problem. If defoliation remains below 10 to 15%, controls are probably not needed.

Cutworms and armyworms. Cutworms are soil dwelling caterpillars having a smooth appearance, 3 pairs of legs and 5 pairs of prolegs. Some species may be up to 5 cm long when mature. The black cutworm, *Agrotis ipsilon* (Hufnagel) (gray to black upper half and distinct greasy appearance), variegated cutworm, *Peridroma saucia* (Hübner) (top line of small, pale spots more distinct on front portion), spotted cutworm, *Xestia* spp. (pairs of black oblique marks on top of last four segments) army cutworm, *Euxoa auxiliaris* (Grote) (body gray with darker top-lateral and spiracle stripes), and the red-backed cutworm, *Euxoa ochrogaster* (Guenée) (top often distinctly reddish bordered with dark bands), feed at night. During the day they can be found under clods of soil or in cracks in the ground near injured plants. The western yellowstriped armyworm, *Spodoptera praefica* (Grote), feeds during the day and like the army cutworm may migrate in mass into potato fields from adjacent crops. Cutworms either cut off stems at or below ground level or strip the foliage during the growing season. They also feed on tubers that are exposed on the surface or accessible through cracks in the soil.

Cutworms spend the winter as partly grown larvae or pupae in the soil. One to several generations occur per season depending upon which cutworm is involved. The adults are dusky-brown to gray miller moths and are commonly observed flying around lights during the warmer seasons.

Control programs are aimed only at seriously damaging infestations because natural enemies generally hold the populations in check. Some defoliation from cutworms can be tolerated. Keeping defoliation between 10 to 15% will generally prevent yield loss. Weed control in previous crops and along field edges also aids in reducing cutworm damage.

Flea beetles. Adult western potato flea beetles, *Epitrix subcrinita* LeConte, and potato flea beetle, *Epitrix cucumeris* (Harris), seldom cause damage severe enough to warrant control but extensive leaf feeding by adults may be an indication of later tuber infestation by larvae. Injury on the surface of the potato tuber consists of rough, winding trails up to 2 mm wide and of varying length. Internal tuber injury consists of shallow, narrow brown subsurface feeding tunnels. These tunnels occur singly or in groups, and are about 1 mm in diameter and up to 6 mm deep. Fungi often fill the tunnels. When potatoes are processed, these injuries must be removed by deep peeling to prevent discolored products.

The adults are about 2 mm in length and metallic greenish-black in color. Adults are active in the spring and feed in weeds until the potato plants emerge. Small, round holes in potato leaves are indicative of adult feeding. Females scatter their eggs in the soil at the base of the potato plants. The eggs hatch in 10 days and the tiny whitish larvae feed on underground stems, roots and tubers for 3 to 4 weeks. There are 1 or 2 generations a year depending on location. Adult western potato flea beetles hibernate under leaves, grass or trash, on margins of fields, along ditch banks and other protected places.

The tuber flea beetle, *Epitrix tuberis* Gentner, occurs in the major potato growing areas of the Northwest except southern Idaho and eastern Oregon. Extensive leaf feeding by adults will cause defoliation of the plants and reduce tuber growth.

Adults chew small holes in leaves, and extensive feeding causes a sieve-like appearance. Severely damaged leaves turn brown and die. Flea beetles can transmit potato diseases, such as spindle tuber and brown rot, and the leaf wounds may allow entry of air or waterborne disease organisms. The appearance and life cycle of the tuber flea beetle are very similar to those of the western potato flea beetle.

Larval damage is much more severe because the larval tunnel goes up to 12 mm directly into the tuber whereas the western potato flea beetle burrows under the peel and seldom penetrates over 6 mm. Some lots of potatoes may be unsuitable for processing when damage by the tuber flea beetle is extensive.

Management. Soil applications of systemic insecticides and foliar applications of insecticides for Colorado potato beetle and green peach aphid control hold flea beetle populations to sub-economic levels.

Garden symphylan

The garden symphylan, *Scutigerella immaculata* (Newport), is not a widespread pest, but it can limit potato production in some localities. Symphylan feeding on the root hairs and rootlets may stunt plant growth before tuber formation. Damage to developing tubers consists of tiny holes in the skin with an undercut cavity lined with hard, dark, corky tissue around each point of injury. Damaged tubers are unmarketable.

Garden symphylans are not insects but are a more primitive centipede-like animal. The adults are white and live below the surface in loose soil where they appear to constantly run in and out among the particles. Symphylans move rapidly away from light, so you must look quickly after exposing the tubers or soil to see them.

Symphylans lay their eggs in the spring or early summer in cavities in the soil. The eggs hatch in 1 to 3 weeks. Under favorable conditions, a new brood develops in 60 days and the adults may live for several years. Optimum temperatures for activity of symphylans are from 10 to 21°C. They readily move up and down in the soil to stay within this range.

Management. Control measures must be very thorough if root crops are to be grown in symphylan-infested soils. Fall fumigation of infested areas can be effective. Insecticides should be broadcast in the spring as close to planting as possible and thorough coverage is essential. Carefully watch the field history for symphylan damage. If damage has not occurred in other crops, damage should be minimal to potatoes.

Grasshoppers

Grasshoppers are pests of potatoes only during years when they migrate out of uncultivated areas. Usually their populations are small and their damage is inconsequential. During outbreak years they can defoliate potatoes and transmit viruses, causing spindle tuber and unmottled curly dwarf.

The several species of grasshoppers that cause the most damage to potatoes are the migratory grasshopper, *Melanoplus sanguinipes* (Fabricius), clearwinged grasshopper, *Camnula pellucida* (Scudder), and the redlegged grasshopper, *Melanoplus femurrubrum* (DeGeer).

Other species may cause local, periodic problems. Grasshoppers lay their eggs in inch-long pods, each containing 10 to 75 eggs, deposited slightly below the surface of the soil in late summer or fall. Each female may lay from 8 to 20 pods.

Hard uncultivated ground is preferred for ovipositing although eggs are sometimes found on the edges of cultivated fields, along ditch banks, in pastures and hay fields. The eggs hatch in the spring depending upon the weather conditions and grasshopper species. The nymphs resemble the adults, but are smaller and without wings. There is one generation per year and the nymphs become mature in summer or early fall.

Management. Control programs need to be initiated only when problems develop. In outbreak years, area-wide programs are more effective than field-by-field treatment for grasshoppers. Also, in outbreak years, watch for blister beetles that may move into the field edge and cause local defoliation.

Leafhoppers

The potato leafhopper, *Empoasca fabae* (Harris), is a North American species considered to be one of the most destructive potato pests in northeastern and midwestern United States. In the West, however, the intermountain potato leafhopper, *Empoasca filamenta* Delong, is important. Nymphs and adults of leafhoppers feed on the undersurface of potato leaves and cause a speckled or white stippled appearance on the lower leaves. Adults are wedge-shaped, green insects with white markings and are about 4 mm long. Nymphs are similar in color but are smaller and lack wings. The intermountain leafhopper does not cause 'hopper burn' or leaf scorching to potatoes like the eastern potato leafhopper. The adults pass the winter in grass and weeds along the field margins and in other areas where they have at least one generation before they move to potato fields.

Management. Control measures specifically for the leafhopper in the West are rarely warranted. Soil applications of systemic insecticides for other pests effectively control the intermountain potato leafhopper. In the East, foliar applications of insecticides are warranted when potato leafhopper is abundant.

Witches' broom and leafhoppers. There are several diseases produced by phytoplasmas in potato. Leafhopper species in the family Cicadellidae have been implied in the transmission of phytoplasmas. Phytoplasmas are pathogens transmitted in persistent form; therefore require long periods of acquisition and an incubation period.

One of these diseases, the disease known as witches' broom, has been reported occurring sporadically in some seed production areas. This disease produces a severe halting of plant growth due to the shortening of stems, and also induces marginal chlorosis of the leaves. The plants do not produce tubers, or only some small ones with enlarged buds.

The control of the diseases caused by phytoplasmas depends exclusively on the use of seed free of these diseases. For this reason, seed must be produced in areas that are known to be free of the vector, and all plants showing some of the described symptoms must be eliminated. Tubers showing proliferation of buds must be eliminated. The control of vectors is not a practical measure and perhaps it only can be used in conditions where the vector remains in the field or forms colonies on the crop.

Leather jackets

Severe damage by the leather jacket, *Tipula dorsimacula* (Walker), may occur in fields planted with potatoes following spring plowing of alfalfa or in low, moist, weedy areas in the field. Larvae feed on tubers, causing round punctures varying from shallow depressions to inch-deep holes.

Leather jackets overwinter in the soil as mature or nearly mature larvae. Adults emerge in the spring and deposit eggs in the vicinity of plant refuse. The larvae initially feed on the decomposing plant tissue in the soil but later transfer to developing tubers.

Mature larvae are about 3–4 cm long, gray to gray-brown, and have characteristic fleshy anal projections. Their skin resembles leather, giving rise to the common name leather jacket. The adult fly is about 2.5 cm long with long, fragile legs that may drop off when the insect is handled. Adults look like giant mosquitoes.

Management. Control consists of avoiding spring incorporation of alfalfa green manure, weed control, and water management to prevent water-soaked areas.

Loopers

The most common loopers found in potato fields are the cabbage looper, *Trichoplusia ni* (Hübner) and the alfalfa looper, *Autographa californica* (Speyer).

Damage is caused by the greenish, white-striped larvae which may be 3 cm long when mature. They differ from cutworms in that they only have three pairs of prolegs. The middle of the larva is characteristically humped up when the insect rests or moves and for this reason they are often called measuring worms. Defoliation usually starts in the middle of the plant. The adult is a gray-brown miller moth that looks like a cutworm adult. There are 2 to 3 generations per year.

Management. Loopers seldom become a serious pest of potatoes even though they may build up high numbers. Damage usually occurs just after the vines have gone into senescence. Loopers often are found with cutworms and are blamed for the cutworm damage because the cutworms are hidden during the day. As long as defoliation remains below 10–15%, control measures are seldom warranted. Foliar sprays applied for Colorado potato beetle will usually control loopers as well.

Lygus bugs

Lygus bugs, *Lygus elisus* Van Duzee and *Lygus hesperus* Knight, are general feeders found on most plants and trees. Damage by lygus is the result of their sucking sap from buds and leaves. Lygus bugs inject a toxin during feeding which kills the area fed upon or causes distorted growth.

Immature lygus bugs (nymphs) are smooth glossy green insects that are similar in size to aphids, but move rapidly when disturbed. Several overlapping generations occur each year that require about 6 weeks each.

Adult lygus are 6 to 7 mm long, green to brown bugs with a yellow triangle on the back. They hibernate in debris in fields or field margins. The insects are strong flyers and move from field to field. They usually move into a potato field just after an adjoining field is harvested or has matured. Damage is most severe on field margins.

Management. Control is seldom necessary because lygus bugs are a sporadic pest. Damage is not often noticed until the lygus bugs have left the field.

Potato aphid

The potato aphid, *Macrosiphum euphorbiae* (Thomas), occurs in most potato growing areas, but infestations are usually not economically important for crops grown for table stock. The aphid is an efficient vector of potato virus Y and PLRV and is therefore important to seed potato production.

The aphid overwinters in the egg stage on rose bushes. Infestations develop on this host in the spring and winged forms are produced after several generations. These fly to various summer hosts that include tomatoes, ground cherry and nightshade in addition to potatoes. Nightshade weeds are apparently the preferred host.

Infestations of potato aphids usually develop on the upper parts of potato plants. This is in contrast to the green peach aphid, which usually occurs on the lower leaves. The potato aphid is green and more elongate and larger (3 to 4 mm) than the green peach aphid. The cauda (tip of the abdomen) is long, extending beyond the tips of the cornicles. The cauda of the green peach aphid is less conspicuous and extends about to the tips of the cornicles. The frontal tubercles of both species are prominent but they are rounded in the potato aphid and angular, appearing boxlike, in the green peach aphid. When exposed to sunlight, potato aphids rapidly move to the opposite side of the leaf while green peach aphid will not move.

Management. Plant certified seed to avoid transmission of Potato Virus Y or plant a variety not susceptible to PVY. Other control measures are seldom warranted. The same management recommendation to reduce virus spread by green peach aphid apply also to this aphid.

Thrips

Two species of thrips are associated with foliar damage to potatoes. These species are the onion thrips, *Thrips tabaci* Lindeman, and the western flower thrips, *Frankliniella occidentalis* (Pergande). Thrips cause damage by severely scarring the undersides of leaves and 'silvering' the tops of leaves.

Extensive damage causes the leaves to become dry and drop. Defoliated plants never recover.

Damage by thrips is usually restricted to the outside 3 to 5 rows adjoining wasteland, grain or dry pastures. Occasionally small fields are totally infested. Thrips move into potato fields when alfalfa is harvested or as grass or grain hosts dry during June or July. Similar damage can be caused by windburn, blown sand or by leaves being hit directly by water from sprinklers.

Thrips are tiny, yellow, brown, or white (nymphs) insects that feed at night or on cloudy days. They rasp and puncture the leaf tissue with their saber-like mouthparts and swallow the sap together with bits of leaf tissue. During the day they are found in cracks in the soil or along leaf veins on the plants.

Management. Thrips are generally kept in check by predators and seldom become a problem. When damage occurs, the thrips population has usually declined before the cause is discovered. If thrips remain in potatoes and halt growth, control measures are warranted.

Twospotted spider mite

Twospotted spider mites, *Tetranychus urticae* Koch, develop some years in potato fields. Problems usually occur downwind from infested bean, corn, alfalfa or clover seed fields or along dusty roads. Spider mites damage plants by puncturing the leaf tissue with their mouthparts to extract the plant juices. Injured cells and those surrounding the injury die, causing loss of chlorophyll. The injury first looks like stippling or small blotching, turning yellowish, then brown. These injury blotches come together causing the leaf to be brittle and brown. In severe infestations brown areas can progress rapidly across the field.

Adult mites are one mm long, yellow, with a dark spot on either side of the body. Nymphs are similar but smaller. Eggs are clear, round spheres found in the feeding areas.

Spider mites overwinter as adults in the soil or in debris in fields, fencerows or field margins. The adult female emerges and lays eggs on the undersides of plant foliage in late spring. A female can lay 20 eggs a day with a total of 300 eggs during her lifetime. The eggs hatch in 3 to 5 days. During hot weather the young develop to adults in 7 to 9 days. The female spins a fine web over the leaf, which apparently protects the eggs and mites from rain and predators. In severe infestations the leaves are tied together with dirty webbing.

When populations become severely crowded, mites climb to the top of a plant or post and secrete a web strand. Some mites are then carried by the wind. This is why sudden infestations commonly develop downwind from previously infested fields.

Management. Sprinkler irrigation washes the foliage, breaking webs and dislodging the mites, thereby reducing populations. There are many predators that attack spider mites, but because insecticides kill many of these, mite outbreaks often follow treatments for aphids or other foliar pests. Since potatoes are usually grown under optimum conditions and spider mites prefer stressed plants, serious problems are not common.

White grubs

The two species of white grubs that frequently damage potatoes are the carrot beetle, *Bothynus gibbosus* (DeGeer), and the ten-lined June beetle, *Polyphylla decemlineata* (Say). Both species are more abundant in sandy soils where grass sod or large quantities of organic matter, such as manure, have been plowed into the soil before potatoes are planted.

The larvae are 2.5 to 3 cm long, C-shaped, dirty white in color with a glossy smooth skin, brown head and 6 prominent legs. Their large abdomen is transparent, allowing the body contents to be seen through the skin. This stage attacks the tubers, causing feeding cavities in the potato which are from 8 to 12 mm in diameter, rough, irregularly shaped and wider than deep. In severe infestations more than half of the tuber may be consumed.

The carrot beetle has an annual life cycle while the ten-lined June beetle spends 2 to 3 years as a grub. Adults of both beetles are awkward flyers. During May and June they feed on leaves of trees at night, and are attracted to lights, thus the name June beetles.

Management. Control of white grubs is difficult because they are found in soils with high organic content, which tend to inactivate insecticides. Currently no insecticides are registered on potatoes for white grubs. But, wireworm materials have been somewhat effective in controlling white grubs. Good weed control may also help reduce grub damage.

Juan M. Alvarez
University of Idaho
Aberdeen, Idaho, USA

Insecticides available for use against potato pests

Class	Commercial name	Common name	Site / Mode of action	Application method	Pest controlled
Biological	Raven	*Bacillus thuringiensis* var. kurstaki	Stomach poison	Foliar	CPB
	Novador, M-Track	*Bacillus thuringiensis* var. tenebrionis	Stomach poison	Foliar	CPB
	Colorado Potato Beetle Beater (Bonide)	*Bacillus thuringiensis* var. san diego	Stomach poison	Foliar	CPB
	Agri-Mek, Avid	Abamectin	Neurotoxin, GABA inhibitor	Foliar	CPB
	Success	Spinosad	Gamma receptor (neurotoxin)	Foliar	CPB
Botanical	Azatin XL Plus BioNeem Margosan-O Neemix	Azadirachtin	Interference with molting, repellent	Foliar	CPB
	Rotenone/Pyrethrin Spray (Bonide)	Rotenone	Respiratory enzyme inhibitors of fish and insects, not mammals	Foliar	CPB
Carbamate	Temik 15G	Aldicarb	Central Nervous System / acetylcholinesterase inhibitor	At planting	Aphids, CPB
	Furadan 4F	Carbofuran	Central Nervous System / acetylcholinesterase inhibitor	Foliar and at planting	Aphids, CPB, wireworms, fleabeetles
	Sevin	Carbaryl	Central Nervous System / acetylcholinesterase inhibitor	Foliar	CPB, leafhoppers, fleabeetles, cutworms, armyworms
	Vydate	Oxamyl	Central Nervous System / acetylcholinesterase inhibitor	In-furrow at planting or foliar	Aphids, CPB, fleabeetles
Chloronicotinyl	Admire 2F	Imidacloprid	Central Nervous System / neurotoxin	In-furrow at planting or seed treatment	Aphids, CPB, leafhoppers, fleabeetles
	Genesis	Imidacloprid	Central Nervous System / neurotoxin	Seed treatment	CPB
	Gaucho-MZ	Imidacloprid	Central Nervous System / neurotoxin	Seed treatment	Aphids, CPB
	Provado	Imidacloprid	Central Nervous System / neurotoxin	Foliar	Aphids, CPB
Chloronicotinyl; pyrethroid	Leverage	Imidacloprid; cyfluthrin	Central Nervous System / interferes with the nicotine type acetylcholine receptor; sodium channel	Foliar	Aphids, CPB, plant bugs, flea beetles, cutworms, loopers, leafhoppers,
Cyclodiene	Thiodan 3EC, and 50WP, Phaser 3EC and 50WSB, Endosulfan 3EC and 50WSB	Endosulfan	Central Nervous System / sodium and potassium balance in neurons	Foliar and chemigation	Aphids, CPB, fleabeetles

(*continued*)

(Continued)

Class	Commercial name	Common name	Site / Mode of action	Application method	Pest controlled
Inorganic	Kryocide, Cryolite	Sodium aluminofluoride	Inhibits enzymes with iron, calcium or magnesium centers	Foliar	CPB
Organochlorine	Methoxychlor	Methoxychlor	Central nervous system depression	Foliar	CPB
Organophosphate	Guthion	Azinphos-methyl	Central Nervous System / acetylcholinesterase inhibitor	Foliar	CPB, Leafhoppers, fleabeetles
	Diazinon	Diazinon	Central Nervous System / acetylcholinesterase inhibitor	Broadcast preplant	Wireworms, fleabeetles
	Di-Syston 15%G	Disulfoton	Central Nervous System / acetylcholinesterase inhibitor	In-furrow at planting	Aphids, CPB
	Di-Syston 8EC	Disulfoton	Central Nervous System / acetylcholinesterase inhibitor	Foliar, may be applied by chemigation	Aphids, CPB
	Lorsban	Chlorpyrifos	Central Nervous System / cholinesterase inhibitor	At planting to postemergence	wireworms
	Monitor	Methamidophos	Central Nervous System / acetylcholinesterase inhibitor	Foliar	Aphids, CPB, Fleabeetles, cutworms, armyworms
	Penncap-M	Methyl parathion	Central Nervous System / acetylcholinesterase inhibitor	Foliar	CPB
	Thimet 15G and 20G, Phorate 20G	Phorate	Central Nervous System / acetylcholinesterase inhibitor	In-furrow at planting or side dress at hilling	Aphids, CPB, wireworms, fleabeetles
	Imidan 70-WSB	Phosmet	Central Nervous System / acetylcholinesterase inhibitor	Foliar	CPB, fleabeetles
Pyridine Azomethine	Fulfill	Pymetrozine	Central Nervous System / interferes with the nicotinic acetylcholine receptor Anti-feeding	Foliar	Aphids
Pyrethroid	Baythroid 2	Cyfluthrin	Central Nervous System / axonic poison, sodium channel disrupter	Foliar	CPB
	Ambush, Pounce	Permethrin	Central Nervous System / axonic poison, sodium channel disrupter	Foliar	Aphids, CPB, leafhoppers, cutworms, armyworms
	Asana XL	Esfenvalerate	Central Nervous System / axonic poison, sodium channel disrupter	Foliar	Aphids, CPB, fleabeetles, cutworms, armyworms
Second-generation neonicotinoids	Actara	Thiamethoxam	Central Nervous System / interferes with the nicotinic acetylcholine receptor Anti-feeding	Foliar	Aphids, CPB, leafhoppers, wireworms
	Platinum	Thiamethoxam	Central Nervous System / interferes with the nicotinic acetylcholine receptor Anti-feeding	Soil applied	Aphids, CPB, leafhoppers, wireworms

Note: For specific rates and time of application for a given product refer to manufacturer recommendations.

References

Casagrande, R. A. 1987. The Colorado potato beetle: 125 years of mismanagement. *Bulletin of the Entomological Society of America* 18: 142–150.

Rowe, R. C. (ed.). 1993. Potato health management. APS Press, St. Paul, Minnesota. 103–115.

Harcourt, D. G. 1971. Population dynamics of *Leptinotarsa decemlineata* (Say) in eastern Ontario. III. Major population processes. *Canadian Entomologist* 103: 1049–1061.

MacGillivray, M. E. 1979. Aphids infesting potatoes in Canada: a field guide. Minister of Supply and Services, Canada. 23 pp.

Zehnder, G. W., M. L. Powelson, R. K. Jansson, and K. V. Raman (eds.). 1997. *Advances in potato pest biology and management.* APS Press, St. Paul, Minnesota. 655 pp.

POTTER, CHARLES. Charles Potter was born in England on January 3, 1907, and rose to become one of the most important workers in chemical insecticides. Beginning in the 1930s with his work on stored product pests at Imperial College, London, Potter made critical observations showing that pyrethrum formulations could be residual under the conditions of a darkened warehouse. To make proper studies of toxicity, laboratory bioassay techniques including precision spray applicators had to be developed. At the Rothamsted Experimental Station, Potter worked vigorously at developing improved techniques to assess potency of pest control materials, resulting in development of an apparatus to become known as the "Potter spraying tower." He served as head of the Insecticides and Fungicides Department, a group that grew and prospered under his direction. Potter's group was active in the introduction of organochlorine and organophosphate insecticides into commercial agriculture, and also was active in solving some of the resulting problems. They greatly improved our knowledge of structure-activity relationships, mode of action, the activities of systemic chemicals, biological and operational properties that affected efficacy, and insecticide resistance. Importantly, he remained convinced, and committed to, the notion that a synthetic material could be developed that had the beneficial attributes of natural pyrethrins. This led to development and introduction of pyrethroid insecticides, a revolutionary class of pest control materials. For his contributions to pesticide research, Potter received numerous awards and honors, including a Congressional Medal from the Third International Congress of Crop Protection, and a team award from UNESCO for development of more photostable pyrethroids. He also served as vice president of the Royal Entomological Society and president of the Association of Applied Biologists. He died on December 10, 1989.

Reference

Needham, P. H. 1990. Charles Potter DSc, DIC, FIBiol, FRES. *Antenna* 14: 57–60.

POTTER WASPS. Members of the family Vespidae (order Hymenoptera). See also, WASPS, ANTS, BEES, AND SAWFLIES.

POULTRY LICE. Members of the family Menoponidae (order Mallophaga). See also, CHEWING LICE.

POUR-ON. A type of pesticide treatment of animals wherein the liquid pesticide formulation is poured on the animal, usually in high volume.

POWASSAN ENCEPHALITIS. See also, TICKS.

POWDERPOST BEETLES (COLEOPTERA: BOSTRICHIDAE: LYCTINAE). The Lyctinae are commonly known as powderpost beetles because of the propensity of the larvae to reduce sapwood, particularly of hardwoods, into a powdery frass. The Lyctinae are worldwide in distribution, each region having an indigenous fauna plus established introduced species. They predominantly are tropical and temperate in distribution. The classification of powderpost beetles is as follows:

Order: Coleoptera
  Suborder: Polyphaga
    Superfamily: Bostrychoidea
      Family: Bostrichidae
        Subfamily: Lyctinae

The two tribes of the Lyctinae are the lyctini and trogoxylini. the major genera of lyctini are *Lyctus, Acantholyctus, Lyctodon, Lycthoplites, Minthea, Lyctoxylon.* The major genera of trogoxylini are *Trogoxylon, Tristaria, Lyctopsis, Lyctoderma, Cephalotoma, phyllyctus*. There are approximately 70 species of powderpost beetles, of which 11 species occur or have become established in the united states. *Lyctus planicollis* lec., *Lyctus linearis* (Goeze), and *Trogoxylon parallelopipedum* (Melsh.) are the most common in the United States.

External morphology

Powderpost beetles are small, 2 to 7.5 mm in length, reddish-brown to black beetles without distinctive spots or markings. They are elongate, body flattened dorso-ventrally, with a prominent, slightly deflexed head constricted behind the eyes. The antennae are 11-segmented, with a two-segmented terminal club, inserted immediately anterior of the eyes. The prothorax is somewhat flattened, and does not form a hood over the head. The tarsi are all five-segmented, with the first segment very small and the fifth segment almost as long as all the preceding segments combined. The tarsal claws are simple.

Life history and habits

Adult powderpost beetles are sexually mature upon emergence. Copulation occurs soon afterward.

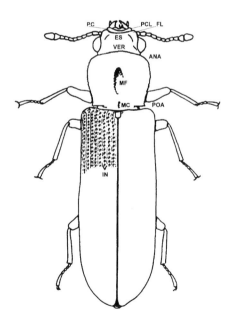

Fig. 833 External morphology of a generalized adult powderpost beetle, dorsal aspect. ANA, anterior angle of pronotum; ES, epicranial suture; FL, frontal lobe; IN, interspace; MC, median canaliculation; MF, median fovea; PC, postclypeus; PCL, postclypeal lobe; POA, posterior angle of pronotum; VER, vertex.

Oviposition takes place 2 to 3 days after mating. The female may feed on the surface of the wood by gnawing the torn fibers, possibly to detect the suitability of the timber for oviposition in relation to food value for the larvae. The female extends the long, flexible ovipositor directly into the lumen of the vessel, tracheae, or pores of the wood. The translucent white cylindrical eggs are laid in depths of from 0.5 to 1 mm. Those woods in which the vessels are most numerous are more liable to heavy attack. The diameter of the vessels in which oviposition occurs is a great importance, as it must be large enough for insertion of the ovipositor. Depending upon temperature, the incubation period may vary from 6 to 20 days. The larva, on hatching, is white and straight bodied, armed at the rear end, and with a pair of small spines. After the first molt, the larva assumes a curved form. The young larva usually tunnels with the grain of the wood. In the later stages, the larval tunnel takes an irregular course. Tunnels approaching the surface of infested wood do not penetrate it but leave a thin, unbroken layer. The larva may remain in the wood for about 10 months, the length of time varying with the temperature, moisture, and condition of the wood. The larva possesses large spiracles on the eighth abdominal segment. When fully matured, it bores its way near the surface of the wood and builds a pupal chamber. The pupal period lasts 12 to 30 days, though this again is variable. When transformation from pupa to adult occurs, the beetle cuts its way to the surface. When emerging, it generally pushes some of the fine dust in front of it, and as a result, small piles of dust often can be seen near new holes. These emergence or flight holes are about 2 or 3 mm in diameter.

The larva feeds mainly on the sapwood of hardwoods, such as oak, ash, hickory, mahogany and bamboo. The chief source of food of the larva is the starch in the cell content of the wood. The cell wall is not digested. Besides starch, certain sugars, disaccharides and a polysaccharide, as well as protein, are necessary constituents of the larval food. As the presence of starch in sapwood is essential for infestation to occur, the greater the starch content, the greater the possible extent of damage. Below a minimum concentration of starch, no attack occurs. Moisture also is essential for the normal development of the larva. It will thrive in wood with a moisture content of between 8 and 30 percent.

Predators and parasites

Predators and parasites are not a factor in the artificial control of the powderpost beetles. Various Hymenoptera have been found to parasitize these beetles. Clerid beetles (Coleoptera) have been reported as predators.

Economic importance

The destructiveness of powderpost beetles to wood and wood products is second only to that of termites. The annual loss of lumber and wood products runs into millions of dollars. As noted previously, powderpost larvae primarily attach to the sapwood of hardwoods. They are more often found in recently dried wood than in old wood. They attack lumber that is used for hardwood floors, crating, furniture, plywood, implement handles, and gun stocks. The damage consists of the destruction of the wood, resulting in a powdery frass as the larvae tunnel their way through the sapwood. When the adult beetles emerge, they further damage the wood by producing exit or flight holes.

Control measures

The control of powderpost beetles in wood may be accomplished by heat treatment, fumigation, chemical treatment, and good lumber yard and saw mill sanitation.

See also, WOOD-ATTACKING INSECTS.

Eugene J. Gerberg
University of Florida
Gainesville, Florida, USA

References

Gerberg, E. J. 1957. A revision of the New World species of powder-post beetles belonging to the family Lyctidae. *U.S. Department of Agriculture Technical Bulletin No. 1157. 55 pp., 14 pls.*

Hickin, N. E. 1963. *The insect factor in wood decay.* Hutchinson and Co., London, United Kingdom. 336 pp., 263 figs.

Ivie, M. A. 2002. 69. Bostrichidae. pp. 233–244 in R. H. Arnett, M. C. Thomas, P. E. Skelley, and J. H. Frank (eds.), *American beetles,* Vol. 2. CRC Press, Boca Raton, Florida.

POWDERPOST TERMITES. A group of termites in the family Kalotermitidae known to attack dry wood not in contact with soil, and reduce it to powder. See also, TERMITES.

POWDERY MILDEW. See also, TRANSMISSION OF PLANT DISEASES BY INSECTS.

PRAYING MANTIDS (MANTODEA). There is some debate about the most appropriate taxonomic position of praying mantids (or mantises). They are considered orthopteroid insects because they appear to have much in common with cockroaches, grasshoppers, crickets, and stick insects, all of which used to be lumped together with mantids within the order Orthoptera. Morphological and molecular evidence suggests that mantids are, in fact, most closely related to Blattodea (cockroaches) and Isoptera (termites), but most authors feel that the 1,900 or so species of mantids are sufficiently different from these groups to be classified separately: Order Dictyoptera, Suborder Mantodea. There are eight families and 28 subfamilies in this suborder:

Amorphoscelididae, containing two subfamilies from Africa and Australia

Chaeteessidae, with only one Neotropical genus, *Chaeteessa*

Empusidae, consisting of eight genera in Africa and Asia

Eremiaphilidae, ground - dwelling mantids of Africa and Asia

Hymenopodidae, with three subfamilies, including tropical flower mimics

Mantidae, the most important family with 21 subfamilies and 263 genera

Mantoididae, with a single Neotropical genus, *Mantoida*

Metallyticidae, with a single Malaysian genus, *Metallyticus*

Distinguishing characteristics

The praying mantis (from the Greek for "prophet") derives both its name and its livelihood from the contradictory morphology and function of the prothoracic legs. When folded against the body, these appendages give the insect the appearance of prayer. However, when an unwary wasp or cricket happens by, the legs unfold at great speed, trapping the prey item in a nearly unbreakable grip between the scimitar-like apical claw of the tibia and the discoidal spines lining the anterior surface of the femur. In addition to the raptorial front legs, the most obvious physical features of mantids are an elongated prothorax and a highly mobile head with which the insect can actually look over its shoulder. Mantids employ accurate binocular vision to assess prey (or enemy) size and distance, and at least some species have an ear that can detect the ultrasound emitted by insectivorous bats hunting the night sky for prey.

Most orthopteroid insects are either herbivores or omnivores, but all mantids are carnivores. Aside from this commonality, mantids are otherwise quite diverse in form and function. There are, for instance, two distinct modes of predation. Most species are sit-and-wait ambush predators such as the leaf mimic, *Phyllocrania illudens*, from Madagascar, the spectacular Malaysian flower mimic, *Hymenopus coronatus*, and the simply cryptic European mantis, *Mantis religiosa*. However, some species are active hunters that run down their prey, such as the North American species, *Yersiniops sophronicum*. Most mantids are winged and sexually dimorphic, but *Brunneria*

borealis from the southern U.S.A. is a wingless, completely parthenogenetic female species. In many winged species, only the male has appreciable flight capability; the wings of females often are too small for flight, but can be used in defense (deimatic displays) or to help disperse sex pheromones. Since males cannot by themselves colonize new habitats, the dispersal of mantid populations within a region is generally slow.

Adult body size varies greatly among species; the smallest known species being *Mantoida tenuis* (1 cm) and the largest is *Ischnomantis gigantas* (17 cm). The genetics of sex determination also differs among mantids: males of some species are XO, while many others are XXY.

Life history

Most species of mantids are relatively rare and narrowly distributed in the tropics, an extreme example of which is *Galapagos solitaria*, which is only found on three of the Galapagos Islands. However, some temperate zone species are both abundant and widely distributed. Thus, in spite of the vastly greater diversity of tropical mantids, we know much more about the biology of a few temperate zone species. The best known, most ubiquitous species undoubtedly is the Chinese mantis, *Tenodera aridifolia sinensis*, which occurs throughout much of Eurasia and (through introduction) eastern North America. The life cycle of this species is closely governed by seasonality, typical of many large-bodied, semelparous insects in temperate regions. Egg hatch for the Chinese mantis occurs in the spring, followed by a long maturation period during the summer and sexual maturity early in the fall. Nymphs develop through six or seven instars, during which they may increase from 10 mm to 10 cm in body length. By the time they are adults, males outnumber females by as much as 2:1. In the fall, adult females emit pheromones to attract males, and perhaps other females as well. The adaptive utility of the first function is obvious; the latter may be explained as a mechanism to increase the total pheromone concentration in the area of grouped females, resulting in a higher probability that males will be attracted.

After mating, a female deposits one or (infrequently) more oothecae on a stalk of vegetation, each ootheca containing from 60 to 300 eggs, and constituting as much as 50% of the female's

Fig. 834 Three male *Tenodera sinensis* competing to mate with the same female. Males typically outnumber females at the beginning of the adult portion of the life cycle, but sexual cannibalism may reduce the discrepancy by the end of the season.

Fig. 835 Female *Tenodera sinensis* with her ootheca. The egg mass weighs about as much as she does after oviposition.

pre-oviposition weight, a considerable parental investment at a time when prey are becoming scarce. The eggs of this and many other species are subject to mortality from parasitoid wasps and, if deposited far enough above the ground, foraging birds.

Adults die with the onset of cold weather, and eggs over-winter. Unlike the European mantis, *Mantis religiosa*, the eggs of which have an obligatory cold diapause, the eggs of the Chinese mantis begin to develop as soon as they are laid, so an extended fall can be lethal if egg hatch occurs before onset of winter. This limits the southern distribution of this species, and extreme cold limits the northern range by desiccating the eggs during winter. Although successive generations of temperate zone species do not overlap, adult females of some tropical species, such as *Cardioptera brachyptera*, live long enough to exhibit parental care by guarding their oothecae and hatchlings against predators.

Mantids in general, and females in particular, are famous cannibals. This trait has led some authors to speculate that a male commits "adaptive suicide" by allowing his mate to consume him, thus contributing to the fitness of his offspring by providing nourishment for egg production. This popular notion is unlikely to be true because females begin to produce their first clutch of eggs before they mate, and the male victim has no assurance of paternity because females may attract and copulate with many males. Furthermore, although sexual cannibalism is frequently observed in the laboratory, under natural conditions males usually escape their first sexual encounter to mate with other females (however, the more mating attempts the greater the probability of being eaten). The simplest explanation for sexual cannibalism is that a very hungry female will eat her suitor before copulation can begin, a mildly hungry female will eat him during or shortly following copulation (if she can catch him), and a female that is less hungry than amorous will let him go. In the first two instances the male does indeed contribute to his mate's nutrition, and therefore to her fitness, but not necessarily to his own because she actually may have been inseminated by a previous male. Considering that females often are food limited at the end of the season and that there are fewer of them than of males, cannibalism of male mantids may be a female's best strategy for producing a healthy ootheca in many instances.

Ecology and economic importance

Praying mantids have a tritrophic niche, i.e., they simultaneously occupy two consumer trophic levels in natural ecosystems by virtue of feeding on herbivores, other carnivores and pollen. Mantids, therefore, can compete with other predator species (e.g., spiders) for food, eat other predators, compete with each other, or cannibalize each other. The fact that all of these processes may be occurring at the same time in the same ecosystem can complicate prediction of the impact of these predators on ecosystem structure and dynamics. Predators of mantids include larger spiders and vertebrates such as birds, lizards, and snakes. Other enemies include chalcoid wasps that are parasitic on eggs, and tachinid flies that parasitize nymphs.

Field experiments have shown that mantids exert both direct (prey reduction) and indirect (prey enhancement) effects on arthropod assemblages. Indirect effects occur because competition with, or predation on, other predators may reduce predation on some herbivorous arthropods. An experiment in

Fig. 836 Adult female *Tenodera sinensis* feeding on a bee. Pollinators such as bees, wasps, and butterflies are frequent prey, especially for adult mantids.

a complex old-field ecosystem revealed that mantids had a positive effect on plant productivity by reducing herbivorous insects, a direct effect known as a trophic cascade because it ramified two trophic levels down from the predators. However, in another study, mantid nymphs indirectly enhanced aphid densities by reducing spider populations. This kind of unpredictability has important implications for one of the most persistent ideas about mantids, that they are good biological control agents. To that dubious end, oothecae of *Tenodera aridifolia sinensis* are sold and distributed through the mail by organic gardening suppliers. Although this has resulted in much of the broad regional distribution of this species in the U.S., the value of these animals to pest control is dubious at best. Because they are generalists, they eat anything that moves within a suitable size range, and this includes spiders, wasps, bees, butterflies, and other desirable species as well as deleterious herbivores. In fact, flower-foraging insects such as bees may constitute a significant portion of their diet at the end of the season when grasshoppers and other herbivorous prey are becoming scarce. Thus, planting mantids in one's garden is not necessarily an efficacious way to control pests. There is also a chance that by spreading this exotic species around the country we may be affecting native arthropod assemblages, including native mantid species such as *Stagmomantis carolina*, which often is eaten by the larger Chinese mantid.

Lawrence E. Hurd
Washington and Lee University, Lexington, Virginia, USA

References
Fagan, W. F., M. D. Moran, J. J. Rango, and L. E. Hurd. 2002. Community effects of praying mantids: a meta-analysis of the influences of species identity and experimental design. *Ecological Entomology* 27: 1–11.
Helfer, J. R. 1963. *How to know the grasshoppers, cockroaches and their allies.* W. C. Brown Co., Dubuque, Iowa. 353 pp.
Hurd, L. E., and R. M. Eisenberg. 1990. Arthropod community responses to manipulation of a bitrophic predator guild. *Ecology* 76: 2107–2114.
Preston-Mafham, K. 1990. Grasshoppers and mantids of the world. *Facts on File*, New York, New York. 192 pp.
Prete, F. R., H. Wells, P. H. Wells, and L. E. Hurd (eds.). 1999. *The praying mantids.* Johns Hopkins University Press, Baltimore, Maryland. 362 pp.

PRECINCTIVE. Native to a specified area, and not occurring elsewhere.

PRECISION. A statistical measure of the repeatability of an estimate relative to a group of estimates from the same population at the same time. Typically measured as the quotient of the standard error of the mean over the mean. Low numerical values indicate high precision, high numerical values indicate low precision. Precision is a key element in developing and evaluating the performance of sampling plans.

See also, SAMPLING ARTHROPODS.

PRECOCENES. Chemical substances from the common bedding plant *Ageratum houstonianum* that affect the development of some insects by causing atrophy of the corpora allata and causing precocious maturity and production of sterile adults.

PREDACEOUS (PREDACIOUS). Animals that attack and feed upon (prey upon) other animals. Ecologists sometimes consider consumption of plants by animals to be predation, but entomologists call such animals herbivores and this process herbivory.

PREDACEOUS DIVING BEETLES. Members of the family Dytiscidae (order Coleoptera). See also, BEETLES.

PREDATION: THE ROLE OF GENERALIST PREDATORS IN BIODIVERSITY AND

BIOLOGICAL CONTROL. The term 'generalist predator' is nearly redundant. The diets of the vast majority of predators are far less specialized than many herbivore-plant associations in which, for example, the life cycle of a phloem-feeding treehopper (Hemiptera: Membracidae) in the temperate zone may be completely dependent on the seasonal phenology of a single species of tree. Although some groups such as ladybird beetles (Coleoptera: Coccinellidae) are relatively specialized on a few related prey species (in this case, mainly aphids or scale insects), the vast majority of predators feed on varied prey. A far more typical group of coleopteran predators, beetles in the family Carabidae, will take nearly anything that moves within a size range they can physically handle, regardless of taxonomic position. Thus, the most important factor determining the niche (ecological role) of insect predators undoubtedly is breadth of diet.

The niche of generalist predators

Predatory insects are fewer in number than herbivores owing to the constraints of the second law of thermodynamics, which dictates that the energy available to higher trophic levels in a food chain diminishes with each transfer. This is why food chains often are represented as pyramids, getting smaller from the ground floor (plants), through the consumer levels, with top carnivores occupying the narrow tip. Most predators are probably generalists because carnivores that feed broadly can sustain larger populations than those specializing on one or a few prey species. A specialist will starve if its prey population gets too low, but a generalist can sustain its population size on alternate prey even if preferred prey species become scarce (see functional response, below). This is a key tenet of niche theory: resource generalists are more buffered against fluctuation in those resources than are specialists.

A voluminous literature suggests that phytophagous insects do not often compete, i.e., are not routinely food limited, whereas most studies that have addressed the question have found that predators usually are limited by their prey. In spite of this apparent thermodynamic disadvantage of the predatory lifestyle, feeding on animal prey does provide one potential advantage over vegetarian fare: higher quality protein. There are even many examples in the animal kingdom of predominately herbivorous animals that are carnivorous as juveniles (when growth is important) or as adults during reproduction (for egg production). After all, adult female mosquitoes (Diptera: Culicidae) feed on blood to make eggs.

Most major insect orders have predaceous members. These may be dedicated carnivores such as dragonflies and damselflies (Odonata), mantids (Dictyoptera), ambush bugs (Hemiptera), and lacewings (Neuroptera), or they may be omnivores such as ants (Hymenoptera), earwigs (Dermaptera), and crickets (Orthoptera). The manner of predation also varies widely: carabid beetles (Coleoptera) chew their victims, while reduviid bugs (Hemiptera) suck them dry as do spiders; water scorpions (Hemiptera) and some mantids, like orb weaving spiders, are sit-and-wait ambush predators, while wasps (Hymenoptera) tend to be active hunters. Some, such as the aquatic naiads of Odonata, will even include vertebrates in their diets. These predators, and others such as diving beetles (family Dytiscidae) and giant water bugs (family Belostomatidae) are particularly adept at capturing small fish and tadpoles.

The amount of research on predatory insects is far exceeded by work on herbivores, for the obvious reason that most insect pests are herbivorous. For that reason, we know much more about plant feeders than about the insects that eat them. Our growing awareness of the environmental hazards of biocide use has contributed to research on some predatory species to investigate their potential for biological pest control. However, we still have much to learn about how predators fit into complex natural communities.

The predator guild and community interactions

A guild is a group of species that use a common resource base. A community of organisms within a habitat generally comprises many guilds using many different, sometimes overlapping, resources. Guilds often consist of species that are not particularly closely related, but that have overlapping resource requirements. Thus, a guild of predatory animals that feed on ground-dwelling arthropods may consist of such invertebrates as wolf spiders, scorpions and tiger beetles, along with vertebrates such as lizards and shrews.

Generalist arthropod predators typically are bitrophic: they simultaneously occupy the third and fourth trophic levels by virtue of feeding both on herbivores and on each other. As a consequence, species within generalist predator guilds may compete with each other for common prey resources, or they may

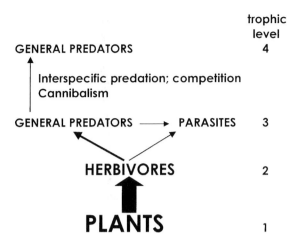

	trophic level
GENERAL PREDATORS	4
Interspecific predation; competition Cannibalism	
GENERAL PREDATORS ⟶ PARASITES	3
HERBIVORES	2
PLANTS	1

Fig. 837 The trophic position of generalist predators is complicated by the varied interactions that can take place among them, as well as by their relationship with the rest of the community. Arrows represent direction and relative amount (line thickness) of energy flow in food webs topped by general arthropod predators, illustrating their bitrophic position as members of both the third and fourth trophic levels.

eat each other. Predation can be either between species or among individuals within the same species, because most generalist predators are cannibals. In fact, cannibalism often has been proposed as a mechanism of population regulation in such predators because the frequency of cannibalism is expected to increase as population density increases, resulting in negative feedback between population size and cannibalism. The selective advantage of cannibalism is simply that a cannibalistic individual can increase its Darwinian fitness by eating members of its own cohort when alternate prey are scarce. The combination of cannibalism and interspecific predation within predator guilds has become known as intraguild predation.

The combination of competition, interspecific predation, and cannibalism not only influences coexistence within this guild, it also complicates the impact of these predators on the rest of the community. Both competition for scarce prey and intraguild predation can occur even among predators of different phyla (e.g., between amphibian larvae and aquatic insects). There is yet another wrinkle in the food web. It has been suggested that predators be thought of as mutualists of plants, because plants supply them herbivore prey and they in turn reduce damage from herbivory. However, predators such as ambush bugs and crab spiders can be viewed as competitors with plants

for a common resource: pollinating insects such as bees and butterflies.

In view of the complex relationship of generalist predators to each other and to the rest of the community, it is not surprising that we do not yet have good predictive models of their impact on diverse insect assemblages. Finding out how generalist predators affect highly diverse natural communities requires careful experimentation in the field, in which predators are either eliminated or their densities augmented in replicated plots, and the results are compared with control plots with normal predator levels. These experiments usually are difficult to perform, and the resultant data (usually in the form of changes in numbers of individuals or biomass among resident species) are nearly always highly variable, necessitating relatively complex statistical analysis. For these reasons, such experiments are not as common in the literature as are laboratory tests of predator feeding, or experiments using low diversity agricultural plots.

Interpretation of the results of experimental predator manipulations on a target community often can be difficult. When a target (prey) population declines, the cause is usually a direct effect of predation. However, frequently predators actually increase populations of prey species, an indirect effect of their feeding on competitors or on other predators of these prey. In a diverse species assemblage, both kinds of results often occur in the same experiment. The potential combinations of positive and negative effects on prey may even serve to cancel out the overall impact of predators on prey assemblages. Our current understanding of these interactions is still too limited to make reliable predictions of which of these effects will be important prior to adding predators to arthropod assemblages. Clearly, more research is needed.

If predators affect herbivore populations, they may as a result affect the plants on which these herbivores feed. This kind of 'top down' effect of predators on successively lower trophic levels is one kind of what is known as a trophic cascade. The other kind, 'bottom up', is when plants are demonstrated to control herbivores (rather than the reverse), and herbivores in turn control predators. Trophic cascades are the subject of a growing literature among theoreticians and experimentalists in ecology, and people involved in biological control are obviously interested in the potential for predators to exert top down control. Unfortunately, there are as yet few experiments in the field demonstrating that arthropod predators can exert top down control. One experiment in a simplified agricultural system showed that spiders could boost plant productivity by reducing herbivores; another experiment demonstrated the same sort of effect for praying mantids in a complex natural community. This may seem promising, but before we can make any kind of general prediction, we need many more studies under various conditions.

Generalist predators and biological control

There are some good examples of biological control of pest species using specialized predators, e.g., the control of scale insects on California citrus by the introduction of the Australian ladybird beetle, *Rodolia cardinalis*. Some evidence suggests that at least some arthropod generalists, such as mites and spiders, can significantly reduce herbivore populations in simplified experimental agricultural system, but very few data have been generated as to the effects of this predation on plants. It is therefore reasonable to ask whether generalist insect predators may be useful in the control of those herbivorous and parasitic species that have exceeded the human economic and nuisance threshold: pests.

Pest species are virtually defined as populations that are out of control, and in order to control a pest population, a predator must track its prey closely. This generally requires two kinds of reactions to changes in the size of a prey population by a predator: numerical and functional responses. Numerical response simply means that a predator can adjust its population size to exploit changes in the prey population. The two ways to do this are through increased reproductive output, and high rates of dispersal into areas of high prey density (and, conversely, reduced reproduction and emigration when prey become scarce, to avoid complete starvation). Both of these mechanisms would entail a time lag in predator response, even if the predator were a specialist; this time lag would likely be much longer for a generalized predator.

Functional response is an increase in per capita feeding rate as prey become more abundant. Ideally, a predator should 'switch' from generalized feeding on many prey species to focus on the rapidly increasing pest population in a frequency-dependent manner. This entails recognition of specific chemical, tactile, or visual cues provided by the pest species from among all the other prey signals in the environment

at the time. This would seem unlikely for a generalist predator, an animal adapted to hunting or ambushing a variety of prey species in a food-limited environment, but even an omnivore may have preference for a specific item from among its dietary choices. The criteria for prey preference in predators vary, but include such factors as ease of capture and nutritional value. The net gain to the predator for an item of prey, when costs of finding and capturing prey are balanced against nutritive gain, are included in the theoretical construct known as optimal foraging theory. This theory predicts that selection will favor those (optimal) behaviors that maximize net return on an animal's foraging expenditure. However, even if a predator did switch to increase its feeding rate on a specific pest, it would reach a satiation level past which further increases in prey density could elicit no faster capture rate per individual predator. Unless this functional response were accompanied by a good numerical response, it would be unlikely to work.

The general unpredictability of community-wide effects of adding predators to experimental systems discussed in the previous section argues against their use in the absence of careful prior study. A predator may feed broadly enough that its diet encompasses a pest species, but once introduced it may exhibit preference for other prey, either benign or beneficial. The fact is that generalist predators may eat as many or more beneficial species as they do pests, and the tradeoff between positive and negative effects of these predators in our gardens can make them unreliable biocontrol agents.

Value of predators to biodiversity

It is becoming increasingly obvious that predators of all kinds play a critical role in the maintenance of biodiversity. So, to the extent that humans are concerned with the variety of life forms in nature, we should pay attention to the carnivores. Human history is rife with examples of human extirpation of vertebrate carnivores, as we have spread ourselves across the habitable world. This has happened largely because tigers, wolves, rattlesnakes, and hawks have been perceived to compete with us for what we eat (including domesticated animals), or are feared because they can harm us. As a result, we have a superabundance of animals like deer, rabbits, rats, and groundhogs, but a general decline in biological diversity in proximity to human settlements.

Among arthropods, predators have been implicated as important selective agents in the evolution of herbivore-plant systems. Evidence of strong natural selection exerted by predators includes the numerous examples of adaptations among herbivorous insects to detect and avoid them. This includes remote sensing to chemically identify the whereabouts of an enemy, clearly a case for considerable evolutionary fine-tuning of a prey to its predator. It seems quite likely that predators play an important role in controlling the diversity of the arthropod portion of the biosphere. This is a large slice of life indeed, considering that about 64% of all species identified so far are insects, so it would not be surprising if much of the rest of global biodiversity depended on them as well.

Adding exotic predators to control pest species may have the undesirable effect of decreasing native insect diversity. Some studies have suggested that introduced predators can interfere with native species, resulting in loss of control over prey populations through release from predation, i.e., making a bad situation worse in terms of levels of pest populations. Or, an introduced predator may simply displace, through competition or intraguild predation, a native predator. An example of this is the Chinese praying mantis, *Tenodera sinensis*, introduced into the United States a little more than 100 years ago. This species has not only become the most abundant and widespread mantid in this country (mainly by human transport of egg cases), it has displaced one of our native species, the Carolina mantid, *Stagmomantis carolina*, from some habitats in North Carolina. In this case, the mechanism is predation: both species lay their eggs on loblolly pines, but the Chinese mantid is larger and finds the Carolina mantid easy prey. It is therefore uncommon to find both species persisting in the same habitat for very long.

Whatever their effects, once introduced insects can be very hard to remove from their adopted home. In archipelagos such as Hawaii and the Galápagos, where entomologists have been keeping score, the proportion of exotic insects making up the species list has been growing at an alarming rate ever since these islands were colonized. Most biologists recognize diet breadth as one indication of how likely an animal is to be invasive. The very fact of their broad diet can make generalist predators ideal invasive species, preadapted for survival in novel habitats with a naive (unadapted) prey assemblage. Given that species invasions are second only to habitat loss as a threat to global biological diversity, and that uncountable introductions have been made inadvertently through global human travel and commerce, we certainly must be especially careful about introducing insects intentionally.

Lawrence E. Hurd
Washington and Lee University
Lexington, Virginia, USA

References
Bernays, E., and M. Graham 1988. On the evolution of host specificity in phytophagous arthropods. *Ecology* 69: 886–892.

Hassell, M. P. 1978. *Arthropod predator-prey systems.* Princeton University Press, Princeton, New Jersey.

Hurd, L. E. 1999. Ecology of praying mantids. pp. 43–60 in F. R. Prete, H. Wells, P. Wells, and L. E. Hurd (eds.), *The praying mantids.* Johns Hopkins University Press, Baltimore, Maryland.

Polis, G. A., C. A. Myers, and R.D. Holt. 1989. The ecology and evolution of intraguild predation: potential competitors that eat each other. *Annual Review of Ecology and Systematics* 20: 297–330.

Sih, A., P. Crowley, M. McPeek, J. Petranka, and K. Strohmeier 1985. Predation, competition and prey communities: a review of field experiments. *Annual Review of Ecology and Systematics* 16: 269–311.

PREDATOR. In entomology, animals that feed on insects, and must eat several or many insects in order to complete their life cycle.

PREDATORY GUILD. A group of different types of organisms that feed on the same resource, such as a developmental stage of an insect.

PREDATORY STINK BUGS (HEMIPTERA: PENTATOMIDAE, ASOPINAE). The subfamily Asopinae belongs to the family Pentatomidae of the suborder Heteroptera (true bugs). About 300 species in 69 genera have been described worldwide.

Order: Hemiptera

  Suborder: Heteroptera

    Infraorder: Pentatomomorpha

      Superfamily: Pentatomoidea

        Family: Pentatomidae

          Subfamily: Asopinae

External morphology

Like other pentatomidae, predatory stink bugs or soldier bugs are of moderate to large size, ranging in length from 7 to 25 mm, and are broadly elliptical in shape. The piercing-sucking mouthparts of predatory stink bugs consist of a four-segmented rostrum or beak (labium) forming a sheath that encloses two mandibular and two maxillary stylets. Whereas the rostrum of phytophagous pentatomids is slender, asopines are characterized by having a thickened rostrum. In asopines, the first segment of the rostrum is markedly thickened and free, which enables the rostrum to swing forward fully, making it easier for the predator to feed on active prey. The appearance of the rostrum can be a useful diagnostic character for distinguishing the beneficial predatory pentatomids from potentially harmful phytophagous pentatomids in the field. The triangular mesothoracic shield, called the scutellum, is usually much shorter than the abdomen, but in some genera it is enlarged, covering most of the abdominal dorsum. The males of Asopinae are unique in combining the presence of genital plates with a thecal shield. The shape of the male claspers or parameres is the most reliable diagnostic characteristic for the identification of species.

Habitat and food

Predatory stink bugs are found in a wide range of natural and agricultural habitats, but many species appear to prefer shrubland and woods. The Asopinae are set apart from the other pentatomid subfamilies by their essentially predaceous feeding habits. It is believed that the Pentatomomorpha have arisen as plant feeders and that only the subfamily Asopinae has secondarily become predaceous. First instars do not attack prey and only need moisture, mainly in the form of plant juices, to develop. Although, for some species, partial development on certain plant-based diets has been reported, nymphs from the second instar on require animal-based diets to complete development. Nymphs and adults, however, are often observed to take up plant juices or free water in addition to feeding on animal prey. Plant feeding primarily provides moisture, but it may also furnish certain nutrients to the bugs. This habit may help predatory stink bugs to sustain their populations in times of prey scarcity. In contrast to some other zoophytophagous heteropterans, plant-feeding asopines have not been reported to injure plants.

Predatory stink bugs attack mainly slow-moving, soft-bodied insects, primarily larval forms of the Lepidoptera, Coleoptera and Hymenoptera. Very few, if any, Asopinae are truly host-specific. Nevertheless, whereas some asopine bugs are generalist predators, attacking a wide array of prey in a diversity of habitats, others appear to be more closely associated with a limited number of insect species and occur in only a few habitats. It has been hypothesized that there may be an evolutionary progression from the drab asopines, which feed rather generally, to the brightly colored asopines, which prefer larvae of Chrysomelidae and, to a lesser extent, Coccinellidae.

Life history

The eggs are usually laid in masses. In some species (e.g., *Podisus*), the eggs bear prominent micropylar processes. There are five nymphal instars. The nymphs take a few weeks to develop, depending on temperature and food. Newly emerged nymphs are highly gregarious, whereas later instars become progressively more solitary with each molt. First instars are not predaceous and take up only water or plant juices; occasionally, they also feed on unhatched eggs. From the second instar on, nymphs begin

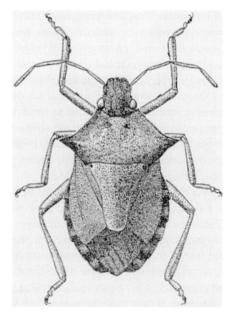

Fig. 838 Line drawing: *Picromerus bidens* (taken from T.J. Henry & R.C. Froeschner, 1988. Catalog of the Heteroptera, or True Bugs of Canada and the Continental United States. E. J. Brill).

searching for prey shortly after molting. The nymphs will feed up to 1 to 2 days before the next molt; at that time, their activity drops and the nymphs will remain resting in a concealed place in preparation for the oncoming molt. Small nymphs tend to attack prey and feed collectively, particularly when the prey is large. Larger nymphs and adults prefer to attack prey individually. Under conditions of food shortage, the nymphs and the adults are highly cannibalistic. The adults are long-lived, with reported longevities of 15 days to more than 3 months, and they lay eggs throughout their entire lifetime. The adults usually move by crawling, but some species (including *Podisus maculiventris*) are also noted to be good flyers. Field surveys suggest that scelionid egg parasitoids, such as *Trissolcus* and *Telenomus* spp., are the most important natural enemies of predatory stink bugs.

Prey location and capture

Predatory stink bugs use visual, chemical and tactile cues to locate and recognize their prey. Vision appears to be the most important sense used by the bugs to locate prey. *Podisus maculiventris* and other asopine bugs react to moving prey at distances up to 10 cm, but their reactive distance to immobile prey is considerably less and detection often seems to occur at antennal or rostral contact. Evidence is accumulating that several asopines can also use airborne chemical cues for prey detection. Both *Podisus maculiventris* and *Perillus bioculatus* are sensitive to systemic volatiles produced by plants in response to prey feeding. *Eocanthecona furcellata* is attracted to its lepidopterous prey based, in part, on a volatile component derived from the chlorophyll ingested by the prey. Further, it has been demonstrated that *Podisus maculiventris* can use vibrations of the substrate as cues for prey location. Prey recognition is based primarily on antennal and rostral contact. Asopines are rather timid predators. After finding the prey and orienting to it, the bugs may spend from several minutes up to an hour stealthily approaching it, often keeping their beak extended forward. The stylets are usually inserted at soft areas of the prey body. During attacking and feeding, the only contact between predator and prey is by the rostrum and stylets. When stylet penetration is perceived by the prey, many prey species will try to escape by vigorous body movements. The harpoonlike structure of the mandibular stylets enables a tenacious holding of the prey during this struggle. Further, predatory pen-

tatomids are presumed to inject a salivary toxin into the prey body that quickly immobilizes it. During feeding, body tissues of the prey are liquefied by the injection of digestive enzymes and by the lacerating action of the stylets. The liquid and partly digested food can then be sucked up through the rostrum. This "solid-to-liquid" feeding process enables the predator to use most of the body tissues of the prey.

Pheromones

Both male and female asopines possess a metathoracic scent gland from which they discharge a disagreeable odor when disturbed, hence the name 'stink bugs.' In addition, many species of predatory stink bugs have discrete pheromone glands, which belong to one of two types: dorsal abdominal glands (in both sexes) or sternal glands (in males only). Species with dorsal abdominal glands are mostly polyphagous predators (e.g., *Podisus*, *Zicrona*), whereas those with sternal glands often appear to be more specialized predators (e.g., *Perillus*, *Stiretrus*).

The best known pheromone system is that of the spined soldier bug, *Podisus maculiventris*. Adult males of this species possess hypertrophied dorsal abdominal glands that produce secretions that function as long-range attractants for adults and immatures of both sexes. Immature predators are thought to use the male pheromone as a cue indicating the presence of prey. In the United States, attractors with a synthetic pheromone are commercially available to lure spined soldier bugs to target areas in early spring when the adults emerge from overwintering. Pheromone-baited traps can also be employed to capture large numbers of predators that can be used to establish mass cultures.

Economic importance

Several predatory pentatomids are believed to have a future for the biological control of various economically important crop pests in different parts of the world. To date, however, the only asopine commercially available for augmentative biological control in North America and Europe is the spined soldier bug, *Podisus maculiventris*. This species is native to North America and has shown good capacity to control a variety of insect pests in field and greenhouse crops, orchards and forests, including the Colorado potato beetle, *Leptinotarsa decemlineata*, and several leaf-feeding caterpillars. The two-spotted stink bug, *Perillus bioculatus*, is another important natural enemy of the Colorado potato beetle in North America. *Podisus nigrispinus* is the most common predatory stink bug in South America. The insect has been the subject of conservation and augmentation biological control programs against leaf-feeding caterpillars in forests, including Eucalyptus stands in Brazil. In Southeast Asia and India, *Eocanthecona furcellata* has received increasing attention for its potential to control outbreaks of lepidopterous and coleopterous defoliators. *Picromerus bidens* was originally a widely distributed Palearctic species, but it has been found in the northeastern United States and eastern Canada since its (accidental) introduction sometime before 1932. Although its biology has been studied to some extent, few studies have attempted to quantify its predatory effectiveness. Given that only 10% of the nearly 300 known species of Asopinae has been studied in more or less detail, obviously an enormous biocontrol potential remains to be investigated.

Patrick De Clercq
Ghent University Ghent, Belgium

References

Aldrich, J. R. 1999. Predators. pp. 357–381 in J. Hardie and A. K. Minks (eds.), *Pheromones of non-lepidopteran insects associated with agricultural plants*. CAB International, Wallingford, United Kingdom.

De Clercq, P. 1999. *Podisus* online. http://allserv.ugent.ac.be/~padclerc/

De Clercq, P. 2000. Predaceous stinkbugs (Pentatomidae: Asopinae). pp. 737–789 in C. W. Schaefer and A. R. Panizzi (eds.), *Heteroptera of economic importance*. CRC Press, Boca Raton, Florida.

Thomas, D. B. 1992. Taxonomic synopsis of the asopine Pentatomidae (Heteroptera) of the Western Hemisphere. *The Thomas Say Foundation Monographs*, vol. 15. Entomological Society of America, Lanham, Maryland.

Thomas, D. B. 1994. Taxonomic synopsis of the Old World asopine genera (Heteroptera: Pentatomidae). *Insecta Mundi* 8: 145–212.

PREDISPOSING FACTORS. Factors which, by their actions, render an organism susceptible to a certain disease; conferring a tendency to disease.

PRE-EMERGENCE TREATMENT. Treatment of a plant, usually with a pesticide, before the plant has emerged from the soil.

PRE-PLANTING TREATMENT. Reference to a treatment applied to a crop before planting.

PREPUPA. A generally quiescent, occasionally active but nonfeeding, stage of insects at the end of the immature development period. The period immediately preceding the molt to the adult stage. In thrips, it is the third instar.

PRESCUTUM. The anterior portion of the meso- and metanotum.
See also, THORAX OF HEXAPODS.

PRESOCIAL BEHAVIOR. Expression of one or two (but not all three) of the following traits of sociality: individuals of the same species cooperate in caring for the young; there is division of reproductive behavior, with more or less sterile individuals working on behalf of the fecund individuals; overlap of at least two generations in life stages contributing to colony labor, so that offspring can assist parents during some period of their life. Thus, presocial behavior is considered to include all the stages intermediate between solitary and eusocial behavior.
See also, SUBSOCIAL, COMMUNAL, QUASI-SOCIAL, SEMISOCIAL, EUSOCIAL BEHAVIOR.

PREVALENCE. The frequency of occurrence. In ecology, the proportion of inhabitable sites or areas inhabited by an organism. With respect to epizootiology, the total number of cases of a particular disease at a given moment of time, in a given population.

PRETARSUS. The terminal portion of the leg. The tarsal claws and associated structures.
See also, LEGS OF HEXAPODS.

PREY. In entomology, insects that are eaten by predatory animals (including other insects). If the insect is killed by a parasitoid or pathogen, the insect is called a host, not the prey.

PRIMARY PARASITOID. A parasitoid of a host, not of another parasitoid.

PRIMARY PRODUCTION. Production by green plants. (contrast with secondary production)

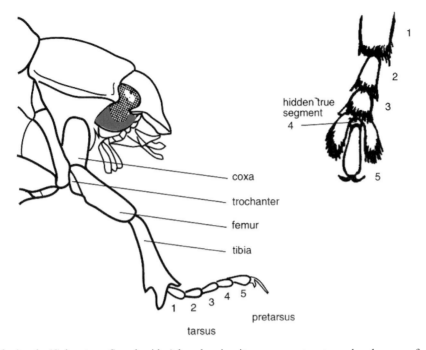

Fig. 839 Leg of a beetle (Coleoptera: Scarabaeidae) leg showing its component parts, and a close-up of one type of beetle tarsus (foot).

PRIMARY PRODUCTIVITY. The rate at which biomass is produced by plants, expressed on a per unit area basis.

PRIMARY REPRODUCTIVES. In social insects, the pair of insects that founds the colony.

PRIMER. In molecular biology, a short oligonucleotide that is attached to a ssDNA molecule in order to provide a site at which DNA replication can begin.

PRIMER PHEROMONE. A pheromone that acts to modify the physiological condition of an animal. (contrast with releaser pheromone)

PRIMITIVE CADDISFLIES. Members of the family Rhyacophilidae (order Trichoptera). See also, CADDISFLIES.

PRIMITIVE CARRION BEETLES. Members of the family Agyrtidae (order Coleoptera). See also, BEETLES.

PRIMITIVE CRANE FLIES. Members of the family Tanyderidae (order Diptera). See also, FLIES.

PRIMITIVE DAMPWOOD TERMITES. Members of the termite family Termopsidae. See also, TERMITES.

PRIMITIVE WEEVILS. Members of the family Belidae (order Coleoptera). See also, BEETLES.

PRIONOGLARIDAE. A family of psocids (order Psocoptera). See also, BARK-LICE, BOOK-LICE, OR PSOCIDS.

PROBABILITY MODEL. In sampling, a mathematical description of the dispersion or distribution of individuals in a population based on numbers per sample unit. Common models include Poisson, Negative-binomial, Binomial, Normal, and Neyman Type A. Such models can form the foundation of a sampling plan.
　　See also, SAMPLING ARTHROPODS.

PROBE. A probe is a molecule labeled with radio-active isotopes or another tag that is used to identify or isolate a gene, gene product, or protein.

PROBOSCIS. The tube-like or beak-like mouth-parts or sucking apparatus of insects that are modified to feed on liquid food.

PROCTODEUM. The hindgut of insects.

PROCTOTRUPIDAE. A family of wasps (order Hymenoptera). See also, WASPS, ANTS, BEES, AND SAWFLIES.

PROCUTICLE. The inner zone of the cuticle, divisible into the hard, outer, dark exocuticle and the soft, inner, light endocuticle, and containing primarily chitin and protein. The procuticle exists during molting, before the inner layers of the cuticle (exocuticle and endocuticle) are sclerotized into distinct layers, but this is a temporary condition.
    See also, CUTICLE, EPICUTICLE.

PRODOXIDAE. A family of moths (order Lepidoptera). They commonly are known as yucca moths.

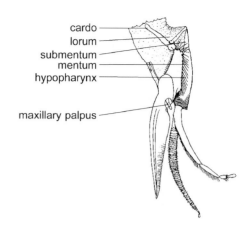

Fig. 841 Left lateral aspect of a proboscis (Diptera).

See also, YUCCA MOTHS, BUTTERFLIES AND MOTHS.

PROFITABILITY CONCEPT. The concept that predators will seek prey that provide the most calories or biomass, as this represent the ''most profitable'' investment of their time by maximizing growth and reproduction. This can also be called optimal foraging.

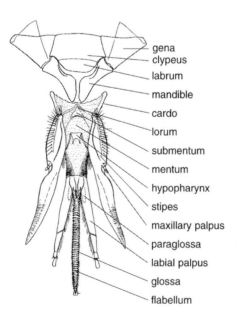

Fig. 840 Anterior aspect of a proboscis (Diptera).

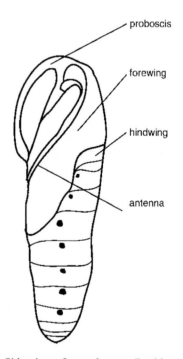

Fig. 842 Side view of a moth pupa (Lepidoptera: Sphingidae) showing the proboscis.

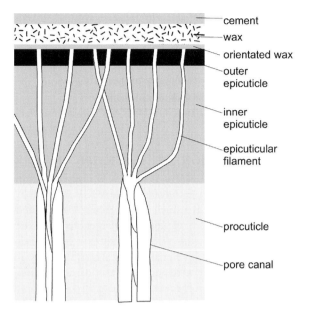

Fig. 843 Cross section of the insect epicuticle (adapted from Chapman, The insects: structure and function).

PROGNATHOUS. A condition in which the head is horizontal and the mouthparts are directed forward. This is particularly common in burrowing and predatory species.

PROGRESSIVE PROVISIONING. Among social insects, the act of provisioning the larva at intervals during its development, as opposed to mass provisioning.

PROHEMOCYTE. Very small hemocytes, possibly giving rise to other types of hemocytes.

See also, HEMOCYTES OF INSECTS: THEIR MORPHOLOGY AND FUNCTION.

PROJAPYGIDAE. A family of diplurans (order Diplura). See also, DIPLURANS.

PROKARYOTE. An organism whose cells lack a distinct nucleus.

PROLARVAE. Newly hatched larvae without completely functional legs, and incompletely sclerotized integument. Such larvae often remain aggregated.

PROLEG. A fleshy, unsegmented appendage serving as a leg and found on the abdomen of some holometabolous insects, particularly caterpillars and sawflies.

PROMINENCE. A section of the body that is raised, elevated, or projecting.

PROMINENT MOTHS (LEPIDOPTERA: NOTODONTIDAE). Prominent moths, family Notodontidae (including processionary moths), total 3,562 species from all faunal regions, most from the Neotropics (1,766 sp.); actual world fauna likely exceeds 4,000 species. The family is in the superfamily Noctuoidea, in the section Cossina, subsection Bombycina, of the division Ditrysia. The subfamily classification varies, but currently involves 8 subfamilies, with segregation into three groups: Oenosandrinina (for Oenosandrinae, with three species in Australia), Thaumetopoeinina (for Thaumetopoeinae, and Notodontinina (for the remaining six subfamilies). Dioptidae (including Doinae) are sometimes included in Notodontidae. Adults small to very large (20 to 124mm wingspan); some with massive bodies. Maculation varied, but many with subdued browns and grays; some white and a few more colorful or with iridescent markings. Adults mostly nocturnal. Larvae are leaf feeders, sometimes gregarious (especially among Thaumetopoeinae) and feeding nocturnally. Host plants include a large variety of plant families, especially for broadleaf forest trees. A number of economic species are

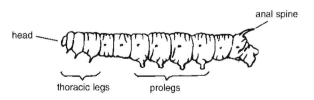

Fig. 844 Lateral view of a 'hornworm' caterpillar (Lepidoptera: Sphingidae).

known, especially among the processionary caterpillars (subfamily Thaumetopoeoinae).

John B. Heppner
Florida State Collection of Arthropods
Gainesville, Florida USA

References

Bender, R. 1985. Notodontidae von Sumatra. In *Heterocera Sumatrana*, 5: 1–101, 14 pl. Heterocera Sumatrana Society, Keltern.
Bryner, R. 2000. Notodontidae–Zahnspinner. In *Schmetterlinge und ihre Lebensräume: Arten-Gefährdung-Schutz. Schweiz und angrenzenden Gebiete*, 3: 403–524, pl. 20–23, 25. Pro Natura-Schweizerische Bund fuer Naturschutz, Basel.
Kiriakoff, S. G. 1964–70. Lepidoptera. Familia Notodontidae. In *Genera Insectorum*, 217(A): 1–213, 11 pl. (1964); 217(B): 1–238, 8 pl. (1967); 217(C): 1–269, 11 pl. (1968); 217(A) (Suppl.): 1–74 (1970); 219: 1–54, 3 pl. (1970). Brussels.
Miller, J. S. 1991. Cladistics and classification of the Notodontidae (Lepidioptera: Noctuoidea) based on larval and adult morphology. *Bulletin of the American Museum of Natural History* 204: 1–230.
Schintlmeister, A. 1992. Die Zahnspinner Chinas (Lepidoptera, Notodontidae). *Nachrichten des Entomologischen Verein Apollo Supplement* 11: 1–343.
Seitz, A. (ed.). 1912–34. Familie: Notodontidae. In Die Gross-Schmetterlinge der Erde, 2: 281–319, pl. 44–49, 56 (1912); 2(suppl.):172–186, 286, pl. 14–15 (1933–34); 6: 901–1070, pl. 143–159 (1931–34); 10: 605–655, pl. 79–84 (1930); 14: 401–444, pl. 68–72 (1928). A. Kernen, Stuttgart.

PROMOTER. A region of DNA crucial to the accuracy and rate of transcription initiation. Usually immediately upstream of the gene itself.

Fig. 845 Example of prominent moths (Notodontidae), *Epicoma melanosticta* Donovan from Australia.

PRONOTAL COMB. In fleas, a row of strong spines on the posterior margin of the pronotum.

PRONOTUM. The upper or dorsal surface of the prothorax, often shield-like in form.

See also, THORAX OF HEXAPODS.

PRONUNCIATION OF SCIENTIFIC NAMES AND TERMS. Many words of Greek and Latin origin are used in entomology and are of two kinds. The first kind is words that have been adopted into English and other modern languages to describe forms, structures and processes. The second kind is the Latin (scientific) names of taxa at all classificatory levels. Their pronunciation has caused confusion.

Once, Latin was a living language. For many hundreds of years after the fall of Rome, Latin was the common language in western Europe, a means by which educated people in various countries communicated. That is why Linnaeus in the 18th century wrote his Systema Naturae in Latin rather than in his native Swedish. It is why Latin was used for binominal nomenclature. Later, French and then English became widely used, and the editions of the International Code of Zoological Nomenclature are written in those two languages although the names of taxa are still nominally Latin and follow the rules of Latin grammar.

Even by the time of Linnaeus, Latin pronunciation had diverged from country to country in western Europe to acquire the characteristics of each native language. By the late 19th century, Latin as pronounced in England (and, by extension, also in the USA, Australia, Canada, and New Zealand) had acquired the vowel and diphthong sounds of English, whereas Latin as pronounced in France had acquired some peculiarities of French. Likewise in Germany and Italy. Although Latin was written identically in countries of western Europe, it no longer was pronounced the same. In recognition of this problem, Latin scholars met to reconstruct pronunciation of Roman Latin, and then to begin its teaching in the early decades of the 20th century. By the mid 20th century, Latin pronunciation as taught in Latin classes in England, the USA, other English-speaking countries, and countries of continental Europe was once again (more or less) uniform. So we may talk about two systems of Latin pronunciation: corrupted Latin as used in the 19th century in English (and differently in other modern languages), and restored Latin as now taught in Latin classes.

But, damage of two forms already had been done. First, many Latin words had been adopted into English and the corrupted form had become the norm in English, as shown for a few words below.

This corrupted English pronunciation is broadly used in biology, medicine and law, and is now engrained in English. We have to accept that such words are English, adopted from Latin, spelled as in Latin, but not pronounced as Latin is now taught. After all, English has adopted and mispronounced many words from other foreign languages. The words beau, bouquet, boutique, lieutenant and lingerie have been adopted from French. They are now used in France, are not relics of a long-defunct language, but yet are commonly mispronounced by English-speakers. The same kinds of changes have befallen other modern languages. French and Spanish too, despite regulation by academies designed to preserve their purity, have adopted, corrupted and mispronounced words from other languages, including English. They have likewise adopted words from Latin, and changed not only their pronunciation but also their spelling.

The International Code of Zoological Nomenclature specifies how scientific names of animals are to be formed. But it does not specify how they should be pronounced.

The most widely used basic entomology textbook in the USA (Borror et al. 1989) provides a few pages about construction of names of scientific names of insects. This explanation for the most part is good and useful. However, as to sounds of vowels, consonants, and diphthongs, it explains only the corrupted English pronunciation as taught in Latin classes in the USA in the 19th century. Its instructions are thus out of line with the way that Latin has been taught for the last 50 years. It promotes an outdated system that makes the pronunciation very hard to understand by anyone who learned Latin and/or pronunciation of scientific names in non-English-speaking countries. Generations of entomologists have relied upon this textbook to explain the latest classification of insects, most without realizing they were being subjected to an outdated concept of Latin pronunciation.

English-speaking botanists have been faced with the same dilemma. However, Stearn (1983) explains the restored Latin pronunciation together with the corrupted English pronunciation.

Restored Latin is the best hope of English-speaking entomologists to be able to communicate names of taxa to entomologists whose native language is something other than English, and to convert international congresses of entomology from something other than towers of Babel. Our objective should be to promote international communication by having a standard system for pronouncing scientific names. It should not be to perpetuate a corrupted, outdated, English mispronunciation of Latin, and even less to foist this on foreigners as being ''Latin.'' Pronunciation of Latin vowels, consonants and diphthongs according to the restored system is explained in the table above. It should not be difficult to learn, especially by anyone who has studied Italian or Spanish. It is further explained in modern textbooks on Latin grammar and in modern Latin dictionaries.

J. Howard Frank
University of Florida
Gainesville, Florida, USA

References

Borror, D. J., C. A. Triplehorn, and N. F. Johnson. 1989. *An introduction to the study of insects* (6th ed.). Saunders College Publishers, Philadelphia, Pennsylvania. xiv + 875 pp.

International Commission for Zoological Nomenclature. 1999. *International code of zoological nomenclature* (4th ed.). International Trust for Zoological Nomenclature, London, England. xxix + 306 pp.

Stearn, W. T. 1983. *Botanical Latin: history, grammar, syntax, terminology, and vocabulary* (3rd ed.). David and Charles, Newton Abbott, Devon, United Kingdom. xiv + 566 pp.

Latin word	Restored Latin pronunciation	Corrupted English pronunciation
Alumnae	alumn.eye	alumn.ee
Alumni	alumn.ee	alumn.eye
Larvae	lar.why	lar.vee
Pupae	poop.eye	pyoo.pee

Pronunciation of Latin vowels, consonants, and diphthongs as now taught in Latin classes

Vowels	Consonants
a (short) as in apple	b (as in English)
a (long) as in father	c as in cat
e (short) as in get	ch as English k (or k-h)
e (long) as in they	d (as in English)
i (short) as in pit	f (as in English)
i (long) as in machine	g as in go
o (short) as in not	h as in hence
o (long) as in note	i (consonant i [= j]) as y in yes
u (short) as in full	k (as in English)
u (long) as in brute	l (as in English)
y as "ew" but without any trace	m as in man
of y sound	n (as in English)
	ph as English p (or p-h)
Diphthongs	q as in quite
ae as y in English try	r always rolled
au as ou in English house	s as in sister
ei as in English rein	t as in tanned
eu as "ay-oo" (stress the "ay")	th as English t
oe as oi in English foil	v as English w
ui as in English we	x as in six
	z as in zero

PROPAGATIVE TRANSMISSION. Transmission of an arthropod transmitted disease wherein the causal organism does not undergo cyclical changes, but multiplies in the body of the arthropod vector.

See also, MECHANICAL TRANSMISSION, CYCLO-DEVELOPMENTAL TRANSMISSION, AND CYCLO-PROPAGATIVE TRANSMISSION.

PROPAGULE. A general term used to describe a reproductive (propagative) stage that will give rise to a new organism, usually in plants (seeds, corms, bulbs) but also in invertebrates (eggs, cysts).

PROPHYLACTIC CONTROL. Preventative control. Control procedures implemented in a pre-emptive manner to avoid the possibility of damage.

PROPLEURON. A lateral portion of the prothorax.
See also, THORAX OF HEXAPODS.

PROPOLIS. The resins and waxes collected by bees and used in construction of nests, and in sealing cracks in the nest wall.

PROPRIORECEPTOR. An internal sensory receptor that senses the internal body condition including relative position of a body's components.

PROPUPA. The first of the quiescent instars in Thysanoptera; this stage lacks functional mouthparts.

PROSOPISTOMATIDAE. A family of mayflies (order Ephemeroptera). See also, MAYFLIES.

PROSTERNUM. The sclerite between the front legs.
See also, THORAX OF HEXAPODS.

PROSTOMIDAE. A family of beetles (order Coleoptera). They commonly are known as juglar-horned beetles. See also, BEETLES.

PROSTOMIUM. A preoral, unsegmented portion of the body, anterior to the first true body segment. This is also known as the acron.

PROTEASE. An enzyme that degrades proteins.

PROTEIN. The polymeric compounds made up of amino acids.

PROTENEURIDAE. A family of damselflies (order Odonata). See also, DRAGONFLIES AND DAMSELFLIES.

PROTENTOMIDAE. A family of proturans (order Protura). See also, PROTURANS.

PROTEOMICS. The science and process of analyzing and cataloging all the proteins encoded by a genome (a proteome). Currently the majority of all known and predicted proteins have no known cellular function. Determining protein function on a genome-wide scale can provide critical clues to the metabolism of cells and organisms. Proteomics involves understanding the biochemistry of proteins, processes and pathways. Two-dimensional gel analyses were used in the late 1970s to identify proteins active (expressed) in different tissues at different times. Now, biological mass spectrometry is a powerful method for protein analysis, involving identification or localization of proteins and interactions of proteins.

PROTHORACIC GLANDS. Endocrine glands found in the thorax and secreting molting hormone (ecdysone) or a closely related ecdysteroid. They are activated by PTTH during the immature stage, and degenerate during or soon after metamorphosis.

See also, ECDYSONE AGONISTS, ECDYSTEROIDS, ENDOCRINE REGULATION OF INSECT REPRODUCTION, METAMORPHOSIS, PROTHORACICOTROPIC HORMONE.

PROTHORACICOTROPIC HORMONE. The growth and molting of insects are cyclical phenomena and are brought about predominantly by two hormones, one produced by neurosecretory cells in the insect's brain, and the other by glands in the prothorax, the prothoracic glands. Periodically, specific neurosecretory cells of the brain synthesize a hormone, the prothoracicotropic hormone (PTTH), that acts on the prothoracic glands, which, in turn, respond to this stimulus by synthesizing and releasing a steroid hormone [ecdysone (E), or 3-dehydroecdysone (3-dE), depending upon the species)] that is ultimately converted at target tissues to 20-hydroxyecdysone (20E), the principle molting hormone of insects. 20E then interacts with various target cells via binding to a high affinity nuclear receptor, which then modulates the expression of specific genes and ultimately, regulates cell growth and differentiation. In the case of the epidermal cells, 20E causes them to deposit a new cuticle and, thus, initiates the molting process. The cyclical synthesis and release of PTTH by the neurosecretory cells of the brain is elicited by environmental factors such as photoperiod, temperature, nutritional state, etc. In the case of moths and butterflies, such as the tobacco hornworm *Manduca sexta*, photoperiod and temperature are important and, indeed, the brain appears to possess an extraretinal photoreceptor, i.e., these PTTH-producing neurosecretory cells may be a self-contained unit directly perceiving the light signal, or perhaps the light signal is transduced elsewhere and the information transferred to these neurosecretory cells.

More than eight decades ago, the Polish biologist Stefan Kopeć first suggested that the control of insect molting was mediated by neurohormones. Using larvae of the gypsy moth, he extirpated the brains of these insects and showed that these 'de-brained' insects lived for several weeks. If the brains were removed ten days or more after the final larval molt, pupation occurred and brainless, but otherwise normal, moths emerged. However, if the brain was removed prior to the tenth day, the caterpillars failed to pupate (i.e., undergo a metamorphic molt) although they survived for a long time. He also showed that if the larvae were divided into two hemolymph-tight compartments by a ligature posterior to the thorax, both the anterior and the posterior portions pupated simultaneously if tied off after the tenth day, but only the anterior portion pupated if ligation occurred prior to this critical period. Kopeć concluded from these studies that the brain liberates a substance into the hemolymph that is essential for pupation, and that it is released on or about the tenth day after the final larval molt. Indeed, not only was this the beginning of the field of insect neuroendocrinology, but it was also the beginning of the science of neuroendocrinology in general, and led to the now accepted dogma that the nervous system of almost all animals functions as an endocrine gland.

About fifteen years after the work of Kopeć, the great British biologist, Vincent Wigglesworth, showed that decapitation of the blood-sucking hemipteran (bug) nymph *Rhodnius prolixus*, within three or four days after feeding, prevented molting, but that decapitation after this period did not. He concluded that there is a factor within the head (i.e., a portion of the brain) that initiates the molting process. These sorts of experiments have been verified consistently in literally hundreds of species over the past seventy years. Further studies in the 1940s showed that regions of the brain containing large neurosecretory cells were the active portions of this neuroendocrine organ. The classic studies of Piepho (Germany), Fukuda (Japan) and Williams (USA) in the 1940s showed dramatically, again via surgical manipulations, that the brain hormone (PTTH) exerted its effect humorally on the prothoracic glands. When the product of the prothoracic glands was identified as ecdysone in the 1950s, it was hypothesized that PTTH elicited, or enhanced, the synthesis and secretion of ecdysone from the prothoracic glands. This finding of a brain-prothoracic gland (neuropeptide-steroidogenesis) axis was of real interest both to insect physiologists examining the control of insect molting and to those interested in evolution, because

this system is quite analogous to the mammalian ACTH-adrenal cortex axis in both a general way, and subsequently, in many details of its transductory biochemistry.

The presence of neurosecretory cells in the nervous system is a very early event in the evolution of animals with data indicating their existence in organisms as primitive as coelenterates (*Hydra*). It should be noted that there are reports that prothoracic glands can also be controlled neurologically by neurons from the brain in some insects, and it has even been suggested that there is direct transport of neurosecretory products by way of channels in the basal lamina, an extracellular matrix that covers all insect tissues. For the most part, however, there is a consensus that PTTH leaves the brain, is stored in a neurohemal organ attached to the brain (the corpus allatum), is released into the hemolymph at specific times and activates the prothoracic gland.

By the 1970s, investigators began to examine the mechanisms by which PTTH might activate the prothoracic glands to synthesize ecdysone. It was believed by some that in order to study the complex transductory mechanisms involved in ecdysone biosynthesis, it would be much too difficult to work with the whole insect. They therefore attempted, and achieved, a model in which the prothoracic glands of *M. sexta* were placed in a culture medium (*in vitro*) in which they can live and function in an almost normal way for several weeks. Such glands were shown to synthesize and release 3-dE which was subsequently converted to E and then 20E. It was

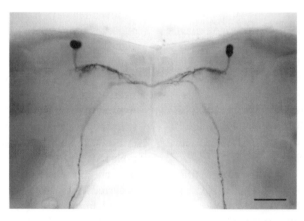

Fig. 846 Immunochemical detection of PTTH in the brain of a day-0 *Manduca sexta* pupa. Note staining in two cell bodies (prothoracicotropes) in each hemisphere of the brain and the crossing of the axon tracts as they move posteriorly to ultimately end in the corpus allatum (not shown) (from Gilbert et al., 2000).

demonstrated that extracts of the brain of the tobacco hornworm could stimulate the synthesis of 3dE in a consistent and predictable way. Of great importance was the observation that the right and left prothoracic glands of this insect secreted the same quantity of 3dE *in vitro* so that one gland of a single insect could be experimentally manipulated, with the other serving as the control. With this *in vitro* assay, a great deal has been learned about how the PTTH signal is transduced into enhanced ecdysteroid synthesis in the prothoracic gland.

The first indication that a mammalian-like signaling mechanism was utilized was the finding that brain extract stimulated the rapid synthesis of cyclic AMP (cAMP) in the prothoracic glands, and that this was followed by 3dE biosynthesis. (In reality, it is the mammal that used an insect-like signaling mechanism because insects evolved several hundred million years before mammals.) With this *in vitro* bioassay, it was also possible to assay particular parts of the *Manduca* brain as noted previously, and by dissecting out specific neurosecretory cells with an eyebrow hair, it was shown that there were two neurosecretory cells in each side of the brain of the insect that contained PTTH. These so-called prothoracicotropes contained varying amounts of PTTH activity during development. It was shown by immunocytochemical means that axons from these neurosecretory cells extend posteriorly and laterally such that they cross and terminate in the neurohemal organs (corpora allata) where PTTH is stored. Thus, the so-called critical period shown by Kopeć, Wigglesworth and many others is actually the time when PTTH is released from these neurohemal organs into the hemolymph due to one of a variety of environmental and physiological signals. It is of interest that cAMP is also a principal component of the signal transductory cascade in the mammalian ACTH (from the pituitary gland)- adrenal cortex axis.

Most of the data discussed here were generated using PTTH-containing brain extracts to study how PTTH activates ecdysteroid synthesis in the prothoracic glands. Consequently, there have been frequent questions as to whether all the changes in prothoracic gland biochemistry that are elicited by brain extracts are due solely to the effect of PTTH. However, very recently, the *Manduca* PTTH has been cloned, the nucleotide sequence determined and purified, recombinant PTTH produced by transformed cells. The use of this pure, recombinant PTTH in the *in vitro*

system has confirmed all observations on the prothoracic gland that were made using the brain extract.

PTTH-prothoracic gland interactions

PTTH has now been purified and cloned, first from the domestic silkmoth *Bombyx mori*, and subsequently from several other moths (Lepidoptera). The analysis of the PTTH protein and gene sequences indicates that this hormone is synthesized as a prohormone that is processed into a shorter homodimeric molecule (25–30 kDa) as it is transported down the axons of the prothoracicotropes to their termini in the corpus allatum. The structure specified by intramonomeric, cystine-cystine bonds is essential for the bioactivity of the hormone. Lepidopteran PTTHs clearly comprise a family of proteins, based on the distribution of cystine-cystine bonds, charged amino acids and hydrophilic regions, but these PTTHs are essentially species-specific in action, indicating that the amino acid sequence (primary structure) is likely important for bioactivity. Outside of the Lepidoptera, the structure of PTTH remains conjectural although the PTTH of the fruit fly *Drosophila melanogaster*, appears to be highly glycosylated and considerably larger than those of the moths. These results suggest two possibilities. First, PTTHs have diverged from an ancestral molecule to such a degree that PTTHs from different taxa have only limited structural similarity with minimal amino acid conservation. Second, PTTHs from different taxa (e.g., flies vs. moths) have evolved from different ancestral proteins, and their only similarity is their ability to elicit ecdysteroid synthesis by the prothoracic gland. Current data cannot resolve this issue.

Periodic increases in ecdysteroid titer are critical to pre-adult insect development, as discussed above, and reflect activation of the prothoracic gland by PTTH. PTTH release occurs in particular daily time "windows," following the integration of a variety of factors, such as time since last molt, nutritional status and physical size. How and where such factors are sensed is unknown, but studies do indicate that PTTH release is acutely controlled by the muscarinic class of acetylcholine-releasing neurons. A neuroendocrine modulation of PTTH release has recently been demonstrated in the cockroach *Periplaneta americana*, involving the neurohormone melatonin, which mediates day-night physiological differences in many organisms. While daily cycles of melatonin and

PTTH levels are not seen in *Periplaneta*, melatonin could be involved in controlling a daily cycle of PTTH release described in *Rhodnius*.

Changes in PTTH release during *Bombyx* development have been investigated using prothoracic glands *in vitro* as well as antibodies against PTTH. Although these two approaches were not in complete agreement, both indicate that major peaks of ecdysteroid secretion by the prothoracic gland are preceded by, and partly overlap with, high PTTH titers, and that some episodes of increased PTTH titer are not associated with obvious ecdysteroid production. This latter observation supports the hypothesis that PTTH may have functions in addition to the regulation of prothoracic gland ecdysteroidogenesis.

PTTH may not be the only molecule that stimulates ecdysteroidogenesis by the prothoracic gland, but our knowledge of the nature of other molecules, their biological significance and their modes of action, is rudimentary and preliminary. A brain-derived "small PTTH" (MW < 10,000) of *Manduca* is the best characterized of three candidate factors. Small PTTH stimulates ecdysteroid synthesis in larval glands *in vitro*, but is only weakly active when applied to pupal glands. Small PTTH appears to stimulate the same second messenger cascades as big PTTH (Ca^{2+} and cAMP: see below), suggesting that it might be an active proteolytic fragment of PTTH, generated during PTTH purification. Conclusive evidence that small PTTH is released into the circulation is lacking. Other candidate ecdysteroidogenic factors that can act upon the prothoracic gland *in vitro* have been partially purified from lepidopteran proctodaea (*Manduca sexta, Lymantria dispar* and *Ostrinia nubilalis*). Given the precedent of the vertebrate gut as an endocrine organ, such findings may not be surprising, but definite roles have yet to be demonstrated *in vivo* for these incompletely characterized insect molecules. It also would not be surprising if there were negative regulators (inhibitors) of prothoracic gland ecdysteroid synthesis and, indeed, at least two candidate molecules have been isolated, one from fly ovaries and a second from a moth brain, but the *in vivo* significance of these data remains conjectural.

The PTTH transductory cascade

In the 1970s, studies of the *Manduca* prothoracic gland showed that levels of cAMP, an important intracellular second messenger, were elevated during periods of increased ecdysteroid synthesis. Further research demonstrated that PTTH stimulated a rapid increase in gland cAMP content *in vitro*, detectable within minutes. However, cAMP generation is not the first intracellular change triggered by PTTH. Experiments revealed that cAMP generation was dependent on the influx of extracellular Ca^{2+}, suggesting that cAMP synthesis required Ca^{2+}-calmodulin dependent adenylyl cyclase activity, which is indeed present in prothoracic glands. A variety of Ca^{2+} channels exist in many cell types and it is not clear which type opens in response to PTTH, let alone the mechanism by which PTTH accomplishes this action. G-proteins, small proteins often associated with peptide hormone receptors, may be involved directly in Ca^{2+} channel opening, but other data indicate that cAMP could be involved in such channel regulation. The observation that PTTH stimulates rapid synthesis of cAMP indicates, based on our knowledge of analogous systems in vertebrates, that the PTTH receptor is likely to be situated in the cell membrane, and that binding of PTTH to the receptor is the first step in PTTH signal transduction. Furthermore, the PTTH receptor is predicted to belong to the family of proteins known as G-protein coupled receptors, and to possess an extracellular region that binds PTTH. Nevertheless, knowledge of the earliest events stimulated by PTTH in prothoracic glands is incomplete.

As intracellular levels of cAMP rise, the important cAMP-dependent kinase, protein kinase A (PKA), is rapidly activated in the prothoracic glands. (Kinases regulate the function of other proteins via the addition of phosphate groups.) However, the natural substrates of PTTH-stimulated PKA activity are not known, and a requirement for PKA activity in PTTH-stimulated ecdysteroidogenesis has not been proven unequivocally. It is clear that analogs of cAMP that enter the prothoracic gland stimulate ecdysteroidogenesis and that an inhibitory, stereoisomeric analog of cAMP blocks PTTH-stimulated ecdysteroidogenesis. However, PKA is not the only protein dependent on cAMP for activity, and, therefore, the possibility remains that proteins other than PKA transduce the PTTH signal into eventual ecdysteroid synthesis. For example, a mammalian cAMP-dependent guanine nucleotide exchange factor has been discovered that activates small G-proteins, leading to phosphorylation-dependent events, including the activation of the 70 kDa S6 kinase, a known target in PTTH action.

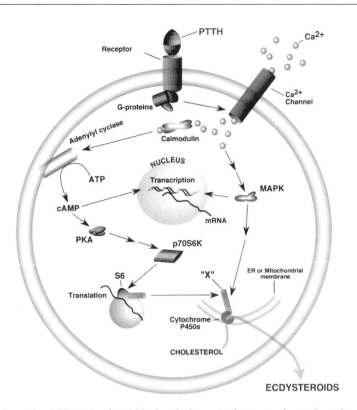

Fig. 847 A model for the major PTTH-stimulated biochemical events that occur in prothoracic gland cells. Proteins and processes connected by single arrows are likely to be directly interacting, while two arrows indicate the probable interposition of one or more additional proteins or events. PTTH, Prothoracicotropic hormone; cAMP, cyclic adenosine monophosphate; p70S6K, 70 kDa molecular weight S6 kinase; S6, ribosomal protein S6; MAPK, mitogen-associated protein kinase; ER, endoplasmic reticulum; 'x', hypothesized, newly synthesized protein that regulates the rate of ecdysteroid synthesis in prothoracic glands stimulated by PTTH.

PTTH stimulates a number of protein phosphorylations in the prothoracic gland via PKA, mitogen-activated protein kinases (MAPKs) and perhaps other kinases. The most prominent and consistent phosphorylated protein is the ribosomal protein S6. The importance of S6 phosphorylation in PTTH-stimulated ecdysteroid synthesis was demonstrated in studies using the drug rapamycin, which blocks S6 phosphorylation by inhibiting the activation of the 70 kDa S6 kinase. In the prothoracic gland, rapamycin inhibits not only S6 phosphorylation, but also PTTH-dependent ecdysteroid and protein synthesis. In vertebrate cells, S6 phosphorylation increases the rate of protein synthesis, especially of mRNAs possessing a polypyrimidine tract at their 5′ transcriptional end, and it is likely that it acts similarly in insect cells. The activation of MAPKs in PTTH-dependent signal transduction has only recently been demonstrated, along with the ability of MAPK inhibi-

tors to inhibit ecdysteroid synthesis. Like the ribosomal protein S6, MAPKs can regulate translation, although by another route, but whether this regulation occurs in the prothoracic glands is currently unknown.

Vertebrate and insect steroidogenic cells share many intracellular mechanisms that function in the regulation of steroid hormone synthesis, as noted previously. In both groups, peptide hormones produced by neurons or brain-associated cells bind to cell membrane receptors associated with G-proteins, with a subsequent generation of cAMP. Increases in intracellular Ca^{2+} occur, as do multiple protein phosphorylations, including that of the ribosomal protein S6. In both taxa, the final transductory step in acutely up-regulating steroid hormone production appears to be the synthesis of one or more short-lived proteins that abolish a rate-limiting bottleneck in the steroid synthesis pathway. In vertebrates, two such proteins are believed to facilitate the movement of cholesterol across the mitochondrial membrane, which is the rate-limiting step in vertebrate steroid hormone production. In insects, the rate-limiting step in ecdysteroid synthesis has not yet been characterized, but recent data from *Drosophila* suggest it may well involve movement of ecdysteroid precursors between intracellular compartments, like the endoplasmic reticulum and the mitochondrion, and that a carrier protein may be involved. Furthermore, it is known that PTTH-stimulated ecdysteroid synthesis in *Manduca* requires protein synthesis, as demonstrated by the ability of translation inhibitors to block PTTH-stimulated ecdysteroid synthesis, and that PTTH stimulates the synthesis of about 10 specific proteins.

PTTH might also regulate other cell types besides those of the prothoracic gland. A thorough discussion of this topic is not within the scope of this review, but it is noteworthy that significant levels of PTTH are present in adult brains, by which time the prothoracic glands have disappeared due to programmed cell death. This observation suggests that PTTH likely regulates cellular processes other than prothoracic gland ecdysteroid synthesis.

A number of the important PTTH-related events in the prothoracic gland is now known. However, knowledge about the events that occur in prothoracic gland cells after PTTH stimulation is still fairly incomplete. For instance, nothing is known about possible PTTH-elicited long term changes in the levels of ecdysteroid-synthesizing enzymes (cytochrome P450s) via either translation or transcription, an effect that might be expected based on vertebrate studies. The PTTH-stimulated activation of a mitogen-activated protein kinase is suggestive in this context, as this family of kinases can migrate into the nucleus to regulate transcription. Furthermore, there are enormous gaps in our knowledge of the events that occur between PTTH contacting a prothoracic gland cell and S6 phosphorylation, and between this phosphorylation and the first steps in the conversion of cholesterol to ecdysteroid. It is likely that continued, integrated molecular and biochemical studies of the PTTH control of ecdysteroid synthesis will fill many of these gaps in our knowledge and that major revisions in our current understanding will surely be necessary in the future.

See also, ECDYSTEROIDS, ECDYSONE AGONISTS, DIAPAUSE.

Lawrence I. Gilbert and Robert Rybczynski
University of North Carolina
Chapel Hill, North Carolina, USA

References
Gilbert, L. I., W. Combest, W. Smith, V. Meller, and D. Rountree. 1988. Neuropeptides, second messengers and insect molting. *BioEssays* 8: 153–157.
Gilbert, L. I., R. Rybczynski, and J. T. Warren. 2002. Control and biochemical nature of the ecdysteroidogenic pathway. *Annual Review of Entomology* 47: 883–916.
Gilbert, L. I., R. Rybczynski, Q. Song, A. Mizoguchi, R. Morreale, W. A. Smith, H. Matubayashi, M. Shionoya, S. Nagata, and H. Kataoka. 2000. Dynamic regulation of prothoracic gland ecdysteroidogenesis: *Manduca sexta* recombinant prothoracicotropic hormone and brain extracts have identical effects. *Insect Biochemistry and Molecular Biology* 30: 1079–1089.
Henrich, V., R. Rybczynski, and L. I. Gilbert. 1999. Peptide hormones, steroid hormones and puffs: mechanisms and models in insect development. pp. 73–125 in G. Litwack (ed.), *Vitamins and hormones*. Academic Press, San Diego, California.
Rybczynski, R., S. Bell, and L. I. Gilbert. 2001. Activation of an extracellular signal-related kinase (ERK) by the insect prothoracicotropic hormone. *Molecular and Cellular Endocrinology* 184: 1–11.
Song, Q., and L. I. Gilbert. 1997. Molecular cloning, developmental expression, and phosphorylation of ribosomal protein S6 in the endocrine gland responsible for insect molting. *Journal of Biological Chemistry* 272: 4429–4435.

PROTHORACIC PLATE. Equivalent to thoracic plate.

PROTHORAX. The most anterior of the three thoracic segments, bearing the first pair of legs.
See also, THORAX OF HEXAPODS.

PROTOCEREBRUM. The largest and most anterior segment of the insect brain, that innervates the compound eyes and ocelli.
See also, NERVOUS SYSTEM.

PROTONYMPH. In mites (Acari), the second instar.

PROTOPODITE. The basal portion of a segmented appendage.

PROTOTHEORIDAE. A family of moths (order Lepidoptera). They also are known as African

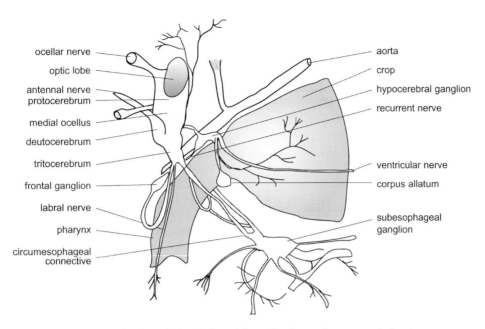

Fig. 848   Diagram of the insect brain, lateral view (adapted from Snodgrass, Insect morphology).

primitive ghost moths. See also, AFRICAN PRIMI-
TIVE GHOST MOTHS, BUTTERFLIES AND
MOTHS.

PROTUBERANCE.   A projection.

PROTURANS (PROTURA).   The order Protura
consists of minute soil-inhabiting hexapods characterized by the lack of eyes and antenna, a 12 segmented abdomen and development by anamorphosis. The
first three abdominal segments have small leg-like
appendages that are capable of movement. The first
pair of legs have enlarged foretarsi that are covered
with many types of setae and sensilla and function
as antenna. The first Proturan species discovered,
Acerentomon doderoi, was described in 1907 by
Silvestri. Some researchers believe that Protura is a
sister group to the Collembola, though Protura may
be a separate class. Here, it is included in the class
Entognatha. Protura have a worldwide distribution
with over 500 described species divided into two distinct suborders: Eosentomoidea and Acerentomoidea
with 9 families:

Class: Entognatha
  Order: Protura

Suborder: Eosentomoidea
  Family: Eosentomidae
  Family: Sinentomidae
Suborder: Acerentomoidea
  Family: Acerentomidae
  Family: Protentomidae
  Family: Hesperentomidae
  Family: Berberentomidae
  Family: Nipponentomidae
  Family: Acerellida
  Family: Antelientomidae

Proturans in the Eosentomoidea possess spiracular
openings on the meso- and metathorax connected to a
primitive tracheal system. Proturans in the Acerentomoidea are without specialized respiratory structures
and respire directly through the cuticle.

Proturans exhibit anamorphosis, a type of development that adds a body segment after a molt. The first
stage, or prelarva, is hatched from the egg with nine
abdominal segments and weakly developed mouthparts. The second stage or Larva I also has nine
abdominal segments but with fully developed mouthparts. Larva II has an additional abdominal segment
added between the eighth segment and the telson,
or last abdominal segment. The next stage is the
maturus junior which has eleven abdominal segments. The maturus junior molts to the adult stage

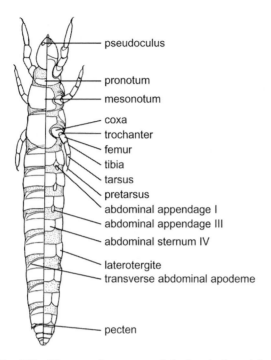

- pseudoculus
- pronotum
- mesonotum
- coxa
- trochanter
- femur
- tibia
- tarsus
- pretarsus
- abdominal appendage I
- abdominal appendage III
- abdominal sternum IV
- laterotergite
- transverse abdominal apodeme
- pecten

Fig. 849 Diagram of a proturan: left, dorsal view; right, ventral view.

except for males in the family Acerentomidae, which have an additional stage known as the pre-imago. The pre-imago has partially developed internal genitalia. It is not known if the adult stage continues to molt throughout the remainder of its life.

The life-history of Protura is poorly understood. Many species can be found in leaf litter, soil that is rich in organic matter, and dead wood. Similarly, the diet of protura is not well known. Their mouthparts are entognathous and most species appear to have modifications for feeding on fungi; however, some species have styletiform or grinding structures. Like most soil arthropods, proturans most likely feed on a variety materials including plants and fungi as well as scavenging on dead arthropods. In culture, Proturans have been observing feeding on mushroom powder, dead mites, and mycorrhizal fungi.

Protura can be collected from soil and leaf litter with Berlese funnels or by the centrifugation sugar flotation technique. They can be stored in 70% ethanol until mounted on permanent microscope slides with Hoyer's or other clearing medium.

Like many other soil arthropod groups, relatively few proturans have been described by taxonomists. Not surprisingly, distribution records for this group are far from adequate to attempt to understand their biogeography. Undoubtedly, there are many hundreds of species yet to be found from the tropics as well as temperate climate areas.

Christopher Tipping
University of Florida
Quincy, Florida, USA

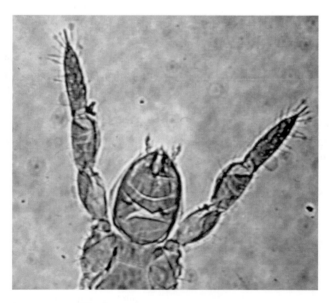

Fig. 850 Foretarsi of *Eosentomon maryae* Tipping.

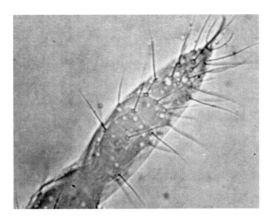

Fig. 851   Close-up of foretarsi showing setae and sensilla.

References

Imadaté, G. 1974. *Fauna Japonica Protura (Insecta)*. Keigaku, Tokyo, Japan. 351 pp.

Nosek, J. 1973. *The European Protura: their taxonomy, ecology and distribution with keys for determination*. Museum d'Histoire Naturelle, Geneva, Switzerland. 345 pp.

Tuxen, S. L. 1964. *The Protura. A revision of the species of the world with keys for determination*. Hermann, Paris, France. 360 pp.

PROVANCHER, (L'ABBÉ) LÉON. Léon Provancher was born at Bécancour, Quebec, Canada, on March 10, 1820. Educated for the Catholic priesthood and later designated Abbot, he eventually gave up this work due to poor health and moved to Cap Rouge, Quebec. There he devoted his time to the natural sciences, including botany, birds, molluscs, worms, and entomology. In 1869 he began publication of the journal "Le Naturaliste Canadien," which was nearly 8,000 pages long before being terminated for lack of support. Also, he commenced publication of the "Petite Faune Entomologique du Canada," which began with a volume on Coleoptera, published in 1877, and eventually included treatises on Orthoptera, Neuroptera, Hymenoptera, and Hemiptera. He labored on this project until 1890, and his most important contributions are on Hymenoptera and Hemiptera. He described hundreds of insects, including 923 species of Hymenoptera alone. This is a particularly remarkable achievement because he was isolated from libraries and collections. He died at Cap Rouge, Canada, on March 23, 1892.

References

Anonymous. 1895. L'Abbe Provancher. *Entomological News* 6: 7.

Essig, E. O. 1931. *A history of entomology*. The Macmillan Company, New York. 1029 pp.

Mallis, A. 1971. *American entomologists*. Rutgers University Press, New Brunswick, New Jersey. 549 pp.

PROVENTRICULUS.   A valve that controls entry of food into the midgut, and located at the terminus of the foregut. It is a muscular organ and capable of some grinding action.

See also, ALIMENTARY SYSTEM, ALIMENTARY CANAL AND DIGESTION.

PROXIMAL.   Pertaining to the part of an appendage closer to the body.

PRUINOSE.   Covered with whitish waxy powder. This condition is common in aphids and scales.

PSEPHENIDAE.   A family of beetles (order Coleoptera). They commonly are known as water-penny beetles. See also, BEETLES.

PSEUDERGATE.   In lower termites, a caste from individuals regressing from nymphs, or derived from

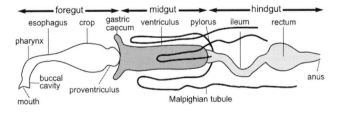

Fig. 852   Generalized insect alimentary system (adapted from Chapman, The insects: structure and function).

larvae. In either case, they comprise the worker caste, but can develop into other castes following additional molts.

PSEUDIRONIDAE. A family of mayflies (order Ephemeroptera). See also, MAYFLIES.

PSEUDOANTAGONISM. An aggressive response by pollinators to orchids that appear to be invaders of the pollinator's territory. The pollinator's response can result in pollination of the orchids.

PSEUDOCAECILIIDAE. A family of psocids (order Psocoptera). See also, BARK-LICE, BOOK-LICE, OR PSOCIDS.

PSEUDOCOCCIDAE. A family of insects in the superfamily Coccoidae (order Hemiptera). They sometimes are called mealybugs. See also, SCALE INSECTS AND MEALYBUGS, BUGS.

PSEUDOCOPULATION. A copulatory response by male pollinators to orchids that mimic females of the pollinator species. The pollinator's response can result in pollination of the orchids.

PSEUDO-CURLY TOP TREEHOPPER, *MICRUTALIS MALLEIFERA* (FOWLER) (HEMIPTERA: MEMBRACIDAE). The pseudo-curly top treehopper or nightshade treehopper, *Micruotalis malleifera* (Fowler), is the only known member of Membracidae to transmit any plant disease agent. This insect was first reported to transmit tomato pseudo-curly top virus (PCTV) in south Florida in 1957. PCTV is now recognized as a member of Geminiviridae based on the inclusion body in the infected cell, serological tests, and sequence evidence. PCTV is transmitted by *M. malleifera* in a semipersistent-circulative manner. Transmission can also be achieved by injecting the treehopper vector with either crude sap or partially purified virus preparation. The transmission efficiency by sap injection is estimated at 30%. Both adults and nymphs are efficient vectors of PCTV. Nymphs retain the virus trans-

missibility after molting. The transmission efficiency reaches 60% after a 6-hr acquisition access period (AAP). The median incubation period of PCTV in *M. malleifera* is estimated at 15 hours after a 6-hour AAP. The average retention period in the vector is 12 days. Pseudo-curly top disease has been a chronic problem in tomato production in south Florida since the 1940s. The incidence of pseudo-curly top on tomato reaches as high as 50% in some years. PCTV host plants include tomato (*Lycopersicon esculentum* Mill.), eggplant (*Solanum melongena* L.), night shade (*S. nigrum* L.), lettuce (*Lettuca sativa* L.), ragweed (*Ambrosia* sp.), tobacco (*Nicotiana glutinosa* L.), chickweed (*Stallaria medea* L.) and jimson weed (*Datura stramonium* L.).

At $25 \pm 1°C$ using eggplant, nighshade, and ground cherry (*Physalis floridana* Rydb.) as rearing hosts, all *M. malleifera* nymphs display five instars. The average developmental times for instars 1 to 5 are 4.6, 4.3, 4.4, 5.4 and 7.8 days, respectively, on eggplant; 4.8, 4.5, 4.6, 5.2, 7.6 days, respectively, on nightshade; and 4.6, 4.3, 4.4, 5.4 and 7.8 days, respectively, on ground cherry. The size of instars 1 to 5 averages as follows: 1.3 mm long and 0.4 mm wide for the first instar, 1.7 mm long and 0.6 mm wide for the second instar, 2.3 mm long and 0.8 mm wide for the third instar, 3.1 mm long and 1.2 mm wide for the fourth instar, and 4.2 mm long and 1.8 mm wide for the fifth instar.

The average adult longevities for females and males are 56.1 and 37.2 days on eggplant, 33.7 and 30.6 days on nightshade and 12.0 and 10.6 days ground cherry, respectively. The adult female is 4.6 mm in length and 2.5 in width. The male is 4.2 mm long and 2.2 mm wide. The eggs are deposited mainly on terminal stems and petioles, but often eggs are found on eggplant leaves. The eggs are imbedded under the epidermal layer with a portion of the egg exposed. The egg is translucent, with one blunt end, and averages 0.9 mm long, 0.3 mm wide. At $25 \pm 1°C$, the average egg incubation period is 13.6 days (range: 12 to 16 days). The mean preoviposition period is 3.6 days. The oviposition period is 51.3 days on eggplant. The average number of eggs per day per female is 2.04. The average number eggs laid per female is 55. Both nymphs and adults are very placid and docile.

Tomato pseudo-curly top is not an economically important problem in tomato production areas. Nightshade plant is the main host for both PCTV and

M. malleifera, which can be effectively controlled in the field and the adjacent areas.

James H. Tsai
University of Florida Ft. Lauderdale, Florida, USA

References

Briddon, R. B., I. D. Bedford, J. H. Tsai, and P. G. Markham 1996. Analysis of the necleotide sequence of the treehopper-transmitted geminivirus, tomato pseudo-curly top virus, suggests a recombinant origin. *Virology* 219: 387–394.

McDaniel, L. L., and J. H. Tsai 1990. Partial characterization and serological analysis of pseudo-curly top virus. *Plant Disease* 74: 17–21.

Simons, J. N. 1962a. The pseudo-curly top disease in south Florida. *Journal of Economic Entomology* 55: 358–363.

Simons, J. N. 1962b. Life-history and behavioral studies on *Micrutalis malleifera*, a vector of pseudo-curly top virus. *Journal of Economic Entomology* 55: 363–365.

Simons, J. N., and D. M. Coe 1958. Transmission of pseudo-curly top virus in Florida by a treehopper. *Virology* 6: 43–48.

Tsai, J. H. 1989. Biology and ecology of treehopper transmission of a geminivirus. 6th International Conference on Comparative and Applied Virology. October 15–21. Banff, Alberta, Canada. Symposium Abstract W7-3.

Tsai, J. H., and L. G. Brown. 1991. Pseudo-curly top of tomato. Plant Pathology Circular No. 344, Florida Department of Agriculture and Consumer Services, Gainesville, Florida.

Tsai, J. H. 2002. Bionomics of *Micrutalis malleifera* Fowler and its transmission of pseudo-curly top virus. pp. 351–361 in K. F. Harris, O. P. Smith, and J. E. Duffus (eds.), *Virus-insect-plant interactions*. Academic Press, New York, New York.

PSEUDOPHASMATIDAE. A family of walking-sticks (order Phasmatodea). They commonly are known as striped walkingsticks. See also, WALKINGSTICKS AND LEAF INSECTS.

PSEUDOPOD. A soft, foot-like appendage, especially on larvae of Diptera.

PSEUDOPOSITOR. A tube-like modification of the tip of the abdomen in certain female insects. See also, ABDOMEN OF HEXAPODS.

PSEUDOPARASITISM. A response by parasitoids to orchids that are apparent potential hosts. The parasitoid's response results in increased pollination of the orchids.

PSEUDOSTIGMATIDAE. A family of damselflies (order Odonata). See also, DRAGONFLIES AND DAMSELFLIES. A family of damselflies (order Odonata). See also, DRAGONFLIES AND DAMSELFLIES.

PSILIDAE. A family of flies (order Diptera). They commonly are known as rustflies. See also, FLIES.

PSILOPSOCIDAE. A family of psocids (order Psocoptera). See also, BARK-LICE, BOOK-LICE, OR PSOCIDS.

PSOCIDAE. A family of psocids (order Psocoptera). See also, BARK-LICE, BOOK-LICE, OR PSOCIDS.

PSOCIDS. Members of the insect order Psocoptera. See also, BARK-LICE, BOOK-LICE, OR PSOCIDS.

PSOCOPTERA. An order of insects, formerly known as Corrodentia. They commonly are known as bark-lice, book-lice, or psocids. See also, BARK-LICE, BOOK-LICE, OR PSOCIDS.

PSOQUILLIDAE. A family of psocids (order Psocoptera). See also, BARK-LICE, BOOK-LICE, OR PSOCIDS.

PSYCHIATRY AND INSECTS: PHOBIAS AND DELUSIONS OF INSECT INFESTATIONS IN HUMANS. Insects are an integral and influential part of our culture, as illustrated by their infiltration of our language, arts, history, philosophy, and religion. However, as human society has become progressively more urbanized, insects have become progressively more estranged. As significant but increasingly intangible elements of our culture, insects now feature prominently in certain psychiatric disorders, much as do religious and extraterrestrial elements. Our perception of insects can range from appropriate apprehension when faced with the possibility of a bee sting, through subclinical and clinical insects phobias, to full blown psychotic delusions of insect attacks and infestations. Here we examine firstly phobias about insects, or entomophobia, which includes acarophobia (scabies) and arachnophobia (spiders). Secondly, we deal with the rarer and more serious delusions about insects which are experienced in some psychiatric disorders.

Most people are at least wary, if not fearful, of certain insects (more correctly arthropods). This may be a reasonable fear based on knowledge or experience (bees, wasps, spiders, mosquitoes), an unreasonable but culturally understandable repulsion (cockroaches or flies), or a misplaced fear resulting from inadequate information (dragonflies, moths, crickets). A true insect phobia, on the other hand, is defined as a 'persistent irrational fear of and compelling desire to avoid insects', which results in significant distress in the sufferer despite recognition that the fear is excessive or unreasonable. The syndrome represents only the tip on an iceberg, with much unnecessary avoidance behavior never reaching a level where medical attention is sought or necessary.

Although insect phobias probably occurred before recorded history, insects are less likely to have been phobic objects in the past. In hygienically urbanized western societies, many people have little first hand experience of insects other than common flies, mosquitoes, cockroaches and ants. Such urban societies are not as mentally or physically prepared for arthropod encounters as are rural communities.

It is not unreasonable to assume that the danger and annoyance insects have caused to man over the millennia has resulted in an ingrained fear of insects in most societies. Bites and stings to humans and domestic animals act not only as stimulators of toxic and allergic reactions, but insects have been the vectors of potentially fatal diseases since prehistoric times (e.g., dengue or malaria). This explanation, however, is likely to account only for some cases at the non-clinical end of the phobic spectrum, not the genuine phobias which satisfy the definition given above. In the latter clinical cases, as in other phobias, the more likely cause is a displacement of diffuse anxiety to an external focus which can be avoided. The choice of insects as the phobic object may be random, symbolic, or perfectly logical. When symbolic the insects often represent filth and soiling. An example of a 'logical' choice of insects as phobic objects is illustrated by the example of a 3-year old girl with an insect phobia, whose symptoms resulted from her being told that her sister with pneumonia had died from a 'bug'!

Treatment for entomophobia (and other phobic disorders) is highly specialized, and is largely determined by the therapist's individual preferences. Methods which have been applied in the past include supportive psychotherapy, desensitization, insight psychotherapy, drug therapy (anxiolytics), modeling, hypnotic regression and reframing, implosive therapy, and combinations of these therapies.

Some detailed information about which insects are dangerous and which are not is probably one of the most useful things that an entomologist can contribute to people unfortunate enough to suffer from this disorder. For example, as most entomologists will know, dragonflies are quite harmless, despite their fearsome name and the fact that they feature as phobic objects in many a person's entomophobia. It is useful to remember that only blood feeders (mosquitoes, fleas, ticks, bedbugs) actively pursue humans. The more common phobic objects (spiders, bees) never bite or sting unless trapped or seriously threatened. The former category comprises insects which are associated with poverty or poor sanitation, but perhaps surprisingly, insects in the latter category are feared the most despite being beneficial to man.

'Delusions of parasitosis' can be defined as an unshakeable false belief that live organisms are present in the skin. The disorder was first described in 1894 by Thibierge, who named it acarophobia. Confusingly the syndrome has also been referred to as dermatophobia, parasitophobia, and entomophobia. In 1946 Wilson and Miller more correctly referred to the disorder as delusions of parasitosis. The 'offending' organisms range from insects to worms and bacteria, the type often depending upon the parasitological knowledge of the patient. Delusions of parasitosis constitute a symptom complex rather than a disease entity, and are found in a variety of physical and mental diseases. The condition can be difficult to differentiate from entomophobia, particularly when there is a phobia of infestation. Patients with delusions of parasitosis actually experience the state of being infested. This is fundamentally different from having a fear of becoming infested, which falls into the category of entomophobia. Even though the syndrome is a psychiatric disorder, patients usually visit a dermatologist rather than a psychiatrist, since they are convinced they have a dermatological problem. Referral to a psychiatrist is nearly always rejected, and the dermatologist has the difficult task of treating these patients.

The syndrome may be preceded by an original and very real arthropod infestation acting as a 'trigger'. Other possible triggers include the itch resulting from systemic disease (diabetes, TB, syphilis) or from alcohol withdrawal ('the DT's' – delirium tremens). The sufferer usually complains about itching, biting, stinging, burning and crawling sensations. Insects are often described as black or white, jumping, and sometimes emerging from cosmetics or toothpaste. The 'matchbox' sign is ominous, where the patient brings unidentifiable specimens to the doctor or pest controller at the first visit. Microscopic examination of the contents usually reveals only lint, scabs, or other household dust. Such negative findings invariably lead to more intense collection and presentation of specimens by the sufferer. One can often elicit a list of attempted treatments including all imaginable varieties of detergents, balms and poisons. Thus there may be excoriations produced by the fingernails on the skin as well as signs of chemical burns as a result of attempts to kill the parasites. The patient is compelled to dig the parasites out, especially before going to bed, and often resorts to the use of a knife, tweezer or other sharp implement, leaving skin lesions consistent therewith.

Because of the variety of diseases in which delusions of parasitosis occur, there is no generally accepted approach to treatment. Psychotherapy and psychoanalysis have been successful in treating delusions of parasitosis associated with repressed conflicts over sexuality and aggression, and drug treatment can provide significant relief of both itch and delusions. A number of other treatments are used less frequently or have fallen out of favor (including ECT - electroconvulsive therapy).

The prognosis is quite variable, and often dependent on those other diagnosed or undiagnosed diseases which contribute to the symptomatology. The itch (formication) of delirium tremens, for example, has an excellent prognosis, but the prognosis is worse in schizophrenia. The prognosis in patients suffering from paranoid conditions is very poor, because these cases are usually not very subjectable to effective treatment. One author describes paranoiacs who would dig into their skins 'up to the time of involuntary parting, and who probably still dig, under somebody else's auspices'!

The details of natural history of the infesting organisms related by patients are often quite complex, and depend upon the patients' previous entomological knowledge. Imagined animals range from fleas, lice and scabies through itchmites, bedbugs and worms, to nondescript 'black things' and insects new to science. The therapist must assess the feasibility of these details by consulting medical entomology texts or local entomologists. It is important to remember that psychiatric patients can be the unwilling hosts of real lice, mites, and bedbugs as easily as can anyone else.

Philip Weinstein and David Slaney
Wellington School of Medicine and Health Sciences
Wellington, New Zealand

References

Bourgeois, M., J. M. Amestov, and J. Durand 1981. Délires d'infestation, dermatozooes et ectoparasitoses délirantes, syndrome d'Ekbom. *Annales Medico Psychologiques* 139: 819–828.

Ekbom, K. A. 1938. Der praesenile Dermatozoenwahn. *Acta Psychiatrica et Neurologica Scandinavica* 13: 227–259.

Olkowski, H., and W. Olkowski 1976. Entomophobia in the urban ecosystem, some observations and suggestions. *Bulletin of the Entomological Society of America* 22: 313–317.

Trabert, W. 1995. 100 years of delusional parasitosis; meta-analysis of 1,223 case reports. *Psychopathology* 28: 238–246.

Thibierge, G. 1894. Les acarophobes. *Revue Generale de Clinique et de Therapeutique* 8: 373–376.

Waldron, W. G. 1962. The role of the entomologist in delusory parasitosis (entomophobia). *Bulletin of the Entomological Society of America* 82: 81–83.

Weinstein, P. 1994. Insects in psychiatry. *Cultural Entomology* 1: 10–15.

Wilson J. W., and H. E. Miller 1946. Delusion of parasitosis (acarophobia). *Archives of Dermatology and Syphilology* 54: 39–56.

PSYCHIDAE. A family of moths (order Lepidoptera). They commonly are known as bagworm moths. See also, BAGWORM MOTHS, BUTTERFLIES AND MOTHS.

PSYCHODIDAE. A family of flies (order Diptera). They commonly are known as moth flies and sand flies. See also, FLIES.

PSYCHOMYIIDAE. A family of caddisflies (order Trichoptera). They (as well as Polycentropodidae) commonly are known as trumpet-net and tube-making caddisflies. See also, CADDISFLIES.

PSYCHOPSIDAE. A family of insects in the order Neuroptera. See also LACEWINGS, ANTLIONS, AND MANTIDFLIES.

PSYLLIDAE. A family of insects in the order Hemiptera. They sometimes are called psyllids or lerp insects. See also, BUGS.

PSYLLIDS. Members of the family Psyllidae (order Hemiptera). See also, BUGS.

PSYLLIPSOCIDAE. A family of psocids (order Psocoptera). See also, BARK-LICE, BOOK-LICE, OR PSOCIDS.

PTEROLONCHIDAE. A family of moths (order Lepidoptera). They are commonly called lance-wing moths. See also, LANCE-WING MOTHS, BUTTER-FLIES AND MOTHS.

PTEROMALIDAE. A family of wasps (order Hymenoptera). See also, WASPS, ANTS, BEES, AND SAWFLIES.

PTERONARCIDAE. A family of stoneflies (order Plecoptera). They sometimes are called giant stone-flies. See also, STONEFLIES.

PTEROPHORIDAE. A family of moths (order Lepidoptera). They commonly are known as plume moths. See also, PLUME MOTHS, BUTTERFLIES AND MOTHS.

PTEROSTIGMA. The dense, often discolored portion of the costal margin of a wing, usually at the end of the radius. This is also known as the stigma.
See also, WINGS OF INSECTS.

PTEROTHORAX. The fused meso- and meta-thorax found in certain winged insects.
See also, THORAX OF HEXAPODS.

PTEROTHYSANIDAE. A family of moths (order Lepidoptera) also known as Parnassian moths. See also, PARNASSIAN MOTHS, BUTTERFLIES AND MOTHS.

PTERYGOTE. An insect bearing wings, or derived from winged ancestors. A member of the Class Insecta, subclass Pterygota.

PTHIRAPTERA. An order name in some classification systems, and comprised of Mallophaga (chewing lice) and Siphunculata (sucking lice). See also, CHEWING LICE, SUCKING LICE.

PTHIRIDAE. A family of sucking lice (order Siphunculata). This family is also known as pubic lice See also, SUCKING LICE.

PTILIIDAE. A family of beetles (order Coleoptera). They commonly are known as feather-winged beetles. See also, BEETLES.

PTILINUM. An inflatable organ on the front of the head, thrust out from a suture just above the base of the antennae, in higher flies. It is expanded when the adult insect is escaping from the puparium.

PTILODACTYLID BEETLES. Members of the family Ptilodactylidae (order Coleoptera). See also, BEETLES.

PTILODACTYLIDAE. A family of beetles (order Coleoptera). They commonly are known as ptilodactylid beetles. See also, BEETLES.

PTILONEURIDAE. A family of psocids (order Psocoptera). See also, BARK-LICE, BOOK-LICE, OR PSOCIDS.

PTINIDAE. A family of beetles (order Coleoptera). They commonly are known as spider beetles. See also, BEETLES.

PTTH. Abbreviation for prothoracicotropic hormone.

PTYCHOPTERIDAE. A family of flies (order Diptera). They commonly are known as phantom crane flies. See also, FLIES.

PUBESCENCE. A covering of setae (hairs).

PUBESCENT. Covered with hair-like structures (setae in insects, trichomes in plants).

PUBIC LICE. Members of the family Pthiridae (order Siphunculata). See also, HUMAN LICE, SUCKING LICE.

PUDDLING BEHAVIOR BY LEPIDOPTERA.
Visitation of mud puddles and patches of moist soil – puddling – is common among many Lepidoptera, as well as other insects such as leafhoppers (Cicadellidae). The phenomenon is most conspicuous among brightly colored butterflies that often form large aggregations at roadside puddles, along streams, or beside pastureland ponds. Among butterflies such as pierids and swallowtails, the presence of one individual often serves as a catalyst for the formation of these aggregations. Puddling is also common among many species of moths and leafhoppers, which are more scattered on the soil at night and do not form the spectacular mud-puddle clubs often seen in diurnal Lepidoptera.

Some puddling species of Lepidoptera pump water through their guts, exuding droplets of water from the tip of the abdomen. Certain species of notodontid moths sit on a film of water and pump large quantities of fluid through their guts, discharging up to 8.5 ml per hour in rhythmic jets up to 30 cm. Even dry patches of soil are visited by some species of butterflies and moths that can moisten the substrate with a bead of saliva passed down the proboscis. Some skippers moisten the substrate by flexing the tip of the abdomen beneath the body and exuding a drop of fluid that they reimbibe.

Although puddling behavior can have different functions such as acquisition of water and thermoregulation, it is believed to play an important role in the procurement of salts, particularly sodium, which is in short supply for many herbivorous insects. Sodium can increase reproductive success significantly for both males and females. Whether the insects are butterflies, moths, or leafhoppers, the sex ratio typically is exclusively or predominantly in favor of males. The few females that visit soil are usually old and worn. The explanation for the biased sex ratio lies in the need to acquire salts for mating. When males pass a spermatophore (packet of sperm) to females, they lose a significant amount of salts that can be reacquired most readily from salt-rich sources, particularly soil and animal products such as dung, sweat, urine, and the exudates of carcasses. The dung of carnivores has a greater attraction than that of herbivorous mammals, presumably because of the richer supply of sodium. Females, although they lose a significant amount of sodium when they lay their eggs, are able to replenish their supplies, in part, from sodium in the male's spermatophore, which is passed to the female during mating. In their first mating, males may transfer a third of their abdominal sodium to a female.

Peter H. Adler
Clemson University
Clemson, South Carolina, USA

References
Adler, P. H. 1982. Nocturnal occurrences of leafhoppers (Homoptera: Cicadellidae) at soil. *Journal of the Kansas Entomological Society* 55: 73–74.
Adler, P. H. and D. L. Pearson 1982. Why do male butterflies visit mud puddles? *Canadian Journal of Zoology* 60: 322–325.
Adler, P. H. 1982. Soil- and puddle-visiting habits of moths. *Journal of the Lepidopterists' Society* 36: 161–173.
Arms, K., P. Feeny, and R. C. Lederhouse 1974. Sodium: stimulus for puddling behavior by tiger swallowtail butterflies, *Papilio glaucus*. *Science* 185: 372–374.
Pivnick, K. A., and J. N. McNeil. 1987. Puddling in butterflies: sodium affects reproductive success in *Thymelicus lineola*. *Physiological Entomology* 12: 461–472.
Smedley, S. R., and T. Eisner 1996. Sodium: a male moth's gift to its offspring. *Proceedings of the National Academy of Science of the USA* 93: 809–813.

PUFFING. A swelling in the giant polytene chromosomes of salivary glands of many dipterans.

PULICIDAE. A family of fleas (order Siphonaptera). They sometimes are known as common fleas. See also, FLEAS.

Fig. 853 Streamside aggregation of puddling pierid butterflies.

PULSATILE ORGAN. This is a pulsating heart-like organ. The function is to maintain circulation through the appendages, including the wings.

PULVILLUS (PL., PULVILLI). Soft pad-like structures found between the tarsal claws, and the short, stiff hairs on the underside of the tarsal joints.

PUNCTATE. Containing impressed points, punctures, or dimples.

PUNKIES. Members of the family Ceratopogonidae (order Diptera). See also, FLIES.

PUPA. The nonfeeding, immobile stage between the larval and adult stages in insects with complete metamorphosis. A stage where major reorganization of the body take place.

PUPARIATION. Formation of the puparium by larval Diptera.

PUPARIUM (PL., PUPARIA). The hardened, thickened integument of the last instar larva of Diptera, in which the pupa is formed.

PUPATION. Formation of the pupal stage in holometabolous insects.

PUPIPAROUS. Giving birth to fully developed larvae that are ready to pupate.

PURPLE SCALE. See also, CITRUS PESTS AND THEIR MANAGEMENT.

PUTOIDAE. A family of insects in the superfamily Coccoidae (order Hemiptera). They are sometimes called giant mealybugs. See also, SCALES AND MEALYBUGS, BUGS.

PUTZEYS, JULES ANTOINE ADOLPHE HENRI. Jules Putzeys, an early Belgian entomologist, was born at Liège, Belgium, on May 1, 1809. He received a doctoral degree from the University of Liège in 1929, and assumed a successful administrative and judicial career. However, he was fascinated by insects, and studied Lepidoptera, Odonata, and especially Coleoptera. In the case of the latter group, he specialized in Cicindelidae, Carabidae, and Pselaphidae. He served as president of the Entomological Society of Belgium from 1874 to 1876, and remained active in the society afterwards. His

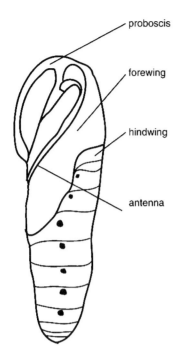

Fig. 854 Side view of a moth pupa (Lepidoptera: Sphingidae).

most important publications were "Prémices entomolgiques" and "Monographie de *Clivina* et des genres voisins." He died on January 2, 1882.

References
De Borre, A. P. 1882. Notice nécrologique sur Jules Putzeys. *Annales de la Société Entomologique de Belgique* 26: iiii–vii.

PYGIDICRANIDAE. A family of earwigs (order Dermaptera). See also, EARWIGS.

PYGIDIUM. The tergum of the last abdominal segment. The supraanal plate. This is the segment often left uncovered by the elytra of beetles.

PYGIOPSYLLIDAE. A family of fleas (order Siphonaptera). See also, FLEAS.

PYGMY BACKSWIMMERS. Members of the family Pleidae (order Hemiptera). See also, BUGS.

PYGMY GRASSHOPPERS. A family of grasshoppers (Tetrigidae) in the order Orthoptera. See also, GRASSHOPPERS, KATYDIDS AND CRICKETS.

PYGMY MOLE CRICKETS. A family of grasshoppers (Tridatylidae) in the order Orthoptera. See also, GRASSHOPPERS, KATYDIDS AND CRICKETS.

PYGMY MOTHS (LEPIDOPTERA: NEPTICULIDAE). Pygmy moths, family Nepticulidae, comprise the most minute moths known, with 868 species described in the world from all faunal regions, although most (over 510 sp.) are from the Palearctic region. The total fauna likely exceeds 1,200 species worldwide. The family, together with Opostegidae, forms the superfamily Nepticuloidea, in the section Nepticulina of division Monotrysia, in the infraorder Heteroneura. There are two subfamilies: Pectinivalvinae and Nepticulinae. Adults minute to small (2.5 to 8 mm wingspan), with head rough-scaled and with a large head tuft of scales, often distinctly colored; haustellum is short and naked (unscaled); labial palpi short, dropping and 3-segmented; maxillary palpi 5-segmented and folded; antennae have an eye-cap developed on the basal segment. Wing maculation is gray or brown, or more brightly colored, sometimes with metallic iridecense. Wing venation is heteroneurous but greatly reduced, with large fringes of hair-like scales on the wing margins and frenular bristles for wing coupling. Adults are diurnally active. Larvae are leafminers, usually blotch mines, although some also mine other plant parts; a variety of host plants are used.

John B. Heppner
Florida State Collection of Arthropods Gainesville, Florida USA

References
Johansson, R., E. S. Nielsen, E. J. van Nieukerken, and B. Gustafson. 1990. The Nepticulidae and Opostegidae (Lepidoptera) of North West Europe. In *Fauna Entomologica Scandinavica*, 231–739 (2 pts). E. J. Brill, Leiden.
Newton, P. J., and C. Wilkinson 1982. A taxonomic revision of the North American species of *Stigmella* (Lepidoptera: Nepticulidae). *Systematic Entomology* 7: 367–463.
Puplesis, R. K. 1994. *The Nepticulidae of Eastern Europe and Asia: Western, Central and Eastern Parts.* Backhuys Publishing, Leiden, 291 pp, 841 fig.

Puplesis, R., and G. S. Robinson. 2000. A review of the Central and South American Nepticulidae (Lepidoptera) with special reference to Belize. *Bulletin of the Natural History Museum, Entomology* 69(1): 3–114.

Scoble, M. J.. 1978. Nepticulidae of southern Africa: a taxonomic revision of the genus *Stigmella* Schrank (Lepidoptera: Monotrysia). *Annals of the Transvaal Museum* 31: 1–129.

PYLORIC VALVE. The valve between the midgut and the hindgut.

PYRALIDAE. A family of moths (order Lepidoptera). They commonly are known as snout and grass moths. See also, SNOUT MOTHS, BUTTERFLIES AND MOTHS.

PYRETHROIDS. Synthetic insecticides that are structurally similar to the toxic components of pyrethrum. See also, INSECTICIDES.

PYRETHRUM. Natural insecticide derived from certain plants in the genus *Chrysanthemum.* Pyrethrum is highly valued for its rapid effects on insects and low toxicity to mammals. See also, BOTANICAL INSECTICIDES, PYRETHRUM AND PERSIAN INSECT POWDER.

PYRETHRUM AND PERSIAN INSECT POWDER. There are numerous examples of plant natural products with interesting agrochemical properties; however, few plant products have had a major role in the development of commercial insecticides. The most important and significant actual application of a plant natural product is pyrethrum or Persian Insect Powder. This powder also has been known as Dalmatian and Japanese powders, or Buhach. The latter powder was produced by the Buhach Producing and Manufacturing Company in California, USA, starting in 1876 from the seeds of Dalmatian pyrethrum.

Pyrethrum powders are made from the dried flowers of several *Chrysanthemum* of the family Compositae. The Persian Insect Powder, obtained from the dried flowers of Persian pyrethrum (*Chrysanthemum coccineum* Willd.), is easily distinguished from those of the Dalmatian pyrethrum (*C. cinerariaefolium*

Vis.) by the purple color of the ray florets. As a result, the Persian Insect Powder is darker in color than the Dalmatian Powder. The plants yielding pyrethrum are also known as pyrethrum. Several other species of *Chrysanthemum* such as *C. achilleae* L., *C. myconis* L., *C. parthenium* Bernh., *C. segetum* L., also are recorded to have toxic properties against insects, but *C. coronarium* L., *C. indicum* L., *C. leucanthemum* L. and *C. frutescens*. L. have only negligible toxicity.

History

The use of powdered dry flower of pyrethrum as an insecticide was a long-established folk practice

Fig. 855 Pyrethrum powders are made from the dried flowers of several *Chrysanthemum* species, including the Persian pyrethrum (*Chrysanthemum coccineum* Willd.), which belong to the family Compositae. Persian pyrethrum is now principally grown for ornamental purposes and exists in many different horticultural forms and colors.

in the Caucasus region and northwest Iran. This powder, known as "Persian Insect Powder" or "Persian Dust," became an article of trade during the 18th century and was exported by caravan from Persia, which is now called Iran. It was introduced into Europe in the 1820s, where it was processed and commercialized. The insecticide's efficacy soon became so apparent that the supply reaching Europe was insufficient to meet the demand. The situation changed, however, when an Armenian merchant discovered the secret of its preparation while traveling in the Caucasus and his son started manufacturing the powder on a large scale in 1828. In spite of this, there were still constraints on the supply. Hand collection of the flowers has always demanded low labor costs and, as a result, has limited its major growing regions to developing countries.

The Persian Insect Powder is reported to have been a mixture of the ground flowers of *C. roseum* Adams and *C. carneum* Bieb. Nevertheless, only the former species is distributed in highlands (up to 2,000 m above the sea level) of Armenia, Caucasus and northern Iran and both species may be synonyms for *C. coccineum* Willd. Commercial cultivation of this species started in Armenia in 1828, and continued until 1840, when the greater insecticidal activity of *C. cinerariaefolium* (known as "Dalmatian Pyrethrum" or "Dalmatian Powder") became known.

The Persian Insect Powder was introduced into Europe in 1828 and into the United States in 1876, and then into Japan, Africa, and South America. Parallel to this development, the "Dalmatian Pyrethrum," which exists naturally along the east coast of the Adriatic Sea extending from Italy to northern Albania and up into the mountainous regions of Croatia, Bosnia and Herzegovina, and Montenegro, was cultivated in 1840 in Dalmatia (a historic region along the Adriatic coasts of Yugoslavia, now part of Croatia). The crop was later introduced into many parts of the world. An interesting story traces the discovery of the effect of the Dalmatian pyrethrum to a German woman of Dubrovnik, Dalmatia, who picked a bunch of flowers for their beauty. When they withered she threw them into a corner where, several weeks later, they were found surrounded by dead insects. She associated the death of the insects with insecticidal property of the flowers and embarked in the business of manufacturing pyrethrum powder.

Use and formulations

The use of pyrethrum spread rapidly because, in the doses necessary to kill household insects, it is nontoxic to humans and domestic animals. The initial uses of pyrethrum was mainly against insects of public health importance. About 1916, kerosene extracts of pyrethrum flowers appeared on the market and were used widely as sprays against flies and mosquitoes. Until the outbreak of World War II, its principal application was as a dust for household use against bed-bugs, fleas, cockroaches and similar pests. It was not until 1918 that pyrethrum sprays were offered on the market for household use. One of the main uses of pyrethrum today is in the preparation of such sprays. Fly sprays for the protection of cattle were also developed. The main supply now is in the form of extracts in hydrocarbon solvents, concentrated to 25%. Furthermore, mosquito coils made from 1.3% pyrethrum powder are widely used in tropical areas. The slow-burning coils are ignited and the smoke acts both as a killing agent and as a repellent. These coils initially irritate the mosquitoes and then motivate them to fly from the source of the stimuli.

Pyrethrum also is one of the most widely used insecticides for controlling insects in stored products because of its low mammalian toxicity, broad spectrum of activity and short residual life. There are many pyrethrum products formulated for stored products insect control. These range from consumer liquid and aerosol products for control of the insects in private homes, to a wide variety of professional products for use by commercial pesticide applicators. In the latter category, the formulations are either used to protect commodities through direct application to grain (including repellent treatments to packaging material), or for space, contact, and spot treatments within food handling establishments. They include, but are not limited to, bakeries, cafeterias, canneries, commercial airplanes, hospitals, mobile caterers, restaurants, schools and supermarkets.

In agriculture, pyrethrum is unique in being exempt from the establishment of tolerances when applied to growing crops. As a result of having a zero-day preharvest interval, it can be used even on the day of harvest. Furthermore, its broad spectrum of activity as well as its renowned capacity for rapid knock-down allows it to be used against a wide range of agricultural insects. However, because of its very low stability, it is not cost-effective on major crops with large acreages where other insecticides, especially photostable pyrethroids, may be used. Nevertheless, the characteristics listed above make it an ideal insecticide for small fields. In addition, one of the traditional agricultural uses of pyrethrum is a tank-mixture at low application rates (about 10 to 20 g pyrethrum per ha) with conventional insecticides. This type of mixture exploits the rapid action of pyrethrum on the insect nervous system because insects which are agitated by the action of pyrethrum are exposed to greater quantities of both pyrethrum and the insecticide in the mixture.

Synergists

It has been known since the 1930s that both the knock-down and the killing effects of pyrethrum and some pyrethroids can be greatly optimized by adding synergists that have little or no toxicity alone. During this period, a number of synergists such as sesamin were discovered. Piperonyl butoxide, the most widely used synergist, was the first highly effective synergist found to potentiate the pyrethrum by 5- to 20-fold. Although the mode of action of synergists is not completely understood, one of their activities is believed to inhibit the oxidative and/or hydrolytic metabolism of the insecticide within the insect.

The pyrethrum plant

There is considerable confusion in the literature about the taxonomy of pyrethrum (i.e., the plant species from which pyrethrum has been obtained). Recent revisions have even changed the genus *Chrysanthemum* to *Tanacetum*. As a result, the two species of plants with historical and commercial importance, namely the Persian Insect Flower (painted daisy, having red flowers) and the Dalmatian Insect Flower (having white flowers) are respectively considered *Tanacetum coccineum* (Willd.) Grierson and *Tanacetum cinerariifolium* (Trevir) Schultz-Bip, though entomologists continue to refer to these plants as *Chrysanthemum* spp. Both are daisy-like herbaceous

Piperonyl Butoxide

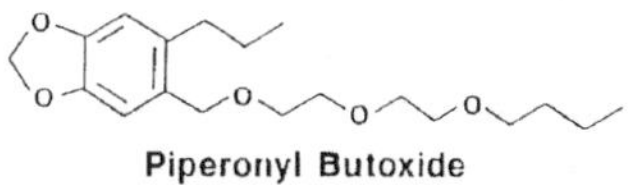

Fig. 856 Piperonyl butoxide, the most widely used synergist, is the first highly effective synergist found to potentiate pyrethrum at the range of 5- to 20-fold.

perennials belonging to the sunflower family (Compositae).

Pyrethrum is a perennial herbaceous plant which, to the casual observer, resembles the ordinary field daisy. When the plant flowers, many shoots originate from the crown and grow on average to 75 cm. The leaves are petioled and finely cut. The shoots branch a few times before terminating into a white (or red, according to the species), daisy-like flower that consists of a few to several hundred flowers per plant. The flower head of pyrethrum, like any other species belonging to the family Compositae, is a compound inflorescence with small flowers, the florets, aggregated together on a convex receptacle (known as the capitulum). The florets are of two kinds: the disc florets and the ray florets. While disc florets with yellow corollas occupy the center of the receptacle, the ray florets with white corollas form the outer rim of the flower head. The disc florets possess both male and female organs, but the ray florets are unisexual (female alone). In contrast to *C. cinerariaefolium* which possesses white ray florets, the ray florets of *C. coccineum* maybe pink, rose, carmine, crimson or, rarely, white. Moreover, the dried flowers of the latter species is easily distinguished from the former by the purple color of the ray florets and the ten-ribbed achenes (instead of five ribs in *C. cinerariaefolium*). *C. coccineum* produces fewer flowering shoots than *C. cinerariaefolium* and is said to be somewhat more resistant to disease and injury. Commercial flowers vary from 6 to 24 mm in width and from 70 to 300 mg in weight.

Pyrethrum production

Pyrethrum propagates easily through different methods including seeds, vegetative splits, stem cuttings rooted under mist and tissue culture. World production of pyrethrum is about 20,000 tons per year. Although the original homes of the commercial flowers of pyrethrum have been Iran, Armenia and Caucasus, a closely related species is now grown in the highlands of East Africa (in particular Kenya, Tanzania, Rwanda and Zaire), Australia (mainly Tasmania), and Papua New Guinea. Eastern Africa, however, remains the main source of supply. For example, from a worldwide production of about 18,000 metric tons in 1992, Kenya alone contributed about 70%. Historically, the major pyrethrum producing countries have been Yugoslavia (until World War I) and Japan (until World War II). By 1941, however, Kenya overtook Japan as the main world producer and Japan ceased to be a significant pyrethrum producer.

Today, the main commercial source of pyrethrum is the mature flower of *Chrysanthemum* (*Tanacetum*) *cinerariaefolium* Vis., cultivated principally in Kenya and Tanzania at elevations over 1,500 m above sea level. In Kenya, pyrethrum is cultivated almost entirely by about 60,000 small-scale farmers. The crop is not only the main source of cash income for these farmers, but it also provides a major source of export revenue for the country. Pyrethrum cultivation is optimal in the temperatures that occur in the highlands of Kenya (1,800 to 2,900 m above sea level). Well-drained soils with moderate organic matter are ideal for crop production. New stands in Kenya are established either from seed or from clonal propagation. In commercial plantations of pyrethrum in Kenya, the first picking of mature flowers occurs shortly after flowering and within 3 to 4 months after transplanting. It is carried out at intervals of 10 to 14 days and continues for ten consecutive months. At the end of this period, the stand is cut back to remove the dead and unproductive plant material. This practice allows the plant to rest and to get ready for growth in the next season. Flower picking is done by hand, then the flowers are dried and delivered to the factory where crude (oleoresin) as well as high-quality refined extract concentrates are produced. About fifteen parts of pyrethrum flowers are required to produce one part of oleoresin, and up to fourteen parts of oleoresin are required to obtain one part of toxic principle of pyrethrum. A small quantity of fine pyrethrum powder is also prepared for sale to manufacturers of insecticide dusts and mosquito coils. It is interesting to note that *C. coccineum* (the origin of Persian Insect Powder) is now principally grown for ornamental purposes and exists in many different horticultural forms.

Active ingredients

The actual investigation on the active components of pyrethrum was carried out by two Swiss chemists (H. Staudinger and L. Ruzicka) at Zurich from 1910 to 1916, but their results were not published until 1924. They discovered that the chief insecticidal components of natural pyrethrum consist of six closely related esters (namely, pyrethrin I, pyrethrin II, cinerin I, cinerin II, jasmolin I, and jasmolin II). While esters of series I are derivatives of chrysanthe-

mic acid and have excellent insect killing activity, esters of series II which are derived from pyrethric acid have high knock-down properties. In commerce, flower heads containing at least 0.7% active constituents are extracted by solvents. The extract is then further processed to get a standard 25% concentrate which consists of about 10% pyrethrin I, 9% pyrethrin II, 2% cinerin I, 3% cinerin II, 1% jasmolin I and 1.1% jasmolin II. The amount of insecticidal constituents of pyrethrum varies, depending on the source. While Kenya flowers are the richest, containing about 1.3% active compounds, some strains developed recently may yield over 2% active compounds. It is interesting to note that pyrethrum continues to compete with synthetic insecticides in the specialized areas where selectivity and low environmental hazard are most important. Furthermore, because of the selective and relatively small-scale uses of pyrethrum during the past centuries, there has been relatively little development of insect resistance. In spite of this, the synthetic pyrethroids may threaten the commercial use of pyrethrum in the future because over-application may lead to a proliferation of pyrethroid-pyrethrum cross-resistant insects.

Pyrethroids

Among the classes of insecticides, the aforementioned six molecules (described under active ingredients) are unique for the intensity of their very rapid action against many species of insects and for their minimal hazard to mammals under normal conditions. Nevertheless, all six esters decompose rapidly and lose their insecticidal activity on exposure to air and light. Therefore, the natural compounds and earlier synthetic analogs are generally suitable only for indoor or protected applications.

In fact, the development of the first commercially competitive pyrethrin analog (resmethrin) by the Rothamsted group, led by Michael Elliott, represents an important milestone in agrochemical research. This discovery subsequently inspired chemical companies to develop a wide spectrum of new synthetic pyrethroids. The discovery of related synthetic pyrethroids stable enough for agricultural applications beginning in the 1970s, has been a revolutionary step toward the production of numerous potent and stable analogs with very low mammalian toxicity. While some pyrethroids are applied with a field rate as low as 2.5 g ha^{-1} (e.g., Deltamethrin), the LD$_{50}$

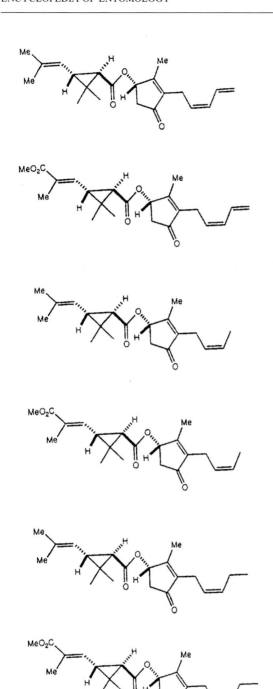

Fig. 857 Structures of chief insecticidal components of natural pyrethrum discovered by Swiss chemists at Zurich (H. Staudinger and L. Ruzicka) during the years 1910 to 1916: From top to bottom: Pyrethrin I, Pyrethrin II, Cinerin I, Cinerin II, Jasmolin I and Jasmolin II.

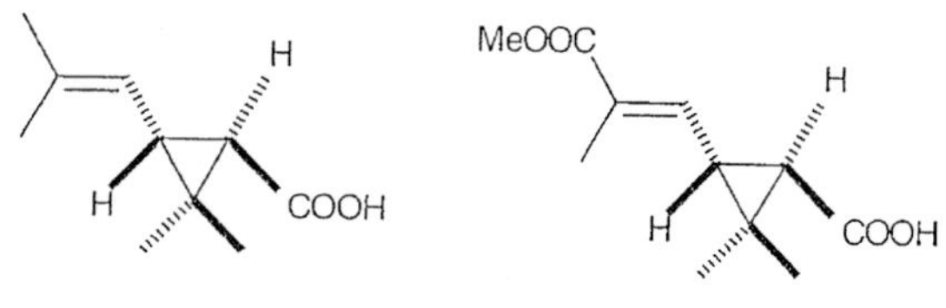

Fig. 858 Structures of chrysanthemic acid (left) and pyrethric acid (right).

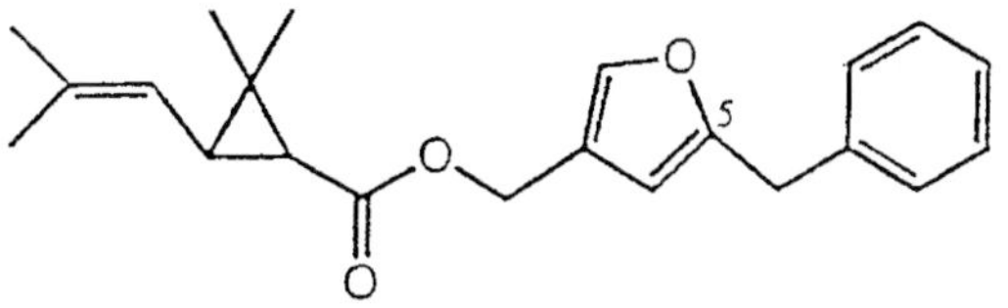

Fig. 859 *Resmethrin*: The first commercially competitive pyrethrin analog developed by the Rothamsted group.

(acute oral toxicity to rats) of some of them may amount to over 4,200 mg kg^{-1} (e.g., Etofenprox). These analogs, which possess favorable properties like those of the natural compounds and at the same time have much greater potency and stability, now constitute about one-quarter of all the insecticides used worldwide. Surprisingly, the small change in structure between the natural and synthetic products has led to a 34-fold increase in its toxicity to houseflies. The addition of synergists in formulations may still lead to much higher activity.

Cyrus Abivardi
Swiss Federal Institute of Technology Zurich, Switzerland

References

Benner, J. P. 1994. Pesticides from nature: Part I. Crop protection agents from higher plants – an overview. pp. 217–229 in L. G. Copping (ed.), *Crop protection agents from nature: natural products and analogues.* The Royal Society of Chemistry, Cambridge, United Kingdom. 501 pp.

Casida, J. E. (ed.). 1973. *Pyrethrum – the natural insecticides.* Academic Press, New York, New York. 329 pp.

Casida, J. E., and G. B. Quistad, G. B. (eds.) 1995. *Pyrethrum flowers: production, chemistry, toxicology, and uses.* Oxford University Press, Oxford, United Kingdom. 356 pp.

Elliott, M. 1994. Synthetic insecticides related to the natural pyrethrins. pp. 254–300 in L. G. Copping (ed.), *Crop protection agents from nature: natural products and analogues*. The Royal Society of Chemistry, Cambridge, United Kingdom. 501 pp.

Holman, H. J. 1940. *A survey of insecticide materials of vegetable origin*. The Imperial Institute, London, United Kingdom. 155 pp.

Zargari, A. 1992. *Medicinal plants*, Vol. 3 (5th ed.). Tehran University Publications, Tehran, Iran. 889 pp. (in Persian).

PYRGOTID FLIES. Members of the family Pyrgotidae (order Diptera). See also, FLIES.

PYRGOTIDAE. A family of flies (order Diptera). They commonly are known as pyrgotid flies. See also, FLIES.

PYROCHROIDAE. A family of beetles (order Coleoptera). They commonly are known as fire-colored beetles. See also, BEETLES.

PYRROLIZIDINE ALKALOIDS AND TIGER MOTHS (LEPIDOPTERA: ARCTIIDAE). Tiger moths (Arctiidae) are known for their bright aposematic coloration meant to repel predators. Their caterpillars often feed on alkaloid-rich host plants, and it is thought that the noxious substances ingested with the leaves determine the toxicity of both larvae and adults. Laboratory rats die when injected with a dose of arctiid alkaloid, and we can safely assume that a natural predator will at least get sick after tasting these brightly colored (and thus memorable) insects, and will avoid eating them in the future. Probably due to these substances, many species of tiger moths were able to become diurnal (daytime) fliers, and many other moths, butterflies, and even beetles form Müllerian mimicry complexes: they all resemble alkaloid-containing arctiid species, so predators should view them all as distasteful.

It has been also suggested that the toxic substances might be synthesized anew by arctiids. In fact, it is the ability of caterpillars to tolerate the noxious compounds secreted for defense against predators that allows them to feed on the toxic plant species. Indeed, the role of pyrrolizidine alkaloids in the biology of arctiid moths is complex.

The significance of pyrrolizidine alkaloids ingested by the larvae in the reproductive biology of arctiids has been studied extensively. It was shown that alkaloids of the arctiid *Creatonotus gangis* L. influence the morphology and chemistry of coremata (large abdominal hair-covered tubes, which are expanded by a male during the courtship, emitting pheromones). The main compound responsible for the scent can be secreted only when pyrrolizidine alkaloids are present in the larval diet. The size and weight of coremata is also proportional to the amount of alkaloids ingested by the larva.

Numerous brightly colored arctiid moths are attracted to the damaged and wilting tissues of Tree Heliotrope (*Tournefortia argentata*, Boraginaceae),

Fig. 860 Arctiid, *Echromia collaris*, and noctuid, *Asota* sp., feeding on alkaloid-rich tissues of a Tree Heliotrope, *Tournefortia argentata*, Solomon Islands.

feeding on sap exuded from leaves and branches. Pyrrolizidine alkaloids that are obtained through feeding on the decomposing plant matter by adult butterflies are used to synthesize danaidone, a compound found in the pheromone secretion of the hair-pencil organ in males that serves as a sexual attractant. Hence, one might assume that alkaloids ingested by the adult tiger moths also are used for pheromone production, or at least for boosting their chemical defense against predators. However, research on several tiger moth species in Florida found that moths instead are seeking pyrrolizidine alkaloids as nitrogen-rich nutrients, which they utilize differently depending on their sex. Egg productivity by females who were mated to alkaloid-deprived males was significantly lower than those of females whose mates had a chance to feed on alkaloid-containing plants. Only young males usually feed on pyrrolizidine alkaloids, subsequently passing nutrients with the spermatophore to a female during the copulation. The spermatophore is then stored in the female's bursa copulatrix and is used up gradually. Unlike males, females seek pyrrolizidine alkaloids after they spend several days on the wing. Alkaloids apparently are metabolized by females, enahancing egg production in their nutrient-depleted abdomens. For an unknown reason, only males or only females of each species in Florida feed on the alkaloids, but on the Solomon Islands, either sex of arctiids feed, with mating also occurring at the congregation sites.

The exact reasons for the phenomenon described above is yet to be understood, but it is clear that multiple biochemical mechanisms underlie the attraction of arctiid moths to various alkaloid-containing plants.

Andrei Sourakov
University of Florida
Gainesville, Florida, USA

References

Boppré, M., and D. Schneider. 1989. The biology of *Creatonotos* (Lepidoptera: Arctiidae) with special reference to the androconial system. *Zoological Journal of the Linnean Society* 96: 339–356.

Goss, G. J. 1979. The interaction between moths and plants containing pyrrolizidine alkaloids. *Environmental Entomology* 8: 487–493.

Hartmann, T., A. Biller, L. Witte, L. Ernst, and M. Boppré. 1990. Transformation of plant pyrrolizidine alkaloids into novel insect alkaloids by arctiid moths (Lepidoptera). *Biochemical Systematics and Ecology* 18: 549–554.

Rotshild, M., R. T. Alpin, P. A. Cockrum, J. A. Edgar, P. Fairweather, and R. Lees. 1979. Pyrrolizidine alkaloids in arctiid moths (Lep.) with a discussion on host plant relationships and the role of these secondary plant substances in the Arctiidae. *Biological Journal of the Linnaean Society* 12: 305–326.

PYRRHOCORIDAE. A family of bugs (order Hemiptera). They sometimes are called red bugs or cotton stainers. See also, BUGS

PYTHIDAE. A family of beetles (order Coleoptera). They commonly are known as dead log bark beetles. See also, BEETLES.

Q

Q-FEVER. See also, TICKS.

QUADRAT. A sampling unit used to assess population density.

QUALITATIVE DEFENSES. (of plants) These are chemical defenses that are toxic, including alkaloids, cardiac glycosides, mustard oils and phytoecdysones, and which are produced by plants to deter herbivory by insects. Qualitative defenses tend to be low molecular weight chemicals that are active at low concentrations, and often are found in unpredictable (short-lived) plant tissues. (contrast with quantitative defenses).

QUANTITATIVE DEFENSES. (of plants) These are chemical defenses that possess protein-complexing and digestibility reducing abilities such as tannins, resins and enzyme inhibitors, and which are produced by plants to deter herbivory by insects. Quantitative defenses are produced in high concentrations in plant tissue, especially in predictable (long-lasting) plant tissues where they are compartmentalized to protect against autotoxicity problems. (contrast with qualitative defenses).

QUARANTINE. Protocols established to prevent entry of an unwanted organism into an area, or out of an area. Also a building constructed to contain organisms, usually for research or breeding.

QUASISOCIAL BEHAVIOR. A level of sociality less than eusocial behavior. A type of presocial behavior. It involves individuals of the same generation sharing a nest and cooperating in brood care.

See also, PRESOCIAL, SOLITARY, COMMUNAL, SUBSOCIAL, SEMISOCIAL, EUSOCIAL BEHAVIOR.

QUATERNARY PERIOD. A geological period of the Cenozoic era, extending from about 2 million years ago to the present.

QUEEN. The primary reproductive in semisocial or eusocial insects.

QUEENSLAND TICK TYPHUS. See also, TICKS.

QUEEN SUBSTANCE. A pheromone component produced by honey bee queens that regulates several aspects of hive function and mating.

QUIESCENCE. A reversible state of inactivity initiated by adverse environmental conditions. Quiescence is also sometimes known as torpor. See also, DIAPAUSE.

QUIET-CALLING KATYDIDS. A subfamily of katydids (Meconematinae) in the order Orthoptera: Tettigoniidae. See also, GRASSHOPPERS, KATYDIDS AND CRICKETS.

R

RACE. A geographically or genetically distinct subgroup of a species.

RADIOLABELING. The attachment of a radioactive atom to a molecule, incorporation of ^{32}P-dNTPs into DNA.

RADIAL CELL. A cell of the wing bordered anteriorly by a branch of the radius.
See also, WINGS OF INSECTS.

RADIAL SECTOR. The lower of the two branches of the radius.

RADIUS. The third of the longitudinal wing veins, between the subcosta and the media. It is subdivided into up to five branches.
See also, WINGS OF INSECTS.

RADOSHKOWSKY, OCTAVIUS JOHN. Octavius Radoshkowsky (Radochkoowsky, Radochkoffsky, Radoszkowskii, etc.) was born in Lomza, Poland on August 7, 1820. He was born into the Polish nobility and educated in Poland. His training also included the military, and he served in the Polish artillery, retiring as Lieutenant-General in 1879. Radoshkowsky had a general interest in entomology, and is considered a systematist and faunist. However, he favored the Hymenoptera, and is best known for his contributions to the Hymenoptera of the Russian Empire. Perhaps his most important contribution, however, is his role in founding and supporting the Russian Entomological Society. He was a founding member, and in his capacity of President he did much to ensure the stability of the Society by acquiring a residence and annual subsidy. He served as president from 1867–1880. Due to serious illness and family circumstances Radoshkowsky relocated from Russia to Poland. He died in Warsaw, Poland, on May 1, 1895.

Reference

Essig, E. O. 1931. *A history of entomology*. The Macmillan Company, New York. 1029 pp.

RAIN BEETLES. Members of the family Pleocomidae (order Coleoptera). See also, BEETLES.

RAIN BEETLES. Members of the family Pleocomidae (order Coleoptera). See also, BEETLES.

RALLIDENTIDAE. A family of mayflies (order Ephemeroptera). See also, MAYFLIES.

RANDOM DISTRIBUTION. A distribution of organisms that lack pattern or order. The distribution is indistinguishable from a pattern based on chance.

RANDOM SAMPLING. A method of allocating sampling units within a sampling universe in which each sample unit has an equal chance of being selected. Random sampling ensures an unbiased estimate, but is rarely used in the practice of arthropod sampling due to time and cost constraints.

See also, SAMPLING ARTHROPODS.

RAPD-PCR. RAPD is derived from the term 'Random Amplified Polymorphic DNA.' It is a type of PCR using single primers of arbitrary nucleotide sequence consisting of 9 or 10 nucleotides with a 50–80% G+C content, and no palindromic sequences. These 10mers can act as a primer for PCR and yield reproducible polymorphisms from random segments of genomic DNA.

RAPHIDIID SNAKEFLIES. Some members of the family Raphidiidae (order Raphidioptera). See also, SNAKEFLIES.

RAPHIDIIDAE. A family of insects in the order Raphidioptera. They commonly are known as raphidiid snakeflies. See also, SNAKEFLIES.

RAPHIDIOPTERA. An order of insects. They commonly are known as snakeflies. See also, SNAKEFLIES.

RAPSMATIDAE. A family of insects in the order Neuroptera. See also LACEWINGS, ANTLIONS, AND MANTIDFLIES.

RAPTORIAL. Suitable for grasping prey. This term usually is used to describe legs.

RASPBERRY CANE BORER. See also, SMALL FRUIT PESTS AND THEIR MANAGEMENT.

RASPBERRY FRUIT WORM. See also, SMALL FRUIT PESTS AND THEIR MANAGEMENT.

RASTER. A complex of hairs, spines and bare areas on the ventral surface of the last abdominal segment of scarab beetle larvae.

RATARDIDAE. A family of moths (order Lepidoptera) also known as Oriental Parnassian moths. See also, ORIENTAL PARNASSIAN MOTHS, BUTTERFLIES AND MOTHS.

RATEMIIDAE. A family of sucking lice (order Siphunculata). See also, SUCKING LICE.

REARING OF INSECTS. Insect rearing is not much different from other forms of animal husbandry and it has similar rewards for a person who is patient, persistent, observant, and reasonably well organized. It is generally safe for humans unless they are hypersensitive to insect proteins, react to the physical irritation of insects or their body parts, work with insects that sting or bite, or expose themselves to toxic substances used in the rearing procedures. Rearing methods can be a simple duplication of nature, such as holding insects and their food plants or animals in cages. In other cases, it can take years to develop a method to successfully rear a very specialized insect, often with breakthroughs based on discovering details about its life history. Methods can also be derived empirically by experimenting with a variety of materials and procedures, until success is achieved. Regardless, the satisfaction in rearing at least the immature stages of insects can be an exciting and fulfilling hobby or vocation.

Once insects have been held to the adult stage, having undergone incomplete (nymphs) or complete (larvae) metamorphosis, the challenge is to induce mating and oviposition of viable eggs. An insect species will mate only in an environment that has the necessary conditions, such as courtship cues, adequate temperature and humidity, appropriate light levels, and often food. Oviposition usually requires a specialized substrate that duplicates or mimics the insect's natural situation. Substrates typically protect the eggs from predators and parasites, and locate them in proximity to larval food. However, given no choice in captivity, some insects will mate and the females will scatter their eggs under completely artificial conditions. First instar larvae of these species travel considerable distances in search of suitable habitats and food.

Taking into consideration the wide range of life histories (e.g., terrestrial, aquatic, subterranean, plant or animal feeding, parasitic, predatory), insect rearing can be approached much the same as raising chickens, rodents or fish. Suitable founders must be collected or purchased, specialized facilities designed and constructed, equipment obtained and in most cases modified, materials and supplies tested, and site-specific rearing procedures established. Both immature and adult insects normally need food, usually with specific nutritional and physical characteristics, presented in certain kinds of containers. Once obtained, fertile eggs may require surface sterilization and incubation in carefully controlled environments. Colonies also must be protected from pathogens and predators, as well as environmental and procedural failures. The rearing processes and insect life history characteristics are checked every generation, e.g., rate of development, pupal weight or adult size, percent emergence and mating, and the number of viable eggs.

Selecting an insect to rear

Virtually any insect can be collected as an egg, larva or nymph and maintained through stages in metamorphoses until it becomes an adult. However, it is considerably more difficult to establish a colony that produces continuous generations. Knowledge is required about the species' habitat, natural food and behavior. Re-creation of soil and aquatic environments, symbiotic relationships, and specialized foods can make rearing difficult. Some insects undergo temperature- and photoperiod-dependent diapause or require host plant cues to terminate multi-year cycles. Feeding by trophallaxis may necessitate maintenance of an entire colony, as in termites, ants and other social insects. Because of these and other peculiar life history characteristics, the easiest insects to rear are relatively small, multivoltine (more than one generation per year), plant-feeding, terrestrial species with wide host ranges and few unusual environmental requirements. Commonly cultured insects include pests of common crops, stored products or landscape plants, as well as abundant species that are not pests. Examples include armyworms, loopers, hornworms, mealworms, weevils, mealybugs, aphids, ants, grasshoppers, and lady beetles.

There are published lists and specialized directories of arthropod cultures, e.g., *Drosophila* spp. strains. "Arthropod Species in Culture" listed colonies of the following taxonomic orders (species, colonies): Acari (41, 77), Anoplura (1, 1), Coleoptera (78, 266), Diptera (168, 301), Hemiptera (90, 206), Hymenoptera (119, 169), Lepidoptera (101, 308), Mallophaga (3, 3), Neuroptera (2, 2), Orthoptera (64, 203), Siphonaptera (2, 7), and Thysanura (4, 11). The most numerous colonies were the two-spotted spider mite, *Tetranychus urticae* (24), confused flower beetle, *Tribolium confusum* (21), yellow fever mosquito, *Aedes aegypti* (21), housefly, *Musca domestica* (28), tobacco budworm, *Heliothis virescens* (25), and American cockroach, *Periplaneta americana* (26). "Suppliers of Beneficial Organisms in North America" is maintained by the California Environmental Protection Agency (http://www.cdpr.ca.gov). This publication lists more than 130 species of beneficial organisms available from 142 suppliers. The species are classified as predatory mites (17 species, 325 suppliers), parasitic nematodes (6, 111), stored product pest parasites and predators (4, 13), aphid parasites and predators (17, 361), whitefly parasites and predators (9, 268), parasites and predators for greenhouse pests (23, 504), scale and mealybug parasites and predators (7, 149), insect egg parasites (12, 238), moth and butterfly larval parasites (6, 30), filth fly parasites (8, 143), other insect parasites (4, 9), general predators (21, 410) and weed controls (36, 50). The most popular species were the lacewings *Chrysoperla carnea* (65 suppliers) and *C. rufilabris* (54), mealybug destroyer, *Cryptolaemus montrouzieri* (52), whitefly parasite, *Encarsia formosa* (54), convergent lady beetle, *Hippodamia convergens* (56), predatory mite, *Phytoseiulus persimilis* (54), and a parasitoid of moth eggs, *Trichogramma pretiosum* (77). Many more species could be reared using the techniques developed for their close relatives.

Founding a colony

Insect colonies are initiated from field-collected specimens or previously established colonies. Any developmental stage can be used, but surface-sterilized eggs are less likely to introduce a pathogen. Eggs are generally used to initiate a colony from another colony because they are durable and easy to ship. However, eggs may be difficult to find in nature, and subsequent larvae or nymphs often suffer high levels of mortality because they are parasitized or, if viable, are not yet laboratory adapted. Thus, it is generally advisable to collect pupae or large nymphs, hold them in individual containers for parasite and pathogen screening, combine the adults in mating cages with food and a suitable oviposition substrate, and collect and treat the eggs for use in establishing a colony. From either source, field or insectary, the degree of success achieved will depend on the quality of colonized insects and skill with which they are reared. Many species that can be colonized and reared for multiple generations are on average much larger, healthier (free of malnutrition, pathogens, parasites and predators), uniform in growth, and more active and fertile than those in nature.

Facilities and equipment

Insect rearing facilities have evolved from so-called 'insectary' buildings with ambient temperature and humidity, to highly controlled facilities that contain insects under high security. Insectaries provide adequate environments for rearing insects in semi-natural conditions on host plants or animals. However, pathogens, parasitoids and predators are not controlled and workers are exposed to potentially dangerous pathogens and allergens. Conversely, temperature, humidity, air quality and quantity, light quality and photoperiod, sanitation, and security are closely maintained in rearing facilities. Insects and supplies are carefully screened for contaminants before they are admitted, and human access and exposure are limited. All openings to a rearing facility are sealed or filtered to prevent insects from entering or leaving, particularly in quarantine facilities. These rearing facilities range from small rooms with single occupants to massive factories with hundreds of workers. Regardless, they must be well insulated and have highly filtered, re-circulated air to be cost efficient. A garage or out-building can easily be modified to serve as an insect rearing facility.

Equipment is required to provide suitable environments, handle the insects, move and store materials and supplies, observe specimens, assure sanitation, and prepare diets. The environment is maintained by air conditioners, fans, humidifiers and de-humidifiers, mechanical or electrostatic air filters, and incandescent and fluorescent lighting. Tables, chairs, a deep sink, and perhaps a dish-

washer are also needed. A dedicated refrigerator is required for storing diet ingredients and a freezer can be used to chill insects for handling. A binocular microscope may be needed to determine sex, screen for parasites and predators, or determine species. Specialized techniques and a high-powered microscope would be required to identify insect pathogens. Laboratory carts with wheels can be very useful for moving insects, materials and supplies. A cooker and blender must be provided to work with gelled diets.

Materials and supplies

Typical insect rearing materials include hardware cloth, screen, or Plexiglas® for making cages; wood or metal framing; shelves and supports, and oviposition substrates. Materials may also be needed to modify rearing containers, support diets, and build small devices for handling insects. Supplies for insect rearing are similar to those found in a family kitchen, including products for preparing and serving diets, protecting tables, and cleaning and sanitizing surfaces. Egg treatment is not much different than cleaning fresh vegetables with a mild bleach solution. Cotton and sponges are used to hold liquids and prevent feeding insects from drowning. All kinds of paper and plastic containers are used as purchased or modified for insect rearing. The most specialized supplies are often dietary ingredients for larvae or nymphs. Diets may be natural, such as certain fruits and vegetables, or formulated from specific nutrients. Medical and veterinary pests are usually maintained on living hosts, although mosquitoes can be reared in trays containing water and food. Most materials and supplies for rearing insects are reasonably inexpensive and specialized diets can be purchased to avoid their preparation.

Insect rearing procedures

Insect eggs are resilient and generally do not require special care unless handled. Newly emerged larvae or nymphs can remain on the plant or animal host until it is no longer suitable for their development. They can then be transferred to new hosts or simply held for pupation, or additional molts if they have incomplete metamorphosis. Eggs are often surface sterilized with 10% bleach solution or another germicidal chemical to eliminate microorganisms, particularly if they are field collected. This procedure requires individual eggs to be removed from bundles

Fig. 861   Rearing a small colony of blister beetles: upper left, adults on desert sunflower; upper right, eggs under a flower; lower left, artificial diet and glass bee cells; and lower right, artificial flower with reared pupae and adults.

and substrates. The bleach solution dissolves the cement and outer layer of the chorion, thus freeing the eggs. However, this treatment renders them susceptible to desiccation or drowning, so they must be handled carefully. Surface sterilization is recommended if eggs are to be placed in containers of artificial diet.

Larvae and nymphs are maintained in containers or cages that provide diet, surface area, appropriate environments, and protection from contamination. The diet must induce feeding, provide adequate nutrition and not deteriorate during the insect's development. As the insects grow, they require more food, space and separation from considerable amounts of frass (feces). Overcrowded larvae often become cannibalistic. The containers and cages buffer the devel-

oping insects from extremes in temperature and relative humidity, as does the rearing facility and holding room. Air circulation is important to balance moisture and minimize vertical stratification in holding areas.

Adult colonies are derived from pupae or nymphs that have been sexed, selected for desired characteristics, and screened for parasites, pathogens, and other contaminants. If closely related insects are being reared, a contaminant can be the wrong species. Food, water and a suitable oviposition substrate are usually provided in a container or cage and, as with the immatures, it is essential to control the environment. A considerable amount of testing may be required to provide the necessary conditions for mating and oviposition. Depending on the kind

Partial list of sources for insects, rearing supplies and associated information.

Company/Institution	Web site addresses (http://)
Assn. of Natural Bio-control Producers	www.anbp.org
Bio Quip Products, Inc.	www.bioquip.com
BioServe	www.bio-serv.com/insect/home.html
Carolina Biological Supply	www.carolina.com
Combined Scientific Supplies	www.overland.net/~insects/
Educational Science	educationalscience.com/breedingsupplies.htm
Entomos	www.entomos.com/
Princ. and Proc. for Rearing Qual. Insects	www.msstate.edu/Entomology/Rearingwksp.html
Southland Products	www.tecinfo.net/~southland/
Suppliers of Beneficial Organisms in N.A.	www.cdpr.ca.gov
Ward's Natural Science Establishment, Inc	www.wardsci.com/
Young Entomologists' Society	members.aol.com/yesbugs/bugclub.html

Examples of simple diets for rearing insects, in this case bollworms (moths). Note that many of the components are diet preservatives, not nutrients.

Vanderzant & Richardson 1962	Shorey & Hale 1965	King & Hartley 1985
Wheat germ	Pinto beans	Wheat germ
Casein	Brewers' yeast	Soybean flour
Agar	Agar	Agar
Sugar	—	Sugar
—	Methyl p-hydroxybenzoate	Methyl paraben
—	Sorbic acid	Sorbic acid
—	Formaldehyde	Aureomycin®
Ascorbic acid	Ascorbic Acid	Vitamin mix
Choline chloride	—	—
B-Vitamins	—	—
Inositol	—	—
—	—	Wesson salt
Water	Water	Water

of insect, these conditions can involve space, lighting, attractants, sex ratios, airflow, host cues, complex courtship behavior, and so forth. Production of viable eggs is usually the greatest problem to overcome in establishing a continuous colony of insects.

Food for immature and adult insects

Insects that feed on plants and animals are often maintained on their natural hosts or associated tissues, such as leaves, fruit, meat or blood. This requires that the hosts be reared continuously or tissues be collected and kept fresh. With insect life cycles extending from a few days to several years, providing host material can be the most time consuming effort in insect rearing. This is particularly true for predatory and parasitic insects because three trophic levels must be synchronized: the

Typical purposes for rearing insects and primary requirements for their production and quality.

Purpose for insects	Primary requirements
Enjoyment	Appearance, Behavior, Longevity
Research	Uniformity, Reliability
Pest Management	Behavior, Reliability, Cost
Display	Appearance, Behavior, Longevity
Education	Appearance, Behavior, Cost
Human Food	Size, Taste, Cost
Animal Food	Size, Behavior, Taste, Cost
Natural Products	Size, Yield, Cost
Fish Bait	Appearance, Behavior, Cost
Biodiversity	Heterogeneity, Behavior

initial plant or animal, second level host and beneficial insect. Consequently, insects are often selected for rearing that feed on human and animal food as immatures and adults, or require only sugar water or nothing as adults. These include crickets, wax moths, cockroaches, stored product pests, certain beetles and grasshoppers, and a wide variety of flies. So, some cockroaches eat dry dog food, grasshoppers eat lettuce, plant bugs eat green beans, flies eat fruit, and so forth. Gelled (often agar-based) diets have been developed to provide water with the food and incorporate preservatives. These kinds of diets can be prepared using published formulas or purchased along with the insects from specialized companies. Diets must be fresh because the ingredients can deteriorate during prolonged storage.

Isolation and sanitation

Insects and their diets attract a host of competitors, predators, parasites and pathogens. Even a very small colony can easily become infested with scavenger or predatory mites, ants, cockroaches, flies and gnats. Most of these kinds of problems can be solved with cleanliness and barriers. Germicidal soaps and a 5% solution of commercial bleach will provide surface sterilization with minimal corrosion. Barriers include sealed walls, electrical outlets and plumbing, reach-in boxes, screened or solid cages, paper and plastic containers, and plastic bags. Food should be stored in a refrigerator or freezer. Rough surfaces, drop ceilings, porous materials and cracks are difficult to sanitize. Positive air pressure helps to keep contaminants from entering and rearing procedures should be accomplished from clean to dirty, with no backtracking. Normally, parasites that penetrate barriers can be eliminated by breaking their life cycles, possible because they develop more rapidly than their hosts. Potentially parasitized insects are isolated from the colony until they are determined to be clean. Pathogens can be more problematic because they are difficult to detect and identify. In practice, diseased insects are carefully discarded and the colony is screened until symptoms cease, often without determining the exact cause. Sanitation and surveillance are increased concomitantly. As in kitchens, some of the bacteria that contaminate insect diets are human pathogens.

Quality of reared insects

Care must be taken to assure the quality of artificially reared insects relative to their intended purpose. Yield, size, sex ratio, rate of development, egg production and mobility are typically monitored depending on the situation. It is also important to keep track of environmental conditions, e.g., temperature, relative humidity, light quality and photoperiod, sanitation, materials and supplies, and work performed. Often check sheets are used for monitoring insect production and quality control charts with

averages and ranges suffice for life history and behavioral measurements. This kind of information is required to "troubleshoot" the rearing process and determine causes of failures to produce acceptable insects. Virtually all severe insect rearing problems result from failure to perform standard operating procedures or defective environmental controls. Once established, rearing operations become routine and individual procedures easy to forget.

As in animal husbandry, insect rearing expertise develops with experience and knowledge gained by trial and error, and by studying information developed by others. It is considered both an art and science, requiring a 'brown thumb' analogous with the green thumb of the gardener. The fastest way to learn new techniques is to visit established insect rearing facilities, talk with colleagues individually and at conferences, participate in an organized course, and read technical publications. Facilities and contacts can be located by Land Grant University faculty members, perhaps at extension and research centers, and state department of agriculture and U.S. Department of Agriculture scientists. Currently, the only comprehensive insect rearing course in the United States is a weeklong, intensive workshop taught twice each year at Mississippi State University. Written information on insect rearing techniques

Fig. 862 Industrial rearing of the Mediterranean fruit fly: upper left, adult fruit fly; upper right, papaya infested with larvae; lower left, trays of larvae on artificial diet; and lower right, rearing facility at Metapa, Mexico.

is scattered mostly in the methods sections of scientific papers or accumulated in difficult to locate, out-of-print books. Although there are many others, the following journals have tended to publish papers on insect rearing: Annals of the Entomological Society of America, Entomologia Experimentalis et Applicata, Florida Entomologist, Journal of Insect Science, BioControl (formerly Entomophaga), and specialized publications, such as the Journal of the Lepidopterists' Society and Journal of Apicultural Research. Commercial suppliers of insects, diets and educational materials are an excellent source of information on rearing (see accompanying table). It can also be advantageous to adapt methods from books on rearing agricultural pests, such as the boll weevil (Curculionidae), screwworm (Calliphoridae), corn earworm (Noctuidae), or tobacco hornworm (Sphingidae). It will require some searching but there is a considerable amount of invaluable information on insect rearing.

See also, NUTRITION IN INSECTS.

Norman C. Leppla
University of Florida
Gainesville, Florida, USA

References

Anderson, T. E., and N. C. Leppla. 1992. *Advances in insect rearing for research and pest management.* Westview Press, Boulder, Colorado. 519 pp.

Leppla, N. C., and T. R. Ashley. 1978. Facilities for insect research and production. U.S. Department of Agriculture Technical Bulletin No. 1576. 86 pp.

Needham, J. G., P. S. Galtsoff, F. E. Lutz, and P. S. Welch. 1937. *Culture methods for invertebrate animals.* Comstock Publishing Co., Ithaca, New York. 590 pp.

Peterson, A. 1964. *Entomological techniques, how to work with insects.* Entomological Reprint Specialists, Los Angeles, California. 435 pp.

Singh, P. 1977. *Artificial diets for insects, mites, and spiders.* Plenum, New York, New York. 594 pp.

Singh, P., and R. F. Moore 1985. *Handbook of insect rearing,* Vols. 1 and 2. Elsevier, Amsterdam, The Netherlands. 488 and 514 pp.

Smith, C. N. 1966. *Insect colonization and mass production.* Academic Press, New York. 618 pp.

Villiard, P. 1969. *Moths and how to rear them.* Funk and Wagnalls, New York, New York. 242 pp.

DE REAUMUR, RENÉ ANTOINE FERCHAUL-T.

René Antoine Ferchault de Réaumur was born in 1683 at La Rochelle, France. He achieved success early in life, being elected to the Academy of Sciences at the age of 24, after publication of some important work on geometry. He worked principally as a naturalist, publishing articles on regeneration of lost appendages by crustaceans, fossils, luminescence by molluscs, the digestive systems in birds, and other topics. Not until 1734 did he begin publishing on insects, but by 1742 he had authored his "Memoires pour servir à l'histoire des insectes" in six volumes, and had begun a seventh volume, on beetles. In this treatise he gave a clear picture of insects and insect habits, and their relationship with nature. In addition to the natural history of insects, he attempted to develop a classification system based on behavior. Thus, he grouped insects, worms, molluscs, and reptiles together. Though this seems strange now, as we are conditioned to classification systems based on structure, classification was still in its formative stages and much more unusual systems had been proposed. This publication gave him prominence among naturalists, and his work remains a classic example of careful observation. However, Réaumur contributed to science and technology in many ways, including invention of a thermometer, the use of porcelain for thermometers, improved steel production and enhanced the economy of tin plate production.

Reference

Pellett, K. L. 1929. Lives of famous beekeepers, René Antoine Ferchault de Réaumur. *American Bee Journal* 69: 291–193.

RECAPTURE TECHNIQUE OF SAMPLING.

An absolute method of sampling that involves capture of insects, marking of captured insects, release of the marked insects, and then recapture of marked and unmarked insects from the natural population. After the recapture effort, the proportion of marked insects is used to estimate the initial population size.

RECOMBINANT DNA TECHNOLOGY.

All the techniques involved in the construction, study and use of recombinant DNA molecules. Often abbreviated rDNA, which can be confused with ribosomal DNA.

RECOMBINATION.

A physical process that can lead to the exchange of segments of two DNA molecules and which can result in progeny from a cross

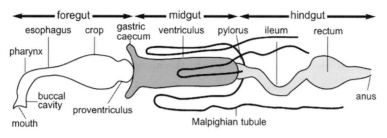

Fig. 863   Generalized insect alimentary system (adapted from Chapman, The insects: structure and function).

between two different parents with combinations of alleles not displayed by either parent.

RECRUITMENT TRAIL.   An odor trail produced by an individual worker that is used to recruit nest-mates to a location where other workers are needed: a food source, nest site, a breach in the nest wall, etc.

RECTAL PAD.   A section of the rectum containing enlarged cells that take up water and ions from the rectum.

RECTUM.   The posterior region of the hindgut leading to the anus.
   See also, ALIMENTARY SYSTEM, ALIMEN-TARY CANAL AND DIGESTION.

RECURRENT VEIN.   A crossvein, often extending obliquely, and usually associated with the costa and subcosta, though this varies among taxa.
   See also, WINGS OF INSECTS.

RED BUGS.   Members of the family Pyrrhocoridae (order Hemiptera). See also, BUGS.

RED FLOUR BEETLE.   See also, STORED GRAIN AND FLOUR INSECTS.

REDI, FRANCESCO.   Francesco Redi was born at Arezzo, Italy, on February 18, 1626. After earning a degree in medicine, he became a tutor for five years, but then moved to Florence where he became physi-

cian to the Grand Dukes Ferdinand II and Cosimo III. He conducted much experimentation aimed at improving medicine, but also was proficient at poetry and music. Among his scientific publications are ''Observations on vipers,'' ''Experiments on the generation of insects,'' and ''Medical consulting.'' Among his literary contributions are the ''Bacco in Toscana,'' and ''Arianna inferma.'' He is best remembered for his work establishing that insects are not produced by spontaneous generation, but rather are derived from other insects. Perhaps more importantly, by working with the development of fly larvae on meat he demonstrated the value of experimentation, and the value of changing one variable at a time in elucidating biology. He died at Pisa, Italy, on March 1, 1698.

Reference

Anonymous. 2002. Franceso Redi – scientific works. Seen at http://galileo.imss.firenze.it/mutli/redi/eopere.html (August 2002).

RED IMPORTED FIRE ANT, *Solenopsis invicta* BUREN (HYMENOPTERA: FORMICI-DAE).   The red imported fire ant, *Solenopsis invicta* Buren (Hymenoptera: Formicidae), is a social insect that produces hills or mounds in open areas where the colonies reside, although colonies occasionally occur indoors, near and within structures, such as utility housings or tree trunks. Fire ant mounds may reach 30 to 40 cm high and 30 to 50 cm in diameter with no holes or central entrance hole on the mound's surface. Inside, mounds have interconnecting galleries that may extend 30 to 40 cm deep, although some tunnels can penetrate to the water table. In response to solar radiation and ambient conditions, fire ants move within the mound seeking optimum temperature for the development of brood (eggs, larvae and pupae), a process called thermoregulation. Foraging worker ants enter and exit through tunnels radiating up to 5 to 10 meters away from the mound. The disturbance of mounds results in a rapid defensive response by worker ants, which quickly run up vertical surfaces to bite and sting any objects encountered.

Two forms of imported fire ant colonies occur: the single queen (monogyne) and the multiple queen (polygyne) forms, the latter containing two or more inseminated, reproductively active queen ants. Worker ants in single queen colonies eliminate additional queens and respond defensively to neighboring colonies to maintain territories. Multiple queen colony worker ants do not display territorial behavior and, consequently, they can produce 3 to 10 times as many ant mounds in a given area of land and result in 500 mounds and 50 million ants per hectare. Areas infested with the single queen form normally have 50 to 75 mounds per hectare.

Description

A mature colony can contain over 200,000 to 400,000 sterile, female worker ants. These ants range in length from 1.5 to 5 mm and are dark reddish brown with black abdomens. Worker ants build up the colony, care for the queen and brood, defend the colony and forage for food. Their functions within the colony are determined by the size and needs of the colony, and by the age of the worker ant. The younger workers serve as nurse ants, which tend and move the queen and brood. Older workers serve as reserves to defend the colony and to construct and maintain the mound. The oldest worker ants become foragers.

Developmental stages include eggs, larvae and pupae, referred to as brood. The eggs are spherical and creamy white. The larvae are legless, cream-colored and grub-like with distinct head capsules. Pupae resemble worker ants and are initially creamy white turning darker before the adult ants emerge (eclose).

Winged, reproductive, male ants develop from eggs that are not fertilized. Fertilized eggs can develop into either sterile, female worker ants, or winged, reproductive females depending upon the nutrition provided to the larval stages and the chemical signals (juvenile hormone level and pheromones) within the colony. Winged, female reproductives are reddish-brown, while males are shiny black with

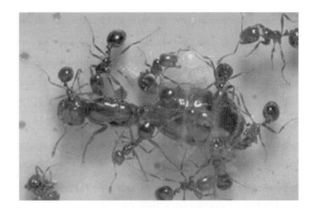

Fig. 864 Fire ant queen and workers.

smaller heads. These ants stay in the colony until conditions exist for their nuptial flight. Queen ants (mated, female reproductives) are larger (9 mm) and have no wings.

Spread

Spread occurs naturally (through mating flights, ground migration and floating colonies on water during floods) and artificially (by humans through shipment of infested articles like nursery potting media, sod, bales of hay, or soil). It is limited primarily by the availability of soil moisture and cold temperatures. Predictions suggest the ant is unable to survive where the minimum yearly temperatures are less than $-12.3°C$ to $-17.8°C$.

Mating flights occur mid-day (10:00 a.m. to 2:00 p.m.) during the first sunny days after a period of rainy weather when temperature are suitable ($24°C$) at any time of the year, but mainly in the spring and the fall. Winged male and female reproductives (also called sexuals or alates) couple in the air (90 to 300 m high). The males die after mating, but the females seeking a nesting site are attracted to shiny surfaces (e.g., water surfaces, shiny truck beds) and land within a mile (1.6 km) or two unless carried farther by the wind.

Colony formation

Newly mated females that survive the nuptial flight and that reach suitable nesting habitats (estimated to be about 1% due to predation and other mortality factors), remove their wings and burrow into the ground. Sealed in a chamber that they dig, these females begin to lay eggs (10 to 20 per day) that hatch into larvae in 6 to 10 days, which are fed from energy produced from the breakdown of flight muscles, infertile (trophic) eggs, young larvae and oil reserves. The first worker ants emerging are uniformly small and are called minums or nanitics.

Life cycle

Each queen ant can produce about 800 eggs per day. The eggs hatch in 8 to 10 days and the larvae develop through four stages (instars) over 12 to 15 days before pupating for a period of 9 to 16 days. Development requires 22 to 37 days, depending on the temperature. Most worker ants live 60 to 150 days with larger ants living longer, but during cooler weather, workers can survive for 8 months or more. Newly established colonies develop winged, reproductive ants after about 6 to 8 months and can produce 4,000 to 6,000 alates per year. Queen ants can live for more than 7 years.

Habitat and food source(s)

Ants communicate through vision (sight), vibration (sound), touch and chemicals (pheromones, including a queen pheromone that attract workers and may suppress dealation and reproduction, and a trail pheromone produced by Dufour's gland associated with the worker ant stinger). They forage when temperatures range from 22°C to 36°C. Upon locating food resources, they develop a trail using a pheromone that directs other worker ants to the site. Fire ants are omnivores, consuming sugars (carbohydrates), certain amino acids, ions in solution and some oils containing polyunsaturated fatty acids. Although they primarily consume other arthropods and the honeydew produced by some types of sucking insects (Hemiptera), they will also consume seeds and other plant parts like developing or ripening fruit, and dead plant and animal tissues.

Worker ant mouthparts are for biting and sipping liquids. Worker ants consume only liquids and particles less than 0.9 microns, storing food in their crops and their postpharyngeal gland (oils only) until feeding it to other worker ants and ultimately, to larvae and queen(s) - a process called trophalaxis. Young larval stages (instars) are fed regurgitated liquid food only. However, the last (fourth) larval stage can ingest solid food particles. Worker ants place bits of solid food in a small depression (called a food basket or praesaepium) just in front of and beneath the larva's mouth (an area called the presturnum). They digest the proteins externally (extraorally) by secret-

Fig. 865 Fire ant mound at base of tree.

ing enzymes and by chewing and swallowing smaller particles.

Pest status

Like a weed species, the aggressive and abundant imported fire ant reduces biodiversity, particularly other ant species, in newly infested areas. These soil-dwelling ant colonies are most commonly found in urban, agricultural and non-agricultural areas and affect people, livestock, wildlife and the flora and fauna of infested lands. Although the exact economic costs of fire ant damage and control are unknown, estimates for the southeastern United States have been more than a half billion to several billion dollars per year. The cost of producing agricultural commodities, such as nursery and sod production, has increased in infested areas. There, quarantine regulations mandate treatment of materials to be shipped to non-infested areas.

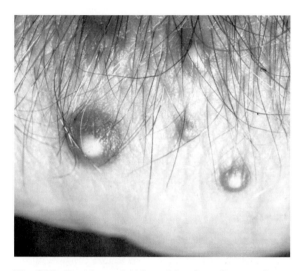

Fig. 866 Pustules on hand resulting from fire ant stings.

Medical importance. For most people, contact with the ants is merely an irritation and a nuisance, but for some people (particularly very young, old or indigent people), the ant's sting can result in medically serious problems or even (rarely) death. Imported fire ant workers are highly defensive and aggressive. When disturbed, they run up vertical objects quickly (8 cm/5 sec) en masse (tens to hundreds) and seem to bite (with their mandibles) and sting (with their modified ovipositor) all at once. The venom injected produces a burning, fire-like sensation for which the ant gets its common name: fire ant. Each worker ant can sting repeatedly. Queen ants do not sting, but they do contain venom. Surveys have documented that more than 50% of the people living in infested areas are stung by the ant annually. Most people can tolerate multiple stings, but may have problems with secondary infections at the sites of the stings. However, about 0.6 to 6.0% of people stung are sensitive to the venom and can experience serious medical problems from even one sting.

The venom produces a fluid-filled pustule within a day or so (10 to 12 hrs) of being stung. The venom consists of aliphatic substituted alkaloids that are cytotoxic (cell-killing), and a small amount of protein, which is responsible for allergic responses (swelling or edema, and anaphylactic shock characterized by swelling, sweating and shortness of breath that can lead to respiratory arrest and heart failure). The pustules persist for several weeks and may leave scars. Ruptured pustules can become infected.

The ants also affect people indirectly by modifying their behavior to avoid contact with the ant. There are also increased health and environmental risks associated with insecticide application or other attempts to control this pest.

Equipment damage. The ants or entire colonies move into buildings or vehicles seeking favorable nesting sites, particularly during flooding and very hot, dry conditions. Fire ant foraging and nesting activities can result in failure of many types mechanical (e.g., hay harvesting machinery, sprinkler systems) and electrical equipment (e.g., air conditioner units, traffic box switching mechanisms).

Agricultural and ecological impact. In addition to being considered medically important pests of people, pets, livestock and wildlife, imported fire ants can also damage crops such as corn, sorghum, okra, potatoes, sunflowers and others by feeding on the seeds, seedlings and developing fruit. Predatory activities of fire ants suppress populations of ticks, chiggers, caterpillars and other insects. Predatory activity attributes to wildlife reductions in some areas. Many animals are also affected by these stinging ants, particularly those that cannot quickly move away from the threat (e.g., very young, old, or confined animals). Ants recruit to moist areas of the body (eyes, genitals), yolk of hatching birds and wounds, and begin stinging when disturbed. Stings result in injury such as blindness, swelling or death. Indirectly, animals avoid infested food, water and nesting areas. The ants can reduce food sources for some insect-eating (insectivorous) animals, such as some birds and lizards, and they may compete with seed-feeding (gramnivorous) animals for food, or by altering the distribution and composition of plant communities.

In addition to direct damage to plants, they also aggravate populations of other insect plant pests like some Hemiptera (aphids, scale insects, mealybugs) by protecting them from natural enemies as the ants consume the sugary honeydew produced by these pests. However, the ants primarily prey on arthropods, such as some species of ticks, many caterpillars and other insects often considered pests. This behavior can provide benefits to producers of cotton, sugarcane and other commodities.

History of spread

The fire ants comprise a group of 18 to 20 insects species native to the New World within the genus,

Solenopsis, which contains about 200 species including "thief ants" (an ant species, most of which are incapable of stinging humans, nesting close to other ant nests and robbing them of stored food and brood). The black imported fire ant, *S. richteri* Forel, was accidentally introduced from the Paraguay River drainage area of South America to the U.S. port of Mobile, Alabama in 1918. The red imported fire ant, *S. invicta*, is believed to have been introduced from northern Argentina or southern Brazil to the same U.S. port around the early 1930s. *S. invicta* spread westward at about 198 km per year.

Several native fire ant species occurred in the U.S. before *S. invicta* and *S. richteri* Forel arrived: *S. xyloni* McCook, the southern fire ant; *S. geminata* (Fab.), the tropical fire ant; and, two desert-adapted species, *S. aurea* Wheeler and *S. amblychila* Wheeler. While *S. xyloni* and *S. geminata* have also been considered to be pest ants, they have been largely displaced in the southeastern U.S. by the red and black imported fire ants. Imported fire ants occur at higher densities than native fire ant species, leading to increased human contact and problems. In the U.S., *S. richeteri* occurs in northern areas of Alabama, Mississippi, Georgia and southern Tennessee, with a sexually reproductive hybrid population of *S. invicta* and *S. richteri* throughout the remainder of these states and part of Georgia.

Solenopsis invicta now occupies 128 million hectares (316 million acres) in nine southeastern states (Alabama, Arkansas, Florida, Georgia, Louisiana, Mississippi, North Carolina, South Carolina, Texas) in the U.S. with limited infestations in Arizona, Oklahoma, Tennessee, New Mexico and California. Infestations also occur in a number of island countries in the Carribean including Puerto Rico, the Bahamas, the British and U.S. Virgin Islands, Antigua and Trinidad. The species has recently been detected in New Zealand and Australia (Brisbane).

History of control efforts in the U.S.

- 1929 – Imported fire ant first detected in the northern Mobile area and Spring Hill, Alabama
- February 1937 – First organized control program in Baldwin County, Alabama (1 to 3 ounces 48% calcium cyanide dust per mound); 2,000 acres or 800 ha treated
- 1948 – Mississippi spends $15,000 to treat ant mounds with chlordane dust; Alabama and Louisiana provide chlordane to farmers free or at cost

- 1957 – Arkansas conducts 12,000 acre eradication project in Union County, and the city of El Dorado applied heptachlor (2 lb/acre) by air; congress appropriates $2.4 million for USDA to begin a federal-state cooperative control and eradication program dedicated to the use of aerial and granular applications of heptachlor and dieldrin (2 lb a.i./acre)
- 1958 – mandated treatments of regulated items initiated to comply with USDA imported fire ant quarantine regulations
- 1960 – heptachlor rate lowered to 0.25 lb a.i./ acre with 2 applications, 3 and 6 months apart
- early 1960s – growing concerns about the detrimental effects of treatments to wildlife, contamination of food and crops ended the program
- 1962 – Lofgren develops conventional bait formulation; mirex developed as an active ingredient: mirex (0.3%) dissolved in soybean oil (14.7%) on impregnated corn grits (85%)
- 1962 to 1978 – 140 million acres treated with mirex bait by ground and aerial equipment, although many treatments consisted of three applications to the same area; thus, approximately 46.6 million acres were actually treated
- 1967 – feasibility studies were initiated by USDA-ARS in Gainesville, Florida, to use mirex to eradicate imported fire ants
- late-1960s to early 1970s - mirex residues found in environment, non-target organisms and to be toxic to estuarine organisms
- 1970–1971 – U.S. Dept. of Interior bans all mirex uses; EPA issues notice of cancellation of mirex
- 1977 – mirex reported to be a potential carcinogen; EPA cancelled registrations 31 December
- 1978 – Amdro[R] (hydamethylnon with oleic or linoleic acid added to increase oil solubility formulated in soybean oil (20%) on a pregelled defatted corn grit carrier) tested
- 1980 Amdro registered by EPA for use on pastures, range grasses, lawns, turf and non-agricultural lands
- 1983 – Prodrone, the first insect growth regulator (IGR) is registered by the EPA
- 1985 – Logic[R] (fenoxycarb), another IGR, registered by the EPA
- 1986 – Ascend[R]/Affirm[R] (abamectin) registered by the EPA
- 1995–2000 – biological control efforts initiated using importation and release of phorid flies

(*Pseudacteon tricuspis* and others) and manipulation of a disease, *Thelohania solenopsae*

- 1998 – Distance®/Spectractide® Fire Ant Bait (pyriproxifen), another IGR, registered by the EPA
- 1998 – Extinguish™ (methoprene), another IGR, registered by the EPA for use in croplands in which other baits can not be applied
- 2000 – Justice®, Eliminator®, Strike® Fire Ant Baits and others containing spinosad (spinosyns A and B combined), a new insecticide class (spinosyn derived from bacterium, *Saccharopolyspora spinosa* Mertz and Yao) registered by EPA
- 2000 – Chipco® FireStar® and TopChoice® (fipronil bait and granular products, a phenyl pyrazole insecticide), approved by EPA

Bastiaan M.Drees
Texas A&M University
College Station, Texas, USA

References

Drees, B. M., C. L. Barr, S. B. Vinson, R. E. Gold, M. E. Merchant, N. Riggs, L. Lennon, S. Russell, and P. Nester, D. Kostroun, B. Sparks, D. Pollet, D. Shanklin, K. Vail, K. Flanders, P. M. Horton, D. Oi, P. G. Koehler, J. T. Vogt. 2000. *Managing imported fire ants in urban areas.* B-6043. Texas A&M University, College Station, Texas. 20 pp.

Drees, B. M., C. L. Barr, D. R. Shanklin, D. K. Pollet, K. Flanders, B. Sparks, and K. Vail. 1999. *Managing red imported fire ants in agriculture.* B-6076. Texas Imported Fire Ant Research & Management Plan. Texas A&M University, College Station, Texas. 20 pp.

Taber, S. W. 2000. *Fire ants.* Texas A&M University Press, College Station, Texas. 308 pp.

Vinson, S. B. 1997. Invasion of the red imported fire ant (Hymenoptera: Formicidae) – spread, biology, and impact. *American Entomologist* 43: 23–39.

Williams, D. F., H. L. Collins, and D. H. Oi. 2001. The red imported fire ant (Hymenoptera: Formicidae): an historical perspective of treatment programs and the development of chemical baits for control. *American Entomologist* 47: 146–159.

RED IMPORTED FIRE ANT TERRITORIAL BEHAVIOR.

The imported fire ant, *Solenopsis invicta* Buren, is a stinging social insect indigenous to parts of Brazil, Paraguay, and Argentina, and it was accidentally introduced to the United States in Alabama in 1929. The fire ant has since spread across the southern United States and Puerto Rico, and because of its venomous sting, it is widely viewed as being an exotic pest. Queens start colonies after nuptial flights that occur year round, but peak flight activity occurs between May and August. Colonies can produce 3,000 to 5,000 queens each year, and fire ant reproduction in a colony can be accomplished by a single queen or by multiple queens in polygynous colonies. Queens are morphologically distinct from other individuals in the colony, but workers range in length from 2 to 6 mm and definite subcastes among workers have not been identified.

Fire ant establish colony mounds with subterranean galleries that can support populations of 50,000 to 230,000 workers. Mound densities of up to 2,500 per hectare, but usually less than 250 mounds per ha have been reported in pastures, cultivated areas, unattended woodlands, and disturbed habitats. Tunnels radiate 2 to 12 mm below ground level from each colony mound. One colony's tunnel system was exhumed and its tunnels were found to collectively extend 84 m. Tunnel exits occur intermittently and they facilitate access to foraging areas. Fire ants are omnivorous and they feed on a wide diversity of arthropods, molluscs, plant material, honeydew produced by hemipteran insects, and nectar. Mass foraging at specific sites is initiated by a trail pheromone secreted by a gland and extruded through the stinger of scout workers inside tunnel systems.

Dyes ingested by fire ants helped to show that, between two different fire ant colonies, fire ants from the two colonies did not forage at the same places, and it was suggested that the fire ants exhibited intraspecific territorial behavior. Another study labeled one fire ant colony with radioactive Zn-65 ingested with molasses and spread throughout the colony by trophallaxis, and a nearby colony was similarly radiolabeled with Mn-54. The territories of both fire ant colonies were shown to be discrete and dynamic over 20 consecutive days. Fire ant colonies in weedy and weed-free sugarcane habitats were labeled with the non-radioactive, stable-activable tracer element samarium. Instrumental neutron activation analyses of the samarium marker revealed that the tagged ants foraged at significantly more sampling stations in weed-free sugarcane habitats than in weedy sugarcane habitats. The smaller territorial size in weedy areas was a result of more dense prey populations, which permitted greater numbers of fire ant colonies per unit area than weed-free sugarcane habitats. It was suggested that a heavily infested area could be completely occupied by a contiguous patchwork of fire ant territories defended by different colonies.

There are nine traits that have contributed to the fire ant's rapid adaptation to the southeastern United States, including a large colony population size, a general omnivorous diet, mass foraging recruitment, massive aerial dissemination of reproductive females, the stinging mechanism. Three traits are related to territoriality: 1) tunneling behavior offers protection

from trail disruption, 2) utilizing foraging areas in water-saturated soil permits access to established territories in areas subject to flooding, and 3) defending temporally dynamic territory boundaries that permit efficient use of foraging areas when specific sites have been depleted.

The fire ant's adaptability and competitiveness have enabled it to become widely established in the southeastern United States despite the presence of other, indigenous ant species. Transect studies have shown that 92–100% of baited sampling sites can

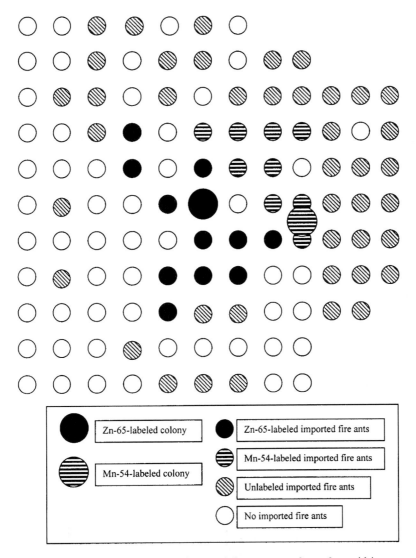

Fig. 867 Day 12 of 20 consecutive days of collecting imported fire ants on a 3 m × 3 m grid in a grassy field in southern Louisiana. Each point on the grid was a 35-ml plastic cup buried to the lip and baited at 1000 h with raw animal fat for 1.5 h. Imported fire ants labeled with Zn-65 and Mn-45 were detected using an autogramma solid scintillation detector with a multichannel analyzer.

be monopolized by fire ants. Since its introduction to the southeastern United States, the fire ant has been a significant competitive force upon the distribution of other formicid species, and species that had previously predominated were frequently displaced. The fire ant has become the dominant formicid species in the southeastern United States. Generally, fire ants can overpower indigenous ant species in combat. Even the Floridian poneroid ant, *Lasius neoniger* Emery, which uses mandibular and chemical defenses to eliminate 2.4 fire ants for every *L. neoniger* killed, eventually succumbs to attrition because of the sheer magnitude of fire ant colony populations. All attempts to control fire ants with mirex applications resulted in the suppression of other ant species, including the tropical fire ant, *Solenopsis geminata* (F.), and *S. xyloni* McCook. Some researchers have suggested that the use of toxins for fire ant eradication has instead contributed to its proliferation in the United States because of fire ant resurgences at the expense of other ant species.

After dominating its niche, the fire ant appears to reach an equilibrium with native ant species. Cohabitation of an area with the competitive fire ant is a result of 1) strategies by other species that exploit other resources in habitat-diverse environments, 2) chemical repellency by some ant species (e.g., *Monomorium minimum* (Buckley)), 3) high mobility (e.g., *Aphaenogaster rudis* Emery can move faster than the fire ant), or 4) proximity to resources. The latter two strategies each contribute to the reduction of recruitment time to essential resources so that opportunities for foraging can be capitalized upon with greater success even within fire ant territories. One study demonstrated that three formicid species sympatric with the fire ant had similarly rapid recruitment times.

The territorial behavior of the fire ant, in combination with the other characteristics that enhance its competitiveness, has permitted this exotic ant to become widely established in the United States. In areas that can support high numbers of fire ants, territories abut each other, and they continuously shift over time to compensate for changes in prey availability (some of which might be caused by predation by the fire ant). The fire ant's offensive, defensive, and foraging capabilities in such conditions have displaced native ant species, but, on the other hand, the fire ant, because of the same characteristics, has emerged as the key predator of some economically important pests, including the boll weevil, *Anthonomus grandis grandis* Boheman, in parts of Texas, and the sugarcane borer, *Diatraea saccharalis* (F.), in Louisiana.

Allan T. Showler
USDA-ARS SARC
Weslaco, Texas, USA

References

Levings, S. C., and F. A. Traniello. 1981. Territoriality, nest dispersion, and community structure in ants. *Psyche* 88: 265–319.

Markin, G. P., J. O'Neal, and J. Dillier. 1975. Foraging tunnels of the red imported fire ant, *Solenopsis invicta* (Hymenoptera: Formicidae). *Journal of the Kansas Entomological Society* 48: 83–89.

Showler, A. T., and T. E. Reagan. 1987. Ecological interactions of the red imported fire ant in the southeastern United States. *Journal of Entomological Science Supplement* 1: 52–64.

Showler, A. T., R. M. Knaus, and T. E. Reagan. 1989. Foraging territoriality of the imported fire ant, *Solenpsis invicta* Buren, in sugarcane as determined by neutron activation analysis. *Insectes Sociaux* 36: 235–239.

Showler, A. T., R. M. Knaus, and T. E. Reagan. 1990. Studies of the territorial dynamics of the red imported fire ant (*Solenopsis invicta* Buren Hymenoptera: Formicidae). *Agriculture, Ecosystems and Environment* 30: 97–105.

Wilson, N. L., J. H. Dillier, and G. P. Markin 1971. Foraging territories of imported fire ants. *Annals of the Entomological Society of America* 64: 660–665.

RED MUSCARDINE. This term has been used to denote various mycoses of insects, caused by species of hyphomycetous fungi and characterized by the appearance of pink to brick-red colors on the body of the dying or dead hosts; certain strains of *Beauveria bassiana* are responsible for a red muscardine of silkworm larvae; *Sorosporella uvella* causes a red muscardine of cutworm larvae (Noctuidae) as well as other insects.

See also, MUSCARDINE, *BEAUVERIA*.

RED PALM WEEVIL, *Rhynchophorus ferrugineus* (OLIVIER) (COLEOPTERA: CURCULIONIDAE). Date palm, *Phoenix dactylifera* L. (Palmales: Palmae), is one of the oldest fruit trees in the world and is mentioned in the Qur'an and the Bible. There are approximately 100 million date palms worldwide, of which 62 million can be found in the Arab region. The origin of the date palm is uncertain. Some claim that the date palm first originated in Babel, Iraq, while others believe that it originated in Dareen or Hofuf, Saudi Arabia, or Harqan, an island on the Arabian Gulf in Bahrain.

The date palm is a perennial and can live about 150 years. The female date palm normally begins to bear dates within an average five of years from the time of planting of the offshoot. The Middle East and North Africa are the major date palm producing areas of the world.

The red palm weevil, *Rhynchophorus ferrugineus* (Olivier) (Coleoptera: Curculionidae), is the most dangerous and deadly pest of the date palm, as well as coconut, oil, sago and other palms.

Information on red palm weevil was first published in 1891 in India. This pest was first described as a serious pest of the coconut palm in 1906, while in 1917 it was described as a serious pest of the date palm in the Punjab, India.

In 1918, red palm weevil caused serious damage to the date palm in Mesopotamia (Iraq), but no insect specimens were collected to confirm it.

Red palm weevil entered and was discovered during the mid-1980s in the Arabian Gulf countries. However, it has become a most destructive pest of date palms in the Middle East.

Distribution of red palm weevil

The red palm weevil occurs in the following countries: Bahrain, Burma, Egypt, India, Indonesia, Iran, Iraq, Japan, Jordan, Kuwait, Oman, Pakistan, Philippines, Qatar, Saudi Arabia, Spain, Taiwan, Thailand, and the United Arab Emirates.

Life cycle

All stages (egg, larva, pupa and adult) are spent inside the palm itself and the life cycle cannot be completed elsewhere. The females deposit about 300 eggs in separate holes or injuries on the palm. Eggs hatch in 2 to 5 days into legless grubs, which bore into the interior of the palm, moving by peristaltic muscular contractions of the body and feeding on the soft succulent tissues, discarding all fibrous material. The larval period varies from 1 to 3 months. The grubs pupate in elongate oval, cylindrical cocoons made out of fibrous strands. At the end of the 14 to 21-day pupation period, the adult weevils emerge. Thus, the life cycle is about 4 months.

All kinds of palms are probably suitable for the development of the red palm weevil, which has been found on the following palms: coconut palm, date palm, nigbong palm, oil palm, ornamental palm, palmyra palm, royal palm, sago palm, sedang palm, sugar palm, talipot palm, and the wild date (toddy) palm.

Eggs. The eggs are creamy white in color, long and oval in shape. The average size of an egg is 2.6 mm long and 1.1 mm wide.

Larva (grub). The full-grown larva is conical in shape and is a legless fleshy grub. It appears yellowish brown, while the newly hatched larva is yellowish white in color, with a brown head. The length of the full-grown larva is 50 mm and the width is 20 mm. The head is brown in color and bent downwards. Mouthparts are well developed and strongly chitinized, which enable the grub to burrow into the trunk. However, the grub requires a moist environment.

Cocoon. When about to pupate, the larva constructs a cocoon of fibers from the palm. The cocoon is oval in shape, with an average length of 60 mm and a width of 30 mm.

Pupa. The pupa is at first cream colored but later turns brown. The head is bent ventrally, the rostrum reaching the tibiae of the first pair of legs. The antennae and eyes are quite prominent. The elytra and wings are brought down ventrally, passing underneath the femora and tibiae of the second pair of legs, overlapping the third pair of legs and meeting in the middle of the abdomen. The average length of the pupa is 35 mm and the width is 15 mm.

Fig. 868 Red palm weevil, *Rhynchophorus ferrugineus*, life cycle: larvae, cocoon, pupa, adult.

Adult. The adult weevil is a reddish brown cylinder with a long prominent curved snout. It varies considerably in size and is about 35 mm in length and 12 mm in width. The head and rostrum comprise about one third of the total length.

The mouthparts are elongated in the form of a slender snout or rostrum, which bears a small pair of biting jaws at the end and a pair of antennae near the base. The rostrum is reddish brown dorsally, and ventrally it is dark brown. In the male, the dorsal apical half of the snout is covered with a pad of short brownish hairs; the snout of the female is bare, more slender, curved and a little longer. The antennae consist of the scape and funicle. The eyes are black and separated on both sides of the base of the rostrum.

The pronotum is reddish brown in color and has a few black spots. These black spots are variable in shape, size and number. The elytra are dark red, strongly ribbed longitudinally, and do not cover the abdomen completely. The wings are brown in color and the weevils are capable of strong flight.

The male weevil has a tuft of soft reddish brown hairs along the dorsal aspect of the snout; this tuft is absent in the female. Also, the male produces an aggregation pheromone "ferrugineol" and/or "ferro-lure" (4-Methyl-5-Nonanol), which a synthetic lure used in the pheromone bucket traps.

Economic importance and damage

Normally, the red palm weevil infests palms below the age of 20 years, where the stem of the young palm is soft, juicy and easily penetrated. The weevils are destructive pests to palms.

The larvae are responsible for damaging the palm, and once they have gained access, the death of the palm generally ensues. The larvae normally never come to the surface, since they begin life inside the palm. Therefore, neither the damage nor the larvae can be seen. However, the trunk of the palm can be infested in any part, including the crown.

The damage caused by only a few larvae is astonishing. Even one larva may cause considerable damage for a young palm (offshoot). It is difficult to assess the actual loss caused by this pest, but undoubtedly it affects the production of date palms.

Methods of control

Integrated pest management (IPM). Since the red palm weevil is difficult to control with just one method, several combinations of control methods should be applied as follows:

Plant quarantine. The transport of offshoots as planting material from infested areas can contribute to the spread of the pest. Strict quarantine at international and national levels should be applied.

Cultural control. Field sanitation and cultural practices are some of the important components to prevent weevil infestation.

- Clean the crowns of the palms periodically to prevent decaying of organic debris in leaf axils.
- Avoid cuts and injuries.
- When green leaves are cut, cut them at 120 cm away from the base.
- Cutting of steps in palms for easy climbing is to be avoided, as this provides sites for egg laying by weevils.
- As palms affected by leaf rot and bud rot diseases are more prone to weevil infestation, they are to be treated with suitable fungicides; after that, application of an insecticide to prevent egg laying by weevils is essential.
- Destroy all dead palms harboring the pest by cutting and burning.

Mechanical control. Dead palms or palms beyond recovery should be split open to expose the different stages of the pest present inside. The debris, including the outer logs and the crowns, should be burned.

Trapping weevils. Trapping and destroying the weevils is another method to control the pest population. Trapping also is used to detect the presence of the pest in the field and also to assess the population.

The aggregation pheromone lure of red palm weevil is used to attract the weevils to the bucket traps which contain pieces of palm stem as food and a solution of insecticide.

Biological control. No effective biological agent has been found that can be employed for the biological control of the pest.

Chemical control.

Preventive. The wounds produced on palms due to cultural practices as well as anther wounds are favorite sites of oviposition by the female weevils. Treatment of such wounds by soaking them with insecticides can be an effective way to prevent red palm weevil entry into palms.

Curative. Once infestation is detected in a palm, curative control must be applied. If the infestation is in the crown, remove the affected and damaged portions and apply insecticide suspension. In case of infestation in the trunk, the infested part should be cleaned and plugged with a mixture of mud and insecticide.

Several insecticides under laboratory and field conditions are tested in order to evaluate the best insecticide that will affect the different stages of the weevil.

Training and education

The cooperation of the farmer is essential in order to successfully implement weevil management. For any large-scale pest management program to succeed, it is imperative that farmers cooperate and become involved at the operational level. This can be achieved by making farmers aware of the seriousness of the problem and training them in various IPM skills.

Abdulaziz M. Al-Ajlan
King Faisal University
Hofuf, Al-Hasa, Saudi Arabia

References

Ajlan, A. M., and K. S. Abdulsalam. 2000. Efficiency of some pheromone traps for controlling red palm weevil, *Rhynchophorus ferrugineus* (Olivier) (Coleoptera: Curculionidae), under Saudi Arabia conditions. Bulletin of the Entomological Society of Egypt, Economic Series. 27: 109–120.

Ajlan, A. M., M. S. Shawir, M. Abo-El-Saad, M. A. Rezk, and K. S. Abdulsalam. 2000. Laboratory evaluation of certain organophosphorus insecticides against the red palm weevil, *Rhynchophorus ferrugineus* (Olivier). *Scientific Journal of King Faisal University (Basic and Applied Sciences)* 1: 15–26.

Hallett, R. H., G. Gries, J. H. Borden, E. Czyzewska, A. C. Oehlschlager, H. D. Pierce, Jr., N. P. D. Angerilli, and A. Rauf. 1993. Aggregation pheromones of two Asian palm weevils, *Rhynchophorus ferrugineus* and *R. vulneratus*. *Naturwissenschafen* 80: 328–331.

Murphy, S. T., and B. R. Briscoe 1999. The red palm weevil as an alien invasive: biology and the prospects for biological control as a component of IPM. *Biocontrol News and Information* 20: 35N–46N.

Nirula, K. K. 1956. Investigations on the pests of coconut palm. *Part IV Rhynchophorus ferrugineus F. Indian Coconut Journal* 9: 229–247.

RED RING OF COCONUT PALMS. See also, TRANSMISSION OF PLANT DISEASES BY INSECTS.

REDTENBACHER, LUDWIG. Ludwig Redtenbacher was a noted Austrian entomologist, and a noted coleopterist. His most important contribution was the Coleoptera treatise of "Fauna Austriaca." He also served as director of the Royal Vienna Zoological Museum. He died in 1876 at he age of 63.

Reference

Anonymous. 1876. Dr. Ludwig Redtenbacher. *Entomologist's Monthly Magazine* 12: 238.

REDUVIIDAE. A family of bugs (order Hemiptera). They sometimes are called assassin bugs, ambush bugs, or kissing bugs. See also, BUGS.

RED WATER FEVER. See also, PIROPLASMOSIS.

REFUGE. (pl., refugia) An area, usually untreated with insecticides, where insects can be preserved. This term usually is used in the context of preserving beneficial insects or pesticide susceptible insects.

REGENERATION. Regrowth of a portion of the body (usually legs in arthropods) that has been lost through injury or autotomy (deliberate shedding of the body part).

REGENERATIVE CELLS. Cells found in the midgut that gradually grow and replace midgut cells that have been lost through age, wear, or other damage.

REGULATION OF SEX PHEROMONE PRODUCTION IN MOTHS. Pheromones are chemicals secreted to the outside of the body to influence the behavior of other individuals, generally within the same species. Sex, aggregation and alarm are some of the common pheromones in insects. Sex pheromones are produced by individuals of one sex, to attract from a distance, members of the opposite sex for mating. On the other hand, contact sex pheromones present in some insects do not elucidate long range attraction, but help in keeping the pairs together thereby facilitating mating. Most species of moths are nocturnal in habit with all the essential activities confined to the dark period of the day. Communication between sexes is almost exclusively mediated through sex pheromones, usually released by the female.

It appears that attraction between sexes in moths had been observed as early as 1690. However, the first insect sex pheromone to be chemically identified was that of the female silkworm, *Bombyx mori*, in 1959. It took 20 years and 500,000 female abdomens to identify this pheromone. With introduction of analytical methods such as high pressure liquid chromatography, gas chromatography and combined gas chromatography and mass spectrometry, it is now possible to identify a pheromone from as few as 1 to 10 moths. Most moth pheromones consist of olefinic aldehydes, alcohols, or acetates with 10 to 20 carbon atoms and one or more sites of unsaturation. The pheromones may consist of a single compound, a specific blend of geometric isomers or a mixture of several components, produced in very small (nanogram) quantities in specialized glands or cells generally located on the telescoping ovipositor of the female. The act of releasing pheromones is commonly referred to as 'calling.'

Most lepidopteran sex pheromones are synthesized de novo from fatty acids corresponding in chain length, double bond position, and stereo chemistry to the type of pheromone produced. Biosynthesis of these acids takes place specifically in the pheromone gland rather than elsewhere in the insect. Typically, the starting material is an acetate that is converted to palmitic acid through fatty acid sythetase; however, preformed acids, such as oleic, linoleic, or linolenic acid, as well as amino acids, may be directly utilized by several species. Subsequent steps are controlled by two key enzyme systems: the microsomal β-oxidation, which results in limited chain shortening by two carbons, and a Δ 11 desaturase, which yields acids with unsaturation at the 11 to 12 position. In the corn earworm, *Helicoverpa zea*, Z-11-hexadecenal the major pheromone component (more than 90%), is produced by oxidation of the corresponding alcohol.

With sexual activity confined to a certain period of the 24 hour day, there is no point in producing a pheromone when it is not needed. Therefore, photoperiod appears to be the major external factor regulating pheromone production in moths. This is particularly true of nocturnal species. However, as we will see later, there are other factors that can impact the regulation of pheromone production. We will use the corn earworm as a model to illustrate the intricacies of pheromone production. Corn earworm, like many moths is a nocturnal species. Sex pheromone is produced only during the night (average amount during peak production is approximately 100 ng), and females have negligible amounts of pheromone in their glands during the day. This diurnal periodicity of pheromone production is controlled by a peptide neurohormone. The hormone is produced in three groups of neurosecretory cells in the subesophageal ganglion and has been named the pheromone biosynthesis activating neuropeptide (PBAN). In the corn earworm and several other species of moths, PBAN is a 33 amino acid residue peptide. It is released into the hemolymph and activates pheromone production at the onset of scotophase. This was demonstrated by a simple bioassay which involved neck-ligating a female to stop pheromone production, followed by injection of either a homogenate of the brain-subesophageal ganglion, or synthetic PBAN into the abdomen of the ligated female. Pheromone titers were then determined by gas chromatography of the pheromone gland extract. One could also conduct the bioassay on a photophase female, without having to ligate it. However, in the gypsy moth the pheromonotropic signal is carried via the ventral nerve cord to the pheromone gland area to activate pheromone production.

The scenario presented for the corn earworm is not universal among moths. For example, species such as the black cutworm and oriental armyworm, which undertake seasonal migration, show delayed sexual maturation as well as delay in initiation of calling. It has been shown that juvenile hormone and PBAN are both involved in regulating pheromone production in these insects. The redbanded leafroller presents a rare case in which it is suggested that PBAN causes the release of a second factor from bursa copulatrix that stimulates pheromone production. In the cabbage looper, females maintain a high titer of pheromone during both day and night. The pheromone glands become competent to produce

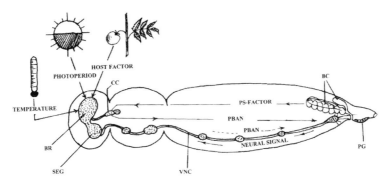

Fig. 869 Extrinsic and intrinsic factors regulating sex pheromone production in a female moth.

pheromone prior to adult eclosion and in response to reduction in ecdysone level. The subesophageal ganglion in this species contains PBAN-like pheromonotropic activity, but its exact role is not clear.

Most neuropeptide hormones interact with surface receptors to activate second messengers which amplify the initial peptide-receptor signal. The second messengers may be small organic molecules such as cAMP, cGMP, ions such as calcium and hydrogen, or inositol phosphate. PBAN has been shown to increase intracellular cAMP levels, and calcium has been reported to be essential for mediating the pheromonotropic response. A G protein-coupled receptor from pheromone glands of *H. zea* was recently cloned. Genes encoding PBAN have been cloned from several insects. Besides PBAN, these genes carry sequences for four other peptides, including one for a diapause hormone.

The females of the corn earworm often mate on the first night after their emergence, and continue to mate throughout their short life, but only once each night. Pheromone production is terminated after mating and resumes during the following night. Continued production of pheromone after mating is a waste of resources and may cause disruption in egg laying, as the males will continue to be attracted to the female. A pheromonostatic peptide (PSP) which is transferred by the male to the female during mating, has been identified. Together with neural signals generated as a result of mating, PSP causes termination of pheromone production. In the corn earworm, pheromonostasis is temporary, and the female can produce pheromone on subsequent nights. However, termination of pheromone production is often permanent in species in which the female mates only once. An example of this is the gypsy moth. In this species,

it is the presence of eupyrene sperm in the spermatheca of the female that is responsible for achieving pheromonostasis.

Another feature in the regulation of pheromone production is the role of host plants. First suggested almost 35 years ago, it was unequivocally demonstrated in 1988 in the corn earworm. Although the corn earworm is a polyphagous insect, it feeds primarily on fruiting parts of all of its hosts. The plant hormone ethylene, associated with fruit ripening, together with other volatile factors from host plants, were shown to induce pheromone production in females of wild or recently colonized corn earworms. This host-factor requirement is not very common among other species of moths, and in the corn earworm it was lost after prolonged laboratory rearing on artificial diet. In addition, temperature has been shown to play a modulating role in control of pheromone production. Thus, pheromone is not produced at very low or very high temperatures.

See also, SEX ATTRACTANT PHEROMONES, PHEROMONES, JUVENILE HORMONE.

Ashok K. Raina
U.S. Department of Agriculture, ARS, SRRC, FSTRU
New Orleans, Louisiana, USA

References

Cardé, R. T., and A. K. Minks. 1997. *Insect pheromone research. New directions*. Chapman & Hall, New York, New York. 684 pp.

Prestwich, G. D., and G. J. Blomquist. 1987. *Pheromone biochemistry*. Academic Press, Orlando, Florida. 565 pp.

Raina, A. K. 1993. Neuroendocrine control of sex pheromone biosynthesis in Lepidoptera. *Annual Review of Entomology* 38: 329–349.

REGULATORY ENTOMOLOGY. Regulatory entomology entails the detection, management and study of non-native or geographically displaced insects in the public interest, principally by governmental agencies and institutions. This branch of entomology began in the 1800s, as governments around the world became aware of the detrimental effects to agriculture and to the public from the unrestricted movement of insect pests associated with trade goods. The United States government's first attempt to intervene in the control of a non-indigenous insect pest occurred in California in 1888. At this time, the vedalia beetle, *Rodolina cardinalis* (Mulsant), and a dipteran parasite, *Cryptochaetum iceryae* (Williston), were imported from Australia to control an accidentally introduced pest, the cottony cushion scale (*Icerya purchasi* Maskell). R.V. Riley, Entomologist to the U.S. Department of Agriculture, coordinated this classical biological control program. Following this successful control effort, interest in an expanded governmental role increased and, in 1904, the Bureau of Entomology was created within the U.S. Department of Agriculture. Other countries, and political subdivisions such as states and provinces, have similarly passed legislation to address the problem of invasion by exotic pests.

A principal role of regulatory entomologists is to prevent the entry of insects to new geographical areas where they may become established as pests. Border inspections of plants began in the U.S. in 1906 to intercept harmful organisms associated with trade goods (agricultural plants and plant products). Inspections have expanded in scope since then to keep pace with increases in trade and travel. In conjunction with border inspections, analyses of future pest status, termed Pest Risk Assessments, associated with the introduction of specific foreign insects are conducted to investigate the potential concerns subsequent to an introduction and to determine appropriate detection and control techniques. Introduction pathways are then examined and targeted for survey. Taxonomic support is critical to the correct targeting and implementation of mitigating measures.

The authority for the U.S. government to establish quarantines was legislated in the Plant Quarantine Act of 1912. Their purpose is to monitor and regulate the movement and spread, both interstate and internationally, of insect pests and trade goods that may harbor them. Regulatory entomologists are responsible for the implementation and enforcement of this act. Early quarantines for insects such as the gypsy moth and the Japanese beetle were effective in limiting the spread of these pests and formed the basis for future efforts. Control measures are also established as components of quarantine programs and can include insecticide treatments, foreign exploration for biological control agents and restrictions on the movement of regulated items such as nursery stock.

Regulatory entomologists also are responsible for determining if exports meet the insect pest requirements of state and foreign governments by certifying that products are free of injurious pests. To assist domestic producers, regulatory entomologists develop protocols designed to limit or prevent the movement of insect pests in or on trade goods. The protocols permit producers to fulfill quarantine regulatory requirements to control insects and to provide shippers with greater flexibility and efficiency in shipping pest-free products. Protocols can include various measures such as heat treatments, fumigation, isolation of production or transit areas, irradiation, insecticide applications, and/or changes in production and handling methods. With increased trade and speed of travel, the likelihood that insects will survive and become established in new geographic areas is increasing. It is the responsibility of regulatory entomologists to identify creative means to limit the number of introductions and to mitigate the impacts of newly established insect pests.

Lloyd Garcia
North Carolina Department of Agriculture and Consumer Sciences
Raleigh, North Carolina, USA

References
Essig, E. O. 1931. *A history of entomology*. Macmillan Publishing Co., New York, New York 1,029 pp.
Metcalf, C. L., W. P. Flint, and R. L. Metcalf. 1993. *Destructive and useful insects, their habits and control* (5th ed.). McGraw-Hill, Inc., New York, New York. 1,087 pp.
Smith, I. M., D. G. McNamara, P. R. Scott, and M. Holderness 1996. *Quarantine pests for Europe* (2nd ed.). CABI/EPPO, CAB International, Wallingford, United Kingdom. 1,425 pp.
USDA, Animal and Plant Health Inspection Service Web site for APHIS history (http://www.aphis.usda.gov/oa/history.html)

REGULATORY SEQUENCE. A DNA sequence involved in regulating the expression of a gene. A promoter or operator.

REHN, JAMES ABRAM GARFIELD. James Rehn was born at Philadelphia, Pennsylvania, USA, in 1881. As a child be became associated with the Academy of Natural Sciences, and then became a staff member and fell under the influence of prominent entomologists. By 1900 he was publishing on Orthoptera, which soon became his life's work. He made collecting trips to Africa and South America, as well as throughout the United States. He worked closely with Morgan Hebard on a great number of projects, and they both became renowned orthopterists. Rehn published over 300 papers on Orthoptera, and several books. Among his most important publications are "The grasshoppers and locusts of Australia" (in three volumes, begun in 1949) and "A monograph of the Orthoptera of North America (north of Mexico), Vol. 1" (with H.J. Grant, Jr.) (1961). He edited the "Transactions of the American Entomological Society" from 1917 to 1924, and served on many important committees of the American Entomological Society. Rehn received many honors, and served as president of the Entomological Society of America. He died on January 25, 1965.

References

Mallis, A. 1971. *American entomologists*. Rutgers University Press, New Brunswick, New Jersey. 549 pp.
Roback, S. S. 1984. James A. G. Rehn and the American Entomological Society. *Entomological News* 95: 163–165.

REITTER, EDMUND. Edmund Reitter was born at Mohelnice in Moravia (Czech Republic) on October 22, 1845. He initially worked as a farm manager, and was responsible for fish pond management on an estate. However, insects were his passion, and he moved to Vienna in 1879 and started a business selling insects. He moved to Mödling, near Vienna in 1881 and expanded his business to include entomological equipment and literature. Mödling became a favorite meeting place for entomologists, but Reitter's wife insisted that they move to Paskov, which was accomplished in 1891. Reitter's house again became a favorite meeting place for entomologists. Reitter is considered a genius among coleopterists because he had an uncanny ability to find useful diagnostic characters, even among species studied extensively by others. He was a prolific publisher, with over one thousand papers to his credit. He worked extensively across the various taxa of the Coleoptera. He also authored "Fauna Germanica" a 5-volume treatise that was the most important book on Coleoptera of Europe for decades. He is credited with describing a phenomenal 1,062 genera and 6,411 species! His personal collection contained 250,000 specimens, including 4,500 types; it is now housed in the Hungarian Natural history Museum. He died at Paskov, Moravia, on March 15, 1920.

Reference

Herman, L. H. 2001. Reitter, Edmund. *Bulletin of the American Museum of Natural History* 265: 126–128

RELAPSING FEVER. See also, TICKS.

RELATIVE HUMIDITY. The percentage saturation of the air with water vapor.

RELATIVE METHODS OF SAMPLING. Techniques used to sample insect populations that provide an estimate of abundance relative to other times or places (e.g., per trap or per unit time of visual search). Types of relative sampling include sex pheromone, blacklight and sticky traps, and amount of plant damage. (contrast with absolute methods of sampling).

See also, SAMPLING ARTHROPODS.

RELEASER PHEROMONE. A pheromone that stimulated the nervous system to produce a quick behavioral response. (contrast with primer pheromone)

REMIGIUM. The rigid anterior portion of the wing, which is directly affected by the wing muscles and is very important in flight.

REMOTE SENSING. Examination of a site from a distance, typically from an airplane but also by satellite.

REMOVAL TRAPPING FOR POPULATION ESTIMATION. An absolute method of insect sampling that involves repeated collection and removal of insects from an area. The rate of decline in abundance as the population is depleted is used to estimate the initial population size.

RENIFORM. Kidney- or bean-shaped. Noctuid (Lepidoptera) moths often bear a reniform spot in the center of the front wings.

REOVIRUSES (CYPOVIRUSES). The family Reoviridae contains members in nine genera that infect animals, plants and fungi. The name of this group is derived from its association with respiratory enteric orphans. All reoviruses are characterized by a segmented dsRNA genome encapsidated within 50 to 70 mm icosahedral particles. Each of the dsRNA segments contains one or two open reading frames. Members of two genera, *Orbivirus* (referred to previously as arboviruses) and *Phytoreovirus*, cause disease in vertebrates and plants, respectively, but are capable of limited replication in their insect vector. In nature, the majority of reoviruses associated with animals cause asymptomatic diseases. Likewise, most of the non-occluded insect reoviruses (unassigned genera) are characterized as being weakly invasive and non-lethal.

A series of reoviruses have been detected in various *Drosophila melanogaster* cell lines and has been reported to contain 10 to 13 segments of dsRNA. These viruses may establish persistent infections and do not appear to adversely affect the host cell. In nature, the reovirus *Drosophila* S virus (DSV) has been isolated from malformed *D. simulins* populations. In *D. simulins*, this virus is maternally transmitted to the epidermal or mesodermal cells of progeny flies. The virus possesses reduced invasive powers; electron microscopy of infected tissues has revealed no evidence of cell lysis, release of the virus, or viral penetration. Infection of the cuticle epidermis results in the S-phenotype characterized by the loss of varying numbers of thoracic bristles. Interestingly, a reovirus of a terrestrial isopod, similar in several respects to DSV, is the causal agent of an inter-sex phenotype. Infection of female isopods by this reovirus induces external pseudohermaphroditism via inhibition of the feminizing endosymbionts affiliated with this isopod host. Other reoviruses related to DSV include the *D. melanogaster* F virus, *Ceratitis capitata* reovirus, and the housefly *Musca domestica*

(HSV) reovirus. Unlike DSV, the HSV is virulent to its host, adult houseflies. The virus replicates in the cytoplasm of hemocytes and appears to undergo partially defective replication in the midgut tissue. In addition to dipteran hosts, reoviruses also have been detected in several hymenopteran parasitoids. The reovirus of the wasp *Diadromus pulchellus* (DpRV) has been well-characterized and is believed to be related to the *Orthoreoviruses*. The DpRV is considered to have a commensal relationship with its wasp host. This virus is distinct in that it produces ploidy-specific genotypes. In haploid male wasps, the DpRV contains 10 dsRNA segments, whereas in diploid female wasps the DpRV contains an additional 3.3 kb supernumerary dsRNA segment.

Members of the genus *Cypovirus* (cytoplasmic polyhedrosis viruses, CPVs) are the best-characterized insect reoviruses. CPVs have been isolated from over two hundred insect species and several non-insect invertebrates. The majority of CPVs have been isolated from the larval stages of Lepidoptera (moths, butterflies). CPVs, like the occluded baculoviruses and entomopoxviruses, are characterized by their ability to produce proteinaceous occlusion bodies. The size and shape of these occlusions may vary according to the viral isolate, the degree of maturation, and the type of host cell. For example, the silkworm *Bombyx mori* is host to a variety of different CPV strains that either produce cuboidal, hexagonal, or tetragonal shaped inclusions. In some cases, more than 10^4 viruses may be occluded within a single occlusion. Cypoviruses produce icosahedral virus particles (65 nm diameter) which encapsidate 10 to 12 dsRNA monocistronic segments (total size 14 to 15×10^6 daltons), an RNA-dependent RNA polymerase, and various enzymes involved in the production of 5' capped, methylated viral mRNAs. The electrophoretic patterns of extracted dsRNA segments have been used to cluster CPVs into twelve electrophenotype groups. Northern blots employing cDNA probes constructed from selected dsRNA segments have demonstrated no homology between the *Cypovirus* and non-occluded reoviruses. Furthermore, little homology was demonstrated among CPV isolates having similar dsRNA electrophenotypes.

In host insects, these viruses preferentially replicate in the midgut columnar epithelium and do not appear to spread to other tissues. However, it has been reported that a CPV of a chironomid is capable of replicating in the fat tissue. The occlusions, ingested by susceptible insects, are disrupted by the conditions in the alimentary tract and release the icosahedral virus particles. These non-enveloped particles contain surface projections that mediate attachment to the microvilli. Within the midgut columnar cells, the virus undergoes replication in a manner similar to that outlined for non-occluded reoviruses. As the name implies, the replication of the CPVs occurs in the cell cytoplasm. During the early phase of the disease, dissected midguts contain regions or zones of infected columnar cells. As the disease progresses, the virus spreads to adjacent cells in a fashion reminiscent of the expansion of viral plaques in cell monolayers. Additional infected cells may undergo extensive hypertrophy and become dislodged from the basement membranes. The sloughing of these cells into the lumen release inocula that can spread the infection to new regions in the midgut. In heavily infected hosts, the large numbers of progeny occlusion bodies generated in columnar cells produce milky colored midgut tissue. The occlusions produced in the midgut cells are released in the feces, aiding in the transmission of this disease. In nature, epizootics of CPV are uncommon. This disease causes problems in insect colonies, especially when insects are mass-reared. The cypovirus disease has several characteristics that make it ideally suited for spread within insectaries. Firstly, these viruses do not cause acute infections; their *in vivo* development is often described as a chronic, debilitating disease that may take weeks to kill infected hosts. During this period, the virus is continuously shed via the feces from infected insects. Secondly, slightly infected hosts may survive and produce chronically infected adults. During both mating and oviposition, the virus can be released from infected alimentary tracts and be passively transmitted to other adults or to the egg surface. At the population level, the presence of a cypovirus may result in reduced food intake, retarded development, and a gradual die-off, resulting in the collapse of the colony.

References

Aruga, H., and Y. Tanada. 1971. *The cytoplasmic polyhedroses virus of the silkworm*. University of Tokyo Press, Tokyo, Japan. 234 pp.

Bellonick, S. 1989. Cytoplasmic polyhedrosis viruses-Reoviridae. *Advances in Virus Research* 37: 173–206.

Nibert, M. L., L. A. Schiff, and B. N. Fields. 1996. Reoviruses and their replication. Pp. 691–730 in B. N. Fields, D. M. Knipe, and P. M. Howley (eds.), *Fundamental virology* (4th ed.). Lippencott-Raven, Philadelphia, Pennsylvania.

REPELLENTS. Chemicals that cause insects to make an oriented movement away from the source of the chemicals.

REPELLENTS OF BITING FLIES. The first use of a repellent was probably by pre-historic humans, sitting in the smoke of a fire to avoid biting flies. This method is still used in many places by remote populations. Some individuals resorted to covering themselves with mud, or grease from the animals they killed, to reduce fly biting. Repellents against insects were first mentioned in classical Roman literature by Pliny, the Naturalist (23 to 79 AD). The Greek physician and pharmacologist, Dioscorides (40 to 90 AD) described the use of wormwood, *Artemisia absinthium*, to repel gnats and fleas. Most of the early repellents consisted of parts of plants or animals, and some were associated with religious beliefs. Native Americans rubbed cedar tree needles, *Juniperus virginiana*, on their bodies to repel insects. Some people wore herbs or plant extracts around their necks or rubbed them on their skins.

Oil of citronella was one of the most widely used early repellents. It was used for human application to repel adult gnats and mosquitoes since 1882, and was initially registered in the U.S. in 1948 as McKesson's Oil of Citronella. Citronella was the standard repellent for many years. Other essential oils that were used as repellents were anise, bay laurel, bergamot, cassia, cedarwood, eucalyptus and wintergreen.

In 1929, dimethyl phthalate was reported as a fly repellent. At Rutgers University, a number of compounds were synthesized, including ethyl hexandiol, and reproducible methods of evaluating repellents were developed. Indalone (dimethyl carbate) was reported as a repellent in 1937.

At the U.S. Department of Agriculture laboratory at Orlando, Florida, and later at Gainesville, Florida, thousands of compounds were tested for repellency. During World War II, the standard military repellent was 6-2-2, a mixture of 6 parts dimethyl phthalate, and 2 parts each of ethyl hexandiol and indalone.

After World War II, the standard military repellent was M-2020, which was a mixture of 40% dimethyl phthalate, 30% ethyl hexandiol and 30% dimethyl carbate. The standard clothing repellent was M-1960, a mixture of 30% benzyl benzoate, 30% *n*-butylacetanilide, 30% 2-butyl-2-ethyl-1,3 propane-diol, and 10% Tween 80.

Diethyl toluamide (DEET) was developed in 1954. DEET was by far the best repellent developed, and is the repellent of choice for the military. Military forces may have to deploy in areas where vector-borne diseases are rampant, and other methods of protection cannot be employed. Pest mosquitoes may exist in such large numbers as to interfere with military operations. A long lasting repellent was required and a new extended duration repellent for-mulation was developed. However, these sorts of conditions normally do not occur in the backyard of suburbia, or on the sunny beaches of the coast. Therefore for normal civilian use, shorter duration repellents may be satisfactory.

In recent years there has been a tendency towards the use of "natural agents" or phytochemicals. Among the so called 'natural' products, or better described as 'herbal' products, are the following:

Citronella. oil of citronella is the volatile oil obtained from the steam distillation of freshly cut or partially dried grass of the genus *Cymbopogon*. Oil of citronella is an essential oil made up of more than 80 compounds of closely related terpenic hydro-carbons, alcohol, and aldeydes.

Geraniol. is a component of citronella. Geraniol *coeur* is a mixture of geraniol, nerol and citronellol.

Quwenling. is derived from the waste distillation after extraction of lemon eucalyptus (*Eucalyptus maculata citriodon*) and used in China as a mosquito repellent. The principal active ingredient is *p*-methane-3,8-diol.

Neem. is a compound found in the leaves, seeds and seed oil of the neem tree (*Azadirachta indica*). The neem tree is a tropical tree, native to East India and Myanmar.

Soy bean oil. is composed of a lecithin base plus other derivatives from soy oil.

Oil of palmarosa. The oil from *Cymbopogon m. martinii* var. *sofia* provides 12 hours of protection against *An. culicifacies*, *An. annularis* and *An. subpictus*.

Castor seed oil. A distillation fraction of castor seed oil was reported as a mosquito repellent.

Thujic acid. Diethylamide of thujic acid distilled from the wood of western red cedar, *Thuja plicata*, was reported as a repellent of *Aedes aegypti*.

Indian privet. The steam distillate of leaves of Indian Privet (*Vitex negundo*) provided 1 to 3 hours of protection against *Aedes aegypti*.

Lemon oil. Lemon oil was similar to DEET in repellency against the sand fly *Lutzomyia youngi*.

Tarweed. An extract of tarweeed (*Hemizonia fitchii*) containing 1,8-cineole (eucalyptol) repelled *Aedes aegypti*.

Eucalyptus camaldulensis. various extracts from this plant, including eucamol, 4-isopropylbenzyl alcohol, *p*-methane 3,8-diol were reported to be repellents to *Aedes aegypti* and *Ae. albopictus*.

(+)-*Eucamolol.* Eucamolol and its 1-epimer showed significant repellent activity against *Aedes albopictus*.

There are numerous reports in the literature of repellency of phytochemicals, but most of these phy-tochemicals have not been developed commercially. Three repellents have recently come to the market, and may be the trend of the future.

Merck 3535. 3-(N-butylacetamino)-propionate has shown strong repellent action against biting flies.

KBR 3023. is a Bayer product, a piperidine deriva-tive, 1-Piperidinecarboxylic acid, 2-(2-hyroxyethyo) 1-methylpropylester.

PMD. is similar to Quwenling, but is derived by a different extraction process. The active ingredient is *p*-methane-3,8-diol with additional isopulegol and citronellol. The product provided 5 hours of complete protection time against *Anopheles funestus* and *An. gambiae*.

Though Permethrin, a pyrethroid, is not techni-cally a repellent, it is used in a manner similar to repellents, to protect a person against biting insects. It is used to impregnate uniforms, clothing and bed nets.

The future repellents may be systemic repellents, or compounds masking the attractiveness of humans.

Compounds may be developed to shield or block the body from bites.

Development of new repellents with increased protection time and effectiveness will likely come from a better understanding of host-finding. Attraction compounds combined with an insecticide may be developed to apply to clothing.

Despite recommendations that repellents be used to prevent mosquito-borne disease transmission, there is very little scientific evidence that repellents can be used to reduce the transmission of vector-borne diseases. It should be recognized that people use repellents to prevent annoyance from mosquito bites, and it would be inappropriate to recommend the use of repellents to prevent disease transmission by biting vectors.

Eugene J. Gerberg
University of Florida
Gainesville, Florida, USA
and
Norman G. Gratz
World Health Organization
Geneva, Switzerland

References

Carlson, D. A. 1997. Insect repellents. pp. 283–297 in D. Rosen, F. D. Bennett and J. L. Capinera (eds.), *Pest management in the subtropics. Integrated pest management – a Florida perspective.* Intercept Ltd., Andover, United Kingdom.

Granett, P. 1940. Studies of mosquito repellents. II. Relative performance of certain chemicals and commercially available mixtures as mosquito repellents. *Journal of Economic Entomology* 33: 566–572.

Gupta, R. K., and L. C. Rutledge. 1994. Role of repellents in vector control and disease prevention. *American Journal of Tropical Medicine and Hygiene* 50 suppl: 82–86.

Gupta, R. K., A. W. Sweeney, L. C. Rutledge, R. D. Cooper, S. P. Frances, and D. R. Westrom. 1987. Effectiveness of controlled-release personal-use arthropod repellents and permethrin-impregnated clothing in the field. *Journal of the American Mosquito Control Association* 3: 556–560.

REPLETE. Individual ants whose crops are greatly distended with liquid food. Repletes serve as living reservoirs, and will regurgitate on demand to supply the colony with food.

REPRODUCTION. As in nearly every other aspect of their lives, insects display great diversity in modes of reproduction. Most insects reproduce in the adult stage by laying eggs, but a few reproduce during an immature stage, a process known as paedogenesis. Some, such as aphids and certain flies, for example, give birth to live young. Although sexual reproduction by union of male and female gametes is typical, certain insects reproduce some or all of the time by laying unfertilized eggs (parthenogenesis).

Oocytes accumulate yolk and cytoplasm and develop in egg chambers or follicles in the ovarioles within the ovary. Follicles are surrounded by a single layer of epithelial cells, the follicular epithelium. The presence or absence of nurse cells associated with ovarioles divides ovaries into two major groups; meroistic ovaries have nurse cells and panoistic ovaries do not. Nurse cells provide nutrients and gene products for the developing oocyte. In panoistic ovaries, similar components probably are provided by other cells, possibly the follicular epithelium. Insects typically produce eggs with a large amount of yolk relative to the cytoplasm, and the yolk is rich in proteins and lipids to be used in forming the new embryo and for energy, respectively. In most insects the yolk is formed from large glycolipoproteins called vitellogenins that are synthesized in fat body cells and transported to the ovaries by hemolymph. In contrast, small molecules called yolk proteins, to distinguish them from large vitellogenins, are incorporated into the yolk by the higher Diptera, such as tephritid fruit flies, drosophilid flies, house flies and related dipterans (but not in mosquitoes, which have vitellogenins). Two different sets of genes are required for the synthesis of yolk proteins and vitellogenins. Some parasitic insects produce small (20 to 200 μm), almost yolkless eggs that are deposited in a host (another insect) where the developing embryo absorbs nutrients from the host through a thin egg shell. In most insects several hormones regulate oogenesis, synthesis of yolk proteins, and additional aspects of reproduction such as pheromone production, mating, and oviposition. Not surprisingly, the complement of hormones is not the same for all groups of insects. When maturation of the oocyte is nearly complete, a vitelline membrane is secreted, followed by secretion of the egg shell, or chorion. After the chorion is secreted onto the egg, it is ready to pass down the oviduct and, in most insects, be fertilized by sperm released from the spermatheca. Hymenoptera and some coccids (Hemiptera) produce haploid males without fertilization and diploid females when the egg is fertilized. The ratio of sex chromosome to autosomes and/or the presence of sex-determining genes are two mechanisms known to determine gender in most insects.

The ovary and egg production

The typical internal reproductive system of females consists of paired ovaries, lateral oviducts, a common oviduct, one or more spermatheca and accessory glands. These structures are located dorsally to the alimentary tract. Generally, several to many ovarioles make up an ovary. Each ovariole contains a string of egg chambers called follicles. The tsetse fly *Glossina* spp. has one ovariole, two other dipterans (*Melophagus* spp. and *Hippobosca* spp.) have two, *Drosophila melanogaster* has from 10 to 30, the blowfly *Calliphora erythrocephala* has about 100, the American cockroach *Periplaneta americana* has eight, and termite queens (Isoptera) may have up to 2,000. The nymphalid butterfly, *Phyciodes phaon*, commonly known as the phaon crescent, has 4 ovarioles in each ovary and about 50 immature oocytes that look like a string of pearls in each ovariole when the adult female emerges from the pupal stage. Just one ovary containing one ovariole occurs in some aphids, and Collembola have sac-like ovaries that do not contain ovarioles. Typically, oocytes in various stages of development occur in each ovariole in many insects, but generally, the most terminal set of oocytes is matured and laid before the other oocytes grow very large, and in some insects at least, the growth of secondary oocytes is under hormonal control. It makes sense that one set of eggs is matured and laid, making room in the abdomen for maturation of the next set.

Nurse cells are associated with developing eggs in meroistic ovaries that occur in most of the Holometabola (except Siphonaptera), and in Hemiptera, Dermaptera, Psocoptera, Siphunculata, and Mallophaga among the Hemimetabola. Panoistic ovaries that do not have special nurse cells likely evolved first, and occur in present-day Thysanura, Odonata, Plecoptera, Blattodea, and Isoptera. Panoistic ovaries evolved secondarily in Ephemeroptera, Orthoptera, and Siphonaptera.

In meroistic ovaries, nurse cells may occur in the same follicle with the developing oocyte (*Drosophila*), in an adjacent follicle (Hymenoptera), or be located in the germarial region with long cytoplasmic

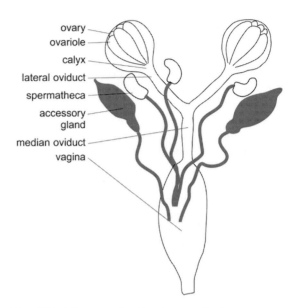

Fig. 870 Diagram of a female reproductive system as found in *Rhagoletis* (Coleoptera) (adapted from Chapman, 1998, The insects: structure and function).

strands projecting to the developing oocyte (some beetles and Hemiptera). Nurse cells transfer nutrients and gene transcripts (mRNA) to developing eggs, and finally are used up in the process. Presumably follicular cells that surround all follicles provide nutrients and gene transcripts in those insects with panoistic ovaries.

Hormones control ovary growth, synthesis of yolk proteins and vitellogenins and their assimilation by the developing oocytes. Juvenile hormone seems to control all reproductive functions in some insects (Orthoptera, for instance), but in others juvenile hormone and ecdysone are important, and in still other groups juvenile hormone, ecdysone, and additional neuropeptide hormones are required. Juvenile hormone is synthesized by the corpora allata in adult insects, as in larvae, but ecdysone is synthesized by the ovaries in adult females. Near the termination of egg growth, the follicular epithelium secretes a thin protein sheet, the vitelline membrane, around the yolk and cytoplasm. Then the follicular epithelium secretes a laminar meshwork of cross-linked proteins, the egg shell or chorion. The follicular epithelial cells, now on the outside of the egg shell, die and are sloughed off the egg. The chorion contains no chitin, and with a few exceptions, it is not mineralized like the egg shell of birds. Sperm, which are released from the spermatheca as the egg passes

Figure labels: ovary, ovariole, calyx, lateral oviduct, spermatheca, accessory gland, median oviduct, vagina

down the common oviduct, traverse the chorion by passing through a small, usually twisted channel, the micropyle. More than one micropyle channel may be present. Although most Diptera have only one, the migratory locust *Locusta migratoria* has 35 to 43 openings.

The male internal system and sperm production

Typically, sperm are produced in paired testes in males, although in some Lepidoptera the testes are fused into a single structure. From the testes they pass into the short vas efferens that leads into the longer vas deferens and finally into the ejaculatory duct. Accessory glands are often, but not universally, part of the male system and when present they are attached by a duct to the vas deferens or to the ejaculatory duct. Secretions from the accessory glands form the spermatophore in some insects, or if no spermatophore is formed, the secretions are added to the sperm prior to transfer to the female. Some of the secretory products provide nutrients and stimulate contractions in the reproductive tract of females, aiding movement of sperm into the spermathecae of the female. Each testis generally consists of a number of tubes or follicles in which spermatozoa are matured. Follicles may vary from one to more than 100 follicles, and may be incompletely separated from each other, as in Lepidoptera, or the testes may consist of several lobes, each with several follicles. In Diptera the testes consist of an elongated and undivided sac. Typically, sperm progress through several devel-

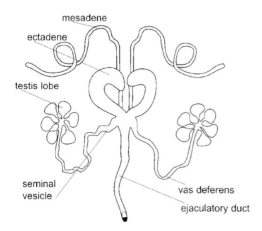

Fig. 871 Diagram of a male reproductive system as found in *Tenebrio* (Coleoptera) (adapted from Chapman, 1998, The insects: structure and function).

opmental stages. Sperm begin as diploid cells secreted by mitotic divisions of cells in the germarium. These initial diploid cells, called spermatogonia, further divide by mitosis into numerous diploid spermatocytes that become enclosed within a cyst or capsule of somatic cells (also diploid). Spermatocytes within a sac or cyst generally arise from the same spermatogonial cell and have synchronized development. Spermatocytes may undergo still more mitotic divisions; there are 5 to 8 divisions in Acrididae and 7 in *Melanoplus*, but eventually in a maturation zone they divide by meiosis and produce haploid spermatids. Spermatids undergo some further maturation in the transformation zone, and usually are bundled together by a thin membrane in this stage, or may be enclosed within a spermatophore. Insect spermatozoa tend to be long and have a slender head region, probably an evolutionary adaptation to navigate the micropyle.

Many males mature sperm during the pupal stage, while others may require several days as an adult to mature sperm. In contrast to the situation in most vertebrates, sperm can survive in the spermatheca of a female for weeks, months, or even years. Honeybee queens have been known to lay fertilized eggs after several years (8 to 9 years in one reported case), and queen ants were reported to contain viable sperm after 15 years.

During most of the 20th century, various experiments failed to show hormonal control of spermatogenesis, but beginning in the 1970s ecdysteroids were found in the testes of flies, crickets, mosquitoes, and various lepidopterans, and experiments showed that they were synthesized there and stimulated spermatogenesis. Neuropeptides also may be involved in a cascade of reactions leading to ecdsyteroid synthesis in testes; physiological levels (approximately 10^{-13} to 10^{-15} M concentrations) of an ecdysiotropic peptide from gypsy moth brain stimulates synthesis of ecdysteroids in testes of larval and pupal gypsy moth males. These observations bring spermatogenesis in insects more in line with well known hormonal control of spermatogenesis in other animals.

An unusual situation evolved in male Lepidoptera. They produce both apyrene sperm without a nucleus and nucleated eupyrene sperm. Only the latter type can fertilize an egg. The two types of sperm are present in some of the earliest families of Lepidoptera that evolved, but one of the most primitive families, the Micropterigidae, do not produce apyrene sperm.

Male Lepidoptera form a spermatophore in the female bursa copulatrix during mating, and both types of sperm are incorporated. As sperm pass down the ejaculatory duct, male silkmoths (*Bombyx mori*) secrete an enzyme, an endopeptidase called initiatorin, from cells in the ejaculatory duct. In the spermatophore, initiatorin begins to act upon the surface coat of both types of sperm, but the coat of the apyrene sperm is more quickly digested, releasing them. Motile apyrene sperm agitate the viscous contents of the spermatophore and aid liberation of the eupyrene sperm. Initiatorin also activates a carboxypeptidase that digests proteins in the seminal secretions and liberates arginine and other amino acids. Arginine is readily converted to glutamate, which sperm metabolize to provide energy for motility.

Pheromones in reproduction

Many insects secrete pheromones, a special class of semiochemicals, to attract mates. Sometimes both sexes produce a pheromone that functions in some aspect of mating. Alternatively, only the female or the male may produce an attractant pheromone, commonly called a sex pheromone. The term 'pheromone' is derived from two Greek words, *pherein*, meaning to carry, and *horman*, meaning to excite. The sex pheromone is usually composed of several chemical components secreted in a particular ratio or blend. Typically, glands at or near the surface of the body secrete the pheromone blend into the air. The total blend generally is considered to be the pheromone. The blend or ratio of components allows more information to be encoded, and allows greater discrimination among closely related species, which often use the same, or some of the same, components. Pheromones have been identified from more than 1,600 species of insects in 90+ families from 9 orders. Lists of Lepidoptera and the chemical nature of their sex pheromones are available on the Internet (Arn et al. 1999). Sex pheromones are sometimes used by predators and parasites to locate a potential host. The sex pheromone of *Dendroctonus frontalis*, a bark beetle, not only attracts a potential mate, but may attract a clerid predator, *Thanasimus dubius*.

Most moths find each other and mate in low light intensity during some part of the evening or night. Perhaps because of activity at low-light intensity, their ability to detect chemicals in the air has evolved to a very high degree of sensitivity. In most Lepidoptera, the female produces the sex pheromone, and males fly to it. The antennae of many male Lepidoptera are plumose, with thousands of small hairs containing pheromone sensitive sensory neurons. The female silkworm moth contains about 1×10^{-8} g pheromone in her body. If this amount of pheromone were released instantly and uniformly distributed into slightly moving air, theoretically a male should be able to sense the pheromone about 4,560 m (2.8 miles) downwind. Any male within that active space, or wandering into its periphery, should be excited to fly upwind. The possibility of a male detecting a single female up to 4,500 m distant stretches the imagination. Several factors probably preclude this theoretical example from being achieved. First of all, females do not release their pheromone instantly, but do so slowly over some period of time each evening. Neither is there uniform distribution of pheromone in the air because of changes in wind current and direction, and buildings, trees, and other objects between the female and the male that disrupt the air flow and create turbulence. In addition, pheromones are adsorbed to the surface of plants and other objects, reducing the amount in the air. Nevertheless, the success of moths in locating each other under semi-dark conditions is impressive and attests to the effectiveness of pheromone in their success.

It is not always the female, however, that produces a sex pheromone. Some male Lepidoptera also produce pheromones, and in most bark beetles (family Scolytidae) both sexes typically produce pheromones. In true fruit flies (family Tephritidae), the male produces the attractant pheromone (with one known exception: the female olive fruit fly, *Bactrocera oleae*).

Sex pheromone synthesis and secretion seem to be under hormonal control in many, perhaps even most, insects, but few specific details have been elucidated. The most thoroughly studied examples of hormonal control of sex pheromone come from moths and cockroaches. A polypeptide hormone that controls the synthesis of the sex pheromone in moths has been named PBAN (Pheromone Biosynthesis Activating Neuropeptide). PBAN is now known from several different moths, and it may have slightly different roles in different moths. For example, in the red-banded leafroller, *Argyrotaenia velutinana*, PBAN regulates pheromone biosynthesis by increasing the supply of octadecanoyl and hexadecanoyl fatty acids needed for pheromone biosynthesis. In several moths, PBAN appears to regulate an enzyme, a Δ11 desaturase, that is necessary to introduce a double bond into the pheromone precursors. In still others, it influences the reduction of fatty acids to the alcohol precursor of acetate pheromones.

Determination of gender

At least three chromosomal systems, with variations, control sex in insects. Probably the earliest to evolve was one in which male sex chromosomes are XY and female XX . This system exists in many different groups today, including *D. melanogaster* and other Diptera. In some insects the Y chromosome has been lost, and those males are XO. Males in the orders Odonata, Orthoptera, and some groups in other orders are XO males. In Lepidoptera and Trichoptera, the female is the heterogametic sex, usually designated as ZW while males are ZZ. In the third system, found in Hymenoptera and some Hemiptera, females are diploid and males are haploid. If an egg is fertilized by a sperm, it becomes a female, and if no fertilization occurs, the egg develops into a haploid male. In honeybees and probably other Hymenoptera, the female can control the release of sperm from the spermatheca, and thus can control the sex of the individuals that will develop from the eggs she lays. In a few insects the sex of individuals is determined by prevailing temperature, as in subarctic mosquitoes, and in some gall midges by available nutrition.

Exactly how the different chromosomal patterns described above lead to sex determination is variable and poorly known for all except a few insects. The ratio of X chromosomes to autosomes (A) in determining sex of offspring has been characterized in *Drosophila melanogaster*. Although the Y chromosome carries genes for factors necessary for the production of motile sperm, it does not carry gender-determining genes in *Drosophila*. The ratio $2X:2A = 1$ (one set of autosomes with an X from each parent) results in normal females, and the ratio of 3X:2A in aneuploids also results in female phenotype. A sex chromosome:autosome ratio of 0.5 (1X:2A, an autosome set from both parents, but an X from the female parent and a Y from the male parent) results in a normal male, and even a smaller ratio (1X:3A) produces a male phenotype. Intermediate ratios can result in mosaic intersexes in which the individual contains both male and female cells. Chromosomal ratios are ultimately expressed in specific gene actions. Since males have only one X chromosome, genes on it that control important functions or synthesis of critical molecules must be twice as active in males as in females, a process known as dosage compensation, but few details are known. Specific sex determining genes control the phenotype of some insects, including the housefly *Musca domestica*, the silkworm moth *Bombyx mori*, and some mosquitoes.

See also, JUVENILE HORMONE, ECDYSTER-OIDS, OOGENESIS, VITELLOGENESIS, ENDOCRINE REGULATION OF INSECT REPRODUCTION, SEX ATTRACTANT PHEROMONES, PHEROMONES, SEX RATIO MODIFICATION BY CYTOPLASMIC AGENTS, STERILE INSECT TECHNIQUE.

James L. Nation
University of Florida
Gainesville, Florida, USA

References

Arn, H., M. Tóth, and E. Preisner. 1999. The pherolist. Internet Edition. www.nysaes.cornell.edu/pheronet

Chapman, R. F. 1998. *The insects: structure and function.* Cambridge University Press, Cambridge, United Kingdom. 770 pp.

Klowden, M. J. 2002. *Physiological systems in insects.* Academic Press, New York, New York. 415 pp.

Nation, J. L. 2002. *Insect physiology and biochemistry.* CRC Press, Boca Raton, Florida. 485 pp.

REPRODUCTIVE RATE. The number of offspring produced per unit of time.

RESERVOIR. A site where organisms can survive, usually in relatively small numbers, and then invade or repopulate an area.

RESILIENCE. An expression of the ability of some species to recover from disturbance. Persistence of a particular level or density despite perturbations. (contrast with stability)

RESILIN. Resilin is a colorless, transparent structural protein of cuticle that has even better elastic properties than natural rubber. A rubber band elongates upon prolonged stretching, but experiments have shown that resilin retains about 97% of its original length even after extended periods of stretching. Resilin is not appreciably sclerotized, or else it could not retain its elasticity. Only small patches of resilin occur in the cuticle, and it mainly occurs at joints where wings or legs are hinged to the thorax. The prealar arm connecting the mesotergum to the first basalar sclerite of the thoracic wall in the wings of the migratory locust *Schistocerca gregaria* contains about 50 μg resilin and 15 μg chitin, and a similar hinge in large *Aeshna* spp. of dragonflies contains 5 to 7 μg resilin. A pad of resilin located near the coxal attachment to the pleural arch is compressed when a jumping flea crouches and prepares for a jump, and as the compressed resilin springs back to its original shape when the leap begins, it helps propel the flea into the air. Resilin is rich in the amino acids glycine and proline, as are some other structural proteins such as collagen, elastin, and silk fibroin. Resilin contains no hydroxyproline, hydroxylysine, tryptophan, or sulfur-containing amino acids, but does contain two unusual amino acids, dityrosine and trityrosine. These latter two amino acids may be involved in light cross-linking of resilin protein chains, enabling them to hold their shape. Studies of resilin biosynthesis are sparce, possibly because of the rather small areas in which it occurs and the difficulty of getting it out of the cuticle since it is insoluble in aqueous solvents. Resilin may be secreted by specialized epidermal cells.

See also, INTEGUMENT: STRUCTURE AND FUNCTION, LOCOMOTION.

James L. Nation
Dept. of Entomology & Nematology
University of Florida, Gainesville, FL 32611-0620

RESISTANCE OF SOLANACEOUS VEGETABLES TO INSECTS.

Tomato (*Lycopersicon esculentum*), potato (*Solanum tuberosum*) and pepper (*Capsicum annuum*) are ''new world'' crops. Tomato and potato originated in the northern regions of South America, and pepper in the tropical regions of Latin America. These crops are thought to be relatively new under cultivation, perhaps 200 to 300 years. Eggplants (*S. melongena*), however, originated in India and have been cultivated for hundreds of years. Wild eggplant species are not known.

The pests and their geographical origin

Almost all the pests are insects belonging to the following orders: Thysanoptera (thrips), Hemiptera (bugs, leaf-hoppers, whiteflies and aphids), Orthoptera (mole-crickets and grasshoppers), Coleoptera (beetles), Lepidoptera (moths and butterflies), and Diptera (flies). Out of the known pest species, only about one-fifth are major pests of primary importance. These may cause severe damage during almost every season and in many countries on different continents. All others are considered as minor pests, restricted to a rather limited area. Within this area, however, they might be severe pests of primary importance. Booklice or psocids (Psocoptera) and the wingless Collembola (springtails) are considered nuisance pests.

The pests affecting these crops either originated from South America, like the crop, and were accidentally reintroduced to the crop in other countries, or are local pests of other plants, which have adapted to the newly introduced crops. The greenhouse whitefly (*Trialeurodes vaporariorum*) and the potato tubermoth (*Phthorimaea operculella*) originated in Central America. The Colorado potato beetle (*Leptinotarsa decemlineata*) originated in North or Central America. The serpentine leafminer (*Liriomyza trifolii*) and the western flower thrips (*Frankliniella occidentalis*) originated in North America. Some other pests, such as the sweetpotato (tobacco) whitefly (*Bemisia tabaci*) and the onion thrips (*Thrips tabaci*), are from Central Asia. The tomato fruitworm (*Keiferia lycopersicella*), the Egyptian cotton leafworm (*Spodoptera littoralis*), the beet armyworm (*Spodoptera exigua*), the *Phytometra* (*Plusia*) *chalcites*, the tomato fruitworm (*Helicoverpa zea*) and the African cotton bollworm (*Heliothis armigera*) seem to be local pests that have adapted to the newly introduced tomato, potato, pepper, and eggplants. Knowing the origin of the pests can assist with the search for resistant plants for breeding.

Mechanisms of resistance

In solanaceous plants, resistance mechanisms are based on exterior and/or interior plant features. The most pronounced are the exterior mechanisms, based on physical means: trichomes or glandular-trichomes and their sticky exudate, which interferes with any small insect like thrips, whiteflies, aphids, leafminer flies or the younger stages of various larvae of other insects. These trichomes cover the plant's surface, leaves and twigs, and provide an exterior defense mechanism. The glandular hair exudate, when dislodged from cell walls, adheres to the legs and mouthparts of the insects and rapidly hardens, entrapping small insects or causing them to fall off the plant. Stuck insects rarely feed, quickly die, and thus are unable to transmit viruses. The interior mechanisms provide resistance that is based on secondary plant substances, which impose repellent, toxic, or hormonal effects, mainly on biting and chewing insects.

Trichomes occur in a multitude of forms in many wild tomatoes. In numerous species, there is a negative correlation between trichome density, and insect feeding and oviposition responses, and the nutrition of larvae. Specialized hooked trichomes may impale adults or larvae as well. Trichomes may also complement the chemical defense of a plant by possessing glands that exude terpenes, phenolics, alkaloids or other substances, which are olfactory or gustatory repellents. In essence, glandular trichomes afford an outer line of chemical defense by advertising the presence of ''noxious'' compounds. These features make them good sources for breeding resistance to virus transmission by insects. This is hereditary in the hybrids of the wild plants with cultivated varieties.

Resistant plants may possess both exterior and interior mechanisms, which then provide simultaneously resistance to a variety of pests. For example, *Lycopersicon* (*Solanum*) *pennellii* (e.g., LA 716) is resistant to a number of insect pests due to the accumulation of acylsugars exuded from type IV trichomes. Thus, it is resistant to *B. tabaci* adults, both causing direct damage and transmitting virus diseases. At the same time, it is also resistant to *S. littoralis*. The small larvae (L_I and L_{II}) become entangled in the sticky exudate of the leaf trichomes, like any small insect. Bigger larvae (L_{III}) generally overcome the stickiness, but they are then confronted with a juvenile hormone mimic in the leaves, which causes high larval mortality, adult morphological deformation, and male sterility. The survivors pupate earlier, and the emerging adults are smaller and less fertile than usual. The large larvae (L_{IV}) are not affected by these mechanisms.

The secondary plant chemicals, present in the ancestors of the domesticated cultivar, provided them with natural resistance to pests. Unfortunately, these chemicals may also be unpalatable or even toxic to people so such plant chemicals were deleted during the breeding process. In this way, the susceptibility to pests of the solanaceous crops was inadvertently increased during their domestication. As a result,

the modern cultivars need to be continuously protected by the grower against pests. The reintroduction of resistance, or partial resistance, to pests into the modern cultivars is an invaluable component of integrated pest management programs.

Vector resistance

Virus diseases are transmitted to these crops by various insect pests (e.g., aphids mainly to potatoes and pepper, and whiteflies and thrips mainly to tomatoes). Because viruses cannot be controlled, and infected plants cannot be cured, the only way (other than by eliminating vector insects) to combat viruses is by preventing their transmission, mainly by promoting resistance.

Hypothetically, two expressions of plant resistance are recognized: resistance, or tolerance, to the virus itself, distinguished as ''virus resistance,'' and resistance to the virus-transmitter, distinguished as ''vector resistance.'' These types of resistance are complementary and can be integrated into the same plant. The advantage of the combined vector-, and virus-resistance is due to the fact that plant resistance to insects is expected to be of greater durability than plant resistance to viruses, and the combination of the two resistance mechanisms will even prolong the durability of plant resistance to virus infection.

Breeding for pest resistance

Tomatoes are attacked by various pests, but within cultivated tomato varieties no sufficient pest resistance has been detected. Wild types such as *L. pimpinellifolium*, *L. peruvianum*, *L. chilense*, and *L. glandulosum* express some resistance to various pests. Nevertheless, adequate levels of resistance were detected mainly in accessions (ecotypes) of *L. pennellii*, *L. hirsutum* (also known as *L. hirsutum* f. *typicum*), and *L. hirsutum* f. *glabratum*. These sources of resistance had been intensively exploited. Some are resistant enough to prevent direct damage, and some, although virus-susceptible themselves, even prevent virus transmission. Glandular trichomes of these plants exude a sticky material and protect the plants by physically trapping and immobilizing of the pests. Others also exude chemicals that are toxic or repellent to the pests. The resistant accessions usually show resistance to various insects.

The potato pests are generally similar to those attacking tomatoes, but most of the sources of resistance had been chosen among wild *Solanum* species,

The incidence of tomato yellow leaf curl virus, expressed in cumulative percent of infected plants, in tomato and in wild *Lycopersicon* spp. in the field. (The most resistant known accessions of the wild species were selected to demonstrate the potential of host plant resistance.)

Weeks From Planting	Tomato cv. "Allround"	*L. hirsutum* f. *glabratum*		*L. hirsutum*		*L. pennellii*	
		(1*)	(2*)	(3*)	(4*)	(5*)	(6*)
2	0.0	0.0	0.0	0.0	0.0	0.0	0.0
3	50.0	14.2	5.0	0.0	0.0	0.0	0.0
4	100.0	47.6	35.7	0.0	0.0	0.0	0.0
5	100.0	57.1	75.5	0.0	0.0	0.0	0.0
6	100.0	92.5	59.3	4.0	0.0	0.0	0.0
7	100.0	92.5	59.3	4.0	0.0	0.0	0.0
8	100.0	100.0	59.3	45.0	46.6	5.0	0.0

L. hirsutum f. *glabratum* (1*) PI 134418 (2*) PI 129157
L. hirsutum (3*) LA 1777 (4*) IVT 771498
L. pennellii (5*) LA 716 (6*) PI 79659

some of which bear tubers. Some of the source plants, e.g., *S. berthaultii*, *S. polyadenium* and *S. taijense*, possess glandular trichomes on the foliage that protect the plants by physically trapping and immobilizing the attacking insects. Other *Solanum* spp. possess a variable degree of secondary plant substances such as glycoalkaloids (e.g., tomatin, demissin, α- and β-solamarie, α- and β-chacinine, and α-solanine), which may have a toxic effect or inhibit feeding. *Solanum berthaultii* releases an aphid alarm pheromone that repels aphids such as *Myzus persicae*.

Aphids

The most severe aphid problem is the transmission of virus diseases in potato. Most of these aphid-borne viruses are non-persistent, or stylet-borne viruses. The agricultural means to prevent this virus transmission are very limited. One of the more efficient ways is the use of aphid vector-resistant plants. Naturally, most breeding efforts in potato have been devoted to aphid resistance. The potato crop is commonly infested by *Myzus persicae*, *Macrosiphum euphorbiae*, and *Aulacorthum solani*. Glandular hairs are quite effective in limiting aphid damage to the potato crop. They occur abundantly on *S. polyadenium*, *S. tarijense* and *S. berthaultii*. The sticky exudate, which is discharged from the trichome glands, precipitates on the aphid's limbs, impedes the aphid's movements and immobilizes it; the aphid finally starves to death. This trait of glandular trichomes is hereditary and expressed in the hybrid progenies of

wild and cultivated plants. *Solanum polyadenium* and *S. berthaultii* are adequate sources for breeding potatoes resistant to aphids. The resistance of *S. chacoense*, *S. stolomfenerum* and *S. demissum*, to *M. persicae*, *A. solani* and *M. euphorbiae* depends on the species of the aphid and on the physiological condition of the leaves. This shortcoming must be considered while using them as sources of resistance.

Resistance in the wild tomato, *L. hirsutum* f. *glabratum* (PI 134417), is due primarily to the presence of the toxic factor that is exuded by the trichomes. A concentrated chloroform extract from the foliage is toxic to *Aphis craccivora*, *A. gossypii* and *M. persicae*.

Tomatoes, to which Mi-1, a *L. peruvianum* gene conferring resistance to the root-knot nematode (*Meloidogyne incognita*) was introgressed, also confer resistance to *M. euphorbiae*. Furthermore, tomatoes that bear the *Mi-1.2* gene, which also provides resistance to nematodes, express resistance to *M. euphorbiae* and to the Q-biotype of *B. tabaci* as well.

Whiteflies

High levels of resistance to the greenhouse (or glasshouse) whitefly, *T. vaporariorum*, are well known in *L. pennellii*, *L. hirsutum* and *L. hirsutum* f. *glabratum* accessions, whereas *L. chilense*, *L. minutum*, and *L. peruvianum* express moderate levels of resistance.

Tomato resistance to direct damage, and to the transmission of the tomato yellow leaf curl virus

(TYLCV), by the cotton (or sweetpotato) whitefly, *B. tabaci*, and resistance to *B. argentifollii*, is known in *L. pennellii*, and to some extent also in *L. hirsutum* and *L. hirsutum* f. *glabratum*, though they are susceptible to the virus itself. The incidence of natural virus infection seems generally to reflect the level of resistance of the plants to *B. tabaci*.

Lycopersicon pennellii accessions LA 716 and PI 79659 are highly resistant to *Bemisia*, whereas the resistance of LA 1277 depends on light intensity and day length. The primary mechanism of resistance is the sticky exudate of the glandular trichomes. Whiteflies that land on these plants become entangled in this sticky exudate and die within a short time. In some accessions, the sticky exudate does not appear on the first 3 to 4 true leaves. These leaves are then unprotected and the whiteflies feed and develop on them readily. When artificially removed exudate is applied on cotton leaves, it protects them from whitefly attack for a long time. A toxic effect, browning and death of whiteflies, can be observed even when the sticky effect of the extracted exudate is neutralized by applying it on filter paper.

Among several *L. hirsutum* accessions, two are known to be rather resistant in respect to population build-up parameters, in comparison to the cultivated tomato: LA 1777 and IVT 771498; however, LA 1363 is rather susceptible. The main mechanisms of resistance are due to the physical interference of the glandular trichomes, and to toxic factors that occur in the trichomes and in the leaf tissues. Among *L. hirsutum* f. *glabratum*, accessions PI 134418, PI 129157 and LA 407 are relatively resistant, and some others are not. In PI 365907, resistance depends on day length and light intensity and probably on some additional environmental factors. The main mechanism of resistance seems to be physical interference of the leaf trichomes and their exudate, which cause immobilization of the insects. It appeared that the exudate also has a repellent effect and, most likely, toxic factors that occur on the leaf surface or within the leaf are involved in plant resistance.

Leafhoppers

Total glycoalkaloid fractions, especially tomatine, provide resistance to the leafhopper in potatoes. *S. hougasii* and *S. bulbocastanum* are among the exploited sources of resistance. *S. berthaultii* (PI 218215) and *S. chacoense* (WRF 888) show less

resistance than expected, probably because of their relatively low tomatine content.

Leafminer flies

Lycopersicon pennellii accessions (LA 1735 and LA 716) possess a very high level of resistance to the leafminer *L. trifollii*. Adequate resistance is also found in *L. cheesmanii* (LA 1401) and in *L. hirsutum* f. *glabratum* (PI 126449). Larval antibiosis also occurs in F₁ hybrids. For breeding purposes, the resistance in *L. cheesmanii* has the most immediate value because it can be reciprocally hybridized with *L. esculentum*, and it produces orange fruit. Oviposition and adult feeding of *L. trifolii* on *L. pennellii* is significantly less than that on the cultivated tomato. This resistance is reduced when the foliage is rinsed with ethanol. Resistant attributes of *L. pennellii* can be transferred to *L. esculentum* through mechanical appression of *L. pennellii* foliage onto *L. esculentum* leaflets. Application of purified 2,3,4-tri-O-acylglucoses (the principal component of type IV glandular trichome exudate of *L. pennellii*) to *L. esculentum* significantly decreased feeding and oviposition on *L. esculentum*. Hence, the principal mechanism of resistance in *L. pennellii* to the leafminer is due to the secretion of acyiglucoses.

Lycopersicon hirsutum (PI 126445 and PI 127826) and *L. hirsutum* f. *glabratum* (PI 126449) possess a high level of resistance to the vegetable leafminer, *L. sativae* and *L. munda*.

Moths and butterflies

Breeding resistance to moths and butterflies has been conducted mainly in tomatoes. The moths and butterflies belong mainly to three families. Members of the Gelechiidae are the tomato pinworm and the potato tuber moth. Both are oligophagous insects, which feed exclusively on Solanaceae under field conditions. High resistance expressed in delayed larval development, high mortality, and reduced leaf consumption to *K. lycopersicella* has been derived from *L. hirsutum* accessions (PI 126445, PI 127826 and LA 361) and from *L. hirsutum* f. *glabratum* accessions (PI 126449 and PI 134417). On these plants, basal cells of type VI trichomes and other epidermal cells as well as most of the palisade mesophyll are not consumed by *K. lycopersicella* larvae. It seems that these structures also contain deterrent chemicals. The resistance of other wild tomatoes — e.g., *L. peruvianum*, *L. peruvianum* var. *humifusum*,

L. cheesmanii f. *minor* and *L. glandulosum* – varies in its intensity. *L. hirsutum* (LA 1777) is highly resistant to *P. operculella*.

Resistance to the tobacco hornworm, *Manduca sexta* (Sphingidae) is expressed in several of the *L. hirsutum* f. *glabratum* accessions (PI 134417 and LA 407).

Several of the moths that attack tomato crops belong to Noctuidae. Resistance of tomato cultivars to the tomato fruitworm *Heliothis zea* varies greatly, indicating that much of the variability can be used to develop less susceptible cultivars. Higher levels of resistance can be obtained from *L. hirsutum* (PI 127826) and *L. hirsutum* f. *glabratum* (PI 126449, PI 129157, PI 134417, PI 134418, PI 251304, PI 251305), whereas the resistance of *L. pimpinellifolium* and *L. peruvianum* is insufficient. The antibiotic factor, which affects *H. zea*, appears to be inherited recessively. *H. armigera* larvae are strongly affected by feeding on leaves of the resistant *L. hirsutum* (LA 1777), *L. hirsutum* f. *glabratum* (LA 407), and *L. pennellii* (LA 716). Highest levels of resistance to *Phytometra chalcites* are found in *L. hirsutum* (LA 1777), *L. hirsutum* f. *glabratum* (LA 407 and LA 1625), and *L. pennellii* (LA 716). Resistance to *S. littoralis* larvae occurs in *L. hirsutum* (LA 1777), *L. hirsutum* f. *glabratum* (LA 407 and LA 1625), and *L. pennellii* (LA 1277). In conclusion, many sources of resistance are common to a various noctuid larvae, and their resistance is based mainly on toxic or antifeedant repellents.

Beetles

Tobacco flea beetle, *Epitrix hirtipennis*, adults are strongly repelled by *L. hirsutum* and *L. hirsutum* f. *glabratum* due to the exudate of their glandular trichomes. Removing this exudate by rinsing the leaves with 75% ethanol eliminates the resistance.

Colorado potato beetle, *Leptinotarsa decemlineata* resistance is due to the four-lobed glandular hairs on the leaves and stems of the wild potato *S. polyadenium*. Sticky material is discharged from these hairs on contact, trapping larvae or encasing their feet. Larvae with encased feet fall off the plants and die.

Feeding rates of *L. decemlineata* are negatively correlated with the occurrence of tomatine, a steroidal glycoalkaloid, in the tomato and in wild tomatoes such as *L. hirsutum*. High tomatine content in leaves virtually inhibits beetle feeding.

M. J. Berlinger
ARO, Gilat Research Center
Beer Sheva, Israel

References

Berlinger, M. J. 1986. Host-plant resistance to *Bemisia tabaci*. *Agriculture, Ecosystems and Environment* 17: 69–82.

Berlinger, M. J., and R. Dahan. 1987. Breeding for resistance to virus transmission by whiteflies in tomato. *Insect Science and Its Application* 8: 783–784.

Berlinger, M. J., M. Tamim, M. Tal, and A. R. Miller. 1997. Resistance mechanisms of *Lycopersicon pennellii* (Corr.) D'Arcy accessions to the *Spodoptera littoralis* (Boisduval) (Lepidoptera: Noctuidae). *Journal of Economic Entomology* 90: 1690–1696.

Juvik, J. A., M. J. Berlinger, Tselila Ben-David, and J. Rudich. 1982. Resistance among accessions of the genera *Lycopersicon* and *Solanum* to four of the main insect pests of the tomato in Israel. *Phytoparasitica* 10: 145–156.

Smith, C. M. 1989. *Plant resistance to insects: a fundamental approach*. John Wiley & Sons, Inc., New York, New York. 286 pp.

RESISTANT. Organisms that are tolerant of conditions that are deleterious to other strains of the same species. In entomology this term usually is applied to plant tolerance of pest damage, or arthropod tolerance of pesticides or pathogens.

RESOURCE CONCENTRATION HYPOTHESIS. The concept that the lack of diversity resulting from a homogeneous plant population limits community stability because herbivores have a ready food supply, favoring herbivore population increase.

RESOURCE PARTITIONING. The differential use of resources by organisms.

RESPIRATORY SIPHON. A tube found in larvae of certain insects, and used for breathing.

RESPIRATORY TRUMPET. Protuberances associated with the prothorax of the pupal stage in certain aquatic Diptera.

RESURGENCE OF PESTS. Increase in abundance of pests, rapidly and sometimes to higher than previous levels, following actions made to suppress them.

RESTRICTED USE PESTICIDE. A pesticide that can be applied only by certified applicators due to its inherent hazard to the applicator, non-target organisms, or environmental quality.

RESTRICTION ENDONUCLEASE. An enzyme that cuts DNA only at a limited number of specific nucleotide sequences. It is also known as a restriction enzyme.

RESTRICTION FRAGMENT LENGTH POLYMORPHISM (RFLP). A polymorphism in an individual, population or species defined by restriction fragments of a distinctive length. Usually caused by gain or loss of a restriction site but could result from an insertion or deletion of DNA between two conserved restriction sites. Differences in DNA RFLPs are visualized by gel electrophoresis.

RETICULATIONS. A netlike structure, usually referring to the pigmented pattern on eyes of Lepidoptera.

RETICULATED BEETLES. Members of the family Cupedidae (order Coleoptera). See also, BEETLES.

RETINA. The light receptive apparatus of an eye.

REY, CLAUDIUS. Claudius Rey, was born at Lyon, France, on September 8, 1817. He was a noted French entomologist who is remembered for his collaborations with Etienne Mulsant on Coleoptera and Hemiptera. Initially he was financially independent due to a family-owned printing business, which allowed him to direct most of his energies to the study of beetles. Though the business went bankrupt in 1847, his uncle offered him a job at a vineyard in southern France. While working at the vineyard he began a collaboration with Mulsant. Together they published almost continuously from 1850 until Rey's death in 1895. He was an honorary member of the Entomological Society of France, and served as president of the Société Francaise d'Entomologie from 1882 until his death.

References

Anonymous. Obituaries, Claudius Rey 1895. *The Entomologist's Monthly Magazine* 31: 122–123.
Herman, L. H. 2001. Rey, Caludius. *Bulletin of the American Museum of Natural History* 265: 129.

RFLP. See restriction fragment length polymorphism.

RHABDOM. The site of the light-absorbing rhodopsin in an ommatidium. The rhabdom typically consists of several rhabdomeres and is situated in the center of the ommatidium.

RHABDOMERE. The part of a retinular cell in an ommatidium that contains the light-sensitive rhodopsin. The rhabdomere membrane is formed into microvilli. The electrical impulses generated by light absorption are transmitted to the optic lobe of the brain.

RHABDOVIRUSES. The rhabdoviruses are a diverse group of more than 200 viruses which can cause disease in both plant and animal hosts. The animal rhabdoviruses are characterized by a bullet-shaped morphology, whereas certain plant rhabdoviruses have bacilliform morphology. The best-known rhabdovirus is the rabies virus, one of the many diseases studied by L. Pasteur in the 1880s. Many of the vertebrate viruses within the genera *Vesiculovirus* (vesicular stomatitus virus, VSV) and *Lyssavirus* (rabies-like viruses) are insect-transmitted and are capable of replicating in mosquitoes, blackflies, midges, sandflies and houseflies, as well as in select insect cell lines. The vesicular stomatitis virus, a relative of the rabies virus, has served as a model for studying the structure and function of rhabdoviruses. The animal rhabdoviruses are composed of a nucleocapsid protein (N) and a matrix protein (M) which form a helical structure encasing the $(-)$ ssRNA. The resulting nucleocapsid or ribonucleic protein core is enveloped by a lipid-bilayer membrane. This envelope contains a viral glycoprotein (G) that anchors the membrane to the M-protein of the ribonucleic protein core. The M-protein (about 220 aa, $pI = 9.07$) is multifunctional, serving various structural roles as well as modulating viral transcription/replication and suppressing certain host cell functions. Extracted $(-)$ssRNA is non-infectious and requires encapsidated RNA transcriptase activity. RNA transcription is dependent upon three proteins, the N protein which complexes to the RNA to form a template, the L-protein which acts as the polymerase, and the polymerase-associated P (or Ns) protein responsible for RNA chain elongation.

The only well-characterized insect rhabdovirus is the Sigma virus of *Drosophila melanogaster*. This virus, discovered in 1937 by L'Heretier and Teissier, is characterized by its ability to induce CO_2 sensitivity (anoxia) and to be vertically transmitted to progeny flies via an extrachromosomal route. Unlike the polydnavirus, the Sigma virus does not integrate its genome into the host chromosome but undergoes limited replication in the cytoplasm of male and female gametes. Interestingly, the CO_2 anoxia, which is temperature dependent, also can be induced in fly and mosquito species that have been challenged with various lassa- and vesiculoviruses. The sensitivity to CO_2 is believed to be due to the presence of virus in the insect nerve ganglia and more specifically to the expression of the viral membrane proteins (G protein). The Sigma virus, unlike other rhabdoviruses, is not cytopathogenic and usually is considered to be harmless to *Drosophila*. This virus, unable to replicate in alternate hosts, undergoes a restrictive replication within *Drosophila* cells. In addition to the Sigma virus, several rhabdovirus-like particles have been detected in other invertebrates. In certain insects, such as the house cricket, *Acheta domestica*, the rhabdovirus VLP is polytropic, causing trembling symptoms and over 80% mortality. In other insects, such as the hymenopteran parasitoid *Biosteres longicaudatus*, a rhabdovirus VLP has been detected in venom apparatus. In this case, the virus does not appear to cause any pathology but may be involved in the suppression of the immune response of the host, the Caribbean fruit fly, *Anastrepha suspensa*. Like the polydnaviruses, this entomopoxvirus virus is associated with the lumen of the venom apparatus of the parasitoid.

References
Coll, J. M. 1995. The glycoprotein G of rhabdoviruses. *Archives in Virology* 140: 827–851.
Fleuriet, A. 1994. Female characteristics in the *Drosophila melanogaster*-sigma virus system in natural populations from Languedoc (southern France). *Archives in Virology* 135: 29–42.
Teninges, D., F. Bras, and S. Dezelee. 1993. Genome organization of the sigma: six genes and a gene overlap. *Virology* 193: 1018–1023.

RHAGIONIDAE. A family of flies (order Diptera). They commonly are known as snipe flies. See also, FLIES.

RHAMMATOCERUS SCHISTOCERCOIDES, A SWARM-FORMING GRASSHOPPER FROM SOUTH AMERICA. The grasshopper *Rhammatocerus schistocercoides* (Orthoptera: Acrididae: Gomphocerinae) causes severe damages in rice, soybeans, maize, sugar cane, and native pastures in Brazil. This species is mainly concentrated between the parallels 12° and 15° south and meridians 52° to 61° west. This area partially occupies the states of Rondonia, Mato Grosso, and Goiás, where since 1983 there have been severe outbreaks (mainly in 1984 to 1988 and 1992 to 1993). This insect also is present in Colombia and Venezuela, where it is reported as one of the main problems in the local agriculture.

Biology

This is a univoltine species. Biological cycles in Brazil and Colombia are approximately the same, with one generation per year, nymphal development during the rainy season and adults during the dry season. In Brazil, females lay their eggs at the beginning of the rainy season, from September to late October. Nymphs (8 to 9 instars) develop during the rainy season, from late October to mid-April. Adults appear in mid-April and remain in the immature state through most of the dry season (May–September). Breeding activity begins in late August, apparently regardless of the rains, followed by the first egg laying in late September. In Colombia, *R. schistocercoides* lays its eggs from February to March. Nymphs are present

Fig. 872 Seventh (penultimate) instar of *Rhammatocerus schistocercoides*.

Fig. 874 Swarm of *R. schistocercoides* in Mato Grosso, Brazil.

Fig. 873 Adult of *R. schistocercoides*.

Fig. 875 Typical savanna-like habitat of *R. schistocercoides*.

from March to August, and adults from August to March.

Behavior

Rhammatocerus schistocercoides has a remarkable gregarious behavior, similar to the behavior of other gregarious acridians, e.g., the desert locust, *Schistocerca gregaria* (Forskål), and the migratory locust, *Locusta migratoria* (L.). To this date, no phase transformation has been demonstrated. Its gregarious behavior is very close to that of the Moroccan locust, *Dociostaurus maroccanus* Thunberg. Hopper bands are formed, spreading over a few hundred to a few thousand square meters, with densities frequently ranging from 5,000 to 10,000 per square meter for the first nymphal instars to 250–500 per square meter for the last instars. Hopper bands have a very standard shape, i.e., generally a frontal formation which

is common in many gregarious species. They move from a few ten to a few hundred meters per day. Swarms are generally small-sized, of a few thousand square meters to a few hectares on the ground. When flying, during the day, these swarms can be much larger, frequently one km or more in length. Density on the ground is around 500 to 1,500 per square meter and 2 to 3 insects per cubic meter when flying. The height of flight is low and does not exceed 5 to 10 meters. Swarms display a typical 'rolling' behavior. Flight activity seems maximal during sexual maturation; it stops temporarily during oviposition. The direction of swarms is mainly determined by the wind. In general, the swarms present a low mobility and move only locally. Distance covered per day is low, even in good conditions, a few hundred meters to 2 to 3 km a day only.

Habitat

In Brazil, *R. schistocercoides* outbreaks occur in the "cerrado" area of Central Brazil (Mato Grosso) and in the "llanos" of Colombia and Venezuela. In Brazil, the vegetation of outbreak areas is mainly pure or bushy savannas. Two main habitats have been recognized: breeding biotopes usually on sandy soils, and dry season dispersion/wandering biotopes on heavier sand-clay soils.

Recent studies revealed that *R. schistocercoides* outbreaks in Mato Grosso cannot be explained by the accelerated agricultural development that has occurred in the affected areas since the early 1980s. Evidence clearly indicates that *R. schistocercoides* outbreaks are a long-standing phenomenon, both in terms of their extent and nature. In fact, the newly cropped and previously uninhabited areas have long been conventional outbreak areas for this species. Much evidence suggests that the outbreaks surely have been the result of irregular interannual rainfall, particularly marked during the dry season and at the onset of the rainy season, just when the grasshopper populations are achieving sexual maturation.

Michel Lecoq
Centre de Coopération Internationale en Recherche Agronomique pour le Développement
Montpellier, France
and
Bonifácio Magalhães
Embrapa Recursos Genéticos e Biotecnologia,
Parque Estação Biológica
Brasília, Brazil

References

Ebratt, E. E., C. E. Correal, M. I. Gómez, L. F. Villamizar, A. M. Cotes, J. C. Gutiérrez, N. B. Triana, E. M. Granja, and G. León. 2000. *La langosta llanera en Colombia. CORPOICA, Ministerio de Agricultura y Desarrollo Rural*. N. Ramirez (ed.). Boletín Técnico. Bogotá, Colombia. 103 pp.

Lecoq, M., and C. V. Assis-Pujol. 1998. Identity of *Rhammatocerus schistocercoides* (Rehn, 1906) forms south and north of the Amazonian rain forest and new hypotheses on the outbreak determinism and dynamics (Acrididae, Gomphocerinae, Scyllinini). *Transactions of the American Entomological Society* 124: 13–23.

Lecoq, M., A. Foucart, and G. Balanca, G. 1999. Behaviour of *Rhammatocerus schistocercoides* (Rehn, 1906) hopper bands in Mato Grosso, Brazil (Orthoptera: Acrididae, Gomphocerinae). *Annales de la Société entomologique de France* 35: 217–228.

Lecoq, M., and I. Pierozzi, Jr. 1995. *Rhammatocerus schistocercoides* locust outbreaks in Mato Grosso (Brazil): a long-standing phenomenon. *The International Journal of Sustainable Development and World Ecology* 2: 45–53.

Lecoq, M., and I. Pierozzi, Jr. 1996. Chromatic polymorphism and geophagy: two outstanding characteristics of *Rhammatocerus schistocercoides* (Rehn 1906) grasshoppers in Brazil (Orthoptera, Acrididae, Gomphocerinae). *Journal of Orthoptera Research* 5: 13–17.

Lecoq, M., and I. Pierozzi, Jr. 1996. Comportement de vol des essaims de *Rhammatocerus schistocercoides* (Rehn, 1906) au Mato Grosso, Brésil (Orthoptera, Acrididae, Gomphocerinae). *Annales de la Société entomologique de France* 32: 265–283.

Miranda, E. E. de, M. Lecoq, I. Pierozzi, Jr., J. -F. Duranton, and M. Batistella. 1996. *Le criquet du Mato Grosso. Bilan et perspectives de 4 années de recherches. 1992–1996*. EMBRAPA-NMA, Campinas, Brésil and CIRAD-GERDAT-PRIFAS, Montpellier, France. 146 pp.

RHAPHIDIPORIDAE. A family of crickets (order Orthoptera). They commonly are known as camel crickets. See also, GRASSHOPPERS, KATYDIDS AND CRICKETS.

RHEOTAXIS. Taxis response with respect to water currents.

RHINOPHORID FLIES. Members of the family Rhinophoridae (order Diptera). See also, FLIES.

RHINOPHORIDAE. A family of flies (order Diptera). They commonly are known as rhinophorid flies. See also, FLIES.

RHINOTERMITIDAE. A family of termites (order Isoptera). They also are called lower subterranean termites. See also, TERMITES.

RHIPICERIDAE. A family of beetles (order Coleoptera). They commonly are known as cicada parasite beetles. See also, BEETLES.

RHIPIPHORIDAE. A family of beetles (order Coleoptera). They commonly are known as wedge-shaped beetles. See also, BEETLES.

RHOPALIDAE. A family of bugs (order Hemiptera). They sometimes are called scentless plant bugs. See also, BUGS.

RHOPALOPSYLLIDAE. A family of fleas (order Siphonaptera). They sometimes are known as club fleas. See also, FLEAS.

RHOPALOSOMATIDAE. A family of wasps (order Hymenoptera). See also, WASPS, ANTS, BEES, AND SAWFLIES.

RHYACOPHILIDAE. A family of caddisflies (order Trichoptera). They commonly are known as primitive caddisflies. See also, CADDISFLIES.

RHYPHAROCHROMIDAE. A family of bugs (order Hemiptera, suborder Pentamorpha). See also, BUGS.

RHYSODIDAE. A family of beetles (order Coleoptera). They commonly are known as wrinkled bark beetles. See also, BEETLES.

RHYTHMS OF INSECTS. Chronobiology is an academic discipline that includes the rhythms of insects. There are several types of rhythms. The most common and prominent are 'circadian rhythms,' which oscillate with a period close to the astronomical (usually between 23 and 25 hours, but an acceptable range is 20 to 28 hours). Furthermore, such a rhythm is evident in every living organism studied. Other rhythms with different frequencies have been determined. The two other major groups are the 'ultradian rhythms' (period of less than 20 hours) and the 'infradian' (greater than 28 hours). They will be presented following the discussion of circadian rhythms. Plants and animals possess accurate time-measuring machinery termed 'circadian clocks.' The circadian system, studied extensively in insects, indicates that a central oscillator is located in the optic lobes serving as a pacemaker, but with neurons associated with neuroendocrine functions. This information is based primarily on studies of *Drosophila*, especially the species *D. melanogaster*. Using larger insects, notably the cockroaches *Periplaneta americana* and *Leucophaea maderae*, it was shown that locomotion was controlled by the clocks in the optic lobes and was entrained solely through compound eyes.

Recent biochemical studies describe the organization of the circadian system. One component is termed the period protein (PER). This appears to colocalize with a peptide pigment dispersing hormone (PDH) in about half of the fruit fly's presumptive neurons. The period protein (PER) is also found in the pacemaker neurons of beetles and moths, but it may have different functions. In moths, the pacemakers are situated in the central brain and are associated with neuroendocrine functions. In other insects, the neurons associated with neuroendocrine functions appear to be closely coupled to optic lobe pacemakers. In some crickets and flies, central brain pacemakers are present in addition to optic lobe pacemakers.

Reliable documentation of a circadian rhythm utilizes the application of a statistically appropriate treatment using the cosine model. This permits one to verify the rhythm and detect possible masking effects when insects have been in unusual environments, or encounter interfering agents. An example illustrating this is shown in a comprehensive study with the caddis fly larvae, *Brachycentrus occidentalis* (Trichoptera), in which an 8-hour feeding schedule in water masked a circadian rhythm determined at 20.3 hours.

Examples of behavioral circadian rhythms

Prominent circadian rhythms include locomotion and flight, foraging, biting habits (especially mosquitoes and other Diptera, cat fleas, among others). Other circadian rhythms include oviposition, hatching, pheromone emission, molting (eclosion in immatures), as well as spermatophore formation and the release of sperm, the secretion of ecdysteroids from prothoracic glands, and cuticle deposition. Oxygen consumption also shows a circadian rhythm as studied in the flour beetle, *Tribolium confusum*, using sterilized adults. Circadian rhythms of oxygen consumption and heart rate have been discovered in the American cockroach. Additional biochemical rhythmic events, described as diurnal, were found with the neuroactive chemicals, acetylcholine and acetylcholinesterase in the cockroach, *Periplaneta americana*.

In summarizing the important role of circadian rhythms, it is evident that insects are best adapted to periodic environments in which the cycle of light and temperature is close to that of a solar day. Cycles with an unnatural period usually result in reduced longevity or impaired physiological function.

'Exogenous' rhythms are those that occur as a direct response to the environmental cycle of light and darkness and temperature. In the absence of these variables, the rhythm does not persist. In contrast, an 'endogenous' rhythm is a periodic system that is a part of the temporal organization of the organism that is self-sustaining, i.e., it 'free runs' in the absence of temporal cues such as the daily cycles of light and temperature.

Developmental processes in insects

Most insects, especially those living at higher latitudes, show seasonally appropriate cycles as indicated by activity, feeding and reproduction during the long days of summer. The winter is passed in a state of dormancy or diapause. Recent experimental studies designed to explain the circadian rhythms and photoperiodism in a diapausing insect have been conducted with the blow fly, *Calliphora vicina*. Both the locomotor activity rhythm and the rhythm system underlying photoperiodic timing are part of a circadian system and are dependent upon the environmental light cycle for entrainment and/or photoinduction. They appear to be brain-centered. Research on the blow fly indicates some differences, however. The locomotor activity in the adult fly shows an average circadian period of 22.5 hours, whereas the photoperiod timing is much closer to 24 hours. The suggestion made is that the locomotor activity is neural, while the photoperiod may have a humoral output, perhaps arising from the ovaries. The conclusion is that circadian rhythms overall are an integral component of insect developmental processes, including diapause.

Studies carried out with the European corn borer, *Ostrinia nubilalis*, led to the conclusion that thermoperiodic and photoperiodic responses involved identical biological clock systems. Similar conclusions have been made with other insects, notably the cockroach, *Leucophaea maderae*, even though the latter insect does not undergo diapause.

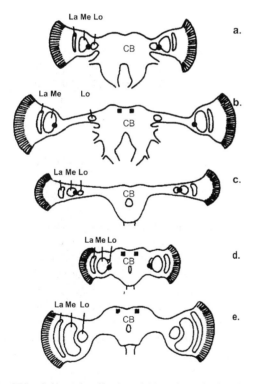

Fig. 876 Schematic of brains of (a) cockroach, (b) cricket, (c) beetle, (d) fly, (e) moth; Within optic lobes are lamina (La), medulla (Me) and lobula (Lo) which process and transfer visual signals from the compound eyes to the central brain (CB). Location of circadian pacemaker centers are indicated by filled circles in the optic lobes and by squares in the central brain. (after Helfrich-Forster et al., Chronobiology International 15: 567–594. 1998.)

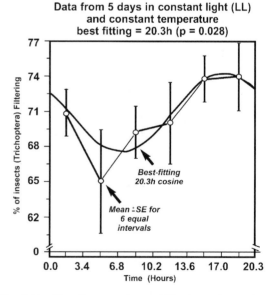

Fig. 877 Circadian waveform of filtering activity of caddis fly larvae (200 larvae) showing best fitting cosine (20.3 hour). (after Sothern et al., Chronobiology International 15: 595–606. 1998.)

Spectrum of rhythms

Rhythm frequencies include ultradian (0.5 to 20 hours), circadian (20 to 28 hours), and infradian (28 hours to 6 days); those with low frequencies include circaseptan (about 7 days), circavigintan (about 20 days), circatrigintan (about 30 days), and circannual (about 1 year).

Circaseptan rhythm deserves a more complete discussion as it was discovered in a primitive insect (springtails, Order Collembola), namely *Folsomia candida*, and can be considered as a model for the study of infradian rhythmicity. This insect is parthenogenetic. Insects were studied in the dark under humid conditions; daily counts of egg laying determined that egg deposition recurred at 7 days, the average number being 46.6, at a constant temperature of 15°C. With higher temperatures (not optimum), the interval between oviposition became circasemiseptan (about 3 days). The lengthening of the egg-laying interval also occurred with aging of the insects. The infradian rhythm of oviposition does not have any obvious environmental counterpart. A study of molting (ecdysis) determined that an infradian rhythm (about 3 days periodicity) occurred at 23°C.

Further studies with *Folsomia candida* determined that life spans could be varied by shifting from an optimal environmental temperature (20°C.) to a non-optimal environmental temperature of 25°C. The temperatures were alternated every 12 hours. The greatest longevity was found to be when the shift intervals were 7 days (circaseptan). A separate study with a Dipteran, the face fly, *Musca autumnalis*, revealed that shifts of the lighting regime every 2 or 9 days shifted at intervals of 7 days apart, had relatively short survival times, whereas some groups that were shifted at intermediate intervals did as well, or better, than the controls. One other documented circaseptan frequency response to the shifts of the lighting regimen of growth was found in an algal species, *Acetabularia mediterranea*.

Lunar rhythms

Lunar rhythms (circatrigintan) occur at about 30 days. Documented examples of lunar rhythms have

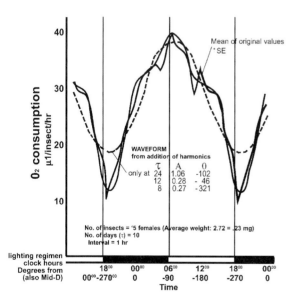

Fig. 878 Oxygen consumption of flour beetles during a 24-hour span using mean values of original data to show fit with a cosine function and a waveform using harmonics. (after Chiba, Y. et al., J. Insect Physiology 19: 2163–2172.1973.)

Fig. 879 *Folsomia candida* adult (Collembola: Isotomidae).

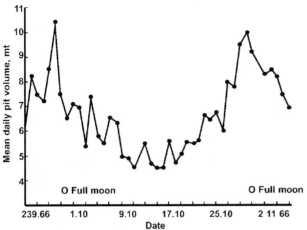

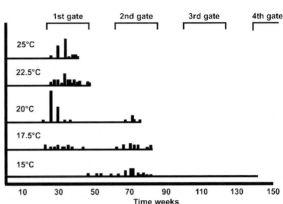

Fig. 880 Daily pit volume of fifty ant lion (*Myrmeleon obscurus*) larvae subjected to normal daylight conditions and each larva fed one ant per day. Times of full moon shown by circles. (after Youthed and Moran, J. Insect Physiology 15: 1259–1271. 1969.)

Fig. 881 Circannual rhythm of pupation in *Anthrenus verbasci* (Oermestidae) using constant conditions of temperature, humidity and darkness. A black square represents the time of pupation to the nearest week, of an individual. Those at higher temperatures utilize the first gate; at lower temperatures an increasing number are required to wait. (after Blake, G. M., Nature, London 183: 126–127.1959.)

been found in a limited number of insects. Studies on the mayfly, *Povilla adusta* (Ephemeroptera), showed a periodicity of emergence from Lake Victoria in West Africa and in a crater lake, Barombi Mbo, with the emergence of adults occurring in greatest numbers after the full moon.

The ant-lion *Myrmeleon obscurus* (Myrmeleontidae), builds pits to capture prey. Using the volume of pits as a measure, it was found that maximum pit-building activity occurred at the full moon (about 28 days). This species also has a solar day activity.

Experiments with a midge, *Clunio marinus* (Chironomidae), living in the intertidal zones of the Atlantic and the Pacific oceans have demonstrated semilunar rhythms. Adult emergence was governed by the superposition of a circadian rhythm that controlled pupal eclosion, and a semilunar rhythm that determined the beginning of pupation. The best documentation was done with the midge population from Normandy, France. The same species showed greater variation of the rhythm in other oceanic areas. In northern latitudes, the brightness of the moon may influence the behavior of the midge.

Circannual rhythms

Circannual rhythms are uncommon. The best documented case is with a carpet beetle, *Anthrenus verbasci* (Dermestidae), which commonly has a life cycle of 2 years, termed 'semi-voltine.' It feeds on material of animal origin and is often found in house sparrow nests. The first winter is spent in diapause as a young larva, and the second winter as a full-grown larva. These larvae pupate and emerge the following spring. The period of the rhythm was between 41 to 44 weeks. When larvae were reared at cooler temperatures, 15 to 20°C, the larvae underwent two cycles of development and did not pupate until about 41 weeks after the first group reared at 25°C. This delay by the group reared at cooler temperatures indicated a phenomenon called 'gated' rather than a continual or gradual sequence of pupation. This 'gated' phenomenon had been previously shown by several workers investigating pupal eclosion in the fruit fly *Drosophila pseudoobscura*. Thus, the circannual rhythm in the carpet beetle was the result of the natural seasonal change, attributed primarily to daylength.

Some experiments with short-lived aphids (Hemiptera) showed that they appear to distinguish spring from autumn. Experiments with *Megoura viciae*, for example, indicated a non-rhythmic interval timer that prevents the aphids from responding prematurely to short days in the spring. No sexual forms were produced for as long as 90 days in a short photoperiod (at 15°C). After this duration, the photoperiod response was restored and reproducing oviparae were produced under the short-day treatment.

Interrelationships of circadian rhythms and insecticide sensitivity

A number of studies have explored the possibility that toxicity to an insecticide or toxin could be related to a critical time in the insect rhythm, particularly with respect to lighting or photoperiod ('photophase' or daylight and 'scotophase' or a dark period). In house flies, the respiratory rate correlated with the conversion of DDT (dichlorodiphenyltrichloroethane) to DDE (a metabolite). Both occurred in the mid-light period. In a separate study, using the flour beetle, *Tribolium confusum*, measurements of their oxygen consumption were related to the sensitivity to a rapid acting organophosphate insecticide, dichlorvos. This species, normally feeding on flour, showed the greatest sensitivity in the middle of the dark period, while the peak in oxygen consumption occurred 2 hours later. Nocturnal species, such as German cockroaches and house crickets, showed the greatest sensitivity a few hours after the onset of a daily dark span. Several diurnal species, including the boll weevil, the pink bollworm and the two-spotted mite, showed the greatest sensitivity shortly after the onset of the light span.

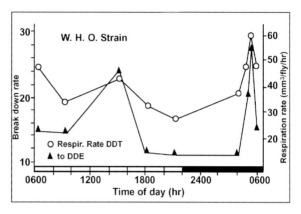

Fig. 882 The effect of time on the treatment of adult flies, *Musca domestica*, with DDT on rate of respiration (open circles) and conversion of DDT to its metabolite, DDE. (after Shipp and Otton, Ent. Exp. et Appl. 19: 235–242. 1976.)

The larvae and adults of the mosquito, *Aedes aegypti*, showed the greatest sensitivity to insecticides at the beginning of the dark period. Biting habits of mosquitoes reveal their greatest occurrence in the early evening, although a second period may

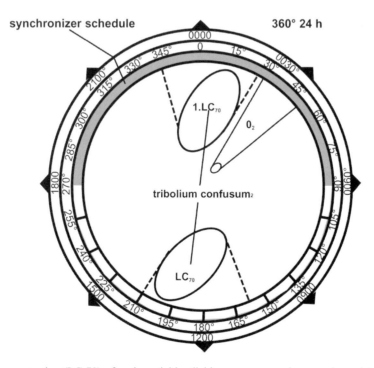

Fig. 883 The lethal concentration (LC 70) of an insecticide, dichlorvos, expressed as a reciprocal to flour beetles, *T. confusum*, indicating a 2-hour earlier lethal effect than peak oxygen consumption. The peak or acrophase of oxygen consumption occurred about 3hours after the middle of the daily dark span. (after Vea, H.H. et al., Chronobiologia 4: 313-323.1977.)

occur before dawn, or late in the dark period. Noteworthy is the fact that both the larvae and the adults had maximum sensitivity to an insecticide at approximately the same time. The insecticides evaluated were dichlorvos on the adult mosquitoes and chlorpyrifos on the mosquito larvae.

Rhythmic events using locomotory activity in relation to sensitivity to an insecticide were studied in the American cockroach. The greatest sensitivity to dichlorvos occurred near the onset of the dark span, with locomotory activity occurring about 3 hours later. In a separate study using house crickets, *Acheta domestica*, a positive relationship occurred with maximum locomotor activity and sensitivity to three anesthetics, namely ether, chloroform and carbon tetrachloride. The peaks occurred during early scotophase. The cricket can be considered to be nocturnal.

A study of the circadian rhythm of the nerve enzyme acetylcholine esterase showed synchronization with sensitivity to two insecticides, carbaryl and malathion. The test insect was the rice grasshopper, *Aiolopus thalassinus*, a diurnal species. The peak or acrophases of both rhythms occurred during the photophase at about the same time. A study conducted much earlier revealed a trimodel (8-hour period) rhythm in the larvae of the mosquito, *Aedes aegypti*. The enzyme Mg^{2+} ATPase was measured in relation to its inhibition by the insecticide gamma chlordane.

Exceptions to the generalization that insecticide sensitivity has circadian rhythm peaks at or near the time of circadian peaks of insect metabolism were found in studies with a nocturnal insect, the cockroach, *Leucophaea madeira*, and the house fly, *Musca domestica*, a diurnal species. Similar peaks were found in sensitivity to pyrethrum, whereas peak activity and metabolism were distinctly different times in a 24-hour period.

The practical application of the circadian sensitivity to toxins has had limited study. Perhaps it deserves greater attention to refine insecticide usage and insect sensitivity.

L.K. Cutkomp
University of Minnesota
St. Paul, Minnesota, USA

References

Blake, G. M. 1959. Control of diapause by an 'internal clock' in *Anthrenus verbasci (L.) (Col. Dermestidae)*. *Nature*, London 183: 126–127.

Cole, C. L., and P. L. Adkisson. 1965. A circadian rhythm in the susceptibility to an insecticide. Pp. 39–313 in J. Aschoff (ed.), *Circadian clocks*. North-Holland Publishers, Amsterdam, The Netherlands.

Cutkomp, L. K., M. D. Marques, R. Snider, G. Cornelissen, J. Wu, and F. Halberg. 1987. Chronobiologic view of molt and longevity of *Folsomia candida* (Collembola) at different ambient temperatures. *Advances in Chronobiology*, Part A: 257–264.

Giebultowicz, J. M. 2000. Molecular mechanism and cellular distribution of insect circadian clocks. *Annual Review of Entomology* 45: 760–793.

Saunders, D. S. 1982. *Insect clocks* (2nd ed.). Pergamon Press, Oxford, United Kingdom. 409 pp.

RIBBED-COCOON MAKER MOTHS (LEPIDOPTERA: BUCCULATRICIDAE).

Ribbed-cocoon maker moths, family Bucculatricidae, total 247 species worldwide, with most species being Nearctic (103 sp.) or Palearctic (86 sp.). Most species are in the genus *Bucculatrix*. The family is part of the superfamily Tineoidea, in the section Tineina, subsection Tineina, of the division Ditrysia. Adults minute to small (5 to 16mm wingspan), with rough-scaled oblique head and an eye-cap on the antennal bases; haustellum naked; labial palpi short and porrect; maxillary palpi minute, 1-segmented or absent. Maculation variable and spotted, often with iridescent markings; venation reduced (especially on hindwings) and long fringes on hindwings. Adults are mostly diurnal. Larvae are leafminers, with some changing to external leaf skeletonizing in later instars, but a few are gall makers or stem miners. Pupation is in a white spindle-shaped cocoon with a ribbed surface, unique to the family. Numerous different host plants are known but many are in Compositae. Some species are economic.

John B. Heppner
Florida State Collection of Arthropods
Gainesville, Florida USA

References

Braun, A. F. 1963. The genus *Bucculatrix* in America north of Mexico (Microlepidoptera). *Memoirs of the American Entomological Society* 18: 1–208, 45 pl.

Friend, R. B. 1927. The biology of the birch leaf skeletonizer *Bucculatrix canadensisella*, Chambers. *Bulletin of the Connecticut Agricultural Experiment Station* (New Haven), 288: 395–406.

Needham, H. B. 1948. A bucculatricid gall maker and its hypermetamorphosis. *Journal of the New York Entomological Society* 56: 43–50.

Scoble, M. J., and C. H. Scholtz. 1984. A new, gall-feeding moth (Lyonetiidae: Bucculatricinae): from South Africa with comments on larval habits and phylogenetic relationships. *Systematic Entomology* 9: 83–94.

Seksjeva, S. V. 1981. Bucculatricidae. In *Identification keys to insects of European Russia*. 4. Lepidoptera, 2: 137–148. Academie Nauk. [in Russian], St. Petersburg.

Selsjeva, S. V. 1994. Review of the mining moths (Lepidoptera, Bucculatricidae) of the fauna of Russia. *Proceedings of the Zoological Institute, Russian Academy of Sciences* 255: 99–120. (1993) [in Russian]

RIBOSOMAL RNA. Ribosomal RNA genes (rRNA genes) are found as tandem repeating units in the nucleolus organizer regions of eukaryotic chromosomes. Each unit is separated from the next by a nontranscribed spacer. Each unit contains three regions coding for the 28S, 18S, and 5.8S rRNAs.

RIBOSOME. A self-assembling cellular organelle comprised of proteins and RNA in which translation of mRNA occurs. Ribosomes consist of two subunits, each composed of RNA and proteins. In eukaryotes, ribosome subunits sediment as 40S and 60S particles.

RICANIIDAE. A family of bugs (order Hemiptera, suborder Fulgoromorpha). All members of the suborder are referred to as planthoppers See also, BUGS.

RICE WEEVIL. See also, STORED GRAIN AND FLOUR INSECTS.

RICHARDIID FLIES. Members of the family Richardiidae (order Diptera). See also, FLIES.

RICHARIIDAE. A family of flies (order Diptera). They commonly are known as richardiid flies. See also, FLIES.

RICHARDS, OWAIN WESTMACOTT. O.W. Richards was born on December 31, 1901, and educated at Oxford University, Oxford, England. In 1927 he moved to Imperial College, and worked at the College Field Station at Slough on stored products pests until 1947. He remained at Imperial College until his retirement in 1967, and served as head of the Zoology and Applied Entomology Department for many years. Richards was principally interested in taxonomy, ecology and the theory of evolution. He had considerable expertise with various groups of flies, but was particularly interested in Aculeate Hymenoptera. His books entitled ''Social insects'' (1953) and ''The social wasps of the Americas'' (1978) reflect this latter interest. Over the course of his career, Richards published over 180 papers and six books, including two revisions (with Gareth Davies) of ''Imms' general textbook of entomology'' (1957, 1977). He served as president of the British Ecological Society and editor of its journal ''Journal of Animal Ecology,'' and was considered one of the most distinguished entomologists of his generation. He died on November 10, 1984.

Reference

Davies, R. G., N. Waloff, and R. Southwood 1985. O. W. Richards. *Antenna* 9: 60-62.

RICINIDAE. A family of chewing lice (order Mallophaga). They sometimes are called hummingbird lice. See also, CHEWING LICE.

RICKETTSIA. Bacteria-like, gram-negative organisms. Insect rickettsia are obligate intracellular pathogens with cell walls, but lack flagella. Some, such as *Wolbachia*, are seldom pathogenic, whereas others, such as *Rickettsiella*, are commonly insect pathogens.

RIFFLE BEETLES (COLEOPTERA: ELMIDAE). The family Elmidae is divided into two subfamilies: Elminae and Larainae. The Elminae are far more abundant and diverse than the Larainae, with about 100 genera and over 1,000 species worldwide, of which 26 genera and about 100 species occur in the United States. This subfamily is represented in every American state but Hawaii and in every bordering province of canada and state of mexico.

Order: Coleoptera
  Suborder: Polypleaga
    Superfamily: Dryopiidea
      Family: Elmidae

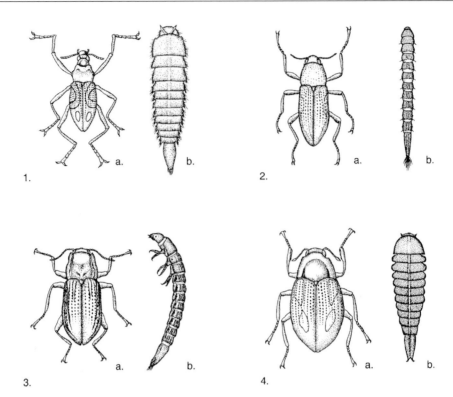

Fig. 884 Riffle beetles: (1) *Ancyronyx variegates*: (a) adult, 2.1–2.6 mm long, conspicuously colored with black and yellow or orange, on or in waterlogged wood in streams of eastern North America; (b) larva, commonly with adults. (2) *Dubiraphia vittata*: (a) adult, 1.8–2.5 mm long, brown or tan, with a broad yellowish longitudinal stripe on each elytron, clings to submerged plants in streams of eastern North America; (b) larva, usually in sediment on the bottom of streams. (3) *Microcylloepus pusillus*: (a) adult, 1.65–2.2 mm long, reddish brown to black, often with reddish lengthwise stripe or spot on each elytron, common on many different substrates in streams of central and eastern North America; (b) larva, side view, often with adults. (4) *Optioservus ovalis*: (a) adult, 2.4-2.6 mm long, plump, brown to black with two elongate yellowish or reddish spots on each elytron, in gravel or among mossy stones in spring-fed streams of eastern states; (b) larva, often with adults.

The Larainae, however, have only about 20 genera and 110 species world wide, with only two genera and no more than four species in the United States. Adult Laraines are usually found near or at the water's edge, sometimes partially submerged. When disturbed, they may fly rapidly. Some tropical species occur in swarms around cataracts or waterfall spray zones. (Big black adults of *Potamophilops* are conspicuous flying about in the spray of the world-famous Iguassu Falls shared by Brazil, Argentina, and Paraguay.) The bodies of Laraines are rather soft, and their lives as adults are much briefer than as larvae. In size, the adults range from about 3 to 9 millimeters.

Elmines are more completely aquatic, adults entering the water soon after pupation and usually never again emerging. They commonly share the habitat and food of their larvae and are rather long-lived (months or years). Although members of a few genera are quite sizable (e.g., some species of *Macrelmis* exceed 5 mm in length), most Elmines are small (1–3 mm) and some are tiny (under 1 mm).

Of the four genera illustrated, *Ancyronyx* is the most unusual. It is represented in the Western Hemisphere only by this one species. Its few known relatives occur in southeastern Asia. The other three genera are represented by a number of species in various regions of the U.S.A. *Microcylloepus*, though abundant in much of the U.S.A., is primarily Neotropical, with many species in South America. *Dubiraphia* is unusual among elmids in that the microhabitat of the larva is so different from that of the adult. *Optioservus* might well be termed the most typical elmid of American and Canadian brooks.

Ecology

Riffles are rippling streams, usually shallows across streambeds where water flows swiftly and is well aerated. In such habitats, elmids are the most typical and abundant beetles, but usually quite inconspicuous. They are mostly bottom dwellers or clinging to submerged wood or debris, and do not swim or come up for fresh air.

The respiratory mechanism of adult Elmines is noteworthy because it enables them to remain permanently underwater without having to come up for fresh air. The key feature, termed a plastron, consists of a thin layer or film of air held by a dense coating of hydrofuge pubescence, or tomentum, on various parts of the legs and body. The plastron is in contact with the air reservoir beneath the elytra, into which the spiracles open. There is normally plenty of dissolved oxygen in the water of stream riffles, and oxygen can readily diffuse into the plastron, and then into the tracheal system and to all body tissues.

Since wetting agents such as soap and detergents destroy plastrons, riffle beetles do not thrive downstream from sites where people bathe and do their laundry. Other types of pollution such as sewage promote excessive bacterial growth and consequent oxygen depletion, and industrial wastes often include toxic substances. Because some species require water of great purity, whereas other species are tolerant of one or another type or quantity of pollutant, riffle beetles are useful as indicators of water quality.

In streams with high mineral content, all sorts of objects may become coated: wood, pebbles, boulders, and riffle beetles. Some elderly Elmines get covered by thick encrustations that outweigh the beetles. Other things than mineral armor often occur on riffle beetles: diatoms and sessile protozoans such as *Vorticella* are common.

Elmids range in color from tan to black, but many exhibit patterns of longitudinal or diagonal stripes or transverse bands, yellow to red. Some have prominent spots and patterns. Such aposematic coloration is probably related to the fact that elmids are not prey for most predators.

All known elmid larvae are aquatic, creeping on, or burrowing in, the substrate—some in mud, sand or gravel, some on, or under rocks, and others on, or in waterlogged wood, leaf-packs or detritus. Their respiratory mechanism employs a three-branched tuft of tracheal gill filaments that may be extruded from a cavity in the last abdominal segment, or withdrawn and covered by a trapdoor-like operculum. They feed chiefly on diatoms and decaying plant materials.

When ready to pupate, mature larvae leave the water to find a suitable place in moist sand, humus, moss, or in a crevice, or beneath a rock, or under loose bark. Some larvae are able to just crawl out of the water in search of pupation sites. Drifting enables mature larvae to get from midstream to snags or boulders that they use to crawl out. Others, in streams with fluctuating water levels, simply wait in shallow water until the level drops so they can pupate in situ. In temperate regions, pupation occurs from late spring through summer, but in humid tropics, it may occur throughout the year. Pupation usually requires a week or two, though a bit longer for larger species.

Most kinds of insects use membranous wings as an important means of dispersal. Beetles are no exception, but their front wings (elytra) are relatively stiff and protective. However, the membranous hind wings are neatly folded beneath the elytra, and can pop out and unfold to serve in flight. Among Elminae, many of the newly emerged adults fly before they enter the water. Other adults never fly because their hind wings do not develop adequately. Usually, neither alates (fliers), nor non-fliers ever leave the water once they have entered it. At relatively low elevations, collecting at lights, or the use of light traps may yield hundreds of elmids. Collectors prefer light-collected specimens because they require so much less cleaning than those taken from streams.

Harley P. Brown
University of Oklahoma,
Norman, Oklahoma, USA

References

Barr, C. B., and J. B. Chapin 1988. The aquatic Dryopoidea of Louisiana (Cleoptera: Psephenidae, Dryopidae, Elmidae). *Tulane Studies in Zoology and Botany* 26: 89–164.

Brown, H. P. 1976. Aquatic drypoid beetles (Coleoptera) of the United States. Biota of Freshwater Ecosystems Identification Manual No. 6. *Water Pollution Control Research Series, USEPA*, Cincinnati, Ohio. 82 pp.

Brown, H. P. 1981. A distributional survey of the world genera of aquatic dryopoid beetles (Coleoptera: Dryopidae, Elmidae, and Psephenidae *sens. lat.*). *Pan-Pacific Entomologist* 57: 133–148.

Brown, H. P. 1983. *A catalog of the Coleoptera of America North of Mexico, Family: Elmidae*. U.S. Department of Agriculture, Agriculture Handbook #529-50. Agriculture Research Service, Washington, DC. i-x, pp.1–23.

Brown, H. P. 1987. Biology of riffle beetles. *Annual Review of Entomology* 32: 253–273.

Leech, H. B., and H. P. Chandler. 1956. Aquatic Coleoptera. Pp. 293–371 in R.L. Usinger (ed.), *Aquatic insects of California*. University of California Press, Berkeley, California. 508 pp.

RIKER MOUNT.   A thin glass-topped exhibition case.

RILEY, CHARLES VALENTINE.   C.V. Riley was born at Chelsea, England, on September 18, 1843. One of the foremost American economic entomologists, he was educated in England, France, and Germany, but at the age of 17 moved to a farm in Illinois, USA. He soon developed an interest in entomology and in 1864 moved to Chicago and began his career in entomology as a reporter for the "Prairie Farmer." In 1868 he was appointed State Entomologist for Missouri. Over the next 9 years he devoted much of his energy to the production of the annual reports on "Noxious, beneficial and other insects of Missouri" for the years 1868-1877. These were landmark publications in the field of economic entomology, and firmly established his reputation. Interestingly, it is not so much the information content of the reports that made such a positive impression at the time, but their appearance. Unlike earlier reports, Riley's were printed in large type and were quite readable, and they had excellent illustrations. Also, he endeavored to keep the reports readable by farmers by keeping the language within the understanding of the readers. During this period Riley also lectured weekly at the University of Missouri. He and B.D. Walsh started the "American Entomologist" in 1868, and Riley became sole editor for volumes 2 to 3. Between 1870 and 1872 Riley lectured at Kansas State University. In 1887 the U.S. Entomological Commission was organized, and consisted of Riley, A.S. Packard, and Cyrus Thomas. He was involved in production of 4 of the 5 reports, and also several bulletins issued between 1877 and 1881. Soon after Riley was appointed to the Entomological commission, the 'locust' plague affecting much of the central and western United States collapsed. Though it was nature, not the Entomological Commission, that caused the disappearance of the grasshoppers, Riley and the Commission received accolades. In 1887, Riley was appointed Entomologist by the United States Department of Agriculture, which led to creation of the Division of Entomology and then the Bureau of Entomology. Riley was adept at identifying qualified entomologists, and soon had a nationwide effort organized. It is largely due to Riley's efforts that the United States Department of Agriculture had such a prominent entomological component, and it is due to his organizational ability that federal entomology was so successful. He founded the entomological collections of the U.S. National Museum with a donation of his own collection of 115,000 specimens. Riley promoted the idea of state experiment stations, which eventually had a profound effect on American agriculture by introducing scientific discipline to the farming process. He also promoted the idea of a federal branch of economic ornithology, which brought great attention to the importance of birds as friends of farmers, and which evolved into the Bureau of Biological Survey and eventually into the Fish and Wildlife Service. Riley also was involved in the formation of the Entomological Society of Washington, in 1884, and the Association of Economic Entomologists, in 1889. Riley edited a 6-volume series called 'Insect life' between 1888 and 1894. Riley played a prominent role in the introduction of natural enemies of cottony cushion scale from Australia into California, and the eventual solution of the grape phylloxera problem on grapes in France. Over the course of his extremely productive career, Riley authored nearly 3,000 papers, and he was a member of nearly all the American and European entomological societies. He died at Washington, DC, on September 14, 1895, after a bicycle accident.

References

Anonymous. 1895. Prof. C.V. Riley, M.A., Ph.D. *Entomological News* 6: 241–243.
Essig, E. O. 1931. *A history of entomology*. The Macmillan Company, New York. 1029 pp.
Mallis, A. 1971. *American entomologists*. Rutgers University Press, New Brunswick, New Jersey. 549 pp.

RING GLAND.   A composite structure encircling the esophagus, and found in some Diptera, that contains both ecdysone-secreting and juvenile hormone-secreting cells.

RING-LEGGED EARWIGS.   Some of the earwigs in the family Carcinophoridae (order Dermaptera). See also, EARWIGS.

RIODINIDAE. A family of butterflies (order Lepidoptera). They commonly are known as metalmark-butterflies. See also, METALMARK BUTTERFLIES, BUTTERFLIES AND MOTHS.

RNA. Ribonucleic acid, one of the two forms of nucleic acids.

RNA POLYMERASE. An enzyme capable of synthesizing an RNA copy of a DNA template.

ROACH STONEFLIES. Members of the stonefly family Peitoperlidae (order Plecoptera). See also, STONEFLIES.

ROBBER FLIES. Members of the family Asilidae (order Diptera). See also, FLIES.

ROBINEAU-DESVOIDY, JEAN BAPTISTE. Jean Baptiste Robineau-Desvoidy was born at St. Sauveur in Puisaye, Department of Yonne, France, on January 1, 1799. He studied in Paris and obtained his doctorate in 1822, and then returned to his native town where he divided his time between medical duties and the study of natural history. Robineau-Desvoidy lived in a marshy area, a generally unhealthy environment. His death in 1857 at the age of 57 is attributed to this environment. At his death, a colleague took on the posthumous publication of Robineau-Desvoidy's work, "Diptères des environs de Paris." He published a large and important work in 1830 that was controversial and not well received. His basic thesis was that the classification of the adults could be based on the mode of life of the immature stage. As an indication of the disapproval by some of his colleagues, Macquart, in his "Histoire Naturelle des Diptères" ignores Robineau-Desvoidy. However, in his 1830 publication "Essai sur les Diptères" Robineau-Desvoidy described many exotic Diptera from South America and the West Indies.

Reference

Papavero, N. 1971, 1973. *Essays on the history of Neotropical Dipterology with special reference to collectors (1750–1905)*. Museu de Zoologia, Universidade de São Paulo.

ROBUST. Within statistics and sampling, the quality of being widely applicable. For example, a sampling plan is robust if it can be used to precisely estimate the density of an arthropod under many different environmental conditions.

See also, SAMPLING ARTHROPODS.

ROBUST MARSH-LOVING BEETLES. Members of the family Lutrochidae (order Coleoptera). See also, BEETLES.

ROCK CRAWLERS (GRYLLOBLATTODEA). Rock crawlers are a small order of primitive orthopteroid insects known only from Siberia, Japan, and North America. The order name is based on the Latin words *gryllus* (cricket) and *blatta* (cockroach). Sometimes this order is called Notoptera. The number of described species numbers about 20 in a single family, Grylloblattidae.

Characteristics

Rock crawlers are elongate, wingless insects. As the name 'Grylloblattodea' implies, they resemble crickets and cockroaches, though the similarity with *Timema* walkingsticks also is apparent. As with other orthopteroids, the mouthparts are of the chewing type. Rock crawlers measure 14 to 30 mm in length, and are elongate and cylindrical. Eyes are small or absent. The head is large, consistent with orthopteroids. The antennae are filiform, moderate in length, and shorter than many crickets and cockroaches. The legs are unspecialized, and the tarsi have 5 segments ending with a pair of claws. The body segments are clearly differentiated, and the abdomen consists of 10 segments. The tip of the abdomen bears a pair of segmented cerci, and in the case of the female, the ovipositor is of nearly equivalent length. The nymphs resemble the adults except for size. There are nine nymphal instars.

Biology

Rock crawlers are found among rocks, often in association with glaciers. Some dwell in caves. They are nocturnal, and apparently feed principally upon organic matter such as dead insects found on snow fields and glaciers. The adult deposits eggs singly in the soil or among moss, and the egg requires an incubation period of about a year. A protracted period of time, probably five years, is required to attain maturity. Rock crawlers are rarely observed, and their biology is poorly known.

References

Arnett Jr., R. H. 2000. *American insects* (second edition). CRC Press, Boca Raton, Florida. 1003 pp.

Gurney, A. B. 1953. Recent advances in the taxonomy and distribution of *Grylloblatta* (Orthoptera: Grylloblattidae). *Journal of the Washington Academy of Science* 43: 325–332.

Kamp, J. W. 1970. The cavernicolous Grylloblattodea of the western United States. *Annals of Speleology* 25: 223–230.

Nickle, D. A. 1987. Order Grylloblattodea (Notoptera). Pages 143–144 in F. W. Stehr (ed.). *Immature insects*, Vol. 1. Kendall/Hunt Publishing, Dubuque, Iowa.

ROCKY MOUNTAIN SPOTTED FEVER.

Rocky Mountain spotted fever (RMSF) is caused by the bacterium *Rickettsia rickettsii*, a member of the spotted fever group of Rickettsiae, and transmitted by several species of ticks belonging to the family Ixodidae. Rickettsiae are intracellular parasites of eukaryotic cells. RMSF is a zoonosis that is cycled in small rodents, rabbits, dogs and other domestic mammals. Humans are accidental hosts and are not part of the natural transmission cycle. The maintenance and persistence of RMSF in domestic and wild animals is complicated. Serological tests have shown that antibodies to *R. rickettsii* exist in many species of ground feeding birds and at least 31 species of mammals. Not all these animals are of equal importance in maintaining the disease in nature. Meadow voles and chipmunks can develop high rickettsemias sufficient to infect more than 50% of ticks feeding on them. Dogs show symptoms of RMSF and often have a high seropositivity in areas where RMSF occurs. They have a low level of rickettsemia which is transitory in nature, and consequently dogs are not believed to be important as a reservoir potential of RMSF. However, dogs are important in bringing infected ticks into contact with humans.

RMSF occurs in most of the continental states of the USA, some parts of Canada, in Mexico, and in some parts of Central America, Brazil and Colombia. The disease is endemic in most of these areas. RMSF, also known as tick-borne typhus, has several local names, including the black measles in the Rocky Mountain region, Mexican spotted fever or fiebre manchada in Mexico, tobia fever in Colombia, and Sao Paulo fever or fiebre maculosa in Brazil. Six species of ixodid ticks are known to harbor and transmit *R. rickettsii* in nature, the Rocky Mountain wood tick *Dermacentor andersoni*, the American dog tick *Dermacentor variabilis*, the lone star tick *Amblyomma americanum*, the brown dog tick, *Rhipicephalus sanguineus*, the cayenne tick *Amblyomma cajennense*, and the rabbit tick, *Haemaphysalis leporispalustris*. The principal vectors of RMSF to humans are the Rocky Mountain wood tick in the western USA and Canada, the American dog tick in the rest of the USA, the brown dog tick in Mexico, and the Cayenne tick in Central and South America. Several species of *Ixodes* from the eastern and southern states of the USA as well as *Dermacentor occidentalis* and *Ixodes pacificus* in California are known to have natural infections of *R. rickettsii*, but their role in the transmission of RMSF is not understood. It is not yet clear what role the lone star tick plays in transmission of RMSF, although it can transmit the disease to humans. The rabbit tick rarely bites man, but it is important in maintaining the rickettsia in animal populations because it is the only vector of *R. rickettsii* found in all regions of the continental USA where the disease occurs. RMSF accounts for over 95% of the reported human rickettsial disease in the United States. Almost all human infections occur through tick bites although infection can occur through the fingers during tick removal, and laboratory workers handling infected ticks or *R. rickettsii* are at special risk.

History of Rocky Mountain spotted fever

RMSF has an important history among the annals of arthropod-borne diseases studied by scientists. It was first described as a disease in 1896 in the Snake River Valley of Idaho, although RMSF is known to have infected pioneers in the Rocky Mountain region since at least 1872. The disease was serious and often fatal. It had a mortality rate of over 90% in the Bitter Root Valley of Montana. The United States Public Health Service sent a physician and scientist, Dr. Howard Taylor Ricketts, to Hamilton, Montana, in the center of the Bitter Root Valley. Here, he started a laboratory in the most primitive of conditions where he carried out classic experiments in vector biology. First he established that the unknown pathogen of RMSF could be transmitted directly to guinea pigs which showed symptoms of the disease after inoculation. Other workers had previously theorized that a tick was the natural transmitter of the disease, but it was Ricketts who experimentally established their role as vectors of RMSF. In his laboratory, he fed a male tick on the ear of a guinea pig previously inoculated with the pathogen and then fed this tick on two other healthy guinea pigs which came down with the disease. Healthy guinea pigs did not contract the disease when placed in close contact with infected guinea pigs, eliminating the possibility of the aerosol route of infection. Ricketts then used infected field-collected ticks to transmit the disease to a guinea pig in his laboratory. He established that any stage of a tick can transmit the pathogen and that the pathogen is transferred from stage to stage in a given tick (transstadial transfer) and to eggs of the infected females (transovarial transmission). It was subsequently found by other scientists that the pathogen was a bacterium which was named *Rickettsia rickettsii* in his honor. Ricketts died in Mexico in 1910 from typhus, a non-tick borne rickettsial disease, shortly after completing his pioneering studies on RMSF.

The laboratory Dr. Ricketts started in Hamilton, Montana, is now the Rocky Mountain Laboratory which is part of the United States National Institute of Allergy and Infectious Diseases of the National Institutes of Health. The laboratory has grown considerably and today is an internationally renowned center for the study of infectious diseases, including RMSF.

Epidemiology

A dramatic shift in the geographical distribution of human cases of RMSF has occurred in the USA. Before the 1930s, most human cases were in the Rocky Mountain region although the disease occurred sporadically throughout the continental USA. Thereafter, human cases of RMSF declined in the Rocky Mountain region, but increased greatly in the southeastern and south-central states. Over 97% of all human cases of RMSF now occur outside the Rocky Mountain region, with some 48% of all cases reported from Oklahoma, Tennessee, North and South Carolina. Why this disease underwent a dramatic decline in the Rocky Mountain region is not known. Its increase in other regions coincides with the expansion of humans from urban areas into suburban and rural areas where people have a greater chance of being bitten by infected ticks. Associated with the change in the distribution of RMSF is a change in who is likely to be infected. In the Rocky

Mountain States most infected persons were males who worked farms and ranches and frequented the woods to hunt, but in the southwestern and eastern USA, females and children as well as males contract the disease. The frequency of reported cases of RMSF in the USA is greatest among males, Caucasians and children. Two thirds of the reported cases occur in children less than 15 years old. The number of RMSF cases reported in the USA has varied from about 250 to 1200 cases per year over the last 50 years. RMSF occurs throughout the year but over 90% of cases occur in the spring and early fall, from April to September with the peak number occurring in June and July.

Ticks can acquire *R. rickettsii* by feeding on a rickettsemic host, or through transovarial or transstadial transmission, and by venereal transmission between ticks. Ticks thus function both as a reservoir and vector of *R. rickettsii*. Venereal transmission is not deemed important in the transmission of *R. rickettsii*. Ticks feeding on an infected host must ingest a "threshold" number of *R. rickettsii* to become infected. After ingestion, rickettsiae disperse out of the midgut, usually within 5 days, to the cells of all tissues and organs of the tick including the ovaries and salivary glands. Feeding has been shown to be a stimulus for rapid multiplication of the pathogen within cells of the infected tick, increasing the chance of passage into an animal host. This rapid multiplication, known as the reactivation phenomenon, greatly increases the chances of transstadial transmission to the next life stage and transovarial transmission to eggs. Transovarial transmission can be as high as 100% of offspring, and offspring are capable of continuing the transmission through as many as 12 generations of ticks. However, in later generations, many infected ticks die soon after completion of feeding, and those ticks that survive lay a smaller clutch of eggs. Transovarial transmission is more effective in females that acquire the pathogen transstadially since *R. rickettsii* cannot penetrate tick eggs after the vitelline membrane develops. *R. rickettsii* are pathogenic to ticks, especially the more virulent strains. The actual percentage of ticks infected with *R. rickettsii* usually varies from 2 to 10% in areas of endemic RMSF but may be less than 1%.

Symptomology

In humans, RMSF has a variable incubation period depending on the severity of the infection. In mild

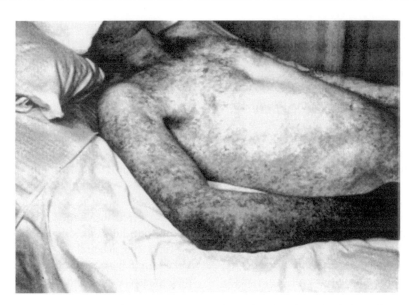

Fig. 885 Photograph of a patient with Rocky Mountain spotted fever showing the entire body and extremities covered with the rash typical of RMSF. (from Sonenshine, 1993, used with permission of Oxford University Press.)

cases, the incubation period is about 3 to 14 days, in severe infections 2 to 5 days. Three symptoms are common in RMSF: a sudden onset of fever, a severe frontal headache, and a rash. Often these three symptoms are accompanied by muscle and joint aches. Patients may also experience vertigo, ringing in the ears and later photophobia. The rash is present in most but not all cases of the disease, but only a little over half of patients have a rash on the palms or soles. Much less common is the additional report of a tick bite. Temperatures can rise to over 40°C. The rash usually appears between the 2nd to 6th day after onset of symptoms, and begins on the wrists and ankles and less commonly on the back then spreads to all parts of the body including the face. The rash starts as small surface spots that later become raised bumps. The spots may form large discolored areas similar to a bruise, which may leave a slight scar after healing. The rash is the result of blood leakage from injured endothelium of small blood vessels. The rash is often delayed in adults for up to 7 days.

Acute cases of RMSF can result in severe damage to vital organs. The heart may suffer from muscle damage, or the patient may become deaf, or develop impaired vision, or partial paralysis. Gangrene can also occur during the disease causing loss of toes or fingers. In fatal cases, death from RMSF usually occurs 9 to 15 days following onset of symptoms.

Death is commonly the result of encephalitis, but may also be caused by renal failure. Strains of *R. rickettsii* with different virulence occur, and several different strains can occur in the same area. Mortality rates have declined from 20% to 30% in the 1940s to anywhere from 3% to 10% presently, no doubt due to the advent of antibiotics effective against the disease and a more educated public. If untreated, recovery from RMSF is slow. Physicians are also now more skilled in early diagnosis of RMSF. Mortality from RMSF is greater when there is a delay in treatment with antibiotics, or when older people are infected, or when people with genetic deficiency of glucose 6-phosphate dehydrogenase are infected. This deficiency can result in a greater hemolysis during the course of the disease.

Treatment

It is important not to delay treatment if RMSF is suspected, and the initial diagnosis and treatment should be made on clinical grounds and must not wait for the result of laboratory tests. Most people who die from the disease have not received prompt treatment after the onset of symptoms. Tetracycline and chloramphenicol are the only antibiotics approved for treatment. The disease can be confused with meningococcemia and human ehrlichiosis. Mortality of untreated RMSF varies from 20% to 80% with an average of 25%. Commercial vaccines have been withdrawn as ineffectual.

Prevention

RMSF is best controlled by avoiding tick-infested areas, or by wearing clothing that covers the ankles and wrists and wrapping the ankles with tape when in tick infested areas. The clothing should be light colored in order to see any crawling ticks. Use of a repellent or acaricide increases protection. Daily examination of the body, especially in areas where ticks are likely to be embedded, should be a routine preventive measure. Proper removal of ticks is important, especially because transmission of the disease can occur during tick removal. For proper examination of the body for ticks and a description of how to properly remove them see the entry on "Ticks." Save the tick for species determination by a physician or scientist. Put it in a plastic bag for storage. Removal of ticks early in feeding lessens the chance of transmission of *R. rickettsii*. Education about symptoms of RMSF is very important.

See also, TICKS.

Lewis B. Coons and Marjorie Rothschild
The University of Memphis
Memphis, Tennessee, USA

References

Burdorfer, W. 1975. A review of Rocky Mountain spotted fever (tick-borne typhus), its agent, and its tick vectors in the United States. *Journal of Medical Entomology* 12: 269–278.

Dumler, J. S. 1994. Rocky Mountain spotted fever. pp. 417–427 in G. W. Beran and J. H. Steele (eds.), *Handbook of zoonoses, Section A: bacterial, rickettsial, chlamydial, and mycotic diseases* (2nd ed.). CRC Press, Boca Raton, Florida.

Sexton, D. J., and E. B. Breitschwerdt. 1998. Rocky Mountain spotted fever. pp. 207–215 in S. R. Palmer, Lord Soulsby and D. I. H. Simpson (eds.), *Zoonoses: biology, clinical practice and public health control*. Oxford University Press, Oxford, England.

Sonenshine, D. E. 1993. *Biology of ticks*, Vol. 2. Oxford University Press, New York, New York. 465 pp.

Sonenshine, D. E., and T. N. Mather (eds.) 1994. *Ecological dynamics of tick-borne zoonoses*. Oxford University Press, New York, New York. 447 pp.

ROCKY MOUNTAIN WOOD TICK, *DERMACENTOR ANDERSONI* STILES (ACARI: IXODIDAE).

The Rocky Mountain Wood Tick, *Dermacentor andersoni* Stiles, is distributed in western Canada east of the Coast Mountains in British Columbia, through Alberta, and Saskatchewan to 105° longitude and north to the 53rd parallel. The tick occurs in the northwestern United States including western Washington, Oregon and eastern California, Nevada, Colorado, Wyoming, Idaho and Montana. The southern limit of its range is northern New Mexico and Arizona, and the eastern limit includes western portions of North Dakota, South Dakota, and Nebraska. Preferred tick habitat in mountainous areas includes rocky slopes, clearing, outcrops and ecotones between grasslands and forest. South-facing slopes are preferred at higher elevations and in northern parts of its range. In drier, prairie habitats, the tick tends to localize in river bottoms and coulees. The presence of shrubby vegetation such as Saskatoon (*Amalanchier*) and rose (*Rosa*) are positive indicators of tick habitat, while the presence of aspen (*Populus*) is a negative indicator of tick habitat.

Rocky Mountain wood tick is a three-host tick and completes its life cycle in 1 to 3 years depending on latitude, elevation, and local environmental conditions. Longer life cycles occur in cooler areas such as higher elevations and more northerly latitudes. Adult ticks are brown with grey patterns and are approximately 3 to 4 mm long and 2 to 4 mm broad. Wood ticks have a rectangular basis capitulum and short, broad palps. The spiracle has a dorsal prolongation and numerous small goblets that distinguish it from the related species *Dermacentor albipictus* and *D. variabilis*. Adult ticks feed on various moderate to large hosts including cattle, sheep, horses, dogs, porcupines, goats, deer and humans. Adults are active in the early spring and quest when temperatures are above 10°C. In the interior of British Columbia, adults are active from mid-March through mid-May, with peak activity in early April. Ticks in prairie regions are active somewhat later, with peak activity in May through June. Adults seek hosts by climbing grasses or low shrubs and waiting for a host to approach. The adults sense carbon dioxide emanations from a potential host, become active, and attach to the host as it passes. Ticks will crawl on the host until they find a suitable site for attachment and feeding. Montane ticks tend to attach and feed on the upper surfaces of cattle near the withers while prairie ticks tend to attach to the undersides of cattle. This may be to avoid excessive exposure to increased solar radiation and host temperatures due to the later periods of activity in prairie regions. Both sexes feed on blood, although male ticks take smaller meals and may redistribute themselves on the host. Ticks that do not find a host in the spring may enter a behavioral diapause, and a small proportion of these can successfully overwinter to seek a host the following spring. Feeding females gain weight slowly during the first few days of feeding, but greatly increase the rate of engorgement after mating. Females require 7 to 10 days to become fully engorged, and can increase their weight from 4 mg to 700 mg. Once engorgement is complete, the females detach, fall to the ground, and seek sheltered areas to oviposit.

The preoviposition period is dependent on temperature, ranging from 5 days at 25°C to 20 to 40 days at 15°C. Females produce from 5,000 to 10,000 eggs over a period of several weeks, and die once oviposition is completed. Eggs hatch in about 4 weeks, depending on temperature. The larvae are six-legged and will attach to smaller mammals such as mice, ground squirrels, and chipmunks. Larval engorgement requires 5 to 10 days, and the fully engorged larvae drop from the host. Molting to the nymphal stage requires 1 to 4 weeks depending on temperature. Larvae are most abundant on hosts during July to August in the interior of British Columbia. Newly molted nymphs may attach to a small mammal host, engorge in 5 to 10 days, drop from the host, and molt to adults in 2 to 4 weeks depending on temperature. Nymphs that are unable to find a host may overwinter, and seek hosts the next spring. Nymphal activity thus occurs over a broader period time than do larvae, from April though August in the interior of British Columbia. Nymphs that are active early in the season are likely those that overwintered, while nymphs active later in the season likely originate from larvae produced the current year. Adults that are produced in the fall do not seek hosts but enter a behavioral diapause that must be terminated by chilling before they will resume normal feeding.

Feeding activity of adult wood ticks on cattle probably does not result in appreciable economic losses due to reduced weight gains of cattle. Adult activity occurs early enough in the season for compensatory weight gain to occur. However, feeding ticks can cause paralysis in cattle in the interior of British Columbia south through Idaho. Paralysis can also affect sheep and humans, as well as dogs and horses. The economic effect of paralysis is difficult to estimate, but results from animal death, underuse of native pasture, and increased costs for surveillance. Currently, tick paralysis is prevented by early season application of synthetic acaricides such as Lindane. However, the availability of such compounds is increasingly limited. Other potential methods for managing ticks based on rangeland management practices have been proposed but not fully investigated.

Rocky Mountain wood ticks can vector several of diseases of humans and livestock. These include anaplasmosis in cattle, and Rocky Mountain spotted fever, Tularemia, and Colorado tick fever to humans. Disease prevention is achieved by removing ticks before feeding, and requires careful inspection after engaging in outdoor pursuits during the season of adult activity. Precautions to prevent ticks from attaching include wearing long pants outdoors, and pulling socks up over the pant legs. Repellents can also be used.

Tim Lysyk
Agriculture and Agri-Food Canada
Lethbridge, Alberta, Canada

References

Gregson, J. D. 1956. The *Ixodoidea of Canada*. Canadian Department of Agriculture Science Serial Publication No. 930.

Rich, G. B. 1971. Disease transmission by the Rocky Mountain wood tick, *Dermacentor andersoni* Stiles, with particular reference to tick paralysis in Canada. *Veterinary Medicine Review* 1: 3–26.

Schaalje, G. B., and P. R. Wilkinson 1985. Discriminant analysis of vegetational and topographical factors associated with the focal distribution of Rocky Mountain wood ticks, *Dermacentor andersoni* (Acari: Ixodidae), on cattle range. *Journal of Medical Entomology* 22: 315–320.

Fig. 886 Wood ticks, unfed and well fed.

Wilkinson, P. R. 1967. The distribution of *Dermacentor* ticks in Canada in relation to bioclimatic zones. *Canadian Journal of Zoology* 45: 517–537.

Wilkinson, P. R. 1968. Pest-management concepts and control of tick paralysis in British Columbia. *Journal of the Entomological Society of British Columbia* 65: 3–9.

RODENT FLEAS. Members of the family Hystrichopsyllidae (order Siphonaptera). See also, FLEAS.

RODENT TRYPANOSOMIASIS: A COMPARISON BETWEEN *Trypanosoma lewisi* and *Trypanosoma musculi.*

Trypanosoma lewisi and *Trypanosoma musculi* are protozoans that belong to the Phylum Sarcomastigophora and Subphylum Mastigophora. The Mastigophora, commonly known as flagellates, typically possess one or more flagella. Although they reproduce asexually by binary fission, sexual reproduction is known to occur. Both *T. musculi* and *T. lewisi* are assigned to the class Zoomastigophorea, family Trypanosomatidae, and order Kinetoplastida because they have flagella, typically with a paraxial rod and an axoneme. In addition, they contain a single mitochondrial kinetoplast located near the flagellar kinetosomes and a Golgi apparatus located close to the flagellar depression.

T. lewisi, a cosmopolitan parasite in the blood of rats (*Rattus* sp.), is transmitted by several species of rat fleas (*Ceratophyllus fasciatus, Nosopsyllus fasciatus*, and *Xenopsylla cheopis*). On the other hand, *T. musculi*, also a cosmopolitan blood parasite, is found in mice and is transmitted primarily by the rat flea, *Nosopsyllus fasciatus*. Both trypanosomes are highly specific for their vertebrate hosts where they live extracellularly, primarily in the bloodstream.

Fleas become infected with these trypanosomes upon ingesting a blood meal from an infected rodent. In the fleas, however, the trypanosomes are present in the alimentary tract, where they undergo morphological changes and then move posteriorly towards the rectum and anus. The trypanosomes are then voided in the feces of the insect vector ('posterior station transmission') and can be ingested by the rodent during grooming. Within the rodent, the trypanosomes now enter the bloodstream by way of the oral mucous membranes. Once in the bloodstream, they undergo further development and are now infective to the adult flea during a blood meal. Rosettes, which usually consist of eight daughter cells, have been found in both *T. lewisi* and *T. musculi* during *in vitro* cultivation. Rodents stressed by starvation, low temperature, multiple infections, or intraspecies fighting do not show an increase in the prevalence of the rosette forms.

Although *T. lewisi was* believed to be primarily monomorphic in appearance with lengths varying from 21 to 36.5 μm and an average of 30.6 μm, two additional forms (type II and III) have been observed. Type II is a morphological variation of the original type and it has a total length of 35.3 to 39.3 μm. The posterior end of this form is extremely elongated. The distance between the posterior end and the kinetoplast (P-K) is 11.9 to 15.0 μm. The average length of type III of *T. lewisi* is 26.7 μm (19.8 to 30.9 μm) and the distance between the posterior end and the kinetoplast is 2.9 to 6.3 μm. The size of *T. musculi* varied from 23 to 30 μm, with an average of 26.5 μm. This parasite has a diameter of 3 to 5 μm. Electron microscopic studies have shown that the kinetoplast of *T. lewisi* appears as a mass of DNA-containing fibrils, about 25 Å in diameter, lying in parallel array and situated within an enlarged portion of the mitochondrion. There are no studies on the kinetoplastic DNA replication of *T. musculi*.

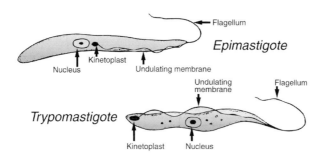

Fig. 887 Epimastigote and trypomastigote forms of mouse trypanosomes.

The two developmental stages of *T. musculi* and *T. lewisi* differ by the position of the kinetoplast. In the epimastigote form, the kinetoplast is anterior to the nucleus, whereas in the trypomastigote form, the kinetoplast is posterior to the nucleus. Epimastigotes have been characterized as dividing, sausage shaped and juvenile forms while trypomastigotes are sometimes referred to as the adult, leaf-shaped stage. During the conversion of the epimastigote stage to the trypomastigote stage, there is a shift in the kinetoplast. The epimastigote form requires iron for its cytochrome oxidase system. When the epimastigote changes to the trypomastigote form, the trypanosomes are believed to utilize the iron from the host.

Both *T. musculi* and *T. lewisi* parasites produce a self-limiting infection comprising of a growth phase, a plateau phase, and an elimination phase during a 2 to 4 week period. In *T. musculi* and *T. lewisi*, upon infection of their specific hosts, there is a latent period which is determined by the size of the parasitic inoculum. Following the latent period, there is a phase of rapidly increasing parasitemia in the infected blood that becomes stabilized after 6 to 7 days in *T. musculi* and 4 to 5 days in *T. lewisi*. This period is followed by a plateau phase in *T. musculi* that lasts about 7 to 10 days. The onset of the plateau phase with the elimination from the blood of all reproductive forms of the parasites is called the first crisis. The antibody ablastin is responsible for this elimination in *T. lewisi*-infected rats. A trypanocidal antibody is responsible for the elimination of reproductive trypanosomes in *T. musculi*-infected mice. A second crisis, at the end of the plateau phase, produces an abrupt fall in the parasitemia, and the parasites disappear from the blood within a few days. An IgM antibody is involved during this second crisis in *T. lewisi* and *T. musculi* infections. *T. lewisi* neither divides in the peritoneal cavity after the first crisis nor remains in the vasa recta of the kidneys after blood recovery from the infection. As opposed to *T. lewisi*, *T. musculi*-infected mice showed both continuous parasitic division in the peritoneal cavity and persistence of dividing forms in the vasa recta of the kidneys of the immune mice. These mice with chronic or latent infections are immune to re-infection and do not experience relapses, although isolated kidney forms and homogenates of kidneys from these animals are infective when inoculated into naive mice.

Trypanosoma musculi and *T. lewisi* infections, although referred to as non-pathogenic, can produce profound changes in the size, cellularity, and histoarchitecture of spleen, liver, lymph nodes, and thymus of the respective hosts. There are obvious pathological changes associated with the elimination of these parasites. They involve splenomegaly, marked enlargement of the lymph nodes, hepatomegaly to a lesser degree, and immunodepression, that are clinical features of the pathogenic trypanosomes in man and animal host.

Life cycle and control of rodent fleas

Fleas are holometabolous insects with an egg, several larval instars, and a pupal stage. Eggs of adult fleas are typically laid in the rodent's nest or bedding, where they hatch within a few days into larvae. The larvae feed on a variety of organic material in the nest or bedding materials of the host. Food of the larvae also may include feces of the adults which contain partially digested blood. The larvae eventually change into a pupa and then either a female or adult male emerges. Both sexes feed on blood.

Because rodent trypanosomiasis generally produces a benign infection in rats and mice, the *T. lewisi*-rat flea system (also *T. musculi*-mouse-flea system) has been used as a laboratory model for studying more virulent trypanosomes that are pathogenic to man and domestic animals. However, it is imperative that flea control strategies (e.g., improved garbage sanitation near homes where rodent populations may build up, as well as interrupting the flea development cycle, or chemicals) be used to reduce the threat or possibility of fleas vectoring the etiological agents of more serious man and animal diseases (e.g., plague, murine typhus, Q fever, etc.).

Clarence M. Lee
Howard University
Washington, DC, USA
and
Earlene Armstrong
University of Maryland,
College Park, Maryland, USA

References

Ashraf, M., R. A. Nesbitt, A. Gugssa, and C. M. Lee. 1999. Development and proliferation of *Trypanosoma musculi* in the presence of adherent splenic cells. *Journal of Parasitology* 85: 129–132.

Bogitsh, B. J., and T. C. Cheng. 1990. *Human parasitology*. Saunders College Publishing, Philadelphia, Pennsylvania.

D'Alesandro, P. A., and A. B. Clarkson 1980. *Trypanosoma*: avidity and adsorbability of ablastin, the rat antibody inhibiting parasite reproduction. *Experimental Parasitology* 50: 384–396.

Shapiro, T. A., and P. T. Englund 1995. The structure and replication of kinetoplast DNA. *Annual Review of Microbiology* 49: 117–143.

Simpson, L. 1987. The mitochondrial genome of kinetoplastid protozoa: genomic organization, transcription replication, and evolution. *Annual Review of Microbiology* 41: 363–382.

Vincendeau, P., B. Guilleman, S. Daulouede, and C. Ripert, C. 1986. *In vitro* growth of *Trypanosoma musculi*: requirements of cells and serum free culture medium. *International Journal of Parasitology* 16: 387–390.

Wechsler, D. S., and P. A. L. Kongshavn 1988. Further characterization of the curative antibodies in *Trypanosoma musculi* infection. *Infection and Immunity* 56: 2379–2384.

ROLLED-WING STONEFLIES. Members of the stonefly family Leuctridae (order Plecoptera). See also, STONEFLIES.

ROMALEIDAE. A family of grasshoppers (order Orthoptera). They commonly are known as lubber grasshoppers. See also, GRASSHOPPERS, KATYDIDS AND CRICKETS.

RONDANI, CAMILLO. Camillo Rondani was an eminent Italian entomologist, and a renowned dipterist. A native of Parma, Italy, he was born on November 23, 1803. After study in France, Rondani served as a professor of natural history at the Royal College, and served as director of the "Instituto Tecnico." An author of many publications on Diptera, his great work "Dipterologiae Italicae Prondromus," of which eight volumes were published from 1856 to 1877, was unfortunately unfinished at the time of his death at the age of 72, on September 18, 1879. Other significant achievements include his "Degli insetti parassiti" and numerous other bulletins, and service to the Italian Entomological Society. He also published on Lepidoptera and Hymenoptera, and described numerous aphids.

References

Anonymous. 1880. Obituaries, Camillo Rondani. *The Entomologist* 13: 120.

Meade, R. H. 1879. Professor Camillo Rondani. *Entomologist's Monthly Magazine* 15: 138–139.

Papavero, N. 1971, 1973. Essays on the history of Neotropical Dipterology with special reference to collectors (1750–1905). Museu de Zoologia, Universidade de São Paulo.

ROOT MAGGOTS. Some members of the family Anthomyiidae (order Diptera). See also, FLIES.

ROOT-EATING BEETLES. Members of the family Monotomidae (order Coleoptera). See also, BEETLES.

ROOT-INFECTING FUNGI. See also, TRANSMISSION OF PLANT DISEASES BY INSECTS.

ROPALOMERID FLIES. Members of the family Ropalomeridae (order Diptera). See also, FLIES.

ROPALOMERIDAE. A family of flies (order Diptera). They commonly are known as ropalomeric flies. See also, FLIES.

ROPRONIIDAE. A family of wasps (order Hymenoptera). See also, WASPS, ANTS, BEES, AND SAWFLIES.

ROSE-GRAIN APHID. See also, WHEAT PESTS AND THEIR MANAGEMENT.

ROSEN, DAVID. David Rosen was born on April 20, 1936, in Tel-Aviv, Israel. He received his education at the Hebrew University of Jerusalem (M.S., 1959; Ph.D., 1965). His thesis research, entitled "Parasites of the Coccoidea, Aphidoidea and Aleurodidea of citrus in Israel" earned him the Jacobsen Prize in 1965. He obtained a postdoctoral appointment with Paul DeBach at the University of California at Riverside, and together they published "Species of Aphytis of the world" (1979). For this important project they received a gold medal from the Filippo Silvestri Foundation in 1980. In 1991 they published a second edition of "Biological Control by natural enemies." David also edited two volumes of "Armored scale insects: their biology, natural enemies and control" (1990), and coedited two volumes of "Pest management in the subtropics" (1994, 1996). Rosen was a powerful voice for practical implementation of biological control, and had a productive career emphasizing population dynamics of pests, particularly biological control of pests, and including the systematics of parasitoids. He served as chairman of the Department of Entomology at Hebrew University of Jerusalem, and was named Vegevani Chair of Agriculture in 1990. He served on the editorial board of several journals, and was well-known and highly respected in international biological control circles. Rosen's career included publication of 140 technical papers, 33 book chapters, and 8 books. He died on January 8, 1997.

References

Hoy, M. A. 2000. The David Rosen lecture: biological control in citrus. *Crop Protection* 19: 657–664.
Gerson, U., and S. W. Applebaum. 1997. David Rosen (1936–1997) In memorium. *Phytoparasitica* 25: 171–174.

ROSS, HERBERT HOLDSWORTH.

Herbert Ross was born in 1908 at Leeds, England, educated in British Columbia, Canada, and received his Ph.D. from the University of Illinois in 1933. Ross began work with the Illinois Natural History Survey in 1927 and was named head of the survey and identification section in 1935. He also held the rank of professor of entomology at the University of Illinois. Ross retired from the University of Illinois in 1969 but then worked for the University of Georgia until 1976. Herbert Ross was an authority on caddisflies (Trichoptera) and published over 200 technical publications, as well as several books. He is best known for a popular introductory text, "A textbook of entomology" (1948 and followed by several editions), but also authored "Understanding evolution" (1966), "Evolution and classification of the mountain caddisflies" (1956), and "Synthesis of evolutionary theory" (1962). Herbert Ross died November 2, 1978, at Athens, Georgia.

Reference

Anonymous. 1979. Herbert Holdsworth Ross, 1908-1978. *Entomological News* 90: 62.

ROSTRUM. In weevils, the snout-like prolongation of the head containing the mouthparts distally. In Hemiptera, this sometimes refers to the beak or piercing-sucking mouthparts.

ROSY APPLE APHID. See also, APPLE PESTS AND THEIR MANAGEMENT.

ROTATION. In agriculture, purposeful alternation of crops grown on the same plot of land. In pest control, purposeful alternation of insecticides used to control a pest population.

ROTENONE. See also, BOTANICAL INSECTICIDES.

ROTOITIDAE. A family of wasps (order Hymenoptera). See also, WASPS, ANTS, BEES, AND SAWFLIES.

ROTTENWOOD TERMITES. Members of the termite family Termopsidae. See also, TERMITES.

ROUBAL, JAN.

Jan Roubal was born at Chudenice, Czech Republic, on August 10, 1880. He was educated at the university in Prague, and in 1905 left the university as a high school teacher. He continued as a gymnasium teacher and director until he retired in 1940, when he directed his entire energy to the study of entomology. Roubal is reputed to have had an encyclopedic knowledge of Coleoptera. This, combined with an outgoing and unselfish personality, made him an important and influential coleopterist in middle Europe. He preferred the Staphylinidae, and influenced many others to work with this large group. He published over 300 papers, mostly on beetles but also on Hemiptera/Heteroptera. His "Katalog Coleoper (brouků) Slovenska a Parkarpatska" was published in three volumes (1930 to 1941) and set a new standard for catalogs because it contained extensive habitat data and comments. Roubal died at Prague on October 23, 1971.

Reference

Herman, L. H. 2001. Roubal, Jan. *Bulletin of the American Museum of Natural History* 265: 129–131

ROUND FUNGUS BEETLES. Members of the family Leiodidae (order Coleoptera). See also, BEETLES.

ROUNDHEADED PINE BEETLE, *DENDROCTONUS ADJUNCTUS* BLANDFORD (COLEOPTERA: CURCULIONIDAE, SCOLYTINAE).

The roundheaded pine beetle, *Dendroctonus adjunctus*, is an important bark beetle that attacks and kills various species of pines from Nevada, Utah, and Colorado south to Guatemala. In the Southwestern United States its primary host is ponderosa pine, *Pinus ponderosa*. It often occurs in trees in conjunction with the western pine beetle, *Dendroctonus brevicomis*, and pine engravers of the genus *Ips*. Like other tree-killing bark beetles, the roundheaded pine beetle causes tree mortality by feeding and developing in the phloem of the tree. Utilization of the phloem for habitat and nourishment results in girdling of the tree so that translocation of nutrients is impeded. The insect also inoculates attacked trees with bluestain fungi *Ophiostoma adjuncti* and *Leptographium pyrinum*, but the ecological role of these in helping the insect kill the tree or providing nutritional benefits to the insect is unclear.

Patterns of tree mortality range from a single tree to groups of tens to hundreds of trees. Occasionally, only a portion of the tree is attacked, commonly referred to as a strip attack, and the tree usually survives. Initial symptoms of attack include an accumulation of brown-reddish boring dust at the base of the tree and in bark crevices. As the insect enters the tree, resin canals, which are the primary defensive mechanism of the tree, are severed and resin exuded. The resin then solidifies on the outer surface of the bark forming a pitch tube. As the tree dies, the foliage changes in color from green to a light yellow and then bright orange within the first year of attack.

In the Southwest, the roundheaded pine beetle has generally a one-year life cycle. The primary emergence and dispersal flight of the insect occurs in late October and early November, with some variation observed among geographical locations. Soon after emergence females seek suitable hosts for attack. Colonization of new hosts involves a complex chemical communication system of pheromones and host chemicals.

Once the adult beetles enter the tree, a small chamber is excavated and then the main egg gallery

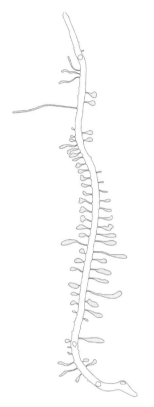

Fig. 888 Roundheaded pine beetle egg gallery (drawing by Joyce VanDeWater).

begins either to the right or left of the chamber for the first 2.5 to 5 cm. The egg gallery then turns vertically moving upward with the grain of the wood and gently meandering to the right or left. The gallery is constructed by the female and packed with frass by the male behind. Galleries as long as 122 cm have been observed but they are normally about 31 cm and located in the inner bark but do score the wood. Ventilation holes are excavated about every 4 to 8 cm. Females oviposit single eggs in niches alternately on both sides of the gallery. Egg niches are about 1.5 mm across and about 1 mm deep and are covered with frass and other debris. Egg niches are about 4 to 8 mm apart. Oviposition begins after the tree is invaded, ceases through the winter, and then resumes in the spring, usually March, and continues into April.

The eggs are pearly white, oblong, and about 1 mm in length and 0.6 mm in width. Some eggs may begin hatching in the fall, but the majority will hatch in the spring. The eggs and egg-laying adults are the primary overwintering stage. The larvae are legless, grublike, mostly translucent but with light reddish-brown coloration. Each larva constructs its own larval gallery, which are most commonly 4 to 15 mm in length. There are four larval instars. Average head capsule width for instars 1 thru 4 is 0.5, 0.7, 1.0, and 1.2 mm, respectively. Average body length for the 4th instar is about 6.3 mm. Larval development through the 3rd instar is rather fast and rather slow in the 4th instar. Pupation occurs at the end of the larval gallery usually between June and August. Average length of the pupae is 5.2 mm. The beetles will usually complete development and become adults by the middle of August but will remain on the tree until the flight period begins again in the fall.

Newly emerged adult beetles are shiny black with a reddish tint in the elytra; parent adults are shiny black. Males exhibit heavy pigmentation along the angular rear margin of the 7th abdominal tergite while females have a transverse elevated ridge on the anterior area of the pronotum, which is lacking in the males. Adult beetles vary widely in total length from 2.9 mm to 6.9 mm but the majority are 4.5 to 6.0 mm. Females tend to be slightly larger than the males.

The roundheaded pine beetle can kill trees in all crown and diameter classes. Studies have indicated average diameter at breast height of beetle-killed of 29 cm in the Sacramento Mountains of south-central New Mexico, 43 cm in the Pinaleno Mountains of Arizona, 48 cm in the Pine Valley Mountain of Utah with dominant, co-dominant, and intermediate trees being killed. In these areas, tree mortality was more common in areas with high stocking and poor tree growth. In an outbreak in Nevada, tree mortality was most common in trees at least 50 cm in diameter at breast height.

Various natural enemies attack the roundheaded pine beetle but little is known about their ecology and impact. The most important appears to be the red-bellied clerid, *Enoclerus sphegeus*. The roundheaded pine beetle is probably the only species in the genus in North America that overwinters mostly in the egg stage and that has a flight period predominantly in the fall.

See also, BARK BEETLES IN THE GENUS *DENDROCTONUS*.

Jose F. Negron
USDA Forest Service
Ft. Collins, Colorado, USA

References

Chansler, J. F. 1967. Biology and life history of *Dendroctonus adjunctus* (Coleoptera: Scolytidae). *Annals of the Entomological Society of America* 60: 760–767.

Lucht, D. D., R. H. Frye, and J. M. Schmid. 1974. Emergence and attack behavior of *Dendroctonus adjunctus* Blandford near Cloudcroft, New Mexico. *Annals of the Entomological Society of America* 67: 610–612.

Massey, C. L., D. D. Lucht, and J. M. Schmid. 1977. Roundheaded pine beetle. *U.S. Department of Agriculture, Forest Service, Forest Insect and Disease Leaflet 155.*

Negrón, J. 1997. Estimating probabilities of infestation and extent of damage by the roundheaded pine beetle in ponderosa pine in the Sacramento Mountains, New Mexico. *Canadian Journal of Forest Research* 27: 1936–1945.

Negrón, J. F., J. L. Wilson, and J. A. Anhold. 2000. Stand conditions associated with roundheaded pine beetle (Coleoptera: Scolytidae) infestations in Arizona and Utah. *Environmental Entomology* 29: 20–27.

Six, D. L., and T. D. Paine. 1996. *Leptographium pyrinum* is a mycangial fungus of *Dendroctonus adjunctus*. *Mycologia* 88: 739–744.

ROVE BEETLES (COLEOPTERA: STAPHYLINIDAE).

The family Staphylinidae belongs to the suborder Polyphaga of the order Coleoptera (beetles). The superfamily Staphylinoidea includes the small families Hydraenidae, Ptiliidae, Agyrtidae, Leiodidae, Scydmaenidae, and Silphidae, and the huge family Staphylinidae.

Order: Coleoptera
  Suborder: Polyphaga
    Superfamily: Staphylinoidea
      Family: Staphylinidae

Four phyletic lines are now (Lawrence and Newton 1995) included in Staphylinidae:

(1) subfamilies Glypholomatinae, Microsilphinae, Omaliinae, Empelinae, Proteininae, Micropeplinae, Neophoninae, Dasycerinae, Protopselaphinae, and Pselaphinae;

(2) subfamilies Phloeocharinae, Olisthaerinae, Tachyporinae, Trichophyinae, Habrocerinae, and Aleocharinae;

(3) subfamilies Trigonurinae, Apateticinae, Scaphidiinae, Piestinae, Osoriinae, and Oxytelinae;

(4) subfamilies Oxyporinae, Megalopsidiinae, Steninae, Euaesthetinae, Solieriinae, Leptotyphlinae, Pseudopsinae, Paederinae, and Staphylininae.

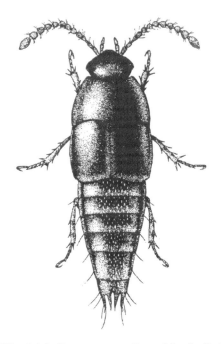

Fig. 889 Adult *Coproprorus rutilus*, subfamily Tachyporinae, 3.8 mm long, West Indies.

The former taxa Brathinidae (now just part of subfamily Omaliinae), Dasyceridae (now subfamily Dasycerinae), Empelidae (now subfamily Empelinae), Glypholomatini (formerly a tribe of Silphidae), Microsilphinae (formerly a subfamily of Silphidae), Pselaphidae (now subfamily Pselaphinae), and Scaphidiidae (now subfamily Scaphidiinae) of earlier authors are here included within the family Staphylinidae. As now constituted, this is one of the largest families of beetles, with over 45,000 species known worldwide and probably over 75% of tropical species still undescribed. In the future, systematists may, however, choose to split the family into the 4 phyletic lines to form 4 families.

Adults

The length of adult Staphylinidae ranges from less than 1 mm to almost 30 mm. Most are under 7 mm long. Most have short elytra, exposing several abdominal segments – but it would be an error to imagine that all have short elytra, or that all beetles with short elytra are Staphylinidae. Typically, they are slender with short elytra and abdominal musculature that renders them very flexible, thus able to enter

narrow crevices. Those that have short elytra trade flexibility for exposure, rendering them subject to desiccation and dependent upon humid habitats. Typically, abdominal segments are bounded by tergites, 1 or 2 sets of paratergites, and sternites, with membranous connections, but in some genera (e.g., of Paederinae and Osoriinae, and partially so in Steninae) the tergites-paratergites-sternites are fused into rings around each segment, probably limiting water loss. The length of adults, which is nevertheless normally used to express size, is an inaccurate determinant of size because abdominal segments typically can be telescoped—making the body appear longer when under moist conditions and alive, but shorter when dried and dead. Eyelessness has evolved in some soil-inhabiting (Leptotyphlinae) and cave-inhabiting species, and winglessness in species occupying environments including mountains, the soil, caves, and seashores.

Non-entomologists sometimes confuse Staphylinidae with Dermaptera. However, the non-opposable valvulae (appendages of the 9th abdominal segment) are not the opposable forcipes of Dermaptera, and the radial wing-folding pattern of Dermaptera is unlike that of Staphylinidae.

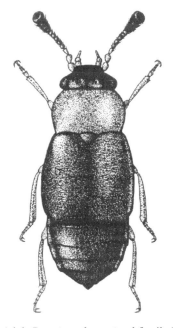

Fig. 890 Adult *Proteinus thomasi*, subfamily Proteininae, 1.5 mm long, North America.

Immature stages

Staphylinid eggs typically are white, spherical, spheroidal, or pyriform. Eggs of some genera within the Staphylininae (e.g., *Philonthus*) have pronounced surface sculpture, allowing identification at least to the species-group level. Larvae are campodeiform (sometimes called staphyliniform). In some subfamilies (Paederinae, Staphylininae, and to a lesser extent in their immediate relatives) the head is relatively more heavily sclerotized and there is a distinct "neck" (nuchal constriction of the head). Prepupae of Aleocharinae, Steninae, and one genus (*Astenus*) of Paederinae spin a silken cocoon in which they pupate. Pupae are obtect, pigmented, and sclerotized in the subfamily Staphylininae, but exarate, white, and unsclerotized in all the other subfamilies. In general, the immature stages develop rapidly, in a few days to a few weeks, and the adults are long-lived.

Habitats

Staphylinidae occupy almost all moist environments throughout the world. Because none of them is truly aquatic, they do not live in open waters; although winged adults may be skimmed from the sea surface far from land, their presence is due to misadventure but attests to their dispersive ability. They live in leaf litter of woodland and forest floors

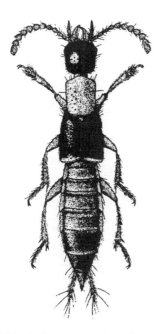

Fig. 891 Adult *Neobisnius occidentoides*, subfamily Staphylininae, 4 mm long, North America.

and grasslands. They concentrate in fallen decomposing fruits, subcortical cavities of fallen trees, drifted plant materials on banks of rivers and lakes, and dung, carrion, and nests of vertebrate animals. Several hundred species live only on seashores. Large numbers of species are specialized to existence in nests of social insects. Many inhabit caves, underground burrows of vertebrate animals, and smaller soil cavities, even of burrows that they (a few of them) excavate. Many live in mushrooms. Adults and even larvae of a few are associated with living flowers. Others climb on plants, especially at night, and hunt for prey. A few seem to live with terrestrial snails, although their role is not understood. Their distribution in arid environments is restricted to moist microhabitats.

Food

A little about the feeding habits of Staphylinidae has been deduced from casual observations by many observers, and from dissections of alimentary canals and from feeding trials and examination of mouthparts by a few. Archetypal staphylinids probably were saprophagous. Saprophagy is still a major feeding mode in Piestinae, Osoriinae, and Proteininae, perhaps with some adaptation to mycophagy. Mycophagy has evolved in Oxyporinae, Scaphidiinae, some Tachyporinae, and a few Aleocharinae. Phytophagy has evolved in some Oxytelinae, to the point where the diet of adults and larvae of *Bledius* consists of diatoms, and at least one species of *Apocellus* has been accused of damaging flowers, one species of *Carpelimus* has (probably wrongly) been accused of damaging cucumbers, and one species of *Osorius* (Osoriinae) has been accused of damaging turf grass. Saprophagy has evolved toward carnivory in other subfamilies (many Tachyporinae, most Aleocharinae, Pselaphinae, Euaesthetinae, Steninae, Paederinae, and Staphylininae), representing the bulk of species in the family, so that it may be said that most Staphylinidae – tens of thousands of species – are facultative predators. Some have specialized, for example *Oligota* (Aleocharinae) as predators of mites, *Erichsonius* (Staphylininae) as predators of soil-inhabiting nematodes, *Odontolinus* (Staphylininae) as predators of mosquito larvae in water-filled flower bracts of *Heliconia* (Heliconiaceae), and *Eulissus* (Staphylininae) on adult dung-inhabiting scarab beetles. One line of Aleocharinae (*Aleochara*) has evolved to become parasitoids in fly puparia.

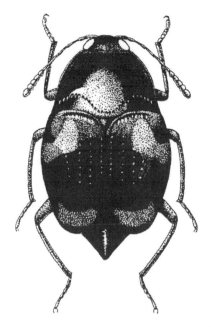

Fig. 892 Adult *Scaphisoma quadriguttatum*, subfamily Scaphidiinae, 4.3 mm long, North America.

Presocial or subsocial behavior

This behavior is known in *Bledius* and *Platystethus* (Oxytelinae) and *Eumicrota* (Aleocharinae). Adults construct chambers in which they deposit and guard their eggs, *Bledius* in sandy or muddy shores, *Platystethus* in the dung of ungulates, and *Eumicrota* in mushrooms. Another dimension of social behavior has arisen in interactions with termites and ants – thousands of species of Aleocharinae, and numerous species of several other subfamilies, are inquilines in the nests of these social insects, with attendant structural and behavioral adaptation. *Emus* (Staphylininae) invades bee nests in Europe, and a species of *Euvira* (Aleocharinae) develops in communal nests of a butterfly in Central America.

Obligate relationships with higher plants

Among those species attracted to flowers, *Pelecomalium testaceum* (Omaliinae) pollinates *Lysichiton americanum* (Araceae) in the mountains of the Pacific coast of the USA and Canada. It is conceivable that *Polyobus* spp. do the same for *Espeletia* (Asteraceae) in the northern Andes of South America. *Charoxus* spp. (Aleocharinae) also have an obligate relationship with plants – the adults are attracted in the Neotropical region to the syconia of *Ficus* spp. (Moraceae) within which they oviposit, but the adults and larvae feed on pollinating wasps (Agaonidae) of those fig flowers.

Relationships with fungi

There are three forms of relationships with fungi. Adults and larvae of many species eat fungi. Others find prey items (fly larvae and other organisms) in fungi. A major association with fungi is that adults of many species are infected by Laboulbeniales (Ascomycetes), and some other fungi, many of which specialize at the level of host genus, tribe, or subfamily.

Relationships with decomposing plants, dung, and carrion

The frequent presence of many staphylinids at decaying plant materials raises the question of whether they arrived there by random movement – and then remained there – or whether they are attracted in much the same way that adults of their prey (fly larvae, etc.) arrived there. Thus, adults and larvae of *Cafius* (Staphylininae) inhabit decaying brown algae (Fucales) on sea beaches and eat fly larvae; adults and larvae of some *Philonthus* (Staphylininae) occur in the dung of ungulates and eat fly eggs and larvae; adults of some *Eulissus* (Staphylininae) are found in the dung of ungulates and there maim and then eat adult scarab beetles; adults and larvae of some *Belonuchus* and *Philothalpus* (Staphylininae) are found in decaying fruits, and eat fly larvae; adults of some *Platydracus* (Staphylininae) are found in carrion and eat fly larvae and adults.

Nests of vertebrates

Some staphylinid species have specialized to live in the nests of vertebrates, especially tortoises, birds, and rodents. Their prey seems to be mainly the larvae of fleas and flies. Thus, in central Asia, where sylvatic plague is endemic, some are credited with suppressing flea populations – and thus help to suppress transmission of plague. Adults of *Amblyopinus* and its close relatives (tribe Amblyopini, subfamily Staphylininae) occur in the fur of some rodents in Central and South America. For years they were suspected of being parasites of these rodents, and taking blood from them. Now, however, they are believed to be phoretic on the rodents, thus being transported from nest to nest. They oviposit in the nests, and larvae feed as predators there of other arthropods.

Secretions and glands

Glandular systems of Staphylinidae are mainly implicated in the production of defensive secretions, of which there is a remarkable array. However, glands of some species that are inquilines in nests of social insects produce substances that appease rather than repel the nest-builders. Further, glands of adult *Stenus* (Steninae) produce a surfactant, stenusin, that enables these beetles to skim over the surface of fresh water into which they have fallen, to regain dry land.

Pederin is a powerful toxin and DNA inhibitor circulating in the hemolymph of all developmental stages of some species of *Paederus* (Paederinae) and close allies in the subtribe Paederina. It is produced by endosymbionts in some, but not all, adult females. Is transferred to eggs at oviposition, and thus to larvae and pupae. Males may obtain it by eating eggs. It is a defensive secretion active against spiders, but seems to have no insecticidal effect. It is a contender for the title of most powerful animal toxin.

Pheromones

A female sex pheromone has been identified in *Aleochara curtula*, but pheromones in 45,000 other species remain unidentified.

Prey capture and pre-oral digestion

The mandibles are the mouthparts typically associated with prey capture by predatory adults and larvae. However, adult *Stenus* (Steninae) use a prey-catching apparatus in which the labium (with its palpi) can be protruded by hemolymph pressure to grip small prey such as Collembola. Oxyporinae, Steninae, Euaesthetinae, Paederinae and Staphylininae have pre-oral digestion. They use mandibles to hold their food, secrete digestive fluids, and imbibe partially-digested food. A consequence is that their digestive systems contain much liquid and few solids, so visual identification of food is difficult. Drops of anal secretions of Neotropical *Leistotrophus* (Staphylininae) adults placed on leaves attract their prey.

Natural enemies

Scattered evidence – needing analysis – suggests that spiders, various insects (including Reduviidae, Carabidae, Asilidae, Formicidae, etc.) amphibia, reptiles, birds, and bats, include Staphylinidae among their diets. Among the parasites, fungi play a major role, and hymenopterous parasitoids, nematodes,

and nemata, a relatively minor role. Non-target effects of chemical pesticides and habitat destruction kill many Staphylinidae.

Biological control

Biological control practitioners have observed predation by various non-specialist Staphylinidae on fly larvae and other invertebrates and have imported various Staphylinidae into Italy, Hawaii, mainland USA, and Easter Island, to capitalize on the perceived benefits – without evident success. Species of *Belonuchus*, *Creophilus*, *Ocypus*, *Philonthus*, *Philothalpus*, and *Thyreocephalus* (Staphylininae) and *Paederus* (Paederinae) have been moved. Other attempts have involved more specialist *Oligota* (Aleocharinae) against tetranychid mites in East Africa, and *Aleochara* against horn fly in mainland USA, but again with little success. Current attempts in Europe involve conservation of native staphylinids, including *Tachyporus* (Tachyporinae) as predators of cereal aphids, and *Aleochara* as predators of root maggots (including augmentative use).

Fossils

Pleistocene fossil Staphylinidae have been reported from northeastern North America from peat bogs, and from Europe. They seem to be of extant species and help to show former distribution of some of these species. Oligocene fossils are amber from the Baltic, the Dominican Republic, and elsewhere, and from shales in the USA (Colorado), France, Germany, and elsewhere. Most of these are recognizably members of modern genera. Deposits from the mid-Cretaceous to lower Jurassic in Eurasia have also yielded fossils. Most of these resemble members of modern subfamilies. The oldest recorded staphylinid, more than 200 million years old, is from the upper Triassic of the USA (Virginia). Species-level identification of present-day staphylinids normally requires dissection, at least of the genitalia - when this cannot be done with fossil specimens they have limited value.

Importance to humans

The importance of staphylinid predation on pests has been demonstrated repeatedly in the literature. They suppress populations of pest insects and mites in numerous crops (agricultural, horticultural and forest entomology), and of biting flies and fleas (medical and veterinary entomology). Their presence in carrion gives them a role in forensic entomology. With one exception (*Paederus* and its close allies) they have trivial importance as pests; but although contact of humans with *Paederus* may cause dermatitis on human skin, the toxin pederin may be harnessed for its therapeutic effects, and some *Paederus* species are valuable predators of crop pests. Finally, Staphylinidae form a substantial part of the world's biodiversity.

J. Howard Frank
University of Florida
Gainesville, Florida, USA
and
M. C. Thomas
Florida State Collection of Arthropods
Gainesville, Florida, USA

References

Frank, J. H. 1991. Staphylinidae. pp. 341–352 in F. W. Stehr (ed.), *Immature insects*, Vol. 2. Kendall-Hunt, Dubuque, Iowa.

Frank, J. H., and M. C. Thomas. 1999. Staphylinidae. "Featured Creatures" creatures.ifas.ufl.edu/misc/beetles/rove_beetles.htm" (27 April 2001).

Herman, L. H. 2001. Catalog of the Staphylinidae (Insecta: Coleoptera). 1758 to the end of the second millennium. *Bulletin of the American Museum of Natural History* 265: 1–4218 [note: this enormous scholarly work includes a bibliography to taxonomic works, together with a catalog of species belonging to most of the subfamilies (excepting only the subfamilies Aleocharinae, Paederinae, Pselaphinae, and Scaphidiinae)].

Lawrence, J. F., and A. F. Newton. 1995. Families and subfamilies of Coleoptera (with selected genera, notes, references and data on family-group names). pp. 779–1006 in J. Pakaluk and S. A. Slipinski (eds.), *Biology, phylogeny, and classification of Coleoptera*. Museum i Instytut Zoologii PAN, Warsaw, Poland.

Newton, A. F., M. K. Thayer, J. S. Ashe, and D. S. Chandler. 2000. Staphylinidae. pp. 272–418 in R. H. Arnett and M. C. Thomas (eds.), *American beetles*. Vol. 1. CRC Press, Boca Raton, Florida.

ROW COVER. A covering, usually consisting of spun-bonded polyester, that is placed over crops to protect them from adverse weather or pests.

ROYAL CELL. In honey bees, a large wax cell constructed by workers and used to rear the queen larvae. In some termites, a cell in which the queen is housed.

ROYAL JELLY. A high-quality food produced by the hypopharyngeal gland of worker honey bees that, if fed to larvae throughout their development, can cause them to develop into queens. Royal jelly also is known as 'bee milk'.

ROYAL MOTHS. Some members of the family Saturniidae (order Lepidoptera). See also, EMPEROR MOTHS, BUTTERFLIES AND MOTHS.

ROYAL PALM BUGS. Members of the family Thaumastocoridae (order Hemiptera). See also, BUGS.

R-STRATEGISTS. Species with life history characteristics making them well suited for exploiting transient environments ('r' is an expression denoting the intrinsic rate of increase). r-selected species represent an extreme on a continuum of life history characteristics, with K-selected species at the other

A comparison of the characteristics of r-selected and K-selected species

	r-selected	K-selected
Habitat type	Unstable, not permanent	More stable
Reproduction	Rapid under favorable conditions; many offspring produced	Usually slower; fewer offspring produced
Development	Rapid; often multivoltine	Slower; often univoltine
Mortality	Often density independent and catastrophic	Often density dependent and more gradual
Population size	Extremely variable in time	Less variable in time
Dispersal capacity & mode	High and random	Lower and oriented
Brood care	Absent	Sometimes present
Body size	Often quite small	Often larger
Competition	Variable, but often low	Often very keen
Ultimate effect	High productivity	Efficient use of resources

end of the continuum. r-selected species can also be said to conform to the 'ruderal strategy.'

RUDERAL. A type of organism, especially plants, found in sites associated with humans and human disturbance. This term is also used with respect to insects that display the characteristics of r-strategists.
See also, r-STRATEGISTS.

RUDIMENTARY. Poorly developed, very small in size, or embryonic.

RUGOSE. Wrinkled or roughened.

RUNOFF. The liquid spray material that drips off the target after a heavy pesticide application is made. The surface water that leaves a field after irrigation.

RUSSIAN WHEAT APHID. See also, WHEAT PESTS AND THEIR MANAGEMENT.

RUST DISEASES. See also, TRANSMISSION OF PLANT DISEASES BY INSECTS.

RUST FLIES. Members of the family Psilidae (order Diptera). See also, FLIES.

RUSTY GRAIN BEETLE. See also, STORED GRAIN AND FLOUR INSECTS.

S

SABROSKY, CURTIS. Curtis Sabrosky was born on April 3, 1910, at Sturgis, Michigan, USA. He graduated from Kalamazoo College in 1931, and obtained a M.S. from Kansas State University in 1933. He taught at Michigan State University from 1936 to 1944, and after military service joined the United States Department of Agriculture at the Systematic Entomology Laboratory. He served in this capacity for 35 years, until retiring in 1980. Widely recognized as an authority on Diptera, Sabrosky published on several families, but considered Chloropidae to be his favorite. He worked diligently on the International Code of Zoological Nomenclature, and served as the president of the Society of Systematic Zoology and the president of the Entomological Society of America. He died on October 5, 1997.

Reference

Crosskey, Roger. 1998. Curtis Sabrosky (1910–1997). *Antenna* 22: 58–61.

SACKBEARER MOTHS (LEPIDOPTERA: MIMALLONIDAE).

Sackbearer moths, family Mimallonidae, total 254 species, all New World and primarily Neotropical (250 sp.). The family is in the superfamily Bombycoidea (series Bombyciformes), in the section Cossina, subsection Bombycina, of the division Ditrysia (some researchers maintain the family in its own monobasic superfamily, Mimallonoidea). Adults medium size (22 to 60 mm wingspan), with head scaling roughened; haustellum short or vestigial; labial palpi short (rarely 2-

Fig. 893 Example of sackbearer moths (Mimallonidae), *Mimallo amilia* (Cramer) from Peru.

segmented); maxillary palpi vestigial; antennae bipectinate; body robust. Wings broad and triangular, usually with a falcate of acute forewing; hindwing rounded. Maculation mostly somber shades of brown or gray, to nearly black, but some more colorful, and with various markings or striae. Adults are nocturnal. Larvae are leaf feeders, with cases having openings that can be plugged by the head and a plate-like structure on the posterior of the body. Host plants are recorded in a number of plant families, including Anacardiaceae, Combretaceae, Fagaceae, Melastomaceae, Myrtaceae, and Rubiaceae, among others. A few can be economic.

John B. Heppner
Florida State Collection of Arthropods
Gainesville, Florida, USA

References

D'Almeida, R. F. 1943. Algunas tipos de generos de orden Lepidoptera. Quarta Nota: Heterocera, Fam. Mimallonidae. *Boletim do Museu Nacional (n.s.) (Zoologie)*, 10: 1–6.

Franclemont, J. G. 1973. Mimallonoidea, Mimallonidae. In R. B. Dominick et al. (eds.), *The moths of America north of Mexicoi Including Greenland, Fasc. 20.1. Bombycoidea [part]*. E. W. Classey, London. 86 pp, 11 pl.

Gaede, M. 1931. Mimallonidae. In *Lepidopterorum Catalogus*, 50: 1–21. W. Junk, Berlin.

Seitz, A. (ed.). 1928–29. Familie: Mimallonidae. In *Die Gross-Schmetterlinge der Erde*, 6: 635–672 (1928); 673, pl. 86–87 (1929). A. Kernen, Stuttgart.

SACBROOD. A disease of honey bee larvae caused by a nonoccluded RNA virus.

SAC SPIDERS (CLASS ARACHNIDA, ORDER ARANEAE, FAMILIES TENGELLIDAE, ZOROCRATIDAE, MITURGIDAE, ANYPHAENIDAE, CLUBIONIDAE, LIOCRANIDAE, AND CORINNIDAE).

The sac spiders are a varied group that at one time was crammed mostly into one family, the Clubionidae. Now, after extensive taxonomic analysis, the nominate family is down to two genera in the United States – *Clubiona* and *Elaver* (formerly *Clubionoides*); all the rest have been transferred to new families. In retrospect, it is obvious that these disparate groups should have family status. In the case of the Tengellidae, Zorocratidae and Miturgidae, the genera are more closely related to the three-clawed spiders than the two-clawed spider families into to which the other 'clubionoid' families fall. The characters used to separate the original family Clubionidae (including Anyphaenidae and the other clubionoid families) from the related Gnaphosidae (Ground Spiders) include: conical, contiguous spinnerets (Gnaphosids, except for the ant-like genus *Micaria*, have cylindrical spinnerets that are usually separated by at least one spinneret diameter), endites lacking a depression (most gnaphosids have a extensive depression in the ventral surface of the endites and *Micaria* has a median depression), and eyes homogeneous in size and shape (the eyes of most gnaphosids are heterogeneous in size and shape).

The Tengellidae, Zorocratidae and Miturgidae are unusual because of their current placement within the three-clawed spiders, although many (but not all) are actually two-clawed. The loss of the third claw (used primarily for web walking) apparently started to occur among non-web builders fairly early, and appears to have occurred more than once.

Based on genitalic characters, these spiders are not closely related to the other 'clubionoid' families, and Griswold has suggested that the tengellids actually are a sister group of the lycosoids. He also implies that the tengellids may turn out to be only a subfamily of the Miturgidae.

The North American genera now placed in the Tengellidae include *Lauricius* and at least four other genera. At least 50 species may be in this family, with about half of them described. *Lauricius hooki* is a flattened, crab spider-like and fast moving Southwestern representative of the family that is commonly found clinging to the underside of rocks. Another North American genus in this family is *Liocranoides*. The cribellate genus *Zorocrates* is now given a separate family – the Zorocratidae, along with the Madagascar genus *Uduba*. *Zorocrates karli* is a wolf spider-like species commonly found in desert scrub in the southwestern US.

The Miturgidae, as now defined, includes several common and also less-known genera. Some (such as *Calamistrula* in Africa) are cribellate and these, plus a few others (such as *Griswoldia*), are three-clawed.

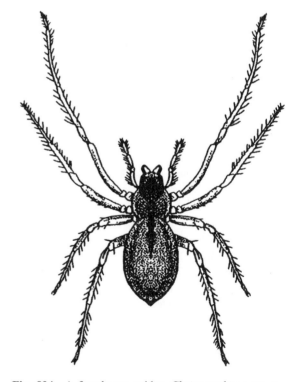

Fig. 894 A female sac spider, *Cheiracanthium punctorium* (Villers).

All of the members of the North American fauna are ecribellate and two-clawed. Several miturgids, such as *Syspira*, resemble wolf spiders. The common sac spider genus *Cheiracanthium* was recently removed from the remains of the Clubionidae and placed in this family. This spider is one of the most studied of the sac spiders in regard to economic impact in agricultural crops. They are known predators of a number of pest species and, like several of the anyphaenids, are known to recognize and eat insect eggs.

The rest of the 'clubionoids' are all two-clawed spiders, apparently more or less related to each other. The first family to be separated from the Clubionidae in most publications was the Anyphaenidae. The members of this family have tracheal spiracles that are from one third to over half the distance from the spinnerets to the epigastric furrow (in most spiders this spiracle is just anterior to the base of the anterior spinnerets.) Another characteristic of the family is the possession of claw tufts of flattened hairs. These are classic sac-building spiders that are probably important members of the natural enemies complex in agroecosystems, especially in tree crops, such as citrus and pecans. They are known to feed on insect eggs and also at extrafloral nectaries, as well as on the more 'normal' moving insect prey, including aphids and other pests. The most numerous anyphaenids in the United States are in the genera *Anyphaena* and *Hibana* (formally *Aysha*).

As noted earlier, the Clubionidae (strictly speaking) is now represented in the United States by only two genera, *Clubiona* and *Elaver*. These common sac spiders are found under rocks and bark, and in sacs inside folded leaves. They have prominent claw tufts on their tarsi.

The Liocranidae are two-clawed, like most of the 'clubionoid' families. This family is probably going to be drastically altered in the near future, as it is analyzed more thoroughly. Many of those initially recognized as liocranids were referred to as phrurolithines in the literature. All of the phrurolithines have multiple macrosetae (spines) on the ventral first tibiae. Phrurolithines are abundant in leaf litter at high elevations in the southwestern United States. North American genera of liocranids, as currently characterized, include *Phrurotimpus*, *Scotinella*, and *Neoanagraphis*. As an example of the fluid taxonomic situation, the phrurolithines, including *Phrurotimpus* and *Scotinella*, were recently transferred by Bosselaers and Jocqué to the Corinnidae.

The Corinnidae is also currently under review. This family includes some ant or velvet ant-like species, as well as some common typical 'sac' spiders in the genera *Trachelas* and *Meriola*. One of the most abundant genera is *Castianeira*. These typically run over the ground much like large ants or female mutillid wasps. Other North American genera in this family are *Mazax* and *Corinna*, with the latter not occurring north of Mexico.

It is sometimes rather disconcerting to the nonspecialist to see so many drastic changes in classification in a common group of organisms. Unfortunately, classification schemes are imperfect because they are based on the data at hand. Whales were once classified with the fishes because it was obvious that, like fishes, they were aquatic, had fins, and were good to eat. The characters of warm-bloodedness, presence of air-breathing lungs, suckling of young, and complicated social interactions resulting from a very unfish-like nervous system were overlooked. Similarly, any 'sac' spider with the eyes in two rows was dumped into the Clubionidae (or even earlier the Drassidae, which also included anyphaenids and gnaphosids). It must be kept in mind that, while more stable than the multitude of common names and classifications, scientific names and classifications are approximations that become (we hope) more accurate as information is added.

Despite the taxonomic muddle of the sac spiders, these creatures are well worth investigating. The probable important roles of sac spiders in the natural enemies complex in agricultural cropping systems is among the best documented of the spiders, although much remains to be done. *Cheiracanthium*, *Hibana*, *Anyphaena* and *Trachelas* are all probably of interest in this regard, as other genera may prove to be in future studies. Several species of *Cheiracanthium*, for example, are common in Egyptian cotton fields, and may be important in controlling various cotton pests. *Hibana velox* has been observed eating the eggs of the sugarcane rootstalk borer weevil, *Diaprepes abbreviatus*, and *Cheiracanthium inclusum* is known to feed on the eggs of soybean loopers.

The life cycle of the miturgid *Cheiracanthium inclusum* (as noted earlier, usually placed in the Clubionidae in the literature) is perhaps the best known of all the sac spiders, primarily because of its agriculture importance. As may be presumed for many, if not most, other sac spiders the cycle is annual, with adults being prominent in the late summer. Adults of the anyphaenid *Hibana incursa*, a common spider in pecan groves in the southwestern U.S., also are found only for a short time in late summer, with immature spiders found throughout the year. During winter they usually are found in silken sacs under bark or in pecan nut hulls. They invade understory plants in the spring, before the pecans leaf out, apparently seeking prey insects, such as aphids.

Of the spiders in these families, only a few are know to have venom that can cause notable effects in humans. All of these are in the genus *Cheiracanthium*, and include both *C. inclusum* and the European species, *C. mildei*. The latter apparently was introduced into the United States a number of years ago. Mild necrotic arachnidism has been reported for both species. North Africa has even larger species of *Cheiracanthium*, but almost nothing is known about their venoms.

David B. Richman
New Mexico State University
Las Cruces, New Mexico, USA

References

Dippenaar-Schoeman, A. S., and R. Jocqué. 1997. *African spiders*. ARC-Plant Protection Research Institute, Pretoria, South Africa.

Griswold, C. E. 1993. Investigations into the phylogeny of the Lycosoid spiders and their kin (Arachnida: Araneae: Lycosoidea). *Smithsonian Contributions to Zoology* 539: 1–39.

Kaston, B. J. 1978. *How to know the spiders* (3rd ed.). McGraw-Hill, Boston, Massachusetts.

Platnick, N. I. 2003. *World Spider Catalog*, version 4.0 http://research.amnh.org/entomology/spiders/catalog81-87/INTRO3.html

Roth, V. D. 1993. *Spider genera of North America* (3rd ed.). American Arachnological Society.

SADDLE.

A chitinous plate on the anal siphon of mosquito larvae.

SADDLE-CASE MAKERS.

Members of the family Glossosomatidae (order Trichoptera). See also, CADDISFLIES.

SAHLBERG, CARL REINHOLD.

Carl Sahlberg was born at Eura, Finland, on January 22, 1779. He entered the Academy in Turku in 1795, and in 1802 obtained the degree of "magister," followed by being appointed the docent of natural history in 1804. In 1810 he obtained a medical license. In 1816 he was named the professor of natural history and economy. After a devastating fire, the University was transferred to Helsinki, and in 1828 he was named professor of zoology and botany. He retired in 1841. Sahlberg did not publish extensively, though the two-volume series "Dissertatio entomologica Fennica enumerans" is noteworthy. He died on October 18, 1860, at Uusinkartano, Finland.

Reference

Herman, L. H. 2001. Sahlberg, Carl Reinhold. *Bulletin of the American Museum of Natural History* 265: 132–133.

SAILER, REECE I.

Reece Sailer was born on November 8, 1915, near Roseville, Illinois, USA. He studied biology at Western Illinois State College from 1934 to 1936, but transferred to the University of Kansas where he obtained his bachelor's degree in entomology in 1938, followed by a Ph.D. in 1942. He worked for a short time as assistant state entomologist in Kansas, then moved to the United States Department of Agriculture, Bureau of Entomology and Plant Quarantine. His principal duties involved identification and research on Hemiptera. A special interest was stink bugs. He also was involved in an extensive and comprehensive assessment of forest fauna treated with DDT, a study that anticipated the eventual problems associated with this popular and widely used insecticide. In 1957 he became Assistant Chief of the Insect Identification Branch. Throughout the 1950s he taught insect ecology at the University of Maryland. He developed a strong interest in biological control, and in 1960 he moved to France to serve as Director of the USDA European Parasite Laboratory. During this time period the laboratory introduced parasitoids of alfalfa weevil, cereal leaf beetle, and European elm bark beetle, as well as predators of balsam woolly adelgid and face fly. In 1966 he returned to Beltsville, Maryland, as Chief of the Insect Identification and Parasite Identification Branch. He retired in 1973 and accepted a graduate research professorship at the University of Florida where he taught biological control and conducted research on citrus whitefly, mole crickets, and other invading pests. Sailer was author of over 100 publications and received numerous honors for his dedicated and effective research and service, including being elected president of the Entomological Society of America. He died on September 8, 1986, while vacationing in Delaware.

Reference

Buckingham, G. R., F. D. Bennett, J. H. Frank, and W. W. Wirth. 1987. Reece I. Sailer 1915 to 1986. *Bulletin of the Entomological Society of America* 33: 1222–23.

SALDIDAE. A family of bugs (order Hemiptera). They sometimes are called shore bugs. See also, BUGS.

SALPINGIDAE. A family of beetles (order Coleoptera). They commonly are known as narrow-waisted bark beetles. See also, BEETLES.

SAMPLE. A collection of sample units from a sampling universe. (contrast with census). See also, SAMPLING ARTHROPODS.

SAMPLE SIZE. The number of sample units collected from the sample universe for a given sampling effort. Ideally, sample size is based on prior information about dispersion and the desired level of precision or accuracy. Sometimes referred to as average sample size, abbreviated ASN.

SAMPLE UNIT. A proportion of the habitable sample universe from which counts of individuals are taken. The sample unit is based on a specific sampling method, should be representative of the behavior and size of the organism, and should strike a balance between cost and variability. Examples include whole leaves, branches or plants, a set number of sweeps along a row, a sticky trap set at canopy height, 10-minute timed count, a 1 m^2 quadrat, etc.

SAMPLE UNIVERSE. The physical area that contains the population of interest. Typical examples include a single crop field, an orchard, a section of stream, the cows in a pasture or barn, etc. Synonymous with the management unit in agricultural production systems.

SAMPLING ARTHROPODS. Sampling is a fundamental component of any experimentally based research program in the discipline of entomology, whether conducted in the laboratory, greenhouse or field. It also is an essential element for describing, measuring and quantifying arthropod population dynamics, whether the goal is to understand the population or community ecology of a given species or group of species, or to develop decision aids for integrated pest management (IPM). Although a population census may be highly desirable for many reasons, rarely does one have the time and resources to exhaustively count every insect, spider or mite within a defined area even if that area is relatively small, say, a small greenhouse or field plot. Instead we must resort to drawing a sample from the population and using the information gained from this sample to estimate variables such as population density or size and to draw inferences that can be applied to the entire population. Fortunately, there is a large body of theory and practical knowledge to guide our efforts in sampling. This short review will attempt to provide a broadly-based, non-mathematical overview of sampling, from discussion of the goals of sampling, to the basic tools for gathering data, to the basic components and mechanics of developing a useful sampling plan.

Samples and sampling are terms with which most entomologists are intuitively familiar. However, as with any field of study, there is a specific vocabulary associated with the study of sampling that may not be as well understood. Thus, brief definitions of some of the more commonly used terms in sampling are provided.

Goals of sampling

The first, and perhaps most important, element in the development of a useful sampling plan or program is to clearly define the goal(s) of sampling. Is the goal to survey for the abundance or presence of one or more insect species over a large geographical region? Or is the goal to intensively study the temporal dynamics and examine rates of mortality of a single insect species in a relatively small area? Perhaps the goal is to develop a simple procedure that a field scout can use to determine the need for remedial pest control in a field crop. A researcher may be interested in testing the effect of a control tactic on populations of a pest insect using replicated field plots. Although some of the information that will be necessary to devise sampling plans is shared among all these and other applications, as we will see below, there are also differences in the sampling methods that may be used and the levels of accuracy and precision that may be required to provide useful data. For example, a quick and simple relative sampling method such as a sweep net or a colored sticky trap may be sufficient for survey detection or for monitoring the relative abundance of one or more insect species over a large region. Here the premium would generally be placed on extensive spatial coverage at the expense of precise estimation of density, which may not be necessary. In contrast, detailed population studies involving the construction of life tables would require an absolute sampling method, such as quadrat or whole plant counts, and fairly large sample sizes so that mortality factors and recruitment to subsequent life stages could be precisely estimated. The development of a decision-aid for pest scouting could use either an absolute or a relative method of sampling depending on the pest and would likely involve the use of a sequential sampling plan that would allow the classification of pest density as either above or below an established economic threshold.

The time and resources that one has to devote to the activity of sampling are almost always in short supply, even in well-funded research programs. Thus, clear delineation of the goal(s) of sampling is the first step towards ensuring that available time and resources will be put to the best use to answer the questions that are being asked of the sample data.

Sampling methods

Many sampling methods have been devised by entomologists for sampling all types of insects, mites and spiders. Each arthropod poses its own unique challenges. In fact, there are probably as many sampling methods as there are entomologists using them. It would be impractical to try and cover all of the available techniques that have been developed. Instead, focus will be placed on relatively few methods that have been widely used to provide a flavor of the diversity. Readers are referred to the excellent and comprehensive treatment of all aspects of sampling provided by Southwood (1978).

For purposes of organizing this discussion, it is convenient to classify sampling methods according to the type of information that they can potentially provide the sampler. These include absolute methods, relative methods, and population indices. Sometimes the difference between an absolute and a relative method is not distinct and a technique that provides an absolute method for one insect will not do so for another in the same habitat. Absolute sampling methods attempt to provide an estimate of density from a specific and quantifiable fraction of the habitat in which the arthropod lives. Generally, these counts have the unit of numbers per square meter or some other unit of ground area or space within structures. Common examples include quadrat sampling in which a square, rectangle or circle of known area is placed on the ground and all arthropods of interest falling within that area are counted *in situ* or the vegetation is collected and later examined in the laboratory. In agricultural, horticultural or rangeland situations the quadrat delineates a volume that may be relatively small if dealing with low-growing grasses or forbs, or relatively large if dealing with corn plants or fruit trees. Quadrats may also be used to sample arthropods within barns, houses or other structures. Often, the quadrat is placed the day before sampling to minimize disturbance. For arthropods that are found only on the plant surface, absolute counts can be made by counting all individuals on a single plant, groups of plants, or even specific parts of plants such as whole leaves, stems, or branches. Sometimes these efforts are aided by the use of various sorts of enclosure devices (e.g., cages or sacks) to confine the arthropods until they can be counted. In many cases, conversion to unit ground area can be accomplished by knowing the number of habitable plants, leaves or branches per unit area. For certain arthropods that are readily dislodged from plants, various kinds of beat cloths and buckets can be used. Suction samplers may provide absolute counts for certain arthropods depending on plant size and growth characteristics. Absolute sampling of soil-dwelling arthropods can be achieved by taking soil cores or digging trenches of a known volume. Surface-dwelling arthropods can be counted directly or collected from the litter within a quadrat. Arthropods can be extracted from collected soil, plant or litter material using a variety of approaches including simple inspection, dry and wet sieving, flotation in concentrated salt solutions, leaf brushing machines, and other methods. Live arthropods can be extracted by chemical fumigation or heat (e.g., Berlese funnel). Emergence traps can be used in some instances to measure the absolute density of adult stages of insects that pupate in or on the soil. Suction traps or aerial nets towed by an airplane or vehicle can be used to obtain absolute samples of flying insects. In this instance the unit is a given volume of air that can be calculated by knowing the volume of air drawn by the motor per unit time (or speed of the vehicle) and other environmental factors such as wind speed.

Fig. 895 Examples of various kinds of sampling methods: A) beat-bucket in which a plant is shaken against the sides of a bucket to dislodge arthropods; B) beat-cloth in which a piece of white cloth is placed on the ground and plants are shaken or beat to dislodge arthropods; C) beat-net (or cloth) in which arthropods are dislodged into a net with a beating stick; D) plant cage that encloses a specific portion of habitat; arthropods are dislodged from the enclosed plant material; E) example of a Berlese funnel in which arthropods are driven from litter or soil into a collection bucket by the application of heat and light from above; F) sweep net; G) D-vac suction sampler; H) high powered suction sampler designed for field row crops; I) trench for sampling soil-dwelling arthropods; J) quadrat; K) visual inspection of a leaf, L) visual inspection of a whole plant. All photographs by the author unless otherwise noted.

Fig. 896 Examples of various kinds of traps for sampling arthropods: A) canopy trap for sampling insects flying vertically out of a crop field; B) emergence trap for sampling adult stages of insects that pupate in the soil; C) window-pane trap for assessing the movement patterns of flying insects; D) modified malaise trap for assessing the movement patterns of flying insects; E) one of a large number of sizes and types of sticky traps; F) suction trap for aerial sampling of flying insects. All photographs by the author unless otherwise noted.

Relative methods are so named because they generally provide estimates of density that are comparable in relation to other estimates made in the same way in the same type of habitat. Relative sampling results in counts per unit of effort and it is not always possible to define the physical units of a method. In contrast with absolute methods, relative methods are generally less costly, easier to perform, and tend to concentrate arthropods. As such they are well-suited to extensive detection and survey work and as components of decision aids in IPM. Relative methods are also widely used in experimental field work where it is often necessary to sample a large number of plots in a short period of time, and only comparative results are sought. In some cases relative estimates can be converted to absolute estimates, but this depends on the specific arthropod and the nature of the sampling method. Common examples of relative methods include the ubiquitous sweep net which is almost the *de facto* standard for estimating densities of arthropods associated with many field crops (e.g., cotton, soybean, small grains, alfalfa). The sweep net can cover a large area of habitat in a short period of time and the particular pattern and number of swings can be standardized across samplers. However, it also samples only a proportion of the population (that inhabiting the tops of plants) and this proportion can change with plant growth, environmental conditions, and time of day. Various methods involve the dislodgement of arthropods from plants into buckets, cloths, nets, trays or other surfaces. Depending on the behavior and biology of the arthropod and the portion of the plant sampled, this may provide either absolute or relative sampling information. In most cases only a small portion of the plant is sampled. The same is true of various suction sampling devices such as the well-known D-vac, modified leaf blowers, or high-powered vacuums attached to carts or tractors. Visual inspections of various plant parts such as leaves or branches are generally considered relative methods, although in many cases, counts on these parts can be related to counts per plant or tree. Timed-counts, which simply involves enumerating all the arthropods that can be counted in a specified period of time are commonly used, especially when the goal it to compare densities of a specific arthropod in a number of different habitats.

By far, traps of various kinds represent the broadest diversity of relative methods. Traps and trapping are the subject of an entry in this volume and so will only be briefly discussed here. Traps can be delineated into those that passively intercept and those that attract actively moving arthropods. Examples of the former include canopy traps, window-pane traps, malaise traps, pitfall traps, and white or clear sticky traps. Examples of the latter include semiochemical- or food-baited traps, traps of various colors (yellow is most common) generally coated with a sticky material, light traps, and sound traps. Mammals and birds also may be used as baits to trap arthropods of medical and veterinary importance. In general, trap counts are very difficult or impossible to convert to absolute counts because they rely on behavior that can be influenced by an array of biological and environmental factors.

Finally, the product or effect of an arthropod population rather than counts of individuals can be used to gauge relative abundance. Common examples of

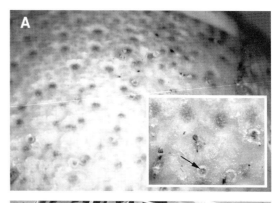

Fig. 897 Examples of methods for estimating population indices: A) counting entrance holes of neonate pink bollworm larvae on the surface of a cotton boll; B) counting honeydew droplets from whiteflies or aphids on water-sensitive paper; C) assessing caterpillar abundance from defoliation. All photographs by the author.

products include the collection of insect frass under-neath plants or trees, the collection of larval or pupal exuviae, or the counting of spider webs. Examples of arthropod effects include the assessment of defolia-tion by foliage-feeding insects, tunneling by stalk-bor-ing insects, or root pruning soil-dwelling insects. Acoustic monitoring can be used to detect the pre-sence and relative abundance of insects inside fruit, grain bins or within the walls of structures. Both pro-ducts and effects can sometimes be related to more quantitative measures of population density. The examination of arthropod effects can be important in their own right in terms of assessing economic damage.

Development of a sampling plan/program

A sampling plan or program is a structured set of rules that guide sampling activities. The sampling plan includes delineation of the sample universe, tim-ing of sampling, the size and nature of the sample unit, how many sample units need to be collected, and how these sample units are spatially allocated among potential sample units in the population. Thus, a sampling plan is distinct from a sampling method or technique (a component of a plan) with which it is sometimes mistakenly confused in the entomological literature. Not all sampling activities in entomology require a sampling plan; for example, collections for systematic or taxonomic purposes. However, if the goal includes the estimation or classification of the density or size of an arthropod population, then the user would be best served by considering, at the very least, selected components of a sample plan to ensure that the sample data are trustworthy. Binns et al. (2000) suggest that a trustworthy sampling plan should adhere to four criteria. First, estimates gener-ated by the plan should be representative of actual pest density thereby avoiding or at least being able to account for any bias. Second, sample estimates should be largely independent of the sampler and of uncontrollable environmental variables such as weather. Third, sample information should be rele-vant to the questions being asked. For example, esti-mates of pest density in a management context should be have an identifiable relationship to crop damage. Finally, the sampling plan must be practical for its intended purpose. For example, it would do little good to develop an elaborate and expensive sampling plan for IPM decision-making that scouts would be unable to implement due to time and resource constraints. These criteria can be met by careful attention to the building blocks of a sample plan.

Sample plan components. Delineation of the sample universe is often dictated directly by the goals of the sampling plan. For example, the population of inter-est for IPM of a particular pest species is generally a single crop field or perhaps a set of crop fields on a single farm that share a similar cultivar and planting date. In a rangeland setting the sample universe might be an entire ranch or perhaps sections or partial sec-tions of the operation. The population of interest in an experimental research program would be a replicate plot representing a given treatment. If the population of interest occurs in an area where physical portions of the habitat are distinct and vary in their suitability for the organism under investigation, several sam-pling universes might be identified or a stratified sampling approach might be employed. This situation will be discussed below.

Likewise, the timing of sampling is often largely dictated by the goals of sampling. Each arthropod species has a distinct phenology and seasonality that will determine the appropriate time to gather samples on the life stage or stages of interest. The timing of occurrence of particular species might be estimated by phenological models, the presence of indicator plants, or historical records. In instances where the timing of occurrence is not well-characterized or highly variable, the use of detection traps may help to determine the need for more intensive sampling by other methods.

Of all the components of a sampling plan, the sam-ple unit is perhaps the most important yet most fre-quently overlooked element. The sample unit defines the unit of habitat represented by the sample and is the foundation for all further development of an efficient sampling plan. Each sample unit should have an equal chance of being selected from the sam-ple universe. It should contain a consistent fraction of the population over time. Thus, definition of the sam-ple unit may include a time component. For example, arthropods may inhabit different parts of a plant over the course of a day due to changes in temperature, light and other factors. The sample unit should be easy to delineate in the sample universe. The sample unit should be appropriately sized relative to the size and behavior of the species in question. For example, a small disk from a leaf may be suitable for spider mites or aphids, but not for large and mobile insects such as assassin bugs or grasshoppers. Finally, the sample unit should strike a balance between variability of counts and the cost of sampling that unit. Using these criteria helps to ensure that the sample unit is representative of the population and help to minimize or eliminate bias.

The sample unit is based on the sampling method from which it is derived. Examples of sample units include 25 sweeps with a sweep net along a single crop row between 0800 and 1000 hours, the lower surface of a leaf five nodes below the terminal apex of a plant, the silks at the tip of one ear of corn, a 8 x 8 cm yellow sticky card oriented horizontally at the top of the crop canopy and exposed for 24 hours, five beats to a tree branch one meter above the ground with a stick over an open 38 cm diameter net, 0.001 cubic meters of soil dug from within a 100 square cm frame, the head or leg of a single cow, or 2 minutes of suction from a D-vac placed over the top third of a crop plant. Very often the sample unit is selected *a priori* based on knowledge of the biology and behavior of the organism and experience, and is not the subject of further investigation. However, careful attention to the size and nature of the sample unit can greatly increase the efficiency of the overall sampling program by minimizing both variability and cost. Established methods are available for comparing various candidate sample units (see Southwood 1978) based on consideration of precision and cost. However, as a guideline, smaller sample units are generally more efficient than larger units. This generalization is based on the typical aggregated dispersion of many arthropod populations and the lower cost of counting individuals on smaller sample units even when a comparatively larger sample size may be required.

Dispersion, or spatial distribution, is a characteristic of populations and can be influenced by a host of biological, behavioral, ecological and environmental factors. Patterns of dispersion are typically categorized as uniform, in which individuals are evenly spaced throughout a habitat, random, in which the location of one individual bears no relation to the location of others, or aggregated (contagious), in which individuals are found in clumps of varying sizes. Many arthropod populations are described as having an aggregated dispersion. Knowledge of dispersion is an important element in the development of sampling plans, influencing both the determination of sample size and the allocation of sample units. The measurement of dispersion is spatially contextual. That is, patterns of dispersion depend on the spatial scale under which they are observed. As a result, the dispersion of a population is greatly influenced by the size of the sample unit. For example a population could be aggregated at the level of a whole plant, but randomly distributed at the level of a single leaf. Thus, in sampling work, dispersion generally refers to the sampling distribution rather than the true spatial distribution. A number of methods are available to explicitly measure dispersion or to provide indices of dispersion. A common index is to simply divide the sample variance by the sample mean. Values < 1, ~ 1 or > 1 indicate uniform, random or aggregated distributions, respectively. For sample plan development it is more typical to characterize dispersion with probability models or empirical mean-variance models. Common probability models include the Poisson for random distributions, and the Negative-binomial for aggregated distributions. The primary limitation of using probability models is that sampling distributions typically vary with changing arthropod density and it is not always possible to account for these changes in the model. Empirical regression models offer a solution to this problem by permitting prediction of sample variance from the sample mean. Such models can provide useful information over a wide range of densities. Common examples include Taylor's power law and Iwao's patchiness regression, both of which can be parameterized with simple linear regression.

Once the sample unit has been selected and the sampling distribution has been characterized by either a probability or empirical model, the sample size can be calculated. How this is done depends on the goals of sampling, the criteria for determining the adequacy of the population estimate or classification, and the resources that can be devoted to sampling activities.

Types of sampling plans. Fixed-sample-size sampling plans, as the name implies, are based on the collection of a set number of sample units. The optimum sample size is the minimum number of sample units that achieves a desired level of performance. Fixed-sample-size sampling can be applied to density classification as well as estimation; however, it is more commonly used for the latter and so consideration will be restricted to estimation here. Several approaches can be used depending on how dispersion is characterized and how precision is quantified. The simplest approach is to calculate the optimum sample size such that the standard error is within a specified proportion of the mean, or that the mean falls within a certain confidence interval of the true mean with a specified probability. Variations on this approach explicitly account for sampling distribution (e.g., Poisson, Negative binomial). A slightly more complicated and robust approach uses a mean-variance model such as Taylor's power law to solve for sample size using the same measures of precision. The main advantage of this approach is that the optimum sample size is dependent on the mean density of the population. Typically, sample size decreases exponentially with increasing density. The reader is referred to Southwood (1978) and Pedigo and Buntin (1994) for mathematical details.

One of the main advantages of a fixed sample size is that sample allocation (see below) can be structured so that sample units are collected from throughout the sample universe. The main disadvantage is that the mean density is unknown for the population under study at any given point in time. One solution would be to calculate sample size for the lowest expected density or the average expected density and then apply that in all cases. This would clearly be inefficient at high densities, or require too many or too few sample units depending on density at any point in time, respectively. The use of double sampling would be an alternative solution. Here a small sample would be collected and used to solve for the required sample size at that point in time.

Sequential sampling is another alternative that solves the problem of unknown density and generally maximizes sampling efficiency. Sequential sampling was developed for military application during World War II and has since become the most widely applied approach for developing sampling plans for arthropods based on both density estimation and density classification. Efficiency is optimized because in sequential sampling the need for further sample information is assessed following the collection of each individual sample unit. Regardless of the mean density, sequential sampling ensures that no more sample units are collected than necessary in order to achieve a predetermined level of precision or classification accuracy. Sequential sampling for estimation of mean density operates by accumulating counts over subsequent sample units and then consulting a 'stop-line'. This stop-line represents the cumulative count as a function of sample size and the desired precision. It can be plotted or presented in table form. Matching the current cumulative count and sample size relative to the stop-line determines whether more sample units are need or whether sampling should be terminated. Once sampling is terminated the mean is calculated by simply dividing the cumulative count by the number of sample units collected. Sequential sampling plans for mean estimation can be developed on the basis of probability or mean-variance models. Frequently, a minimum and maximum sample size is specified as part of the sequential plan. The reader is referred to Nyrop and Binns (1991) and Pedigo and Buntin (1994) for mathematical details.

A second application of sequential sampling involves the classification of population density. Most often this approach is applied to the development of decision-aids for IPM where one simply wants to know whether pest density is above or below the economic threshold. Operationally, sequential sampling

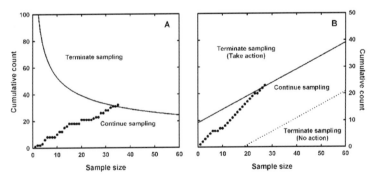

Fig. 898 Hypothetical examples of A) sequential sampling plan for estimating mean density with a fixed precision and B) sequential sampling plan for classifying mean density relative to an economic threshold. The lines denote the stop-lines and the data points represent a hypothetical sample that terminates after crossing the stop-line.

for classification is similar to that described above for mean estimation except that a critical density (e.g., economic threshold) and upper and lower decision boundaries must be specified. The calculation of the stop-lines is also more complex. Several approaches have been developed. One goes by the daunting name of the sequential probability ratio test (abbreviated SPRT) and can be based on various probability models including the Poisson, negative-binomial, normal or binomial. A second technique termed Iwao's confidence interval method is based on the normal distribution, and a third called the sequential interval procedure is based on the binomial distribution. All three approaches result in pairs of stop lines rather than a single one, delineating three distinct decision zones. More complex plans can be developed involving multiple stop lines designating even finer decision zones. Such sampling plans are popular for use in IPM because of their high degree of efficiency. Very few sample units are required when densities are much smaller or larger than the critical density and the maximum number of sample units is required only when the density is very near the critical density. Sequential sampling plans for classification are developed and evaluated on the basis of two criteria, the operating characteristic (OC) and the average sample number (ASN). The OC estimates the probability of classifying density below the critical density as a function of the true mean density. Various parameters of the sequential plan can be adjusted to increase or decrease the level of classification accuracy and the associated sample size.

Up to this point the discussion has focused on enumerative sampling in which all individuals inhabiting a sample unit are counted. However, depending on program goals, the general efficiency of sampling can be potentially improved by determining only the presence or absence of individuals in the sample unit. This approach, referred to as binomial sampling has been widely used in the development of decision aids for IPM and is typically implemented as a sequential classification sampling plan as described above. Generally, the foundation for this approach is the development of a relationship between mean density and the proportion of sample units containing at least T individuals. T is known as the tally threshold and traditionally has a value of 1 (true presence/absence); however, it can take on any value and proper selection of T can often improve the accuracy of classification. This mean density-presence/absence relationship can be determined using various probability models or it can be represented by one of several empirical models. In some instances it is more convenient to base the economic threshold simply on binomial counts and, thus, there is no need to develop a relationship with mean density. A well-devised and tested binomial sampling plan can significantly reduce the cost of sampling without sacrificing accuracy of density classification in IPM decision-making. Binomial sampling can also be an efficient approach for estimating population density, although its application for this purpose has been limited. The reader is referred to Nyrop and Binns (1991), Pedigo and Buntin (1994), and Binns et al. (2000) for mathematical details on sequential sampling for classification using both enumerative and binomial sampling.

A final sampling approach, termed variable-intensity sampling, is a classification method that shares some of the attributes of both fixed-sample-size and sequential sampling plans. It was developed primarily to address a potential deficiency in sequential sampling having to do with adequate coverage of the sample universe. As a result of its high efficiency, sequential sampling often terminates sampling after only a small portion of the sampling universe is observed. Variable intensity sampling solves this problem by delineating a transect (see allocation below) across the sample universe, dividing that transect into equal lengths, and requiring at least one sample unit in each section. Thus, the sample is more representative, especially in situations where pest density might be highly variable across a field. The overall number of sample units taken in each transect section ultimately depends on prior sampling information much as in a traditional sequential sampling plan. Variable intensity sampling can be based on enumerative or binomial counts.

Allocation of sampling units. The final element of a sampling plan is the collection of sample units from the sample universe. Random sampling assumes that every possible sample unit within the sample universe has an equal chance of being selected. As such, random sampling ensures that the sample will provide unbiased estimates (assuming that the proper sample unit is used and sample size is adequate). Random sampling can be accomplished by enumerating every sample unit and then selecting them using a random number table, drawing numbers from a hat,

Definitions of some common terms associated with arthropod sampling.

Absolute method	A sampling method that allows the estimation of numbers per unit of habitat, generally per unit of ground area.
Accuracy	A measure of the closeness of an estimate to the true mean or variance of a population.
Bias	An unidirectional deviation of an estimate from the true mean or variance of a population.
Binomial sampling (presence/absence sampling)	Sampling based on determining the presence or absence of one or more individuals in the sample unit in lieu of complete counting of all organisms. Commonly used for pest management decision-making application. Contrast with enumeration sampling.
Census	Complete enumeration of every individual within a defined sample universe. Contrast with sample.
Classification	A sampling plan that classifies population density as being either above or below some predetermined level (e.g., economic threshold), or belonging within some density class (e.g., low, medium, high). Commonly used in pest management decision making application. Contrast with estimation.
Dispersion/distribution	The spatial patterning of individuals in a population in their habitat. This pattern can be broadly described as uniform, random or, most commonly, aggregated. Dispersion can be quantified by explicit spatial indices, or more typically by probability models (e.g., Poisson, Negative-binomial) or empirical models (e.g., Taylor's power law). The measurement of dispersion is dependent on the sample unit size and population density. Most often the term is used to denote the sampling dispersion/distribution and not true spatial patterning. Dispersion is an important element in the development of a sampling plan.
Double sampling	A sampling approach in which an initial sample (usually small) is drawn and used to determined the necessary sample size for a subsequent sample within the same time period.
Efficiency	A measure of the level of precision or accuracy per unit of cost (time or currency).
Enumeration sampling	Sampling based on the complete counting of all individuals in the sample unit. Contrast with binomial sampling.
Estimation	A sampling plan that numerically estimates population density or intensity. Commonly used for detailed population dynamic and experimental studies. Contrast with classification. The term can also denote the process of calculating various statistical parameters such as variance.
Fixed-sample size sampling plan	A sampling plan in which a predetermined number of sample units is collected based on a prescribed level of desired precision, classification accuracy, or level of confidence. Contrast with sequential sampling.
Mean-variance model	A model, generally a regression model, which predicts the variance of a sample from an estimate of the sample mean. Such models can characterize dispersion over a large range of population densities and are commonly used in developing sampling plans. Common examples are Taylor's power law and Iwao's patchiness regression.
Operating characteristic	A measure of the accuracy of a classification sample. An operating characteristic curve shows the probability of classifying density below a critical value (e.g., economic threshold) as a function of true mean density. Ideally, probabilities should be near 1 at densities far below the critical value, near 0.5 at densities very near the critical value and near 0 at densities far above the critical value. Often abbreviated as OC.
Population index	A sampling method that attempts to provide an indirect estimate of population density based on an associated product (e.g., frass) or effect (e.g., defoliation).
Precision	A statistical measure of the repeatability of an estimate relative to a group of estimates from the same population at the same time. Typically measured as the quotient of the standard error of the mean over the mean. Low numerical values indicate high precision, high numerical values indicate low precision. Precision is a key element in developing and evaluating the performance of sampling plans.

Probability model	A mathematical description of the dispersion or distribution of individuals in a population based on numbers per sample unit. Common models include Poisson, Negative-binomial, Binomial, Normal, and Neyman Type A. Such models can form the foundation of a sampling plan.
Random sampling	A method of allocating sampling units within a sampling universe in which each sample unit has an equal chance of being selected. Random sampling ensures an unbiased estimate, but is rarely used in the practice of arthropod sampling due to time and cost constraints.
Relative method	A sampling method that results in numbers per unit effort. Generally much faster and easier than absolute methods. Sometimes possible to convert counts to absolute density. Common examples are sweep nets and sticky traps.
Robust	Within statistics and sampling, the quality of being widely applicable. For example, a sampling plan is robust if it can be used to precisely estimate the density of an arthropod under many different environmental conditions.
Sample	A collection of sample units from a sampling universe. Contrast with census.
Sampling method/technique	A particular tool or technique used to gather information on population density from the sample universe. Examples include sweep nets, beat clothes, visual counts, suction devices, and various kinds of attractive and passive traps.
Sampling plan/program	A structured set of rules for collecting a sample that is based on delineation of the sample universe, knowledge of dispersion, a specific sample unit, a predetermined sample size (but see sequential sampling), time of sampling, and a given allocation of sample units throughout the sample universe (e.g., random, stratified, systematic).
Sample size	The number of sample units collected from the sample universe for a given sampling effort. Ideally, sample size is based on prior information about dispersion and the desired level of precision or accuracy. Sometimes referred to as average sample size, abbreviated ASN.
Sample unit	A proportion of the habitable sample universe from which counts of individuals are taken. The sample unit is based on a specific sampling method, should be representative of the behavior and size of the organism, and should strike a balance between cost and variability. Examples include whole leaves, branches or plants, a set number of sweeps along a row, a sticky trap set at canopy height, 10-minute timed count, a 1 m^2 quadrat, etc.
Sample universe	The physical area that contains the population of interest. Typical examples include a single crop field, an orchard, a section of stream, the cows in a pasture or barn, etc. Synonymous with the management unit in agricultural production systems.
Sequential sampling plan	A sampling plan in which "stop-lines" continually assess the need for additional sample units during the sampling effort on a given day or time interval. These stop lines are typically based on a prescribed level of desired precision or classification accuracy. Sequential sampling is efficient because no prior knowledge of density is required and no more sample units than necessary are collected. Contrast with fixed-sample size sampling.
Stratified sampling	A method of allocating sampling units in which the sample universe is subdivided into 2 or more sections and sample units are randomly collected from each subdivision in proportion to their size or relative variability. Examples include the subdivision of a crop field into border and interior sections, the subdivision of a plant along vertical strata, or different rooms within a structure.
Systematic sampling	A method of allocating sample units within a sample universe in which sample units are collected at fixed intervals along a predetermined pattern. Examples include sampling along transects or along V or X shaped patterns in crop fields. Often, the starting point of the pattern is determined randomly. Systematic sampling is probably the most common method for allocating sample units in most agricultural systems because it is simple and time-efficient.
Tally threshold	In binomial sampling, the number of individuals required to be present per sample unit to consider the sample unit infested. Proper selection of the tally threshold can improve the accuracy of classification sampling.
Variable-intensity sampling plan	A sampling plan that shares the characteristics of fixed-sample size and sequential sampling plans in which prior sampling information is used to evaluate the number and allocation of subsequent sample units, but ensures that sample units are collected throughout the sample universe.

or using a computer random-number generator. This assumes that the spatial location of each sample unit can be correctly identified which may be extremely difficult. In field crops, orchards or greenhouses with evenly spaced plants, sample unit location can be simplified somewhat by defining a coordinate system and randomly selecting a random row and a random number of steps or plants along the row from the edge. A variation of random sampling is stratified sampling. This approach may be used if there is some a priori knowledge that the population being sampled is heterogeneously distributed. Common examples are the differential distribution of insects in border rows compared with interior rows of a crop field, the differential distribution of insects within the vertical strata of a plant, or differences in distributions of arthropods within structures. In instances such as these a more precise estimate may be achieved by allocating sample units to different strata in proportion to the size of each strata, or in relation to variability within each strata. More sample units would be collected in strata that are known to be more variable and vice versa. The allocation of sample units within a stratum is random, preserving the benefits of random sampling, but different formulae are needed to estimate mean density and variance.

Despite the desirable qualities of random sampling, it is rarely used in practice mainly because proper implementation is costly. It is much more common for sampling plans to be implemented using some sort of systematic sampling. In systematic sampling, some predetermined path is chosen and sample units are collected at fixed intervals along this path. Typical examples are diagonal transects, or X-, V-, or Z-shaped patterns laid out across the sample universe. Aside from being easy and inexpensive to implement, systematic sampling generally allows sample units to be collected from throughout the entire sample universe, resulting in a more representative sample. A degree of randomness can be added by randomly selecting the starting point of the transect and the interval between sample units. A disadvantage to systematic sampling is that it can produce biased estimates if there is some underlying pattern in the population that closely mimics the pattern of sample unit collection. Further, there are no exact formulae for calculation of variance and so estimates based on random sampling are often assumed. The overall effect of these limitations is not well-known, however, proper testing and evaluation (see below) can help to ensure that the sample obtained is reliable and provides the necessary levels of precision and accuracy.

Testing and evaluation of sample plans. Ideally, sampling plans should be developed from observations that encompass the geographic area and range of environmental conditions that future users of the sampling plan are likely to encounter. In reality, however, sampling plans are often developed from a more restricted range of observations and then applied to similar or novel situations. In addition, regardless of the extent of data collection, any sampling plan is based on observations that are measured with some amount of error. Thus, it is important that the performance of a sampling plan be evaluated so that its limitations and strengths can be better defined. This evaluation process is known as validation. In the past, generally little attention was paid to sample plan validation, however, this situation is changing and several validation approaches have been formalized in the last 10–15 years. The simplest involves the use of Monte Carlo simulation. In this method the sampling plan is applied to sample units counts that are randomly drawn on a computer from a probability distribution (e.g., Poisson, Normal, Negative binomial) that is representative of the sampling distribution of the arthropod in question. This procedure is repeated a large number of times (usually 500 or more) to represent a large number of possible sampling outcomes and average precision or classification accuracy (and their variances) are calculated. An alternative approach uses real sample data independently collected at the time of sample plan development or collected from an area where the sample plan is to be implemented. These real data are then resampled on a computer much as in the Monte Carlo approach. The advantage of the resampling method is that the data rather than a probability distribution are used to represent the sampling distribution. The result is a more robust test of the sampling plan and the assumed probability model upon which it may be based. A disadvantage is that larger amounts of data collection are required. With both approaches, the validation data can have an explicit spatial arrangement so that issues of sample unit allocation can be tested as well. The reader is referred to Naranjo and Hutchison (1997) and Binns et al. (2000) for further detail and available computer software.

Steven E. Naranjo
USDA-ARS, Western Cotton Research Laboratory
Phoenix, Arizona, USA

References

Binns, M. R., J. P. Nyrop, and W. van der Werf. 2000. *Sampling and monitoring in crop protection: the theoretical basis for developing practical decision guides.* CABI Publishing, Wallingford, United Kingdom.

Cochran, W. G. 1977. *Sampling techniques* (3rd ed.). John Wiley and Sons, New York, New York.

Flint, M. L., and P. Gouveia. 2001. IPM in practice: principles and methods of integrated pest management. University of California, Statewide Integrated Pest Management Project, Agriculture and Natural Resources Publication 3418.

Morris, R. F. 1960. Sampling insect populations. *Annual Review of Entomology* 5: 243–264.

Naranjo, S. E., and W. D. Hutchison. 1997. Validation of arthropod sampling plans using a resampling approach: software and analysis. *American Entomologist* 43: 48–57.

Nyrop, J. P., and M. Binns. 1991. Quantitative methods for designing and analyzing sampling programs for use in pest management. pp. 67–132 in D. Pimentel (ed.), *Integrated pest management.* CRC Press, Boca Raton, Florida.

Pedigo, L. P., and G. D. Buntin (eds.) 1994. *Handbook of sampling methods for arthropods in agriculture.* CRC Press, Boca Raton, Florida.

Southwood, T. R. E. 1978. *Ecological methods: with particular reference to the study of insect populations* (2nd ed.). Chapman and Hall, London, United Kingdom.

SAMPLING PLAN. A structured set of rules for collecting a sample that is based on delineation of the sample universe, knowledge of dispersion, a specific sample unit, a predetermined sample size (but see sequential sampling), time of sampling, and a given allocation of sample units throughout the sample universe (e.g., random, stratified, systematic).

SANDERSON, DWIGHT. Dwight Sanderson was born at Clio, Michigan, on September 25, 1878. He earned B.S. degrees from both Michigan State University and Cornell University. He then worked in entomology in Maryland, Delaware, and at Texas A&M and the University of New Hampshire. At New Hampshire he became Director of the Agricultural Experiment Station and was elected president of the American Association of Economic Entomologists. In 1910 he became dean of the College of Agriculture at West Virginia University and then director of the experiment station. He was awarded a doctorate in sociology from the University of Chicago in 1921. Sanderson was interested in insect develop-ment and in insect pest control. He authored "Insects injurious to staple crops" (1902), and "Insect pests of farm, garden, and orchard" (1911). He was a founder of the Journal of Economic Entomology and was instrumental in getting the Federal Insecticide Act of 1910 passed by the United States Congress. He also wrote four books on rural sociology. He died at Ithaca, New York, on September 27, 1944.

Reference

Mallis, A. 1971. *American entomologists.* Rutgers University Press, New Brunswick, New Jersey. 549 pp.

SAND FLY. A name applied to various small biting flies (Diptera), including members of the family Psychodidae, Simuliidae, and Ceratopogonidae, though this name refers most correctly to members of the family Psychodidae. See also, FLIES.

SANITATION. The practice of eliminating pests by removal or destruction of materials or sites that might harbor pests.

SAP BEETLES (COLEOPTERA: NITIDULIDAE). The family Nitidulidae belongs to the superfamily Cucujoidea in the suborder Polyphaga of the order Coleoptera (beetles).

Order: Coleoptera

Suborder: Polyphaga

Superfamily: Cucujoidea

Family: Nitidulidae

Six subfamilies of Nitidulidae are recognized, although the classification at both the generic and subfamily levels is still in a state of flux. One subfamily, Brachypterinae, was recently elevated to family status. The subfamily Cybocephalinae is often placed in its own family Cybocephalidae, and sometimes Cateretidae, Brachypteridae, Cybocephalidae and Smicripidae are included in the Nitidulidae.

The Nitidulidae are a small family of mostly small beetles. There are 165 species in North America and about 3000 described worldwide, although many more remain undescribed.

Adults

The length of adult Nitidulidae ranges from about 1 mm to about 12 mm. Bodies are often depressed, usually oval shaped, but some elongate. The adults often bear red or yellow spots. The head is

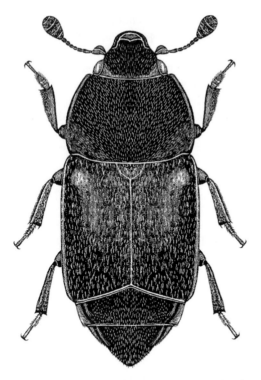

Fig. 899 Adult dusky sap beetle, *Carpophilus lugubris*.

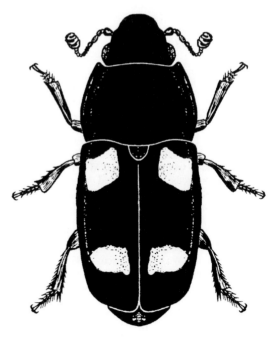

Fig. 900 Adult fourspotted sap beetle, *Glischrochilus quadrisignatus*.

prognathous. The antennae are 11-segmented, with the apical three segments forming a club. The antenna is often received in a groove on the underside of the head. The tarsi are five-segmented on all legs, but with only four segments in the Cybocephalinae.

Larvae

The larvae are elongate, and 2 to 20 mm long, but most less than 12 mm, parallel sided and frequently flattened. In many larvae, the mandibles bear a well-developed prostheca. Most larvae with paired, fixed appendages (urogomphi), but they are lacking in the Meligethinae and Cybocephalinae. The spiracles are two openings at the end of tubes (except in Cybocephalinae). The pupae are exarate.

Food

Nitidulidae display a wide range of feeding habits, though they are generally considered to be saprophagous and mycetophagous. As the common name (sap beetles) implies, many (especially *Carpophilus* and *Glischrochilus*) are found at sap flows on injured trees. Cybocephalinae are predatory on scale insects, and *Glischrochilus* have been reported to be predators of wood-boring larvae. Many species (e.g., *Pallodes*, *Pocadius*, *Thalycra*, *Cychramus*, *Aphenolia*) feed on fungi. Pollen and nectar-feeding occur in the *Meligethes*, most *Conotelus* and some *Carpophilus*. *Nitidula* and *Omosita* are carrion feeders and *Amphotis* are associated with ant colonies at least part of the year.

Economic importance

Aethina tumida is a serious pest in bee hives. *Carpophilus dimidiatus* is a minor pest of stored grains, and *C. lugubris* is a pest of sweet corn. Several species in the genera *Carpophilus*, *Glischrochilus*, *Epuraea* and *Colopterus* are involved in transmitting the pathogenic fungus causing oak wilt.

See also, BEETLES.

Dale Habeck
University of Florida
Gainesville, Florida, USA

References

Habeck, D. H. 2002. Nitidulidae. In R. H. Arnett and M. C. Thomas (eds.), *American beetles*, Vol. 2. CRC Press, Boca Raton, Florida.

Lawrence, J. F. 1991. Nitidulidae (Cucujoidea) including Brachypteridae, Cateretidae, Cybocephalidae, Smicripidae. pp. 456–460 in F. W. Stehr (ed.), *Immature insects*, Vol. 2. Kendall/Hunt, Dubuque, Iowa.

Lawrence, J. F., and Newton, Jr. 1995. Families and subfamilies of Coleoptera (with selected genera, notes, references and data on family-group names). pp. 779–1006 in J. Pakaluk and S. A. Slipinski (eds.), *Biology, phylogeny, and classification of Coleoptera.* Museum: Instytut zoologii PAN, Warsaw, Poland.

Parson, C. T. 1943. A revision of Nearctic Nitidulidae (Coleoptera). *Bulletin of the Museum of Comparative Zoology* 92: 121–278.

SAPROPHAGOUS. Feeding on dead or dying plant or animal tissue.

SAPYGIDAE. A family of wasps (order Hymenoptera). See also, WASPS, ANTS, BEES, AND SAWFLIES.

SARCOPHAGIDAE. A family of flies (order Diptera). They commonly are known as dung flies. See also, FLIES.

SARCOPTIC MANGE. A disease caused by mites of the genus *Sarcoptes*; scabies.

SARCOSOMES. Large mitochondria found in muscle, especially thoracic muscle.

SATURNIIDAE. A family of moths (order Lepidoptera). They commonly are known as giant silkworm moths, emperor moths, or royal moths. See also, EMPEROR MOTHS, BUTTERFLIES AND MOTHS.

SATYRS. Some members of the family Nymphalidae, subfamily Satyrinae (order Lepidoptera). See also, BUSH-FOOTED BUTTERFLIES, BUTTERFLIES AND MOTHS.

SAUCER BUGS. Members of the family Naucoridae (order Hemiptera). See also, BUGS.

SAUNDERS, WILLIAM. William Saunders was born in England in 1820, and moved from Devonshire to Canada in 1832, where he became known as a pioneer in horticulture and entomology, and Canada's first economic entomologist. Though lacking much formal education he received some technical training in chemistry and became a druggist in London, Ontario. However, he loved nature and took to the study of plants and insects. He became a specialist in medicinal plants, and became well-known as a supplier of quality products to the medical profession. He also became a professor at the University of Western Ontario. Saunders was co-founder (with C.J.S. Bethune) of the Entomological Society of Canada in 1863, and also co-founder of the Canadian Entomologist, in 1868. Interestingly, the co-founders were also the only two authors of the first two editions of this journal! Saunders managed the journal for five years, ending in 1886 when he moved to Ottawa. In 1883 he published an important treatise, "Insects injurious to fruit." He was also appointed Director of Experimental Farms of the Dominion in Ottawa, in 1886, and helped establish the first experimental farm. He died on September 13, 1914, in London, Ontario.

Reference

Mallis, A. 1971. *American entomologists.* Rutgers University Press, New Brunswick, New Jersey. 549 pp.

SAUSSURE, HENRI LOUIS FREDERIC DE. Henri Saussure was born at Geneva, Switzerland, on November 27, 1829. A noted specialist in Hymenoptera and Orthoptera, Saussure initially studied social wasps, but is best known for work on the grasshoppers. He received his education in Switzerland and Paris, and also studied with the entomologist François Jules Pictet de la Rive. Saussure traveled to the West Indies, Mexico and the United States, returning to Europe with valuable collections of animals. His interests were quite broad, and in 1858 he founded the Geographical Society of Geneva. In Geneva, he was also associated with the Natural History Museum of Geneva. Saussure died at Geneva on February 20, 1905.

Reference

Essig, E. O. 1931. *A history of entomology.* The Macmillan Company, New York. 1029 pp.

SAWFLIES (HYMENOPTERA: SYMPHYTA). Sawflies are members of the order Hymenoptera, suborder Symphyta. Adult sawflies can be distinguished from the other members of that order (Suborder Apocrita: the parasitic Hymenoptera, wasps, bees, and ants) by their saw-like ovipositor, lack of constriction at the base of the abdomen, fly-like appearance and more extensive wing venation. Sawfly larvae are eruciform (caterpillar-like) and phytophagous (except members of the family Orussidae) rather than grub-like and parasitic or predatory. They are often confused with lepidopteran larvae, but have numerous distinguishing characteristics, including six or more pairs of abdominal prolegs, a lack of chrochets on the abdominal prolegs, one lateral ocellus on each side of the head, and distinctly segmented maxillary and labial palpi. Furthermore, sawfly larvae generally lack long hairs, spines, or tubercules, common characteristics of many lepidopteran larvae.

There are approximately 8,000 species of sawflies worldwide, with most of the diversity found in the Northern Hemisphere. Diversity appears to be relatively low in the tropics. However, this may reflect a lack of significant sampling and study of sawflies in this region. Most sawflies, whose hosts are woody plants, are external leaf or needle feeders, and are relatively monophagous. In northern forest ecosystems, sawflies are the second most dominant group of defoliators behind the Lepidoptera. A number of introduced species have major pest status and many can periodically cause severe localized damage. Because studies of this group have not been extensive, they are not as well known to the public and it is likely that sawfly damage has often been attributed to other groups, particularly the Lepidoptera.

There are thirteen families within the Symphyta. The Diprionidae and the Tenthredinidae contain most of the economically important species, with the latter family being by far the most speciose group. The table

Family	Approximate No. of species	Distribution	Hosts / Feeding habits
Xyelidae	75	Holarctic	Pollen of *Pinus*, bud and twig borers of *Abies*, external leaf feeders on *Ulmus* and *Carya*.
Pamphiliidae	300	Holarctic	Web-spinning and leaf-rolling species on hardwoods, conifers, and shrubs.
Megalodontidae	50	Palearctic	Web spinners on herbaceous plants only.
Pergidae	500	Worldwide, most abundant in neotropics, dominant in Australia	Known hosts include species of *Eucalyptus, Quercus, Carya,* and *Nothofagus*. Hosts and habits are unknown for most species.
Argidae	800	Worldwide, most abundant in neotropics, dominant in Africa	All genera but one feed on angiospermous trees, shrubs, and herbaceous plants. One Australian genus feeds on *Cupressus*.
Diprionidae	125	Holarctic, with extensions into the northern tropics	Feed exclusively on conifers, primarily *Pinus*.
Cimbicidae	150	Holarctic	Defoliators of woody plants.
Blasticotomidae	6	Palearctic	Stem borers of ferns.
Tenthredinidae	5000	Worldwide, rare in Australia	Many subfamilies, feed on a variety of plants including horsetails, ferns, grasses, sedges, hardwoods, and conifers.
Xiphydriidae	100	Worldwide, most diverse in sub-tropical Asia	Wood borers of hardwoods.
Siricidae	100	Mostly holarctic, some tropical	Wood borers of conifers and hardwoods.
Cephidae	100	Holarctic	Twig and stem borers of grasses, serious grain crop pests.
Orussidae	75	Worldwide, most diverse in tropics and subtropics	Only non-phytophagous group of Symphyta, parasitic on wood borers, especially Coleoptera.

describes the species diversity, distribution, major hosts, and feeding habits of each sawfly family.

Adult sawflies are sexually mature upon emergence. Following mating, females of most species use their saw-like ovipositor to create a slit in the host plant tissue, into which they insert their eggs. Like other Hymenoptera, sawflies have haplodiploid sex determination, in which unfertilized eggs become males and fertilized eggs become females. Larvae of many species feed gregariously, particularly during the early instars. In most species, male larvae pass through five instars, females six. Mature larvae that have finished feeding usually drop off from the host plant, burrow into leaf litter or soil, and enter dormancy as a pre-pupae, often within a cocoon. For most species, the pre-pupae is the overwintering stage, and pupation occurs in the spring.

There is considerable variation in life history and voltinism (number of generations per year) within the Symphyta. Voltinism can even vary within a species depending on latitude and within a population depending on local microclimate. Most species in the Northern Hemisphere are univoltine (one generation per year), with an increase in multivoltinism towards the tropics and semivoltinism (more than one year required to complete life cycle) towards the poles. The red-headed pine sawfly, for example, has one to five generations per year throughout its range from Ontario, Canada to Florida, respectively.

A handful of sawfly species have major economic importance, particularly those that have been introduced from other regions. Some important conifer pests that were introduced into North America from Europe include the European pine sawfly, *Neodiprion sertifer*, the introduced pine sawfly, *Diprion similis*, the European spruce sawfly, *Gilpinia hercyniae*, and two European strains of the larch sawfly, *Pristiphora erichsonii*. Other native species in North America that frequently plague conifer plantations are the red-headed pine sawfly, *Neodiprion lecontei*, and the yellow-headed spruce sawfly, *Pikonema alaskensis*. The European woodwasp, *Sirex noctilio*, along with a symbiotic fungus, has been a substantial problem in the Southern Hemisphere where it was introduced into the highly profitable radiata pine plantations of Australia. It poses a similar threat to pine plantations in other regions such as South America. It is likely that new sawfly pests will emerge as species continue to be introduced into new areas in the absence of their natural enemies.

See also, WASPS, ANTS, BEES, AND SAWFLIES, and SAWFLIES (HYMENOPTERA: TENTHREDINIDAE).

Christopher Asaro
University of Georgia
Athens, Georgia, USA

References

Coulson, R. N., and J. A. Witter. 1984. *Forest entomology: ecology and management*. John Wiley and Sons, New York, New York.

Drooz, A. T. 1985. *Insects of eastern forests*. USDA, Forest Service, Miscellaneous Publication No. 1426.

Furniss, R. L., and V. M. Carolin. 1977. *Western forest insects*. USDA, Forest Service, Miscellaneous Publication No. 1339.

Smith, D. R. 1993. Systematics, life history, and distribution of sawflies. pp. 3–32 in M. R. Wagner and K. F. Raffa (eds.), *Sawfly life history adaptations to woody plants*. Academic Press, San Diego, California.

Wilson, L. F., R. C. Wilkinson, and R. C. Averill. 1992. *Redheaded pine sawfly: its ecology and management*. USDA, Forest Service, Agriculture Handbook No. 694.

SAWFLIES (HYMENOPTERA: TENTHREDINIDAE).

The Tenthredinidae belong to the Hymenoptera, suborder Symphyta. 'Sawflies' is a generic term generally designating the superfamily Tenthredinoidea. In this section, the term 'tenthredinid' will be used for clarity.

Systematics and geographic distribution

The superfamily comprises the Argidae, Blasticotomidae, Cimbicidae, Diprionidae, Pergidae and Tenthredinidae. The oldest fossil record of the tenthredinids is from the Early Cretaceous, whereas the other families evolved much later, namely from the Middle Tertiary.

Class: Insecta

Order: Hymenoptera

Superfamily: Tenthredinoidea

Family: Tenthredinidae

Over 5,000 tenthredinid species are described worldwide and distributed approximately in 430 genera. Major genera that contain over 50 species are *Amauronematus* (approximately 170 species described), *Athalia* (80), *Dolerus* (200), *Nematus* (240), *Macrophya* (170), *Pachyprotasis* (110),

Pristiphora (180), *Tenthredo* (700), *Tenthredopsis* (60). Today, seven or sometimes eight subfamilies are recognized, as follows: Allantinae (represented approximately by 110 genera; e.g., *Athalia*, *Eriocampa*), Blennocampinae (105; e.g., *Blennocampa, Monophadnus, Periclista*), Heterarthrinae (40; e.g., *Caliroa, Fenusa, Profenusa*), Nematinae (55; e.g., *Cladius, Euura, Hemichroa, Hoplocampa, Nematus, Phyllocolpa, Pontania, Priophorus, Pristiphora, Trichiocampus*), Selandriinae (including Dolerinae; 75; e.g., *Dolerus*), Susaninae (1), and Tenthredininae (50; e.g., *Macrophya, Pachyprotasis, Tenthredo, Tenthredopsis*).

Tenthredinids occur as native species on almost all continents, being absent in Antarctica and New Zealand, and rare in Australia and smaller islands. They are the most diverse and abundant in the temperate zones of the northern hemisphere, preferring humid biotopes. However, some subfamilies are the most diverse in (sub)tropical zones (Selandriinae, Tenthredininae), while the Nematinae are virtually the only sawflies in the subarctic zone. The geographic distribution of tenthredinids is limited by the fact that adults are not strong fliers and by the distribution of the larval host plant. In contrast, the distribution of some species has been expanded in the last centuries by accidental introductions, mainly from Europe to North America.

The monophyly of the tenthredinids remains under debate since no unambiguous apomorphies could be found so far.

Morphology

Adults vary in size from 2.5 to 15 mm. Their color pattern consists of black as well as pale orange, yellow or green (that fade to straw coloration in dead specimens). Most tenthredinids can be identified by a combination of the following characters: antennae filiform and composed of 9 flagellomeres; thorax with mesoscutellar appendage distinct; pronotum markedly concave along the posterior margin; foretibiae with two apical spurs, hind tibiae without subapical spurs; forewing with subcosta absent or as cross-vein; ovipositor saw-shaped (and that gave the name 'sawflies' to the group). The subfamilies are separated mainly on the basis of wing morphology.

Tenthredinid larvae are eruciform, meaning subcylindrical in shape and resembling caterpillars. The body is depressed only in some species. The larva possesses 7 or 8 pairs of pseudopods, or prolegs,

on abdominal segments 2 to 8 and 10 (in most subfamilies), 2 to 7 and 10 (Nematinae), or 2 to 8 (e.g., *Caliroa*), except in the mining taxa where pseudopods are reduced or absent. A tenthredinid larva can be distinguished easily from that of a lepidopteran since the latter never possess more than 5 pairs of pseudopods, and never a pair on the second abdominal segment. The abdomen of tenthredinid larvae consists of 10 segments, each one composed of 3 to 7 annulets. The head capsule is well developed and antennae are short with 3 to 5 segments. The nervous, tracheal, muscular, circulatory and reproductive systems are roughly similar and comparable to other insects, whereas the digestive and excretory systems, fat body and integument show interspecific differences of systematic interest.

Life history and habits

The biology of the Tenthredinidae is highly diversified, of course by comparing one species with another. However, all species of a genus or subfamily can show a similar biological trait. Examples of such taxa are given in brackets from now on throughout the text. Another preliminary remark is that general characteristics are given, while exceptions usually are omitted.

The emergence of adults is strongly linked in non-(sub)tropical regions to the phenology of the host plant (of the larvae, see later). In holarctic regions, Dolerinae and many *Amauronematus* emerge early, at the very beginning of the spring, but most tenthredinid species occur as adults in May and June.

After its molt from pupal to adult stage, the tenthredinid will remain during 1 to 7 days in the cocoon. Newly emerged sawflies will drink rain or dew droplets. Later, they can feed on honeydew, but they feed most typically on pollen and nectar, thus helping somewhat in pollination, whereas several species are carnivorous (*Tenthredo*) and some others do not feed at all. On a still, sunny day, one might commonly find adults, especially on flowers of the Apiaceae, Ranunculaceae, Rosaceae, Saxifragaceae, Compositae, and on catkins of Salicaceae, Betulaceae, etc. Tenthredinid adults generally do not fly very much, and usually only for a short distance.

In captivity, females live 1 to 2 weeks, males only a few days. The effective occurrence of adults in the field might spread over weeks or months. Male emerge somewhat earlier than females. Individuals of both sexes will find each other by optical and

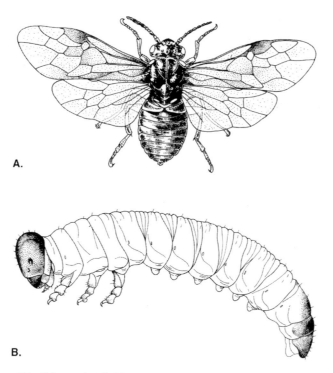

Fig. 901   Adult (a) and larva (b) of the tenthredinid *Hoplocampa flava* (drawings by Marylise Leclercq).

chemical cues. During copulation, lasting generally 1 to 2 minutes, the two bodies point in opposite directions, facing away from each other (i.e., stophandry).

Tenthredinids reproduce sexually or asexually. Parthenogenesis is a typical and frequent feature. It is generally either facultative (arrhenotokous, i.e., unfertilized eggs produce males, whereas fertilized eggs produce females), or obligatory (thelytokous, i.e., eggs always produce females, males being unknown). For arrhenotokous species, males can be obviously absent or rare in given populations and geographic areas. In most tenthredinid species, the number of chromosomes is n = 7 to 10, extreme values being n = 5 to 22. A trend seems to be that the chromosomes become more numerous during evolution.

Normally a female possesses from 40 to 250 eggs. Eggs that will be laid are generally already mature in the emerging female. Egg sizes range from 0.4 to 3.9 mm in length, and from 0.2 to 0.9 mm in width. The female cuts with a saw-like ovipositor such plant tissue as a petiole, leaf vein, or fertilized flower. Then, eggs are deposited in the created slit, more often singly than in small clusters. In the particular case of leaf galling species, the female will lay eggs and concur-

rently a secretion that will start the development of the leaf gall (later on, the larva will continue to stimulate this process). In a few other species, it has been observed that females lay their eggs on plants different from the larval host plant. Once deposited, the egg will mature (into an embryo) by which its volume will increase, swelling, thereby, the surrounding tissue. Egg eclosion will occur 1 to 3 weeks after eggs are laid, depending on temperature; eggs also are particularly sensitive to a moisture deficiency.

All tenthredinid larvae feed on plant material. Their host plants are horsetails and ferns (Pteridophyta), resinous trees (Coniferae) as well as flowering plants (Angiosperms), namely herbs, shrubs and deciduous trees, but mosses, Compositae, Lamiaceae and Apiaceae are rarely attacked. Compared to other herbivorous insects, tenthredinid larvae have a relatively narrow diet breadth; most species are 'specialists' feeding on one given plant genus or even one plant species. Exceptions to this rule are found among the Tenthredininae. The host range observed today in tenthredinid species suggests, to some extent, a radiation of the insects parallel to the evolution of their host-plants.

Larvae generally feed freely on leaves, but some species groups are leaf miners (Heterarthrinae), leaf rollers (*Phyllocolpa*, *Blennocampa*), leaf gallers (*Pontania*), shoot gallers (some *Euura*), fruit borers (*Hoplocampa*), etc. Within the Nematinae, it is likely that from free-feeding larvae evolved the habit to live more and more deeply in the plant (tissue), namely from leaf rollers through leaf gallers to shoot gallers.

There are 4 to 8 larval instars, males showing often one instar fewer than females. Larval growth can be best followed, beside weight, by measuring the head capsule width. This will allow the instar to be determined. The feeding larval stages together require approximately 10 to 30 days. This period and the feeding rhythm are influenced by species, sex, climatic conditions, etc.

The location of a larva feeding freely on a leaf can be the leaf edge (Nematinae). The leaf edge generally corresponds to the external border of the leaf, but it can be an internal border when a larva makes a hole on the edge of which the larva will continue to feed. Furthermore, the location of a larva can be on one of both sides of a leaf. The larva then either only feeds on the parenchyma (*Caliroa*) or from the edge of the leaf while it remains with its body on the underside of the leaf (*Trichiocampus*). A case of acoustic communication has been reported in the gregarious species *Hemichroa crocea*. By scratching the leaf surface with the abdomen, the larvae of a group remain together and are informed on food quality.

The last larval molt leads to a so-called prepupa. This instar either feeds (most Nematinae), or not (Blennocampinae). The prepupa can be different in color and external morphology compared to the previous instars (Blennocampinae, Tenthredininae). It will drop off and crawl into the soil (litter) in order to build a cocoon. Some species not going into the soil will bore into a fruit or bark of a plant that is, generally, not the food plant, and some species (in the Tenthredininae, Selandriinae) do not spin a cocoon. The prepupa undergoes an apolysis leading (without a molt) to a form that is somewhat retracted. This is generally the resting, overwintering stage. Then, at the end of the winter the prepupa molts into the pupa that is an instar lasting only a few weeks.

Most species have one generation per year (univoltine), but they can have up to three or four. Conversely, within one species, a part of a population can overwinter more than one winter. Most northern species enter diapause that then needs to be broken by a specific temperature and by light.

The influence of abiotic parameters on the life cycle of sawflies can be listed as follows: light (photoperiod) on diapause; light, relative humidity and air pressure on emergence timing and on activity of adults; temperature on development time of eggs and larvae as well as on larval and pupal weight. Moreover, larva and pupa are highly sensitive to dryness, but a very moist soil will enhance risk for the (pre)pupa to be attacked by fungi and micro-organisms.

Natural enemies and defense strategies

Numerous arthropods are known to prey upon tenthredinids, as follows: Acari (on the egg stage); Insecta: Hemiptera, Pentatomidae (larva); Cimicidae (egg); Thysanoptera (egg, larva); Dermaptera (larva); Coleoptera, Elateridae (prepupa and pupa); Mecoptera (adult); Diptera, Asilidae (adult); Hymenoptera, Tenthredinidae (adult), Vespidae (larva), Formicidae (larva). Spiders seem to avoid tenthredinids due to the occurrence of chemical defenses (see later).

Birds such as tits, woodpeckers, cuckoos, partridges, chaffinches and starlings have been observed preying sometimes heavily on tenthredinid larvae. Avian predation seems to be especially important in the regulation of prey populations when these populations are at low densities. Mammals such as shrews and voles are well known to destroy overwintering prepupae and pupae, and are important as well in the regulation of populations. Deleterious compounds originating from a toxic host plant and which the feeding larva previously sequestered can deter shrews from attacking tenthredinid cocoons. Amphibians and reptiles, which can feed on tenthredinid adults and larvae, are probably less important predators than birds and mammals.

Many tenthredinid adults drop off when disturbed. Then they feign death for a few seconds with the antennae and legs pressed to the body as at the pupal stage. On the contrary, other species are rather aggressive, and can use their strong mandibles to bite an aggressor (for instance, your finger) as well as predators. Most tenthredinid adults are cryptic in appearance, but some species do mimic dangerous insects (e.g., *Tenthredo vespa* resembling a wasp).

Deposition within the plant tissue protects tenthredinid eggs. One tenthredinid species is known to defend the eggs after oviposition: the female of

Pachynematus itoi remains with stretched wings upon her cluster of eggs until their hatch.

Larvae living within plant tissue are protected mechanically against some predators, but when attacked, their escape capabilities are of course reduced. In contrast, some free-feeding larvae can drop off the plant when disturbed. The stimuli eliciting this escape behavior are visual or mechanical (by vibration, touching). On the ground, they will often curl their body so that the abdomen protects the head. A regurgitated droplet can be emitted simultaneously (*Dolerus*).

The visual appearance of the larvae is generally one of the following patterns. Most free-living larvae are greenish (cryptic), resembling the leaf color, whereas others are brightly colored (e.g., black head and black spots and lines on a yellowish background of the body). To these opposite trends are linked a different level of gregariousness, larvae being solitary or gregarious, respectively.

The integument of tenthredinid larvae often presents adaptations acting as a form of protection (that is passive) and/or a defense (that is linked to a given behavior). The larvae of most Nematinae species raise up their abdomen once disturbed. Gregarious larvae can perform such movements of the body simultaneously, which can visually frighten birds at distance. The movement of a single larva can deter an approaching parasitoid. Larvae can also dislodge, using their abdomen, an attacking predator. If Nematinae larvae with an already raised up abdomen are continued to be disturbed, then ventral glands of the abdomen can be evaginated. The glandular secretion will evaporate and will be able to repel arthropods such as ants. In some Nematinae species, the ventral glands are developed enough to allow a chemical protection against birds. Larvae can be covered by developed setae (*Cladius*) or spines (*Periclista*), by a wax secretion (*Eriocampa*) or a slimy liquid (*Caliroa*). The hemolymph of many tenthredinid species is distasteful to ants. Larvae can sequester in their hemolymph toxins originating from plants. The hemolymph can be released by bleeding, spitting, or upon wounding.

A majority of tenthredinid parasitoids belong to the Ichneumonidae (Hymenoptera), which are believed to be often specialized on one host species. Tenthredinids are also frequently parasitized by Chalcidoidea (Hymenoptera) and Tachinidae (Diptera). Most parasitoid species attack the tenthredinid at the larval stage, but some species are either egg, or cocoon (i.e., prepupa) parasitoids. The parasitoid female lays one or several egg(s) per tenthredinid individual. The egg is laid either on the body surface or directly in the body of the larva. When a parasitoid egg is in the body of a larva, hemolymph cells of the latter can encapsulate the parasitoid egg, killing it. This is a form of counter-adaptation of the sawflies that become, consequently, resistant to a parasitoid.

Tenthredinids can be attacked by Nematodes (Mermetidae) as well as by fungal, bacterial, protozoan and viral agents. Pathogenic micro-organisms are unknown from the galling sawflies *Pontania* and *Euura*, so that inhabiting the plant tissue might constitute a protection towards diseases (and more generally to natural enemies such as parasitoids).

Economic importance

Tenthredinid pests are especially important as forest defoliators. Pests are known on deciduous trees (e.g., *Fenusa* on birch, Nematinae species on willow, poplar, birch), their economic effect being probably stronger on conifers (e.g., *Pristiphora abietina* on spruce, *P. erichsonii* on larch). However, there exist only a few real outbreak species among tenthredinids. On the other hand, tenthredinid species are sometimes regional pests in orchards (*Hoplocampa* spp. on apple, plum and pear; *Nematus ribesii* and *Pristiphora appendiculata* (= *pallipes*) on red-current and gooseberry), in agriculture (e.g., *Athalia rosae* on several Brassicaceae such as turnip) and on ornamental plants (e.g., *Blennocampa phyllocolpa* (= *pusilla*) on rose, *Pristiphora rufipes* (= *aquilegiae*) on columbine).

Tenthredinids can be beneficial. Monophagous species are used sometimes as biological control agents of weeds (e.g., *Monophadnus spinolae* introduced in New-Zealand and feeding on *Clematis recta* that is invasive there; *Priophorus brullei* (= *morio*) in Hawaii on *Rubus*). At a local level, tenthredinid richness and/or the occurrence of given 'indicator' species can inform us on the biodiversity of a studied biotope.

See also, WASPS, ANTS, BEES, AND SAWFLIES, and SAWFLIES (HYMENOPTERA: SYMPHYTA).

Jean-Luc Boevé
Royal Belgian Institute of Natural Sciences
Brussels, Belgium

References

Benson, R. B. 1950. An introduction to the natural history of British sawflies (Hymenoptera Symphyta). *Transactions of the Society for British Entomology* 10: 45–142.

Goulet, H. 1992. *The insects and arachnids of Canada*. Part 20. The genera and subgenera of the sawflies of Canada and Alaska: Hymenoptera: Symphyta. Agriculture Canada Publication #1876. 235 pp.

Lacourt, J. 1999. Répertoire des Tenthredinidae ouest-paléarctiques (Hymenoptera, Symphyta). *Mémoires de la Société Entomologique de France no. 3*. Paris, France. 432 pp.

Lorenz, H., and M. Kraus. 1957. *Die Larvalsystematik der Blattwespen (Tenthredinoidea und Megalodontoidea)*. Akademie Verlag, Berlin, Germany. 339 pp.

Schedl, W. 1991. *Hymenoptera Unterordnung Symphyta. Handbuch der Zoologie. Band IV Arthropoda: Insecta*. Walter de Gruyter, Berlin, Germany. 136 pp.

Smith, D. R. 1979. Suborder Symphyta. pp. 3–137 in K.V. Krombein, P.D. Hurd, Jr., D.R. Smith, and B.D. Burks (eds.), *Catalog of Hymenoptera in America North of Mexico, Vol. 1*. Smithsonian Institution Press, Washington, DC.

Taeger, A., and S. M. Blank (eds.) 1998. *Pflanzenwespen Deutschlands (Hymenoptera, Symphyta) Kommentierte Bestandsaufnahme*. Verlag Goecke & Evers, Keltern, Germany. 364 pp.

Viitasaari, M. 2002. *Sawflies (Hymenoptera, Symphyta). I. A review of the suborder, the Western Palearctic taxa of Xyeloidea and Pamphilioidea*. Tremex Press, Helsinki, Finland. 516 pp.

Wagner, M., and K. F. Raffa (eds.) 1993. *Sawfly life history adaptations to woody plants*. Academic Press, New York. 581 pp.

SAWTOOTHED GRAIN BEETLE. See also, STORED GRAIN AND FLOUR INSECTS.

SAY STINK BUG. See also, WHEAT PESTS AND THEIR MANAGEMENT.

SAY, THOMAS. Thomas Say was born at Philadelphia, Pennsylvania, USA, on July 27, 1787, and is considered by most to be the 'father of American entomology,' and the first great American systematist. Others argue that F. V. Melsheimer was the father of American entomology, but none dispute that Say was the first American of equal stature with the European entomologists of the era. He was educated as a pharmacist in Philadelphia, where his father was an apothecary and physician. He lacked interest in medicine, however, favoring natural history. In 1812 he became a member of the Academy of Natural Sciences of Philadelphia, and became influenced by this enthusiastic group of naturalists who were in the very early stages of building a fabulous museum collection. The Academy became his home as well after a business failure, and he slept and ate at the Academy. Say began to publish in 1817. He participated in collecting trips to the Georgia and Florida coasts, the Rocky Mountain region, and western Canada, eventually taking up residence at a utopian Quaker commune in New Harmony, Indiana. At New Harmony, Say became superintendent for literature, science and education for the community. Say is well known to economic entomologists because he described many of the economically important pests found in the United States. He published a 3-volume treatise, ''American Entomology,'' during the period of 1817 to 1828, as well as numerous papers. He was a member of all the important entomological societies. Say is also credited with being the 'father of American conchology,' publishing six papers on the subject while living at New Harmony. Say died in New Harmony, Indiana, on October 10, 1834.

References

Essig, E. O. 1931. *A history of entomology*. The Macmillan Company, New York. 1029 pp.

Mallis, A. 1971. *American entomologists*. Rutgers University Press, New Brunswick, New Jersey. 549 pp.

SCABIES. A disease caused by mites of the genus *Sarcoptes*; sarcoptic mange. See also, MITES.

SCALE. A modified, flattened seta on the surface of an insect.

SCALE AND HIERARCHY IN INTEGRATED PEST MANAGEMENT. Integrated pest management (IPM) has been defined as a control system that, in the context of the associated environment and the population dynamics of the pest species, utilizes all suitable techniques and methods in the most compatible manner possible and maintains pest populations at levels below those causing economic injury. Conceptually, IPM operates at the interface of two multidimensional systems: the ecological and the socioeconomic systems, which are hierarchically ordered in ascending levels of complexity and expanding spatial scales.

From a socioeconomic standpoint, IPM schemes are designed to maximize the benefits from control measures, minimize adverse environmental impact and maximize social benefits. Presumably, these objectives are easier to meet on industrialized agricultural holdings than on subsistence farms. African agriculture is characterized by a large majority of small-scale farmers who cultivate small landholdings of less than one hectare to a few hectares and practice intercropping in horticulture-based production systems. Often, farmers do not have titles to the land, and ownership of the harvest is not restricted to the ones cultivating it, which is important when it comes to implementing IPM.

From an ecological standpoint, IPM schemes integrate compatible pest control methods by considering the population dynamics of a pest and its interaction with the commodity. Even the most basic level of population dynamics involves the problem of multiple scales. The problem of multiple scales permeates the study of ecological process and pattern, uniting aspects of space, time and organizational complexity.

Concepts of scale and hierarchy

Scale has numerous definitions in the ecological literature. One common definition is that it denotes the resolution within the range of a measured quantity. Scale, therefore, is the unit measure (resolution) relative to the largest multiple (range). This definition applies to space, time and mass. Scale differs from hierarchical levels in that it is recorded as a quantity and involves (or at least implies) measurements and measurement units. Scale-dependent patterns can be defined as a change in some measure of pattern with change in either the resolution or the range of measurement. Scale also refers to power law relations between variables. Specifically, a quantity Q scales with another measured quantity Y (e.g., area, duration, mass) as follows.

$$\frac{Q}{Q_{ref}} = \left(\frac{Y}{Y_{ref}}\right)^{\beta} \qquad [1a]$$

Hence

$$Q = (Q_{ref}Y_{ref}^{-\beta})\,Y^{\beta} \qquad [1b]$$

The exponent of these power laws can sometimes be obtained by dimensional analysis. In ecology, power laws are typically obtained empirically, by regressing the logarithm of Q against the logarithm of Y. Examples of power law scaling in ecology include body size allometry (Y = mass), species area curves (Y = area), and Taylor's Power Law (Y = mean density).

Like the word 'scale,' hierarchy has multiple meanings in ecology. One common definition is that it refers to discrete levels in hierarchically organized systems, where levels are distinguished by rates that differ by one or more orders and magnitude. Rate differences cause structural differences among levels. Hierarchical organization means that higher level processes can steer and constrain lower level processes, while, at the same time, higher level features might emerge from lower level dynamics. Hierarchical organization can be described a two step procedure: top-down across two or three levels of organization at a time by analysis, then bottom up across the same levels by synthesis. Also, natural laws do not apply equally at all scales, but are formulated at particular levels of space, time and complexity. At each level of organization phenomena exist that require new laws and principles which still cannot be predicted from those at more fundamental levels. As one proceeds away from those levels at which a law was cast, the explanatory capability wanes. Hierarchy theory takes into account the problem of aggregation errors by identifying levels of nested systems within larger systems. From this perspective, it is natural to analyze data across multiple scales to seek discontinuities that define changes in the levels. Translating hierarchical concepts into equations that can be used in applied ecology has proven difficult. Hierarchical levels are useful in labelling different levels of human organization as it affects ecosystems including agroecosystems where IPM applies.

Application of scale and hierarchy concepts to IPM

Human institutions and their economic activities introduce levels of field, farm and village and beyond. Clearly, the hierarchy levels are identified on organizational rather than observational ground. Often, the IPM literature refers to the hierarchical organization of decision-making in IPM ranging from farmers to extension agents and policy makers. To simplify the treatment of the topic, this discussion is restricted to a single-species pest population and does not elaborate on its interaction with the commodity or with social institutions and policy.

Recognize that there is a connection between temporal and spatial scales. Nevertheless, for the sake of simplicity, they are treated separately.

An adequate understanding of spatio-temporal dynamics is a prerequisite for planning and undertaking control measures within an IPM framework. This understanding is often obtained by the analysis of age-structured populations changing in time and space, via statistical inference or mathematical modelling. Both methods require scale considerations. Statistical inference in IPM relies on spatial design considerations of plot size and layout, which introduce considerations of spatial scale. As shown below, changes in the sample unit size and in the distribution of the samples in the sampling universe also require scaling considerations. Inference from a sample to a larger area introduces questions of scale that bear on population density estimation used for decision-making in threshold-based IPM.

Spatial scales

In homogenous fields, there is no standing pattern in underlying physio-chemical environment, individual sites may vary from one another at any particular time due to history, but will exhibit similar statistical features over long periods of time. For example, the overall dynamics of an acarine predator-prey system in a greenhouse is composed of local dynamics. Homogeneous sampling universes enable the use of simple random sampling procedures, i.e., a method of selecting units so that every one of the samples has an equal chance of being drawn. In heterogeneous sampling universes it is advantageous to use stratified sampling procedures and work with subpopulations having known sampling probabilities.

The design of a sampling plan requires the definition of the physical size of the sampling unit and recommends cost-benefit procedures. There are significant advantages of a large number of samples with small units over a small number with large units. Also, sample size should be chosen according to the nature of the organisms, and with consideration of the objective of the study.

Homogeneous sampling universe. The range of measurement, or sampling universe, is given by the dimension of a field. In weed control, there is no doubt of the importance of field-level management, while this is not always true for other pests. Many invertebrates show few differences between field

and farms, and seemed to be changing evenly across the whole area in response to factors operating at higher levels.

The design of efficient agricultural experiments requires minimizing residual variability in order to detect small effects with the least effort. To overcome spatial heterogeneity, the efficiency of many small plots versus a few big plots needs to be assessed. The number of samples required to meet reliability criteria depends on the spatial distribution of the pest population. Frequency distributions, or variance δ^2 to mean μ relationships, are generally used for the design of sampling plans. Selected here as an example is Taylor's power law even though other descriptions as well as frequency distributions could be used with equal justification. According to this law, the variance in density of organisms $\delta^2 (\mu)$ scales as a power law of the mean μ.

$$\delta^2(\mu) = a\mu^b \qquad [2]$$

The parameter a be definition depends on the units of μ and is thus largely, possibly wholly, a sampling or computing factor. The scaling exponent b ranges typically from 2 to 4 and was interpreted as an index of aggregation. There are publications with the explicit or implicit assumption of a constant species-specific b. In simulation studies, however, b varies with the sample unit size. This result has important implications both for the design of sampling plans and for the interpretation of spatial distributions. Namely, b describes a consistent relationship between sample variance and sample mean over a range of densities, on a spatial scale connected to the sample unit size. This is consistent with observation that different sampling unit sizes detect different spatial patterns.

The dependence of Taylor's Power law on sample unit size can be estimated from data or by deriving it from assumptions regarding population processes and models as well as frequency distributions. A diffusion-limited aggregation model to derive [3] and yields:

$$\delta^2(q) = \delta^2(1)q^b \qquad [3]$$

That is, the variance for a unit size quadrate q is scaled up by a factor q^b when the quadrate size is increased by the factor q. Similar scaling relations describe fractal symmetry, then from this proposes that the exponent b can be interpreted as a fractal dimension.

Heterogeneous sampling universe. Consider the case of an inhomogeneous field and the extension of the range of measurement to neighboring fields. In statistical terminology, the measurement is strata and, therefore, reliance is on stratified sampling procedures for population estimation. Sawyer and Haynes use [1] to represent the spatial distribution of cereal leaf beetle trap catches in each stratum (field) and for the region, with mean density v

$$\delta^2(v) = a_s v^b \qquad [4]$$

These authors recognize that the intercept contains strata-specific information. However, Sawyer and Haynes stress the need to validate the assumption of constant proportions, e.g., density invariant, distribution of individuals in the different strata. Stratification addresses discontinuities in the sampling universe. To the degree that it is effective, it indicates a shift from the homogeneous to heterogeneous distributions. Stratification can be based on physical conditions (e.g., soil moisture). It also can be based on imposed levels of organization of land use practice at the scale of the field, farm and village.

Spatial range extension

A wide spatial range is addressed when pest movements and control operations require that density estimates are done on village levels. Mass trapping is an efficient technique for the control of the savannah tsetse flies *Glossina pallidipes* and *G. morsitans submorsitans*. The minimum range for control operations covers an area of several km², and typically includes many farms and villages. The scale is related to the area of attraction of an odor-baited trap that may attract flies within a 50 m to 100 m radius. For monitoring purposes, traps are deployed on a 0.25×1.0 km grid. Control operations are carried out with 4 to 12 traps per km², depending on the fly density. In both cases, the sampling universe is assumed to be homogeneous.

In a highly structured sampling universe, subsampling procedures may be appropriate for density estimation. Thereby, a relatively small scale sample is cast over two or more levels or stages. Subsampling refers to the fact that a unit on a given level is not measured completely but is itself sampled.

Temporal scale

A temporal range covering a crop growing period may often be too short for a complete pest generation to develop. Often, however, the crop may be exposed to either single generation or even multi-generation infestation patterns. In any case, the infestation patterns influence the temporal scale for density estimates.

Models on the dynamics of physiologically structured populations often represent infestation patterns adequately. For example, the outcome of weed-crop plant interactions and thus, yield loss, is determined by temporal coincidence: a scale of 12 hours may affect yield.

Temporal range extension

The temporal range is extended to cover a series of growing periods at the field, or at higher levels. For example, the immature life stages of the European cockchafer *Melolontha melolontha*, develop in one or more crops during an often 3-year period, and soil samples with surface of 0.5×0.5 m and adequate depths are taken at the field level. The adults operate in areas of several km², hence, adult flight patterns are studied during the time of adult occurrence, at the level of villages.

In general, the extension of the temporal range and frequent measurements allow the use of spectral analyses, where the measure of association, plotted against the frequency of measurement, depends on the time scale used. Hence, the results obtained by spectral analyses are specific to the time scale used, and for generalization purposes, should be completed by scaling up work.

Mass scale

In many cases, the amount of food consumed by a pest can be linked to yield losses. In a small range of pest organism masses, it may be considered as proportional to the body mass, while scaling up for higher body mass may require nonlinear functions. Likewise, respiration costs do not increase proportionally to an increase in individual body mass. Often, the relationship

$$r(m) = r(1)m^c \qquad [5]$$

is used. That is, the respiration r for a unit mass organism is scaled up by a factor m^c. This is important when explaining the population dynamics on the basis of the supply-demand theory.

Concluding remarks

The IPM literature often refers to concepts of hierarchy and scale, but does not rely on a coherent terminology adopted in this paper. Often, the IPM literature refers to spatial levels of fields, farms and villages and beyond as well as to levels in hierarchical decision-making organization, ranging from farmers to policy makers. Presumably, this is because human institutions and economic considerations allow a convenient organizational identification of levels. There are strong indications, however, that these levels may not have the same importance in pest population dynamics, hence their relevance for studies and management should be supported by data. Scale related problems including the implications of scale selection have received little attention in IPM, but we expect an increasing interest in the treatment of scaling and in the design of field experiments. These studies might confirm that, from an ecological standpoint, scaling functions (variants of equation [1]) are easier to define from data than hierarchies.

IPM schemes remain often narrowly designed for field-specific measures and do not appear to fully consider the ecological knowledge obtained in the past decade. When addressing hierarchies, bottom-up approaches receive more attention than top-down approaches. This could be corrected by putting IPM into a landscape context. This approach could complement field- and species-specific IPM programs and become useful when designing IPM schemes at the community and ecosystem level. When dealing with landscape concepts, however, the distinction between the level of organization and the level of observation should be taken into account.

Johann Baumgärtner
International Centre of Insect Physiology and Ecology (ICIPE) Addis Ababa, Ethiopia
and
David C. Schneider
Memorial University of Newfoundland,
St. John's, Newfoundland, Canada

References

Allen, T. F. H., and T. B. Starr. 1982. *Hierarchy*. University of Chicago Press, Chicago, Illinois.
Kogan, M., B. A. Croft, and R. F. Sutherst. 1999. Applications of ecology for integrated pest management. pp. 681–736 in C. B. Huffaker and A. P. Gutierrez (eds.), *Ecological entomology (2nd ed.)*. Wiley, New York, New York.
Levin, S. A. 2000. Multiple scales and the maintenance of biodiversity. *Ecosystems* 3: 498–506.
Sawyer, A. J., and D. L. Haynes. 1978. Allocating limited sampling resources for estimating regional populations of overwintering cereal leaf beetles. *Environmental Entomology* 7: 63–66.
Schneider, D. C. 1994. Quantitative ecology, spatial and temporal scaling. Academic Press, San Diego, California.
Schneider, D.C. 1998. Applied scaling theory. pp. 253–269 in D. L. Peterson and V. T. Parker (eds.), *Ecological scale: theory and applications*. Columbia University Press, New York, New York.
Schneider, D. C. 2001. Scale, concept and effects of. pp. 245–254 in S. A. Levin (ed.), *Encyclopaedia of biodiversity*. Academic Press, San Diego, California.

SCALED FLEAS. Members of the family Leptopsyllidae (order Siphonaptera). See also, FLEAS

SCALE INSECTS AND MEALYBUGS (HEMIPTERA: COCCOIDEA).

Scale insects, including mealybugs, comprise a diverse group of insects within the superfamily Coccoidea in the suborder Sternorrhyncha containing more than 7,500 species worldwide. Scale insects initially were called coccids, a name derived from *Coccus* meaning berry. These small, often cryptic plant-feeding insects are considered pests with several species capable of inflicting substantial economic damage. Scale insects are distinguished from other hemipterans by several morphological and behavioral traits. The adult females are always wingless and usually have a powdery covering or waxy test protecting their body. Their mouthparts arise between the front coxae. When legs are present, they have one (rarely two) segmented tarsi with a claw. The first instars always are mobile, while later instars often are sedentary. Most females deposit their eggs either beneath their tests or in ovisacs. The adult males may be winged or wingless. When present, a pair of wings is located on the mesothorax and a pair of hamulohalteres often occurs on the metathorax. The head of the male is distinguished by two or more pairs of various types of eyes and non-functional mouthparts. Because females are present longer throughout the year, the taxonomy of the group is based on their morphology.

The scale insects usually are arranged in 22 families worldwide that include: the Aclerdidae (grass scales), Asterolecaniidae (pit scales), Beesoniidae,

Carayonemidae, Cerococcidae (ornate pit scales), Coccidae (soft scales or coccids), Conchaspididae (false armored scales), Cryptococcidae (bark crevice scales), Dactylopiidae (cochineal scales), Diaspididae (armored scales), Eriococcidae (felted scales), Halimococcidae, Kermesidae (gall-like scales), Kerriidae (lac scales), Lecanodiaspididae (false pit scales), Margarodidae (giant scales), Micrococcidae, Ortheziidae (ensign scales), Phenacoleachiidae, Phoenicococcidae (date scales), Pseudococcidae (mealybugs), and Stictococcidae. Six families (Electrococcidae, Grimaldiellidae, Inkaidae, Jersicoccidae, Kukaspididae, and Labiococcidae) are represented by extinct species found in amber. In the U.S., 856 species are known representing 16 families. The families Diaspididae, Pseudococcidae, Coccidae, and Asterolecaniidae, respectively, contain the highest numbers of species. Adult females in the more primitive families often have well-developed legs providing them more mobility than adult females in the more advanced families that often are sessile.

Economic impact of scale insects

Many species are important pests of fruit, nut, forest and ornamental trees, greenhouse and house plants, as well as several agricultural crops. Losses due to plant death and increased production costs are estimated to exceed five billion dollars annually worldwide with losses in North America approaching 500 million dollars annually. The most economically important infestations of scale insects are in cultivated plantings, nurseries, greenhouses, and other urbanized areas where ecological disturbances have occurred. These minute scale insects often go unnoticed on the branches and leaves until a heavy infestation is present, which results in loss of plant vigor, dieback, leaf drop, fruit damage, or eventual death of the host plant. Over an extended period, the tests or scale covers from successive generations encrust the stems and limbs that not only damage the plant, but diminish their aesthetic appeal. Some scale insects can injure the plant surface by injecting their needle-like stylets into the tissue, producing open wounds, which may allow plant pathogens to enter into the plant. For example, damage caused from feeding by the beech scale is believed to be responsible for the outbreak of beech bark disease that now threatens the beech groves in the Great Smoky Mountains National Park. In addition, several species of mealybugs are known vectors of pathogens that may lead to the death of the plant.

In some instances, the saliva injected during feeding may be toxic to the host plant resulting in gall formation, deformed areas, or discoloration around the feeding site. Species of several families (e.g., Eriococcidae, Asterolecaniidae, and Beesoniidae) induce gall formation on their hosts. Where the mouthparts are inserted, normal growth of the host is disrupted and eventually the tissue forms a closed or open gall around the scale. Some scales (mostly in the Australian region) develop inside these galls on leaves, stems, and twigs. Those species that develop within closed galls often exhibit morphological changes that may include an enlarged head and more posteriorly located thoracic spiracles, with the spiracles in Beesoniidae positioned near the anal opening. Species of the gall-forming genus *Apiomorpha* develop an elongated and constricted abdomen. Open galls in the form of slight depressions or pits are induced by species of Asterolecaniidae and Lecanodiaspididae. While golden oak scales produce distinctive open galls on oaks, galls formed by sweetgum scales appear as reddish to whitish specks on the ventral surface of the leaf of sweetgum. Damage around the feeding sites often is manifested in the formation of chlorotic spots. The feeding activity by San Jose scales may produce discolorations in the form of red rings on apples, while Florida red scales produce light circular areas on citrus. The birch margarodid is found embedded within the bark of birch with only a hollow, waxy tube-like filament linking it to the external environment.

Secondary injury to the plant host by most soft scales and mealybugs results from exuding a liquid excretion (often referred to as honeydew), which is a by-product of the large volume of sap they ingest. The excreted honeydew that falls onto the lower leaves and branches serves as a substrate for the development of sooty mold, which inhibits growth, lowers fruit yields and sugar content, and reduces the aesthetic value of the host plant. Honeydew also attracts ants that collect and feed on the sugary residue. In return, the ants often protect the scales from natural enemies.

Several species have been beneficial to cultures over the past centuries. Species of *Kermes* and cochineal scales were used as a dye source from antiquity to the present. Carminic acid is a natural red dye from the cochineal scale (*Dactylopius* spp.), a family of scale insects that feed on cacti. This dye has been used commercially worldwide since the 16th century and was the dominant dye source used in cosmetics, food, medicine, and textile products until its replacement by aniline dyes. The white waxy tests of the Chinese wax scale were used to make wax candles and figurines in China. Also, candles made from the tests of this species were used in religious ceremonies as a symbol of purity. Ground-dwelling margarodids, known as ground pearls, are collected and strung into bracelets and necklaces in several South American countries. For centuries, the secretions from lac scales have been processed into shellac and used to protect a variety of items including furniture, machinery, etc. Also, adult females of the giant scale, *Llaveia axin*, are processed for their 'fat' content that is used as a lacquer coating on wood products, especially art and sculpture creations, in Mexico and Central America. The ornate pit scale, *Cerococcus quercus*, was used as a glue source and as a chewing gum by native American Indians of the southwestern U.S. In addition, a few species are effective in suppressing invasive weeds, such as some species of cochineal scales on cacti.

General morphology

Dramatic sexual dimorphism is particularly exhibited within the scale insects. The adult male's appearance is so different from the adult female that pairing them as a species can be difficult. Because of the diverse differences between the males and females, taxonomists use a different morphological nomenclature to describe the two sexes.

Female development in the more primitive species consists of the egg, first instar (crawler), second instar, third instar, in some species, and adult stage. Although exceptions occur (e.g., Pseudococcidae), females of the more advanced species differ by having the egg, first instar (crawler), second instar, and adult. Adult females are pyriform, subcircular or elliptical, wingless, and usually without any constriction between the head and thoracic regions. Ocelli often are present in species of the more primitive families (e.g., Margarodidae, Ortheziidae, and Phenacoleachiidae). However, the ocelli may be reduced to a slight protuberance, pigmented spots, or completely absent in adult females in the remaining families.

Antennal segments range from zero in Beesoniidae, one-segmented in several families (e.g., Aclerdidae, Asterolecaniidae, Cerococcidae, Diaspididae, and Phoenicococcidae), to as many as 13 segments in some species of Margarodidae (except Margarodinae) and Phenacoleachiidae. Mouthparts consist of a clypeolabral shield and a labium with species within scale insect families exhibiting a one to four-segmented labium. Two pairs of thoracic spiracles are present in all families; however, species of Margarodidae and Ortheziidae also have submarginal abdominal spiracles. Legs are absent in adult females of several families (e.g., Aclerdidae, Asterolecaniidae, Beesoniidae, Phoenicococcidae, and Diaspididae), but range from one-segmented stubs (e.g., most Cerococcidae, species of Cryptococcidae with metathoracic plate) to five-segmented with a tarsal claw in other families. Leg development may be variable among species within some families ranging from absent to five-segmented legs (e.g., Lecanodiaspididae and Pseudococcidae). When present, the leg types vary from digging in Margarodidae to walking or running in most other families.

Body segmentation is pronounced in the more primitive families (e.g., Margarodidae and Ortheziidae) and in some species of a few advanced families (e.g., Pseudococcidae and Eriococcidae). Segmentation of the cephalothorax is indistinct in most species, but the presence of marginal setae often is used to distinguish segments. Ventrally, the thoracic segments are demarcated by the attachment of the legs, when present, and the location of the spiracles on the mesothorax and metathorax. Segmentation is more visible on the ventral surface of the abdomen than on the dorsum. Segments often are delineated by transverse creases and marginal setae. However, the number of discernible segments often varies among species with two to three segments visible in Phoenicococcidae to eight or nine in Cerococcidae. In diaspidids, the last four segments of the abdomen are fused into one pygidial structure with plates and small lobes on the margin. Apparent segmentation also is reduced in Aclerdidae, Asterolecaniidae, Conchaspididae, Crytococcidae, Kermesidae, and Kerriidae. The vulva is located on the eighth segment, and an anal ring often denotes the terminal segment. The anal ring is covered and protected by a pair of anal plates in Coccidae, while a single plate occurs in Aclerdidae and Cerococcidae. A pair of anal plates with an associated arched anal plate surrounds the anal ring in Lecanodiaspidae and some species of Asterolecaniidae.

Male development consists of the egg, first and second instars, prepupa, pupa, and adult stages. Males are nearly always linear in appearance, possessing prominent cephalic, thoracic, and abdominal regions. Adult males have 9- to 10-segmented antennae, a non-functional mouth, a pair of compound eyes is present in Margarodidae and Ortheziidae, and dorsal and ventral ocelli occur in most species in other families. The thorax of most species contains a pair of mesothoracic wings, metathoracic hamulohalterae (except Aclerdidae), and 5-segmented legs. In some species, wings may be present in one generation and reduced or absent in others. On the abdomen, the aedeagus is distinguished by a distinctive penile sheath with 0–3 pairs of caudal filaments. Adult males do not feed and live only one to three days. Adult males resemble aphids and small flies.

First instar nymphs (crawlers) are mobile with well-developed 5-segmented legs, 5- to 6-segmented antennae, and a variety of wax pores and ducts. Each anal lobe terminates with a long anal lobe seta. Sexual dimorphism is visible in the first instar nymphs of several species (e.g., Kuno scale and flat grass scales) representing various families. These differences are evident by the number and arrangement of setae and wax-secreting pores and ducts. Sexual differences in the second instars are generally more pronounced than in crawlers with males being more elliptical than the more ovoid females. The presence or absence of legs among scale insect species is variable. In some species, legs may be present in the males, but absent in the sedentary females. Males often are distinguished by the presence of various ducts and glands, which are usually absent or reduced in the females or appear later in the adult stage.

Protective test or scale cover

One of the more prominent features of the scale insects, including mealybugs, is the development of a protective covering over their bodies. The covering may vary from a powdery secretion to a complete soft or hard waxy coating over the body. The wax is produced by an assortment of pores and ducts on the derm, and the covering is often referred to as a test or scale cover. In some families (e.g., Pseudococcidae), a variety of protective coverings occur ranging from a coating of white waxy powder to long wax strands. In species of Ortheziidae, the body is covered by a series of waxy plates. The Kerriidae aggregate and secrete a flexible lac compound for protection. Members of the genus *Cerococcus* also tend to aggregate close together resulting in wax production completely covering their bodies and sections of adjacent scales. The species within the families Asterolecaniidae and Lecanodiaspididae secrete a papery to waxy test with the individual located in a pit or depression on its host. Many species of Eriococcidae secrete a felt-like sac completely enclosing the female. The Coccidae have a variety of test forms ranging from felt-like in *Eriopeltis*, papery in *Lecanium*, waxy in *Ceroplastes*, to cottony in *Pulvinaria*. Females of Beesoniidae develop within galls and no longer produce tests to cover their bodies. In several species of Margarodidae, the waxy excretions that flow over and cover the body aid in the development of a cyst-like covering.

The protective test or waxy covering over the body allows most species to maintain a favorable humidity level, while preventing rapid temperature changes. The test also functions as a protective barrier against predators and parasitoids. The tests of several species provide protection by blending in with the color and texture of the bark on the plant host and are most often in shades of gray or brown. However, some species display ornate shapes and colors. A stellate appearance is common with waxy extensions projecting from the margins or from the complete test. A number of species in various families are covered by individual wax strands presenting the appearance of a cotton ball. The colors of the waxy scale covers range from hues of brown to a brilliant white to rosy pink to vibrant yellows. The tests of several species display a disruptive pattern. For example, some soft scale species are vivid red to crimson with white zebra-like markings (e.g., calico scale), and others have dark backgrounds with series of irregular white spots to break up the outline of the body, perhaps making them less conspicuous to parasitoids and predators. Some mealybugs aggregate in small groups and appear berry-like, while species of Kermesidae resemble galls or plant buds.

Life cycle and behavior

The life cycle is quite varied among the Coccoidea depending upon the family and species. Development also differs between the female and the male. Most species have one or two generations per year, but some have multiple generations. A number of polyphagous species have two or more overlapping generations outdoors, but reproduce continuously in greenhouses. Generally, all stages of development are present throughout the year in species with multiple generations. Developmental time for the various stages is usually influenced by temperature and humidity. However, the development time for a number of species may take several years. For example, the spruce mealybug has a four-year life cycle that involves multiple migrations annually from the foliage to bark crevices or the duff at the base of the tree. The Chilean margarodid, *Margarodes vitium*, may remain in a cyst-like state for seven or eight years and only after heavy rains emerges as adult females. This adaptation may provide this species the ability to survive periods of extended droughts.

Although most scale species reproduce bisexually, some Coccoidea exhibit parthenogenetic and hermaphroditic reproduction. Adult females that possess protective tests covering their bodies usually have a sedentary nature and a limited active courtship. Some species (e.g., San Jose scale) are known to produce sex pheromones to attract males. The mobile males must locate and mate with females on the host plant. Upon finding a receptive female, the male of Diaspididae inserts the stylus under the pygidial region with the hind legs assisting in the insertion. Although adult males are ephemeral, living only about one to three days upon emergence, they can fertilize several females during this period. Virgin females of the San Jose scale are known to exert their pygidia from beneath their tests and slowly move it back and forth in a short arc. During this activity, females probably release sex pheromones to attract males.

Some females (e.g., Pseudococcidae and Dactylopiidae) actively seek out males for mating. In a species of *Pseudococcus*, the two sexes live apart during the winter with the female inhabiting the smaller branches and the males, still in the pupal stage, on the trunk of the host tree. The adult females become active during early spring and travel down the branches to the trunk where they continue their search for males. Adult females usually precede the emergence of males by several days. After mating, females return to the branches, construct their ovisacs, and lay their eggs. In the fall, the nymphs separate and the males descend to the tree trunks to pupate. A similar behavior is found in northern red oak kermes. Males travel down to the lower levels of the tree trunk, and occasionally leave the tree to pupate, while females remain on branches in the canopy. Upon adult emergence, males fly into the tree canopy to locate and mate with females.

Females protect the eggs during embryonic development, and also the newly hatched crawlers, by laying them under the scale cover or in an ovisac. During the course of egg deposition, the bodies of females in several families (e.g., Asterolecaniidae and Diaspididae) shrivel up to eventually occupy only the anterior one-third of the waxy test. In several species of Coccidae and Kermesidae, the female's exoskeleton hardens providing protection for the eggs against adverse environmental changes and many natural enemies. Most species lay eggs, producing from less than a few dozen to over five thousand eggs. A variety of species in several families (e.g., Margarodidae, Ortheziidae, and Coccidae) deposit their eggs in an ovisac. The type of ovisac and the number of eggs produced vary with the species. The ovisac may completely cover the insect's body, be reduced to cover only part of the abdomen, or be positioned ventrally. Females of Ortheziidae produce a long, cylindrical ovisac for egg deposition. Several species of Eriococcidae and some Coccidae make a fine structured felted sac to retain the eggs until hatching. One of the strangest means of protecting the eggs in some species (e.g., Margarodidae) is the development of a deep invagination of the integument forming a cavity. Females that produce eggs that hatch within their bodies, such as the pineapple mealybug, hairy mealybug, and tuliptree scale, shield the crawlers for a short period by covering them with their abdomens and tests. Most species overwinter in the egg stage (e.g., oystershell scale), while others overwinter as second instars (e.g., European fruit lecanium), or in the adult stage (e.g., golden oak scale).

Parthenogenesis has been documented in several species (e.g., golden oak scale and brown soft scale). It has been speculated that parthenogenesis favors a higher reproductive potential for sedentary species in harsh environments. Some species exhibit life cycles that include a combination of bisexual and parthenogenetic reproduction (e.g., oleander scale and oystershell scale). One of the rarest forms of reproduction is hermaphroditism that occurs in a few species, such as the cottony cushion scale.

A mutualistic relationship exists among species of several scale insects and ants. In exchange for honeydew, a soft scale or mealybug population may be attended and protected by ants, or physically taken within the ant's tent or nest where they are not only protected from parasitoids and predators, but also against adverse environmental changes and fungal infestations. For example, tuliptree scales often are found in association with ants that construct a paper-like tent over a group of adult females protecting them from parasitoids and predators. When leaving the nest on the nuptial flights, females of the ant *Acropyga paramaribensis* always carry a fertilized rhizoecus coffee mealybug female in their mandibles. These mealybugs are maintained within a special chamber in the nest where they are cared for, and in return, provide honeydew to the ants when stroked. Another remarkable way of dispersal occurs in species of *Hippeococcus* that have long raptorial legs and digitules for clinging to attending *Dolichoderus* ants. When disturbed, they cling to the ants and are carried into their nests. The presence and active associations of ants with scale insects and mealybugs result in an increase in their numbers, and a subsequent reduction in the scale insect populations following the removal of the ants. Because ants are so protective, some predators have adapted by mimicking the appearance of their prey to move freely about and feed on mealybugs, such as the larvae of the lady beetle, *Cryptolaemus montrouzieri*.

Host selection and distribution

The age of the Coccoidea is placed in the Cretaceous Period which basically coincides with the time angiosperms became the dominant plant form. This adaptation to the angiosperms was so successful that the majority of present day species cannot leave the host plant and survive. Only a few species are found on gymnosperms. Species may be monophagous (e.g., sweetgum scale), oligophagous feeding on plants in one genus (e.g., kermes oak scales on *Quercus*) or one plant family (e.g., flat grass scales primarily on Gramineae grasses), or polyphagous (e.g., juniper scale on Cupressaceae and Taxodiaceae). Members of some families (e.g., Pseudococcidae) feed on both deciduous and conifer hosts with some being polyphagous, while others are oligophagous or monophagous. Most species feed on a variety of plant species within a given generic or family level. Although most coccid species are polyphagous, members of several families have specific plant associations. These include members of Dactylopiidae on cacti, Halimococcidae on palms, most Kermesidae

that feed on oaks, and Phoenicococcidae on palms. Also, most species of Aclerdidae, some Coccidae and Pseudococcidae feed on grasses, while the eriococcid genus *Apiomorpha* feeds exclusively on *Eucalyptus*. A number of species prefer particular habitats. Several members of Ortheziidae are found in wet meadows, mosses, and in the soil of slopes and forest areas, while the boreal ensign scale commonly is found in leaf litter.

Species of several families actively seek out the host in many instances. For example, species of Margarodidae, Ortheziidae, and Pseudococcidae are mobile throughout their various life stages, and are able to walk from area to area on a given host or even to different hosts. Several species overwinter in the detritus or in cracks on the trunk of a tree and migrate the following spring to the stems and fruits to feed. For those sedentary adult females, the problem of finding food is resolved by the female ovipositing on the host plant.

The mobile crawler is the usual stage that finds a suitable host and feeding site. Dispersal takes place primarily through the movement of the crawlers from tree to tree, by wind, or on the bodies of other insects, spiders, mites, birds, and mammals. Within a day or two after hatching, crawlers migrate, either by walking or with the aid of wind, to new feeding sites on the host plant or adjacent plants. Upon locating a suitable site, crawlers settle, insert their stylets, begin to feed, and continue to mature. Once they establish their position on the host, an adequate food supply is available throughout their development. Species that infest annual plants are unique because their hosts are present for only a short period of the year. Several species have adapted some unusual development patterns. For example, the male crawlers often settle on the leaves, while females tend to settle more often on the twigs (e.g., black willow scale). A common habit exhibited by species in a number of families is for a cluster of male crawlers to settle and develop around a female or group of females increasing the potential of finding a mate upon maturity (e.g., walnut scale). In those species where the crawler settles and develops on the leaves, the leaf veins often are the preferred sites. There is usually a progressive migration back to the stems or twigs by members of the overwintering generation in species with multiple generations. The distance covered on the host plant may vary from species to species depending upon the number of generations per year. While some species settle on specific areas of the host, others (e.g., euonymus scale) do not appear to differentiate between the leaves or stems as preferred sites. Many species commonly infest the underside of limbs and stems of host plants that may protect them from adverse weather conditions, exposure to direct sunlight, and predators. Successive generations force specimens to disperse to other areas to obtain feeding sites that eventually result in a heavy encrustation of the branches.

Several species are known to select multiple locations on the host. The position depends upon the time of year, available space, and their stage of development. The solenopsis mealybugs feed in the crown of their host in the summer, but migrate onto the roots before winter. Members of Eriococcidae may be found on the upper or lower leaf surfaces of one or several hosts, in the axils of leaves and smaller stems, on the surface of smaller stems, on the bark of larger branches, or on the main tree trunk. Sugarcane mealybug crawlers actively walk to the tops of stalks where the developing new tissue allows for easy insertion of their stylets. The females aggregate around the nodes of the stalk. In early spring, they may go to the leaf axil, but development time is slow until they later migrate to the stalks. Orientation of first instars on the underside of leaves and on stems does not appear to be at random; for instance, cottony cushion scales position themselves with their heads upward on the stems. Aside from the numerous species that feed above ground, an enormous fauna of subterranean species exists that feed on the roots of plants. These under-explored niches hold the potential for the discovery of new species, even in North America.

Morphological changes in body shape and size often are influenced by population density and site location. In crowded conditions, the tests usually develop abnormally around other females or around projections of their host (oystershell scale). The position and location on the host often have an effect on the shape of the individual. Species that settle on or between the leaf sheaths and stems of grasses exhibit a dorso-ventrally flattened appearance (e.g., *Aclerda*, *Antonina*, and *Chaetococcus*). The immature azalea bark scale, *Acanthococcus azaleae*, often settles between the petiole and branch at the axillary bud, and frequently acquires a wedge-shaped appearance. Kermesids that reside on the limbs of their host are gall-like or spherical. Host-induced variation results in shape, size, and color variations among populations (e.g., European fruit scale).

Population density is affected by several biotic and abiotic factors. Seasonal abundance is highest in the summer and fall, and declines during the winter months. Mortality is attributed to adverse weather conditions, such as heavy rains washing crawlers off the host, high relative humidity, fungal development on the test and insect, or hot and dry conditions. In some instances, drier environments produce more females, while more moist environments produce more males (e.g., citrus mealybug). Distributional patterns of species in several families imply macroenvironmental preferences as to the elevation, regions inhabited, and climatic conditions. The number of species and family taxa tend to be fewer as the elevation increases. Microenvironmental preferences are demonstrated by the host plants selected and the area and position taken by the species on the host. It is postulated those increases in the population size of endemic species in disturbed habitats are in part due to air pollution and lower levels of natural enemies.

For many species, favorable environmental conditions often result in several generations of scales within one season. The movement of the crawlers is directly affected by temperature with increased activity as the temperature rises. The length of time the crawlers remain under the female ranges from a few hours to two or three days depending on the temperature. Newly emerged crawlers remain under the test on cool, cloudy days, but leave immediately during warmer days. The ephemeral adult males are feeble fliers and more often rely on walking over the host plant for a successful encounter with a female. Their range is generally believed to be restricted to that of the population; however, some males in flight are carried to other parts of the host plant or to other plants. In those sessile stages where individuals retain their legs, a short mobile period may occur between molts. The degree of mobility is dependent on the flexibility of the test that extremely limits the distance they can travel.

Control

The transport of exotic species from one country to another is enhanced by the international trade in plants and fruits. As a result, several exotic species have entered the United States on imported plants. Many of these species are destructive and pose substantial threats to native trees and shrubs, as well as plants grown in commercial greenhouses. Monitoring

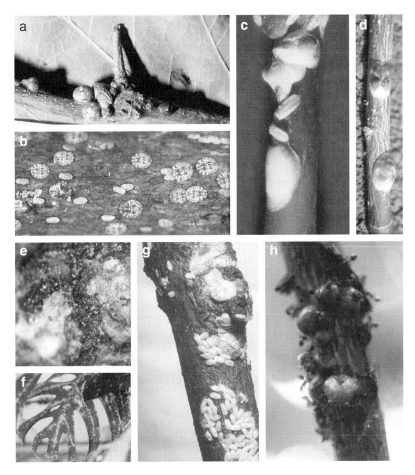

Fig. 902 Some scale insects: a, *Allokermes kingi* (Cockerell) on *Quercus rubra* L. (Kermesidae); b, *Lecanodiaspis prosopidis* (Maskell) on *Aescules* sp. (Lecanodiaspididae); c, *Chionaspis salicisnigrae* (Walsh) on *Salix nigra* Marsh. (Diaspididae); d, *Neolecanium cornuparvum* (Thro) on *Magnolia acuminata* L. (Coccidae); e, *Quadraspidiotus juglansregiae* (Comstock) on *Cornus floridensis* L.: male and female (Diaspididae); f, *Pseudococcus longispinus* (Targioni Tozzetti) on *Dieffenbachia* sp. (Pseudococcidae); g, *Acanthococcus quercus* (Comstock) male and female clusters on *Quercus borcalis maxima* Ashe. (Eriococcidae); h, *Toumeyella liriodendri* (Gmelin) attended by ants on *Liriodendron tulipfera* L. (Coccidae).

plants for ant activity, sooty mold growth, or abnormalities in color and structure of stems and leaves can help locate scale insect infestations in landscape and plant production systems. Several chemicals are available for use against pest populations with the most effective results obtained when applied against the emerging and unprotected first instars (i.e., crawlers). Infestations of house plants and shrubs around the house may be controlled by handpicking or spraying with soapy water, insecticidal soaps, or oils. Several natural enemies including parasitic wasps primarily in the families Encyrtidae and Eulophidae, and predators in the families Coccinellidae, Anthribidae, and Nitidulidae, also are effective at maintaining pest populations at low levels. To obtain the most effective control of pest species, identification of the species or family groups is essential. A key to identify the families in the U.S. is provided.

Key to Families of Coccoidea in the U.S. (Modified from Kosztarab, 1996)

1 Abdominal spiracles present 2
 – Abdominal spiracles absent 3
2(1) Anal ring distinct, with band of pores and 6 setae; eyes usually stalked; antennae 3- to 8-segmented .... Ortheziidae

- Anal ring reduced, with no pores or setae; eyes rarely stalked; antennae 1- to 13-segmented .... Margarodidae

3(1) Anal opening covered by 2 triangular anal plates (except *Physokermes* with one plate); abdomen with a well-developed anal cleft Coccidae

- If anal plate present, only one cover over anal opening (although sclerotized plates lateral of anal opening may be present); anal cleft, if present, not as well developed, as in Coccidae 4

4(3) Caudal margin with furrows; usually under leaf sheaths of grasses and reeds .. Aclerdidae

- Caudal margin without furrows and ridges; habitat and host variable 5

5(4) Cluster pore plate present below each posterior thoracic spiracle; anal ring surrounded by short and stout setae .... Cryptococcidae

- Cluster pore plate absent; anal ring not surrounded by setae 6

6(5) Dorsal spine on abdomen; with dorsal brachial plate and tube; anterior spiracle larger than posterior ... Kerriidae

- Dorsal spine on abdomen absent; without dorsal brachial plate and tube; spiracles similar in size 7

7(6) 8-shaped pores present on dorsum 8

- 8-shaped pores absent from dorsum ... 11

8(7) 8-shaped pores on dorsum and in a submarginal band on venter; ventral tubular ducts scattered over entire body; antennae 1- to 9-segmented; on various hosts 9

- 8-shaped pores restricted to dorsum; ventral tubular ducts form a submarginal band around body margin; antennae 5-segmented; on Fagaceae only .... Kermesidae

9(8) Tubular ducts with inner filaments; antennae 1- to 9-segmented; with cribiform plates .. 10

- Tubular ducts without inner filaments; antennae 1-segmented; without cribiform plates .... Asterolecaniidae

10(9) Antennae 1-segmented, with an associated cluster of 5- to 7-locular pores; with sclerotized triangular plate lying over anal opening; tubular ducts without constriction on outer ductile .... Cerococcidae

- Antennae 7- to 9-segmented, without associated 5- to 7-locular pore cluster; with arched and anal plates around anal

opening; tubular ducts with constriction on outer ductile .... Lecanodiaspididae

11(7) Usually with dorsal ostioles, cerarii, ventral circuli, and trilocular pores; anal ring with inner and outer rows of pores .... Pseudococcidae

- Without dorsal ostioles, cerarii, ventral circuli, and trilocular pores; anal ring variable; tubular ducts invaginated ... 12

12(11) Terminal abdominal segments fused into a pygidium; anal opening simple; body covered with thin, hard shieldlike test, with exuviae of previous instars 13

- Terminal abdominal segments not fused to form a pygidium; anal opening often setiferous; test not as above or body covered by waxy powder or strands ... 14

13(12) Antennae 1-segmented; labium 1-segmented; legs absent; without dermal slits on venter .... Diaspididae

- Antennae 3- or 4-segmented; labium 2-segmented; legs present; with dermal slits on venter .... Conchaspididae

14(12) 8-shaped tubular ducts on dorsum and ventrum; anal ring reduced, without pores but with 2 minute setae; antennae 1-segmented; legs absent .... Phoenicococcidae

- 8-shaped tubular ducts absent; anal ring normally with pores and setae; antennae well developed; legs present 15

15(14) Without dorsal clusters of 3- to 5-locular pores; cruciform pores usually present; on hosts other than Cactaceae ... Eriococcidae

- With dorsal clusters of 3- to 5-locular pores; cruciform pores absent; only found on Cactaceae .... Dactylopiidae

Paris Lambdin
University of Tennessee
Knoxville, Tennessee, USA

References

Ben-Dov, Y., and D. Miller. 2001. ScaleNet. http://www.sel.barc.usda.gov/scalenet/scalenet.htm

Borror, D. J., C. A. Triplehorn, and N. F. Johnson 1989. *An introduction to the study of insects* (6th ed.). Saunders College Publishers, New York, New York. 875 pp.

Kosztarab, M. 1996. *Scale insects of northeastern North America: identification, biology, and distribution.* Special Publication #3. Virginia Museum of Natural History, Martinsville, Virginia. 650 pp.

McKenzie, H. L. 1967. *Mealybugs of California, with taxonomy, biology and control of North American species (Homoptera: Coccoidea: Pseudococcoidae)*. University of California Press, Berkeley, California. 526 pp.

Poole, R. W., and P. Gentili (eds.) 1996. *Nomina Insecta Nearctica: a check-list of the insects of North America. Vol. 4: Non-holometabolous orders*. Entomological Information Service, Rockville, Maryland. 731 pp.

SCALY CRICKETS. A subfamily of crickets (Mogopistinae) in the order Orthoptera: Gryllidae. See also, GRASSHOPPERS, KATYDIDS AND CRICKETS.

SCAPE. The basal segment of the antenna. See also, ANTENNAE OF HEXAPODS.

SCARABAEIFORM LARVA. A larval body form resembling a scarab beetle larva (white grub). Characteristics include hairy, thick bodied, well-developed thoracic legs, the absence of prolegs, and C-shaped in overall body form. This also is known as scarabaeoid. It is found in the beetle families of the superfamily Scarabaeoidea.

SCARAB BEETLES. Members of the family Scarabaeidae (order Coleoptera). See also, BEETLES.

SCARABAEIDAE. A family of beetles (order Coleoptera). They commonly are known as scarab beetles. See also, BEETLES.

SCATOPSIDAE. A family of flies (order Diptera). They commonly are known as minute black scavenger flies. See also, FLIES.

SCAVENGER MOTHS (LEPIDOPTERA: BLASTOBASIDAE). Scavenger moths, family Blastobasidae, total over 296 species worldwide, with many known from North America and Europe, in three subfamilies; actual fauna probably exceeds 600 species. There are two subfamilies: Holcocerinae and Blastobasinae (Symmocinae have been included as well but now are a tribe of Autostichinae, Oecophoridae). The family is part of the superfamily Gelechioidea in the section Tineina, subsection Tineina, of the division Ditrysia. Adults small (5 to 35 mm wingspan), with head smooth-scaled; haustellum scaled; maxillary palpi 4-segmented. Maculation mostly somber shades of gray with few markings. Adults nocturnal as far as is known. Larvae are scavengers or detritus feeders, sometimes feeding on plant fruits, flowers, or seeds, among a number of plant families, but few are known biologically. At least one species lives with coccids (Hemiptera) but predation on the coccids has not been confirmed.

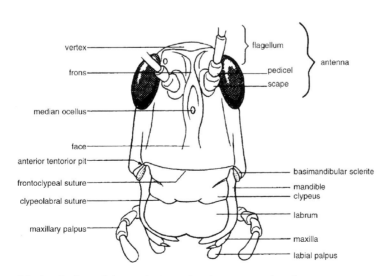

Fig. 903 Front view of the head of an adult grasshopper, showing some major elements.

Fig. 904 Example of scavenger moths (Blastobasidae), *Blastobasis huemeri* (Sinev) from Italy.

John B. Heppner
Florida State Collection of Arthropods
Gainesville, Florida USA

References

Adamski, D. 1995. Review of the Blastobasidae of the Republic of the Seychelles (Lepidoptera: Gelechioidea). *Proceedings of the Entomological Society of Washington* 97: 489–499.

Adamski, D. 2002. *A synopsis of described neotropical Blastobasinae (Lepidoptera: Gelechioidea; Coleophoridae)*. Entomological Society of America, Lanham, Maryland (Thomas Say Monograph). 150 pp.

Adamski, D., and R. L. Brown. 1989. Morphology and systematics of North American Blastobasidae (Lepidoptera: Gelechioidea). *Mississippi Agricultural and Forestry Experiment Station Technical Bulletin* 165: 1–70.

Adamski, D., and R. W. Hodges. 1996. An annotated list of North American Blastobasinae (Lepidoptera: Gelechioidea: Coleophoridae). *Proceedings of the Entomological Society of Washington* 98: 708–740.

Buszko, J. 1978. Blastobasidae. In *Klucze do Oznaczania Owadów Polski. 27. Motyle – Lepidoptera*, 36:22–32. Polskie Towardzystwo Entomologiczne [in Polish].

SCELIONIDAE. A family of wasps (order Hymenoptera). See also, WASPS, ANTS, BEES, AND SAWFLIES.

SCENOPINIDAE. A family of flies (order Diptera). They commonly are known as window flies. See also, FLIES.

SCENT GLAND. A gland producing an odor.

SCENTLESS PLANT BUGS. Members of the family Rhopalidae (order Hemiptera). See also, BUGS.

SCHEERPELTZ, OTTO. Otto Scheerpeltz was born near Olomouc, Czech Republic, on July 16, 1888. He enrolled in the Technische Hochschule, in Vienna, to become an engineer, but instead became a teacher. He continued at this profession until his retirement in 1945. In 1922, while still employed as a teacher, he entered the University of Vienna to study zoology and botany, and graduated in 1930. There he became interested in the Staphylinidae. Scheerpeltz amassed one of the largest collections of Staphylinidae, containing 300,000 specimens and 10,000 types. He authored 286 papers describing 1,405 species and 181 genera, and is considered to be an important contributor to our knowledge of this family. He died at Vienna, Austria, on November 10, 1975.

Reference

Herman, L. H. 2001. Scheerpeltz, Otto. *Bulletin of the American Museum of Natural History* 265: 137–139.

SCHIZOPTERIDAE. A family of bugs (order Hemiptera). See also, BUGS.

SCHNEIDERMAN, HOWARD ALLEN. Howard Schneiderman was born at New York, New York, USA, on February 9, 1927. He received a bachelor's degree from Swarthmore College in 1948, and a M.S. and Ph.D. from Harvard University in 1949 and 1952, respectively. Early in his career he became recognized as an authority on juvenile hormone chemistry. He was a member of the faculty at Cornell University from 1953 to 1961. Schneiderman helped found the Developmental Biology Center at Case-Western Reserve University, where he also served a chairman of the Biology Department. During this period he was co-discoverer of JH II. In 1969 Schneiderman moved to the University of California at Irvine, where he chaired the Department of Developmental and Cell Biology, and later became dean of the School of Biological Sciences. During this period Schneiderman's focus shifted from the developmental biology of Lepidoptera to *Drosophila*. In 1979 he joined the Monsanto Company, and eventually was chief scientist and senior vice president of research and development. His activities with Monsanto and collaborative arrangements with universities earned him national prominence as one of the architects of industrial biotechnology research, and one of the leading figures in physiology, development, and genetics of insects. Schneiderman published over 200 papers during his career, and was awarded many honors, including election to the National Academy of Sciences. He died on December 5, 1990, in St. Louis, Missouri.

Reference

Oberlander, H. 1991. Howard Allen Schneiderman. *American Entomologist* 37: 124.

SCHIZOPODID BEETLES.

Members of the family Schizopodidae (order Coleoptera). See also, BEETLES.

SCHIZOPODIDAE.

A family of beetles (order Coleoptera). They commonly are known as schizopodid beetles. See also, BEETLES.

SCHMIDT'S LAYER.

A zone of new cuticle deposition within the procuticle.
See also, CUTICLE.

SCHRECKENSTEINIIDAE.

A family of moths (order Lepidoptera). They are commonly known as bristle-legged moths. See also, BRISTLE-LEGGED MOTHS, BUTTERFLIES AND MOTHS.

SCHWARZ, EUGENE AMANDUS.

Eugene Schwarz was born in Liegnitz, Germany, on April 21, 1844. He was educated at the University of Breslau and University of Leipzig. He moved to the United States in December of 1872. He left Germany because he had deceived his father, who expected his son to study philosophy when, in fact, he studied zoology and entomology. Schwarz joined the Museum of Comparative Zoology at Harvard University in 1872, and initially worked as a preparator. He moved to Detroit where he co-founded the Detroit Scientific Association and helped initiate the Hubbard-Schwarz collection of Coleoptera. In association with this effort many collecting trips were made in the Great Lakes region and to Florida. Beginning in 1878 he was employed by the Division of Entomology of the United States Department of Agriculture, and he remained in this capacity for most of the period of employment, which ended with retirement in 1926. During the early years of his employment he studied cotton pests in the southern United States, and fig pests in California. His principal interest, however, remained the systematics of Coleoptera, and he amassed a large collection in the U.S. National Museum, where he was named curator of Coleoptera. He was one of the founders of the Entomological Society of Washington and reportedly had significant influence in the Department of Agriculture and National Museum. Schwarz published 395 papers over the course of his career. He died at Washington, DC, on October 15, 1928.

References

Essig, E.O. 1931. *A history of entomology*. The Macmillan Company, New York. 1029 pp.
Mallis, A. 1971. *American entomologists*. Rutgers University Press, New Brunswick, New Jersey. 549 pp.

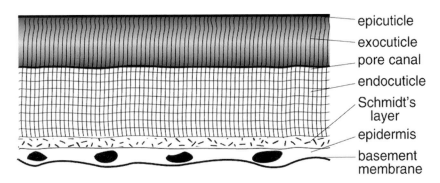

epicuticle
exocuticle
pore canal
endocuticle
Schmidt's layer
epidermis
basement membrane

Fig. 905 Cross section of the insect cuticle and epidermis (adapted from Chapman, The insects: structure and function).

SCIADOCERIDAE. A family of flies (order Diptera). See also, FLIES.

SCIARIDAE. A family of flies (order Diptera). They commonly are known as dark-winged fungus gnats. See also, FLIES.

SCIENTIFIC NAME. A Latin or Latinized name given to all biological organisms and consisting of two parts, a genus and species. The first of the two names is the genus and is capitalized. The second is the species and is not capitalized. Both words are italicized. The author's name (the person who first provided a technical description of the species) often follows the scientific name. If the genus name has been changed since the organism was named by the author, then the author's name is placed in parentheses to indicate that a change has been made. When several species in the same species are discussed together, then 'species' may be abbreviated spp. For example, in discussing the genus *Spodoptera*, we might refer to the members as *Spodoptera* spp. On the other hand, a single species is indicated as 'sp.' See also, SCIENTIFIC NAMES AND OTHER WORDS FROM LATIN AND GREEK.

SCIENTIFIC NAMES AND OTHER WORDS FROM LATIN AND GREEK. These words are often misused by people who do not understand the rules governing them. The rules are explained here simply.

Words from Latin and Greek adopted into English – singular and plural forms

Words such as larva, sensillum, and elytron appear to be Latin or Greek. Many of them were invented by scientists and were never part of classical Latin or Greek. They were invented because new words were needed to name newly described structures. For hundreds of years they had to be in Latin (or Greek) because books about biology and medicine were written in Latin (with a few entries in Greek), which was the international language of science. When such books began to be written in English, those invented words were not changed. Instead they were adopted into English. More were added later, also with a Latin or Greek origin, because by then the tradition was to use words of that form.

Most English nouns form the plural by adding -s (or -es) to the singular. There are exceptions (man, ox, goose, mouse, etc.) but these are relatively few.

However, almost all words adopted from Latin or Greek still form their plural in the way this is done in Latin or Greek, not by adding -s. Exceptions are the words abdomen and trochanter, whose English plurals are now spelled by adding -s.

Singular and plural forms of some words from Latin

Singular (is)	Plural (are)
alumnus	alumni
cercus	cerci
fungus	fungi
ocellus	ocelli
palpus	palpi
tarsus	tarsi
alga	algae
alumna	alumna
coxa [1]	coxae
	exuviae
lamella	lamellae
larva	larvae
maxilla	maxillae
pupa	pupae
seta	setae
tibia	tibiae
agendum [2]	agenda
bacterium	bacteria
cilium	cilia
datum [2]	data
flagellum	lagella
labium	labia
medium	media
labrum	labra
ommatidium	ommatidia
operculum	opercula
ovum	ova
sensillum	sensilla
sternum	sterna
tergum	terga
tympanum	tympana
foramen	foramina
corpus	corpora
femur	femora
genus	genera
[3]	faeces
	feces (American)
axis	axes
penis	penes
species	species

[1] the singular does not exist
[2] the singular is rarely used in English
[3] the singular is not used in English

Singular and plural forms of some words from Greek

Singular (is)	Plural (are)
chrysalis	chrysalids
	(or chrysalides)
proboscis	proboscides
criterion	criteria
elytron	elytra
ganglion	ganglia
protozoon[1]	protozoa
stemma	stemmata
stigma	stigmata
stoma	stomata
thorax	thoraces
	thoraxes (American)

---

[1] protozoan is an English word formed from Protozoa

If you have studied Latin you will know that you need to remember only a small subset of that grammar to use the correct singular or plural form of each word. Some of those words are shown in the table below, in two columns, giving the singular and plural forms that may be memorized. They are grouped (by their last letters) according to their placement in Latin grammar, which helps to place others of similar form.

The form of these words is not fixed because poorly trained news writers commonly use the plural forms algae, bacteria, data, larvae and media as if they were singular. This misuse confuses a large audience and may influence the future direction of English.

Words that are considered to be Latin (not English)

Scientific (Latin) names of insects and other animals are either singular or plural, not both. Whether they are singular or plural depends upon the rank (species, genus, tribe, family, etc.).:

Names of species are singular

Danaus plexippus is (not are) a species of butterfly

The name of each and every species consists of two words. These two words are conventionally printed in italic font. When written by hand or typed on a typewriter, they are underlined, which indicates to a publishing company that they should be set in italics. The two words are normally written or printed using upper and lower case (large and small letters),

and then the first letter of the first word should be capitalized, and the first letter of the second word should not be capitalized. In titles, all letters of both words may be capitalized.

When upper and lower case letters are used, it is incorrect to capitalize the first letter of the second word of the name of a species of insect or any other animal (according to current and recent editions of the International Code of Zoological Nomenclature). The rules of botanical nomenclature differ, so that it is permitted for names of certain plants, and it once was permitted to do the same for some animals.

The convention of putting these words into italics came about to indicate that these words are Latin, not English. Neither of the two words may have any diacritical mark (accent) because Latin has none.

The first word of a species name is the name of the genus to which it belongs. The second word of a species name has variously been called the specific epithet, the trivial epithet, or even the "specific name" although this last is a confusing term (because it is not the name of a species).

Most if not all scientific journals require that the name of a species be spelled out in full at its first mention in a text. In subsequent uses of the name in the same text, it may be abbreviated by reducing the generic name to a single capital letter followed by a period (full stop), as in:

D. plexippus

Some entomologists have adopted other conventions for abbreviating generic names of insects (for example, mosquitoes) in the names of species.

Names of genera, too, are considered to be latin, are conventionally italicized, and are singular:

Names of genera are singular

Danaus is (not are) a genus of butterfly

Names of all higher categories are plural:

Names of tribes, families, orders, classes and phyla are plural

---

Tribe:
   Brachinini are (not is) the bombardier beetles
Subfamily:
   Cicindelinae are (not is) the tiger beetles

Family:
    Culicidae are (not is) the mosquitoes
Order:
    Plecoptera are (not is) the stone flies
Class:
    Insecta are (not is) the insects
Phylum:
    Arthropoda are (not is) a phylum of invertebrates

The Latin (or Greek) terminations of these scientific names make it plain that they must be treated as singular (species and genera) or plural (all higher categories). Article 11.7 of ICZN (1999) declares that "family-group names" (names of tribes to families) must be in the nominative plural (see a Latin grammar for further explanation), so cannot be construed as singular.

To follow the rules, we must write "Coleoptera are a huge order" and "Culicidae have many species that transmit disease." Why, then, do some people write "Coleoptera is a huge order" or "Culicidae has many species that transmit disease?"

There seem to be three reasons why some writers treat these words as singular:

(1) They have not learned that the words are plural. This is quite common among students. Even some more experienced writers thoughtlessly copy the erroneous style of earlier writers.

(2) They argue that when they write "Coleoptera is speciose", what they really mean is "the order Coleoptera is speciose", but scientific writing should have no hidden meanings, so this excuse is not valid.

(3) They perversely refuse to accept what they are told are the established rules, or want to flaunt their ignorance of Latin and Greek and disdain for the International Code of Zoological Nomenclature, and want to make their own rules.

It is of course acceptable to write "the order Coleoptera is speciose."

Another error is made by students in such expressions as "A Culicidae was seen feeding" by which they mean a mosquito (not the whole family!) was seen feeding. This may be written as "a culicid was ..." or "a mosquito was ..." or "an individual of the family Culicidae was..." or "individuals of a species of Culicidae were ..."

Although the names of higher categories (tribe and above) are considered to be Latin, they are not conventionally printed in italic font. Each must have a capital initial letter.

J. Howard Frank
University of Florida
Gainesville, Florida, USA

References

Brown, R. W. 1956. Composition of scientific words (2nd ed.). Published by the author. Baltimore, Maryland. 882 pp.

International Commission on Zoological Nomenclature (ICNZ). 1999. International code of zoological nomenclature. International Trust for Zoological Nomenclature, London, United Kingdom. xxix + 306 pp.

SCIOMYZIIDAE. A family of flies (order Diptera). They commonly are known as marxh flies. See also, FLIES.

SCIRTIDAE. A family of beetles (order Coleoptera). They commonly are known as marsh beetles. See also, BEETLES.

SCLERITE. A rigid section or plate of cuticle forming a component of the insect's body wall, and bounded by sutures or membranous areas.

SCLEROGIBBIDAE. A family of wasps (order Hymenoptera). See also, WASPS, ANTS, BEES, AND SAWFLIES.

SCLEROTIN. Cuticular protein that has been irreversibly hardened and darkened by chemical cross-linking of the molecules.

SCLEROTIZATION. The process of sclerotization, or tanning, results in the hardening of the cuticle after an insect molts. Sclerotization involves the cross-linking of proteins, aided by quinones, which typically contribute to the darkening of the cuticle.

SCOLEBYTHIDAE. A family of wasps (order Hymenoptera). See also, WASPS, ANTS, BEES, AND SAWFLIES.

SCOLIIDAE. A family of wasps (order Hymenoptera). See also, WASPS, ANTS, BEES, AND SAWFLIES.

SCOLOPIDIUM. (pl., scolopidia) A mechanoreceptor sensillium located beneath the cuticle that perceives vibration, stretch or pressure.

SCOLUS. (pl. scolli) An extension of the integument, in the form of a projection or tubercle, that bears stout setae. These are found on some highly spinose insects, such as the caterpillars of Saturniidae.

SCOLYTIDAE. Considered by some to be a family of beetles (order Coleoptera). They commonly are known as bark beetles or ambrosia beetles. Here they are treated as a subfamily (Scolytinae) of Curculionidae. See also, BEETLES.

SCOPA. Patches of branched hairs occurring on the body of bees, and used to collect and hold pollen. The location of the scopa varies among bees.

SCOPURIDAE. A family of stoneflies (order Plecoptera). See also, STONEFLIES.

SCORPIONFLIES (MECOPTERA). This is a small, primitive order of holometabolous insects.

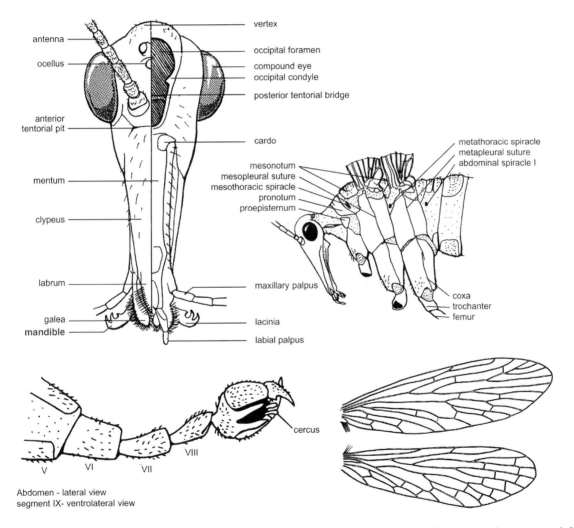

Fig. 906 Morphology of adult scorpionfly (Mecoptera: Panorpidae): above left, the head, showing anterior aspect on left, and posterior aspect on right; above right, lateral view of the thoracic region; below left, lateral view of the tip of the male abdomen; lower right, the front and hind wings.

The order name is based on the Greek words *mecos* (length) and *pteron* (wing). The anterior region of the head usually is prolonged into a beak, and in some the tip of the abdomen is curved upward, scorpion-like.

Classification

The fossil record suggests that these once were very abundant insects, but they now number only about 480 species from around the world. Normally they are placed in nine families

Class: Insecta

Order: Mecoptera
Family: Nannochoristidae
Family: Bittacidae – hangingflies
Family: Boreidae – snow scorpionflies
Family: Meropeidae – earwigflies
Family: Eomeropidae
Family: Apteropanorpidae
Family: Choristidae
Family: Panorpodidae – short-faced scorpionflies
Family: Panorpidae – common scorpionflies

Characteristics

These small or moderate-sized holometabolous insects bear two pairs of wings that are nearly equal in size. They measure 3 to 30 m in size, but many are quite small. The wings are membraneous, and often spotted, though the Boreidae are wingless. The antennae are filiform. The anterior portion of the head is drawn out into a beak-like structure that bears biting mouthparts. The compound eyes are well developed, and three ocelli are usually present. The legs are long and slender. The abdomen is elongate, and in the male Panorpidae it is drawn out into a scorpion-like, upcurved tip. Ten abdominal segments generally are evident. The larvae resemble caterpillars or larvae of scarab beetles, but some differ in that they have small (3 to 20 ommatidia) compound eyes.

Biology

Mecopterans are terrestrial insects, and most are predatory though some are scavengers or phytophagous. Some take insects trapped in spider's webs. They most often feed on flies (Diptera), but also take aphids, caterpillars, moths, and others. They inhabit dense vegetation and shaded situations such as the edges of wooded areas. In the snow scorpionflies, adults are active in the winter and can be observed on the snow, but also frequent mosses, on which the larvae feed. In some species, males emit a pheromone to attract females, and then provide a prey item to induce her to allow copulation. The caterpillar-like larvae live above-ground, but most pupate in the soil.

References

Arnett Jr., R. H. 2000. *American insects* (second edition). CRC Press, Boca Raton, Florida. 1003 pp.

Byers, G. 1987. Order Mecoptera. Pages 246–255 in F.W. Stehr (ed.). *Immature insects, Vol. 1*. Kendall/Hunt Publishing. Dubuque, Iowa.

Byers, G. W., and R. Thornhill 1983. Biology of the Mecoptera. *Annual Review of Entomology* 28: 203–228.

Willmann, R. 1987. The phylogenetic system of the Mecoptera. *Systematic Entomology* 12: 519–524.

SCORPIONS (CLASS ARACHNIDA, ORDER SCORPIONES). The scorpions are an ancient group of arachnids that date back at least as far as the Silurian, over 400 million years ago. They began as aquatic arthropods, as evidenced by the lack of tarsal claws on the earliest fossil scorpion such as *Palaeophonus*. Some of the earliest forms have also been demonstrated to have had gills. While scorpions may have invaded the land during the Devonian, by the Carboniferous (350 million years ago) some had book lungs, were certainly terrestrial, and many were almost indistinguishable from modern scorpions. At least some (including the earliest) appear to have had compound eyes, but these have been lost except for a small cluster of 'simple eyes' at the front corners of the carapace. Scorpions are now considered to be closely related either to the harvestmen (Opiliones) or to the extinct Eurypterida (often placed with the horseshoe crabs in the Class Merostomata). The exact relationships are still being worked out, and both claims are found in the literature.

Scorpions are primarily nocturnal and have adapted to live in very dry deserts, as well as tropical and subtropical forests and scrub, high mountains, grasslands, caves, intertidal zones of the oceans and other habitats. In deserts, they survive largely by living in deep burrows or under rocks, and only coming out at night to feed and mate. Like most arachnids, scorpions are predators and use their claw-like pedipalps to capture and hold prey while stinging it with the stinger at the end of their long curved postabdomen.

Fig. 907 Scorpions bear claw-like pedipalps, and a stinger at the tip of the abdomen.

The anatomy of scorpions is unusual among the Arachnida. In addition to the claw-like pedipalps, five-segmented postabdomen and stinger (telson) just mentioned, scorpions also have four pairs of walking legs, a pair of jointed and claw-like chelicerae, a pentagonal (most), triangular (family Buthidae and Microcharmidae) or split (family Bothriuridae) sternum, a genital plate and a pair of odd comb-like sensory structures called pectines. The exact function of the pectines is still debated, but they seem to function in habitat scanning and in chemoreception. Other important sense organs include the eyes (both median and lateral), which function in vision, light level perception and entrainment of daily biological rhythms, and the trichobothria, which are primarily air-borne vibration and touch receptors.

The biology of scorpions is much better known now than in J. Henri Fabre's time. Fabre, working in the latter half of the nineteenth century, did not discover the male's production of a spermatophore at the end of the 'promenade a deux' of scorpion courtship. He was under the mistaken impression that the mating actually took place under a rock or other object, after which the female ate the male. In reality, the walk or dance of the scorpions starts when male and female contact each other, and the male grasps the female's pedipalps. The spermatophore is produced and the male pulls the female over the package of sperm, which is then picked up in her genital opening. Then he either escapes (probably in most cases in the wild), or gets caught and eaten (as Fabre observed in the close confines of his scorpion cages.) A few

species, notably a few *Tityus* from northern South America, reproduce parthenogenically, and have few or no males.

Young scorpions are produced alive by viviparity. The young of some species actually are attached to the mother at their mouth, forming a sort of placenta. At birth the baby scorpions clamber onto the mother's back and ride until the first molt, after which they typically disperse. Scorpions may live for at least several years, even after reaching adulthood. Some, such as *Hadrurus*, as well as some of the diplocentrids, ischnurids, and scorpionids have been reported to live as long as 20 years, about the same at that reported for some tarantulas.

The taxonomy of scorpions has been undergoing great changes in the last few years. Sissom recognized nine families in 1990 (Bothriuridae, Buthidae, Chactidae, Chaerilidae, Diplocentridae, Ischnuridae, Iuridae, Scorpionidae and Vaejovidae). By 2000 this had nearly doubled to 16 families (Bothriuridae, Buthidae, Chactidae, Chaerilidae, Diplocentridae, Euscorpiidae, Heteroscorpionidae, Ischnuridae, Iuridae, Microcharmidae, Pseudochactidae, Scorpionidae, Scorpiopidae, Superstitioniidae, Troglotayosicidae and Vaejovidae). Two more families, the Hemiscorpiidae and the Urodacidae, have recently been proposed by Prendini, the new families being separated from the Ischnuridae and the Scorpionidae, respectively. A recent cladistic analysis published by Soleglad & Sissom has now placed the Scorpiopidae as a subfamily under the Euscorpiidae. These changes are not especially surprising because they are the result of new analytical techniques that have included more parameters, including biochemical and morphological characters.

The North American species of scorpions are primarily in the families Buthidae, Chactidae, Diplocentridae, Iuridae and Vaejovidae. The family Superstitioniidae is represented in the United States by the southwestern species *Superstitionia donensis*. The family Euscorpiidae includes several species in tropical Mexico and Guatemala and South America, but is otherwise all Old World, including Asia. Most scorpion species in the United States are found in the southern tier of states, with the largest number in Texas, New Mexico, Arizona and California. One species reaches British Columbia in the west and another reaches Virginia in the east. There are no scorpions in the northeastern United States and most of Canada.

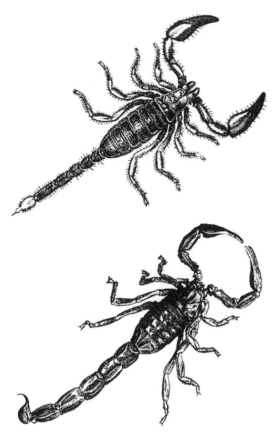

Fig. 908 *Anuroctonus* sp. scorpion (family Iuridae) (above), found in the western United States; and *Buthus* sp. scorpion (family Buthidae) (below), found in Europe, Africa and the Middle East.

Of the buthids, only the genus *Centruroides* is found in the United States. There are five known species, with three (*C. gracilis*, *C. hentzi* and *C. guanensis*) occurring in Florida and two (*C. exilicauda* and *C. vittatus*) in the midwestern and southwestern United States. The genus *Centruroides* is more complex in Mexico, where it contains a number of dangerous species, such as *C. noxious*, *C. limpidus*, *C. infamatus*, *C. elegans*, and *C. suffusus* (the so-called Durango Scorpion). All are very slender scorpions, with the metasoma especially elongate in the male, and most have a subacular tubercle just anterior to the stinger on the ventral side of the telson. Most are straw-colored, except for *C. gracilis*, which is dark and gets as large at 11 cm in length. A similar large and dark-colored species, *C. margaritatus*, is found in Mexico and Central America, south into South America. *Centruroides gracilis* has also been

introduced to the Canary Islands and *C. margaritatus* is known from the Cape Verde Islands, Sierra Leone and Gambia. These are the only members of the genus found outside the New World.

In the rest of the world there are a number of buthid genera. In South America and nearby islands the genus *Tityus* includes several dangerous species. In the Middle East and North Africa there are several genera including dangerously venomous species. These include *Leiurus* and *Androctonus*. These scorpions are large (8 to 10 cm) in length and possess highly potent neurotoxins. There is a record of a healthy adult dying in four hours after being stung by *Androctonus australis*. Oddly, the effects of the venom differ from population to population and some local populations are more venomous than others. In the rest of Africa there are other genera that may be dangerous, such as *Hottentotta*. Some members of *Buthus*, which is widespread from southern Europe into Asia and Africa, can also be dangerous. Common, but apparently much less dangerous widely ranging genera include *Isometrus* and *Lychas*. *Isometrus maculatus* is common in Hawaii, where it is likely introduced. Although readily available through the exotic pet trade, live specimens of the more venomous buthids, such as *Androctonus australis*, should only be kept by experienced researchers.

Chactids occur from Mexico south into Central and South America. Only the genus *Nullibrotheas* is known from North America. The family is closely allied to the Vaejovidae and was initially difficult to separate. All chactids are now known to have many accessory trichobothria (slender sensory hairs), while almost all vaejovids have the basic trichobothrial pattern without accessories.

Diplocentrids are represented in the United States by the genus *Diplocentrus*, which ranges south into Mexico. This genus consists of chunky, highly polished rock-dwelling scorpions with large pedipalps and pronounced subacular tubercles. None are known to be dangerous. Other genera occur in Mexico, Central America, South America, the Caribbean and parts of Asia.

Iurids include the largest of North American scorpions in the genus *Hadrurus*. These desert-dwelling scorpions can reach over 12 cm in length and make an impressive sight when out foraging after dusk. The genus *Anuroctonus* (only recently removed from the Vaejovidae) contains only one species, which is distinguished from all other scorpions by the male having a bead-like swelling on the stinger. It lives primarily in burrows under rocks in desert and coastal mountains from Utah and northern Arizona through California. Of the remaining four genera, *Hadruroides* is known from Ecuador and Peru, including the Galapagos Islands, and *Iurus* is found in Turkey. None are known to be dangerous.

Vaejovids make up the largest part of the North American fauna. The genus *Vaejovis* is by far the largest, with over 60 species known from the United States and Mexico. *Vaejovis spinigerus* is the common 'devil scorpion' of the southwestern United States. *Vaejovis coahuilae* replaces it in the Chihuahuan Desert. *Vaejovis carolinianus* is found in the Southeast, north of Florida. A number of other species are known from Arizona, New Mexico and Texas, with many more in Mexico. Other common genera include *Paruroctonus*, *Serradigitus*, *Smeringurus* and *Uroctonus*. Most of these were in the old genus *Vaejovis*, which is still huge despite the deletions.

Currently recognized families not found in North America include Bothriuridae (South America, Africa, India and Australia), Chaerilidae (one genus from south and southwest Asia), Euscorpiidae (Europe, Africa and Asia), Heteroscorpionidae (all from Madagascar), Ischnuridae (Europe, Asia, Africa, South America, and Australia), Microcharmidae (Africa), Pseudochactidae (Asia), Scorpionidae (Africa and Asia), Urodacidae (Australia), and Hemiscorpiidae (Asia and northeastern Africa). Some of the Ischnuridae and Scorpionidae genera are commonly sold in the exotic pet trade, especially *Scorpio* and *Pandinus*. The latter and the Ischnurid genus *Hadogenes* include the largest living scorpions known-reaching around 20 cm in length. These giants have venoms with very low toxicity and thus are not really dangerous. Several species of *Pandinus* are now on the CITES list, and so trade in these giant scorpions is monitored for wild-caught individuals.

David. B. Richman
New Mexico State University
Las Cruces, New Mexico, USA

References

Fet, V., W. D. Sissom, G. Lowex, and M. E. Braunwalder. 2000. *Catalog of the scorpions of the world* (*1758–1998*). New York Entomological Society, New York, New York.

Keegan, H. L. 1980. *Scorpions of medical importance.* University Press of Mississippi, Jackson, Mississippi.

Kjellesvig-Waering, E. N. 1986. A restudy of the fossil scorpions of the world. *Paleontographica Americana* 55: 1–287.

Polis, G. A. (ed.) 1990. *The biology of scorpions.* Stanford University Press, Stanford, California.

Prendini, L. 2000. Phylogeny and classification of the superfamily Scorpionoidea Latreille 1802 (Chelicerata, Scorpiones): an exemplar approach. *Cladistics* 16: 1–78.

Rein, J. O., and K. McWest. 2001. The scorpion files. www.ub.ntnu.no/scorpion-files/

Soleglad, M. E., and W. D. Sissom. 2001. Phylogeny of the family Euscorpiidae Laurie 1896 (Scorpiones): a major revision. pp. 25–111 in V. Fet and P.A. Selden (eds.), *Scorpions 2001.* In Memoriam Gary A. Polis. British Arachnological Society, Plymouth, Devon United Kingdom.

SCOUTING. Systematic, regular monitoring of plants or animals, normally as part of an effort to determine the need for suppression..

SCRAMBLE COMPETITION. Competition for limited resources, often intraspecific competition, wherein all individuals have equal access and a free-for-all occurs, with resources being dissipated among the many individuals, and in some cases inadequate supply for any individuals. This also is known as exploitation competition.

SCRAPER. In some Orthoptera (katydids and crickets), the anal angle of the tegmen, which rubs against a file; part of the stridulatory apparatus.

SCRAPERS. Insects in aquatic communities that feed by grazing on algae growing on rocks.

SCRAPTIIDAE. A family of beetles (order Coleoptera). They commonly are known as false flower beetles. See also, BEETLES.

SCREWWORM FLIES. See also, FLIES (DIPTERA), MYIASIS, AREA-WIDE PEST MANAGEMENT, STERILE INSECT TECHNIQUE.

SCRUB TYPHUS. A rickettsial disease transmitted by some chigger mites.

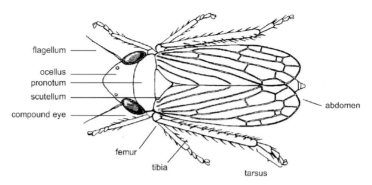

Fig. 909 Adult leafhopper (Hemiptera: Cicadellidae).

SCUDDER, SAMUEL HUBBARD. Samuel Scudder was born at Boston, Massachusetts, USA, on April 13, 1837. He is known for his pioneering work in the Orthoptera and Lepidoptera, but especially in insect paleontology, and is considered to be the founder of insect paleontology in the United States. He was educated at Williams College and Harvard University, in Massachusetts. He was appointed custodian of the Boston Society of Natural History in 1864, and assistant librarian of Harvard University in 1879. Scudder was one of the founders of the Cambridge Entomological Club, founder of the journal Psyche, and editor of Psyche from 1883 to 1885. He served as president of the Boston Society of Natural History from 1882 to 1887, and was employed as paleontologist with the United States Geological Survey from 1886 to 1892. Scudder described over 2,000 fossil insects and published the ''Index to the known fossil insects of the world, including myriapods and arachnids,'' in 1891. Scudder's interest in Orthoptera began in 1862 and continued until about 1900; during this period he developed the reputation as being America's foremost orthopterist. He described about 630 species and 106 genera of grasshoppers. Among his important publications on Orthoptera were ''Catalogue of the Orthoptera of North America described previous to 1867,'' and ''Catalogue of described Orthoptera of the U.S. and Canada.'' Scudder's work on Lepidoptera began in 1889 when he published a 3-volume treatise ''Butterflies of the Eastern United States and Canada.'' Over the course of his career, Scudder published 791 papers and was a member of numerous scientific societies. He died in Cambridge, Massachusetts, on May 17, 1911.

References
Essig, E. O. 1931. *A history of entomology*. The Macmillan Company, New York. 1029 pp.
Mallis, A. 1971. *American entomologists*. Rutgers University Press, New Brunswick, New Jersey. 549 pp.

SCUTELLERIDAE. A family of bugs (order Hemiptera). They sometimes are called shield-backed bugs. See also, BUGS

SCUTELLUM. In Hemiptera, the triangular mesothoracic region between the base of the wings; in other winged insects, a small posterior section of the meso- and metanotum, separated from the scutum by a suture.
See also, THORAX OF HEXAPODS.

SCUTUM. The middle region of the meso- and metanotum, set off from the scutellum by a suture.
See also, THORAX OF HEXAPODS.

SCUTTLE FLIES. See also, MYIASIS.

SCYDMAENIDAE. A family of beetles (order Coleoptera). They commonly are known as antlike stone beetles. See also, BEETLES.

SCYTHRIDIDAE. A family of moths (order Lepidoptera). They commonly are known as flower moths. See also, FLOWER MOTHS, BUTTERFLIES AND MOTHS.

SEAL LICE. Members of the family Echinophthriidae (order Siphunculata). See also, SUCKING LICE.

SEASIDE EARWIGS. Some of the earwigs in the family Carcinophoridae (order Dermaptera). See also, EARWIGS.

SEAWEED FLIES. Members of the family Coelopidae (order Diptera). See also, FLIES.

SECONDARY PESTS. Pests that normally are not considered to be important, but which become important after a change in pest control or cropping practices.

SECONDARY PLANT SUBSTANCE. Material found in plants that has no apparent role in basic plant metabolism, but which functions to deter herbivory or herbivore survival in some manner.

SECONDARY PRODUCTION. Production by herbivores, carnivores and detritus feeders. (contrast with primary production)

SECTORIAL CROSS VEIN. Cross veins that connect radius veins at the base of the branches. See also, WINGS OF INSECTS.

SEDGE MOTHS (LEPIDOPTERA: GLYPHIPTERIGIDAE). Sedge moths, family Glyphipterigidae, total 431 species from all regions, mostly in the genus *Glyphipterix* (in the past often misspelled as *Glyphipteryx*), with the largest number from the Australian-New Zealand region; actual world fauna probably exceeds 800 species. There are two subfamilies: Orthoteliinae (only a single species from Europe) and Glyphipteriginae. The family is part of the superfamily Yponomeutoidea in the section Tineina, subsection Tineina, of the division Ditrysia. Adults small (5 to 35 mm wingspan), with head smooth-scaled; haustellum naked; labial palpi tufted and somewhat flattened; maxillary palpi 2 to 4-seg-

mented. Wings elongated and forewings typically with a somewhat falcate apex in many species. Maculation generally dark with various white and metallic-iridescent spots and markings. Adults are diurnal, usually in proximity of the host plants. Larvae are mostly borers in seeds, stems, or leaf axils, and a few are leafminers, but most tropical species are unknown biologically. Host plants are mostly sedges (Cyperaceae), rushes (Juncaceae), and grasses (Gramineae), plus a few other plant families.

John B. Heppner
Florida State Collection of Arthropods
Gainesville, Florida USA

References
Arita, Y. 1987. Taxonomic studies of the Glyphipterigidae and Choreutidae (Lepidoptera) of Japan. *Transactions of the Shikoku Entomological Society* 18: 1–244.
Arita, Y., and J. B. Heppner. 1992. Sedge moths of Taiwan (Lepidoptera: Glyphipterigidae). *Tropical Lepidoptera*, 3 (Suppl. 2): 1–40.
Diakonoff, A. N. 1986. Glyphipterigidae auct. sensu lata (sensu Meyrick, 1913). In H. G. Amsel et al. (eds.), *Microlepidoptera Palaearctic*. Vol. 7. G. Braun, Karlsruhe. 436 pp, 175 pl.
Heppner, J. B. 1982. Synopsis of the Glyphipterigidae (Lepidoptera: Copromorphoidea) of the world. *Proceedings of the Entomological Society of Washington* 84: 38–66.
Heppner, J. B. 1985. *The sedge moths of North America (Lepidoptera: Glyphipterigidae)*. Flora & Fauna Publishing: Gainesville 254 pp.
Kyrki, J., and J. Itämies. 1986. Immature stages and the systematic position of *Orthotaelia [sic] sparganella* (Thunberg) (Lepidoptera: Yponomeutidae). *Systematic Entomology* 11: 93–105.

SEED BEETLES. Members of the family Bruchidae (order Coleoptera). See also, BEETLES.

SEED BUGS. Members of the family Lygaeidae (order Hemiptera). See also, BUGS.

SEED CHALCIDS. Members of the family Eurytomidae (order Hymenoptera). See also, WASPS, ANTS, BEES, AND SAWFLIES.

SEED PREDATION BY INSECTS. Seeds represent an important food resource. Nearly 70% of all human food comes from seeds or their products, and a large portion of the remainder is produced from animals fed on seeds. Insects also consume large quantities of seeds. Seed predation, also referred to as granivory, represents a particular type of herbivory in which a mobile predator consumes a sessile prey (the seed). This prey constitutes a high food quality, but is usually too small to allow full development of the predator with the consumption of only a single prey item. Because of its magnitude and ubiquity, entomologists and plant ecologists have long recognized the important role that seed predation plays in population and community dynamics, and in evolution. Impacts of seed predation can be found in natural habitats including temperate and tropical forests, grasslands, wetlands, mangroves, prairies, and deserts. Insect seed predation has also been documented in human dominated ecosystems such as forest plantations, orchards, and herbaceous annual crop fields.

Conceptually, and due to differences in mechanisms, seed predation can be divided into predispersal and postdispersal phenomena. Predispersal seed predation refers to the consumption of seeds on the parent plant prior to being shed. These seeds represent a spatially and temporally aggregated resource with defense mechanisms provided by the parent plant. Small and sedentary specialist feeders belonging to the insect orders of Diptera, Lepidoptera, Coleoptera, and Hymenoptera are mostly responsible for predispersal seed predation.

Despite the wide occurrence of predispersal seed predation, it is difficult to generalize about the impact that feeding on flowers, ovules, and seeds has on plant recruitment and population dynamics. Several studies have shown that compensatory flowering or seed development may reduce any consequence of predispersal predation. Nevertheless, insect feeding on flowers and developing seeds has been found to be a key factor determining the spatial distribution and density of certain plant species. For example, experimental exclusion of Diptera, Lepidoptera, Hymenoptera, and Thysanoptera has shown that interactions between insect herbivory and seedling survival could determine variation in plant density along a 100-km gradient from the coast to the inland mountains of San Diego County, California, USA.

Although relatively little is known about the extent and impact of predispersal seed predation by insect in herbaceous crop systems, several research-

ers have reported its occurrence. For example, in soybean fields, the joint effect of *Niesthrea louisianica*, a scentless plant bug that feeds on seeds, and fungi of the genera *Fusarium* and *Alternaria*, has been proposed as a tool to control velvetleaf (*Abutilon theophrasti*), a common weed of agricultural systems. The combined effect of seed predation and fungal infection significantly reduced the density of viable velvetleaf seeds. Other examples of predispersal seed predation of agricultural weeds include *Orellia ruficauda* (Diptera: Tephritidae) eating large quantities of seeds of Canada thistle (*Cirsium arvense*), as well as fruit fly, weevils, and moths eating giant ragweed (*Ambrosia trifida*) seeds.

Seeds are inconspicuous once they have been dispersed from the parent plant. These seeds are primarily fed upon by mobile and generalist herbivores in a process referred as postdispersal seed predation. Post-

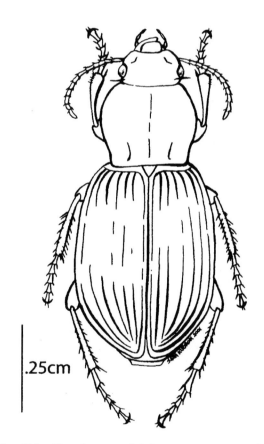

Fig. 910 *Harpalus pensylvanicus*, a common carabid beetles of many agricultural and natural habitats known to eat large number of seeds (Drawing by S. Kudrom).

dispersal seed predators include vertebrates such as birds and rodents, as well as insects of the order Hymenoptera and Coleoptera. In particular, carabid beetles of the genera *Amara*, *Zabrus*, and *Harpalus* are conspicuous seed-eaters. Because of this, considerable research has been conducted on the physiological, behavioral, and morphological adaptation of carabid beetles to seed predation. Carabid beetle larvae consume seeds on the soil surface and accumulate them in below-ground burrows, but seed predation by adult beetles occurs mostly on the soil surface. Seed predation habits of carabids are related to the shape of the mandibles. Specialist granivores of subfamily Carabinae have squat, slender and pointed mandibles, whereas true granivores of Amarini and Harpalini have short, stout, quadrate mandibles with large mandibular muscles, and generalist granivores have intermediate shapes.

Ants are another important group of seed predators found in many ecosystems. In the Sonoran and Chihuahuan deserts of Arizona, USA, colonies of the specialized genera *Pogonomyrmex*, *Veromessor*, and *Pheidole*, as well as the omnivorous genera *Novomessor* and *Solenopsis* contain thousands of foraging workers. During periods of warm temperature, high humidity, and food availability these workers collect large quantities of seeds and store them in underground galleries. Similarly, in tropical agroecosystems of Mexico the native fire ant, *Solenopsis geminata*, harvests large amounts of weed and stores them underground. Despite the large number of seeds attacked by ants, the implications of this behavior for plant population and community dynamics are not straightforward. On one hand, ant nests may provide microsites rich in essential nutrients such as phosphorus and nitrogen that may boost seed germination and seedling growth. On the other hand, seeds captured by ants suffer high rates of destruction and are buried several centimeters below ground, which reduces seedling establishment and growth.

Much less is known about the seed feeding habits of the field cricket, *Gryllus pennsylvanicus*. This species is one of the most abundant and widely distributed field crickets in the northeastern United States, occurring in various grassy habitats, such as crop fields, pastures, weedy areas, roadsides, and lawns. Recent laboratory and greenhouse studies indicated that *G. pennsylvanicus* actively locates and feeds on seeds of several weed species. In a 24-h period under laboratory conditions, a female cricket ate an average of 223 redroot pigweed (*Amaranthus retroflexus*) seeds.

In contrast with predispersal seed predation, the parent plant can not respond to seed death once seeds have been shed. The lack of a physiological feedback implies that the mother plant cannot compensate for an increase in postdispersal seed predation by producing more seeds. Therefore, changes in the intensity of postdispersal seed predation generate responses that occur at the evolutionary level. Defense strategies against postdispersal seed predation include physical mechanisms (e.g., thorns, hard seed covers), camouflage, toxicity through high concentrations of a variety of chemicals, and mast seeding, i.e., the production of seeds in a bimodal distribution with years of large seed outputs and years of small seed outputs.

In recent years, there as been a growing interest in the potential of pre- and postdispersal invertebrate seed predators to act as biological control agents of invasive plants and agricultural weeds. The probability of success of seed-feeding insects in reducing plant abundance depends not just on the establishment and reproduction of large insect populations, but also on characteristics of plant populations. If seed production is high, seedbank density is large, and there are few 'safe sites' for germination and establishment, then the density of these sites, not the number of seeds in the soil, determines seedling densities. It is also necessary to consider the number of seedling plants that reach adult stages of development. Nonetheless, there are several cases in which successful control of invasive plants has been achieved through seed feeding insects. For example, the apionid *Trichapion lativentre* and the weevil *Ryssomatus marginatus* feed on seeds prior to their dispersal. These two insects have been successfully employed in South Africa to reduce the abundance of *Sesbania punicea*, an invasive shrub from South America. Also, two seed-head attacking flies, *Urophora affinis* and *U. quadrifasciata*, are promising biological control agents for control of spotted knapweed (*Centaurea maculosa*), which is a serious problem in rangelands, meadows, and roadsides, especially in sandy soils. Under experimental conditions, the joint effect of these two flies can reduce spotted knapweed seed production up to 95%.

Despite the importance of seed feeding insects as biological control agents, non-target effects may be serious potential problems. For example, the Eurasian weevil *Rhinocyllus conicus* has been released in

Canada and the United States since 1968 for the biological control of introduced thistle species. This weevil feeds on thistle seedheads and attacks not just the targeted invasive species, but several species of native thistles.

In herbaceous crop systems, invertebrate seed predation appears to be a relatively safe strategy to control agricultural weeds. The relatively large size that crop seeds have in comparison with agricultural weeds, the short time span between crop planting and emergence, the tendency of invertebrate seed predator populations to peak when crops are harvested rather than when they are planted, the ground-dwelling habits of invertebrate seed predators, and their inability to fed on buried seeds suggest that invertebrate crop seed predation should not be a concern to farmers.

In conclusion, insect seed predation is a particular type of plant/insect interaction occurring in both natural habitats as well as agricultural systems. Because of the extremely diverse nature of insect-seed relations, it is difficult to make generalizations about the dynamics and consequences of insect seed predation. However, several patterns emerge: 1) seed predation involves plant-insect interactions that have coevolved in chemical, spatial, and temporal dimensions; 2) invertebrate seed predation involves processes spanning from the individual energy budget to dynamics of entire communities; 3) due to several biotic and abiotic factors, seed predation by insects is highly variable in time and space, 4) because insects predators are usually selective, seed predation has a strong potential to affect plant population and community dynamics; and 5) the cumulative effects of insect seed predation coupled with other sources of plant mortality, such as seedling herbivory and seed decay, can determine spatial patterns of plant abundance. Finally, entomologists interested in developing a weed biological control program based on pre- or postdispersal seed predation should carefully study the potential for non-target effects prior to the release of natural enemies.

See also, FOOD HABITS OF ARTHROPODS, GROUND BEETLE FEEDING ECOLOGY.

Fabián D. Menalled
U.S. Department of Agriculture,
National Soil Tilth Laboratory
Ames, Iowa, USA
and

Matt Liebman
Iowa State University
Ames, Iowa, USA

References

Crawley. M. J. 1992. Seed predators and plant population dynamics. pp. 157–191 in M. Fenner (ed.), *Seeds: the ecology of regeneration in plant communities.* CAB International, Wallingford, Oxon, United Kingdom.

Janzen, D. H. 1971. Seed predation by animals. *Annual Review of Ecology and Systematics* 2: 465–492.

Louda, S. M. 1989. Predation in the dynamics of seed regeneration. pp. 25–51 in M. A. Leck, V. T. Parker, and R. L. Simpson (eds.), *Ecology of soil seed banks.* Academic Press, New York, New York.

Menalled, F., M. Liebman, and K. Renner. In press. The ecology of weed seed predation in herbaceous cropping systems. In D. Batish (ed.), *Handbook of sustainable weed management.* Food Products Press, Binghamton, New York.

SEED TREATMENT. A pesticide application to the seed prior to planting to protect the seed, and sometimes the seedling, from injury.

SEGMENT. A major subdivision of the body or appendage, separated from other segments by areas of flexibility.

SELECTIVE PESTICIDE. Pesticides that are toxic principally to the target pest. Pesticides having few adverse effects on nontarget (beneficial) organisms.

SELECTIVE PRESSURE. A force acting on populations that results in some individuals leaving more progeny (descendents, genes) than others, contributing to natural selection.

SELF-REGULATION. The concept that deterioration in the intrinsic quality of individuals regulates populations rather than some external factor such as predators or weather.

SELYS-LONGCHAMPS, MICHEL EDMOND DE. Michel Selys-Longchamps was born at Paris, France, on May 25, 1813. Though lacking formal education, he was a man of wealth, rank, and political influence. He represented various communities in the Belgium parliament and later served as senator and then president of the senate. In addition to such public service, Selys-Longchamps was interested in Odonata, as well as neuropteroid and orthopteroid insects, and in ornithology. His wealth and influence allowed him to amass one of the great insect collections of Neuroptera and Orthoptera. He is now considered to be the father of the study of Odonata. Selys-Longchamps authored about 250 publications, including "Revue des odonates ou libellules d'Europe" in 1850, "Catalogue des insectes Lépidoptères de la Belgique," (1857) and "Memoirs of Odonata from New Guinea, Philippines, Japan, palearctic regions, Europe, Sumatra, and Burma" (1878–1891). He died at Luttich, Belgium, on December 11, 1900.

Reference

Essig, E. O. 1931. *A history of entomology*. The Macmillan Company, New York. 1029 pp.

SEMATURIDAE. A family of moths (order Lepidoptera). They commonly are known as American swallowtail moths. See also, AMERICAN SWALLOWTAIL MOTHS, BUTTERFLIES AND MOTHS.

SEMELPARITY. When organisms produce all their progeny in a single reproductive event, or over a short period of time.

SEMILOOPER. Caterpillars, usually of the order Lepidoptera, family Noctuidae, with a reduced number of prolegs, usually one or two pairs fewer than normal. These loopers move in a looping motion, but the looping movements are small. (contrast with looper)

SEMINAL VESICLE. An expansion of the vas deferens of the male which stores sperm.

SEMIOCHEMICAL. Any chemical involved in the chemical interaction between organisms, particularly signaling chemicals that allow transfer of a message. These also sometimes are called 'infochemicals.' They may be intraspecific or interspecific. Examples

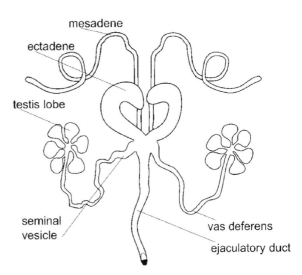

Fig. 911 Diagram of a male reproductive system as found in *Tenebrio* (Coleoptera) (adapted from Chapman, The insects: structure and function).

include pheromones, kairomones, allomones, and synomones. See also, CHEMICAL ECOLOGY.

SEMISOCIAL BEHAVIOR. A level of sociality less than eusocial behavior. A type of presocial behavior. As in quasisocial behavior, semisocial behavior involves individuals of the same generation sharing a nest and cooperating in brood care. However, it also includes division of reproductive behavior, with a worker caste caring for the young of the reproductive caste.

See also, PRESOCIAL, SOLITARY, COMMUNAL, SUBSOCIAL, QUASISOCIAL, EUSOCIAL BEHAVIOR..

SENESCENCE. The process of aging in organisms.

SENESCENT. The period of life after maturity.

SENSILLUM. (pl., sensilla) A simple sense organ associated with the integument, often taking the form of a hair or bristle.

SENSORIUM. (pl. sensoria) A transparent sensory pit on the antennae, usually near the tip.

SENSORY NEURON. Nerves that connect a peripheral sensory location with the insect's central nervous system. Types of sensory neurons include: mechanoreceptors (for touch, position and sound), chemoreceptors (for taste and smell), hygroreceptors (for humidity), photoreceptors (for light), and thermoreceptors (for heat).

SENTINAL HOST. A method of early detection of viruses transmitted by mosquitoes. Various birds, most often chickens, are caged in areas where mosquitoes are active. Blood from the birds is sampled periodically and tested for antibodies specific to arboviruses of interest. Antibody presence can trigger warning of mosquito transmitted disease in human and animal populations.

SEPSIDAE. A family of flies (order Diptera). They commonly are known as black scavenger flies. See also, FLIES.

SEPTICEMIA. A morbid condition caused by the multiplication of bacteria in the blood, and accompanied by production of toxins. *Serratia* and *Streptococcus* species commonly are implicated.

SEPTOBASIDIUM. *Septobasidium* (Order: Septobasidiales, Phylum: Basidiomycota, anamorph: *Harpographium*) is parasitic on colonies of scale insects. *Basidiospores* germinate on the host insect, and hyphae penetrate the cuticle and underlying tissues. Additionally, an extensive hyphal mat forms over the insect as it adheres to the host plants. In *Septobasidium*, this hyphal mat may encase an entire colony of insects with each insect located in an individual chamber. Hyphal mats are often easily recognizable and range in color from brown to yellow, red, or purple. These fungi usually do not kill host scales, but do cause sterility in infected individuals. They cannot survive in nature without the host insects, depending upon them for nutrients and dispersal. In turn, the fungi form an umbrella-like structure that provides a protective environment for any uninfected scales that exist under the mycelial umbrella. At the level of the colony, the relationship

between Septobasidiales and scale insects may therefore be considered to be mutualistic.

Reference
Couch, J. N. 1938. *The genus Septobasidium.* University of North Carolina Press, Chapel Hill, North Carolina.

SEQUENTIAL SAMPLING. A sampling plan in which "stop-lines" continually assess the need for additional sample units during the sampling effort on a given day or time interval. These stop lines are typically based on a prescribed level of desired precision or classification accuracy. Sequential sampling is efficient because no prior knowledge of density is required and no more sample units than necessary are collected. (contrast with fixed-sample size sampling).

See also, SAMPLING ARTHROPODS.

SERICOSTOMATIDAE. A family of caddisflies (order Trichoptera). See also, CADDISFLIES.

SERRATE. Notched, like the teeth of a saw.

SERRATIA ENTOMOPHILA. Several species within the genus *Serratia*, including *Serratia marcescens*, *S. entomophila* and *S. proteamaculans*, have been identified as entomopathogenic. These species can be isolated using the selective caprylate thallous agar. Many *S. marcescens* strains produce characteristic bright red colonies. Insects infected with *S. marcesens* succumb within one to three days and often

Fig. 912 Healthy (top) and infected (bottom) *Costelytra zealandica* grubs. Note the absence of the dark underlying midgut in the grub infected with *Serratia entomophila.* These amber grubs have voided their guts and have ceased feeding.

exhibit a reddish tinge. Whether or not *S. marcesens* is able to actively invade healthy insects is unclear. In many cases, this disease has been associated with poor sanitation and crowded rearing conditions. Certain parasitic hymenoptera have been reported to vector this bacterium on contaminated ovipositors. Very likely, both the pigmented and non-pigmented *S. marcescens* strains are opportunistic pathogens.

Strains of *S. entomophila* and *S. proteamaculans* are true pathogens of the New Zealand grass grub, *Costelytra zealandica*, and are the causal agents of amber disease. *Serratia* strains that induce amber disease possess a common large 140 kb plasmid (pADAP) which, when removed by heat-curing cells, resulted in loss of virulence. These pADAP plasmid bearing strains of *Serratia* exhibit a high degree of host specificity and have been found to infect only *C. zealandica*. Larvae infected with pathogenic *S. entomophilia* and *S. proteamaculans* rapidly cease feeding. It has been shown that infection by pathogenic *Serratia* strains causes a reduction in the trypsin activity in the gut, disrupting the normal digestive process. The gut contents, which are normally black, are clarified, conferring an amber coloration to infected larvae. The amber larvae are unable to feed and assimilate nutrients and undergo a starvation period that may last from one to four months. During this period, the bacteria colonize the gut and become attached to cuticular surfaces including the external cuticle. Late in infection, bacteria may penetrate the hemocoel and cause a generalized septicemia. Alternatively, the colonizing population of this pathogen may be displaced by secondary bacteria prior to host death. Interestingly, the pathology induced by *S. entomophila* is similar to that caused by *S. proteamaculans*.

References

Jackson, T. A., A. M. Huger, and T. R. Glare. 1993. Pathology of amber disease in the New Zealand grass grub *Costelytra zealandica* (Coleoptera: Scarabaeidae). *Journal of Invertebrate Pathology* 61: 123–130.

Jackson, T. A., D. G. Boucias, and J. -O. Thaler. 2001. Pathobiology of amber disease caused by *Serratia* spp. in the New Zealand grass grub *Costelytra zealandica*. *Journal of Invertebrate Pathology* 48: 232–243.

SERRITERMITIDAE. A family of termites (order Isoptera). See also, TERMITES

SESIIDAE. A family of moths (order Lepidoptera). They commonly are known as clearwing moths. See also, CLEARWING MOTHS, BUTTERFLIES AND MOTHS.

SESSILE. Immobile; incapable of moving.

SETA. (pl., setae) A hair or bristle.

SETACEOUS. Bristle-like. This term usually is used to describe antennae.
See also, ANTENNAE OF HEXAPODS.

SETAL MAP. A diagrammatic representation of the setae found on the body of larvae.

SEX ATTRACTANT PHEROMONES. Pheromones are chemicals emitted and secreted by an individual and received by another individual of the same species, in which they induce a species specific reaction. By far the greatest variety of pheromone-mediated systems occurs among insects. Sex pheromones (chemicals, or blends of chemicals, emitted by an individual of one sex for the purpose of inducing sexual behaviors by members of the opposite sex) are the only pheromone mediated system in common usage by members of all insect orders. This undoubtedly stems from the universal need to reproduce and the development of chemosensory organs early in the evolutionary history of Hexapoda, perhaps even before the development of light-sensitive organs. In general, sex pheromones act as sexual attractants when they induce the receiving sex to move toward the releasing sex or as copulatory stimulants, or aphrodisiacs, when released or perceived in close proximity to the receiving sex to induce a receptivity or copulatory behaviors. However, it should be noted that attractant pheromones also may act as copulatory stimulants at close range. Although known sex pheromones act to induce a behavioral change in the receiving sex (releaser pheromones) there is no reason not to assume that sex pheromones could have primer effects and induce physiological changes associated with sex. Indeed, it is quite conceivable that volatile pheromones could induce release of neuropeptides or other rapidly dispersed hormones or cause nervous signals that would induce physiological changes required for effective mating. Quite simply, we have not looked extensively for such effects.

Historically, the use of conspecific attractant pheromones by insects was documented as early as 1690 when John Ray reported several male *Biston betularia* (Lepidoptera) flying around a caged female. This knowledge of the attractive capacity of female Lepidoptera was used by such great naturalists as Fabré for collection of rare specimens; the procedure used was essentially the same as that of Ray. The use of live females, and naturally produced sex pheromones for population monitoring of the gypsy moth, *Lymantria dispar* (L.), began as early as 1914, but by 1920 females had been replaced by organic solvent extracts of abdomen tips, which remained attractive for longer periods of time than did females. Attempts to isolate the chemicals responsible for induction of sexual attraction and behavior also were begun in the 1920s. However, the methods available for chemical analysis were inadequate to obtain concrete identifications and required such tremendous amounts of sample that little headway was made. Butenandt and his colleagues identified the first sex pheromone, produced by females of the silkworm moth, *Bombyx mori* (L.), in 1959. The elucidation of this pheromone, bombykol ((*E,Z*)-10,12-hexadecadiene-1-ol), had taken 20 years and required sacrifice of 500,000 females. With the development of improved analytical chemistry techniques, methods for collection of volatile chemicals released by insects, highly sensitive electrophysiological techniques like the electroantennogram detector and sophisticated bioassay techniques, the isolation and identification of sex pheromone components has become much easier, rapid and requires far fewer insects than those used for earlier work. In fact, sex pheromones have now been identified for at least 2,000 species representing more than 90 families of insects, and the list grows every day.

Chemistry

Sex pheromones used by insects are all lipidic in nature, and most are composed of acetogenins and mevalogeninsins. Apart from this, there is no general theme associated with the types of lipids that act as pheromones. However, relatively low molecular weight compounds (20 carbons or less) generally serve as long distance attractants because of their volatility. Most of these contain oxygenated functional groups like alcohols, esters, ketones, aldehydes, epoxides, and lactones, but non-oxygenated monoterpenes, sesquiterpenes and hydrocarbons also serve as long distance attractants. Long chain wax esters and hydrocarbons that are not volatile serve as close range stimulants and require contact by the receiver in order to induce sexual behavior. Often, closely related groups of insects use pheromones synthesized in a similar fashion, as is evidenced by the use of long chain cuticular hydrocarbons as contact sex pheromones by muscid flies, the acetates and propionates of monoterpene and sesquiterpene alcohols used by scale insects, and the blends of 12 to 16 carbon saturated, and unsaturated alcohols, aldehydes and acetates used by noctuid moths. These moth sex pheromones are formed by Δ-11-desaturation of palmitic (or octadecanoic) acid followed by varying degrees of chain shortening via β-oxidation and subsequent modification of the acid moiety to form the oxygenated functional group. Nonetheless, there are many insects that use blends of compounds that are structurally unrelated for sexual signaling. For example, males of the Mediterranean fruit fly release a blend of compounds containing straight chain alcohols, monoterpenes, sesquiterpenes, lactones, and nitrogen containing compounds including 1-pyrroline and 2-ethyl-3,5-dimethyl pyrazine.

Pheromone function

The key to effective sexual signaling with pheromones is the use of chemicals, or chemical blends, that are species specific because it is metabolically very expensive to produce or respond to pheromones if no mating reward results. In fact, sex pheromones of insects are the primary means by which many closely related species achieve reproductive isolation. Although there are instances in which insects use a single component for sexual communication, these cases are rare and, more often than not, involve the use of chiral compounds in which one enantiomer is active and the other inhibitory. For example, the sex pheromone of the Japanese beetle is (*R,Z*)-5-(1-decenyl)-dihydro-2(3H)-furanone and the (*S,Z*) enantiomer is inhibitory. Thus, a significant decline in trap capture has been documented when as little as 1% of the (*S,Z*) enantiomer is incorporated into lures containing the (*R,Z*) enantiomer. The most common method for imparting species specificity is the use

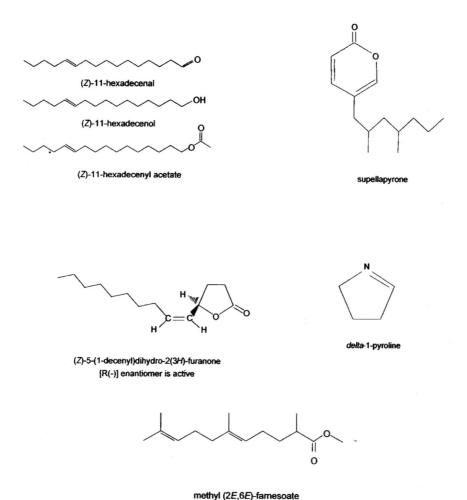

Fig. 913 Examples of insect sex pheromones. Sex pheromones common to noctuid moths include (*Z*)-11-hexadecenal, (*Z*)-11-hexadecenol and (*Z*)-11-hexadecenyl acetate. That of the Japanese beetle is (*Z*)-5-(1-decenyl)dihydro-2(3*H*)-furanone. Supellapyrone is the pheromone of the brown banded cockroach, *Delta*-1-pyroline is from the Mediterranean fruit fly and methyl (2*E*,6*E*)-farnesoate is produced by male stink bugs.

of blends of chemical compounds. As such these pheromones are best equated with perfumes because, as with perfumes used by humans, very minor differences in the blend of chemicals results in significant differences in the attractive odor of the pheromone. Blend specificity can be imparted by a number of simple changes in chemistry of the compounds including the use of: 1) compounds having different oxygenated functional groups (aldehydes, acetates, alcohols or ketones); 2) compounds having different numbers of carbons in the skeleton; 3) compounds having different degrees of unsaturation; 4) compounds having different geometries of double bonds;

5) blends of compounds having different ratios of the same components; 6) blends of compounds containing different numbers of components.

Despite the fact that sex pheromones have been identified from a large number of insect species, detailed studies on perception and signal processing have been conducted on surprisingly few insect species. Thus, we are just beginning to understand the mechanisms involved in peripheral perception and central signal processing. Peripheral perception of pheromones occurs at the antenna, where specialized sensilla act to filter out pheromone molecules from the air. The cuticular wall of the sensilla contains numerous pore tubules through which odorant molecules diffuse into the lumen of the sensilla. Once in the lumen, the pheromone molecules are bound to pheromone binding proteins that solublize the pheromone in the aqueous receptor lymph and carry them to specialized receptors on the surface of dendrites in the sensillum. Binding of pheromone molecules with receptors on the dendrites leads to a G-protein-mediated activation of phospholipase C and generation of inositol triphosphate, which activates inositol triphosphate-gated Ca^{++} channels in the dendritic membrane, and results in electrochemical signal transduction. After release from the dendritic receptor, pheromone molecules must be degraded rapidly. This is accomplished by aggressive pheromone degrading enzymes in the receptor lymph. These degrading enzymes do not distinguish between pheromone molecules that have bound with receptors, or those that have not, so there is competition between pheromone binding proteins and degrading enzymes in the receptor lymph. Pheromone-responsive axons exit the antenna and enter specialized structures in the antennal lobe of the brain called the macroglomerular complex. The macroglomerular complex is sex specific, being found only in the sex that perceives and responds to sex pheromone, and it is here that efferent axons form synapses between interneurons and afferent neurons. The actual mechanisms that translate electrical signals generated by pheromone responsive neurons in the macroglomerular complex into behavioral events associated with response to sex pheromones are, as yet, not known.

Use of sex pheromones for insect management

Sex pheromones have been targets for development of strategies for insect control because of their species specificity. Initially, sex pheromones were primarily used as monitoring tools to determine population densities of pests in agricultural crops or storage facilities and this is still the major use of sex pheromones. When deployed in traps, sex pheromones can be used as detection tools to provide: 1) an early warning system about the presence of a pest, 2) a survey tool to determine the population distribution of insects, and 3) a way to determine if quarantine pests have been introduced to an area. Sex pheromone trapping systems are also effective for determining a population threshold for: 1) the timing of control treatments, 2) timing of other sampling methods, for example, searching for larvae, and 3) risk

assessment. Population monitoring also provides ways to assess population trends, the effects of control programs and dispersion of pests. Additionally, sex pheromones of insects can be effectively used to determine the distribution and spread of entomophagous insects released as biological control agents.

Sex pheromones also are used for direct control of insect pest populations. Direct control using sex pheromones can be achieved by mass trapping, attraction of pest insects to baits containing pesticide (attract-and-kill protocols) and disruption of mating. Although mass trapping is conceivably possible with sex pheromones alone, the fact that mated females are not normally captured in sex pheromone baited traps significantly reduces the efficacy of this technique. However, sex pheromones are used in combination with other attractants for mass trapping. For example, mass trapping of the Japanese beetle with lures containing the female produced sex pheromone along with the feeding attractants phenethyl propionate, eugenol and geraniol has been shown to be effective. Males are attracted to the sex pheromone in large numbers and inclusion of the feeding lure results in capture of great numbers of females. Additionally, the use of pheromones that act in dual roles has been shown to be effective for mass trapping. For example, the pheromone of the boll weevil acts as an aggregation pheromone in the spring and fall, which results in capture of both sexes, but in summer only females are attracted. Capture of both sexes in spring and fall reduces the populations significantly prior to infesting cotton, and capture of females in the summer keeps the population in check during the growing season.

A second method of population suppression with sex pheromones is the attract-and-kill system in which pheromone dispensers also contain a conventional pesticide. The key to such systems is the use of a pheromone system that is as effective, or preferably more effective, in attracting insects than is the releasing sex.

Mating disruption has become the most successful method for direct control of insect pests using sex pheromones. In this technique, large amounts of pheromone are released throughout control areas, either by strategic distribution of a relatively small number of pheromone dispensers, each of which releases substantial amounts of pheromone, or by broadcasting large numbers of small dispensers which release relatively small amounts of pheromone throughout the entire control area. Although the exact mechanisms responsible for mating disruption with pheromones are not well understood, five mechanisms have been proposed to account for mating disruption. These mechanisms include: 1) sensory fatigue in which constant exposure to high levels of pheromone causes sensory adaptation of pheromone receptors on the antenna or an habituation to pheromone in the central nervous system which, in either case, inhibits the receiving sex from responding to natural pheromone; 2) competition between natural and synthetic sources (false trail following) in which the number of pheromone plumes in a field generated from synthetic sources is far greater than the number of the signaling sex and, thus, the receivers spend all of their time finding synthetic sources; 3) camouflage of natural sex pheromone plumes could occur when the amount of pheromone released is so great that it forms a fog permeating the entire atmosphere which does not allow the receiver to track down a natural point source of sex pheromone (i.e., the emitter); 4) sensory imbalance can occur when an 'off blend', not so different from the natural blend that insects fail to respond to it, is completely superimposed over the natural point sources of pheromone and results in the disorientation of the insect; and 5) mating disruption also occurs when compounds that inhibit the response of insects to the natural pheromone blend are included in the synthetic disruptant, as is evidenced by disruption of *Eupoecilia ambiguella* using technical grade pheromone which contains the pheromone *(Z)*-9-dodecenyl acetate plus as little as 0.1% of the inhibitory isomer *(E)*-9-dodecenyl acetate.

See also, REPRODUCTION, PHEROMONES, REGULATION OF SEX PHEROMONE PRODUCTION IN MOTHS.

Peter E. A. Teal
U.S. Department of Agriculture, Center for Medical, Agricultural and Veterinary Entomology
Gainesville, Florida, USA

References

Carde, R. T., and A. K. Minks (eds.) 1997. *Insect pheromone research new directions*. Chapman and Hall, New York, New York.
Millar, J. G., and K. F. Haynes (eds.) 1998. *Methods in chemical ecology, Vol. 1. Chemical methods*. Chapman and Hall, New York, New York.
Prestwich, G. D., and G. J. Blomquist (eds.) 1987. *Pheromone biochemistry*. Academic Press, Orlando, Florida.
Ridgeway, R. L., R. M. Silverstein, and M. N. Inscoe (eds.) 1990. *Behavior-modifying chemicals for insect management*. Marcel Dekker, New York, New York.

SEX CHROMOSOME. A chromosome which is involved in sex determination.

SEX PHEROMONE. Pheromones that attract the opposite sex for mating. These are often used, in conjunction with traps, to monitor abundance of insects. See also, CHEMICAL ECOLOGY.

SEX RATIO MODIFICATION BY CYTOPLASMIC AGENTS. The normal sex ratio of progeny produced by mated diploid insects is one female to one male (1:1); typically, unmated diploid females cannot produce progeny. That normal 1:1 sex ratio is modified when unmated insects reproduce by parthenogenesis (called arrhenotoky or thelytoky). With arrhenotokous (haplo-diploid) species, such as bees and wasps in which both haploid males and diploid females occur, the sex ratio of progeny produced by mated females commonly is biased towards females, often with 2 or 3 females for each male. Unmated arrhenotokous females can produce only haploid male progeny. Thelytokous species commonly have only females, although sometimes rare males are produced. These sex-determining systems are based on the chromosomes in the nuclear genome of insects. However, normal sex ratios in insects can be modified by a number of organisms (bacteria, viruses, protozoans) and by genetic elements called meiotic drive elements, which are genes on the chromosomes that modify the normal distribution of chromosomes to egg and sperm during meiosis. See also, MEIOTIC DRIVE IN INSECTS.

Cytoplasmic agents distort normal sex ratios

Most cytoplasmic agents (bacteria, viruses, spiroplasmas) are inherited primarily through the mother and thus are cytoplasmically or maternally inherited. Cytoplasmic agents that can manipulate their insect host's sex ratio (usually increasing the proportion of females) and promote their own spread are called cytoplasmic sex ratio distorters. The spread of a cytoplasmic sex ratio distorter often reduces the fitness of its host and can drive insect populations to extinction.

Sex ratio distorters are usually suspected if single-pair crosses produce a heavily biased sex ratio in the progeny, although meiotic drive agents are also possible mechanisms.

At least 50 cases have been identified in which cytoplasmic agents alter sex ratios in insects or mites. Examples are found in the Diptera, Heteroptera, Coleoptera, Lepidoptera or Acari (mites). Cytoplasmic sex ratio distorters may be widespread, but undetected, in other arthropods. Maternal transmission rates of these agents typically are high, although a few daughters may fail to become infected. The altered sex-ratio conditions are found in natural populations at frequencies ranging from low to high. The infections may reduce fitness of the hosts, with egg hatch or larval survival reduced in the progeny of infected females. Specific examples are described below.

Spiroplasmas. The sex ratio condition of *Drosophila willistoni*, and related neotropical *willistoni* group species, is due to a spiroplasma. Spiroplasmas are maternally inherited and transovarially (transmitted in the egg) transmitted. The spiroplasmas are lethal to male embryos. Spiroplasmas can be transmitted between species by injecting hemolymph containing them. Spiroplasmas from different species are different and a virus associated with the spiroplasmas also is different for each spiroplasma. When spiroplasmas from different species are mixed, they clump because the viruses lyse the spiroplasmas of the other species.

L-form bacteria. The *Drosophila paulistorum* complex contains six semispecies (subgroups derived from a single species that are thought to be in the process of speciation) that do not normally interbreed in the field. When they are crossed in the laboratory, fertile daughters and sterile sons are produced. Streptococcal L-form bacteria that are associated with the male sterility have been isolated and cultured in artificial media. The L-forms are transferred through the egg cytoplasm and each semispecies is associated with a different L-form bacterium. The L-forms can be microinjected into females and can produce the expected male sterility in the progeny.

Wolbachia. These rickettsia-like organisms are one of the most commonly described microorganisms found in arthropods. *Wolbachia* are intracellular and found in the reproductive tissues, as well as in the

somatic cells of some species. *Wolbachia* have different effects on their arthropod hosts, but one that is relevant to this entry is their ability to alter sex ratio. The role of *Wolbachia* as endosymbionts was discussed in the section on Symbionts of Arthropods.

Wolbachia are gram-negative rods that cannot be cultured easily outside their hosts and they are widespread, having been found in up to 76% of all insect species surveyed. In addition to insects and mites, *Wolbachia* have been found in Crustacea and nematodes. Knowledge of the physiological and phenotypic effects of *Wolbachia* on most of their hosts remains limited.

In the Crustacea, *Wolbachia* infect isopods, including *Armadillidium album*, *Ligia oceanica*, *A. nasatum*, *Porcellionides pruinosus*, *Chaetophiloscia elongata* and *Spaeroma rugicauda*. Some *Wolbachia*-infected isopods regularly produce female-biased broods because the *Wolbachia* change genetic males (homogametic ZZ individuals) into functional females. Such individuals are chromosomally male (ZZ) but phenotypically appear and function as females. These 'daughters' of infected mothers produce all-female or highly female-biased progeny, resulting in isopod lineages that are chromosomally males but are functional females. Interestingly, there has been speculation that some *Wolbachia* genes have been transferred to the isopod nuclear genome–reminiscent of the movement of genes over evolutionary time from the mitochondria (originally a microbial symbiont) to the nuclear genome of eukaryotes.

Wolbachia can cause thelytoky (production of all females), male killing, and female mortality in insects and mites. Some *Wolbachia* improve fertility or vigor while others appear to decrease these traits in their hosts. Some species appear to have *Wolbachia* only in their germ line (ovaries and testes) while others have *Wolbachia* in somatic tissues, as well. *Wolbachia*-induced thelytoky (production of females only) in the Hymenoptera has been found in the Tenthredinoidea, Signiforidae and Cynipoidea, as well as at least 70 species of parasitoids (Aphelinidae, Encyrtidae, Eulophidae, Pteromalidae, Torymidae, Trichogrammatidae, Cynipidae, Eucoilidae, Braconidae, Ichneumonidae, Proctotrupoidae). Many hymenopteran parasitoids have both arrhenotokous and thelytokous strains, with thelytoky associated with the presence of *Wolbachia*.

In thelytokous populations of parasitoids, unfertilized eggs give rise to females. A number of thelytokous parasitoids (*Ooencyrtus submetallicus*, *Pauridia peregrina*, *Trichogramma* sp., *Ooencyrtus fecundus*) produce a few males, usually less than 5%, when reared at temperatures over 30°C. Heat treatments are known to reduce or eliminate *Wolbachia* infections in some species, so it is thought that the low rates of male production are due to reduction or elimination of *Wolbachia* from some females. Sometimes, the rare males produced by cured females have been shown to mate and transfer sperm, indicating that these males have normal vigor and fertility. In other cases, the rare males are infertile, suggesting that the *Wolbachia* infection had existed for a long time in the population, which could have allowed selection for essential fertility genes to be relaxed over evolutionary time. In addition to heat treatments, several antibiotics (tetracycline hydrochloride, sulphamethoxazole and rifampin) can induce the production of males in some thelytokous parasitoid populations.

The cytogenetic changes that occur during meiosis to restore an unfertilized haploid egg to diploidy (thus permitting thelytoky) has been studied in *Trichogramma*. In the eggs of *Wolbachia*-infected *Trichogramma* females, meiosis progresses to the stage of a single haploid pronucleus but the diploid chromosome number is restored during the first mitotic division. This happens because, during anaphase, the two identical sets of chromosomes do not separate and the result is a single nucleus containing two copies of the same set of chromosomes, resulting in a female that is completely homozygous at all loci.

Wolbachia infection can influence mating behavior of insects, as well as kill males. For example, populations of the butterfly *Acraea encedon* are found across Africa. In many populations, females produce only female progeny while other populations produce both males and females in a normal 1:1 sex ratio. The production of all female progeny is caused by a *Wolbachia* strain that kills males. *A. encedon* typically deposit clutches of 50 to 300 eggs and newly hatched larvae often cannibalize unhatched eggs, only gradually dispersing into smaller groups.

The evolution of male-killing by *Wolbachia* may be favored when the behavior and ecology of a species makes antagonistic interactions between host siblings or sib cannibalism likely. Under field conditions, the *Wolbachia* infection rates in *A. encedon* females may result in populations with a serious shortage of males. As a consequence, the mating behavior of *Wolbachia*-infected *A. encedon* has been altered.

Normally males seek out and compete for individual females that are located near larval food plants. However, when male-killing *Wolbachia* are present in high frequency in a population, *A. encedon* females instead form dense aggregations in grassy areas near trees, perhaps to attract rare males as mates.

B chromosomes and cytoplasmic factors. Sex ratio in the haplo-diploid (arrhenotokous) parasitoid wasp *Nasonia vitripennis* can be altered by at least two different mechanisms. Some natural populations of *N. vitripennis* carry a supernumerary or B chromosome that causes a condition called paternal sex ratio (PSR). B chromosomes are found in many plant and animal species and are small non-vital chromosomes mostly consisting of heterochromatin (DNA that is noncoding for proteins). Most B chromosomes have few genes but they often cause a small fitness cost to their host, making them selfish genetic elements. Some B chromosomes are thought to be derived from normal chromosomes and are transmitted at higher rates than expected, thus exhibiting 'drive'.

The PSR chromosome is carried only by haploid male *N. vitripennis* and is transmitted via sperm to fertilized eggs. After an egg is fertilized by a PSR-bearing sperm, all normal paternally derived chromosomes condense into a chromatin mass and are lost, leaving only the maternal chromosomes. The PSR chromosome itself survives. Thus, the PSR chromosome changes fertilized diploid (female) eggs into haploid males that carry the PSR chromosome.

Where did the PSR chromosome come from? Molecular analyses suggest the PSR chromosome has sequences that are homologous with sequences found in *Nasonia giraulti*, *N. longicornis* and *Trichomalopsis dubius*, but not with *N. vitripennis*. Thus, the PSR chromosome could have been present prior to the divergence of the genera *Trichomalopsis* and *Nasonia*. Alternatively, PSR may have crossed the species barrier more recently (horizontal transfer) through a series of transfers between species capable of mating. Experimental interspecific transfer of PSR was successful after these species were cured of *Wolbachia* which causes cytoplasmic incompatibility between them. The transferred PSR chromosome continued to function in both recipient species.

The sex ratio of *Nasonia vitripennis* can be modified by cytoplasmic factors, including 'Son-killer', a maternally transmitted bacterium that prevents development of unfertilized male eggs and 'Maternal Sex

Ratio', a cytoplasmically inherited agent that causes female wasps to produce nearly 100% daughters.

Male killing in Coccinellidae

The Coccinellidae appear particularly prone to invasion by male-killing organisms, with four different groups (*Rickettsia*, *Spiroplasma*, Flavobacteria and *Wolbachia*) having been associated with this phenomenon. Coccinellids may be especially susceptible to invasion by male-killing microbes due to their behavior; coccinellids typically feed on aphid populations that are patchy and deposit eggs in tight batches. This behavior could promote egg cannibalism by newly hatched larvae when normal prey populations are low. Such cannibalism could provide an additional mechanism by which the bacteria are transmitted to new individuals.

The evolution of the male killing ability may have evolved in the bacteria because they are almost exclusively transmitted vertically from mother to eggs; thus, bacteria in male hosts are at an evolutionary dead end, so male-killing has a fitness cost of zero from the bacterial point of view. Furthermore, the death of male embryos could augment the fitness of the remaining female brood members by providing food to those females carrying the relatives of the male-killing bacteria. A full understanding of the evolutionary dynamics of male-killers and their hosts remains elusive.

See also, REPRODUCTION.

Marjorie A. Hoy
University of Florida
Gainesville, Florida, USA

References

Bourtzis, K., and S. O'Neill. 1998. *Wolbachia* infections and arthropod reproduction. *BioScience* 48: 287–293.

Cook, J. M., and R. D. J. Butcher. 1999. The transmission and effects of *Wolbachia* bacteria in parasitoids. *Researches on Population Ecology* 41: 15–28.

Hurst, G. D. D., and F. M. Jiggins. 2000. Male-killing bacteria in insects: mechanisms, incidence and implications. *Emerging Infectious Diseases* 6: 329–336.

Jiggins, F. M., G. D. D. Hurst, and M. E. N. Majerus 1998. Sex ratio distortion in *Acraea encedon* (Lepidoptera: Nymphalidae) is caused by a male-killing bacterium. *Heredity* 81: 87–91.

O'Neill, S. L., A. A. Hoffmann, and J. H. Werren (eds.) 1997. *Influential passengers. inherited microorganisms and arthropod reproduction.* Oxford University Press, Oxford, United Kingdom. 214 pp.

Wrensch, D. L., and M. A. Ebbert (eds.) 1993. *Evolution and diversity of sex ratio in insects and mites.* Chapman and Hall, New York, New York.

Relevant website

The European *Wolbachia* Project: Towards Novel Biotechnological Approaches for Control of Arthropod Pests and Modification of Beneficial Arthropod Species by Endosymbiotic Bacteria. http://wit.integratedgenomics.com/GOLD/Wolbachia.html
http://www.ncbi.nim.nih.gov

SEXUAL RECOMBINATION. The process by which DNA is exchanged by homologous chromosomes by pairing and crossing over during meiosis.

SEXUAL REPRODUCTION. Reproduction involving the union of gametes. (contrast with parthenogenetic reproduction)

SEXUALES. In aphids, the sexual forms of male and females.

SEXUAL SELECTION. Sexual selection is the class (i.e., form, section, axis) of natural selection that is associated with genetic differences among individuals in their ability to (a) compete for, or (b) select, fertile partners in reproduction (i.e., mates). The former is typically associated with males, and the latter, with the female sexual role, but such distinction is continually eroded as ever more is learned of the details of mating behavior in different kinds of organisms.

Competition and mate attraction on the one hand, and mate choice on the other, can involve: direct physical combat among rivals; physical appearance reflecting health (e.g., symmetry or conformity to "type," as these reflect nutritional and disease history); vigor or strength as demonstrated via "ritualized" displays; the ability to provide or defend resources for partners, and many other comparative and competitive categories and situations.

It can be useful to think of natural selection and sexual selection as the agents of genetic changes that promote individual adaptation (any genetically controlled attribute – e.g., structure, physiological process, or behavior – that increases an organism's probable genetic contribution to succeeding generations; a characteristic that enhances an organism's chances of perpetuating its genes, usually by leaving descendants) in genetic lineages along two axes: the first axis, ecological/somatic adaptation is that connected with growth, development, maintenance, and survival; and the second, reproductive/genetic adaptation is that directly connected with procreation, the successful production of competitive, fertile offspring. Such a separation in a thought experiment quickly reveals the strong connection yet important distinction that exists between natural and sexual selections, and that must be recognized by entomologists in a number of fields – taxonomy, behavioral ecology, ecology, and pest control and management, to mention a few. For example, a taxonomist who wishes to understand the process of speciation and the genetic divergence of populations needs to give special attention to the mate choice aspects of sexual selection and its connection with classical "reproductive isolation" theory.

Examples of sexual selection phenomena in insects include weapons used by males to control resources and gain access to females. If a resource, something that females lay their eggs into or eat is relatively rare, then males that wait on or near the resource are likely to encounter potential mates among the females that come to oviposit or feed. If this resource is relatively small, then a male may be able to keep it and its associated females to himself by defeating intruding males. Under these circumstances males may have evolved weapons that intimidate or help physically remove sexual rivals. In certain wood and dung-feeding beetles (Scarabaeidae), males sometimes have enormous horns arising from the head and prothorax that are used as levers or pinchers to fling or carry other males from branches, logs, and feces.

Examples of male display and female choice are seen when resources are relatively common, and males cannot profitably wait by an oviposition or feeding site to encounter females. In such situations, should females encounter male interference at a resource site they can easily find another, one without males, where they are not forced to fend off or submit to sexual advances. Under these circumstances males tend to evolve courtship displays that persuade females that they are suitable mates. For example, certain Tephritidae, such as *Ceratitis capitata* Wiede-

mann (the Mediterranean fruit fly), are among the most serious pests of fruits and limit exports wherever they occur. Males produce a buzzing sound ("song") with their wings as they attempt to mate, and within limits, females select as mates those with louder songs, even when a male's effort is artificially amplified by playing a recorded song at a higher volume than one could otherwise produce. The "meaning" of (i.e., relevant information within) such displays is often difficult to decipher, but it may be that the song is a display of vigor that advertises an underlying genetic quality.

Sex role reversals sometimes occur in species with male "nuptial gifts." Males sometimes provide females with a food item before mating, and in these circumstances females appear to choose mates based on the quality of the gift presented. It may be that females are primarily interested in the food, but they may also judge the ability of the male, i.e., his underlying genetic quality, by his capacity to find and capture or produce certain items. In a number of katydids (Tettigoniidae), including the agricultural pest the Mormon cricket (*Anabrus simplex Haldeman*), males produce a very large spermatophylax. This protein-containing wad of cheese-like material is associated with the ejaculate and can constitute as much as 30% of a male's body weight. Females use this substance as food and those that receive larger "gifts" have greater fecundity (achieved reproduction). When other foods are scarce, and in species where males produce an exceptionally large gift, there sometimes occurs a sex-role reversal. That is, it is the females that seek out and display toward choosy males. Male Mormon crickets estimate the weight of a potential mate, hence the numbers of eggs she is likely to produce in the future as she mounts him (orthoptera fashion) prior to copulation. Males reject light-weight females in favor heavier ones, and in this way promote the favored outcome, that their resources will be used in the production of greater numbers of offspring.

See also, VISUAL MATING SIGNALS.

James E. Lloyd
University of Florida
Gainesville, Florida, USA
and
John Sivinski
U.S. Department of Agriculture,
Agricultural Research Service
Gainesville, Florida, USA

References

Blum, M., and N. Blum. 1979. *Sexual selection and reproductive competition in insects*. Academic Press, New York, New York.

Choe, J. C., and B. J. Crespi. 1997. *Mating systems in insects and arachnids*. Cambridge University Press, Cambridge, United Kingdom.

Gwynne, D. 2001. *Katydids and bush-crickets: reproductive behavior and evolution of the Tettigoniidae*. Cornell University Press, Ithaca, New York.

Lloyd, J. E. 1979. Mating behavior and natural selection. *Florida Entomologist* 62: 17–34.

Sirot, L. The evolution of insect mating structures through sexual selection. *Florida Entomologist*. (in press)

Sivinski, J., M. Aluja, G. Dodson, A. Freidberg, D. Headrick, K. Kaneshiro, and P. Landolti. 2000. Topics in the evolution of sexual behavior in the Tephritidae. pp. 751–792 in M. Aluja and A. Norrbom (eds.), *Fruit flies (Tephritidae): phylogeny and evolution of behavior*. CRC Press, Boca Raton, Florida.

Thornhill, R., and J. Alcock. 1983. *The evolution of insect mating systems*. Harvard University Press, Cambridge, Massachusetts.

SEXUPARA. (pl., sexuparae) In aphids of the family Pemphigidae, winged viviparous parthenogenetic females that produce both males and females. See also, APHIDS.

SHADE TREE ARTHROPODS AND THEIR MANAGEMENT.

There are unique issues associated with arthropod management of shade trees. These include: the high value of the plant material, including replacement costs; the siting of plants in areas of high human traffic; the unique management considerations associated with the above; and the aesthetic issues that can be important in management decisions.

Almost invariably, shade trees are a blend of native species and, increasingly, exotic species that have desirable features for the site. Concurrently, associated arthropod pests are also a mixture of native and exotic species and in many areas, the latter predominate. For example, among the more serious exotic shade tree pest species present in the United States are the gypsy moth, *Lymantria dispar* (L.), oystershell scale, *Lepidosaphes ulmi* (L.), the European elm bark beetle, *Scolytus multistriatus* (Marsham), the twospotted spider mite, *Tetranychus urticae* (Koch), and the Japanese beetle, *Popillia japonica* (Newman).

Hundreds of species of insects and mites seriously affect shade trees worldwide. These are often differentiated using a combination of taxa and feeding habits.

ARTHROPODS WITH SUCKING MOUTHPARTS.

Aphids. These may be considered in a broad sense to include 'true aphids' (Aphididae), as well as 'woolly aphids' associated with deciduous plants (Aphididae: Eriosomatinae) and adelgids (Adelgidae). Aphids primarily extract sap from the phloem. This injury is often well tolerated by healthy plants, but may be a serious stress. Species that colonize branches and trunks, such as the hemlock woolly adelgid (*Adelges tsugae* Annand), can be particularly damaging. On some plants, aphids colonize during periods of new growth and produce serious curling distortions of leaves or needles that can adversely affect the growth and form of plants. Aphids (Aphididae) also excrete honeydew, a significant aesthetic problem with shade trees. Among the more important aphids associated with North American shade trees are the tuliptree aphid, *Illinoia liriodendri* (Monell), the apple aphid, *Aphis pomi* De Geer, the giant willow aphid, *Tuberolachnus salignus* (Gmelin), the Norway maple aphid, *Periphyllus lyropictus* (Kessler), the leafcurl ash aphid, *Prociphilus fraxinifolii* (Riley), and the woolly apple aphid, *Eriosoma lanigerum* (Hausmann).

Soft scales. Several families of scales within the superfamily Coccoidea are associated with shade trees. Most share habits that include: phloem feeding with associated honeydew production; mobility after the first instar (crawler stage), often involving migration with early instars on foliage and later stages on small branches and twigs; and the production of hundreds or even a couple thousand eggs. Damage can include loss of vigor, dieback of branches and nuisance problems associated with honeydew.

The most important family is Coccidae, the soft scales. Among the more damaging species in North America are the cottony maple scale, *Pulvinaria innumerabilis* (Rathvon), the striped pine scale, *Toumeyella pini* (King), the European fruit lecanium, *Parthenolecanium corni* (Bouché) and the calico scale, *Eulecanium cerasorum* (Cockerell). Related families of scales of importance to shade trees include giant scales or the margarodids (Margarodidae), gall-like scales (Kermesidae), eriococcids

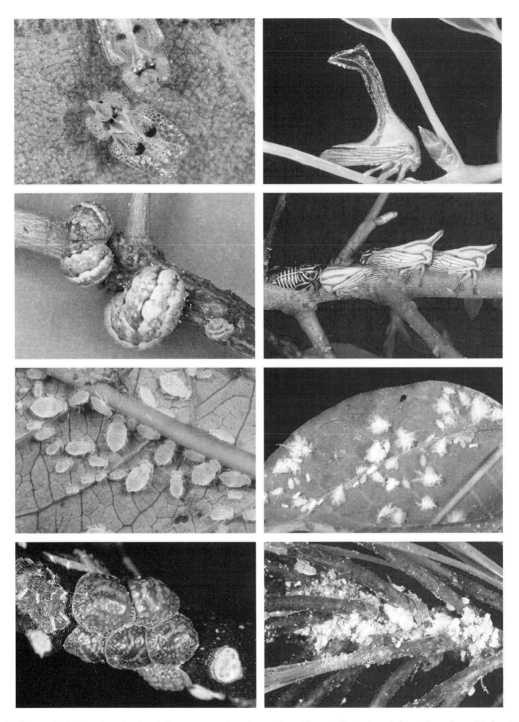

Fig. 914 Some shade tree insects: top left, sycamore lace bug, *Coryuthuca ciliata*; top right, a thorn bug, *Umbonia crassicornis*; second row, left, kermes scale, *Kermes* sp.; second row, right, a treehopper, *Platycotis vittata*; third row, left, crapemyrtle aphid, *Tinocallis kahawaluokalani*; third row, right, Asian woolly hackberry aphid, *Shivaphis celti*; bottom left, tuliptree scale, *Toumeyela liriodendri*; bottom left, acute mealybug, *Oracella acuta* (photo credits: left column by James Castner, right column by Lyle Buss).

(Eriococcidae), false pit scales (Lecanodiaspididae) and pit scales (Asterolecaniidae).

Armored scales. Armored scales (Diaspididae) include some of the most damaging scale insects associated with shade trees. Unlike the soft scales, armored scales do not feed on the phloem, but confine feeding to the mesophyll and parenchyma cells. Honeydew is not produced, but cells around the feeding site are often killed and sustained infestations can cause foliage loss, branch dieback, and increased susceptibility to canker producing fungi. Typically, armored scales produce no more than 2 to 3 dozen eggs that hatch within the cover (test) of the female. The first instar nymphs (crawlers) are active for only a few days and then they lose their legs at the first molt. Some important armored scales in North America include the obscure scale, *Melanaspis obscura* (Comstock), the walnut scale, *Quadraspidiotus juglansregiae* (Comstock), the pine needle scale, *Chionaspis pinifoliae* (Fitch) and the California red scale, *Aonidiella aurantii* (Maskell).

Leafhoppers. Many leafhoppers (Cicadellidae) are associated with shade trees. Most feed on the phloem, typically producing minor injuries in the form, loss of vigor and honeydew production. Others feed on the parenchyma, causing white flecking (stippling) injuries. A leafhopper of particular concern in much of eastern North America is the potato leafhopper, *Empoasca fabae* (Harris), that seriously damages the phloem when it feeds, disrupting photosynthesis and producing an injury known as 'hopperburn.'

One subfamily of leafhoppers (Cicidellinae), known as the sharpshooters, feed on the xylem. Some of these are capable of transmitting certain xylem-inhabiting bacteria that produce diseases of shade trees. Lethal yellowing of coconut, Pierce's disease (bacterial scorch) and ash yellows are shade tree diseases spread by leafhoppers.

Minor insect families

Bugs (Hemiptera). Several families of true bugs can be damaging to shade trees. Plant bugs (Miridae) are primarily important when they feed on new growth, killing the area around the feeding site. This produces distortions of new growth and necrotic spotting; in heavy infestation, some plant bugs may temporarily kill emergent growth. Lacebugs (Tingidae) often produce pale-colored spotting of leaves, caused by the destruction of the mesophyll, which can adversely affect the appearance and vigor of plants.

Some true bugs associated with shade trees are most important not for the injuries they produce to plants, but because of the subsequent nuisance problems they produce when they move to adjacent buildings for winter shelter. The boxelder bug, *Boisea trivittata* (Say), and the western conifer seed bug, *Leptoglossus occidentalis* Heidemann, are notorious for this habit in parts of North America.

Several additional families of Hemiptera suck sap from the foliage and/or woody parts of shade trees including certain whiteflies (Aleyrodidae), mealybugs (Pseudococcidae) and psyllids (Psyllidae). These insects tend to be more important on shrubs and ornamental plants.

Spider mites. Several species of spider mites (Tetranychidae) are important pests of shade trees. Feeding damage is caused by breaking into the surface cell layers and removing the sap. This results in small, pale colored or necrotic spots that may coalesce and produce a generalized yellowing or bronzing, often with associated premature leaf drop.

Spider mite problems can sometimes be aggravated due to pesticide use practices unfavorable to natural enemies. Most species also thrive under relatively dry conditions. Many spider mites tend to be most abundant during warmer periods of the year, such as the twospotted spider mite, *Tetranychus urticae* (Koch) and the honey locust spider mite, *Platytetranychus multidigituli* (Ewing). Others attain peak population during cooler periods of spring and fall, such as the spruce spider mite, *Oligonychus ununguis* (Jacobi) and the European red mite, *Panonychus ulmi* (Koch).

Eriophyid mites. Eriophyid mites are commonly associated with the foliage of most shade trees, but few cause visible injury. A few, known as 'rust mites,' can cause bronzing of the foliage often associated with increased thickening and brittleness of the foliage. Other eriophyid mites, discussed below, produce galls.

Gall-making arthropods

Galls are some of the most conspicuous injuries associated with shade trees. Galls are abnormal growths caused by feeding and other activities of insects and mites. Galls on shade trees can range from being fairly simple swellings to bizarre alterations of plant tissue. Galls frequently attract considerable attention and concern, but few are damaging to the health of shade trees. Several families of arthropods produce galls on shade trees with the primary ones summarized below:

Gall-making aphids. Some aphids (Aphididae, subfamily Eriosomatidae) produce galls that usually take the form of a simple swelling of the leaves. These insects have complex life cycles that typically involve a spring form that produce galls on the twigs, petioles or leaves of trees followed by a summer form that develops on the roots of a different host plant. The most important gall-making aphids found on shade trees in North America are found in the genus *Pemphigus*.

Adelgids. Adelgids (Adelgidae) are associated with conifers and may produce galls that involve stunting and/or thickening of terminal growth. Typical are the Cooley spruce gall adelgid, *Adelges cooleyi* (Gillette) and the eastern spruce gall adelgid, *Adelges abietis* (L.). Most gall-producing adelgids show alternation of hosts and do not produce galls on the primary host plant.

Phylloxerans. Phylloxerans (Phylloxeridae) are uncommon pests of shade trees, limited to certain leaf and petiole gall-forming species that attack *Carya* spp. (pecan). However, other phylloxerans can be seriously damaging to other woody plants, notably the grape phylloxera, *Daktulosphaira vitifoliae* (Fitch), that produces, alternately, leaf galls and root galls on grape vines.

Psyllids. Several genera of psyllids (Psyllidae) produce galls on the foliage or on the woody parts of plants. Leaf galls on foliage take the form of blisters, pouches, or nipples. Slight swellings are produced on twigs and small branches. A very common genera in North America is Pachypsylla, which includes over a half dozen species all of which are associated with *Celtis* spp.

Eriophyid mites. The eriophyid mites (Eriophyidae) associated with shade trees make an extremely diverse range of gall forms on shade trees. Simple eruptions of blisters on leaf surfaces (blistergalls) are common on some Rosaceae. Other eriophyid mites make more elaborate galls in the form of pouchgalls, or are further extended into elongate fingergalls. A proliferation of plant hairs producing small, felt-like mats is characteristic of other eriophyid mites. These are known as an erineum (erinea, pl.). Other eriophyid mites produce gross disorganization and cell proliferation of buds and flowers.

Gall midges. Hundreds of species of gall midges (Cecidomyiidae) are associated with woody plants. Galls typically are fairly simple, involving stunting and/or swelling of foliage. Leaves and needles that are galled usually shed prematurely.

Gall-making sawflies. Some sawflies in the family Tenthredinidae make galls in the form of simple balls or swellings in leaves or stems, primarily of willow.

Gall wasps. The largest family of gall making insects are gall wasps (Cynipidae). Hundreds of species produce galls of very determinate form on leaves and stems. These galls may involve ball like growths, sometimes with bizarre projections and spines, that have a very different appearance from the tissues from which they are differentiated. Some galls produced by gall wasps exude honeydew.

Life cycles of many gall wasps can involve alternating forms that produce different types of galls at different periods in the life cycle. Despite their abundance, essentially, all galls produced are limited to *Quercus* spp. and *Rosa* spp. Only a very few, limited to gall production on stems and twigs, cause significant plant injury.

Defoliators

Leaf feeding on shade trees is a habit widespread among the orders Lepidoptera, Coleoptera and Hymenoptera. Phasmida and Orthoptera are minor orders associated with shade tree defoliation.

Defoliation injuries to established shade trees are often well tolerated in terms of plant health. Usually only sustained, high levels of defoliation over several years significantly increases risks to plant health. However, defoliation stresses can favor increased incidence of wood borers, bark beetles and certain fungal pathogens that act more aggressively in weakened plants.

Defoliation of shade trees can have serious aesthetic effects. In addition to leaf injuries and loss, some defoliating insects may also produce large amounts of silk, a habit particularly common among some Lepidoptera in the families Lasiocampidae and Arctiidae. Dropping excrement can also be a nuisance issue with defoliating insects of shade trees.

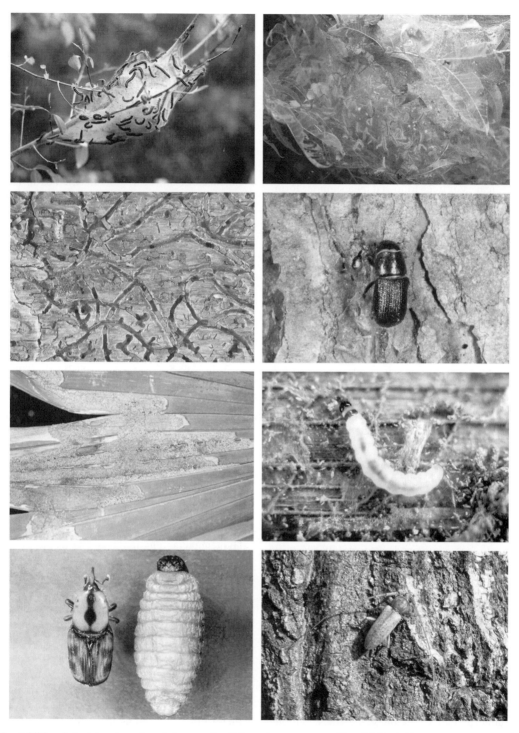

Fig. 915 Additional shadetree insects and damage: top left, eastern tent caterpillar, *Malacosoma americanum*; top right, fall webworm, *Hyphantria cunea*; second row, left, galleries caused by southern pine beetle, *Dendroctonus frontalis*; second row, right, black turpentine beetle, *Dendroctonus terebrans*; third row, left, damage caused by palm leafskeletonizer, *Homaledra sabella*; third row, right, palm leafskeletonizer, *Homaledra sabella*; bottom left, giant palm weevil, *Rhynchophorus cruentatus*, adult (left) and larva (right),; bottom right, red oak borer, *Enaphalodes rufulum* (photo credits: left column and top right, James Castner; others by Lyle Buss).

LEAFMINERS

Insects that develop between the leaf surfaces are known as leafminers; species with this habit that are associated with conifers are known as needleminers. Common leafminers that are associated with shade trees are found in three orders: Coleoptera (primarily Chrysomelidae), Hymenoptera (primarily Tenthredinidae) and Lepidoptera (several families). Some Diptera, primarily Agromyzidae, are occasionally associated with shade trees, although they are more important on shrubs and ornamentals. Leafminers tend to be classified by the pattern of the mine that they produce; blotch, serpentine, or tentiform are terms commonly used in their description. Leafminers on shade trees can seriously affect the aesthetic appearance of foliage, but are very rarely significantly damaging to the health of established shade trees.

BARK BEETLES

Bark beetles (Scolytidae or Curculionidae: Scolytinae) develop under the bark, scoring the cambium during their production of larval galleries and producing girdling wounds. Although bark beetles sometime can kill shade trees, they often are considered secondary pests that largely limit infestations to trees that are wounded, diseased, recently transplanted or otherwise weakened. Almost all bark beetles have an association with fungi, primarily in the genera *Ceratocystis* or *Ophiostoma*. The transmission of fungi can greatly increase the importance of bark beetles. The most spectacular example of this involves Dutch elm disease. This devastating disease of the American elm has largely eliminated this formerly very popular shade tree in much of North America. Dutch elm disease is produced by infection with *Ophiostoma ulmi*, which is primarily spread by the smaller European elm bark beetle.

WOOD BORERS

Most insects that develop within wood are collectively known as wood borers. This general term includes several families from three insect orders.

In the Coleoptera, the metallic wood borers (Buprestidae), known as flathead borers in the larval stage, are particularly damaging to shade trees. The larvae make meandering tunnels beneath the bark, sometimes causing girdling of the plants. Members of the genus Agrilus are particularly important as shade tree pests in North America. Many longhorned

beetles (Cerambycidae), known as roundhead borers in the larval stage, are also important. Larval injuries tend to involve tunneling of the interior of the trunks and branches, sometimes producing significant weakening and predisposing plants to breakage. A few species of weevils (Curculionidae), also develop as wood borers.

The most aggressively damaging wood boring insects are found in the order Lepidoptera. The clearwing borers (Sesiidae), include dozens of important species of woody plants. Larvae make irregular gouging wounds under the bark, with most species concentrating feeding around the crown of the trunk. In the Pyralidae, the genus *Dioryctria* includes several species that tunnel the trunks and branches of pines. Carpenterworms (Cossidae), is quite a large species that is associated with a wide variety of trees and shrubs.

Relatively few Hymenoptera develop as borers in shade trees. Most prominent are the horntails (Siricidae). Although they primarily limit attacks to trees that are in serious decline or recently killed, the ability of some to transmit white rot fungi can cause premature tree death.

Tip and terminal borers

Several insects develop by tunneling tips and terminals. This habit is particularly common among certain families of Lepidoptera and Coleoptera; Hymenoptera are a minor order with species of this habit. A particularly common group that girdles the twigs and the new growth of pines are known as 'tip moths,' primarily found in the lepidopteran genera *Rhyacionia* and *Eucosoma*. Some insects, such as the weevils in the genus *Pissodes*, only develop within the terminal growth of conifers.

Root damaging species

Roots are consumed by the larval stages of several families of Coleoptera and some Lepidoptera. The adults of many root-feeding species typically feed above ground on leaves and twigs and may be important in both stages. For example, several root-feeding weevils (Curculionidae) in the genus *Otioryhnchus*, consume roots as larvae and later chew on foliage as adults. The larvae of root-feeding species in the family Scarabaeidae also commonly feed on foliage as adults.

Fig. 916 Additional shade tree insects and damage: top left, oriental scale, *Aonidiella orientalis*; top right, dictospermum scale, *Chrysomphalus dictyospermi*; second row, left, magnolia white scale, *Pseudaulacaspsis cockerelli*; second row, right, an oak gall, insect species unknown; third row left, a bagworm, insect species unknown; third row, right, walnut caterpillar, *Datana intergerrima*; bottom left, yellow-necked caterpillar, *Datana ministra*; bottom right, forest tent caterpillar, *Malacosoma disstria* (photo credits: top row and second row, left, by Lyle Buss; others by James Castner).

Whitney Cranshaw
Colorado State University
Ft. Collins, Colorado, USA

References

Johnson, W. T., and H. H. Lyon 1988. *Insects that feed on trees and shrubs* (2nd ed.). Cornell University Press, Ithaca, New York. 556 pp.

Pirone, P. P. 1978. *Diseases and pests of ornamental plants* (5th ed.). John Wiley and Sons, New York, New York. 556 pp.

Howard, F. W., D. Moore, R. M. Giblin-Davis, and R. G. Abad. 2001. *Insects on palms*. CABI Publishing, Wallingford, United Kingdom. 400 pp.

SHARP, DAVID. David Sharp was born at Towcester, England, on October 18, 1840. He studied at St. Bartholomew's Hospital and the University of Edinburgh, and received a medical degree in 1866. He was torn between his interest in insects and his career as a physician, and sought positions that would allow him the leisure to also study insects. In 1885 he moved to Cambridge to serve as Curator of the University Museum, effectively ending his medical career. Sharp made many important contributions to our knowledge of Coleoptera, including "Fauna Hawaiiensis" (1899, 1908), "Beetles of Central America" (1894), "Insecta" (1895, 1899), a large portion of the Cambridge Natural History (1895 and 1899), and many other voluminous works. He authored more than 250 publications. He also worked tirelessly on behalf of "Entomologist's Monthly Magazine," "The Entomologist," and "Zoological Record". Sharp served as president of the Entomological Society of London, and was fellow of several societies. He died August 26, 1922, at Brockenhurst, England.

Reference

Lucas, W.J. 1922. David Sharp, M. A., M.B., F.R.S. etc., 1840–1922. *The Entomologist* 55: 217–121.

SHARPSHOOTERS. Some members of the family Cicadellidae (order Hemiptera). See also, LEAFHOPPERS, BUGS.

SHEATH. A structure enclosing others cells or structures.

SHELF LIFE. The period of time that a pesticide can be stored without losing its effectiveness.

SHELFORD, VICTOR ERNEST. Victor Shelford was born at Chemung, New York, USA, on September 22, 1877. He received his bachelor's degree in 1903 and his Ph.D. in 1907, both from the University of Chicago. He remained on the staff of the University of Chicago until 1914 when he moved to the University of Illinois at Urbana. Although he retired in 1946, he remained active and continued to publish for many years. Shelford's contributions to entomology included several important and long-term studies on the relationship of insect abundance to weather and climate. However, he also worked with vertebrate populations, and made major contributions to characterization of ecosystems in North America. Some of his important publications included "Animal communities in temperate America" (1913), "Laboratory and field ecology" (1929), "Bio-Ecology" (with F.E. Clements, 1939), and "The ecology of North America" (1963). Shelford was a pioneer in the ecological sciences, was named Eminent Ecologist by the Ecological Society of America in 1968, and has been called the 'father of animal ecology in the United States.' He died on December 27, 1968.

Reference

Buffington, J. D. 1970. Victor Ernest Shelford. 1877–1968. *Annals of the Entomological Society of America* 63: 347.

SHIELD-BACKED BUGS. Members of the family Scutelleridae (order Hemiptera). See also, BUGS.

SHIELD-BACKED KATYDIDS. A subfamily of katydids (Tettigoniinae) in the order Orthoptera: Tettigoniidae. See also, GRASSHOPPERS, KATYDIDS AND CRICKETS.

SHIELD BUGS. Members of the family Pentatomidae (order Hemiptera). See also, BUGS.

SHIELD BEARER MOTHS (LEPIDOPTERA: HELIOZELIDAE). Shield bearer moths, family Heliozelidae, total 106 species from all regions, with more than half the species split between North America (31 sp.) and Australia (36 sp.); actual fauna probably exceeds 200 species. The family is in the superfamily Incurvarioidea, in the section Incurvariina, of division Monotrysia, infraorder Heteroneura.

Adults minute to small (3 to 9 mm wingspan), with very smooth head scaling (one sp. has roughened head scales in the Neotropics); haustellum scaled and scales usually merging with head scaling to form a triangular-appearing face; labial palpi porrect; maxillary palpi minute, 5-segmented. Maculation usually dark with shining metallic-iridescence, plus some light markings. Adults are diurnal. Larvae make serpentine leaf mines at first, then make blotch mines in later instars. Host plants include a variety of hardwood trees and bushes.

John B. Heppner
Florida State Collection of Arthropods
Gainesville, Florida USA

References

Kuroko, H. 1961. The genus *Antispila* from Japan, with descriptions of seven new species (Lepidoptera, Heliozelidae) *Esakia* 3: 11–24, pl. 4–9.

Lafontaine, J. D. 1973. Eastern North American species of *Antispila* (Lepidoptera: Heliozelidae) feeding on *Nyssa* and *Cornus*. *Canadian Entomologist* 105: 991–994.

Snodgrass, R. E. 1922. The resplendent shield-bearer and the ribbed-cocoon-maker: two insect inhabitants of the orchard. *Smithsonian Reports* 1920: 485–509, 3 pl.

Wojtusiak, J. 1976. Heliozelidae. In *Klucze do Oznaczania Owadów Polski. 27. Motyle – Lepidoptera* 94:1–18. Polskie Towarzystwo Entomologiczne [in Polish], Warsaw.

SHINING FLOWER BEETLES. Members of the family Phalacridae (order Coleoptera). See also, BEETLES.

SHINING LEAF CHAFERS. Members of the subfamily Rutelinae, family Scarabaeidae (order Coleoptera). See also, BEETLES.

SHINY HEAD-STANDING MOTHS (LEPIDOPTERA: ARGYRESTHIIDAE). Shiny head-standing moths, family Argyresthiidae, include 160 species, mostly from Holarctic regions; actual fauna probably exceeds 450 species. The family is part of the superfamily Yponomeutoidea in the section Tineina, subsection Tineina, of the division Ditrysia. The family is placed as a subfamily of Yponomeutidae by some researchers. Adults small (6 to 15 mm wingspan), with head slightly roughened; haustellum naked; labial palpi upcurved; maxillary palpi minute, 1-segmented. Wings elongated and hindwings usually more linear and with long fringes. Maculation various

shades of tan with white and darker spots, often with forewings lustrous. Adults mostly crepuscular, but many may be diurnal. Larvae are leafminers and needleminers, and some mine in various plant parts. Numerous plants are recorded as hosts, with many records in Cupressaceae, Ericaceae, Fagaceae, Pinaceae, and Rosaceae, among others. Adults typically angle their bodies with the head down when at rest, with the wings held tightly onto the body. Several species are economic in the genus *Argyresthia*.

John B. Heppner
Florida State Collection of Arthropods
Gainesville, Florida USA

References

Busck, A. 1907. Revision of the American moths of the genus *Argyresthia*. *Proceedings of the United States National Museum* 32: 5–24, pl. 4–5.

Freeman, T. N. 1972. The coniferous feeding species of Argyresthia in Canada (Lepidoptera: Yponomeutidae). *Canadian Entomologist* 104: 687–697.

Friese, G. 1969. Beiträge zur Insekten-Fauna der DDR: Lepidoptera–Argyres thiidae. *Beiträge zur Entomologie* 19: 693–752, pl. 1–2.

Gershenson, Z. S. 1974. Moli gornostajevi (Yponomeutidae, Argyresthiidae). In *Fauna Ukraine* 15: 1–129. Kiev.

Moriuti, S. 1969. Argyresthiidae (Lepidoptera) of Japan. *Bulletin of the University of Osaka Prefecture (B)* 21: 1–50.

SHIP-TIMBER BEETLES. Members of the family Lymexylidae (order Coleoptera). See also, BEETLES.

SHORE BUGS. Members of the family Saldidae (order Hemiptera). See also, BUGS.

SHORE FLIES. Members of the family Ephydridae (order Diptera). See also, FLIES.

SHORT-FACED SCORPIONFLIES. Members of the family Panorpodidae (order Mecoptera). See also, SCORPIONFLIES.

SHORTHORNED GRASSHOPPERS. A family of grasshoppers (Acrididae) in the order Orthoptera. See also, GRASSHOPPERS, KATYDIDS AND CRICKETS.

SHORT-HORNED SPRINGTAILS. A family of springtails (Neelidae) in the order Collembola. See also, SPRINGTAILS.

SHORT-WINGED FLOWER BEETLES. Members of the family Brachypteridae (order Coleoptera). See also, BEETLES.

SHREDDERS. Insects living in aquatic communities that feed by chewing up large particles such as leaves.

SIALIDAE. A family of insects in the order Megaloptera. they commonly are known as alderflies. See also, ALDERFLIES AND DOBSONFLIES.

SIBLING SPECIES. Two or more species that are nearly identical in their appearance, yet reproductively isolated, are referred to as sibling or cryptic species. Although they have been discovered in organisms from protozoans to elephants, they are probably best known and abundantly documented among the insects. As small organisms, insects can exploit a multitude of niches, changing physiologically and behaviorally during speciation while retaining their structural similarity. Two or more sibling species constitute a species complex, and in some groups of insects, the existence of five or more sibling species is common.

Differences in habitat, host preferences, reproductive behavior, pheromones, seasonality, or other attributes within a so-called species may first suggest the presence of sibling species. What appears to be a generalist species using a broad range of hosts or occupying a diversity of habitats is often a group of sibling species, each a specialist. For example, the treehopper *Enchenopa binotata* originally was considered a single polyphagous species, but is actually a group of at least six species, each developing on a different species of plant.

Confirmation, or even discovery, of sibling species often comes from detailed behavioral, chromosomal, ecological, or molecular studies. These studies demonstrate conclusively that populations on different hosts or in different habitats are not merely host races or varieties. Once sibling species are revealed and studied, they can be formally named. With appropriate scrutiny of pure material, they often can be distinguished morphologically from one another. Often, however, only one or two life stages offer distinguishing morphological characters. In some species complexes of mosquitoes, for instance, only the eggs are morphologically distinct.

The frequency of sibling species is unknown for most insect groups because many are still being discovered. Perhaps half of the crickets in North America have been discovered by the sounds that they make. Among North American black flies, more than one quarter was first discovered through studies of the giant polytene chromosomes in their larval salivary glands.

Sibling species often overlap broadly in their distributions and have been used to illustrate probable examples of sympatric speciation. In the *Rhagoletis pomonella* species complex, sympatric speciation might have produced the various sibling species when these fruit flies shifted and adapted to new host plants in the same geographic area.

The recognition of sibling species is critical to meaningful biological research and effective pest management. The *Simulium damnosum* (black fly) species complex consists of as many as 40 sibling species, each biologically distinct but only some of which actually transmit the filarial nematode that causes human onchocerciasis, or river blindness. Many sibling species of mosquitoes also differ in their ability to transmit agents of disease. Successful management of pest species, many of which are members of species complexes, relies on precisely targeting them in the appropriate habitats. Natural enemies, especially parasitoids, used in biological control are often complexes of sibling species. The success or failure of biological control programs can rest on the recognition of sibling species because parasitoids are often highly host specific.

Peter H. Adler
Clemson University
Clemson, South Carolina, USA

References
Adler, P. H. 1988. Ecology of black fly sibling species. pp. 63–76 in K. C. Kim and R. W. Merritt (eds.), *Black flies: ecology, population management, and annotated world list*. Pennsylvania State University Press, University Park, Pennsylvania.
Bush, G. L. 1993. Host race formation and sympatric speciation in *Rhagoletis* fruit flies (Diptera: Tephritidae). *Psyche* 99: 335–357.
Henry, C. S. 1985. The proliferation of cryptic species in *Chrysoperla* green lacewings through song divergence. *Florida Entomologist* 68: 18–38.
Mayr, E. 1999. *Systematics and the origin of species from the viewpoint of a zoologist*. Harvard University Press, Cambridge, Massachusetts. 368 pp.

Rosen, D. 1977. The importance of cryptic species and specific identifications as related to biological control. pp. 23–35 in J. A. Romberger (general ed.), *Biosystematics in agriculture. Beltsville Symposia in Agricultural Research*, Vol. 2. Allanheld, Osmun & Co., Montclair, New Jersey.

Wood, T. K., and S. I. Guttman 1983. *Enchenopa binotata* complex: sympatric speciation? *Science* 220: 310–312.

SIDEDRESSING. Application of an agricultural chemical to a growing crop, often in a strip or band along the side of a row of plants.

SIEROLOMORPHIDAE. A family of wasps (order Hymenoptera). See also, BEES, and WASPS, ANTS, BEES, AND SAWFLIES.

SIEVE PLATE. A perforated covering over, or just inside, the opening of a spiracle.

SIGMOID CURVE. An S-shaped curve (e.g., the logistic curve).

SIGN. Any aberration or manifestation of disease indicated by a change in structure. (contrast with symptom)

SIGNAL WORD. Words required to appear on most pesticide labels that denote its toxicity: 'danger-poison' for highly toxic materials, 'warning' for moderately toxic materials, and 'caution' for slightly toxic or nontoxic.

SIGNIPHORIDAE. A family of wasps (order Hymenoptera). See also, WASPS, ANTS, BEES, AND SAWFLIES.

SIGNORET, VICTOR ANTOINE. Victor Signoret was born at Paris, France, on April 6, 1816. This distinguished French entomologist is known as one of the first great authorities on Hemiptera, with particular expertise on Coccidae. He conducted important work on Phylloxera, and several scales are named after him. He died at Paris, France, on April 3, 1889.

Reference

Essig, E. O. 1931. *A history of entomology*. The Macmillan Company, New York. 1029 pp.

SILENT SLANTFACED GRASSHOPPERS. A subfamily (Acridinae) of grasshoppers in the order Orthoptera: Acrididae. See also, GRASSHOPPERS, KATYDIDS AND CRICKETS.

SILKEN FUNGUS BEETLES. Members of the family Cryptophagidae (order Coleoptera). See also, BEETLES.

SILKEN-TUBE SPINERS. Members of the family Philopotamidae (order Trichoptera). See also, CADDISFLIES.

SILK. Silk usually is the hardened labial or salivary secretion of Lepidoptera, though certain Neuroptera also produce a similar material from rectal glands. It is best known as a product of silk moths, *Bombyx* spp.: the cocoon filament spun by the fifth instar larva of *Bombyx mori* and other silk moths. Each cocoon filament contains two cylinders of fibroin, each surrounded by three layers of sericin. Fibroin is secreted by the cells of the posterior portion of the silk gland. The fibroin gene is present in only one copy per haploid genome, but these silk gland cells undergo 18 to 19 cycles of endomitotic DNA replication before they begin transcribing fibroin mRNAs. The sericin proteins are named because they contain abundant serines (over 30% of the total amino acids). Sericins are secreted by the cells from the middle region of the silk gland.

SILKWORMS. This name refers to the family Bombycidae of the order Lepidoptera, but more commonly is used to designate the Chinese silkworm, *Bombyx mori* L. Chinese silkworm is the most important of the silkworms because it is the principal source of silk used by humans to construct silken cloth. With the possible exception of the European honey bee, *Apis mellifera*, it is the most beneficial insect to humans. For centuries, Chinese silkworm has been the basis of large industries in both Asian and European countries, where raw silk is produced. In turn, it is manufactured into articles of clothing in many countries. The production of silk, called sericulture, is very labor intensive, so increasingly it is concentrated in Asian countries where labor costs are lower. A single cocoon produces a single thread measuring from 300 to 900 meters in length. About 3000 cocoons are required to manufacture a pound of silk.

A by-product of silk production is the silkworm pupae. The pupae are used as animal feed. Production is estimated at 50,000 to 100,000 tons annually in India and China, and is fed to pigs, chickens, rats and fish. In addition, ground silkworm pupae sometimes are used as a bait in food-based baits for pest ants.

As the common name suggests, *Bombyx mori* is native of Asia, where it has been domesticated for so many centuries (approximately 5000 years) that, like many domestic animals, it has lost the ability to survive without human intervention, and cannot fly. Selective breeding has resulted in races that differ in the quality, quantity, and color of silk produced by the insect. It apparently no longer occurs in the wild.

The adult moth has a wing span of 4–6 cm. They are heavy bodied and creamy white, with dark wing veins and sometimes several weak dark transverse lines crossing the wings. Adults do not feed. Each adult female produces about 300 nearly spherical eggs. The eggs are white initially but soon darken to bluish gray. Larvae normally feed on white or black mulberry leaves (*Morus* sp.), though they will accept other foliage. They are whitish in color, and bear a pronounced anal horn similar to that found in the hornworms of the Sphingidae. There are five instars. Larvae complete their development in about 45 days, attaining a length of 45 to 55 mm. At pupation, they spin a large, oval, white or yellow cocoon. If allowed to complete their development, the adult will emerge from the cocoon in 12 to 16 days, and they can produce up to 16 generations per year. If allowed to emerge, however, the silk is damaged, so in silk production systems the silkworm pupae are killed in hot water. Silkworms are readily reared in containers if mulberry foliage or artificial diet is provided. Because they are easy to culture, they are popular for use in schools. However, silkworms are quite susceptible to disease.

See also, ERI SILKWORM, SILKWORM MOTHS.

SILKWORM MOTHS (LEPIDOPTERA: BOMBYCIDAE).

Silkworm moths, family Bombycidae, total 166 described species, all Old World and primarily Oriental (146 sp.), with only five species known for Africa. Two subfamilies are involved: Bombycinae and Prismostictinae. The family is in the superfamily Bombycoidea (series Bombyciformes), in the section Cossina, subsection Bombycina,

Fig. 917 Example of silkworm moths (Bombycidae), *Bombyx mandarina* (Moore) from Taiwan.

of the division Ditrysia. Adults medium size (19 to 64 mm wingspan), with head scaling roughened; haustellum absent; labial palpi reduced, 1-segmented; maxillary palpi absent; antennae bipectinate (serrate or filiform in females); body robust. Wings broadly triangular, usually with acute and falcate forewings; hindwings rounded. Maculation is mostly subdued browns and grays, with various striae or other markings; or white as in the domesticated silkworm moth. Adults are nocturnal. Larvae are leaf feeders. Host plants predominate in Moraceae. The silkworm (*Bombyx mori*), used for silk production, is related to other native species in China (probably *Bombyx mandarina*), but is now completely white after centuries of breeding in culture.

John B. Heppner
Florida State Collection of Arthropods
Gainesville, Florida USA

References

Holloway, J. D. 1987. Family: Bombycidae. In *The Moths of Borneo*, 3:74–90, pl. 9. Malayan Nature Society (Malayan Nature Journal, 41).

Jost, B., J. Schmid, and H. -P. Wymann. 2000. Bombycidae – Seidenspinner. In *Schmetterlinge und ihre Lebensräume: Arten - Gefährdung - Schutz. Schweiz und angrenzenden Gebiete*, 3:399–402, pl. 17. Pro Natura-Schweizerische Bund fuer Naturschutz, Basel.

Mell, R. 1951. Der Seidenspinner. In *Die Neue Brehm-Bücherei*, 34. A. Ziemsen, Wittenberg. 45 pp.

Rougeot, P. C. 1971. Bombycidae. In *Les Bombycides (Lepidoptera-Bombycoidea) de l'europe et du Bassin Méditerranéen*. In *Faune de l'Europe et du Bassin Méditerranéen,* 5: 42–46. Masson, Paris.

Seitz, A. (ed.). 1910–27. Familie: Bombycidae. In *Die Gross-Schmetterlinge der Erde*, 2: 189–192, pl. 29–30, 35 (1911); 10: 433-442, pl. 57 (1910); 14: 283–285, pl. 41 (1927). A. Kernen, Stuttgart.

SILPHIDAE. A family of beetles (order Coleoptera). They commonly are known as carrion beetles. See also, BEETLES.

SILVANID FLAT BARK BEETLES. Members of the family Silvanidae (order Coleoptera). See also, BEETLES.

SILVANIDAE. A family of beetles (order Coleoptera). They commonly are known as silvanid flat bark beetles. See also, BEETLES.

SILVERFISH (THYSANURA). This is a primitive order of insects known as Thysanura or Zygentoma. The order name is derived from the Greek words *thysanos* (bristle) + and *oura* (tail). Because these insects are wingless and none of their ancestors appear to be winged, this order is placed (with the order Archeognatha) in the subclass Apterygota; in this regard apterygotes differ from all other insects.

Classification There are about 320 species found throughout the world.

Class: Insecta

    Subclass: Apterygota
      Order: Thysanura
         Family: Lepidotrichidae
         Family: Nicoletiidae
         Family: Lepismatidae

Characteristics

Silverfish are small, measuring 2 to 20 mm. They are elongate and somewhat flattened, with three tail-like appendages. The body is soft, and may or may not bear scales. They are whitish, gray, or brownish. The median appendage is only slightly longer than the lateral cerci. The antennae are variable in length, but often long. The compound eyes are small and separate, or absent, and ocelli are absent. The more advanced form of the mouthparts (mandibles) is the basis for separation of the Thysanura from the Archeognatha. There is no evidence of external metamorphosis; the immatures and adults differ only in size.

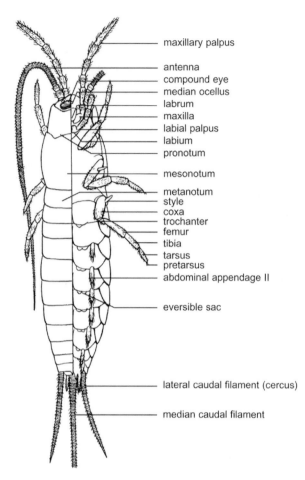

maxillary palpus
antenna
compound eye
median ocellus
labrum
maxilla
labial palpus
labium
pronotum
mesonotum
metanotum
style
coxa
trochanter
femur
tibia
tarsus
pretarsus
abdominal appendage II
eversible sac
lateral caudal filament (cercus)
median caudal filament

Fig. 918 A diagram of a silverfish showing a dorsal view (left) and a ventral view (right). Note that the medial caudal filament is abbreviated, and normally is about twice the length of the lateral caudal filaments.

The number of molts is considerable; 20 to 70 have been reported. They can live from 1 to 4 years.

Ecology

Silverfish are found in cryptic (beneath bark) and subterranean (animal burrows, caves, ant and termite nests) habitats. Often they require high humidity, though some can survive within buildings, where they are minor pests. Silverfish feed on plant materials, and those that inhabit buildings feed on starchy material such as book bindings and wall paper glue.

References

Arnett Jr., R. H. 2000. *American insects* (second edition). CRC Press, Boca Raton, Florida. 1003 pp.

Paclt, J. 1963. Thysanura fam. Nicoletiidae. *Genera Insectorum de P. Wytsman*, Fasicle 216e. 58 pp.

Paclt, J. 1967. Thysanura fam Lepidotrichidae, Maindroniidae, Lepismatidae. *Genera Insectorum de P. Waytsman*, Fasicle 218e. 86 pp.

Remington, C. L. 1954. The suprageneric classification of the order Thysanura (Insecta). *Annals of the Entomological Society of America* 47: 277–286.

Wygodzinsky, P. 1987. Order Thysanura. Pages 71–74 in F.W. Stehr (ed.). *Immature insects*, Vol. 1. Kendall/Hunt Publishing, Dubuque, Iowa.

SILVERLEAF WHITEFLY, *BEMISIA ARGENTIFOLII* BELLOWS AND PERRING (HEMIPTERA: ALEYRODIDAE).

The silverleaf whitefly, *Bemisia argentifolii* Bellows and Perring, is also known as B-type of *B. tabaci* Gennadius. *B. tabaci* is commonly known as the cotton, sweetpotato or tobacco whitefly, and is believed to have originated in the Orient. To date there are over 1,100 species of whiteflies recognized worldwide. Among these, only three (*B. tabaci*, *Trialeurodes vaporariorum* (Westwood), and *T. abutilonea* [Hald.]) are known as vectors of plant viruses. *Bemisia tabaci* is the most common and important whitefly vector of plant viruses. This species is multivoltine and polyphagous, attacking at least 506 dicot and monocot species in 74 families. However, host specialization is commonly observed for certain *B. tabaci* populations. It is commonplace that populations raised on one host species may be difficult to establish on another, although the populations may eventually become well adapted to the new host. In the last decade, *B. argentifolii* has become prominent because of its ability to cause silverleaf symptom on squash and other *Cucurbita* spp.; irregular ripening in tomato and stem whitening in *Brassica* spp., and to transmit numerous plant viruses. The etiology of silverleaf of squash and irregular ripening of tomato is not known. It is believed that this disorder is toxicogenic derived from the whitefly saliva, because symptoms increase in severity with increasing whitefly numbers, and removal of whiteflies after silverleaf development results in non-symptomatic new growth. The whitefly-transmitted plant viruses are divided into three categories based on their particle morphology, and nucleic composition: carlavirus-like virus, closterovirus, and geminivirus. The latter is the largest group, and all geminiviruses have a paired particle morphology. They either have a monopartite DNA genome of 2.6 kb. or bipartite DNA genome of 5.2 kb. Transmission of geminiviruses by whitefly vectors are similar to that of leafhoppers and aphids; virus acquisition and inoculation efficiency increases with increasing aquisition access times, a minimum of several hours latent period is required, and the virus persists after molting and for life. The acquisition

Fig. 919 Adult and immature silverleaf whitefly, *Bemisia argentifolii* (photo J. Tsai).

efficiencies of a Florida isolate of bean golden mosaic virus (BGMV) (a geminivirus) are 27.1, 27.3, 38.7, 48.8, 67.1, and 72.9% after 2, 4, 8, 24, 48 and 72-hour acquisition access periods, respectively. The respective inoculation efficiencies by single adults after 1, 4, 8, 24 and 48-hour inoculation access periods are 4.3, 3.3, 21.9, 38.2 and 67.6%. The minimum inoculation time by individual adults is 15 minutes. Viruliferous adults can retain BGMV until the death of insect.

Morphological similarities between *B. tabaci* A and B types are evident. *Bemisia argentifolii* was described as a distinct species in 1993 based on pupal cases, PCR-based DNA analysis, allozymic frequency, crossing experiments and mating behavior differences. In retrospect, it is quite possible that many plant viruses reported to be transmitted by *B. tabaci* prior to 1993 might well be transmitted by *B. argentifolii*.

The length of the life cycle of *B. argentifolii* varies with temperature, host plant and whitefly population. The average egg incubation period decreases as temperature increases from 15 to 30°C, ranging from 4 to 26 days. Within 15 to 30°C, the developmental time for four instars decreases significantly as temperature increases, ranging from 79 to 9 days. However, the average nymphal development time at 35°C is significantly longer than that at 25 to 30°C.

The adult longevity of females averages from 44 days at 20°C, to 22 days at 35°C. The oviposition of *B. argentifolii* averages 324 eggs at 20°C, to 22 eggs at 35°C. The average generation time ranges from 46 days at 20°C, to 18 days at 30°C.

Host plants affect nymphal development times. At 25°C, there is a uniform egg incubation period of 6 days when *B. argentifolii* are reared on eggplant, tomato, sweet potato, cucumber and garden bean. The average developmental time from egg to adult emergence on these host plants varies significantly, ranging from 20 days on garden bean, to 17 days on eggplant. The body lengths of second, third and fourth instars are not different when reared on these different hosts, but the males are always smaller than females, regardless of their rearing hosts. The average female longevity on these plants is 24, 21, 17, 10, and 13 days, respectively. The average number of eggs laid per female on these host plants are 234, 168, 78, 66 and 84 eggs, respectively. The mean generation time of *B. argentifolii* on these hosts ranges from 23 to 27 days.

Whiteflies are difficult to control with insecticides, and frequently they develop resistance to pesticides. Barriers such as row covers and reflective mulches have been effective in reducing disease incidence when the vector populations are low. Biological controls, and cultural controls such as removing volunteer crops and weeds within and around the field, planting trap crops, and practicing crop free periods, are commonly used with various degrees of success. However, the classical plant breeding technique incorporating engineered types of resistance into a single cultivar or variety would probably have a greater potential for a long term control strategy.

See also, SWEETPOTATO AND SILVERLEAF WHITEFLIES, WHITEFLIES.

James H. Tsai
University of Florida
Gainesville, Florida, USA

References

Harrison, B. D. 1985. Advances in geminivirus research. *Annual Review of Phytopathology* 23: 55–82.

Perring, T. M., A. D. Cooper, R. J. Rodriguez, C. A. Farrar, and T. S. Bellows, Jr. 1993. Identification of a whitefly species by genomic and behavioral studies. *Science (Washington, DC)* 259: 74–77.

Tsai, J. H., and K. H. Wang 1996. Development and reproduction of *Bemisa argentifolii* (Homoptera: Aleyrodidae) on five host plants. *Environmental Entomology* 25: 810–816.

Wang, K., and J. H. Tsai 1996. Temperature effect on development and reproduction of silverleaf whitefly (Homoptera: Aleyrodidae). *Annals of the Entomological Society of America* 89: 375–384.

Yokomi, R. K., K. A. Hoelmer, and L. S. Osborne 1990. Relationship between the sweetpotato whitefly and the squash silverleaf disorder. *Phytopathology* 80: 895–900.

SILVESTRI, FILIPPO. Filippo Silvestri was born at Bevagna, Italy, on June 22, 1873. He was educated at the University of Palermo, and worked at the University of Rome and the Buenos Aires Museum. In 1904 he was appointed director of the Zoological Station at Portici, and he held this position until his death. Silvestri was a remarkable entomologist, publishing 470 papers, of which 320 were on systematics. He established the orders Protura and Zoraptera, and had considerable interest in myrmecophilous and termitophilous insects. He died at Bevagna, Italy, on June 1, 1949.

Reference

Herman, L. H. 2001. Silvestri, Filippo. *Bulletin of the American Museum of Natural History* 265: 143–144.

SIMPLE EYES.

The eyes of insects other than compound eyes: stemmata and ocelli.

See also, OCELLI, STEMMATA.

SIMPLE METAMORPHOSIS.

A type of metamorphosis characterized by slight changes in body form during the developmental period, and the presence of the egg, nymphal and adult stages. The pupal stage is lacking. This is also known as incomplete metamorphosis or hemimetabolous development. Some of the aquatic insects displaying simple development (Odonata, Ephemeroptera, Plecoptera) differ in the degree of difference between the mature and immature stages, and so are sometimes said to have gradual metamorphosis or paurometabolous development.

See also, METAMORPHOSIS, COMPLETE METAMORPHOSIS, GRADUAL META-MORPHOSIS.

SIMULIDAE.

A family of flies (order Diptera). They commonly are known as black flies or buffalo gnats. See also, BLACK FLIES, BLACK FLIES AFFECTING LIVESTOCK, FLIES.

SIMULIUM SPP., VECTORS OF *ONCHOCERCA VOLVULUS*: LIFE CYCLE AND CONTROL.

Members of the dipteran genus *Simulium* are of significant medical and veterinary importance. They belong to the suborder Nematocera and family Simuliidae, usually called black flies. These flies have long segmented antennae of about 11 segments, and all immature stages are aquatic, or at least live in very moist environments. Members of the family have a reduced mesothoracic segment giving them the appearance of having a humped back hence their alternate name humped back flies. Some members are also called buffalo gnats. Most of the species belong to the genera *Cnephia*, *Simulium*, and *Prosimulium*.

Simulium damnosum (the most widely known), and *S. neavei* vector the filarial parasite *Onchocerca volvulus* that causes human onchocerciasis ('oncho') or river blindness in several East and West African

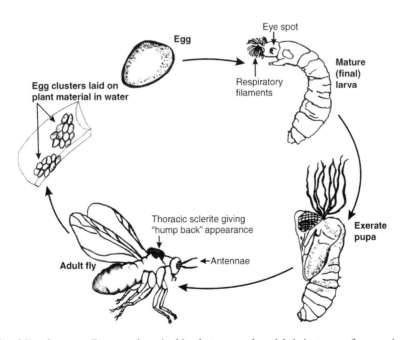

Fig. 920 Life cycle of *Simulium* spp. Eggs are deposited in clusters on plant debris in or near fast-running rivers. Each egg is ovoid to wedge-shaped. There are three larval instars and the third stage (mature) larva has two well-defined eye spots and a delineated head, thorax, and abdomen. The pupa is exerate with well-developed eyes and elongate cephalic respiratory filaments. Adult flies have a characteristic 'hump back' because of the modified thoracic sclerite.

countries. *S. ochraceum*, a smaller species, is the most common vector of the parasite in Mexico and several Central and South American countries. Other species of *Simulium*, and those of *Cnephia* and *Prosimulium*, attack livestock and birds worldwide and in some cases cause toxemia and anaphylactic shock. Their bites also are irritating to man. *S. meridionale* are irritating pests of poultry, while *S. vittatum* and *S. articum* in the United States and *S. colombaschense* in Canada attack cattle. In Australia, several other species of *Simulium* transmit the filarial worm *Onchocerca gibsoni* to cattle.

The female is haematophagous, using the protein from the blood meal for egg development, but feeds on plant juices at other times while the male feeds only on plant juices. When the short piercing-sucking mouthparts of a female fly previously infected with the *O. volvulus* parasite is inserted into the skin of the victim, the infective L3 (third instar) nematode enters the wound. Molting, maturation and offspring production of *O. volvulus* occur in the connective tissues of the host, inducing symptoms of onchocerciasis.

Simulium damnosum is actually a complex of different strains, and are known to use aggregation and oviposition pheromones to mate in swarms followed by mass oviposition in riverine habitats. A female may oviposit up to 800 eggs in her lifetime. Eggs are attached to rocks or plant material just below the water surface. The larva is elongate, with a distinct head, and adheres itself by means of a hook on its abdominal segment to a silken pad spun from its salivary glands on a rock, stick, or other submerged vegetation. The larva is thereby suspended in the fast running water from which it extracts microscopic organisms using a special filtering apparatus around its mouth. Fast running, well-oxygenated water is necessary for larval development.

There are 6 to 7 instars if conditions are favorable. Prior to pupation, the last instar spins a silken cocoon around itself. Long filaments that serve as gills facilitate respiration. The pupal stage may last up to four weeks but *S. damnosum* adults usually emerge after 8 to 12 days and live for up to four weeks. Chromosomal variation and morphological keys have been used to identify adults, larvae, and pupae of the Simuliidae in various European, African, and South American countries.

A major *Simulium* and onchocerciasis control effort was successfully initiated in West Africa by the Onchocerciasis Control Programme (OCP) in 1974 by the World Health Organization, the Food and Agricultural Organization (FAO), the United Nations Development Program, and other international and non-governmental agencies. The OCP employed several strategies to decrease the incidence of 'oncho' and improve economic conditions in the affected regions. In addition to educating local people and administering chemotherapy with the antifungal drug ivermectin against the nematode infection, the OCP significantly reduced the *Simulium* population with aerial spraying of the large riverine habitats. They used a mixture of several larvicides, at minimum concentrations, and biological control agents including *Bacillus thuringiensis* and mermithid nematodes that parasitize the insects. The OCP utilized computers and satellites at hydrological stations to predict the flow of the river and estimate the appropriate dose of larvicides needed. Thus, the program successfully controlled black fly populations with minimal negative effects on the environment.

In Central America as in Africa, in addition to the spraying of breeding habitats, the *Simulium* populations have been controlled by clearing vegetation around housing, and decreasing irrigation that causes fast flowing water.

Pauline O. Lawrence
University of Florida
Gainesville, Florida, USA

References

Anonymous. 1994. *Onchocerciasis in a nutshell*. OCP, World Health Organization. Onchocerciasis Control Programme, 1974–1994.

Cheng, T. C. 1986. *General Parasitology* (2nd ed.). Academic Press, New York, New York. 827 pp.

Davies, J. B., L. Oskam, R. Lujan, G. J. Schoone, C. C. M. Kroon, L. A. Lopez-Martinez, and A. J. Paniagua-Alvarez. 1998. Detection of *Onchocerca volvulus* DNA in pools of wild-caught *Simulium ochraceum* by use of the polymerase chain reaction. *Annals of Tropical Medicine and Parasitology* 92: 295–304.

McCall, P. J. 1995. Oviposition/aggregation pheromones in the *Simulium damnosum* complex. *Medical and Veterinary Entomology* 9: 101–108.

Schmidt, G. D., and L. S. Roberts 1989. *Foundations of Parasitology* (4th ed.). Times Mirror/ Mosby Publishers, St. Louis, Missouri. 750 pp.

Wigg, D. 1993. *And then forgot to tell us why. A look at the campaign against river blindness in West Africa*. World Band Development Essays. The World Bank, Washington, DC. 44 pp.

SIPHLAENIGMATIDAE. A family of mayflies (order Ephemeroptera). See also, MAYFLIES.

SIPHLONURIDAE. A family of mayflies (order Ephemeroptera). See also, MAYFLIES.

SIPHON. The breathing apparatus of a mosquito larva. Less often, this term is used to describe the tube-like mouthparts of certain insects, particularly Hemiptera.

SIPHONAPTERA. An order of insects. They commonly are known as fleas. See also, FLEAS.

SIPHUNCULATA. An order of wingless ectoparasitic insects, also known as order Anoplura, and sucking lice. It sometimes is treated as a suborder in the order Phthiraptera. See also, SUCKING LICE.

SIPHUNCULUS. (pl., siphunculi) The paired tubular structures on the back of an aphid; cornicles.

    See also, APHIDS, CORNICLES, ABDOMEN OF HEXAPODS.

SIRICIDAE. A family of wood wasps (order Hymenoptera, suborder Symphyta). They commonly are known as horntails. See also, WASPS, ANTS, BEES, AND SAWFLIES.

SISYRIDAE. A family of insects in the order Neuroptera. They commonly are known as spongillaflies. See also LACEWINGS, ANTLIONS, AND MANTIDFLIES.

SIXSPOTTED SPIDER MITE. See also, CITRUS PESTS AND THEIR MANAGEMENT.

SKELETONIZE. To remove the tissue of a leaf except for the veins, leaving a 'skeleton'.

SKIFF BEETLES. Members of the family Hydroscaphidae (order Coleoptera). See also, BEETLES.

SKIN BEETLES. Members of the family Trogidae (order Coleoptera). See also, BEETLES.

Fig. 921 Example of skipper butterflies (Hesperiidae), *Hesperia leonardus* Harris from Florida, USA.

SKIN BEETLES. Members of the family Dermestidae (order Coleoptera). See also, BEETLES.

SKIPPER BUTTERFLIES (LEPIDOPTERA: HESPERIIDAE). Skipper butterflies, family Hesperiidae, include about 4,100 species from all faunal regions; most are Neotropical, with 2,338 species. The actual world fauna probably exceeds 4,500 species. Seven skipper subfamilies are recognized: Megathyminae, Coeliadinae, Pyrrhopyginae, Pyrginae, Trapezitinae, Heteropterinae, and Hesperiinae. The family is in the superfamily Papilionoidea (series Hesperiiformes), in the section Cossina, subsection Bombycina, of the division Ditrysia. Adult small to medium size (16 to 82 mm wingspan); antennae with clubs mostly elongated and hooked distally (club more compact in Megathyminae). Wings mostly triangular with somewhat acute forewing apexes; hindwings mostly rounded (rarely with tails); body robust. Maculation is primarily subdued shades of brown, often with various pale or hyaline spotting; rarely very colorful. Adults are diurnal, usually with very rapid flight, but a few tropical species are crepuscular. Larvae are leafrollers or borers, and typically have a large head capsule followed by a narrow prothorax (or 'neck'). Host plants are primarily grasses (Gramineae) and other monocots, but some other hosts are also utilized; the primitive Nearctic Megathyminae are borers in Agavaceae. A few economic species are known, particularly on rice.

John B. Heppner
Florida State Collection of Arthropods
Gainesville, Florida, USA

References

Evans, W. H. 1937–53. *A catalogue of the Hesperiidae.* British Museum, London. 5 volumes.

Freeman, H. A. 1970. Systematic review of the Megathymidae. *Journal of the Lepidopterists' Society* 23 (Suppl. 1): 1–59.

Hayward, K. J. 1948–50. *Genera and species animalium Argentinarum.insects, Lepidoptera. familia Hesperiidarum.* G. Kraft, Buenos Aires. 2 volumes.

Lindsey, A. W., E. L. Bell, and R. C. Williams, Jr 1931. The Hesperioidea of North America. *Journal of the Science Laboratory, Denison University Bulletin* 26: 1–142.

Seitz, A. (ed.). 1909–31. Familie: Hesperiidae. In *Die Gross-Schmetterlinge der Erde,* 1:330–355, pl. 84–89 (1909); 1(suppl): 307–326, 353; pl. 16 (1931); 5: 833–1056, pl. 160–193 (1921–24); 9:1 027–1107, pl. 163–175 (1927); 13: 505–588, pl. 75–80 (1925). A. Kernen, Stuttgart.

Sonderegger, P. 1997. Hesperiidae – Dickkopffalter. In *Schmetterlinge und ihre Lebensräume: Arten - Gefährdung - Schutz. Schweiz und angrenzenden Gebiete,* 2:69–164, pl. 1-3. . Pro Natura-Schweizerische Bund fuer Naturschutz, Basel.

SKIPPER FLIES. Members of the family Piophilidae (order Diptera). See also, FLIES.

SLAVERY. A phenomenon among ants wherein slave-making ants capture larvae and pupae of another ant species and take them (the future slaves) to their nest where they are raised and become part of the slave-making ant colony, carrying out their normal functions but for the benefit of the slave-making ant species.

SLENDER SPRINGTAILS. A family of springtails (Entomobryidae) in the order Collembola. See also, SPRINGTAILS.

SLUG CATERPILLAR MOTHS (LEPIDOPTERA: LIMACODIDAE). Slug caterpillar moths, family Limacodidae, total 1,104 known species worldwide, the largest family of Cossoidea, mostly tropical and especially biodiverse in the Oriental tropics; likely world total is near 1,600 species or more. There is no established subfamily classification thus far. The family is in the superfamily Cossoidea (series Limacodiformes) in the section Cossina, subsection Cossina, of the division Ditrysia. Adults small to medium size (9 to 80 mm wingspan) (typically 20 to

Fig. 922 Example of slug caterpillar moths (Limacodidae), *Monoleuca erectifascia* Dyar from Florida, USA.

30 mm), with head small and relatively rough-scaled; haustellum and maxillary palpi vestigial or reduced; labial palpi short; antennae short, usually bipectinate in males (typically becoming filiform near apex). Body robust. Wings usually quadrate and rounded, although some with forewings more pointed and elongated; hindwings usually ovoid. Maculation mostly various shades of brown, but some are colorful (rarely hyaline or with hyaline patches or spots), with few markings but for many with diagonal lines, various markings (some with bright green patches); hindwings usually unicolorous to match forewings or lighter (rarely hyaline). Adults perhaps only nocturnal; many with unique resting postures. Larvae slug-like (often colorfully marked over green base color) and mostly polyphagous leaf feeders, with vestigial prolegs and usually an extensive array of poisonous stinging spines dorsally. Large numbers of host plants utilized. Few species are economic, other than medically as stinging caterpillars, but palm defoliators can be a problem in the tropics.

John B. Heppner
Florida State Collection of Arthropods
Gainesville, Florida, USA

References

Cock, M. J. W., H. C. J. Godfray, and J. D. Holloway (eds.) 1987. *Slug and nettle caterpillars: the biology, taxonomy and control of the Limacodidae of economic Importance on palms in South-East Asia.* CABI, Wallingford. 270 pp, 36 pl.

Guenin, R. 1997. Limacodidae – Schneckenspinner (Asselspinner, Schildmotten). In *Schmetterlinge und ihre Lebensräume: Arten - Gefährdung - Schutz. Schweiz und angrenzenden Gebiete*, 2: 441–446, pl. 11. Pro Natura-Schweizerische Bund fuer Naturschutz. Basel.

Holloway, J. D. 1986. Family: Limacodidae. In *Moths of Borneo*, 1:47–150, pl. 5–4. Malayan Nature Society (Malayan Nature Journal 40:4 7–150, pl. 5-9).

Holloway, J. D. 1990. The Limacodidae of Sumatra. pp. 9–77 In *Heterocera Sumatrana*. Heterocera Sumatrana Society, Göttingen.

Janse, A. J. T. 1964. Limacodidae. In *The moths of South Africa*, 71–136, pl. 1-44. Pretoria.

Seitz, A. (ed.). 1912–37. Familie: Limacodidae. In *Die Gross-Schmetterlinge der Erde*, 2: 339–347, pl. 49–50 (1912); 2(suppl.): 201–209, pl. 15 (1933); 6: 1104–1136, pl. 164–167 (1935–37); 10: 665-728, pl. 85–89 (1931); 14: 447–472, pl. 73–75 (1928). A. Kernen: Stuttgart.

SMALL DUNG FLIES. Members of the family Sphaeroceridae (order Diptera). See also, FLIES.

SMALL FRUIT FLIES. Members of the family Drosophilidae (order Diptera). See also, FLIES.

SMALL FRUIT PESTS AND THEIR MANAGEMENT. In the United States, the production of small fruit is becoming increasingly important as growers diversify from traditional staple crops to specialty fruits. Blueberries, cranberries, strawberries, and grapes comprise the major small fruit crops with combined annual sales exceeding $3 billion dollars. Despite these figures, further growth and development of small fruit industries will depend on producers' ability to develop effective pest management programs that specifically target key pests within these systems. The passage of the 1996 Food Quality Protection Act (FQPA) and public concern over food safety, as well as pressure from environmentalists, are forcing producers and pest management specialists to develop programs aimed at reducing pesticide residues in food and improving agricultural worker safety.

The goal of this section is to provide biological information on some of the major pests within small fruit systems in the United States. It is difficult to cover all of the economic pests that affect small fruit crops in a single section; therefore, a special effort has been made to highlight key insect pests that cause serious economic damage within the various small fruit systems. We attempted to provide up-to-date information on distribution, biology, and damage symptoms and focused our management efforts on effective monitoring and cultural techniques. A special effort was made to provide useful IPM strategies that can be adopted to minimize the use of pesticides. The information that is provided can be used as a basis to provide growers, consultants, and pest management specialists' options for managing key pests in these systems. Although there is substantial information in the literature on the use of reduced-risk pesticides, we did not make any spray recommendations because pesticide information changes frequently as more research becomes available.

Blueberries

In the U.S., commercial blueberries are grown on approximately 41,000 acres of land. In 2001, the total market value of blueberries produced was $165 million dollars. Michigan leads the U.S. in producing over 40% of the northern highbush blueberries, *Vaccinium corymbosum* L., while Maine produces more than 90% of the lowbush blueberries, *V. agustifolium* Aiton. Rabbiteye blueberries, *V. ashei* Reade, are grown mostly in the southeastern United States, with the highest production coming from Georgia. The recently developed southern highbush blueberries, *V. corymbosum* X *V. darrowi* Camp are grown primarily in Florida, although Georgia and some areas in the western U.S., including Oregon and California also produce small quantities of southern highbush blueberries. Blueberries are a rich source of antioxidants as well as other nutrients that are associated with lowering the risk of cardiovascular diseases and cancer. Blueberries are attacked by several pests including blueberry maggot, *Rhagoletis mendax* Curran, cranberry fruitworm, *Acrobasis vaccinii* Riley and cranberry tipworm, *Dasineura oxycoccana* (Johnson).

Pests of blueberries, *Vaccinium* spp. (family Ericaceae)

Blueberry maggot, *Rhagoletis mendax* Curran (Diptera: Tephritidae).

Distribution. Blueberry maggot, *Rhagoletis mendax* Curran, is an important pest of commercially grown blueberries, *Vaccinium* L. spp., in the eastern and Midwestern United States. Only one generation per year has been reported throughout its distribution. Temperature, soil type, moisture and the presence of wild host plants may affect rate of emergence. Deerberry, *V. stamineum* L., and huckleberries,

Fig. 923 Adult blueberry maggot.

Gaylussacia Humboldt, Bonpland, and Kunth spp., have been identified as principal host species in the southeast.

Biology. Adults emerge from overwintering puparia when berries begin to ripen, and flies will continue to emerge until all cultivars have fully ripened. Adults are black with a white spot on the thorax and are approximately 3 to 4 mm in length. The abdomen of females is black and pointed with four white cross bands. Males are slightly smaller than females and have only three white cross bands on their abdomen. Both males and females have a characteristic F-shaped marking on their wings. Mature females oviposit eggs singly into maturing berries. The larva is a creamy white, legless maggot approximately 4 to 6 mm in length. Larval development requires 18 to 20 days, at which time mature larvae drop to the ground to pupate in the soil. Puparia are smooth brown colored capsules and usually are located about 2 to 3 cm below ground level.

Damage. Larvae feed on the internal tissues of the fruit causing major destruction of the tissues. There is a no tolerance for maggots in fruit, and *R. mendax* infestations may render shipments of blueberries unmarketable.

IPM strategies. Blueberry maggot can be controlled effectively with insecticides if applications are properly timed. Three principle types of traps should be considered when monitoring blueberry maggot: 1) yellow sticky boards, 2) green sticky spheres and 3) sticky red spheres. Baiting traps with ammonium acetate can increase monitoring efficiency. In highbush blueberries, a minimum of two sticky boards or spheres is needed for every two hectares (about 5 acres). One trap should be placed near (10 to 20 m) the border of the planting adjacent to wild host plants and the other trap should be placed in the center of the two-hectare block. Yellow sticky boards should be hung 15 cm (6 inches) above the bush and positioned in a "V" orientation with the sticky surface facing downward. Traps should be monitored weekly and recoated or replaced after three weeks of exposure to maintain efficiency. Spraying for blueberry maggot should commence when two or more flies are found on any one trap during a sampling period, or when the total number of flies over all traps in a field equals or exceeds five in one week. In the field, *Daichasma alloeum*, an opiine parasitoid, is known to attack third instar blueberry maggot larvae.

References

Gaul, S. O., K. B. McRae, and E. N. Estabrooks. 2002. Integrated pest management of *Rhagoletis mendax* (Diptera: Tephritidae) in lowbush blueberry using vegetative field management. *Journal of Economic Entomology* 95: 958–965.

Liburd, O. E., S. R. Alm, R. A. Casagrande, and S. Polavarapu. 1998. Effect of trap color, bait, shape, and orientation in attraction of blueberry maggot (Diptera: Tephritidae) flies. *Journal of Economic Entomology* 91: 243–249.

Cranberry fruitworm, Acrobasis vaccinii Riley (Lepidoptera: Pyralidae).

Distribution. Cranberry fruitworm is a pest of *Vaccinium* spp. throughout the eastern United States and Canada, from Nova Scotia southward to Florida. Typically, cranberry fruitworm is a univoltine pest, although a second generation is possible under favorable climatic conditions.

Biology. Cranberry fruitworm adults are brownish-gray moths with a blend of light and dark coloration. Adult wingspan is approximately 18 mm, with females having a slightly larger and more orange-colored abdomen. Adult moths are nocturnal, with female activity preceding that of males at dusk/after sunset. Mated females lay eggs near the calyx-cup of expanding cranberries and blueberries. Females often lay eggs in berries located near bog or planting edges. Cranberry fruitworm eggs are white and slightly elliptical, approximately 1 mm in diameter. Newly hatched larvae exit the fruit near the oviposition site and crawl over the berry surface before re-entering the same berry at either the stem end or the calyx to begin feeding. Mature larvae are pale green with a darkened head capsule, approximately 15 mm in length. Larvae pupate in hibernacula composed of sand, leaves and debris, and emerge as adults the following spring.

Damage. Damage caused by cranberry fruitworm is characterized by internal feeding. Larvae consume the pulp of approximately 5 to 8 berries, webbing berries together at their point of contact. Infested berries may exhibit premature color change, and larval frass often becomes entangled in the webbed cluster. Early instar larvae sometimes construct a silken enclosure over the entry site. Webbed fruit containing frass is a characteristic of *A. vaccinii* damage that is useful for differentiating between damage caused by cherry fruitworm and sparganothis fruitworm.

IPM strategies. Current IPM strategies employ the use of pheromone-baited, (*E,Z*)-8,10-pentadeca-dien-1-ol and (*E*)-9-pentadecen-1-ol acetate, winged traps to monitor male *A. vaccinii* activity. Traps are available commercially, although their effectiveness may depend on several factors, including height with respect to host plant and geographic position within a bog or planting. Traps placed in adjacent habitats within close proximity to blueberry planting or cranberry bogs may provide additional information regarding male moth flight throughout the season. Monitoring for cranberry fruitworm also can be accomplished by examining individual clusters or plants for the presence of eggs. This strategy requires the aid of a hand lens, and since infestations of cranberry fruitworm may be aggregated, an accurate assessment may require a large sample size. Monitoring tactics will be useful for predicting peak oviposition and for timing insecticide applications. One cultural technique for managing cranberry fruitworm in cranberries involves late-water flooding because hibernacula cannot survive in water exceeding 16°C for more than two weeks.

Selected list of arthropods that affect blueberries, *Vaccinium* spp., in the United States

Common name(s)	Scientific name	Order	Family
Blueberry bud mite	*Acalitus vaccinii* (Keifer)	Acari	Eriophyidae
Japanese beetle[2]	*Popillia japonica* Newman	Coleoptera	Scarabaeidae
Blueberry flea beetle	*Altica sylvia* Malloch	Coleoptera	Chrysomelidae
Blueberry leaf beetle	*Pyrrhalta vaccinii* Fall	Coleoptera	Chrysomelidae
Plum curculio[2]	*Conotrachelus nenuphar* (Herbst)	Coleoptera	Curculionidae
Blueberry stem borer	*Oberea myops* Hald	Coleoptera	Cerambycidae
Cranberry weevil[1]	*Anthonomus musculus* Say	Coleoptera	Curculionidae
Cranberry tipworm[1]	*Dasineura oxycoccana* (Johnson)	Diptera	Ceciomyiidae
Blueberry maggot	*Rhagoletis mendax* Curran	Diptera	Tephritidae
Blueberry aphid	*Illinoia pepperi* McGillivray	Hemiptera	Aphididae
Terrapin scale	*Mesolecanium nigrofasciatum* (Pergande)	Hemiptera	Coccidae
Sharpnosed leafhopper	*Scaphytopius magdalensis* Provancher	Hemiptera	Cicadellidae
Blueberry stem gall wasp	*Hemadas nubilipennis* Ashmead	Hymenoptera	Pteromalidae
Blueberry leafminer	*Gracilaria vacciniella* ely	Lepidoptera	Gracillariidae
Blueberry tip borer	*Hendecaneura shawiana* Kearfott	Lepidoptera	Tortricidae
Read-banded leafroller	*Argyrotaenia velutinana* Walker	Lepidoptera	Tortricidae
Fruittree leafroller	*Archips argyrospilus* Walker	Lepidoptera	Tortricidae
Oblique-banded leafroller	*Choristoneura rosaceana* Harris	Lepidoptera	Tortricidae
Blueberry spanworm	*Itame argillacearia* Packard	Lepidoptera	Geometidae
Fall webworm	*Hyphantria cunea* (Drury)	Lepidoptera	Arctiidae
Yellow-necked caterpillar	*Datana ministra* (Drury)	Lepidoptera	Notodontidae
Cranberry fruitworm[1]	*Acrobasis vaccinii* Riley	Lepidoptera	Pyralidae
Cherry fruitworm[1]	*Grapholita packardi* (Zeller)	Lepidoptera	Tortricidae
Blueberry thrips	*Frankliniella vaccinii* Morgan	Thysanoptera	Thripidae
Blueberry thrips	*Catinathrips kainos* O'Neill	Thysanoptera	Thripidae

[1] also infest cranberries. [2] also infest grapes.

References

Mallampalli, N., and R. Isaacs 2002. Distribution of egg and larval populations of cranberry fruitworm (Lepidoptera: Pyralidae) and cherry fruitworm (Lepidoptera: Tortricidae) in highbush blueberries. *Environmental Entomology* 31: 852–858.

McDonough, L. M., A. L. Averill, H. G. Davis, C. L. Smithhisler, D. A. Murray, P. S. Chapman, S. Voerman, L. J. Dapsis, and M. M. Averill 1994. Sex pheromone of cranberry fruitworm, *Acrobasis vaccinii* Riley (Lepidoptera: Pyralidae). *Journal of Chemical Ecology* 20: 3269–3279.

Cranberry tipworm, Dasineura oxycoccana (Johnson) (Diptera: Cecidomyiidae).

Distribution. Cranberry tipworm is distributed throughout the eastern United States and Canada. In the north, cranberry tipworm is a major pest of cranberries. In the south, cranberry tipworm infests rabbiteye and southern highbush blueberries and is often referred to as 'blueberry gall midge.'

Biology. Tipworms are small insects, approximately 3 mm in size with long, slender legs, delicate wings (with microtrichia) and globular antennae. Female tipworms are slightly larger than males and have a distinctively orange-colored abdomen. A single female is capable of ovipositing more than 20 eggs into the developing tissue of terminal tips (cranberry) or buds (blueberry). Eggs hatch within 2 to 3 days, and the emerging larvae develop through three or four instars that are characterized by color: transparent (first instar), white (second instar), and orange (third instar), respectively. Larvae drop to the ground and pupate in the upper layers of the planting soil. There are several generations per year. Climatic conditions and latitude influence the number of generations within a particular region.

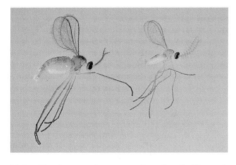

Fig. 924 Cranberry tipworm female (left) and male (right).

Damage. Tipworm damage is the result of larval feeding within the tips/buds, which ultimately kills the developing tissue and terminates growth of the affected shoot. Vegetative buds also may be affected after fruits are harvested. Curled, cupped leaves characterize cranberry tipworm damage in a blueberry planting. In general, cranberry tipworm prefers young shoots rather than older growth. Cranberry tipworm damage often is misdiagnosed as frost damage in blueberries. In both blueberries and cranberries, post-harvest damage may limit flowering potential for subsequent seasons.

IPM strategies. Monitoring for cranberry tipworm can be accomplished by destructively sampling floral and vegetative buds during and after bud swell, when scales have begun to separate. A dissecting microscope is necessary to view eggs and early larval instars. In blueberries, applications of reduced-risk insecticides may be most effective if targeted during the egg-hatching stage of the first generation. Cultural control practices include the use of bed sanding and alteration in pruning and fertilization practices.

References

Sampson, B. J., S. J. Stringer, and J. M. Spiers 2002. Integrated pest management for *Dasineura oxycoccana* (Diptera: Cecidomyiidae) in blueberry. *Environmental Entomology* 31: 339–347.

Cockfield, S. D., and D. L. Mahr. 1994. Phenology of oviposition of *Dasineura oxycoccana* (Diptera: Cecidomyiidae) in relation to cranberry plant growth and flowering. *Great Lakes Entomologist* 27: 185–188.

Cranberries

Cranberries are produced mainly in Wisconsin, Massachusetts and New Jersey. In 2001, the U.S. produced 500 million pounds of cranberries valued at $115 million dollars. Wisconsin leads the country in producing 60% of the nation's cranberries, followed by Massachusetts and New Jersey, which produced 142 and 57 million pounds, respectively. Cranberries are a rich source of vitamins, potassium and fiber, and have been reported to play a significant role in lowering the incidence of cystitis and other urinary tract infections. Recent studies also have indicated that cranberries are a good source of antioxidants. Like other plants belonging to the genus *Vaccinium*, cranberries are susceptible to several pests including Sparganothis fruitworm,

Sparganothis sulfureana (Clemens) Riley, black-headed fireworm, *Rhopobota naevena* (Hübner) and cranberry fruitworm, *Acrobasis vaccinii* Riley (discussed under blueberry section).

Pests of cranberries, *Vaccinium macrocarpon* Aiton (family Ericaceae)

Sparganothis fruitworm, Sparganothis sulfureana (Clemens) (Lepidoptera: Tortricidae).

Distribution. Sparganothis fruitworm is a major pest of cranberries throughout the eastern United States and Wisconsin. This species is not known to affect production of *Vaccinium* spp. on the west coast of the United States. Sparganothis fruitworm also has been reported on strawberry.

Biology. Sparganothis fruitworm has no approved common name. Adults are approximately 15 mm in length and are characterized by having yellowish wings with a distinctly brownish "X" marking on the dorsal surface. Larvae have a yellow-green body with paired white spots on the dorsal surface and a yellow head. Sparganothis fruitworm has two generations per year. The first generation larvae are indirect pests, feeding on young foliage and webbing together terminal vine tips. Second generation larvae feed internally, causing direct damage to the fruit. Sparganothis fruitworm overwinter as larvae.

Damage. Fruit entry sites are generally larger and more tattered compared with cranberry fruitworm. Direct injury result from internal feeding and consumption of fruit pulp, with no evidence of frass inside affected berries.

IPM strategies. Sparganothis fruitworm may be difficult to manage due to organophosphate resistance. There may be some promise for mating disruption with sprayable micro-encapsulated pheromones. The pheromone (E)-11-tetradecenyl acetate (E11-14: Ac) has been known to significantly reduce mating among virgin females in cranberry plots.

References
Polavarapu, S., G. C. Lonergan, H. Peng, and K. Neilsen 2001. Potential for mating disruption of *Sparganothis sulfureana* Clemens (Lepidoptera: Tortricidae) in cranberries. *Journal of Economic Entomology* 94: 658–665.

Cockfield, S. D., S. L. Butkewich, K. S. Samoil, and D. L. Mahr. 1994. Forecasting the flight activity of *Sparganothis sulfureana* Clemens (Lepidoptera: Tortricidae) in cranberries. *Journal of Economic Entomology* 87: 193–196.

Cranberry girdler, Chrysoteuchia topiaraia (Zeller) (Lepidoptera: Pyralidae).

Distribution. Cranberry girdler, classified as a sod webworm because of the characteristic webbing at the feeding site, is a pest of cranberries throughout the United States and Canada. This species also

Selected list of arthropods that affect cranberries, *Vaccinium macrocarpon* Aiton, in the United States

Common name(s)	Scientific name	Order	Family
Cranberry weevil[1]	*Anthonomus musculus* Say	Coleoptera	Curculionidae
Cranberry rootworm[1]	*Rhabdopterus picipes* (Oliver)	Coleoptera	Chrysomelidae
Cranberry white grub	*Phyllophaga anxia* LeConte	Coleoptera	Scarabaeidae
Black vine weevil	*Otiorhynchus ovatus* L.	Coleoptera	Curculionidae
Cranberry tipworm[1]	*Dasineura oxycoccana* (Johnson)	Diptera	Cecidomyiidae
Cranberry aphid	*Ericaphis scammelli* Mason	Hemiptera	Aphidae
Cranberry girdler	*Chrysoteuchia topiaraia* (Zeller)	Lepidoptera	Pyralidae
Blackheaded fireworm	*Rhopobota naevena* (Hübner)	Lepidoptera	Tortricidae
Cranberry fruitworm[1]	*Acrobasis vaccinii* Riley	Lepidoptera	Pyralidae
Green spanworm	*Itame sulphurea* Packard	Lepidoptera	Geometridae
Brown spanworm	*Ematurga amitaria* Gueneé	Lepidoptera	Geometridae
Sparganothis fruitworm[2]	*Sparganothis sulfureana* (Clemens)	Lepidoptera	Tortricidae
Gypsy moth	*Lymantria dispar* L.	Lepidoptera	Lymantriidae
Spotted fireworm[1]	*Choristoneura parallela* (Robinson)	Lepidoptera	Tortricidae
Cranberry blossomworm	*Epiglacea apiata* Graote	Lepidoptera	Noctuidae
False armyworm	*Xylena nupera* Lintner	Lepidoptera	Noctuidae

[1]also infest blueberries. [2]also infest strawberries.

affects grasses that are common to many cranberry bogs.

Biology. Adult moths are approximately 9 mm in length and silvery, with light brown coloration on the outer edge of the forewing. Small black dots may also be visible on the outer margin of the forewing, although adults lose their coloration soon after emerging. Females lay eggs randomly within the bog floor. Eggs (0.5 mm) appear white when laid and attain a pinkish color when mature prior to hatching. Larvae have a dark brown head and cream-colored abdomen with fine hairs. Girdler larvae are concealed largely in the leaf litter on the bog floor and feed on the bark and internal vascular system of low-lying cranberry vines. Mature larva is approximately 13 mm in length. Prior to pupation, larva prepares a cocoon made from leaf litter for overwintering in the soil. Pupation occurs in the spring and there is one generation per year.

Damage. Larval feeding results in 'girdled' or completely severed vines, which interrupts water and nutrient transfer in affected plants. Damage symptoms (including foliage browning) may appear in the late fall, although more extensive damage may not be visible until the spring. The occasional appearance of many dead or dying uprights in late summer and early fall could be a sign of girdler damage. Damage may be more abundant in bogs with accumulated trash.

IPM strategies. Commercially available pheromone traps are useful for monitoring adult populations. There are some indications that certain plant species, including foxtail, *Alopecurus pratensis* L., can be used as monitoring tools in cranberry bogs. Cultural control practices include sanding and flooding. Parasitic wasps, spiders, and ground beetles may offer some degree of control in some areas.

References

Cockfield, S. D., and D. L. Mahr 1994. Prediction models for flight activity of the cranberry girdler (Lepidoptera: Pyralidae) in Wisconsin. *Great Lakes Entomologist* 27: 107–109.
Roland, J. 1990. Use of alternative plant species as a monitoring tool for the cranberry girdler (Lepidoptera: Pyralidae). *Environmental Entomology* 19: 721–724.

Blackheaded fireworm, *Rhopobota naevena* (Hübner) (Lepidoptera: Tortricidae).

Distribution. Blackheaded fireworm is a major pest of cranberry production in North America.

Biology. Adults are darkly colored small moths, approximately 7 mm, with light and dark brown-banded forewings. Mature larvae have shiny black heads with greenish-yellow abdomens and measure approximately 8 mm in length. Young larvae burrow into the green tissue of developing leaves eventually moving to the growing tips. Larvae often tunnel into swollen buds and web terminal leaves together with silk. Larvae feed for about 3 to 5 weeks, during which time a single larva can construct more than five silken tents. Blackheaded fireworm overwinters in the egg stage on the underside of cranberry leaves. Overwintering eggs are yellow, disc-shaped, and approximately 0.5 mm in length. There are two or three generations per year. First generation egg hatch begins in the spring. Pupation occurs either within tents or in the trash layer on the bog floor. Females lay eggs on the underside of cranberry leaves in the late spring/early summer. Development of the second generation proceeds similarly to the first except that second generation larvae cause direct damage to blossoms and berries. In addition, second generation adults fly in mid/late summer. A third generation is possible under favorable conditions.

Damage. Damage symptoms caused by first generation larvae appear similar to leafminer damage. First generation larvae also mine terminal buds, which may reduce fruit set. Second generation larvae cause direct damage to the flowers and fruit, often hollowing the fruit. Affected bogs exhibit a scorched appearance.

IPM strategies. Monitoring for larvae often is conducted by viewing runner tips or sweepnet sampling. Particular attention should be paid to areas of previous infestation and bed edges. Pheromone traps are commercially available for monitoring adults. Regional recommendations for trap maintenance and spacing may vary.

References

Fitzpatrick, S. M., J. T. Troubridge, C. Maurice, and J. White 1995. Initial studies of mating disruption of the black headed fireworm of cranberries (Lepidoptera: Tortricidae). *Journal of Economic Entomology* 88: 1017–1023.
Averill, A. L., and M. M. Sylvia 1998. *Cranberry insects of the northeast: a guide to identification, biology, and*

management. University of Massachusetts, Cranberry Experiment Station Bulletin.

GRAPES

Among the small fruit crops, grapes have the highest market value in the U.S., averaging more than 2 billion dollars in annual sales. Grapes are produced on almost a million acres of land in the U.S. About 85% of the grapes produced are processed for wine production. A small amount of grapes is dried for raisins, or processed for jams and jellies, and less than 10% are used for juice production. California leads the U.S. in grape production, generating more than 90% of total sales, followed by Washington and New York, which produce 4 and 2%, respectively. Most American grape cultivars are hybrids of native American species and European grapes, *Vitis vinifera* L. Hybridization combines the excellent wine quality of European grapes with the insect and disease resistance qualities of the wild American grape species. American hybrids were used during the latter part of the nineteenth century to combat a major pest, grape phylloxera, *Daktulosphaira vitifoliae* (Fitch). A few European cultivars are grown in the colder regions of the U.S. Muscadine grapes, *Vitis rotundifolia* Michx., are grown on small acreage in the southeastern regions of the U.S. Muscadine grapes have a long history in commercial and backyard culture in eastern U.S. and were the first American grape to be cultivated. Muscadine grapes usually are sold in groceries as fresh pack. Like other small fruits, grapes are susceptible to many insect and disease pests. The glassy-winged sharpshooter, *Homalodisca coagulata* (Say), recently has become established in California and is a major threat to the state wine industry because it vectors the bacterium *Xylella fastidiosa*, which causes Pierce's disease in grapes.

Pests of grapes, *Vitis* spp. (family Vitaceae)
Grape berry moth, Endopiza vitana Clemens (Lepidoptera: Tortricidae).

Distribution.   Grape berry moth is a key pest of grapes throughout the United States and Canada east of the Rocky Mountains. Prior to grape cultivation in this region, the grape berry moth infested wild grapes.

Biology.   Adults are brown moths with grayish markings, with a 9 to 12 mm wingspan. There are two or three generations per year. First generation

adults emerge from pupation in late spring. Females lay eggs singly on buds or stems, while the later generation(s) lay eggs directly on the fruit. Eggs, which are approximately 0.7 mm, must be viewed using a hand lens. Eggs hatch into creamy white larvae within a week in most regions. Larvae have a brown head and dark thoracic shield. The abdomen changes color from white to green to purple as the larvae mature. Full grown larvae measure about 10 mm in length. Larvae generally construct cocoons from leaves and leaf litter, although some larvae pupate within the fruit cluster in which they once fed. Pupae are light brown to green in color and are 5 mm long.

Damage.   First generation larvae feed on tender stems, blossom buds, and berries. Later generations cause direct damage to the fruit, often boring into several grape berries at their point of contact and webbing the cluster together. Up to seven berries may be damaged by a single larva.

IPM strategies.   Pheromone lures are available commercially for monitoring the emergence of male moths within vineyards and in adjacent locations. These pheromone baited traps are useful for properly timing insecticide applications where grape berry moth is an annual pest. Cultural control methods include removing infested clusters, generally when infestations are mild. Leaf litter containing cocoons can be destroyed after harvest. In the spring, applying a layer of compact soil on top of the leaf litter may reduce adult emergence.

References
Hoffman, C. J., T. J. Dennehy, and J. P. Nyrop 1992. Phenology, monitoring, and control decision components of the grape berry moth (Lepidoptera: Tortricdae) risk assessment program in New York. *Journal of Economic Entomology* 85: 2218–2227.
Trimble, R. M., D. J. Pree, and P. M. Vickers 1991. Potential of mating disruption using sex pheromone for controlling the grape berry moth, *Endopiza viteana* (Clemens) (Lepidoptera: Tortricidae), in Niagara peninsula, Ontario vineyards. *Canadian Entomologist* 123: 451–460.

Grape root borer, Vitacea polistiformes (Harris) (Lepidoptera: Sesiidae).

Distribution.   The grape root borer is a major pest of grapes in the eastern United States, posing the most serious threat in southern regions. The life cycle of grape root borer requires two to three years.

Fig. 925   Grape root borer.

Biology.   Adults are dark brown with thin yellow bands across the abdomen and resemble paper wasps. The forewings are dark brown, while the hind wings are transparent. Adult moths are daytime fliers and live for approximately seven days. Females lay up to 500 dark brown eggs in clusters on the soil surface, on grape leaves, or on nearby weeds. Eggs hatch in about two weeks, and first instar larvae drop to the ground and tunnel into vine roots. Larvae are cream colored with a brown head. Studies suggest that only 3% of neonate larvae survive threats of parasitism, predation and desiccation. However, once larvae become established in the vine roots, mortality is very low. Mature larvae can be up to 40 mm in length. Second year larvae leave the vine roots to pupate in cocoons near the soil surface. The pupal stage lasts approximately 35 to 40 days.

Damage.   Grape root borer damage is often difficult to diagnose since the damaging larvae are hidden within the vine roots. However, evidence such as protruding pupal skins left by emerging adults may indicate that an infestation is present. Larval feeding results in reduced vine vigor and a reduction in grape yield. Only two or three larvae are necessary to destroy an entire plant.

IPM strategies.   Cultural methods for control of grape root borer include mounding soil under vines just after pupation in order to reduce adult emergence. Proper weed management is also important for reducing potential egg laying sites and to increase larval mortality due to desiccation. Two species of *Heterorhabditis* nematodes have shown promise as biological control agents against grape root borer

larvae. Pheromone-baited traps are recommended for monitoring adult moths. These traps should be placed about 100 m apart inside the vineyard and along adjacent woodland boundaries. Monitoring information may be useful for timing insecticide applications, although contact insecticides are ineffective once larvae reach the root system.

References

Anonymous. 1992. *Grape pest management*, 2nd ed. Publication 3343. Division of Agriculture and Natural Resources, University of California, Oakland, California.

Liburd, O. E., and G. Seferina 2003. *Grape root borer life stages and IPM strategies in Florida*. Institute of Food and Agricultural Sciences, University of Florida, Gainesville, Florida.

Glassy-winged sharpshooter, Homalodisca coagulata (Say) (Hemiptera: Cicadellidae).

Distribution.   The glassy-winged sharpshooter, *Homalodisca coagulata* (Say), is native to the southeastern United States. Glassy-winged sharpshooter recently has become established in California where it vectors the bacterium *Xylella fastidiosa*, which causes Pierce's disease in grapes.

Biology.   Adults measure up to 2 cm in length and have large membranous wings with red markings. There are distinguishing small yellow dots on the head and thorax. The glassy-winged sharpshooter overwinters as an adult and emerges under favorable climatic conditions. Generally, there are two generations per year in semi-tropical climates. Females lay eggs in groups of 10 to 20 on the underside of leaves, just under the epidermis. After oviposition, females protect their eggs with 'brochosomes,' which are white hydrophobic particles produced in specialized locations within the Malphigian tubules.

Damage.   The glassy-winged sharpshooter feeds on vines, not leaves. Early symptoms of Pierce's disease include wilting, the appearance of water loss, which is caused by bacterial growth that blocks the flow of xylem in affected plants. Subsequent damage includes discolored leaf margins, shriveled fruit, leaf drop, and irregular maturation of new canes. Several other sucking insects may transmit the *X. fastidiosa* bacterium, including the blue-green leafhopper, *Graphocephala atropunctata* (Signoret), although the common grape leafhoppers (discussed later) are not known to be vectors.

IPM strategies. Host plant resistance methods, which focus on planting vines resistant to *X. fastidiosa* may reduce the incidence of Pierce's disease. In general, muscadine grapes are resistant to *X. fastidiosa*, which may be responsible for the absence of Pierce's disease in some places, including Florida.

References

Brlansky, R. H., L. W. Timmer, W. J. French, and R. E. McCoy. 1983. Colonization of the sharpshooter vectors, *Oncometopia nigricans* and *Homalodisca coagulata* by xylem limited bacteria carriers of xylem limited bacterial disease. *Phytopathology* 73: 530–535.

Phillips, P. A. 2001. Protecting vineyards from Pierce's disease. *California Grower* 25: 19–20.

Grape leafhoppers, Erythroneura spp. (Hemiptera: Cicadellidae).

Distribution. There are several species of Cicadellidae that are referred to as grape leafhoppers. In

Selected list of arthropods that affect grape production in the United States

Common name	Scientific name	Order	Family
Grape erineum mite	*Colomerus vitis* (Pagenstecher)	Acari	Eriophyidae
Two-spotted spider mite[1]	*Tetranychus urticae* Koch	Acari	Tetranychidae
Pacific spider mite	*Tetranychus pacificus* McGregor	Acari	Tetranychidae
Willamette spider mite	*Eotetranychus willamettei*	Acari	Tetranychidae
Grape curculio	*Craponius inaequalis* (Say)	Coleoptera	Curculionidae
Plum curculio[2]	*Conotrachelus nenuphar* (Herbst)	Coleoptera	Curculionidae
Grape bud beetle	*Glyptoscelis squamulata* Crotch	Coleoptera	Chrysomelidae
Grape flea beetle	*Altica chalybea* (Illiger)	Coleoptera	Chrysomelidae
Grape colaspis	*Colaspis brunnea* (Fabricius)	Coleoptera	Chrysomelidae
Southern grape rootworm	*Fidia longipes* (Melsheimer)	Coleoptera	Chrysomelidae
Grape cane girdler	*Ampeloglypter ater* LeConte	Coleoptera	Curculionidae
Grape cane gallmaker	*Ampeloglypter sesostris* LeConte	Coleoptera	Curculionidae
Japanese beetle[2]	*Popillia japonica* Newman	Coleoptera	Scarabaeidae
Grape trunk borer	*Clytoleptus albofasciatus* (Laporte & Gory)	Coleoptera	Cerambycidae
Grape blossom midge	*Contarinia johnsoni* Felt	Diptera	Cecidomyiidae
Grape sawfly	*Erythraspides vitis* (Harris)	Hymenoptera	Tenthredinidae
Grape seed chalcid	*Eroxysoma vitis* (Saunders)	Hymenoptera	Eurytomidae
Grape whitefly	*Trialeurodes vittata* (Quaintance)	Hemiptera	Aleyrodidae
Sharpshooters	*Homalodisca coagulata* (Say)	Hemiptera	Cicadellidae
Grape mealybug	*Pseudococcus maritimus* (Ehrhorn)	Hemiptera	Pseudococcidae
Grapevine aphid	*Aphis illinoisensis* (Shimer)	Hemiptera	Aphididae
Grape leafhopper[1]	*Erythroneura comes* (Say)	Hemiptera	Cicadellidae
Grape leafhopper[1]	*Erythroneura elegantula* Osborn	Hemiptera	Cicadellidae
Variegated grape leafhopper	*Erythroneura variabilis* Beamer	Hemiptera	Cicadellidae
Grape phylloxera	*Daktulosphaira vitifoliae* (Fitch)	Hemiptera	Phylloxeridae
Grape scale	*Diaspidiotus uvae* (Comstock)	Hemiptera	Diaspididae
Grape berry moth	*Endopiza vitana* Clemens	Lepidoptera	Tortricidae
Omnivorous leafroller	*Platynota stultana* Walshingham	Lepidoptera	Tortricidae
Orange Tortrix	*Argyrotaenia citrana* (Fernald)	Lepidoptera	Tortricidae
Grape leaffolder	*Desmia funeralis* (Hübner)	Lepidoptera	Pyralidae
Grapevine looper	*Eulythis diversilineata* (Hübner)	Lepidoptera	Geometidae
Grape plume moth	*Pterophorus periscelidactylus* Fitch	Lepidoptera	Pterophoridae
Grape root borer	*Vitacea polistiformes* (Harris)	Lepidoptera	Sesiidae
Grapeleaf skeletonizer	*Harrisina americana* (Guérin-Méneville)	Lepidoptera	Zygaenidae
Western grapeleaf skeletonizer	*Harrisina brillians* Barnes & McDunnough	Lepidoptera	Zygaenidae
Western flower thrips[1,2]	*Frankliniella occidentalis* (Pergande)	Thysanoptera	Thripidae
Eastern flower thrips[1,2]	*Frankliniella tritici* (Fitch)	Thysanoptera	Thripidae
Florida flower thrips[1,2]	*Frankliniella bispinosa* (Morgan)	Thysanoptera	Thripidae

[1] also infest raspberries and strawberries. [2] also infest blueberries.

general, members of the genus *Erythroneura* cause the most serious damage to grapes. The grape leafhopper, *Erythroneura comes* (Say), is the most abundant species in the northeastern United States. In Florida, *E. vulneurata* is the most problematic species, while *E. elegantula* and the variegated leafhopper, *E. variabilis*, occur in California. These species are also believed to vector Pierce's disease.

Biology. Leafhoppers of the genus *Erythroneura* overwinter as adults, emerging from nearby hibernation sites when temperatures exceed 16°C. Adults are approximately 3 mm in length and are pale yellow in the beginning of the season. As the season progresses, adults obtain a deeper color, often with distinct markings or spots. Eggs are laid singly on the underside of the leaf just below the epidermis. Nymphs hatch within 7 to 10 days. Newly hatched nymphs are nearly transparent and have distinct red eyes. There are five nymphal instars, and with each molt thorax markings and wing pads become more prominent. Both nymphs and adults are very active.

Damage. Adults and nymphs bore small holes into the underside of leaves and extract cell sap and chlorophyll, generally when plant nutrients are abundant. Affected leaves have a mottled appearance and may lose their color. Severe injury to the leaf tissue may result in leaf drop and reduced fruit quality. Excreta (honeydew) from leafhoppers may accumulate on the foliage and berries making them susceptible to fungal attack, particularly sooty molds.

IPM strategies. Monitoring strategies for leafhoppers include the use of yellow sticky boards or counting the number of nymphs and adults relative to the number of leaves. Removal of alternate host plants and weed management in areas adjacent to vineyards may reduce food sources and potential overwintering sites. Coverage on the undersides of leaves is important when insecticide treatments are warranted for leafhopper control.

References

Martinson, T. E., and T. J. Dennehy. 1995. Varietal preferences of *Erythroneura* leafhoppers (Homoptera: Cicadellidae) feeding on grapes in New York. *Environmental Entomology* 24: 550–558.

Martinson, T. E., R. Dunst, A. Lakso, and G. English-Loeb. 1997. Impact of feeding injury by Eastern grape leafhopper (Homoptera: Cicadellidae) on yield and juice quality of Concord grapes. *American Journal of Enology and Viticulture* 48: 291–302.

Grape leaffolder, Desmia funeralis (Hübner) (Lepidoptera: Pyralidae).

Distribution. The grape leaffolder is a common pest of grapes throughout the United States. It is native to eastern North America but it is now distributed from southern Canada to northern Mexico. Although damage is typically minor, outbreaks of grape leaffolder have been problematic occasionally in both California and Florida. There are several generations per year in the southeastern United States and California, although damage appears to be most problematic during the second and third generations.

Biology. Adult moths are nearly black and have slightly iridescent wings. Wingspan varies from 2 to 2.5 cm. There are two white, oval-shaped spots on the forewings of both sexes. Males have a white band on the hindwings, while a similar band may be partitioned into two spots in females. Mated females lay elliptical-shaped eggs singly on the underside of leaves, often near the veins. Emerging larvae feed on the leaves of grapevines. At maturity, larvae are 15 to 20 mm long and nearly transparent; larvae also may appear green as a result of consuming leaf material. The head of the larva is light brown and the abdominal segments have fine yellow hairs. If larvae are disturbed, they wriggle from their silken enclosure and drop to the ground. Pupation generally occurs within the leaf-roll.

Damage. Larvae feed for about two weeks before rolling the edge of a leaf to complete development. The leaf folds are constructed with silk threads. Rolled leaves have their undersides exposed, and often the upper surface is skeletonized. Third generation larvae may completely defoliate a vineyard, resulting in sun-scorched berries. In some instances, larvae may feed directly on the berries.

IPM strategies. In general, populations of grape leaffolder are balanced by the presence of several larval parasites, including *Bracon cushmani* (Muesebeck). However, growers could monitor vineyards with a history of leaffolder outbreaks. Adult moths can be monitored with tent-shaped traps baited with terpinyl acetate or by using blacklight traps. If insecticide treatment becomes necessary, sprays are generally most effective against early instars.

References

Millar, J., J. S. McElfresh, and F. de Assis Marques 2002. Unusual acetylenic sex pheromone of grape leaffolder (Lepidoptera: Pyralidae). *Journal of Economic Entomology* 95: 692–705.

Aliniazee, M. T., and E. M. Stafford. 1973. Sex pheromone of the grape leaffolder, *Desmia funeralis* (Lepidoptera: Pyralidae): laboratory and field evaluation. *Annals of the Entomological Society of America* 66: 909–911.

Strawberries

In the U.S., California is the largest producer of strawberries, generating more than 50% of the nation's strawberries on over 25,000 acres of land. Florida ranks second in the country in terms of quantity of strawberries produced, followed by Oregon and the state of Washington. During 2001, the total value of strawberries produced in the U.S. exceeded $1 billion dollars. In the south, strawberries are grown as an annual crop and are cultivated between October and May. In the northeast, strawberries are grown as a perennial crop. Some of the cultivars commonly grown in the north include Kent, Annapolis and Honeoye. In the south and western U.S., Camarosa, Sweet Charlie, Chandler, and Aromas are fairly common cultivars. Each variety has its own unique characteristics with distinct advantages and disadvantages. Strawberries are a rich source of vitamin C and folic acid. They are relatively low in calories and some varieties are rich in antioxidants. Strawberries are susceptible to several mite and insect pests including the two-spotted spider mite, *Tetranychus urticae* Koch, plant bugs belonging to the genus *Lygus*, tarnished plant bug, *Lygus lineolaris* (Palisot de Beauvois), in the east, and the western plant bug, *L. hesperus* Knight. The strawberry clipper, *Anthonomus signatus* (Say), once was considered a major pest of strawberries, but recent reports have indicated that the floral buds removed early in the season by strawberry clipper are compensated with more vigorous bud growth later in the season, resulting in larger more dense fruit.

Pests of strawberries, *Fragaria* spp. (family Rosaceae)

Two-spotted spider mite, *Tetranychus urticae* Koch (Acari: Tetranychidae).

Distribution. The two-spotted spider mite is distributed widely across North America, and is known to feed on more than 180 species of plants in both greenhouse and outdoor environments. The two-

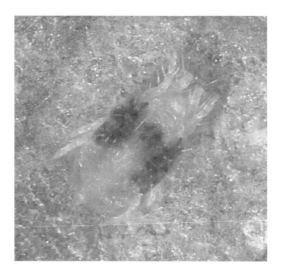

Fig. 926 Two-spotted spider mite.

spotted spider mite occurs year-round in the greenhouse, but its presence in the field requires warm temperatures.

Biology. The two-spotted spider mite develops through five phases: egg, larvae, proto-nymph, deuto-nymph and adult. The larva can be distinguished from the other stages because it has only six legs and is the size of the egg. The protonymph and deutonymph resemble the adult. Adults are approximately 0.5 mm long and range in color from light green to dark red, usually with two dark spots on the dorsal surface of the abdomen. Adults overwinter under vegetative cover, and mated females begin feeding and laying eggs when temperatures increase. Maturity from egg to adult requires at least five days, but may approach one month, depending on environmental conditions. Dispersal of two-spotted spider mites can occur by either ambulatory or aerial means. The vast majority of dispersal takes place aerially due to the small size of mites and the amount of time it would take for a spider mite to travel from one plant to another. There are several generations per year.

Damage. Two-spotted spider mite feeds on the underside of strawberry leaves by inserting their mouthparts into the leaf cells and extracting cellular fluid from the plant. The result is chlorosis, which leads to a decrease in photosynthetic activity by the plant and a subsequent reduction in fruit yield. Infested leaves are often mottled brown and the leaf surfaces may be covered with fine webbing. Mite colonies tend to be localized rather than evenly distributed and are more abundant when the weather is dry.

IPM strategies. Sampling for two-spotted spider mite can be accomplished by collecting leaves from the field and counting the number of mites on individual leaves, generally with the aid of a microscope. Heavy infestations can be treated with reduced-risk miticide. An alternative to treatment with miticides is the inoculative release of predatory mites, primarily in the family Phytoseiidae. Monitoring for the presence of phytoseiid mites in the fields should be conducted in conjunction with two-spotted spider mite sampling.

References
Easterbrook, M. A., J. D. Fitzgerald, and M. G. Solomon 2001. Biological control of strawberry tarsonemid mite *Phytonemus pallidus* and two-spotted spider mite, *Tetranychus urticae* on strawberry in the UK using species of *Neoseiulus* (*Amblyseius*) (Acari: Phytoseiidae). *Experimental and Applied Acarology* 25: 25–36.

Croft, B. A., and L. B. Coop 1998. Heat units, release rate, prey density, and plant age effects on dispersal by *Neoseiulus fallacis* after inoculation into strawberry. *Journal of Economic Entomology* 91: 94–100.

Strawberry sap beetle, Stelidota geminata (Say) (Coleoptera: Nitidulidae).

Distribution. There are several species of sap beetles that affect strawberries throughout the United States, including the strawberry sap beetle, *Stelidota geminata* (Say). Other sap beetles, including the picnic beetle, *Glischrochilus quadrisignatus* (Say), and the dusky sap beetle, *Carpophilus lugubris* Murray, also may be problematic in many areas.

Biology. Adult beetles are brown and somewhat mottled, approximately 3 mm in length with capitate antennae. The elytra are shortened, leaving the terminal abdominal segments exposed. Larvae range in color from white (early instar) to pale yellow (mature) and have a light brown head. The body is elongate and the mouthparts are exposed. Larval development is much shorter compared with the adult lifestage. *Stelidota geminata* pupae are tan colored prior to adult emergence and are about 4 mm in length. Adults emerge in the spring from decaying vegetation and mate and lay eggs near decomposing plant material. Larvae develop in decomposing material in or near the soil before pupating. The cycle from egg to adult takes about 30 to 35 days.

Damage. The sap beetle is also a contaminant of strawberries, which may cause large batches of fruit to be rejected by processors. In some cases the adults bore into the strawberry fruit, rendering it unmarketable.

IPM strategies. The sap beetle may be monitored with traps baited with synthetic volatile compounds. Since the beetles are attracted to ripe fruits, insecticide treatments are usually not recommended. One cultural tactic is to harvest berries before they become over-ripe. In addition, damaged, decaying or diseased fruit should be removed from the field.

References
Peng, C., and R. N. Williams 1991. Effect of trap design, trap height, and habitat on the capture of sap beetles (Coleoptera: Nitidulidae) using whole-wheat bread dough. *Journal of Economic Entomology* 84: 1515–1519.

Miller, K. V., and R. N. Williams. 1982. Seasonal abundance of *Stelidota geminata* (Say) in selected habitats strawberry sap beetle, USA. *Journal of the Georgia Entomological Society* 17: 112–117.

Strawberry bud weevil (clipper), Anthonomus signatus (Say) (Coleoptera: Curculionidae).

Distribution. Strawberry bud weevil is a pest of several small fruits in the northeastern United States and Canada, including strawberries, blueberries, and brambles. This species often is called the 'strawberry clipper' because clipped buds are characteristic damage resulting from weevil infestation.

Biology. Strawberry bud weevil overwinters as an adult, emerging from protected areas such as mulch and fence lines when temperatures become favorable. Adults are approximately 3 mm in length. They are reddish brown beetles with a pronounced snout. Mated females lay eggs singly inside flower buds and then partially clip the blossom below the bud. The infested bud may remain on the plant or drop to the ground, where egg and larval development occurs. Eggs hatch within 1 week. The larvae are white and develop within the bud, reaching maturity after 3 to 4 weeks. Adults emerge in early to mid-summer, depending on region.

Damage. The major damage caused by strawberry bud weevil occurs during egg lay, as infested buds wilt and fail to produce fruit. Prior to egg lay, adults also may feed on pollen from unopened buds. Brownish-purple holes in the sepals may indicate pollen feeding, although this type of damage does not necessarily disrupt normal fruit development.

IPM strategies. Monitoring tactics for strawberry bud weevil should focus on counting the number of clipped buds in a particular area. One method involves placing a 30×60-cm frame over plants and counting the number of clipped buds. Growers should pay particular attention to rows located near potential overwintering sites. Insecticide applications may be warranted if the number of clipped buds exceeds 13 per sampling area (30×60 cm). Sampling should continue after insecticide applications are made in order to assess the need for further applications.

References

Mailloux, G., and N. J. Bostanian 1993. Development of the strawberry bud weevil (Coleoptera: Curculionidae) in strawberry fields. *Annals of the Entomological Society of America* 86: 384–393.

English-Loeb, G., M. Pritts, J. Kovach, R. Rieckenberg, and M. J. Kelly 1999. Compensatory ability of strawberries
to bud and flower removal: implications for managing the strawberry bud weevil (Coleoptera: Curculionidae). *Journal of Economic Entomology* 92: 915–921.

Tarnished plant bug, *Lygus lineolaris* (Palisot de Beauvois) (Hemiptera: Miridae).

Distribution. Tarnished plant bug feeds on wild and cultivated plants throughout North America, although it is a pest primarily in temperate regions. Regarding small fruit production, tarnished plant bug often affects strawberries and brambles (raspberry and blackberry).

Biology. Tarnished plant bug adults exhibit a greenish-brown coloration that is mottled with red. Adults are approximately 6 mm in length and are identified by a characteristic yellow triangle at the junction of the wings anteriorly. Tarnished plant bugs overwinter as adults under bark, leaves, and other ground covers that offer protection. Adults become active during the blossoming period of fruit production and feed on nutritive plant juices. Females lay 1 mm long, cream-colored eggs in the plant tissue, either directly on the fruit crop or within adjacent ground cover. There are five nymphal stages, characterized by a progressive color change from green to

Selected list of arthropods that affect strawberry production in the United States.

Common name	Scientific name	Order	Family
Strawberry spider mite	*Tetranychus turkestani* Ugarov & Nikolski	Acari	Tetranychidae
Two-spotted spider mite[1,2]	*Tetranychus urticae* Koch	Acari	Tetranychidae
Cyclamen mite	*Phytonemus pallidus* (Banks)	Acari	Tarsonemidae
Strawberry bud weevil[1]	*Anthonomus signatus* (Say)	Coleoptera	Curculionidae
Strawberry root weevil[1]	*Otiorhynchus ovatus* L.	Coleoptera	Curculionidae
Strawberry rootworm	*Paria fragariae* Wilcox	Coleoptera	Chrysomelidae
Strawberry crown borer	*Tyloderma fragariae* (Riley)	Coleoptera	Curculionidae
Black vine weevil[1]	*Otiorhynchus sulcatus* (Fabricius)	Coleoptera	Curculionidae
Rough strawberry root weevil[1]	*Otiorhynchus rugosostriatus* (Goeze)	Coleoptera	Curculionidae
Strawberry sap beetle	*Stelidota geminata* (Say)	Coleoptera	Nitidulidae
Tarnished plant bug[1]	*Lygus lineolaris* (Palisot de Beauvois)	Hemiptera	Miridae
Strawberry aphid	*Chaetosiphon fragaefolii* (Cockerell)	Hemiptera	Aphididae
Strawberry root aphid	*Aphis forbesi* Weed	Hemiptera	Aphididae
Strawberry whitefly	*Trialeurodes packardi* (Morrill)	Hemiptera	Aleyrodidae
Meadow spittle bug	*Philaenus spumarius* (L.)	Hemiptera	Cercopidae
Potato leafhopper[1,2]	*Empoasca fabae* (Harris)	Hemiptera	Cicadellidae
Strawberry leafroller	*Ancylis comptana* (Froelich)	Lepidoptera	Tortricidae
Strawberry crown miner	*Aristotelia fragariae* Busck	Lepidoptera	Gelechiidae
Strawberry crown moth	*Synanthedon bibionipennis* (Boisduval)	Lepidoptera	Sesiidae
Western flower thrips[3]	*Frankliniella occidentalis* (Pergande)	Thysanoptera	Thripidae
Eastern flower thrips[3]	*Frankliniella tritici* (Fitch)	Thysanoptera	Thripidae
Florida flower thrips[3]	*Frankliniella bispinosa* (Morgan)	Thysanoptera	Thripidae

[1]also infest brambles. [2]also infest grapes. [3]also infest blueberries and grapes.

Fig. 927   Tarnished plant bug.

brown. The entire lifecycle requires 25 to 40 days, yielding a varying number of generations, depending on seasonal weather conditions.

Damage.   Fruit affected by tarnished plant bug feeding exhibits 'catfacing,' which is the result of tissue development around a centralized wound. Damage caused by egg lay may cause more intense fruit distortion and may be accompanied by one or more blemishes (scabs). The severity of damage resulting from tarnished plant bug injury may vary with cultivars.

IPM strategies.   There are several techniques that can be used for monitoring tarnished plant bug populations. Direct plant examinations or visual inspections, sticky traps, sweep sampling, and fruit injury counts are only some of the techniques used for monitoring tarnished plant bug activity in the field. Any one of these tactics may be preferred in one setting over another. Sweep netting is a simple way to sample for nymphs in nearby vegetation. Visual examination is a direct method for visualizing insect activity, although adults may fly if disturbed, and this method may vary considerably with weather conditions. Ground cover management, including mowing and tilling, may be useful for reducing migrating tarnished plant bug populations.

References

Panneton, B., A. Belanger, C. Vincent, M. Piche, and M. Khelifi 2000. Effect of water volume rates on spray deposition and control of tarnished plant bug (Hemiptera: Miridae) in strawberry crops. *Phytoprotection* 81: 115–122.

Handley, D. T., and J. E. Pollard 1993. Tarnished plant bug (*Lygus lineolaris*) behavior on cultivated strawberries. *Acta Horticulturae* (Aug): 463–468.

Brambles

Brambles are defined as plants belonging to the genus *Rubus* and include blackberries and raspberries. Tayberries are brambles resulting from a cross between blackberry and raspberry. Blackberries and raspberries are the two most common bramble crops grown in the U.S. In 2001, more than 6,000 acres of blackberries were produced in the U.S. with an estimated value exceeding $14 million dollars. Oregon leads the country in blackberry production with more than 90% of the U.S. sales. Among the brambles, raspberries are cultivated on the largest acreage. The three most commonly types of raspberry grown in the U.S. are red, black and purple. In 2001, approximately 16,000 acres of raspberries were produced in the U.S. with more than 50% of the production coming from the state of Washington. Brambles are susceptible to several key pests including raspberry cane borer, *Oberea bimaculata* (Olivier), rednecked cane borer, *Agrilus ruficollis* (F.), and blackberry psyllid, *Trioza tripunctata* (Fitch).

Pests of raspberries and blackberries (brambles), *Rubus* spp. (family Rosaceae)
Raspberry cane borer, Oberea bimaculara (Olivier) (Coleoptera: Cerambicidae).

Distribution.   The raspberry cane borer is native to North America and it is found throughout the northeastern United States and Canada.

Biology.   The posterior region of the head (prothorax) is bright yellow with two or three black dots. Raspberry cane borer requires a two-year cycle to complete its development. Females lays eggs on the canes and then girdle a hole approximately 6 mm in diameter around the eggs. After eggs hatch, larvae bore further into the cane. They overwinter in the soil. During the following season, larvae continue to bore into the canes until they reach the raspberry crown. During the second spring mature larvae continue to feed, eventually pupating in the hollow

Selected list of arthropods that affect brambles (raspberry and blackberry) production in the United States.

Common name	Scientific name	Order	Family
Raspberry cane borer	*Oberea bimaculara* (Olivier)	Coleoptera	Cerambicidae
Rednecked cane borer	*Agrilus ruficollis* (F.)	Coleoptera	Buprestidae
Raspberry fruitworm	*Byturus unicolor* (Say)	Coleoptera	Byturidae
Eastern raspberry fruitworm	*Byturus rubi* Barber	Coleoptera	Byturidae
Western raspberry fruitworm	*Bytunus bakeri* Barber	Coleoptera	Byturidae
Raspberry cane maggot	*Pegomya rubivora* (Coquillett)	Diptera	Anthomyiidae
Blackberry psyllid	*Trioza tripunctata* (Fitch)	Hemiptera	Psyllidae
Blackberry leafhopper	*Dikrella californica* (Lawson)	Hemiptera	Cicadellidae
Blackberry gallmaker	*Diastrophus nebulosis* (O.S)	Hymenoptera	Cynipidae
Raspberry sawfly	*Monophadnoides geniculatus* (Hartig)	Hymenoptera	Tenthredinidae
Blackberry leafminer	*Metallus rubi* Forbes	Hymenoptera	Tenthredinidae
Blackberry skeletonizer	*Schreckensteinia festaliella* (Hübner)	Lepidoptera	Heliodinidae
Raspberry bud moth	*Lampronia rubiella* (Bjerkander)	Lepidoptera	Incurvariidae
Raspberry crown borer	*Pennisetia marginata* (Harris)	Lepidoptera	Sesiidae
Raspberry leafroller	*Olethreutes permundana* (Clemens)	Lepidoptera	Tortricidae

surface of the cane. Adults begin to emerge in early summer and produce eggs shortly after.

Damage. Heavy infestation results in wilted shoots and toppled canes.

IPM strategies. Canes should be examined weekly for hollow stems and signs of wilting. One cultural technique, which may reduce total infestation of raspberry cane borer, involves the removal of hollow canes and wilted shoots that show signs of injury. After canes are removed, end surfaces should be re-inspected to ensure that no hollow surfaces remain in the existing canes.

References
Bessin, R. 2001. *Rednecked and raspberry cane borers*. Extension publication, University of Kentucky, College of Agriculture http://www.uky.edu/Agriculture/Entomology/entfacts/fruit/ef209.htm.
Eaton, A. 1994. *Raspberry cane borer, small fruit pest*. Extension fact sheet #53, Cooperative Extension Unit, University of New Hampshire, Durham, New Hampshire.

Raspberry fruitworm, Byturus unicolor (Say) (Coleoptera: Byturidae).

Distribution. Raspberry fruitworm is distributed throughout the eastern United States wherever raspberries are grown.

Biology. Adult beetles are yellowish brown, hairy and about 8 mm long. Adults lay grayish-white eggs on swollen, unopened blossom buds, which may be deposited at the base of the developing fruit. Eggs hatch into larvae within a few days, and the emerging larvae feed on the developing fruit for five to six weeks. Larvae pupate in the soil inside of a cocoon. Raspberry fruitworm overwinters as an adult, and emerges the following year when blossoms begin to develop.

Damage. Raspberry fruitworm larvae feed on raspberry receptacles, causing berries to dislodge from the plant stems. In some cases, larvae feed on internal tissues of immature berries. Adults feed on blossom and foliage but rarely cause significant damage. High populations of adult fruitworm may result in characteristic longitudinal holes in the foliage.

IPM strategies. Collecting and inspecting known numbers of blossoms and looking for eggs can be used as a strategy for monitoring populations of raspberry fruitworm. Developing berries also should be examined for larvae. Cultural strategies include cultivating the soil to bury raspberry fruitworm pupae and exposing pupae to natural predators. Infested plant debris also should be removed from the soil. The use of some selected reduced-risk insecticides may also play a role in regulating raspberry fruitworm populations.

Reference
Baker, W. 1949. Biology and control of the western raspberry fruit worm in western Washington. Bulletin of the State College of Washington. Agricultural Experiment Station #497. 63 pp.

Occasional pests of small fruits

Japanese beetle, Popillia japonica Newman (Coleoptera: Scarabaeidae).

Distribution. The Japanese beetle is native to Japan and was first detected in the United States in New Jersey in 1916. Since then, the Japanese beetle has migrated southward into Alabama and Georgia. Specimens also have been collected in Florida, although the Japanese beetle is not considered a pest in that region. High populations of Japanese beetles have been recorded in Michigan and Ohio, whereas isolated populations have been detected in some western states.

Biology. Japanese beetle adults are metallic green with an amber hue on their forewings. Adults are about 10 mm long and 7 mm wide with characteristic rows of hair along each side of the abdomen. Adults are active by day and spend the nighttime hours within tunnels in the soil. Within 10 days of emergence females begin to lay eggs in the soil of short grasses and continue to lay eggs for several weeks. Larvae (grubs) are usually cream colored just after hatching. Larvae develop through three instars and turn darker with age as they mature. During the winter, larvae move downward into the soil to hibernate when temperatures are cool. After hibernation, grubs tunnel up through the soil to feed on grasses to complete their development. There is usually one generation per year.

Damage. Adult beetles are generalist feeders that skeletonize the leaves of more than 400 broadleaf plants. Feeding damage is characterized by brown lacy leaves. However, adults also may attack flower buds and fruit, including blueberries and grapes. When populations are high, the adult becomes a contaminant in the harvest of blueberries and grapes, thus interfering with the marketability of the fruit. Larvae generally are not injurious to fruit crops but their existence in the ground cover serves as inoculum for next year's adult population.

IPM strategies. The strategies for controlling adults and grubs are quite different. Traps baited with a combination of sex and aggregation pheromones are available commercially for monitoring adult populations of Japanese beetle. However, traps may attract adults from adjacent areas, which may increase the potential for damage to fruit crops. When chemical control is warranted, several applications of insecticide may be necessary to provide adequate control since adults continue to emerge all season long. In the past, grubs have been controlled using the bacterial milky diseases *Bacillus popilliae* Dutky and *B. lentimorbus* Dutky. This method has been relatively successful in the eastern United States. Recently, certain species of entomophagous nematodes have been shown to be effective in suppressing larval populations. Cultural methods of controlling Japanese beetle include decreasing irrigation to young grasses between rows and mowing of ground cover.

References

Stewart, C. D., S. K. Braman, B. L. Sparks, J. L. Williams-Woodward, G. L. Wade, and J. G. Latimer 2002. Comparing an IPM pilot program to a traditional cover spray program in commercial landscapes. *Journal of Economic Entomology* 95: 789–796.

Cappaert, D. L., and D. R. Smitley 2002. Parasitoids and pathogens of Japanese beetle (Coleoptera: Scarabaeidae) in southern Michigan. *Environmental Entomology* 31: 573–580.

Flower thrips, Frankliniella spp. (Thysanoptera: Thripidae).

Distribution. There are several species of flower thrips that affect small fruit production throughout the United States. The eastern flower thrips, *Frankliniella tritici* (Fitch), is found east of the Rocky Mountains, and western flower thrips, *F. occidentalis* (Pergande), occurs throughout the United States and Canada. In the northeast, the blueberry thrips *F. vaccinii* and *Catinathrips kainos* are minor pest problems in lowbush and highbush blueberries, causing leaf distortion and discoloration in small isolated patches in fields. In Florida, a species of flower thrips, *F. bispinosa* (Morgan), locally called Florida flower thrips, causes significant damage to southern highbush and rabbiteye blueberries. *Frankliniella bispinosa* also has been detected as far north as Georgia.

Biology. In the absence of microscopic examination, flower thrips species are difficult to distinguish. Members of the genus *Frankliniella* are approximately 1 to 1.3 mm in length and yellow to brown in color, often with gray blotching across the abdomen. Males are generally smaller and paler than females. Eggs are cylindrical and laid within the plant tissue, making them difficult to see. Thrips development progresses through several nymphal instars (latter stages sometimes are referred to as prepupae and

Selected list of arthropods that affect small fruit production in the United States

Common name	Scientific name	Order	Family
Plum curculio[1]	*Conotrachelus nenuphar* (Herbst)	Coleoptera	Curculionidae
Picnic beetle[2]	*Glischrochilus* spp.	Coleoptera	Nitidulidae
Japanese beetle[1]	*Popillia japonica* Newman	Coleoptera	Scarabaeidae
Rose chafer[3]	*Macrodactylus subspinosus* (Fabricius)	Coleoptera	Scarabaeidae
Green June beetle[2]	*Cotinis nitida* (L.)	Coleoptera	Scarabaeidae
Leaffooted bugs[3]	*Leptoglossus* spp.	Hemiptera	Coriedae
Stink bugs[3]	*Acrosternum hilare* (Say)	Hempiptera	Pentatomidae
Gypsy moth	*Lymantria dispar* L.	Lepidoptera	Lymantriidae
Oblique-banded leafroller[2]	*Choristoneura rosaceana* (Harris)	Lepidoptera	Tortricidae
Redbanded leafroller[3]	*Argyrotaenia velutinana* Walker	Lepidoptera	Tortricidae
Western flower thrips[3]	*Frankliniella occidentalis* (Pergande)	Thysanoptera	Thripidae
Eastern flower thrips[3]	*Frankliniella tritici* (Fitch)	Thysanoptera	Thripidae
Florida flower thrips[3]	*Frankliniella bispinosa* (Morgan)	Thysanoptera	Thripidae

[1] also infest blueberries and grapes. [2] also infest strawberry and brambles. [3] also infest blueberries, grapes, strawberries and brambles.

pupae), each of which resembles the adult morph without wings. Adults and nymphs have rasping mouthparts. Most flower thrips migrate over long distances and are known to have an extensive host range.

Damage. Thrips are capable of causing damage in at least two different stages of fruit development. During the blooming period, thrips feed within the flowers by inserting their mouthparts into developing tissue and remove fluids. The resulting effect can be either aborted or distorted fruit. Thrips also may feed on pollen, causing additional floral abortion. Thrips also may feed on foliage, removing cell sap from plant tissues. If populations of thrips are high during the fruiting period, significant damage to the fruit can be observed, particularly in areas where the fruits are touching or closely clustered. This may result in silvery blemishes resembling a halo. Some species of thrips may be of particular concern for their potential to vector various plant viruses.

IPM strategies. Because thrips are very mobile, their management can be very difficult. There are several methods available for monitoring flower thrips populations, including the use of sticky traps (white or blue), tapping infested flowers over a white surface, and dipping infested flowers into alcohol. Grower preference and crop specifics may dictate which monitoring method works best in a particular area. It is also helpful to know which species is being targeted. Several predatory insects may be successful for suppressing flower thrips populations, including the minute pirate bug, *Orius insidious* (Say). The fungal pathogen *Beauveria bassiana* has shown potential for control of *F. occidentalis*, and may be a potential agent for control of other thrips species in the future. Cultural methods for control of thrips species include elimination of alternate hosts near the main crop, particularly if the blooming periods overlap.

Oscar E. Liburd and Erin M. Finn
University of Florida
Gainesville, Florida, USA

References
Moritz, G., D. Morris, and L. Mound 2001. *Thrips ID: pest thrips of the world*. CSIRO Publishing, Collingwood, Victoria, Australia.
Childers, C. C., and J. K. Brecht 1996. Colored sticky traps for monitoring *Frankliniella bispinosa* (Morgan) (Thysanoptera: Thripidae) during flowering cycles in citrus. *Journal of Economic Entomology* 89: 1240–1249.

SMALL GREEN STINK BUG, *Piezodorus guildinii* (WESTWOOD) (HEMIPTERA: HETEROPTERA: PENTATOMIDAE).

The small green stink bug, *Piezodorus guildinii* (Westwood), is a Neotropical pentatomid found from the southern U.S. to Argentina. It first was described from the island of St. Vincent, and frequently has been reported from Central and South America. It is a major pest of soybean in South America. In Brazil, this stink bug was seldom found on soybean until the early 1970s. Subsequently, it has become more common, ranging from Rio Grande do Sul (32° S latitude) to Piau (5° S latitude). With the expansion of soybean production to the central, west, and northeast regions of the country, *P. guildinii* is, perhaps, the most important pest of soybean in Brazil.

The list of food plants includes some economically important plants in addition to soybean, mostly legumes such as common bean, pea, and alfalfa. It is reported occasionally on sunflower, cotton, and guava, but it is not believed to be a serious pest of these crops. Native host plants include species of indigo legumes, *Indigofera* spp., in the southern U.S., Colombia, and Brazil. It also feeds on legumes of *Sesbania* and on *Crotalaria*.

Eggs are blackish and laid in two parallel rows. First, second, third, fourth, and fifth instars are 1.30, 2.25, 2.58, 4.60, and 7.87 mm long, respectively. Adults are light green to yellowish, with a red-band at the base of the scutellum, particularly on females.

The number of egg masses per female will vary from approximately 3 on soybean up to 37 in some indigo species. Total number of eggs per female will vary from approximately 28 on soybean to approximately 200 on *Indigofera hirsuta* to approximately 500 on *I. truxillensis*. Adults will have a total longevity of approximately 50 days on soybean (mean of female and male) up to approximately 90 days on some of the indigo legumes. Despite its relatively low performance on soybean compared to native legumes, *P. guildinii* is highly detrimental to this crop, as noted previously.

Piezodorus guildinii, like most of the phytophagous pentatomids associated with soybean, feeds primarily on pods. The nature and extent of the damage to this crop is similar to that reported for southern green stink bug, *Nezara viridula*. *Piezodorus guildinii* usually is the first species to appear in soybean fields during flowering or even earlier. Apparently, *P. guildinii* is better adapted to feeding on flowering plants than other pentatomid species. However, it does have to feed on reproductive structures to thrive.

Chemical insecticides, such as monocrotophos and endosulfan, are the major weapons used to control *P. guildinii*. Natural enemies include several species of egg parasitoids such as *Telenomus mormideae* Lima, *Telenomus* sp., *Trissolcus basalis* (Wollaston), *T. scuticarinatus* Lima, *Ooencyrtus submetallicus* (Howard), and *Ooencyrtus* sp. Nymphs and adults of the pentatomid *Tynacantha marginata* Dallas were reported as predators of fifth instars, and adults are reported to be a substantial part of the diet of some birds in Argentina.

See also STINK BUGS, HEMIPTERA.

Antônio R. Panizzi
Embrapa
Londrina, Brazil

References

McPherson, J. E., and R. M. McPherson. 2000. *Stink bugs of economic importance in America North of Mexico.* CRC Press, Boca Raton, Florida. 253 pp.

Panizzi, A. R. 1997, Wild hosts of pentatomids: ecological significance and role in their pest status on crops. *Annual Review of Entomology* 2: 99–122.

Panizzi, A. R., and F. Slansky,Jr. 1985. Review of phytophagous pentatomids associated with soybean in the Americas. *Florida Entomologist* 8: 184–214.

Panizzi, A. R., and J. G. Smith. 1977. Biology of *Piezodorus guildinii*: oviposition, development time, adult sex ratio, and longevity. *Annals of the Entomological Society of America* 70: 35–39.

Schaefer, C. W., and A. R. Panizzi (eds.) 2000. *Heteroptera of economic importance.* CRC Press, Boca Raton, Florida. 828 pp.

Fig. 928 Adult of small green stink bug.

SMALL RICE STINK BUG, *OEBALUS POECI-LUS* (DALLAS) (HEMIPTERA: HETERO-PTERA: PENTATOMIDAE). The small rice stink bug, *Oebalus poecilus* (Dallas), also known as *Solubea poecila* (Dallas), is perhaps the most important pest of rice in South America. Although showing preference for Gramineae such as rice, barley, oat, corn and wheat, it also is associated with soybean, cotton, and guava. There are about 42 species of host plants for this pest within the state of Rio Grande do Sul, Brazil.

Adults are rust to black dorsally, with yellowish spots; the venter usually is darker in females than males. Males are 6.9 to 8.3 mm long and females are 7.4 to 9.5 mm long. The small rice stink bug has distinctive morphs that distinguish non-diapausing and diapausing adults. The adult key characters are the pronotal shape and color of the body. In the non-diapausing summer morph, the lateral angles of the pronotum are spinose and the predominant dorsal coloration is dark brown, almost black; in the over-wintering diapausing morph, the lateral angles are rounded and the predominant dorsal coloration is light brown.

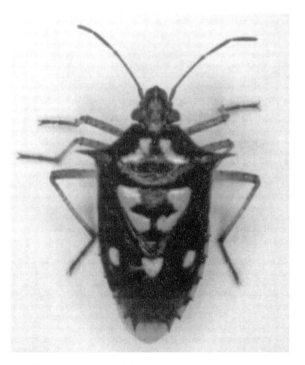

Fig. 929 Small rice stink bug.

Despite the importance of *O. poecilus* as a pest, little has been published on its biology. The egg color changed due to the embryonic development, which allowed the age of the egg to be estimated. *Oebalus poecilus* is reared successfully in the laboratory using panicles of the weed *Polygonum punctatum*.

Nymphs and adults fed on the developing grains of rice, and the nature and extent of damage depend on the stage of grain development at the time of attack. Florets fed upon during early endosperm formation (milk stage) resulted in either empty glumes or severely atrophied kernels. Feeding during later stages of endosperm development (dough and hard stages) resulted in a chalky discoloration around the feeding site. Rice with this damage, known as 'pecky rice,' is weakened structurally and often breaks under mechanical stress during milling. Furthermore, the pecky rice that escapes breakage is of inferior quality as a finished product because of the discoloration. Other types of damage include withering of young plants, reduction of tillers, and grain emptiness due to feeding activity.

In spite of being a major pest, no effective method to manage *O. poecilus* on rice fields has been developed. However, a pest management strategy for *O. poecilus* in southern Brazil based on planting time has been recommended. This method consists of delaying and restricting the time of rice planting. Planting the crop in the first half of December results in a synchrony between the pest and the crop because the developing grain is available in March when bugs entering the adult stage are diapausing morphs that migrate to hibernating sites instead of remaining in the crop. The result is that damage is considerably reduced.

Natural enemies of *O. poecilus* include the egg parasitoid *Telenomus mormideae* Lima, and the nymph and adult parasite *Beskia cornuta* Brauer & Bergenstan (Diptera: Tachinidae).

See also STINK BUGS, HEMIPTERA.

Antônio R. Panizzi
Embrapa
Londrina, Brazil

References
Albuquerque, G. S. 1993. Planting time as a tactic to manage the small rice stink bug, *Oebalus poecilus* (Hemiptera, Pentatomidae), in Rio Grande do Sul, Brazil. *Crop Protection* 12: 627–630.

Schaefer, C. W., and A. R. Panizzi (eds.) 2000. *Heteroptera of economic importance*. CRC Press, Boca Raton Florida. 828 pp.

Silva, C. P. 1992a. Aspectos biológicos básicos de *Oebalus poecilus* (Dallas, 1851) (Heteroptera: Pentatomidae) por ataque de parasitóides de ovos na cultura de arroz. *Anais da Sociedade Entomológica do Brasil* 21: 225–231.

Silva, C. P. 1992b. Mortalidade de *Oebalus poecilus* (Dallas, 1851) (Heteroptera:) Pentatomidae) por ataque de parasitóides de ovos na cultura de arroz. *Anais da Sociedade Entomológica do Brasil* 21: 289–296.

van Halteren, P. 1972. Some aspects of the biology of the paddy bug, *Oebalus poecilus* (Dallas), in Surinam. Surinaam. *Landb*. 2: 23–33.

SMALL RNA VIRUSES OF INVERTEBRATES.

Baculoviruses are the best-known viruses associated with insects. They have large, rod-shaped particles that measure around 30 to 60 nm by 250 to 300 nm which contain double-stranded DNA genomes. Baculovirus particles are most often embedded in large proteinaceous occlusion bodies that usually measure between 0.5 and 2 µm and thus are visible by light microscopy. However, during the last 30 to 40 years a large number of non-occluded, small (25 to 40 nm in diameter), icosahedral viruses have been reported in a variety of insects and other invertebrates. With their small size, spherical appearance, and genomes comprised of RNA, they have frequently received the general epithet picornaviruses. This name derives from the combination of terms pico (= small) + RNA + viruses. However, to avoid confusion with the recognized taxonomic group of vertebrate viruses known as the *Picornaviridae*, the insect viruses are best referred to by the more general term small RNA viruses (SRVs). Wherever possible we have used the terminology and nomenclature recognized by the International Committee for Taxonomy of Viruses (ICTV) in their VIIth report.

There are three recognized families of invertebrate SRVs: *Nodaviridae*, *Tetraviridae* and *Dicistroviridae* (previously termed the 'cricket paralysis-like viruses'). In addition, there is an emerging grouping of picorna-like viruses that show superficial similarity to the *Picornaviridae*. The basic characteristics of these four groups are presented below. Moreover, while a large number of SRVs still remain unclassified and show no affinities with other recognized groups of viruses, we also provide a brief description of some of the more novel virus-types to be found among this more ad hoc assemblage. In the final sections of this chapter, we also provide a brief summary of the general pathology of insect SRVs and of their potential for use as biocontrol agents and biopesticides.

Nodaviridae

This is a unique and well-studied group with virus representatives from both vertebrate and invertebrate hosts. The members of the family from the genus *Alphanodavirus* generally infect insects while those from the genus *Betanodavirus* are those isolated from fish species. The nodavirus genome is comprised of two separate single-stranded RNAs: RNA1 (approximately 3,100 nucleotides long) and RNA2 (approximately 1,400 nucleotides long). The 30 nm diameter virus particles are primarily assembled from a single protein of 40,000 to 45,000 molecular weight (MW) whose precursor is coded by RNA2. The protein(s) that form the structural shell (the capsid) which surrounds the viral nucleic acid generally are referred to as the capsid proteins. The type virus of the family, Nodamura virus (NV), was isolated from *Culex* mosquitoes but is able to replicate in other invertebrate hosts (e.g., *Galleria melonella* larvae and honey bees, *Apis mellifera*) as well as in suckling mice. There are also nodavirus isolates from Lepidoptera (Boolara virus (BoV) from the hepialid, *Oncopera inticoides* and gypsy moth virus from *Lymantria ninayi*), and from Coleoptera (black beetle virus (BBV) from *Heteronychus arator*; Flock House virus (FHV) and Manawatu virus (MwV) from *Costelytra zealandica*). Surprisingly, FHV has been shown to replicate and assemble infectious virus particles in some plants, although these infections only occur if the viral genome is introduced directly into the plant cells (by a process known as transfection). Some of the alphanodaviruses (e.g., FHV, BBV) generate extraordinarily large yields (milligram quantities of virus) during laboratory infections of two *Drosophila* cell lines (i.e., Schneider's Line 1 and Line 2 cells). They also are capable of establishing persistent infections in these cell lines under certain conditions and one nodavirus, New Zealand virus (NZV; also known as *Drosophila* line 1 virus, DLV) is most likely to have resulted from a chance laboratory infection of this continuously cultured cell line.

Tetraviridae

Tetraviruses have been isolated so far only from, or shown to infect, Lepidoptera, mainly saturniid, limacodid and noctuid moths. The particles of tetraviruses are relatively large (38 to 40 nm in diameter) and exhibit a lower density than most of the other SRVs of insects (i.e., approximately 1.30 g/ml in caesium chloride cf. 1.34 to 1.43 g/ml for the dicistroviruses and picorna-like viruses). The group name is derived from the unusual T = 4 (tetra) symmetry of their particles which contrasts to the T = 1 or T = 3 symmetry common to most other small RNA viruses. The major tetravirus capsid protein is large, around 60,000 MW, although a minor protein of 7,000 to 8,000 MW can also be present in some types. The tetraviruses are divided into two genera: Betatetravirus (the *Nudaurelia β*-like viruses) and Omegatetravirus (the *Nudaurelia ω*-like viruses), although the majority of isolates have yet to be assigned to either group. Members of both genera contain a major single-stranded RNA approximately 5,300 to 6,500 nucleotides long. However, Omegatetraviruses also contain a second, smaller strand of RNA approximately 2,500 nucleotides long that has been shown to code for the capsid protein. The betatetraviruses *Darna trima* virus (DtV) and *Setothosea asigna* virus (SaV) (*Thosea asigna* virus) have been successfully used in biological control programs directed against their primary limacodid hosts, which are pests of oil palms. Field trials with an Omegatetravirus, *Heliothis armigera* stunt virus (HaSV), have indicated that it has potential for the control of the cotton bollworm (*Helicoverpa armigera*) even though no natural epizootic of the virus has been recorded.

Picorna-like viruses

The vertebrate *Picornaviridae* include such well-known members as poliovirus, hepatitis A virus (HAV) and foot-and-mouth disease virus (FMDV), as well as the plethora of viruses whose infections result in the common cold (rhinoviruses). Although the derivation of the family name simply alludes to being small RNA viruses, the true picornaviruses all have three major capsid proteins which encase a ssRNA genome around 7,500 nucleotides long. Many picornaviruses also have a fourth, internal capsid protein. During infection, the genomic RNA is translated to produce a single, long polypeptide (greater than 200,000 MW). This is cleaved to produce the various non-structural proteins (involved in RNA replication and protein processing) observed during the replication cycle as well as the capsid proteins which eventually are assembled into mature virions which contain the RNA genome.

Although a number of the invertebrate small RNA viruses are remarkably similar to picornaviruses with respect to their particle diameters, densities and protein compositions, it is at the molecular level of the genome where important differences have recently become apparent. Thus, while these recent genomic studies have produced evidence that some insect SRVs produce a single polypeptide from the genomic RNA and have their genomes organized like the picornaviruses with their structural proteins at the 5′ end of the genome, the specific arrangement of the capsid proteins is more like the dicistroviruses (see below). Three viruses have, to date, been found to have such characteristics, namely Infectious flacherie virus (IFV) of silkworms, sacbrood virus (SBV) of honeybees and *Perina nuda* picorna-like virus (PnPV) from the ficus transparent wing moth, *Perina nuda*.

Other picorna-like SRVs are either insufficiently characterized or, like the type member, Cricket paralysis virus (CrPV), have been shown to have a genome structure and replication characteristics that warrant them being placed in a separate group, the so-called Cricket paralysis-like viruses or *Dicistroviridae*.

Dicistroviridae (cricket paralysis-like viruses)

A number of other insect SRVs originally isolated in the 1960s and 70s, such as CrPV, *Drosophila* C virus (DCV) and IFV, also were thought to be picornaviruses since they produced large precursor polypeptides which were then post-translationally cleaved to produce three major capsid proteins and the non-structural proteins. However, there were a number of discrepancies which suggested that these well-studied insect viruses, as well as others, represented a novel category of SRVs. The proteins observed during infection, for example, were produced in unequal amounts, i.e., some were present at significant molar excess whereas equivalent amounts would be expected from the true picornavirus strategy. Also, partial sequence analysis indicated that the 3′ end of the genome (where picornaviruses code for the non-structural proteins) appeared to be directing the synthesis of the virus capsid proteins. The sequencing and analyses of the complete genomes of DCV and several other viruses, including CrPV, has since revealed that these viruses actually translate their RNA in two separate parts in stark contrast to the true picornaviruses. The existence of two separate cistrons (or Open-Reading Frames, ORFs) is the derivation of the name, *Dicistroviridae*. Interestingly, these viruses also use a novel method of translating the polyprotein; the more 3′ of the two ORFs has dispensed with the initiation codon used for starting the translation, of virtually every other eukaryotic protein. Other confirmed dicistroviruses are the lepidopteran *Plautia stali* intestine virus (PSIV); several viruses isolated from hemipterans (*Triatoma* virus, TrV; *Rhopalosiphum padi* virus, RhPV; Himetobi P virus, HiPV); and the hymenopteran isolates Acute bee paralysis virus (ABPV) and Black queen cell virus (BQCV). As more viruses are sequenced, the *Dicistroviridae* will undoubtedly accumulate more members. Recently, a virus that infects penaid shrimp (Taura syndrome virus) has been found to have all of the characteristics of a discistrovirus, demonstrating that these viruses are not restricted to the Insecta.

Unclassified viruses

Most of the insect SRVs isolated to date remain unclassified. These include observations made of presumably RNA-containing virus-like particles in insect extracts examined by electron microscopy or crystalline arrays noted in thin sections. There are also virus isolates which are reasonably well characterized biophysically but which still defy classification because they do not easily fit into pre-established taxonomic categories. Some have unique features, like the lemon-shaped particles of Chronic bee paralysis virus (CBPV) and *Drosophila* RS virus or the knobbed capsid structure of kelp fly virus (KFV). At present, there is insufficient data to confidently classify these viruses. Other unclassified examples include the icosahedral *Acyrthosiphum pisum* virus, which produces a single major capsid protein and several smaller capsid proteins from an array of overlapping ORFs, and Bee virus X and Bee virus Y that have only single capsid proteins of around 50,000 MW.

Pathology of SRVs

Despite the size of their single-stranded RNA genomes, which comprise only about 3 to 15 genes, SRVs often are capable of rapidly killing or severely debilitating their host. In other instances the SRVs appear to be benign, existing as latent or unapparent infections in their hosts who show no obvious symptoms of infection. Thus, the presence of SRVs may remain undetected for years until their hosts display obvious symptoms which in many instances seem to be associated with increased stress in the host. For example, during studies on the Australian field cricket, *Teleogryllus commodus*, in the 1960s, apparently healthy young crickets were collected and brought into the laboratory. Subsequently, in these large overcrowded laboratory colonies some crickets became paralyzed and died. The paralytic disease, which spread rapidly and devastated the colony, was found to be caused by the now well-known Cricket paralysis virus.

Similar outbreaks of other virus diseases in laboratory colonies of insects are not unheard of even in the best-maintained rearing facilities. However, the symptoms of disease often are difficult to define. They may range from reduced longevity of the host insect (*Drosophila* A virus and *Drosophila* C virus), general lethargy (*Setothosea asigna* virus), through to the more obvious symptoms of paralytic disease (CrPV and Aphid lethal paralysis virus) and inability to molt or developmental stunting (*Helicoverpa armigera* stunt virus). Many insect SRVs are readily transmitted from individual to individual but many also are transmitted through the germ-line to the progeny of infected individuals. In the latter case, surface sterilization of eggs using low concentrations of hypochlorite often can eliminate the offending SRV but re-infection from laboratory sources can occur readily.

Potential for insect SRVs as biocontrol agents and biopesticides

SRVs replicate in the cytoplasm of infected cells and may produce on the order of 250,000 to 1 million particles within each infected cell during their 12 to 24 hour replication cycle. Thus, their capacity to increase in number and spread to uninfected individuals is enormous. This, coupled with the often severe pathological effects of infection, makes them candidates for use as biological control agents or biopesticides of insect pests. There are several examples of their successful use in the field. In the 1960s in Uganda, the lasciocampid moth *Gonometa podocarpi* was a serious pest of pine plantations. Some larval cadavers were found to contain a small RNA virus and a crude extract of these, applied in areas where no diseased larvae had been detected, resulted in effective control of the pest. Other examples are in the Indonesian states of Sabah and Sarawak in the early 1970s with the control of *Darna trima*, a pest of oil palms and in the 1990s with *Epicerura perigrisea*, a lepidopteran defoliator of two economically important tree species in the Ivory Coast.

There has been some reluctance to seriously consider the insect SRVs as biocontrol agents/ biopesticides primarily because of similarities to the vertebrate picornaviruses. Safety testing, as with any biological agent, would be required. Host range considerations also would be important since some of the SRVs can affect a wide range of insect hosts. CrPV, for example, has been shown to multiply in a large number of species from five different insect orders (Lepidoptera, Orthoptera, Diptera, Hymenoptera and Heteroptera) and arguably has the widest host range of all known viruses. Some SRVs (e.g., the tetraviruses) have a highly restricted host range and may infect only a few lepidopteran species. Others, like kelp fly virus (KFV), may infect only a few species but from different orders (Diptera and Lepidoptera). The SRVs are particularly attractive control agents for pests where no baculoviruses are available.

The host range of any SRV contemplated as a biological control agent would require specific tests with reference to the ecological considerations of the areas in which their use was contemplated. The use of selective application strategies or baits may be advantageous. Economic concerns may limit commercial production of SRVs although large quantities, in some cases, can be produced directly in insect hosts or in available insect cell lines. However, local cottage industry production and application as in the cases in Uganda, Sabah and Sarawak, and the Ivory Coast may be the most feasible alternative in many situations.

Paul D. Scotti
The Horticulture and Food Research
Institute of New Zealand
Mt. Albert Research Centre
Auckland, New Zealand
and
Peter D. Christian
National Institute for Biological Standards and Control
Potters Bar, Hertfordshire, United Kingdom

References

Miller, L. K., and L. A. Ball (eds.) 1998. *The insect viruses.* Plenum Press, New York, New York. 413 pp.

Granoff, A., and R. Webster (eds.) 1999.*The encyclopaedia of virology (2nd ed.).* Academic Press, New York, New York. 2,000 pp.

Longworth, J. F. 1978. Small isometric viruses of invertebrates. *Advances in Virus Research* 23: 103–157.

SMALL WINTER STONEFLIES. Members of the stonefly family Capniidae (order Plecoptera). See also, STONEFLIES.

SMALL-HEADED FLIES. Members of the family Acroceridae (order Diptera). See also, FLIES.

SMICRIPIDAE. A family of beetles (order Coleoptera). They commonly are known as palmetto beetles. See also, BEETLES.

SMINTHURIDAE. A family of springtails in the order Collembola. They commonly are known as globular springtails. See also, SPRINGTAILS.

SMITH, JOHN BERNHARDT. John B. Smith was born at New York City, New York, USA, on November 21, 1858. Though admitted to the bar in 1879, after practicing for a few years he abandoned the legal profession to pursue a career in entomology. He had many interests and proficiencies, and became known as both an economic entomologist and a taxonomist. The members of the Brooklyn Entomological Society were likely responsible for switching Smith from a legal to an entomological career. Initially he worked for the United States Department of Agriculture, but in 1886 he became assistant curator in entomology at the U.S. National Museum. There he published some excellent works including ''Monograph of the Sphingidae of America North of Mexico,'' ''A preliminary catalogue of the Arctiidae of temperate North America,'' ''A revision of the lepidopterous family Saturniidae,'' and ''Contributions toward a monograph of the family Noctuidae.'' In addition to this work on Lepidoptera he also studied Coleoptera while at the Museum, and in 1889 moved to Rutgers University. In 1894 he became state entomologist for New Jersey. Though taxonomy was Smith's first interest, he applied his considerable energy and enthusiasm to solving the state's pest problems, and so became involved in insecticide studies aimed at vegetable, fruit, shade tree, livestock and medical pests. A particular success was demonstration of the effectiveness of fish-oil soap for suppression of San Jose scale. However, he is best known for the development of mosquito management techniques for the New Jersey salt marshes, which emphasized ditching and draining. His vision of ridding New Jersey of salt marsh breeding mosquitoes was met with ridicule by the state legislature, but Smith was both intelligent and persistent, so he persevered and eventually the legislature appropriated funds to support this project. One of the important concepts that he promoted was that mosquitoes did not honor municipal boundaries, and that mosquito control had to be handled on a large geographic scale. Smith's approach was quite effective at reducing salt marsh mosquitoes, and eventually became adopted all along the Atlantic coast of the United States. Smith authored more than 600 titles, including several important books, including "Mosquitoes occurring within New Jersey, their habits, life history, etc.," "Catalogue of the Lepidopterous superfamily Noctuidae found in boreal America," "Explanation of terms used in entomology," and "Economic Entomology." Smith belonged to many societies, often serving in a leadership role, and was a fellow of the American Association for the Advancement of Sciences, and the New York Academy of Sciences. He died March 12, 1912.

References

Grosbeck, J. A. 1912. Professor John Bernhardt Smith, Sc.D.. *Entomological News* 23: 193–196.

Mallis, A. 1971. *American entomologists*. Rutgers University Press, New Brunswick, New Jersey. 549 pp.

Crans, W. J. 2002. John B, Smith, Pioneer for organized mosquito control in New Jersey. *Wing Beats* 13(2): 6–7, 13.

SMITH, HARRY SCOTT.

Harry Smith was born at Aurora, Nebraska, on November 29, 1883. He obtained his A.B. (1907) and M.S. (1908) degrees from the University of Nebraska. From 1909 to 1913 he worked for the United States Bureau of Entomology, studying natural enemies of important pests. In 1913 Smith was appointed superintendent of the California State Insectary in Sacramento, and in 1923 was transferred to the University of California at Riverside where he was put in charge of the Department of Biological Control. He remained in this capacity until his retirement in 1951, and did much to foster implementation of biological control in California and the western United States. Smith was largely responsible for the large-scale use of the coccinellid *Cryptolaemus moutrouzieri* for citrus mealybug suppression, and the wasp *Macrocentrus ancylivorus* for Oriental fruit moth suppression. He also pioneered the use of biological suppression of Klamath weed with insect herbivores. Smith was elected president of the American Association of Economic Entomologists, and received an honorary doctorate from the University of Nebraska. He died at Riverside on November 28, 1957.

Reference

Mallis, A. 1971. *American entomologists*. Rutgers University Press, New Brunswick, New Jersey. 549 pp.

SMITH, RAY F.

Ray Smith was born at Los Angeles, California, USA, on January 20, 1919. He received all his degrees at the University of California-Berkeley. He joined the faculty there in 1946 , and became Department Chairman in 1959. Smith authored over 300 publications, and championed the ecological approach to pest management, and the minimization of pesticides in agricultural ecosystems through integration of pesticides with other approaches to pest suppression. A paper he co-authored on integrated control was cited as one of the most important publications on crop protection, and Smith is considered by some to be the 'father of pest management.' Among his many honors and awards are the C.W. Woodworth Award, Fellow and President of the Entomological Society of America, and member of the National Academy of Sciences. He also was co-recipient of the 1977 World Peace Prize. He died August 23, 1999, in Lafayette, California.

Reference

Scalise, K. 1999.UC Berkeley pest control scientist Ray F. Smith has died at age 80. University of California, Berkeley, news release, 9/30/99. Seen at http://www.berkeley.edu/news/media/releases/99legacy/9-3-99.html (August 2002).

SMOKY MOTHS.

Some members of the family Zygaenidae (order Lepidoptera). See also, BURNET MOTHS, BUTTERFLIES AND MOTHS.

SMOOTH SPRINGTAILS. A family of springtails (Isotomidae) in the order Collembola. See also, SPRINGTAILS.

SNAKEFLIES (RAPHIDIOPTERA). The Raphidioptera, or snakeflies, are virtually restricted to the Holarctic region. They are easily identified by their general appearance, have an elongate head and prothorax, with the legs arising posteriorly. The females have a long ovipositor. There are two families which have similar biologies. The larvae usually live in the crevices in the bark of trees or shrubs, or in the superficial stratum of the soil around the roots of shrubs. The pupae are strange in that they can move about in their environment. Most of the taxonomy is based on the male and female terminalia.

The family Inocelliidae is a small family of about 20 species in about five genera. They differ from the Raphidiidae in lacking ocelli. Two genera occur in the New World, *Negha* Navás (also in Spain) and the endemic *Indianoinocellia* Aspöck & Aspöck.

The family Raphidiidae consists of about 168 species in 26 genera. The majority of genera and species are found in the Palaearctic Region. There are only two genera found in North America, *Alena* Navás, which is the southernmost occurring genus in the world with two species living in the pine forests of Oaxaca, Mexico, and *Agulla* Navás which is restricted to the western half of North America. Larvae of *Alena* were found under pine bark.

Lionel Stange
Florida Department of Consumer and Agricultural Services, Division of Plant Industry
Gainesville, Florida, USA

Fig. 930 A snakefly, *Agulla bicolor* (Albarda) (Raphidioptera: Raphidiidae) (photo L. Buss).

Reference

Aspöck, H., Aspöck, U., and H. Rausch. 1991. *Die Raphidiopteren der Erde. Eine monographische Darstellung der Systematik, Taxonomie, Biologie, Ökologie und Chorologie der rezenten Raphidiopteren der Erde, mit einer zusammenfassenden Übersicht der fossilen Raphidiopteren (Insecta:Neuropteroidea)*. Goecke & Evrs: Krefeld. 2 vols., 730 and 550 pages.

SNIPE FLIES. Members of the family Rhagionidae (order Diptera). See also, FLIES.

SNODGRASS, ROBERT EVANS. Robert Snodgrass was born at St. Louis, Missouri, on July 5, 1875. His family relocated to Kansas and then California, and Robert entered Stanford University in 1895. He majored in zoology, and due to his interest in birds he was encouraged by V.L. Kellogg, the entomology instructor, to study bird lice. This culminated in two publications on louse morphology. He accepted a teaching position at Washington State University when he graduated in 1901, but his practical jokes were not fully appreciated and he returned to Stanford. There he caused problems by stripping the mulberry trees to feed his silkworms, so he moved on to San Francisco and eventually Washington, DC, where he commenced work for the Bureau of Entomology. Not satisfied with that employment, Snodgrass moved to New York to study art. He work in entomology off and on for brief periods, then settled in with the Bureau of Entomology in 1917 and taught entomology at the University of Maryland from 1924 to 1947. Snodgrass authored over 80 publications, but is best known for his expertise in morphology, exemplified by his books: "Anatomy and physiology of the honeybee" (1925), "The principles of insect morphology" (1935), and "Textbook of arthropod anatomy" (1952). After nearly 70 years, his 'principles' book remains the definitive work on the subject. He died in Washington, DC, on September 4, 1962.

References

Mallis, A. 1971. *American entomologists*. Rutgers University Press, New Brunswick, New Jersey. 549 pp.
Schmidt, J. B. 1963. Robert Evans Snodgrass (1875–1962). *Entomological News* 74: 141–142.

SNOUT BEETLES. Members of the family Curculionidae (order Coleoptera). See also, BEETLES.

Fig. 931 Example of snout butterflies (Libytheidae), *Libytheana bachmanii* (Kirtland) from Pennsylvania, USA.

SNOUT BUTTERFLIES (LEPIDOPTERA: LIBYTHEIDAE).

Snout butterflies, family Libytheidae (also called beaks), are a small family of only 12 species but with at least one species in each faunal region. Alternate classifications, either as a separate family, or as a subfamily within Nymphalidae, continue to plague specialist views on the group, but the family basically represents the lineage to Nymphalidae and are very unique if included within the latter family. The family is in the superfamily Papilionoidea (series Papilioniformes), in the section Cossina, subsection Bombycina, of the division Ditrysia. Adults medium size (25 to 55 mm wingspan), with very long porrect labial palpi; forelegs somewhat reduced and not used for walking. Wings triangular, typically with a falcate forewing apex, often somewhat truncated; hindwing relatively rounded with emarginate margin. Maculation mostly shades of brown with orange or red suffusion and some light and dark spotting. Adults are diurnal. Larvae are leaf feeders. Host plants are in Ulmaceae.

John B. Heppner
Florida State Collection of Arthropods
Gainesville, Florida USA

References

Heppner, J. B. 2003. Libytheidae. In *Lepidopterorum Catalogus*, (n.s.). Fasc. 99. Association for Tropical Lepidoptera, Gainesville. 12pp.

Hering, E. M. 1921. Die geographische Verbreitung der Libytheiden. *Archiv der Naturgeschichte (Berlin)*, 87A 4: 248–296, 2 pl.

Pagenstecher, A. 1901. Libytheidae. In *Das Tierreich* 14: 1–18.

Seitz, A. (ed.). 1908-19. Unterfamilie: Libytheini. In *Die Gross-Schmetterlinge der Erde*, 1: 251–252, pl. 71 (1908); 5: 622-623, pl. 120D (1916); 9: 767-771, pl. 139 (1914); 13: 293-294, pl. 61 (1919). A. Kernen, Stuttgart.

Shields, A. O. 1979. Zoogeography of the Libytheidae (snouts or breaks [sic]). *Tokurana* 9: 1–58.

SNOUT MOTHS (LEPIDOPTERA: PYRALIDAE).

Snout moths, family Pyralidae, comprise the third largest family of Lepidoptera, with about 16,500 described species, but a probable fauna of at least 25,000 species worldwide. There are 19 subfamilies in the classification, divided into two groups: group Crambinina, with 14 subfamilies (Crambinae, Schoenobiinae, Cybalomiinae, Linostinae, Scopariinae, Musotiminae, Midilinae, Nymphulinae, Odontiinae, Noordinae, Wurthiinae, Evergestinae, Glaphyriinae, and Pyraustinae), and group Pyralinina, with 5 subfamilies (Pyralinae, Chrysauginae, Galleriinae, Epipaschiinae, and Phycitinae). The group names are sometimes elevated to separate families, as was already done over 100 years ago, but they can equally be maintained within the single family Pyralidae as has long been the practice. By far the largest subfamily is the Pyraustinae, with about 7,500 species worldwide. The family is in the superfamily Pyraloidea in the section Tineina, subsection Tineina, of the division Ditrysia. Adults small to large (6–95 mm wingspan), with head mostly rather smooth-scaled; haustellum scaled; labial palpi usually porrect and prominent; maxillary palpi 4-segmented (rarely 2 to 3-segmented). Wing shape usually has rather elongated and pointed forewings. Maculation extremely varied, often subdued and with various markings, but also many very colorful species; hindwings usually unicolorous and pale. Adults mostly nocturnal, but some are crepuscular and a few are diurnal. Larvae are mostly leafrollers or leaf webbers, but many are borers, root feeders, detritus feeders (including stored products pests), and a few are leafminers, plus rare myrmecophilous species, and even some aquatic groups making cases (Nymphulinae). Host plants are in a large number of plant families. A large number of economic species are in this family, including pests on virtually all crops and forest trees. Most Crambinae and Schoenobiinae larvae feed on grasses, including crops such as rice and other grains. Pyralinae and Phycitinae

Fig. 932 Examples of snout moths (Pyralidae): top left, (subfamily Crambinae), *Diatraea crambidoides* (Groe) from Florida, USA; top right, (subfamily Galleriinae), *Galleria mellonella* (Linnaeus) from Italy; bottom left, (subfamily Phycitinae), *Cactoblastis cactorum* (Berg) from Florida, USA; bottom right, (subfamily Pyraustinae), *Diathrausta harlequinalis* Dyar from Florida, USA.

include a number of stored products pests, now mostly worldwide in distribution due to their association with dry stored foods. Galleriinae have beehive pests.

John B. Heppner
Florida State Collection of Arthropods
Gainesville, Florida USA

References

Balinsky, B. I. 1994. *A study of African Phycitinae in the Transvaal Museum*. Johannesburg. 208 pp.

Bleszynski, S. 1965. Crambinae. In H. G. Amsel, H. Reisser and F. Gregor (eds.), *Microlepidoptera Palaearctica*. Vol. 1. Vienna: G. Fromme. 533 pp, 133 pl. [in German]

Goater, B. 1986. *British pyralid moths: a guide to their identification*. Harley Books, Colchester. 175 pp (8 pl.).

Heinrich, C. 1956. American moths of the subfamily Phycitinae. *Bulletin of the United States National Museum* 207: 1–581.

Munroe, E. G. 1972–76. Fasc.13.1, Pyraloidea (in part), In R. B. Dominick, et al. (eds.), *The moths of America north of Mexico*, 13.1: 1-304, 9 pl. (1972–73); 13.2: 1-150, 8 pl. (1976). E. W. Classey and R. B. D. Publishing, London

Neunzig, H. H. 1986–97 Pyraloidea, Pyralidae (part). Phycitinae (part). In R. B. Dominick, et al. (eds.), *The moths of America north of Mexico*, 15.2:1-113, 6 pl. (1986); 15.3 :1-165, 5 pl. (1990); 15.4:1-157, 4 pl. (1997). Wedge Entomological Research Foundation: Washington.

Roesler, R.U. 1973–93. Phycitinae [part]. In *Microlepidoptera Palaearctica*, 4(1):1-752, 4(2): 1-137, 170 pl. (1973); 8: 305, 82 pl. (1993).

Slamka, F. 1997. *Die Zünslerartigen (Pyraloidea) Mitteleuropas: Bestimmen – Verbreitung – Flugstandort – Lebensweise der Raupen*. 2nd ed. Bratislava. 112 pp (13 pl.).

Zimmerman, E. C. 1958. Lepidoptera: Pyraloidea. In *Insects of Hawaii*, 8: 1-456. University of Hawaii Press, Honolulu.

SNOW, FRANCIS HUNTINGTON. Francis Snow was born at Fitchburg, Massachusetts, on June 29, 1840. He was an industrious young man from the start, and in the course of his life developed an astounding number of interests and competencies, among them ornithology, botany, geology, paleontology, mineralology, and entomology. Professionally, he was at various times a minister, teacher, and chancellor of the University of Kansas. He grew up in the exciting days just before the American Civil War, and the antislavery views of his father strongly influenced young Francis Snow. He excelled at Williams College, from which he graduated in 1862, and was most interested in natural history. He taught high school briefly, then entered Andover Theological Seminary. An avowed pacifist, he assisted chaplains and aided the wounded during the war. In 1866 Snow moved to Kansas to accept a professorship in mathematics and natural science. Snow quickly gained a reputation for being a knowledgeable, sincere, and patient teacher, greatly admired by the students at the University of Kansas. Snow was quite impressed with the grasshoppers in Kansas; at times they were so numerous that the train could not move because the crushed bodies of the grasshoppers prevented traction of the wheels on the tracks. A swing of an insect net would collect up to 190 grasshoppers. However, Snow was interested in everything, and he and his students collected insects, birds, plants, minerals and other elements of natural history earnestly, an activity that would eventually result in a fine museum. Williams College bestowed on Snow an honorary Ph.D. in 1881. In 1882 he was named State entomologist, and in 1885 the legislature named the Snow Hall of Natural History in his honor. In 1890 he accepted the position as chancellor. Though Snow was a collector, he was not a taxonomist. He devoted much of his energies to economic entomology, a fact that encouraged the practical farmers of Kansas to support him as chancellor. As chancellor he strengthened the faculty of both science and humanities. In 1889 Snow's son died an accidental death, and in the wake of this unfortunate event Snow suffered nervous disorders.

He never completely recovered, and he died on September 20, 1907, while visiting Wisconsin.

Reference

Mallis, A. 1971. *American entomologists.* Rutgers University Press, New Brunswick, New Jersey. 549 pp.

SNOW SCORPIONFLIES. Members of the family Boreidae (order Mecoptera). See also, SCORPIONFLIES.

SOCIAL INSECT PHEROMONES. Social insects are unique in that they have overlapping generations in which adult workers (normally sterile) assist their mother in rearing sisters and brothers. In addition, there are reproductive and non-reproductive castes constituting a division of labor. Our discussion of social insect pheromones will be restricted to the social Hymenoptera (Formicidae, ants; Apidae, bees; Vespidae, wasps) and Isoptera (Termitidae, termites), although there are examples of social thrips and cockroaches.

Regardless of the size of the social insect colony, which can range from tens of individuals to millions, social interactions are required for effective food retrieval, brood and queen care, regulation of caste (sexuals/workers), recognition and exclusion of non-nestmates, and other tasks. There are several sensory mechanisms available to social insects for communication, e.g., tactile and vibratory, however, chemical communication has evolved to a high level of complexity in these insects.

There are advantages to the use of chemical signals as a means of information transfer. The chemicals are relatively small, volatile structures that are energetically inexpensive to biosynthesize and easy to release into the surrounding airspace. In addition, because the signal is detected through space by the sense of smell, it is functional under various environmental conditions. Some disadvantages are that it is difficult to direct the signal at a single recipient unless detection is through direct contact. Also, once released, the signal cannot be changed rapidly or in some cases the signal dissipates before the information has been transferred to all intended recipients. Specific terminology has been developed to define chemical interactions between organisms. We will restrict ourselves to intraspecific chemical communication.

Pheromones are chemicals released by an individual that have an effect on members of the same species. The word pheromone is derived from the Greek 'pheran' meaning to transfer and 'horman' meaning to excite. There are two pheromone subcategories, a) releaser pheromones – produce an immediate response in the recipient individual, e.g., a male moth orienting toward the sex pheromone released by the female moth; and b) primer pheromones – perception triggers the initiation of a complex physiological response that is not immediately observable. Both pheromone types are extremely important in maintaining colony social structure.

Unlike hormones, pheromones must be secreted outside the insect's body via exocrine glands. It is a testament to the importance of pheromones to the success of social insects that they have evolved a very diverse repertoire of exocrine glands – over 70, with 45, 21, 14, and 11 in ants, bees, wasps, and termites, respectively. Some of these glands function to produce wax, other types of building materials, or defensive compounds; however, many exocrine glands have become specialized for the production of pheromones that play major roles in communication and the maintenance of colony social structure. Pheromone-producing glands can be found anywhere on the social insect – from the pretarsal glands of bees to the frontal and labial glands of termites. In addition, the ovipositor of solitary insects evolved into the sting in the aculeate Hymenoptera, bees, wasps, and ants. Most important for this discussion is that the accessory glands associated with the ovipositor evolved into the Dufour's and poison glands. Both of these glands have become very important in pheromone communication in social insects.

The pheromone products have diverse uses among social insects. Recruitment pheromones help guide workers to food materials and in a different context can lead the entire colony to new nesting sites. When a worker ant, wasp, bee, or termite detects an intruder it will release an alarm pheromone that acts to excite and attract nearby workers, which then respond aggressively toward the intruder. Other compounds act as sex pheromones, queen recognition pheromones, territory markers, and brood pheromones, among others. Additional layers of complexity are added because pheromone structure and function varies within the morphological/functional caste system in social insects. Not only is there variability between queens and workers different, but also between the different worker castes as well. Sometimes, a single chemical plays different pheromonal roles, depending on the caste releasing it and the context of the situation. With such a plethora of glandular sources and elicited behaviors there appears to be no pattern in pheromone chemical structure or the glandular source relative to behavior elicited. To illustrate the unpredictability of glandular source versus function, trail pheromones in ants have been reported from the Dufour's gland, venom gland, hindgut, pygidial gland, Pavan's gland, and the postpygidial gland. Thus, the glandular source of a pheromone that elicits a particular biological behavior must be determined through bioassay.

Pheromone glands

Pheromone-producing glands are specialized for the distribution of their secretory products. Besides the biosynthesis of the active pheromone components, the glands must be able to regulate the release of the secretion. The mechanisms of release are not well understood; however, glands that have reservoirs may have specialized musculature at the reservoir opening that controls the release of the gland contents. Interestingly, the act of opening the mandibles initiates the release of alarm pheromones from man-

dibular glands of some ant species. Where there is no reservoir, glandular products may be released directly to the outside as they are produced. Release may be continuous or synthesis and release could be triggered by exogenous stimuli.

The chemistry of social insect pheromones is very diverse. Releaser pheromones elicit an immediate response in the recipient and are generally detected by the antennae in the air and therefore must be volatile. Primer pheromones elicit physiological change and may or may not be volatile, because they can be distributed to colony members by grooming and/or trophallaxis (passage of regurgitated material from one individual to another). We will briefly go through the major pheromone behavioral categories and their associated chemistry.

Alarm pheromones

Alarm pheromones represent an important evolutionary development for eusocial insects because they enable colony worker resources to be focused at specific colony needs, e.g., nest or food defense. The alarm signals only act to alert other members of the colony. If the alerted workers subsequently find an intruder they will likely respond aggressively by attacking. If they find food they may be stimulated

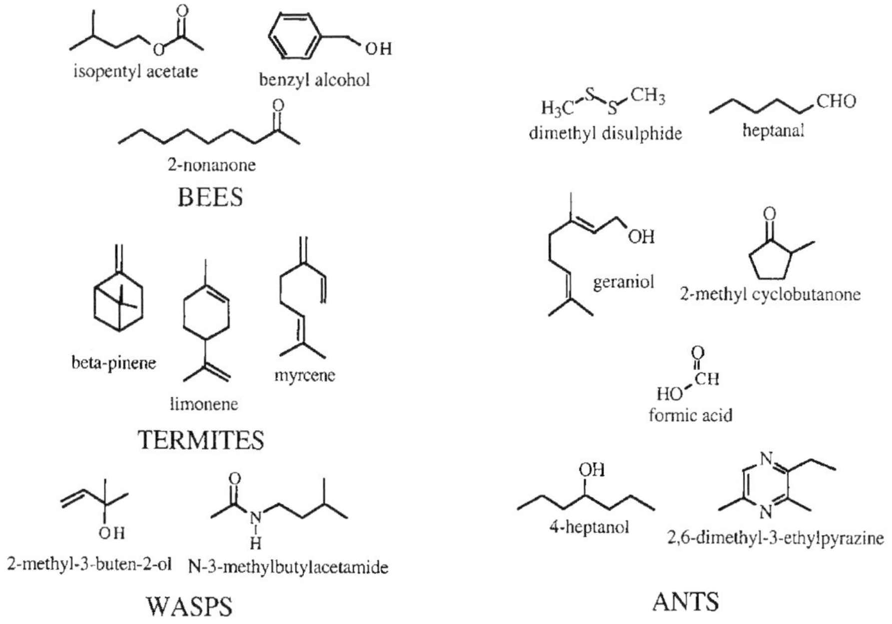

Fig. 933 Representative alarm pheromones identified from bees, termites, wasps and ants, illustrating the diversity of structural types.

to ingest the food. So, while the alarm reaction is usually associated with defensive behavior, it is subsequent stimuli that often dictate what happens next. The release of alarm pheromones may be initiated by the physical disturbance of an individual worker or by the chemical recognition of an intruder. Once the alarm response is initiated, its release may be 'telegraphed' to other nearby workers. This helps explain anecdotal accounts that hundreds of nestmate bees or wasps respond to the accidental disturbance of a single bee or wasp. Similarly with fire ants, people report that worker ants will sneak up their leg, then when in position all of them will sting in unison. Actually the ants simply go undetected for a period of time, until a movement disturbs one ant that then releases an alarm pheromone, quickly activating nestmates in the area. While the mandibular gland is the most common source of alarm pheromones they have also been reported from the pygidial gland, Dufour's gland, Koschevnikov's gland (associated with the sting apparatus of bees), and other glands.

There are many reported behaviors associated with alarm pheromones. In ants for example, attraction to the alarm pheromone source, increased speed of movement, frenzied running, aggression, raised head and more have all been reported as alarm behaviors. More complicating is that the context in which the alarm pheromone is released can affect worker response, e.g., colony disturbance, an encountered intruder, age of the colony, etc. Consequently, it is important for the researcher to clearly define the alarm behavior being investigated.

The requirement for a quick response to the alarm pheromone, as well as for its rapid dissipation after the perceived need has passed suggests that alarm pheromones should be small volatile compounds. While in some instances only one component is reported, e.g., 2-methyl-3-buten-2-ol for the hornet (*Vespa crabro*), it is expected that many more potentially active compounds will be isolated and identified. For example, initially only isopentyl acetate was isolated from Koschevnikov's gland in honey bees (alarm pheromone source); however, subsequent analysis has identified at least 22 additional compounds. Many ant alarm pheromones have been isolated and identified. Considering the necessity that alarm pheromones must be highly volatile, the variety of structural types pays homage to the biosynthetic versatility of the social insects and ants in particular.

Future work in this area should focus on the precise definition of the alarm behavior, development of a behavior-specific bioassay, and then the bioassay-driven isolation of the active compounds.

Recruitment pheromones

Recruitment pheromones are compounds involved in bringing colony workers to a particular location

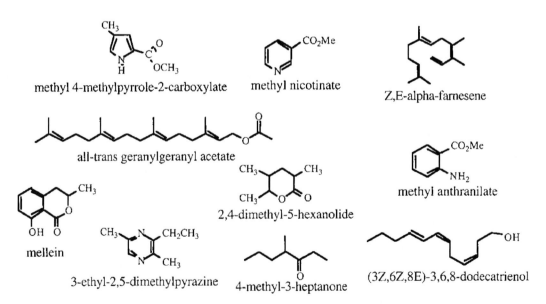

Fig. 934   Representative recruitment pheromone structures illustrating the diversity of structural types.

where they are needed. An ant trail between the colony's nest to a dead insect or other food material is a familiar sight to most people. The trail is the end result of a complex process; but since it is easily observed this pheromone system is sometimes referred to as the trail pheromone. Nestmates may be recruited to gather food resources, or to defend their territory against invasion, or recruitment pheromones may be used to guide the migration of the colony to a better nest site.

The terrestrial ants, of necessity, have evolved a wide variety of recruitment mechanisms and glandular sources of recruitment pheromones. In ants the Dufour's gland, poison gland, the pygidial glands, and sternal glands, hindgut, and rectal gland have been reported as sources of recruitment pheromones. Pheromones from these sources can be readily envisioned being deposited on a solid substrate. Similarly the sternal gland is used in several genera of termites as a source of trail pheromones. Thus far this gland is the only known source of termite trail pheromones. The flying social insects also have a need to recruit a worker force to food supplies and/or to guide colony members to new nesting sites and hence recruitment pheromones can also play a role. Social wasps do not commonly recruit to food sources but do recruit to new nest sites. During this process workers drag their gasters on a substrate to deposit sternal gland secretions, which then guides the migrating colony to a new nest site. Stingless meliponine bees deposit mandibular gland products at intervals on the ground while the Nasanov gland in honeybees is a source of marking pheromones. Most investigations of recruitment pheromones have focused on ants, primarily because they form readily observed trails on the surface and bioassays have been relatively easy to develop.

The red imported fire ant, *Solenopsis invicta*, is a good model to illustrate the behavioral components of the recruitment process. The process begins when a foraging scout worker discovers a food source too large for it to carry back to the colony. During the foraging process the scout keeps track of its position relative to a light source (sun, street light, etc.) and is able to navigate pretty much straight back to the colony. As it makes its way back, it periodically deposits minute amounts of Dufour's gland products to the substrate. The Dufour's gland is attached to the base of the sting apparatus and the products exit through the sting. Thus, in this case, the sting is periodically extended to touch the substrate. When the scout returns to the nest Dufour's gland products are used to attract other workers toward the scout and activate them to follow the very weak initial trail. The attracted and activated workers detect and respond to the trail by following it to the food source. The movement back and forth along the trail is called orientation. The newly recruited workers ingest some food, stimulating them to reinforce the trail with additional recruitment pheromone as they return to the nest. More workers are recruited until the food source is covered with fire ants and additional workers cannot get to the food and do not reinforce the trail. As the food source diminishes fewer workers reinforce the trail and because of its volatility, the concentration of the deposited recruitment pheromone weakens until the food is gone and the trail evaporates completely. The recruitment process can be broken down into several sub-categories. a) Initial trail laying by the scout ant; b) attraction of additional workers to the scout ant; c) activation (induction) of the workers to follow the trail; d) trail orientation. Separate behavioral bioassays were developed for each of the categories and used to guide the isolation of the responsible pheromone components. Z,E-α-farnesene appears to be solely responsible for orientation, but does not attract worker fire ants. A mixture of Z,E-α-farnesene and a bicyclic homofarnesene account for 100% of the Dufour's gland activity in an olfactometer (attraction) bioassay. Remarkably, when this mixture is presented to foraging ants as a trail they do not orient along the artificial trail! This illustrates that the workers first need to be activated or put into a more alert state. In some ant species the scout worker will physically agitate perspective recruits, whereas the fire ant can do the same job with Dufour's gland products. This activation requires virtually a reconstitution of the complete chemical profile of the Dufour's gland.

The fire ant provides one example; however, in some ant species the scout attracts and escorts a single nestmate back to the food source and in others the scout physically activates nestmates who are then able to independently follow the trail. Social insects occupy virtually all ecological niches and have evolved a myriad of mechanisms to accomplish tasks such as recruitment to bring nestmates to a point in space where extra legs and mandibles are needed. The chemistry of many recruitment pheromones has been determined.

undecane

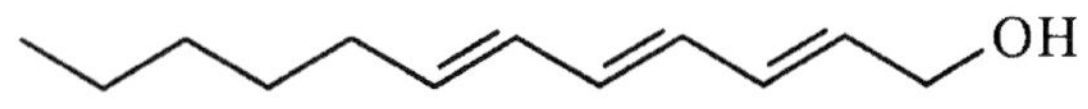

3,6,8-dodecatrien-1-ol

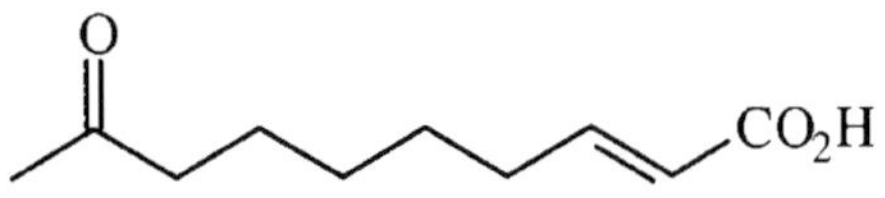

9-keto-2(E)-decenoic acid

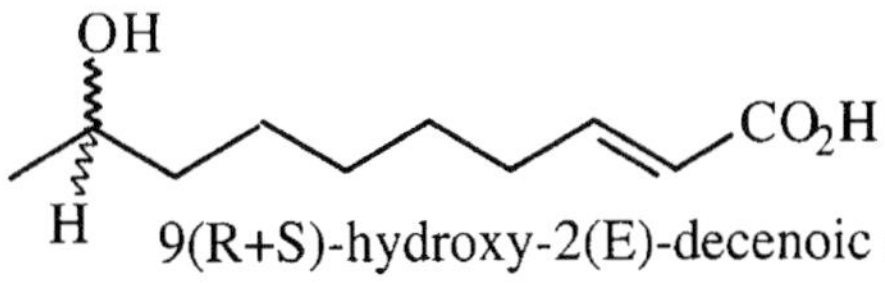

9(R+S)-hydroxy-2(E)-decenoic acid

Fig. 935 The structures of known social insect sex pheromones.

Sex pheromones

Sex pheromones are substances that are emitted by males or females that function to attract the opposite sex for the purpose of mating. Most commonly the female attracts the male(s) to her. There are many examples among solitary insects, especially lepidopteran species, of both the associated behaviors and the chemistry of the pheromones involved. However, few sex pheromones have been reported in social insects, probably because of the difficulty in developing appropriate bioassays. Most social insects, including the terrestrial variety (ants and termites), mate during 'mating flights'. The mating flights are usually

triggered by specific environmental conditions, so laboratory assays are difficult to establish and observations of the natural phenomenon have been rare. For example, the red imported fire ant, *Solenopsis invicta*, may send off hundreds of thousands of male and female winged sexuals, with males taking flight approximately 30 minutes before the females. The males and females find each other, presumably through sex pheromones, and mate about 100 to 200 meters in the air, then the females come down to attempt to start a colony. No one has yet seen a mating pair of fire ants and no one has been able to get them to mate successfully in the laboratory!

The honey bee 'queen substance' produced by the mandibular glands is used in the context of an established colony to control the production of female sexuals. These same glandular products are used by virgin queens to attract male drones for mating. While there are many compounds produced by the mandibular glands, only two compounds appear to play a sex pheromone role: (E)-9-oxodec-2-enoic acid and (E)-9-hydroxydec-2-enoic acid. The other two examples of sex pheromones in social insects are also pheromones that have different behavioral effects depending on the context of their use. In termites 3,6,8-dodecatrien-1-ol is a trail pheromone, but has also been implicated as a sexual attractant for two termite species: *Pseudacanthotermes spiniger* and *Reticulitermes santonensis*. The one and only sex pheromone from an ant species was isolated from *Formica lugubris*. The pheromone undecane was isolated from the mandibular glands and accounted for 100% of the sex pheromone bioassay activity. This compound is also used by workers as an alarm pheromone.

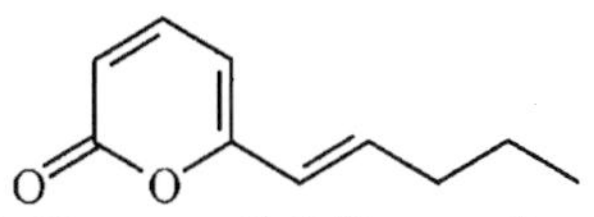

(E)-6(1-pentenyl)-2-H-pyran-2-one

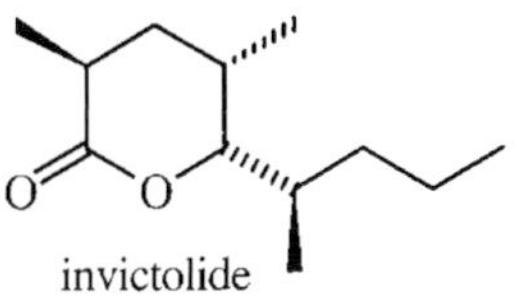

invictolide

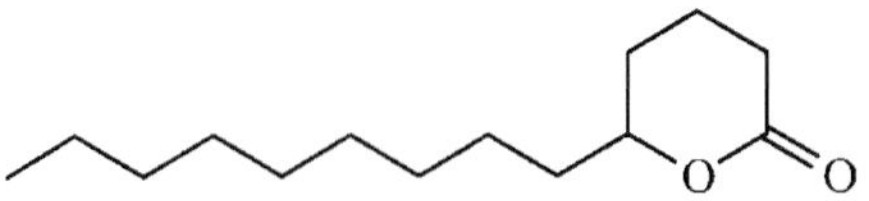

d-n-hexadecalactone

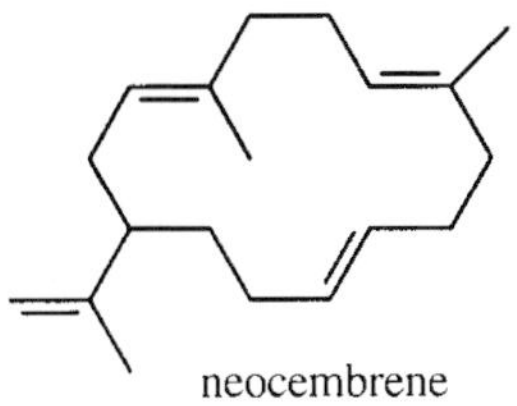

neocembrene

Fig. 936 The structures of queen recognition pheromones.

Queen pheromones

Queen recognition pheromones or more simply 'queen pheromones' are exocrine gland products released by the queen that usually attract workers to her, eliciting care and protection. Most queens of social insects have the ability to attract workers, so the behavior is well documented. Again, various glands serve as sources for the queen pheromone, from the mandibular glands in the head of honey bees to the poison sac in the abdomen of fire ants. The queen benefits from the attention of the workers, and the workers may also use the pheromone signal to gain information about the queen. For example, the release of poison sac contents is directly linked to queen egg laying, thus workers may be able to assess the fecundity of their queen based on the amount of pheromone released into the colony.

The chemistry of these pheromone systems has been elucidated in only a few systems. In the fire ant *Solenopsis invicta* the source of attractant pheromones was found to be the queen's poison sac. Two lactones have been identified as the active pheromones. The pharoah's ant, *Monomorium pharaonis*, produces a macrocyclic compound, cembrene A, in its Dufour's gland that elicits worker attraction. Ejection of the sting and possible deposition of Dufour's gland products on the eggs has also been reported for this species. The honey bee queen pheromone is produced by the mandibular gland and has turned out to be a very complex mixture of compounds that by themselves are only slightly active but when taken together act as synergists, reproducing the activity from the mandibular gland itself. Unusual for pheromones, most of the compounds are found in tens of μgs. In spite of the large quantity of pheromones

produced, the synergistic effects of the components complicated the isolation of the active compounds. No queen pheromones have been isolated and identified yet from termite species. Only one compound, δ-n-hexadecalactone, has been isolated as a queen pheromone from a wasp species. It was isolated from head extracts and affect worker behavior.

Releaser and primer pheromones

As mentioned earlier there are two broad categories of pheromones, releaser and primer pheromones. The above discussion centers around releaser pheromones whose detection yields an immediate observable behavioral change. Thus, bioassay development is easier and results can be obtained quickly. This area of social insect pheromone communication has foraged ahead rapidly, while primer pheromones have lagged far behind. This is because they trigger the initiation of complex physiological responses that are not immediately observable. It may take days or more to obtain results. Bioassays have been developed that demonstrate the existence of primer pheromones, but the chemistry has not been forthcoming, except for the honey bee.

The honey bee, *Apis mellifera*, is one of the most extensively studied social insects, in part because of its economic importance and the ready availability of colonies. Interestingly, the queen's mandibular pheromone blend of five components acts as both releaser and primer pheromones. The blend attracts workers to the queen as in 'queen recognition' and therefore has releaser pheromone effects. It acts as a primer pheromone by inhibiting queen rearing by workers. Additionally, it retards the build up of

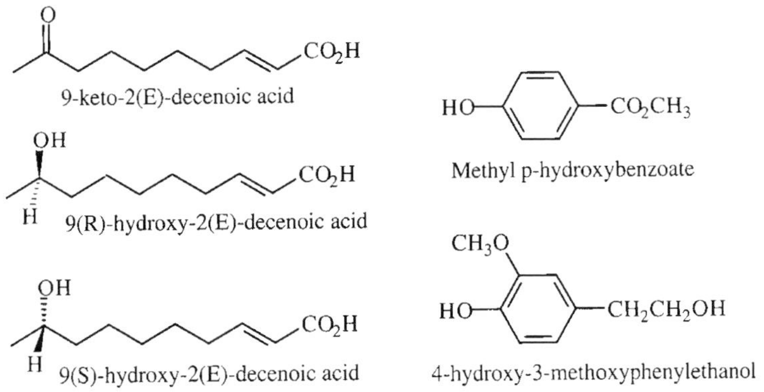

Fig. 937 The five components that constitute the queen produced honey bee primer pheromone.

juvenile hormone titer in worker bees, which effectively acts to slow the developmental progression of workers from brood tending to foraging. These effects clearly represent changes in the physiology of honey bee workers.

Most, if not all, social insect primer pheromones are produced by the queen and function in some way to prevent reproductive competition. Another example comes from the fire ant. The fire ant queen has been demonstrated to produce several primer pheromones that function to: (a) inhibit ovary development in female sexuals; (b) suppress egg production by mature queens in polygyne colonies; (c) inhibit the production of female sexuals; (d) regulate nestmate recognition sensitivity, such that newly mated queens are executed by workers. Unfortunately, the structures of these interesting primer pheromones are still unknown.

We have discussed the main behavioral categories of pheromones in social insects; however, there are other categories, such as territorial marking and brood pheromones, a discussion of which can be found in the references. In addition, there are nestmate recognition cues that are on the surface of individuals that allow members of one colony to distinguish members of another colony of the same species. While the effect is conspecific, the behavioral response is different for nestmate and non-nestmate; acceptance or attack, respectively. Nestmate recognition cues may be acquired from the environment, inherited, or derived from a combination of the two sources. Thus, nestmate recognition cues do not quite fit into the realm of pheromones.

We have demonstrated the complexity of social insect pheromones – their sources, associated behaviors, and chemistry – and hope that readers will be stimulated to delve deeper into the literature to further uncover the beauty and intricacies of this fascinating subject.

See also, PHEROMONES, SOCIALITY IN INSECTS.

Robert K. Vander Meer and Catherine A. Preston
United States Department of Agriculture,
Agricultural Research Service
Gainesville, Florida, USA

References

Traniello, J. F. A., and S. K. Robson 1995. Trail and territorial communication in social insects. pp. 241–286 in R. Cardé and W. J. Bell (eds.), *The chemical ecology of insects* 2. Chapman & Hall, New York, New York.

Vander Meer, R. K., M. Breed, and M. L. Winston. 1998. *Pheromone communication in social insects: ants, wasps, bees, and termites*. Westview Press, Boulder, Colorado.

Winston, M. L. 1992. Semiochemicals and insect sociality. pp. 315-333 in B. D. Roitberg and M. B. Isman (eds.), *Insect chemical ecology*. Chapman & Hall, New York, New York. 359 pp.

Vander Meer, R. K., and L. Morel. 1988. Brood pheromones in ants. pp. 491-513 in J. C. Trager (ed.), *Advances in myrmecology*. E.J. Brill, New York, New York. 551 pp.

Hölldobler, B., and E. O. Wilson. 1990. *The ants*. Harvard University Press, Cambridge, Massachusetts.

SOCIALITY OF INSECTS.

The social insects represent a major evolutionary success story. Their cooperative nest-building, prey capture, and recruitment for food have captured the attention and fueled the imaginations of humans. This cooperation also has contributed to the numerical abundance and ecological dominance of social insects in many habitats. For example, in tropical forests and savannahs, ants and termites outweigh the impressive and very conspicuous mammalian fauna. In all but the coldest environments, ants are the predominant predators of other insects, and they are also major dispersers of seeds of numerous plant species. Social bees pollinate a large proportion of the flora throughout the world, both because they are abundant and because many of them have efficient recruitment tactics such as the remarkably complex dance language of honey bees. Termites are the primary decomposers of wood and other materials containing cellulose in the tropics. Termites and ants together move more soil than earthworms in all tropical habitats, and even in temperate regions, have a major role in mixing nutrients in the soil and creating topsoil. From these examples, it is evident that the social insects figure prominently in nutrient cycles and energy flow in the biosphere.

The majority of all insects are solitary. For solitary species, the most social point in their life cycle is often the interaction between mates, after which females go about their business of reproduction without cooperating with other members of their species. Parasitic wasps lay eggs in their hosts, grasshoppers oviposit in the soil, and butterflies seek out larval host plants on which to lay their eggs. The common thread among these diverse solitary groups of insects is that their offspring develop into adults without interaction with their mothers.

Eusocial insects

At the other extreme are the eusocial insects. Eusociality is a term that refers to a set of three very specific criteria: overlap of generations (i.e., mothers live sufficiently long to interact with their adult offspring); cooperative brood care (i.e., females provide some care to offspring that are not their own, and individual larvae are provisioned by more than one female); reproductive division of labor (i.e., some individuals of a group leave behind more offspring than others).

Most of the insects we think of as highly social – honey bees, yellowjacket wasps and hornets, ants, and termites – meet the criteria of eusociality. In fact, most eusocial species belong to the order Hymenoptera, with all the species of ants (about 9,000 species), about ten percent of bees (more than 2,000 species), and some wasps (about 1,000 species) being eusocial. The other main group of eusocial insects contains the termites, order Isoptera, with all 2,200 species being eusocial. There are only a handful of other eusocial organisms by the above set of criteria: a bark beetle in Australia, various gall-forming aphids and thrips, sponge-inhabiting snapping shrimp, and naked mole rats in Africa.

Between the extremes of solitary and eusocial insects are numerous species in many different taxa that exhibit some, but not all, of the criteria for eusociality. These insect species often are described by the nonspecific term, presocial. Because eusocialty evolved from presocial societies, considerable attention has been directed at presocial insects in an attempt to elucidate those factors that favor the evolution of more complex social systems. Presociality can be further divided into parasociality and subsociality. The definitions and examples of these two major categories of subsocial species will be discussed here, and their central role in the evolution of sociality will be considered later in this article.

Parasocial insects

Parasocial species are those in which the colonies are comprised of individuals of the same generation, and are usually groups of sisters that nest together. Thus, in parasociality, the first criterion of eusociality (overlap in generations) is not met. For example, females of the carpenter bee (*Xylocopa virginica*), that is common over much of eastern North America, construct galleries in wood. The young adult bees emerge after their mother has died and often they remain in the gallery in a communal society over winter. Once spring arrives, they disperse and initiate nests individually. Although the offspring share the nest for a period of time, there is little or no cooperation between individuals. In some parasocial colonies (e.g., some neotropical orchid bees, *Euglossa* spp.), although all adult females have developed ovaries and apparently are equally capable of laying eggs (i.e., no reproductive division of labor), there are more adult bees than cells with larvae being provisioned at any point in time. This is suggestive of cooperative brood care. There also may be some slight specialization in nest guarding, but overall cooperation in such parasocial bee societies usually is rudimentary. Tropical parasocial spiders (e.g., *Anelosimus eximius*, *Agelena consociata*) are able to construct huge webs in clearings and thereby encounter more and larger prey than their solitary relatives can. Within their societies, these spiders exhibit a variety of relatively complex behaviors, but they lack reproductive division of labor.

Subsocial insects

The other broad category of presociality is subsociality. Subsocial insects are those in which females show extended care of their offspring that enhances offspring survival. The specific selective forces that favor subsociality vary widely, but generally are linked to either very harsh conditions or very favorable but ephemeral conditions. The European rove beetle, *Bledius spectabilis*, provides an example of a species that lives under extremely unfavorable conditions. Adult females construct vertical burrows in tidal mudflats. After feeding on algae on the mud at low tide, the beetles retreat to their burrows and plug the entrances. Without the adult beetle present when the burrows are submerged at high tide, the younger immatures are unable to maintain the tunnels and they die from lack of oxygen. At the opposite extreme, burying beetles (*Nicrophorus* spp.) reproduce on small rodents that represent abundant, high quality, ephemeral resources. In order to reproduce successfully they must find, then defend, a dead rodent from potential competitors such as fly larvae, fungi and bacteria, and even other *Nicrophorus* beetles. Consequently, a reproductive pair of beetles usually fights other *Nicrophorus* to gain control of the rodent, after which they shave it and treat it with mandibular secretions. The female, and sometimes the male, remain with the larvae, protecting and even directly feeding them, until they pupate in the soil. At that time, the continued presence of the adults cannot enhance offspring survival, and the beetle(s) disperse to seek new cadavers upon which they can reproduce.

The evolutionary 'breakthrough' that enabled some presocial species to evolve more complex eusocial societies is reproductive division of labor. In most cases, that translates into morphologically distinct queen and worker castes, with queens doing most or all of the reproduction. In the *Origin of Species*, Darwin wrote that the sterile workers and soldiers of social insects presented "one special difficulty, which at first appeared to me insuperable, and actually fatal to my whole theory." Why should some individuals "give up" their own reproduction in order to assist the queen of their colony? The evolution of sterile individuals seems contrary to the principles of natural selection, in which those individuals with the highest fitness as measured by number of offspring contribute most to succeeding generations. Darwin believed he solved this puzzle by treating the colony as the unit of selection. If specialization into distinct worker and soldier castes enhances colony survival and functioning, then the colony will be able to invest greater energy into the rearing of reproductives (new gynes [potential queens] and males) and as a result will experience enhanced reproductive success. As a consequence, colonies could evolve various specializations of queens, workers, and soldiers that contribute to colony-level fitness. This view of social insect colonies probably helped to usher in the early-1900s view of social insect societies as 'superorganisms,' in which colonies were seen to have attributes or abilities that were distinct from the sum of the behaviors of the individuals that comprise those colonies.

In 1964, W.D. Hamilton III proposed a different view of sociality based on kinship. He pointed out that in the Hymenoptera (bees, wasps and ants), males arise from unfertilized eggs. A consequence of this is that all sperm from a single male are identical. Female nestmates (sisters) that have the same father share three quarters of their genes in common, more than they share with their own offspring (one half). Hamilton pointed out that in hymenopteran colonies in which the queen has mated only once, it is genetically beneficial for females to remain in their nest to cooperate in rearing sisters rather than to risk leaving to start a new nest and raise daughters with which she would be less closely related and would share fewer genes. Hamilton extended his theory of 'inclusive fitness' to all kin, both direct offspring and others that share some genes through shared ancestry (e.g., sisters, cousins, etc.).

Hamilton's theory of inclusive fitness ushered in a new wave of research on the role of kinship (i.e., the proportion of genes shared through common ancestry) on social systems, a field termed 'sociobiology.' One of the first steps was to determine whether organisms could recognize kin. Now, four decades later, there is ample evidence that within all the taxa of social insects, kin recognition abilities have been demonstrated. Recognition is mediated by chemical 'signatures,' both intrinsic, or genetically determined, and environmentally influenced, on the surface of the insect that are learned by nestmates. But simply having the ability to recognize kin does not mean that social insects also choose to direct aid towards individuals that are more closely related to them. Again, after numerous contributions from many researchers, the evidence has mounted that proves that, in most cases, social insects can act in ways to enhance their inclusive fitness. A classic demonstration of the operation of kin recognition comes from 'worker policing' in honey bees. Worker bees have the ability to lay unfertilized eggs that develop into males, but when the queen is present in a normally functioning colony, worker-laid eggs usually are cannibalized by other workers. One can predict mathematically that it is in the workers' best interests to attempt to reproduce, because they are more closely related to their own sons (relatedness $= 1/2$) than to their brothers (i.e., queen-derived unfertilized eggs; $r = 1/4$). However, because queens naturally mate with about ten males, when a worker in the colony encounters a worker-laid egg, roughly nine times out of ten it will share only $1/8$ of its genes with it as a result of common ancestry. When faced with eggs laid by workers (the majority of which are related by $1/8$) and queen-laid male eggs ($r = 1/4$), it is in the best interests of worker bees to destroy the eggs laid by other workers. In fact, when tested empirically, it was found that after 24 hours all worker-laid eggs had disappeared from colonies, whereas approximately 50% of queen-laid male eggs remained. As another example, paper wasps (*Polistes* spp.) have been shown to preferentially join with nestmates (e.g., sisters) rather than non-nestmates when more than one foundress wasp cooperate to initiate nests in spring.

On the surface it seemed that W.D. Hamilton had discovered the major factor leading to insect sociality, namely the genetic asymmetry resulting from the haplodiploid sex determination system (females diploid, males haploid) of Hymenoptera. The simple fact that females could gain more inclusive fitness by staying in their nest to cooperate in rearing sisters rather than leaving to initiate new nests to rear daughters seemed to answer the question. However, several pieces of information suggest that the situation is not so simple. As early as 1976, Robert Trivers and Hope Hare pointed out that the inclusive fitness gained by rearing their sisters ($r = 3/4$) is lost in rearing brothers ($r = 1/4$) rather than sons ($r = 1/2$) if colonies rear equal numbers of queens and males. Secondly, there are many haplodiploid taxa that have failed to evolve eusociality. Even most taxa in which offspring are produced clonally by their mothers, so that sisters and brothers share all of their genes in common, are not eusocial. Thirdly, the termites are an extremely diverse and successful group of insects, but their males derive from fertilized eggs and there is no genetic asymmetry between daughters and sisters, or sons and brothers. Although Hamilton's ideas have dramatically changed the perspective of biologists with respect to the concept of fitness and the role of kin in animal sociality, other considerations have eliminated the 3/4 relatedness hypothesis as the sole factor favoring the evolution of sociality.

In place of a unifying theory, most scientists now recognize that each taxonomic group of insects has its own set of conditions that have influenced the evolution of sociality. As an example, paper wasp (*Polistes*) larvae have the ability to eat proteinaceous prey, while adults with the narrow petiole connecting their head and thorax to the abdomen are restricted to a liquid diet. In times of inclement weather and the ensuing food shortages that regularly occur early in the nesting cycle, the adults solicit and receive food from the larvae to prevent their own starvation. While this behavior allows the queens and their brood to survive into warmer, less threatening times of year, it results in the offspring produced early in the nesting cycle receiving suboptimal nutrition. It has been suggested that these nutritionally compromised adults gain higher inclusive fitness by remaining on their natal nest rather than initiating new nests on their own. In the case of sweat bees (family Halictidae) that nest in aggregations in soil, brood parasitism often is reduced by sharing of the nest with another female. When either of the females is away foraging, the other is usually in the nest defending it from intruders. This mutualism enhances the overall reproductive success of both females and has played a significant role in the repeated evolution of sociality in sweat bees. In termites, symbiotic gut microorganisms are passed between colony members. When nymphs molt, they must become re-inoculated with these symbionts from their nestmates. The exchange of these symbionts has undoubtedly been a factor in the evolution of termites from their closest relatives, the cockroaches. However, Hamilton's work made it clear that no matter what the specific set of circumstances surrounding a particular taxonomic group, loss of reproductive capacity will not evolve unless the interactions involve related individuals (i.e., kin).

Bernard Crespi has examined in detail the life history traits of the less typical eusocial organisms, as a way to elucidate the suite of traits that are most likely to enable eusociality to evolve. He has suggested that three conditions are necessary for the evolution of eusociality: coincidence of food and shelter, strong selection pressure for defense of nests, and the ability to defend nests. In this context, the occurrence of the sting in Hymenoptera probably has played a large role in the evolution of sociality in bees, wasps, and ants. Crespi suggests that additional eusocial insect taxa may be found among long-lived bark beetles that have substantial cohabitation of adults and their offspring, book lice (order Psocoptera) and webspinners (order Embiidina) that occupy long-lived webs, and other gall-inhabiting taxa.

Because evolution cannot be viewed directly, we must infer the evolutionary pathways to eusociality followed by various groups of social insects. In the past, this was achieved by logical reconstruction of the stepwise changes that could have led to eusociality using the biologies of living species. An excellent example involves the evolution of sociality in wasps. Howard Evans used the comparative approach to relate social systems of various wasp species to each other. He developed a widely accepted framework, from solitary species to primitive mother-offspring (subsocial) associations, to more complex subsociality, and finally eusociality with reproductive division of labor as seen in temperate hornets and yellowjackets (subfamily Vespinae). With the advent of cladistic methods to construct phylogenies (evolutionary histories), and statistical analyses that allow the certainty of those phylogenies to be tested, we have made important advances in the understanding of the evolution of sociality. Returning to wasp evolution, by 'mapping' behavioral states onto the phylogeny of the subfamilies of vespid wasps, it has been discovered that the most basal wasp subfamily in the phylogeny that exhibits sociality contains only parasocial associations of sisters, not subsocial mother-daugher associations. The sweat bees provide another example of the approach of combining phylogenetic analyses with behavioral data. It has long been recognized that parasocial associations of females have been involved in the repeated evolution (and loss) of eusociality within the family. By mapping states of sociality onto phylogenies, it is possible to identify those species for which further study will best illuminate the factors affecting the evolution of sociality, by finding closely related species that differ in their degree of social complexity. In a similar way, this approach of combining behavioral traits with phylogenies has enhanced our understanding of the evolution of sociality in termites.

The superorganism concept was mentioned earlier. Early in the 20th century, the concept that insect colonies were analogous to whole organisms complete with 'germ plasm' (queens and males) and 'soma' (workers and soldiers) gained enormous popularity. Despite the importance of the superorganism concept as a paradigm for studying the biology of social insects for a significant period of time, it had largely disappeared by the 1960s. As ethological approaches to the study of animal behavior gained in force, so too did the application of the scientific method to field studies. Modern scientists entered an era of hypothesis formulation and testing. While the superorganism concept provided interesting analogies to other organisms, it failed to provide concrete hypotheses that could be tested empirically. Interestingly, the concept has seen a revival since the development of powerful personal computers. Researchers can quantify simple rules of behavior by observing individual insects, then combine them through computer simulations to recreate behavioral patterns of self-organized behavior with remarkable accuracy. This approach has been used to model, among other behaviors, pattern formation on honey bee combs, nest construction by termites, and army ant raids.

The study of insect sociality has led to many profound discoveries about insects. Some areas of research that should provide new insights in the future include the study of molecular and genetic changes that occur as complex societies evolve, the specific conditions that favor sociality, the relative importance of parasocial and subsocial routes to eusociality, more complex integration of systems of self-organized behavior, and enhanced understanding of the role of social insects in ecosystem functioning.

Gard W. Otis
University of Guelph
Guelph, Ontario, Canada

References
Brockmann, H. J. 1984. The evolution of social behaviour in insects. pp. 340-361 in J. R. Krebs and N. B. Davies (eds.), *Behavioural ecology: an evolutionary approach (2nd ed.)*. Sinauer Associates Inc., Sunderland, Massachusetts.

Crespi, B. J. 1994. Three conditions for the evolution of eusociality: are they sufficient? *Insectes Sociaux* 41: 395–400.

Hamilton, W. D. 1964. The genetical evolution of social behaviour, I and II. *Journal of Theoretical Biology* 7: 1–52.

Hölldobler, B., and E. O. Wilson 1990. *The ants*. Belknap Press of Harvard University Press, Cambridge, Massachusetts.

Michener, C. D. 1974. *The social behavior of the bees*. Belknap Press of Harvard University Press, Cambridge, Massachusetts.

Ross, K. G., and R. W. Matthews. 1991. *The social biology of wasps*. Comstock Publishing Associates, Ithaca, New York.

Trivers, R. L., and H. Hare 1976. Haplodiploidy and the evolution of the social insects. *Science* 191: 249–263.

Wilson, E. O. 1971. *The insect societies*. Belknap Press of Harvard University Press, Cambridge, Massachusetts.

SOCIAL PARASITES. Insects that obtain their sustenance at the expense of a colony of insects from another species, without destroying the colony.

SOCIOBIOLOGY. (in insects) The study of social behavior, and population characteristics among social insects.

SOD WEBWORM. See also, TURFGRASS INSECTS AND THEIR MANAGEMENT.

SOFT-BODIED PLANT BEETLES. Members of the family Artematopididae (order Coleoptera). See also, BEETLES.

SOFT SCALES. Some members of the family Coccidae, superfamily Coccoidae (order Hemiptera). See also, BUGS.

SOFT-WINGED FLOWER BEETLES. Members of the family Melyridae (order Coleoptera). See also, BEETLES.

SOIL MITES (ACARI: ORIBATIDA AND OTHERS). Soil is a complex of living and non-living components which are present in different combinations. Small arthropods, including several groups of mites, contribute to the humus fraction and permit complexes of soil organisms to exist. Even though the role of mites in soil mixing may be small in comparison with that of larger invertebrates such as earthworms, insects, crustaceans and millipedes, mites exercise an important function in mineral turnover, vegetation succession and as decomposers of organic matter. Densities of 50,000 to 250,000 or more mites per square meter may be found in the upper layers of soil. Dozens of species may be found in a small area when soil is rich in organic material such as decaying vegetation, dung or animal remains. In combination with microflora, which the mites may disperse, soil mites help in decomposing organic matter which they cannot digest. Much of the work on soil mite taxonomy, biology and ecology, including their role in nutrient cycling and fertility-humification processes, has been carried out in Europe.

Mites in the Suborder Oribatida (= Cryptostigmata), also called moss mites or beetle mites, are distinctive soil mites, ranging in size from 200 to 1400 μm. They are common in soils that are relatively undisturbed and have a high organic content. Some species of Oribatida may be found on the

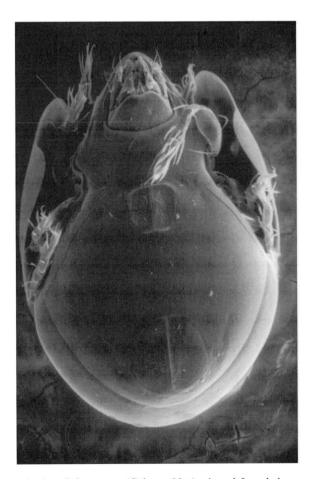

Fig. 938 A soil-dwelling oribatid mite, *Galumna* sp. (Galumnoidea), viewed from below.

trunks of trees or on low-growing vegetation. Oribatida species can even be found associated with roots of plants very deep in the soil. There are an estimated 150 families in the Oribatida in 10,000 species in 1,100 genera. The name "beetle mites" is a result of the fact that many have hard, dark brown exoskeletons and somewhat resemble miniature beetles. The term moss mites is due to the fact that they are often found in mosses.

A summary of their biology is as follows: Sperm transfer is indirect, and mating takes place when males deposit spermatophores that are taken up by the female (indirect sperm transfer). Spermatophores are stalked and often deposited in groups. Many are destroyed by being eaten. Copulation has never been observed in oribatids and parthenogenesis (reproduction without mating) is known in some families.

Females typically deposit one to twelve eggs at a time. After mating, females deposit oval eggs singly or in small clumps in old exuviae, the axils of moss bracts, under detritus, and in the soil pores. Often the female retains a number of eggs until all are mature and she deposits them at once. There is a prelarva, larva, three nymphal stages and the adult. The length of time to reach adulthood depends on temperature, but is relatively slower than that of other mite suborders; smaller oribatid species have a shorter development time and life span than larger species. Rate of development also is influenced by whether the diet is adequate and whether the mites are crowded. In general, small species may have several generations a year (compared to one generation a week for the two-spotted spider mite, a plant-feeding pest in the suborder Prostigmata). Larger species may

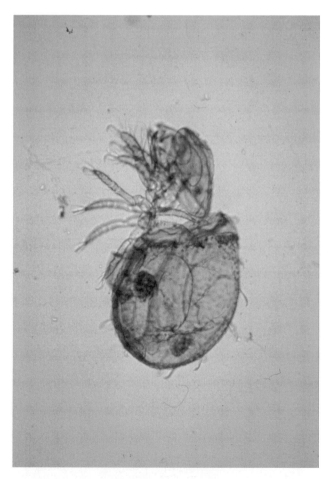

Fig. 939 A soil-swelling oribatid mite, *Haplophorella cucullata* (Ewing), side view (photo by Roy Norton, SUNY, Syracuse).

have one generation per year and some oribatid speces may require several years to complete their life cycle in cool climates.

Oribatids chose optimal temperatures (or at least avoid unfavorable temperatures). Humidity requirements are species specific; some are able to survive for days at low relative humidities while others need higher relative humidities to survive. Salinity and soil pH are important; soils rich in organic matter have a high pH.

Oribatid mites are microphytophages, feeding on pollen, small pieces of algae, lichens, highly decomposed humus, and fungal hyphae. They can also be macrophytophagous, feeding on dead plant tissues, parenchyma of needles, or directly on dead wood. Some are nonspecialized, feeding on leaf tissue and on fungal mycelia. The larger species are generally macrophytophagous or nonspecialized and thus play the main role in litter decomposition. Some oribatids prefer animal food and can be maintained on cricket powder in the laboratory. Some oribatids feed on the numerous nematodes in the soil. Laboratory rearing of some species can be accomplished with dried mushrooms, chopped lichens, decomposing leaves, or an artificial diet of dextrose and casein or brewer's yeast.

Only a few soil mite species have a direct role in altering soil morphology. Most exercise indirect effects through stimulation of microbial activity and distribution of spores. Pesticides can have a negative effect on nutrient cycling by permitting energy to remain bound up in undecomposed plant debris.

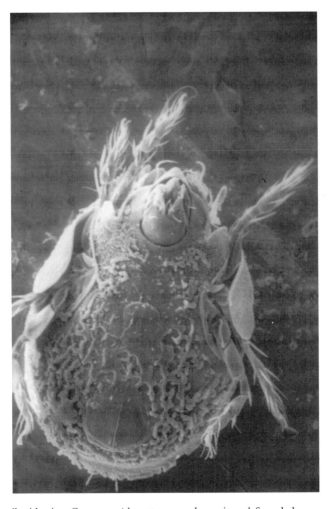

Fig. 940 A soil-dwelling oribatid mite, *Ceratozetoidea pteromorphae*, viewed from below.

Pesticides can positively affect nutrient cycling by killing predators of saprophagous arthropods, thus releasing detritus-feeding populations.

Species in the Oribatida are thought to be especially important decomposers of plant debris. On mulched plots, they make up 60% of the total arthropod species, compared to 8.5% on mowed or fallow plots. Distribution of microarthropods is irregular under annual crops, in part due to the effects of cultivation. Fewer mites are found in grasslands than in forest communities. Microarthropod distribution is influenced by water, pore space, oxygen saturation deficits, temperature variations, ground cover, fungi, flooding, cropping, cultivation, organic matter, disturbance, soil compaction, soil type and texture, predation, and feeding habits.

Oribatida serve as food for centipedes, symphyla, diplurans, spiders, pseudoscorpions, opilionids, ground beetles, and ants. The Oribatida also are fed on by beetles in the families Pselaphidae and Ptiliidae. Some birds, salamanders, frogs, skinks and other small lizards also may feed on oribatid mites. Oribatid mites may function as vectors of anoplocephaline tapeworms; moss mites are intermediate hosts for the transmission of cysticercoids of these tapeworms and 12 families, 25 genera and 32 species are confirmed vectors.

Mites other than species in the Oribatida also are found in the soil; Actinedida, Gamasida and Acaridida are present, although are not as numerous. Much remains to be learned about the role of the Oribatida, and other mite groups, in soil biology and ecology.

Marjorie A. Hoy
University of Florida
Gainesville, Florida, USA

References

Balogh, J. 1972. *The Oribatid genera of the world.* Akademiai Kiado, Budapest, Hungary.

Evans, G. O., J. G. Sheals, and D. MacFarlane. 1961. *The terrestrial Acari of the British Isles. An introduction to their morphology, biology and classification.* British Museum of Natural History Press, London, United Kingdom.

Walter, D., and H. Proctor. 1999. *Mites. Ecology, evolution and behavior.* CABI Publishing, Wallingford, United Kingdom.

SOLDIER BEETLES. Members of the family Cantharidae (order Coleoptera). See also, BEETLES.

SOLDIER FLIES. Members of the family Stratiomyidae (order Diptera). See also, FLIES.

SOLENOPHAGES. Arthropods that feed at blood vessels, and specifically from small veins. (contrast with telmohages)

SOLITARY BEHAVIOR. Showing none of the following traits of sociality: individuals of the same species cooperate in caring for the young; there is division of reproductive behavior, with more or less sterile individuals working on behalf of the fecund individuals; overlap of at least two generations in life stages contributing to colony labor, so that offspring can assist parents during some period of their life.

See also, PRESOCIAL, SUBSOCIAL, COMMUNAL, QUASISOCIAL, SEMISOCIAL, EUSOCIAL BEHAVIOR.

SOLITARY MIDGES. Members of the family Thaumaleidae (order Diptera). See also, FLIES.

SOLITARY PARASITOID. A parasitoid that has nutritional requirements limiting the ability of the host insect to support but a single parasitoid.

SOMA. (pl., somata) The cell body of a nerve cell. Also known as the perikaryon. See also, NERVOUS SYSTEM.

SOLDIER. A member of the nonreproductive caste in social insects that contributes to the welfare of the colony through defense of the colony. Soldiers often bear an enlarged head and mandibles.

SOMABRACHYIDAE. A family of moths (order Lepidoptera) also known as Mediterranean flannel moths. See also, MEDITERRANEAN FLANNEL MOTHS, BUTTERFLIES AND MOTHS.

SOMATIC CELLS. All the eukaryotic body cells except the germ line cells and the gametes they produce.

SOMATIIDAE. A family of flies (order Diptera). See also, FLIES.

SOOTY MOLD. A dark-colored fungus growing on the honeydew secreted by hemipterous insects, usually aphids or scales. See also, TRANSMISSION OF PLANT DISEASES BY INSECTS.

SOROSPORELLA. *Sorosporella* was first described from a coleopteran, *Cleonus punctiventris*, in 1886 by Krassilstschik, and then two years later from a lepidopteran by Sorokin. Krassilstschik named the fungus *Tarichium uvella*, while Sorokin called it *Sorosporella agrotidis*. In 1889, Giard suggested that these two genera were the same and combined the two names into *Sorosporella uvella*. Speare, in the early 1900s, wrote the only detailed descriptions of *Sorosporella* to date; included were his observations on phagocytosis of the hyphal bodies by hemocytes in cutworms, which constitute one of the first reports of cellular defense response in insects invaded by a fungal pathogen.

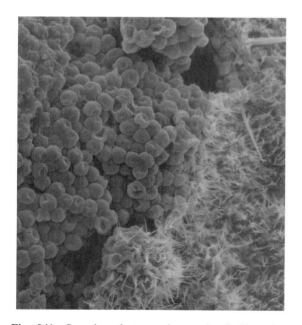

Fig. 941 Scanning electron micrograph of chlamydospores produced in mole crickets infected with *Sorosporella*. The cuticle of the mole cricket has been peeled away exposing internal chlamydospores.

Sorosporella exists endogenously as globose, brick-red, thick-wall chlamydospores which are 6 to 10 μm in diameter. These structures form within the insect hemocoel from the yeast-like vegetative hyphal bodies and cohere into solid masses that, at maturity, may become dry and powdery, and then separate. *Sorosporella* germinates on water or various types of media. On Sabouraud maltose agar, the germinating chlamydospores produce a white mycelial mat with a synnematous-type growth sometimes evident. The mycelia generate conidiophores and phialides, which produce hyaline, ellipsoidal conidia approximately 1 to 1.5 μm in diameter. This conidial stage is known as the alternate state of *Sorosporella* and was identified as *Syngliocladium*. This is not a teleomorph-anamorph association since both genera represent asexual states. The laboratory-produced conidia were infectious to several different insects, but *Sorosporella-Syngliocladium* has never been seriously targeted as a biocontrol agent, especially in comparison to other more promising Hyphomycetes.

Drion G. Bouicas
University of Florida
Gainesville, Florida, USA

References

Pendland, J. C., and D. G. Boucias. 1987. The hyphomycete *Sorosporella-Syngliocladium* from mole cricket, *Scapteriscus vicinus*. *Mycopathologia* 99: 25–30.

Petch, T. 1942. Notes on entomogenous fungi. *Transactions of the British Mycological Society* 25: 250–265.

Speare, A. T. 1920. Further studies of *Sorosporella uvella*, a fungous parasite of noctuid larvae. *Journal of Agricultural Research* 18: 399–451.

SOUND PRODUCTION IN THE CICADOIDEA. One of the characteristic sounds of summer is the calling of cicadas. The sounds cicadas produce are used in the attraction of mates, courtship, and as an alarm to startle potential predators. These sounds can be as intense as 109 dB at 50 cm and choruses can be heard for more than a mile. Each species has a unique calling song and there are significant differences between the frequency and temporal patterns of species that share the same environment. Cicadas use three main mechanisms to produce sound.

The primary sound production mechanism in cicadas is the timbal organ. The timbal is a chitinous

Fig. 942 Right timbal of *Beameria venosa* (Uhler). The operculum can be seen extending under the ventral surface of the abdomen. The wings have been removed for clarity.

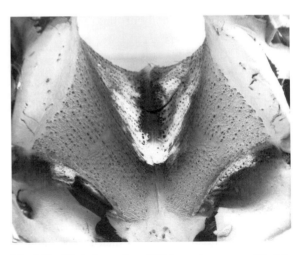

Fig. 943 Timbal muscles of *Tibicen winnemanna* (Davis). The ventral attachment is to the Chitinous V in the lower center of the micrograph.

membrane located on the dorso-lateral surface of the first abdominal segment of most male cicadas. The timbal contains a flattened area, termed the timbal plate, and several ribs that help to strengthen the timbal. The timbal produces sound pulses when it is deformed by a timbal muscle. The timbal muscles are located within the abdomen attached to a support structure (the Chitinous V) on the ventral side and is attached dorsally to the timbal plate by an apodeme. To produce sound, the timbal muscle contracts pulling on the timbal. Eventually the muscle force is sufficient to buckle the timbal plate and continued muscle force leads to buckling of the timbal ribs. Sound pulses are produced when the timbal plate and ribs buckle. In addition, some timbals will produce sound when they return to their resting position as the timbal muscle relaxes. Rebound of the buckled timbal to a resting state is facilitated by the rubber-like protein resilin that is incorporated into the timbal. The rate of timbal muscle contraction determines the sound pulse repetition rate for each species. The frequency of the sound that is produced by a particular species is primarily determined by the resonance of the timbal organ. There is a good inverse correlation between body size and the frequency of the song produced in cicadas where larger species have lower frequency calls. However, there are several body structures that can alter the frequency of the sound produced.

Multiple accessory structures have been shown to affect the sound produced by the timbal organs. The timbal covers protect the timbals from damage in members of the family Cicadidae but they also can act as resonating structures to increase the amplitude of the cicada call. The opercula are plate like structures on the ventral surface of the abdomen that can also resonate or modify the frequency of the timbal. The abdominal air sacs transform the abdomen into an amplifier to increase the intensity of the call. The tympana are thin membranes located dorsolaterally at the base of the abdomen that can dampen the sound produced and also act as the main site of sound radiation, so cicadas radiate sound through their eardrums. The timbal tensor muscle acts to put tension on the timbal that increases the amplitude of the emitted sound. The dorsal and ventral abdominal muscles act to elevate and depress the abdomen. The changes in abdominal position can bring the abdomen into resonance with the timbals, and also produce frequency and amplitude modulations in more complex calls.

Stridulation is a second method of sound production in cicadas. A stridulatory apparatus has been described in various cicada genera from around the world. A typical stridulatory apparatus is composed of a file and scraper similar to the system found in crickets, grasshoppers, and some heteropterans. The cicada stridulatory apparatus is normally found associated with the tegmina and the mesonotum. However, one stridulatory apparatus has been described as a modification of the genitalia although it has not been seen being used in the field. Stridulation is

Fig. 944 The timbal is attached to the timbal muscle by an apodeme. A timbal cover can be seen lying over the surface of the timbal.

used primarily as a secondary acoustic system in cicadas. The stridulating cicadas that have been studied use stridulation in addition to the timbal call. In addition, stridulation has permitted some females to respond to male timbal calls.

A final method of sound production in cicadas is crepitation. The North American genera *Platypedia* and *Neoplatypedia* have replaced timbal songs with crepitation but other genera use crepitation in addition to timbal song production. A crepitating cicada will snap the wings together or bang the wings on vegetation to produce a sound pulse. Individual sound pulses are produced when the wings are snapped against the body. There does not appear to be a cost in terms of the sound intensity to the cicadas that use a crepitation signal. It has been found that a crepitating species does not differ from the expected sound intensity based on the relationship between body size and timbal song intensity. A benefit to crepitation appears to be the two-way communication between males and females that it permits. The two-way communication between males and females in crepitating and stridulating species can ensure proper species identification when it comes to selecting a mate.

Allen Sanborn
Barry University
Miami Shores, Florida, USA

References

Bennet-Clark, H. C. 1997. Tymbal mechanics and the control of song frequency in the cicada *Cyclochila australasiae*. *Journal of Experimental Biology* 200: 1681–1694.

Bennet-Clark, H. C., and D. Young. 1992. A model of the mechanism of sound production in cicadas. *Journal of Experimental Biology* 202: 3347–3357.

Moore, T. E. 1993. Acoustic signals and speciation in cicadas (Insecta: Homoptera: Cicadidae). pp. 269–284 in D. R. Lees and D. Edwards (eds.), *Evolutionary patterns and processes*. Linnean Society Symposium Series 14. Academic Press, London, United Kingdom.

Pringle, J. W. S. 1954. The physiological analysis of cicada song. *Journal of Experimental Biology* 31: 525–560.

Sanborn, A. F., and P. K. Phillips. 1999. Analysis of acoustic signals produced by the cicada *Platypedia putnami* variety *lutea* (Homoptera: Tibicinidae). *Annals of the Entomological Society of America* 92: 451–455.

SOUTHERN BACTERIAL WILT. See also, TRANSMISSION OF PLANT DISEASES BY INSECTS.

SOUTHERN BLOTTING. A technique developed by E. M. Southern for transferring DNA fragments isolated electrophoretically in an agarose gel to a nitrocellulose filter paper sheet by capillary action. The DNA fragment of interest is then probed with a radioactive nucleic acid probe that is complementary to the fragment of interest. The position on the filter is determined by autoradiography. The related techniques for RNA and proteins have been dubbed 'Northern' and 'Western' blots, respectively.

SOUTHERN CHINCH BUG. See also, TURF-GRASS INSECTS AND THEIR MANAGEMENT.

SOUTHERN GREEN STINK BUG, *Nezara viridula* **(L.) (HEMIPTERA: HETEROPTERA: PENTATOMIDAE).** The southern green stink bug, *Nezara viridula* (L.), has a worldwide distribution, occurring throughout the tropical and subtropical regions of Europe, Asia, Africa, and the Americas. To this distribution should be added California in the United States and some areas in South America, wherein the bug is expanding its distribution, in particular in Paraguay, south Argentina, and central-west toward the north-east of Brazil. The expansion in

South America is the result of increased acreage for soybean production.

Nezara viridula is highly polyphagous, and more than 150 species of plants of over 40 families have been recorded as hosts. Despite its polyphagy, this stink bug shows preference for leguminous and brassicaceous plants.

This bug completes several generations per year, depending on the favorability of the environment. For instance, in the United States, adults leave overwintering shelters during spring and begin feeding and ovipositing in clover, small grains, early spring vegetables, corn, tobacco, and weed hosts, where they complete the first generation. A second generation is completed on leguminous weeds, vegetables and row crops, cruciferous, and okra, which are typical midsummer hosts. Third generation adults move into soybean where they complete the fourth and fifth generations. Another example of host plant sequence for *N. viridula* was studied in Paraná state, southern Brazil. The bug concentrates on soybean during the summer, which is very abundant at this time; two to three generations are completed on this crop. During fall, adults move to wild hosts, which include the star bristle, *Acanthospermum hispidum*, and the castor bean, *Ricinus communis*, where they feed but do not reproduce. They also move to wild legumes, such as the beggar weed, *Desmodium tortuosum* and *Crotalaria* spp., where they feed and reproduce, with its resulting offspring completing a fourth generation. During late fall and early winter, *N. viridula* completes a fifth generation on radish, *Raphanus raphanistrum*, and mustards, *Brassica* spp. During winter, it may feed but will not reproduce on wheat, *Triticum*

aestivum. During spring, a sixth generation is completed on Siberian motherwort, *Leonurus sibiricus*. Many other cases of *N. viridula* host plant sequences have been studied elsewhere, indicating the importance of knowing these sequences to manage this pest species better. Both pods and seeds of legumes generally provide a suitable food source for nymphs, whereas survival is poor on leaves and stems.

Adults are green and 12 mm long. Eggs are firmly glued together in rows within compact polygonal clusters. At the time of deposition, eggs are cream colored, turning pinkish or reddish as development progresses. Nymphs in the early instars are strongly gregarious, and this behavior tends to disappear as the nymphs grow older.

Significant reductions in seed yield, quality, and germination can result from feeding by this pest; damaged seeds can have an increased incidence of pathogenic organisms. The feeding punctures (flanges) on seeds cause minute darkish spots, and generally chalky-appearing air spaces are produced when the cell contents are withdrawn.

The effects of *N. viridula* on corn are more severe on younger plants, and yield reductions are attributed to total ear loss rather than reduction in kernel weight. On tomato, its feeding is associated mostly with reduced quality of the fruit. *N. viridula* also can attack nut crops. It is one of the most damaging pests of macadamia, causing nut abortion in small nuts and quality damage in larger nuts in the trees or on the ground. In recent years, *N. viridula* has been reported as a pest of flue-cured tobacco. Both adults and nymphs extract plant fluids from younger tender growth, causing the leaves to wilt and flop over due to loss of turgor pressure.

Several tactics have been studied and implemented to manage *N. viridula* attacking many crops. The use of chemical control still is the major tool used by growers, but many other tactics within the context of cultural practices and biological control have been investigated and used. For chemical control, chlorinated hydrocarbon, organophosphates and organochlorophosphate insecticides, such as methyl parathion, monocrotophos, trichlorphon, dimethoate, and endosulfan have been recommended. Some of these insecticides are recommended to control *N. viridula* on soybean using reduced dosages mixed with sodium chloride, which increase their efficacy. Also, the use of a natural

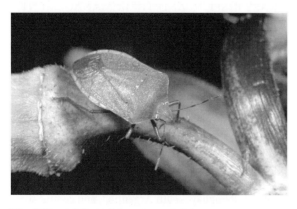

Fig. 945 Adult southern green stink bug.

insecticide, such as an extract of neem seed, has been reported to decrease the scars on pecan nuts caused by *N. viridula* feeding.

Trap cropping has been utilized to suppress *N. viridula* and other pentatomid species on soybean. It takes advantage of the fact that stink bugs colonize soybeans during the pod-set and pod-filling stages of plant development. Thus, early maturing or early planted soybeans are highly attractive to stink bugs and can be used to exploit this behavioral response by attracting large populations of the bugs into small areas containing these trap crops (usually less than 10% of the total area). The concentration of ovipositing females and the subsequent nymphal population can be controlled effectively with insecticides before the next generation of adults disperses to the surrounding fields. The key to success of this cultural practice is to control nymphs before they become adults and disperse to adjacent fields. If insecticide controls are not timed properly, then this management strategy is useless and, in fact, actually intensifies the stink bug problem in the main crop. Inoculative releases of the egg parasitoid *Trissolcus basalis* (Wollaston) (Hymenoptera: Scelionidae) in early-maturing soybean, used as trap crops, caused a reduction and delay in the population peak of this stink bug.

Some progress has been made in the development of host plant resistance to *N. viridula* feeding on some crops. But, so far, cultivars with variable levels of resistance have been little used, mainly because the potential production of these cultivars is usually lower than other commercial cultivars.

Nezara viridula is attacked by numerous natural enemies including parasitoids, predators, and entomopathogens. As many as 57 parasitoid species are recorded for *N. viridula*. The most important of these are the egg parasitoid *Trissolcus basalis* (Wollaston) and, in the New World, the adult parasitoid *Trichopoda pennipes* (F.) in North America, and *Trichopoda nitens* (Blanchard) in South America.

Several arthropod predators have been reported to feed on *N. viridula*. There is a complex of nonspecific predators, and the fire ant, *Solenopsis invicta* Buren, is mentioned as an effective predator of eggs and early instars in the United States.

Antônio R. Panizzi
Embrapa
Londrina, Brazil

References

Jones, W. A. 1988. World review of the parasitoids of the southern green stink bug, *Nezara viridula* (L.) (Heteroptera: Pentatomidae). *Annals of the Entomological Society of America* 81: 262–273.

McPherson, J. E., and R. M. McPherson. 2000. *Stink bugs of economic importance in America North of Mexico*. CRC Press, Boca Raton, Florida. 253 pp.

Panizzi, A. R. 1997. Wild hosts of pentatomids: ecological significance and role in their pest status on crops. *Annual Review of Entomology* 42: 99–122.

Panizzi, A. R., and F. Slansky Jr., 1985. Review of phytophagous pentatomids associated with soybean in the Americas. *Florida Entomologist* 68: 184–214.

Schaefer, C. W., and A. R. Panizzi (eds.). 2000. *Heteroptera of economic importance*. CRC Press, Boca Raton, Florida. 828 pp.

Todd, J. W. 1989. Ecology and behavior of *Nezara viridula*. *Annual Review of Entomology* 34: 273–292.

SPARGANOTHIS FRUITWORM. See also, SMALL FRUIT PESTS AND THEIR MANAGEMENT.

SPARKLING ARCHAIC SUN MOTHS (LEPIDOPTERA: ERIOCRANIIDAE). Sparkling archaic sun moths, family Eriocraniidae, are a Holarctic family of 25 species, of which about half the species are in Europe and half in North America. The family, plus the related Acanthopteroctetidae, are the only members of the superfamily Eriocranioidea, which form the infraorder Dacnonypha of the suborder Glossata. Adults small (6 to 13 mm wingspan), with roughened head scaling; haustellum reduced; mandibles present but vestigial (haustellum absent); labial palpi 3-segmented; maxillary palpi 5-segmented and folded. Maculation rather colorful with various iridescent spots of gold on darker forewings. Adults are diurnally active. Hosts are mostly in Betulaceae and Fagaceae, and the egg is inserted by the female into tender host leaves in early spring, whereupon the young larvae feed before the leaves turn hard. Larvae are leafminers with serpentine mines that become blotch mines as the larvae grow.

John B. Heppner
Florida State Collection of Arthropods
Gainesville, Florida USA

References

Busck, A., and A. Böving. 1914. On *Mnenonica auricyanea* Walsingham. *Proceedings of the Entomological Society of Washington* 16: 151–163.

Davis, D. R. 1978. A revision of the North American moths of the superfamily Eriocranioidea with the proposal of a new family, Acanthopteroctetidae (Lepidoptera). *Smithsonian Contributions to Zoology* 251: 1–131.

Heath, J. 1976. Eriocraniidae. In J. Heath, *The moths and butterflies of Great Britain and Ireland*, 1: 156–165. Harley Books, London.

Sutter, R. 2000. Beiträge zur Insektenfauna Ostdeutschlands: Lepidoptera-Eriocraniidae (Insecta). *Faunistische Abhandlungen* 22: 49–67.

SPATULATE. Broad at the tip and narrow at the base. The tip is flattened or sometimes recessed and spoon-like.

SPEAR-WINGED FLIES. Members of the family Lonchopteridae (order Diptera). See also, FLIES.

SPECIALIST. Insect occupying a narrow niche, or consuming a narrow range of food.

SPECIATION PROCESSES AMONG INSECTS.

The development of molecular genetics has led to a renewed attempt to define the concept of the species. Originally, 'species' was defined by Ernst Mayr as an interbreeding group of organisms (= population) that is reproductively isolated from other such groups (typically referred to as the biological species concept or BSC). However, the increasing frequency of reported viable hybrids produced by crosses of 'parental species' in natural systems (often recognized as distinct species until molecular analyses revealed that they are intermediate crosses of various degrees between the parentals and/or hybrids), suggests that Mayr's definition was inadequate. As a result, other species concepts were developed such as the phylogenetic species concept (PSC), which identifies species (and other taxonomic categories) based on the presence of shared derived characters. In other words, taxa that share recently derived morphological or genetic characters are assumed to be more closely related than lineages that do not share these derived traits. While this definition circumvented the repro-ductive isolation requirement imposed by the BSC, it has its own unique set of problems. For instance, the phylogenetic (i.e., evolutionary) relationships inferred from a cladistic analysis can be affected by the choice of characters used in the assessment (this is also true for studies that rely on the use of shared derived genetic sequences).

Regardless of the functional definition of the term, the evolution of new species is traditionally believed to occur through allopatry or, less frequently, through the more controversial pathway of sympatry. Allopatric speciation, which involves a geographic separation and isolation of segments of a formerly contiguous population, is firmly established as the primary mechanism by which new species evolve. Sympatric speciation also predicts the development of new species under certain restrictive conditions, but this mechanism does not require speciating subpopulations to be geographically isolated from one another. It is precisely this lack of geographic isolation that has rendered sympatric speciation so contentious because, without spatial isolation among speciating subpopulations, it is difficult to imagine a scenario in which interbreeding (i.e., gene flow) among diverging groups can be greatly reduced or eliminated (a requirement for speciation).

Insects in general, and phytophagous insects in particular, provide some of the strongest evidence for the sympatric pathway of speciation. Several recent studies have found evidence suggesting that host range expansion by phytophagous insects may result in the development of new species without geographical isolation of host-specific subpopulations. The tenability of this mechanism is directly tied to the problem of fitness trade-offs and potential gene flow among diverging subpopulations. According to the generalized model, females mistakenly oviposit on an atypical (i.e., outside its normal host range) plant species and, if progeny survive on this novel host (to which they are, by definition, non-adapted), temporal or seasonal differences in host plant phenology may effectively isolate this subpopulation from conspecifics on the natal host species. However, for this scenario to be feasible, insects from the atypical host species must show a high degree of fidelity for this novel host even though they are likely to suffer decreased survival, fitness and performance compared to insects from the original host.

How can such a paradox be resolved? Many phytophagous insects are specialists on one or a few hosts and they often exhibit host-specific assortative mating. Although the reasons for this behavior are unclear, plant-feeding insects, especially endophages such as leafminers or gallers, may become chemically marked by the host plant during development. As adults, they may rely on these chemical cues when searching for mates and/or oviposition sites. Additionally, host-specific mating behavior may simply provide the most reliable source for locating a potential mate, a likely possibility if the host shift is to a plant with a different phenology than the ancestral host species. For instance, in one of the best-documented cases of nascent speciation, host-associated races of the apple maggot fly (*Rhagoletis pomonella*) complex, undergo courtship and mating on their individual host plants, often on the fruits. This is also true for the tephritid sawfly (*Eurosta solidaginis*), which consists of two host-associated races that attack the goldenrods, *Solidago altissima* and *S. gigantea*, in which mating and oviposition occur on the natal host plant. Mating of the soapberry bug, *Jadera haematoloma*, which has expanded its host range to include several introduced species, is also limited to the host plant.

Host-specific associative mating is likely to result in reduced gene flow between subpopulations or races that utilize different host species if insect development time is variable and host fidelity among subpopulations is high. For example, spring-summer gall persistence (measured as LD_{90} and correlated with larval development time) of host-associated populations of the gall midge, *Asphondylia borrichiae*, which galls the terminals of three plants in the aster family, ranges from approximately 29 days on *Iva imbricata* to 70 days on *Borrichia frutescens*. Since these midges only live for a couple of days as adults after emergence from the gall, it is very likely that host shifts have resulted in semi-isolated subpopulations of *A. borrichiae* because adults may be limited to mating with conspecifics from the same host species owing to different host-specific emergence times. Host-associated populations of *A. borrichiae* also exhibit a high degree of host fidelity (at the level of plant genus). Midges that emerged from sea oxeye daisy (*Borrichia frutescens*) attacked the natal host approximately 97% of the time, whereas midges that emerged from the other two hosts (*Iva* spp.) attacked these hosts exclusively. Similar allochronic isolation owing to variable host plant phenology occurs among host-specific races of *R. pomonella* that feed on the fruits of several plants including blueberry, domestic apple and hawthorn. Prior to pupation, larvae emerge from the fruit of the host plant and drop to the ground where they pupate in the soil. Larvae from blueberry pupate 1 to 2 weeks earlier than those from apple and approximately four weeks earlier than those from hawthorn. Synchronization of adult eclosion also mirrors differences in host-specific fruit phenology; adults from the blueberry race eclose first, followed by those from apple and hawthorn, respectively. Gene flow among the ancestral host, hawthorn, and the derived host, domestic apple, has been estimated at 4-6% per generation.

Moreover, larvae that find themselves on a novel host plant may experience reduced levels of attack by natural enemies such as parasitoids. For the apple maggot fly, which has apparently expanded its host range numerous times from its ancestral host, hawthorn (*Crataegus* spp.), including expansion to the commercially important domestic apple (*Malus pumila*) introduced from Europe, both interspecific competition and parasitism are greatly reduced on the derived host. *Asphondylia borrichiae* also appears to experience lower levels of parasitism on two species of *Iva* compared midges on *B. frutescens*. Indeed, reduced levels of parasitism and/or competition on derived hosts may offset lower rates of survival and fitness on sub-optimal hosts, thereby aiding the establishment and maintenance of host-choice polymorphisms.

Although phytophagous insects provide some of the best evidence for the sympatric pathway of speciation, it does not appear to be limited to herbivorous insects. For instance, the predatory lacewings *Chrysoperla carnea* and *C. downesi*, which differ in color, are believed to have originated in sympatry relatively recently. Most likely a color polymorphism arose in the ancestral population, which lead to habitat-specific assortative mating among subpopulations. *Chrysoperla carnea* is light green-brown and prefers deciduous woodlands, while *C. downesi* is dark green and prefers coniferous forests. Differences in habitat-specific survivorship/fitness and habitat preference has favored fixation of different color morphs and asynchronous breeding seasons among the subpopulations that would eventually become *C. carnea* and *C. downesi*.

While phytophagous insects provide model systems for investigating sympatric speciation, some plant-feeding insect guilds such as leafminers and gallers, which are intimately associated with their hosts, may possess characteristics that facilitate the formation of host-associated races or species in sympatry including: (1) sessile larvae that are restricted to feed on the host plant selected by the ovipositing female; (2) endophagous larvae that feed exclusively within the tissue of the host plant; (3) significant host-specific variation in larval survival and performance (including differential attack by natural enemies); (4) differences in host plant phenology (and host-specific insect development) that result in allochronic isolation of host-specific subpopulations (i.e., plant-mediated reduction in gene flow); and (5) host fidelity of ovipositing females. While many biologists still harbor doubts about sympatric speciation, several recent studies provided strong evidence for this evolutionary pathway. After all, Charles Darwin clearly stated in Chapter 4 of his *magnum opus*, 'On the Origin of Species': "I can see no means to agree...that migration and isolation are necessary elements for the formation of new species."

Anthony Rossi
University of North Florida
Jacksonville, Florida, USA

References

Berlocher, S. H., and J. L. Feder. 2002. Sympatric speciation in phytophagous insects: moving beyond controversy? *Annual Review of Entomology* 47: 773–815.

Darwin, C. 1859 (reprinted 1958). *On the origin of species by natural selection or the preservation of favored races in the struggle for life*. Mentor, New York, New York. 479 pp.

Jaenike, J. 1990. Host specialization in phytophagous insects. *Annual Review of Ecology and Systematics* 21: 243–273.

Mopper, S., and S. Y. Strauss (eds.). 1998. *Genetic structure and local adaptation in natural insect populations*. Chapman and Hall, New York, New York. 449 pp.

Rossi, A. M., P. Stiling, M. V. Cattell, and T. I. Bowdish. 1999. Evidence for host-associated races in a gall-forming midge: trade-offs in potential fecundity. *Ecological Entomology* 24: 95–102.

Thompson, J. 1988. Evolutionary ecology of the relationship between oviposition preference and performance of offspring in phytophagous insects. *Experimentalis et Applicata* 47: 3–14.

Via, S. 1990. Ecological genetics and host adaptation in herbivorous insects: the experimental study of evolution in natural and agricultural systems. *Annual Review of Entomology* 35: 421–446.

Wheeler, Q. D., and R. Meier (eds.). 2000. *Species concepts and phylogenetic theory: a debate*. Columbia University Press, New York, New York. 230 pp.

SPECIES. Populations with individuals that are similar in structure and behavior, and capable of mating and producing fertile offspring. See also, SPECIATION PROCESSES AMONG INSECTS.

SPECIES AREAS-RELATIONSHIP. The concept that the number of herbivore species associated with a plant is positively related to the area occupied by the plant. This is an outgrowth of island biogeography theory, in which larger islands support larger faunas; alternative explanations include the length of time a plant has been available for colonization (with more insects accumulating over time) and habitat heterogeneity (with plants living over a wider range being exposed to greater environmental and faunistic diversity).

SPECIES COMPLEX. A group of closely related species, or a species that appears to be variable, and is thought to consist of more than one undefined species.

SPECIES DIVERSITY. An index of community diversity that considers both species richness and the relative abundance of species.

SPECIES ODOR. The odor associated with the body of social insects It is unique to a species, though it may be a component of the colony odor.

SPECIES RICHNESS. The number of species present in an assemblage, community, or other defined ecological unit.

SPERM PRECEDENCE. In insects with multiple matings, the sperm of the last male to copulate with the female is most likely to father the offspring. The last sperm to be stored in the spermatheca are the first to be released.

SPERMATHECA. A small cuticular storage chamber that stores sperm and is associated with the median oviduct of females. Impregnated females

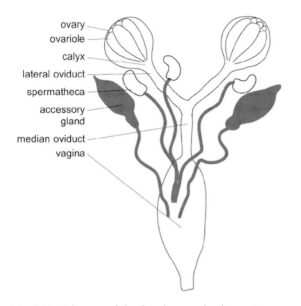

ovary
ovariole
calyx
lateral oviduct
spermatheca
accessory
gland
median oviduct
vagina

Fig. 946 Diagram of the female reproductive system, as found in *Rhagoletis* (Diptera) (adapted from Chapman, The insects: structure and function).

can store sperm for weeks, months or even years depending on the species, and release sperm as an egg passes through the oviduct, resulting in fertilization.

SPERMATID. (pl., spermatozoa) A germ cell in the testis that eventually becomes sperm. Developmentally, the spermatid develops from the spermatocytes by mitosis, and mature into sperm.

SPERMATOCYTE. A cell in the testis that divides, eventually forming the sperm. Developmentally, spermatocytes occur in multicellular cysts created by mitosis of spermatogonia, and develop into spermatids.
See also, REPRODUCTION, ENDOCRINE REGULATION OF REPRODUCTION.

SPERMATOGENESIS. The production of sperm cells in the testis.
See also, STERILE INSECT TECHNIQUE, REPRODUCTION.

SPERMATOGONIA. Cells developing in the testis and forming spermatocytes mitotically. The first step in the production of sperm. See also, REPRODUCTION, ENDOCRINE REGULATION OF REPRODUCTION.

SPERMATOPHORE. A covering or capsule around the sperm.

SPHAERITIDAE. A family of beetles (order Coleoptera). They commonly are known as false clown beetles. See also, BEETLES.

SPHAEROCERIDAE. A family of flies (order Diptera). They commonly are known as small dung flies. See also, FLIES.

SPHAEROPSOCIDAE. A family of psocids (order Psocoptera). See also, BARK-LICE, BOOK-LICE, OR PSOCIDS.

SPHECIDAE. A family of wasps (order Hymenoptera). See also, WASPS, ANTS, BEES, AND SAWFLIES.

SPHECOLOGY. The scientific study of wasps.

SPHECOPHILE. An organism that spends part of its live cycle in wasp colonies.

SPHERULOCYTE. A type of hemocyte of intermediate size.
See also, HEMOCYTES OF INSECTS: THEIR MORPHOLOGY AND FUNCTION.

SPHINDIDAE. A family of beetles (order Coleoptera). They commonly are known as dry-fungus beetles. See also, BEETLES.

SPIDER BEHAVIOR AND VALUE IN AGRICULTURAL LANDSCAPES. Spiders are one of the largest groups of predatory organisms in the animal kingdom with more than 30,000 species distributed over 60 families worldwide. They are so diverse they are found almost everywhere on earth, from arctic islands to dry desert regions. They are particularly abundant in areas of rich vegetation. It is no exaggeration to say that spiders have conquered all possible ecological niches on land.

Even though spiders are found in abundance all over the world, they have not been fully used in pest management programs. The widespread, undeserved reputation of spiders as being dangerous to people and their animals is one reason why they are not given proper credit as important natural pest control agents. With the exception of one group under the family *Uloboridae,* all spiders have poison glands. However, not all spiders are poisonous. There are many more useful spider species than harmful ones. Of over 30,000 known species, only about two dozen are considered to be of any medical hazard to humans. The venom produced by spiders is intended to affect insects and the effect of the venom on humans is usually only slight. Spider venom was not designed to kill humans. Human death by spider venom is an unfortunate reaction, much the same as adverse human reaction to some wasp and bee stings.

Some of the most well-known, dangerous spider species in the world are shown below.

Most of the dangerous spiders are shy and only become aggressive and bite when provoked. There is controversy over spiders as to whether they are friends or foes in agricultural landscapes. Generally, they are friends of farmers and only a very few, those that can harm humans, are considered foes. Knowing how to recognize the harmless spiders and how to prevent and control the dangerous ones can avert needless concern and reduce the chances of harm to humans.

Importance of spiders in agriculture

In reality, spiders are a component of the natural enemy complex in nearly all agricultural landscapes. To date, over 300 studies on spiders in different agricultural crops are documented. In general, spiders are extremely beneficial because they are important predators of pests such as thrips, caterpillars, aphids, plant bugs, leafhoppers, flies, and other arthropod pests in home gardens and crop fields. Spiders, along with other predators, are important in agriculture because they regulate and balance arthropod pest populations. They are beneficial in keeping insect populations in check, which far outweighs the hazard posed by the few spiders that occasionally bite humans and pets.

Some well-known spiders of importance.

Common name	Species name	Location
Australian whistling tarantula	*Selenocosmia sp.*	Australia
Black widow spider	*Latrodectus mactans*	Southern USA and other warm regions around the world
Brown recluse spider	*Loxosceles reclusa*	S and Central N America
Cellar spider	*Pholcus phalangioides*	Europe, Asia, N and S America
False widow spider	*Steatoda grossa*	S Europe, N Africa, W Asia
Greek trap-door spider	*Cyrtocarenum sp.*	Western Australia
Horned baboon spider	*Ceratogyrus darlingi*	E Africa, Zimbabwe, Mozambique
Katipo	*Latrodectus katipo*	New Zealand
Red-back spider	*Latrodectus hasselti*	Gulf region and Australia, New Zealand and Japan
Spanish funnel-web spider	*Macrothele calpeina*	Spain and NW Africa
Sri Lankan ornamental tarantualas	*Poecilothera fasciata*	Sri Lanka
Sydney funnel-web spider	*Altrax robustus*	Australia, New South Wales and Victoria
Tube-web spider	*Segestria florentina*	Europe, Argentina, New Zealand
Violin spider	*Loxosceles rufescens*	S Europe, N Africa, Japan, N America, Australia, New Zealand
Woodlouse spider	*Dysdera crocota*	Europe, Japan, N and S America, S Africa, Australia, New Zealand

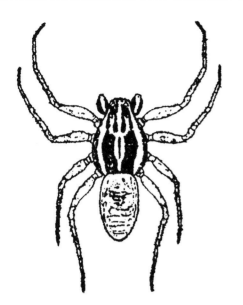

Fig. 947   Beneficial spider, *Lycosa pseudoannulata*.

The importance of spiders as pest control agents is not a recent discovery. Several ancient books in agriculture mention their importance. In China, the use of spiders for pest control dates back 2,000 years and even today, many elderly people in Chinese villages consider the number of spiders in a field as a measure of its potential agricultural productivity. The scientific literature amply demonstrates the biological control potential of spiders. A classic example of a successful biological control system using spiders was reported on rice. A wolf spider, *Lycosa pseudoannulata*, successfully controlled the brown planthopper, *Nilaparvata lugens*, which is the major pest of rice in Southeast Asia. In Britain, theridiids (comb-footed spiders) and erigonids (dwarf spiders) suppress orchard mites. In Australia, another species of small theridiid, *Acahearaneae veruculata*, successfully controls the tortricid moth. A study in an unsprayed peach orchard in the Niagara Peninsula of Canada shows other species of theridiids, *Theridion murarium* and *Philodromus praelustris*, regulate the density of phytophagous mites.

Feeding behavior

Great differences exist among spiders concerning the ways in which they capture their prey. While some spiders stalk their prey and others ambush it, most spiders trap their prey by snares and a few live by eating the prey captured by other spiders (these are

called commensals). Spiders paralyze their prey by injecting poison secreted by a pair of poison glands in the chelicerae (front jaws). The ducts from these glands open on each side through a minute pore located near the tip of the fang.

Prey catching strategies have been thoroughly studied in web-building spiders. For instance, the sheet web spider (family *Agelenidae*) hides in its funnel-shaped nest and rushes out only when the prey has been trapped in the web. Members of the family *Linyphiidae* construct a horizontal sheet web with vertical threads that serve as trapping lines for insects. Once the prey is trapped by the threads, the spider shakes the web until the victim falls onto the sheet, then bites through the web and pulls down the victim. Some web-building spiders (e.g., theridiids) produce threads with glue droplets that trap and glue insects to the broken threads. The insects become progressively more entangled while attempting to escape. The alerted spider quickly climbs down and throws more sticky threads over the victims before biting them.

The various groups of orb-weaving spiders exhibit three related strategies of prey capture: strategy 1 – seize with chelicerae, pull out of web, carry in chelicerae, wrap at hub and secure to web, then feed (e.g., *Nephila* spp.); strategy 2 – bite, wrap in web, cut out of web, carry on thread or carry on chelicerae, wrap at hub and secure to web, then feed (e.g., *Nephila*

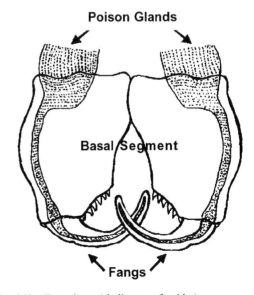

Fig. 948   Front jaws (chelicerae of spider).

spp.); and strategy 3 – wrap, bite, cut out of web and carry on thread, or, cut out of web and carry on chelicerae, wrap at hub and secure to web, then feed (e.g., *Araneus* and *Argiope)*.

The wandering spiders do not rely on snares. Most locate and overpower their prey directly. Among the more familiar examples of wandering spiders are the wolf spiders (*Lycosidae*), the jumping spiders (*Salticidae*), most of the crab spiders (*Thomisidae*), and the two-clawed foliage spiders (*Clubionidae, Anyphaenidae*). These spiders run in search of their prey and pounce upon it when an opportunity arises. Wandering spiders, exemplified by lycosids or wolf spiders, actively hunt for their prey. Wandering spiders spend some of their time stationary adopting a sit and wait strategy. Wandering spiders catch their prey in a different manner than web-builders. First, they locate the prey. Then they turn toward the victim and grasp it with the tips of the front legs, pulling it to the chelicerae and then biting (venom injection). Next, they release their grasp with the legs and hold the prey with the chelicerae instead, fastening some silk thread over the immobilized victim, and then commence feeding. Wandering spiders may locate their prey either by visual cues or by vibration, depending on the species. Classical examples of visually guided predators are the jumping spiders. They have highly developed main eyes that can analyze shapes, and can even recognize motionless prey. Blind spiders, like the clubionids and related species, rely on the vibration of the substrate, or immediate contact with a victim to elicit direct catching movements.

Three species of sac spiders, [*Chiracanthium inclusum* (Clubionidae), *Hibana velox* (Anyphaenidae), and *Trachelas volutus* (Corrinnidae)], belong to the wandering spider group. Studies show they do not capture prey in webs, but actively hunt their prey, usually at night. During the day, they hide in tubular silken capsules that they construct, which give them the common name sac spiders. They are important in agriculture because they feed on the larvae and prepupae stages of citrus leafminer (*Phyllocnistis citrella* Stainton), one of the major pests of citrus. These spiders belong to the group of spiders with poor vision and they are nocturnally active hunters. Their prey capture depends greatly on substrate vibration. They detect their prey through the vibration of the substrate where the prey is concealed. Movement of citrus leafminer larvae and prepupae creates vibrations on the leaf substrate, which serves as a cue for the spiders to locate the positions of the prey. The prey capture sequence for these three species of spiders follows a similar pattern. During the searching period, the spider moves around and then stops for a while to locate sources of vibration. Once the spider touches the prey with its legs, the spider turns rapidly toward the prey and grasps it.

Two methods of prey attack are exhibited by the three species of sac spiders. In one method, the spider punctures the mine, immobilizes the larva then it bites and sucks the larval body fluid. In the second behavioral pattern, the spider makes a slit on the mine, and then pulls the larva or prepupa out of the mine using its forelegs. The first gentle touch of the forelegs quickly changes into a powerful grip aided by the special hairs on the spider's forelegs. For *Cuppiennius*, another genus of wandering spider, the forelegs are able to improve their secure hold by means of the adhesive hairs or scopulae on the tarsi. This may also be true for *Chiracanthium inclusum*, *Hibana velox*, and *Trachelas volutus* since these species possess tarsal scopulae and dense claw tufts. After securing a hold, the prey is pulled quickly toward the spider's body, then the chelicerae of their fangs move apart and are inserted rapidly into the nearest part of the victim's body. Immediately after the bite, the tips of the legs release their grip and the prey is held in the air only with the chelicerae, thereby minimizing any danger to the spider from the prey. This is advantageous because the victim has no contact with the substrate and therefore cannot apply any direct force to free itself. Only after the prey has become immobilized by the venom does actual feeding (the chewing and the exuding of digestive juice) begin.

Studies show that for *Chiracanthium inclusum*, *Hibana velox*, and *Trachelas volutus*, start to feed on citrus leafminer larvae at their first nymphal stage. This is not surprising, since spiders in general, after their first nymphal molt, are self-sufficient. At this stage, they have developed their sensory hairs, their legs are equipped with the typical claws, the eyes have the bulging lenses, and the mouthparts are already differentiated for hunting and feeding. The percent of citrus leafminer consumption for all the species of hunting spiders varies among the different nymphal stages and the adult stage. Consumption increases as they develop to later nymphal stages, especially for *H. velox*. Consumption peaks at the third nymphal stage for all of the species, halfway to the adult stage. For all of the species, feeding slows down and sometimes stops for one or two days before molting and then resumes right after molting. Most spiders that are preparing to molt withdraw into their retreat for several days and stop feeding. This phenomenon occurs naturally in all spiders.

Conservation

The diversity and voracious appetite of spiders in almost all agricultural systems suggests the importance of spiders as predators of insect pests and other arthropods. Thus, they should be considered as an important component of integrated pest control programs. Conservation and augmentation of spiders in fields is a simple and efficient method of pest control. There are several ways to conserve and augment the populations of spiders in the field. One way is to avoid the use of harmful pesticides. Studies show that there are more spiders and fewer insect pests in unsprayed fields than in sprayed ones. Another way is to manipulate their habitat by providing natural mulches (ground cover), which serve as refuge areas for spider breeding or by naturally increasing habitat diversity in the field. A more diverse agroecosystem in terms of habitat structure results in increased environmental opportunities for biological control agents like spiders, and consequently improved biological control. One way to enrich the vegetation structure of cropping systems is through weed management. The impact of weed diversity in the form of weed borders, alternate rows, or by providing weeds in certain periods of the crop growth is important in keeping the populations of the existing beneficial natural control agents, including spiders.

Divina Amalin
USDA, APHIS
Miami, Florida, USA

References

Bristowe, W. S. 1971. *The world of spiders*. New Naturalist, London, United Kingdom.
Chant, D. A. 1956. Predaceous spiders in orchards in southeastern England. *Journal of Horticultural Science* 31: 35–46.
Foelix, R. F. 1996. *Biology of spiders (2nd ed.)*. Oxford University Press, Oxford, United Kingdom.
Jackson, R. R. 1996. Predatory behavior of jumping spiders. *Annual Review of Entomology* 41: 287–308.

SPIDERS. Spiders differ from insects, crustaceans and other members of the phylum Arthropoda by having two body parts rather than the three of insects and crustaceans,and the multiple body parts of other arthropods. They also completely lack antenna – the only arthropod group to lack this sensory organ. Spiders use structures called chelicerae in feeding rather than the mandibles found in other arthropod groups.

Classification and evolution

Spiders share the class Arachnida with other organisms such as scorpions (order Scorpiones), pseudoscorpions (order Pseudoscorpiones), camel spiders (order Solifugae), vinegaroons (order Uropygi), tailless whipscorpions (order Amblypygi), daddy longlegs (order Opiliones) and mites and ticks (order Acari). All these arachnid groups are thought to have evolved from an ancestor similar to eurypterids (order Eurypterida). These marine chelicerates were common in the seas from 500 to 245 million years ago (Ordovician to the Permian). Some eurypterids grew to two meters long. These giant aquatic arthropods were formidable predators of the ancient seas and had similar morphological characteristics as extant scorpions. In fact, some of the earliest scorpion fossils resemble more recent eurypterid fossils. Scorpion fossils from the Silurian period probably were aquatic because gills are apparent on these fossils. The arachnids that evolved from a eurypterid-like ancestor moved from the marine environment to the terrestrial environment but, largely, maintained their predatory nature.

The earliest fossil arachnids include certain scorpions that date back to the Silurian period between 440 to 410 million years ago. The earliest true spiders appear in the fossil record in the Devonian Period about 380 million years ago. Early terrestrial arthropods had certain adaptations that gave them an advantage in the transition from an aquatic to a terrestrial environment including jointed legs to support the body against the pull of gravity and an exoskeleton to protect them from the desiccating effects of the air. The respiratory structures of the early aquatic chelicerates were probably similar to the book gills found on horseshoe crabs. These book gills are, in turn, similar in structure to the internal book lungs of extant terrestrial arachnids. It is tempting to imagine that as the early chelicerates were making the evolutionary transition from an aquatic to a terrestrial existence, the external respiratory structures became internalized as the book lungs.

Following is a current taxonomic classification system for spiders, and the currently accepted families and numbers of genera and species within each family can be found at: http://research.amnh.org/entomology/spiders/catalog81–87/INTRO2.html.

Kingdom: Animalia

  Phylum: Arthropoda
    Subphylum: Chelicerata
      Class: Arachnida
        Order: Araneae
          Suborder: Mesothelae
            Family: Liphistiidae
          Suborder: Opisthothelae
            Infraorder: Mygalomorphae
              Family: Atypidae
              Family: Antrodiaetidae
              Family: Mecicobothriidae
              Family: Hexathelidae
              Family: Dipluridae
              Family: Cyrtaucheniidae
              Family: Ctenizidae
              Family: Idiopidae
              Family: Actinopodidae
              Family: Migidae
              Family: Nemesiidae
              Family: Microstigmatidae
              Family: Barychelidae
              Family: Theraphosidae
              Family: Paratropididae
            Infraorder: Araneomorphae
              Family: Hypochilidae
              Family: Austrochilidae
              Family: Gradungulidae
              Family: Filistatidae
              Family: Sicariidae
              Family: Scytodidae
              Family: Periegopidae
              Family: Drymusidae
              Family: Leptonetidae
              Family: Telemidae
              Family: Ochyroceratidae
              Family: Pholcidae
              Family: Plectreuridae
              Family: Diguetidae
              Family: Caponiidae
              Family: Tetrablemmidae
              Family: Segestriidae
              Family: Dysderidae
              Family: Oonopidae

Family: Orsolobidae
Family: Archaeidae
Family: Mecysmaucheniidae
Family: Pararchaeidae
Family: Holarchaeidae
Family: Micropholcommatidae
Family: Huttoniidae
Family: Stenochilidae
Family: Palpimanidae
Family: Malkaridae
Family: Mimetidae
Family: Eresidae
Family: Oecobiidae
Family: Hersiliidae
Family: Deinopidae
Family: Uloboridae
Family: Cyatholipidae
Family: Synotaxidae
Family: Nesticidae
Family: Theridiidae
Family: Theridiosomatidae
Family: Symphytognathidae
Family: Anapidae
Family: Mysmenidae
Family: Pimoidae
Family: Linyphiidae
Family: Tetragnathidae
Family: Araneidae
Family: Lycosidae
Family: Trechaleidae
Family: Pisauridae
Family: Oxyopidae
Family: Senoculidae
Family: Stiphidiidae
Family: Neolanidae
Family: Zorocratidae
Family: Psechridae
Family: Zoropsidae
Family: Zoridae
Family: Ctenidae
Family: Agelenidae
Family: Cybaeidae
Family: Desidae
Family: Halidae
Family: Amphinectidae
Family: Cycloctenidae
Family: Hahniidae
Family: Dictynidae
Family: Amaurobiidae
Family: Phyxelididae

Family: Titanoecidae
Family: Nicodamidae
Family: Tengellidae
Family: Miturgidae
Family: Anyphaenidae
Family: Liocranidae
Family: Clubionidae
Family: Corinnidae
Family: Zodariidae
Family: Cryptothelidae
Family: Chummidae
Family: Homalonychidae
Family: Ammoxenidae
Family: Cithaeronidae
Family: Gallieniellidae
Family: Trochanteriidae
Family: Lamponidae
Family: Prodidomidae
Family: Gnaphosidae
Family: Selenopidae
Family: Sparassidae
Family: Philodromidae
Family: Thomisidae
Family: Salticidae

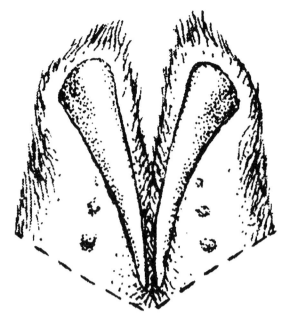

Fig. 949   Chelicerae of a spider in the subfamily Mygalomorphae (drawing by Eric Parrish, University of Colorado).

The suborder Mesothelae contains only the single family Liphistiidae. Species of liphistiids are found only in China, Japan, Southeast Asia, and Sumatra. Liphistiids are the only spiders that have a segmented abdomen. Abdominal segmentation has been lost in all more recently evolved lineages. Liphistiids also have four distinct pairs of spinnerets instead of the three pairs found in all other spiders.

The suborder Opisthothelae contain all remaining spider families which are divided into two infraorders: the Mygalomorphae and the Araneomorphae. In North America, the mygalomorphs are commonly called tarantulas and include the quintessential large, hairy pet store spiders. The mygalomorphs are united as a single lineage based largely on spinneret and male genitalic characters. They have only two functional pairs of spinnerets and the bulb of the male pedipalp is fused in most families of mygalomorphs. Most mygalomorphs also are characterized as having paraxial chelicerae that are positioned more or less parallel to one another like the tongs of a garden rake. These robust spiders also are characterized as having two pairs of book lungs.

The infraorder Araneomorphae contains the majority of spider species worldwide. This infraorder is sometimes referred to as the 'true' spiders even though liphistiids and mygalomorphs are as true as they come! This infraorder contains the best known and most familiar spiders such as the orb weavers

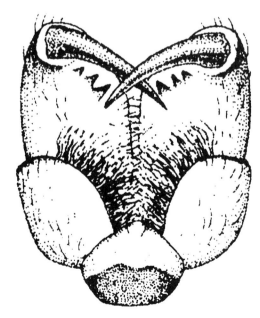

Fig. 950   Chelicerae of a spider in the subfamily Araneomorphae (drawing by Eric Parrish, University of Colorado).

in the family Araneidae, the wolf spiders in the family Lycosidae, the jumping spiders in the family Salticidae, and the crab spiders in the family Thomisidae. Araneomorphs typically have opposing chelicerae that move somewhat like the blades of scissors, and they typically have either one pair of book lungs or lack book lungs altogether. They also have three pairs of spinnerets, although some families of araneomorphs have a residual pair of anterior spinnerets that has evolved into a flattened spinning plate called the cribellum. Ecribellate spiders have a non-functional structure called the colulus.

Arachnid characteristics. Most arachnids share certain characteristics including: (1) a body divided into two parts – the cephalothorax or prosoma and the abdomen or opisthosoma; (2) no antennae; (3) chelicerae; (4) a unique pair of appendages in the front of their bodies called pedipalps involved in feeding and/or in mating; (5) four pairs of walking legs; (6) external digestion; (7) indirect sperm transfer; (8) elaborate courtship behaviors; and (9) a strict predatory lifestyle. However, not all arachnids share all these characteristics. For example, the cephalothorax and abdomen of opilionids and acarines have become fused giving them the appearance of having only a single body part rather than the two characteristic of the other arachnid groups. Some species in these two orders also differ in having direct sperm transfer. Opilionid males have a penis that delivers the sperm directly into the female's body. Some opilionids can digest solid particulate food and, therefore, do not rely solely on pre-digested, or externally digested, material. Opilionids and acarines also differ in that neither group is a strict predator. Opilionids are, for the most part, scavengers, whereas acarines can be predators, ectoparasites, or plant feeders.

Spider phylogeny. The order Araneae currently includes over 38,000 species. This number includes only those species known to science. New species are being discovered every year, and the true species diversity is probably closer to 80,000 according to Dr. Norm Platnick of the American Museum of Natural History. Spiders are found in 110 families divided into three suborders: the Mesothelae, the Mygalomorphae, and the Araneomorphae. The suborder Mesothelae includes a single Asian family, the Liphistiidae. The liphistiids are distinct from all other spiders because they retain the abdominal segmentation common to more ancestral arachnid orders such as the scorpions, the camel spiders (or solifugids) and the vinegaroons (or uropygids). The suborder Mygalomorphae includes those spiders referred to in the new world as tarantulas and referred to elsewhere as bird spiders. These spiders tend to be large, robust animals and are differentiated from the araneomorphs by having chelicerae that move more or less parallel to one another, somewhat like the tongs of a garden rake and by having two pairs of book lungs instead of just one pair. The Mygalomorphae includes 16 families worldwide. The suborder Araneomorphae includes the majority of the world's spiders found in 94 families. All these spiders have chelicerae that move in opposition to one another somewhat like salad tongs. It is thought that the Mesothelae are the evolutionarily most primitive (earliest evolved) lineage of spiders and that the Araneomorphae are the most recently evolved lineage.

The majority of araneomorph spiders are further divided into two groups known as the Haplogynae and Entelegynae. In the haplogyne araneomorphs, the terminal part of the female's genitalia – that portion that opens to the outside – serves as both a copulatory duct and a fertilization duct. In other words, the external portion of the genitalia connects directly with the seminal receptacles, or spermathecae. Haplogyne females have no complex sclerotized structures associated with their genital opening. The entelegyne spiders, on the other hand, have females with very complex structures associated with their genital openings. These sclerotized structures are collectively referred to as the epigynum. In the entelegyne spiders, side ducts lead from the fertilization duct to the spermathecae. Seventeen families make up the Haplogynae branch of araneomorph spiders and 74 families make up the Entelegynae branch. Three araneomorph families – the Hypochilidae, the Gradungulidae, and the Austrochilidae – are not considered either haplogynes or entelegynes. Hypochilidae form a sister group to all the other Araneomorphae taxa. The hypochilids, like the Mesothelae and Mygalomorphae, are considered a more 'primitive' (i.e., earlier evolved) group. Remnants of abdominal segmentation are present in hypochilids and hypochilids have a cribellum (see next section). The Gradungulidae and Austrochilidae together form a sister group to the entelegyne/haplogyne branch of the evolutionary tree. The species in these two families are found primarily in South America and Australia, probably due to a Gondwana distribution.

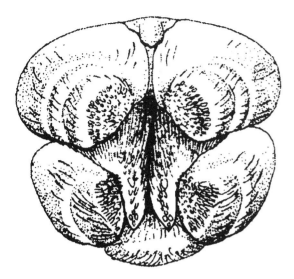

Fig. 951 Spinnerets of a spider (drawing by Eric Parrish, University of Colorado).

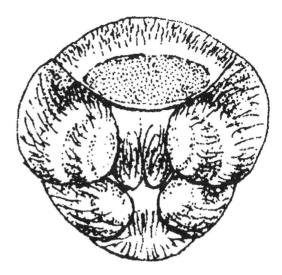

Fig. 952 Cribellum located anterior to the spinnerets (drawing by Eric Parrish, University of Colorado).

Silk production. Silk is characteristic of all spiders. Spider silk emerges from structures at the posterior end of the spider's abdomen called spinnerets. The proteinaceous material is produced in silk glands in the abdomen. As the silk protein is exuded through the tiny opening of the silk spigot on the surface of the spinneret, most of the water is re-absorbed. The protein configueration changes from liquid to solid. The solid proteinaceous silk is further strengthened and solidified when it is stretched or put under tension as it emerges from the spinneret.

Most spiders have three pairs of spinnerets – the anterior, the median and the posterior pairs. However, it is thought that ancestral spiders had four pairs: an anterior lateral pair, an anterior median pair, a median pair and a posterior pair. Spiders in the suborder Mesothelae still have four pairs of spinnerets, but one pair (the anterior median) is nonfunctional and residual. Many mygalomorph spiders have only two pairs of functional spinnerets having lost the anterior pairs (i.e., the pairs closest to the cephalothorax). Among araneomorph spiders, most species have three pairs of functional spinnerets. The anterior median pair has been lost and a rudimentary, non-functional structure called the colulus remains, positioned between the anterior lateral pair of spinnerets. Other species of araneomorphs, called cribellate spiders, have a flattened spinning plate, called the cribellum, positioned between the anterior lateral pair of spinnerets. It is thought that the cribellum and the colulus are homologous structures and both are thought to be evolutionary derivatives of the anterior median pair of spinnerets. The cribellum is also thought to be an ancestral, or plesiomorphic, characteristic among the araneomorph spiders. The cribellum is covered with upto tens of thousands of individual silk spigots. In some cribellate species, the cribellum is divided into two small plates; in others, it is a single continuous plate. Cribellate spiders also have a specialized comb, called a calamistrum, on the metatarsal segment of the last leg. The calamistrum is used to comb the cribellate silk out of the cribellum. Both cribellate and ecribellate spiders are found in the Haplogynae and in the Entelegynae. Therefore, it is currently thought that the cribellum is a polyphyletic characteristic.

On the surface of each spinneret are different types of silk spigots each of which leads to a silk gland in the abdomen of the spider. Spiders can possess up to six different types of silk glands each secreting a different kind of silk for different functions. These glands include ampullate glands, aciniform glands, tubuliform glands, piriform glands, aggregate glands and flagelliform glands. Ampullate glands produce dragline silk and spider web frame threads. Draglines are continually produced by spiders as they move around in the environment and serve as safety lines. Aciniform glands produce the silk used to wrap prey as well as silk used by male spiders to construct sperm webs and silk used by

Fig. 953 Calamistrum located on the metatarsal segment of the last leg of a cribellate spider. Used to comb the silk out of the cribellum (drawing by Eric Parrish, University of Colorado).

females in constructing the outer walls of the silken egg sacs. Tubuliform glands are only found in female spiders and are also involved in the construction of the egg sacs. Piriform glands are used in the production of prey capture silk. Flagelliform and aggregate glands are found only in orb weaving spiders. Aggregate glands produce the glue that covers the prey capture silk of ecribellate spiders' webs. Flagelliform glands produce the axial thread of the sticky spiral. Adult male spiders lose their aggregate and flagelliform glands when they mature and, thus, lose the ability to build prey capture webs.

Spider silk is remarkable for its strength and elasticity. In terms of tensile strength, it is about as strong as the best steel. In terms of its tenacity and elasticity, dragline silk would have to be 80 km long before it broke under its own weight and it is more elastic than the stretchiest nylon. Many spiders recycle silk proteins by eating their webs. Scientists have radioactively labeled silk to determine how the silk proteins are used by such spiders. The results of these studies indicate that the silk proteins appear again as new silk in just 30 minutes – spiders are clearly the masters of recycling. In World War II, spider silk was used for the cross hairs in gun sights because of its strength and durability. Natives of New Guinea used spider webs as fishing nets. The real Dr. Thomas Muffet of Little Miss Muffet fame was an actual practicing physician who had a fondness for spiders. One of Dr. Muffet's treatments for cuts or abrasions was to wrap the wound in fresh spider webs. Some spider silk has an acidic pH (around pH 4) and is, therefore, resistant to bacterial decomposition. Because of this, there seems to be some benefit to Dr. Muffet's treatment of wounds.

Spiders use silk for a variety of purposes. Dragline silk protects the spider walking around the habitat from plummeting to its death when a sudden gust of wind knocks it off a leaf; egg sac silk provides the spider eggs some protection from desiccation and from egg parasites and predators; non-sticky silk provides the framework for webs and serves to line burrows and retreats of non-web building spiders; sticky silk allows ecribellate web building spiders to capture prey by gluing the prey to the web; cribellate silk allows cribellate web building spiders to capture prey by tangling the hairs of the prey in the fine sheaths of silk combed from the cribellum by the calamistrum. Silk production is at the core of a spider's day-to-day existence.

Prey capture. When spiders encounter prey, most either bite the prey, injecting it with venom and then wrap the prey in silk, or they wrap the prey to immobilize it and then inject the wrapped prey with venom. Whatever immobilization technique is used, venom plays a crucial role. Venom is produced by all families of spiders except those in the families Uloboridae and Holerchaeidae. All other spiders have venom glands in the cephalothorax that are linked to the fangs by ducts running through the chelicerae. One of the common misconceptions about spiders is that the venom of spiders acts not only to immobilize the prey, but also to digest the prey. This is incorrect. The venom is not involved in digestion and contains only trace amounts of digestive enzymes if any at all.

Spider venoms of medical importance. Another common misconception about spiders is that the venom of many spiders is harmful to humans. Of the over 38,000 species of described spiders worldwide, only 160 or fewer have venom harmful to humans – not even 1% of the species. In the United States, the most common spiders with venom of medical importance are the black widow species in the genus *Latrodectus* (family Theridiidae) and the brown spiders, or recluse spiders in the genus *Loxosceles* (family Sicariidae). These spiders have proteinaceous toxins in their venom. Both black widows and brown spiders are fairly timid and only bite when severely provoked. People get bitten by black widows primarily when they accidentally press down on a resting spider hidden beneath a stone, under a log, or under the seat of a privy in an outhouse. In areas where *Loxosceles* is common inside homes, people usually get bitten when the spider seeks shelter in clothing dropped on the floor. As a person puts on the clothing, the spider is threatened and may bite in defense.

The venom of black widows is neurotoxic and can cause severe systemic pain by disrupting the transmitter substances at the neural synapses. Toxicity of

Latrodectus venom varies between different vertebrates. Rabbits and sheep, for example, are little affected by bites from black widows, whereas horses and camels are very susceptible to injections of the venom. The effect of black widow bites on humans varies between individuals depending on the amount of venom injected, the age and the health of the individual. Rarely do bites from black widows cause death, but they can cause severe abdominal pains, cramps and pain in the joints and other systemic effects. Antivenin is now readily available in areas where black widows are common. However, because the antivenin is in horse serum and because many people have allergic reactions to the serum, other treatments may be attempted first to ease the pain caused by the bite. Injection of calcium salts has shown positive results.

Loxosceles (or recluse) venom is proteolytic and hemolytic, primarily affecting tissues at the bite site. Bites from these spiders can cause deep, ulcerous wounds that heal very slowly, sometimes taking up to three weeks. These lesions can become extensive. They are typically treated topically with corticosteroids. Antibiotics often are given to prevent secondary bacterial infection of the wound site. In extreme cases, surgical excision of necrotic tissue is necessary, often followed by skin grafts.

One of the more infamous spiders with medically significant venom is the Sydney funnel web spider, *Atrax robustus* (Mygalomorphae, family Hexathelidae). This spider is found only in Australia. The venom of male *A. robustus* acts upon neuromuscular endings, stimulating the release of acetylcholine, adrenaline and noradrenaline from the autonomic nervous system. Males are much more venomous than females and much more likely to be encountered. Females tend to remain in the funnel webs, whereas adult males are found wandering in search of females and may, in their wanderings, have a face to face encounter with a human. When only a little venom is injected, the bite site is very painful for hours to days but, otherwise, only mild reactions to the venom are evident. A more serious bite can result in nausea, vomiting, abdominal pain, arrhythmia and other systemic symptoms. In severe cases, confusion, coma and death by asphyxiation can result from an *A. robustus* bite. Antivenin is now readily available in areas of Australia where the Sydney funnel web spider is found. This antivenin is extremely effective.

Digestion of prey. Once the prey has been immobilized using venom, the spider regurgitates digestive enzymes onto the prey through the mouth. The mouth is located between three plates situated on the ventral side of the cephalothorax: a central plate called the labium and two lateral plates called the endites. These enzymes that are regurgitated onto the prey break down, or pre-digest the body of the prey until the body has been completely liquefied. When a spider bites its prey, the fangs create holes in the exoskeleton of the prey that serve as entryways for the digestive enzymes. Some spiders create further holes in the body of the prey using teeth lining the fang furrows on the chelicerae. Spiders without such teeth on the chelicerae rely on the holes created by the fangs to digest and suck up the pre-digested prey. With these spiders, all that is left when they are done is the intact exoskeleton of the prey – a hollow husk of its former self.

A spider's stomach is the main organ used to imbibe the pre-digested food. The stomach is located in the dorsal portion of the cephalothorax. It is connected to the spider's exoskeleton by a series of muscle bands. The apodeme, or muscle attachment, of the dorsal band can be seen clearly on the cephalothorax of a tarantula as a median indentation – somewhat like a dimple. Circular muscle bands surround the stomach. When the muscle bands connecting the stomach to the body wall contract, the stomach lumen expands; when these muscles relax, the circular bands reduce the volume of the stomach. The spider's stomach thus acts as a pump, drawing fluid into the body when the long bands contract, or pumping fluid out of the stomach and either forward as regurgitant, or laterally and backward into the remaining tubes of the digestive system. The stomach of spiders, appropriately, is called a sucking stomach.

A spider cannot ingest solid particulate matter, it can only suck into its body pre-digested liquids. Hairs are often found on the lateral borders of the endites and on the anterior border of the labium. These hairs function to filter out particulate material. The pharynx is lined with downward pointing ridges that also act to filter out larger particles. These larger particles are spit out of the mouth by an anti-peristaltic stream of digestive fluid and are concentrated as a solid pellet, which is then cleared away from the mouthparts using the pedipalps. While feeding, the spider continually vomits digestive fluid and sucks up the resulting liquified prey until all that is left is the empty exoskeleton, or bits of exoskeleton. When larger spiders feed on vertebrate prey, the pre-digestive process is readily apparent as the mouse, or small bird becomes a glistening, liquefied ball of material. All that is left when a vertebrate is eaten is a collection of bones, cartilage, feathers and other indigestible material.

Extensions of the digestive system, called diverticula, extend out into all areas of the spider's body including the legs, the eyes and the abdomen. These extensions create an efficient and effective transport system for the pre-digested food. Partly because of these extensive diverticula and partly because of the low metabolism of spiders, these animals can go a very long time without eating. Pet tarantulas can be fed once a month or less and be perfectly content (or so it seems).

Excretion. The posterior portion of the digestive system ends in a stercoral pocket, or cloacal chamber. The excretory products of the main excretory organs, the Malpighian tubules, empty into this sac-like stercoral pocket. The main excretory products secreted by spiders are guanine, adenine, hypoxanthine and uric acid – all of which are largely insoluble in water. Therefore, spiders lose very little water through excretion. These excretory products are stored in the stercoral pocket until the pocket is periodically emptied and the material excreted through the anus. In some spiders, waste products from digestion are stored as crystals of guanine just below the transparent cuticle where they are seen as white markings on the abdomen.

Respiratory and circulatory systems. When talking about the circulatory system of spiders, it makes sense to talk about the respiratory system as well because the primary function of the circulatory system is oxygen transport. Oxygen enters the body of a spider through openings in the body called spiracles. For mygalomorph spiders, these spiracles lead to two pairs of book lungs. Araneomorph spiders have either one pair or no book lungs. The posterior pair of book lungs in most araneomorph spiders is thought to have evolved into a pair of tubular tracheae.

The spiracle opening into the book lung is a passive opening – in other words, it cannot be closed. The air flows between thin plates of tissue called lamellae. In between these lamellae are evaginations of the body cavity filled with hemolymph, or blood. Oxygen diffuses through the lamellae into the hemolymph. The tubular tracheae of spiders are similar to the tracheae of insects, except that insects have multiple spiracular openings into the tracheae, whereas spiders have either one or two spiracles opening into their tracheae. Also, in spiders, the tracheoles are open ended unlike the closed system of tubes found in insects. Therefore, in spiders, oxygen carried by the tracheal tubes is dumped directly into the hemolymph.

For both the book lungs and tracheae, oxygen enters the body passively with little help from muscular contractions. The tracheal system varies in different araneomorph families. In some, the tubes branch out only into the abdomen; in others, the tubes branch all over the spider's body including into the cephalothorax. In some very tiny spiders, they have lost the book lungs completely and rely on an extensive system of tracheal tubes to get their oxygen.

The hemolymph of spiders contains oxygen-carrying molecules called hemocyanin. This molecule functions very similarly to hemoglobin found in mammals. However, hemoglobin has iron atoms as part of its structure and is contained in blood cells, whereas spider hemocyanin has copper atoms and is not contained within blood cells. Whereas the iron atoms gives mammalian blood its red color when it becomes oxygenated, the copper atoms gives spiders' blood a blue-green color. Hemocyanin is nearly as effective as hemoglobin in binding to oxygen molecules. The hemolymph of spiders also contains cells that are involved in blood clotting, wound healing, fighting off infections and storing different types of compounds for transport throughout the body.

After the hemocyanin molecules pick up the oxygen from the book lungs and from the open-ended tracheae, this oxygenated blood is carried throughout the body. This transport is carried out via the circulatory system. As with all arthropods, spiders have an open circulatory system — the hemolymph bathes the internal organs and tissues and is not contained in blood vessels. Nevertheless, a circulatory pathway does exist. A spider's heart is located in the dorsal portion of the abdomen. As the heart muscle relaxes, holes in the sides of the heart, called ostia, open and the oxygenated blood is sucked into the heart. The heart then contracts, which causes the ostia to close and forces the blood anteriorly and posteriorly into a system of open-ended arteries. The oxygenated blood thus gets circulated throughout the body in a directional manner such that as the hemolymph loses its oxygen, it flows back toward the abdomen and the respiratory organs before returning to the heart. The circulation of the hemolymph follows a gradient of decreasing blood pressure. This carries the deoxygenated blood back to the abdomen and over the book lungs or tracheal openings.

In addition to its other functions, the hemolymph of spiders also serves another very important function. Spiders have both flexor and extensor muscles in all their leg joints except the femur-patella joint and the tibia-metatarsal joint, which both lack extensor muscles. Spiders extend their legs at these joints by increasing the blood pressure in their limbs. This increased pressure is caused by a contraction of muscles connecting the carapace (or top of the cephalothorax) with the sternum (or bottom plate of the cephalothorax) thus reducing the volume of the first body segment and increasing blood flow into the extremities. If spiders lose too much water, they can no longer increase hemolymph pressure sufficiently to extend the legs. Spiders caught inside homes away from any source of free water are often found with their legs curled up underneath them. If found in time, a droplet of water can revive such a hapless spider. Although spiders can survive long periods without food, most cannot survive long without some source of moisture.

Sensory hairs and pits.   Hairs on a spider's body serve as mechanoreceptors, responding to touch, vibrations, air currents and positions of the spider's joints. In some mygalomorph species, the hairs on the abdomen also serve for defense. These hairs are often barbed and, when provoked, the spider brushes them off using a rear leg. Some wandering spiders, such as mygalomorphs and jumping spiders – spiders that hunt without a web – have scopula and claw tufts on the tarsi and metatarsi. These hairs are more structural than sensory. They are very fine hair fringes that allow these spiders to climb and keep a foothold on very smooth surfaces. Any smooth surface is covered by a fine film of water. The thousands of small hairs

making up the claw tufts and scopula allow these spiders to walk across these surfaces surefootedly due to capillary forces.

On the leg segments of spiders are long, fine hairs called trichobothria. These hairs are sunk into sockets with multiple dendritic nerve endings at the base. They are found on the legs and are extremely sensitive to air currents – even those produced by the wings of an insect flying nearby. They probably serve in localizing prey or enemies. Patterns of trichobothria are often diagnostic of certain families. Sensory structures called slit sense organs are located near the leg joints. These slits are also connected to dendrites. Clusters of slit sense organs are called lyriform organs because they look like a stringed instrument called a lyre. Slit sense organs, or slit sensilla serve multiple functions. Some slit sensilla are sensitive to airborne sounds, some probably serve as gravity receptors, some respond to vibrations within a certain frequency, others respond to movements of the spider's own extremities.

Open-ended hairs at the ends of the legs and pedipalps serve as taste receptors. Spiders will avoid pungent or acidic substances upon contact. An adult spider can have over 1,000 contact chemoreceptors, or taste hairs, on the tarsal segments of its legs. These contact chemoreceptors probably enable the spider to tell good-tasting prey from bad and enable males to detect female pheromones that may be deposited in their silk. It is clear that spiders can also respond to airborne odors, or smells. It is thought that the 'noses' of a spider are small, open pits on the dorsal side of each tarsus. These pits are called tarsal organs. Spiders use these tarsal organs to respond to airborne cues from prey and mates. Although the majority of spiders are solitary predators often living and hunting far away from conspecifics, when a female matures, males are able to quickly locate her. This localization is almost certainly done via airborne pheromones or chemical cues that females deposit either on the substrate or in their silk.

Vision.   Most spiders have eight eyes arranged as two rows, although some families are characterized by spiders with only six or two eyes and some cave dwelling species have lost their eyes altogether. Often, the eye arrangement is diagnostic for a particular family. For example, jumping spiders (Salticidae) and wolf spiders (Lycosidae) can be identified to the family level solely by their eye pattern. The

Fig. 954   Scopula and claw tufts of a mygalomorph spider (drawing by Eric Parrish, University of Colorado).

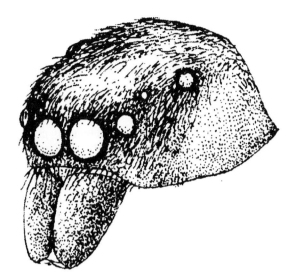

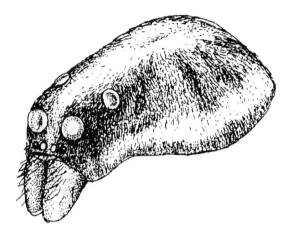

Fig. 956 Eye pattern of a wolf spider in the family Lycosidae (drawing by Eric Parrish, University of Colorado).

Fig. 955 Eye pattern of a jumping spider in the family Salticidae (drawing by Eric Parrish, University of Colorado).

structure of a spider's eyes is similar to the ocelli of insects. Spiders have simple eyes with a single outer lens, an inner cellular vitreous body and a layer of visual and pigment cells which, together, make up the retina. For the vast majority of spiders, vision is not the primary sense they rely upon in their daily lives. Most spiders are nocturnal and, in the low light levels of the night, they rely primarily on tactile and chemical cues. Web building spiders, in particular, have what, by vertebrate standards, would be considered poor vision. They respond to changes in light intensity, or to changes in the polarization of incoming light, but probably cannot see fully formed images of objects. However, wandering spiders or cursorial hunters, such as wolf spiders, jumping spiders and crab spiders are much more reliant on vision both to catch prey and to court mates.

Jumping spiders, in particular, have exceptional eyesight. They can probably see in color and can perceive images of objects 8 to 10 cm from them. A jumping spider will respond to movement of objects up to 40 cm away. These spiders are diurnal and can often be found hunting on the sides of houses, fences, barns or other structures. When a human moves close to a jumping spider, the spider often responds to the movement and orients its body in order to face the person. Their large main eyes, or anterior median eyes, and their responsiveness to the slightest movement gives these spiders a very per-

sonable countenance (at least to arachnophiles). Salticids have elaborate courtship behaviors involving zigzag dances and specific movements of the front legs and pedipalps. The bright colors of many salticids probably serve a role in mate choice.

Wolf spiders also have keen vision. Most wolf spiders are crepuscular, or nocturnal. These spiders have a light-reflecting crystalline structure called a tapetum behind the retinal cells of the secondary eyes (secondary eyes include all eyes except the anterior median eyes). It is thought that the tapetum functions to gather incident light in low light conditions. The tapetum also allows arachnologists to hunt wolf spiders at night using headlamps. The light from the headlamp reflects off the tapetum so spider hunters can find wolf spiders by their eyeshine. Diurnal hunters, such as salticids, lack tapeta. Wolf spiders also have keen vision and actively pursue their prey and their mates. They also have elaborate courtship behaviors involving visual signals. In addition, some wolf spiders incorporate vibratory cues in their courtship rituals by tapping their legs and pedipalps against the substrate.

Central nervous system. Whereas insects have interconnected ganglia throughout their bodies, spiders have only two major ganglia. Both nerve centers are located in the cephalothorax: the supra- and the subesophageal ganglia. The subesophageal ganglion consists of nerves controlling the legs and pedipalps. A nerve bundle called the cauda equina arises from the subesophageal ganglion and enters the abdomen. The supraesophageal ganglion, located, as the name

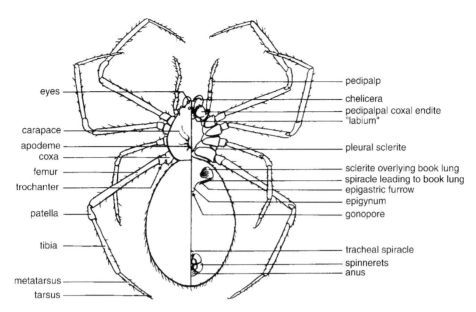

Fig. 957 Composite drawing of typical spider with principal body parts labeled. Left half is dorsal view; right half is ventral view.

implies, above the spider's esophagus, includes the nerves to the chelicerae, the pharynx, the poison glands and the eyes. The supraesophageal ganglion is considered the spiders 'brain.'

Although some aspects of a spider's behavior are 'pre-programmed' or instinctive, in many respects, spiders show a wide range of behavioral plasticity in their response to environmental cues – spiders, in other words, can think. Perhaps they do not think in the way that humans define that term, but they certainly show more behavioral flexibility than one would expect of an animal showing purely instinctive responses to cues.

A good example of this behavioral plasticity is seen with the jumping spider in the genus *Portia* found in New Zealand. Although most jumping spiders hunt primarily insects, *Portia* specializes in hunting other species of spiders as well as spider egg sacs and insects. *Portia*, as do all jumping spiders, has very keen vision. If it is stalking an insect, its behavioral hunting strategy is similar to that of any other jumping spider. Most insects that *Portia* hunts cannot fight back very effectively so the spider uses stealth, but not caution, in its hunt. However, if *Portia* is stalking another spider, the tables can turn quickly and hunter can become hunted. *Portia*'s hunting strategy changes commensurate with the type of prey. If the prey is a web building spider, *Portia* will approach the web and pluck it to mimic an insect caught in the silk. If the web builder responds by approaching the vibrations, *Portia* pounces. However, if the web builder does not respond, or if the web builder is larger than *Portia*, the jumping spider will try different vibratory cues or different hunting tactics until its behavior is effective in luring the web builder close enough to kill. *Portia* is even known to detour around a particularly dangerous prey in order to approach from a direction from which *Portia* itself cannot be attacked. A spider's brain may only be the size of a pinhead, but the integration of electrical signals from the spider's appendages and eyes can result in extraordinarily complex behaviors.

Sex among the spiders. The internal sexual organs of spiders – the testes and ovaries – are located in the abdomen. However, male spiders have no intromittent organs associated with the genital opening. As with most arachnids, spiders have indirect sperm transfer. With primitive groups of arachnids, such as scorpions, the male deposits a stalked sperm packet, or spermatophore, on the substrate and, using complex courtship behaviors, manipulates the female over the spermatophore. Male spiders have evolved a more direct strategy for ensuring that their spermatophore makes it into the female.

Fig. 958 Some common spiders: top row left, black widow, *Latrodectus mactans* (family Theridiidae); top row right, brown recluse, *Loxosceles reclusa* (family Sicariidae); second row left, golden silk, *Nephila clavipes* (family Tetragnathidae); second row right, green lynx, *Peucetia viridans* (family Oxyopidae); third row left, a jumping spider, *Plexippus paykulli* (family Salticidae); third row right, a long-jawed orb-weaver, *Tetragnathus* sp. (family Tetragnathidae); bottom row left, a tropical orb-weaver, *Eriophora ravilla* (family Araneidae); bottom row right, a spiny orb-weaver, *Gasteracantha cancriformis* (family Araneidae). All photos by J. Castner, University of Florida.

Male spiders can be distinguished from females by the shape of their pedipalps. Male spiders look like they are wearing boxing gloves on the end of their pedipalps. When a male spider matures, he builds a small platform of silk called a sperm web. In some groups, the sperm web consists of just a few lines of silk. Onto this sperm web, the male deposits a droplet of sperm from his genital opening. He then presses a tube associated with the pedipalp called the embolus into this droplet of sperm and sucks the sperm into the bulb of the pedipalp where it is stored until the male mates. The pedipalps of entelegyne spiders are much more complex than those of haplogynes. However, mating in all groups of spiders requires the expulsion of the sperm stored in the pedipalp into the female's genital opening. In entelegyne spiders, complex structures on the male pedipalp correspond to structures located on the female's epigynum.

The majority of spiders are solitary predators and will opportunistically feed on whatever prey is available, even if the prey is of its own species. A male spider must, therefore, be cautious when approaching a female and must effectively communicate to her that he is a mate and not a meal. This is particularly important among web building spiders because there is often dramatic dimorphism between males and females with males being considerably smaller than their female conspecifics. When a male web building spider locates a female, probably via airborne pheromones, he sings a silk song to her communicating that he is a mate and not an insect caught in the web. If the female is mature and is willing, she will respond to his vibrations in a species specific manner. If she is not responsive to mating, she may try to attack the male.

Male wandering spiders, such as wolf spiders or jumping spiders, dance for their lady loves. The courtship behaviors of male jumping spiders are particularly striking in that they often involve zigzag dances, and waving of the legs and pedipalps. Wolf spiders may add a vibratory component by tapping their pedipalps on the ground. Some male spiders have stridulatory structures on their bodies that are probably involved in courtship.

When a male's courtship behaviors have been accepted by a female, he will approach her cautiously and may stroke her. Finally, he will insert his embolus into her genital opening to inject the sperm. The female has seminal receptacles where the male's sperm is stored. Some females will mate multiple times with the same or with different males. In some species, depending on the location of the seminal receptacles in relation to the oviducts, the first male to mate with the female fertilizes most of her eggs, whereas in other species, the last male's sperm takes precedence. In some species, the male will deposit a mating plug into the female's genital opening in an attempt to prevent her from mating with other males. Sometimes, the mating plug is the broken tip of the male's embolus.

When male web building spiders mature, the silk glands responsible for creating prey capture silk become non-functional. In other words, the aggregate

Fig. 959 Trapdoor spider (above) and burrow with door closed (middle) and open (below) (photos by author).

and flagelliform glands of male ecribellate spiders become non-functional and the glands associated with the cribellum of cribellate spiders become non-functional. When male web building spiders mature, they are no longer thinking about food, but only of true love. Partly as a consequence of this, male web building spiders have shorter life spans then their conspecific females. This difference in life span holds true for the majority of spiders including mygalomorphs.

Brood care, hatching and dispersal. Within a few weeks after copulating, female spiders are ready to lay their eggs. The stored sperm is released and fertilization occurs in the oviduct just prior to egg emergence from the genital opening. The female builds a silken egg sac to protect the eggs. In some spiders, the egg sac is an elaborate structure with thick walls of silk. However, some spiders, such as those in the daddy longlegs or cellar spiders family (Pholcidae) (which share a common name with the arachnids in the order Opiliones) attach the eggs together with just a few silk strands. Most females can produce multiple egg sacs, each containing from a few to over 1,000 eggs depending on the species. In many species, the female guards the egg sac against egg sac parasitoids and predators until hatching. Female spiders in the family Uloboridae, for instance, remain in contact with the suspended egg sac, brushing off any insect that might land or walk on it. Female wolf spiders (Lycosidae) attach the silken, disc-shaped egg sacs to their spinnerets and carry them around until the egg sacs hatch. Female fishing spiders (Pisauridae) carry their egg sacs in front, holding onto the silk with their chelicerae. Female jumping spiders (Salticidae) remain in silken retreats with their egg sacs until they have hatched.

Within a few weeks, the egg sacs hatch except in those species that overwinter. In some cases, as with wolf spiders, the mother assists her young by cutting a hole in the egg sac to facilitate emergence. In most cases, the offspring cut their own way out of the silken structure. After hatching, some females will continue to guard and care for their young. For example, wolf spiderlings, upon hatching, crawl up onto their mother's back. Wolf spider females have specialized knobbed hairs on the back of the abdomen to which the babies cling. The female carries her offspring on her back until the yolk supply from the egg stage is depleted and they are old enough to hunt for themselves. In a very few species, the mother

actually provides food for her newly hatched offspring either in the form of prey the mother has killed, or in the form of regurgitated food.

After one or two weeks, the spiderlings of most species are ready to disperse. Ballooning is one common method of dispersal seen among araneomorph spiderlings. These spiderlings crawl to the top of a blade of grass, a twig, a branch or some other structure and release a strand of silk. Air currents catch the silk, often called gossamer, and lift the tiny spider up, up and away. Although some of these 'flying spiders' may land only a meter or two away from the take-off point when their silk gets tangled in the branches of a nearby tree, others can travel truly extraordinary distances by this means of aerial dispersal. Some ballooning spiders have been found on ships far out at sea and have been collected from airplanes. Because of this ability to travel long distances by ballooning, spiders are often one of the first pioneer species to establish themselves on distant oceanic islands. However, lack of food and water probably makes life difficult if not impossible for these pioneer predators. In a few families of very tiny spiders, even the adults will sometimes move to a new habitat by means of ballooning. In other spider families, including some mygalomorphs, dispersal mechanisms are much more mundane. The spiderlings, when they are old enough to hunt, simply wander away from natal ground and establish themselves in a new area.

Spiderlings look much like the adult stages except that they are often paler and have no fully developed sexual organs. The complex structures of the male pedipalp do not appear until the final molt, nor do the structures associated with the female's epigynum. The palpal and epigynal characters are the structures used by arachnologists to identify spiders to species. Tiny spiders require about five molts to reach maturity, whereas larger species can require up to 10 molts. Araneomorph spiders do not undergo any further molts after reaching maturity; mygalomorph females, however, continue to molt all their lives. The molting process in spiders is similar to that in insects. As with insects, all chitinous structures, including such internal structures as the trachea, the book lungs and, with mygalomorph females, the linings of the seminal receptacles, are molted. For adult mygalomorph females, this means that any sperm stored in the seminal receptacles when they molt is lost and the female becomes newly virginal (the one group of animals that can actually regain its virginity!).

Fig. 960 Some spider webs: funnel web (upper left), theridiostomatid web (lower left), and hanging web (right) (photos by author).

Most araneomorph spiders either drop down on a silken thread, or enter a silken retreat or burrow when they are getting ready to molt. Mygalomorphs turn over onto their backs prior to molting. Many a pet tarantula owner has thrown away their pet thinking it was dead when it was just getting ready to put on a new skin. As with all arthropods, molting is a dangerous time for the spider. The spider is vulnerable to predators during the molting process and just after completing the molt when the new exoskeleton is not completely sclerotized.

Most male spiders, araneomorphs and mygalomorphs, die shortly after maturation. In temperate regions, most species live only one or two years. The majority of adults die soon after the first hard frost. In tropical regions, lifespans can vary and there can be multiple generations of one species at any time. The longest lived spiders are some mygalomorph species that have recorded lifespans in the laboratory of up to 30 years for females and up to about 10 years for males. In temperate regions, some spiders overwinter in the egg sac stage – the development arrested until warmer temperatures in the spring. In other species, the spiderlings overwinter by seeking protection in warmer microhabitats in cracks, crevices, or in the leaf litter layer.

Spiders as predators. Spiders are found in every terrestrial ecosystem on earth except Antarctica.

Although a few spiders can obtain some nutritional value from pollen on occasion, all species of spiders, including the occasional pollen eater, are overwhelmingly predatory in their lifestyle. They are one of the top invertebrate predators in terrestrial communities. An arachnologist, William S. Bristowe, once estimated that a one acre field (0.4 hectare) in England could be home to over two million spiders. Although no volunteers have been found to empirically verify this extraordinary number, spiders are in high enough numbers to affect the populations of their prey species. Most spiders are generalist predators and, because of their euryphagous diet, have not been considered useful biocontrol agents in agroecosystems. However, more and more studies have indicated that spiders can, in fact, suppress pest populations. Exploring the usefulness of spiders as biocontrol agents in integrated pest management is a burgeoning field of research.

See also, SPIDER BEHAVIOR AND VALUE IN AGRICULTURAL LANDSCAPES, PREDATION: THE ROLE OF GENERALIST PREDATORS IN BIODIVERSITY AND BIOLOGICAL CONTROL, CONSERVATION BIOLOGICAL CONTROL.

Paula E. Cushing
Denver Museum of Nature and Science
Denver, Colorado, USA

References

Coddington, J. A., and H. W. Levi 1991. Systematics and evolution of spiders (Araneae). *Annual Review of Ecology and Systematics* 22: 565–592.

Foelix, R. F. 1996. *Biology of spiders* (2nd ed.). Harvard University Press, Cambridge, Massachusetts.

Hillyard, P. 1994. *The book of the spider: a compendium of arachno-facts and eight-legged lore.* Avon Books, New York, New York.

Nentwig, W. (ed.). 1987. *Ecophysiology of spiders.* Springer-Verlag, Berlin, Germany.

Platnick, N. I. 2002. The world spider catalog, version 2.5. American Museum of Natural History, online at http://research.amnh.org/entomology/spiders/catalog81-87/index.html.

Shear, W. A. 1986. *Spiders: webs, behavior, and evolution.* Stanford University Press, Stanford, California.

Wise, D. H. 1993. *Spiders in ecological webs.* Cambridge University Press, Cambridge, Massachusetts.

SPINDLE-SHAPED. Elongate, cylindrical, thickened at the middle and tapering at the ends.

SPINE. A large, stout seta or thorn-like process extending from the integument.

SPINED SOLDIER BUG, *Podisus maculiventris* (HEMIPTERA: PENTATOMIDAE). The spined soldier bug, *Podisus maculiventris* (Say), is the most common predatory stink bug in North America. It is naturally distributed from Mexico, the Bahamas and parts of the West Indies into Canada. The insect occurs in a variety of natural and agricultural ecosystems, such as shrubs, woods, streambanks, orchards and several field crops.

Adult spined soldier bugs are pale brown to tan in color and are 10 to 14 mm long, with the females somewhat larger than the males. They are shield-shaped and have a prominent spine on each ''shoulder.'' The adults have a distinct dark spot on the membranous tip of each forewing. The adults, as well as the nymphs, have a thickened rostrum or beak which they can extend fully forward to feed on prey, and which they keep folded under their body when not feeding. The females start ovipositing about 1 week after emerging and usually lay eggs throughout their entire lifetime. The adults are long-lived, with reported longevities ranging from 1 to 4 months under laboratory conditions. Fecundities measured in the laboratory range from 300 to over 1,000 eggs per female, mainly depending on the prey supplied.

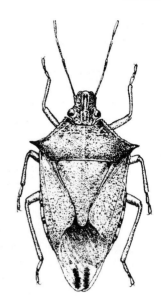

Fig. 961 *Podisus maculiventris.*

The eggs are deposited in loose oval clusters, each containing on average 15 to 30 eggs. The eggs are about 1 mm in diameter and 1 mm long, are cream-colored to black in color and have a ring of about 15 long micropylar processes around the operculum on top. The duration of the egg stage is 5 to 8 days, depending on the temperature.

There are five nymphal instars. The nymphs are round to oval rather than shield-shaped. Young nymphs are red and black in color, whereas older nymphs are marked with black, red, yellow-orange and cream bands and patches. The wing pads become clearly visible from the fourth instar on. The length of the respective instars averages 1.3 to 1.5 mm, 2.5 to 3 mm, 3.5 to 4 mm, 5 to 6 mm and 7 to 9 mm. First instars are highly gregarious and hardly move about. In later instars, the nymphs become progressively more solitary with each molt. First instars are not predaceous and usually only take up moisture, although they have been observed feeding on unhatched eggs from the same mass. The nymphs start actively searching for insect prey shortly after molting to the second instar. The nymphs (and adults) become cannibalistic when food is short. The nymphal stage requires 20 to 30 days for complete development, depending on the temperature and on the availability and quality of food.

Development from the egg to the adult stage typically takes 4 to 5 weeks at 20 to 25°C. A lower threshold temperature for immature development is approximately 12°C. In Canada and in central and northern parts of the United States, the spined soldier bug usually has 2 to 3 generations per year and hibernates as an adult from October to April. Overwintering takes place in litter, under tree bark, under stones, etc. In warmer climates, like in the southern United States, the insect is active all year.

Although in natural habitats it is primarily found in association with soft-bodied larvae of plant-feeding Lepidoptera, Coleoptera and Hymenoptera, the spined soldier bug is a strongly generalist predator, reportedly attacking more than 90 insect species from 8 orders. Nonetheless, there may be important differences in developmental and reproductive success depending on the prey species or stage supplied. The nymphs and the adults of the spined soldier bug have frequently been observed feeding on plant juices. This plant-feeding habit appears to be particularly important when prey is scarce. Plant-feeding primarily provides moisture, but it may also furnish certain nutrients to the bugs at critical times. Plant-feeding by spined soldier bugs has not been reported to cause plant damage.

Several studies have favorably assessed the predation capacity of the spined soldier bug on various insect pests, including the Colorado potato beetle, *Leptinotarsa decemlineata*, the Mexican bean beetle, *Epilachna varivestis*, the pine sawfly, *Diprion similis*, the fall webworm, *Hyphantria cunea*, the cotton leafworm, *Alabama argillacea*, the cotton bollworm, *Helicoverpa zea*, the tobacco budworm, *Heliothis virescens*, and the beet armyworm, *Spodoptera exigua*. From the 1930s up to the 1980s, numerous attempts have been made to introduce the spined soldier bug into different parts of Europe for the control of the Colorado potato beetle, but the predator never became established, probably because of its inability to overwinter. The predator is currently being used in augmentative biological control of leaf-feeding caterpillars (mainly belonging to Noctuidae) in European and North American greenhouse crops. For augmentative releases, the nymphs are preferred to the adults, because the latter tend to fly quickly from the treated plots. Two releases of 0.5 fourth-instars per square meter in commercial glasshouses in The Netherlands proved to be effective in keeping noctuid caterpillars in check on sweet peppers, eggplants and some ornamentals. Hot spots of aggregating caterpillars have been treated successfully at densities of 5 to 10 nymphs per square meter. In field crops in North America, natural populations of the spined soldier bug are often too low to effectively suppress outbreaks of pests like the Colorado potato beetle, particularly early in the season. Natural populations may be augmented by releasing laboratory-reared individuals. In small field plots, releases of about 3 second- to third-instars per potato plant reduced high densities of Colorado potato beetle eggs and larvae and provided good foliage protection. However, the cost of rearing and releasing the predator on a large scale may be prohibitive. Alternatively, a synthetic pheromone can be used to lure spined soldier bugs to target areas in early spring, as adults emerge from overwintering. The pheromone can also be employed to capture large numbers of predators that can be used to establish mass cultures.

Podisus maculiventris is easily reared in the laboratory on different factitious prey, including larvae of the yellow mealworm, *Tenebrio molitor*, and the greater wax moth, *Galleria mellonella*. Wax moth larvae appear to be a particularly good food, yielding good survival, rapid development and high fecundity. However, given the great voracity of the predator, extensive parallel cultures of prey insects are needed, rendering its mass production less economical. An artificial diet based on beef is available to support the production of consecutive generations of the bug, but development is somewhat prolonged and fecundity is about half of that obtained when wax moth larvae are the primary food.

Although the spined soldier bug has been commercially available in Europe and North America since the 1990s, wider adoption of this beneficial in augmentative biological control is hindered mainly by its high production cost. In Europe, there is also some concern that releases of this and other exotic generalist predators may have detrimental effects on nontarget, native fauna.

Patrick De Clercq
Ghent University
Ghent, Belgium

References

Couturier, A. 1938. Contribution à l'étude biologique de *Podisus maculiventris* Say, prédateur américain du Doryphore. *Annales Epiphyties Phytogénétique* 4: 95–165.

De Clercq, P. 2000. *Podisus* online. http://allserv.uget.be/padclerc/

De Clercq, P. 2000. Predaceous stinkbugs (Pentatomidae: Asopinae). pp. 737–789 in C. W. Schaefer and A. R. Panizzi (eds.), *Heteroptera of economic importance*. CRC Press, Boca Raton, Florida.

O'Neil, R. J. 1995. Know your friends: spined soldier bug. Midwest Biological Control News, vol. II, no. 8. http://www.entomology.wisc.edu/mbcn/kyf208.html

Weeden, C. R., A. M. Shelton, and M. P. Hoffmann 1998. Biological control: a guide to natural enemies in North America. *Podisus maculiventris* (Hemiptera: Pentatomidae). http://nysaes.cornell.edu/ent/biocontrol/predators/podisus.html

SPINY SHORE BUGS. Members of the family Leptopodidae (order Hemiptera). See also, BUGS.

SPHINGIDAE. A family of moths (order Lepidoptera). They commonly are known as sphinx, hawk or hummingbird moths, or hornworms. See also, HAWK MOTHS, BUTTERFLIES AND MOTHS.

SPINNERET. A structure that produces silk in immature insects. The spinneret is often a finger-like structure.

SPHINX MOTHS. Some members of the family Sphingidae (order Lepidoptera). See also, HAWK MOTHS, BUTTERFLIES AND MOTHS.

SPIDER BEETLES. Members of the family Ptinidae (order Coleoptera). See also, BEETLES.

SPIDER MITES. See also, MITES.

SPIDER WASPS. Members of the family Pompilidae (order Hymenoptera). See also, WASPS, ANTS, BEES, AND SAWFLIES.

SPINE-TAILED EARWIGS. Members of the earwig family Forficulidae (order Dermaptera). See also, EARWIGS.

SPINOSE EAR TICK. See also, TICKS.

SPINY RAT LICE. Members of the family Polyplacidae (order Siphunculata). See also, SUCKING LICE.

SPIRACLE. (pl., spiracles or spiraculae) An external opening of the system of ducts (trachea) used to transfer atmospheric gases into, and out of, the body of arthropods. They are commonly found along each side of the body.
See also, TRACHEAL SYSTEM AND RESPIRATORY GAS EXCHANGE.

SPIRACULAR BRISTLES. Bristles found in close proximity to a spiracle.

SPIRACULAR GILL. A gill formed by an extension of a spiracle, or a plastron around a spiracle.
See also, TRACHEAL SYSTEM AND RESPIRATORY GAS EXCHANGE.

SPIRACULAR PLATE. A plate found adjacent to or surrounding a spiracle.

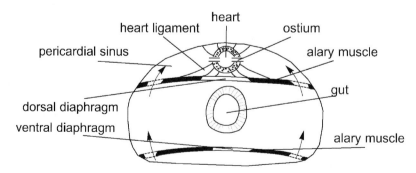

Fig. 962 Cross section of the abdomen of an insect, showing the principal trachea and tracheal connections (adapted from Chapman, The insects: structure and function).

SPIREA APHID. See also, CITRUS PESTS AND THEIR MANAGEMENT.

SPITTLEBUGS (HEMIPTERA: CERCOPOI-DEA). The superfamily Cercopoidea belongs to the suborder Auchenorrhyncha of the order Hemiptera. It includes four taxa, the family Cercopidae, Aphrophoridae, Clastopteridae and Machaerotidae, which encompass the spittlebugs proper and the tube-dwelling machaerotids. Some taxonomists still include all four groups as subfamilies within the traditional unified family Cercopidae.

The Clastopteridae (about 85 described species in one genus) and Machaerotidae (about 105 species in 27 genera) are sister groups, limited to the New and Old World, respectively. The Machaerotidae are predominantly tropical, with a handful of temperate species in China and Japan. The Clastopteridae are abundant from Canada to Argentina, all in the single speciose genus *Clastoptera* (the status of an additional isolated Philippine genus is problematic). The Aphrophoridae (about 820 species in 150 genera) probably represent a polyphyletic group, in which the Cercopidae constitute one monophyletic branch. The Aphrophoridae are best represented in north temperate habitats, but also occur widely in the tropics. There are approximately 1,360 described species in the Cercopidae in 140 genera, and a recent revision lists 416 of those species distributed in the New World. This group is predominately tropical. Only two species, *Prosapia bicincta* and *P. ignipectus*, occur north of Mexico in the Americas, and only two genera, *Cercopis* and *Haematoloma*, occur in Europe.

Adult spittlebugs, known as froghoppers for their quick jumps and putative resemblance to tiny frogs, are named in parallel to other Auchenorrhyncha such as the leafhoppers, planthoppers and treehoppers. They are distinguished from these groups by 1 or 2 strong spines on the metatibia, a crown of spines at the distal end of the tibia, and nymphs that live in spittle masses or calcareous tubes filled with liquid.

Biology and behavior

Nymphs and adults of the Cercopoidea have sucking mouthparts and feed on the xylem sap of host plants. Compared to phloem sap, xylem sap is a challenging food resource because it is normally under negative pressure and offers a very dilute source of organic nutrients. Special adaptations to this resource include a filter chamber shunt to manage high volumes of fluid and an inflated clypeus and reinforced apodemes to support the enlarged sucking muscles. Other consequences include a relatively long life cycle and unique defensive opportunities for the nymphs compared to phloem-feeding Auchenorrhyncha.

The Cercopoidea have five nymphal instars. Nymphs in the Clastopteridae, Aphrophoridae and Cercopidae construct spittle masses at their feeding sites and gain protection from desiccation and most natural enemies. The spittle is composed of bubbles added to excreted, attenuated xylem sap fortified with mucopolysaccharides that are secreted by the Malphigian tubules. Nymphs form the bubbles one-by-one with repeated thrusts of the tip of the abdomen out of the excreted liquid to pick up air in a special pocket formed by the ventral abdominal plates. The *Clastoptera* tend to have very aqueous spittle masses, while those of the Aphrophoridae are frothy and those of the Cercopidae intermediate. Typically there is a single nymph per spittle mass, but sometimes two or more nymphs cohabit one mass. On occasion, nymphs of two species share a common spittle mass. The widespread Neotropical aphrophorid species *Cephisus siccifolius* forms large communal spittle masses with dozens of nymphs. Machaerotid nymphs live in liquid-filled calcareous tubes constructed at the feeding site. They form a spittle mass, at the mouth of the tube, only in conjunction with the adult molt.

Adult spittlebugs do not make spittle masses. Like nymphs, they excrete large volumes of sap, but flick it away in small drops, so it does not accumulate. Large species may drip like faucets when actively feeding. In Africa, species of the aphrophorid genus *Ptyelus* drip so copiously that it appears to be raining beneath their dry season 'raintree' hosts.

Adult Aphrophoridae, Clastopteridae and Machaerotidae depend on camouflage or disruptive coloration and jumping as defensive mechanisms. The Cercopidae exhibit a unique shared character, reflex bleeding. When threatened, adult Cercopidae emit odoriferous orange hemolymph from the tips of their feet, helping to repel attackers. This defense is accompanied by warning coloration in a large proportion of taxa. Many species exhibit striking red, yellow, orange or black markings, making them conspicuous in their natural surroundings.

Substrate communication plays an important role in courtship and other behaviors among adult Cercopoidea. Tymbals located on the first abdominal segment produce vibrations that are transmitted through the host plant. In the Cercopidae, male courtship calls vary in pulse structure (simple to compound), duration, peak frequency and pulse repetition frequency. A receptive female transmits her response via the substrate to guide males to copulation. There is also evidence for other classes of vibrational communication, such as distress calls.

Most temperate spittlebugs are univoltine. A few have two generations per year. Most tropical spittlebugs are multivoltine, with life cycles tied to seasonal rainfall patterns.

Ecology

Spittlebugs feed on an unusually wide range of plants. Very generally speaking, the Cercopidae tend to feed on herbaceous monocots, the Clastopteridae and Machaerotidae on flowering trees and shrubs, and the Aphrophoridae on conifers and herbaceous dicots. The are many exceptions to these generalizations, however, and some species feed on an extraordinarily wide range of hosts. *Philaenus spumarius*, the aphrophorid meadow spittlebug, may have more recorded hosts than any other phytophagous insect. Presumably, wide host range is a consequence of xylem feeding, xylem sap being chemically less defended and rather similar in composition across plant groups. Within this diversity of hosts, many spittlebugs have an affinity for nitrogen-fixing plants that may provide a richer and more constant source of xylem sap nutrition. The Cercopidae, for instance, have a predilection for plants that have associative nitrogen fixation through root zone bacteria, while many Aphrophoridae prefer legumes, and many *Clastoptera* prefer actinorhizal hosts. Other factors that influence host plant or feeding site selection include concentration of amino acids, tissue toughness, depth of xylem elements, presence of trichomes and growth habit or architecture.

Two species of the genus *Mahanarva* in Central America have aquatic nymphs that live in the flowers of *Heliconia* plants, the only known truly aquatic spittlebugs. Nymphs of several species of Aphrophoridae and *Clastoptera* as well as one machaerotid have associated Diptera (Drosophilidae: *Cladochaeta*, *Leucophenga*, *Paraleucophenga*) larvae that share their spittle masses. The precise nature of these associations, whether parasitic or commensal, is not fully understood.

In the temperate zone several aphrophorid species (primarily of the genera *Aphrophora* and *Peuceptyelus*) are serious pests of conifers. *Philaenus spumarius* is a pest of alfalfa and other legume forage crops, as well as strawberries. *Clastoptera* species attack cacao, pecans and citrus but are only rarely of serious concern. The cercopid species *Haematoloma dorsatum* is a pest of pines in Southern Europe and, as elaborated below, the grass-feeding Cercopidae as a group are serious pests in Latin America.

As xylem feeders, spittlebugs are infrequent disease vectors, but all tested species transmit the xylem-limited bacterium *Xylella fastidiosa*, which causes Pierce's disease of grape and diseases of several other plants. *Xylella fastidiosa* is a serious concern in California and other areas where this bacterium poses a threat to commercially important crops. Spittlebugs also transmit the bacterium that causes Sumatra disease of cloves and the mycoplasma that causes stunt disease of *Rubus*.

Grass-feeding Neotropical spittlebugs

As pests, the Neotropical grass-feeding spittlebugs of the cercopid subfamily Ischnorhininae are by far the most important Cercopoidea. This economically critical subset includes dozens of species from at least 11 genera (*Aeneolamia, Deois, Isozulia, Kanaima, Mahanarva, Maxantonia, Notozulia, Prosapia, Sphenorhina, Tunaima, Zulia*). These insects are destructive pests of forage grasses and sugar cane, and occasionally other cultivated graminoids, such as turfgrass and rice. The following sections relate specifically to this well-studied group.

Life cycle. Eggs go through four generalized developmental stages, completing development in 2 to 3 weeks. In the case of diapause and quiescence, time to hatching can be prolonged up to 530 days. Although soil is the most common oviposition site, litter and plant stem surfaces are used by some species and favored in preference to soil by others.

The five nymphal instars can be reliably differentiated by morphological measurements, particularly the width of the head capsule. Recently emerged first instar nymphs usually establish spittle masses on roots at the soil surface or in cracks and pores of the upper layer of soil. While the nymphs of most species remain near the soil, others migrate upwards into leaf axils on more erect hosts. In sugar cane, for instance, *Aeneolamia postica* nymphs emerge from eggs in the soil and complete development on surface roots, while *Mahanarva andigena* nymphs emerge from eggs on litter or the base of old leaf sheaths and favor aerial spittle masses in leaf whorls. Development time varies with species and climate, ranging from 4 to 9 weeks. Duration of the adult phase ranges from 1 to 3 weeks.

Adults usually exhibit sexual dimorphism, with males smaller and often more brightly colored than females. Some species also exhibit multiple color forms within one or both sexes, while unrelated species often share common color patterns, complicating species identification. The styles, plates and aedeagus of the male genitalia are the most diagnostic taxonomic features.

Agricultural ecology. Nymphs and adults of grass-feeding spittlebugs occur during the wet season. Most species pass the dry season as drought tolerant diapause eggs. Depending on species, habitat and duration of the wet season, they achieve 1 to 6 generations per year. In highly seasonal sites, the return of the wet season promotes synchronous hatching and damaging early season outbreaks. Population dynamics are characterized by high population fluctuations and population synchrony. In continuously humid sites, all life stages are present throughout the year, fluctuations and synchrony are reduced, and yield loss can be less severe.

The most widely reported natural enemies are fungal entomopathogens, predaceous larvae of the pan-Neotropical syrphid fly *Salpingogaster nigra* that attack nymphs in their spittle masses, parasitic nematodes (Mermithidae), and generalist robber flies (Asilidae). Egg parasites (Eulophidae, Mymaridae, Trichogrammatidae), parasitic flies (Pipunculidae) and parasitic mites (Erythraeidae) also have been reported.

Economic impact. Grass-feeding spittlebugs are widely distributed in the Neotropics and are economically important in lowlands and highlands from the southeastern U.S. to northern Argentina. They attack all economically important genera of tropical forage grasses, particularly species of African origin. Adult feeding causes a phytotoxemia identified by a chlorosis that spreads from the feeding site. Nymphs cause damage similar to water stress, identified by browning spreading from the leaf tip. Severe outbreaks result in the yellowing off of the entire aboveground portion of the plant. This damage lowers forage production,

quality, and palatability, reduces milk and beef production, inhibits the establishment and persistence of improved pastures, and increases soil degradation.

Pest management. Management of spittlebugs in forage grasses and sugar cane is still haphazard, given the challenges posed by the diverse situations in which they occur. Due in part to incomplete information on the biology and habits of the specific pest species, there has been a tendency to over-generalize the spittlebug/host/habitat interactions to the detriment of control efforts. In pastures, controlled grazing, effective fertilization, burning, host plant selection and diversification of forage species offer possibilities for control. A long-term breeding program is achieving successes for spittlebug resistance in *Brachiaria* grasses, the most widely planted forage grasses in South America and the most susceptible to spittlebug attack. In sugar cane, control measures include cultural control tactics and applications of commercial fungal entomopathogens. Management successes in both systems are limited by poorly developed IPM decision tools such as yield loss assessments, sampling schemes and thresholds, in addition to the fundamental challenges inherent to IPM in extensive and low value agroecosystems such as rangelands.

In systems with highly seasonal precipitation, the recommended management strategy is temporal and spatial identification of the early season nymph outbreaks to target control tactics that will suppress population development before the invasion of uninfested areas and the propagation of future generations. In less seasonal systems, the recommended management strategy will depend on cultural control techniques such as host plant selection and controlled grazing, and there will be more options for biological control.

Daniel C. Peck
International Center for Tropical Agriculture
Cali, Colombia
and
VintonThompson
Roosevelt University
Chicago, Illinois, USA

References

Carvalho, G. S., and M. Webb. Review of the New World spittlebugs (Hemiptera, Auchenorrhyncha, Cercopidae). (In preparation)

Hamilton, K. G. A. 1982. *The spittlebugs of Canada (Homoptera: Cercopidae).* The insects and arachnids of Canada, Part 10. 102 pp.

Peck, D. C. 1999. Seasonal fluctuations and phenology of *Prosapia* spittlebugs (Homoptera: Cercopidae) in upland dairy pastures of Costa Rica. *Environmental Entomology* 28: 372–386.

Thompson, V. 1994. Spittlebug indicators of nitrogen-fixing plants. *Ecological Entomology* 19: 391–398.

Valério, J. R., S. L. Lapointe, S. Kelemu, C. D. Fernandes, and S. J. Morales 1996. Pests and diseases of *Brachiaria* species. pp. 87–105 in J. W. Miles, B. L. Maass, and C. B. do Valle (eds.), *Brachiaria: biology, agronomy and improvement*. CIAT, Cali, Colombia.

SPLENETIC FEVER. See also, PIROPLASMOSIS.

SPONGILLAFLIES (NEUROPTERA: SISYRIDAE). The insect order Neuroptera is primarily a terrestrial group. Unique among the neuropterans, however, is the family Sisyridae that have aquatic larvae. The common name of this family is spongillaflies and is derived from the predaceous, and arguably parasitic, feeding of the larvae on freshwater sponges (Porifera: Spongillidae). Spongillafly larvae occur in various types of freshwater habitats including streams, lakes and impoundments. Comparatively little is known about the biology of sisyrids in relation to other aquatic insect groups largely due to their unique larval habitats and their relative rarity. The classification is as follows:

Order: Neuroptera

    Suborder: Planipennia
        Superfamily: Hemerobioidea
            Family: Sisyridae

Taxonomy and distribution

The known genera of spongillaflies include *Climacia*, *Sisyra*, *Sisyrina* and *Sisyborina*. The genus *Climacia* is distributed primarily in the New World tropics although three species are represented in temperate regions. The genera *Sisyrina* and *Sisyborina* consist of only three described species and is restricted to India and Asia, respectively. *Sisyra* occurs throughout the world from boreal to tropical environments. Only about 60 described species of spongillaflies among all genera occur worldwide.

Adult spongillaflies superficially resemble brown lacewings (Family Hemerobioidae). However, because of the branching pattern of the forewing venation, brown lacewings appear to have two radial sector veins while the spongillaflies have only one clearly

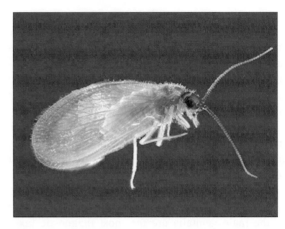

Fig. 963 *Sisyra vicaria* (Walker). Female, lateral view.

defined radial sector vein. In addition, most of the coastal crossveins of brown lacewings are branched, but those of sisyrids are not branched. Most spongillaflies are uniformly brown in color, but some species in the genus *Climacia* have colorless wings marked with black and brown patches or streaks.

Sisyrid larvae are robust, odd-looking insects having fairly long antennae, and slender legs each bearing a single tarsal claw. Their bodies bear numerous stout setae, and they are variably colored ranging from yellowish-brown to green. In addition, the second and third instar larvae bear two- or three segmented, transparent ventral gills that are folded beneath the abdominal segments. The mouthparts of larval spongillaflies are modified into elongate, unsegmented stylets used for sucking fluids from their sponge hosts.

Fig. 964 *Climacia areolaris* (Hagen). Female, lateral view.

Fig. 965 *Climacia chapini* Parfin & Garneg. Larva, dorsal view.

Natural history

Spongillafly larvae most often are found crawling about the surface of their host sponge, but occasionally they may be found within the cavities and recesses of the sponge. Several genera of freshwater sponge are parasitized by spongillaflies and all are widely distributed in North America. These genera include *Anheteroeyenia, Dosilia, Ephydatia, Eunapius, Heteromeyenia, Radiospongilla, Spongilla* and *Trochospongilla*. Some sponges do not form large, robust colonies and may appear only as a thin encrustation on woody debris or stones. The latter forms are more difficult to locate and collect than the former, but they still can support spongillaflies. Rarely, spongillaflies have been reported feeding on bryazoans and algae. Adult spongillaflies often can be found on flowering plants and other riparian vegetation during daylight. The feeding habits of adult spongillaflies are not well known, but they have been observed feeding on insect eggs, and nectar also may be a food source.

Spongillaflies have complete life cycles similar to that of other neuropterans. Female spongillaflies typically lay eggs in masses of 2–5 on emergent vegetation or other objects overhanging the water. The eggs are oval, whitish to yellowish in color, and covered with a web of white silk. The eggs hatch in about one week and the first instar larvae fall into the water and immediately proceed to search for a host sponge. First instar larvae "jump about" presumably to aid in their distribution. Jumping is accomplished by tucking the various appendages underneath the body and then rapidly straightening them out causing the body to lift off the substrate and move forward. Mature larvae swim in a vertical position and gain forward movement by snapping the body. There are three larval instars in the life cycle with the final instar larvae ranging from 4 to 8 mm in length.

Prior to pupation, final instar larvae leave the water and climb onto objects along the bank such as plants, tree limbs and other solid structures often some distance from the water. Pupae are housed in hemispherical, usually double-walled, silken cocoons. The inner wall of the pupal cocoon consists of tightly woven mesh and the outer wall is more loosely constructed.

Spongillafly larvae in North America occasionally are attacked by the parasitoid wasp *Sisyridivora cavigena* Gahan (Hymennoptera: Chalcidoidea: Pteromalidae). The female wasp pierces the pupal cocoon of the spongillafly using a long ovipositor, stings the pupa, and lays its eggs. In some instances, the female wasp may use a feeding tube constructed from their own ovipositors to feed on the body fluids of the host. The extent and effects of such parasitism on spongillafly populations are unknown.

Collecting methods

Adult spongillaflies are strongly attracted to a variety of lights including incandescent, ultraviolet, and mercury-vapor. Such lights provide an excellent means by which to collect adult spongillaflies. Adult spongillaflies also can be collected by sweeping the vegetation bordering aquatic habitats. Larvae can be collected by hand from freshwater sponges, and by using a variety of benthic collecting devices and

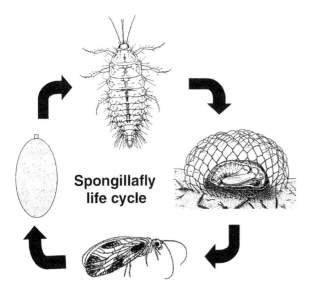

Spongillafly life cycle

Fig. 966 Generalized life cycle of a spongillafly. The pupal cocoon is cut-away to show the pupa inside. Illustrations are redrawn and modified from those of Brown (1952) and Pupedis (1987).

techniques such as kick-nets, Surber and Hess samplers, and Eckman and Ponar grabs.

Preservation

Larval and pupal spongillaflies ideally should be fixed in Kahle's solution (parts by volume: 15–95% ethyl alcohol, 6–10% formalin, 1-glacial acetic acid, 30-distilled water) for approximately 24-hours when possible, and then permanently stored in 70% ethyl or isopropyl alcohol. However, 80% ethyl or isopropyl alcohol will serve this purpose satisfactorily when Kahle's solution is unavailable. Adult specimens ideally should be pinned or pointed, but preservation in 70% ethanol or isopropyl also is suitable.

See also, LACEWINGS, ANTLIONS, AND MANTIDFLIES.

David E. Bowles
Texas Parks and Wildlife Department
Austin, Texas, USA

References

Brown, H. P. 1951. *Climacia areolaris* (Hagen) parasitized by a new pteromalid (Hym.: Chalcidoidea). *Annals of the Entomological Society of America* 44: 103–110.

Brown, H. P. 1952. The life history of *Climacia areolaris* (Hagen), a neuropterous "parasite" of freshwater sponges. *American Midland Naturalist* 47: 130–160.

Parfin, S. I., and A. B. Gurney. 1956. The spongilla-flies, with special reference to those of the Western Hemisphere (Sisyridae, Neuroptera). *Proceedings of the United States National Museum* 105: 421–529.

Pupedis, R. J. 1978. Tube feeding by *Sisyridivora cavigena* (Hymenoptera: Pteromalidae) on *Climacia areolaris* (Neuroptera: Sisyridae). *Annals of the Entomological Society of America* 71: 773–775.

Pupedis, R. J. 1980. Generic differences among new world spongilla-fly larvae and a description of the female of *Climacia striata* (Neuroptera: Sisyridae). *Psyche* 87: 305–314.

Pupedis, R. J. 1987. Foraging behavior and foods of adult spongilla-flies (Neuroptera: Sisyridae). *Annals of the Entomological Society of America* 80: 758–760.

SPORE. A reproductive stage of fungi, usually somewhat resistant to adverse environmental conditions, that is capable of growing into a new organism.

SPOTTED TENTIFORM LEAFMINER. See also, APPLE PESTS AND THEIR MANAGEMENT.

SPOT TREATMENT. The application of pesticides to the area where pests actually are found rather than to the entire field.

SPREADER. An adjuvant added to a pesticide to enhance the ability of the pesticide to spread over a larger area of the foliage, and often combined with another adjuvant to produce a combined 'spreader-sticker'.

SPREAD-WINGED DAMSELFLIES. A family of damselflies in the order Odonata: Lestidae. See also, DRAGONFLIES AND DAMSELFLIES.

SPRING STONEFLIES. Members of the stonefly family Nemouridae (order Plecoptera). See also, STONEFLIES.

SPRINGTAILS (COLLEMBOLA). Springtails are a primitive entognathous order of hexapods that sometimes (along with Protura and Diplura) are considered to be insects. They are among the most widespread and abundant terrestrial arthropods, and are found in Antarctica, where arthropods are scarce. Generally they are considered to be useful as they assist in the decomposition of organic materials, and few are pests.

Classification

There are over 6000 species of springtails, and they are found throughout the world, including Antarctica. The order generally is subdivided into two suborders: Arthropleona, the elongate-shaped species, and Symphypleona, the globular species.

Class: Entognatha

Order: Collembola
   Suborder: Arthropleona
      Family: Neanuridae
      Family: Odontellidae
      Family: Brachystomellidae
      Family: Poduridae – water springtails
      Family: Hypogasturidae – elongate – bodied springtails
      Family: Onychiuridae – blind springtails
      Family: Isotomidae – smooth springtails

Family: Entomobryidae – slender springtails
Family: Cyphoderidae
Family: Paronellidae
Family: Oncopoduridae
Family: Tomoceridae
Family: Coenaletidae
Family: Actoletidae
Family: Microfalculidae
Family: Protentomobryidae
Suborder Symphypleona
Family: Neelidae – short – horned springtails
Family: Sminthuridae – globular springtails
Family: Mackenziellidae – Mackenzie globular springtails
Family: Dicyrtomidae

Characteristics

These are small animals, measuring only 0.25 to 10 mm in length, and usually less than 5 mm long. They are either elongate or globular in body form. Their color varies greatly, and though often obscure, some are brightly colored. The antennae are short to medium in length, and consist of 4 to 6 segments. The compound eyes are small, with only a few facets per eye. The mouthparts are basically the biting (chew-

ing) type, but sometime extensively modified, and somewhat enclosed by the head. The legs are small, and lack extensive modifications. Wings always are absent. The abdomen consists of 5 or 6 segments, though some may be obscure. Springtails often are equipped with a jumping apparatus that serves as the basis for the common name. This apparatus consists of a fork-like furcula originating near the tip of the abdomen that is flexed ventrally and held by a catch, the tenaculum. When the tenaculum releases, the furcula springs with a snap that propels the insect forward. In addition, springtails possess a small ventral tube-like structure on the first abdominal segment, called a collophore or ventral tube. The collophore has various functions, including water absorption and excretion, and possibly adhesion to smooth surfaces. They lack cerci. Metamorphosis may be lacking or incomplete. Collembolans often display 6 to 8 molts during their life, with sexual maturity attained after the fifth molt. Thus, unlike insects, they continue to molt as adults. The immatures usually resemble the adults in external morphology.

There are three major body types found among springtails: globular, elongate, and grub-like. In globular species, the abdominal segments are fused to

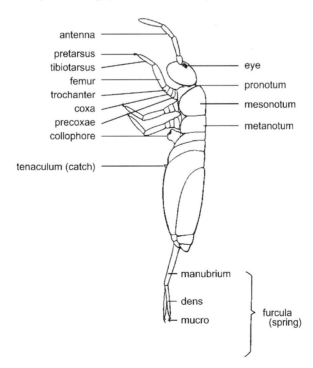

Fig. 967 A diagram of a springtail showing a lateral view.

form a globe-shaped abdomen. Such species typically inhabit open, grassy habitats. In elongate species, the first thoracic segment is reduced in size relative to the other thoracic segments, producing a "neck"-like structure. Also, the legs and antennae, and furcula are elongate, and the eyes well developed. Sometimes these insects are will marked with patterns or colors, and the body is covered with scales or hairs that help reduce the rate of drying when the springtails forage over the surface of the soil. This body form is called entomobryomorphoid. The grub-like springtails have short legs, antennae, and furcula, and poorly developed eyes. The first thoracic segment is the same size as the other thoracic segments. Such springtails typically lived deep in the litter or soil, and the body form is called podomorphoid.

Biology

Springtails occur in numerous habitats, including the water surface of fresh water, the tidal region of salt water, in soil, leaf debris, bird nests, beneath bark, and occasionally on foliage. They feed primarily on lichens, pollen, fungi, bacteria, and carrion, though a few feed on seedling plants or are carnivorous. In general, the above-ground species are in the suborder Arthropleona, are slender in shape, brightly colored, relatively large, and have long antennae and legs, and relatively well-developed eyes. The below-ground species are small, pale colored, with short antennae and legs, and with reduced eyes and the jumping apparatus. The species in the suborder Symphypleona, which are globular in shape, dwell in moist habitats or caves, and are active jumpers. Like the other above-ground species, their eyes are well developed, and their body pigmented. Springtails generally lack trachea, with gas exchange occurring through the integument. Springtails often aggregate in large groups, and this can be observed on the surface of water, snow, or on organic material. The purpose or cause of aggregations is unknown.

The reproductive behavior of springtails is consistent with other entognathous insects. Sperm transfer is indirect, with sperm deposited in stalked droplets on the ground. In most cases, sperm uptake by the female is a passive process, but in some of the more advance species, the male guides the female to his sperm.

References

Arnett, Jr., R. H. 2000. *American insects (second edition)*. CRC Press, Boca Raton, Florida. 1003 pp.

Christiansen, K. 1964. Bionomics of Collembola. *Annual Review of Entomology*. 9: 147–178.

Christiansen, K. and P. F. Bellinger. 1998. *The Collembola of North America north of the Rio Grande, a taxonomic analysis*. Grinnell College, Grinnell, Iowa. 1520 pp.

Hopkin, S. P. 1997. *The biology of the springtails*. Oxford university Press, Oxford, United Kingdom. 330 pp.

Snider, R. J. 1987. Class and order Collembola. Pages 55–65 in F. W. Stehr (ed.). *Immature insects*, Vol. 1. Kendall/Hunt Publishing, Dubuque, Iowa.

SPUR. A moveable spine. Spurs are often located on a leg segment, usually at the tip of the segment.

SPURIOUS VEIN. A vein-like thickening of the wing membrane. The presence of a spurious vein is an important diagnostic feature in Syrphidae (Diptera).

See also, WINGS OF INSECTS.

SPURTHROATED GRASSHOPPERS. Subfamilies of grasshoppers in the old world (Catantopinae) and new world (Melanoplinae) in the order Orthoptera: Acrididae. See also, GRASSHOPPERS, KATYDIDS AND CRICKETS.

SQUASH BUG, *ANASA TRISTIS* (DEGEER) (HEMIPTERA: COREIDAE). The squash bug attacks cucurbits (squash and relatives) throughout Central America, the United States, and southern Canada. In some areas of North America it is the most serious pest of cucurbits.

Description and life cycle

The complete life cycle of squash bug commonly requires 6 to 8 weeks. Squash bugs have one generation per year in northern climates and 2 to 3 generations per year in warmer regions. In intermediate latitudes the early-emerging adults from the first generation produce a second generation whereas the late-emerging adults go into diapause. Both sexes overwinter as adults. The preferred overwintering site seems to be in cucurbit fields under crop debris, clods of soil, or stones but sometimes adults also are found in adjacent wood piles or buildings.

Eggs. Eggs usually are deposited on the lower surface of leaves. The elliptical egg is somewhat flattened and bronze in color. The average egg length is about 1.5 mm and the width about 1.1 mm. Females deposit about 20 eggs in each egg cluster. Eggs may be tightly clustered or spread a considerable distance apart, but an equidistant spacing arrangement is commonly observed. Duration of the egg stage is about 7 to 9 days.

Nymph. There are five nymphal instars, collectively requiring about 33 days for complete development. The nymph is about 2.5 mm in length when it hatches, and light green in color. The second instar is initially about 3 mm long, and its color is light gray. The third, fourth, and fifth instars initially are about 4, 6–7, and 9–10 mm in length, respectively, and darker gray. The youngest nymphs are rather hairy, but this decreases with each subsequent molt. In contrast, the thorax and wing pads are barely noticeable at hatch, but get more pronounced with each molt. Young nymphs are strongly gregarious, a behavior that dissipates slightly as the nymphs mature.

Adult. The adult measures 1.4 to 1.6 cm in length and is dark grayish brown in color. In many cases the edge of the abdomen is marked with alternating gold and brown spots. Adults are long-lived, surviving an average of about 75 to 130 days, depending on availability and quality of food.

Host plants

Squash bug has been reported to attack nearly all cucurbits, but squash and pumpkin are preferred for oviposition and support high rates of reproduction and survival. There is considerable variation among species and cultivars of squash with respect to susceptibility to damage and ability to support growth of squash bugs. New World varieties are preferred. Studies conducted in the United States reported survival of squash bug to be 70, 49, 14, 0.3, and 0% when nymphs were reared to the adult stage on pumpkin, squash, watermelon, cucumber, and muskmelon (cantaloupe), respectively.

Damage

The squash bug causes severe damage to cucurbits because it secrets highly toxic saliva into the plant. The foliage is the primary site of feeding but the fruit is also fed upon. The foliage wilts, becomes blackened, and dies following feeding. Often an entire plant or section of plant perishes while nearby plants remain

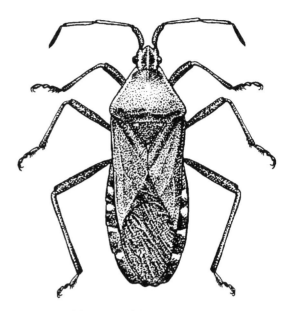

Fig. 968 Adult of squash bug, *Anasa tristis.*

healthy. The amount of damage occurring on a plant is directly proportional to the density of squash bugs.

Natural enemies

Several natural enemies of squash bug are known, principally wasp egg parasitoids (Hymenoptera: Encyrtidae and Scelionidae). Up to 30% parasitism among eggs collected in Florida, USA has been reported. The best known natural enemy is a common parasitoid of several hemipterans, *Trichopoda pennipes* (Fabricius) (Diptera: Tachinidae). The brightly colored adult fly is easy to recognize, having a gold and black thorax and an orange abdomen, with a prominent fringe of feather-like hairs on the outer side of the hind tibia. Flies develop principally in the adult bug, initially castrating the female, and then killing her when the fly emerges. In Connectcut, USA about 20% on the squash bugs have been found to be parasitized in late summer.

Management

Squash bug adults are unusually difficult to kill with insecticides. Although adult control can be accomplished if the correct material is selected, it is advisable to target the more susceptible nymphs. Squash bugs are not often considered a severe pest of large-scale cucurbit production, probably due to the absence of suitable overwintering sites in well managed crop fields and because the bug's effects are diluted by the vast acreage. However, small fields and home gardens are commonly damaged.

Pollinators, particularly honeybees, are very important in cucurbit production, and insecticide application can interfere with pollination by killing honeybees. If insecticides are to be applied when blossoms are present, it is advisable to use insecticides with little residual activity, and to apply insecticides late in the day, when honeybee activity is minimal.

Adult squash bugs preferentially colonize larger, more mature plants. Thus, early-planted crops may be especially prone to attack. Numbers are also highest on plants during bloom and fruit set. Use of early-planted crops as a trap crop has been proposed, but due to the high value of early season fruit most growers try to get their main crop to mature as early as possible. The use of squash or pumpkin as a trap crop to protect less preferred host plants such as melons and cucumbers is reported to be effective.

The tendency of squash bugs to aggregate in sheltered locations can be used to advantage by home gardeners. Placement of boards, large cabbage leaves, or other shelter for squash bugs induces the bugs to congregate there during the day where they are easily found and crushed. Row covers and netting delay colonization of squash, but bugs quickly invade protected plantings when covers are removed to allow pollination.

Removal of crop debris in a timely manner is very important. Squash bugs will often be found feeding on old fruit or in abandoned plantings, so clean cultivation is essential to reduce the overwintering population.

See also, VEGETABLE PESTS AND THEIR MANAGEMENT.

References

Beard, R. L. 1940. The biology of *Anasa tristis* DeGeer with particular reference to the tachinid parasite, *Trichopoda pennipes* Fabr. *Connecticut Agricultural Experiment Station Bulletin* 440: 597–679.

Bonjour, E. L. and W. S. Fargo. 1989. Host effects on the survival and development of *Anasa tristis* (Heteroptera: Coreidae). *Environmental Entomology* 18: 1083–1085.

Bonjour, E. L., W. S. Fargo, and P. E. Rensner. 1990. Ovipositional preference of squash bugs (Heteroptera: Coreidae) among cucurbits in Oklahoma. *Journal of Economic Entomology* 83: 943–947.

Bonjour, E. L., W. S. Fargo, A. A. Al-Obaidi, and M. E. Payton. 1993. Host effects on reproduction and adult longevity of squash bugs (Heteroptera: Coreidae). *Environmental Entomology* 22: 1344–1348.

Capinera, J. L. 2001. *Handbook of vegetable pests.* Academic Press, San Diego. 729 pp.

SQUIRREL LICE. Members of the family Enderleinellidae (order Siphunculata). See also, SUCKING LICE.

SSDNA. Single-stranded DNA.

STABILITY. An expression of constancy. In population biology it refers to population densities and age structures that remain fairly constant over long periods of time. (contrast with resilience)

STABLE FLY, *STOMOXYS CALCITRANS* (DIPTERA: MUSCIDAE). The stable fly (*Stomoxys calcitrans* L.), sometimes referred to as the dog fly or the biting house fly, is primarily a pest of cattle. This species closely resembles a house fly, but possesses piercing-sucking instead of sponging mouthparts. Both sexes of the stable fly feed on blood 2 to 3 times per day from a variety of warm-blooded animals including humans. The act of feeding by several flies can produce quite an annoyance to cattle. Fly avoidance behavior often includes constant stomping of feet, tail switching, and 'bunching,' whereby the herd gathers or 'bunches' together in a tight circle with heads facing inward to protect their front legs from feeding flies. When fly avoidance behavior takes up most of an animal's daily activity, it is not eating. Stable flies have been reported to reduce weight gain in beef cattle and lower milk production in dairy cattle. Research has estimated an economic threshold of less than 1 to 5 feeding flies per front leg depending on animal size, age, breed and production system. Because stable flies feed on blood, they are capable of transmitting pathogenic organisms that cause anthrax, brucellosis, swine erysipelas and equine infectious anemia to a variety of their animal hosts.

Blood-seeking stable flies can also affect recreational areas. In some areas of the U.S. (e.g., Florida panhandle beaches, coastal areas of New Jersey, Great Lakes Region and Tennessee Valley Authority lakeshores) stable flies often disperse from cattle producing areas and congregate along the shorelines.

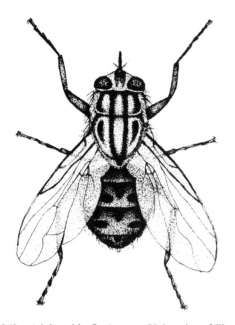

Fig. 969 Adult stable fly (source: University of Florida).

Weather systems associated with the passage of cold fronts often move the flies to these areas. It is currently unknown how or why this happens.

Although most stable flies seldom disperse farther than needed to obtain a blood meal, studies have shown that this pest can routinely travel distances up to 8 km and as far as 225 km on storm fronts to Florida panhandle beaches. Suburban areas that have expanded into rural cattle production areas have experienced problems with stable flies. Oftentimes, this phenomenon occurs when stable fly populations

are not adequately controlled on surrounding farms and they disperse into the surrounding environs. However, a study in Kansas indicated that a suburban stable fly problem was not the result of nearby rural livestock operations. Rather, local developmental areas for stable flies, such as numerous compost heaps located within the suburban landscape, contributed greatly to the problem. In any event, legal issues have arisen with many livestock operations going out of business.

Biology

Stable flies go through complete metamorphosis. The eggs are laid on decayed vegetation, or vegetation mixed with cattle feces and/or urine. Improperly composted vegetative material, or decomposing grass clippings that remain wet can provide adequate developmental medium for stable flies in residential areas.

After oviposition, the eggs hatch in 1 to 5 hours and the larvae go through three successive molts called instars. When the larva is mature (i.e., third instar), it ceases to feed and often migrates from the developmental area to pupate in drier conditions. Sometimes larvae will enter the ground to the depth of about 2 to 4 cm to pupate. At the time of pupation, the exoskeleton from the last larval instar will harden. This 'shell' forms the puparium in which the pupa will reside. The pupa is the transitional stage from larva to adult fly. Depending upon temperature and other environmental parameters, the cycle from egg to adult can take as little as 10 days, but can be extended to two months or longer under unfavorable

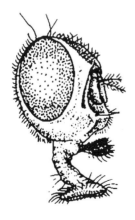
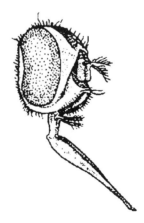

Fig. 970 Comparison of nonbiting and biting mouthparts: left, sponging mouthparts of the adult house fly (source: Novartis Corporation); right, piercing-sucking mouthparts of the adult stable fly (source: Axtell, R.C. 1986. Fly control in confined livestock and poultry production. CIBA-GEIGY Agric. Div., Greensboro, NC.).

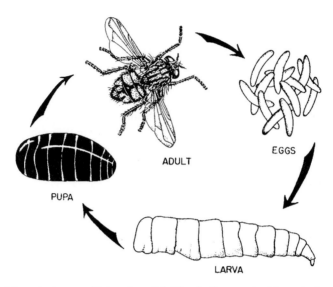

Fig. 971   Generalized fly life cycle (source: University of California Cooperative Extension).

conditions. Several generations are produced annually and tremendous numbers of biting adults can develop if not held in check.

Adult flies can live about 20 days under favorable conditions. Adult stable fly seasonal peaks appear to change with climatic zone in the U.S. Populations in the more temperate areas (e.g., Minnesota, Michigan, and New York) primarily peak in mid-summer, whereas further south (e.g., Texas and Florida) the population peak is bimodal, occurring in late spring and again in late summer to early fall.

Control

The most effective method of reducing stable flies is to eliminate the habitat for fly development. The removal of accumulated feed, silage, manure, grass clippings, etc., in a timely manner (e.g., weekly) will help to reduce the fly population. If removal is not feasible, then it should be piled in rows and turned at least once a week to facilitate drying. In lieu of this, heavy black plastic can be placed over the piles (especially compost heaps). Oftentimes, the temperature produced by anaerobic decomposition of the pile will be high enough to kill any fly larvae present. Edges of the plastic should be covered with soil to create firm contact with the ground and prevent adult flies from ovipositing underneath the plastic.

Stable fly pupae have a variety of minute wasp parasitoids that parasitize them as a form of natural control. The most common parasitoids are referred to as microhymenoptera, and belong to the family Pteromalidae. These tiny wasps (about 2.0 mm in length) parasitize stable flies by ovipositing one or many eggs (depending on species) into a pupa. The egg(s) hatch and the larva(e) consume the developing fly, eventually killing it. The immature wasp then pupates and emerges from the stable fly puparium as an adult. As a result, wasps in this family are considered to be of potential use for the control of stable flies in animal facilities. Research has shown that native wasp parasitoids have the best potential for colonizing and impacting native stable fly populations. Currently, knowledge on how many parasitoids should be released, and the optimum number of releases to obtain control is poorly understood.

Application of insecticides to organic materials is not recommended because the quantity needed would be greater than the labeled rate to achieve any degree of control. Also, organic substrates often bind up insecticides making them ineffective against the targeted pest. Stable fly control on animals usually involves spraying labeled insecticides onto the animal's legs. Because stable flies often rest on the sides of buildings, a residual insecticide applied to such surfaces may provide short-term control of adult flies. Currently, repellents do not give long-term relief for animals or humans against stable fly biting annoyance.

James E. Cilek
Florida A & M University
Panama City, Florida, USA

References

Foil, L. D., and J. A. Hogsette 1994. Biology and control of tabanids, stable flies and horn flies. *Revue Scientifique et Technique. Office International des Epizooties, Paris* 13: 1125–1158.

Petersen, J. J., and G. L. Greene. 1989. *Current status of stable fly (Diptera: Muscidae) research*. Miscellaneous Publication of the Entomological Society of America. Number 74. 53 pp.

Scholl, P. J., and J. J. Petersen. 1985. Biting flies. pp. 49–63 in R. E. Williams, R. D. Hall, A. B. Broce, and R. E. Williams (eds.), *Livestock entomology*. John Wiley and Sons, New York, New York. 335 pp.

Steelman, C. D. 1976. Effects of external and internal arthropod parasites on domestic livestock production. *Annual Review of Entomology* 21: 155–178.

Thomas, G. D., and S. R. Skoda (eds.) 1993. *Rural flies in the urban environment*. Agricultural Research Division, Institute of Agriculture and Natural Resources, University of Nebraska-Lincoln Research Bulletin 317. 97 pp.

STABLE TRANSFORMATION. Transformation that alters the germplasm of an organism so that the progeny transmit the trait through subsequent generations.

STAG BEETLES. Members of the family Lucanidae (order Coleoptera). See also, BEETLES.

STADIUM. The time interval between molts in a developing arthropod.

STÅL, CARL. Carl Stål was born at the castle of Carlberg, in Sweden, on March 21, 1833. He was the greatest Swedish hemipterist and orthopterist, and one of the world's greatest authorities on Hemiptera. He began his education at Uppsala, Sweden, including medical studies, and obtained further education at Stockholm and Jena. Stål became an assistant to the Swedish coleopterist C.H. Boheman at the National Zoological Museum, and eventually supervised the entomological section of the Museum. Although he favored Hemiptera, Stål was an authority on Coleoptera, Orthoptera and Hymenoptera. Some of his noteworthy publications include "Hemiptera Fabriciana" (two volumes: 1868, 1869), "Enumeratio Hemipterorum" (five volumes: 1870 to 1876), "Revisio Pentatomidarum, Coreidarum, Lygaeidarum, Reduviidarum et Tingitidarum Europae," and "Recensio Orthopterorkum" (1873 to 1876). At his death, Stål was the world's authority on Hemiptera, and he left the Museum with the finest collection of Hemiptera in the world. He died at Trösundavik, Sweden, on June 13, 1878.

References

Anonymous. 1878. Obituary, Carl Stål. *Entomologist's Monthly Magazine* 15: 94–96.

Essig, E. O. 1931. *A history of entomology*. The Macmillan Company, New York. 1029 pp.

STALK OR STEM-INFECTING FUNGI. See also, TRANSMISSION OF PLANT DISEASES BY INSECTS.

STALK-EYED FLIES. Members of the family Diopsidae (order Diptera). See also, FLIES.

STAPHYLINIDAE. A family of beetles (order Coleoptera). They commonly are known as rove beetles. See also, BEETLES.

STEBLIDAE. Formerly a family of flies (order Diptera). They (and Nycteribiidae) were commonly are known as bat flies, but are now included in the family Hippoboscidae. See also, FLIES.

STEINHAUS, EDWARD ARTHUR. Edward Steinhaus was born at Max, North Dakota, USA, on November 7, 1914. H graduated from North Dakota State University in 1936 with a degree in bacteriology, and then moved to Ohio State University where he received a Ph.D. in 1939. In 1940 he joined the staff of the U.S. Public Health Service Rocky Mountain Laboratory, and then moved to the University of California at Berkeley in 1944. He organized and chaired the Laboratory of insect pathology in 1945, which became the Department of insect pathology in 1960, and then the Division of invertebrate pathology in 1963. Later he moved to the University of California at Irvine as dean of Biological Science. Edward Steinhaus was the most distinguished insect pathologist of his era, and created the world's foremost program in insect pathology at Berekeley. He authored 200 publications on insect pathology and the first comprehensive publications on the subject: "Insect microbiology" (1946), "Principles of insect pathology" (1949), and "Laboratory exercises in insect microbiology and insect pathology" (1961). In addition, he edited a two-volume treatise "Insect pathology, an advanced treatise" (1963). Steinhaus founded and edited an important scientific journal, "Journal of invertebrate pathology," and founded the Society for Invertebrate Pathology. He was honored by the Entomological Society of America by being selected to present a Founders Memorial Lecture, and also served as president of the Society. He died at Newport Beach, California, on October 20, 1969.

References

Linsley, E. G., and R. F. Smith. 1970. Edward Arthur Steinhaus (1914-1969). *Journal of Economic Entomology* 63: 689–691.
Pristavko, V. P. 1970. In memory of Professor E. A. Steinhaus (1914-1969). *Entomological Review* 49: 306–307.

STEM CELLS. Stem cells are able to self-renew and generate cell populations that differentiate to maintain adult tissues. There are about two stem cells in the ovary of *Drosophila* that maintain oocyte production.

STEMMATA. The small, simple eyes found on the side of the head of larvae in holometabolous insects. The number of stemmata varies from one to six in different taxa. They are similar to ommatidia of compound eyes in structure and function. A single simple eye of this type is called a stemma.

STEMMOCRYPTIDAE. A family of bugs (order Hemiptera). See also, BUGS.

STEM SAWFLIES. Members of the family Cephidae (order Hymenoptera, suborder Symphyta). See also, WASPS, ANTS, BEES, AND SAWFLIES.

STENOCEPHALIDAE. A family of bugs (order Hemiptera, suborder Pentamorpha). See also, BUGS.

STENOPELMATIDAE. A family of crickets (order Orthoptera). They commonly are known as Jerusalem crickets. See also, GRASSHOPPERS, KATYDIDS AND CRICKETS.

STENOPHAGOUS. Feeding on a small number of hosts. Also called oligophagous.

STENOPSOCIDAE. A family of psocids (order Psocoptera). See also, BARK-LICE, BOOK-LICE, OR PSOCIDS.

STENOPSOCIDAE. A family of psocids (order Psocoptera). See also, BARK-LICE, BOOK-LICE, OR PSOCIDS.

STENOSTRITIDAE. A family of wasps (order Hymenoptera). See also, WASPS, ANTS, BEES, AND SAWFLIES.

STENOTRACHELIDAE. A family of beetles (order Coleoptera). They commonly are known as false long-horned beetles. See also, BEETLES.

STEPHANIDAE. A family of wasps (order Hymenoptera). See also, WASPS, ANTS, BEES, AND SAWFLIES.

STEPHANOCIRCIDAE. A family of fleas (order Siphonaptera). See also, FLEAS.

STEREOKINESIS. Kinesis response with respect to contact with surfaces.

STERILE INSECT CONTROL. A pest control method using sterilized insects to interfere with normal mating behavior in a wild population, thereby disrupting reproduction of the wild (non-sterile) insects.

STERILE INSECT RELEASE METHOD (SIRM). A technique used to control pest insects. Large numbers of mass produced males are given nonlethal but sterilizing doses of radiation or chemical mutagens and then released in nature. The natural populations are so overwhelmed by releases of large numbers of these males, that females are almost always fertilized by them. The resultant matings produce inviable progeny and a new generation is not produced. Used to eradicate the screwworm from the USA.

See also, STERILE INSECT TECHNIQUE.

STERILE INSECT TECHNIQUE. The sterile insect technique (SIT) is a form of birth control imposed on a population of an insect pest to reduce its numbers. Thus far, this has involved rearing large numbers of the target insect pest species, exposing them to gamma rays to induce sexual sterility and releasing them into the target population of the pest on an ecosystem-wide or area-wide basis. The concept of releasing insects of pest species to introduce sterility into wild populations, and thus to control them, was independently conceived by three scientists on three different continents. The first scientist to propose genetic control of insect species was A.S. Serebrovskii, a geneticist in the Institute of Zoology, Moscow State University, who proposed the use of chromosomal translocations for population suppression. The second was Dr. F.L Vanderplank at a British Overseas Service tsetse research field station in rural Tanganyika (now Tanzania). He showed that the sterility of interspecific hybrids of two tsetse species could strongly suppress a field population of one of the species. The third was E.F. Knipling, a scientist with the U.S. Department of Agriculture. The sterile insect technique was conceived by Knipling in 1937 and first developed for use against the screwworm, *Cochliomyia hominivorax* (Coquerel), a deadly parasite of warm-blooded animals and man in the Americas. This article is limited to a discussion of the latter technique.

General overview

The sterile insect technique harnesses the sex drive of insects. The idea of doing this first occurred in 1937 to Edward F. Knipling, an entomologist with the U.S. Department of Agriculture at Menard, Texas. Knipling did not know of a method to induce sexual sterility. However, in 1926 Herman Müller at the University of Texas had found that X-rays cause numerous changes or mutations in the genes of the vinegar fly, *Drosophila melanogaster* Meigen, and that at high radiation doses, dominant lethal muta-

tions are induced in the sperm or the eggs, which cause the embryos to die. After World War II, both the United States and the Soviet Union were testing hundreds of atomic weapons in the atmosphere. Müller was concerned that the radioactive fallout would induce numerous deleterious mutations in people, so in 1950 Müller publicized his findings that ionizing radiation induces mutations in the vinegar fly, and at high radiation doses, the insects are completely sterilized. Immediately thereafter, under Knipling's leadership, the sterile insect technique was developed and used against the screwworm fly, a major insect parasite of warm-blooded animals and man. Over a period of 43 years, this technique was used to eradicate the screwworm from the United States, Mexico, and Central America to Panama.

Considerable research has been conducted to develop the use of the sterile insect technique against a variety of major pests. For example, Dr. M.D. Proverbs initiated such research on the codling moth, *Cydia pomonella* (L.), in 1955 at Summerland, British Columbia, Canada, but more than two decades lapsed before a practical program was launched against this major pest of temperate tree fruits. Since radiation affects the chromosomes of moths quite differently than those of flies, and since the large scale rearing of moths also is more difficult, the work on moths has lagged behind similar work with flies.

The sterile insect technique has been used with great success against various species of tropical fruit flies. The Japanese used the sterile insect technique to eradicate the melon fly, *Bactrocera cucurbitae* Coquillett, from Okinawa and all of Japan's southern islands, and this has opened the main markets in Japan to fruits and vegetables produced on Okinawa and the islands.

Chile used the sterile insect technique to rid the entire country of the Mediterranean fruit fly, *Ceratitis capitata* (Wiedemann). By 1980 the entire country of Chile had become a medfly-free zone, and since then Chilean fruits in huge volumes have entered the U.S. market without the need for any quarantine treatments. This has dramatically strengthened the economy of Chile. Now Argentina, Peru and other countries have sterile insect technique programs that they hope will enable them to become fly-free zones with free access to markets in southern Europe, Japan and the United States.

Also, Mexico has used the sterile insect technique to get rid of the Mediterranean fruit fly. Indeed

Fig. 972   Edward F. Knipling in 1946.

Mexico is ridding large sections of its territory of all fruit fly species of economic importance. The Mexican states of Baja California, Chihuahua and Sonora have been freed of all economically important species of fruit flies, so that citrus, stone fruits, apples and vegetables are being exported from these states without any postharvest treatment. In other parts of Mexico, low prevalence fruit fly areas are being established by means of a systems approach.

To prevent the establishment of medflies that enter the United States with smuggled fruit, sterile medfly males are being released continuously over the Los Angeles Basin and over Tampa and Miami. This has obviated the need to spray these urban areas with the insecticide malathion to suppress outbreaks.

Tsetse flies, which transmit trypanosomes to people and to livestock, are regarded as the root cause of rural poverty in Sub-Saharan Africa because they prevent mixed farming. Food is produced with hoes and spades because African sleeping sickness kills draft animals. Milk cannot be produced and manure

is not available to fertilize the worn-out soils. The conquest of sleeping sickness would allow the use of cattle for draft and for milk production, and this is key to facilitating rural development and well being in Sub-Saharan Africa. In 2001 the African Heads of State committed their countries to a campaign to rid Africa of sleeping sickness. This undertaking is known as the "Pan African Tsetse and Trypanosomosis Eradication Campaign." Eradication of the tsetse fly, *Glossina austeni* Newstead, and sleeping sickness from Zanzibar Island was completed in 1997 by means of the sterile insect technique. Previously tsetse populations had been eradicated by this means in small areas in Nigeria and Burkina Faso. Obviously the conquest of African sleeping sickness will require decades of concerted effort.

The sterile insect program against the codling moth is of great global significance, since it is pioneering the practical application of this technique against a species of moth. There are a number of very damaging moth species, and these include the gypsy moth, various cutworm moths, the corn earworm, the European corn borer, the diamondback moth which plagues vegetable production and is resistant to almost all chemical insecticides, rice stem borers, a number of species that destroy stored products, and many others. For a number of years, sterile pink boll-worm moths, *Pectinophora gossypiella* Saunders, have been released over the cotton fields in the San Joaquin Valley of California to prevent the establish-ment of pink bollworms that migrate from southern California. However, the prevention of incipient infestations is not nearly as demanding as coping with a well-entrenched pest population, such as the codling moth in the Okanagan Valley of British Columbia, Canada. Success against the codling moth in the Okanagan is likely to provide the confidence needed to initiate SIT programs against moth pests in other countries.

Clearly over the past four decades the SIT has come to play a progressively greater role in overcom-ing poverty, and in expanding free, fair and safe trade between countries, and this is being accomplished with minimal pollution and ecological disruption.

The theoretical basis of the SIT and screwworm eradication

The New World screwworm is adapted to range throughout the tropics and subtropics. Before it was eradicated from the USA, the parasite survived the winters in southern Arizona, California, Florida and Texas, and ranged far to the north during the sum-mers. The conceptualization of the SIT arose out of the tremendous difficulty of coping with infestations of wounds of livestock by what was believed to be the carrion-feeding blow fly, then known as *Cali-troga macellaria* (Fabricius). (The common name of this species was later changed to the secondary screwworm.) However, Emory Cushing, an employee of the Bureau of Entomology and Plant Quarantine, U.S. Department of Agriculture (USDA) at Menard, Texas, suspected that a complex of species was involved. Consequently, Cushing took specimens of the various life stages collected from wounds, car-casses and traps to the Liverpool School of Tropical Medicine for investigation under the guidance of Pro-fessor Walter S. Patton, an expert on the taxonomy of Diptera. Cushing and Patton discovered that in addi-tion to *C. macellaria*, a second species was involved in wound infestations, which they named *Calitroga*

Fig. 973 Female screwworm laying a mass of eggs on the edge of a wound.

americana Cushing and Patton. This second species, an obligate parasite, proved to be the major species involved in causing wound infestations, but its iden-tity had been mistaken for *C. macellaria*, which feeds on necrotic tissue and not on living flesh. Thus, it was said that Cushing and Patton had ''unmasked the great insect imposter.'' Clearly, efforts to suppress screwworm attack through the practices of burying carcasses and mass trapping *C. macellaria* adults had been utterly futile. The name of this New World screwworm was later changed to *Cochliomyia homi-nivorax* (Coquerel), since the screwworm had first been described by Coquerel, a French naval physi-cian, who in 1854 described this parasite as the cause of the deaths of prisoners held on Devil's Island, French Guyana.

Of course, the biology and ecology of this new spe-cies needed to be elucidated, and this was facilitated by an unfortunate development. In 1933 a drought visited the Great Plains, and ranchers sought to rescue cattle by shipping them to Georgia and Florida where water and grass were plentiful. This program inadver-tently introduced the screwworm to the southeastern USA, whereas previously it had not existed east of the Mississippi. The climate in Florida permitted the pest to reproduce year round, and its population grew explosively. The USDA sought to ameliorate the pro-blem by establishing a small facility at Valdosta, Georgia, with the goal of educating cattle owners on ways of managing their herds in order to minimize the number of screwworm cases.

One of the scientists assigned to the Valdosta laboratory was Edward F. Knipling, who had grown up on a farm at Port Lavaca, Texas, where coping

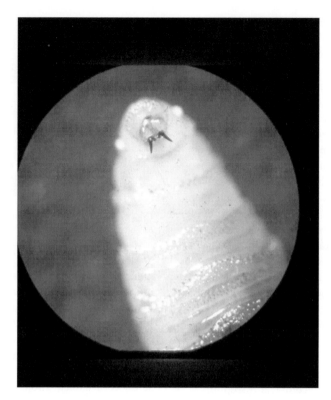

Fig. 974 Anterior end of a screwworm larva showing the mouth hooks used to tear living flesh.

with screwworm-infested wounds in farm animals was an unrelenting chore. The USDA scientists at Valdosta taught cattle owners to control the insemination of cows so that they would calve only during the winter, and also to restrict any operations that result in wounds (branding, castration, dehorning) to the winter when the screwworm populations were depressed. Knipling found time to investigate the biology of the newly described screwworm. Unlike the common blowfly, which can complete its development on carcasses, the screwworm proved to be an obligate parasite of warm-blooded mammals. The female fed on the serum oozing from wounds and laid her eggs at the edges of wounds, and not on carcasses. In addition, the density of screwworms per square mile was quite low. These and later studies led Knipling to estimate the number of screwworms that survive the winter to be roughly 10 to 20 per square mile, and that peak densities rarely exceeded 1,000 screwworms per square mile.

Meanwhile, Raymond C. Bushland, a young scientist at Menard, Texas, had the duty of rearing the screwworm species on rabbits in order to provide screwworm larvae for evaluating candidate toxicants for inclusion in wound treatments. Bushland devised an economical method of rearing the screwworm on a mixture of hamburger meat, citrated blood and formalin. Bushland was readily able to rear about 5,000 screwworms on one pound of hamburger meat.

In 1937, Knipling was transferred to Menard, Texas, and for the first time he was able to observe the mating behavior of screwworms in the cages of adults maintained by Bushland. Knipling noted that the flies mated on the second and third day of adult life, and that the females mated only once, but that the males remained highly aggressive sexually. Thus, the great idea was born in Knipling's mind that if sexual sterility could be induced in the males, and if vast numbers could be reared, sterilized and released into nature, then the screwworm population could be suppressed strongly. Knipling reasoned that if a natural population could be over-flooded overwhelmingly with sexually sterile males for several successive generations, then most of the wild females would be mated with sterile males and the population would decline towards extinction. Knipling developed very

Rates of mortality and survival required to prevent population growth and decline in relation to various intrinsic rates of increase.

Intrinsic rate of increase	Progeny per female	Number that must survive to prevent population from declining (fraction)	To prevent population increase			
			Number that must die	Fraction that must die	Percent that must die	Percent that may survive
2-fold	4	2 (1/2)	2	2/4 = 1/2	50	50
3-fold	6	2 (1/3)	4	4/6 = 2/3	67	33
4-fold	8	2 (1/4)	6	6/8 = 3/4	75	25
5-fold	10	2 (1/5)	8	8/10 = 4/5	80	20
10-fold	20	2 (1/10)	18	18/20 = 9/10	90	10
20-fold	40	2 (1/20)	38	38/40 = 19/20	95	5

simple mathematical models to simulate the dynamics of screwworm populations, and to assess the effects of the sterile male method and of insecticides on the dynamics of screwworm populations.

Knipling recognized that the level of suppression that is required to prevent further growth (stabilize) a population depends on its actual rate of increase. This can be summarized as follows:

Knipling estimated that the overwintering screwworm population typically increases at roughly five-fold between generations, and that this same rate of increase is likely to occur for the next two or three generations. For example, in an area of 100,000 square miles, if 1,000,000 screwworms overwinter,

this population will increase to 5,000,000, 25,000,000 and 125,000,000 in the F1, F2 and F3 generations, respectively. In any subsequent generations, the rate of increase would be less than five-fold so that the population would not greatly exceed 1,000 per square mile. In order to calculate the consequences of releasing sterile screwworms into a population of fertile screwworms, Knipling listed the various types of matings, calculated the frequency of each type of mating and assigned ten progeny to each mating of a normal (fertile) female (NF) with a normal (fertile) male (NM). When both sterile males (SM) and sterile females (SF) are released, there are four types of matings. Thus, the above wild population has 500,000

Method of calculating frequencies of various types of matings and resultant progeny when sterile males (SM) and females (SF) are released in to a population of normal males (NM) and females (NF).

Type of mating	Number of matings	Progeny per mating	Number of progeny
NF × NM	$\dfrac{500,000 \times 500,000}{5,000,000} = 50,000$	10	500,000
NF × SM	$\dfrac{500,000 \times 4,500,000}{5,000,000} = 450,000$	0	0
SF × NM	$\dfrac{500,000 \times 4,500,000}{5,000,000} = 450,000$	0	0
SF × SM	$\dfrac{4,500,000 \times 4,500,000}{5,000,000} = 4,050,000$	0	0

Trend of an insect population subjected to sterile insect releases when the normal increase rate is five-fold.

| Generation | Uncontrolled natural population (5X increase rate) | Controlled population | | Ratio sterile to fertile |
		Natural population	Sterile population	
1	1,000,000	1,000,000	9,000,000	9:1
2	5,000,000	5,00,000	9,000,000	18:1
3	25,000,000	1,31,625	9,000,000	68:1
4	125,000,000	9,535	9,000,000	942:1
5	625,000,000	50	9,000,000	180,000:1

normal males (NM) and 500,000 normal females (NF), and 4,500,000 sterile males (SM) and 4,500,000 sterile females (SF) are released. The frequencies of the various matings and the progeny produced are shown in the following table:

Using this method, Knipling constructed a model showing the trend of this hypothetical screwworm population subjected to sterile insect releases when the normal increase rate is five-fold.

Knipling's model indicated that if we assume a given insect has a net capacity to increase five-fold each generation, the ratio of fully competitive sterile to fertile insects will have to be 4:1 in order to keep the natural population stable. However, when the increase rate is five-fold, a sustained release rate sufficient to achieve an initial sterile to fertile ratio of 9:1 will have the effect shown in the table above. The first column of figures in that table shows the relative trend of an uncontrolled population increasing at a five-fold rate. Theoretically, an initial ratio as low as five sterile to one fertile will be adequate to start a downward trend in the natural population when the net increase rate is only five-fold. Actually, starting with this lower over-flooding ratio, theoretical elimination of the population will be achieved with fewer insects than will be required with a 9:1 ratio. However, in these models we are concerned with the overall population. In actual practice, the density of insects will vary in different parts of the environment. Moreover, the distribution of sterile insects will never be uniform. Therefore, in control operations, the initial ratio should be sufficiently high to be certain that an overall reduction in the population will occur in all parts of the environment from the start. In some instances, the sterility procedures might reduce the vigor and competitiveness of the organism. Allowance must be made for this factor.

The ratio of sterile to fertile insects increases asymptotically as the density of the wild population declines to low levels. Thus, in order to take advantage of the tremendous power of the SIT against sparse populations of pests, Knipling advocated that releases of sterile insects should be initiated when the wild population was at a seasonal low, or immediately after adverse weather events, such as freezes and hurricanes, had decimated it. In addition, Knipling designed pest management systems in which insecticides, biocontrol agents, etc., were used to reduce the density of the target population to a level at which the SIT could manifest its great suppressive power.

Over a period of several years, Knipling and Bushland continued to delve into screwworm behavior and to discuss Knipling's notion on finding a way to harness the sex drive of the screwworm to combat the parasite in a preventive manner rather than making rescue wound treatments after the livestock had become infested. However, neither Knipling nor Bushland knew of a method of inducing sexual sterility in insects. The quest to harness the screwworm's sex drive was interrupted in 1940 with Knipling's transfer to Portland, Oregon, to conduct investigations on mosquitoes and then by World War II when in 1942 both Knipling and Bushland were assigned to Orlando, Florida, to develop methods to combat insect vectors of diseases of concern to military personnel.

In 1946, Knipling was appointed Chief of the USDA's research on insects affecting livestock, man, households, and stored products. Knipling again turned his attention to the possibility of eliminating the deadly screwworm by means of the production and release of sterile flies. An absolutely essential technology was needed to induce sexual sterility in insects. However, such a method had already been

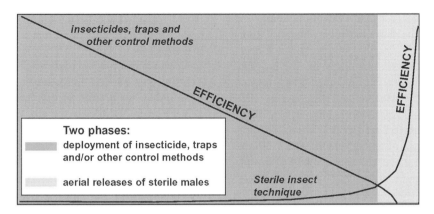

Fig. 975 Schematic representation of the sequential use of methods of suppression that efficiently decimate a dense population followed by use of the sterile insect technique, which becomes progressively more suppressive as the pest population declines. Source: FAO/IAEA.

discovered in 1926 at the University of Texas by Herman J. Müller, who used X-rays to induce mutations in the vinegar fly, *Drosophila melanogaster*. In 1950, Müller, now a Nobel laureate, published a popular article in the American Scientist with the intent of explaining to the scientific community that radioactive fallout from atmospheric testing of atomic bombs produced both lethal and non-lethal mutations in the sperm and eggs of living organisms. Müller had found that at elevated doses, X-rays caused complete sterility in both male and female flies. In correspondence between Müller and Knipling, Müller assured Knipling that X-rays and gamma rays would probably be effective in sexually sterilizing the screwworm. Accordingly, Knipling assigned Bushland to investigate the effects of X-rays on the screwworm. In brilliantly executed laboratory experiments, Bushland and Hopkins identified doses of X-rays that completely sterilized both sexes of the screwworm. They found that somatic damage did not occur when pupae were irradiated shortly before the adults emerged, because at this late stage all divisions of somatic cell required to form the adult had been completed. Also, Bushland showed that in laboratory cages, sexually sterile males, in competition with normal males, reduced the number of progeny of normal females, in accordance with Knipling's theoretical calculations.

Next, in 1953 Bushland's team evaluated the sterile male release concept on Sanibel Island, two miles off the coast of Florida. By over-flooding the native population with sexually sterile flies, the numbers of screwworm cases were, at times, reduced to zero. However, the continuous influx of flies from the Florida peninsula prevented the total elimination of the screwworm.

It was Knipling's incredibly good fortune that in 1953, B. A. Bitter, an animal health officer on the Dutch island of Curaçao, contacted him about screwworm control forty miles off the coast of Venezuela. Bitter requested the assistance of the USDA in dealing with a major screwworm infestation that was taking a major toll on goats and dairy cows. The USDA negotiated a cooperative agreement with the Government of The Netherlands to conduct a pilot eradication experiment on this island of 176 square miles. Bushland's team at Orlando produced sterile flies, and A. H. Baumhover and W. D. New were dispatched to conduct the operational program on Curaçao. In 1954, about 150,000 sterile screwworm flies per week were released over Curaçao, and within three months – the time needed for four generations to mature – the screwworm had been eradicated from the island. The Curaçao venture entailed considerable political risk, since failure surely would have resulted in severe criticism from the Congress for wasting money on such an unheard-of scheme.

The Florida livestock industry soon demanded that the USDA conduct an eradication program in the southeastern United States, and a plan to do so was developed. However, Knipling was hesitant to take this giant step, and he argued for an additional two years to develop more robust technologies, which could reduce program costs by $2 million. However, Florida Governor Thomas Leroy Collins asked Knipling, ''Why wait two years to save $2 million when losses are $10 million per year?'' Knipling often cited the Governor's astute reasoning in encouraging subordinates to take well thought out risks. The Florida Legislature appropriated $3 million for the program and the U.S. Congress matched this. The program started late in 1957 when a cold air mass invaded Florida and killed all screwworms southward to a line running from Tampa to Vero Beach, Florida. The program took advantage of this development by initially releasing sterile flies over the northern half of Florida to prevent reinfestation. An aircraft hangar in Sebring, Florida, was converted quickly into a giant 'fly factory' and, by the fall of 1958, it produced 50 million sterile flies per week. A fleet of 20 aircraft dropped sterile flies over the infested area, while livestock owners treated all wounds with insecticide smears. The University of Florida Cooperative Extension Service trained cattle producers in management practices needed to minimize opportunities for screwworms to infest animals, to report all cases of infestations, to collect samples of larvae from infested wounds and submit them for identification at the program's headquarters in Sebring. No screwworms could be found in Florida after the end of June 1959.

Cattle producers in the states bordering on Mexico insisted on a program to eradicate the pest from the entire southwest. The Southwest Animal Health Research Foundation raised $3 million from producers to initiate the program. The Texas Legislature had appropriated $2.8 million to match the federal appropriation. A mass rearing facility was built by the Foundation at an air base in Mission, Texas. The plan was to let the cold weather destroy the screwworms to the north of the overwintering area, and to release sterile flies only on the overwintering area. As soon as screwworms had been eradicated from the overwintering area, the sterile flies would be deployed to create a barrier zone along the U.S.-Mexico border to protect against reproduction by invading flies. In this way, the parasite was eliminated from Texas, New Mexico, Arizona and California, but the parasite population in the adjacent area of northern Mexico was managed to minimize the number of flies that would enter the United States.

During 1966 no screwworms could be found in the United States for several months, and the state governors persuaded the U.S. Secretary of Agriculture to declare that the parasite had been eradicated from the United States. This declaration caused the screwworm to be considered a foreign pest, and thus the federal government became solely responsible for costs incurred when screwworms reappeared in the United States.

Eradication of the screwworm to the Isthmus of Tehuantepec was the long-range goal, but the program could not be moved into Mexico because the Governments of Mexico and the United States took six years – until 1972 – to negotiate an agreement. The strategy of eradication was replaced with area-wide population management as a static holding action along the entire border with Mexico. During this static decade many difficulties arose. The sterile fly barrier proved to be narrower than the flight range of the screwworm, so that infestations occurred north of the international boundary. The deer population in Texas exploded, and many ranchers reduced the number of cowboys needed to treat wounds. Screwworm cases began to arise as far as 300 miles north of the Mexican border. Serious difficulties were encountered in 1968 when almost 10,000 cases were recorded in the United States, and 1972 was a disaster with more than 95,000 cases. Prestigious scientists criticized the program and decried the strategy of eradication.

The Mexico-American Screwworm Eradication Agreement was signed in 1972 by the two Secretaries of Agriculture with the purpose of removing the parasite to the north of the Isthmus of Tehuantepec and then establish there a sterile fly barrier. For this purpose, a Joint Mexico-U.S. Screwworm Eradication Commission was created. Field operations began in 1974 and a newly constructed screwworm factory near Tuxtla Gutierrez came into full production in January 1977 with a capacity of 500 million flies per week. By 1982, the Commission employed 2,031 people.

The approaches and technologies employed were the latest versions of those that had been pioneered in the U.S. campaigns with the addition of SWASS (Screwworm Adult Suppression System), and a pelletized form of attractants and toxicant developed under the leadership of Wendell Snow for broadcast application. SWASS was effective in attacking high population densities, a necessity to obtain an overwhelming ratio of sterile to wild flies.

The initial deployment of sterile flies was in the Mexican states bordering the U.S., Baja California, and regions of Tamaulipas and Vera Cruz, which provide very favorable screwworm habitat and propitious routes for flies migrating toward Texas.

Heavy emphasis was placed on examining and, if it was necessary, treating every animal. Some livestock inspectors traveled routes by horseback that required 15 days to traverse. About 11% of the larvae recovered from wounds were secondary species. By 1984, the Commission had achieved the goal of eradicating to the Isthmus of Tehuantepec.

Dr. Peneda-Vargas, who served as the Mexican Director of the Commission during this period, wrote, "I feel compelled to state that this accomplishment, eradication, was the result of hard work by thousands of people both in the Commission and outside of it, but it would be impossible to overstate the contribution of the Mexican rancher. It was the livestock owner who first saw the need for the program, who initiated and supported it for a period of almost 25 years, and who was in the final analysis most responsible for its success. It was his attention to his animals and his cooperation that made it possible to locate and eliminate every infestation from this large country with all its extremes of terrain and ecology."

In 1986, operations were extended to the Yucatan Peninsula and the countries bordering Mexico. Mexico was declared free of screwworm in February 1991. In order to further reduce the threat of the screwworm to the United States, the decision was made to move the sterile fly barrier to Panama. Thus, agreements were signed between the U.S. Department of Agriculture and the various Central American governments. By 2001, the screwworm had been cleared from Central America north of Panama and a barrier of sterile flies was established across Panama from the Atlantic to the Pacific Ocean.

It is important to keep in mind that with modern travel and transportation, *C. hominivorax* could easily be re-introduced into areas that have been cleared of this plague. The defenses against this parasite would be improved if all of the Caribbean countries could be freed of the parasite, and eventually the problem in South America must be addressed. However, sufficient progress has been made to enable us to recognize the triumph of an original idea and the benefits nuclear science can add to our quest to assure a reliable and adequate food supply.

Physiological and cytogenetic bases of sterility
Reproductive physiology of male insects. To understand the physiological nature of sexual sterility in insects, familiarity with fertility and reproduction in insects is important. The insect male's mating drive is scarcely affected by castration, and spermatogenesis is under the control of the frontal ganglion and the prothoracic gland. The paired testes become visible in the early instars, and in many species sperm are produced before adult emergence. Nevertheless, the development of the germ cells is not correlated closely enough with morphogenesis to estimate the degree of sex-cell maturation based on the individual's life stage. In many species, all types of germ cells are present in the late immature instars and in adults. However, in other species, most notably those belonging to the Lepidoptera, only one or two types of germ cells are present at a given life stage. The spermatozoa pass from the testis through the vasa deferentia into paired seminal vesicles for temporary storage. Typically, a pair of accessory glands flanks

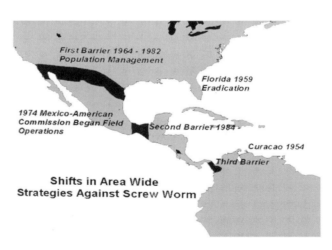

Fig. 976 Progress in eradication of the screwworm.

Types of germ cells present in the testes of many species of male insects at a given life stage.

Life stage	Cell stage in testes
Hatching to penultimate level	Spermatogonia
Pupa	Spermatogonia, primary and secondary spermatocytes
Pupa and adult	Spermatogonia, primary and secondary spermatocytes, spermatids and mature sperm

Types of germ cells present at given life stages of the silkworm (Sado, 1961).

Life stage	Cell stage in testis
Hatching to first instar	Primary spermatogonia
Second to third instar	Secondary spermatogonia
Third to fourth instar	Spermatocytes in early meiotic prophase
Fourth to fifth instar	Spermatocytes in early and late meiotic prophase
Fifth instar to spinning	Same as fourth instar plus a few early spermatids
Pupa	Spermatids and fully formed spermatozoa
Emerged adult	Only fully formed spermatozoa

the seminal vesicles, and their secretions are transferred with sperm during copulation.

Each testis consists of tubules or follicles, and each tubule is divided into a number of zones. Most distal in the tubule lies the germarium comprised of the primary spermatogonia. Usually, each spermatogonium detaches from the germarium and becomes enveloped by cyst cells. The spermatogonium then undergoes mitotic divisions to form 256 primary spermatocytes. Next, in the zone of maturation and reduction, each of these primary spermatocytes divides into two haploid secondary spermatocytes, and then each secondary spermatocyte divides to produce two spermatids. In the zone of transformation, the spermatids (still enclosed in the cyst) elongate to form sperm. Finally the sperm rupture the cyst and accumulate in the seminal vesicles as densely packed bundles. In Diptera, during mating the sperm

are transferred in semen from the seminal vesicles to the female's spermatheca, while in Lepidoptera the male synthesizes a spermatophore, fills it with sperm bundles, and positions the opening of spermatophore to allow the sperm bundles to pass down the seminal duct to the spermatheca.

The amount of time required for a spermatogonium to develop into a bundle of mature sperm varies from species to species, but in most species this time period probably ranges from 5 to 15 days. In *Drosophila melanogaster*, this period is roughly 5 to 6 days, and another two days of maturation are required before the sperm are transferred during mating. Newly formed sperm bundles continuously enter the elastic seminal vesicle to increase its volume and to mix with sperm that had matured previously.

Male insects produce much less sperm than do male mammals. An inseminated *Drosophila* female stores 500 to 700 sperm in her spermatheca, and all of these sperm may be utilized, because a female may lay more than 500 eggs. When the *Drosophila* female remates, the sperm of the second mating may displace the majority of the stored sperm. In contrast, when the female boll weevil remates, the sperm received from the second mating are used much more frequently than the sperm from the first mating.

The frequency of copulation by males, and their ability to transfer sperm in successive matings, varies between species. *Habrobracon* males can mate 14 times in rapid succession and transfer sperm in every mating. Boll weevil males mate 2.55 ± 1.50 times per day. Male *Aedes aegypti* have been observed to mate 30 times in 30 minutes, however, sperm is transferred only to the first 3 or 4 females. Also, in

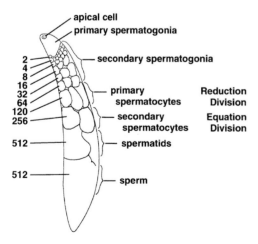

Fig. 977 Schematic diagram of a testis tubule as found in Diptera and Coleoptera. The numbers on the left indicate the number of cells in a cyst or sperm bundle.

Schedule of spermatogenesis in *Drosophila melanogaster* (Martin, 1965).

Late spermatogonia	Primary spermatocytes	Secondary spermatocytes	Spermatids		Sperm
0 hours	24 hours	48 hours	72 hours	96 hours	120 hours

Drosophila, few sperm are transferred after the 3rd closely spaced mating. In *Drosophila* this inability to transfer sperm is not caused by the lack of sperm, but by the exhaustion of the accessory gland secretion, without which sperm transfer is impossible. Regeneration and replenishment of accessory gland secretion in *Drosophila* requires more than 2 hours, and 24 hours in *Aedes aegypti*.

Reproductive physiology of female insects. Each female possesses a pair of ovaries for egg production. Each ovary consists of ovarioles varying in number from one to several hundred depending on the species. Primitive orders such as Odonata and Orthoptera have panoistic ovarioles without special cells for yolk production. In such insects, nutrients absorbed from the hemolymph are converted into yolk by an enlarged oocytes nucleus or germinal vesicle.

In other orders, special nutritive cells or trophocytes for yolk production are present in the meroistic ovarioles. Depending on the location of the nurse cells in relation to the oocytes, meroistic ovarioles are categorized as either telotrophic or polytrophic. In telotrophic ovarioles the trophocytes are distant from the oocytes, and they deliver yolk via cytoplasmic strands. In polytrophic ovarioles, the trophocytes lie adjacent to the oocytes and form part of the egg follicle. Telotrophic ovarioles are characteristic of Coleoptera, while polytrophic ovarioles are characteristic of Diptera and Hymenoptera. The following

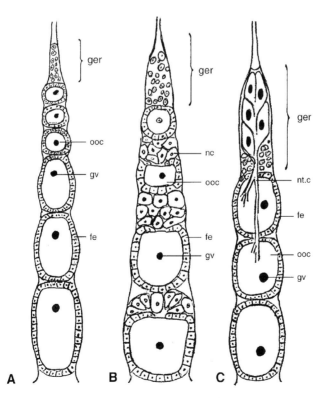

Fig. 978 Diagrams of the principal types of ovarioles based on R.E. Snodgrass (1935): (A) panoistic (Odonata and Orthoptera); (B) meroistic and polytrophic (Diptera and Hymenoptera); and (C) meroistic and telotrophic (Coleoptera). fe = follicular epithelium, ger = germarium, gv = germinal vesicle, nc = nurse cell, nt.c = nutritive chord, ooc = oocyte.

description pertains to species with polytrophic ovarioles.

Egg production depends on mitotic division of oogonial cells. These oogonial divisions produce a cyst of interconnected cells. Depending on the species, each oogonium gives rise to one oocyte and 1, 7, 15, 31 or more trophocytes. For egg maturation to commence, important changes first must occur in the nuclei of the trophocytes. These cells must nurse the growing oocytes. In preparation for this, the nurse-cell chromosomes undergo endomitosis, which entails the repeated replication of trophocyte chromosomes without cell division.

During oogenesis, the volume of a trophocyte cell nucleus increases an estimated 2,000-fold. Thereafter, the oocyte proceeds to grow rapidly. In *Drosophila*, the volume of the oocytes increases by over 100,000-fold in three days. After the oocytes have matured, the trophocytes degenerate and disappear. After vitellogenesis has been completed, the follicle cells secrete the chorion, but leave an opening, the micropyle, to facilitate entry of sperm. Finally, the egg ruptures through the follicle, passes down the vagina, receives sperm from the spermatheca, and is deposited. The presence of the sperm in the egg stimulates its nucleus to divide into four haploid nuclei. One nucleus becomes the female pronucleus and unites with the sperm (syngamy) to form the zygote.

Causes of sterility. Sterility is defined as the inability to produce offspring. Conversely, fertility is the ability to reproduce. Fecundity is the number of progeny produced per female. Thus, only females can be infecund. Sterility may be caused by (1) inability of females to lay eggs (infecundity), (2) inability of males to produce sperm (aspermia), or inability of sperm to function (sperm inactivation), (3) inability to mate or (4) dominant lethal mutations in the reproductive cells of either the male or the female. All of these mechanisms may be induced by exposure of insects to gamma rays or X-rays or to certain chemicals. In addition, sterility may be induced by insect growth regulators, which can be transferred from a treated male to an untreated female during mating, and subsequently disrupt the development of the embryo by interfering with endocrine mechanisms.

Dominant lethal mutations are of foremost importance in the effective use of sterility by means of the sterile insect technique (SIT). In many instances, the successful use of the SIT also requires that infecund-

ity has been induced. For example, in the use of the SIT against the New World screwworm, both males and females are irradiated in the late pupal stage with a dose of gamma rays that induces dominant lethal mutations in all of the sperm of the male and that induces infecundity in all females. In addition, this dose must destroy the spermatogonia of the male so that the male cannot recover fertility.

Infecundity. A reduction in fecundity is expressed as a depression of egg production. Depressed egg production is almost always observed following the treatment of female insects with certain chemicals (antimetabolites and alkylating chemicals), various forms of ionizing radiation (X-rays, gamma rays, and neutrons), ingestion of radioisotopes and intense ultraviolet radiation.

Egg production is dependent on the differentiation and development of oocytes from oogonia and their supply of nutrients from properly functioning nurse cells (trophocytes). Both the oogonia and the trophocytes are readily impaired by the above treatments. Severe damage to these two cell types results in permanent infecundity. In Diptera, the trophocytes attain a high degree of polyploidy during development. Before the trophocytes have fully differentiated (during endomitosis), they are extremely sensitive to the effects of ionizing radiation and to antimetabolites and alkylating agents. On the other hand, after the trophocytes have differentiated they are extremely resistant to impairment by these agents. For example, when the yellow fever mosquito, *Aedes aegypti*, females are irradiated four hours after ingesting a blood meal, only 10 krad are required to induce infecundity, yet 100 krad are required to induce the same effect when the irradiation occurs 42 hours after the blood meal.

Inability or failure to mate. Irradiation of either male of female insects with very high doses of ionizing radiation or of chemosterilant, or the application of these mutagenic agents when many somatic cells are dividing, can prevent the treated insects from mating, or at least from mating readily. For example, weevils are debilitated at substerilizing doses of ionizing radiation or by most alkylating chemosterilants. This phenomenon was studied in detail in the boll weevil, *Anthonomus grandis* Boheman. In the midgut of the adult boll weevil, the secretory cells are sloughed to release the digestive enzymes into the lumen of the gut, and these secretory cells are continuously replaced by dividing cells. The latter are destroyed with doses of ionizing radiation or alkylating agents that are only a fraction of the doses required to induce dominant lethal mutations in the sperm. The damage to the midgut prevents the boll weevil from digesting food, and also makes the gut vulnerable to penetration by microorganisms. The capacity of male weevils to mate falls off sharply several days after exposure to ionizing radiation or most alkylating agents.

In the Lepidoptera, the large doses required to induce sterility in the males may impair the transfer of sperm from the male to the spermatheca of the female. During coitus, lepidopteran males enclose their sperm in a spermatophore and position the opening of the spermatophore against the opening of the seminal duct so that the sperm can traverse from the spermatophore down the seminal duct to the spermatheca. Irradiated males in some species experience difficulty in placing the spermatophore opening against the seminal duct opening. Consequently the females from such pairings do not lay eggs, and they call for another mate.

Sperm inactivation. In most insects of economic importance, the frequency of sperm inactivation is not easily ascertained because lack of egg hatch can be the result of dominant lethal mutations, aspermia or sperm inactivation. However, sperm inactivation by mutagenic agents is studied most readily in the Hymenoptera in which fertilized eggs develop into females, and nonfertilized eggs develop into males. Any egg fertilized with a sperm in which a dominant lethal mutation has been induced will not hatch because the mutation will cause the embryo to die. Generally, inactivation of a significant portion of the sperm requires higher doses than are needed to induce dominant lethal mutations. Careful studies have shown that in some species, some sperm are inactivated at fairly low doses of radiation. However, sperm inactivation has not been found to be an important factor affecting applications of the sterile insect technique.

Dominant lethal mutations. Ionizing radiation induces dominant lethal mutations. A dominant lethal mutation is a change in the nucleus of the sperm or the egg that usually causes the death of the zygote, even when such mutation is introduced by only one of the two germ cells that unite at fertilization. (Of course there are also dominant lethal mutations that

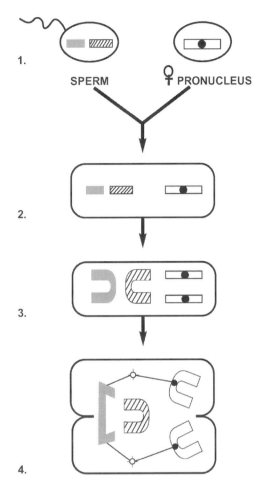

Fig. 979 Schematic representation of the effects of a broken chromosome in a sperm on cell division in the zygote or early cleavage stage. (1) Irradiated sperm with broken chromosome unites with normal oocyte pronucleus; (2) and (3) after syngamy, each chromosome produces sister chromatids. Broken ends of chromatids stick together and heal to form acentric and dicentric isochromosomes. (4) At the cleavage division, a chromosome bridge is formed and the acentric fragment is lost. The bridge may terminate cell division or it may break. In any case the daughter cells will be genetically unbalanced.

are expressed at later stages of development and they prevent reproduction of the bearer of the mutation.) A dominant lethal mutation does not impede the maturation of the treated germ cell into a gamete or the subsequent participation of the gamete in fertilization to form the zygote. However, the dominant lethal mutation prevents the zygote from developing into a reproducing adult. Lethal mutations are not lethal to the treated cell. Rather, lethal mutations are lethal to a descendant of the treated cell. Because only one dose of a dominant allele is needed for expression, dominant lethal mutations are expressed in the first generation of the treated cell. However, recessive lethal mutations may be passed on to many generations before they appear in homozygous form required for expression.

Numerous studies that involved insect species in diverse orders and families have shown that dominant lethal mutations are characterized by the presence of chromosome fragments and bridges between dividing nuclei in the zygote or young embryo. The vast majority of dominant lethal mutations induced by ionizing radiation are the result of chromosome breaks. A broken chromosome, in which the two segments do not rejoin and heal prior to cell division, is the simplest kind of dominant lethal mutation. The chromosome fragment, which lacks a centromere, may not be included in the daughter nucleus, and this would be a lethal deficiency. Moreover, the broken end of a chromosome is sticky and readily joins with any other broken end of a chromosome. Chromosome breaks that rejoin, but not in the original fashion, invariably have a lethal effect. Such asymmetrical exchanges result in acentric fragments (fragments without a centromere) and dicentric chromosomes. Dicentric chromosomes form bridges between the two daughter nuclei formed during cell division. Such a bridge may block cell division or the bridge may break to form genetically unbalanced chromosome fragments that are lethal. In the eggs of the screwworm formed in matings of irradiated males and untreated females, numerous chromosome aberrations occur in the first two cleavage divisions, and that embryonic development seldom progressed beyond the second cleavage division. However, when an irradiated female was mated to an untreated male, numerous chromosome aberrations were found during the first two meiotic divisions. These meiotic divisions occur after the egg is laid and before the cleavage divisions can occur. Again, numerous chromosome aberrations occurred during the cleavage divisions and the embryos invariably died before blastoderm formation.

Aspermia. Aspermia is the lack of a supply of sperm. Either mature sperm are not produced or their supply becomes exhausted and their replenishment is prevented. Aspermia can be induced by exposure of immature insects to a dose of a mutagenic agent sufficiently high to inhibit the spermatogenic cycle (kill

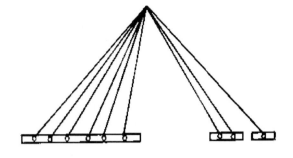

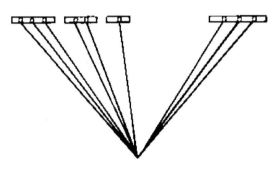

Fig. 980 Diagrammatic representation of a one holokinetic chromosome at late metaphase in a dividing nucleus. The breaks have been induced with ionizing radiation. In the F_1, broken ends of chromosomes join to form a variety of translocations between chromosomes. These translocations cause lethal genetic imbalances in the gametes, and are the primary cause of sterility in the F_1 generation.

the spermatogonia). Irradiation of an immature whose testes contains only spermatogonia may destroy them, so the resulting adult is devoid of sperm. Irradiation of the adult male also can result in aspermia because the spermatogonia are killed at substerilizing doses. Thus, such an adult male may exhaust his supply of sperm after several matings.

In order to achieve permanent sterility by means of irradiation, it is necessary to kill all of the spermatogonia. Any spermatogonia that survive irradiation will continue to divide and to repopulate the germarium. Sperm derived from such surviving spermatogonia lack chromosome breaks, and hence they largely lack dominant lethal mutations.

Transfer of a chemosterilant during mating. A compound with hormonal effects can be transferred to females by treated males in the process of mating in sufficient amounts to prevent the development of progeny. This was first shown by Massner et al. with *Pyrrocoris apterus* (L.). As pointed out by Knipling, such a chemical sterilant, when applied to individuals in the wild populations has the following two effects: (1) it prevents the treated insect from reproducing and (2) it converts the treated insect into a biological agent that can nullify the biological potential of untreated individuals with which it mates. For example, if 90% of a population of 1,000,000 were simply killed, the 100,000 survivors would reproduce normally. However, if 90% of the population was effectively treated with the sterilant, then the 900,000 treated insects would compete for mates with the 100,000 untreated insects. Thus only 10,000 insects would succeed in reproducing. Clearly, if this level of treatment were continued for several more generations, the population would become extinct. Pheromone-baited traps fitted with contaminating devices were developed to dispense the mutagenic chemosterilant bisazir to flies. Subsequently, an insect growth regulator was used as the sterilant, and it provided encouraging results in a field trial. The latter compound arrests development during the pupal stage. More recently, this approach was used against tsetse on the Buvuma Islands, Uganda. Unfortunately, only 25 to 30% of wild tsetse populations are attracted into the contaminating devices, though models indicate that the effect of this level of sterilization of untreated individuals by mating with chemosterilant-contaminated individuals is considerable. This autosterilization approach is entirely compatible with the release of sterile insects, and is especially applicable as a component of barriers to reinvasion of zones from which the pest has been completely eliminated.

Inherited sterility in Lepidoptera

Because the very high doses of gamma rays (40,000 krads) required to induce complete sterility in codling moth males is quite debilitating in most species, the effects of partially sterilizing doses of radiation on the competitiveness of the codling moth were investigated. When the treated males were crossed to untreated females, most of the progeny were males. Further, F_1 males crossed to untreated females produced F_1 males with a higher level of sterility than their irradiated fathers. Also, the F_1 females had a high level of sterility. Simulation models showed that released males with inherited sterility

would suppress the wild population to a greater extent than the release of equal numbers of fully sterile males.

The phenomenon of inherited sterility was found to occur in species whose chromosomes have a 'diffuse centromere' with multiple spindle fiber attachments that are characteristic of the Lepidoptera, Hemiptera and Acari. Such chromosomes have a diffuse centromere and have been referred to as holokinetic or holocentric. Some have argued that these terms are not appropriate, since the kinetochore plates do not extend the full length of the chromosome, but only about 70% of the length, as in the milkweed bug, *Oncopeltus fasciatus* (Dallas). In any case, these multiple spindle fiber attachments insure that most radiation-induced chromosome breaks will not lead to the immediate loss of chromosome fragments. Indeed, these fragments persist and are transmitted through the germ cells to the next generations. The chromosome fragments may unite and form various types of chromosomal translocations, which result in genetically unbalanced gametes in the F_1, and serve as the primary cause of the enhanced level of sterility.

In spite of the existence of inherited sterility, its application for suppressing lepidopteran pests has been slow to develop because ionizing radiation harshly affects (1) the formation and functioning of the two types of sperm, eupyrene and apyrene, and (2) the ability of the male to properly position the spermatophore opening adjacent to the opening of the seminal duct and thus facilitate the transfer sperm to the spermatheca.

Eupyrene sperm occur in bundles and possess nuclei and large mitochrondrial derivatives. These nucleate sperm are needed to fertilize the egg. Apyrene sperm do not occur in bundles after they leave the testis, and they lack nuclei. Apyrene sperm appear to be necessary for the transport of the eupyrene sperm from the spermatophore down the seminal duct into the spermatheca, and to prevent the inseminated female from calling for another mate. Eupyrene sperm are susceptible to damage by irradiation, and damage to the flagellum and apex of the sperm cell are frequent. Moreover, in the F1 male progeny, the number of eupyrene sperm per sperm bundle appears to be reduced, and the many sperm exhibit ultrastructural abnormalities. However, only normal appearing sperm are found in the spermathecae of females that mated with an F_1 male.

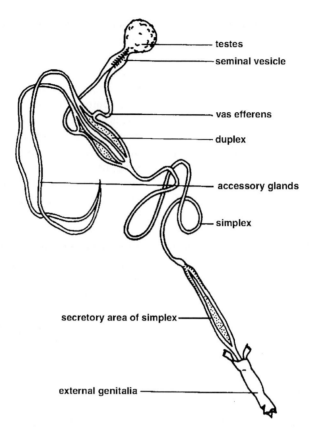

testes
seminal vesicle
vas efferens
duplex
accessory glands
simplex
secretory area of simplex
external genitalia

Fig. 981 Diagram of the male reproductive system of the tobacco budworm, *Heliothis virescens* (Fabricius). Courtesy of H.M. Flint.

Even though F_1 males have not been irradiated, they often fail to transfer sperm to the spermathecae of females. For example, when P_1 adult *Anagasta kuehniella* (Zeller) males received 15 to 20 krads of gamma rays, roughly 50% of F_1 males failed to transfer eupyrene sperm. Nevertheless, season-long releases of partially sterile males, which resulted in production of F_1 progeny on the field, strongly suppressed wild populations of *Heliothis virescens* in North Carolina and of *Cydia pomonella* in Washington.

Appropriate applications and some limitations of the SIT

The sterile insect technique is a tactic almost always used in association with other control tactics as part of a pest management system designed to accomplish a certain strategy of pest management. Thus, a 'tactic' is a method for detecting, monitoring or controlling a pest. A 'system' is an assemblage of tactics that are applied simultaneously or sequentially so that the effects on the pest population of individual tactics are either additive or mutually potentiating, and so that counter-productive (negative) interactions are avoided or minimized. A pest management 'strategy' is a broad overall plan for employing pest control tactics to minimize negative ecological, economic, sociological and/or political impacts of a pest problem, both in the short term and in the longer term. Successful strategies should strongly protect agriculture on a sustained basis.

Pest management strategies include (a) prevention or containment, (b) temporary alleviation, (c) management of local populations, (d) management of total populations in an ecosystem or area-wide pest management, and (e) eradication of a pest, i.e., the destruction of every individual of a species in an area surrounded by natural or man-made barriers sufficiently effective to prevent re-invasion except through the intervention of man.

Each control tactic has characteristics that need to be considered in the design of a system to implement a strategy effectively and efficiently. Ecological

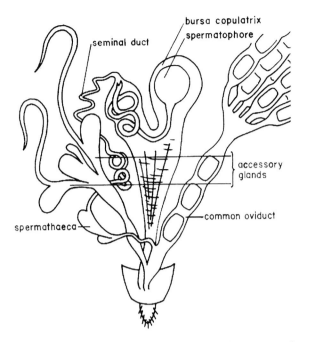

Fig. 982 Diagram of the female reproductive system of the tobacco budworm, *Heliothis virescens* (Fabricius). Note the position of the open end of the spermatophore at the entrance of the seminal duct. Courtesy of H.M. Flint.

selectivity is important to avoid the destruction of natural enemies needed to prevent the resurgence of the pest population following a control operation. Most conventional insecticides are relatively nonselective. Highly selective tactics include the SIT and other genetic techniques, resistant crop varieties, certain insect pathogens, parasitoids and predators and certain artificial and naturally occurring attractants. Light traps, some attractant baits, 'general' predators, parasitoids and pathogens, and certain insecticides are only moderately selective. These characteristics are displayed in the following chart:

Ideally pest management systems are based on an appreciation of the selectivity and efficiency of the available tactic. Whenever possible, highly selective measures should be chosen. When suppression of populations to very low levels is required, methods that are effective against high populations, and methods that are effective against low populations, should be integrated in such a way that the former potentiate the latter. When the economic threshold of the pest is moderately high, several tactics that are effective against dense populations often may be combined to give much more reliable suppression than a single method.

Because the SIT is not efficient in combating dense pest populations, it is almost always used in concert with a method that is effective against dense populations. Conversely, since the SIT is extremely efficient in suppressing or eradicating low density pest populations, it is used to eradicate incipient infestations of major pests. By analogy, the SIT is useless against the 'roaring blaze' but extremely effective in extinguishing the sparks. Thus, the SIT is used effectively as a preventive measure to prevent the establishment of immigrant populations of the Mediterranean fruit fly in the vicinities of Los Angeles, Tampa and Miami, and of the pink bollworm in cotton growing areas of California. Similarly, the SIT was used to prevent the spread of the screwworm throughout Africa and to eradicate it from its foothold along the Mediterranean coast of Libya.

The SIT is extremely useful for combating the leading edge of a pest population invading uninfested territory. To be effective, the width of the sterile insect barrier must be greater than the range of dispersal of the wild females. Thus, the MOSCAMED Program has used the SIT as an important component of the barrier to contain the Mediterranean fruit fly in

	Constant efficiency at all pest densities	Efficient at high pest densities	Efficient at low pest densities
Highly selective ecologically	Resistant varieties; sex pheromones used to confuse; synthetic attractants	Selective insect pathogens, parasitoids and predators	SIT and other genetic techniques; sex pheromone baited traps
Moderately selective ecologically	Light traps; food baits; systemic insecticides	General predators, parasitoids and pathogens	
Nonselective ecologically	Most chemical insecticides; cultural measures; harsh weather events		

Guatemala and prevent its movement into Mexico. Since 1957 the SIT has been used in barriers to contain populations of the screwworm.

Pest-free zones and pest-free fields of production are area-wide approaches to facilitating exports of agricultural commodities without the need for post-harvest or quarantine treatments. Both the International Plant Protection Convention and the Sanitary and Phytosanitary Agreement are structured to accept and encourage area-wide pest management as a tool for promoting safe trade and contributing as much as possible to the complementary goals of food security and economic security for all countries. In Florida, sterile Caribbean fruit flies, *Anastrepha suspensa* Loew, are part of a system used each year to create fly-free fields of citrus production as required to permit the export of fresh citrus to Japan.

The effectiveness of the SIT is affected by many variables. For pest species, which have contagious or clustered distributions, it is the over-flooding ratio in the clusters that determines the effectiveness of the SIT. Seasonal patterns of abundance must be taken into account, since a high overflooding ratio can be established most easily when the wild population is low, and since the SIT has tremendous impact when imposed on a wild population with a low or negative intrinsic rate of increase. The longevity of gravid females is important, because they tend not to mate. Thus, it is often cost effective to make one blanket treatment with an insecticide to kill gravid females. The overlapping of generations prolongs the period during which sterile insects must be released. Advantage should be taken of climatic events that synchronize insect development. For example, in the program to eradicate the screwworm from Libya, releases of the sterile flies were begun on December 18, 1990, when the cold rainy weather had halted all reproduction and the wild population consisted entirely of pupae in the soil. Thus, early in February 1991, when wild flies emerged synchronously from beneath the soil, they encountered numerous sterile males. Consequently, the wild population was decimated, and no wild flies could be found after the end of April 1991. The induction of close to 100% sterility is critically important in combating pest species with high intrinsic rates of increase. Also, rearing, sterilization, and release procedures may diminish the quality of the released flies. Clearly no other method of pest suppression is influenced by as many factors as is the SIT.

The SIT is an underutilized and widely misunderstood technology. There is a need for additional data to correlate the frequencies of matings of sterile males with wild females with the dynamics of the pest population.

Economic analysis of SIT programs

Apart from clearly objective measures such as technical effectiveness or cost efficiency, there are several subjective measures that come into the evaluation of SIT and other area-wide programs. These include risk of failure of containment or eradication, the boundaries of externalities (for example, variable probabilities of pesticide drift under different conditions, or concerns of people who feel they have no economic stake in a program, as is usually the case when an agricultural pest must be combated in an urban setting), and time preferences for returns on capital investments (such as insect rearing facilities). Such subjective issues may cause great difficulties in reaching agreed decisions, even with a consensus on the data used. It is useful to analyze three general classes of economic problems: (1) annual comparisons of costs and benefits, (2) initial capital costs with delayed projected benefits, and (3) initial costs with delayed and uncertain future benefits.

The two basic principles governing the selection of area-wide management over individual local pest management are effectiveness and efficiency. Containment and eradication can be achieved only on an area-wide basis, and total effectiveness is required. Questions of efficiency subsequently arise in deciding which area-wide techniques to employ, not in comparing them to uncoordinated local control. Nevertheless, economic efficiency must be considered at an early stage to decide whether containment and eradication are affordable goals, and if so, whether the resources can be assembled.

The SIT also may be applied on an area-wide basis for population suppression and not for eradication, and this is certainly the case when the SIT is used to create pest-free fields of production. Suppression may or may not be more effective by area-wide management than by management on a local basis, but the main issue is one of cost efficiency. Is area-wide pest suppression more cost efficient than the sum of local pest management? This is a challenge, because local pest management often yields private returns of 4:1.

Size, coordination and economies of scale. Legal authority to conduct a coordinated SIT program may be based on existing laws, or it may require a referendum of all affected parties. Usually, in the USA such referenda must pass with a majority of two thirds.

The optimal size of an area or a sub area for a SIT program depends on the mobility of the pests, the uniformity of their densities and the costs of coordination, and whether pre-existing decision-making and administrative institutions exist. Total detection plus suppression costs per hectare of crop usually will decline as the size of the area increases. However, suppression costs usually decline with increasing size, but they may increase in very large programs if expensive items as helicopters or airplanes must be purchased. Per hectare organizational costs (management, meetings, solving financial problems, etc.) tend to rise sharply with project size.

Approaches to dealing with problems of coordination and economies of scales include: centralized handling of area-wide surveillance with suppression left to individual producers, and division of the area into zones and sub-zones. For example, a zone may correspond to a political subdivision, and sub-zones may be established for areas of uniform pest density. In some eradication programs, growers in areas where the pest is only a minor problem are unwilling to pay as much per hectare as growers in areas were the pest is devastating.

Benefit-cost assessment and discounting net returns. Investment in large-scale SIT programs usually is made with the expectation that program benefits will accrue over a multi-year time horizon. Therefore, we must discount future benefits to balance them against present or near term expenditures. The stream of discounted annual benefits and costs for many years of an undertaking can be summed up and expressed as a single value, known as the present value net benefits (PVNB). The formula for calculating the PVNB for a 15-year project is as follows:

$$PNVB = NB_1 + w_2NB_2 + w_3NB_3 + w_4NB_4$$
$$+ \cdots w_tNB_t \cdots w_{15}NB_{15},$$

where NB_t represents the net benefits in year t, and w_t represents the weighting factor for year t. The weighting factors are a function of a discount rate (r):

$$w_t = \frac{1}{(1+r)^t}$$

The discount rate is the 'opportunity cost' of the money, or the interest value that money could earn if allocated to the best alternative use. This rate may be established by subtracting the national inflation rate from the bank interest rate for savings. In normal times, this procedure generally will produce a figure around 4 or 5% in developed countries. This represents the 'reasonable' person's discount on the future, since people put their money in the bank to gain this premium, and otherwise they would spend it now. So, the benefit of eradication next year is worth 5% less if it is brought back to the present. Benefits in 20 years are only worth 37% of their face value when brought back to the present. In riskier economic environments, discount rates will be much greater, so the calculated net present value of future benefits may be insignificant. However, for SIT programs involving vectors of human diseases, the futures of groups of people are at stake, and it does not seem appropriate to discount benefits in the manner appropriate for private investments. The health of the human population 30 years in the future seems just as important as the health of the population at present. Nevertheless, investments in vector control programs must be subjected to critical analysis in the interest of efficient and sound management. However, if high discount rates (e.g., 25%) are selected commensurate with economic risk in many developing countries, then it seems unlikely that vector control programs in these countries would ever be launched.

With respect to the screwworm eradication in Jamaica, discount of the benefits and costs at 11% and produce the following data:

The benefit-cost ratio for screwworm eradication in Mexico was in the order of 10:1.

Discounted value of	After Year 3		After Year 10	
	Scenario 1 (US$ million)	Scenario 2 (US$ million)	Scenario 1 (US$ million)	Scenario 2 (US$ million)
Benefits	18.3	18.3	48.0	48.0
Program costs	4.5	8.2	4.5	8.2
Net savings	13.8	10.1	43.5	39.8
Benefit-cost ratio	4.1:1	2.2:1	10.7:1	5.9:1

Who should pay for SIT programs? Or how much does each group of beneficiaries gain?. In developing a concept of how the cost of an area-wide project should be partitioned between producers and consumers, it is helpful to consider the market effects on the market price of the commodity produced. An effective area-wide program will have the immediate effect of increasing supply of the commodity. The relative benefit to producers and consumers can be determined easily by analyzing the relationship of unit price to quantity of production.

If producers profit significantly from the SIT program (increased yield and lowered costs), they may be expected to expand the area under production. If the price of the commodity is elastic (because extra production is easily stored and exported), then an increase in supply will not greatly reduce the market price. In this case, consumers are likely to benefit more than producers, and they should shoulder part of the program cost. On the other hand, for a commodity with an inelastic demand any program that greatly stimulates production will cause the unit price to fall, and this could be damaging to producers.

SIT programs have good potential for capturing returns through commercial sales or privatized management. Privatization is likely to become common for SIT where it is used for suppression rather than eradication. In eradication programs, the period in which sterile insects are used may be too short to recover the cost of private investment in a rearing facility. However, in suppression programs, profits realized over a much longer period can repay private investment in facilities and avoid the need for publicly initiated or coordinated control programs. In particular, the easy and efficient intercontinental transport of sterile insects by air can be expected to allow commercial competition based on price and quality and opportunities for specialization and economies of scale in production facilities.

Waldemar Klassen
University of Florida
Homestead, Florida, USA

References

Bloem, S., K. A. Bloem, J. E. Carpenter, and C. O. Calkins 2001. Season-long releases of partially sterile males for control of the codling moth, *Cydia pomonella* (Lepidoptera: Tortricidae) in Washington apples. *Environmental Entomology* 30: 763–769.

Graham, O. H. 1985. Symposium on eradication of the screwworm from the United States and Mexico. *Miscellaneous Publications of the Entomological Society of America* 62: 1–68.

Klassen, W. 1989. Eradication of introduced arthropod pests: theory and historical practice. *Miscellaneous Publications of the Entomological Society of America* 73: 1–29.

Klassen, W., D. A. Lindquist, and E. J. Buyckx 1994. Overview of the Joint FAO/IAEA Division's involvement in fruit fly sterile insect technique programs. pp. 3-26 in C.O. Calkins, W. Klassen, and P. Liedo (eds.), *Fruit flies and the sterile insect technique*. CRC Press, Boca Raton, Florida. 272 pp.

Krafsur, E. S. 1998. Sterile insect technique for suppressing and eradicating insect population: 55 years and counting. *Journal of Agricultural Entomology* 15: 303–317.

LaChance, L. E., D. T. North, and W. Klassen 1968. Cytogenetic and cellular basis of chemically induced sterility in insects. pp. 99–157 in G.C. LaBrecque, and C.E. Smith (eds.), *Principles of insect chemosterilization*. Appleton-Century-Crofts, New York, New York.

STERNITE. The lower or ventral plate of a body segment, or a portion of this region.

STERNUM. The ventral portion of any segment, often used to describe the ventral portion of the thorax.

STICK INSECTS. Members of an order of insects (order Phasmatodea). See also, STICK AND LEAF INSECTS.

STICK AND LEAF INSECTS (PHASMIDA). Phasmida are terrestrial, nocturnal, phytophagous insects found in nearly all temperate and tropical ecosystems. Scientists have described over 3,000 species, yet this figure is uncertain since some taxon names are synonyms, and many new species have not been formally described.

They are variable in appearance, ranging from relatively plain forms, to some that are wonderful mimics of sticks and/or leaves. They display varying degrees of brachyptery, and can be winged or wingless. The tarsi have three articles in *Timema* Scudder and five in other Phasmida. Cerci are composed of one article, except for adult males of *Timema* which have a lobe on the right cercus.

Sexual dimorphism is usually extreme: the males are smaller and more gracile than the females.

Reproduction is typically sexual, but parthenogenesis occurs frequently. The egg capsule is distinctively shaped, possessing a lid called the operculum and a micropylar plate. Eggs are large and oftentimes highly sculptured resembling plant seeds. They are laid singly, and are either dropped, flicked, buried, glued to a surface, or riveted to a leaf. Some species that drop the eggs rely on ants to disperse them in a process analogous to myrmecocory. The entire life cycle from egg to adult can take from several months to several years, depending on the species.

The order consists of two extant monophyletic groups: the genus *Timema* and the remaining species of Phasmida termed the Euphasmida. These two clades diverged at least 50 million years ago and are phenotypically distinct.

Timema are small, wingless Phasmida that lack elongation of the body segments. They have three tarsal articles, unlike other Phasmida which have five. Using her large paddle-shaped cerci, the female coats each egg with defecated soil before oviposition. This forms a coating that may protect the egg from fire.

Timema are found only in the mountains of the western United States and are primarily associated with the Chaparral biome. They live on their host plants and are highly cryptic due to their coloration and habits while resting. When disturbed they release a spray from the prothoracic exocrine glands, but it has not been demonstrated to have a defensive function. They are unusual in that they feed on a variety of plants, including both Gymnosperms and Angiosperms.

Euphasmida are commonly referred to as stick and leaf insects. Exquisitely camouflaged, many species look like twigs and may have the appearance of being covered by lichens, mold, bird feces, or moss. The leaf-like forms usually bear a striking resemblance to foliage, exhibiting leaf veins, mildew spots and even apparent insect feeding damage.

Their primary means of avoiding predators is crypsis. If discovered, they either play dead (catalepsy), or they try to scare the predator with a startle display which can include wing flashing, leg kicking, or spastic motion. Some species also release an irritating 'tear gas-like' spray when disturbed. In addition, many Euphasmida can purposely lose some of their legs to help free them from a predator's grasp, or free them from the exuvia while molting. If this occurs during the immature stages, they can regenerate the lost limbs during successive molts.

Compared to other insects, Euphasmida are large, and a few are gigantic. Several species measure over 200 mm in length and the world's largest extant insect is the Euphasmida *Phobaeticus serratipes*, with one documented female measuring 555 mm.

Fig. 983 Some stick insects: top left, *Diapheroma femorata*; top right, *Anisomorpha buprestoides* (Stoll); lower left, *Exactosoma tiaratum*; lower left, *Heteropteryx dilatata* (top photos by Lyle Buss, lower photos by Mike Turco).

Since Euphasmida are wonderful looking insects, and are relatively easy to rear in captivity, they are popular as pets and for displays at zoological gardens.

Erich Tilgner
University of Georgia
Athens, Georgia, USA

References

Bragg, P. E. 1995. The phasmid database version 1.5. *Phasmid Studies* 3: 41–42.

Crespi, B. J., and C. Sandoval 2000. Phylogenetic evidence for the evolution of ecological specialization in *Timema* walking-sticks. *Journal of Evolutionary Biology* 13: 249–262.

Seow-Choen, F. 1995. The longest insect in the world. *Malay Naturalist* 48: 12.

Tilgner, E. 2001. Fossil record of Phasmida (Insecta: Neoptera). *Insect Systematics and Evolution* 31: 473–480.

Tilgner, E. H., T. G. Kiselyova, and J. V. McHugh. 1999. A morphological study of *Timema cristinae* Vickery with implications for the phylogenetics of Phasmida. *Deutsche Entomologische Zeitschrift* 46: 149–162.

STICK KATYDIDS. A subfamily of katydids (Saginae) in the order Orthoptera: Tettigoniidae. See also, GRASSHOPPERS, KATYDIDS AND CRICKETS.

STICKY TRAP. A method of insect sampling that uses an adhesive material to capture insects. Sticky traps normally use visual attraction based on color, or chemical attraction based on sex pheromone, to attract the insect to the trap. Some sticky traps depend instead on interception of flying insects with transparent material such as glass to obtain an unbiased assessment of population density because they do not concentrate the insects, as do lures.

STICTOCOCCIDAE. A family of insects in the superfamily Coccoidae (order Hemiptera). They are sometimes called false soft scales. See also, SCALES AND MEALYBUGS, BUGS.

STIGMA. The dense, often discolored portion of the costal margin of a wing, usually at the end of the radius. This is also known as the pterostigma.
    See also, WINGS OF INSECTS.

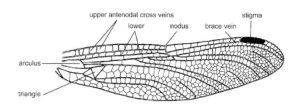

Fig. 984 Front wing of a dragonfly (Odonata).

STILETTO FLIES. Members of the family Therevidae (order Diptera). See also, FLIES.

STILT BUGS. Members of the family Berytidae (order Hemiptera). See also, BUGS.

STILT-LEGGED FLIES. Members of the family Micropezidae (order Diptera). See also, FLIES.

STINK BUGS. Members of the family Pentatomidae (order Hemiptera). See also, BUGS.

STINK BUGS (HEMIPTERA: PENTATOMIDAE), EMPHASIZING ECONOMIC IMPORTANCE. This is one of the largest groups within the suborder Heteroptera, the so-called true bugs. This family is the third largest of the suborder; it is surpassed by the Reduviidae, the predatory assassin bugs, and the Miridae, the plant bugs or capsids, which is by far the largest family of Heteroptera. Of over 36,000 described species of true bugs, more than 4,000 species belong to the family Pentatomidae. The classification of the family is as follows:

Order: Hemiptera

  Suborder: Heteroptera
    Superfamily: Pentatomorpha
      Family: Pentatomidae

Within the family pentatomidae there are eight subfamilies: Asopinae, Cyrtocorynae, Discocephalinae, Edessinae, Pentatominae, Phyllocephalinae, Podopinae, and Serbaninae. Members of the subfamily Asopinae are distributed worldwide, and are predatory. Because of their feeding habits they are important biological control agents, particularly those of the genera *Podisus* and *Perillus*. Members of the subfamily Cyrtocorinae are found only in the Neotropical Region, and are cryptically colored, resembling tree bark. They are of minor economic importance, feeding on branches of weeds, shrubs or trees. Eventually they may be found feeding on soybean. The Discocephalinae are flat and colored dark, typical from the Neotropics and of no economic importance. The Edessinae are large stink bugs, being abundant in the Neotropics. They are pests of several crops, including Leguminosae and Solanaceae. The Pentatominae contains most of the species in the family and are plant feeders. They are cosmopolitan. These phytophagous pentatomids are characterized by being round or ovoid, with five-segmented antennae, three-segmented tarsi, and a scutellum that is short, usually narrowed posteriorly, and more or less triangular. Several species are pests of major crops, feeding mostly on legumes and brassicas, but by being extremely polyphagous, also damage fruit trees, nut trees, palms, etc. Members of the subfamilies Phyllo-

cephalinae, Podopinae, and Serbaninae are, in general, flat, feeding on bark of trees. They are mostly present in the Neotropics. With very few exceptions, they are of no economic importance.

Pentatomids are called stink bugs because they produce a disagreeable odor by means of scent glands which open in the region of the metacoxae. Some nymphs have scent glands located on the dorsum of the abdomen.

Stink bugs feed by inserting their stylets into the food source to suck up the nutrients. By doing this, they cause injury to plant tissues, resulting in plant wilt and, in many cases, abortion of fruits and seeds. During the feeding process, they also may transmit plant pathogens, which increase their damage potential. Because they feed on several plant species of economic importance, they are regarded as major pests.

The economic importance of these insects varies greatly from species to species and within a species depending on the plant attacked. From the subfamily

Fig. 985 Some stink bugs (Pentatomidae) of economic importance: top left, Neotropical brown stink bug, *Euschistus heros*; trop right, small green stink bug, *Piezodorus guildinii*; lower left, small rice stink bug, *Oebalus poecilus*; lower right, southern green stink bug, *Nezara viridula*.

Edessinae, the species *Edessa meditabunda* (F.) is a pest of many vegetable crops, particularly Solanaceae such as tomato and potato, and Leguminosae, such as peas, soybean, and alfalfa. It also feeds on cotton, eggplant, tobacco, sunflower, papaya, and grapes. Of the many host plants, soybean is perhaps the most important. Among the Pentatominae, the pest species injure a wide range of plants, from vegetables to trees. The generalist southern green stink bug, *Nezara viridula* (L.), is a pest of grain, legumes and vegetables worldwide; *Piezodorus* spp. feed mostly on legumes, being a severe pest of soybean in South America, in Africa, and in the Orient. Members of the genus *Euschistus* are major pests of several crops, mostly damaging legumes such as soybean in the New World; *Eurydema* spp. in the Orient also attack grain legumes and vegetables. Members of the genera *Oebalus*, *Mormidea*, and *Aelia* are major pests of Gramineae, particularly on rice and wheat. Pests of trees include the spined citrus bug, *Biprorulus bibax* Breddin, on *Citrus* spp. in Australia; *Lincus* spp. on coconut in South America; *Bathycoelia thalassina* (Herrich-Schaeffer) on cocoa in Africa; and *Plautia* spp. on orchard plants in the Orient. The wide ranges of bug species and the host plants fed upon, with frequently detrimental effects to plant production, make this perhaps the most economically important group of insects among the Heteroptera. As an example of this, total losses from Hemiptera during 1985, including stink bugs, were estimated at 3.5 million dollars for the pecan industry in Georgia. In addition, pentatomids may disturb human beings by invading houses in large numbers for overwintering. Finally, because they are polyphagous, pentatomids feed on an array of wild plants.

After copulating, which may last for several hours in an end-to-end position, stink bugs lay eggs in clusters. First instars remain on the top of egg shells, apparently acquiring symbionts, and do not feed. During the second and third stadia, they are gregarious and feed, in general, on immature seeds of their host plants. As nymphs develop through the fourth and fifth stadia, the gregarious behavior decreases, and nymphs disperse. Fifth instars molt into adults, which usually disperse further to colonize other host plants. Several generations (1 to 7) are produced per year, depending on the favorability of abiotic (e.g., temperature, photoperiod, humidity) and biotic (e.g., food availability) factors.

Stink bugs have several natural enemies, the most common being the egg parasitoids (Hymenoptera) and tachinid flies (Diptera: Tachinidae), which are parasites of adults. Birds also prey on stink bugs.

See also, SOUTHERN GREEN STINK BUG, SMALL GREEN STINK BUG, NEOTROPICAL BROWN STINK BUG, SPINED SOLDIER BUG, PREDATORY STINK BUGS, HEMIPTERA.

Antônio R. Panizzi
Embrapa
Londrina, Brazil

References

McPherson, J. E., and R. M. McPherson. 2000. Stink bugs of economic importance in America north of Mexico. CRC Press, Boca Raton, Florida. 253 pp.

Panizzi, A. R. 1997. Wild hosts of pentatomids: ecological significance and role in their pest status on crops. *Annual Review of Entomology* 42: 99–122.

Schaefer, C. W., and A. R. Panizzi (eds.). 2000. *Heteroptera of economic importance*. CRC Press, Boca Raton, Florida. 828 pp.

Schuh, R. T., and J. A. Slater. 1995. *True bugs of the world (Hemiptera: Heteroptera). Classification and natural history*. Cornell University Press, Ithaca, New York. 336 pp.

STINK BUGS, PREDATORY (HEMIPTERA: PENTATOMIDAE, ASOPINAE).

The subfamily Asopinae belongs to the family Pentatomidae of the suborder Heteroptera (true bugs). About 300 species in 69 genera have been described worldwide.

Order: Hemiptera

    Suborder: Heteroptera
      Infraorder: Pentatomomorpha
        Superfamily: Pentatomoidea
          Family: Pentatomidae
            Subfamily: Asopinae

External morphology

Like other pentatomidae, predatory stink bugs or soldier bugs are of moderate to large size, ranging in length from 7 to 25 mm, and are broadly elliptical in shape. The piercing-sucking mouthparts of predatory stink bugs consist of a four-segmented rostrum or beak (labium) forming a sheath that encloses two mandibular and two maxillary stylets. Whereas the rostrum of phytophagous pentatomids is slender, asopines are characterized by having a thickened rostrum. In asopines, the first segment of the rostrum is markedly thickened and free, which enables the rostrum to swing forward fully, making it easier for the predator to feed on active prey. The appearance of the rostrum can be a useful diagnostic character for distinguishing the beneficial predatory pentatomids from potentially harmful phytophagous pentatomids in the field. The triangular mesothoracic shield, called the scutellum, is usually much shorter than the abdomen, but in some genera it is enlarged, covering most of the abdominal dorsum. The males of Asopinae are unique in combining the presence of genital plates with a thecal shield. The shape of the male claspers or parameres is the most reliable diagnostic characteristic for the identification of species.

Habitat and food

Predatory stink bugs are found in a wide range of natural and agricultural habitats, but many species appear to prefer shrubland and woods. The Asopinae are set apart from the other pentatomid subfamilies by their essentially predaceous feeding habits. It is believed that the Pentatomomorpha have arisen as plant feeders and that only the subfamily Asopinae has secondarily become predaceous. First instars do not attack prey and only need moisture, mainly in the form of plant juices, to develop. Although, for some species, partial development on certain plant-based diets has been reported, nymphs from the second instar on require animal-based diets to complete development. Nymphs and adults are, however, often observed to take up plant juices or free water in addition to feeding on animal prey. Plant feeding primarily provides moisture, but it may also furnish certain nutrients to the bugs. This habit may help predatory stink bugs to sustain their populations in times of prey scarcity. In contrast to some other zoophytophagous heteropterans, plant-feeding asopines have not been reported to injure plants.

Predatory stink bugs attack mainly slow-moving, soft-bodied insects, primarily larval forms of the Lepidoptera, Coleoptera and Hymenoptera. Very few, if any, Asopinae are truly host-specific. Nevertheless, whereas some asopine bugs are generalist predators, attacking a wide array of prey in a diversity of habitats, others appear to be more closely associated with a limited number of insect species and occur in only a few habitats. It has been hypothesized that there may be an evolutionary progression from the drab asopines, which feed rather generally, to the brightly colored asopines, which prefer larvae of Chrysomelidae and, to a lesser extent, Coccinellidae.

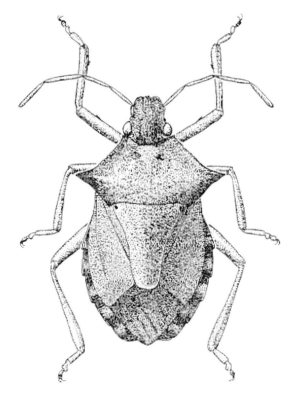

Fig. 986 *Picromerus bidens* (taken from T.J. Henry and R.C. Froeschner, 1988. Catalog of the Heteroptera, or true bugs of Canada and the continental United States. E.J. Brill, Leiden, The Netherlands).

Life history

The eggs are usually laid in masses. In some species (e.g., *Podisus*), the eggs bear prominent micropylar processes. There are five nymphal instars. The nymphs take a few weeks to develop, depending on temperature and food. Newly emerged nymphs are highly gregarious, whereas later instars become progressively more solitary with each molt. First instars are not predaceous and take up only water or plant juices; occasionally, they also feed on unhatched eggs. From the second instar on, nymphs begin searching for prey shortly after molting. The nymphs will feed up to 1 to 2 days before the next molt; at that time, their activity drops and the nymphs will remain resting in a concealed place in preparation for the oncoming molt. Small nymphs tend to attack prey and feed collectively, particularly when the prey is large. Larger nymphs and adults prefer to attack prey individually. Under conditions of food shortage, the nymphs and the adults are highly cannibalistic. The adults are long-lived, with reported longevities of 15 days to more than 3 months, and they lay eggs throughout their entire lifetime. The adults usually move by crawling, but some species (including *Podisus maculiventris*) are also noted to be good flyers. Field surveys suggest that scelionid egg parasitoids, such as *Trissolcus* and *Telenomus* spp., are the most important natural enemies of predatory stink bugs.

Prey location and capture

Predatory stink bugs use visual, chemical and tactile cues to locate and recognize their prey. Vision appears to be the most important sense used by the bugs to locate prey. *Podisus maculiventris* and other asopine bugs react to moving prey at distances up to 10 cm, but their reactive distance to immobile prey is considerably less and detection often seems to occur at antennal or rostral contact. Evidence is accumulating that several asopines can also use airborne chemical cues for prey detection. Both *Podisus maculiventris* and *Perillus bioculatus* are sensitive to systemic volatiles produced by plants in response to prey feeding. *Eocanthecona furcellata* is attracted to its lepidopterous prey based, in part, on a volatile component derived from the chlorophyll ingested by the prey. Further, it has been demonstrated that *Podisus maculiventris* can use vibrations of the substrate as cues for prey location. Prey recognition is based primarily on antennal and rostral contact. Asopines are rather timid predators. After finding the prey and orienting to it, the bugs may spend from several minutes up to an hour stealthily approaching it, often keeping their beak extended forward. The stylets are usually inserted at soft areas of the prey body. During attacking and feeding, the only contact between predator and prey is by the rostrum and stylets. When stylet penetration is perceived by the prey, many prey species will try to escape by vigorous body movements. The harpoonlike structure of the mandibular stylets enables a tenacious holding of the prey during this struggle. Further, predatory pentatomids are presumed to inject a salivary toxin into the prey body that quickly immobilizes it. During feeding, body tissues of the prey are liquefied by the injection of digestive enzymes and by the lacerating action of the stylets. The liquid and partly digested food can then be sucked up through the rostrum. This "solid-to-liquid" feeding process enables the predator to use most of the body tissues of the prey.

Pheromones. Both male and female asopines possess a metathoracic scent gland from which they discharge a disagreeable odor when disturbed, hence the name 'stink bugs.' In addition, many species of predatory stink bugs have discrete pheromone glands, which belong to one of two types: dorsal abdominal glands (in both sexes) or sternal glands (in males only). Species with dorsal abdominal glands are mostly polyphagous predators (e.g., *Podisus*, *Zicrona*), whereas those with sternal glands often appear to be more specialized predators (e.g., *Perillus*, *Stiretrus*).

The best known pheromone system is that of of the spined soldier bug, *Podisus maculiventris*. Adult males of this species possess hypertrophied dorsal abdominal glands that produce secretions that function as long-range attractants for adults and immatures of both sexes. Immature predators are thought to use the male pheromone as a cue indicating the presence of prey. In the United States, attractors with a synthetic pheromone are commercially available to lure spined soldier bugs to target areas in early spring when the adults emerge from overwintering. Pheromone-baited traps can also be employed to capture large numbers of predators that can be used to establish mass cultures.

Economic importance

Several predatory pentatomids are believed to have a future for the biological control of various economically important crop pests in different parts of the world. To date, however, the only asopine commercially available for augmentative biological control in North America and Europe is the spined soldier bug, *Podisus maculiventris*. This species is native to North America and has shown good capacity to control a variety of insect pests in field and greenhouse crops, orchards and forests, including the Colorado potato beetle, *Leptinotarsa decemlineata*, and several leaf-feeding caterpillars. The two-spotted stink bug, *Perillus bioculatus*, is another important natural enemy of the Colorado potato beetle in North America. *Podisus nigrispinus* is the most common predatory stink bug in South America. The insect has been the subject of conservation and augmentation biological control programs against leaf-feeding caterpillars in forests, including Eucalyptus stands in Brazil. In Southeast Asia and India, *Eocanthecona furcellata* has received increasing attention for its potential to control outbreaks of lepidopterous and coleopterous defoliators. *Picromerus bidens* was originally a widely distributed Palearctic species, but it has been found in the northeastern United States and eastern Canada since its (accidental) introduction sometime before 1932. Although its biology has been studied to some extent, few studies have attempted to quantify its predatory effectiveness. Given that only 10% of the nearly 300 known species of Asopinae has been studied in more or less detail, obviously an enormous biocontrol potential remains to be investigated.

Patrick De Clercq
Ghent University
Ghent, Belgium

References

Aldrich, J. R. 1999. Predators. pp. 357–381 in J. Hardie and A. K. Minks (eds.), *Pheromones of non-lepidopteran insects associated with agricultural plants*. CAB International, Wallingford, United Kingdom.

De Clercq, P. 1999. *Podisus* online. http://allserv.rug.ac.be/~padclerc/

De Clercq, P. 2000. Predaceous stinkbugs (Pentatomidae: Asopinae). pp. 737–789 in C. W. Schaefer and A. R. Panizzi (eds.), *Heteroptera of economic importance*. CRC Press, Boca Raton, Florida.

Thomas, D. B. 1992. Taxonomic synopsis of the asopine Pentatomidae (Heteroptera) of the western hemisphere. *The Thomas Say Foundation Monographs*, Vol. 15. Entomological Society of America, Lanham, Maryland.

Thomas, D. B. 1994. Taxonomic synopsis of the Old World asopine genera (Heteroptera: Pentatomidae). *Insecta Mundi* 8: 145–212.

STIPES. A portion of the maxilla. In Diptera, it is sometimes modified into a piercing structure or a lever for flexing the proboscis.

See also, MOUTHPARTS OF HEXAPODS.

ST. LOUIS ENCEPHALITIS.

St. Louis encephalitis is the most important mosquito-transmitted disease in the United States. It was first detected in the city after which it is named in 1939 where it claimed 220 lives. It is distributed throughout most of the continental United States although epidemics have only occurred in the Midwest and Southeast. The Centers for Disease Control (CDC) report more than 4,000 confirmed cases in the United States since 1964. The disease is also found in parts of Central and South America and the Caribbean. The pathogen that can cause the disease is an arbovirus (arthropod-borne virus). These viruses are maintained in nature through transmission between vertebrate hosts and

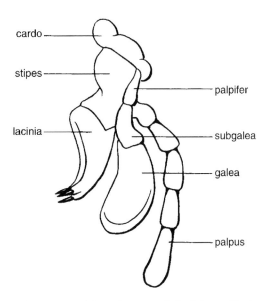

Fig. 987 External lateral aspect of the left maxilla in an adult grasshopper, showing some major elements.

blood-feeding arthropods (mosquitoes, flies, ticks, etc.). The St. Louis encephalitis virus belongs to the family Flaviviridae and is related to the Japanese encephalitis virus.

The virus transmission cycle

The transmission cycle of St. Louis encephalitis is not completely known, but we know that wild bird populations are of paramount importance in the cycle. When a wild bird is infected by the bite of a carrier mosquito, it produces large amounts of the virus in its bloodstream that can then infect other mosquitoes that subsequently bite the bird. This process is known as 'amplification' and is essential for virus transmission. The hosts, in this case wild birds, are known as 'amplification hosts'. Birds do not show any symptoms of the disease and become immune after infection. In an infected mosquito, the virus replicates and invades other parts of the body, including the salivary glands. From there, the virus can be transferred to other hosts because when mosquitoes bite they inject a small amount of saliva to act as an anti-coagulant.

Mosquitoes remain infective for life, but the virus disappears from the amplification host in a few days. As more and more birds become immune after recovering from the infection, transmission to uninfected mosquitoes diminishes. Therefore, as infected mosquitoes die off (they usually live a week or two) the

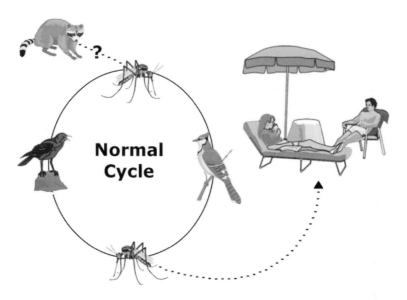

Fig. 988 Transmission cycle of St. Louis encephalitis.

epidemic subsides. It is unclear what roles other vertebrates such as raccoons and squirrels play in the transmission cycle of St. Louis encephalitis. Mosquitoes known to play a role in transmission of the St. Louis encephalitis virus include *Culex nigripalpus* in Florida, *Culex pipiens* and *Culex quinquefasciatus* in the Midwest and Gulf Coast, and *Culex tarsalis* in western states.

Although humans are not part of the natural transmission cycle, they can become infected by the bite of a vector mosquito. However, humans are 'dead-end hosts', meaning that sufficient amplification of the virus does not occur in humans and therefore, uninfected mosquitoes cannot acquire the virus from a human. Also, St. Louis encephalitis is not transmitted directly from human to human.

Symptomology and treatment

There have been more than 4,000 human cases of St. Louis encephalitis reported in the United States since 1964. However, less than 1% of St. Louis encephalitis infections are clinically apparent and the vast majority of cases go unreported. Symptoms of infection range from mild headaches and low fever to full meningoencephalitis symptoms including severe headaches and high fever, impaired motor skills, drowsiness, seizures, and central nervous system malfunction. Neurological symptoms may persist for some time after the disease has subsided. Symptoms are usually milder in children than in adults and are

most severe in the elderly. Mortality rates for the disease range from 5 to 15%.

All resident of areas where active cases of the disease have been confirmed are considered at risk of contracting the disease. Disease risk increases in those persons that are frequently exposed to mosquito bites. The elderly are particularly prone to exhibit severe symptoms and death. Diagnosis of the disease is confirmed by a four-fold rise in antibody titer, or detectable immunoglobular M antibody in sera or cerebrospinal fluid, or by isolation of the viral agent from the patient. Other techniques such as enzyme-linked immunosorbent assay (ELISA) and polymerase chain reaction techniques can increase the speed and accuracy of diagnosis, and are increasingly being used in clinical settings. There is no specific treatment for the disease and no vaccines are available for humans. Affected persons are given supportive treatment for the symptoms and after-effects.

Preventing infection

As in most arthropod-transmitted diseases, prevention involves personal protection and control of the vector populations to break the transmission cycle in nature. Personal protection includes the use of clothes that protect against mosquito bites (long-sleeved shirts and long pants), the use of repellents containing DEET, and avoidance of areas where mosquitoes are present, particularly after dark, when the *Culex* mosquitoes are most active. Vector population control includes the use of approved pesticides to kill adult and/or immature mosquitoes, and reduction or elimination of sites known to breed mosquitoes.

Surveillance programs are important parts of disease prevention. An effective surveillance program can provide an early warning of when the risk of disease transmission to humans is high. During those times, local health departments may alert the public that personal protection measures against mosquito bites are advisable. It can also trigger increased mosquito control efforts in the affected area. Surveillance programs monitor virus activity in nature, usually by monitoring virus activity in captive chicken flocks ('sentinel chickens'). Weather, particularly rainfall, is an important factor in prediction of vector abundance and is thus an integral part of successful surveillance programs. Direct monitoring of mosquito abundance, behavior, and movement is obviously of paramount importance in predicting risk for human populations, and a variety of techniques are routinely used by mosquito control and health agencies to monitor these vector-related variables. Finally, if disease transmission to humans does occur, it is important to monitor the distribution of cases in time and space, to determine the demographic characteristics of infected persons, and to try to determine the likely place of exposure for each case.

Jorge R. Rey
University of Florida
Vero Beach, Florida, USA

References

Chamberlain, R. W. 1988. History of St. Louis encephalitis. In T. P. Monath (ed.), *The arboviruses: epidemiology and ecology*. CRC Press, Boca Raton, Florida.

Eldridge, B. F. 1987. Strategies for surveillance, prevention and control of arbovirus diseases in western North America. *American Journal of Tropical Medicine and Hygiene* 37: 775–885.

Lazoff, M. 1998 *Encephalitis*. In Emergency medicine - an online reference. http://emedicine.com/EMERG/topic163.htm.

Tsai, T. F., and C. J. Mitchell. 1988. St. Louis encephalitis. In T. P. Monath (ed.), *The arboviruses: epidemiology and ecology*. CRC Press, Boca Raton, Florida.

STOCHASTIC MODEL. A mathematical model that is based on probabilities. Thus, the prediction of the model is based not on fixed numbers, but a range of possible numbers. (contrast with deterministic model)

STOMA. (pl., stomata) A pore in the epidermis of plants through which gas exchange occurs. They are most abundant on leaves.

STOMACH POISON. A toxicant that acts only after it has been eaten.

STOMATOGASTRIC NERVOUS SYSTEM Several small ganglia, including the frontal ganglion and hypocerebral ganglion, that control the foregut muscles.

STOMODEUM. The foregut of insects.

See also, ALIMENTARY SYSTEM, ALIMENTARY CANAL AND DIGESTION.

STONE BROOD. A disease of larval and adult bees, caused by the fungi *Aspergillus flavus.*

STONEFLIES (PLECOPTERA). The Plecoptera (stoneflies), more than any other order of insects, are typical inhabitants of running waters. Nearly all species occur exclusively in streams, and most are restricted to running water habitats of mountainous regions of the world. Usually water temperatures of these streams are below 25°C and with high dissolved oxygen levels. These requirements make them an excellent indicator of water quality.

Stoneflies display incomplete metamorphosis (hemimetabolous development), though it is sometimes called gradual metamorphosis (paurometabolous development).

The order name Plecoptera is derived from the Greek 'pleco' or folded and 'ptera' or wings.

Adult stoneflies can be distinguished from other insects that undergo hemimetabolous or incomplete metamorphosis by the long, thin antennae; membranous wings, when not used, lying flat over the abdomen; and the front wings straight and about the same length as the body. The front wings usually have a 'fish skeleton-like' venation pattern. The hind wings usually have an expanded posterior area, that folds like a fan, hence Plecoptera, or 'pleated' wing insects. Some adult stoneflies have reduced wings, or even lack wings. Additionally, most species have two thin tails or cerci that project from the end of the abdomen.

The nymphs usually have cylindrical or flattened bodies; the head with long, thin antennae; wing pads

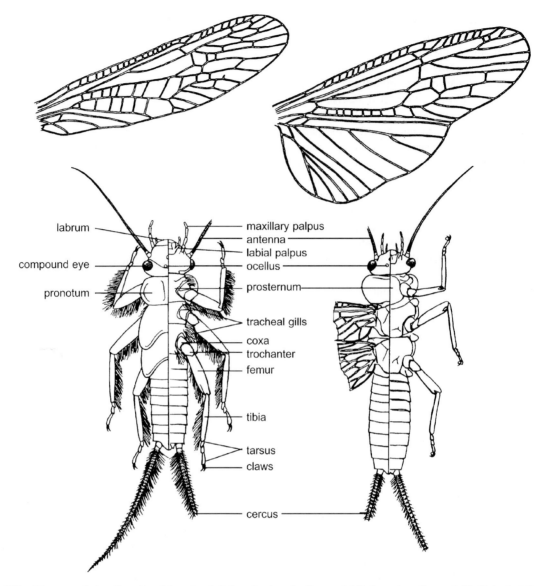

Fig. 989 Diagram of stonefly wings (above); adult female stonefly (lower right); immature (lower left). Note that the adult and immature are divided into a dorsal perspective (left portion of each diagram) and ventral perspective (right portion).

usually present on the thorax in more mature individuals; three pairs of segmented legs, each with two terminal claws; and two long, thin tails or cerci filaments at the end of the abdomen. Some species have single or branched gills on the thorax or abdomen.

Most stonefly nymphs are herbivores, feeding principally on plant detritus. Even the young nymphs of many carnivorous species feed on fine detritus before switching to animal prey. The Perlodidae, Per-lidae, and Chloroperlidae are predominantly predators as older instars. Mature nymphs range in length from a few millimeters (some capniids) to over five centimeters (Pteronarcyidae). Although some species emerge in autumn, most stoneflies transform to adults in spring or early summer. Mature nymphs often congregate at emergence sites such as piles of rocks, bridge abutments, and woody debris. The cast skins or exuviae of the nymphs are often abundantly seen attached to these sites.

Stonefly nymphs occur most commonly in lotic habitats with rocky bottoms, and with high dissolved oxygen concentrations. Some species are restricted to seeps and springs, others to high gradient coldwater streams. One species is known only from Lake Tahoe. Many stonefly nymphs, when subjected to low dissolved oxygen concentrations, will exhibit a 'push up' behavior, that apparently increases the rate of water movement over the body and gills.

A remarkable behavior of adult stoneflies is the two-way communication between sexes called drumming. Males tap, rub or scrap their elongate abdomens upon various substrates such as branches and leaves. The female adult detects this vibration, and she will answer, allowing for an eventual location of each other. The drumming pattern in amplitude, frequency, or duration is species specific.

Numerous 'winter stoneflies' occur in North America. These species typically emerge as soon as openings appear in the ice (late winter, early spring); the dark-colored adults are seen on snow-covered stream banks, bridges, fence posts or tree trunks. The remarkable aspect of these stoneflies is the reversal of the typical development pattern of aquatic insects, in that growth and maturity occur during the coldest part of the year. Eggs laid by adult females during the winter or early spring hatch quickly, and the nymphs migrate into the interstitial spaces of the loose, rocky streambed (known as the hyporheic zone) and begin a state of inactivity or diapause throughout the warmer months. As water temperatures begin to cool in the fall and early winter, nymphs move back to the water body substrate surface and complete development.

The classification of the order is relatively well established, with over 2,000 species in the world placed in fifteen families. Nine of these families occur in North America, including about 626 species. A web site maintained by B. P. Stark provides a listing of all North American species by province or state: http://www.mc.edu/campus/users/stark/sfly0102.htm

Order: Plecoptera

  Suborder: Arctoperlaria
    'group' Euholognatha
      Family: Capniidae
      Family: Leuctridae
      Family: Nemouridae
      Family: Notonemouridae
      Family: Taeniopterygidae
      Family: Scopuridae

    'group' Systellognatha
      Family: Chloroperlidae
      Family: Peltoperlidae
      Family: Perlidae
      Family: Perlodidae
      Family: Pteronarcyidae
  Suborder: Antarctoperlaria
      Family: Austroperlidae
      Family: Diamphipnoidae
      Family: Eustheniidae
      Family: Gripopterygidae

A short synopsis of the families found in North America follows:

Family Capniidae (small winter stoneflies)

This family is composed primarily of small species. It contains many 'winter stoneflies,' with adults typically emerging from January to April. The numerous species are most common in small streams, including seeps and springs. Nymphs are usually found in leaf packs, accumulations of woody debris, and develop during the coldest water temperatures of winter. The early instars diapause in the hyporheic zone of the stream. Most species of capniids are detritivore shredders. Common genera include the eastern *Allocapnia* and the western *Capnia*.

Family Leuctridae (rolledwinged stoneflies)

Nymphs occur in lotic habitats from intermittent streams, springs to small rivers, and are associated primarily with leaf packs, woody debris or mineral substrates. Nymphs are considered shredder-detritivores. Adults of some species emerge during late winter to early spring (*Paraleuctra*), some during the entire summer (*Leuctra*), and others in the fall (*Despaxia*). Some species are very tolerant of low pH.

Family Nemouridae (nemourid stoneflies)

Nymphs occur in a wide range of aquatic habitats, from seeps, springs, to lakes. Some species also are found in large rivers. Nymphs typically occur in leaf packs, woody debris, but also rocky areas. Most species are restricted to higher elevations. Most adults are recognized by one-segmented cerci and the X-pattern of veins of the front wing. The life cycles vary from one to two years for completion. Emergence of adults occurs typically in the spring, but some species emerge in the winter, and others in the fall. Several genera are common, *Amphinemura* in eastern North America and *Malenka* in western North America.

Family Taeniopterygidae (winter stoneflies)

Another well-known group of 'winter stoneflies.' Adults in warm-weather regions emerge during the winter, and in cool climates and at high elevations they emerge through the spring and early summer. The nymphs are characterized by their stout, robust habitus with divergent wing pads, and can be found in almost all types of permanent streams, small and large. Nymphs usually are associated with leaf packs, wood debris, but also coarse mineral substrates. They are primarily detritivore shredders. The common genera in the east are *Strophopteryx* and *Taeniopteryx*, and *Taenionema* in the West.

Family Chloroperlidae (green stoneflies)

Chloroperlid nymphs are small, generally elongate, and lack gills. Nymphs typically inhabit the hyporheic zone, the interstitial spaces between substrate particles in the streambed. Some species reside at considerable depths in the substrate and laterally under the banks. Most species are found only in clean, cool streams of mountainous areas. Adults are recognized by slender bodies and yellow or green coloration. The western genus *Paraperla* is a collector-gatherer and scraper. Other genera are predaceous on other aquatic insects (often chironomid midges), but also act as collector-gatherers.

Family Perlidae (common stoneflies)

The distinctively patterned yellow and brown nymphs can be found in all types of streams from intermittent to large rivers. Some species can even occur along wave-swept shores of lakes. Members of this family inhabit warmer waters than most other stoneflies. Nymphs are usually associated with large loose rocks, but also leaf packs, woody debris, and submerged logs.

Several genera *Hesperoperla* and *Claassenia* in the West and *Acroneuria* and *Paragnetina* in the East are relatively large and conspicuous stoneflies. The nymphs are predatory, feeding largely on other aquatic insects. The life history varies from 1 to 3 years, with most adults emerging from May to August. Adults often can be found attracted to lights near streams.

Family Perlodidae (perlodid stoneflies)

Perlodid nymphs are medium to moderately large-sized and predatory. If gills are present, they are simple (unbranched) and are restricted to the thorax or neck region. Nymphs of most species are restricted to cool, swift, mountain streams with rocky bottoms. Some species are found only at splash zones of seeps. Others are restricted to large silty western rivers. All species apparently have a single generation per year. Emergence is generally in the spring, but some species are active as adults in the fall. Several species have egg diapause as long as 5 to 7 months. *Isoperla* is the most common and widespread genus in North America.

Peltoperlidae (roachlike stoneflies)

The 'roach like' appearance of the nymphs makes them one of the most recognizable of all North American stoneflies. Nymphs can be abundant in leaf packs of small woodland streams of the southern Appalachians. Some species are restricted to splash zones of seeps, springs, and even intermittent streams. All species are shredders-detritivores, and are important in the breakdown of leaves. Adults generally emerge in the spring and early summer.

Family Pteronarcyidae (giant stoneflies)

Nymphs of this family are usually found in cool steams and rivers of small to medium size. Nymphs can be especially abundant in accumulations of woody debris, or among large rocks in the swifter reaches. They are shredders that typically feed on submerged leaf material, but sometimes take in animal material. The life cycle requires one to four years to complete, depending on geographic location. Adult emergence typically occurs in May or June for most species. The adults are the largest stoneflies found in North American, reaching 5 to 6 centimeters.

The genus *Pteronarcys* is known as salmonfly to anglers, and widespread over most of North America, whereas the other genus *Pteronarcella* is restricted to the West. Giant stonefly adults use reflex bleeding to repel potential predators.

Boris C. Kondratieff
Colorado State University
Fort Collins, Colorado, USA

References

Baumann, R. W., A. R. Gaufin, and R. F. Surdick. 1977. The stoneflies (Plecoptera) of the Rocky Mountains. *Memoirs of the American Entomological Society* 31: 1–207.

Stark, B. P., and B. J. Armitage (eds.). 2000. Stoneflies (Plecoptera) of Eastern North America. Volume I. *Bulletin of the Ohio Biological Survey New Series* 14: 1–98.

Stewart, K. W., and B. P. Stark. 2002. *Nymphs of North American stonefly genera (Plecoptera)*. 2nd ed. Caddis Press, Columbus, Ohio, 510 pp.

Zwick, P. 2000. Phylogenetic system and zoogeography of the Plecoptera. *Annual Review of Entomology* 45: 709–746.

STOP CODON. One of the three mRNA codons (UAG, UAA, and UGA) that prevents further polypeptide synthesis.

STORAGE PROTEIN RECEPTORS. During metamorphosis of holometabolous insects, most larval structures are broken down, while adult tissues are formed. Many new proteins need to be synthesized, requiring vast mounts of amino acid precursors. These amino acid resources, however, must be present prior to pupation, because pupae are not able to feed and take up nutrients from external sources. Specific larval serum proteins, often called 'storage proteins', provide these amino acid resources. These proteins accumulate in the hemolymph of last instar larvae, and move shortly before pupation into the fat body, where most new proteins are synthesized. Storage proteins enter the fat body in small vesicles, which fuse with one another to form larger, often crystalline protein granules. As development progresses, these granules are hydrolyzed to provide the needed amino acids.

Most storage proteins are members of the hexamerin family; these proteins are composed of 6 subunits of 80 kDa each, and are evolutionarily related to hemocyanin. The most common storage protein is arylphorin, a protein rich in aromatic amino acids, but most species contain other storage proteins as well. In one lepidopteran family (Noctuidae), an additional, structurally unrelated storage protein has been found: a very high density lipoprotein (VHDL) composed of four 150 kDa subunits. VHDL is colored brightly blue, due to biliverdin that is bound non-covalently to the protein.

The uptake of storage proteins into fat body cells is selective. Storage proteins are present at very high concentrations in the hemolymph of last instar larvae in both lepidopteran and dipteran species, but prior to pupation they are efficiently removed from the hemolymph, and sequestered by the fat body. Other hemolymph proteins, such as lipophorin, remain in the hemolymph. Storage proteins enter the fat body via endocytosis; to be selective, the uptake is mediated by specific endocytotic receptors, proteins that are associated with the fat body membrane. Such receptors have been found in Diptera and Lepidoptera.

Dipteran storage protein receptors

Storage proteins receptors have been identified in two dipteran species, namely the fleshfly *Sarcophaga peregrina* and the blue bottle fly *Calliphora vicina*. These proteins bind arylphorin in a saturable manner, with dissociation constants between 2 and 800 nM. Binding generally requires Ca^{2+} ions and is pH dependent, with optimal binding at the slightly acidic pH of the hemolymph (pH 6.5). While the reported molecular weights for the arylphorin receptor vary between 50 and 140 kDa, it is now clear that the observed variability is the result of the cleavage of an initial protein of around 140 kDa. The protein is encoded by a gene that is evolutionary related to its own ligand, arylphorin, and other members of the hexamerin superfamily. The genes of the arylphorin (or storage protein) receptors of *Calliphora vicina* (Genbank ID 630903) and *Sarcophaga bullata* (Genbank ID 984655) show high sequence homology with the previously identified fat body P1 protein from *Drosophila melanogaster* (Genbank ID 544281) which appears to be the storage protein receptor in that insect species. The transcription of the latter gene is under the control of the hormone 20-hydroxy ecdysone. While the exact mechanism of action of the receptor remains unclear, proteolytic cleavage of the initial gene product is necessary for the formation of the active receptor. Following the removal of the signal peptide, a protease removes a C-terminal fragment of ~50 kDa. The remaining receptor protein is then moved to the outside of the cell, where it can bind to its ligand. The uptake of the bound arylphorin, however, cannot occur until the receptor is further processed by another protease, which appears to be under the control of the hormone 20-hydroxy ecdysone, however at the post-translational level.

Interestingly, the active receptor does not contain a typical membrane spanning domain, and is likely a peripheral membrane protein that is attached to the membrane through another transmembrane protein.

Lepidopteran storage protein receptors

In contrast to the activation of the dipteran storage proteins receptor, the receptor identified in Lepidoptera does not seem to be regulated by proteolytic cleavage. The apparent receptor is a basic 80 kDa protein

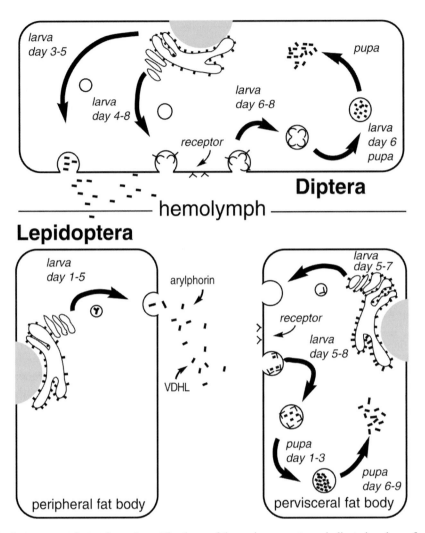

Fig. 990 Model of storage protein uptake and use. The times of the various events are indicated as days after hatching for the dipteran species, and as days after the preceding molt for Lepidoptera.

that is glycosylated. In *Helicoverpa zea*, the protein binds arylphorin as well as VHDL, with similar binding constants (Kd˜100 nM). Binding of its ligands occurs between pH 6.5 and 7.5, and requires at least 4 mM Ca^{2+}. The receptor protein can be found only during the second half of the last larval instar, and only in those regions of the fat body that take up storage proteins, located centrally surrounding the gut (perivisceral fat body). In contrast, the storage proteins themselves are synthesized in different regions of the fat body (peripheral fat body). Electron microscopy revealed that the uptake process occurs through classical receptor mediated endocytosis, via coated pits that are internalized as coated vesicles into the perivisceral fat body, where they fuse to form endo-

somes. Fusion with primary lysosomes leads to the digestion of membrane components and of the receptor protein, so that the electron dense protein granules containing arylphorin and VHDL can be formed.

The receptor protein is generally associated with the cell membrane fraction, and cannot be found in the hemolymph. It does not bind to antibodies against arylphorin or VHDL, and antibodies raised against the receptor do not cross-react with these or other cytosolic proteins. Its sequence was recently reported; the receptor appears to be also related to members of the hexamerin superfamily, namely to the basic juvenile hormone suppressible proteins identified in another noctuid species. It is not entirely clear how it is associated with the fat body cell membrane. Although it contains one potential trans-membrane region, it lacks other motifs typically found in endocytotic receptors, like adaptin binding sites that link the receptor to the coat protein clathrin. Hence, another integral membrane protein may be required to serve as an intermediary of the uptake process, as it has been suggested for dipteran species.

See also, METAMORPHOSIS.

Norbert H. Haunerland
Simon Fraser University
Burnaby, British Columbia, Canada

References

Burmester, T., and K. Scheller. 1999. Ligands and receptors: common theme in insect storage protein transport. *Naturwissenschaften* 86: 468–474.

Burmester, T., and K. Scheller. 1997. Developmentally controlled cleavage of the *Calliphora* arylphorin receptor and posttranslational action of the steroid hormone 20-hydroxy ecdysone. *European Journal of Biochemistry* 247: 695–702.

Haunerland, N. H. 1996. Insect storage proteins: gene families and receptors. *Insect Biochemistry and Molecular Biology* 26: 755–765.

Haunerland, N. H., and P. D. Shirk. 1995. Regional and functional differentiation in the insect fat body. *Annual Review of Entomology* 40: 121–145.

Wang, Z., and N. H. Haunerland. 1994. Receptor mediated endocytosis of storage proteins by the fat body of *Helicoverpa zea*. *Cell and Tissue Research* 278: 107–115.

STORED GRAIN AND FLOUR INSECTS AND THEIR MANAGEMENT.

Hundreds of insects are found in stores of grain and milled cereal, but only a few species cause direct damage to products that are in sound condition. Most pest species can feed only on processed or damaged grain and mold, and their presence results in food contamination with their waste, caste skins, webbing, and body parts. Insects may also contribute to the conditions that cause grain to heat and mold. Approximately fifty species are deemed pests of economic concern. Nearly all of them are beetles or moths.

Environmental factors

Insect infestation in stored grain is fostered by moisture, moderate heat, and damaged kernels. Not only does high moisture content in grain and processed food provide a favorable physical environment for development of many pest species, it also fosters the development of the molds on which some insects feed. The most favorable grain moisture range for stored grain insects is from 12 to 18 percent. In many instances, insect infestation amplifies mold problems in grain by exposing otherwise hidden endosperm surfaces to molds, transporting mold spores to new areas and encouraging mold germination in microhabitats made moist by insect metabolic activity. Indeed, insect and mold metabolic activity can raise grain temperatures.

At grain temperatures below $60°F$ ($15°C$) insect development and activity are minimal and pest populations can be expected to remain static. Temperatures a few degrees above $35°F$ ($90°C$) are detrimental or lethal to stored-product insects, and brief treatments of $122°F$ ($50°C$) are used for stored-product pest control.

Grain insects move within the grain mass at a rate that is determined by the season and grain temperature. During the summer and fall, insect infestations are usually on the surface of the grain. In cold weather, insects congregate at the center and lower portions of the grain and may escape detection until high population numbers are reached.

The presence of damaged grain, dust, and chaff, collectively known as dockage, is quite beneficial to the survival and reproduction of insects that feed externally. For example, flour beetles are reportedly unable to maintain populations in dockage-free wheat. Accordingly, grain stored without cleaning is subject to more severe insect infestation than is cleaned grain.

Damage

Some insects, designated primary pests, damage grain by developing inside kernels (egg, larvae, pupae), feeding on the inner endosperm, and chewing holes in the kernels through which the adult insects exit. The five primary pests of stored grain in the U.S.A. are the lesser grain borer, *Rhyzopertha dominica* (Fabricius), the rice weevil, *Sitophilus oryzae* (L.), the granary weevil, *Sitophilus granarius* (L.), the maize weevil, *Sitophilus zeamais* Motschulsky, and the Angoumois grain moth, *Sitotroga cerealella* (Olivier) (Gelechiidae), all of which deposit eggs in undamaged kernels. Other insect species, designated secondary pests, do not develop within the kernels, although they may hide inside cracked grain, making detection very difficult. Some secondary pests, including most of those designated as grain beetles, feed primarily on mold. Other species such as the sawtoothed grain beetle, *Oryzaephilus surinamensis*

(L.), the flour beetles, and moth larvae feed on damaged grain or fines.

Insect damage can reduce grain quality through loss of weight or nutritional quality, spreading and encouraging mold germination, adding to the fatty acid content of the grain, and leaving quantities of uric acid that cause grain rancidity. Insects also create fines and broken kernels when feeding that reduce air flow through grain and prevent proper aeration. In addition, the presence of insects in a grain sample can cause price discounts and shipping restrictions.

In the U.S.A., the presence of two insects of any kind in 1 kg of wheat, rye, or triticale causes the grain to be graded the lowest quality category, sample grade. Insect tolerances in finished commodities such as flour or cornmeal are stricter. Perhaps the most difficult problems associated with insect infestation of grain are quarantine and export restrictions. Many nations require the fumigation or return of grain shipments that harbor the Khapra beetle, *Trogoderma granarium* Everts. Under the most restrictive export laws, detection of a single insect in grain precludes its shipment.

Monitoring and control

Grain processing facilities are monitored by light or pheromone traps and by visual inspections. In stored grain, plastic pitfall traps are the dominant method, but sticky traps are useful for headspaces. Also, grain can be inspected by screening or sieving and searching in the screenings for insects, examining kernels for damage, checking grain for webbing, and investigating off-odors. Monitoring is especially important in summer and early fall when temperatures are optimum for rapid development. Grain temperature should be monitored as an indicator of grain condition and to locate areas of bulk grain in which conditions are favorable for insects. The number of insects found in traps is recorded and charts constructed so that changes in population size can be easily noticed and management tactics can be implemented as needed. Traps are used following control interventions to determine whether the treatments are effective.

The first defense against stored-grain insects is exclusion by means of pest-resistant construction, protective packaging, and sanitation. Empty storage and processing facilities can be treated with chemical insecticide sprays. Residual grain protectants are sometimes added directly to grain at the beginning of storage. These are usually organophosphates, but synergized pyrethrum or diatomaceous earth may be used. The dominant method for control of insects in stored grain is fumigation with phosphine. Nonchemical controls include temperature modification, either by heat treatment or lowering temperature through aeration. Atmosphere modifications to reduce oxygen are used primarily outside the U.S.A.

Grain weevils

The three economically important species of the weevil family Curculionidae that infest cereal grains in storage are the rice weevil, *S. oryzae*, the maize weevil, *S. zeamais*, and the granary weevil, *S. granarius* (L.). Adult weevils can be distinguished from other grain-infesting insects by their long snouts with chewing mouth parts at the tip and elbowed antennae. Both the rice and maize weevils have two light areas on each elytron and dense small, nearly round pits on the pronotum, giving a rough appearance. The granary weevil is uniformly colored and has less dense elliptical pits on the pronotum, giving the prothorax a glossier appearance than that of the other two species. Adults of all three species are dark brown, with the rice and maize weevils being slightly darker than granary weevils. Adult maize and granary weevils are

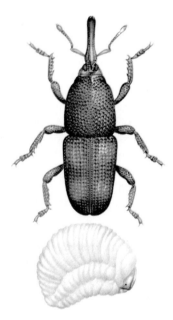

Fig. 991 Rice weevil, *Sitophilus oryzae* (L.): adult above, larva below.

3 to 4.5 mm long, and the slightly smaller rice weevils are about 3 mm long. Egg, larval, and pupal stages are spent within the grain kernel. The larvae of all three species are nearly identical, being legless and crescent shaped. As with the adults, larvae of maize and granary weevils are slightly larger than larvae of rice weevils.

After mating, a female weevil uses her long snout to chew a small cavity in a grain kernel into which she places an egg. She then seals the hole with a gelatinous egg plug. From this time until the new adult chews its way out of the kernel, the infestation is inapparent. Each female can lay as many as 300 to 400 eggs in her lifetime. Under similar conditions, rice weevils generally lay more eggs than maize or granary weevils. Two rice weevils may develop in a wheat kernel, one on either side of the crease, but usually only one maize or granary weevil develops in a kernel. The four larval instars feed on the endosperm of the grain kernel. Pupation and adult emergence take place within the kernel. The adult chews its way out of the kernel, leaving visible damage behind. About one third of a wheat kernel is consumed by rice weevils and over half by maize or granary weevils. Under favorable conditions, granary weevils require about 40 days to complete their development and maize and rice weevils about 35 days. Almost immediately after emerging as adults, the females are mated and the developmental cycle is repeated. The adult weevils live as long as five to eight months.

The rice weevil is the most destructive pest of stored grain. It is found in all parts of the world where grain is used and is particularly abundant in warm climates where it breeds continuously. Although wings are present some question the ability of the rice weevil to fly. Infestations of grain field origin lend credibility to the presumption of flight ability. The rice weevil's habitat is almost identical to that of the granary weevil. The granary weevil is one of the oldest known insect pests. It has been carried to all parts of the world by commerce. It prefers temperate climates and, in the U.S., is more frequently found in the northern states than the southern states. The granary weevil cannot fly and must reside in accumulations of grain or be carried from place to place in infested lots of grain. Maize weevils are capable of flight, and in warm areas, they commonly fly to fields and infest corn before it is harvested. They can easily move from one lot of stored grain to another.

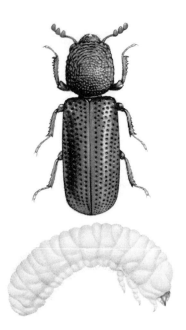

Fig. 992 Lesser grain borer, *Rhyzopertha dominica* (F.): adult above, larva below.

Grain borers

The family Bostrichidae, which comprises mainly wood-boring beetles, has two species that are important primary pests of grain. The lesser grain borer, *R. dominica*, is one of the most damaging insects found in grain, especially in wheat. The larger grain borer *Prostephanus truncatus* (Horn) is a more serious pest of maize in warm tropical areas, especially where maize is stored on the cob.

Adults of the lesser grain borer are cylindrical and about 3 mm long. The prothorax forms a hood over the head, which is directed downward. The hood has distinct, sharp tubercles on the top and near the front margin of the prothorax. The wing covers have rows of indentations, but they are glossy. Each antenna has a distinctive loose, three-segmented club at the end. Individual segments of the club are frequently found in flour milled from wheat that has been infested with the lesser grain borer. The adult has functional flight wings and is a strong flyer. Most larvae develop inside whole grain kernels and thus are rarely visible. The larvae have three pairs of small legs located on the segments immediately behind the head. The head is a partially withdrawn into the thorax. The body of the larva is grub-like with the thorax and last segments of the abdomen enlarged.

The female lays as many as 300 to 400 eggs, placing them among, but not inside the grain kernels, sometimes in clusters of up to 45. Larvae chew into kernels or penetrate through cracks in the pericarp. Inside the kernel, the borers pass through three to five larval instars and a pupal stage and emerge as adults. In some instances, where broken kernels and much dust are present, larvae may feed and develop externally. Adult insects live an average of seven to eight months and often return to kernels from which insects have emerged to feed and more extensively damage the grain.

Lesser grain borer infestations are characterized by the presence of grain dust and a characteristic sweetish, musty odor. The dust is mostly fecal matter. In localized heavy infestations, the dust and heavily damaged kernels form compacted areas that can cause uneven aeration or fumigant penetration.

The lesser grain borer is better able to survive in drier grain than the weevils, which probably accounts for its being a more serious pest of farm-stored wheat in the central and southern plains areas of the United States.

Larger grain borers are similar to lesser grain borer in development, habits, and appearance. The adults also have the typical cylindrical borer shape, but they are somewhat larger than the lesser grain borer, measuring 3 to 4 mm in length. In contrast to those of the lesser grain borer, the posterior portions of the elytra are steeply flattened with many small tubercles on the surface. The head is turned under and covered by the prothorax. The antennae terminate in flattened, three-segmented clubs similar in appearance to those of the lesser grain borer. The larvae are similar in appearance to the lesser grain borer larvae, but larger.

The larger grain borer is a serious pest of corn, as well as dried cassava. It is seldom found in the United States but is common in Central and South America. It was introduced into Kenya in the 1970s and has rapidly spread across sub-Saharan Africa, where it has become an extremely serious pest. The climate and the prevalence of farm storage corn on the cob in parts of Africa are particularly favorable for larger grain borer infestations. Adult borers infest corn shortly before and after storage. Shelled corn is a marginal medium for development, but unhusked corn is particularly vulnerable to infestation. Females bore into the attached kernels and lay eggs in side tunnels. Larvae feed mainly on dust, much of which

Fig. 993 Sawtoothed grain beetle, *Oryzaephilus surinamensis* (L.): adult above, larva below.

is created by adult feeding. The developmental cycle can be completed in about 27 days under optimum conditions and about a week longer in dry grain.

Grain beetles

Often the most abundant insects in stored grain are those belonging to a guild of dorso-ventrally-flattened beetles that feed on damaged grain, dust and mold. None of these grain beetles attack undamaged grain. The larvae of grain beetles do not stay within a single grain, but crawl about actively to feed. The development times are typically about a month under favorable conditions, and populations often build up rapidly.

The family Laemophloeidae includes eight species in the genus *Cryptolestes* that are secondary pests of cereals and cereal products. Two are prominent, particularly in moist environments. The rusty grain beetle, *Cryptolestes ferrugineus* (Stephens), is a cold-tolerant species and is often the most abundant insect in stored grain in temperate climates. The flat grain beetle *Cryptolestes pusillus* (Schönherr) is a minute beetle of about 1.5 mm in adult length. It is cosmopolitan, but less tolerant of low temperature and humidity than the rusty grain beetle and is more common in warm moist climates such as Southeast Asia.

The members of the family Silvanidae are generally larger (2 to 4 mm adults) and more active than

Cryptolestes species, but they lack the flying ability of *Cryptolestes*. The sawtoothed grain beetle, *O. surinamensis*, derives its common name from the tooth-like projections on the sides of its thorax. It is among the best-known cosmopolitan pests in stored grain and grain products and is considered the most serious stored grain pest in the United Kingdom. The merchant grain beetle, *Oryzaephilus mercator* (Fauvel), is often confused with the sawtoothed grain beetle, but it prefers oilseed products, including nuts. Several other Silvanidae can infest cereals but are of minor importance.

Flour beetles

In the family Tenebrionidae, *Tribolium castaneum* (Herbst), the red flour beetle, and *Tribolium confusum* Jacquelin du Val, the confused flour beetle, are the most intractable insect pests of grain and flour processing areas. They are notoriously difficult to control with chemical insecticides. The adults are 3 to 4 mm long, reddish brown to brown and rather glossy. The light, honey-colored larvae have darkened heads and darkened forked processes at the tips of their abdomens that distinguish them from other grain-infesting larvae. Like other grain insects that feed externally, the larvae develop best in the presence of grain dust and broken kernels. Where such material is abundant, flour beetles can develop in grain with as little as 8% moisture content. The larvae will pass through 5 to 11 larval instars depending on growing conditions, and pass from egg to adult in less than a month under ideal conditions. The adults are known to live for up to five years.

Stressed or agitated flour beetles secrete quinones that have a pungent odor and can give a pink hue to heavily infested flour. There is concern that the quinones may be carcinogenic.

Mealworms

The mealworms comprise several large members of the family Tenebrionidae that are nocturnal and frequent dark, damp places. Three species are common, the yellow mealworm *Tenebrio molitor* L., the dark mealworm *Tenebrio obscurus* Fabricius, and the lesser mealworm, *Aphitobius diaperinus* Panzer. The yellow and dark mealworms are the largest insects found in grain and grain products. The larvae are 26 to 30 mm long and resemble wireworms. Under natural conditions, the yellow and dark mealworms complete only one generation per year and,

consequently, are not serious pests. The lesser mealworm is similar to the yellow and dark mealworms, but reaches only 5.5 to 7 mm in length. All three species prefer damp, out of condition grains and grain products, but they also feed on various deteriorating food materials. Lesser mealworms are especially common in poultry feed.

Dermestid beetles

Dermestidae is a small but important family whose members scavenge and feed on animal matter. It includes some 55 species that have been reported as stored-product pests. The black carpet beetle, *Attagenus megatoma* (Fabricius), is a cosmopolitan species that is often found in cracks in warehouses, but most of the dermestid pests are *Trogoderma* species. *Trogoderma* adults are reddish-brown to black, ovoid and convex, mainly 2 to 4 mm in length, and covered with scales. The larvae are very hairy, with setae of various characteristic shapes. *Trogoderma inclusum* LeConte, *T. glabrum* (Herbst), and *T. variable* Ballion are common pests of grain, seed, and other stored products. The adults of these beetles live only 1 to 3 weeks, but development from egg to adult can take from one month to as long as a few years. The ability of the larvae to remain in a quiescent state for several months in the absence of food makes *Trogoderma* species especially dangerous pests.

The Khapra beetle is the only truly phytophagous dermestid beetle. It feeds on many stored foodstuffs in addition to grain. Worldwide, it is the most destructive member of this group. The Khapra beetle is believed to have been eradicated from the U.S. in 1961 and has been a quarantine target since that time.

Other beetles

The cadelle, *Tenebroides mauritanicus* (L.), is a member of the family Trogositidae that infests flour, meal, and grain worldwide. Larvae and adults have the destructive habit of moving from kernel to kernel, devouring the germs. Both stages are omnivorous and will eat other insects.

The cigarette beetle, *Lasioderma serricorne* (Fabricius) and the drugstore beetle, *Stegobium paniceum* (L.) are in the family Anobiidae. Both are found in all temperate and subtropical regions. As its common name indicates, the cigarette beetle infests tobacco, but it also attacks various food products, including grain. Its ability to penetrate packaging materials makes it a threat to high value stored products. The drugstore beetle has an extremely diverse diet, and takes its name from its ability to feed on pharmaceutical drugs.

Lepidoptera

Among the Lepidoptera, there is only one important primary feeder in grain, the Angoumois grain moth. It is a pest of stored grain throughout the tropics and subtropics, primarily in on-farm stores. Each larva eats out the inside of a single grain and remains in the grain through pupation, leaving no visible signs of infestation. The complete life cycle requires approximately one month at 30°C. Adults are 5 to 7 mm long, pale brown moths with fringed wings and long up-turned labial palpi. The female adult lives only 5 to 12 days, during which time she lays about 200 eggs. Angoumois grain moths are thought to compete with *Sitophilus* spp. for larval habitat and may dominate in drier habitats that are stressful for the weevils.

In the family Pyralidae, the subfamily Phycitinae or knot-horn moths, comprises the most important secondary lepidopterous pests of stored grains. Among them, the Indianmeal moth *Plodia interpunctella* (Hübner) is the most important pest of stored foodstuffs worldwide. It is especially common on corn, hence its common name.

In grains the larvae feed primarily on embryos. During the development of the 5 to 7 instars, Indianmeal moth larvae spin webs that are often dense enough to attract attention. The last instar spins a silken cocoon. Adults are 5 to 10 mm in length and about 16 mm in wingspan. They are easily recognized by the distinctive appearance of the forewing, which is light gray basally and bright copper red distally.

The females lay up to 400 eggs that can develop into adults in as few as 28 days under optimal conditions. In the tropics, there may be 6 to 8 generations; in temperate zones 1 to 2 generations are usual.

In addition to the Indianmeal moth, several other pyralids are pests in stored grain. The Mediterranean flour moth, *Ephestia kuhniella* Zeller, a native of Europe, is a general feeder that was once the most troublesome pest in U.S. flour mills. Its pest status was primarily due to the particle matting created by the larva-produced silken thread. Other occasional pyralid pests include the meal moth, *Pyralis farinalis* (L.), the tobacco moth *Ephestia elutella* (Hübner), the raisin moth, *Cadra figulilella* (Gregson), and the almond moth, *Cadra cautella* (Walker). Among these the almond moth is the most serious pest especially in rice and sorghum grain in tropical and subtropical climates.

Jeffrey C. Lord
U.S. Department of Agriculture,
Grain Marketing and Production Research Center
Manhattan, Kansas, USA

References

Cotton, R. T. 1963. *Pests of stored grain and grain products*. Burgess Publishing Co., Minneapolis, Minnesota.

Mueller, D. K. 1998. *Stored product protection... a period of transition*. Insects Limited, Inc., Indianapolis, Indiana.

Pederson, J. R. 1992. Insects: identification, damage and detection. pp. 435–489 in D. B. Sauer (ed.), *Storage of cereal grains and their products*. American Association of Cereal Chemists, St. Paul, Minnesota.

Subramanyam, Bh., and D. W. Hagstrum. 1996. *Integrated pest management of insects in stored products*. Marcel Dekker, New York, New York.

STRAIGHT-SNOUTED WEEVILS. Members of the family Brentidae (order Coleoptera). See also, BEETLES.

STRATIFIED SAMPLING. A method of allocating sampling units in which the sample universe is subdivided into two or more sections and sample units are randomly collected from each subdivision in proportion to their size or relative variability. Examples include the subdivision of a crop field into border and interior sections, the subdivision of a plant along vertical strata, or different rooms within a structure.

See also, SAMPLING ARTHROPODS.

Fig. 994 Indian meal moth, *Plodia interpunctella* (Hübner): adult above, larva below.

STRATIOMYIDAE. A family of flies (order Diptera). They commonly are known as soldier flies. See also, FLIES.

STRAWBERRY BUD WEEVIL. See also, SMALL FRUIT PESTS AND THEIR MANAGEMENT.

STRAWBERRY SAP BEETLE. See also, SMALL FRUIT PESTS AND THEIR MANAGEMENT.

STREPSIPTERA. An order of insects. They commonly are known as twisted-wing parasites or stylopids. See also, STYLOPIDS.

STRESS-INDUCED HOST PLANT FREE AMINO ACIDS AND INSECTS. Various researchers have demonstrated that individual amino acids differ in their effects on insect growth and development. Subsets of amino acids are the mechanistic bases of relationships between total plant nitrogen and insect vitality. For example, artificial diets with the amino acid distribution found in anthers were found to be successful for rearing tobacco budworms, *Heliothis virescens* F. At least to some extent, nutritional strength has been linked with insect preference for host plants and higher levels of readily available (unbound) amino acids presumably improve insect development. Similarly, plants with accumulations of certain free amino acids (FAAs) are, in some cases, preferred by some herbivorous insects.

Abiotic plant stress factors and FAAs

Higher levels of some FAAs, especially free proline, in association with water deficit stress occur in many plants, including cotton, *Gossypium hirsutum* L.; soybeans, *Glycine max* (L.) Merr.; wheat, *Triticum aestivum* L. Pers.; cowpea, *Vigna unguiculata* L.; tomato, *Lycopersicon esculentum* Mill.; and jack pine, *Pinus banksiana* Lamb. In corn, *Zea mays* L., seedlings, drought treatment caused an increase in total FAA content and a consistent rearrangement of the amino acid pool with increased accumulation of free alanine, arginine, valine, aspartate, serine, threonine, and tyrosine. Lower amounts of free glutamic acid, glycine, and methionine likely resulted from a strong decrease in glutamate. In cotton, increasing degrees of

water deficit stress increased foliar free alanine, arginine, glycine, histidine, isoleucine, phenylalanine, proline, and valine, but associations between the weekly volume of water provided and the quantity of individual FAAs were not always linear.

Researchers have most often focussed on the relationship between water deficit stress to plants and free proline accumulations. Suspension culture cells of water deficit adapted tomatoes to polyethylene glycol had a 300-fold bigger free proline pool than untreated cells, the result of a 10-fold rise in proline synthesis via the glutamate pathway. Use of N-15 labels in tomato plants determined that ornithine was an unlikely precursor to proline, and it has been suggested that the rate of proline synthesis in unstressed cells only slightly exceeds the rate required to sustain both protein synthesis and free proline pool maintenance. Mechanisms restricting free proline oxidation, and oxidation of other FAAs, can operate at different rates governed by the degree to which water deficit stress is imposed. Some researchers have interpreted the accumulation of large pools of free proline and other FAAs in water deficit stressed rice as being an indication of protein synthesis pathway disruption, or an increase in protein degradation. Others concluded that the higher free proline levels in *Arabidopsis thaliana* (L.) Heynh tissues with low water content resulted from increased biosynthesis. Osmotic stress in *A. thaliana* induced a rapid accumulation of free proline through *de novo* synthesis from glutamate, and during recovery from osmotic stress, accumulated free proline is oxidized to glutamate and the first step of this process is catalyzed by proline oxidase. A study using water deficit stressed bermudagrass, *Cynodon dactylon* (L.) Pers., showed that 10 to 100 times more free proline accumulated than in controls and it was suggested that, because proline turns over more slowly than other FAAs during water deficit stress, free proline might function primarily as a nitrogen storage compound. Alternatively, in hydroponic potatoes, *Solanum tuberosum* L., subjected to water deficit stress induced by polyethylene glycol, plants were able to grow in 'dry' conditions by accumulating organic solutes for use in osmoregulation. Osmoregulation in the tubers occurred in two phases: 1) during the first two days hexoses were accumulated, and 2) after 7 days osmotic adjustment was mostly due to the accumulation of FAAs, especially proline (150 times the amount found in unstressed controls). Free proline might act as a compatible solute, a readily utilizable source of energy, and a sink for soluble nitrogen. Plants originating in dry climates were found to have higher levels of free proline than plants in wetter climates, and it has been postulated that the high free proline levels accompany survival and growth rather than being a result of damage. One study demonstrated that free proline sprayed on cotton plants counteracted the effects of water deficit stress. Leaf relative water content, dry matter content, and chlorophyll content and stability to heat increased. FAA accumulations associated with disease and nematode infections of plants might result from osmotic disruption, but also from other reasons, including xylem injury repair, syncytial wall formation in nematode-induced root galls, cell wall formation, and blockage of protein synthesis. Other abiotic stresses that have been associated with significantly increased foliar free amino acids accumulation include high soil salinity and soil nitrogen levels (each results in increased free proline in particular), and shade (particularly free arginine).

Biotic plant stress factors and FAAs

Weed competition. Competition with weeds has been associated with increased free arginine accumulations, but not proline, in sugarcane and cotton. In cotton, shade from weeds was attributed to be the cause of the stress. Sugarcane canopy closure shaded out weed growth after weeds had already caused a significant sugarcane stand reduction, and the higher free arginine levels in weed-free sugarcane habitats were attributed to shade caused by denser stands of the crop itself.

Fungi and bacteria. Plants infected with various pathogens have been found in association with increased accumulations of FAAs. *Fusarium solani* (Mart.) altered host plant metabolic activity in citrus, and it was suggested that the higher levels of free arginine, asparagine, glutamic acid, glycine, and proline were caused by a fungal toxin that enhanced xylem membrane permeability. Pecan trees, *Carya illinoinensis* (Wangenh.) K. Koch, are more susceptible to fungal leaf scorch, *Phomopsis* sp. and *Glomerella cingulata* (Ston.), when under water deficit stress, and FAAs were more abundant in scorch-infected tree leaves. It was suggested that foliage with higher FAA levels might favor the growth and development of leaf scorch. Tobacco, *Nicotinia tabacum* L., infected with crown gall, caused by the bacteria *Agro-*

bacterium tumefasciens, had 70-fold higher accumulations of free proline than healthy controls. Similarly, crown gall infected tomatoes, *Lycopersicon esculentum* Mill., had 22 times more free proline as compared to healthy controls. Non-pathogenic fungi can also influence FAA accumulations in plants. For example, the vesicular-arbuscular mycorrhizal fungus, *Glomus faciculatum* (Thax. Sensu Gerd.) caused ammonia-fed corn to accumulate more FAAs than control plants. Major components of the FAA pool were glycine, glutamine, alanine, serine, asparagine, and GABA. These FAAs were between 2- and 3-fold more concentrated in *Glomus*-colonized roots.

Viruses. Tomato plants infected with tomato leaf curl virus had higher concentrations of soluble amino acids than controls, and it was suggested that virus infection stimulates nitrogen metabolism and mobilization of proteins. The rate of turn-over of FAAs into bound amino acids is enhanced by virus infection. FAAs are not directly converted to amino acids in viral protein, but they are incorporated into viral protein by way of bound amino acids. Aspartic acid, glutamic acid, serine, alanine, phenylalanine, and other amino acids are actively turned over and used for the synthesis of viral protein (these amino acids comprise much of the viral protein). FAAs have also been shown to increase in association with a variety of viruses infecting different plants, including barley yellow dwarf virus infection of barley, *Hordeum vulgare* L., and wheat, *Triticum aestivum* L.; cassava mosaic virus infection of cassava, *Manihot esculenta* Crantz; pigweed mosaic virus infection of pigweed, *Amaranthus* sp.; southern bean mosaic virus infection of cowpea, *Vigna* sp.; common bean mosaic virus infection of mung bean, *Vigna radiata* (L.) Wilczek; and soybean mosaic virus and bean pod mottle virus infection of soybean, *Glycine max* (L.) E. Merill.

Plant stress, FAAs, and phytophagous nematodes

Some phytophagous nematodes have been shown to be associated with higher host plant FAA levels than healthy controls, including *Radopholus similis* (Cobb) Thorne infected grapefruit, *Citrus* x *paradisi* Mcfad., seedlings; reniform nematode, *Rotylenchulus reniformis* Linford & Oliveira, in cotton, *Gossypium barbadense* L.; green gram, *Phaseolus aureus* Roxb.; black gram, *P. mungo* (L.); and cowpea ; and stem nematode, *Ditylenchus dipsaci* (Kuhn) Filipjev, infecting alfalfa, *Medicago sativa* L. Augmentation of root knot nematodes, *Meloidogyne* spp., was associated with lower levels of free cysteine, histidine, proline, and serine concentrations in sugarcane, *Saccharum officinarum* L. Shading in dense stands of sugarcane elevated levels of some FAAs compared to less dense stands of sugarcane, and the increases were correlated to populations of stunt nematodes, *Tylencorhynchus annulatus* Cobb. Concentrations of FAAs were unchanged when roots of *Beta patellaris* (Moq.), a resistant species of beet, were infected with cyst nematodes, *Heterodera schactii* (Schmidt). However, in the susceptible sugar beet, *B. vulgaris* L., concentrations of aspartic acid, glutamic acid, and glutamine, and total FAAs were significantly higher when infected with cyst nematodes. Another study determined that infection of tomato plants by *Meloidogyne javanica* (Treub) Chitwood resulted in accumulation of free proline in galls. The concentration of free proline increased with increasing density of nematodes, but the highest concentration occurred at the time of egg production, and the concentration of free proline was highest in *M. javanica* eggs and egg sacs compared to portions of the plant. Proline is reported as being a major constituent of the egg shells of phytophagous nematodes, including *M. javanica* and *Heterodera rostochiensis* Wellenweber. Other researchers have suggested that proline might stimulate xylem production following injury induced by phytophagous nematodes.

Although the previous examples of research have indicated an association between nematode feeding and elevated accumulations of FAAs in the host plant, other studies have shown that host plant resistance to nematodes is conferred through high levels of FAAs. For example, quantities of FAAs in segments of cotton roots resistant and susceptible to *Meloidogyne incognita* (Kofoid and White) were compared. Following infection, the susceptible cultivar had greater percentage increases in certain FAAs than a resistant cultivar, but the sum total of FAAs was greatest in the resistant cultivar. Another study, using ring nematode, *Criconemella xenoplax* (Raski) Luc & Raski, showed that nematode resistant cultivar peach, *Prunus persica* (L.) Batsch, seedlings accumulated significantly more FAAs in the root tissue in association with phytophagous nematode populations.

Virus infected host plants, presumably with greater accumulations of FAAs, have been shown to be associated with higher numbers of certain phytophagous nematodes than healthy control plants. *Meloidogyne javanica* juveniles entered bean, *Phaseolus* sp., plant roots, infected with tobacco ring spot virus more than uninfected plants, though the growth of the nematodes was unaffected. The root knot nematode grew more rapidly on tobacco mosaic virus-infected beans than on controls, but the virus did not affect the number of nematodes entering the root. Root exudates can attract nematodes, and it is possible that the exudates are attractive at least in part because they contain FAAs or other constituents of the exudate signal high FAA levels in the host plant.

Plant stress, FAAs, and phytophagous insects

One study demonstrated that mildly water deficit stressed soybean plants were preferred by Mexican bean beetle, *Epilachna varivestris* Mulsant, larvae, possibly as a result of higher levels of FAAs in the foliage. Beet armyworm, *Spodoptera exigua* (Hübner), eggs have been reported as being most concentrated on water deficit stressed cotton in fields and this observation might be because water deficit stressed cotton plant foliage has greater accumulations of FAAs. However, water deficit stress has been shown to decrease aphid populations because the insects are more dependent on water pressure in the plant than other insects.

Insects that depend on host plant water pressure for efficient ingestion of nutrients have been shown, in some cases, to prefer virus infected host plants over non-infected plants, presumably because the nutritional value of the plant is enhanced by higher accumulations of FAAs. Pigeon pea, *Cajanus cajan* (L.) Millsp., plants infected with a mosaic virus were infested with more potato leafhoppers, *Empoasca kerri* Pruthi, than healthy plants. The virus-infected leaves had higher accumulations of total nitrogen, peptides, and FAAs than the controls. Sweetpotato whitefly, *Bemisia tabaci* (Gennadius), populations were higher on pumpkin, *Cucurbita pepo* L., infected with squash leaf curl virus than healthy pumpkin plants. Sugar beet, *Beta vulgaris* L., leaves with a mosaic virus were more suitable food for aphids than uninfected leaves. Grasshopper, *Melanoplus* sp., preferred to feed on leaves of sunflower, *Helianthus* sp., infected with rust, *Puccinia* sp. Cotton aphids, *Aphis gossypii* Glover, lived longer and produced more offspring on pumpkin plants infected with zucchini yellow mosaic virus than non-infected plants. Aphids, *Rhopalosiphum padi* (L.) and *Sitobion avenae* (F.), reared on oats, *Avena byzantina* (Koch), infected by various isolates of barley yellow dwarf virus were much more likely to mature as winged adults than were aphids reared on healthy oats. However, sweet-potato whitefly populations were lower on squash, *Cucurbita* sp.; tomato; and cotton plants infected with squash leaf curl virus, chino del tomate virus, and cotton leaf crumple virus, respectively, than non-infected controls. In some studies, no significant differences were observed between herbivorous insect populations and virus-infected host plants.

Insects absorb nitrogen through the gut primarily in the form of FAAs or small peptides. Thus, the initial cost of proteolysis is saved if amino acids are ingested in this form. The distinction between FAAs and those bound in proteins may be important for insects such as aphids that cannot ingest large molecules or are not exposed to large molecules because of the composition of the plant's phloem sap on which insects like aphids feed, but this is separate from the costs of proteolysis. In terms of optimizing insect development, it is probable that the balance between different amino acids is particularly important. For example, pink bollworm, *Gossypiella pectinophora* (Saunders), larvae raised on a diet in which one of the essential amino acids was omitted grew slowly and failed to survive beyond the second instar. Increased glutamic acid caused growth stimulation. In some cases, amino acids such as glycine have been shown to be essential. Beet armyworms prefer to oviposit on pigweed, *Amaranthus hybridus* L., over cotton, *G. hirsutum*, and third instars prefer to feed on pigweed foliage over cotton foliage. Beet armyworm larvae also grow and develop better on pigweed than on cotton, and it has been suggested that the lack of gossypol and other deterrent or toxic compounds found in cotton leaves, and the higher accumulations and greater diversity of FAAs in pigweed leaves make pigweed the superior host. However, in some insect-plant relationships, preferences for the plant host that most favors phytophagous insect growth and development is not always evident.

At least in some cases, the increased availability of readily available nitrogen appears to not only improve the chances of some disease organisms becoming established, but might also increase the likelihood of survival of insects feeding on the infected plants. Environmental stresses, whether biotic or abiotic, seem to affect the survival, fecundity, morphology, and vitality of some phytophagous insects through physiochemical processes mediated by corresponding changes of FAA accumulations in the host plant.

Allan T. Showler
USDA-ARS SARC
Weslaco, Texas, USA

References

Brodbeck, B., and D. Strong. 1987. Amino acid nutrition of herbivorous insects and stress to host plants. pp. 346–364 in P. Barbosa and J. C. Schultz (eds.), *Insect outbreaks*. Academic Press, London, United Kingdom.

Costa, H. S., J. K. Brown, and D. N. Byrne. 1991. Life history traits of the whitefly, *Bemisia tabaci* (Homoptera: Aleyrodidae) on six virus-infected or healthy plant species. *Environmental Entomology* 29: 1102–1107.

Dadd, R. H. 1968. Dietary amino acids and wing determination in the aphid, *Myzus persicae*. *Annals of the Entomological Society of America* 61: 1201–1210.

Helms, J. A., F. W. Cobb, Jr., and H. S. Whitney. 1971. Effect of infection by *Verticicladiella wagenerii* on the physiology of *Pinus ponderosa*. *Phytopathology* 61: 920–925.

Kennedy, J. S. 1951. Benefits to aphids from feeding on galled and virus-infected leaves. *Nature* 168: 825–826.

Showler, A. T. 2002. Effects of water-deficit stress, shade, weed competition, and kaolin particle film on selected foliar free amino acid accumulations in cotton, *Gossypium hirsutum* L. *Journal of Chemical Ecology* 28: 615–635.

Showler, A. T., T. E. Reagan, and K. P. Shao. 1990. Nematode interaction with weeds and sugarcane mosaic virus in Louisiana sugarcane. *Journal of Nematology* 22: 31–38.

STRIA. Grooves or indented lines.

STRIATE. With groves or indented lines.

STRIDULATE. To produce a noise by rubbing together two surfaces.

STRIDULATING SLANTFACED GRASSHOPPERS. A subfamily (Gomphocerinae) of grasshoppers in the order Orthoptera: Acrididae. See also, GRASSHOPPERS, KATYDIDS AND CRICKETS.

STRIGILATION. The oral removal of secretions by one insect from another.

STRIP CROPPING. Cultivation or harvesting crops in strips rather than in large expansive fields, with the strips sometimes following the natural contours of the land.

STRIPE. A line that runs horizontally or lengthwise on an insect. This term often is confused with a "band" or transverse line.

STRIPED EARWIGS. Members of the earwig family Labiduridae (order Dermaptera). See also, EARWIGS.

STRIPED WALKINGSTICKS. A family of walkingsticks (Pseudophasmatidae) in the order Phasmatodea. See also, WALKINGSTICKS AND LEAF INSECTS.

STRUCTURAL GENE. A gene that codes for an RNA molecule or protein other than a regulatory gene.

STRUCTURAL GENOMICS. The study of protein structure based on DNA sequences.

STRUCTURAL PEST. A pest that affects the structural wood in buildings.

STYLET. The elongated needle-like portions of the piercing-sucking type of insect mouthparts in Hemiptera.
See also, MOUTHPARTS OF HEXAPODS.

STYLET-BORNE VIRUS. A virus that does not persist in the vector, but which is a mechanical contaminant of the insect's mouthparts (stylets). Also known as nonpersistent virus.

STYLOPIDAE. A family of insects in the order Strepsiptera. See also, STYLOPIDS.

STYLOPIDS (STREPSIPTERA). This small group of parasitic insects, also known as twisted-wing parasites, has sometimes been considered to be part of Coleoptera. The order name is based on the Greek words *strepti* (twisted) and *pteron* (wing). They appear to be related to the beetle families Meloidae and Rhipiphoridae because they display hypermetamorphosis, but this may simply be an example of convergent evolution, and they are best

considered to be a separate order. Indeed, some recent molecular studies have suggested that stylopids may be more closely related to Diptera than to Coleoptera. Clearly this is an unresolved issue.

Classification
This group is not well known, and only about 600 species have been described. In most systems of classification there are about eight families. Bohartillidae and Callipharixenidae are limited in distribution, but the others are found widely

Class: Insecta

   Order: Strepsiptera
      Family: Callipharixenidae
      Family: Bohartillidae
      Family: Stylopidae
      Family: Myrmecolacidae
      Family: Elenchidae
      Family: Halictophagidae
      Family: Corioxenidae
      Family: Mengenillidae

Characteristics
The stylopids exhibit extreme sexual dimorphism. Females of this order spend their entire live parasitizing other insects. The tip of the females abdomen often are observed protruding from between the abdominal segments of parasitized (sometimes called stylopized) hosts. There is not much structure to the females. Although all females are wingless, some even lack external adult characteristics such as eyes, legs and antennae. Thus, the female is primarily a sac capable of ingesting food and producing eggs. Although the females lack wings, and retain a larviform appearance, the males are free-living and winged. Interestingly, the front wings of males are reduced to small club-like structures while the hind wings are fan-shaped. The antennae are heavy and pronounced. The eyes are pronounced. The mouthparts are a degenerate biting type. The abdomen consists of seven segments. The number of tarsal segments is variable among species, ranging from two to five per tarsus. Stylopids range in size from 0.4 to 4.0 mm.

Biology
Stylopids parasitize Hymenoptera, Orthoptera, Thysanura, and Hemiptera. Each species tends to

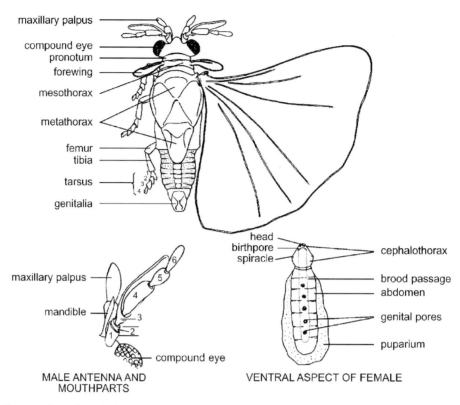

Fig. 995 A diagram of a male stylopid (above) showing a dorsal view with the left wing removed. The lower diagrams show a close-up of the male antenna and mouthparts (left) and a ventral view of the wingless parasitic female (right).

confine its activities to a few closely related hosts. They display a type of hypermetamorphosis combined with a complex life cycle. The immatures (called triungulin larvae) often hatch within the body of their mother and exit from her genital canal. They escape from the body of the host insect as possible, and must locate a new host. In the case of species parasitizing wasps, they likely attach to a wasp when it visits flowers, hitch a ride back to the wasp nest, and then locate and parasitize wasp larvae. The short-lived, winged males are attracted to the sexually mature female stylopids, whose abdomen protrudes from the host insect's body, where they mate. The female spends most of her life attached to the host, and sealed in a puparium.

References

Arnett Jr., R. H. 2000. *American insects (second edition)*. CRC Press, Boca Raton, Florida. 1003 pp.

De Carvalho, E. L. and M. Kogan. 1991. Order Strepsiptera. Pages 659–673 in F. W. Stehr (ed.). *Immature insects*, Vol. 2. Kendall/Hunt Publishing, Dubuque, Iowa.

Kathirithamby, J. 1989. Review of the order Strepsiptera. *Systematic Entomology* 14: 41–92.

Kinselbach, R. 1990. The systematic position of the Strepsiptera (Insecta). *American Entomologist* 36: 292–303.

Whiting, M. F., J. C. Carpenter, Q. S. Wheeler, and W. C. Wheeler. 1997. The Strepsiptera problem: phylogeny of the holometabolous insect orders inferred from 18S and 28S ribosomal DNA sequences and morphology. *Systematic Biology* 46: 1–68.

STYLUS. A finger-like process.

SUBALARE. A small, upper portion of the pleuron, posterior to the wing, to which wing muscles are inserted.

SUBCOSTA. The longitudinal vein running parallel to the costa. It usually turns and connects to the anterior margin of the wing before reaching the wing apex, and usually is unbranched.

See also, WINGS OF INSECTS.

SUBCOSTA FOLD. The depression between the costa and the radius.

SUBCUTICULAR SPACE. The space created between the epidermal cells and the endocuticle during molting.

SUBDORSAL. A region between the dorsal and lateral areas.

SUBESOPHAGEAL GANGLION. A ganglion located in the head that is found beneath the digestive tract, and which innervates the mouthparts.

See also, ENDOCRINE REGULATION OF INSECT REPRODUCTION, NERVOUS SYSTEM.

SUBFAMILY. In classification, a major division of the family unit consisting of related tribes or genera. Subfamily names end in -inae. Although not one of the major taxonomic units, subfamilies sometimes are quite useful in describing clusters of like taxa.

SUBGALEA. A maxillary segment associated with the stipes.

SUBGENITAL PLATE. A plate covering the gonopore from below.

SUBIMAGO. The first of two winged instars of mayfles (Ephermeroptera) after they emerge from the water.

SUBMEDIAL ARC. Pigmentation in the form of an arc occurring on the face or top of the head in caterpillars.

SUBMENTUM. The basal portion of the insect labium, by means of which it attaches to the head.

See also, MOUTHPARTS OF HEXAPODS.

SUBSOCIAL BEHAVIOR. A level of sociality less than eusocial behavior. A type of presocial behavior. It involves cooperation within a family, usually expressed as a mother caring for her offspring. However, the cooperation does not extend to include to cooperation between sisters.

See also, PRESOCIAL, SOLITARY, COMMUNAL, QUASISOCIAL, SEMISOCIAL, EUSOCIAL BEHAVIOR.

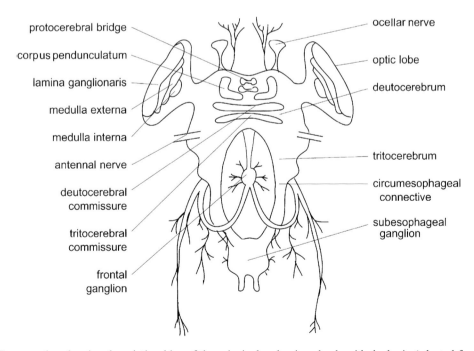

Fig. 996 Cross section showing the relationships of the principal endocrine glands with the brain (adapted from Chapman, The insects: structure and function).

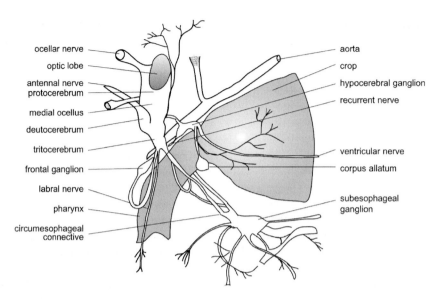

Fig. 997 Lateral view of the insect brain(adapted from Snodgrass, Insect morphology).

SUBSPECIES. A subdivision of species, usually a geographic race and differing slightly or greatly in appearance but capable of interbreeding. Subspecies generally carry little importance, but may be indicative of speciation in progress. In some taxa, particularly with butterflies (Lepidoptera, in part) subspecies are widely recognized.

SUBSPIRACULAR. The area immediately below the spiracles.

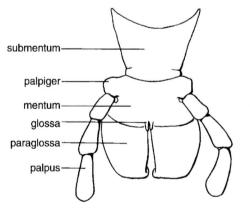

Fig. 998 External aspect of the labium in an adult grasshopper, showing some major elements.

SUBTERRANEAN TERMITES. A group of termites in the family Rhinotermitidae that maintain contact with the soil as they attack wood. See also, TERMITES.

SUCCESSION. A progressive series of changes in the vegetation of a site, starting with colonization and ending with establishment of a climax community. Also called ecological succession.

SUCKING LICE (SIPHUNCULATA). This order of wingless ectoparasitic insects is also called Anoplura, and sometimes is grouped together with the chewing lice (Mallophaga) into a single order, Phthiraptera. The order name is based on the Greek word *sipho* (tube) and the Latin *cula* (little). Sucking lice are closely related to chewing lice, but their evolution is poorly understood.

Classification

There is a total of about 15 families of sucking lice found throughout he world. At this point, it is a relatively small order, however, containing only about 500 species. Many species probably remain to be described. Most species are quite host specific.

Class: Insecta

Order: Siphunculata
    Family: Polyplacidae – spiny rat lice

Family: Linognathidae – pale lice
Family: Enderleinellidae – squirrel lice
Family: Hoplopleuridae
Family: Neolinognathidae
Family: Hamophthiriidae
Family: Ratemiidae
Family: Microthoraciidae
Family: Echinophthiriidae – seal lice
Family: Pthiridae – pubic lice
Family: Pedicinidae
Family: Pecaroecidae – pecarry lice
Family: Hybothiridae
Family: Haematopinidae – ungulate lice
Family: Pediculidae – body lice

Characteristics

Sucking lice are found on the bodies of their hosts. They are small, measuring only 0.5–5 mm in length, and oval in shape. Their head is narrow, a distinguishing character relative the chewing lice. Their eyes are reduced or absent; ocelli are absent. The antennae are short, and 3- to 5-segmented. The mouthparts, which are about as long as the head are adapted for piercing and sucking, and are retracted into the head when not in use. The body is dorsoventrally flattened. The segments of the thorax are fused, and the thoracic spiracles located dorsally. The abdominal segments are distinct. The tarsi consist of only one segment, and there is only one large claw on each tarsus. Cerci are absent. Metamorphosis is incomplete (hemimetabolous development).

Biology

Sucking lice feed only on blood of mammals. Two (or three depending on how *Pediculus humanus* is treated) species affect humans, and about 12 species affect domestic animals. The families tend to contain lice with very similar feeding habits. For example, the echinophthiriids feed on seals, sea lions, walruses and river otter; the enderleinellids on squirrels; the haematopinids on ungulates such as pigs, cattle, horses and deer; the hoplopleurids on rodents and insectivores; the linognathids on even-toed ungulates such as cattle, sheep, goats, reindeer and deer, and on canids such as dogs, foxes and wolves; the pecaroecids on peccaries; the pediculids on the head and body of humans; the polyplacids on rodents and insectivores; and pthirids on gorillas and humans. The eggs generally are cemented to the hairs of the host. There are three nymphal instars in nearly all species.

See also, HUMAN LICE.

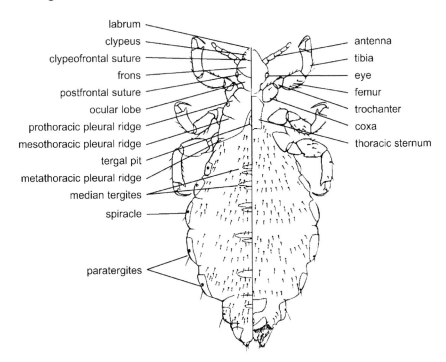

labrum
clypeus
clypeofrontal suture
frons
postfrontal suture
ocular lobe
prothoracic pleural ridge
mesothoracic pleural ridge
tergal pit
metathoracic pleural ridge
median tergites
spiracle
paratergites

antenna
tibia
eye
femur
trochanter
coxa
thoracic sternum

Fig. 999   A diagram of a sucking louse showing a dorsal view (left) and a ventral view (right).

References

Arnett Jr., R. H. 2000. *American insects* (second edition). CRC Press, Boca Raton, Florida. 1003 pp.

Durden, L. A. and G. G. Musser. 1994. The sucking lice (Insecta, Anoplura) of the world; a taxonomic checklist with records of mammalian hosts and geographical distributions. *Bulletin of the American Museum of Natural History* 218: 1–90.

Hopkins, G. H. E. 1949. The host-associations of the lice of mammals. *Proceedings of the Zoological Society of London* 119: 387–604.

Kim, K. C. and H. W. Ludwig. 1978. The family classification of Anoplura. *Systematic Entomology* 3: 249–284.

Kim, K. C. 1987. Order Anoplura. Pages 224–245 in F. W. Stehr (ed.). *Immature insects*, Vol. 1. Kendall/Hunt Publishing, Dubuque, Iowa.

Kim, K. C., H. D. Pratt, and C. J. Stpjanovich. 1986. The sucking lice of North America. An illustrated manual for identification. Pennsylvania State University Press, University Park, Pennsylvania.

Price, M. A. and O. H. Graham. 1997. Chewing and sucking lice as parasites of mammals and birds. *U.S. Department of Agriculture, Technical Bulletin 1849*. 309 pp.

SUGARCANE ROOTSTOCK BORER WEEVIL.

See also, CITRUS PESTS AND THEIR MANAGEMENT.

SUGAR-FEEDING IN BLOOD-FEEDING FLIES.

In the order Diptera, there are several groups of haematophagous (blood-feeding) flies that also feed on carbohydrates. Many a hiker can attest to the fact that when you get swarmed by black flies and accidentally crush a few between your teeth, the black flies taste quite sweet. This is due to the presence of sugars within the crop, the sugar-storage organ.

In contrast to some blood-sucking flies (e.g., stable fly, family Muscidae; tsetse fly, family Glossinidae), in which both sexes feed on blood, it is only the females that blood-feed in the following dipteran families: Simuliidae (black flies); Culicidae (mosquitoes); Tabanidae (deer flies, horse flies); and Ceratopogonidae (biting midges). Depending on the species, sugars are required by both sexes for increased longevity and/or flight. Females benefit from the blood meal because they use the extra protein for egg development; however, in certain species fecundity is increased following sugar feeding.

Sugar meal sources

The major sources of sugar meals are floral nectar, extrafloral nectaries, and hemipteran honeydew; minor sources include tree sap, rotting or damaged fruit, and leaf wounds. Floral nectar contains three major sugars, namely fructose, glucose and sucrose. A few floral nectars also contain maltose, melebiose and/or raffinose. Most studies on the sugar-feeding biology of biting flies have used the cold-anthrone test to detect fructose and fructose-containing sugars, under the assumption that most biting flies use floral nectar as their main source of sugar. Recently, several studies have employed other techniques such as gas chromatography, high performance liquid chromatography and thin layer chromatography in which the trisaccharide melezitose (and in some instances, the trisaccharide erlose or the tetrasaccharide stachyose) is used as an indicator sugar to show when flies have fed on honeydew. Using these techniques, the incidence of honeydew feeding in biting Diptera has been reported to be from approximately 5 to 55% in mosquitoes (with most species less than 15%), 0 to 5% in biting midges, 20 to 50% in black flies (with an average of 35%), 45 to 85% in deer flies (with an average of 65%) and 50% in horse flies.

Black flies and sugar sources

A long-held view has been that black flies preferentially take nectar from plants with small white, yellow or green flowers. There have been several records of black flies feeding on ivy and willow. Folklore has it that on the Canadian Shield black flies are a major pollinator of lowbush blueberries (*Vaccinium angustifolium* and *V. myrtilloides*), but it has been shown experimentally that although they will enter the flowers and feed avidly on blueberry nectar, black flies do not assist in pollination. The exact plants on which black flies nectar-feed can sometimes be determined by examining the crop contents for pollen grains and identifying the pollen microscopically. They have been observed feeding directly on the honeydew produced by adelgids found on tamarack.

Mosquitoes and sugar sources

Depending on the species, early spring *Aedes* mosquitoes have been recorded nectar-feeding on Canada plum, pin cherry and a variety of woodland plants. Even in the high Arctic, mosquitoes have been observed feeding on floral nectar from the flower *Dryas integrifolia*. A relatively high incidence of crops tested positive for honeydew sugars in *Anopheles* mosquitoes (about 55%) but other genera seem less inclined to feed on hemipteran honeydew.

Fig. 1000 Mosquito feeding on nectar from plant nectary. (Photo by Doug Burkett, University of Florida.)

Tabanids and sugar sources

There have been only a few observations of tabanids feeding on floral nectar, whereas several researchers have observed tabanid flies feeding on aphid honeydew (e.g., on the leaves of marsh elder, *Iva frutescens*). For salt marsh horse flies it has even been suggested that honeydew represents a more important source of sugar than floral nectar.

Role and timing of sugar-feeding

The role of the sugars in the biology of biting flies varies according to the species. For example, it is known that a mosquito species that feeds almost exclusively on human blood, *Aedes aegypti*, does not feed often on sugars in the wild although it will nectar-feed when flowering plants are abundant. In fact, in *Aedes aegypti*, feeding on sugars results in a decrease in longevity and fecundity.

Most other biting Diptera (including most mosquitoes) are thought to sugar-feed soon after emergence for it is from sugar-meals that energy from general maintenance and for flight is acquired. Most haematophagous Diptera will also sugar-feed before mating, host-seeking and oviposition. As a result, it is generally thought that sugar-feeding occurs more often than blood-feeding.

Almost all mosquitoes, black flies, biting midges and tabanids survive significantly longer in captivity when provided with sugar-meals (such as 10% sucrose) as opposed to water alone. The composition of the sugar-meal can affect fly longevity as well as other traits such as flight distance and duration, parasite transmission and/or egg development.

The optimal sugar-feeding regime (i.e., source and frequency of sugar-feeding) for a given species and gender will depend on a number of factors including the amount of larval fat body reserves carried forward into the adult stage, the amount of flying to be done, the parasite load, and the parous state (for females).

Following sugar-feeding and digestion, circulating monosaccharides are removed from the hemolymph, assembled into large macromolecules of glycogen and stored in the fat bodies. When sugars are required, there is hydrolysis of individual glucose molecules from the glycogen reserves. The non-reducing disaccharide, trehalose, is then synthesized and released into the hemolymph. Subsequently, when flight energy is required, the trehalose is transported to the flight muscles (where it is broken back down into monosaccharides which fuel the actual flight).

In the females of some *Culex* mosquitoes lipids accumulate in the fat body after sugar-feeding. This is correlated indirectly with a behavioral switch to host-seeking (i.e., becoming attracted to blood-host odors over floral odors). Furthermore, it has been shown that the switch to host-seeking mode following a sugar-meal is mediated by juvenile hormone (JH) activity (caused by follicular development that, itself, can be triggered by a sugar-meal). Diapausing females in the genus *Culex* sugar-feed more frequently and for longer periods than do non-diapausing mosquitoes in order to build up adequate fat reserves. Nocturnal sugar-feeding behavior is also greater in diapausing vs. non-diapausing mosquitoes.

Historical considerations

Hemiptera appear in the fossil record in the Permian, Diptera in the Triassic and flowering plants (angiosperms) in the Cretaceous. There are three ancestral traits among Diptera that facilitate honeydew usage. First, the presence of a pseudotracheate labellum allows flies to secrete saliva onto dried sugars and then suck up the liquefied sugars. This type of labellum seals off the dissolved sugars from the air, thereby reducing evaporative water loss. Second, the presence of tarsal sensilla sensitive to sugars allows the flies to 'taste' sugars with their feet. This is convenient for testing the surface of leaves for honeydew (but less convenient for tasting sugars within floral nectaries). Third, flies tend to be attracted to glossy spots on leaf surfaces; this helps foraging flies flying over vegetation to locate honeydew droplets.

It is not surprising, therefore, that black flies (which appeared in the Jurassic) are capable of using honeydew sugars; effectively, nectar-feeding was a more recent development in this family. Interestingly, tabanids appeared in the fossil record at about the same time as the flowering plants and apparently maintained an ancestral preference for honeydew over floral nectar. Mosquitoes, in contrast, did not appear in the fossil record until the late Cretaceous (Senonian) or the Paleocene of the Tertiary, long after the angiosperm radiation had begun; this probably explains why mosquitoes seem to preferentially exploit nectar sources. In addition, mosquitoes have olfactory chemosensilla on their palps which respond to nectar odors and their long mouthparts are well adapted for probing flowers for nectar.

Fig. 1001 Example of sun moths (Heliodinidae), *Heliodines galapogoensis* Heppner & Landry from Galapagos Islands.

Fiona F. Hunter
Brock University
St. Catharines, Ontario, Canada

References

Clements, A. N. 1992. *The biology of mosquitoes*, Vol. 1. *Development, nutrition and reproduction*. Chapman & Hall, London, United Kingdom. 509 pp.

Crosskey, R. W. 1990. *The natural history of blackflies*. John Wiley & Sons Inc., London, United Kingdom. 711 pp.

Foster, W. A. 1995. Mosquito sugar feeding and reproductive energetics. *Annual Review of Entomology* 40: 443–473.

SULCUS. (pl., sulci) A groove or furrow formed by the infolding of the body wall.

SULFURS. Some members of the family Pieridae (order Lepidoptera). See also, YELLOW-WHITE BUTTERFLIES, BUTTERFLIES AND MOTHS.

SUN MOTHS (LEPIDOPTERA: HELIODINIDAE).

Sun moths, family Heliodinidae, are a small family of 56 species, mostly Neotropical (31 sp.); actual fauna probably exceeds 100 species. The family is part of the superfamily Yponomeutoidea in the section Tineina, subsection Tineina, of the division Ditrysia. Adults small (7 to 15 mm wingspan), with head smooth-scaled; haustellum naked; labial palpi porrect; maxillary palpi minute, 1-segmented. Wings elongated, with lanceolate hindwings with long fringes. Maculation usually brilliantly colored, with metallic-iridescent markings and sometimes raised scale tufts. Adults are diurnal. Larvae are mostly leaf skeletonizers, but some are borers in fruit racemes. Several plant families are used as hosts, particularly Chenopodiaceae, Nyctaginaceae, Portulacaceae, and Scrophulariaceae, among others.

John B. Heppner
Florida State Collection of Arthropods
Gainesville, Florida USA

References

Emmet, A. M. 1985. Heliodinidae. In J. Heath and A. M. Emmet (eds.), *The moths and butterflies of Great Britain and Ireland*, 2: 410–411, pl. 11. Harley Books, Colchester.

Falkovitsh, M. I. 1981. Heliodinidae. In *identification keys to insects of European Russia. 4. Lepidoptera*, 2: 529–531. Academie Nauk: St. Petersburg. [in Russian]

Le Marchand, S. 1937. Les Heliodinidae. *Amateur de Papillon* 8: 217–221.

Powell, J. A. 1991. A review of *Lithariapteryx* (Heliodinidae), with description of an elegant new species from coastal sand dunes in California. *Journal of the Lepidopterists' Society* 45: 89–104.

Wester, C. 1956. Comparative bionomics of two species of *Heliodines* on *Mirabilis* (Lepidoptera, Heliodinidae). *Proceedings of the Entomological Society of Washington* 58: 43–46.

SUNN PEST. See also, WHEAT PESTS AND THEIR MANAGEMENT.

SUPERGENE. Mutually advantageous genes clustered together on the same chromosome by selection for this beneficial association. Separation of the genes is so infrequent that they function as if they were a single gene.

SUPERNUMERARY. (larvae and nymphs) The occurrence of extra (additional beyond the normal number) instars in immature insects. Additional juvenile hormone stimulates this effect.

SUPERORDER. Closely related orders, sharing common traits and thought to have evolved from one another or from a common ancestor. This level of classification is not often used, but is useful in understanding phylogeny.

See also, CLASSIFICATION

SUPERORGANISM CONCEPT. In ecology, the concept that communities consist of some species that are tightly linked together now and in their evolutionary history. In social insects, the concept that the different castes perform functions equivalent to the physiological properties of a single organism.

SUPERPARASITISM. When more parasites (parasitoids) are developing on or in a host insect than can survive to maturity.

SUPERSEDURE. The replacement of an old or sickly queen by a new one reared by the workers.

SUPPLEMENTARY REPRODUCTIVE. A queen or male termite that takes over the reproductive functions of the colony after removal of the primary reproductive.

SUPRASPIRACULAR. The area immediately above the spiracles.

SURFACTANT. A pesticide adjuvant that affects surface tension of the pesticide, and enhances penetration and retention. An abbreviated form of 'surface active agent'.

SURVIVORSHIP. The probability of a newly born individual surviving to a particular age, usually to maturity or reproduction.

SURVIVORSHIP CURVE. A plot of the declining numbers of a cohort or population as the individuals die over time. Survivorship is usually expressed as a percentage of the initial population.

SUTURE. A seam or impressed line that indicates the juncture of two body plates.

SWALLOWTAIL BUTTERFLIES (LEPIDOPTERA: PAPILIONIDAE). Swallowtail butterflies, family Papilionidae (including birdwings, parnassians, and kites), total about 589 species worldwide, with about 250 species being Indo-Australian. Three subfamilies are recognized: Baroniinae (a single relict species in Mexico), Parnassiinae, and Papilioninae. The family is in the superfamily Papilionoidea (series Papilioniformes), in the section Cossina, subsection Bombycina, of the division Ditrysia. Hundreds of local races have been given subspecies names, and form names without validity, for European *Parnassius* (Parnassiinae). Adults large (35 to 285 mm wingspan); head sometimes roughened (Parnassiinae); antennae mostly with elongated clubs (more compact in Baroniinae and Parnassiinae). Wings triangular and mostly with relatively acute apexes (rounded in Baroniinae and most Parnassiinae; hindwings usually with tails (some species tailless, including Baroniinae and most Parnassiinae); body mostly robust. Maculation very varied, with

Fig. 1002 Example of swallowtail butterflies (Papilionidae), *Byasa polyeuctes* (Doubleday) from India.

many color combinations and sometimes with iridescence, but Parnassiinae mostly pale (even hyaline) with various spotting on both fore- and hindwings. Adults diurnal; usually slow and gliding fliers. Larvae leaf feeders; with an osmeterium defensive gland behind head. Host plants include many plant groups, especially Crassulaceae, Lauraceae, Leguminosae, Rutaceae, and Saxifragaceae, among others. Birdwings use various Aristolochiaceae vines. Some economic species are known, mainly citrus feeders.

John B. Heppner
Florida State Collection of Arthropods
Gainesville, Florida USA

References

D'Abrera, B. 1975. *Birdwing butterflies of the world.* Hill House, Melbourne. 216 pp.

Igarashi, S. 1979. *Papilionidae and their early stages.* Kodansha, Tokyo. 2 volumes.

Rothschild, W. 1895. A revision of the Papilios of the Eastern Hemisphere, exclusive of Africa. *Novitates Zoologicae* 2: 167–463, pl. 6.

Scriber, J. M., Y. Tsubaki, and R. L. Lederhouse (eds.). 1995. *Swallowtail butterflies: their ecology and evolutionary biology.* Scientific Publishing, Gainesville. 459 pp (32 pl.).

Seitz, A., and H. Stichel. 1906–31. Familie: Papilionidae. In *Die Gross-Schmetter linge der Erde,* 1: 7–38, pl. 1–16 (1906); 1(suppl.): 7–91, 327–332, pl. 1–6 (1929–31); 5: 11–51, 1012–1014, pl. 1–17 (1907–24); 9: 11–118, pl. 1–50 (1908–09); 13: 11–28, pl. 1–9 (1908–10). A. Kernen, Stuttgart.

Tsukuda, E., and Y. Nishiyama. 1980. *Butterflies of the South East Asian islands.* I. Papilionidae. Plapac, Tokyo. 459 pp, 166 pl.

Tyler, H. A., K. S. Brown, Jr., and K. A. Wilson. 1994. *Swallowtail butterflies of the Americas. a study in biological dynamics, ecological diversity, biosystematics and conservation.* Scientific Publishers: Gainesville 376 pp (157 pl.).

Weiss, J.-C. 1992–94. *The Parnassiinae of the world.* Venette: Sci. Nat. 3 volumes.

SWALLOWTAIL MOTHS (LEPIDOPTERA: URANIIDAE).

Swallowtail moths, family Uraniidae, comprise about 120 species from all tropical regions, mostly Indo-Australian (85 sp.); one species strays into the United States-Mexican border region (mainly in Texas). Two subfamilies are known: Microniinae, for the smaller mostly white species with very short tails (all Old World), and Uraniinae for the more well known larger and long-tailed species. The family is in the superfamily Uranioidea,

Fig. 1003 Example of swallowtail moths (Uraniidae), *Urania leilus* (Linnaeus) from South America.

in the section Cossina, subsection Bombycina, of the division Ditrysia. Adults medium to large (31 to 160 mm), with head somewhat roughened; haustellum naked; labial palpi often long with short apical segment; maxillary palpi small, 1-segmented; antennae often serrate and thickened. Wings triangular, with acute forewings and mostly tailed hindwings (margins of hindwings usually with some emarginations). Maculation usually dark browns to black, with green or pale striae, but some smaller species mostly white with very short tails on hindwings; one colorful African species (*Chrysiridia*) has three tails on each hindwing. Adults nocturnal or diurnal, with some of the larger diurnal species known to migrate (*Urania*). Larvae are leaf feeders. Host plants are known in Asclepiadaceae and Myrtaceae for the larvae in Microniinae, and Euphorbiaceae for the larvae in Uraniinae.

John B. Heppner
Florida State Collection of Arthropods
Gainesville, Florida USA

References

Dalla Torre, K. W. von. 1924. Uraniidae. In *Lepidopterorum Catalogus,* 30: 1–57 [part]. W. Junk, Berlin.

Holloway, J. D. 1998. Family Uraniidae. In *The moths of Borneo,* 8: 77–78, 82–92, pl. 5–6. Malayan Nature Society (Malayan Nature Journal, 52).

Regteren Altena, C. O. van. 1953. A revision of the genus *Nyctalemon* Dalman (Lepidoptera, Uraniidae) with notes on the biology, distribution, and evolution of its species. *Zoologische Verhandlingen* 19: 1–57, 4 pl.

Seitz, A. (ed.). 1912–33. Familie: Uraniidae. In *Die Gross-Schmetterlinge der Erde. 2. Die palaearktischen Spinner und Schwärmer*, 2: 275–276, pl. 22, 48 (1912); 2(suppl.): 171–172, pl. 15 (1933); 6: 829–837, pl. 67, 138 (1930); 10: 93–103, pl. 69–72 (1929); 14: 387–394, pl. 67, 48 (1928). A. Kernen, Stuttgart.

Smith, N. G. 1992. Reproductive behaviour and ecology of *Urania* (Lepidoptera: Uraniidae) moths and their larval food plants, *Omphalea* spp. (Euphorbiace ae). Pages 576–593 In D. Quintero and A. Aiello (eds.), *Insects of Panama and Mesoamerica: selected studies*. Oxford University Press, Oxford.

SWAMMERDAM, JAN.

SWAMMERDAM, JAN. This Dutch naturalist played an important role in the development of biology during the 1660s and 1670s. This was an important period in the development of science because it marked the re-ascendency (dormant since the Greek scholars) of scientific observation and experimentation, and the consequent diminution of divine interpretation. Swammerdam was born in Amsterdam on February 12, 1637; his father was an affluent apothecary who, though frustrated by Jan's lack of a 'proper' vocation (medicine), apparently provided support to Jan throughout his life. Indeed, the elder Swammerdam is likely responsible for the younger Swammerdam's interest in insects, as he had a notable insect collection. Jan enrolled in the medical program at the University of Leiden at the age of 24 and received his M.D. in 1667, though there is no evidence that he ever practiced medicine. Jan Swammerdam made several important anatomical discoveries, including the presence of valves in lymph ducts, respiration and nerve-muscle function; the presence of human ovarian follicles; and the mechanism of penile erection. In 1669 he published a book: "Historia Generalis Insectorum," in which he put forth a revolutionary classification of insects based on their mode of development. His work supported that of Francesco Redi, who maintained (contrary to popular belief) that insects were not the product of spontaneous generation, but originated from eggs deposited by females. Stimulated by the work of the physiologist Marcello Malpighi, in 1670 Swammerdam began to study insects under a microscope. In this effort, he greatly advanced our knowledge of insect anatomy, particularly that of the honey bee. Unfortunately, Swammerdam fell under the influence of the religious mystic Antoinette Bourignon, and abandoned his scientific activities for several years, only to become disillusioned with Bourignon and return to science in 1676. The result of his latter work, particularly his magnificent drawings, were published much later by the Dutch physician Herman Boerhaave in 1737 to 1738 as "Bibliae Naturae," revealing a wealth of detail about insects and insect anatomy. Swammerdam died prematurely at the age of 43 in Amsterdam on February 17, 1680.

References

Cobb, M. 2002. Eric Weisstein's world of biography. Seen at http://scienceworld.wolfram.com/biography/Swammerdam.html (August 2002).

Pellett, K. L. 1929. Lives of famous beekeepers, John Swammerdam. *American Bee Journal* 69: 130–131.

Westfall, R. S. 2002. Catalog of the scientific community. Seen at http://es.rice.edu/ES/humsoc/Galileo/Catalog/Files/swamrdam.html (August 2002).

SWARMING.

SWARMING. The exodus of the reproductive (or reproductives) from the original nest, with the intent to form a new colony. In bees, this is normally accomplished by a single female accompanied by a large number of workers, whereas in ants and termites, usually large numbers of reproductives disperse but new colonies are founded by only a pair of individuals.

SWEETPOTATO AND SILVERLEAF WHITEFLIES (HEMIPTERA: ALEYRODIDAE).

SWEETPOTATO AND SILVERLEAF WHITEFLIES (HEMIPTERA: ALEYRODIDAE). The whitefly (Hemiptera: Aleyrodidae) species *Bemisia tabaci* (sweetpotato whitefly or cotton whitefly) was described by Gennadius infesting tobacco in Greece in 1889. In the mid 1980s, following outbreaks of a similar, but not identical, insect (initially called *B. tabaci* biotype B) was recognized and in 1994, Bellows and Perring described it as *B. argentifolii* (silverleaf whitefly). To date, the validity of the 'B' biotype as a separate species is still disputed. In addition, about 10 biotypes have since been recognized using molecular and biochemical methods. They differ in host plant relationships and their capacities to transmit plant disorders and/or viral diseases. In the present discussion, the whole *B. tabaci* complex will be treated under the name *B. tabaci*.

Following its description in 1889, *B. tabaci* was recorded from the United States in 1894, and shortly thereafter in several African and Asian countries. It was considered to have originated in India or Pakistan until recently, when molecular studies pointed to an African origin. Moreover, some of the different biotypes probably arose locally; e.g., the ''B'' and the ''Q'' biotypes originated in the Mediterranean basin, E is from Benin, H from India and K from Pakistan. *Bemisia tabaci* was first reported as a pest from India, The Sudan, and several mideastern countries. In the United States, damage was first noted when cotton leaf crumple virus was found to be transmitted by *B. tabaci* in 1954. During the last 20 years, with the spread of the B biotype, *B. tabaci* became a cosmopolitan pest. It attacks vegetable, ornamental and field crops in tropical and subtropical regions, and greenhouse crops in colder climates.

Plant injury

Bemisia tabaci damages plants through direct removal of plant assimilates, honeydew contamination, virus transmission and the induction of physiological plant disorders. Direct damage may range from slight plant injury to plant death. The honeydew of both nymphs and adults, and the sooty mold fungi that develop on the honeydew, contaminate plant leaf surfaces, thereby reducing their photosynthetic efficiency. The insects also contaminate the produce, reducing its market value. The most severe contamination-related economic damage is caused to cotton, where stickiness may sharply reduce lint value. The transmission of more than 70 different kinds of viral plant diseases by *B. tabaci* is a major agricultural problem. Most viruses belong to the Begomoviruses (Geminiviridae), but recently Carlaviruses, Closteroviruses, Criniviruses, Luteoviruses, Nepoviruses and Potyviruses also were recorded. Some Begomoviruses, such as the Tomato Yellow Leaf Curl Virus (TYLCV) and the East African Cassava Mosaic Virus (EACMV), drastically curtail crop production and raise great difficulty in devising an integrated or biological control program for the pest. Physiological plant disorders have mainly been associated with type B of the pest The best known disorder is leaf silvering of squash, in which the plant cuticle separates from the underlying cells giving the leaf a silvery appearance, and tproviding a basis for its new name (silverleaf whitefly or *B. argentifolii*). Another disorder, ''uneven tomato ripening'' appears when heavy whitefly populations are present during tomato ripening season.

Biology

Like all whiteflies, the *B. tabaci* female inserts its eggs into the leaf tissue by means of a pedicel. On most plants, development occurs on the underside of the leaves. They hatch into crawlers that settle close to the point of hatching, insert their mouth parts into the phloem, molt into the second instar, and loose their ability to walk on the leaf. The second to fourth instars are spent in the same location, with the developing nymphs growing from ca. 0.2 mm to 0.8 mm in length depending upon growing conditions and plant species. The nymphs may have a glabrous or spiny dorsum, in direct correspondence to the degree of pubescence of the leaf upon which they develop. The nymphs feed by inserting their 100–300 μm long mouthparts (stylets) into medium to small phloem tubes and imbibing the exuded fluid. The fourth instar includes a developing stage during which it continues to grow, and a later sessile stage that can be distinguished through appearance of larger, red eye spots. This sessile stage corresponds to the pupal stage of the Holometabola when larvae undergo metamorphosis to adults. Therefore, the red-eyed stage of the fourth instar is often referred to as a 'pupa'. The adults emerge through the dorsum of the 'pupal' case, leaving a T-shaped split behind. The body of the emerging adults is covered with wax and is completely white. Both sexes are winged, and their body is about 12 mm. long, with males somewhat shorter than the females.

Bemisia tabaci is arrhenotokous, having haploid males and diploid females. Their longevity, fecundity and developmental duration may vary greatly depending the temperature, the host plant and the whitefly biotype. Typically, females live 20 to 30 days in outdoor conditions during which they often lay 50 to 150 eggs per female. However, the number of eggs may vary, and as many as 500 eggs per female have been recorded. Developmental duration from egg to adult ranges between two weeks in the summer and over two months in the winter, in subtropical and temperate regions.

Emerging adults may stay on the leaf of emergence, mate there and lay their eggs near the point of emergence. Often, however, they migrate to the top of the canopy where they settle on younger leaves of the same or of neighboring plants. In addition to trivial, short-distance flight, long distance migration (up to 7 kilometers) has been recorded. Individuals that migrate have somewhat different body proportions, a smaller egg load and less investment in vitellogenins than the trivial flyers.

Host range

The full host plant range of *B. tabaci* is probably not known. Recent reports record over 500 different plant species with some reports reaching as many as 900 hosts. Given the ease at which this pest adopts new host plants, it is useful to consider any plant, especially of the dicotyledons, as a potential host, rather than to adhere to a fixed host plant list. Host plant finding is a random process until the plant itself is reached and selection takes place following a short probing. *B. tabaci* interacts with plant physiology, and plant nutrition, including nitrogen levels, influences the insect's developmental success. Feeding by *B. tabaci* can change the source-sink relationships in young melon plants, redirecting the flow of amino acids and thus enriching its feeding sites. Induced plant resistance, in which previously infected plants are relatively less suitable for insect growth than unifested plants, does not seem to affect *B. tabaci*, thus differing from other phytophagous insects such as the leaf miner *Liriomyza trifolii* or from several noctuid larvae.

Bemisia tabaci adults produce several compounds that apparently render them better adapted to the high summer temperatures in which they live and to the variation in osmotic pressures they may experience while feeding on a wide variety of plants. In addition to the presence of heat-shock proteins, which are produced in response to short-term heat stress, diurnal increases in environmental temperature induce *B. tabaci* adults to accumulate a high level of sorbitol (15–27-fold increase to ca. 0.5 M) in their body fluids which apparently protect the insect's proteins from heat damage. The presence of the recently discovered sugar isobemisiose in the adult's body fluids, and its absence from the excreted honeydew, have led to the conclusion that the creation of this trisaccharide and/ or it decomposition into smaller moieties facilitates adjustment of the osmotic pressure in the haemolymph, enabling the insect to withstand the relatively wide fluctuations in sugar concentration of the imbibed phloem.

Management

Because it is a very serious pest of both greenhouse and outdoor field crops, this whitefly is often treated with insecticides. This brought about widespread insecticide resistance that ranges from 5 to 10-fold, to over 1000-fold, to both conventional materials such as organophosphates and pyrethroids, and to the newer insect growth regulators (IGRs) such as buprofezin and pyriproxyfen. IPM programs, including the insecticide resistance management (IRM) strategies recently established in Israel and the southwestern USA, depend on alternating key insecticides and limiting the numbers of yearly applications of IGRs, thus reducing the buildup of resistance. The use of insect pollinators such as bumblebees in tomatoes and in peppers, and the high incidence of resistance, brought about a widespread search for other control solutions. These include plant varieties that are resistant to the transmitted viruses or to the whitefly, the use of cultural methods such as enclosing the crop in 50-mesh screenhouses, the use of UV absorbent sheets plastic (that apparently affect the insect's orientation) and screens in greenhouses and the utilization of natural enemies: parasitoids, predators and fungal diseases. Many of the natural enemies appear spontaneously in the field crops, but may not do so early enough or in large enough numbers to control the pest by themselves. Therefore, their augmentation through timely releases in concert with utilization of other control measures, are often recommended.

Natural enemies

The principal species of *B. tabaci* parasitoids belong to the genera *Eretmocerus* and *Encarsia*, whereas the predators may belong to numerous families and vary greatly according to the geographic location. In recent years, predators that invaded Spanish greenhouses in the spring, mainly Miridae and Anthocoridae, were found to be effective and are artificially reared and sold as control in agents in addition to the parasitoids. The use of fungal diseases to control *B. tabaci* is not widespread, mainly due to their cost and their lack of efficacy under low humidity conditions.

Organizations, literature and web sites

A *Bemisia* website is available: http://rsru2.tamu.edu/BIRU/sweetpot.html. It directs readers to the various United States-based agencies that work and publish on this insect. In addition several international organizations are devoted to the study and application of *Bemisia* control. The main ones are:

European Whitefly Studies Network (EWSN) at:

http://www.jic.bbsrc.ac.uk/hosting/eu/ewsn/What.html

Sustainable Integrated Management of Whiteflies as Pests and Vectors of Plant Viruses in the Tropics, an SP-IPM Task Force led by CIAT at:

http://www.cigar.org/spipm/tf/wgv.html

Action Network for Whitefly and Geminivirus Management in Ibero America and the Caribbean, at:

http://www.catie.ac.cr/moscablanca

The quarterly newsletter *Mosca Blanca al Día* (*Whitefly Update*), which has been published since 1993, is included in this site.

In addition, over 300 publications about *B. tabaci* appear each year, listed by Dr. S. Naranjo, T. Henneberry and colleagues (Western Cotton Research Laboratory of the USDA, Phoenix, AZ) at:

http://www.wcrl.ars.usda.gov

Until the end of 2000 a *Bemisia Newsletter* was published and can be obtained at:

http://207.43.217.12/biru/BEMISIA13.htm

From 2001 on, it has been united with the EWSN newsletter that appears about every 3 months since 1998 and can be seen at:

http://www.jic.bbsrc.ac.uk/hosting/eu/ewsn/EWSN_News04.htm

See also, WHITEFLIES (HEMIPTERA: ALEYRODIDAE).

Dan Gerling
Tel Aviv University
Ramat Aviv, Israel

References

Ohnesorge, B., and D. Gerling. 1986. *Bemisia tabaci* - ecology and control. *Agriculture, Ecosystems and Environment* 17. (Special issue).

Gerling, D., and R. T. Mayer. 1996. *Bemisia* 1995: taxonomy, biology, damage, control and management. Intercept Ltd., Andover, United Kingdom. 702 pp.

Naranjo, S. E., M. R. V. Oliveira, P. C. Ellsworth, and O. A. Fernandes (eds.). 2001. Challenges and opportunities for pest management of *Bemisia tabaci* in the new century. *Crop Protection* 20.

SWEETPOTATO FLEA BEETLE, *CHAETOCNEMA CONFINIS* (COLEOPTERA: CHRYSOMELIDAE: ALTICINAE).

Chaetocnema (*Tlanoma*) *confinis* Crotch, 1873 is a very small (1.4 to 1.8 mm) and light species. Although thought to have originated in North America (USA), it has become ubiquitous in the tropics of the Old and New World. It is becoming the most widely spread species of leaf-beetle. It has two known synonyms: *C. flavicornis* J. LeConte and *C. etiennei* Jolivet. *Chaetocnema perplexa* Blake from Central America is probably also a junior synonym.

Morphology

Chaetocnema confinis is a very small species; even the female is less than 2 mm. The adult is black to dark bronze, moderately glossy, with the antennae pale brown, the ventral surface black to dark brown, the femora brown to dark brown, the tibiae brown to pale brown, the tarsi pale brown. The striae on the elytra are parallel and the punctures moderate. The punctures are deep on the pronotum.

The male has the fore and middle tarsus with the first segment somewhat enlarged. The aedeagus relatively small and sinuate. *Chaetocnema perplexa* Blake seems to differ by only a small variation of the shape of the aedeagus. The eggs are deposited in small groups and do not differ from those of related species (0.8 × 0.2 mm). The larvae are eruciform, straight, cylindrical, legged and measure 6 mm at the final stage. They live on plant roots.

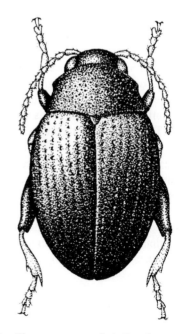

Fig. 1004 *Chaetocnema confinis* Crotch.

Biology

Normally, *C. confinis* larvae feed inside Convolvulaceae roots, but they also attack the collar at the limit between stem and root. Long narrow grooves are eaten in the leaves, especially on the upper surface along the veins, during May and early June in North America, and throughout the year in the tropics. When these channels are numerous, the leaf may wilt. In the tropics there are many generations per year and no winter diapause. Otherwise in the US the species is univoltine.

In the US, *C. confinis* overwinters as adult under trash and various sheltered places. When they come out of hibernation, adults attack the plants when they are set out from seedbeds. By the end of June, they leave the sweet potatoes, migrating to bindweed where they lay their eggs and die. The larvae hatch after 3 weeks incubation. The new generation of beetles will feed and enter diapause in the fall. In the US and Canada, both sexes are present. In the Old World tropics only thelytokous parthenogenetic females are present, and so far as is known this is the condition in Central and South America.

Distribution

Chaetocnoma confinis originated in the US and Canada where both sexes are present. The beetle walks, flies and jump easily. Its recent dispersal in Africa and Asia-Pacific is probably due to typhoons and hurricanes because the beetle is light, small and winged. The first specimen was discovered on *Ipomoea aquatica*, a vegetable grown in Reunion, in 1976. It was then described under the name of *C. etiennei* Jolivet and ten years later M. L. Cox synonymized it with *C. confinis*. Having only females in the tropics makes identification rather difficult.

Then the beetle invaded Madagascar, Mauritius and spread over most of tropical and southern Africa. *C. confinis* is also known from Brazil and Galapagos, Central America, Oceania (Palau), New Caledonia, Asia (India, Vietnam, Thailand, Ryukyu islands, etc.). At the speed that it spreads, New Guinea, Indonesia and Australia are the next targets. It is very probable that the species is already in New Guinea and several Indonesian islands. Probably it will soon cover all the tropical lands of Indian Ocean and Pacific.

Food plants

The beetle feeds on Convolvulaceae. It feeds on *Convolvulus arvensis* (bindweed) in the US and Canada, but it has adapted to many other Convolvulaceae: sweet potato, *Ipomoea aquatica*, *I. pandurata*, *Calystegia sepium*, *Pharbitis purpurea*, *P. cathartica*, etc.

The beetle has become secondarily polyphagous, and adapted mostly in North America to corn (*Zea mays*), sugarbeet, tomato, and many other crops and weeds. In the tropics, so far it remains a major pest of sweet potato and related plants.

Conclusion

Few parasites have been recorded on this beetle and related species: fungi (Laboulbeniales, *Beauveria*), nematodes (*Howardula*), bacteria (*Xanthomonas*) and one parasitoid (Ichneumonidae).

Except in the US, where the beetle can become a serious pest on various crops, sweet potato seems to be the only host in the tropics. Several other *Chaetocnema* species and other alticines are also parthenogenetic and that seems to help in the expansion of the species. *Chaetocnoma confinis* is invading the whole of the tropics, mainland and islands together, and is becoming the most ubiquitous chrysomelid. Exchanges between Europe and the US have been common in the past (Colorado Potato Beetle, *Plagiodera* and many others) but only due to passive importations. This time the beetle seems to have dispersed by wind. Strict quarantine measures are certainly necessary in places like Australia when importing sweet potatoes, but aerial dispersion is very probable in the near future.

Pierre Jolivet
Paris, France

References

Biondi, M. 2002. Checklist of Afrotropical species of the genus *Chaetocnema* Stephens (Col. Chrys.: Alticinae): synonymies and geographical distribution. *African Entomology* 10: 265–284.

Cox, M. L. 1996. *Parthenogenesis in the Chrysomeloidea.* pp. 133–151 in P. Jolivet and M. L. Cox (eds.), *Chrysomelidae biology. 3. General studies.* SPB Academic Publishing, Amsterdam, The Netherlands.

Jolivet, P. 1979. Réflexions sur l'écologie, l'origine et la distribution des chrysomélides (Col.) des îles Mascareignes, avec la description de deux espèces nouvelles. *Bulletin Société Linnéenne de Lyon* 48: 524–528; 48: 606–608 and 641–649.

Jolivet, P. 1998. Les nouveaux envahisseurs ou les chrysomélides voyageurs (Col.). *L'Entomologiste, Paris* 54: 33–44.

White, R. E. 1996. A revision of the genus *Chaetocnema* of America, north of Mexico (Col. Chrys.). *Contribution of the American Entomological Institute* 29: 1–156.

SWEETPOTATO WEEVILS AND THEIR ERADICATION PROGRAMS IN JAPAN.

The sweetpotato weevil, *Cylas formicarius* (Fabricius) (Coleoptera: Brentidae), is the most destructive pest of the sweet potato, *Ipomoea batatus* (Linnaeus). The insect occurs circumglobally in tropical and sub-tropical regions of Asia, the Pacific, Africa, the Caribbean and the United States. *Cylas* spp. are believed to have originated in Africa and/or Madagascar; only *C. formicarius* occurs world-wide as a pest species. Larvae and adults damage fleshy roots and stems of sweet potatoes. Larvae cause serious damage because they tunnel through the roots. Furthermore, the roots respond to weevil larvae feeding by producing ipomeamarone, a toxic furanoterpenoid with an unpleasant odor. The entire roots are rendered inedible, even though they are damaged in part, and because of their bitter taste, they cannot be eaten by livestock such as pigs, either. Because most of the larvae are cryptic in the roots below the soil surface, it is difficult to control the weevil populations effectively by the application of chemical insecticides. In 1986, the sex pheromone of *C. formicarius* was isolated, identified and synthesized in the United States. The synthesized pheromone strongly attracts the male adults of the weevil. Pheromone-trap monitoring systems have been developed in the field, and the monitoring trap is also used as a regulatory tool to detect an invasive organism in quarantined areas.

Fig. 1005 Adult sweetpotato weevil. 7 mm (photo by S. Moriya).

The West Indian sweetpotato weevil, *Euscepes postfasciatus* (Fairmaire) (Coleoptera: Curculionidae), is another major pest of sweet potato in the tropical and sub-tropical islands of the Pacific and the Caribbean and a part of Central and South America. The range of its distribution is narrower than that of *C. formicarius* as a whole, but the two species coexist in some places such as the Pacific and the Caribbean regions. The adults of both species are quite different in shape; that is, *Cylas* is a snout beetle of ant-like appearance, while *Euscepes* is inconspicuous and looks like a soil particle, especially when it feigns death. However, their immature stages are indistinguishable by the naked eye, as is the damage they do to sweet potato roots. Unlike *C. formicarius*, no sex pheromone has been found in *E. postfasciatus* up to now, and no evidence has been obtained either in the laboratory or the field that this weevil can actually fly, even though the adults possess well-developed hind wings. These facts may limit available information on the management of *E. postfasciatus*.

In Japan, these two weevil pests are exotic and distributed only in the Ryukyu Islands and Bonin Islands, the southernmost part of Japan. Because the weevils are not found on the main islands of Japan, transporting host plants from infested areas to uninfested areas is prohibited by quarantine regulations, as with quarantine programs for *C. formicarius* in the United States and *E. postfasciatus* in the islands of the South Pacific. To solve the problems caused by this restriction, eradication programs were initiated by the Kagoshima and Okinawa Prefectural Governments in 1988 and 1990, respectively, under the supervision of the National Government of Japan.

A combination of two unique methods was adopted on *C. formicarius* to eradicate it from the target areas. One is to use sex pheromone traps to annihilate males; the other is to release into the field masses of laboratory-reared males sterilized by gamma irradiation, thereby depriving wild females of their chance to mate with wild males. Because of the absence of effective attractants for *E. postfasciatus*, only the sterile insect technique (SIT) could be applied to their case.

There are three steps in the course of the eradication programs, as there were with the melon fly, *Bactrocera cucurbitae* (Coquillett), which was successfully eradicated from the same area of the Ryukyu Islands in 1993. The first step was essentially

Fig. 1006   Adult West Indian sweetpotato weevil. 3–4 mm (photo by S. Moriya).

a research phase to develop basic techniques required for the procedures, including mass rearing, a sterilization and marking method, and pilot eradication experiments started on two small islands (21 to 35 ha) in 1988 and 1990. The second step was to confirm the feasibility of the programs; thus, in 1994, experimental eradication projects were initiated on two "medium sized" islands of about 57 and 63 km^2. The sweetpotato weevil populations were almost eradicated in the target areas of both islands. The final step of the programs was put into practice in 2001 to eradicate the two weevil pests from all the distribution areas (ca. 3,500 km^2) in the Ryukyu Islands. Since this program constitutes the first attempt at area-wide eradication of sweetpotato weevils in the world, it might take longer for it to reach its goals than the melon fly eradication project, which took 22 years.

Seiichi Moriya
National Agricultural Research Center
Tsukuba, Japan

References

Chalfant, R. B., R. K. Jansson, D. R. Seal, and J. M. Schalk. 1990. Ecology and management of sweet potato insects. *Annual Review of Entomology* 35: 157–180.

Jansson, R. K., and K. V. Raman (eds.). 1991. *Sweet potato pest management: a global perspective.* Westview Press, Inc., Boulder, Colorado. 458 pp.

Moriya, S. 1995. The possibility of eradicating two sweet potato pests, *Cylas formicarius* and *Euscepes postfasciatus*, in Japan. *Food & Fertilizer Technology Center (FFTC) Extension Bulletin* 403: 1–7.

Moriya, S., and T. Miyatake. 2001. Eradication programs of two sweetpotato pests, *Cylas formicarius* and *Euscepes postfasciatus*, in Japan with special reference to their dispersal ability. *Japan Agricultural Research Quarterly (JARQ)* 35: 227–234.

Sutherland, J. A. 1986. A review of the biology and control of the sweetpotato weevil *Cylas formicarius* (Fab). *Tropical Pest Management* 32: 304–315.

SWEZEY, OTTO HERMAN.   Otto Swezey was born near Rockford, Illinois, USA, on June 7, 1869. He attended Lake Forest College (B.A. 1896) and then obtained an M.A. from Northwestern University in 1897. He taught biology at Northwestern until 1903. He gained employment with the Hawaiian Sugarcane Planters Association in 1904, apparently the first American entomologist to be employed by a private company. In Hawaii, Swezey was instrumental in establishing biological control for sugarcane pests. He was a founder of the entomological section of the Bishop Museum in Honolulu, and pursued his interest in Lepidoptera, though he was more interested in describing life histories than in describing insects. Nevertheless, he described over 100 Lepidoptera. Swezey was editor of the Proceedings of the Hawaiian Entomological Society for 39 years, and also served as president of that society. A major publication was "Forest insects in Hawaii" (1954). Swezey died at San Jose, California, on November 3, 1959.

Reference

Mallis, A. 1971. *American entomologists*. Rutgers University Press, New Brunswick, New Jersey. 549 pp.

SWIFTS.

Members of the family Hepialidae (order Lepidoptera). They also are known as ghost moths. See also, GHOST MOTHS, BUTTERFLIES AND MOTHS.

SWORD-TAIL CRICKETS.

A subfamily of crickets (Trigonidiinae) in the order Orthoptera: Gryllidae. See also, GRASSHOPPERS, KATYDIDS AND CRICKETS.

SYLVATIC.

A disease that is found in, or acquired in, a wooded or other undeveloped natural habitat.

SYLVEIRA CALDEIRA, JOÃO DA.

João da Sylveira Caldeira was born at Rio de Janeiro, Brazil, on June 28, 1800. After studying medicine at the University of Edinburgh, Scotland he worked for a time as a preparator at the ''Jardin des Plantes'' in Paris. Returning to Rio de Janeiro, he and others were commissioned to revise and publish the ''Flora Brasiliensis.'' He also served as professor of chemistry and director of the National Museum. Following the proclamation of Brazilian independence in 1822, the holdings of the National Museum were greatly increased. He also established exchanges with European museums. Sylveira contributed materially to the collections of Macquart, and many species described by the latter are attributed to Sylveira, and several species were named after him. He died in Rio de Janeiro on July 4, 1854.

Reference

Papavero, N. 1971. 1973. *Essays on the history of Neotropical Dipterology with special reference to collectors (1750–1905)*. Museu de Zoologia, Universidade de São Paulo.

SYMBIONT.

An organism living symbiotically with another organism.

SYMBIONTS OF INSECTS.

Symbiosis has been defined as mutualism (a beneficial interdependence of different species), and as parasitism (a form of interaction that is beneficial only to one partner and detrimental to the other). Today, symbiosis is sometimes equated with the concept of mutualism, in which one partner is considered the host and the other the symbiont. A difficulty with this definition is that it can be difficult to resolve relationships because the boundary between symbiosis and parasitism may be fluid. Symbionts may be intracellular or extracellular.

Mitochondria are intracellular symbionts

All arthropods have a nuclear genome with chromosomes consisting of DNA and proteins. In addition, insect cells also contain multiple copies of mitochondria, which are involved in basic functions of the cell. Mitochondria are generally accepted to be microbial symbionts that were modified after a long process of evolution within eukaryotic cells. Mitochondria retain a distinctive genome that is replicated and expressed, but all mitochondria are incapable of independent existence. Most of the genes in mitochondrial DNA were transferred from the mitochondrion to the nuclear genome of its eukaryotic host over the course of evolution. Only a few genes are left in the mitochondria which are located in the cytoplasm.

Multiple symbionts may be present

In addition to mitochondria, insects have intimate relationships with a diverse array of other symbionts, including viruses, bacteria, yeasts, and rickettsia. The details of the relationship between the host and the microorganism usually are unknown. There are intracellular symbionts (such as viruses and bacteria) as well as the extracellular symbionts.

The rice weevil *Sitophilus oryzae* (Rhynchophoridae), for example, has four genomes important to the weevil's biology. These are: nuclear, mitochondrial, principal gut endosymbiont, and *Wolbachia*. The principal gut endosymbiont is found (3×10^3 bacteria/cell) in specialized cells called bacteriocytes. A total of 3×10^6 of these bacteria are found in each weevil. The principal symbionts induce the specific differentiation of the bacteriocytes and increase mitochondrial oxidative phosphorylation. Their elimination impairs many physiological traits.

The multiple genomes found in this weevil support the serial endosymbiotic theory, which states that

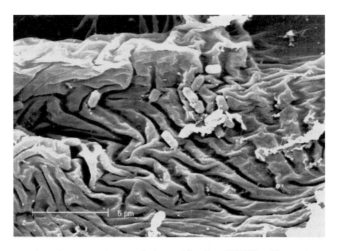

Fig. 1007 Gut symbionts: scanning electron micrograph (magnification 6000X) of bacteria attached to the convolutions of the peritropic membrane in a *Ceratitis capitata* intestine (photo provided by Carol Lauzon, University of California, Hayward).

endosymbiosis did not occur just once in eukaryotic evolution with the origin of a nucleus, or even twice, when an anaerobic single-celled organism acquired a respiring bacterium which ultimately became the mitochondrion. Molecular data suggest the gut symbiont of *Sitophilis* was established prior to acquisition of *Wolbachia*. One scenario suggests that the colonization of cereal plants by *Sitophilus* was facilitated by the acquisition of the gut symbiont, a vitamin supplier.

Symbiont function(s)

In many cases, symbionts possess metabolic capabilities that the insect host lacks and the insects use these capabilities to survive on poor or unbalanced diets. The insect and its microbes often require their mutual association. Many insects freed of symbionts grow slowly and produce few progeny; many symbionts cannot grow outside their insect host. The amazing diversity of relationships and organisms involved in symbiotic relationships with insects has raised many questions, but provided few clear cut answers because many symbionts cannot be cultured outside their hosts. Many symbionts are contained in special structures and transmitted by a highly-specific method, including transovarial transmission (transmission through the egg), to progeny. Transmission also can occur when larvae feed on contaminated egg shells or feces.

Some insect species contain several different types of symbionts in different tissues, including the gut, malpighian tubules, salivary gland, fat body, or gonads. Usually these symbionts are neutral or beneficial to their hosts. For example, symbionts of some aphids (Hemiptera) synthesize essential amino acids, sterols, and vitamins and may function as genetic elements distinct from the host nuclear genome. Symbionts found in scale insects (Hemiptera) are particularly diverse, with almost twenty different types of associations described so far.

Antlions (Myrmeleontidae) prey on other insects by sucking out the body fluid after first paralyzing their prey with a toxin produced by salivary gland secretions produced by salivary bacteria. The paralyzing toxin produced by the bacterial endosymbionts is a homologue of GroEL, a protective heat-shock protein that functions as a molecular chaperone in the common bacterium *Escherichia coli*. In the antlion, the GroEL protein may act on receptors in insects to induce paralysis, perhaps having evolved this non-chaperone function to establish a mutually beneficial antlion-bacterium relationship. It is unknown if such insecticidal proteins are produced by endosymbionts to help other fluid-feeding predatory insects.

Insects lacking some symbionts are apparently completely normal. For example, in the beetle family Cerambycidae, all of which live in wood, some species have symbionts while others lack them. Thus, the hypothesis that symbionts always supply a deficiency in the insect's diet appears to be simplistic.

Gut symbionts

The primary habitat for the majority of micro-organisms associated with insects is the digestive tract, especially the hind gut. The termite gut is one of the better studied examples of symbiosis, and molecular tools are improving our ability to resolve the taxonomy of these complex relationships.

The hindguts of termites can be compared to small bioreactors where wood and litter is degraded, with the help of symbiotic microorganisms, to provide nutrients. The hindgut of termites is a structured environment with distinct microhabitats. The dense gut microbiota includes such organisms as Bacteria, Archaea, Eukaryotes, and yeasts. These organisms do not occur randomly within the gut, but may be suspended in the gut contents, located within or on the surface of flagellates, or attached to the gut wall. The identity, exact number, and location of most is inadequately known because these organisms are difficult to culture outside their hosts.

Molecular tools will provide significant new information. For example, the microbiota of termites includes spirochaetes, which account for as many as 50% of the organisms present in some termite species. Spirochaetes are a distinct phylum within the bacterial domain, but relatively little is known about them. One molecular analysis of spirochaetes in the termite *Reticulitermes flavipes* suggested there are at least 21 previously unknown species of *Treponema*, suggesting that the long-recognized and striking morphological diversity of termite gut spirochaetes is paralleled by their genetic diversity, which may reflect substantial physiological diversity.

Omnivorous cockroaches also have microbial communities within their guts, but the associations are less interdependent than those of termites. The gut microbial communities in cockroaches anaerobically degrade plant polymers and include hydrogen-consuming bacteria, especially methanogens (bacteria that produce methane gas). The densities of

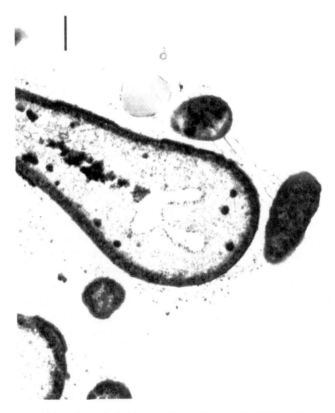

Fig. 1008 Gut symbionts: transmission electron micrograph (magnification 12,000X) of bacterial cells attached directly to the peritrophic membrane in the gut of *Anastrepha ludens*. Data bar 500 nm (photo provided by Carol Lauzon, University of California, Hayward)

these microorganisms can be enormous; for example 5×10^{12} bacteria per ml were found in the hindgut of the cockroach *Periplaneta americana*.

Tsetse flies (Glossinidae) are vectors of African sleeping sickness disease in humans and animals. Microorganisms associated with these flies, that are blood feeders, are responsible for nutrients that the flies are unable to synthesize. Different micro-organisms have been found in midgut, hemolymph, fat body and ovaries. Until molecular techniques were used, their taxonomic status was unresolved. Now we know that at least three different microorganisms are present: the primary (P) symbiont *Wigglesworthia glossinidia* is an intracellular symbiont residing in specialized epithelial cells that form a U-shaped organ (bacteriome) in the anterior gut. The secondary gut symbiont *Sodalis glossinidius* is present in midgut cells. The third, *Wolbachia*, is found in reproductive tissues.

Tsetse flies are viviparous, with females retaining each egg within her uterus, where it hatches. One young larva matures at a time and is born as a fully developed third instar larva. During its intrauterine life, the larva receives nutrients and both of the gut symbionts from its mother via milk-gland secretions. The *Wolbachia* are transmitted transovarially. Efforts to eliminate tsetse symbionts with antibiotics result in retarded growth and a decrease in egg production; because it is impossible to eliminate only one gut symbiont at a time, it is difficult to decipher the role each plays. However, they play a role in metabolism, supplying B-complex vitamins. *Sodalis* also produces a chitinase, which appears responsible for increasing the susceptibility of its host to the sleeping sickness trypanosome. An analysis of the *Wigglesworthia* and *Sodalis* genomes indicate that they each form a distinct lineage in the Proteobacteria. The phylogenetic data suggest that a tsetse ancestor was infected with a *Wigglesworthia* and from this ancestral pair tsetse species evolved along with the different *Wigglesworthia* strains existing today. No evidence exists that horizontal transfer of *Wigglesworthia* symbionts occurred between tsetse species. However, the *Sodalis* symbiont might have been acquired recently by each tsetse species or there may have been multiple horizontal transfers between tsetse species.

Perhaps the best studied gut endosymbiont of insects is *Buchnera aphidicola*, an associate of aphids. The complete genome of this symbiont has been sequenced. *Buchnera* is found in huge cells (bacteriocytes) in most of the 4,400 aphid species, supplying its host with essential amino acids. In return, *Buchnera* is given a stable and nutrient-rich environment. Aphids become sterile or die if their symbionts are eliminated. This relationship has been stable for 200 to 250 million years and *Buchnera* and aphids have co-evolved over a long period. About 9% of the *Buchnera* genome is devoted to producing essential amino acids for use by the aphid. Genes for nonessential amino acids are absent in *Buchnera* and thus this symbiont depends on its host for these, making *Buchnera* and the aphid co-dependent.

Analyses of different aphid species and their *Buchnera* symbionts indicate that vertical transmission of the symbionts has occurred from the time of the common ancestor of aphids, approximately 150 to 250 million years ago. Thus, the aphids and their symbionts appear to have co-speciated and there is no evidence of horizontal transfer, even within a single aphid species. In many *Buchnera* lineages, genes involved in tryptophan and leucine biosynthesis are present on plasmids rather than in the *Buchnera* genome. The location of these genes on plasmids would allow increased expression and, thus, increased benefit to their hosts. The number of copies of the plasmids appear to vary across *Buchnera* in different aphid lineages, perhaps reflecting coordinated, adaptive adjustment to the nutritional needs of the different aphid hosts. The genome of *Buchnera* is unusual when compared to the free-living bacterium *E. coli*. First the sequences are very AT-biased (about 28% GC) and second, DNA sequences evolve faster in *Buchnera* than in free-living relatives. Finally the genome of *Buchnera* from *A. pisum* is reduced, consisting of only about 650 kb (about one-seventh of the genome size of *E. coli*). It appears to contain only a subset of about 600 of the 4500 genes present in an *E. coli*-like ancestor.

Remarkably, it appears that each *Buchnera* contains 50 to 200 chromosomal copies, with the number of copies varying with the life cycle stage of the host, suggesting chromosome amplification is used to vary the contribution of the symbiont to its host's nutrition. The amplification of genome copy number to 200 copies/cell is very unusual in the microbial world; *E. coli* typically has one or two chromosomes per cell. The dramatic reduction in genome size of *Buchnera* and the extraordinary increase in genome copy number make this intracellular symbiont resemble eukaryotic cell organelles such as mitochondria and chloroplasts—which are considered to be evolutionary descendants of symbiotic bacteria. *Buchnera* resemble these organelles also in that they are transmitted maternally between host generations.

A less intimate relationship between gut microbes and insects is that between *Enterobacter agglomerans*, found in the gut of the apple maggot *Rhagoletis pomonella*. Enterobacteriaceae are the most common microorganisms associated with the apple maggot in the gut and female reproductive organs, and there are suggestions the flies use the bacteria for some vital function(s). In addition to *E. agglomerans*, *Klebsiella oxytoca* is found in the gut of *R. pomonella* and both are most abundant in the esophageal bulb, crop and midgut. These bacteria are found on host plants and other substrates in the environment. It appears that the bacteria provide usable nitrogen for *R. pomonella* and other tephritids by degrading purines and purine derivatives, making them facultative symbionts. The relationship between the *Enterobacter* and *Klebsiella* species is probably complex. *Rhagoletis* gut symbionts exist as a biofilm in the gut. A biofilm is a complex, structured community of microbes attached to surfaces. Biofilms function as a cooperative consortium in a complex and coordinated manner.

A full understanding of the genetic and evolutionary roles played by symbionts remains to be determined. In the few cases that have been well studied, a genetic interplay between insect host and symbiont occurs, factors are supplied to each from the other, and the microorganism has specific means of movement and relocation within the insect. Insects commonly control movement and multiplication of the symbiont, and the symbiont often influences growth and reproduction of the insect. A symbiont must not be subject to suppression or elimination by the insect immune system. Our understanding of how microorganisms were incorporated into insect tissues and cells and have evolved remains fragmentary, but advancing with the use of molecular tools.

Bugs within bugs within mealybugs

Perhaps most amazing is the recent discovery that there are bugs within bugs within mealybugs. Mealybugs (Pseudococcidae) have endosymbionts that live within the cytoplasm of large, polyploid host cells of a specialized bacteriome. The symbionts provide nutrients to their hosts and the relationship between hemipteran insects and these primary endosymbionts is ancient, perhaps dating to the origins of the families or superfamilies 100 to 250 million years ago.

The mealybug *Planococcus citri* packages its intracellular endosymbionts into mucus-filled spheres which surround the host cell nucleus and occupy most of the cytoplasm. These spheres are structurally unlike eukaryotic cell vesicles, and it was recently demonstrated that the mealybug cells actually harbor two types of Proteobacteria. The bacteria are not co-inhabitants of the spheres. Rather, the spheres themselves are β-proteobacteria with γ-proteobacteria living inside. This is the first report of an intracellular symbiosis involving two species of bacteria and the authors hypothesized that the internalization of the γ-proteobacterium by the β-proteobacterium may facilitate the exchange of genes and gene products to slow or reverse the genetic degradation that is common to long-term intracellular symbionts over evolutionary time.

Wolbachia

The genus *Wolbachia* is found in the α-proteobacteria. These bacteria are one of the most commonly found intracellular microorganisms found in arthropods. *Wolbachia* are gram-negative rods that cannot be cultured easily outside their hosts and are widespread, with estimates of infection frequency ranging to as many as 76% of all arthropod species.

Wolbachia have been implicated as both the cause of alterations in sex ratio (resulting in thelytoky and male killing), and cytoplasmic incompatibility in arthropods. Some *Wolbachia* improve fertility or vigor in their hosts, while other strains of *Wolbachia* appear to decrease these traits in their hosts.

The molecular mechanism(s) by which reproductive incompatibility are induced by *Wolbachia* are hypothesized to be due to *Wolbachia*'s ability to modify sperm. This hypothesis suggests that paternal chromosomes are modified during spermatogenesis by *Wolbachia* and this modification is 'rescued' in eggs of females infected with the same strain of *Wolbachia* during fertilization. If, however, the female is not infected with *Wolbachia* and mates with an uninfected male or male infected with a different strain of *Wolbachia*, then the embryos die. Some *Wolbachia* strains have been identified that fail to modify sperm but can rescue the modification in eggs of other *Wolbachia* strains.

Species in which *Wolbabchia*-induced cytoplasmic incompatibility has been documented include mosquitoes (*Culex pipiens, Aedes scutellaris, A. albopictus*), flies (*Drosophila similans, D. melanogaster*), the moth *Ephestia cautella*; beetles (*Tribolium confusum* and *Hypera postica*), the parasitoid *Nasonia*; and the hemipteran *Laodelphax striatellus*. Mites in the families Tetranychidae and Phytoseiidae also exhibit cytoplasmic incompatibility due to infection with *Wolbachia*.

Cytoplasmic incompatibility caused by *Wolbachia* may be partial or complete. Sometimes incompatibility is found in both reciprocal crosses (A × B <u>and</u> B × A, bidirectional incompatibility), perhaps due to the presence of different strains of *Wolbachia* in each population. Incompatibility more often is found in one reciprocal cross (A × B <u>or</u> B × A, unidirectional incompatibility). Cytoplasmic incompatibility typically is incomplete (less than 100%), perhaps due to inefficient transfer of *Wolbachia* to all progeny or to differences in the titer of *Wolbachia*. Such differences in titer could occur naturally if the infected insects encounter antibiotics in their environment or if they experience high temperatures (typically greater than 30°C).

Some insects appear to have *Wolbachia* only in their germ line tissues (ovaries and testes) while others have *Wolbachia* in somatic tissues. Large numbers of *Wolbachia* have been found in ovaries and testes of populations with cytoplasmic incompatibilities. Incompatible strains have been converted to compatible by treating the colonies with heat or antibiotics to eliminate (or reduce the titer of) the *Wolbachia*.

Wolbachia can be transferred to new populations experimentally by microinjecting infected egg cytoplasm into uninfected eggs. Transinfected strains of *D. simulans* and *D. melanogaster* with high titers of *Wolbachia* exhibited cytoplasmic incompatibilities at high levels, but those with low titers exhibited low levels of incompatibility, suggesting that a threshold level of infection is required and that host factors may determine the density of the *Wolbachia* in the host.

Wolbachia have been identified in at least 70 species of parasitic Hymenoptera, including species in the Aphelinidae, Encyrtidae, Eulophidae, Pteromalidae, Torymidae, Trichogrammatidae, Cynipidae, Eucoilidae, Braconidae, Ichneumonidae, Proctotrupoidae and in three dipteran parasitoids (Tachinidae). Both cytoplasmic incompatibility and induction of parthenogenesis occur in these parasitoids. Many hymenopteran parasitoids have both bisexual (arrhe-

notokous) and unisexual strains consisting only of females (thelytoky), probably due to the presence of *Wolbachia*.

Phylogenetic analysis suggests that the *Wolbachia* common ancestor evolved between 80 and 100 million years ago, whereas the arthropod common ancestor occurred at least 200 million years earlier. Thus, *Wolbachia* probably invaded arthropods through horizontal transmission. In fact, the molecular phylogenies of *Wolbachia* and arthropods do not match, supporting the hypothesis that horizontal transmission is important in the distribution of different strains of *Wolbachia* among arthropods.

Several methods have been proposed as mechanisms for horizontal transfer, including the movement of *Wolbachia* from host arthropods to their parasitoids and vice versa. Experimental microinjection (artificial horizontal transfer) of *Wolbachia* from the parasitoid *Muscidifurax uniraptor* into its host *D. simulans* resulted in a temporary infection, but no specific phenotypic effects were observed. These results suggest that host-symbiont interactions are important for successful establishment of a *Wolbachia* infection in a new host. The reasons for failure to experimentally transfer *Wolbachia* remain unknown, but it is clear that *Wolbachia* has successfully bridged large phylogenetic distances in its horizontal movements over evolutionary time. Some arthropods have been found to have double or even triple infections of *Wolbachia*. The effects of these multiple infections are diverse, but usually are unknown.

The availability of genetic information about *Wolbachia* has fostered increased knowledge of *Wolbachia*. The molecular technique called the polymerase chain reaction (PCR) allows scientists to study *Wolbachia* even though *Wolbachia* cannot be cultured outside their arthropod hosts. The availability of PCR primers for *Wolbachia* genes revolutionized the study of the distribution and evolution of *Wolbachia*. The *Wolbachia* genome project will further revolutionize such studies. Based on a phylogeny developed using DNA sequences from the *ftsZ* gene, *Wolbachia* infecting arthropods have been divided into Groups A and B, which are estimated to have diverged from each other 58 to 67 million years ago. Phylogenies based on the *wsp* gene sequences have yielded more groups, indicating considerable genetic variation exists among different *Wolbachia* strains.

Wolbachia may have a role in the speciation of arthropods by generating reproductive isolation, although some argue that the role of *Wolbachia*'s in this important process remains unproved. Typically, *Wolbachia* cause unidirectional cytoplasmic incompatibility when a *Wolbachia*-infected male mates with an uninfected female. The eggs or embryos of such matings die, resulting in a fitness cost to uninfected females, which over time results in the infected cytotype becoming fixed in the population. A problem with this speciation hypothesis is that *Wolbachia* females transmit less than 100% of the time to progeny, so some progeny will be produced that are compatible. Secondly, incompatibility is not completely expressed (incomplete penetrance of the trait) when infected males and uninfected females mate in natural populations, perhaps due to differences in the titer of the *Wolbachia* within the different individuals. Furthermore, selection on both the host and *Wolbachia* may favor reduced penetrance of the incompatibility phenotype or loss of *Wolbachia*. This could lead to a situation in which there is no gene flow to some gene flow, reducing reproductive isolation. Thus, unidirectional incompatibility caused by *Wolbachia* may be insufficient to cause the reproductive barriers that could lead to speciation, although such reproductive isolation could assist in the process of speciation. Additional factors, such as hybrid sterility (sterility of the hybrid when crossed with either of the parental species) and hybrid breakdown (the inviability or sterility of progeny resulting from a backcross of hybrid progeny with either of the parental species) may also be involved in the speciation process.

A second speciation mechanism possibly associated with *Wolbachia* may be by the induction of thelytoky (reproduction by females only), as has been found in the hymenopteran *Encarsia formosa*. Some populations of *Encarsia* no longer have males, so that they essentially become clonal and over time could differentiate genetically. A third potential *Wolbachia* speciation mechanism is by bidirectional incompatibility; if a population is infected with two different strains of *Wolbachia* that are incompatible with each other, then the incompatibility could act as a postzygotic reproductive barrier.

How *Wolbachia* are maintained in populations has considerable theoretical and practical importance. *Wolbachia* have been proposed as vectors for genetically transforming their host arthropod, as well as mechanisms for driving genes into arthropod populations in genetic manipulation projects for improved pest control.

The interest in the biology and evolution of *Wolbachia*, with its fascinating effects on reproductive isolation (thus potentially having effects on speciation), sex ratio, feminization, and male killing, has led to the development of a *Wolbachia* genome project, with four different groups of *Wolbachia* as targets.

Despite the wealth of information obtained about *Wolbachia* within the past few years, our understanding of the role of *Wolbachia* in arthropod biology and evolution probably remains fragmentary. For example, some *Wolbachia* in arthropods were recently shown to contain bacteriophages named WO. A phylogenetic analysis of different WOs from several *Wolbachia* strains yielded a tree that was not congruent with the phylogeny of the *Wolbachia*, suggesting that the phages were active and horizontally transmitted among the various *Wolbachia*. Because all *Wolbachia* strains examined had WO, the phage might have been associated with *Wolbachia* for a very long time, conferring some benefit to its microbial hosts.

Polydnaviruses as symbionts

A particularly interesting example of an intimate relationship between insects and symbionts is illustrated by the relationship between polydnaviruses and some parasitoids. Polydnaviruses are relatively newly-recognized viruses that are found only in the Braconidae and Ichneumonidae among the parasitic Hymenoptera. Polydnaviruses are symbiotic proviruses that have double-stranded circular DNA genomes; they are literally poly-DNA-viruses, having segmented genomes composed of several circular DNA molecules. For example, the genome of the virus within the parasitic wasp *Campoletis sonorensis* consists of 28 DNA molecules ranging in size from approximately 5.5 to 21 kb, with the total genome size approximately 150 kb.

Polydnaviruses are important in ensuring that some species of braconids and ichneumonids are able to successfully parasitize their insect hosts. At least fifty species of parasitic wasps have been shown to contain polydnaviruses and over 30,000 species are thought to carry polydnaviruses. Polydnaviruses alter the host insect's neuroendocrine and immune responses, prevent encapsulation of wasp eggs and larvae by host hemocytes, and influence development of the host to benefit the wasp. Genera of parasitoids containing polydnaviruses appear to have more species and have broader host ranges than sibling groups lacking them, suggesting the viruses contribute to the evolutionary success of their hosts. The two polydnaviral groups, Ichnoviridae and Brachoviridae, are phylogenetically and morphologically distinct and use different mechanisms to inhibit host immunity and development. The association between braconid parasitoids and their viruses appears to have lasted at least 60 million years. Thus, the viruses and braconid parasitoids appear to have a long association and have evolved a variety of interactions with their lepidopteran hosts.

Polydnaviruses replicate only in braconid or ichneumonid wasp ovaries and are secreted into the oviducts from where, during oviposition, they are injected into host lepidopteran larvae. The viruses appear to be vertically transmitted and integrated into the chromosome of the wasp. Each wasp species appears to carry a polydnavirus characteristic of that species. If one species within a particular genus carries a polydnavirus, they all are likely to do so.

Insects possess immune mechanisms that protect them from microorganisms, other invertebrates, and abiotic materials. Protection occurs through constitutive factors or by inducible humoral and cellular responses. Many parasitoid wasps are internal parasites and spend part of their lives in the bodies of other insects. Many behavioral, morphological, nutritional, and endocrine factors determine whether the interactions between a host and a parasitoid will lead to development of the parasitoid or to its destruction. The polydnavirus influences the immune system of the insect host, which allows the parasitoid eggs and larvae to survive. The virus replicates asymptomatically in the parasitoid but causes a pathogenic virus infection in the wasp's lepidopteran host. The virus alone can induce altered immune responses in some hosts, but in other hosts the venom injected by the wasp also must be present for the full effect of the virus to occur. Parasitoid wasps thus appear to benefit significantly from the polydnaviruses that replicate in their reproductive tracts. The virus also clearly benefits if the wasp is able to reproduce, because polydnaviruses are known to replicate only within their wasp hosts.

The polydnavirus-parasitoid-lepidopteran host system provides an unusual example of an obligate mutualistic association between a virus and a parasitoid that functions to the detriment of the parasitoid's lepidopteran host. The origin of polydnaviruses is unknown, as is how they became established in the parasitoid genome. It has been speculated that polydnaviruses may have potential value in agricultural pest management programs if genetically engineered pathogens (viruses, bacteria, fungi) containing polydnavirus genes could produce products that immunosuppressed the target pest, making the pathogens more effective. Alternatively, genetically engineered parasitoids could be developed that exhibit a modified host range, making them more effective in controlling pests.

Marjorie A. Hoy
University of Florida
Gainesville, Florida, USA

References

Aksoy, S. 2000. Tsetse-a haven for microorganisms. *Parasitology Today* 16: 114–118.

Bandi, C., B. Slatko, and S. L. O'Neill. 1999. *Wolbachia* genomes and the many faces of symbiosis. *Parasitology Today* 15: 428–429.

Baumann, P., N. A. Moran, and L. Baumann. 1997. The evolution and genetics of aphid endosymbionts. *BioScience* 47: 12–20.

Beckage, N. E. 1998. Parasitoids and polydnaviruses. *BioScience* 48: 305–310.

Brune, A., and M. Friedrich. 2000. Microecology of the termite gut: structure and function on a microscale. *Current Opinion in Microbiology* 3: 263–269.

Cazemier, A. E., J. H. P. Hackstein, H. J. M. Op den Camp, J. Rosenberg, and C. van der Drift. 1997. Bacteria in the intestinal tract of different species of arthropods. *Microbial Ecology* 33: 189–197.

Cook, J. M., and R. D. J. Butcher. 1999. The transmission and effects of *Wolbachia* bacteria in parasitoids. *Researches in Population Ecology* 41: 15–28.

Douglas, A. E. 1992. Symbiotic microorganisms in insects. *Encyclopedia of Microbiology* 4: 165–178.

Gray, M. W. 1989. Origin and evolution of mitochondrial DNA. *Annual Review of Cell Biology* 5: 25–50.

Hales, D. F., J. Tomiuk, K. Wohrmann, and P. Sunnucks. 1997. Evolutionary and genetic aspects of aphid biology: a review. *European Journal of Entomology* 94: 1–55.

Heddi, A., A. M. Grenier, C. Khatchadourian, H. Charles, and P. Nardon. 1999. Four intracellular genomes direct weevil biology: nuclear mitochondrial, principal endosymbiont, and *Wolbachia*. *Proceedings of the National Academy of Sciences of the USA* 95: 6814–6819.

Hurst, G. D. D., and M. Schilthuizen. 1998. Selfish genetic elements and speciation. *Heredity* 80: 2–8.

Jeyaprakash, A., and M. A. Hoy. 2000. Long PCR improves *Wolbachia* DNA amplification: *wsp* sequences found in 76% of 63 arthropod species. *Insect Molecular Biology* 9: 393–405.

Kaufman, M. G., E. D. Walker, D. A. Odelson, and M. J. Klug. 2000. Microbial community ecology and insect nutrition. *American Entomologist* 46: 173–184.

Krell, P. J. 1991. The polydnaviruses: multipartite DNA viruses from parasitic Hymenoptera. pp. 141–177 in E. Kurstak (ed.), *Viruses of invertebrates*. Marcel Dekker, New York, New York.

Lauzon, C. R., R. E. Sjogren, S. E. Wright, and R. J. Prokopy. 1998. Attraction of *Rhagoletis pomonella* (Diptera: Tephritidae) flies to odor of bacteria: apparent confinement to specialized members of Enterobacteriaceae. *Environmental Entomology* 27: 853–857.

Lilburn, T. G., T. M. Schmidt, and J. A. Breznak. 1999. Phylogenetic diversity of termite gut spirochaetes. *Environmental Microbiology* 1: 331–345.

Moran, N. A., and P. Baumann. 2000. Bacterial endosymbionts in animals. *Current Opinion in Microbiology* 3: 270–275.

O'Neill, S. L., A. A. Hoffmann, and J. H. Werren (eds.). 1997. *Influential passengers. inherited microorganisms and arthropod reproduction*. Oxford University Press, Oxford, United Kingdom. 214 pp.

Schwemmler, W., and G. Gassner (eds.). 1989. *Insect endocytobiosis: morphology, physiology, genetics, evolution*. CRC Press, Boca Raton, Florida.

Webb, B. A. 1998. Polydnavirus biology, genome structure and evolution. pp. 105–139 in: L. K. Miller and A. Ball (eds.), *The insect viruses*, Plenum, New York.

Werren, J. H. 1997. Biology of *Wolbachia*. *Annual Review of Entomology* 42: 587–609.

SYMBIOSIS. A close association of two different organisms in which some benefit is derived from the association. A special type of symbiosis is mutualism, in which both organisms benefit.

SYMBIOSIS BETWEEN PLANTHOPPERS AND MICROORGANISMS. Symbiosis is a rather common biological phenomenon and is defined as living together intimately between two or more dissimilar organisms. At least nine orders of insects possess symbiotes, which may be different in forms and are located in various organs (mycetoms) or cells (mycetocytes). Many plant-sucking hemipterans harbor intracellular symbiotes. For example, aphids have symbiotic bacteroids, leafhoppers have rickettsia-like organisms in addition to bacteroids, and planthoppers harbor mostly the yeast-like symbiote. The brown planthopper, *Nilaparvata lugens* Stål, is one of the most destructive pests of rice in Asia, and is extensively studied on its symbiotic association with microorganisms.

The yeast-like symbiote of *N. lugens* is located in fat body cells of the abdomen, but never in the head or thorax. It measures 2.5×8 to $10\,\mu m$ in size. This symbiote is bacilliform and unicellular, has a two-layered cell wall, and reproduces by budding, forming a convexity at the budding site. The fat body in the abdomen is full of yeast-like symbiote, as observed by histological sectioning. Ultrastructural observations showed that the yeast-like symbiote has a large nucleus, ribosomes, rough endoplasmic reticulum, Golgi apparatus, and many mitochondria in cytoplasm. Its morphology resembles cellular structure of the yeast. Because it is not cultured *in vitro*, this microorganism is named the yeast-like symbiote. However, a phylogenetic study based on direct sequencing of its 18S rRNA genes showed that the yeast-like symbiote in three planthoppers (*N. lugens*, *Sogatella furcifera*, and *Laodelphax striatellus*) is placed in the class Pyrenomycetes of the subphylum Ascomycotina. Therefore, the taxonomic position of the yeast-like symbiote is at present uncertain.

The population of yeast-like symbiote increases through nymphal development of *N. lugens*. The total number reaches a peak before oviposition in female adults and then declines, while the maximal number in males is at the fifth nymphal instar. The females harbor more yeast-like symbiote than the males, especially at the fifth nymphal and adult stages. The population declines drastically by incubating the neonate nymphs at $32°C$, or above, for 3 days or longer, resulting in subsymbiotic or aposymbiotic insects (insects with less symbiotes or without symbiotes). Incubation at $32°C$ causes some cellular changes, such as disintegration of its cell wall, degeneration of nucleus, and loss of cytoplasm. The reduction in yeast-like symbiote number caused by high-temperature treatment (at $32°C$) could be due to these cellular changes. In addition, treatment of yeast-like symbiote with some antibiotics also may result in lowering the symbiote number. For example, 0.3% Polyoxin S and 0.1% chlamphenicol as well as 0.1% cycloheximide may reduce yeast-like symbiote number, and thus kill the insects due to loss of the symbiotes.

The physiological roles of yeast-like symbiote can be studied by comparing normal with subsymbiotic or aposymbiotic insects. The brown planthopper eggs treated with high temperature on day 1, 4 or 7 after oviposition display reduced hatchability only on day 7, indicating that the symbiote is more significant

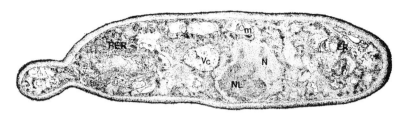

Fig. 1009 Schematic diagram of a yeast-like symbiote in the brown planthopper, *Nilaparvata lugens* (after Cheng and Hou, 1996). Abbreviations: N, nucleus; NL, nucleolus; m, mitochondrium; r, ribosome; RER, rough endoplasmic reticulum.

in the late embryonic development than the earlier stages. The eggs within the high-temperature treatment harbor only a few of the symbiotes, and their symbiote ball in eggs through embryonic development is free of symbiotes. The embryos of subsymbiotic eggs cannot undergo blastokinesis and dorsal closure, and fail to hatch due to lack of differentiation of the abdominal segments. Electrophoretic profile of the eggs laid by the subsymbiotic females showed the absence of several minor proteins, which are usually found in the fat body of normal females. A protein of 131 kDa is barely detectable in the subsymbiotic insects, and is not found in the ligated eggs in which

the symbiote ball is completely separated from the developing germ band. Therefore, the symbiote seems to supply its host with proteins for normal embryonic and postembryonic development. The yeast-like symbiote number in females may be reduced after injection with lysozyme solution, and some of the eggs are unable to hatch due to failure in blastokinesis. The embryos of ligated eggs could complete segmentation and differentiation normally before 110 h of embryonic development, but the abdominal segments fail to differentiate after dorsal closure, forming the head embryos. Partially ligated eggs harbor some symbiotes and may produce normal

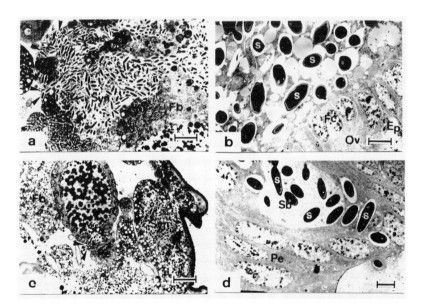

Fig. 1010 Histological micrographs showing transovarial transmission of yeast-like symbiote in *Nilaparvata lugens* (rearranged from Cheng and Hou, 2001). (a) The yeast-like symbiote moves out of the syncytium by exocytosis. Bar = 10 μm; (b) The free yeast-like symbiote in hemolymph approaches to the ovarioles near pedicel. Bar = 10 μm; (c) The yeast-like symbiote enters follicle cell around the primary oocyte by endocytosis (arrow heads) at epithelial plug. Bar = 10 μm; (d) The yeast-like symbiote aggregates at the posterior end of the mature egg after entering, forming a symbiote ball. Bar = 10 μm. Abbreviations: Ep, epithelial plug; Fb, fat body; Fc, follicle cell; Oo, oocyte; Ov, ovariole; Pe, pedicel; S, symbiote; Sb, symbiote ball.

larvae. Hence, the yeast-like symbiote is significant in abdominal segmentation and differentiation of the planthopper embryo.

The yeast-like symbiote of *N. lugens* passes to progeny by transovarial transmission. The symbiotes in mycetocytes move out of the syncytium, which is formed from a layer of fat body cells, by exocytosis, and are released into hemocoel in females. Then, the free yeast-like symbiote in hemolymph approach to the ovarioles near the pedicel and are enclosed by follicle cells. They enter the follicle cells around the primary oocyte by endocytosis at epithelial plug of the ovariole. The yeast-like symbiote aggregates at the posterior end of the mature egg after entering, forming a symbiote ball, and finally complete the transovarial transmission. Therefore, the yeast-like symbiote is intimately associated with the brown planthopper through generations and is indispensable for its host life.

Roger F. Hou
National Chung Hsing University
Taichung, Taiwan

References

Chen, C. C., L. L. Cheng, and R. F. Hou. 1981. Studies on the intracellular yeast-like symbiote in the brown planthopper, *Nilaparvata lugens* Stål. II. Effects of antibiotics and elevated temperature on the symbiotes and their host. *Zeitschrift fur angnewandte Entomologie* 92: 440–449.

Chen, C. C., L. L. Cheng, C. C. Kuan, and R. F. Hou. 1981. Studies on the intracellular yeast-like symbiote in the brown planthopper, *Nilaparvata lugens* Stål. I. Histological observations and population changes of the symbiote. *Zeitschrift fur angnewandte Entomologie* 91: 321–327.

Cheng, D. J., and R. F. Hou 1996. Ultrastructure of the yeast-like endocytobiont in the brown planthopper, *Nilaparvata lugens* (Stål) (Homoptera: Delphacidae). *Entocytobiosis and Cell Research* 11: 107–117.

Cheng, D. J., and R. F. Hou. 1998. Cellular changes of the yeast-like endocytobiont in the brown planthopper, *Nilaparvata lugens* (Stål), after high-temperature treatment. *Endocytobiosis and Cell Research* 12: 177–183.

Cheng, D. J., and R. F. Hou. 2001. Histological observations on transovarial transmission of a yeast-like symbiote in *Nilaparvata lugens* Stål (Homoptera, Delphacidae). *Tissue and Cell* 33: 273–279.

Hou, R. F., and Y. H. Lee. 1984. Effect of high-temperature treatment on the brown planthopper, *Nilaparvata lugens*, with reference to physiological roles of its yeast-like symbiote. *Chinese Journal of Entomology* 4: 107–116.

Lee, Y. H., and R. F. Hou. 1987. Physiological roles of a yeast-like symbiote in reproduction and embryonic development of the brown planthopper, *Nilaparvata lugens* Stål. *Journal of Insect Physiology* 33: 851–860.

Noda, H., N. Nakashima, and M. Koizumi. 1995. Phylogenetic position of yeast-like symbiotes of rice planthoppers based on partial 18S rDNA sequences. *Insect Biochemistry and Molecular Biology* 25: 639–646.

SYMPATRIC. Possessing the same geographic range as a related organism without loss of genetic identity. (contrast with allopatric)

SYMPATRIC SPECIATION. Speciation that occurs in the same place, due to such factors as consumption of different food (allophagic speciation) or occurrence at different times (allochronic speciation).

SYMPHYLANS (CLASS SYMPHYLA). Symphylans are insect relatives in the subphylum Atelocerata (Myriapoda), and are related to centipedes (Chilopoda) and millipedes (Diplopoda). They possess a single pair of Malpighian tubules, no median simple eyes, and have tomosvary organs (humidity and/or chemoreceptors). They also have musculated antenna and diffuse nervous systems. There are approximately 160 species within the single order.

Phylum: Arthropoda

Subphylum: Atelocerata
Class: Symphyla
Order: Scolopendrellida

The apterygotes (wingless insects) may be derived from ancestral symphylans via paedomorphosis (juvenile characteristics are retained in the adult.).

Symphyla are less than 10 millimeters long, slightly dorsoventrally compressed, and white or colorless with a slender, soft body. They resemble small centipedes without the fangs found in Chilopoda. The head is distinct and well-developed, and the mouthparts are endognathous (mouthparts are concealed by the labium or cranial folds). Symphylans resemble the apterygotes with a single pair of long antenna with many segments and a single pair of spiracles are open on the head. They lack eyes and are blind. There are 15 to 22 body segments present, with an average of 15. Each body segment has one pair of hook-like legs. Immatures begin with 6 to 7 pairs of legs. All legs on one side move simultaneously, alternating side to side.

There are two legless caudal segments at the tip of the body, which may posses a pair of spinnerets for

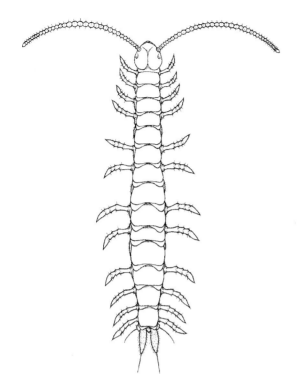

Fig. 1011 An adult symphylan (original by M. L. Baker, Louisiana State University-Shreveport).

silk production. The number of dorsal tergites (14 to 24) exceeds the number of true segments. The fourth segment has a gonopore. In an unusual fertilization ritual, the egg is taken from the gonopore and stored in the mouth, where it is exposed to sperm from a spermatophore. Spermatophores are generally deposited on the ground. Ten to twenty eggs are laid in clusters on the soil. Symphylans do not dig tunnels, but take advantage of naturally occurring cracks and worm burrows for shelter. Symphylans inhabit soil that is rich in organic material, particularly humus and decaying wood. They feed on plant material, especially the roots. They can become a pest in greenhouses and in the field. They are cosmopolitan, yet concentrated in the tropics. Garden symphylan, *Scutigerella immaculata* (Newport), is the best known of the pests species, and has a wide host range. Though likely originating in Europe, it is now widely distributed.

Beverly Burden
Louisiana State University-Shreveport
Shreveport, Louisiana USA

References
Barnes, R. S. K. (ed.). 1998. *The diversity of living organisms*. Blackwell Science Ltd., Oxford, United Kingdom. 345 pp.
Borror, D. J., C. A. Triplehorn, and N. F. Johnson. 1989. *An introduction to the study of insects (6th ed.)*. Harcourt Brace College Publishers, Stamford, Connecticut.
Michelbacher, A. E. 1938. The biology of the garden centipede, *Scutigerella immaculata*. *Hilgardia* 11: 55–148.

SYMPTOM. Any aberration in function (including behavior), indicating disease. (contrast with sign)

SYNANTHROPIC. Living in association with humans, without actually feeding on humans.

SYNAPSE. (pl., synapses) The narrow gap between two nerves, or between a nerve and a muscle, that is bridged by a chemical neurotransmitter. Acetylcholine is a major synaptic transmitter in the insect central nervous system. See also, NERVOUS SYSTEM.

SYNCHROA BEETLES. Members of the family Synchroidae (order Coleoptera). See also, BEETLES.

SYNCHROIDAE. A family of beetles (order Coleoptera). They commonly are known as synchroa beetles. See also, BEETLES.

SYNCHRONOUS MUSCLE. Muscle in which the frequency of contraction is controlled directly by nervous impulses, with each contraction caused by nervous stimulation. (contrast with asynchronous muscle)

SYNDROME. A group of signs and symptoms characteristic of a particular disease.

SYNECOLOGY. Study of groups of organisms in relation to their environment (population, community, and ecosystem ecology).

SYNERGISM. Chemical interactions may occur when two or more toxic chemicals are given to a living organism simultaneously or sequentially. Such interactions may result in a toxicity that is additive of individual toxicities of the chemicals or that is greater or less than the additive toxicity of these chemicals if administrated separately. Generally, the effects of chemical interactions can be divided into four types, including additive effects, synergism, potentiation and antagonism.

Both synergism and potentiation refer to the action or phenomenon in which two or more chemicals (e.g., insecticides) administered simultaneously or sequentially achieve a greater toxicity than the sum of the individual toxicities of the chemicals administered separately. However, the term synergism is reserved for the case in which one chemical has little or no toxicity if administered alone, whereas in the case of potentiation both chemicals have appreciable toxicities.

A relatively non-toxic chemical substance applied with an insecticide to increase the toxicity of the insecticide against insect pests is known as a synergist. Different synergists may confer their synergistic effects on an insecticide through different mechanisms. They may inhibit the metabolic detoxification of the insecticide, increase the penetration of the insecticide through the insect cuticle, or act at the binding site(s) of the insecticide on the receptor proteins in target insects.

The most commonly used insecticide synergists are metabolic inhibitors that block certain detoxification enzymes so that insecticide detoxification in target insects is abolished or significantly reduced. Such synergists may serve as research tools to help determine the mechanisms of insecticide detoxification in a living organism and as control agents to increase the efficacy of insecticides in a pest management program. For example, the insecticide synergist piperonyl butoxide (PB or PBO) is a rather potent inhibitor of cytochrome P450-dependent polysubstrate monooxygenases (PSMOs, formerly known as mixed-function oxidases or MFOs), a major group of detoxification enzymes in living organisms. Both DEF (S,S,S-tributyl phosphorotrithioate) and triphenyl phosphate (TPP) inhibit various esterases that may be involved in detoxification of various insecticides including organophosphates, carbamates and pyrethroids. Currently, there is no specific inhibitor available for glutathione S-transferases (GSTs). However, diethyl maleate (DEM), which conjugates with and rapidly depletes glutathione, has been commonly used as a synergist to suppress the GST activity.

The degree of synergism is often determined by two parallel but separate bioassays for an insecticide. One bioassay treats insects with mixtures of the insecticide and synergist or pretreats the insects with synergist and then with insecticide. The other bioassay treats the insects with insecticide only. If both bioassays determine the LD_{50} (lethal dose for 50% of the target insect population) values for the insecticide with and without the synergist, the level of synergism, which often is presented as the synergism ratio (SR), can be calculated:

$$SR = LD_{50}(\text{without synergist})/LD_{50}(\text{with synergist}).$$

Insecticide bioassays with synergists can help researchers learn possible detoxification mechanisms for a particular insecticide in a target population. For example, Yang et al. (2001) evaluated the susceptibility and possible detoxification mechanisms of the Banks grass mite (BGM), *Oligonychus pratensis* (Banks), and the two-spotted spider mite (TSM), *Tetranychus urticae* Koch, to three selected insecticides with three synergists (TPP, DEM and PBO). Their study suggests that esterases, GSTs, and PSMOs all play important roles in the detoxification of the pyrethroid bifenthrin in the two-spotted spider

Effects of chemical interactions on chemical toxicity

Chemical effect	Characteristics of chemical effect
Additive toxicity	The combined toxicity of two chemicals administered together is equal to the sum of the toxicities of the chemicals administered separately.
Synergism	The combined toxicity of two chemicals administered together is greater than the sum of the toxicities of these chemicals, one of which has little or no intrinsic toxicity, if administered separately.
Potentiation	The combined toxicity of two chemicals administered together is greater than the sum of the toxicities of these chemicals; both have appreciable toxicities if administered separately.
Antagonism	The combined toxicity of two chemicals administered together is less than the sum of the toxicities of these chemicals administered separately.

Effect of three synergists (TPP, DEM, and PBO) on bifenthrin toxicity against the two-spotted spider mite using glass vial residual contact bioassay. Data adapted from Yang et al. (2001).

Chemical	LC_{50} (95% CI) (μg/ml)	DF	Slope + SE	$P(\chi^2)$	SR
Bifenthrin	114.0 (92.3–144.0)	5	0.42±0.03	0.35	–
Bifenthrin + TPP	18.5 (13.9–23.6)	5	0.37±0.03	0.32	6.2
Bifenthrin + DEM	27.5 (20-5-36.8)	4	0.34±0.03	0.45	4.1
Bifenthrin + PBO	25.1 (19.8–31.0)	5	0.42±0.03	0.20	4.5

The probability for Chi-square. The $P(\chi^2)$ value larger than 0.05 indicates a significant fit between the observed and expected regression lines at the 0.05 level. Synergism ratios (SR) were calculated as LC_{50} (bifenthrin without synergist) $\div LC_{50}$ (bifenthrin with synergist); All the synergism ratios are statistically significant ($P < 0.05$) based on the non-overlapping 95% confidence intervals (95% Cis) of the LC_{50} values.

mite because TPP, DEM and PBO enhanced the toxicity of bifenthrin by 6.2-, 4.1- and 4.5-fold, respectively.

Kun Yan Zhu
Kansas State University
Manhattan, Kansas, USA

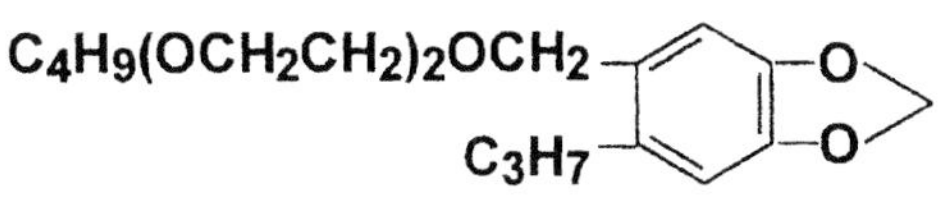

$$C_4H_9(OCH_2CH_2)_2OCH_2$$
$$C_3H_7$$

Piperonyl butoxide (PBO)

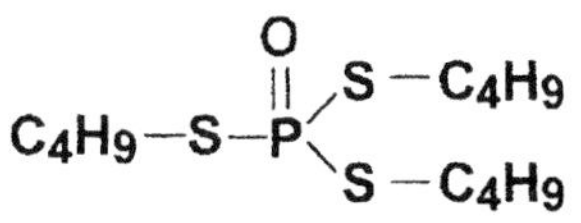

$$C_4H_9-S-P \begin{matrix} O \\ \| \\ \\ \end{matrix} \begin{matrix} S-C_4H_9 \\ \\ S-C_4H_9 \end{matrix}$$

DEF

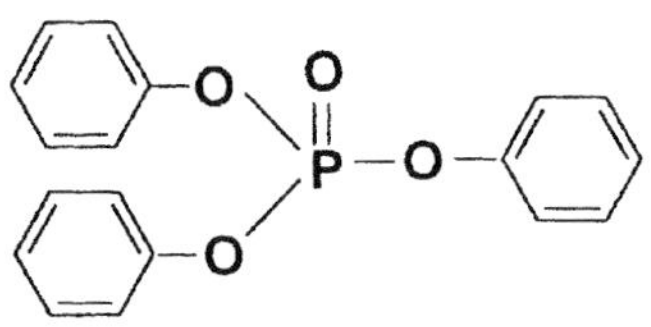

Triphenyl phosphate (TPP)

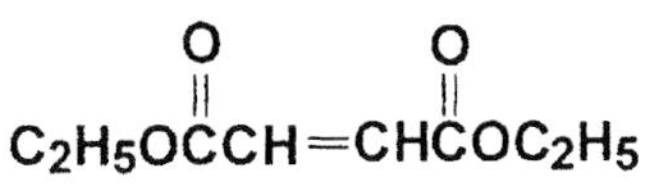

$$C_2H_5OCCH=CHCOC_2H_5$$

Diethyl maleate (DEM)

Fig. 1012 Chemical structure of some commonly used insecticide synergists.

References

B-Bernard, C., and J.R. Philogène. 1993. Insecticide synergists: role, importance, and perspectives. *Journal of Toxicology and Environmental Health* 38: 199–223.

Eaton, D.L. and C.D. Klaassen. 2001. *Principles of toxicology.* Pp. 11–34 in C.D. Klaassen (ed.), Cassarett and Doull's toxicology: the basic science of poisons, 6th ed. McGraw-Hill, New York, New York.

Gordh, G., and D.H. Headrick. 2001. *A dictionary of entomology.* CABI Publishing, New York, New York. 1032 pp.

Hodgson, E. 1994. *Chemical and environmental factors affecting metabolism of xenobiotics.* pp. 154–175 in E. Hodgson and P.E. Levi (eds.), Introduction to biochemical toxicology. Appleton & Lange, Norwalk, Connecticut.

Parkinson, A. 2001. *Biotransformation of xenobiotics.* pp. 133–224 in C.D. Klaassen (ed.), Cassarett and Doull's toxicology: the basic science of poisons, 6th ed. McGraw-Hill, New York, New York.

Raffa, K.F., and T.M. Priester. 1985. Synergists as research tools and control agents in agriculture. *Journal of Agricultural Entomology* 2: 27–45.

Yang, X., K.Y. Zhu, L.L. Buschman, and D.C. Margolies. 2001. Comparative susceptibility and possible detoxification mechanisms for selected miticides in banks grass mite and two-spotted spider mite (Acari: Tetranychidae). *Experimental and Applied Acarology* 25: 293–299.

SYNERGIST. A chemical added to insecticides that increases the toxicity of the active ingredient.

SYNLESTIDAE. A family of damselflies (order Odonata). See also, DRAGONFLIES AND DAMSELFLIES.

SYNOMONE. A chemical that benefits both the producer and perceiver. A synomone is a type of semiochemical. See also, CHEMICAL ECOLOGY.

SYRINGOGASTRIDAE. A family of flies (order Diptera). See also, FLIES.

SYRPHID FLIES. Members of the family Syrphidae (order Diptera). See also, FLIES.

SYRPHIDAE. A family of flies (order Diptera). They commonly are known as flower flies, hover flies, or syrphid flies. See also, FLIES.

SYSTEMATICS. In entomology, an abbreviated form of 'biosystematics,' the relationships of taxa based on study of their genetics and evolutionary history.

SYSTEMATIC SAMPLING. A method of allocating sample units within a sample universe in which sample units are collected at fixed intervals along a predetermined pattern. Examples include sampling along transects or along V or X shaped patterns in crop fields. Often, the starting point of the pattern is determined randomly. Systematic sampling is probably the most common method for allocating sample units in most agricultural systems because it is simple and time-efficient.

SYSTEMIC. Absorbed and translocated through the vascular system of plants or animals.

SZENT-IVÁNY, JÓZSEF GYULA HUBERTUS. József Szent-Ivány was born on the 3rd of November 1910 in Budapest, Hungary. He matriculated in Rimaszombat and from 1928 he continued his studies in Vienna, Austria. Between 1933 and 1936, he was a student at the Budapest University of Science, where he completed his studies and received his doctoral degree of zoology, geography and mineralogy. In 1936, he joined the staff of the Hungarian Natural History Museum, where he primarily worked on Lepidoptera. In 1938, he established *Fragmenta Faunistica Hungarica*, a zoological journal, and from 1937 until 1944 he was the editor of *Folia Entomologica*, the official publication of the Hungarian Entomological Society. He collected huge numbers of Lepidoptera for the museum and during World War II, at his own cost, he translocated the entire collection from Budapest to Tihany, a small town where it was safe from bombing raids. The war disrupted his scientific work and as the Russians occupied the country, he fled to the West. In 1950, he migrated to Australia but it took another 4 years before he could continue with his entomological work. He gained employment as government entomologist in Papua New Guinea. He spent 12 years there, building an insect collection of more than 100,000 specimens, which formed the basis of the Papua New Guinea Entomological Institute. His entomological activities were not restricted to collecting and taxonomy alone. He took active part in running the Institute and conducting numerous studies in the field of crop protection against insect pests, projects of medical and forest entomology, environmental studies and a whole host of other scientific activities. He published a great number of papers, mostly in English, German and Hungarian. After his retirement in 1966, he remained active as an entomologist, taking on many assigments, consultation projects and honorary positions. From 1972–74 he was the president of the committee which investigated and found the lost grave of Sámuel Fenichel at Stephansort. In 1975, the newly established entomological research station in Wau was named Szent-Ivány Laboratory. In 1985, he received from Queen Elizabeth II the Order of Australia. In 1988, he was admitted into the Hungarian Academy of Science. A few days after this honor, on the 8th of June 1988, József Szent-Ivány passed away.

George Hangay
Narrabeen, New South Wales, Australia

References
Gressitt, J. I. 1968. Szent-Ivány, *J. Bibliography of New Guinea Entomology*. Honolulu, Hawaii.
Balázs, D. 1989. *Búcsú Szent-Ivány Józseftől*. Földrajzi Múzeum Tan. 6.

T

TABANIDAE. A family of flies (order Diptera). They commonly are known as horse flies or deer flies, and some are called yellow flies or greenheads. See also, FLIES.

TACHINID FLIES. Members of the family Tachinidae (order Diptera). See also, FLIES.

TACHINIDAE. A family of flies (order Diptera). They commonly are known as tachinid flies. See also, FLIES.

TACHINISCIDAE. A family of flies (order Diptera). See also, FLIES.

TAENIDIUM. (pl., taenidia) Cuticular ridges that strengthen and support the walls of the trachae.

TAENIOPTERYGIDAE. A family of stoneflies (order Plecoptera). They sometimes are called winter stoneflies. See also, STONEFLIES.

TAGMA. (pl., tagmata) One of the major body regions. In insects, this is the head, thorax and abdomen. In many other arthropods it is the cephalothorax and abdomen.

TAIGA TICK, *Ixodes persulcatus* **SCHULZE (ACARI: IXODIDA: IXODIDAE).** This is one of the most well-studied species among hard ticks. The intensive study of *Ixodes persulcatus* was initiated in the late 1930s when this tick was found to be the main vector and reservoir of a virus causing a severe human disease, tick-borne encephalitis, that is spread all over the range of the tick. Two other viruses responsible for human pathogenicity were isolated from the taiga tick: Powassan virus in the southern Russian Far East, and Kemerovo virus in Western Siberia and the European part of Russia. Later, *I. persulcatus* was found to be the main vector of Lyme borrelioses caused by *Borrelia garinii* and *B. afzelii*. Recently, *I. persulcatus* appeared to be a vector of human ehrlichiosis. This tick may also play some role in mechanical or biological transmission of agents of some other human diseases. Transstadial and transovarial passage of many pathogens has been proven. Thus, this tick is one of the most important vectors and reservoirs of human pathogens from various groups. Mixed infections in the same populations of ticks, as well as in the same specimens of ticks were documented. A double infection of humans was recorded, and this event may dramatically change the clinical course of human illness.

The range of *I. persulcatus* covers a huge territory of boreal coniferous (taiga) forests, from the Baltic Sea to the Pacific Ocean, including not only the territory of the former Soviet Union (mainly Russia), but also some spots in Poland and parts of Northern China, Mongolia and Korea, as well as a part of Hokkaido Island and several smaller nearby islands in Japan. The spread of the taiga tick to the north

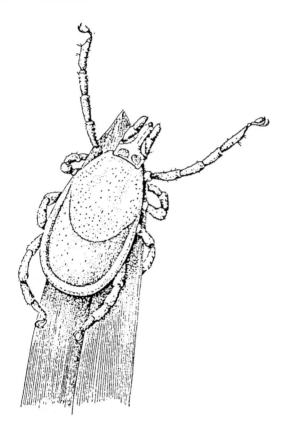

Fig. 1013 Unfed tick on foliage, seeking a host.

of its reproductive range is limited by a temperature factor, and to the south by a humidity factor. The single findings of the tick to the north of its range are explained by its introduction by birds during spring migrations to their breeding sites. Evolutionarily, the taiga tick is one of the youngest representatives of the closely related species in the subgenus *Ixodes* that are united in the *I. persulcatus* group or *I. ricinus* complex, but its range is the largest one among all ticks of this group.

As most species of ticks living in the temperate zone, *I. persulcatus* has a well-defined seasonality. The unfed ticks search for hosts (questing or host seeking) during the spring-summer season of the year. They use a so-called 'ambush strategy' for seeking hosts: unfed ticks are positioned on the vegetation, from the litter layer to the tips of stems or branches, scanning nearby spaces with their chemoreceptors (part of Haller's organs, located on the dorsal surface of tarsus I). They spend from several hours to several days in questing position depending

on particular environmental conditions. Adult ticks sense the host stimuli from a distance of 5 to 10 m. They can detect paths of regular host migration which explains their concentration nearby. Questing ticks either catch onto a passing host or, sensing it from a distance, move in its direction. Ticks cannot constantly be in a questing position because their water reserves would become depleted. To maintain their water balance, ticks must regularly migrate down to the litter and soil where the temperature is lower and humidity higher. During the day, unfed adults have two peaks of activity: in the morning (8 to 10 a.m.) and in the late afternoon (after 4 p.m.). Nymphal activity is maximal at dusk and the beginning of the night. However, cloudy weather, rain or warm nights may significantly change this pattern.

The capacity of *I. persulcatus* for active migrations is limited. The questing larvae can commonly migrate upwards up to 20 cm, nymphs up to 40 to 50 cm, and adults up to 1.0 m. The height of vertical migrations is directly correlated with the air humidity. Ticks also are unable to carry out long horizontal migrations: 5 m (up to 10 m) for adults and 0.5 m (up to 1.5 m) for larvae. The rate of adult movement may be as high as 30 cm/min at 23°C but is generally lower, decreasing with decreasing temperatures. The taiga tick has no eyes, though it can distinguish between more or less illuminated areas, preferring the latter. Photosensitive cells were found approximately in the same area where eyes might be, and their response to light is no less pronounced than in some ticks having eyes.

The season of adult activity lasts for 3 to 6 months, from the time of snow melting (April–May) until July–September, depending on the conditions of a particular region. The season of activity for unfed subadults (larvae and nymphs) mainly corresponds to that for adult ticks though it may be rather longer. The activity season does not correspond to the life-span of individual ticks. After hatching or molting, unfed subadults live about 14 to 15 months and unfed adults live about 12 months, including behavioral diapause during hibernation. The active (host-seeking) life itself is much shorter, from 1 day to about 75 days for unfed adults. The unfed ticks do not become active immediately after post-molting development or diapause but gradually, during the main part of the activity season. Most adults become active during the first 15 to 30 days of the season. The adults that become active earlier in the season live longer than ticks that become active later. The abundance of active adults sharply increases during the first third of the season (often during a shorter period) and then gradually diminishes till the end of the season. Unfed ticks that do not find a host and cannot feed during the activity season die due to complete expenditure of their nutritional reserves.

The taiga tick is a typical exophilic three-host tick. Each parasitic stage feeds on a vertebrate host only once: the larva feeds for 3 to 5 days, the nymph for 3 to 6 days and an adult female for 6 to 10 days. The duration of feeding strongly depends on the host species. The obligatory condition for normal female engorgement is its insemination. The female *I. persulcatus* can be inseminated not only on a host but also in nature before finding a host. In some populations, about 50% of unfed females collected from vegetation were inseminated. Males feed several times during their active life for 15 to 30 minutes each time. The larvae and nymphs increase their weight during feeding 15 to 30-fold, while engorged females are heavier than unfed females 100 to 150-fold. The mass of an unfed larva is 0.03 to 0.045 mg, whereas the mass of fully engorged females reaches 250 to 470 mg and they lay from 2,000 to 4,000 eggs. The minimal weight of engorgement, after which females are capable of laying single eggs, is about 20 to 35 mg. The heavier the engorged female, the greater the ratio of the number of laid eggs/mg of females weight, until the female weight is 140 to 200 mg, after which the dependence of egg number/mg of engorged females becomes linear and equals approximately 8.5 to 9.5 eggs/mg of the female weight. Values of all these parameters differ for tick populations from different parts of the range.

The engorged ticks drop off the host into the litter where they develop before molting, or where the females oviposit and the eggs develop. The engorged subadults feeding in the late summer have no time to undergo the necessary metamorphosis by the autumn; they enter the morphogenetic diapause and molt into the next stage only in the second part of the summer of the following year. The molted unfed specimens hibernate, being in behavioral diapause and become active next season. Only eggs and engorged adults cannot hibernate. The general duration of the entire life cycle (from eggs to eggs) may fluctuate from two years under most favorable conditions (the southern part of the Russian Far East), to five years. In most cases, the tick populations consist of ticks of different cohorts (having life cycles of different durations).

The taiga tick is one of the most opportunistic (generalist) tick species, using as hosts nearly 300 species of different vertebrates (about 100 species of mammals, more than 175 species of birds and a few species of reptiles). Larvae mainly feed on small mammals (rodents and insectivores), nymphs on medium-sized mammals (rodents, lagomorphs, some carnivores) and birds, and adults on large mammals. The questing height of ticks of each stage coincides with the size and location of their main hosts. In years of depression of the abundance of small mammals, or under specific environmental conditions, subadults may change their hosts for larger ones or for hosts less typical for them under normal conditions. Adults actively attack people, and are the main source of human infection by different pathogens. *Ixodes persulcatus* adults are more aggressive than their close relative, the sheep tick *I. ricinus*. Their efficient attachment to the host body is provided by a long hypostome covered with numerous recurved teeth preventing tick removal.

A number of parasites and predators of the taiga tick are known. However, the prospects of their use for tick control are rather slim. The taiga tick is a member of some rather complicated communities formed long ago and there is little chance of destroying a well developed balance of such communities by introducing new agents. The high stability of adult tick abundance from year to year (the difference is not more than one order of magnitude) indicates the extreme fitness of this species. This stability is provided for by a number of factors, such as a complicated life cycle, existence of different types of diapauses, numerous hosts from various groups, and high aggressiveness of adults towards potential hosts.

The simplest mode of human protection from taiga tick attacks and bites is proper clothing when in forested areas and regular self- and cross-inspections followed by tick removal. There are data on the protective effect of some repellents, which can be used on clothing or applied directly to the skin, to reduce tick infestation. Large-scale treatment of tick populated forest areas by persistent acaricides, which were used in the 1950s to 1970s, are now banned. The only target for treatment by non-persistent acaricides might be the popular sites of human resting during the summer.

See also, TICKS.

Igor Uspensky
The Hebrew University of Jerusalem
Jerusalem, Israel

References

Filippova, N. A. (ed.) 1985. *The taiga tick, Ixodes persulcatus Schulze (Acarina, Ixodidae).* "Nauka", Leningrad, Russia. 416 pp. (in Russian).

Korenberg, E. I. 2000. Seasonal population dynamics of *Ixodes* ticks and tick-borne encephalitis virus. *Experimental and Applied Acarology* 24: 665–681.

Uspensky, I. 1993. Ability of successful attack in two species of ixodid ticks (Acari: Ixodidae) as a manifestation of their aggressiveness. *Experimental and Applied Acarology* 17: 673–683.

Uspensky, I. 1995. Physiological age of ixodid ticks: aspects of its determination and application. *Journal of Medical Entomology* 32: 651–664.

Uspensky, I. 1996. Tick-borne encephalitis prevention through vector control in Russia: an historical review. *Review of Medical and Veterinary Entomology* 84: 679–689.

Uspensky, I., and I. Ioffe-Uspensky. 1999. The relationship between engorged female weight and egg number in ixodid ticks: a biological interpretation of linear regression parameters. *Acarologia* 40: 9–17.

TALLY THRESHOLD. In binomial sampling, the number of individuals required to be present per sample unit to consider the sample unit infested. Proper selection of the tally threshold can improve the accuracy of classification sampling.

See also, SAMPLING ARTHROPODS.

TANAOSTIGMATIDAE. A family of wasps (order Hymenoptera). See also, WASPS, ANTS, BEES, AND SAWFLIES.

TANDEM RUNNING. A from of communication among some ants, wherein ants trail one another, with each ant maintaining contact with the preceding individual by means of its antennae.

TANGLE-VEINED FLIES. Members of the family Nemestrinidae (order Diptera). See also, FLIES.

TANNING. The process of sclerotization (hardening) of the cuticle after an insect molts; tanning involves the cross-linking of proteins, aided by quinones. Darkening may accompany tanning. Transparent cuticle can be tanned (hardened), as occurs over the compound eye, though in the absence of darkening.

TANOCERIDAE. A family of grasshoppers (order Orthoptera). They commonly are known as desert longhorned grasshoppers. See also, GRASSHOPPERS, KATYDIDS AND CRICKETS.

TANYDERIDAE. A family of flies (order Diptera). They commonly are known as primitive crane flies. See also, FLIES.

TANYPEZID FLIES. Members of the family Tanypezidae (order Diptera). See also, FLIES.

TANYPEZIDAE. A family of flies (order Diptera). They commonly are known as tanypezid flies. See also, FLIES.

TARNISHED PLANT BUG, *Lygus lineolaris* PALISOT DE BEAUVOIS (HEMIPTERA: MIRIDAE). The tarnished plant bug, *Lygus lineolaris* (Palisot de Beauvois) is a widely distributed mirid, known from Alaska to southern Mexico. It is a pest of a wide range of seed, vegetable, fruit, fiber and forage crops. More than 600 different plant species have been reported as potential hosts for this insect. Its ability to cause economic losses is often related to the migratory behavior of the adult and the phenology of the host plant. Depending on the host plant species, feeding may cause any of the following injuries: malformation of fruit, abnormal growth habits, necrosis, abscission of fruiting structures, or production of shriveled or embryoless seeds.

Life stages

Egg. A freshly laid egg is translucent. Later it is yellowish. It is oval and slightly curved on one side. One end is obtuse and broadly rounded, whereas the other end is almost squarely truncated. The hardened egg shell (chorium) is smooth. Each is 0.85–1.6 mm long and 0.22–0.28 mm wide.

Nymphs. The five instars are very similar in form and all are yellowish-green. They walk rapidly and drop readily when disturbed.

First instar. The yellowish-green body has a pale orange spot in the middle of the caudal margin of the third abdominal segment. The body length is 0.85–1.10 mm and the body width is around 0.40 mm. The width of the head at the eyes is 0.34–0.36 mm.

Second instar. The body is yellowish to yellowish pea-green. The third abdominal segment exhibits a bright orange-yellow spot with a slightly smaller spot at the posterior margin. The body length is 1.30–1.65 mm and the body width is about 0.60 mm. The width of the head at the eyes is 0.45–0.52 mm.

Third instar. The body is green. The abdominal gland is indicated by a black spot. Toward the end of this stage, four dark thoracic spots begin to appear.

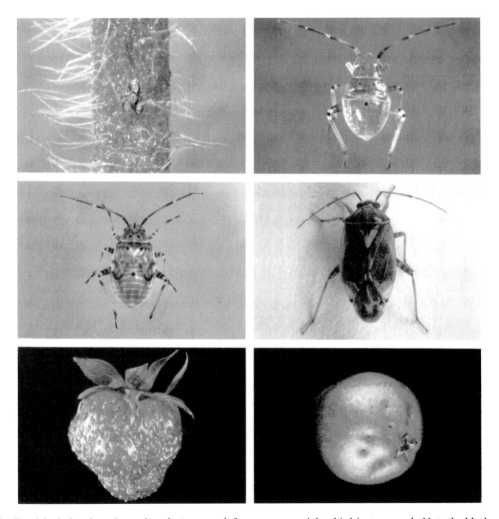

Fig. 1014 Tarnished plant bug, *Lygus lineolaris*: upper left, eggs; upper right, third-instar nymph. Note the black spot of the abdominal gland; second row left, fifth-instar nymph. Note the four black spots on the thorax. The wing pads are yellowish with irregularly marked brown lines and extend to the fifth and sixth abdominal segments; second row right, the adult, dark brownish; lower left, a berry injured by tarnished plant bug nymphs. Note the closeness of the achenes where the bug has fed; lower right, a pear injured by tarnished plant bug. Note the 'pit.'

Wing pads also appear and extend on to the second abdominal segment. The body length is 1.7–2.2 mm and the body width is about 1 mm. The width of the head at the eyes is 0.58–0.60 mm.

Fourth instar. A great variation of color exists and green, red, white and black predominate. The four black thoracic spots are more prominent. The caudal margin of the third abdominal segment bears a large black spot. Two prominent reddish bands appear on each femur. The wing pads extend to the third abdominal segment. The body length is 2.1–2.7 mm and the body width is about 1.5 mm. The width of the head at the eyes is 0.78–0.80 mm.

Fifth instar. The color is variable, but greenish in general. The yellowish head has five longitudinal brownish stripes converging behind, but not attaining, the posterior margin of the vertex. The thorax and wing pads are yellowish with irregularly marked brown lines. The abdomen is yellowish or greenish yellow. The wing pads extend onto the fifth or sixth abdominal segment. The veins begin to appear in the pads. Four black spots are conspicuous on the thorax. The dorsal abdominal gland is indicated by a conspicuous black spot on the caudal margin of the third abdominal segment. The legs are variable in color. The body length is 3.2–4.2 mm and the body width is about 2 mm. The width of the head at the eyes is 0.94–0.96 mm.

Adult. The color is variable, usually greenish or brownish, with reddish brown markings on the wings. The thorax has five longitudinal dark stripes. The scutellum has two medium and two lateral black or reddish lines. Eleven segments comprise the abdomen, although some of these are considerably modified. The first ventral segment is reduced to an elastic membrane that joins the abdomen to the thorax. In the female, the seventh segment is named the subgenital plate, as it extends posteriorly in the mid-ventral

Annotated list of crop plants attacked by the tarnished plant bug

Forest trees (nurseries)			
Bare-root pine and container nurseries	Conifers	Douglas-fir	Larch
Fruit trees			
Apple	Cherry	Prune	Grape
Peach	Pear	Pecan	Plum
Quince			
Small fruits			
Blackberry	Currant	Grape	Raspberry
Strawberry			
Commercially grown flowers			
Aster	Bachelor's button	Bleeding heart	Calendula
Carnation	Chrysanthemum	Cosmos	Dahlia
Garden balsam	Geranium	Gladiolus	Golden glow
Hollyhock	Impatiens	Iris	Marigold
Nasturtium	Peony	Poppy	Rose
Sage	Salvia	Shasta daisy	Snapdragon
Stock	Strawflower	Sunflower	Sweet pea
Verbena	Zinnia		
Garden crops			
Asparagus	Beet	Broccoli	Cabbage
Carrot	Celery	Cucumber	Eggplant
Endive	Escarole	Horseradish	Lettuce
Lima beans	Mustard	Onion	Pea
Pepper	Potato	Radish	Snap beans
Spinach	Squash	Swiss chard	Tomato
Turnip			
Forage crops			
Alfalfa	Birdsfoot trefoil	Clover	Soybeans
Other crops			
Oilseed rape	Tobacco	Sunflower	Sweet corn
Seed heads of wheat and other grasses			

region to cover the base of the ovipositor. The body length is 4.9–5.7 mm for males and 5.2–6.0 mm for females. The body width is 2.4–2.8 mm for males and 2.4–3.0 mm for females. Moreover, around the base of the ovipositor on the ventral surface of the abdomen, there is usually a dark brown patch that may extend to the thorax.

Life cycle

Tarnished plant bugs overwinter as diapausing adults either beneath plant litter or duff ground cover, or between the leaves of plants and long dry grasses. In piled leaves and rubbish, winter survival is about 29%, whereas in orchard sod it is only 6%. Overwintering adults start emerging in mid-April at temperatures as low as 8°C. These adults feed first on the opening buds of trees such as apple, peach and shrubs. A little later they migrate to early appearing annual plants. The favorite spring food plants include black current (*Ribes nigrum* L.), wild currents (*Ribes* spp.), common mullein (*Verbascum thapsus* L.), sheep sorrel (*Rumex acetosella* L.), yellow rocket (*Barbarea vulgaris* R.Br.), and strawberries (*Fragaria* spp.).

Overwintering adults gradually disperse on these spring plants as the temperature increases and then migrate to other favorable emerging weeds or cultivated plants. The bugs usually feed on the sap of growing tips or the reproductive parts of a plant, such as the buds of flowers, or the rapidly growing meristematic tissue of such plants as asparagus, alfalfa and other forage crops.

Tarnished plant bugs generally migrate to a host only when the plant enters its reproductive growth stage. There is a seasonal succession of hosts as the bugs feed on different plants from spring to fall. Nymphs can disperse long distances (15–20 m) within a short time and adults can travel at least 5.1 km in sustained flight. Most adults fly no higher than 1 m from the ground, although some adults have been collected at elevations as high as 1,500 m. Multiple matings occur, but a single mating is sufficient for a female to lay viable eggs throughout her entire life. In Canada, overwintering females oviposit from the first week of May to the third week of June (about 50 days). Optimal oviposition occurs between 21 and 27°C and oviposition does not take place below 16°C. Eggs can sustain low temperatures of 10°C for 15 days without adverse effect. Eggs are embedded in the stem, in the petioles, or in the mid-

ribs of leaves. They may be deposited singly or sometimes in groups of two or three in close proximity.

Following eclosion, the nymphs feed immediately on the succulent parts of the host plant and molt five times. The choice of the host plant dictates significantly the duration of each stage. Laboratory studies, with the temperature fluctuating irregularly between 17–30°C, showed that female longevity ranged from 31 to 68 days, whereas male longevity ranged from 19 to 41 days. In southern Quebec, Canada, adults of the first generation appear about mid-June and reach maximum abundance around mid-July. During this time, adults migrate in great numbers from early season hosts, e.g., June bearing strawberries toward other host plants such as the following: raspberry (*Rubus* spp), garden beans (*Phaseolus vulgaris* L.), potato (*Solanum tuberosum* L.), pepper (*Capsicum annuum* L.), turnip (*Brassica napus* L.), sugar beet (*Beta vulgaris* L.), red clover (*Trifolium pratense* L.), celery (*Apium graveolens* L), wild mustard (*Sinapsis arvensis* L.), brown knapweed (*Centaura jacea* L.), and lamb's-quarters (*Chenopodium album* L.). This list is by no means complete. Eggs laid by summer adults produce adults 20 to 25 days later. For a given area, the seasonal sequential presence of tarnished plant bugs on wild plant hosts is practically the same year after year. The first generation adults mate when they are 4 to 6 days old. The pre-oviposition period varies between 9 and 13 days, with individual ranges from 5 to 29 days. The fecundity may reach up to 140 eggs per female with an average oviposition of 0–3.4 eggs per female per day. The population dynamics of the second generation nymphs is a function of the immigration time of the adults, the oviposition period of each individual and the nutritive quality of the different host plants. With the season progressing, more and more overlapping occurs between the developmental stages and the different generations of the tarnished plant bug. The overlap results in a smooth exponential growth pattern as opposed to a stepwise pattern. From the end of July to the second week of September, all stages of the bug can generally be found in the field. First generation adults that emerge in mid-June oviposit on several hosts and lead to a continuous emergence of second generation summer adults from the beginning to the end of August. A third generation of nymphs could occur on some cultivated hosts such as alfalfa and asparagus, and on wild plants such as common ragweed (*Ambrosia artemisiifolia* L.),

Annotated list of weeds attacked by the tarnished plant bug

Spring

Annual fleabane	*Erigeron annuus* (L.) Pers.
Common mullein	*Verbascum thapsus* L.
Common ragweed	*Ambrosia artemisiifolia* L.
Garden sorrel	*Rumex acetosa* L.
Wild strawberries	*Fragaria* spp.
Yellow rocket	*Barbarea vulgaris* R. Br.

Summer

Black mustard	*Brassica nigra* (L.) Koch
Boneset	*Eupatorium perfoliatum* L.
Brown knapweed	*Centaurea jacea* L.
Lamb's-quarters	*Chenopodium album* L.
Narrow-leaved hawk's beard	*Crepis tectorum* L.
Ox-eye daisy	*Chrysanthemum leucanthemum* L.
Pineappleweed	*Matricaria matricarioides* (Less.)
Redroot pigweed	*Amaranthus retroflexus* L.
Shepherd's-purse	*Capsella bursa-pastoris* (L.)
St. John's-wort	*Hypericum perforatum* L.
Tansy	*Tanacetum vulgare* L.
White mustard	*Sinapsis alla* L.
White sweet-clover	*Melilotus alba* Desr.
Yellow sweet-clover	*Melilotus officinalis* (L.) Lam.
Wild carrot	*Daucus carota* L.
Wild mustard	*Sinapsis arvensis* L.

Autumn

Canada fleabane	*Erigeron canadensis* L. (= *Conyza canadensis* (L.) Crong)
Canada goldenrod	*Solidago canadensis* L.
Canada thistle	*Cirsium arvense* (L.) Scop.
Common ragweed	*Ambrosia artemisiifolia* L.
European stinging nettle	*Urtica dioica* L. subsp. *dioica*
False ragweed	*Iva xanthifolia* Nutt.
Narrow-leaved goldenrod	*Solidago graminifolia* (L.) Salisb.
Red-stemmed aster	*Aster puniceus* L.
Rough goldenrod	*Solidago rugosa* Mill.
Tall white aster	*Aster simplex* Willd.

Canada golden rod (*Solidago canadensis* L.), rough goldenrod (*Solidago rugosa* Mill.), tall white aster (*Aster simplex* Willd.), stinging nettle (*Urtica dioica* L.), Canada fleabane (*Erigeron canadensis* L.), and the red-stemmed aster (*Aster puniceus* L.). In the fall, 25 to 30 days are needed for the tarnished plant bug to complete its life cycle. In southeastern Canada there are two generations per season whereas in the cotton belt in the southern USA, there are four to seven generations. Past the second generation, it is difficult to distinguish the generations. For example, on lamb's quarters characterized by continuous germination, second and third generation adults and larvae are found on the same plant. Therefore, the extended vegetative and flowering period of this weed contributes to the build-up of tarnished plant bugs in adjacent fields.

Despite some variations, the sex ratio is 1:1 irrespective of the host plant. The bugs are mainly phytophagous; however, they can sometimes feed as facultative predators on soft-bodied arthropods such as aphids and mites, or as scavengers on dead nymphs and adults of their own species.

As fall progresses, the testes and ovaries atrophy and the adults enter diapause. Exposure of nymphs of the first four stages (the photosensitive stages) to photoperiods of 12.5 hours or less induces diapause in the adult stage. Rearing these nymphs to a photoperiod of 13.5 hours or more will prevent the adults to enter diapause. Continuous light prevents diapause in young adults and terminates diapause in diapausing adults. Fifth-instar nymphs and adults are not sensitive to diapause-inducing photoperiods.

Known parasitoids of *Lygus* bugs

Parasitoid name	Family	Target	Comments
Indigenous species			
Polynema pratensiphagum Walley	Mymaridae	Egg	72% *Lygus* mortality in Quebec
Anaphes iole Girault	Mymaridae	Egg	
Anaphes ovijentanus Crosby & Leonard	Mymaridae	Egg	50–85% *Lygus* mortality widespread
Erythmelus miridiphagus Dozier	Mymaridae	Egg	
Telenomus spp.	Scelionidae	Egg	
Peristenus pallipes (Curtis)	Braconidae	Larvae & adults	First generation May–June; second generation Aug.–Sept. combined 15–20% *Lygus* mortality in Quebec; 38% in Ontario and 22% in Saskatchewan-Alberta
Alophora opaca (Coquillett)	Tachinidae	Adults	7% *Lygus* mortality in Quebec
Alophorella sp.	Tachinidae	Adults	0.8% *Lygus* mortality in Ontario
Imported species			
Peristenus stygicus Loan (USA, Sask., & Alta.	Braconidae	Larvae	Diapausing race from France and a nondiapausing race from Turkey; polyvoltine
Peristenus rubricollis (Thomson) (USA)	Braconidae	Larvae	From Poland; univoltine
Peristenus digoneutis Loan	Braconidae	Larvae	From France (potato, alfalfa & rye fields); released in alfalfa field in New Jersey 1979–1988; 36% parasitism of first generation and 29% parasitism of second generation; released in seed alfalfa, Saskatchewan, and strawberries, Quebec, in 1991, 1992; Currently, descendents from the New Jersey releases are collected at St. Clothilde Qc. It is bivoltine; population at maximum abundance about same time as *Lygus* populations; attacks early stage nymphs of *Lygus*.

Attractants

There is now some evidence that virgin female tarnished plant bugs release an attractant to males. However, the attractant has not yet been isolated and identified. The presence of an aggregation pheromone that attracts both sexes also has been suggested. In any case, there are still no practical applications for these findings.

Pest host relationships

In southeastern Canada and northeastern United States, adults are subject to late fall temperatures that induce the tarnished plant bug to hibernate until the following spring. Management practices to control this bug in fields must consider plant hosts as reservoirs for the build-up of adults and nymphs in the vicinity of cultivated fields. Any flowering plant seems to attract the tarnished plant bug. Nevertheless, there are a few exceptions such as dandelions, sowthistle, milkweed and dogbane, which are all sap latex-type plants. Goldenrod is one of the rare latex-type sap plants which is very attractive in the fall to field populations of this bug.

Plant hosts range from highly attractive to highly resistant to the tarnished plant bug. Resistance to *Lygus* has been reported in some cultivars of celery, alfalfa, cotton, beans and carrots. Some plants do not attract the bug at all. For example, outbreaks have never been reported on wheat, rye and turf grass.

Although the tarnished plant bug is able to develop on a wide range of hosts, field observations show that some plants consistently support much higher populations than others. Management on selected hosts at particular growth stages over a short time, e.g., before the nymphs become adults, may result in lowering bug populations in crop agro-ecosystems. Sites with ephemeral, lush growth that dries out quickly provide a release of tarnished plant bugs and are of immediate concern during the cropping season. In uncultivated fields, weeds growing in sequence that are acceptable to tarnished plant bug development will retain their bug populations throughout the year without any activity of dispersion toward the surrounding crop lands, as long as the site is not disturbed or dried up.

Natural enemies

Presently, it seems that natural enemies, other than a few species of insect parasitoids, are not capable of checking a *L. lineolaris* population once it begins to build up. Endemic populations of *L. lineolaris* are under tremendous environmental pressure that checks their populations. In such instances, natural enemies play an important role. We do not yet have quantitative analyses of the intricate relationships between prey (*L. lineolaris*) and their natural enemies. In addition to parasitic Diptera and Hymenoptera, an identified species of nematode has been reported to rarely parasitize *L. lineolaris* adults in Quebec. Several predators are known to attack *L. lineolaris*, including damsel bugs (*Orius* spp., Anthocoridae), leaffooted bugs (Coreidae), stink bugs (Pentatomidae), lacewings (Chrysopidae), solitary wasps (Sphecidae), robberflies (Asilidae), lady beetle larvae (Cocinellidae), and spiders (Oxyopidae, Tetragnathidae and Thomisidae).

Damage

In southeastern Canada and northern United States, overwintering adults are pests of fruits such as apple, peach and pear. The adults injure the buds causing a circular depression referred to as ''pit'' in apples and pears. In peaches, the fruit appears as if it has been gouged when small and the injury is called ''cat facing.'' Furthermore, a certain percentage of fruit drop occurs following *Lygus* attack. The nymphs that will comprise the first generation attack strawberries causing malformed berries where the achenes are very close to each other and straw-colored. Similar to peaches, the injury is called ''cat facing.'' Several vegetables such as celery, tomato, eggplant and pepper are also attacked by first generation adults and second generation nymphs and adults, causing flower abscission, which reduces yield in some years. In celery, early injuries to the stalk are of no economic consequence. However, near harvest, injury to petioles and necrosis of leaflets leads to secondary infections by bacteria. The damage is called black joint. It is serious and causes unacceptable economic losses. Though cotton is not its preferred host, the tarnished plant bug is a serious pest of cotton. Migration occurs when the wild hosts have dried along cotton fields. Following the completion of the second generation, the bugs enter the cotton fields and damage all fruiting forms of cotton. Early to mid-June pinhead squares (immature flower buds) are attacked and turn yellowish, dry up and eventually drop off the plant. In addition to feeding on pinhead squares, tarnished plant bugs also feed on terminal shoots. This activity results in several physiological changes caused by the toxins injected into the meristematic tissue of the plant terminal. The tall, fruitless plant is characterized by shortened internodes, swollen nodes and excessive lateral branches. This condition is often referred to as ''crazy cotton.'' Later, damage to larger squares shows up as darkened anthers in white blooms resulting in poor pollination and deformed bolls. These are called ''dirty blooms.'' Eventually, the tarnished plant bugs even attack the bolls causing small, dark, sunken spots on the outside of the boll and brownish discoloration inside the boll.

See also, SMALL FRUIT PESTS AND THEIR MANAGEMENT.

N. J. Bostanian
Agriculture and Agri-Food Canada
Horticultural Research and Development Center
St. Jean-sur-Richelieu, Quebec, Canada

References

Bostanian, N. J. 1994. The tarnished plant bug and strawberry production. *Technical Bulletin 1994-1E*. Agriculture and Agri-Food Canada, St. Jean-sur-Richelieu, Canada.

Day, W. H. 1987. Biological control efforts against *Lygus* and *Anthocoris* spp. infesting alfalfa, in the United States, with notes on other associated mirid species. pp. 20–39 in R. C. Hedlund and H. M. Graham (eds.), *Economic importance and biological control of Lygus and Adelphocoris in North America*. U.S. Department of Agriculture, Washington, DC. ARS-64.

Tingey, W. M., and E. A. Pillemer. 1977. *Lygus* bugs: crop resistance and physiological nature of feeding injury. *Bulletin of the Entomological Society of America* 23: 277–287.

Young, O. P. 1986. Host plants of the tarnished plant bug, *Lygus lineolaris* (Heteroptera: Miridae). *Annals of the Entomological Society of America* 79: 747–762.

TARSAL CLAW. The claws at the tip of the tarsus.

See also, LEGS OF HEXAPODS.

TARSOMERES. The major subdivisions of the tarsus (the foot), the distal segment of which is called the pretarsus, and bears claws and sometimes a pad or arolium.

See also, LEGS OF HEXAPODS.

TARSUS. (pl., tarsi) The jointed portion of the insect leg distal to the tibia, consisting of tarsomeres and often bearing claws on the distal segment (pretarsus). The 'foot' of an insect.

See also, LEGS OF HEXAPODS.

TASTE AND CONTACT CHEMORECEPTION. The sense of taste provides animals with the ability to identify potential food by the perception of certain nutrients, and also to detect potentially toxic materials. Insects are no exception, but unlike most other animals their taste receptors are not restricted to the region around the mouth and they may be able to recognize oviposition cues, and, occasionally, intraspecific signals as well as food. In addition, whereas in vertebrates the taste receptors are stimulated by chemicals in solution, insects have the capacity to perceive chemicals on dry surfaces. For this reason, it is referred to as "gustation" or "contact chemoreception" rather than "taste."

Insect contact chemoreceptors are usually in the form of hairs, or conical projections from the cuticle, with a pore at the tip. The pore permits chemicals to pass through the cuticle to the sense cells beneath. Each hair, or cone, contains the sensitive endings (dendrites) of (commonly) four sensory cells, and each of these responds to a different range of chemicals. The ranges of chemicals reflect the habits of the insect and also the position and specific function of a particular hair.

Very little is known about why an insect chemosensory cell responds to one, or some chemicals and not to others, but it is probable that, as in other animals, this is determined by receptor molecules in the cell membrane of the dendrite just inside the pore in the cuticle. A taste receptor cell may have only one type of receptor molecule, in which case, its response is limited to one, or a few, structurally similar

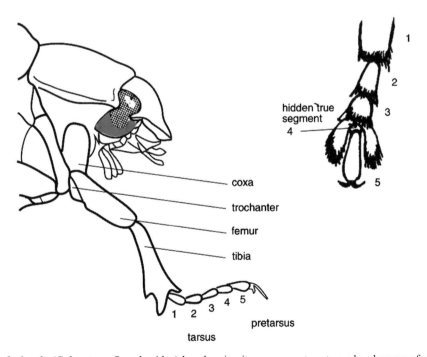

Fig. 1015 Leg of a beetle (Coleoptera: Scarabaeidae) leg showing its component parts, and a close-up of one type of beetle tarsus (foot).

chemicals that interact with the receptor. Other cells have more than one type of receptor molecule, and so have the capacity to respond to a range of structurally different chemicals. Some cells that are presumed to have these different characteristics are described below.

The interaction of the appropriate chemical with a receptor molecule leads to changes in the electrical potential across the cell membrane and an electrical signal is sent along the axon that connects each sensory cell to the central nervous system. Connections within the central nervous system determine whether the insect responds positively or negatively to the stimulating chemical, that is, whether it accepts or rejects food, or an oviposition site. In practice, how the insect responds nearly always depends on the integration of signals from a number of receptors.

Contact chemoreceptors are found primarily on the mouthparts, the labrum, maxillae, labium, and hypopharynx, and they may be especially abundant on the maxillary and labial palps when these are present. A small number are present on the epipharynx (the inside surface of the labrum) just outside the mouth in all insects that have been investigated. Many insects also have contact chemoreceptors on the tarsi, and on the antennae, although most of the chemoreceptors on the antennae are olfactory receptors. Females of some Orthoptera, Diptera, Hymenoptera and Lepidoptera have contact chemoreceptors on their ovipositor, but this is not a general characteristic of females even within a group.

The numbers of contact chemoreceptors associated with the mouthparts is very large in cockroaches, grasshoppers and related insects. An adult of the migratory locust, *Locusta migratoria*, for example, has about 3,000 contact chemoreceptors on the mouthparts. In these insects, additional receptors are produced at each molt. By contrast, much smaller numbers are present in sucking bugs. Aphids have no contact chemoreceptors on the exposed parts of the proboscis, but they have a few on the epipharynx. As a result, aphids must draw fluid into the cibarial cavity before they are able to taste it. This is not true of all sucking bugs, however, many of which have contact chemoreceptors on the tip of the labium, which is the first mouthpart structure to contact the food prior to feeding. Holometabolous insects also generally have few mouthpart contact chemoreceptors, especially in larvae. Caterpillars

have only 16–20 contact chemoreceptors and this number does not change throughout development. In grasshoppers, species with more restricted diets tend to have smaller numbers of contact chemoreceptors on the mouthparts than polyphagous species. Among caterpillars, however, this is not true. The numbers are similar irrespective of feeding habit.

Feeding

The primary function of contact chemoreceptors on the mouthparts is the selection of food. Once the insect bites into food, some receptors become immersed in the juices released from the food and so function very much like taste receptors in vertebrates by detecting compounds in solution. However, the receptors on the maxillary and labial palps often come into contact with food before the insect bites. They can detect compounds on the dry surface of a leaf, for example. In grasshoppers, crickets and cockroaches, the palps are vibrated rapidly so that the contact chemoreceptors at their tips are brought into a rapid series of brief contacts with the surface. As many as 15 contacts per second may be made, and each period of contact probably lasts less than 20 milliseconds. This behavior, known as palpation, probably serves two functions. It allows the insect to receive a more sustained flow of information from the receptors than would be possible if contact were maintained because the receptors become adapted (as our taste receptors do after the first mouthful of a sweet drink), and it also allows the insect to sample a larger surface area than if the palps remained stationary. The information provided by palpation before an insect starts to feed enables it to make feeding decisions more rapidly and also, perhaps, to avoid the possible intake of noxious compounds if the food contains toxins.

Insects, like other animals, can taste the major nutrients essential for their development, survival and reproduction: sugars (as a source of energy), amino acids (the building blocks for protein), and inorganic salts and water. There is no direct evidence that insects are able to taste proteins. The specific chemicals to which an insect's taste receptors respond, however, vary with the feeding behavior of the insect. It is common for one of the sense cells in a contact chemoreceptor to be sensitive to some sugars, and another to be sensitive to some amino acids. Some insects have also been shown to possess a "water cell," that is, a cell that responds to water and very dilute salt solutions, and it is possible that this type of cell is of widespread occurrence. All insects appear to have a cell responding to inorganic salts, but this cell exhibits increased activity as the salt concentration increases (unlike the "water cell" whose activity declines with increased salt concentration), and it is probably a cell that inhibits further feeding, ensuring that the insect does not ingest excessive amounts of salt. The other nutrients such as sterols, fatty acids and vitamins appear to be acquired as a result of their widespread occurrence in the insect's food and do not require specific taste receptors.

A cell responding to sugars does not necessarily respond to all sugars, nor does an amino acid-sensitive cell respond to all amino acids. In the woolly bear caterpillar of the moth, *Grammia geneura*, for example, a sugar-sensitive cell in one hair only responds to the fruit sugar, fructose. This cell probably has only one type of receptor molecule that only reacts with fructose. However, a cell in another hair close by responds to sucrose and glucose although not to fructose. This cell probably also has a single type of receptor molecule that reacts with both sucrose and glucose because of similarities in their molecular structure. In the same way, cells in different hairs may respond to different ranges of amino acids. This probably allows an insect to detect food of different qualities. Even so, not all different types of sugar or amino acids are tasted by any one insect even though some of them are essential for its development. It obtains a balanced diet mainly because a food with some nutrients, which it can taste, will generally contain most of the others, which it cannot taste.

In addition to nerve cells that respond to nutrient compounds, many plant-feeding insects possess sensory cells that respond to plant secondary compounds. These compounds, alkaloids, terpenoids, and many others, are produced by plants outside the normal primary metabolic pathways (hence "secondary") and are important in the plant's ecology. Many of them inhibit feeding by herbivorous animals, including insects, and the human interpretation is that they taste "bitter." Sometimes they are also toxic. In a majority of plant-feeding insects, stimulation of a cell by these secondary compounds inhibits, or deters feeding, and the cells are usually called "deterrent cells." A deterrent cell usually responds to several different secondary compounds, but by no means all the different types. This almost certainly reflects that fact that it has several different types of receptor molecules.

Because all plants contain secondary compounds, whether or not an insect feeds on a particular plant and the amount it eats is dependent on the balance between information received from taste cells signaling the presence of nutrients, and deterrent cells signaling the presence of secondary compounds. However, in some insect species that feed only on particular plant species or groups of plants, characterized by a specific chemical, it is common for the deterrent cells to have lost any sensitivity to that chemical so the plant is no longer "distasteful." Correspondingly, some now have a sensory cell that responds only to that compound or class of compounds and which indicates "acceptability" rather than "distastefulness" to the insect. The best-known example of this is the response to glucosinolates. These are sulfur-containing compounds characteristic of the cabbage family. For most plant-feeding insect species, these compounds stimulate deterrent cells and inhibit feeding, but in many species that feed habitually on plants in the cabbage family, the deterrent cells are not affected. Instead, another sensory cell responds to the glucosinolates, indicating acceptability so that the insects are stimulated to feed or oviposit.

In species that feed on vertebrate blood, the contact chemoreceptors have a sense cell that responds to adenine nucleotides such as ATP and ADP. These compounds are released when the insect probes into the host's blood vessels and damages blood cells. They provide the insect with an unequivocal signal that appropriate food is available.

Not only the chemoreceptors on the mouthparts are involved in feeding decisions, however. This is most obvious in fluid feeding insects like flies and bees, where extension of the proboscis, which is necessary before the insect can feed, is induced if chemoreceptors on the tarsi are stimulated with sucrose. In the tsetse fly, *Glossina fuscipes*, similar receptors respond to some of the common components of human sweat, such as uric acid and the amino acids leucine and valine. Stimulation with these compounds causes the insect to probe with its proboscis and is presumably part of the normal host recognition process. Tarsal receptors are also involved in host recognition by plant-feeding insects.

Oviposition

It is common for contact chemoreceptors to be involved in oviposition, although attraction to the oviposition site often depends on the sense of smell.

Contact chemoreception in this context has been most thoroughly investigated in some flies such as the cabbage fly, *Delia radicum*. The tarsal receptors on this fly respond to glucosinolates on the leaf surface of the host plant. Other insects, like the imported cabbage butterfly, *Pieris brassicae*, and the cabbage looper, *Trichoplusia ni*, that lay their eggs on cabbage plants also have tarsal receptors responding to glucosinolates. Contact with these compounds induces the insects to lay eggs. A number of butterflies drum on the leaf surface with their fore tarsi when selecting a leaf for oviposition. This appears to be a process analogous to palpation, bringing the contact chemoreceptors on the tarsi into a series of brief contacts with the leaf surface. There is no evidence that the leaf surface is broken by these activities, so the insect does perceive chemicals on the surface, not from within the plant.

Although contact chemoreceptors do occur on the ovipositors of some insects, there is not yet clear evidence of their roles in oviposition.

Pheromone detection

Contact chemoreceptors are sometimes important in the perception of pheromones involved in sexual recognition, prevention of oviposition, and trail following. In the tsetse flies, *Glossina* species, the male recognizes a female by touching her with contact chemoreceptors on his fore tarsi. This contact also enables him to determine whether or not she has mated through the perception of specific molecules in the wax on the surface of the female's cuticle. Changes in the proportions of some components occur at the time of mating because wax from the male contaminates the female, and the male is able to detect the difference with his tarsal receptors. A similar process occurs in *Drosophila* fruit flies, except that in this case, the female wax alters as a result of changes in synthesis in the female following mating. Females of the German cockroach, *Blattella germanica*, also have sex specific compounds in their cuticular wax which function as a sex recognition pheromone. At the start of courtship, the male touches the female with his antennae, presumably detecting the compounds with contact chemoreceptors on the antennae.

Contact chemoreceptors are used by the adults of some flies to detect oviposition deterrent pheromones laid down by previously ovipositing females to reduce the likelihood of competition with their larvae. This is known to be the case in adults of the apple maggot fly, *Rhagoletis pomonella*, and the

cabbage butterfly. The cabbage seed weevil, *Ceutorhynchus assimilis*, however, detects its oviposition deterrent pheromone with contact chemoreceptors on the club of the antenna. Although contact chemoreceptors are common on insects' antennae, this is one of the few instances where their function is known. Many insects are known to make rapid vibrations of the antennae when they encounter a potential food or prey item. This process is known as antennation, and it is probably analogous to palpation with the insect bringing contact chemoreceptors at the tip of the antenna into brief contacts with the substrate. Parasitic Hymenoptera, for example, are commonly seen doing this when they encounter a potential host.

Tent caterpillars follow chemical trails deposited by conspecifics. The trails are detected by contact chemoreceptors on their maxillary palps.

It is almost certain that there are many situations in which insects use contact chemoreceptors on their antennae or tarsi for the perception of intraspecific signals. See also, ULTRASTRUCTURE OF INSECT SENSILLA.

Reg Chapman
University of Arizona
Tucson, Arizona, USA

References
Chapman, R. F. 1995. Chemosensory regulation of feeding. pp. 101–136 in R. F. Chapman and G. de Boer (eds.), *Regulatory mechanisms in insect feeding*. Chapman & Hall, New York, New York.
Glendinning, J. I., N. Chaudhuri, and S. C. Kinnamon 2000. Taste transduction and molecular biology. pp. 315–351 in T. E. Finger, W. L. Silver, and D. Restrepo (eds.), *The neurobiology of taste and smell*. Wiley-Liss, New York, New York.

TAXIS. (pl., taxes) A directional movement in response to an environmental stimulus, directed toward or away from a stimulus. A movement toward the stimulus is considered to be a positive taxis, a movement away is a negative taxis. Types of taxes include anemotaxis, astrotaxis, chemotaxis, geotaxis, hygrotaxis, phonotaxis, phototaxis, rheotaxis, and thermotaxis.

TAXONOMY. The principles and procedures according to which species are named and assigned to taxonomic groups.

TEACHING AND TRAINING ENTOMOLOGY: INSTITUTIONAL MODELS. Research on insects, whether to understand basic biology or to provide management strategies, most often occurs within thef context of an institutional paradigm. Precisely the same is true for teaching/training programs and extension. The structure of an organization in which researchers, trainers or extension specialists work is crucial to understanding the goals, motivations and successes of those human resources. Understanding the institutional framework within which programs are developed and delivered is therefore an important consideration when evaluating program effectiveness.

There are five principal models of institutional and procedural frameworks within which research, teaching and/or extension programs associated with entomology occur. Three are fairly widely used and two have the potential to become global. The very framework chosen within which to pursue entomology may well affect the types and quality of information developed. Evaluating the effectiveness of entomological programs cannot be complete without recognizing the institutions within which such programs are developed and reside.

The degree to which entomological programs developed under one paradigm can function successfully when transferred to another is therefore an important and an intriguing issue, but one that has been virtually ignored on a global basis. Each of the five major models currently used has advocates and critics, and each has relative strengths and weaknesses. However, discussion of program strengths and weaknesses most often fails to consider the institutional context in which they have developed historically, and thus fails to provide a comprehensive analysis of when, how and why these programs succeed or fail.

The land grant system: USA

During Abraham Lincoln's presidency, the United States Congress was moved to provide legislation directly aimed at assisting farmers. This made both practical and political sense, as some 65% of the US population was involved in agriculture in one way or another. Under Lincoln's guidance, the United States Department of Agriculture was established in 1861. Then, four successive Federal Acts (Morrill Land Grant Act, Hatch Act, second Morrill Act and Smith-Lever Act), dating from 1862 to 1914, set aside lands for agricultural research and put the

infrastructure and human resources in place to conduct that research. The second Morrill Act (1890) set aside a second set of lands and infrastructure for the development of what are often termed Historically Black Colleges and Universities (HBCUs), and the Smith-Lever Act (1914) created the Federal Extension System.

The Land-Grant system is based on the concept of combining teaching, research, and extension within a single institution, and it took about 50 years to establish all three functions. All 50 states and US protectorates and territories employ the system. America has led the world in agricultural production since the Land Grant system was established.

Proponents of the Land-Grant model point to the productivity of American agriculture and the associated research infrastructure. Indeed, due in part to the Federal and state investment in this institutional model, the USA is today the world's leading producer of food with only about 1.8% of the population involved in food production. The USA grants more graduate degrees in the agriculture-related disciplines than any other nation, and more international students seeking advanced degrees in agriculture choose to study in the United States than anywhere else. Including extension in the model indicates the importance of a forward and backward link between farmers and researchers. Finally, this model ensures that advances in agricultural technology to find their way into the classroom very quickly, ensuring that future generations of agricultural professionals and farmers are able to apply the new technologies.

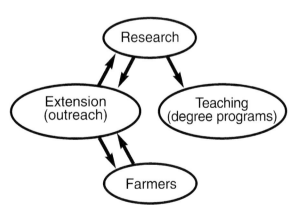

Fig. 1016 Basic structure and flow of information in a Land-Grant Institution as originally conceived.

Not everyone agrees that this model is ideal. Critics of the Land-Grant model point out several weaknesses in the system.

1. Over the years, extra-mural grants have supplanted Federal and state direct support for agricultural research and extension. Grants programs are fickle and trendy. Land-grant institutions have a long history of "following" extra mural priorities (integrated pest management in the 70s, sustainable agriculture in the 80s, and genetic engineering now). The wholesale commitment to these types of trends by significant percentages of land-grant scientists is indicative of how susceptible this model is to abrupt shifts in priorities in the quest to acquire funds and publish. Research is driven more by granting agency priorities than by farmer needs.

2. Due in part to the emphasis on extra-mural funding, Land-Grant institutions tend to recruit faculty with ever-narrowing expertise so they can compete successfully for grants. As a result, the system lacks scientists who have either the skills or the incentives to integrate knowledge across disciplines. Individuals who are capable of critically analyzing the potential impacts of new technologies in a broader social, economic and ecological framework are rare.

3. The departmental and college management structure of Land-Grant institutions is, in many ways, the antithesis of the type of holistic program structure needed to resolve real problems. University units are rewarded for their individual productivity, not for their collective cooperation. While universities constantly discuss the need for interdisciplinary approaches, their very structure often prevents them from implementing effective cross-disciplinary solutions.

4. The ability of the USA to capitalize on the high productivity of Green Revolution agriculture is as much attributable to good fortune as to agricultural research, and it is likely not sustainable. Green Revolution agriculture depends on high levels of off-farm inputs that make successful farmers more dependent on a few transnational corporations that supply seed, fertilizers, equipment, and chemicals while also increasing the farmers' debt. Other farmers simply cannot obtain these inputs. There appears to be an internal conflict of priorities in the Land-Grant Model.

One priority is to feed the masses at any cost. The other is to keep small farmers in business for the social good.

5. Teaching in Land-Grant institutions often is the single least rewarded activity. Reward and professional advancement depend largely on the amount of grant dollars obtained and the number of publications produced. Relevance to real farm problems may not rank high among the factors taken into consideration for faculty promotion.

6. Extension often does not work as it was conceptualized. The fundamental idea was for extension to serve as the voice of the farmers and to vocalize the research needs seen by farmers as priorities. That extension voice would then explain those priorities to university researchers who, using their research expertise, would investigate the issues, develop solutions and have the extension voice deliver those solutions back to the farmers. In reality, researchers are driven by grant priorities, rather than extension-relayed farmer needs.

7. Finally, critics point out that the USA has moved from an original farm policy designed to keep farmers on the farm and farming profitable, to a "cheap food" policy designed to feed urban populations who, by and large, elect Federal officials. The Land-Grant infrastructure evolved to pay little attention to the promotion of agricultural procedures that would make farming sustainable and conserve natural resources.

While the debate continues, the sheer volume of output from Land-Grant programs is impressive. In the USA, the Land-Grant system encompasses a wide range of farming enterprises. Large scale, corporate farming often becomes a source of information for Land-Grant scientists as corporations have the capital to invest in staff and research on their crops and pests that tax-based Land-Grant institutions do not. Small farmers often cannot afford to adopt the technologies proposed by Land-Grant scientists and tend to utilize a mix of modern science and traditional farming practices.

The system has not gained acceptance outside the U.S. and, in fact, has really not ever been replicated elsewhere, although components of the system have been implemented in other places. This model depends greatly on a collaborative relationship among peers for its success. Peer review, for example, is a major factor in evaluating the quality of teaching, research and extension. Similarly, funding for the system involves Federal, state and local support, and farmers must be empowered to ensure that their problems and priorities are imposed on the entire structure. All of these characteristics are typical of the social and political structure in the United States. The poor "transportability" of the U.S. system, especially to non-European based cultures and to nations where democracy is not the norm, may be due, in large part, to the absence of an appropriate social, cultural and political context for the system. The Land-Grant model may be culturally insensitive, but not necessarily wrong. Agricultural productivity appears to be like every other increasingly globalized activity: one cannot compete if one doesn't play the top-level game. This, of course, is not highly acceptable to politically sensitive people who want to maintain local culture and merely improve agricultural productivity. Global competition likely will demand local cultural changes in the way farmers and governments spend their time and money.

Training and visit (T&V): World Bank

T&V is a child of the World Bank, and this model was conceptualized and first utilized in Asia in the 1970s. This hierarchical system began in India, Turkey, Burma, Nepal, Sri Lanka and Thailand, and today exists in more than 40 nations. The basic premise of the T&V system is that much more information and technologies exist than are used by farmers, and the principal reason for this is the ineffectiveness of extension delivery systems. T&V thus places major emphasis on training and outreach. Research is most often limited to site-specific adaptive research of existing technologies. While more recent T&V programs have tended to re-emphasize the research component, the approach continues to focus heavily on outreach.

T&V is a top-down approach. A subject matter specialist trains a limited number of regional extension agents who, in turn, provide training to local extension agents who, in turn, train farmers. This approach emphasizes the timeliness of actions, and the progress of agricultural productivity at the farm and regional levels is "tracked" by the training process. For example, training on improved seedbed preparation will be delivered just before the time when seedbed preparation is to occur. Frequent, regular reporting from the bottom up is a key feature of the T&V model. The Land-Grant system has "field days" and "demonstration plots" where farmers can come to witness innovations. While not hierarchical in nature, as is T&V, the idea is somewhat the same.

Proponents of T&V argue that the system is ideal because it focuses on those human resources that are native to indigenous cultures, provides a "filter" to take existing technology and adapt it to local conditions, and it is both relevant and timely. T&V takes complete advantage of cultural and social hierarchies already in place. The fact that T&V is currently utilized in over 40 nations, so say the proponents, is indicative of its relevance. Simply put, it works.

Critics of T&V point to several shortcomings.

1. Methods are geared to the top, best-educated, most articulate facilitators (regional extension personnel). Most T&V programs end up with a paltry number of highly capable facilitators. The methods that are emphasized both confuse and leave behind the other participants in the system. One result is that the less capable extension agents train their local agents poorly, and the farmers who rely on them, in essence, receive poor advice.

2. In the same vein, T&V historically has not sanctioned facilitators for not attending all training sessions. One result is poorly equipped facilitators, and the problem progresses down the hierarchy.

3. This approach, in many ways, "force fits" the T&V timetable into an often-unreceptive farmer schedule resulting in poor adoption of advice.

4. T&V programs historically are narrow in perspective and not linked to broader, yet relevant, issues of great concern to farmers. Timely seedbed technology is useless if not linked to socially relevant local schedules and to lending periods and priorities of local banks.

5. T&V programs were developed for irrigated systems where water was controlled and never attempted to develop procedures exportable to rain fed agriculture.

6. Many analyses have shown T&V approaches to be overly costly and non-sustainable.

T&V differs from the Land-Grant system in its almost sole focus on a hierarchical extension-training approach. An interesting note is that virtually all the countries where T&V is employed are in World Bank-funded efforts in Asia. T&V does not appear to have been widely adopted elsewhere. However, the essence of T&V sometimes is practiced in other countries like Jamaica despite no formal funding of projects actually called T&V.

Once again, cultural and social factors may play an important role in determining how well this system "travels" from its Asian heartland to other parts of the world. The very hierarchical model that assumes that knowledge flows "downward" from more educated and knowledgeable experts to practitioners (farmers) would not, for example, be very acceptable to many farmers in the United States or perhaps in most societies where egalitarian norms are strong.

Farming systems: United States Agency for International Development

The farming systems model originated from programs in Latin America and, over the last two decades, has been promoted by a variety of international development agencies. The single largest advocate of this model in the 1980s was the United States Agency for International Development (USAID). Many of the CGIAR centers (notably CIP, CIMMYT and IITA) also have used this model.

The farming systems model differs widely from T&V in that farming systems primarily emphasizes research, not extension. Underlying the farming systems model is the assumption that most existing technology has not been adapted to specific regional or local conditions. Farmers are, therefore, unable to adopt and utilize what would be useful methods. The major emphasis of farming systems is on site-specific, adaptive research. While there is episodic invention of "new technology," the major emphasis focuses on local tests of existing technologies. Proponents of this model argue that there are major differences between the farming systems model and the Land-Grant model. In practice, those differences are often hard to discern. Both models depend on strong linkages between research and extension and are difficult to implement in places where the two functions are not housed in the same institution. Further, many advocates of the Land-Grant system argue that both models emphasize features such as on-farm research, site-specific testing and multidisciplinary research teams. There is, in practice, one major difference between the two models. The farming systems model emphasizes research much more than extension, and farming systems projects that demonstrate mass dissemination of information are rare.

Fig. 1017   Farming systems training effort in The Gambia, West Africa.

A basic tenet of the farming systems model is that many global farmers have been by-passed by the Green Revolution. Their land holdings or their economies were too poor to take advantage of high-yielding agriculture and its associated inputs. Thus, the advocates say the most relevant approach is the use of a multidisciplinary "rapid rural appraisal" team to ask farmers' opinions, followed by research aimed at tailoring existing technology to individual farmers' fields.

Critics of the farming systems model point to a number of constraints on its effectiveness.

1. The disciplinary make-up of the initial rapid rural appraisal team can severely bias what the team determines is pertinent research.

2. Farming systems teams have often been unable to define their tasks and, consequently, served only to "spread the gospel" of farming systems rather than to provide holistic solutions for real problems. Farming systems failed in Africa for precisely this reason.

3. Many farming systems projects to date have focused on crop variety and/or fertilizer trials — the very things that these farmers supposedly could not afford, thus giving rise to the farming systems model in the first place.

4. To date, adherents of the farming systems model have been primarily economists, anthropologists and a few agronomists. The broad spectrum of agricultural disciplines apparently has not bought into the basic premise of farming systems.

5. It is an expensive system, relying on a highly trained and educated body of researchers becoming very familiar with local conditions, problems and farmers. The central role of the multi-disciplinary team, along with the emphasis on the site-specificity of knowledge, means that any institution using this approach must make a relatively large investment in the local farming systems team and cannot spread the effect of that investment across a very large geographic area. Not surprisingly, like T&V, most places using the farming systems model also have been the recipients of funding from the principal donor advocating the model.

The basic argument for the origination of the farming system model would seem pertinent, and the approach has been tried in a number of locations with mixed results. Like the Land-Grant model, which it resembles in many other aspects, the farming systems model depends on mutual respect among peers and between agricultural professionals and farmers. In societies where there are strong biases against effective communication between individuals of differing social, economic or cultural status, the all-important communication between researcher and farmer is likely to fail. Again, the success of this system may well depend on the social and cultural context in which it exists, and its distinct lack of success in some parts of the world may reflect cultural and social realities rather than simply the ineffectiveness of the system itself in solving farmers' problems.

Interestingly, the World Bank funded both T&V and Farming Systems projects in Africa. T&V came first as an extension effort and was followed by farming systems as a research effort (Malawi, Tanzania, Ethiopia). World Bank saw these two models as complementary. Yet, African successes at solving real problems in a cost effective manner are hard to find.

Training-driven research: Indonesian Integrated Pest Management

This model is relatively new and has not been implemented in many places. Basically, it has been developed as part of an Asia-wide integrated pest management effort funded originally by the Australian government and is now managed by the Global Integrated Pest Management Group within FAO in Rome, Italy. It offers exciting possibilities to overcome many of the flaws inherent in the other three models.

Training-driven research is based on two fundamental premises: (1) do not conduct research unless it is needed and (2) use training as the tool to identify what research is needed. This model is like T&V in that it trains personnel who, in turn, train others and farmers. It is like farming systems in that it focuses only on "relevant" research, and it is like the Land-Grant model in that it combines training, extension and, when needed, research.

Farmers and those who assist and advise them need to be trained. Thus, experts establish a rigorous, field-based training program for the farmer support system, then for the farmers. Beginning with field preparation, each and every step in the agricultural production annual cycle is demonstrated to the trainees, both in the field and on site. Experience of

Fig. 1018 Farming systems project to teach proper goat-dipping techniques for tick control in Kenya. Slide provided by Dr. Sandra Russo, International Programs, University of Florida.

the trainers, literature and formal knowledge possessed by the trainers are incorporated into the training course. As long as the information exists to proceed in relevant fashion through all the steps of production (from field preparation, to fertilization, to pest management, to harvesting, to marketing), there is no need for research.

If, at any step, the trainer recognizes that the information to train appropriately does not exist, there is a call for "research." This may take the form, however, of a demonstration rather than a full-blown, randomized experiment. In any case, the product of this research is not a journal article, but rather feedback provided directly to the training program. This is an iterative process resulting, sooner or later, in an ability to appropriately train farmers and their assistants without conducting research until a new, unknown situation arises.

This model differs markedly from the Land-Grant model where funded research is the top priority. Further, the first priority of Land-Grant research is not training, but publication and the volume of such publications produced is germane to continued employment in the Land-Grant system. Training-driven research also offers a paradigm to regiment discipline expertise. Often, Land-Grant multidisciplinary teams attacking some aspect of agriculture will find,

once the problem is truly understood, they have the wrong expertise on the team (e.g., hiring a bacteriologist when a virologist was needed). Training-driven research avoids this issue by first identifying the particular need, then seeking the appropriate expert.

Training-driven research, to date, has not been attempted in enough places to have many proponents or critics. One criticism to date focuses on TDR's inability to demonstrate any significant research accomplished after identification of the problem in training. Indonesian rice has been the principal target to date. As it expands and is attempted, similar to the first three models, inherent strengths and weakness will be identified.

Farmer participatory research: CGIAR Network

Concerns over the lack of appropriate technology transfer have given rise to a more recent institutional paradigm. Farmer participatory research (FPR) has its origins within the CGIAR system and, from all appearances, has had its greatest impact to date in Asia. The conceptual foundation for FPR is a belated appreciation of mainstream agricultural scientists for the importance of indigenous knowledge to agricultural systems management, especially with resource-poor, marginal farmers.

Fig. 1019   In-field training for rice farmers and support staff at a farmer field school, Indonesia.

Operationally, FPR begins with four fundamental steps: (1) community-based dialogue with individuals or small groups used to identify local changes in agriculture over a generation or two; (2) community-based analysis used to identify high-priority problems that need resolution; (3) community-based inventory of local methods attempted to solve the key agricultural problems; and (4) community-based assessment of how well methods tried in (3) have performed. The purpose of FPR is to initially learn how farmers think and assess agricultural productivity and damage. Only then can new technology be injected in a socially and culturally relevant fashion.

A recent FPR project in Viet Nam illustrates how the approach functions. Vietnamese farmers were invited to test the heuristic , "We do not need to spray insecticides in the first 30 days after transplanting [rice]". Volunteer rice farmers were invited to participate, and each reported results based on individual assessment of perceived damage. Analytical variables used to determine success or failure of the project included farmers' beliefs, intentions, spray frequencies, timing and targets, yields, inputs and other management practices.

Such experiments, if managed properly, usually are inexpensive and easy to conduct, and they facilitate farmer learning by pragmatically "testing" a new idea. The process provides researchers the opportunity to learn how farmers think and perceive success or failure. The FPR approach can be quite expensive if attempted over large areas. The presence of well-educated scientists will affect farmers differentially, and care must be taken to minimize this influence by including farmers at all levels of planning and decision-making. Reductionist scientists often view FRP as scientifically "weak," however, those using this approach must remember that the idea is to evaluate farmer responses and learn how they perceive success and failure. The idea is not to conduct well-controlled, reductionist experiments. Adoption rates for new varieties of plants and improved technology are well-documented in the literature, however, much less is known about the adoption or adaptation of information into farmer decision-making.

Fundamental to FPR is a concerted effort to train local agriculturists to proceed through the above four steps so that, after a few iterations, local farmers and their support network work independently of outside influence. Unlike reductionist science, FPR begins by assuming that indigenous peoples have usable solutions to many agricultural problems or they would not have survived. The effort is to extract those solutions up front and, by doing so,

Fig. 1020 Viet Nam farmer participatory training session on evaluation of insecticides during first 30 days after sowing. Photo provided by Dr. Kong Luen Heong, International rice Research Institute, Los Baños, Philippines.

convince local farmers that they will not be treated simply as passive recipients of modern, science-based information. Rather, they will be partners in the experimentation and delivery phases of agricultural development. There is growing evidence that the most effective extension programs in Land-Grant models also operate this way. Unfortunately, this is not yet the norm.

FPR is a unique approach and quite different from station-based research and traditional extension-delivery systems. It is, in many ways, the antithesis of the Land-Grant model. Unlike farming systems, FPR does not have to "adapt" station-based research because the blend of indigenous knowledge and modern science is inherent in FPR. Unlike the T&V approach, there is no hierarchy involved and no loss of information between tiers. FPR most closely resembles the training-driven research model, but has been tried in more places to date than TDR. There are on-going efforts in Asia, Africa and Latin America that use FPR as the paradigm. In Ecuador, small groups of farmers form committees and decide on a theme they want to research and then conduct that research under the guidance of an extension specialist. Colombia, under the auspices of CIAT, has

adopted the same approach. There is increasing awareness that talking with farmers is an improvement over talking at farmers.

Unlike the Land-Grant, T&V and farming systems models, FPR is a new approach. In essence, it has been developed out of the perception that the other models have failed to resolve real problems with marginal farmers. The concept of formalized farmer field schools appears unique to TDR and FPR, although, Land-Grant extension experts would argue their "turn row conversations" with farmers serve precisely the same purpose. As TDR and FPR gain maturity, the global agricultural community can begin to assess their real impact and their relative value as compared to the other three institutional paradigms.

In agricultural societies of relative wealth where individuals, not social groups, make decisions, and where egalitarian norms prevail, systems resembling, at least in part, the Land-Grant system have prevailed. Land holdings and economies conducive to high-yielding agriculture seem to be requirements for the success of Land-Grant style entomological programs, probably because of the emphasis on the ability of individual farmers to make critical

Characteristics of five institutional models of agricultural (including entomology) research, training and extension

Characteristic	Land-Grant	T&V	Farming systems	Training driven research	Farmer participatory research
Major focus on research	Yes	No	Yes	No	Yes (with an altered definition of research)
Major focus on publication	Yes				
Major focus on extension	Yes	Yes	No	Yes	No
Focus on use of indigenous knowledge	No	No	No	Yes	Yes
Major focus on farmer training	No	Yes	No	Yes	Yes
Targets users with relatively good resources	Yes	No	No	No	No
Targets marginal users	No	Yes	Yes	Yes	Yes
Regiments technical expertise	No	No	No	Yes	Yes
Puts farmers in active role	No	No	No	Yes	Yes
Combines research, training and extension	Yes	No	No	Some	Inherently
Sufficiently tested outside area(s) of origin	No	No	No	No	No
Area(s) targeted to date	USA	Asia; some Africa; some Latin America	Latin America; some Africa	Asia	Asia; some Latin America; some Africa

decisions about how they will manage their resources. The other four models have both been conceptualized and implemented with farmer groups who have virtually none of these characteristics. The latter groups consist of farmers that are relatively poor, often make group or community decisions, typically have poor access to credit and are labor intensive. These models appear to be more successful in places where cultural norms play a stronger role in determining acceptance of new practices than in places where decisions about technological innovation are less subject to social and political control.

By far, the Land-Grant system is the model that adheres most rigorously to reductionist experimentation. Large numbers of farmers under the Land-Grant, T&V, and farming systems approaches are today still treated as passive recipients of information, not as active participants in the identification and generation of new information. The farming systems approach does ask farmers what they think is important as a way to initiate the model. Training-driven research and Farmer Participatory Research truly involve the farmer and the associated support system

as active participants in the process. Both these latter paradigms make use of indigenous knowledge systems and thus strive to have as end products agricultural programs that are blends of traditional approaches and modern science.

All five models have inherent strengths and weaknesses, but only three have been attempted to date on a fairly broad scale. Clearly, the questions asked about insects, the methods chosen for investigation, the methods used in training and the consequences of entomological programs can vary markedly from one model to the next. Recognition of which model predominates in any effort is fundamental to success, and global entomologists need to address, perhaps in a side-by-side comparative fashion, just how transportable information developed under one model is to farmers living under another. This could be one of the reasons that "technology cannot be simply exported." International donor organizations clearly need to be attentive to matching the institutional model being promoted with appropriate cultural, social and economic norms.

An example

How might the use of an exotic biological control agent in a particular crop be implemented under each of the five models? For argument, the crop is soybean and, among its insect pests, is one that pesticides have failed to control. It is the key pest and it limits yield annually. The idea is to try a biological control approach.

Under the land-grant model. Entomologists with biological control expertise, either individually or in small groups, would survey the literature for what is known about biological control of the target pest. Two avenues would be explored initially: the potential use of indigenous natural enemies and the importation of exotic ones. As the indigenous complex was not maintaining the pest's population density below acceptable levels, the choice might be made to search for exotic natural enemies. The USDA/APHIS protocol would be utilized for permits, and travel likely would be to China, the indigenous home of the soybean plant. If the pest has been subjected to biocontrol previously through the efforts of others and the results were published, travel to China might not be necessary.

Collections of potential natural enemies would likely result in the importation of each and every one found to inflict mortality on the pest in China. In quarantine, each potential natural enemy would be checked for hyperparasites, and then reared to sufficient densities to test any deleterious effects a release might have. Later, progeny of the imports would be released and evaluated for effectiveness. Results would be published in the refereed journal literature. At this point, extension agents and farmers would best get access to the information if they had been working in teams of individuals from the various agencies involved. Otherwise, those responsible for implementation would need to be keenly aware of the refereed journal literature.

Practical use of the new imported natural enemy(ies) might demand a mass rearing facility. Thus, further development could occur if, and only if, someone procured the resources for that facility and dedicated it to the rearing and release of natural enemies in soybean. If inoculative releases worked, the problem would be less costly. A key point here is the fragile relationship between research and extension. Researchers perceive that they get "rewarded" for publishing the results, not implementing them.

Extension specialists have to be alert to know the new technology has been developed, especially if they have not been included as part of the overall team. Viewed as too simplistic a solution by some, one way around this dilemma is to have scientists who have formal, evaluated appointments in both research and extension.

Under the T&V model. Once someone else did the research, experts who understood biological control could bring the imported biocontrol agents to a regional training session where extension specialists were being trained. Those extension agents could be shown how to identify the natural enemies, what they did to the pests and how to tell if the biological control worked. They also could be taught how to establish a cottage rearing operation. This process would continue down the training hierarchy, provided there were enough natural enemies to go around and provided the extension specialists had enough formal education to grasp some basic entomology and the concept of biological control. A lot of useful information likely would get lost between training tiers, leaving the ultimate farmer user wondering what all the fuss was about.

Under the farming systems model. The entire issue would likely never even come up unless there was an entomologist or other biological scientist on the rapid rural appraisal team who was aware of the biological control agents that might be tested on local farms. Given the presence of the expertise, biocontrol agents could be imported, released on the farm and evaluated, with the farmers observing. If the biological control was effective, farmers would of course want to adopt it and the entire farming systems effort would have to become more akin to a T&V effort to train the farmers about the conservation of natural enemies and perhaps a cottage rearing and release operation.

Under the training-driven research model. Like farming systems, this model would demand that trainers be up to date on the latest research literature and recent progress to even know there was a new biological control agent that might be used locally. If so, use of the agent in field training exercises would be quick and would not have to pass through as many tiers of training as with the T&V model. The same problem outlined above exists in this model, too. It does little good to educate farmers about a new technique and entice them to adopt it if nobody teaches them how to implement it in a self-sufficient fashion.

Under the farmer participatory research model. If the CGIAR or other staff working with local farmers were able to identify the indigenous knowledge including experiences with natural enemies, there likely would be receptivity to the importation and use of an exotic biocontrol agent. In fact, such indigenous knowledge might even identify a better candidate for use than the one potentially coming from China. If there was no evidence that indigenous knowledge systems reflected previous exposure to natural enemies, training in the concepts and uses of natural enemies would, in essence, have to occur. Techniques to sustain the biocontrol effort, if accepted in local culture, would become part of the training.

Some general conclusions

Institutional models for the research-extension-training activities would appear to best serve and reflect the societies in which they originate. The Land- Grant model likely could never develop in a strong hierarchical society. Much of the success of T&V in Asia and Israel comes from the fact that the organization of T&V "fits" those societies. The farming systems approach or Land-Grant approach do not work well in places where people are simply afraid of, or are unused to, expressing their opinions (e.g., Haiti and Cameroon).

There does appear to be a single model potentially usable in the wide array of situations that exist in today's world: farmer participatory research. All farmers and farming communities, save large industry systems, have an indigenous culture. This is especially true for resource poor farmers anywhere in the world. They have survived to date with, to one extent or another, a blend of traditional approaches and modern science. FPR and, to almost as great an extent, TDR offer what appear to be extremely high probabilities for solutions to agricultural problems that are acceptable to farmers almost by definition, that include farmers in the process and that avoid the inherent loss of information quality that occurs in the transfer of information.

There is debate over the extent to which the Land-Grant system involves farmers as active participants. No doubt, the better extension systems do just that, however, the extent to which farmer involvement permeates the breadth of Land-Grant efforts has not, to our knowledge, been gauged. FPR has been harder to implement in relatively wealthy nations like the USA where farmers see their tax revenues as providing agricultural services and resources they can tap "free of charge" without having to participate themselves.

Very few people outside the USA really understand the role of the extension specialists in the Land-Grant Model. To make this model work as designed, this is a key position, yet, some Land-Grant universities only recently have begun to employ individuals with formal, evaluated appointments in BOTH extension and research. Others utilize the inter-agency committee approach, and still others do not address the problem in any overt fashion. Joint appointments are still the exception, not the rule. The whole idea of state specialists grew up informally in the Land-Grant model as a need, but has over time become detached from formal research. Europeans do not employ extension advisory groups at all.

Three of the aforementioned systems have a rather disappointing track record on environmental issues. The reason is that all three fail to really focus on the farmer and long-term sustainability. It is hard to think of the off-site impacts of items like synthetic, organic pesticides if the farm is not viewed as part of a larger ecological and cultural system. Farming Systems appears to have the worst track record of the group on environmental and pest issues, with T&V a close second. The Land-Grant model is a classic example of how the system can derail on environmental issues (e.g., total commitment to pesticides for almost 25 years). FPR and TDR try overtly to avoid this problem by beginning their paradigm with farmer knowledge. Many modern Land-Grant extension personnel argue that current extension programs involve farmers in active decision-making more now than at any time in history.

Relationships among research, extension and training, independent of institutional models, demand choices be made on various levels. Some cultural, social, and economic systems clearly are more amenable to one choice versus another.

** Public versus private
** Government versus non-government
** Top-down (bureaucratic) versus bottom-up participatory)
** Profit versus nonprofit

** Free versus cost-recovery
** General versus targeted sector
** Multipurpose versus single purpose
** Technology-driven versus need-oriented

The role of publicly funded research and extension is changing. In the U.S., large-scale farmers depend less and less upon information generated and delivered through Land-Grant models. Part of the reason is that more and more of total agricultural productivity comes from corporate, rather than individual farmer, sources. Corporations have their own staff. The agricultural chemical companies provide all sorts of information and technical advice, but are single-minded in approach and profit motivated. Even in developing nations, where farming systems, T&V, FPR and TDR have predominated, they have been sponsored mostly by international donors and have regressed once funding ceased. So, just what is the role for public sponsored research-extension-training? There is a role if such programs will re-define their clientele base from richer corporate-industrial programs to poorer, smaller farming enterprises and beginning producers. With this new clientele definition must come a commitment to long-term sustainability of both the farmer and the farm.

Carl S. Barfield and Marilyn E. Swisher
University of Florida
Gainesville, Florida, USA

References

Barfield, C. S., and M. E. Swisher 1994. Integrated pest management: ready for export? Historical context and internationalization of IPM. *Food Reviews International* 10: 215–267.
Benor, D., and M. Baxter. 1984. *Training and visit extension.* World Bank, Washington, DC. 199 pp.
Shaner, W. W., P. F. Phillips, and W. R. Schmehi 1982. *Farming systems research and development: guidelines for developing countries.* Westview Press, Boulder, Colorado. 405 pp.

TEACHING ENTOMOLOGY: A REVIEW OF TECHNIQUES.

Both positive and negative interactions with insects have long been a part of the human experience. However, people have not always had an opportunity to study insects in any formal, systematic fashion. Insects as biological organisms and various institutions where they can be studied are both part of the discussion on teaching entomology.

With the Neolithic Revolution (ca. 8,000 BCE), humans began the process of sedentary agriculture and civil development. Social scientists have provided us with at least anecdotal information on some of the roles played by insects in the human quest for survival and expansion. Early in human existence, insects vectored diseases, competed for food, consumed structures and were general irritants. They also were subjects of curiosity and sources of food and food products like honey.

Insects are the single largest and most diverse group of living organisms on Earth, and humans begin noticing and interacting with them almost from birth. Insects are among the first living organisms that children notice, and today's elementary education includes a host of activities that center on insects – butterfly gardens, games, music and use of the Internet being but a few examples. Students can choose to study entomology formally as part of their college education. A bit of historical context on the university as an institution where entomology is taught is relevant.

Entomology and higher education

Virtually none of the biological sciences existed as logical, written disciplines until Aristotle, and not until the ancient Greeks did the study of insects become characterized by the delight in observational opportunity insects provided. The 11th Book of the Roman Pliny's *Historia Naturalis* (77 AD) was the most comprehensive treatise on insects to date; however, it contained almost no original observations. Pliny's time and writings marked the beginning of a body of knowledge that would allow university degree programs to arise in the 19th century.

Higher education in Europe, from antiquity until the end of the middle ages, was maintained in the monastic and cathedral schools, enriched by a huge influx of knowledge from the Islamic world. These institutions eventually developed into universities as products of cross-cultural influence. Modern universities thus treasure a multi-cultural faculty and student body and a plethora of teaching styles. Those early European institutions were places where ideas spread, and where local and national authority was engaged. Entomology was not an area of formal study in these early institutions. Insects were engaged as plentiful examples of the biological world and, fairly often, as 'pests'. Indeed, the first real concerns about insects as a group were more in the context of 'sources of products' (e.g., honey) and 'competitors' (e.g., agricultural pests) than just for the need to understand their biology and history. The Catholic Church would often put 'bad' insects on trial and hire them defense attorneys, ultimately 'banishing' them from select provinces. These practices occurred fairly frequently from the middle ages right up to the start of the 19th century. Both the Bible and the Torah refer frequently to insects, paying special attention to the parables that can be extracted from insect life and to whether or not select insects are kosher. Entomology was important because most everyone had personal experiences with insects, not because it was a recognized area of study.

The study of entomology was nested within the general biological sciences in Europe and America throughout the 17th and most of the 18th centuries. Things began to change toward the start of the American Civil War. In 1862, President Abraham Lincoln established the United States Department of Agriculture and set in motion what would ultimately be three key Federal Acts to institutionalize the research and teaching (Morrill Land Grant Act and Hatch Act) and extension (Smith-Lever Act) functions by which all Land-Grant Universities are recognized today. While entomology is taught at many higher education institutions, the vast majority of American entomology majors are housed within colleges of agriculture and life sciences at Land-Grant institutions. Part of the reason for this situation is that, to many in the general public, insects are viewed as pests and colleges of agriculture were developed, at least in part, to help solve the public's pest problems.

After WWII, departments of entomology grew at almost every Land-Grant institution. Today, many of these institutions offer undergraduate degrees in entomology, and most offer graduate degrees. An undergraduate major in entomology, depending on the institution, will take 18–30 hours of entomology courses offered in sequence. Graduate entomology degrees take 30–90 additional hours, the majority of which are insect-related. Graduate degrees usually include a research experience.

Teaching entomology at the university level

The present treatise will focus very little on the content of entomology courses; rather, the focus will be aimed at the style and methods through which any given content may be delivered. Clearly, different styles are applicable to different audiences, and the professor of any single course must understand the background and motivation of his/her audience before the most appropriate style can be selected. Interestingly, this is not how it usually works. Most often, any professor has his/her long-adopted style, and it is left to the clientele (students) to make the adjustments. There are possible consequences of this long-adopted practice, as illustrated in the associated table.

These various professor roles translate in myriad ways to an individual style used in the professor-student classroom experience. The entire point is that there is not a single best way to teach. The capabilities of both professor and students are basic ingredients in the choice of teaching style. An assessment of the students' optimum learning style early in the term is ideal if the professor has both the desire and the capability to shift teaching styles to align with learning styles. Here are a few possibilities:

The Socratic, discussion approach. This approach to teaching entomology is used most often in small classes and can be applicable to either entomology students or general students simply curious about insects. It is commonly used in undergraduate honors classes at large universities, and small colleges and universities often use this approach throughout their curriculum. Students who take these classes typically are interested, motivated, and engaging. They are willing to read and discuss a plethora of historical, social and current issues related to insects and then perhaps, via a series of essays, analyze and critique what they have read. In this context, entomology becomes merely the biological medium in which to have students develop their interpersonal, analytical, critical thinking, oral presentation, and English grammar skills. Such students may become enticed to explore entomology as a major and an ultimate career, but most often these students, if they are interested in the biological sciences at all, are focused on medicine or veterinary medicine.

Success in this approach demands specific criteria on the part of both students and professor. The professor must be knowledgeable of a wide array of entomological issues and literature. Even more important, he/she must be charismatic, enga-

Five models for the presence of an instructor in the classroom: (summarized from Grasha 1966).

	Expert	Formal authority	Personal model	Facilitator	Delegator
Professor Role	Source of knowledge; status as expert; gets students prepared	Status symbol; reinforces correct, acceptable and standard methods; establishes learning goals	Role model; teaches students to emulate him as model	Emphasizes personal student-professor interactions; guides learning; creates independent students; supports and encourages	Focuses on student independence; serves as a resource
Advantages	Copious information; development of skills	Clear expectations; identification of acceptable approaches	Emphasis on direct observation and following the role model	Personal flexibility; focus on students' needs and goals; open-mindedness	Students become independent learners
Disadvantages	Volume of information; students may be intimidated; lack of focus on underlying processes that led to information	Rigidity in approach to students	Feeling of inadequacy if students feel they cannot live up to model	Time-consuming; often lacks positive reinforcement	Student anxiety; student may be misperceived as ready for independent work

ging and able to entice bright students to develop curiosity about entomology. The professor must be able to write and speak well, as these students will discern quickly between what is said and what is done in critique of written and oral presentation performances. The students must be engaging and willing to participate in class discussions. The Socratic Approach is, in many ways, the antithesis of more traditional presentation of copious facts about insect morphology, physiology, behavior and control. The idea is to entice students to explore, on their own, issues presented incompletely in discussion periods, thus developing skills that will serve them well during their university tenure. This approach will simply not work if either the students or the professor are not engaging or if the class resorts to a host of information-laden slides, films, blackboard text or, more frequently now, Power Point or Internet presentations.

Topics typical of this style of teaching may include many that do not appear, on the surface, to have anything to do with entomology. However, discovering their links to entomology is part of the desired outcome. For example:

1. Pliny's *Historia naturalis*
2. The Idea of "Food Insurance"
3. Social attitudes on pesticides versus pharmaceuticals
4. What was The Green Revolution?
5. In your supermarket, what foods are indigenous to the contiguous 48 states?
6. Medieval higher education?
7. Colonial American living?
8. What is social Darwinism?
9. Who is Norman Borlaug and what did he do?
10. What is the role of empiricism in science?

For example, in a classroom of 20 or so students, the professor initiates a discussion about the diet of Colonial Americans. Initially, the professor seeks to understand what the students already know about this subject. He discovers this by asking them and expecting them to engage in dialogue with him and with each other. The conversation expands to include the crops grown and the problems the colonists had with food production. Analogies are drawn periodically to modern agriculture and its problems. As some of the problems were insect related, the dialogue eventually permits the students to discover the linkages between Colonial American diet and the study of insects, as well as the various means of controlling those insects under social and technological conditions existing in colonial times. The result is a group of students who are better able to place insects and entomology into appropriate historical context and then draw inferences to the importance of entomology today.

This teaching style assuredly contributes to the student's knowledge of biological science and thus meets both professorial and student objectives in this context. There is a decent chance this approach will entice such students to become interested in entomology as a major and a career − something the vast majority of them have never before considered. Many will not even know it is possible to major in entomology.

The lecture/lab approach. By far, the most common way entomology is taught in modern universities is as a 3 or 4 credit hour course. This involves a series of lectures (2–3 per week) and 1–2 laboratory sessions per week. Lectures are used to present facts about entomology, and the lab is experientially based and involves dissection and study of the various physical and physiological features of insects. Interestingly, what actually occurs during the lecture period can follow any of the teaching styles presented in this paper; however, by far the most common is using the professor as the expert whose role it is to pass entomological facts and knowledge on to the students. The labs, if scheduled as typically desired, mirror the lecture periods and provide visual, hands-on experience for the topics discussed that week during lecture.

Lecture approaches vary from professor to professor, but there is concern under this model to ensure sufficient information is imparted to justify the student moving on to the next course in the sequence. General undergraduates who simply want biological science credits do take these type courses just because they want to learn about organisms they have experienced daily. It is the entomology majors, however, who professors feel the need to get ready for continued study of entomology. Lectures are replete with facts, drawings, information; labs follow

suit. The accompanying diagrams illustrate typical information presented under this style.

The emphasis is often on learning facts about insects. Professors pursue these facts in various ways. Some choose a systematic approach that focuses on the principal orders and a comparison of their morphologies, physiologies, behaviors and control. Others use a type of comparative biology approach that may compare insect adaptations with those of other animals. While various entomological issues may be discussed, students spend time hearing the expert impart expertise, then committing that expertise to memory. Labs will have students dissecting, searching for, and identifying these morphological structures, and a test may include a lab practical where students must identify parts labeled. This is precisely the approach used in Colleges of Medicine.

Entomology lecture/lab courses do not have to unfold in any pre-determined fashion, and many professors use creativity in what they impart as entomological knowledge. The point here, however, is that introductory entomology courses are typically taught for entomology majors or at least biological science majors, not for non-science students, despite the fact that diverse students often enroll. Under this approach, more emphasis is placed on facts about insects than on issues associated with insects. This is not a criticism, but rather an observation.

Introductory entomology textbooks offer a range of approaches to teaching the subject. Most will begin with chapters that discuss the general importance of insects to science and to humans. From there, the authors' approaches vary greatly. Likely, the most common approach is to begin immediately with external and internal anatomy and physiology. Sensory systems, reproduction, development, life history, and systematics follow. Then, insects as part of special habitats (e.g., aquatic and terrestrial systems) are presented, followed by subjects like insect societies, insects and plants, predation, parasitism, medically important insects, insect pest management and collection/preservation methods. Several texts will place collection and preservation methods, plus insect systematics, immediately following introductory chapters. In any case, most existing entomology texts will cover the above subjects, independent of which order of presentation is chosen.

The teaching approach used is tied more to the bent and capability of the individual professor than

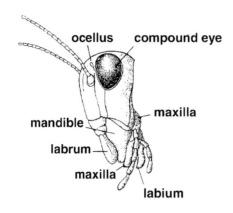

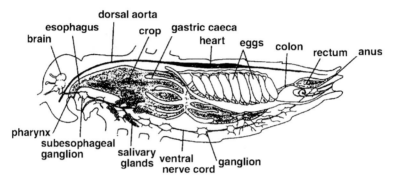

Fig. 1021   Elements of insect morphology and anatomy.

to the subject matter. One teaching strategy is more of a comparative biology approach that might compare, for example, the insect leg and locomotion to analogous structures and functions in mammals, birds, reptiles, etc. This approach is effective in placing insects and their adaptations in perspective among all animals and is depicted in accompanying illustrations.

Another strategic approach is to compare insects among the various orders and explore metamorphosis, wing type, locomotion, habitat, life history, etc. as a means to compare-and-contrast the different orders of insects. This approach is illustrated in the accompanying pictures.

There is no shortage of information (journals, books, Internet sources) from which to supplement the basic text to expand student knowledge of insects both as biological organisms and as social entities interacting with humans. Indeed, social scientists (especially anthropologists) often use insects to illustrate daily life, problems and indigenous approaches indicative of specific cultures. Some instructors of basic entomology will include sections on indigenous cultures and the role insects played therein.

Comparison of five entomology teaching styles

	Socratic	Lecture/lab	Land lab	FFS	Internship
Experientially based	NO	NO	YES	YES	YES
Focus on insect facts	NO	YES	YES (limited)	YES (limited)	YES (limited)
Focus on student analytical skills	YES	NO	YES	YES	YES
Demands large resources	NO	NO	YES	YES	NO
Teacher focused	NO	YES	NO	NO	NO
Student focused	YES	NO	YES	YES	YES
Ideally requires supplemental funding	NO	NO	YES	YES	YES

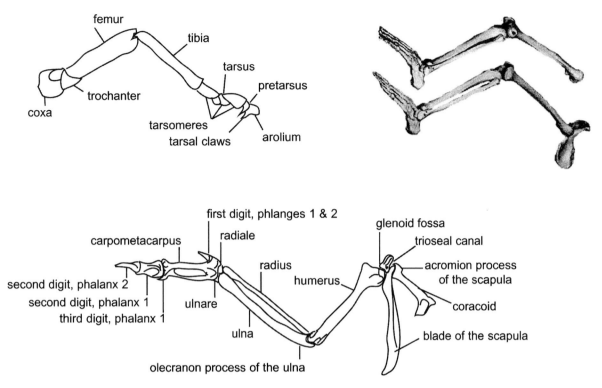

Fig. 1022 Comparative biology approach to teaching insect leg morphology and its analogous structure and function in other animals. Human leg unlabeled; an exercise might be for students to locate analogous structures on human anatomy.

Increasingly, entomology departments are offering courses to non-science majors. This is both as a way to educate undergraduates to the possibility of entomology as a major and as a mechanism to capitalize on the general student's previous experience with insects and their required enrollment in a specific number of science courses. Perusal of university catalogs usually will reveal courses such as: The Insects, Bugs and People, Pesticides and Pills, Ecology and Human Intervention, Insects and Society, The Insect World and You. These type courses typically are taught in one of two diverse formats. First, they frequently appear as honors courses taught to small-sized classes consisting of very intelligent students. In this context, the aforementioned Socratic style is most prevalent. Such students are often searching for creative majors, as their academic credentials permit them to select most any major they desire. The second prevalent delivery format is a lecture, replete with detailed syllabus and copious illustrations, to a large class. This delivery style is the aforementioned lecture/lab, but without the lab portion in most cases. Most every academic unit, including entomology, in

a large university now offers such courses for non-majors. Strategically, these courses expose searching students to possibilities of a major they had not theretofore considered. Further, these courses allow students more flexibility in fulfilling general education requirements.

Internships. Most professional disciplines in the biological sciences require extensive internship experiences prior to certification. One cannot imagine a licensed Medical Doctor or Veterinarian without such experiences before they are sanctioned to ply their trade. The basic biological sciences in general, and entomology as one example therein, do not typically make such requirements. However, those parts of entomology related to agricultural plant protection often do require internships as part of graduation requirements. While wonderful in their intent and educational possibilities, internships are not without their potential problems.

An internship typically occurs toward the end of an undergraduate degree program. The idea is to match a student's need for hands on experience with

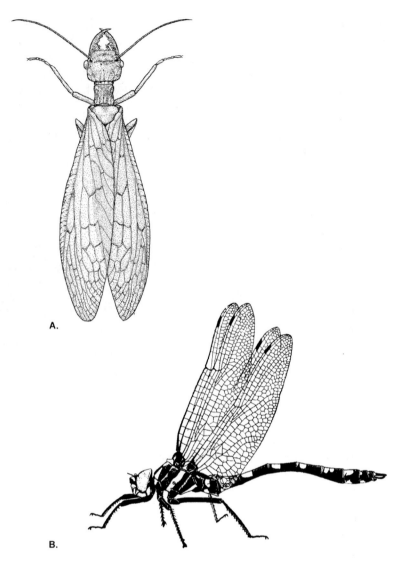

Fig. 1023 Neopterous (A) versus Paleopterous (B) wing types characteristic of different orders of insects.

a public or private sector business. The student would typically leave the university and locate at the business site for 1-2 semesters. Under the supervision of the business contact, the student would work daily in several of the tasks in which the business is involved. This might include plot preparation, fumigation, sampling, pest control, assessment of pest damage, interaction with the public, identification of insect pests, bookkeeping, and seminars. The entire experience would be aimed at permitting the student to translate academic knowledge obtained to marketplace demands. The university professor in

charge of internships would make periodic visits to assess the student's progress and to witness first-hand the range of activities in which the student was involved. The student would register for a university course and pay associated tuition, and he/she may or may not receive salary compensation for the work.

Matching undergraduates with appropriate public or private internship opportunities often means asking them to relocate or commute for 1-2 semesters during their undergraduate studies. Many students are reticent to encumber the additional travel and per diem costs. Parents express similar concerns. If an internship is a requisite for graduation, care must be taken to ensure the intern's experience does not resort to a more trivial exercise based on student economics rather than on learning needs. Care also must be taken to ensure the business contact understands the needs of the student intern and is not simply looking for a short-term source of cheap labor. Medical and Veterinary interns accept the requisite; however, they are older, graduate students, away from parental influence and knowledgeable that the economic pay-offs of certification are potentially great. Often, none of these issues are the same for undergraduates. The more successful undergraduate intern programs provide financial support, if justified, to offset additional costs to students.

Field-based courses. This section refers to courses taught totally in the field, not to courses where sporadic field trips are part of a lecture/lab style. If field-based courses are targeting agricultural entomology, they require facilities: land, equipment, etc. There is debate over the cost needs for such facilities. Some feel adequate field-based courses are limited only by the creative imagination of professors teaching them. Others feel adequate land, equipment and staff are crucial to the success of such courses. As universities often do not consider such teaching resources high-priorities, these courses often piggyback onto research efforts, using plots developed for some professor's research. This can work, but often encounters problems as the timing associated with teaching conflicts with the timing of needed research activities.

Many universities have natural areas that can, within certain guidelines, be used to teach various aspects of entomology. In any case, the fundamental idea in a field-based course is to provide students with experiential learning. Under ideal conditions, such a course would be offered on facilities set aside exclusively for teaching and would be under the control of teachers so that the timing of activities could occur in an orderly fashion consistent with educational paradigms.

Also known as the Land-Laboratory Approach, this style is somewhat analogous to a teaching hospital where doctors and nurses get experiential learning under real situations – all supervised by professorial experts. Where else is it possible to give developing professionals the opportunity to gain practical experience in a supervised learning environment? Many educators feel such practical experience should be a mandatory part of science education.

Fig. 1024 Zamorano students in land-laboratory exercise, Pan American School of Agriculture, Honduras, Central America.

Joliet Junior College in Joliet, Illinois, exemplifies the intent and structure of the Land-Laboratory concept. As stated in their advertisements, they exist to "serve as a teaching tool for instructors to use hands-on learning as a means to reinforce classroom instruction." The Land-Laboratory exists as 101 acres of land donated for the exclusive purpose of teaching. The Pennsylvania State University has a similar facility dedicated to land analysis. In most cases, Land-Laboratories exist because the land and perhaps the infrastructure needed to make it operational were donated. Rarely do universities encumber the costs of a Land-Laboratory from their operational budgets.

The Escuela Agrícola Panamericana (Pan American School of Agriculture, EAP) in Honduras, Central America, provides an example of both experiential learning and the Land-Laboratory style. Students attend from throughout the Caribbean Basin and beyond and spend 0.5 day in class and 0.5 day in the field in a wide array of agricultural experiential learning activities including land preparation, planting, harvesting, processing, marketing, plant protection, outreach and with both plant and animal systems. Entomology plays a prominent role in the experiences of the "Zamoranos".

Teaching at the International Rice Research Institute (Philippines) exemplifies the same approach and has been in place much longer than the EAP. The general trend is for international research institutes that also teach to place a higher premium on Land-Laboratory approaches than do U.S. universities. This is true likely because they have the land and labor to make this approach work and because they understand that hands-on experience for their student clientele is of maximum importance.

Extra-university models. The international arena offers a variety of teaching approaches aimed directly at farmers and professionals who provide technical support to farmers. In this context, the stakes (possible food self-sufficiency) are much higher than the pursuit of an undergraduate college degree. These programs offer both challenge and opportunity to university-based degree programs in entomology.

The farmer field school approach. The principal sponsor and promoter of this approach is the United Nations Food and Agriculture Organization, Rome, Italy. Principally, the Global Integrated Pest Manage-

ment Group within the UN/FAO developed this model based on experiences in Rice Integrated Pest Management programs in Indonesia. This approach serves as a model that universities might want to emulate in those aspects of entomological education that demand experiential learning.

The FFS was established to provide training to farmers and their support network (researchers and extension specialists) on the problems and opportunities associated with specific crops. The idea is to follow crop development from land preparation to marketing of the harvest, teaching at every step the techniques demanded, the pests encountered, the sampling needed, the pest management approaches needed, etc. Those same trained people obtain the multiplier effect by training others. The idea is to teach, in a hands-on fashion, the details for any specific crop. As the crops change, so do the issues and thus additional training may be needed. In 2002, FAO is assisting Korea in the establishment of a junior college level curriculum of study based exactly on the FFS concepts. Students would hold night sessions on the more general aspects of, say, entomology. Unlike the lecture/lab approach typical of American Universities, the FFS approach begins with the field and then moves to general classroom experience.

Fig. 1025   Farmer Field School training in corn-based agriculture, Honduras.

The FFS approach served as the basis for development of The Farmer Centered Agricultural Resources Management (FARM) Program that, until funding declined, consisted of the member countries of China, India, Indonesia, Nepal, Philippines, Sri Lanka, Thailand and Viet Nam. The entire FARM program was supported by the UNDP and implemented by FAO. In 2002, there were 19 sites in 8 nations covering some 10,000 Asian households. FARM, and its parent idea, the FFS, was one of the most rapidly expanding teaching resources in the world. It offered practical hands-on training that could be supplemented with more traditional academic education in a wide array of subjects, including entomology. Sadly, politics overtook needs and funding was cut to the point the program vanished in its original form.

The purpose of the FFS is not pursuit of academic degrees. Training is focused on the specificities of individual crops or cropping systems and is increasingly supplemented with "after hours" academic work on subjects like entomology. Students are adults, not undergraduates in pursuit of degree requirements. Results of this educational style are improved chances of food self-sufficiency, not a B.S. degree and employment. However, in the few places where the fundamental concepts of the FFS or its close relative, the Land-Laboratory, have been directly combined with academic programs, the results have been astonishing. These are the rare situations that make agricultural and entomological education the most analogous to professional school education in the USA.

Cross style comparisons

Three things limit what style is used to teach entomology: (1) imagination, (2) budget, and (3) either professor or student capability. The most appropriate criterion for a comparison among styles is whether or not students achieved course objectives. The reality is that various styles have different intents and expected outcomes. The problem occurs when a style of teaching is used that has a low probability of generating the desired outcomes.

Use of the lecture/lab approach where the lecture is used to present copious facts about insects which are to be memorized and repeated on a multiple choice exam may not be the most appropriate for a group of students who have selected beginning entomology for its general education function and who are simply curious about insects. Not providing the

Fig. 1026 Farmer-field school classroom training, Indonesia.

future field practitioner of agricultural entomology a field-based experience will not achieve desired outcomes in the most efficient manner.

If the professor desires students to become independent thinkers, capable of analysis and presentation, playing the role of "expert" who imparts facts about insects that are to be learned and demonstrated on an exam is probably not the most ideal approach. Tailoring teaching style to a particular student audience is hard, not easy. It takes effort and does not occur automatically. Just like any other aspect of the human experience, some professors can do it and others cannot. This is not a criticism, but a fact. Ideal learning experiences tend to occur in entomology or any other subject when care is taken to match the style of teaching to the needs and learning styles of students in that class that term. Adoption of a single teaching style to be used throughout one's

Fig. 1027   Farmer Field School plot facilities for rice-based agriculture, Indonesia.

Fig. 1028   In-field training on rice insects, Indonesia.

tenure is, quite literally, a ''hit and miss'' proposition where desirable and less desirable outcomes will occur among semesters.

The key to any professor adjusting to dynamic learning styles of students is an ability to gauge quickly their response to a variety of styles. Literally, what works well one term and gets students involved and responsive is not automatically what will work best next term, given a new group of students. Many, if not most, university professors of basic entomology courses offered to entomology or science majors, do not attempt to adjust to perceived student learning styles. Such adjustments are much more common in entomology courses offered to non-science majors.

Carl S. Barfield
University of Florida
Gainesville, Florida, USA

References

Evans, E. P. 1998. *The criminal prosecution and capital punishment of animals.* Richard Clay, Ltd., Bungay, Suffolk, United Kingdom. 336 pp.

Grasha, A. 1996. *Teaching with style.* Alliance Publishers, Pittsburgh, Pennsylvania.

Smith, R. F., T. E. Mittler, and C. N. Smith (eds.). 1973. *History of entomology.* Annual Reviews, Inc., Palo Alto, California. 517 pp.

TEAK MOTHS (LEPIDOPTERA: HYBLAEI-DAE).

Teak moths, family Hyblaeidae, are a small tropical family of 18 species, mostly Indo-Australian and in the genus *Hyblaea* (one pantropical species is also established in southern Florida). The family is in the superfamily Pyraloidea in the section Tineina, subsection Tineina, of the division Ditrysia (sometimes placed in its own superfamily, Hyblaeoidea). Adults medium size (25 to 49 mm wingspan), with head relatively smooth-scaled; haustellum naked; labial palpi porrect and prominent; maxillary palpi 3 to 4-segmented. Maculation mostly shades of brown, with colorful spotted hindwings. Bodies are usually robust. Adults are diurnal or perhaps crepuscular. Larvae are leaf rollers. Host plants are in Bignoniaceae and Verbenaceae. One economic species: the teak leafroller. Due to their robust form, these moths were often associated with Noctuidae in the past.

John B. Heppner
Florida State Collection of Arthropods
Gainesville, Florida USA

Fig. 1029 Example of teak moths (Hyblaeidae), *Hyblaea puera* (Cramer) from Taiwan.

References

Common, I. F. B. 1990. Superfamily Hyblaeoidea. pp. 334–337 In I. F. B. Common, *Moths of Australia.* Melbourne University Press, Melbourne.

Dalla Torre, K. W. von. 1928. Die Hyblaeinen (Noktuiden). *Entomologische Jahrbücher* 37: 162–164.

Koning, H. S. de, and W. Roepke. 1949. Remarks on the morphology of the teak moth, *Hyblaea puera* Cr. (Lep. Hyblaeidae). *Treubia* 20: 25–30.

Singh, B. 1956. Description and systematic position of larvae and pupa of the teak defoliator, *Hyblaea puera* Cramer (Insecta, Lepidoptera, Hyblaeidae) *Indian Forest Records* (n.s.). *Entomology* 9: 1–16.

Viette, P. E. L. 1962. Les Noctuidae Hyblaeinae de Madagascar (Lep.). *Bulletin Mensuel de la Société Linneenne de Lyon,* 30: 191–194.

TECHNICAL GRADE.

A chemically pure preparation. This is often used to describe research-grade pesticides as opposed to commercial formulations.

TEGMEN.

(pl., tegmina) The thickened front wing of Orthoptera and related insects.
See also, WINGS OF INSECTS.

TEGULA.

A small flap or lobe at the anterior edge of the forewing of some insects. It is also known as the patagium.

TELEGEUSIDAE.

A family of beetles (order Coleoptera). They commonly are known as long-lipped beetles. See also, BEETLES.

TELEPHONE-POLE BEETLES. Members of the family Micromalthidae (order Coleoptera). See also, BEETLES.

TELOGANELLIDAE. A family of mayflies (order Ephemeroptera). See also, MAYFLIES.

TELOGANODIDAE. A family of mayflies (order Ephemeroptera). See also, MAYFLIES.

TELOMERE. Telomeres are the physical ends of eukaryotic chromosomes. They protect the ends of chromosomes and confer stability. Telomeres consist of simple DNA repeats and the non-histone proteins that bind specifically to those sequences.

TELOPODITE. The portion of the limb beyond the base; the shaft.

TELMOPHAGES. Arthropods that feed at blood vessels, and specifically from pools of blood created by lacerating vessels. (contrast with solenophages)

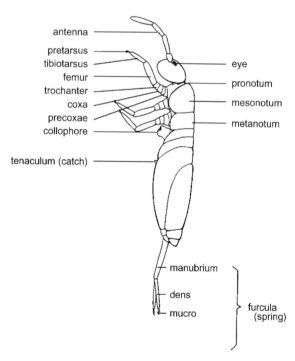

Fig. 1030 Lateral view of a springtail (Collembola).

TEMPLATE. A macromolecular mold for synthesis of another macromolecule. Duplication of the template takes two steps; a single strand of DNA serves as the template for a complementary strand of DNA or mRNA.

TENACULUM. In Collembola, a small structure on the third abdominal segment that serves as a clasp for the furcula.
    See also, ABDOMEN OF HEXAPODS.

TENEBRIONIDAE. A family of beetles (order Coleoptera). They commonly are known as darkling beetles. See also, BEETLES.

TENERAL. The condition of an insect after molting but before the new cuticle has hardened.

TENT CATERPILLARS. Some members of the family Lasiocampidae (order Lepidoptera). See also, LAPPET MOTHS, BUTTERFLIES AND MOTHS.

TENTHREDINIDAE. A family of sawflies (order Hymenoptera, suborder Symphyta). They commonly are known as common sawflies. See also, WASPS, ANTS, BEES, AND SAWFLIES.

TENTORIUM. The internal invaginations of the exoskeleton occurring in the head tergite.
    See also, HEAD OF HEXAPODS.

TEPHRITIDAE. A family of flies (order Diptera). They commonly are known as fruit flies (but not to be confused with small fruit flies, Drosophilidae) See also, FLIES.

TERATEMBIIDAE. A family of web-spinners (order Embiidina). See also, WEB-SPINNERS.

TERATOMYZIDAE. A family of flies (order Diptera). See also, FLIES.

TERGUM. The dorsal section of a body segment. Also called tergite. They sometimes are named after their body segment (e.g., mesotergum is the tergum of the mesothoracic segment).

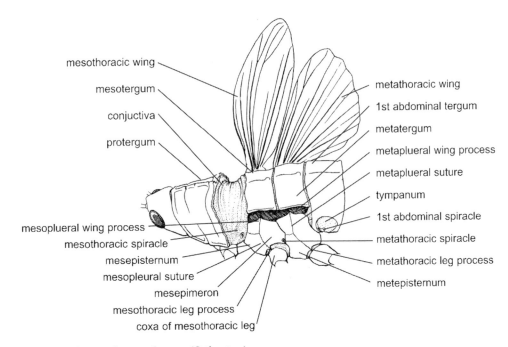

Fig. 1031 Head and thorax of a grasshopper (Orthoptera).

TERMEN. In Lepidoptera, the edge of the wing in the distal (lateral) area.

TERMINAL ARBORIZATION. An extensive branching of the dendrites at the end of a nerve cell.

TERMINAL FILAMENT. Long, slender projections from the last abdominal segment.

TERMITAPHIDIDAE. A family of bugs (order Hemiptera, suborder Pentamorpha). See also, BUGS.

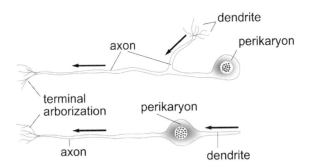

Fig. 1032 Diagrams of insect nerve cells showing direction of nervous impulse (adapted from Chapman, The insects: structure and function).

TERMITARIUM. A termite nest, or the artificial net used to house termites in a laboratory.

TERMITES (ISOPTERA). Termites are small to medium-sized orthopteroid insects that are cryptic in habit. All species live in eusocial colonies and feed primarily on cellulose. Although referred to in older literature as "white ants", termites are unrelated to ants. The name of the order is derived from the Greek words "iso" (equal) and "ptero" (wing) that describe the similar length and shape of both the fore and hind wings of the reproductive alates. Mature colonies are composed of task-specific castes that typically include one or more pairs of reproductives, about 0 to 25% soldiers, and a majority of immature or sterile workers. During part of the year, colonies may also contain some maturing or fully winged reproductives (alates, imagos) destined to leave their colony in brief, but often intense, dispersal flights. The order Isoptera is divided into seven families. The family Rhinotermitidae is divided into seven small and closely allied subfamilies and the Termitidae into four large and diverse subfamilies. By historical convention, all but 2% of termite genera end in the suffix "*termes*", the Latin word for termite.

ORDER: Isoptera, the termites	No. genera	No. species
Family: Hodotermitidae, Old World Harvester Termites	3	19
Family: Kalotermitidae, Drywood and Dampwood Termites	21	452
Family: Mastotermitidae, Giant Northern Australian Termite	1	1
Family: Rhinotermitidae, Lower Subterranean Termites	14	350
Subfamily: Coptotermitinae	1	79
Subfamily: Heterotermitinae	3	143
Subfamily: Prorhinotermitinae	1	18
Subfamily: Psammotermitinae	1	8
Subfamily: Rhinotermitinae	7	56
Subfamily: Stylotermitinae	1	43
Subfamily: Termitogetoninae	1	3
Family: Serritermitidae	2	2
Family: Termitidae, Higher Termites	240	2021
Subfamily: Apicotermitinae, Soldierless Termites	42	215
Subfamily: Macrotermitinae, Fungus-Growing Termites	14	365
Subfamily: Nasutitermitinae, Nasutiform Termites	93	674
Subfamily: Termitinae	91	771
Family: Termopsidae, Primitive Dampwood or Rottenwood Termites	5	21
TOTAL	286	2870

Phylogeny and fossils

Termites are closely related to cockroaches and mantids. Based on molecular and morphological characters, the Mastotermitidae are the most primitive termite family, followed by the Termopsidae, Hodotermitidae, Kalotermitidae, Serritermitidae, Rhinotermitidae, and finally the "higher" termites of the family Termitidae. The earliest fossil records of termites are from the Cretaceous period (144–65 million years before present) and include the families Mastotermitidae, Hodotermitidae, and Termopsidae. Cretaceous fossils consist of wings or imago bodies in sedimentary deposits or amber. Excellent soldier and alate fossils have been found for the Mastotermitidae, Kalotermitidae, Rhinotermitidae, and Nasutitermitinae in Dominican amber from the Oligocene to early Miocene (20–30 million YBP). Over one hundred extinct termite species have been identified worldwide except from Africa.

Description of important taxa

The family Mastotermitidae consists of a single primitive species, *Mastotermes darwiniensis*, a large subterranean nester that attacks sound wood in

Occurrence of termite families and subfamilies of Termitidae by zoogeographic region

Family	Region							
	Australasian	Ethiopian	Oriental	Malagasy	Nearctic	Neotropical	Palaearctic	Papuan
Hodotermitidae		X					X	
Kalotermitidae	X	X	X	X	X	X	X	X
Mastotermidae	X							
Rhinotermitidae	X	X	X	X	X	X	X	X
Serritermitidae						X		
Termopsidae	X	X	X		X	X	X	
Termitidae	X	X	X	X	X	X	X	X
SUBFAMILY								
Apicotermitinae		X	X	X	X	X		
Macrotermitinae		X	X	X				
Nasutitermitinae	X	X	X	X	X	X		X

northern Australia. The alates of *M. darwiniensis* have some roach-like characters and queens lay eggs in batches that resemble cockroach ootheca. The Termopsidae are large to very large primitive wood-nesting termites found in damp logs in temperate forests of both hemispheres. The Hodotermitidae are an Old World family consisting of large termites that nest in the soil and openly forage in grasslands and savannas for herbaceous growth. Unlike other termites, the eyes of Hodotermitidae workers are rather well developed. The Kalotermitidae include wood-dwelling species found throughout the tropics and subtropics. Although some species require wood with moderate to high moisture content, other kalotermitids thrive in low moisture conditions and are called drywood termites. Drywood termites produce characteristic six-sided fecal pellets that are stored in galleries and periodically ejected from the wood through surface holes. One drywood species in particular, *Cryptotermes brevis*, is a worldwide pest known only from structural lumber and furniture.

The family Rhinotermitidae is worldwide in distribution and contains primarily wood-feeding subterranean nesters and includes some of the most important pest genera attacking buildings: *Coptotermes* (Coptotermitinae), *Heterotermes*, and *Reticulitermes* (Heterotermitinae). A few *Coptotermes* are known to build mounds in Australia. The Prorhinotermitinae (*Prorhinotermes*) typically locate their nests inside and under damp logs near oceanic coastlines. The Psammotermitinae constitute one genus (*Psammotermes*) of subterranean termites that occurs in arid lands of Africa and the Middle East. Some genera of the Rhinotermitinae, such as the neotropical *Rhinotermes* and the paleotropical *Schedorhinotermes* have dimorphic soldiers, one with large mandibles and the other with an extended labrum, that bear little or no resemblance to one another. The monogeneric subfamilies Stylotermitinae and Termitogetoninae are limited to central and Southeast Asia, respectively. The Serritermitidae are an obscure family known only from a few collections of two species in Brazil.

By far the largest termite family is the Termitidae inclusive of all evolutionarily advanced or ''higher termites''. The soldierless subfamily Apicotermitinae includes grass feeders and humivores. Most apicotermitine species live in diffuse underground gallery systems but some African genera, such as *Apicotermes*, build elaborate geometrically proportioned nests. The Macrotermitinae are an Old World subfamily of large mound-building termites. These medium to very large termites eat fungus grown on special combs in their nest. Many species provision the fungus with herbaceous vegetation that is gathered by workers from forests, grasslands, and agroecosystems. The massive eight meter-high chimney mounds of *Macrotermes* have classically defined the landscape of the African savanna. The Macrotermitinae, which also include the diverse genera *Microtermes* and *Odontotermes*, range east to the Philippines. The Nasutitermitinae are a tropicopolitan subfamily of arboreal, mound, and subterranean nesting termites that are characterized by long-snouted soldiers. The snout, or nasus, is the armature from which a volatile and sticky secretion is exuded as a defensive mechanism. Some Nasutitermitinae, primarily in the genus *Nasutitermes*, feed on wood and can be serious pests of structures, while others are humivores. In some Neotropical genera, e.g., *Armitermes*, the soldiers have both a nasus and long sickle-shaped mandibles. The subfamily Termitinae is the most diverse of all the termitid subfamilies. Included is the genus *Cubitermes* of sub-Saharan Africa which is characterized by mushroom-shaped mounds. The pantropical *Termes* and related genera have soldiers with snapping mandibles and arboreal or rotten log nests. The *Amitermes* and related genera like *Microcerotermes* occur in the tropics and subtropics worldwide where they feed on wood, grasses, and cellulosic debris. *Amitermes* is subterranean nester in the Nearctic Region while species from northern Australia build large epigeal mounds. *Microcerotermes* tend to be arboreal or subterranean in nesting habit.

Diversity and distribution

Almost 2,900 termite species in 286 genera have been described and it is likely that more than 4,000 species may ultimately be recorded. Termites are found in all zoogeographic regions of the world and many oceanic islands between latitudes of about 50 degrees north (southern British Columbia) and 45 degrees south (southern Chile and South Island, New Zealand). The greatest diversity, measured by numbers of genera and species, occurs in the tropical non-arid regions of Africa, Asia, and the Americas. Some regions show remarkable speciation by single genera such as the *Cryptotermes* from the circum-

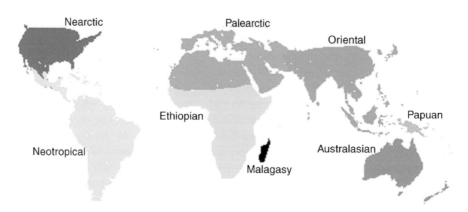

Fig. 1033 Worldwide occurrence of termites in the major zoogeographic regions. Unshaded areas (northernmost North America, Eurasia, southernmost South America) do not support termites.

Caribbean with 25 species, the *Amitermes* of Australia with 61 species, the *Nasutitermes* of the Neotropics with 72 species, and the *Odontotermes* of tropical Asia with nearly 100 species. Certain genera, such as *Reticulitermes* (Rhinotermitidae) and all those of the Termopsidae are characteristic elements of more temperate climates. In turn, *Heterotermes* and *Neotermes*, respectively, occupy similar niches in the tropics. Numerous genera are found in all tropical regions of the world. These include the *Neotermes*, *Glyptotermes*, *Kalotermes*, *Procryptotermes*, *Cryptotermes* (Kalotermitidae), *Coptotermes*, *Heterotermes*, and *Prorhinotermes* (Rhinotermitidae), and *Anoplotermes*, *Microcerotermes*, *Amitermes*, *Termes*, and *Nasutitermes* (Termitidae).

International trade has facilitated the spread of "weedy" species from native habitats across natural barriers to non-native locations worldwide. Human transport, especially by boat, has facilitated the spread and establishment of about a dozen major pest species. Among them are the drywood termites (Kalotermitidae) *Cr. brevis* (West Indian drywood termite), *Cr. havilandi*, *Cr. dudleyi*, *Cr. domesticus*, *Cr. cynocephalus*, and *Incisitermes minor* (Western drywood termite). Invasive species among the lower subterranean termites include *Coptotermes formosanus* (Formosan subterranean termite), *Co. gestroi*, and *Reticulitermes flavipes* (Eastern subterranean termite). Among the Termitidae, only one species, *Nasutitermes costalis*, has become established in a non-native habitat.

Morphology

All termite castes are soft-bodied. Only the soldier head-capsules, the heads and bodies of some imagos, and the mandibles of all castes have moderate to heavy sclerotization. Termite workers and imagos have chewing mouthparts, filiform or moniliform antennae, and a shield-like pronotum behind the head. Soldier mandibles are highly developed to atrophied and variable in shape depending on the defenses employed. Important morphological characters used to classify termites include wing venation and mandible dentition in imagos, head capsule and mandible shape in soldiers, and gut configuration and mandible dentition in workers. Alates have compound eyes and ocelli while eyes of other castes are reduced to spots or are completely absent. The tarsi of most termites are not well suited for climbing

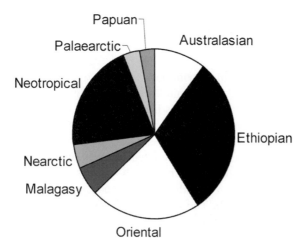

Fig. 1034 Proportion of termite genera found in the eight zoogeographic regions.

smooth surfaces. Worker and soldier bodies are usually light colored but some have darker pigmentation. Alate bodies and wings range in color from pale yellow to black. Soldier head capsules range in color from light yellowish to orange, reddish-brown, or black. Among the smallest of termite soldiers is *Atlantitermes snyderi* (Nasutitermitinae) from Trinidad and Guyana with a total length of 2.5 mm while one of the largest is *Zootermopsis laticeps* (Termopsidae) from Arizona and Mexico at 22 mm long. The largest termite alates are the African *Macrotermes* measuring up to 45 mm in length with wings while the smallest is *Serritermes serrifer* (Serritermitidae) at 6 mm with wings. Alates of some *Incisitermes* and *Glyptotermes* (Kalotermitidae) and Apicotermitinae are less than 7 mm long with wings.

Life cycle and behavior

The inception of a termite colony occurs when a male and female alate pair after a dispersal flight. Flights are typically associated with weather conditions superimposed over annual cycles. Rainfall is a prerequisite for flights by many subterranean species. The beginning of a rainy season coincides with flights by many species. After separation from their wings along specialized suture lines, the reproductive pair runs in tandem until a suitable nesting site is encountered and selected. Once in the protected site, e.g., a crevice in a branch, a tree hole, or underneath a log or stone, the new king and queen build a small nuptial chamber in which the first eggs are deposited and hatch into larvae. The first batch of developing workers and first soldier are cared for by the

Fig. 1035 Termite castes. A. Eggs and larvae of *Cryptotermes brevis* (Kalotermitidae), Florida. B. Soldier (L), worker, nymph, and alate of *Amitermes floridensis* (Termitinae), Florida. C. King (top) and queen of *Incisitermes minor* (Kalotermitidae), California. D. Queen (center) of *Nasutitermes rippertii* (Nasutitermitinae), Bahamas.

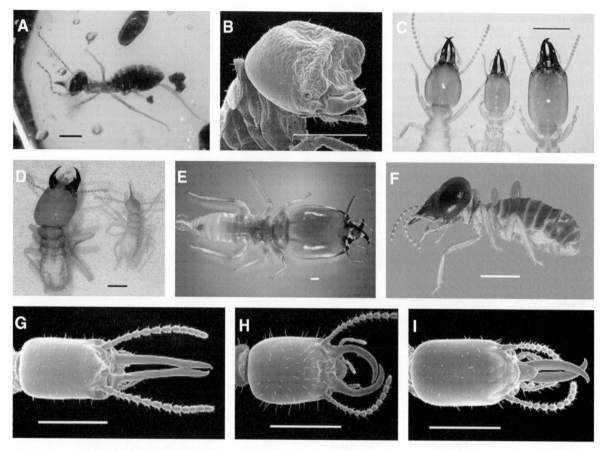

Fig. 1036 Termite soldiers. A. *Constrictotermes* sp. in amber, Dominican Republic. B. *Cryptotermes brevis* (Kalotermitidae), Florida. C. Rhinotermitidae: *Coptotermes formosanus* (L), Heterotermes sp., and *Reticulitermes flavipes*, Florida. D. *Rhinotermes marginalis* (Rhinotermitidae) major (L) and minor, Dominican Rep. E. *Macrotermes* sp. (Macrotermitinae) major, Kenya. F. *Nasutitermes costalis* (Nasutitermitinae), Trinidad. G. *Termes hispaniolae* (Termitinae), St. Croix. H. *Amitermes amicki* (Termitinae), Aruba. I. *Microcerotermes arboreus* (Termitinae), Trinidad. Scale bars = 1 mm.

reproductives. The workers, usually blind and composed of both sexes, in turn begin to forage and feed the royal pair, soldiers, and younger sibling larvae. Feeding is accomplished by trophallaxis in which food is passed mouth-to-mouth from workers to their dependent nestmates. As foraging and food resources increase, the size and reproductive output of the queen grow accordingly. The abdomens of mature queens in the Rhinotermitidae and Termitidae, filled with egg-swollen ovaries, reach a state of near immobility called physogastry. In the Kalotermitidae, castes develop along a single line with both soldiers and alates molting from the immature worker (pseudergate) line. In the higher termites, each caste differentiates along separate developmental pathways after hatching with the workers becoming sterile adults.

Unlike social Hymenoptera, the king is retained and periodically fertilizes the queen. After a number of years when a large crop of workers and soldiers are present, the colony matures to produce its first batch of alates, completing the reproductive cycle. Colonies will produce alates for many years after maturation. Termites have some complex symbiotic relationships with a community gut-inhabiting microorganisms. Various bacteria (prokaryotes) and protists (eukaryotes) are important for cellulose digestion, nitrogen fixation, and metabolism. Protists are absent from the Termitidae. Termites also produce their own cellulytic enzymes. Proctodeal feeding, the transfer of gut contents via anus to mouth, facilitates the exchange of gut biota to freshly molted nestmates.

Fig. 1037 Termite colonies. A. *Zootermopsis angusticollis* (Termopsidae), California. B. *Neotermes castaneus* (Kalotermitidae), Florida. C. *Anoplotermes* n. sp. (Apicotermitinae), Bahamas. D. *Nasutitermes costalis*, (Nasutitermitinae), Florida. Scale bar = 5 mm.

Defense mechanisms and natural enemies

Termites are an ideal food source for other animals. They are relatively slow moving, soft-bodied, and rich in fat and protein. Termites passively avoid casual predation because of their cryptic habits of living in nests, feeding mostly under cover, and foraging in narrow galleries constructed in soil or wood. Most termite species, however, have a highly specialized soldier force to defend the colony from specialized invaders, especially ants. Soldiers of the Kalotermitidae, Termopsidae, and Hodotermitidae rely solely on mechanical defenses including powerful crushing mandibles. Some kalotermitid genera, e.g., *Cryptotermes* and *Calcaritermes*, have soldiers with plug-shaped or phragmotic heads used to seal off nest galleries in wood from ants. Termites in other families use various combinations of biting, piercing, or slashing mandibles and repellent, sticky, and/or toxic chemical secretions which they exude, brush, or squirt onto ants. In most

species of Nasutitermitinae, soldier mandibles are altogether lacking and instead, the soldiers use their conical nasi to squirt repellent secretions over several body lengths.

Many animals, especially birds, bats, rodents, reptiles, and amphibians opportunistically gorge themselves on termite alates during dispersal flights. Some large mammals, such as the aardvark and aardwolf of Africa, pangolins of the paleotropics, anteaters of the neotropics, the echidna of Australia, and the sloth bear of Asia are specialized predators of termites and use claws and long tongues to excavate and feed inside hard nest structures. Humans, especially in Africa, trap and eat *Macrotermes* alates during seasonal dispersal flights from their mounds.

Some arthropods have evolved an association with termites that includes living in termite nests. Those arthropods that have an obligate relationship with the termite colony are called termitophiles. Usually the termitophiles rely on termites for both food and shelter. The great majority of termitophiles are beetles. Some staphylinid beetles even resemble their hosts. Other termitophile taxa include isopods, millipedes, phorid flies, and silverfish.

Ecology

Termite nests consist of a network of galleries that interconnect all colony members and foraging sites. Nest types include single-piece wood nests, (Termopsidae, Kalotermitidae), diffuse subterranean nests (Mastotermitidae, Rhinotermitidae, Termitidae), hollow tree, log, and stump nests connected to other food resources via subterranean galleries (Rhinotermitidae, Termitidae), on-soil (epigeal) mounds (Hodotermitidae, Rhinotermitidae, Termitidae), and arboreal nests (Termitidae). Arboreal nests are constructed of feces while epigeal nests are composed mostly of soil. Foraging is the means by which termite colonies locate and exploit new food resources. If such resources are abundant, secondary nest structures may be constructed nearby. Usually workers, but in some cases soldiers, follow specific search patterns to locate food. If food is encountered, the foragers will recruit additional nestmates to exploit the food source for the colony. When on exposed surfaces, foraging trails are usually protected from predators by tube or sheet-like enclosures made of feces or soil. Recruitment to food locations is enforced and oriented by chemical trails released from the sternal gland located on the ventral surface of the abdomen. Large immovable food sources may be covered with a protective fecal or soil sheeting. Smaller food items such as leaves and grasses are either covered with soil sheeting or are subdivided and carried back to the nest for consumption, storage, or fungal gardens. Foraging territories for single termite colonies may extend 100 meters or more from the nest. Water, either in food or substrata, can have a profound influence on foraging. Most termites forage in the direction of moisture gradients and prefer moist foods and nesting sites. In savannas and arid lands, surface objects caste heat shadows that are often investigated by termites foraging near the soil surface.

Although cryptic for the most part, the absolute numbers of termites in most tropical and semitropical forests, deserts, and savannas ranges from several hundred to thousands of individual termites per square meter. Colony populations of termites vary from several hundred (Kalotermitidae) to several million (Rhinotermitidae, Termitidae) individuals.

The biomass of termites often rivals that of terrestrial vertebrates and ranges from less than 1 g to over 10 g per cubic meter. Food consumption by termites can be greater than that of large ungulates. Termite foods include dry, moist, wet wood, bark, live wood, leaf litter, soil (humus), live and dried grasses and herbs, and leaves, lichens, fungi, and algae. Termites are also significant, albeit, relatively minor sources of global atmospheric methane and carbon dioxide. Termites, along with earthworms, create large turnover and aeration of organic and mineral content in soil and thus benefit plant growth and nutrient cycling.

Termites as pests

About 10% of the termite species worldwide attack structural lumber, wood products of a broad array including paper, and agricultural and forage crops. About 1–2% of all species are major pests and account for the bulk of the untold billions of dollars in damage caused by termites around the world each year. The vast majority of termites is a cryptic and essential component of the environment but has very little direct impact on humans. Termite control in developed countries relies on chemical applications in the form of specialized residual soil termiticides, wood preservatives, baits, dusts, and fumigants often applied by professional pest control operators. People in developing countries may treat termite infestations with unsafe or unproven methods or simply replace damaged and infested wood.

Collecting

Because termites are delicate and can desiccate rapidly, they should be preserved in alcohol (85% ethanol is best) while still alive. A hatchet, trowel, and aspirator are essential collecting tools. Obvious places to collect termites include nests, foraging tubes, and dead limbs and logs. Subterranean species also congregate under boulders, stones, and other surface debris. Herbivore dung, under which some subterranean termites forage, is another surface feature to collect from. Small dried twigs on trees, tree holes, leaf litter, and disturbed areas such as road, trail sides, or even garbage dumps are good collecting places. A few species even infest branches, stems, and roots of live trees and shrubs. Termite alates that fly at dusk or at night are attracted to lights where

Fig. 1038 Termite workings. A. Fecal pellets of *Cryptotermes brevis* (Kalotermitidae), Florida. B. Exposed damage and aerial nest of *Coptotermes formosanus* (Rhinotermitidae), Florida. C. Nest galleries of *Neotermes jouteli* (Kalotermitidae) in structural lumber, Florida. D. Arboreal nest of *Nasutitermes rippertii* (Nasutitermitinae), Bahamas. E. Soil mound of *Macrotermes* sp. (Macrotermitinae), Ivory Coast. F. Foraging tubes of *Microcerotermes arboreus* (Termitinae), Trinidad.

they can be collected by aspirator or net. Because identification relies heavily on characters of soldiers or imagos, one should take multiple samples of each caste present. When alates are not found, one might be lucky to find a queen, king, or both. Nestmates from one colony should be combined as a single sample to confirm their relationship.

See also, WOOD-ATTACKING INSECTS.

Rudolf H. Scheffrahn
University of Florida
Ft. Lauderdale, Florida, USA

References
Abe, T., D. E. Bignell, and M. Higashi (eds.). 2000. *Termites: evolution, sociality, symbiosis, ecology.* Kluwer Academic Publishers, Dordrecht, The Netherlands. 466 pp.
Edwards, R., and A. E. Mill. 1986. *Termites in buildings. Their biology and control.* Rentokil Ltd., East Grinstead, United Kingdom. 261 pp.
Krishna, K., and F. M. Weesner (eds.). 1969. *Biology of termites,* Vol. 1. Academic Press, New York, New York. 598 pp.
Krishna, K., and F. M. Weesner (eds.). 1970. *Biology of termites,* Vol. 2. Academic Press, New York, New York 643 pp.
Pearce, M. J. 1997. *Termites: biology and pest management.* CAB International, New York, New York. 172 pp.

TERMITIDAE. A family of termites (order Isoptera). See also, TERMITES.

TERMITOLOGY. The scientific study of termites.

TERMITOPHILE. An organism that spends at least part of its life cycle with termites.

TERMOPSIDAE. A family of termites (order Isoptera). They also are called primitive dampwood termites or rottenwood termites. See also, TERMITES.

TERRITORIAL PHEROMONE. Pheromones that serve as markers, delineating territory used by an insect and causing others of the same species to avoid that space.

TERTIARY PERIOD. A geologic period extending from about 65 to 2 million years ago. The beginning of the Cenozoic era.

TESSARATOMIDAE. A family of bugs (order Hemiptera, suborder Pentamorpha). See also, BUGS.

TESTICULAR FOLLICLES. Tubules in the testes that form sperm.
See also, TESTIS, REPRODUCTION.

TESTIS (PL., TESTES). An assemblage of testicular follicles, each of which produces sperm (spermatozoa). Each testis empties into a tube called the vas deferens, the seminal vesicles, and then into the ejaculatory duct. The seminal vesicles may be simply an extension of the vas deferens, or may be expanded for sperm storage. Associated with the ejaculatory duct is the accessory gland, which produces secretions that aid in sperm transfer. Most noteworthy of the secretions is the spermatophore, a sac containing sperm.
See also, REPRODUCTION.

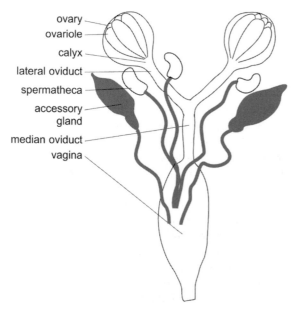

Fig. 1039 Diagram of a male reproductive system as found in *Tenebrio* (Coleoptera) (adapted from Chapman, The insects: structure and function).

TETHINID FLIES. Members of the family Tethinidae (order Diptera). See also, FLIES.

TETHINIDAE. A family of flies (order Diptera). They commonly are known as tethinid flies. See also, FLIES.

TETRACAMPIDAE. A family of wasps (order Hymenoptera). See also, WASPS, ANTS, BEES, AND SAWFLIES.

TETRATOMIDAE. A family of beetles (order Coleoptera). They commonly are known as polypore fungus beetles. See also, BEETLES.

TETRIGIDAE. A family of grasshoppers (order Orthoptera). They commonly are known as pygmy grasshoppers or grouse locusts. See also, GRASSHOPPERS, KATYDIDS AND CRICKETS.

TETTIGARCTIDAE. A family of bugs (order Hemiptera, suborder Cicadomorpha). See also, BUGS.

TETTIGOMETRIDAE. A family of bugs (order Hemiptera, suborder Fulgoromorpha). All members of the suborder are referred to as planthoppers See also, BUGS.

TETTIGONIIDAE. A family of katydids (order Orthoptera). They commonly are known as long-horned grasshoppers. See also, GRASSHOPPERS, KATYDIDS AND CRICKETS.

TEXAS BEETLES. Members of the family Brachypsectridae (order Coleoptera). See also, BEETLES.

TEXAS CATTLE FEVER. See also, PIROPLASMOSIS.

TEXAS CITRUS MITE. See also, CITRUS PESTS AND THEIR MANAGEMENT.

THAUMALEIDAE. A family of flies (order Diptera). They commonly are known as solitary midges. See also, FLIES.

THAUMASTELLIDAE. A family of bugs (order Hemiptera, suborder Pentamorpha). See also, BUGS.

THAUMASTOCORIDAE. A family of bugs (order Hemiptera). They sometimes are called royal palm bugs. See also, BUGS.

THEILERIA. See also, PIROPLASMOSIS.

THELYTOKOUS PARTHENOGENESIS. A type of parthenogenesis wherein females produce only diploid females from unfertilized eggs.

THEOCHODAEID SCARAB BEETLES. Members of the family Ochodaeidae (order Coleoptera). See also, BEETLES.

THEREVIDAE. A family of flies (order Diptera). They commonly are known as stiletto flies. See also, FLIES.

THERMOREGULATION IN INSECTS. Temperature is a physical component of the environment that affects animals from the subcellular to the population level. Because cells are units of chemical reactions and chemical reactions are temperature dependent, animal activity is determined by temperature. Temperature has been shown to be an important influence on all aspects of insect life. Numerous studies have shown that temperature influences cellular activity by changing the efficiency of enzymes, changes the physiology of tissues such as nerves and muscles, alters the rate of development, determines when species emerge, limits the biogeography of species, and determines when a species can be active. There are two strategies for any animal to deal with temperature: they can be a thermoconformer or a thermoregulator.

A thermoconformer is an animal that permits its body temperature to fluctuate with and is about equal to environmental temperature. For many insects being a thermoconformer is determined by the size or activity patterns of a species. For example, a very small insect will exchange heat with the environment so rapidly that its body temperature will always approximate ambient temperature or a nocturnal insect with a low metabolism may not be able to elevate body temperature above ambient conditions. The major problem with being a thermoconformer is that enzymes must be able to work over a greater temperature range than the enzymes of a thermoregulator. This means the enzymes are less efficient at any particular temperature in the thermoconformer. The benefit to being a thermoconformer is that an animal spends no time or metabolic energy attempting to regulate body temperature.

A thermoregulator is an animal that maintains body temperature within a limited range regardless of changes in ambient temperature. A thermoregulator has the benefit of keeping its enzymes and physiological systems within the temperature range where it operate most efficiently. The cost to a thermoregulator is the use of time or metabolic energy needed to maintain its body temperature within a specific temperature range. Thermoregulators can be classified as either ectotherms or endotherms based on the source of the energy used in the regulation of body temperature (the terms poikilotherm or cold-blooded and homeotherm or warm-blooded are no longer used due to ambiguities).

An ectotherm is an animal that uses energy from the environment to regulate body temperature. These animals generally are heliotherms, using the radiant energy of the sun to regulate body temperature, but there are also thigmotherms, animals that use the energy of the substrate to regulate body temperature. Ectothermy is a metabolically inexpensive form of thermoregulation. The major metabolic expense is the cost of transporting the animal from sun to shade. The problem with ectothermy is the animal is still dependent on the environment for a heat source and thus activity. This problem was solved by the evolution of endothermy.

An endotherm is an animal that produces metabolic heat specifically for thermoregulation. Endotherms have been identified in many insect orders. The source of heat in endothermic insects is the flight musculature. Endothermic insects will increase heat generation through muscular activity to elevate body temperature to the range necessary for activity. The heating can occur without flight or wing movements but shivering can be observed in many night flying moths or bees at flowers on cool days prior to take-off. The body temperatures of endothermic insects have been recoded more than 35°C above ambient temperature. The use of metabolic energy for thermoregulation frees the endotherm from possible environmental constraints on activity.

Behavior

Thermoregulation can occur through behavioral and/or physiological mechanisms. The first option used by any organism to thermoregulate is behavior. Behavioral mechanisms are metabolically inexpensive and produce immediate results. Very simple forms of behavioral thermoregulation are changing body orientations and shuttling between the sun and shade. Changing body orientation to the sun is another common behavior that alters the heat exchange of an insect. When animals have low body temperatures, they will orient their body axis perpendicular to the sun to maximize radiant heating. Insects will orient their body parallel to the sun when they have an elevated body temperature to minimize the exposed surface area and radiant heat uptake. When an animal has a low body temperature, it will sit in an exposed position to gain radiant heat from the sun. When an animal has an elevated body temperature, it will seek a shaded location to decrease the radiant gain and lose heat to the environment. One benefit an insect has in thermoregulating is its small size. The high surface to volume ratio means that heat exchanges quickly with the environment and it can take advantage of small microclimates within the habitat.

Microhabitat selection is another important behavioral mechanism of thermoregulation. For example, a desert insect can change the ambient temperature to which it is exposed by more than 30°C simply by selecting a particular microhabitat within the environment. Insects will also employ vertical migration as a means to optimize the ambient conditions in which they are found. Desert animals will move away from the ground as ground temperature (and the boundary layer above the earth) becomes warmer. Similarly, if it is cool and windy, insects will move toward the ground in an effort to find a warmer microclimate. Stilting is a similar behavior seen in some beetles and grasshoppers. When their body temperature is low, the insects will press their body against the ground to increase the rate of heat uptake by conduction and keep their body in the warm boundary layer next to the ground. When their body temperature becomes elevated, they will extend their legs lifting their body as high as possible above the ground. These types of behavioral mechanisms are used by endotherms as well as ectotherms.

Physiology

Physiological mechanisms of thermoregulation can occur through endothermy, evaporative cooling, or thermal adaptation. Endothermic insects generally use the heat generated by the flight musculature to raise body temperature. The flight musculature is a good source of heat because it makes up a significant portion of body mass and it is a highly aerobic tissue. Electrical recordings from the muscles of moths and bees show that the moths shiver (simultaneously activating wing elevator and depressor muscles) to elevate heat production. As the body temperature of the insect increases, the activity of the wing elevator and wing depressor muscles become more out of phase and flight is then initiated. Overheating is prevented by varying the rate of heat production, through circulatory adaptations that increase heat loss, or through the cessation of activity. The functional significance of endothermy has been described for many insect species and represents the diversity of insect behavior. Endothermy permits such diverse behaviors as flight, foraging, acoustic activity, dung ball rolling, maintenance and defense or territories, the use of habitats unavailable to ectothermic species, predator avoidance, defense, brood incubation, and hive temperature regulation.

Evaporative cooling represents a significant avenue of heat loss for animals. The problem with using evaporative cooling is that there must be a water source for the animal to evaporate. The small size of insects means they have a relatively small water reservoir in their bodies but several evaporative cooling mechanisms have evolved in insects. Evaporative cooling by extruding a bubble from the mouth has been described in moths and bees. A bubble of saliva is extended, heat is lost as water evaporates, the bubble is withdrawn back into the mouth to pick up more heat, and the process is repeated until the animal is sufficiently cool. This system is very good at cooling the head but is limited in its ability to cool the entire body. Locusts evaporatively cool through an abdominal pumping mechanism while some caterpillars will spread rectal fluid on their ventral surface to cool through evaporation.

Another group of insects in which evaporative cooling has been studied extensively is cicadas. Cicadas can avoid the potential water balance problems since they feed on xylem fluid. Thus, the desert cicadas that evaporatively cool have access to a water source that other animals cannot use. As a result, desert species like *Diceroprocta apache* (Davis) and *Okanagodes gracilis* Davis can continue their activity while other animals in the environment have sought shelter from the extreme desert heat. In addition, these animals can survive water loss of over 40% of their total body mass. Water is evaporated through pores in the cuticle of the thorax and abdomen of cicadas. The evaporative response of cicadas appears to be energy dependent as toxins can eliminate the response and the response can change dramatically by altering ambient temperature. Evaporative cooling responses are not universal in insects and appear to be restricted to desert species or species whose metabolism may cause dangerous increases in body temperature.

A final physiological mechanism of thermoregulation is thermal adaptation. Insects will adapt their enzymes to work best under particular conditions. The enzymes will show optimal activity temperatures that are related to their environment. In addition, the membrane composition of insects will change dependent upon where the insect lives and even the daily fluctuations in ambient temperature. These changes in membrane composition are necessary to maintain the fluidity of the membrane and thus the integrity of the cell.

Finally, there have been several morphological adaptations that assist insects in thermoregulating. Light coloration in hot environments or dark coloration in cool environments changes the rate of heating and the maximum temperature that can be attained by an insect which increases the time a species can be active in their particular environment. In fact, some dragonflies undergo a temperature dependent color change to alter the amount of radiation uptake. Animals from warmer climates may also reflect more infrared radiation. There are desert beetles that have highly convex elytra forming a large subelytral space that acts to decrease heat transfer to the body while the animal is active in the sun. Some desert tenebrionid beetles have legs that are much longer than their forest relatives. These long legs elevate the animal above the boundary layer and significantly increase the time of activity for the species. Many endothermic and ectothermic species that live in cool habitats are covered with pile. The pile acts as insulation to help conserve heat within the body.

Allen Sanborn
Barry University
Miami Shores, Florida, USA

References

Heinrich, B. (ed.). 1981. *Insect thermoregulation.* John Wiley & Sons, New York, New York. 328 pp.

Heinrich, B. 1993. *The hot-blooded insects: strategies and mechanisms of thermoregulation.* Harvard University Press, Cambridge, Massachusetts. 601 pp.

May, M. L. 1979. Insect thermoregulation. *Annual Review of Entomology* 24: 313–349.

May, M. L. 1985. Thermoregulation. pp. 507–552 in G. A. Kerkut and L. I. Gilbert (eds.), *Comprehensive insect physiology, biochemistry, and pharmacology.* Pergamon Press, New York, New York.

Sanborn, A. F. 1998. Thermal biology of cicadas (Homoptera: Cicadoidea). *Trends in Entomology* 1: 89–104.

THERMOTAXIS. Taxis response with respect to temperature.

THICK-HEADED FLIES. Members of the family Conopidae (order Diptera). See also, FLIES.

THIRD GENERATION INSECTICIDE. Organic insecticides derived from knowledge of insect's hormones. By mimicking hormones, the insecticides provide great selectivity, and are less disruptive to nontarget organisms.

THOMAS, CYRUS. Cyrus Thomas was born at Kingsport, Tennessee, USA, on July 27, 1825. A lawyer and Lutheran minister by training, he became one of America's foremost systematists and economic entomologists. Thomas had little formal education and did not attend college; nevertheless, he learned science, mathematics, and practiced law. He sought an area of science where he could be recognized, and which could be mastered with little expense and the materials at hand; entomology met those needs. While still practicing law, he began to publish entomology papers. Thomas served with the Hayden Geological Survey of the territories of the West and Southwest from 1869 to 1874. From 1874 to 1876 he taught natural history at the University of Illinois, and was state entomologist for Illinois, publishing six annual reports. Thomas served on the United States Entomological Commission for five years. One of his most significant achievements was to author a comprehensive treatment of Aphididae, included in the "Third annual report of the state entomologist of Illinois on noxious and beneficial insects," published in 1879. He is also remembered for authorship of the "Synopsis of the Acrididae of North America," published in 1873. In his taxonomic work he named chiefly aphids and grasshoppers. Despite a successful career in entomology, he devoted the last 25 years of his life to archaeological and ethnological studies, and became an authority on the Cherokee, Shawnee, and Maya people. Thomas died at Washington, DC, on June 27, 1910.

References

Anonymous. 1910. Obituary, Cyrus Thomas, Ph.D. *Entomological News* 21: 387–388.

Essig, E. O. 1931. *A history of entomology*. The Macmillan Company, New York. 1029 pp.

Mallis, A. 1971. *American entomologists*. Rutgers University Press New Brunswick, New Jersey. 549 pp.

THOMSON, CARL GUSTAV.

Carl Gustav Thomson was born at Skaane, Sweden, on October 13, 1824. He became a student at Lund in 1843, and attained his Ph.D. in 1850. Beginning in 1850 he worked in various capacities at the Lund Zoological Museum. Thomson tackled difficult taxonomic problems successfully. He studied principally Hemiptera, Hymenoptera, and Coleoptera, but also Diptera. A prolific author, he published nearly 9,000 pages during his lifetime, though his work on Diptera is not highly regarded. He died on September 20, 1899, in Lund, Sweden.

References

Papavero, N. 1971, 1973. *Essays on the history of Neotropical Dipterology with special reference to collectors (1750–1905)*. Museu de Zoologia, Universidade de São Paulo.

Herman, L. H. 2001. Thomson, Carl Gustav. *Bulletin of the American Museum of Natural History* 265: 150–151.

THORACIC PLATE.

In caterpillars, the shield-like dorsal covering or plate on the body segment immediately behind the head, usually dark in color. Also known as the cervical shield.

THORAX.

The second or middle of the three major body regions of insects, and the section bearing wings (if present) and jointed (true) legs.

See also, THORAX OF HEXAPODS.

THORAX OF HEXAPODS.

The arthropods, and particularly the insects, present a basic model of body organization constructed on a metameric base, about which is produced the process of tagmosis. This process consists of the grouping of a variable number of segments to form a suprasegmentary unit called the tagma.

Tagmosis is probably produced to establish a more operative model in the development of certain functions important to the life of those animals. In this case, in the hexapods, the thoracic tagmata have come about by the grouping of three segments, called the prothorax, mesothorax and metathorax, which have developed principally for locomotion.

The formation of the thorax by three segments was pointed out in the nineteenth century by Adouin and later recognized by the majority of authorities, except by some like Verhoeff, who at the beginning of the twentieth century considered that the intersegmentary regions were authentic segments, whereby five segments would form the thorax. This theory was rapidly rejected, as it was demonstrated through comparative studies of morphology and embryology that the hexapod thorax consists of only three segments.

Except in the apodous larvae of certain orders, each thoracic segment carries a pair of legs, and in addition the winged forms present a pair of wings on the mesothorax and on the metathorax, with both segments constituting a functional 'subtagma' called the pterothorax. The presence of legs and wings on the thorax demonstrates that it is a tagma specialized for locomotory function, developed and controlled by muscles and ganglia situated in the thorax itself. It is probable that this locomotory function originated the differentiation of the thoracic tagma from the other two that form the body of the hexapods: the head and the abdomen. The fact that the thorax is found basically formed in the most primitive hexapod groups and that it appears perfectly individualized

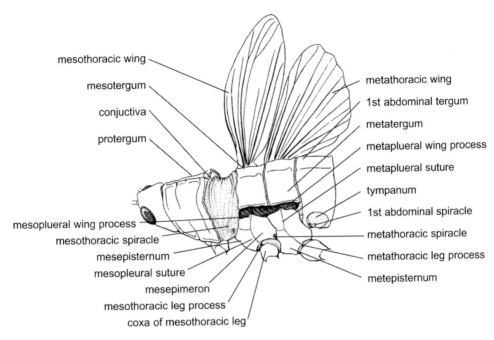

mesothoracic wing

mesotergum

conjuctiva

protergum

metathoracic wing

1st abdominal tergum

metatergum

metaplueral wing process

metaplueral suture

tympanum

1st abdominal spiracle

metathoracic spiracle

metathoracic leg process

metepisternum

mesoplueral wing process

mesothoracic spiracle

mesepisternum

mesopleural suture

mesepimeron

mesothoracic leg process

coxa of mesothoracic leg

Fig. 1040 Lateral view of a grasshopper head and thorax (Orthoptera: Romaleidae).

in the first stages of ontogenetic development makes one think that the differentiation of this tagma ought to have occurred in early periods of insect evolution. As much from the structural as functional point of view, the hexapod thorax is the most specialized and centralized tagma if we compare it with analogous tagmata of other arthropods.

Before beginning a description of the thorax itself, it is necessary to emphasize the existence of an anterior region of the body called the neck or cervical region, which is of mixed origin: labial and prothoracic. The morphological, embryological, musculature and neurological data demonstrate that the neck results from the fusion of the dorsal and ventral parts of the labial and prothoracic segments, respectively. The cervical sclerites can be found in the dorsal position (dorsal cervical or dorso-cervical sclerites), lateral (lateral cervical or latero-cervical sclerites) and ventral (ventral cervical or ventro-cervical sclerites). The lateral cervical sclerites can originate from the anterior part of the propleura [some Thysanura (Zygentoma) and the majority of holometabolous insects] or from the presternal region (hemimetabolous insects and Coleoptera). The dorsal cervical sclerites are considered structures of secondary formation and are united anteriorly to the dorsal

posterior margin of the occiput. The same occurs with the ventral ones.

Modern studies of the hexapod thorax began with the decade of the 1940s, particularly such authors such as Snodgrass, Weber, Barlet, Carpentier, Francois, La Greca, and Matsuda. As in each arthropod segment, each hexapod segment presents a dorsal region called the notum or tergum, another ventral region named the sternum, and between the two, a zone that constitutes the pleuron.

The tergal region presents variations in the number of sclerites and sutures, whose interpretation in Pterygota should be analyzed with a previous understanding of what occurs in the Apterygota. The Protura display a more elaborate tergum whose homologization with the rest of the hexapods is difficult. In Diplura, Archaeognatha and primitive Thysanura, an anterior zone exists, which, according to Matsuda, is named the pseudoprescutum. It is separated from the rest of the tergum by a pseudoprescutoscutal suture that produces toward the interior a suture named the pseudophragma. The study of *Nicoletia* is interesting in that it explains the passage of a primitive tergum to another evolved in the Pterygota by the loss of the pseudophragma, which would have no functional significance on developing a phragma

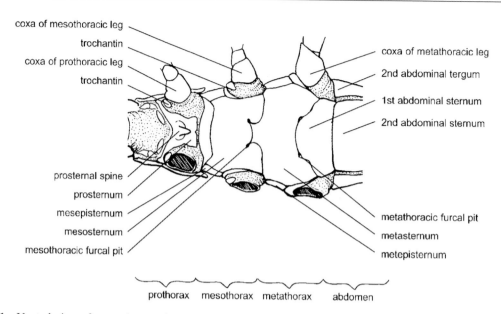

Fig. 1041 Ventral view of a grasshopper thorax (Orthoptera: Romaleidae).

with the incorporation of an acrotergite; the pseudo-prescutoscutal suture would be transformed into the prescutoscutal suture and the pseudoprescutum into the prescutum. In the case of *Lepisma*, a narrow antenotum (delimited by an antecosta), a prescutum and a scutum (separated by a prescutoscutal suture) and a postnotum united laterally to the pleuron are differentiated; these sclerites are already homologous to those of the Pterygota.

The tergal region of the pterothorax of Pterygota (meso and metathorax of winged insects), presents sclerites delimited by two types of sutures: some homologous to those of the Apterygota and other, new ones that appear by the wings. The wings, in addition, produce modifications in the zone of insertion that fundamentally affect the lateral margins of the tergum.

Among the sutures homologous to those of the Apterygota are the antecosta, which delimited anteriorly a zone that corresponds to the acrotergite and gives rise to a phragma toward the interior. The prescutoscutal suture is situated behind the antecosta, delimiting a sclerite named the prescutum, whose lateral zones form the prealar arms.

Among the sutures not homologous to those of the ''Apterygota'' stands out the scutoscutellar suture, which has a sinuous form and reaches the axillary ligament. This suture separates two sclerites, an anterior (scutum) from another posterior (scutellum). The proper parapsidal lines and the lateral parapsidal lines are differentiated as intrascutal formations; both arising from the prescutoscutal suture, the first in a more or less median position and the second lateral. Two marginal lines exist within the scutum, one that delimits the anterolateral angle of the scutum, forming the suralare, a sclerite that includes the anterior and antemedian notal wing processes. The other marginal suture delimits the posterolateral angle of the scutum, which includes the posterior notal wing process. In primitive forms of certain orders a tergal fissure is differentiated, which transversally unites the sides of the scutum at the level of the anterior or antemedian notal wing process. The recurrent scutoscutellar suture originates from or near the center of the posterior margin of the scutellum, diverging toward the anterior part. The last sclerite is the postnotum.

The presence of the wings brings with it a series of modifications in the lateral margins of the tergum, differentiating a series of processes, some of which have been mentioned previously. Thus, the most anterior process is called the prealar arm or branch, which in the majority of orders is a prolongation of the prescutum, although in others like the Hymenoptera it does not exist. Posteriorly, five points of articulation with the wings are differentiated, which are called notal wing processes. From the anterior part to the posterior, they are the anterior and antemedian

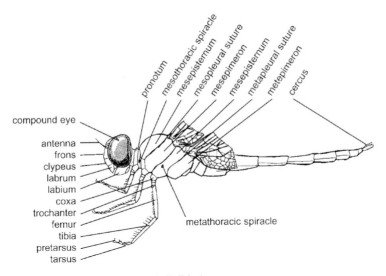

Fig. 1042 Lateral view of a dragonfly (Odonata: Libellulidae).

(already mentioned), median, postmedian, and posterior (also mentioned). In the case where the five processes are evident, the first four are joined with the first axillary sclerite and the last with the third axillary sclerite.

The notal region of the thorax can present different modifications. Among the most customary and conspicuous is the prolongation of the mesothoracic scutellum, which in certain cases projects toward the back, on top of the metanotum (for example, Hemiptera: Scutelleridae), and in other cases adapts to its conformation.

The sternum, as already mentioned, corresponds to the ventral zone. In generalized Diplura,

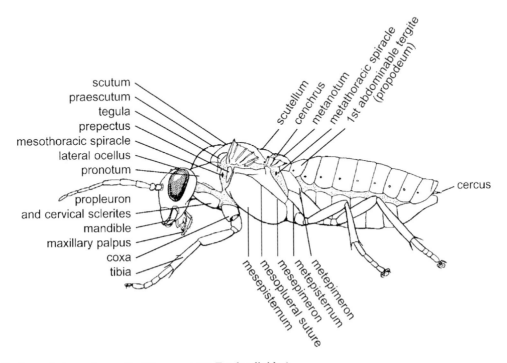

Fig. 1043 Lateral view of a sawfly (Hymenoptera: Tenthredinidae).

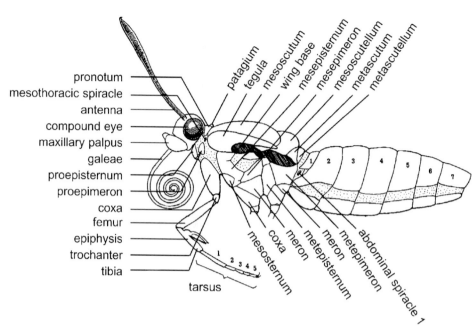

Fig. 1044 Lateral view of a moth (Lepidoptera: Sphingidae).

Archaeognatha and some Thysanura, five parts or apotomes are distinguished, which from the anterior to the posterior part are: the presternum, the basisternum, the furcasternum, the spinasternum and the poststernum. In principle, the question is posed of whether the poststernum, instead of representing the final portion of the sternum, constitutes the more anterior of the following sternum that would have been incorporated. In any case, the different apotomes of the sternum of Aptergygota have a relatively well-defined position, with the poststernum being a posterior segmentary structure, although it does not appear in Ptergygota.

The association of the mentioned sclerites to the sternum, and the reductions that occur during evolution toward advanced hexapod groups, are justified through the study of endosternal formations. In the Apterygota, a non-cuticular endoskeletal complex is distinguished, in addition to spinasternal processes in Diplura and pleural apodemes in Archaeognatha. Cuticular formations are well defined in Ptergygota, corresponding to the furca and the spina.

As a consequence of losses, modifications and fusions of certain zones, the original division of the sternum in Pterygota has suffered changes that do not permit delimiting a clear position from its apotomes. Thus, the lack of a presterno-basisternal suture makes the differentiation of a presternum very difficult. The posternum is lost in the Pterygota, with this loss affecting the two anterior apotomes (spinasternum and furcasternum). The reduction affects, in the first place, the most posterior apotome, which is the spinasternum, being reduced in certain cases to the spine, as occurs in the mesosternum of the Orthoptera. The spine is also lost in more evolved groups like Coleoptera, Diptera and Hymenoptera, causing reductions to the furcasternum, which can reach the point of disappearing. The anterior part of the metasternum then occupies its place, shifting toward the front and situating itself close to the furcal bases. When this occurs, obliteration is produced on the ventral border between the meso and metathorax. This obliteration, which exists in different levels of development in advanced insects, appears to directly affect the formation of a functional subtagma, the pterothorax, to which reference has already been made. In this process, a tendency for the reduction of the anterior and posterior parts of the sternum can be observed, always leaving a well-developed apotome, the basisternum.

All the studies carried out on the origin of the pleuron demonstrate that this zone of the thorax, situated between the tergum (or notum) and the sternum in Archeognatha, Thysanura (Zygentoma) and

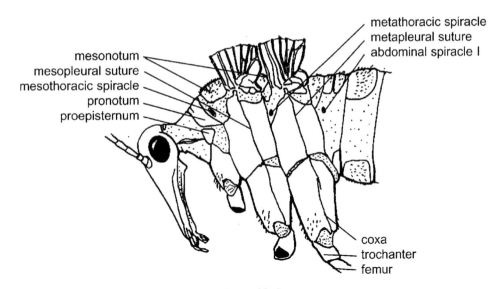

Fig. 1045 Lateral view of a scorpion fly (Mecoptera: Panorpidae).

Pterygota, consists of a ventral subcoxal area and another dorsal or pleural area, which have become indistinguishable because of the sclerotization. In this sense, an evolutionary sequence can be established from the pleuron of Archaeognatha to that of the pterothorax, considering the increase in sclerotization tied to the development of the wings. In primitive Pterygota like Blattodea, the sclerotization is not very pronounced, and the sclerites that surround the base of the coxa have not been obliterated, so clear homologies can be established with Archaeognatha.

If one considers the pleural formations existing in the segments of the pterothorax, there are distinguished, in the first place, a pleural suture and its corresponding internal costa. In principle, this suture ought to present a vertical path as is observed in certain orders like Neuroptera, with a tendency to present an oblique path toward the front. The most dorsal part constitutes the pleural wing process. This suture has been obliterated in some orders of advanced hemimetabolous insects like Hemiptera (Heteroptera) while conserved in secondarily apterous orders like Siphunculata, Mallophaga and Siphonaptera, and in orders like Psocoptera. Internally, it is distinguished by way of a pleural branch that is united to the furca through a tendinous structure or a muscle. This pleural suture divides the pleuron into two regions: an anterior one (episternal region or episternum) and a posterior one (epimeral region or epimeron).

Another important suture is the paracoxal, which divides the pleuron into two rings, an external anapleural ring, or anapleurite, and an internal ketapleural ring, or katapleurite. Its presence in Archaoegnatha indicates that it is a primitive suture.

Considering the two pleural sutures mentioned up to now, the pleuron remains divided into four regions: anepisternum, katepisternum, anepimeron and katepimeron. In species of certain orders like Neuroptera, Hemiptera and Hymenoptera, another suture exists, named the transepimeral, which is a continuation of the paracoxal suture in the episternal part. The anepisternum remains divided by the anapleural suture in one part dorsally (the anepisternum proper), and another part ventrally (the preepisternum). In species of some orders like Psocoptera, the anapleural suture is situated more ventrally, with another suture being differentiated in those cases named the transepimeral suture, which divides the anepisternum proper into two portions (dorsal and ventral). In primitive orders, beginning from the anepisternum a small sclerite called the basalare is differentiated, situated just in front of the pleural wing process. When the area that surrounds the basalare is membranous, it can extend posteriorly, originating a basalar incision that in some orders like Dermaptera forms the episternal suture. Behind the pleural wing process the subalar sclerite is differentiated, which in reality is a sclerite differentiated from the tergum.

Three bridges exist. The prealar bridge or prealare, which has already been mentioned in the section dedicated to the tergum. The precoxal bridge or precoxale, which is formed by the preepisternum uniting with the basisternum and, are separated by the pleurosternal suture. This suture is obvious in Apterygota and in the majority of hemimetabolous Pterygota but does not exist in certain holometabolous insects, in those in which the preepisternum and basisternum are fused. This fusion is not produced in other holometabolous insects, since the preepisternum extends forward, forming the lateral cervical sclerite. The third bridge is the postalar bridge or postalare, a structure similar to the prealare, but formed by the connection between the dorsal margin of the anepimeron and the lateral extension of the postnotum.

The trochantin is a sclerite differentiated from the katepisternum, which can present four points of articulation in Thysanura, one dorsal with the katapleuron and three ventral with the coxa. In Ptergygota, only the anterior coxal-trocantinal articulation is conserved, with various modifications of the trochantin being observed. It can be isolated (prothorax of Psocoptera), partially fused to the katepisternum, conserving the anterior coxal-trocantinal articulation (eutrochantin), and it can even be undifferentiated due to the strong sclerotization of the katepisternum, as occurs in advanced orders.

In the more primitive groups, in the posterior zone of the katepisternum the sternopleurite is found, which appears to correspond, according to Matsuda (1970) to a part of the ventral zone of the katapleuron. This ventral pleurite in holometabolous insects is not appreciated, and it is possible that it has been transformed in the ventral articulation process of the coxa.

It can be concluded that the evolution of the hexapod thorax has been a continual process, between the thorax of the groups of Apterygota and that of the Pterygota, a process based on the loss and modifications of different structural elements. See also, WINGS OF INSECTS.

Severiano F. Gayubo
Universidad de Salamanca
Salamanca, Spain

References

Bitsch, J. 1994. The morphological groundplan of Hexapoda: critical review of recent concepts. *Annales de la Societé Entomologique de France* 30: 103–129.

Kristensen, N. P. 1998. The groundplan and basal diversification of hexapods. pp. 281–293 in R.A. Fortney and R. H. Thomas (eds.), *Arthropod relationships*. Systematics Association, Special Volume Series 55.

Manton, S. M. 1977. *The arthropoda. Habits, functional morphology and evolution*. Clarendon Press, Oxford, United Kingdom.

Matsuda, R. 1970. Morphology and evolution of the insect thorax. *Memoirs of the Entomological Society of Canada* 76: 1–431.

Snodgrass, R. E. 1952. *A textbook of arthropod anatomy*. Comstock Publishing, Ithaca, New York.

THRIPIDAE. A family of thrips (order Thysanoptera). They commonly are known as common thrips. See also, THRIPS.

THRIPS-PARASITIC NEMATODES. Several species of nematodes belonging to the genus *Thripinema* (= *Howardula*) (Tylenchida: Allantonematidae) are known to naturally parasitize thrips (Thysanoptera).

Taxonomy, host range and distribution

The genus *Thripinema* was erected by Siddiqi in 1986 during a taxonomic revision of the species. To date, there are five described species of *Thripinema*: *T. reniraoi*, *T. aptini*, *T. nicklewoodi*, *T. khrustalevi* and *T. fuscum*. Together these nematodes have been recovered from twelve species among eight genera of thrips: *Thrips physapus* L., *T. trehernei* Prisner, *T. physopus* L., *Aptinothrips rufus* Gmelin, *Frankliniella vaccinii* Morgan, *F. occidentalis* Pergade, *F. fusca* Hinds, *Taeniothrips vaccinophilus*, *Stenothrips graminium*, *Catinathrips vaccinophilus*, *Heliothrips* sp., *Megaluriothrips* sp. The biogeographical range of thrips parasitic nematodes is known to include the UK, Germany, USA, Canada, Russia and India. However, the extent of this distribution probably reflects where surveys have been made. Given *Thripinema* spp. hosts include introduced agricultural pest species found worldwide, thrips-parasitic nematodes are probably widely distributed.

Biology and life cycle

Unlike soil-dwelling nematodes of the family Rhabditidae, *Thripinema* spp. are not mutually associated with pathogenic bacteria and do not appear to kill their host. Rather, *Thripinema* spp. have evolved a parasitic lifestyle and develop through a single heterosexual generation in the live host. Although infected thrips show no obvious physical signs of the internal parasite, a consequence of infection is that embryos do not develop and adult female thrips are effectively sterile. The cause of host sterility is unknown, but nematodes may deprive the host of the protein required for oogenesis or secrete a toxin, which damages the reproductive organs. Male thrips are similarly parasitized, although the effect of infection on male fertility is unknown. During the infection stage, parasitic female nematodes penetrate a thrips host through intersegmental membranes. Following infection, the female nematode swells to a sac-like organism in which her reproductive organs become the only visible structure. Nematode eggs are laid into the host hemocoel. Upon hatching, the resulting vermiform juveniles feed on fluids within the thrips' abdominal cavity. When mature, both male and female nematode progeny penetrate into the lumen of the hosts' gut, and are continually released for the lifetime of infective thrips via the anus in the frass or via the ovipositor. Relatively little is known about the biology of *Thripinema* spp. following its emergence from the host. However, an intriguing aspect of the life cycle of the free-living stage is that they appear to attack their hosts in the above-ground parts of the plant. While all free-living nematodes require high humidity, *Thripinema* spp. appears to exploit the moist microclimate found within a leaf gall, flower perianth or developing foliage terminal. Since mature thrips nematodes exit the host via the anus or ovipositor, these free-living forms naturally accumulate in and around thrips foraging sites where defecation and oviposition are pronounced and where susceptible hosts may be found. Fertilization of parasitic female nematodes is thought to occur outside the host despite the fact that survival of this 'free-living' stage may only be a few hours; thus the cycle is complete.

Thripinema spp. are not currently mass produced commercially. However, the ability of these nematodes to attack thrips in their preferred feeding sites, areas that are often impenetrable to insecticides and natural enemies, have led several investigators to speculate that thrips parasitic nematodes have potential for thrips management in agriculture.

Steven Arthurs
Texas A&M University
College Station, Texas, USA

Reference

Loomans, A. J. M., T. Murai, and I. D. Greene 1997. Interactions with hymenopterous parasitoids and parasitic nematodes. pp. 355–397 in T. Lewis (ed.), *Thrips as crop pests*. CAB International, Wallingford, United Kingdom.

THRIPS (THYSANOPTERA). The Thysanoptera, thrips, are a diverse insect order with worldwide distribution. There are approximately 5800 species described from 9 families. The order is divided into two distinct suborders: Tubulifera and Terebrantia. These two suborders can be distinguished by the shape of the last abdominal segment of the adult stage which is short and pointed in the Terebrantia, or long and tubular in the Tubulifera. Nearly all described species are less than 5 mm in length, and can be yellow, green, black, or red colored. The name Thysanoptera, derived from the Greek words, 'thysanos' meaning fringe and 'ptera' meaning wings, refers to the 2 pairs of slender wings which have few or no veins and bear a dense fringe of long hairs. These hairs allow for greater wing area and increased flight efficiency. Thrips are thought to be closely related to Hemiptera and Psocoptera.

Order: Thysanoptera
   Suborder: Tubulifera
      Family: Phlaeothripidae
   Suborder: Terebrantia
      Family: Aeolothripidae

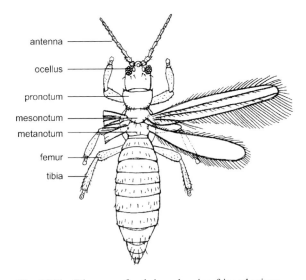

Fig. 1046 Diagram of a thrips, showing fringed wings.

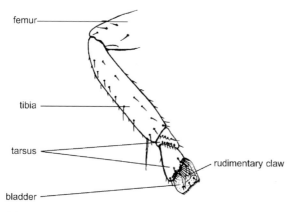

Fig. 1047 Leg of a thrips; note that claws are not apparent.

Family: Merothripidae
Family: Heterothripidae
Family: Thripidae
Family: Melanthripidae
Family: Uzelothripidae
Family: Adeheterothripidae
Family: Fauriellidae

Thrips are holometabolous insects with complete metamorphosis. Development includes the egg, 2 larval instars, 2 to 3 pupal stages and the adult. The pupal stages do not feed but are capable of limited movement. Females of terebrantia have a curved

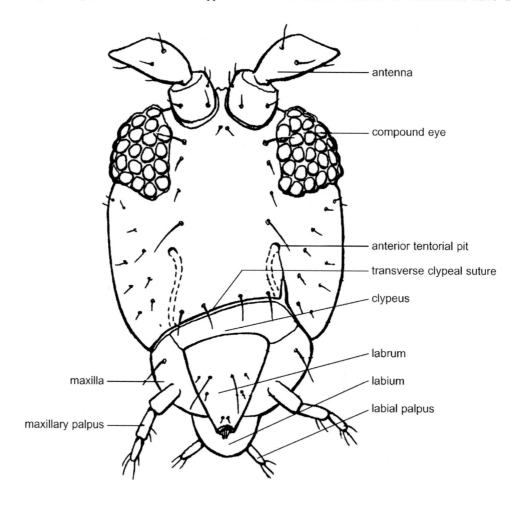

Thrips Head

Fig. 1048 Head of a thrips; note lack of symmetry due to absence of the right mandible.

Some economically important thrips and their distribution

Common and species name	Present geographic distribution	Native range
Western flower thrips *Frankliniella occidentalis* (Pergande)	World-wide; Mediterranean climates, greenhouses in cooler climates	Central California
Tobacco thrips *Frankliniella fusca* (Hinds)	North America, Mexico,and Puerto Rico	North America, Mexico, and Puerto Rico
Flower thrips *Frankliniella tritici* (Fitch)	Eastern Temperate North America and Mexico	Eastern Temperate North America and Mexico
Melon thrips *Thrips palmi* Karney	Tropics, southern Florida	Southeast Asia
Onion thrips *Thrips tabaci* (Lindeman)	World-wide	Mediterranean climates
Cotton bud thrips *Frankliniella schultzei* (Trybom)	World-wide	Africa
Florida flower thrips *Frankliniella bispinosa* (Morgan)	Southeastern U. S.	Southeastern U.S.
Yellow tea thrips or chillie thrips *Scirtothrips dorsalis* Hood	Southeast Asia, India, Africa, Australia	Southeast Asia
Thrips setosus Moulton No common name	Japan	Japan
Frankliniella intonsa (Trybom) No common name	Europe, Russia, Britain, Spain, Mongolia, Japan, British Columbia, USA(Washington state).	Europe, Russia, Britain, Spain, Mongolia, Japan,

ovipositor which is used to insert eggs into plant tissue. The antenna are short with 6 to 10 segments. Thrips have asymmetrical sucking mouthparts, possessing only the left mandible. Thrips are weak flyers and short directed flights are called thripping. Longer range dispersal is dependent on wind currents. Within the same species, populations may have individuals with reduced wings (brachypterous) or no wings (apterous) depending on environmental conditions such as density, food quality, and season.

Thrips are important members of the ecosystem as both herbivores and predators. Many species live in leaf litter or dead wood and feed nearly exclusively on fungus and will often supplement their diet with plant pollen. Other thrips species are gall formers and display primitive eusocial behavior. Predatory thrips are often beneficial species in agronomic situations and can help regulate populations of mites and other small insect pests including other thrips. Thrips are often associated with disturbed growth areas where large numbers can occur quickly on new plant growth.

The vast majority of the described species of thrips are herbivorous, with several being destructive pests of grain crops, fruits, vegetables and ornamentals. Certain species are important pests of plants grown in greenhouses. Feeding activities can result in plant deformities, scarring, loss of yield, and in some cases, transmission of plant pathogens. Plant-feeding thrips pierce and suck juices from the outer layer of cells, causing stippling, or small scars, on leaves, flowers and fruit. This feeding damage may result in stunting of the plant, premature leaf drop and aborted fruit. Flowers that have been damaged by thrips feeding may be deformed and fail to open properly. As many as 90 species of thrips are of economic importance, including 9 species capable of vectoring plant viruses in the genus *Tospovirus*. These viruses are the most important disease of agronomic crops in many regions of the world today.

Christopher Tipping
University of Florida
Quincy, Florida, USA

References
Lewis, T. 1998. *Thrips as crop pests*. Oxford University Press, Oxford, United Kingdom.

Moritz, G., and L. Mound. 2001. Thrips ID: pest thrips of the world. Interactive CD. ACIAR, CSIRO, Collingwood, Victoria, Australia.

Stannard, L. J. 1968. *The thrips, or Thysanoptera of Illinois.* Illinois Natural History Survey, Urbana, Illinois.

THROSCIDAE. A family of beetles (order Coleoptera). They commonly are known as false metallic wood-boring beetles. See also, BEETLES.

THUNBERG, CARL PETER. Carl Thunberg was born at Joenkoepping, Province of Smaaland, Sweden, on November 11, 1743. He studied at the University of Uppsala under Linnaeus, and served as a medical doctor in South Africa for several years before returning to Sweden. While in South Africa, he also visited the Dutch colonies of the Far East, and was able to investigate the fauna of Ceylon, Java and Japan. He replaced Carl Linnaeus, Jr., at the University of Uppsala, and transformed the Royal Gardens into botanical gardens to honor Linnaeus. Though known principally as a botanist, Thunberg published numerous entomological papers, including some descriptions. A prolific publisher, he authored over 150 scientific publications. He died August 8, 1828.

Reference

Papavero, N. 1971, 1973. *Essays on the history of Neotropical Dipterology with special reference to collectors (1750–1905).* Museu de Zoologia, Universidade de São Paulo.

THURINGIENSIN. A soluble, heat-stable beta-exotoxin produced by *Bacillus thuringiensis* during the vegetative growth stage of the bacteria. It has several properties, including insecticidal, feeding deterrent, and teratological effects.

THYATIRIDAE. A family of moths (order Lepidoptera). They commonly are known as false owlet moths. See also, FALSE OWLET MOTHS, BUTTERFLIES AND MOTHS.

THYREOCORIDAE. A family of bugs (order Hemiptera). They sometimes are called negro bugs. See also, BUGS

THYRETIDAE. A family of moths (order Lepidoptera) also known as African maiden moths. See also, AFRICAN MAIDEN MOTHS, BUTTERFLIES AND MOTHS.

THYRIDIDAE. A family of moths (order Lepidoptera). They commonly are known as picture-winged leaf moths. See also, PICTURE-WINGED LEAF MOTHS, BUTTERFLIES AND MOTHS.

THYSANOPTERA. An order of insects commonly are known as thrips. See also, THRIPS.

THYSANURA. An apterygote order of insects. They commonly are known as silverfish. This order is also known as Zygentoma. See also, SILVERFISH.

THYSANURIFORM LARVA. These are active, flattened, chitinous, free living, and principally predaceous larvae. Subcategories of thysanuriform larvae include campodeiform, caraboid, trianguloid, naupliiform and planidiiform.
See also, CAMPODEIFORM, CARABOID, TRIANGULOID, NAUPLIIFORM, PLANIDIIFORM.

TIBIA. (pl., tibiae) The section of the insect leg between the femur and the tarsus, usually one of the largest sections and often bearing spines or spurs. See also, LEGS OF HEXAPODS.

TICK-BORNE ENCEPHALITIS. See also, TICKS.

TICK PARALYSIS. Adult ticks inject considerable quantities of saliva into the hosts during feeding. In some species of ticks, proteins in the saliva can paralyze the vertebrate host. At least 43 species of ticks have been reported to cause paralysis, but some records are doubtful. Paralysis caused by *Dermacentor andersoni* Stiles in North America and *Ixodes holocyclus* Neumann in Australia have been studied most extensively. *Dermacentor andersoni*, the Rocky

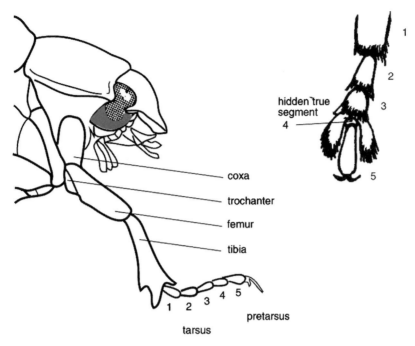

Fig. 1049 Leg of a beetle (Coleoptera: Scarabaeidae) leg showing its component parts, and a close-up of one type of beetle tarsus (foot).

Mountain Wood Tick, causes paralysis in a variety of hosts including sheep, cattle, horses, dogs and humans. Paralysis has been reported in various wildlife species, but this is rare and may be because affected animals are not readily visible. Wood ticks can paralyze cattle at doses of 25 to 83 mg tick per kg host body weight, and in sheep at doses of 37 to 70 mg tick per kg host. Increasing numbers of ticks per host may increase the incidence of paralysis, but this has only been demonstrated for laboratory animals. Humans may be paralyzed by a single tick. There are at least 10 cases of human paralysis in children aged 3 to 8 years that were caused by single ticks ranging in weight from 60 to 300 mg.

Although *D. andersoni* is widely distributed in western North America, paralysis is mainly associated with tick populations in a relatively small portion of its range. Paralysis occurs in the interior of British Columbia, and extends into Idaho, Oregon and Montana. Paralysis is relatively rare in the prairie regions of its distribution. Even though paralysis is associated with a particular geographic location, the ability of individual ticks to cause paralysis varies considerably within areas that paralysis occurs. Only a small proportion of ticks may be capable of causing paralysis

when placed singly on susceptible laboratory animals.

Paralysis is typically associated with unmated female ticks that have fed for at least 4 to 5 days on the host. Males do not cause paralysis as they do not engorge to the extent that females do, and not inject the copious amounts of saliva into the host that females do. Female ticks are most likely to cause paralysis after they have attained a minimum weight of 40 mg. Although paralysis will occur in humans of all ages, it is less common in adults compared with young children as adult humans are more likely to detect and remove feeding ticks. In cattle, yearlings are the most susceptible age class as older cattle may develop immunity, and calves can have ticks removed by maternal grooming. Paralysis typically occurs in the early spring when adult ticks are actively searching for hosts. Cattle placed on pasture in the early spring are most at risk for paralysis.

Tick paralysis is a progressive condition and worsens the longer that ticks are attached. Regardless of the host, paralysis begins with loss of hind limb function (posterior ataxia), proceeding to full ataxia. The victims lose the ability to stand or sit. Cattle

Fig. 1050 Cow immobilized by tick feeding.

will become sternally recumbent, and as paralysis proceeds, will become laterally recumbent. If ticks are not removed, death due to respiratory failure will occur. Prompt removal of ticks from paralyzed animals can result in rapid recovery, even as the ticks are being removed. The earlier ticks are removed, the more rapid is the recovery. Paralysis is most common in cattle. Several thousand cattle have been reported paralyzed, with some occurrences involving entire herds of over 100 head. Death rates vary, but may range upwards of over 20 to 25% of an affected herd. Several hundred cases of human paralysis have occurred in the interior of British Columbia since the early 1900s.

Tick paralysis is believed to be caused by a toxin produced in the salivary glands and released during feeding. Injection of artificially collected saliva can cause paralysis. The rapid recovery of the host following tick removal suggests that infectious agents are not directly involved. Increasing incidence of paralysis with increasing doses of ticks further suggests that a toxin is involved. To date, the nature and identity of the toxin has not been identified. The mode of action of the toxin is not well understood, but is associated with reduced nerve conduction, possibly by attacking the nerve membrane. It is thought to

involve motor polyneuropathies, with only limited participation of the afferent pathways.

Paralysis caused by *Ixodes holocyclus* is similar in some respects to that caused by *D. andersoni*, and markedly different in others. *Ixodes holocyclus* is among the most virulent of paralyzing ticks, and is common in moist areas along the eastern coast of Australia. Approximately 10 females can paralyze young calves 30 to 40 kg in weight, and 20 to 25 can paralyze calves 80 to 160 kg. The main hosts of the tick are various species of bandicoots, but the tick will cause paralysis in dogs, cats, domestic livestock and humans. Paralysis is associated with feeding by the adult females, and these are active from June through December. Toxins are secreted by the tick after the third day of attachment, and paralysis occurs after feeding for 4 to 5 days. Variation in virulence has not been found as has with *D. andersoni*. Symptoms are similar to those caused by *D. andersoni*, including an ascending flaccid paralysis, however, the victim may become acutely ill, and vomiting may occur. The symptoms may worsen following removal of the tick, and symptomatic treatment or administration of a canine hyperimmune antiserum. Immunity to paralysis can develop through previous exposure, through mother's milk, or by immunization.

The toxin produced by *Ixodes holocyclus* appears to have a different mode of action compared to other tick paralysis toxins, and may act to inhibit acetylcholine release at the neuromuscular junction. Considerable research has been conducted to isolate and identify the toxin produced by *Ixodes holocyclus*. Initial work suggested the toxin was a 40 to 80 kDa protein, but more recent studies suggest the toxin is of a lower molecular weight and related to scorpion neurotoxins. Toxins from other tick species, such as *Rhipicephalus evertsi evertsi,* are approximately 68 kDa.

The evolutionary significance of the toxins remains unclear. Toxins may have evolved to reduce mortality caused by host grooming, may have been conserved from an ancestor that used venom to immobilize prey, or may have other functions associated with tick-host interactions and paralysis is accidental.

Tim Lysyk
Agriculture and Agri-Food Canada
Lethbridge, Alberta, Canada

References

Gothe, R. 1981. Tick toxicoses of cattle. pp. 587–598 in M. Ristic and I. McIntyre (eds.), *Diseases of cattle in the tropics. Current topics in veterinary medicine and animal science*, Vol. 6. Martinus Nijhoff, The Hague, The Netherlands.

Gothe, R., K. Kunze, and H. Hoogstraal. 1979. The mechanisms of pathogenicity in the tick paralyses. *Journal of Medical Entomology* 16: 357–369.

Gregson, J. D. 1973. *Tick paralysis: an appraisal of natural and experimental data.* Monograph No. 9. Agriculture Canada.

Stone, B. F., B. M. Doube, and K. C. Binnington. 1979. Toxins of the Australian paralysis tick *Ixodes holocyclus*. pp. 347–356 in J. G. Rodriguez (ed.), *Recent advances in acarology*, Vol. 1. Academic Press, New York, New York.

Wilkinson, P. R. 1982. Tick paralysis. pp. 275–282 in J. H. Steele, G. V. Hillyer, C. E. Hopla (eds.), *CRC handbook series in zoonoses.* CRC Press, Boca Raton, Florida.

TICKS. Ticks are obligatory blood feeding ectoparasites of reptiles, birds and mammals. They belong to the subclass Acari, order Ixodida. Ticks are the largest mites, ranging in length from about 2 mm to 30 mm depending on the species and life stage. All ticks pass through an egg and three active stages, a six legged larva also known as a "seed tick," an eight legged nymph and an adult. Each active stage molts into the next and, with few exceptions, each requires a blood meal before molting. The complex relationship of the tick to its vertebrate host that may date back to the dinosaurs, and the need for the tick to obtain a blood meal have been the principle driving forces for their evolution.

Classification of ticks: the order Ixodida

About 850 species of ticks are arranged in three families:

Class: Insecta
  Order: Ixodidae
    Family: Agasinae
      Subfamily: Argasinae
      Subfamily: Ornithodorinae
      Subfamily: Otobinae
      Subfamily: Antricolinae
      Subfamily: Nothoaspinae
    Family: Nuttallielinae
    Family: Ixodidae
      Subfamily: Ixodinae
      Subfamily: Rhipicephalinae
      Subfamily: Amblyomminae
      Subfamily: Haemaphysalinae
      Subfamily: Hyalomminae

Argasidae. The argasidae with some 159 species is divided into five subfamilies each with one genus: argasinae (*Argas* with seven subgenera and 56 species), Ornithodorinae (*Ornithodoros* with 100 species), Otobinae (*Otobius* with two species), Antricolinae (*Antricola* with about eight species), and Nothoaspinae (*Nothoaspis* with one species).

Nuttallielidae. The Nuttallielidae has a single species *Nuttalliella namaqua* known only from females and nymphs. It is found in South Africa and Tanzania where it is believed to be a parasite on small mammals such as rodents and the rock hyrax. What little is known about this species suggests that it has many characteristics in common with both of the other two tick families.

Ixodidae. The Ixodidae, with some 650 species, is divided into five subfamilies. The subfamilies are grouped into two divisions, the Prostriata and the Metastriata. The Prostriata has a single subfamily, the Ixodinae, with one genus, *Ixodes*, and some 245 species. The Metastriata contains four subfamilies. The largest of these, the Rhipicephalinae, has eight

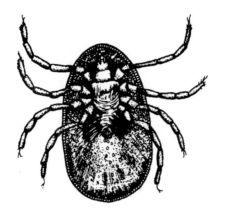
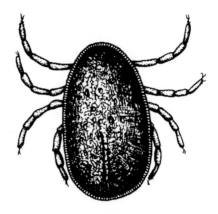

Fig. 1051 Adult of *Argas persicus*, the blue bug. The left side is a ventral view and the right side is a dorsal view. (from Marquardt et al., 2000, used with permission of Harcourt Academic Press.)

genera, *Dermacentor* with 30 species, *Cosmiomma* with one species, *Nosomma* with one species, *Rhipicephalus* with 70 species, *Anomalohimalaya* with three species, *Rhipicentor* with two species, *Boophilus* with five species, and *Margaropus* with five species. The other three subfamilies are smaller, Amblyomminae has two genera, *Amblyomma* with 102 species, and *Aponomma* with 24 species; Haemaphysalinae has one genus, *Haemaphysalis* with 155 species, and Hyalomminae has one genus, *Hyalomma* with 30 species.

External morphology of ticks

Ticks have a typical acarine body, but one that has been adapted to an ectoparasitic life. The body is divided into a movable head region, the capitulum or gnathosoma and the idiosoma, which makes up the rest of the body. There is no visible segmentation. The small dorso-ventrally flattened body of unfed ticks makes it difficult for the host to remove them.

The capitulum, or gnathosoma, consists of the basis capituli, the palps, and the chelicerae, and the hypostome. The gnathosoma is anteriorly situated in the larvae of soft ticks and in all instars of hard ticks. In postlarval instars of soft ticks, it is not visible from above, being mostly hidden by the overlapping anterior part of the idiosoma. The gnathosoma is connected with the idiosoma by a cavity, the emargination in hard ticks, or the camerostome in soft ticks. The connection is via a soft articulation membrane that allows the gnathosoma to be flexed or extended.

The basis capituli, an integumental ring that encircles the mouthparts, contains the shafts of the chelicerae, the salivary ducts and the pharynx. Its dorsum bears the area porosa in female hard ticks. The paired palps have four segments, and resemble legs in soft ticks, but have a more flattened shape than legs in hard ticks. In soft ticks, the terminal segment is normal in appearance but, in hard ticks, it is much shorter and can be retracted or protruded. The palps bear several types of setiform sensilla.

The paired two-segmented chelicerae have been modified as cutting organs that are unique among the mites. Chelicerae consist of a bulbous base, an elongated shaft, and the cutting digits, or articles. In many species, a cheliceral hood covers the articles. The cutting edges of the internal and external articles are laterally oriented. The internal article is moved from side to side by the tendons attached to the powerful muscle masses in the bulbous cheliceral bases. The external article moves with it. This movement is used by the tick to tear into the hosts skin prior to embedding its mouthparts. Sensory structures occur on the external and internal articles.

The hyopostome, located on the ventral gnathosoma, is a dorso-ventrally flattened protrusion with recurved teeth on its ventral side, and a preoral canal on its dorsal side. The preoral canal leads to the mouth and the muscular pharynx. The action of the pharynx sucks up blood through the preoral canal. The preoral canal and the space between the dorsal surface of the hypostome and the ventral surfaces of the chelicerae are the common channels for blood intake and saliva outflow. In hard ticks, the intake of

HYPOTHETICAL MALE AND FEMALE IXODIDAE (HARD TICKS)
WITH KEY CHARACTERISTICS LABELED

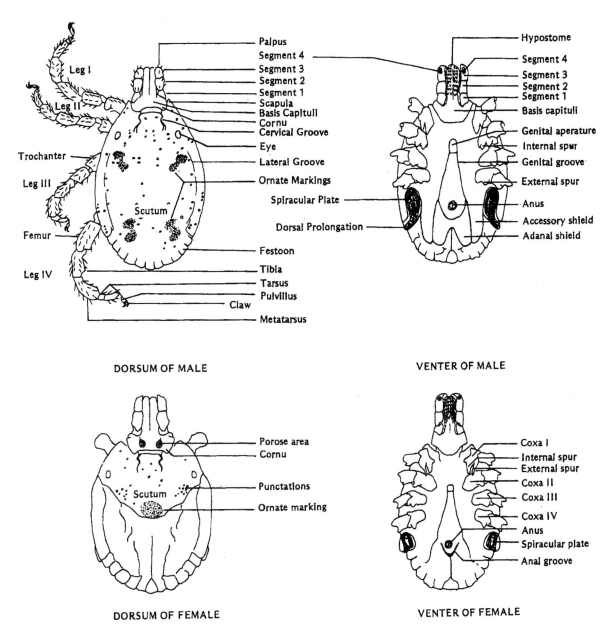

Fig. 1052 Diagrams illustrating the external anatomy of a hypothetical male and female ixodid tick with key characteristics labeled. Note that the chelicerae have been removed (compare to following figure). (from Marquardt et al., 2000, used with permission of Harcourt Academic Press.)

blood alternates with saliva flow with some pauses between the two events. At the junction of the preoral canal with the pharyngeal valve, a short and pointed flap-like labrum exists.

The mouthparts of hard ticks vary in length. Those species with longer mouthparts are referred to as longirostrate, and those with smaller mouthparts as brevirostrate.

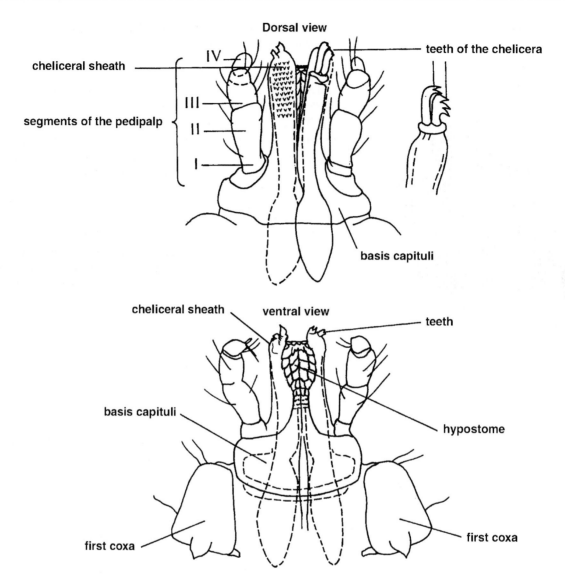

Fig. 1053 Capitulum and mouthparts of a tick. (from Marquardt et al., 2000, used with permission of Harcourt Academic Press.)

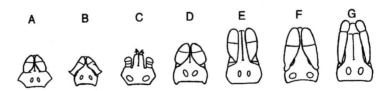

Fig. 1054 Capituli of the genera of hard ticks, family Ixodidae. (A) *Rhipicephalus*; (B) *Haemaphysalis*; (C) *Boophilus*; (D) *Dermacentor*; (E) *Ixodes*; (F) *Hyalomma*; (G) *Amblyomma*. (from Marquardt et al., 2000, used with permission of Harcourt Academic Press.)

Fig. 1055 Female *Ixodes persulcatus* in questing posture, waving its first pair of legs. (from Marquardt et al., 2000, used with permission of Harcourt Academic Press.)

The idiosoma is divided into the podosoma which bears the walking legs and, in adults, the genital opening, and the opisthosoma that lies posterior to legs IV and bears the spiracular plate and the anal opening. Some hard ticks have patterns in the cuticle, usually in the scutum or the dorum of the basis capituil, which are referred to as markings or ornamentations.

The idiosoma of soft ticks is tough and leathery and, with the exception of some larval forms, has no sclerotized plates or shields. Eyes, when present, are found laterally above the fourth pair of legs in soft ticks and, in hard ticks, on the lateral surface of the scutum. The cuticle of all Argasidae, but *Argas*, have tiny ridges (mammillae) and depressions (discs) to which muscles attach. Some hard ticks have

grooves, or festoons, along portions of the ventral posterior body margin.

The idiosoma of hard ticks bears several sclerotized plates or shields. In female hard ticks, an anterior shield, the scutum, occurs on the dorsum. The alloscutum covers the remainder of the idiosoma. The alloscutum is not sclerotized and is folded in unfed ticks. This allows the alloscutum to greatly enlarge during feeding to accommodate the large blood meal. Male hard ticks take small blood meals, the alloscutum is lacking, and the scutum covers the whole dorsum. Various shields such as the accessory shield and adanal shield occur only in male hard ticks. Sexual dimorphism is well developed in hard ticks. Soft ticks are so-called because they lack the scutum. There is little visible external difference between males and females in soft ticks, except the shape of the genital opening.

The podosoma contains the walking legs. Each leg is divided into six segments or podomeres, from proximal to distal: the coxa, trochanter, femur, tibia, metatarsus and tarsus. Muscles, hydrostatic pressure from hemolymph and bending stresses are used to produce movement. Muscles and tendons run between the joints. Joints between the body and the legs, between the podomeres, and between the tarsus and the apotele allow movement of the legs. The direction of movement is determined by the type of articulation, the extent of the flexible arthrodial membranes, and the insertion points of the muscles and the position of the tendons. The tarsus of each leg has an apotele, or pretarsus, which includes the claws and, in hard ticks, only the pulvillus. The pulvillus is a flap-like structure with a central lumen containing lipid compounds that are secreted onto the surface and may act as an adhesive that allows the tick to walk on almost any surface and to move vertically.

The opisthosoma contains the opening of the respiratory tract and the anus. Adult and nymphal ticks respire through tracheae, most larvae and all eggs respire through their integument. However, the larvae of some species of *Argas* and *Ornithodoros* have a simple tracheal system that opens through minute apertures between coxae I and II. Tracheae in other ticks open to the outside through a complex structure, the stigmata (or spiracle). This is associated with a sclerotized elevated region of the cuticle, the spiracular plate. A pair of spiracular plates occur near coxae IV in all adult and nymphal ticks. The spiracles of soft and hard ticks are similar, although the spiracular plate is often less conspicuous in the Argasidae. The exchange of gases occur through pores that perforate the spiracular plate and lead to a labyrinth of chambers that in turn lead to the trachea. Spiracular pores help prevent dust and other debris from entering the trachea. Pillars of cuticle (pedicles) run from the floor to the top of these chambers. A valve-like structure, the atrial valve, occurs beneath the spiracular plate. High carbon dioxide concentrations stimulate the atrial valve to open.

The tick respiratory system is designed, in part, to conserve water. Ticks that live in arid habitats have the most efficient spiracles to reduce water loss. The pores and the elaborate labyrinth system beneath the sieve plate function to reduce water transpiration during respiration. When stigmata remain open, loss of water increases, which explains why a discontinuous ventilation cycle exists in hard ticks A discontinuous ventilation cycle coupled with a low metabolic rate is probably the reason unfed adults can survive long periods of starvation and desiccation off the host. Engorged hard ticks ventilate continuously with little spiracular control.

Host seeking and tick feeding

Most ticks are ambushers, seeking their hosts in a passive manner. The exceptions are some species of *Hyalomma*, which are active hunters that will crawl several meters toward a host after perceiving their odors and movements. Ticks have evolved specialized structures to sense the presence of a host. The most important of these is Haller's organ, a specialized area of the dorsal surface of the first pair of legs found on all ticks. Haller's organ, along with other setae, can sense CO_2, ammonia found in sweat and urine, hydrogen sulfide found in the breath, belches or flatulence, nearby movement, and a rapid rise in temperature. When seeking a host ticks extend their first pair of legs and assume a questing position. Ticks process sensory information through their central nervous system, or synganglion, which is described in the section on mites. Host seeking behavior is rhythmic in some ticks. *Rhipicephalus appendiculatus*, which mostly infests cattle, has a bimodal diurnal periodicity, that is, a peak of host seeking that occurs twice during each day. The peaks shift depending on the time of year.

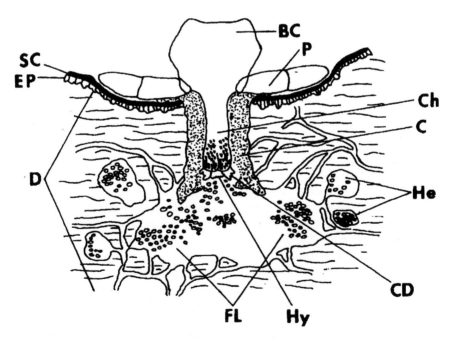

Fig. 1056 Diagram of the feeding lesion that develops beneath the attached female ixodid. The lesion is filled with blood and inflammatory infiltrates. BC, basis capituli; C, cement cone; CD, cheliceral digit; Ch, outer cheliceral sheath; D, dermis of host skin; EP, epidermis of host skin; FL, feeding lesion; He, hemorrhage; Hy, hypostome; P, palps; SC, stratum corneum of host skin. (from Coons and Alberti, 1999, used with permission of John Wiley and Sons, Inc).

Once on the host, a tick selects a feeding site then embeds its mouthparts into the skin. First, the tick uses its legs to elevate the body to a sharp angle, then it cuts the skin with its chelicerae and inserts the hypostome and chelicerae into the wound using a rocking motion of the body. The palps are splayed out on the surface of the hosts skin. Ticks are pool feeders, that is, they feed on blood that flows into a wound site in the case of soft ticks, or a feeding lesion in the case of hard ticks. Soft ticks create wounds by using their chelicerae to tear into small blood vessels and capillaries. Soft ticks are rapid feeders compared to hard ticks, finishing their blood meal in 20 to 60 minutes depending on the species. Exceptions to this rapid feeding are the larvae of many species of *Argas* and *Ornithodoros*, which require several days to complete a blood meal.

After insertion of the hypostome, all hard ticks, with the exception of some species of *Ixodes*, secrete a substance from their salivary glands that hardens into a cone-like structure. This ''cement'' cone covers the mouthparts but not the palps. This cone has multiple functions. It cements the tick onto the host. It creates a feeding tube that make the mouthparts more effective at taking up blood. It limits contact with the host to a small opening at the apex of the cone. Hard ticks do not create a wound at the feeding site similar to that of soft ticks. Instead, a specialized feeding lesion develops beneath the cement cone. This lesion does not develop without the host's initial inflammatory reaction. Both the bite wound of soft ticks, and the feeding lesion of hard ticks are maintained by saliva from the tick's salivary glands.

The paired salivary glands of ticks are complex, multifunctional organs that are very important in off-host physiology as well as feeding. Salivary glands consist of grape-like acini connected to a system of salivary ducts. The ducts lead to a preoral cavity, the salivarium, which is isolated by a trap door mechanism. When the salivarium is opened, saliva flows into the host; when it is closed, blood is sucked into the digestive system by the action of the muscular pharynx. Both soft and hard ticks alternate blood sucking with saliva production. Acini contain specialized cells that produce the many bioactive compounds found in the saliva. One type of acini produces a salt that takes up atmospheric water when secreted on the mouthparts of nonfeeding ticks. This is used to replenish the water lost during the off host part of the tick life cycle. The

salivary glands of hard ticks undergo a great developmental change during feeding to become organs of osmoregulatiori that return excess water and ions from the bloodmeal back into the host via the saliva. Tick salivary glands are innervated with a catecholaminergic-like synapse.

Tick saliva compounds have anti-hemostatic, anti-inflammatory and immnuosuppresant properties. These compounds help the tick to obtain an adequate bloodmeal and avoid rejection by the host.

Antihemostatic compounds in tick saliva inhibit platelet adhesion, activation and aggregation, blood coagulation, and vasoconstriction. Some compounds promote vasodilation. The importance of preventing platelet activation at the tick feeding site is apparent by the number of compounds that *Ornithodoros moubata* has evolved to prevent platelet activation that include at least five different compounds. Moubatin, a 17 kDa polypeptide, Tick Adhesion Inhibitor, a 15 kDa polypeptide, and disagregin, a 6 kDA peptide, prevent exposed collagen on damaged blood vessel walls from activating platelets. Disagregin also prevents the formation of fibrinogen and ADP. Apyrase, an enzyme that hydrolyzes adenosine diphosphate (ADP) is also present in the saliva of *Ornithodoros moubata*. ADP can activate platelets and is released from injured cells and activated platelets. Apyrase is found in the saliva of a wide variety of ticks in addition to *O. moubata*. An exception is *Ambylomma americanum*, which has no apyrase. In this tick, inhibition of platelet aggregation is carried out by high concentrations of prostaglandins (PGE_2 PGD_2). *Ixodes scapularis* saliva also contains high levels of prostaglandins (PGI_2, PGE_2). The prostaglandins are the most powerful known inhibitor of platelet aggregation known and can actually cause desegregation of aggregated platelets. Prostaglandins, PGI_2, PGE_2 and PG_2, are also potent vasodilators that can help increase blood flow during feeding.

Blood coagulation is initiated by either the extrinsic system or the intrinsic system. The extrinsic system is initiated by exposure of blood to subendothelial components such as collagen, which then activates tissue factor. The intrinsic system is initiated by the release of tissue thromboplastin from injured cells. Each system leads to a common reaction that involves factor Xa (the Stuart factor), catalyzing prothrombin to thrombin. Thrombin catalyzes the conversion of soluble fibrinogen to an insoluble fibrin mesh that forms a blood clot. Most tick anticoagulants inhibit either the serine protease factor Xa, or thrombin. An exception to this is a compound from *Dermacentor andersoni* saliva that inhibits the serine protease VIIa, and a compound from *Dermacentor variabilis* salivary glands that inhibits tissue factor.

The vertebrate host's inflammatory and immune systems are closely linked and many anti-inflammatory compounds of tick saliva also inhibit or suppress the host's immune system. Compounds in tick saliva can inhibit both the innate and the acquired immune response of the vertebrate host. Innate immune responses include those mounted by the host immediately without the requirement of previous contact with the foreign invader. These include the process of inflammation and a series of soluble proteins (complement) that can destroy or damage foreign objects. Other components of the innate response are natural killer cells and the interferons. Studies have shown that some factors in tick saliva can inhibit complement, suppress the activity of natural killer cells and also suppress the anti-viral action of some interferons. This includes the production of nitric oxide by host macrophages, which is used to destroy foreign objects. The acquired immune response is highly specific and requires prior exposure to the foreign object which must be antigenic. Acquired immunity can be divided into humoral immunity carried out by antibodies, which are bloodborne proteins of the Immunogloblin superfamily manufactured by B lymphocytes or B cells, and cell-mediated immunity, which is carried out by T-lymphocytes or T cells. Compounds in tick saliva suppress antibody production and responses of host T-lymphocytes. The cement cone, which is known to be antigenic, binds host immune response factor IgG. This may act as a molecular mask to prevent recognition of the cement cone as a foreign object by the host's acquired immune response. A novel family of immunoglobulin-binding proteins (IGBPs) may also play an important role in evading the host's acquired immune response. IGBPs have been found in all ixodid species studied to date, and have been isolated from both saliva and hemolymph. IGBPs secreted by males co-feeding with their mates have been shown to enhance the female's ability to complete her blood meal and lay a larger clutch of eggs.

After formation of the feeding lesion early in feeding, hard ticks suppress the host's inflammatory response. The high levels of prostaglandins in tick saliva inhibit neutrophil function and suppress the release of inflammatory mediators from mast cells. Bradykinin, an inflammatory mediator of vertebrates that potentiates pain, is deactivated by the saliva of *Ixodes scapularis*. Histamine binding compounds and host immunoglobulin binding compounds have been identified in tick saliva. Histamine is an important mediator of the inflammatory response, and immunoglobulins mediate the host's immune response.

Tick drop-off from the host after feeding is not uniform with respect to day/night cycles. Completion of feeding at certain times has evolved to ensure that ticks drop off into habitats that are the most advantageous for their reproduction and for the availability of hosts for the next life stage. Photoperiod is the most common external factor that affects drop-off timing. Individual ticks of the same species feeding on the same host may finish at different times.

Physiology of ticks

Digestive system

Ticks have a typical acarine digestive system consisting of a foregut, midgut and hindgut. The foregut is divided into a mouth, a muscular pharynx that is designed to suck up blood, and an esophagus. The midgut is divided into a ventriculus, or stomach, and several ceaca. Digestion takes place in the midgut, which is the largest division of the digestive system. A peritrophic membrane occurs in several species of ticks. A short hindgut leads from the midgut to the anal opening. Digestion starts when the bloodmeal is taken up by the midgut cells using receptor-mediated endocytosis into coated vesicles. These are sorted into endosomes that fuse with lysosomes. The bloodmeal is digested in the lysosomal system in a process termed heterophagy. The end products are residual bodies that accumulate in the cell as digestion proceeds. The products of digestion are released into the hemolymph. A pulsating organ, the heart, is located dorsally and helps circulate the hemolymph throughout the body cavity. Soft ticks digest only a portion of the bloodmeal as needed. This probably contributes to their ability to live for relatively long periods without an additional bloodmeal. Hard ticks digest the entire bloodmeal, but do so in stages. The first stage occurs before rapid engorgement. Cells in the epithelial lining of the midgut fill up with residual bodies during the digestive process and are sloughed off into the midgut lumen. They are replaced by new cells that repeat the digestive process, slough off and are replaced. The second stage of digestion occurs during rapid engorgement. To enter rapid engorgement, a female must be inseminated. During rapid engorgement, midgut cells become greatly distended and are filled with endosomes containing the bloodmeal. At this time, digestion is minimal. Following completion of the bloodmeal, the female drops off the host and the third stage of digestion occurs. During this time, all the bloodmeal is utilized and the midgut cells fill up with residual bodies. Cells do not slough off during second or third stage of digestion.

Osmoregulatory and excretory organs

Osmoregulatory organs are used to eliminate the excess water and ions of the blood meal. In hard ticks, salivary glands develop into osmoregulatory organs and return water and ions to the host as saliva. In soft ticks, coxal glands are the osmoregulatory organs and eliminate excess water and ions to the outside through the coxal fluid, which appears as a colorless drop between the base of legs I and II during, or shortly after, feeding. Coxal glands are absent in hard ticks. In hard and soft ticks, paired Malpighian tubes accumulate guanine, the end product of nitrogenous metabolism. Malpighian tubes empty into the lower digestive tract near the junction of the midgut and hindgut. Guanine is then expelled through the anus as a white paste-like substance along with the undigested portion of the bloodmeal, which appears as small red pellets.

Water loss

Unfed ticks must prevent desiccation to survive. This is accomplished by physical barriers and physiological and behavioral adaptations. The most important physical barriers are the impermeable cuticle and a discontinuous respiratory cycle in which the spiracle remains closed except during gas exchange. Physiological adaptations include the excretion of guanine which is a dry waste product, and dry fecal material. An important behavioral adaptation is that ticks move down close to the ground into microenvironments with increased humidity following questing. Here, the tick can actively reabsorb water from the environment using a salt solution secreted onto its mouthparts by the salivary glands.

Reproduction

Male ticks inseminate their mates by transferring packets of sperm in a structure termed a spermatophore. Parthenogenesis is rare, but occurs in both hard and soft ticks and involves thelytoky where females produce only females. In female ticks, a large amount of the blood meal is converted to the female-specific glycolipoheme protein vitellogenin, which is secreted into the hemolymph and taken up by the eggs to become vitellin. Vitellin is the yolk of the tick egg and is the main energy source of the developing embryo.

Soft and hard ticks wax their eggs using Gene's organ which is located in the anterior body cavity. The Gene's organ turns inside out during oviposition. Unwaxed eggs dehydrate and do not develop. In female hard ticks, secretions from the porous area on the dorsal surface of the basis capituli are also added to the eggs during waxing.

Semiochemicals

Semiochemicals are informational chemicals. Two types of semiochemicals have been identified in ticks. Pheromones regulate beneficial functions between the same species, and allomones repel predators. Four different types of pheromones occur in ticks: assembly pheromones, aggregation-attachment pheromones, sex pheromones and a primer pheromone.

Assembly pheromones are common in many species of soft and hard ticks and induce clustering of ticks. These pheromones are interspecific and effect all life stages. Perception of assembly pheromones cause free living ticks to cluster. Guanine, the major excretory product of nitrogenous metabolism in ticks, acts by contact as a non-specific assembly pheromone in the species of *Argas*, *Ornithodoros*, *Amblyomma*, and *Rhipicephalus*.

Aggregation-attachment pheromones induce attachment of ticks to areas where males are feeding. They are produced by males of *Amblyomma* and affect the behavior of adults and, in some cases, nymphs. Females of *A. maculatum*, *A. hebraeum*, *A variegatum*, and *A. marmoreum* will not embed in hosts unless feeding males are present.

Sex pheromones elicit behavior patterns that lead to copulation. A pheromone in the coxal fluid of female *Ornithodoros*, which is most active a few days following feeding, elicits courtship behavior in sexually active males. In all hard ticks except *Ixodes*, there is a complex, but similar, courtship behavior. Three different types of sex pheromones are involved. A volatile attractant sex pheromone, 2,6-dichlorophenol, is produced by the foveal glands in most of these species. These glands are located on the dorsal alloscutum just posterior to the scutum in all metastriate ticks. After contact with the female, males must identify a second pheromone, the mounting sex pheromone, or they will abandon the female. A third sex pheromone, the genital sex pheromone, must be recognized by males of some species before a spermatophore is formed and copulation completed.

A primer pheromone, in this case the fecundity reducing pheromone, has been identified in *Argas arboreus*, where its effect is most noticeable when ticks are crowded. The large wax glands of *Dermacent variabilis* and *Amblyomma americanum* secrete an allomone that repels predators.

Life cycles and habitats of ticks

Many aspects of the life cycles and habitats of soft and hard ticks differ. Differences also occur between two divisions of hard ticks, the Prostriata and Metastriata.

Postembryonic development

The Argasidae have multiple nymphal stages with different species having different numbers of nymphal molts. Following a final blood meal, the last instar nymph molts into the adult stage. The Ixodidae have a single nymphal stage.

Feeding

Depending on the species, soft ticks feed on the host once to many times in a given life stage. Feeding is much shorter for soft ticks (15 minutes to several hours) than for hard ticks, although some larvae require several days to complete feeding. Soft ticks ingest a relatively small amount of blood compared to hard ticks. In a few species of soft ticks, the larval stage molts to a nymph without feeding. Hard ticks take a much larger blood meal (up to or greater than 100 times their unfed body weight in some species) than soft ticks, and feed for a long time (9 to 13 days depending on the species). Hard ticks feed only once in a given life stage. Larvae complete feeding in the shortest time, adult females in the longest time. Females pass through a feeding stage known as rapid engorgement, while the males do not. It is during this feeding stage, which occurs 24 to 48 hours prior to the completion of feeding that most of the blood meal is taken up. Depending on the species, each ixodid

tick can consume up to 15 ml of blood during feeding. This is made possible by the tick returning excess water and ions from the blood meal to the host through its saliva. Compared to females, males produce only about one-twentieth the amount of saliva.

Mating and reproductive cycles

Ticks either mate on, or off, the host. Almost all soft ticks mate off the host. Two exceptions are *Argas* (*Microargas*) *transversus*, and *Otobius megnini*. The former is a small tick that spends its entire life on the Galapagos giant tortoise, and is the only known tick to oviposit on its host. The latter tick is the spinose ear tick, an economically important species. A blood meal is not necessary to initiate gametogenesis in species of Prostriata. They can mate before feeding, and can mate on, or off, the host depending on the species. In many of the nidicolous (nest or burrow dwelling) species of *Ixodes*, males cannot feed and do not occur on hosts. Instead, they seek out females and usually mate off the host. This life cycle may have evolved in nidicolous, or nest dwelling species, to ensure that females can survive long periods without feeding and still oviposit. An exception occurs in the subgenus *Ixodiopsis* where, despite a nidicolous life, males feed. All species of Metastriata mate on the host.

Soft and hard ticks have different reproductive cycles. Soft ticks go through multiple egg laying (gonotrophic) cycles. Females lay a clutch of eggs during each cycle. Each clutch of eggs is usually progressively smaller. Autogeny, the ability to lay eggs without a bloodmeal, is obligatory in genera where females have nonfunctional mouthparts, i.e. *Otobius*, *Antricola*, and possibly *Nothoaspis*; and facultative in *Argas persicus*, and several species of *Ornithodoros*, including *O. lahorensis*. Female hard ticks go through one gonotrophic cycle, lay a single clutch of eggs, then die.

Hosts

Ticks either have multiple hosts, or have a one-, two-, or three-host life cycle. Almost all soft ticks have multiple hosts although the hosts may be the same species. Two exceptions are *Argas transversus*, and *Otobius megnini*.

All species of ixodids have either a one, a two- or a three-host life cycle. In a one-host life cycle, ticks infest the host as larvae then remain on the same host as nymphs and adults. Adults mate following a blood meal. Mated females drop off the host and lay eggs in a protected niche such as a crack in the ground, or under vegetation litter. About 12 species of metastriate ticks have a one-host life cycle. In a two-host life cycle, the larvae infest a host then remain on the host to feed as nymphs after molting. Following completion of feeding, the replete nymph drops off to molt into an adult. The adult then infests a second host to feed and mate, after which the female drops off to oviposit. Only a dozen or so species of metastriate ticks have a two-host life cycle. Two examples are *Hyalomma anatolicum excavatum* and *Rhipicephalus evertsi*. By far the majority of metastriate ticks have a three-host life cycle where each life stage infests a host. Usually, each host is a different species and each is larger than the previous host. Molting from one stage to another occurs off the host, and mating occurs on the penultimate host then the replete female drops off to oviposit.

In their preference of hosts, ticks are categorized as either host-specific or opportunistic. Host-specific ticks are less common, and are considered specialists. An example is the cattle tick *B. microplus*. These ticks respond dramatically to the odors of cattle, but not to human odors. Opportunist ticks are generalists and have a wide range of hosts. Two examples are *Ixodes scapularis*, whose hosts include a number of species of birds and mammals, and *Amblyomma americanum*, whose host list includes reptiles, birds and mammals.

Many species of ticks have repeated feeding success on their natural hosts, but not on unnatural laboratory hosts. Repeated infestations of *Dermacentor variabilis* on their natural hosts, such as deer mice, results in successful feeding, but when larvae and nymphs are fed on guinea-pigs, feeding success is reduced. Successful molting to the next stage by ticks that have completed feeding is also reduced.

Habitats

Ticks can be divided into nidicolous and non-nidicolous. Nidicolous ticks either live in, or in close association with, the host's dwelling. That is, the host's nest, burrow, cave or shelter. Non-nidicolous ticks live throughout the host's range. Almost all species of Argasidae and many Prostriata are nidicolous. Most Metastriata are non-nidicolous. The behavior of nidicolous ticks is adapted to specialized niches. Such behavior can include (1) the ability to survive for years in the absence of a host. (2) negative phototropism, an avoidance of light, (4) thigmotropism, a posi-

tive reaction to contact with a solid object, so these ticks prefer small cracks or crevices in their environments that contact both surfaces of their body, (5) a narrow tolerance of temperature and humidity, often the optima is that of the host's residence.

Non-nidicolous ticks live throughout the host's range, but most are not distributed in a uniform fashion throughout this range. Some are found in several different habitats of their host's range, while others have a single preferred type of habitat, for example a deciduous forest, where most of their population is found. Important factors that determine this non-uniform type of distribution are climate, vegetation, availability of hosts, presence or absence of diapause, and the ability of the tick species to withstand adverse environmental conditions. No single ecological factor can account for this non-uniform type of distribution.

The presence of diapause enables many tick species to synchronize their populations with the presence of hosts or with favorable climatic conditions. Diapause is a state of low metabolic activity in an arthropod that is mediated by neurohomone(s). Two types exist in ticks, behavioral diapause, or developmental diapause. Ticks in behavioral diapause will not feed when offered hosts and often fail to quest for hosts. Behavioral diapause is the most common type. Developmental diapause involves delayed completion of development in an egg, or fed immature. Developmental diapause is often an important regulator of the developmental cycle in nidicolous ticks that use migratory birds or bats for hosts. Usually this diapause occurs as delayed oviposition, which ensures that expansion of the tick population is related to the presence of the host. Great numbers of nidicolous ticks can occur in a given nest or burrow. Bird nests can contain up to 20,000 ticks per square foot, and warthog burrows have contained 250,000 ticks per burrow. In some cases, seabirds have abandoned their nests and fledglings due to the presence of large numbers of ticks. Diapause has not been reported in nidicolous ticks that have non-migratory hosts. Diapause is widespread in non-nidicolous ticks. In *Ixodes rincinus*, developmental diapause exists in the egg and all active stages, which results in separate spring and fall feeding populations. Delayed oviposition (developmental diapause) occurs in *Dermacentor marginatus*. Behavioral diapause occurs in *Dermacentor albipictus*, *Dermacentor andersoni* and *Dermacentor variabilis* to name a few species.

Disperson

Ticks move slowly over short distances and must rely on their hosts for dispersal. Most widespread tick species have hosts that, in turn, are widely dispersed, or are migratory. An example is the brown or red dog tick, *Rhipicephalus sanguineus*, which is found throughout the world. Many ticks have at least one life stage that infests migratory birds that can carry them between continents. Using hosts as mechanisms of dispersal has resulted in the spread of tick transmitted diseases through large distances, and has made many of the diseases transmitted by ticks multifocal.

Ticks of medical and veterinary importance

Three genera of soft ticks, *Argas*, *Ornithodoros* and *Otobius* have species that are medical or veterinary pests. The other genera, *Antricola* and *Nothoaspis*, are parasites on bats.

The genus *Argas* is divided into seven subgenera, two of which, *Argas* and *Persicargas* infest birds. The other subgenera are parasites on bats and small mammals. Almost all species in the genus *Argas* are nocturnal and occur in arid habitats with long dry seasons. *Argas* (*Persicargas*) *persicus*, the fowl tick or blue bug, is the most wide spread soft tick on poultry. Larvae of this tick feed on the host for 2 to 10 days depending on the situation. Fed larvae drop off and molt in 4 to 16 days. Nymphs seek hosts and complete feeding within 30 minutes. There are three nymphal stages. Adults complete feeding within 45 minutes. Females feed before laying each clutch of eggs, and this cycle may occur daily. Several hundred eggs are laid in the early clutches, and only a couple of dozen in the later ones. *Argas* (*Persicargas*) *persicus* originated in the Palearctic region but, following domestic poultry, this tick has spread throughout the world with the possible exception of the Neotropical region. In addition to domestic poultry, it is found on wild birds. The fowl tick transmits two diseases to poultry, fowl spirochetosis and a rickettsial disease caused by *Aegyptianella pullorum*. Three other closely related ticks, *Argas* (*Persicargas*) *walkerae*, *Argas* (*Persicargas*) *persicus*, and *Argas* (*Persicargas*) *arboreus*, are also parasites on birds. In the southern U.S.A., *Argas persicus*, *Argas sanchezi*, and *Argas radiatus* are parasites on poultry, but are not problems in large commercial poultry houses.

Three species of *Ornithodoros*, *Ornithodoros moubata*, *Ornithodoros savignyi*, and *Ornithodoros lahorensis*, are important parasites of man and domestic animals. The African tampan, or eyeless tampan, *Ornithodoros moubata*, is a complex of at least four species. *Ornithodoros moubata* sensu strictu (in the narrow sense) occurs throughout semi arid, or arid Africa from Kenya southward to South Africa. The life cycle is relatively short, about four months from egg to adult. The larvae are non-feeding and molt directly into a nymph. There are three or more nymphal stages. Females deposit several hundred eggs per batch. More than seven clutches are laid, which can total over 2,500 eggs. Large infestations of this tick can kill pigs. The tick can become a parasite of humans that live in huts or in dwellings with cracks. *Ornithodoros moubata* is the vector of *Borrelia duttoni*, which causes endemic relapsing fever. Other species in this complex include *Ornithodoros moubata porcinus*, found in Africa, *Ornithodoros tholozani* and *Ornithodoros erraticus*, found in the Middle East. *Ornithodoros moubata porcinus*, the African warthog tick, and *Ornithodoros erraticus*, transmit the causative agent of African swine fever. The eyed tampan, *Ornithodoros savignyi*, occurs in the arid regions of Africa, the Middle East, India and Sri Lanka where it is a parasite on domestic stock and humans. It does not transmit disease, but heavy infestations can cause severe damage. *Ornithodoros lahorensis* is a parasite of domestic stock throughout its range in Asia and the southern republics of the former Soviet Union. *Ornithodoros hermsi*, *Ornithodoros turicata* and *Ornithodoros parkeri*, feed on humans along with other hosts, and can vector *Borrelia* species that cause relapsing fever.

The spinose ear tick, *Otobius megnini*, is a parasite on cattle, horses and companion animals. It occurs mostly in stables and in other animal shelters. The spinose ear tick has spread from America to southern Africa and India, and has been reported from humans in the later region. The larvae have a capitulum that is over one-third its body length. Larvae invade the ear and feed for 5 to 10 days then molt to a nymph. Nymphs reattach within the ear. The integument of the nymph has numerous spines from which the tick gets its common name. The second stage nymph drops off the host and seeks small, hidden areas such as cracks where they molt to a fiddle-shaped adult. The adult does not feed. Its hypostome is small and has no teeth. Mated females lay up to 1,500 eggs in cracks in the walls of the shelters at a height suitable for transfer to a large host. Larval spinose ear ticks die within four months, but adults can live up to two years without hosts.

In the Ixodidae, the genus *Ixodes* has the largest number of species. All are three-host ticks without eyes and festoons. The capitulum is much longer in the female than in the male. The most important species group is the *Ixodes ricinus-persulcatus* complex which has a wide distribution from North America south into Mexico, and in Europe south to the Sahara, and in Asia south to the Himalayas. This complex is important in vectoring pathogenic viruses of man, the *Borrelia* that causes Lyme disease, and the protozoa that cause piroplasmosis. Important species of this group are the blacklegged tick, or American deer tick, *Ixodes scapularis* (= *dammini*), found in North America, the European sheep tick, or European castor bean tick, *Ixodes ricinus*, and the tiaga tick, *Ixodes persulcatus*. *Ixodes scapularis* occurs in North America. *Ixodes ricinus* is found along the western parts of the British Isles, and Norway southward to Iran and Turkey, Bulgaria, Italy and the Pyrenes. *Ixodes persulcatus* has a wide distribution from Japan westward into Germany. All three are vectors of the causative agent of Lyme disease. *Ixodes persulcatus* is also the main vector of the virus that causes Russian spring-summer encephalitis. This tick is more tolerant of temperatures and is more cold hardy than *Ixodes ricinus*. Two other species, the Karoo paralysis tick *Ixodes rubicundus* in southern Africa, and *Ixodes holocyclus* in Australia, cause paralysis in mammals. Lyme disease is transmitted by *Ixodes rincinus* in Europe, *Ixodes scapularis* in the Eastern U.S. and *Ixodes pacificus* in the Pacific coast states and intermountain west of the U.S. *Ixodes pacificus* is a major cause of tick paralysis.

The European sheep tick *Ixodes ricinus* is a 3-host tick with a 2 to 6 year life cycle depending on location. Immatures attack birds, but sheep can be the hosts for all three life stages. This tick has separate spring and fall feeding populations. The larvae, seek a host in the spring or fall, feed for 3 to 6 days, then drop off and molt. After molting, the larvae ascend grass and twigs to assume a questing position but, here, they tend to lose body water causing them to descend to a microclimate on or near the ground where water is at or near saturation. They rehydrate in this microclimate. The nymphs seek a host the next spring or fall, engorge for 3 to 5 days, then drop off the host and molt. The adults seek a host the next spring or fall. Females take from 5 to 14 days to complete feeding. Fertilized females lay from 500 to 2,000 eggs that hatch by late spring or late fall, but the larvae usually do not feed until the following spring or fall.

Ixodes scapularis is found from Maine south to Florida and west into central Texas and possibly Mexico, and from Maine west into Minnesota and Iowa. It is a small 3-host tick that attacks a wide variety of birds and mammals including man. Larvae and nymphs attack small rodents especially the white footed mouse *Peromyscus leucopus*. Adults occur on larger animals with the white-tailed deer *Odocoileus virginianus* being the most common. A large population of deer is an important factor in the presence of large populations of *Ixodes scapularis*. Changes in agriculture and patterns of human use in the eastern U.S. have resulted in humans coming more into contact with areas supporting populations of *Ixodes scapularis*. The tick population in these areas has increased dramatically due to the increase in the deer population. This tick is an important vector of the causative agent of Lyme disease, human babesiosis and human granulocytic ehrlichiosis. The Western black legged tick, *Ixodes pacificus*, is found on the pacific coast states and British Columbia. It is closely related to *Ixodes scapularis*. Immatures of *Ixodes pacificus* feed on small rodents and lizards, the adults feed on deer, horses and man. Throughout its range, this tick occurs in regions of higher mositure. *Ixodes pacificus* transmits the causative agent of Lyme disease and equine granulocytic ehrlichiosis.

Species of *Dermacentor* are mostly large, ornate, brevirostrate ticks with eyes and festoons. This genus is most common in the New World. *Dermacentor marginatus* transmits Siberian tick typhus and *Dermacentor reticulatus* infests livestock. Both occur throughout Europe and Asia. *Dermacentor albipictus*, the winter tick, occurs from the west coast to the east coast of Canada and far north to almost 60° latitude. This one-host tick does not feed during the summer. Heavy infestations on large horned animals such as moose can result in death. The Pacific Coast tick, *Dermacentor occidentalis*, is found from Oregon to California where it transmits anaplasmosis. This species also causes paralysis in livestock and deer.

The American dog-tick *D. variabilis* occurs in the central and eastern United States (U.S.) northward into southern Canada and Nova Scotia. The American dog tick vectors the *Rickettsia* bacteria that causes Rocky Mountain spotted fever, the bacterium that causes tularemia and anaplasmosis. *D. variabilis* is an important pest of domestic animals and man throughout its range. It can cause canine paralysis. In the southern U.S., a life cycle can be completed in 3 months under favorable conditions. However, in the northern regions of its distribution, a life cycle can take 2 years. *D. variabilis* has expanded its range in recent years. It is found in the mid-western U.S. and Pacific states where local populations have been established, mostly along river valleys. Northward expansion is probably limited by cold temperature because the tick has not established breeding populations beyond the mean winter (December to February) $0°C$ isotherm. Westward expansion is probably limited by the lack of rainfall and deciduous forests or brushy habitats. The American dog-tick prefers field-forest ecotones. *D. variabilis* is a three-host tick. Replete females drop from the host and, within 4 to 10 days, lay a clutch of 4,000 to 6,500 eggs preferably in cracks and crevices on the ground. In the summer, the eggs begin to hatch in about 35 days depending on the temperature. The larvae feed mostly on wild mice and voles for 3 to 12 days then drop off and molt to nymphs. The nymphs also prefer wild mice and voles as hosts. They feed for 3 to 11 days then drop off to molt on the ground and emerge as adults. The adults prefer dogs and other larger mammals including man. The females engorge in 6 to 13 days. Mating takes place on the host. Unfed adults may live for more than 2 years.

The Rocky Mountain wood tick, *D. andersoni*, occurs in the western U.S. and British Columbia. It is a three-host tick with a similar life cycle to that of *D. variabilis*. The adults prefer large mammals, the nymphs smaller mammals and the larvae feed mostly on wild mice. *D. andersoni* transmits the causative agents of Rocky Mountain spotted fever, Colorado tick fever and anaplasmosis. The tick causes paralysis in mammals.

The genus *Anocentor*, has a single species, the Neotropical ear tick of horses *A. nitens*, found in the southern U.S. westward to Texas and southward to Brazil. This one-host tick has no ornamentation and the eyes are obsolescent. Some workers place this tick in the genus *Dermacentor*.

Nosomma has a single species, *N. monstrosum*, which has a wide range of hosts including cattle, buffalo, humans, boar, bear, horses and dogs, with the larvae infesting rodents. This three-host tick occurs in India and Southeast Asia.

Species of *Rhipicephalus* are brevirostrate, small, mostly inornate ticks. This genus is found in the Old World, mainly in Africa south of the Sahara and southern Arabia. One species, the brown or red dog tick *R. sanguineus*, is now cosmopolitan with a greater geographical distribution than any other species of tick. *R. sanguineus* has a wide host range preferring dogs and other carnivores, some large herbivores, lagamorphs, rodents, bats, reptiles, and some primates including humans. This tick is a warm-climate species that probably originated in Africa. Its ability to colonize shelters in cold climates has significantly contributed to its distribution. The brown dog tick vectors *Babesia canis* and *Rickettsia conorii*. The latter pathogen causes boutonneuse fever. In some areas of Mexico, this tick transmits *Rickettsia rickettsii* that causes Rocky Mountain spotted fever.

Other important species of *Rhipicephalus* on domestic animals are the brown ear tick *R. appendiculatus*, *R. evertsi* and *R. simus*. *R. appendiculatus* is a three-host tick that occurs in the eastern and southern parts of Africa south of the equator. This tick is found on goats and sheep but is mostly a parasite of cattle. Its distribution reflects an interaction between climate, vegetation and hosts. It is thought that the presence of cattle is necessary for the tick to become established. Temperatures around $4°C$ will kill all stages of engorged *R. appendiculatus*. A dry climate prevents eggs from hatching through desiccation. Under ideal conditions, a life cycle in this tick can be completed in 3 or 4 months. The number of cycles per year depend on the local conditions. In southern Africa, there is one generation per year, but in Tanzania, two generations per year can occur under optimal conditions. *R. appendiculatus* vectors *Theileria parva* which causes east coast fever in cattle, the Nanovirus that causes Nairobi sheep disease, and *Babesia bigemina* that causes babesiosis in cattle.

The two-host African red tick *R. evertsi*, infests domestic cattle, goats, sheep and wild ungulates. The eggs of this tick hatch on the ground and the larvae seek a host animal where they attach to the inner surface of the ear. After feeding to repletion, larvae detach and molt into nymphs that attach in the same area of the ear and feed to repletion then drop off the host to molt on the ground. The adults seek a second host, attaching mostly in the perianal region under the base of the tail or less commonly on the teats, the base of the legs, or the scrotum. Here they mate and females feed to repletion then drop off to lay eggs on the ground.

All five species of *Boophilus* are brevostriate, one-host ticks and all are parasites on large hoofed mammals, especially cattle. No festoons, or eyes are present. Three species are important vectors of babesiosis to cattle: *B. microplus*, the cattle tick is found in the Neotropical, Afrotropical and Australian regions; *B. decoloratus* occurs in tropical Africa; and *B. annulatus* is found in North America. The life cycle of *B. microplus* is completed in about 5 weeks under favorable conditions and can require up to several months under less favorable conditions. Females lay a single clutch of 2,000 to 3,000 eggs that hatch in about 2 weeks at 70% humidity. Larvae quest on the tips of grass and twigs but do not move down to rehydrate in a microenivronment. Larvae feed for about 4 days on the host then molt into a nymph after a quiescent period of about 2 days. The nymphs may move about the host before attaching. They feed for about a week and then molt into an adult. This molt is also preceded by a short quiescent period. The adults mate on the host. The females take about 3 weeks to feed to repletion, then drop off to lay eggs. The oviposition period lasts some 10 days and is preceded by a short preoviposition period of several days. *B. microplus* must have a high rainfall, and cannot be found in dry areas with a low humidity. This tick is widespread throughout the warmer climates of the world. In the tropics, *B. microplus* are found on cattle throughout the year. In subtropical regions, it has a seasonal cycle. A commercial vaccine has been developed against "concealed antigens" of the tick midgut. This vaccine has reduced tick fertility by as much as 70% in some cases. Booster shots are necessary.

Some species of *Margaropus*, infest giraffes in the Sudan and East Africa, but *M. winthemi*, the winter horse tick, is a parasite on horses in southern Africa. *Margaropus* are one-host ticks that are considered relic boophilids.

Amblyomma are mostly large, highly ornamented, longirostrate ticks with eyes and festoons. Most species occur in the tropics. All have a three-host life cycle. Their long mouthparts are especially damaging to cattle hides. Species from the southern U.S. are the only members of this genus from the temperate region, although some nymphs occur on migratory birds from Africa northward through Europe and Asia. The African brown ear tick, *A. appendiculatus*, and the tropical bont tick *A. variegatum*, transmit *Cowdria ruminantium*, the causative agent of heartwater fever in cattle. *A. nuttali* has the distinction of producing the largest clutch of eggs ever recorded from a single tick, over 22,000.

The lone star tick, *Amblyomma americanum*, is found across the southeastern U.S. from central Texas to the Atlantic coast and north to New York. It prefers a forested habitat. The questing activity of this tick is most active from sundown to the late evening hours when its principle host, white-tail deer, forage. Large numbers of ticks occur in, or near, the bedding of hosts. Active stages have no host preference. Immatures are found on birds and all sizes of mammals, but adults occur mostly on medium to large mammals. All stages of this tick attack man. *Amblyomma americanum* is well adapted to forest communities. Two factors that must be present to support large populations of *Amblyomma americanum* are suitable hosts, and a moist microenvironment to protect the ticks from desiccation. An area with a forest canopy and lots of vegetative ground cover is ideal habitat for *Amblyomma americanum*. Male ticks are found on deer all year. Engorgement of the females may be a photoperiod response as there is a small drop-off period that begins in May, and is over in late August. The larvae first occur on deer in late June and continue until mid-November. The nymphal activity, which occurs from March to October, is the longest of any stage. The lone star tick is a vector of Rocky Mountain spotted fever and Q fever.

The Cayenne tick *Amblyomma cajennense*, is also found in the southwestern U.S. south through Mexico and Central and South America and the West Indies. It commonly attacks man and many other animals. It transmits Rocky Mountain spotted fever from Mexico south to Brazil. The Gulf Coast tick *A. maculatum*, is a parasite on deer and cattle. The deer population has dramatically increased, and this has resulted in an increase in the population of *A. maculatum*.

Some or all species of *Hyalomma* are longirostrate, medium-sized ticks with eyes that may or may not have festoons and scutal ornamentations. This genus most likely originated in the dry deserts of Kazakhstan and Iran in the Palaearctic. Some or all species of *Hyalomma* are hardy ticks.

Several economically important species exist. The *H. marginatum* complex and *H. anatolicum anatolicum* are major vectors of the arbovirus that causes Crimean-Congo haemorrhagic fever. The *H. marginatum* complex extends from India and Indochina westward throughout southern and southeastern Europe into the Near East and North Africa. Populations are scattered throughout the drier areas of Africa south of the Sahara from the Red Sea to the Atlantic Ocean. Hosts of immature ticks in this complex include wild birds, small mammals, hedgehogs and hares. The adults attack any domestic animal especially cattle, horses and camels. Migrating birds are important in spreading these ticks. A typical *H. marginatum* life cycle is completed in 116 days at 18 to 19°C. All stages of this tick complete feeding in about 6 days. The females lay from 4,300 to 15,500 eggs in a single clutch. Unfed adults can live for over a year.

Species of *Haemaphysalis* are small, inornate, brevirostrate ticks. The genus is most common in the tropics where it probably originated. *H. punctata* in the Australian region, and *H. longicornis* in the Oriental region, are important parasites of cattle. The latter tick is also found on red deer in New Zealand. *H. leachi*, the yellow dog tick of Africa, is a common parasite of dogs and carnivores throughout the tropics and subtropics of Asia and Africa. It transmits the protozoan pathogen of malignant jaundice in dogs. In the Oriental region, *H. spinigera* transmits the arborvirus that causes Kyasanur Forest Disease. Two species of *Haemaphysalis* occur in North America. The rabbit tick, *H. leporispalustris*, occurs in the New World from Alaska south to Argentina. It has a large host list that includes horses, cats, dogs and birds. It rarely attacks man, but is responsible for the transmission of Rocky Mountain spotted fever and tularemia among wild animals. The bird tick, *H. chordelis*, is common on upland game birds in North America and is an important parasite of turkeys. *H. mageshimaenis* is found in Japan, and is rare in that it has both bisexual and parthenogenetic reproduction.

Economic importance of ticks

Ticks are of great economic importance. Many disorders are caused directly by the interaction of the host to the tick. These can occur locally at the tick bite area, but some are systemic such as toxic reactions, or toxicoses and host paralysis produced by tick toxins. Ticks have a high vector potential and transmit more varieties of serious diseases to vertebrates than any other blood feeding arthropod. They are second behind mosquitoes in transmitting diseases to humans, and first in veterinary importance. Some species of ticks have a toxin in their saliva that can cause death through paralysis. Another of the many afflictions imposed on their host has to do with blood loss through heavy tick burdens. Large mammals can harbor enormous numbers of ticks. A moose can have more than 50,000 ticks feeding on it at any given time, as can a small giraffe. The record is probably a single caribou found with more than 400,000 ticks. Heavy tick infestations also occur in livestock and companion animals where loss of blood can result in the development of anemia, or in extreme cases, death.

Several superficial local disorders such as dermatosis, inflammation, itching, swelling and ulceration can occur at the tick feeding site. In some individuals, tick feeding can produce a hypersensitivity reaction that can be local, or in severe cases, systemic, and can even result in anaphylactic shock. These reactions usually occur early in the course of tick feeding. The feeding site can become a means of secondary infections from pathogens not transmitted by the tick. The infestation of the auditory canal by ticks, otocariasis, can cause serious secondary infections. Proper removal of ticks is important to minimize secondary infections because skin ulceration and lesions can result from improper or partial removal of tick mouth parts.

Tick toxins introduced into the host through the saliva cause paralysis or death in domestic animals, some wildlife, and humans. Tick paralysis manifests itself as a motor paralysis that spreads from the lower limbs to the upper limbs and head region within hours. Paralysis rescinds following removal of the tick, or ticks, except in the case of *Ixodes holocyclus*, which is discussed below. Paralysis is most likely to occur when a tick embeds on the neck, especially near the base of the skull. Many different ticks can cause paralysis, but five species are notorious. These are *Dermacentor andersoni* and *Dermacentor variabilis* in North America; *Ixodes rubicundus*, *Rhipicephalus evertsi evertsi* and *Argas walkerae* in South Africa; and the Australian paralysis, or scrub tick, *I. holocyclus* in Australia. The toxin of the latter tick, known as holocyclotoxin, has been isolated, characterized and an antitoxin developed. The scrub tick is the most virulent paralysis tick in the world. Most cases of paralysis are caused by females because larvae and nymphs have much less toxin, and the adult males feed only by inserting their mouthparts into females. Removal of *I. holocyclus* does not rescind the paralysis, which must be treated by an antitoxin given intravenously.

The high vector potential of ticks is a direct result of the following characteristics: (1) Ticks are persistent blood feeders that stay attached to the host for long periods allowing ample time for the transfer of pathogens. Hard ticks feed for 5 to 14 days depending on the species. (2) Ticks have great longevity that enhances the chances of acquiring and transmitting pathogens. (3) Ticks have a high reproductive potential. Hard ticks lay a single clutch of eggs that can number over 20,000 depending on the species. (4) Ticks have few natural enemies due to heavy body sclerotization. (5) Pathogens in ticks can persist through transovarial transmission to the next generation, and from one life stage to another through transstadial transmission. (6) Ticks transmit pathogens through several routes. Most are transmitted through the saliva, but some *Borrelia* are transmitted through the coxal fluid. Stercoral transmission occurs through the feces and requires pathogens that can survive in the dry excrement of ticks, so this route is uncommon. However, *Coxiella burnetti* and *Rickettsia conori* are transmitted in this fashion. (7) Some pathogens are transmitted directly from an infected tick to a non-infected tick while both are feeding on the same host. This phenomena is termed co-feeding and may be an important means of maintaining tick-borne pathogens. Anti-hemostatic, anti-inflammatory and immnuosuppresant actions of tick saliva facilitate blood-feeding and may indirectly enhance pathogen transmission. This is termed saliva-activated transmission (SAT). In some diseases, such as tick-borne encephalitis, uninfected ticks can acquire the pathogen when co-feeding with infected ticks on hosts that do not exhibit a viremia, or systematic infection. This phenomenon is known as non-viremic transmission and may be due to SAT. Bacterial pathogens may be acquired by continuing to feed at a localized site where previously infected ticks have fed.

Tick transmitted pathogens include arboviruses (arthropod-borne viruses), rickettsiae, other bacteria, and protozoans. Most tick-borne diseases are zoo-noses, that is they occur in wild and or domestic animals that act as a natural reservoir of infection. Man is only an incidental host in many of these diseases. Some zoonotic diseases are transmitted within their natural reservoirs by a species of tick that does not attack man. A different tick species then transmits the disease to man. If the zoonoses includes a large wild animal reservoir with multiple vector species of ticks it becomes difficult to control.

Tick-borne arboviruses are members of three families, the Flaviviridae, the Bunyaviridae and the Reoviridae. The Flaviviridae includes the genus *Flavivirus* whose virons are spherical, about 40 to 50 nm in diameter with a lipoprotein envelope, and a genome that consists of a single molecule of single-stranded RNA. The Bunyaviridae includes the genus *Nairovirus* in which all species are transmitted by ticks. The *Nairovirus* are spherical in shape, 80 to 100 nm in diameter, with a lipid envelope that has glycoprotein projections. Their genome consists of three molecules of single-stranded RNA. The Reoviridae contain the genus *Orbivirus* which causes Colorado tick fever. This virus is icosahedral in shape, 60 to 80 nm in diameter, with an outer protein coat and a genome of 12 pieces of RNA. All tick-borne arboviruses replicate in intermediate vertebrate hosts as well as in the tick. The major arboviral diseases transmitted by ticks include Tick-Borne Encephalitis, Louping Ill, Omsk Hemorrhagic Fever, Kyasanur Forest Disease, Powassan Encephalitis, Crimean-Congo Hemorrhagic Fever, Nairobi Sheep Disease, Colorado Tick Fever and African Swine Fever. Other lesser known tick-borne arboviral diseases occur in birds, especially seabirds. Treatment for human arbovirus diseases is often relegated to supportive measures. Hospitalization can help especially if a high fever or other symptoms can be treated.

Tick-Borne Encephalitis (TBE) is caused by a *Flavivirus* and can affect humans, sheep, monkeys, mice and hamsters. It is transmitted by species of *Ixodes*. In Europe and Northern Asia, it is transmitted mostly by *Ixodes ricinus*, and in areas of the far east by *I. persulcatus*. TBE can also be transmitted through fresh milk or cheese from infected goats or sheep. TBE is most common during spring and summer. The disease is maintained in nature by infections of small mammals such as rodents. Agricultural and forest workers are most at risk of infection, but urban residents that spend time in the forests and countryside are also at risk. Two subtypes of TBE are recognized, Russian Spring-Summer encephalitis (RSSE), which is found in Siberia, the southern republics of the former Soviet Union and in north-eastern China; and Central European Encephalitis (CEE), which occurs in Russia west of the Ural mountains and Europe. RSEE is more serious than CEE. RSSE is a life-threatening disease with a mortality rate of from 8% to 54%. However, an effective vaccine exists. RSSE is characterized by high fever, headache, nausea followed by the symptoms of developing encephalitis, which include paralysis especially in the upper body regions. Paralysis may persist in some patients that have recovered. The milder European form is biphasic with a brief remission after 4 to 6 days of illness followed by renewed fever. Brain dysfunction and meningeal irritation are common. Mortality in the milder CEE subtype is from 1% to 5%.

Louping Ill (LI) is caused by a *Flavivirus* and is found in Great Britain where it has been known for centuries among sheepherders in Scotland and Ireland. LI infects a variety of mammals including man, and some birds, especially upland game birds. The virus is transmitted by *I. ricinus*. Transstadial, but not transovarian, transmission occurs. The disease is so named because it causes an erratic "louping" gait in sheep. The mortality in sheep is severe, often as high as 60%, and has been known to reach 100%. LI causes a severe, fatal encephalitis that often results in permanent neurologic damage in animals that recover. In humans, LI also produces an encephalitis that can be severe and even result in death. Those at risk include farmers, veterinarians, and animal husbandry workers. A vaccine for sheep exists, but sick animals are destroyed. Lambs are protected by a maternal antibody that disappears after weaning.

Omsk Hemorrhagic Fever (OHF) is caused by a *Flavivirus* and affects humans, rats, mice, muskrats and wild rodents. OHF occurs in Siberia and is transmitted by *D. reticulatus* and *I. apronophorus*. Immature *D. reticulatus* ticks bite mostly water voles, which are found in a forest-steppe habitat. The cyclic nature of the vole population can produce a huge population of ticks, many of whom are infected with the virus. Adult *D. reticulatus* ticks infect several different large mammals and humans. The tick is the reservoir, which is maintained by transovarial transmission. *I. apronophorus* does not transmit the virus

to humans, but is thought to help maintain the virus in nature by feeding on many different small mammals especially voles and muskrats. Muskrats have a high incidence of infection (up to 30% of the population). Transmission of OHF to humans can occur through handling infected muskrat carcasses, or by drinking infected water contaminated by muskrats or voles. Muskrats are amplifying hosts for the disease. Hunters of muskrats are at severe risk. The disease occurs during the spring and summer seasons. OHF is characterized by hemorrhagic symptoms. Mortality is less than 5%. No vaccine is available, but some cross-protection occurs with the Tick-Borne Encephalitis vaccine.

Kyasanur Forest Disease (KFD), caused by a *Flavivirus* affects humans and monkeys. KFD occurs in the Kyasanur State Forest and a surrounding area in India where it is transmitted by *H. spinigera*. Increased human activity in the forest has increased the opportunity of this tick to attack humans. The pathogen is maintained within a given tick by the transstadial route, but is not transmitted transovarially. KFD has a sudden onset after a short incubation period (2 to 7 days). Symptoms include coughing, diarrhea, vomiting, a severe fever for up to 12 days and muscle pain. Mortality is 8% to 10%. Monkeys are very susceptible to KFD. They are viremic for days, during which time they have many ticks feeding on them. In this fashion, monkeys act as an amplifying host for the disease. Human cases of this disease are increasing. Individuals working in the forest are especially at risk, however, an effective vaccine exists.

Powassan Encephalitis (PE) is caused by a *Flavivirus*. It occurs in Canada, Russia and northeastern parts of the U.S. where it has been isolated from a variety of vertebrates and causes a sickness in humans, horses, and foxes. PE is transmitted by several species of *Ixodes*, *Dermacentor* and *Haemaphysalis*. *I. cookei* is an important vector of PE in North America. No vaccine for PE exists.

Crimean-Congo Hemorrhagic Fever (CCHF) is caused by a *Nairovirus* that is transmitted by a wide range of tick species. It was first described in 1100 AD. CCHF is widely distributed throughout Europe, Asia and Africa, and is most common in arid or semiarid regions of these areas. CCHF is transmitted mostly by ticks of the genus *Hyalomma*, or by contact with blood or tissues from human patients or infected livestock. Person to person spread can occur through respiratory secretions and excreta, which can cause serious outbreaks of the disease in hospitals. In Eurasia, *Hyalomma marginatum marginatum* and *D. marginatus* are the principal vectors. The clinical disease has been found only in humans, but the virus occurs in a wide variety of mammals. CCHF is found in isolated enzootic foci throughout its range. An enzootic disease occurs in a population of animals at all times. Ground-feeding birds are important in maintaining the disease and migratory birds are important in its spread. Incubation of CCHF takes from 3 to 7 days. The disease has a sudden onset with fever, chills, photophobia and severe headache. Muscle pain occurs mostly in the legs and back. The hemorrhagic state of the disease involves bleeding from the mucous membranes and the appearance of a round red spot, which is due to intradermal hemorrhage. Mortality is from 10% to 50%. Ticks are the reservoirs of CCHF. The rate of infection in ticks is amplified by feeding on infected mammalian hosts, and by transstadial and transovarial transmission. A vaccine exists in Russia and parts of Eastern Europe. Controlling this disease is very difficult due to its widespread distribution and the large number of different tick species involved.

Nairobi Sheep Disease (NSD), also known as Ganjam virus disease, is caused by a *Nairovirus*. NSD is found in sheep, goats, and wild ruminants in East Africa. The principle tick vector is *R. appendiculatus*. There is no evidence for a wildlife reservoir. Onset of the disease is sudden with a dramatic rise in temperature as high as 41°C. A nasal discharge and diarrhea is common, along with abdominal pain. Mortality is high. Humans exposed to the virus develop an antibody, but no symptoms, with the exception of one case. Treatment of NSD involves destroying sick animals. In areas of enzootic NSD, sheep are protected by a maternal antibody. Epizootics, diseases that spread rapidly through an animal population, are related to large increases in the vector populations and require the use of acaricides.

Colorado Tick Fever (CTF) is caused by an *Orbivirus*, and is found in the Western U.S. where it infects humans and many other mammals. The principle vector to humans is *Dermacentor andersoni*. Transovarian transmission does not occur. The disease is enzootic in wild rodents. Wherever the principle vector occurs with infected reservoirs of wild rodents and humans, local outbreaks will occur.

CTF has a short incubation period in humans of 1 to 4 days. A biphasic fever occurs in about 50% of the cases. Symptoms include chills, nausea, sore throat and retroorbital pain. Sometimes meningitis occurs, and leukopenia is common. However, mortality is low, less than 0.2%. The peak incidence of CTF occurs from April to July. Treatment is supportive and no vaccine exists. Prevention of tick infestation in focal areas of the disease is the best method of control.

African Swine Fever (ASF) occurs in Europe, Africa, South America and the Caribbean. This icosahedral-shaped virus is about 200 nm in diameter with a lipoprotein coat and a single molecule of double-stranded DNA. It is the only known DNA virus to be transmitted by an arthropod. Originally, it was placed in the family Iridoviridae, but is now unassigned to any family. Wild swine, including warthogs, harbor the virus. Domestic pigs are at risk with the mortality being as high as 100% in herds. Several species of *Ornithodoros* have been shown to be vector-competent in the laboratory, but the only natural vectors are *O. erraticus* and *O. m. porcinus*. Transstadial, transovarial and sexual transmission of ASF has been shown to exist in *O. moubata*. Many of the vector-competent species exist outside the known distribution of ASF suggesting that the disease has a good possibility of spreading. ASF is characterized by sporadic epizootics. Ticks are not the only way of spreading the disease, direct contact with animals is also a factor. Three forms of the disease are known in domestic pigs - acute ASF, subacute ASF and chronic ASF. Acute ASF causes a high fever (up to 41°C) about 3 days following infection. Then the fever subsides and the animal dies. In the subacute form of ASF, the fever follows an irregular course for 3 to 4 weeks, then the sick pigs either die, or recover and become carriers of the virus. In chronic ASF, the main symptoms are stunted growth and emaciation. In chronic ASF, sick pigs often remain carriers and can live for a long time. They eventually die from a secondary illness.

The *Rickettsia* are bacteria that are obligate, intracellular parasites of vertebrates. Unlike viruses, they have both DNA, RNA and bacterial cell walls. During evolution, the *Rickettsia* have lost several enzymes and cell components necessary to live outside the cell. Some 15 genera of *Rickettsia* are recognized, seven of which are transmitted by ticks: *Rickettsia*, *Coxiella*, *Ehrlichia*, *Cowdria*, *Anaplasma*, *Aegyptianella*, and *Haemobartonella*.

Rickettsia are divided into three groups, each containing serologically related species. Two groups, the Spotted Fever Group (SFG), and the Typhus Group (TG) are transmitted by ticks. The third group, the Scrub Typhus Group, is transmitted by a mite. The only species of TG transmitted by ticks is *R. canada*, which was first isolated from rabbit ticks in Ontario. *R. canada* has not yet been shown to cause a human disease despite a single case of possible human infection where the patient presented with Rocky Mountain Fever-like symptoms. The SFG is made up of the following species: *R. rickettsii*, *R. montana*, *R. belii*, *R. rhipicephali*, *R. parkeri*, *R. conorii*, *R. australis*, *R. sibirica*, *R. slovaca*, *R. helvetica*, and *R. akari*. In general, species of the SFG cause severe diseases characterized by headache, chills and fever. These rickettsiae grow mainly in the cytoplasm of the host's cells, but can also grow in the nucleus. Infected host cells are seldom killed by SFG rickettsiae. They have an optimal growth temperature of 32°C to 34°C. They do not cause hemolysis. The associated rash in humans is the result of damage to capillaries that causes blood to leak out. SFG rickettsia can cause collapse of the cardiovascular system.

Several SFG rickettsiae are not associated with human disease. *Rickettsia parkeri* occurs in *A. maculatum* and *Amblyomma americanum*, from Texas through Mississippi and Georgia. The principle hosts are domestic animals. *Rickettsia montana* is found in *D. andersoni* and *D. variabilis* in at least 13 states in the U.S. and is maintained naturally by small mammals. *Rickettsia rhipicephali* is found in *R. sanguineus*, *D. andersoni*, *D. variabilis*, and *D. occidentalis* in the southeastern U.S., and is maintained by small mammals. It is not pathogenic in dogs. *Rickettsia belii* is found in *D. andersoni*, *D. variabilis*, *D. occidentalis*, *D. albipictus*, *H. leporispalustris*, *A. cooleyi*, and *O. concanensis* in eight states in the U.S. Again, it is maintained in nature by small mammals.

A species of tick can harbor more than one serotype of *Rickettsia*. However, experimental evidence shows that a serotype present in a given tick interferes with the establishment of another serotype of *Rickettsia*. This phenomena is termed rickettsial interference. Ticks are considered the reservoir of infection in rickettsial diseases. This is due to the fact that infected animals remain rickettsemic for just several days and horizontal transmission of rickettsiae from an infected animal to a feeding tick is not as efficient as transstadial and transovarial transmission. Venereal transmission of rickettsiae in ticks during mating does not occur.

Treatment of rickettesial diseases is greatly dependent on the proper early diagnosis of the disease. Several effective antibiotics exist including tetracyclines, doxycyclines, chloramphenical, chloromycetin and cotrimexazol.

The most severe disease caused by SFG rickettsiae is Rocky Mountain Spotted fever (RMSF) caused by *R. rickettsii*. This disease is covered in a separate entry. The other human diseases caused by SFG rickettsiae are discussed below.

Boutonneuse Fever (BF) (= Mediterranean spotted fever, Fievre Boutonneuse, Marseilles fever, Kenya tick typhus, South African tick bite fever, Indian tick typhus) is caused by *Rickettsia conorii*, which is closely related to *R. rickettsii*. BF is widely distributed in Southern Europe, Africa, Western and Central Asia and India. It can be transmitted by many species of hard ticks. The principle vector in the Mediterranean is *R. sanguineus*, while in southern Africa, *R. evertsi*, *A. hebraeum*, *H. leachi* and *R. sanguineus* transmit the disease. In India, *I. ricinus*, *H. leachi* and *R. sanguineus* are vectors. BF is maintained in nature by a wide variety of smaller mammals. It is principally a seasonal disease. Endemic areas exist in Israel and throughout Africa. The disease is characterized by chills, fever, lymphadenitis, headaches and joint and muscle aches following a 5- to 7-day incubation period. Fever can reach 40°C. A distinctive ulcer, known as a tache noire, is covered with a black crust and appears at the site of the tick bite. Untreated cases recover, but some virulent strains can result in death. The pathology is similar to RMSF, but milder. Vaccines are not available. Control of ticks on dogs is the most effective method of preventing the disease.

North Asian Tick Typhus (NATT), is caused by *Rickettsia siberia*, and is transmitted by species of *Dermacentor*, *Hyalomma* and *Haemaphysalis*. The most common are *D. marginatus*, *D. silvarum*, *D. reticulatus*, *D. nuttalli*, *Hy. asiaticum*, *Hy. japonicum*, *H, punctata* and *H. concina*. NATT occurs from Siberia to Mongolia, and from Central Asia to Eastern Europe. Natural foci of this diseases exist in populations of small mammals. Ticks can harbor *R. siberia* for long periods of time. The pathogen is moved through the tick's life cycle by transstadial and transovarial transmission. In the former Soviet Union, NATT most commonly occurs in farm workers. It occurs from Spring to Fall. After an incubation period that lasts about a week, the disease manifests as chills and a fever that becomes intermittent after about a week. A small lesion may develop at the tick bite. A rash appears on the extremities and spreads to the trunk.

Queensland Tick Typhus (QTT) is caused by *Rickettsia australis* and transmitted by *I. holocyclus*, and possibly other species of this genus. The disease occurs in Queensland, Australia, mostly in a savanna habitat with intermittent rain forest. A large population of rodents and marsupials is necessary to support the tick population. QTT is a mild disease. The incubation period varies from a few days to over a week. QTT manifests as a general malaise, headache and a mild fever pattern that is often remittent. The tick bite site shows an eschar, a lesion similar to that of the tache noire seen in Boutenneuse fever. Nearby lymph nodes are enlarged and painful. A variable rash appears.

Q-Fever (QF), also known as nine mile fever or the Balkan grippe, is caused by the rickettsiae *Coxiella burnetii*, and is transmitted by many species of hard and soft ticks. The disease was first recognized in Australia in 1936, where the Q in Q-fever came from the word "Query." The disease is world wide in distribution, except in Antarctica. The pathogen is found in small mammals, reptiles, birds and domestic animals. *Rickettsia* are shed in the feces of infected animals. Unlike other members of the *Rickettsia*, *C. burnetii* can survive long period outside the host cells. It is believed to be able to survive up to six years in tick feces. This hardiness may be due to the presence of a sporelike cell in its life cycle. *Coxiella burnetii* is only secondarily transmitted by ticks. Most commonly, it is transmitted through airborne transmission, consumption of infected milk, handling of contaminated wool or hides, infected animal feces, animal bedding and contaminated clothing. The most susceptible people are agriculture and laboratory workers where sheep and goats are used for scientific experiments. The pathogen can enter the body through abrasions in the skin, inhalation in the lungs, mucous membranes, the gastrointestinal tract and possibly placental transfer. The disease has an incubation period of about 20 days. It starts with a sudden fever of 38–40°C which

may last for up to two weeks and can have a biphasic pattern. Headache, diarrhea, sore throat, sweats and chills also occur. A rash is usually absent, but when it occurs, is found on the trunk and shoulders. Pneumonia can occur in some areas with acute Q-fever. Mortality rates usually are less than 1%, and a vaccine exists.

Ehrlichia are obligate, intracellular parasites of white blood cells, especially monocytes. They cause ehrlichiosis, a serious and sometimes fatal blood disease in animals and humans. Canine ehrlichiosis, caused by *Ehrlichia canis*, and transmitted by *R. sanguineus*, is found world wide. Following an incubation period of about two weeks, the infected dog has a fever that can reach 40.5°C, edema, anorexia, conjunctivitis and pancytopenia. Dogs lose weight. In most breeds, the disease has a mild form, but in German shepherds, it may produce a severe hemorrhagic condition known as tropical canine pancytopenia. Ehrlichiosis can persist in dogs for years without clinical symptoms. *E. ewingii* causes canine granulocytic ehrlichiosis, a similar but less severe disease. *E. (= Cytoecetes) phagocytophila*, is transmitted to cattle and to sheep by *I. ricinus*. The disease, known as tickborne fever, manifests itself in weight loss, reduced milk production and sometimes abortion. It occurs in Ireland, Great Britain and is widely distributed in Europe. This disease can make young animals, especially lambs, susceptible to other more serious diseases. *E. chaffeensis* causes human ehrlichiosis. The pathogen is believed to be transmitted by *Amblyomma americanum* and *D. variabilis*, but the vector has not been proven. The disease in man is similar to RMSF. However, in human ehrlichiosis, a rash occurs in less than a 33% of patients, and a rash on the palms and feet occurs in less than 5% of patients, which differs from the rash found in RMSF. *E. equi*, which is transmitted by *I. pacificus*, causes equine granulocytic ehrlichiosis in horses and a wide variety of other mammals in California. Several *Ehrlichia* have unknown vectors, but all are assumed to be transmitted by ticks. Examples are *E. risticii*, which causes equine monocytic ehrlichiosis (= Potomac horse fever); *E. bovis*, from cattle; *E. platys*, which infects canine blood platelets; and *E. senetsu*, which causes ehrlichiosis in humans in Japan.

Two important livestock diseases are heartwater, which is caused by *Cowdria ruminantium*, and anaplasmosis, which is caused by three species of

Anaplasma. Two other species of rickettsiae are transmitted by ticks. *Aegyptianella pullorumn* is transmitted by species of *Argas*, especially *A. persicus*, and causes a disease of fowl that can be severe in young chickens. *Haemobartonella canis*, transmitted by *R. sanguineus*, is found in the red blood cells of dogs, but is not associated with any pathology. *H. canis* is widely distributed throughout the world. It is transmitted by *R. sanguineus*.

Ticks transmit pathogens from four genera of bacteria other than *Rickettsia*. These are *Borrelia*, *Francisella*, *Klebsiella* and *Staphylococcus*. *Borrelia* are helically coiled, Gram-negative, motile spirochetes that cause Lyme disease, several types of relapsing fever, epizootic bovine abortion, bovine borreliosis and fowl (avian) spirochetosis. *Borrelia* are similar to the spirochetes that cause syphilis and Leptospirosis. Lyme disease is covered in a separate entry.

Several species of *Borrelia* cause tick borne relapsing fever, a disease known since ancient times. Each of these species of *Borrelia* is transmitted by a given species of soft tick. Some species of *Borrelia* are transmitted in the coxal fluid of their vectors, but other vector species do not produce coxal fluid until they leave the host and, therefore, transmission must follow a different route. Some vectors, such as *O. hermsi* and *O. turicata*, transmit the spirochetes in their saliva. In Africa south of the Sahara, *B. duttoni* is transmitted by *Ornithodoros moubata*. In Spain, Portugal and northern Africa exclusive of Egypt, *B. hispanica* is transmitted by *O. erraticus* and causes Hispano-African relapsing fever. In Morocco and Libya, *B. crocidurae* is transmitted by *O. erraticus*, and causes North African relapsing fever. *O. erraticus* also transmits *B. merionesi* in Egypt and Senegal, *B. microti* in Kenya and Turkey, and *B. dipodilli* in Iran. All cause relapsing fever. *B. persica* is transmitted by *O. tholozani*, from China through India and Iran to areas of the former USSR into Egypt. It causes Asiatic-African relapsing fever. *B. caucasica* is transmitted by *O. verrucosus* from the Caucasus to Iraq, and causes Caucasian relapsing fever. *B. latyschewii* is transmitted by *O. tartakovskyi* in Iran and central Asia, and causes relapsing fever. American tickborne relapsing fever is caused by at least six different species of *Borrelia*: *B. hermsii* is transmitted by *O. hermsi* in the Western U.S.; *B. turicatae* is transmitted by *O. turicata* in the southwestern U.S.; *B. parkeri* is transmitted by *O. parkeri*

in the western U.S.; *B. mazzotti* is transmitted by *O. talaje* in the southern U.S.; and *B. venezuelensis* is transmitted by *O. rudis* in Central and South America. Relapsing fevers are zoonoses that involve the circulation of the *Borrelia* between reservoir hosts and vector species of *Ornithodoros*. With one exception, the primary reservoir hosts of all species of *Borrelia* that cause tick borne relapsing fevers are species of rodents, chipmunks and squirrels. The exception is *B. duttoni*, which has humans as its primary reservoir host. This form of relapsing fever is endemic in Kenya and other East African countries because the vector *O. moubata* has adapted to human dwellings, especially the huts made of mud or straw. Studies on the DNA of the relapsing fever *Borrelia* suggest that all species are closely related.

All tick-borne relapsing fevers have similar clinical features and pathology. Relapsing fever spirochetes migrate rapidly from the bite site into the hosts circulatory system. An incubation period lasts from 2 to 18 days. Symptoms appear abruptly and include headache, fatigue, chills and fever that can be as high as 41°C. Subsequent periods of fever are relapses. The first attack lasts about 3 days, followed by another after 7 days, and one or more attacks after that. The virulence of the disease abates with secondary relapses, which tend to be shorter and milder. Relapses are probably due to the pathogen's change in its antigenicity thus evading the host's immune system. Relapsing fevers are treated with antibiotics. Because all *Ornithodoros* that transmit relapsing fever *Borrelia* are nidicolous, humans are incidental hosts and are not involved in the zoonotic cycle. Relapsing fevers are enzootic and occur in humans as scattered local outbreaks.

Epizootic bovine abortion is caused by *B. coriacease*, and is transmitted by *O. coriaceus*. It is a serious problem in the western U.S. Bovine borreliosis, caused by *B. theleri*, is transmitted by the *Boophilis* species and *R. evertsi*. Fowl or avian spirochetosis is caused by *B. anserine*, and is transmitted by *A. persicus*, *A. reflexus*, and *A. miniatus*. All are species of the *Argas* subgenus *Persicargas*. The disease is pathogenic for most domestic fowl, but less so for guinea fowl and pigeons. It occurs world wide mostly as an endemic disease.

Francisella tularensis causes tularemia, which has a worldwide distribution. The most common mammals with the disease are cottontail rabbits, muskrats and rodents. Many species of ticks can transmit the disease. The pathogen undergoes transovarial transmission in ticks. In the U.S., the disease is most commonly associated with rabbit hunting. Transmission can occur in three ways: through the skin, through inhalation of the pathogen and through contaminated water or meat. Each method of transmission produces a distinct form of the disease. Entry of the pathogen through the skin from the bite of a tick results in the ulceroglandular form of tularemia. Following a two-day incubation period, the patient has a fever of up to 41°C, accompanied by chills and shaking. The fever plus enlarged lymph nodes (buboes) and severe headache can last up to a month in untreated patients. A vaccine exists, but must be administered every 3 years and is not always protective. Streptomycin is effective. *Klebsiella paralytica*, transmitted by *D. albipictus*, causes moose disease in North America. The disease is of limited importance. *Staphylococcus aureus*, transmitted by *I. ricinus*, causes tick pyaemia in sheep in Britain. This disease is not widespread or very pathogenic, and ticks may not be necessary for its transmission.

The protozoan class Piroplasmea contains the genera *Babesia* and *Theileria*. All of whose species are transmitted by ticks. These are parasites of vertebrate blood cells and cause piroplasmosis such as human babesiosis and east coast fever in cattle.

Control of ticks

The high reproductive rate, host variability, wide dispersion, secretive habits and longevity of many ticks make control difficult. Strategies for tick control involve the reduction of transmitted diseases and the reduction of numbers of ticks on animals to an acceptable economic level. Surveillance of ticks in a given area, which involves determining the species, their population density and any pathogens present, is an important part of any control strategy. Control measures have included one or more of the following: use of pesticides, vaccination of susceptible animals, environmental management, and protection of individual humans. The latter, coupled with education about the species of tick present and its life cycle, is probably the most effective measure to prevent transmission of tick-borne diseases to humans. Biological control, which has proven valuable in controlling many insect pests, has so far not been of much

use in controlling ticks. Some parasitic wasps have been released in an attempt to control *D. variabilis* in the northeastern U.S., but this has not proven effective. Oxpeckers are natural agents of biological control and, in Africa, consume large numbers of ticks from ungulates. Likewise, fire ants will eat ticks. How these natural enemies can be used by man to control ticks is unclear.

Pesticides

Pesticides (= acaricides) are applied by spraying a given area, or more commonly, by dipping or spraying individual domestic animals. Other methods of application use a systemic acaricide, or involve controlled delivery systems such as collars impregnated with a slow releasing acaricide. The use of acaricide impregnated cotton fibers that are taken by wild mice to their nests has been successful in reducing populations of *Ixodes scapularis*; or the use of baited stations that dispense the acaricide to the animal when it takes food from the device has been successful in controlling tick populations on domestic livestock in some regions and on deer in the southwestern U.S. The development of resistance to acaricides in recent years and their adverse environmental effects have prompted the use of other means such as vaccines and environmental management to control ticks. Other strategies use pheromones or CO_2 to attract ticks to stations containing pesticides.

Vaccines

Several commercial vaccines against ticks now exist. These all take advantage of the fact that host IgG in the midgut of the tick is immunologically competent, and is transported through the midgut as such into the hemolymph of the tick. The most effective vaccines use concealed antigens. These are antigens from the tick that the host has never been exposed to before being vaccinated. Concealed antigens usually require booster shots of the vaccine.

Environmental management

Ticks need a suitable habitat and suitable hosts for an abundant population to develop. Habitat modification involves controlled burning, use of herbicides, mowing to destroy vegetation, and plastering to seal cracks and crevices in human dwellings. The latter is an effective control on some species of nidicolous ticks such as *Ornithodoros moubata*. Indigenous herdsman in Africa have used controlled burning for centuries to limit ticks that attack cattle. In some cases, removal of vegetative litter and opening up the ground to intense sunlight by altering the forest canopy has been effective. Reducing or denying hosts is most effective when the target area is isolated and has little possibility of their reintroduction. If a wildlife host exists, control of ticks in an area is very difficult, although fencing has been used with some success to exclude large wildlife hosts such as the Cervidae. It is possible to keep domestic animals out of tick infested areas by rotating pastures.

Protecting humans

Protecting humans against ticks involves prevention of infestation, and proper removal of attached ticks. Within the urban environment, measures to prevent infestation include strategies to kill and repel ticks inside the home and on companion animals. In the home, foggers and sprays for the house and yard are important. Topical repellants and sprays, anti-tick collars and shampoos are effective means of killing and repelling ticks on dogs and other companion animals. Measures that prevent infestation outside the urban environment involve avoidance of heavily infested areas, especially during high tick activity. If this is not possible, then tucking trousers into exposed socks using some sticky tape and the use of repellents to clothes and or exposed skin (follow the directions on the label). Complete inspection of one's person immediately after leaving the area is essential. Ticks prefer the area around the waist, the axillary region, the genital and perianal region, the neck and the head. It is important to know how to remove ticks because improper removal can result in secondary infections and, in some cases, the transmission of pathogens from the tick to the person removing the tick. Attached ticks should be removed by placing tweezers between the tick and the skin of the host and pulling slowly and gently up until the tick is removed. Once removed, the tick should be examined to make certain that the mouthparts are not still in the host. Once removed, the tick should be saved. Preserve the tick in a container with 70% ethanol or rubbing alcohol. Preserved in this manner, a physician or scientist can identify the tick, and also determine if it is capable of transmitting a disease. Following removal, the bite area should be cleaned, an antiseptic applied and the area covered with a small dressing. See also, MITES.

Lewis B. Coons and Marjorie Rothschild
The University of Memphis
Memphis, Tennessee, USA

References
Coons, L. B., and G. Alberti. 1999. The Acari-ticks. pp. 267–514 in F. W. Harrison and R. Foelix (eds.), Microscopic anatomy of invertebrates, Vol. 8B, Chelicerate Arthropoda. Wiley-Liss, New York, New York.

Sonenshine, D. E. 1991. *Biology of ticks*, Vol. 1. Oxford University Press, New York, New York. 447 pp.

Sonenshine, D. E. 1993. Biology of ticks, Vol. 2. Oxford University Press, New York, New York. 465 pp.

Sonenshine, D. E., and T. N. Mather (eds.) 1994. *Ecological dynamics of tick-borne zoonoses*. Oxford Unversity Press, New York, New York. 447 pp.

Marquardt, W. C., R. S. Demaree, and R. B. Grieve. 2000. *Parasitology and vector biology* (2nd ed.). Harcourt Academic Press, New York, New York. 702 pp.

TICKS AS VECTORS OF PATHOGENS. Worldwide, ticks transmit an exceptional number and diversity of micro-parasites (viruses, bacteria, rickettsia, protozoan parasites, even filarial worms) that cause disease in humans and their livestock. In temperate regions, ticks surpass insects in importance as vectors, and a number of 'new' tick-borne pathogens recently have been recognized as causing human disease in the northern hemisphere. These include the spirochete bacteria (*Borrelia burgdorferi* s.l.) that cause Lyme disease and the rickettsia *Erhlichia* spp. In reality, however, the burden of these infections pales into insignificance beside the medical and veterinary impact of ticks in the tropics. Added to the ticks' role as vectors, the direct damage they do as parasites is a major brake on livestock productivity. Together, this imposes huge economic burdens where they can least be afforded.

Like most true vectors, ticks are blood feeders, for which they are superbly designed; they cut through the host's skin with a pair of toothed chelicerae and suck up body fluids from the sub-dermal lesion through a hypostome. Yet as vectors, ticks appear to rather poorly designed, lacking the high mobility and frequent feeding habits characteristic of most insect vectors. Ticks have no wings, nor do they jump. Nidiculous (nest-dwelling) Argasid (soft) ticks live in semi-permanent, or seasonally repeated, close association with their hosts. Most Ixodid (hard) ticks are not nidiculous; typically they climb to some vantage point on the vegetation from where they contact a passing host, a procedure that exposes them to considerable moisture stress. Intermittently, they must return to the ground where they can absorb water from the moist air, but this is energetically expensive.

Ticks minimize the costs of achieving contact with their host by taking very few, very large meals. This is taken to extreme by Ixodid ticks, which feed once per life stage, as larvae, nymphs and adults, and reproduce once after the adult meal. Most ticks drop to the ground after each meal, where they develop to the next stage. Some species take both the larval and nymphal meals from the same host, and *Boophilus* ticks, worldwide vectors of cattle babesias, take all three meals from the same individual host before dropping back to the ground. Each meal is enormous; even after concentrating the blood by returning 30 to 70% of the imbibed fluid to the host via salivary secretions, on average immature tick stages increase their body weight by about one order of magnitude, and adult females by two orders of magnitude. To accommodate such a large volume of blood, ticks secrete new endocuticle during a long phase of very slow blood intake, before engorging rapidly towards the end of the meal that typically takes from about four (immature stages) to 14 (adult females) days to complete. This prolonged feeding itself incurs a cost as hosts mount strong hemostatic and immunological defenses; in response, tick saliva contains an impressive cocktail of pharmacologically active components. At the same time, the large volume of saliva can transport large numbers of infective parasites, whose infectivity is enhanced by their entering the host at an immuno-modulated site.

To exploit as a vector an hematophage that feeds only once per life stage, a parasite must survive trans-stadially. It is acquired from an infected host by a tick of one stage, maintained through the tick's development and moulting processes, and transmitted to a new host by the following tick stage. The parasite's transmission cycles are thus determined by the tick's stage-specific host relationships (larvae that acquire infections from one host species may later feed as nymphs on hosts of a different species) and rates of development, survival and reproduction. An infection in one host acquired by many feeding larvae may be retained and transmitted to several new hosts by individuals of one or both of the succeeding stages, nymphs and adults. This achieves horizontal amplification between vertebrate hosts, but as only about 10% of ticks survive from one stage to the next, this route is more limited than it may at first appear. On the other hand, those parasites that are passed trans-ovarially, from females via the eggs to larvae of the next generation, can exploit the tick's impressive fecundity (several thousand eggs) to achieve considerable potential for vertical amplification of prevalence in addition to any horizontal amplification. Even though trans-ovarial transmission is usually inefficient, with less than 20% of an infected tick's larval progeny being infected, nevertheless the abundance of larvae (typically 100 times more numerous than adult ticks and 10 times more than nymphs) can make this route quantitatively significant.

Due to these biological peculiarities of ticks, the quantitative framework (i.e., model) for estimating the transmission potential of tick-borne pathogens differs from that for insect-borne pathogens. The tick's feeding pattern makes the concept of an individual's daily biting rate inappropriate, but, most importantly, it introduces a long delay between acquisition and transmission of an infection. The off-host inter-stadial period, comprising both post-engorgement development and questing for the next host, is functionally equivalent to a very long extrinsic incubation period as the tick is not infective until it is ready to feed again. This period is specific to the vector rather than to the transmitted parasite, and may last from one month to more than a year depending on the tick stage, the temperature-dependent rate of development and whether diapause occurs.

Ticks are renowned for their longevity. Engorged ticks must survive the long, temperature-dependent inter-stadial development periods, and unfed ticks must survive the long period while they quest for hosts, which may not always be available just when and where the tick needs them. Unlike insect vectors, however, a tick's vectorial capacity does not increase proportionally with a prolonged life-span. This is because however long an individual tick survives, it does not feed on and infect more than one host per life stage. Although tick longevity (which may exceed that of smaller host species such as rodents) ensures an enduring reservoir of infection, it slows

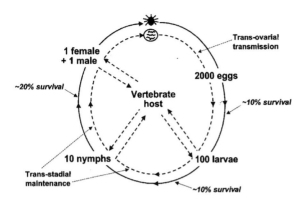

Fig. 1057 *Ixodes ricinus* life cycle with typical reproduction and mortality rates and pathogen transmission cycle.

the pace of transmission, because long survival as any one life stage increases the delay between acquisition and transmission of infection.

An important limiting factor on the transmission potential of any parasite is the period of host infectivity, determined by rates of host mortality and recovery from infection. If the transmitted parasites are highly virulent, high daily mortality rates of infected hosts may significantly reduce transmission potential. If a host dies prematurely, the feeding ticks will not complete their blood meals. The risk of killing their hosts is avoided by those tick-borne viruses that limit their infection to a non-lethal, non-systemic form (i.e., limited to certain parts of the host's body) that is, nevertheless, highly transmissible to ticks over short periods of time. The quantitative impact of this transmission route arises from the large numbers of infectible ticks that co-feed with an infected tick at sites of localized infection. This is a natural feature of tick-host interactions, as prolonged meals taken on certain preferred parts of the host's body result in very large aggregations of ticks feeding close together on some host individuals. The immuno-modulatory effects of saliva secreted by so many co-feeding ticks further facilitate this transmission route. Although first recognized for tick-borne viruses, this route has now been identified for *B. burgdorferi* s.l., and shown to be sufficient to allow sheep to support natural cycles of Lyme borreliosis in the absence of other, systemically infected hosts.

Whereas insect vectors are viewed as the bridge between reservoir vertebrate hosts, the identity of reservoir host and bridge is reversed for the many parasites that survive in their long-lived tick vectors rather than in vertebrates. A multitude of specific parasite-vector-host interactions have clearly evolved to allow many micro-parasites to exploit the transmission potential offered by ticks, even though ticks are not endowed with the features we normally associate with insect vectors.

Sarah Randolph
University of Oxford
Oxford, United Kingdom

References

Nuttall, P. A. 1998. Displaced tick-parasite interactions at the host interface. *Parasitology* 116: S65–72.

Randolph, S. E. 1998. Ticks are not insects: consequences of contrasting vector biology for transmission potential. *Parasitology Today* 14: 186–192.

Randolph, S. E., L. Gern, and P. A. Nuttall. 1996. Co-feeding ticks: epidemiological significance for tick-borne pathogen transmission. *Parasitology Today* 12: 472–479.

Sonenshine, D. E., and T. N. Mather (eds.) 1994. *Ecological dynamics of tick-borne diseases*. Oxford University Press, Oxford, United Kingdom. 447 pp.

TIGER BEETLES. Some members (tribe Cicindelini) of the family Carabidae, sometimes treated as a separate family, Cicindelidae (order Coleoptera).

See also, GROUND BEETLES (COLEOPTERA); TAXONOMY, BEETLES.

TIGER MOTHS (LEPIDOPTERA: ARCTIIDAE). Tiger moths, family Arctiidae (including flag moths and wasp moths), total 11,155 species worldwide, primarily Neotropical (about 6,000 sp.);

Fig. 1058 Example of tiger moths (Arctiidae), *Grammia virgo* (Linnaeus) from USA.

actual world fauna likely exceeds 14,000 species. The family is in the superfamily Noctuoidea, in the section Cossina, subsection Bombycina, of the division Ditrysia. There are five subfamilies among three groups: group Pericopinina (with Pericopinae), group Arctiinina (for Lithosiinae and Arctiinae), and group Ctenuchinina (for Ctenuchinae and Syntominae). Various specialists use other classifications. Adults small to large (8 to 115 mm wingspan); tympanal organs absent or vestigial in Syntominae; hindwings greatly reduced in some groups (wasp moths). Maculation extremely varied, but mostly very colorful, especially among the wasp moths (Ctenuchinae and the Old World Syntominae) which mimic wasps in many cases. Adults mostly nocturnal but many are crepuscular or diurnal (Pericopinae, Ctenuchinae, and Syntominae). Larvae are leaf feeders. Host plants are varied among numerous plant families, including mosses and lichens. A few are economic. Among the largest adults are females of *Aglaomorpha histrio*, from China, while the smallest are some of the Lithosiinae.

John B. Heppner
Florida State Collection of Arthropods
Gainesville, Florida USA

References

Bryk, F., E. Strand, and H. Zerny. 1912–37. Arctiidae. In *Lepidopterorum Catalogus*, 7: 1–179 (1912); 22: 1–416 (1912); 26: 501–900 (1922); 45: 1–57 (1931); 82: 1–126 (1937). W. Junk, Berlin.

Fang, C. L. 2000. Lepidoptera Arctiidae. In *Fauna Sinica (Insecta)*. Vol. 19. Science Press, Beijing. 589 pp, 20 pl. [in Chinese]

Holloway, J. D. 1988. Family Arctiidae. In *The moths of Borneo*, 6:1–101, 19 + 6 pl. Malayan Nature Society (Malayan Nature Journal, 42).

Jacobson, N L., and S. J. Weller. 2001. *A cladistic study of the Arctiidae (Lepidoptera) using characters of immatures and adults.* Entomological Society of America (Thomas Say Monograph) Lanham. 130 pp.

Marmet, P., and J. Schmid. 2000. Arctiidae–Bärenspinner. In *Schmetterlinge und ihre Lebensräume: Arten-Gefährdung-Schutz. Schweiz und angrenzenden Gebiete*, 3: 581–744, pl. 26–34. Pro Natura-Schweizerische Bund fuer Naturschutz, Basel.

Seitz, A. 1909–34. Familie: Arctiidae. In *Die Gross-Schmetterlinge der Erde*, 2: 37–108, pl. 9–18 (1909–10); 2(suppl.): 53–94, 278–290, pl. 1–8 (1931–34); 6: 33–230, 293–455, 469–497, pl. 9–31, 38–67 (1915–25); 10: 61–92, 105–290, pl. 10–30 (1912–15); 14: 41–122, pl. 3–5, 8–19 (1926). A. Kernen, Stuttgart.

Watson, A., and D. T. Goodger 1986. Catalogue of the Neotropical tiger-moths. *Occasional Papers on Systematic Entomology*, 1: 1–71 (4 pl.).

TILLYARD, ROBIN JOHN.

Robin Tillyard was born January 31, 1881, at Norwich, England. He was educated at Dover College and Queen's College, University of Cambridge, and received a Sc.D. in 1920. He taught math and science in Sydney, Australia, from 1905 to 1913, and began publishing on dragonflies in 1905. He was named a lecturer in zoology at the University of Sydney in 1917, and received several honors from that University and from the Linnean Society of London. He also served as Chief of the Biological Department of Cawthron Institute in New Zealand, and Chief of the Division of Economic Entomology, CSIRO, Canberra, Australia. Tillyard published nearly 100 scientific papers, on many aspects of entomology and most orders of insects, but was especially interested in their evolution, in fossil insects, and in dragonflies. He also provided numerous contributions to "The illustrated Australian encyclopedia" (1925) and "The insects of Australia and New Zealand" (1926). Tillyard died in 1937.

Reference

Musgrave, A. 1932.p. 316–321 in *Bibliography of Australian Entomology, 1775–1930.* Royal Zoological Society of New South Wales.

TIMEMA WALKINGSTICKS.

A family of walkingsticks (Timemidae) in the order Phasmatodea. See also, STICK AND LEAF INSECTS.

TIMEMIDAE.

A family of walkingsticks (order Phasmatodea). They commonly are known as Timema walkingsticks. See also, STICK AND LEAF INSECTS.

Timarcha LATREILLE (COLEOPTERA: CHRYSOMELIDAE, CHRYSOMELINAE).

The leaf-beetle genus *Timarcha* Latreille comprises about four subgenera, 125 species and 30 subspecies, spread in eastern North America and around the Mediterranean basin. They are absent from Syria, Lebanon, Israel, and Egypt, where very probably European and African species once met, but were eradicated during the Pleistocene desertification. In Libya, it survives along the coast on the western side and in

Cyrenaica, but also survives on some oases, 80 km south. *Timarcha* does not reach or survive in the central Sahara (Hoggar, 2918 m) where some *Chrysolina* live.

Fossil *Timarcha*, before the Pleistocene, are unknown, but *Timarcha* is a very old genus, perhaps related to the Upper Jurassic *Timarchopsis*, a Siberian fossil. The genus *Timarcha* combines plesiomorphic characters (very primitive nervous system, primitive aedeagus and tegmen, etc.) with apomorphic ones, like the welding of the elytra in tenon and mortise and a complete aptery. How long has the beetle been apterous? Probably very early, during the Cretaceous, because it is found with *Meloe* (Meloidae), one of the rare beetles that is apterous at the pupal stage. Aptery and a subelytral cavity are forms of protection against heat and water loss in many tenebrionids living in desert and semidesert areas. Such structures reduce transpiration, act as a thermal buffer for heat flow and allows the beetle to store the maximum amount of water possible to compensate for the loss of liquid through reflex bleeding. Lack of flight muscles and shortening and widening of metasternum are a direct consequence of aptery. *Pimelia* spp. (Tenebrionidae), also totally apterous, often mimic the *Timarcha* spp. and there seems to be a certain concordance in northern Africa between the local *Timarcha* and *Pimelia* species. It could be more Müllerian mimicry than Batesian, however, because *Pimelia* regurgitates liquid when disturbed.

The archaic characteristics of the adult, and morphological structures of the larvae, probably warrant for *Timarcha* a subfamily of its own, the Timarchinae, situated between primitive ones (Aulacoscelinae, Sagrinae) and more evolved ones (Chrysomelinae). That status has already been proposed by various authors, but many others remain hesitant and prefer the status of tribe (Timarchini) at the beginning of the Chrysomelinae. Timarchini would be a monogeneric tribe with four subgenera. All other genera formerly included among the Timarchini are now listed among the Entomoscelina, a subtribe. No other Chrysomelinae has a ring piece around the aedeagus, divided at its base, with a ciliated cap-piece on the top. Several rare chrysomeline genera show a ring piece (tegmen) devoid of any cap piece on the top. Normally Chrysomelinae genera and species have a v or Y-shaped tegmen, a more evolved form of aedeagus. Farrell's molecular analysis of the Chrysomelinae unfortunately missed *Timarcha*, a key genus.

The genus *Timarcha* probably originated in the steppes of Central Asia, from where it has been eliminated by the Pleistocene glaciations. It adapted through a complicated system of egg or adult diapause to Middle Europe and North America, including middle-sized mountains. It still does not occupy areas that were glaciated in the Pleistocene in the US and Canada (except Vancouver island) and in Europe (below Scotland, the Baltic States and Denmark). Species of *Timarcha* can survive moderate cold, but in Normandy, for instance, one species (*T. goettingensis* (Linné)) becomes active in February when the sun shines. Most of the species are diurnal, but some can be crepuscular or entirely nocturnal. *Timarcha punctella* Marseul in Tunisia and Libya, and *T. laevigata* (L.) in Morocco, are active during the day when other species in the North African mountains are mostly crepuscular. Through the Middle European mountains, adults of the subgenus *Metallotimarcha* Motschulsky (*T. metallica* Laicharting, *T. hummeli* Faldeman, etc.) seem to be mostly crepuscular or nocturnal in activity. In the United States and southern Canada, adults of the subgenus *Americanotimarcha* Jolivet are entirely nocturnal. During the summer, they climb over *Rubus* plants at around 9 pm and start to go down around 5 am to hide under the trash, fallen leaves, and near the roots of brambles. They do the same on strawberry plants. European and African *Timarcha* show abundant reflex bleeding (haemorrhage) around the mouth and between femur and tibia. American species, being nocturnal and in this way being protected from most of the predators, show a very discrete bleeding. Not only does *Timarcha* blood taste bitter, but it is toxic in its contents (anthraquinones). Also, the purely nocturnal species of *Timarcha* (the American species, and also some of the *Metallotimarcha*) do not have the elytra fused. Probably, problems of water loss are not acute during the night.

Timarcha species fed originally on *Plantago* species (*P. albicans* L., *P. maritima* L., etc.) and this choice is maintained with most of the African and southern European species (no data are known on Asian *Timarcha*). This host selection behavior is absent among north European and North American species. However, the dual choice (Rubiaceae/Plantaginaceae) is still shown along the French Atlantic coast with *Timarcha maritima* Perris which feeds alternatively on *Galium arenarium* Lois and *Plantago*

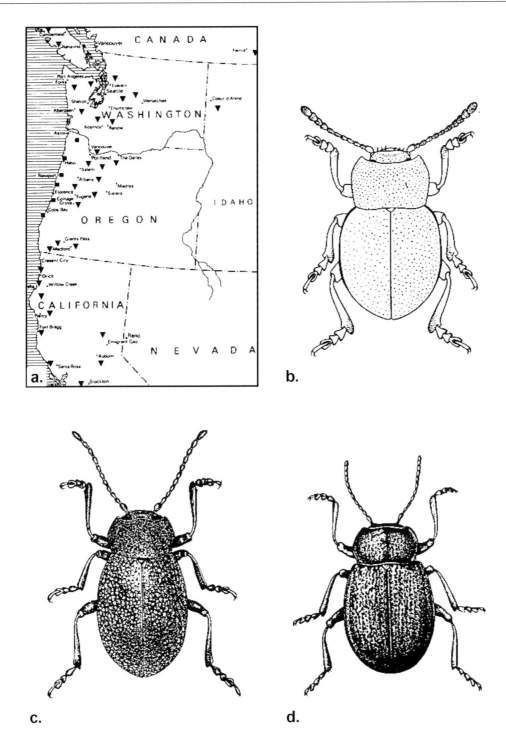

Fig. 1059 Timarcha: (a) Distribution of the genus *Timarcha* in the USA and Canada. Localities in Montana are not visible on the map; (b) *Timarcha* (*Metallotimarcha*) *metallica* Laicharting from Belgium; (c) *Timarcha* (*Americanotimarcha*) *intricata* Haldeman, Oregon, USA; (d) *Timarcha* (*Americanotimarcha*) *cerdo* Stål, Oregon, USA.

maritima L. Many Iberian or Moroccan species have adapted to the same diet, Rubiaceae/Plantaginaceae, related taxonomically and chemically, but they eventually switch to other related plant families: *Veronica* (Scrophulariaceae), *Scabiosa* (Dipsacaceae) or not related: *Launaea* (Asteraceae), *Carrichtera*, *Iberis*, *Alyssum* (Brassicaceae). Southern species (Sicily, Spain, Morocco) display such adaptations, but generally Rubiaceae (*Galium*, *Rubia*, *Crucianella*, *Asperula*, *Sherardia*) along with various species of *Plantago* form part of the diet. There are also reports of *Timarcha metallica* feeding on *Vaccinium* spp. (Ericaceae) and other Ericaceae. The difference in food plants between Old and New World stocks of *Timarcha*, and the big difference in chromosome formula between the two groups, indicate a long isolation, probably since the separation of the northern land masses of the Atlantic in the early Eocene period. American *Timarcha* feed on Ericaceae and Rosaceae.

Timarcha is a K-strategist and lays few big eggs protected by a primitive ootheca mostly made of buccal secretions. Aptery and suppression of flight muscles likely allows the female to increase its egg production. Aptery is a serious handicap in case of fragmentation of the habitat. Once *Timarcha* is eliminated from a habitat, repopulation is unlikely.

Timarcha is distributed in middle and southern Europe, northern Africa and western North America, and new species are described from time to time in Turkey, Italy and Corsica. However, *Timarcha* has completely disappeared from central Asia, and its ephemeral and possible existence in Teneriffe, Japan and elsewhere (Iceland) seems to be due to accidental introduction with forage not followed by survival. *Timarcha* has survived in some big islands (Corsica, Sardinia, Sicily, Mallorca, Minorca) and small ones (Channel islands, Chausey archipelago, Aegades, small islands off the coast of eastern Spain, etc.), but has disappeared or has never existed in the eastern Mediterranean islands, including Cyprus and Crete. Malta remains *Timarcha* free, but probably lost the beetle due to intense urbanization since the Greek and Roman times. It was connected with Sicily during the Cenozoic. Stranger is the lack of *Timarcha* in three small Balearic islands, perhaps due to geological history. *Timarcha* has disappeared completely from the eastern and central USA, probably during the Pleistocene, though host plants are present in the Apalachian Mountains and elsewhere. It is also difficult to understand why the *Timarcha* did not cross the Sahara when it was vegetated, and did not colonize the Hoggar and the East African mountains as *Chrysolina* did. In some ways, the original distribution of *Timarcha* could have been similar to that of *Pimelia* and Pimeliinae in the West before the desertification of the Middle East, with a wider distribution towards Mauritania and the western Sahara, plus Egypt, Sudan, eastern Mediterranean and western Asia.

Being protected by their toxic fluid (by regurgitation) and prebuccal and tibio-femoral reflex bleeding (by anthraquinones) and eventually also by a nocturnal life, *Timarcha* spp. have few parasites and parasitoids (Ichneumonidae and Braconidae), mites under the subelytral cavity (Canestriniidae) and microbial commensals: gregarines. Practically no predators (birds, lizards) feed on them. When resting during the cold season in the Mediterranean, adults of *Timarcha* often gather together under tufts of plants, showing a tendency to gregarism.

American and European species of *Timarcha* differ in number of chromosomes, the American species having a much greater number. The basic formula in Europe is: $2n = 12$ and in America: $2n = 44$.

The black body color of *Timarcha* and the red blood extruded abundantly likely have an aposematic effect because of the contrast with the green background of the food plant. However, aposematism is of no use for nocturnal species.

The size of *Timarcha* adults varies from 5 to 8 mm (*T. cerdo*) to 18 to 23 mm (*T. tangeriana* Bechyne). Generally a size of 10 to 12 mm (*T. goettingensis* (L.)) is common.

Conclusions

The genus *Timarcha* remains rather enigmatic. In certain areas it seems to be evolving very quickly. Isolated valleys and mountains, acting as islands, like the Pyrenees in France and Spain, and the Atlas in Morocco, seem to be the focus of strong variation. Along the Moroccan coast and in the Pyrenees, small morphological differences, probably linked with interbreeding, sometimes show small variations in food habits. In the Pyrenees, each river, each valley seems to have a small variation.

Separation of the genus between America and the Old World has always puzzled entomologists. Very few other cases are similar among the arthropods.

Transpacific migration remains a possibility, but the genus is absent in the Far East.

Extinction is caused by urbanization, fragmentation of the habitat, use of insecticides and herbicides, general pollution and many other reasons. The survival of many species is actually in jeopardy in Europe and in the USA. Probably several species will soon be extinct, and many more are endangered.

Pierre Jolivet
Paris, France

References

Farrell, B. D. 1998. 'Inordinate fondness' explained: why are there so many beetles? *Science* 281: 555–559.

Gomez-Zurita, J., C. Juan, and E. Petitpierre. 2000. The evolutionary history of the genus *Timarcha* (Col Chrys.). Inferred from mitochondrial COII gene and partial 16S rDNA Sequences. *Molecular Phylogeny and Evolution* 14: 304–317.

Jolivet, P. 1994. Remarks on the biology and biogeography of *Timarcha* (Chrysomelidae Chrysomelinae). pp. 85-97 in D.G. Furth (ed.), *Proceedings 3rd International Symposium on Chrysomelidae, Beijing*, 1992. Backhuys Publishers, Leiden, The Netherlands.

Jolivet, P. 1995. A status report on the species of *Timarcha* (Col. Chrys.). *Insecta Mundi* 9: 153–154.

Jolivet, P., and E. Petitpierre. 1973. Plantes-hôtes connues des *Timarcha* Latreille (Col. *Chrys.*). Quelques considérations sur les raisons possibles du trophisme sélectif. *Bulletin Société Entomologique de France* 78: 9–25.

TINEID MOTHS.
Members of the family Tineidae (order Lepidoptera). See also, FUNGUS MOTHS, BUTTERFLIES AND MOTHS.

TINEIDAE.
A family of moths (order Lepidoptera). They commonly are known as tineid moths, fungus moths, or clothes moths. See also, FUNGUS MOTHS, BUTTERFLIES AND MOTHS.

TINEODIDAE.
A family of moths (order Lepidoptera) also known as false plume moths. See also, FALSE PLUME MOTHS, BUTTERFLIES AND MOTHS.

TINGIDAE.
A family of bugs (order Hemiptera). They sometimes are called lace bugs. See also, LACEBUGS, BUGS.

TIPHIIDAE.
A family of wasps (order Hymenoptera). See also, WASPS, ANTS, BEES, AND SAW-FLIES (HYMENOPTERA), and TIPHIID WASPS (HYMENOPTERA: TIPHIIDAE).

TIPHIID WASPS (HYMENOPTERA: TIPHIIDAE).
The family Tiphiidae belongs to the informal group (series) Aculeata, which includes all stinging wasps, of the order Hymenoptera (ants, wasps and bees). The Tiphiidae were previously placed in the superfamily Scolioidea and later Tiphioidea. The current classification of the Hymenoptera which recognizes only three aculeate superfamilies, Chrysidoidea, Apoidea and Vespoidea, places the Tiphiidae in the superfamily Vespoidea.

Order: Hymenoptera
  Suborder: Apocrita
    Informal Group: Aculeata
      Superfamily: Vespoidea
        Family: Tiphiidae

Most members of the Tiphiidae are believed to be parasitoids of the edaphic (soil inhabiting) larval stage of certain beetles (Coleoptera). Primarily a tropical family, the Tiphiidae are represented worldwide by seven subfamilies containing about 1,500 species. These subfamilies are the Anthoboscinae (six genera), Diamminae (one genus), Thynninae (50 genera), Tiphiinae (nine genera), Brachycistidinae (13 genera), Myzininae (12 genera) and Methochinae (two genera). The Thynninae is the largest subfamily of the Tiphiidae, present mainly in the Neotropical regions and Australia. The five subfamilies represented in North America include the Tiphiinae, Myzininae, Methochinae, Anthoboscinae and Brachycistidinae. Of these subfamilies, the Tiphiinae is the largest containing about 140 species in North America. The most common group of Tiphiinae are the members of the genus *Tiphia*, represented by about 100 North American species.

External morphology

The adult Tiphiidae range from 5 to 30 mm in length. The wasps are typically black but may have red, yellow or white markings. The antennae are 12-segmented in females and 13-segmented in males. The pronotum and mesonotum are separated by a suture and not fused. The middle and hind legs are often heavily spined. Sexual dimorphism ranges from slight to pronounced and in some species, the wings

Fig. 1060 Tiphiid wasps: A. Adult female *Tiphia pygidialis*, subfamily Tiphiinae, 12 mm long, North America. B. Masked chafer, *Cyclocephala* sp. with egg of *T. pygidialis*. C. Fifth instar *T. pygidialis* devouring *Cyclocephala* sp. host. D. Cocoon of *T. pygidialis*.

of the female are reduced or absent. Males often have a modification of the eighth sternite (or hypopygidium) forming an upward curved spine that may be mistaken for a sting. Two morphological features which help to distinguish the Tiphiidae from other families are a mesosternum with two posterior lobes and the separation of abdominal segments by well-defined constrictions.

Life history and habits

The Tiphiidae are all solitary wasps and develop as ectoparasitoids on their soil-dwelling hosts. Most tiphiids are parasitoids of larval Scarabaeidae, commonly referred to as white grubs. The Thynninae, Tiphiinae and Myzininae contain species that, for the most part, are parasitoids of white grubs. The exception is the Myzininae, in which a few of the species parasitize larval Tenebrionidae, Cicindellidae or Cerambycidae (all Coleoptera). The biology of the Anthoboscinae and Brachycistidinae has not been studied but it is believed that members of these two subfamilies are also parasitoids of scarabaeid larvae. The Diamminae, an Australian group, parasitizes mole crickets (Orthoptera: Gryllotalpidae) and the

Methochinae parasitize the larvae of tiger-beetles (Coleoptera: Cicindellidae).

Adult Tiphiidae feed primarily either on honeydew produced by other insects, or on the nectar of flowering plants. Female wasps may also feed by biting a paralyzed host during an oviposition event and then imbibing the hemolymph exuding from the wound. Females lay about 50 eggs over their 3 to 4 week lifespan, but in a few species may lay as many as 100 eggs. As in most members of Hymenoptera, fertilized eggs become females and males are produced with unfertilized eggs. Females control the sex of their offspring, with male eggs placed on smaller hosts and female eggs placed on larger hosts.

The best-studied group of Tiphiidae is the genus *Tiphia*. *Tiphia* are mostly host specific, i.e., each wasp species parasitizes larvae of only one species of grub. In cases where more than one grub species is parasitized, all host grub species are usually of the same genus. The emergence of adult wasps is synchronized with the presence of the third instar of grubs with a 1-year life cycle or the second instar of grubs with a 2- or 3-year life cycle. After mating, the female burrows into the soil and uses kairomones from grub frass (feces) and body odor trails to locate a host. Once a host is found, the wasp stings the grub ventrally between the first and second thoracic segments causing temporary paralysis. An egg is then laid on the host. The position of the egg on the host differs among the species of *Tiphia*. The larval *Tiphia* hatches from the egg in 3 to 7 days. Upon hatching, the wasp larva pierces its host's cuticle with its mouthparts and feeds on its body fluids. After a period of about 21 days, the fifth instar *Tiphia* devours the remaining non-sclerotized body parts of its host and then spins a silken cocoon. In temperate regions, *Tiphia* have one generation per year and will overwinter in the cocoon. Tropical species of *Tiphia* have multiple generations each year.

Predators and parasites

The impact that natural enemies have on tiphiid populations is not known. Bombyliid flies (Diptera) and rhipiphorid beetles (Coleoptera) are common parasites reared from the cocoons of various Tiphiidae. Sphecid and mutillid wasps (Hymenoptera) and nematodes are less commonly observed as natural enemies of tiphiids.

Importance in biological control

During the 1920s and 1930s, 14 species of *Tiphia* were introduced into the U.S. from Japan and China to control the spread of the Japanese beetle, *Popillia japonica* (Coleoptera: Scarabaeidae). Only two of them, *Tiphia popilliavora* and *T. vernalis*, became established. Only *T. vernalis* is known to still be established throughout much of the range of the Japanese beetle. Currently, efforts are underway to move *Tiphia vernalis* to areas of the U.S. where the Japanese beetle has recently spread but *Tiphia vernalis* is not yet established. Of the native Tiphiidae, *Tiphia pygidialis* and *T. relativa* are common parasitoids of masked chafer, *Cyclocephala* spp., grubs and *T. berbereti*, *T. intermedia*, *T. tegulina*, *T. transversa* and *T. vulgaris* parasitize June beetle, *Phyllophaga* spp., grubs.

See also, WASPS, ANTS, BEES, AND SAWFLIES (HYMENOPTERA).

Michael E. Rogers and Daniel A. Potter
University of Kentucky
Lexington, Kentucky, USA

References

Brothers, D. J. 1999. Phylogeny and evolution of wasps, ants and bees (Hymenoptera, Chrysidoidea, Vespoidea and Apoidea). *Zoologica Scripta* 28: 233–249.

Clausen, C. P. 1940. *Entomophagous insects*. McGraw-Hill, London, United Kingdom. 688 pp.

Clausen, C. P., H. A. Jaynes, and T. R. Gardner. 1933. Further investigations of the parasites of *Popillia japonica* in the far east. *Technical Bulletin No. 366*, United States Department of Agriculture, Washington, DC.

Kimsey, L. S. 1991. Relationships among the tiphiid subfamilies (Hymenoptera). *Systematic Entomology* 16: 427–438.

King, J. L., and J. K. Holloway. 1930. *Tiphia popilliavora* Rohwer, a parasite of the Japanese beetle. *Circular No. 145, United States Department of Agriculture*, Washington, DC.

Rogers, M. E., and D. A. Potter. 2002. Kairomones from scarabaeid grubs and their frass as cues in below-ground host location by the parasitoids *Tiphia vernalis* and *Tiphia pygidialis*. *Entomologia Experimentalis et Applicata* 102: 307–314.

TIPULIDAE. A family of flies (order Diptera). They commonly are known as crane flies. See also, FLIES, CRANE FLIES (TIPULIDAE AND OTHERS).

TISCHERIIDAE. A family of moths (order Lepidoptera). They commonly are known as trumpet leafminer moths. See also, TRUMPET LEAFMINER MOTHS, BUTTERFLIES AND MOTHS.

TOAD BUGS. Members of the family Gelastocoridae (order Hemiptera). See also, BUGS.

TOBACCO. See also, BOTANICAL INSECTICIDES.

TOE BITERS. Members of the family Belostomatidae (order Hemiptera). See also, BUGS.

TOGOSSITIDAE. A family of beetles (order Coleoptera). They commonly are known as barkgnawing beetles. See also, BEETLES.

TOLERANCE. The ability of a host to grow and reproduce normally while supporting an insect population that would normally be damaging.

TOMATO BIG BUD. See also, TRANSMISSION OF PLANT DISEASES BY INSECTS.

TOMATO RUST MITE. See also, FOUR-LEGGED MITES.

TOMOCERIDAE. A family of springtails in the order Collembola. See also, SPRINGTAILS.

TONIC RECEPTORS. Sensory neurons that are slow to adapt to continuing steady stimuli, with the receptor potential remaining elevated. (contrast with phasic receptors)

TOOTH-NECKED BEETLES. Members of the family Derodontidae (order Coleoptera). See also, BEETLES.

TORRE-BUENO, JOSÉ ROLLIN DE LA. José de la Torre-Bueno was born at Lima, Peru, on October 6, 1871. At the age of 14 he and his family moved to the United States. He attended Columbia University and graduated in 1894. He obtained employment with the General Chemical Company of New York, where some of his principal responsibilities involved editorial work. He was an active member of the Brooklyn Entomological Society, and helped revive the society's publications as the Bulletin of the Brooklyn Entomological Society, New Series, and of Entomologica Americana, New Series. He served as editor of both until he died. Torre-Bueno was a hemipterist, and published "Synopsis of the North American Hemiptera Heteroptera" in three parts (1939, 1941, 1946). In 1937 the Brooklyn Entomological Society published Torre-Bueno's "Glossary of Entomology." Though the "Glossary" was a revision of J.B Smith's "An explanation of the terms used in entomology," it became a classic publication and remains available to this day in reprinted form. He died in Tucson, Arizona, on May 3, 1948.

Reference

Mallis, A. 1971. *American Entomologists*. Rutgers University Press, New Brunswick, New Jersey. 549 pp.

TORTOISE BEETLES. Members of the subfamily Cassidinae of the beetle family Chrysomelidae. The lateral expansion of the pronotum and elytra hide the legs and head, giving a tortoise-like appearance to the insect. See also, BEETLES.

TORTOISE SCALES. Some members of the family Coccidae, superfamily Coccoidae (order Hemiptera). See also, SCALE INSECTS AND MEALYBUGS, BUGS.

TORTRICID MOTHS. Members of the family Tortricidae (order Lepidoptera). See also, LEAFROLLER MOTHS, BUTTERFLIES AND MOTHS.

TORTRICIDAE. A family of moths (order Lepidoptera). They commonly are known as leafroller moths or tortricid moths. See also, LEAFROLLER MOTHS, BUTTERFLIES AND MOTHS.

TORYMIDAE. A family of wasps (order Hymenoptera). See also, WASPS, ANTS, BEES, AND SAWFLIES.

TOWNES, JR., HENRY K.. Henry Townes was born at Greenville, South Carolina, USA, on January 20, 1913. He received B.S. and B.A. degrees from Furman University in 1933, and then attended Cornell University where he attained the Ph.D. in 1937. He taught at Cornell and Syracuse Universities until receiving a fellowship in 1940 to work at the Philadelphia Academy of Natural Sciences in Philadelphia, where he prepared a catalog on the Nearctic Ichneumonidae. He also worked at the U.S. National Museum, where he was employed by the United States Department of Agriculture as a taxonomist. In 1949 he moved to North Carolina State University, where he worked on tobacco pests, and traveled to the Philippines to advise the government on pests of rice and corn. In 1956 he moved to the University of Michigan to work on Ichneumonidae. Henry and his wife Marjorie founded the American Entomological Institute in Ann Arbor, Michigan, in 1962 as an independent, nonprofit research institute. Henry also worked briefly at Michigan State University and Carleton University during the early stages of the institute, but it soon became a mecca for parasitic hymenopterists, and a full-time job. The Institute relocated to Gainesville, Florida, in 1985. Henry was passionate about the study of Ichneumonidae, became known as the world's foremost authority, and amassed the world's best collection. He published nearly 140 articles or books during his career, of which nine were at least 500 pages long. Along with colleagues, he established a firm knowledge of this group, one of the largest families of insects known. Henry Townes died at Gainesville, Florida, on May 2, 1990.

Reference

Buckingham, G. R., V. K. Gupta, and M. C. Townes 1991. Henry K. Townes, Jr. *American Entomologist* 37: 252–253.

TOXEMIA. A condition produced by the dissemination of toxins in the blood, though the bacteria are confined to the gut (e.g., brachytosis).

TOXICANT. A toxic substance, often the active ingredient in pesticide formulations.

TOXICITY. The ability of a toxin to harm or kill an organism.

TOXICOGENIC INSECTS. Insects capable of producing disease in plants due to injection of saliva or other secretions, and in the absence of microbial pathogens.

TRACHEA. (pl., tracheae) An internal tube that delivers air to tissues, and part of a system of large and small tubes. The external openings of the system are called spiracles, and the smaller tubes are called tracheoles. Air sacs, or dilated areas of trachea, are often found in the system.

See also, TRACHEAL SYSTEM AND RESPIRATORY GAS EXCHANGE.

TRACHEAL GILL. An extension of the insect's body that contains numerous tracheae and serves as a site for oxygen extraction from water. The heavily tracheated flaps or filaments of the body wall are usually lateral or ventral expansions of the thorax or abdomen, and gas exchange occurs across the body surface.

See also, TRACHEAL SYSTEM AND RESPIRATORY GAS EXCHANGE.

TRACHEAL MITE, *Acarapis woodi.* The tracheal mite, *Acarapis woodi* (Rennie), is an internal parasite of honey bees. It feeds and reproduces inside

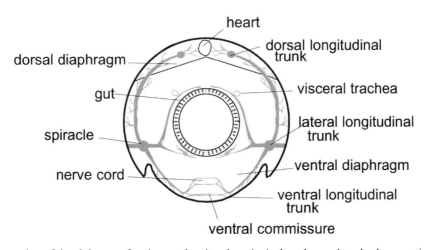

Fig. 1061 Cross section of the abdomen of an insect, showing the principal trachea and tracheal connections (adapted from Chapman, The insects: structure and function).

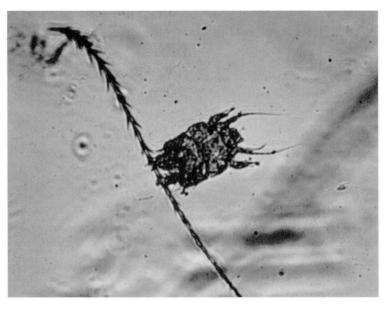

Fig. 1062　Female tracheal mite clinging to a bee hair, in wait for a new host bee.

the prothoracic tracheae of the adult bee. If a bee is heavily infested, its tracheae become scarred and filled with mites. Bees from heavily infested colonies may be seen crawling on the ground nearby, unable to fly.

Young bees are more susceptible to infestation than older bees. The female mite enters the tracheal tube at the spiracle, and lays eggs within several days of entry. The egg is followed by a larval and a pharate nymphal stage. The immature stages last 11 to 12 days for males and 14 to 15 days for the larger females. Immature and mature mites feed by piercing the host bee's tracheal wall with stylet mouthparts, and consuming the hemolymph. Male and female offspring mate in the trachea. Only mated females exit the spiracle to find another host bee.

The means by which tracheal mites kill bees is not entirely clear. Some bees found crawling near infested hives have no mites or any apparent damage to their tracheae. Possibly, pathogen outbreaks in a bee colony are stimulated or mediated by the mite.

Dispersal from bee to bee is most rapid in winter when the bees cluster tightly. Since no bees are reared during winter, the proportion of bees which are infested rises considerably in winter. Consequently, a heavy infestation is often discovered on the first warm days of early spring, when many bees are found crawling on the ground near the hive. Tracheal mites spread from hive to hive when infested bees drift to, or rob honey from, other hives nearby.

An earlier name, the acarine mite, has been abandoned because it is meaningless and redundant. Two closely related mites, *A. externus* and *A. dorsalis*, live externally on the bees. They are not known to be harmful to their hosts.

History

The tracheal mite was first identified in 1919, when it was found in Scotland. Originally the mite was implicated in the 'Isle of Wight disease', which had plagued the British Isles earlier in the century. However, more recent studies suggest that the disease was not caused simply by mite infestations.

Since then, the mite has been discovered in many other parts of the world. It is very well adapted to life as a honey bee parasite and is not found on hosts other than *Apis* species. Consequently, the host-parasite relationship must be very old.

Mexico was found to have the mite around 1980. In 1984 it was discovered in the United States, having apparently entered Texas from Mexico. Infestations were confirmed in Canada in 1986. By the late 1980s many thousands of hives in the United States and Canada were dying yearly. This parasite was rapidly dispersed by the shipment of queen bees, package bees, and migratory beekeeping operations.

By the late 1990s, however, tracheal mite infestations had declined dramatically. Now the mite continues to be endemic in North America as it is elsewhere. The decline of tracheal mite infestations is still not well understood.

Diagnosis and control

Tracheal mites are identified by removing the prothoracic tracheae alone, or within the anterior portion of the bee's thorax. Thoracic muscle tissue is cleared with potassium hydroxide solution. Tracheal mites and scarring of the tracheae are easily seen by low power microscopy. Positive identification is not possible for those without a microscope.

Menthol, formic acid and chlorobenzilate have been used effectively as fumigants to control the mites inside the hive. Vegetable oil also has been used with success, mixed with sugar and placed inside the hive. Bees that walk over the oil preparation acquire a fine coating which seems to inhibit mite dispersal among bees in the hive.

Some bees groom themselves of mites. This trait is heritable, so selection for grooming behavior can be part of a breeding program for tracheal mite resistance.

Thomas C. Webster
Kentucky State University
Frankfort, Kentucky, USA

References

Bailey, L., and B. V. Ball 1991. *Honey bee pathology* (*2nd ed.*). Academic Press, New York, New York.

Webster, T. C., and K. S. Delaplane (eds.). 2001. *Mites of the honey bee*. Dadant and Sons, Hamilton, Illinois.

Wilson, W. T., J. S. Pettis, C. E. Henderson, and R. A. Morse 1997. Tracheal mites. In R.A. Morse and K. Flottum (eds.), *Honey bee pests, predators and diseases* (*3rd ed.*). A.I. Root, Medina, Ohio.

TRACHEAL SYSTEM AND RESPIRATORY GAS EXCHANGE.

Insects breathe through a complex network of tubules, the tracheae. Large tracheae connect to spiracles opening at the surface of the body, where air enters and carbon dioxide exits. Spiracles usually occur on the pleural surface of the body, typically one on each side of each segment, but numerous variations have evolved. Airflow may be tidal, in and out of the same spiracles, or directed flow with inflow through anterior spiracles and outflow through posterior abdominal ones. Interconnecting longitudinal and transverse tracheal trunks make directed flow possible and more efficient than tidal flow because the system is constantly flushed and incoming air is not mixed with used air. Larger tracheal tubes send off branches that become smaller in diameter as they ramify to all tissues. The smallest diameter tracheoles (tubes 1 μm diameter or less) end blindly on the surface of most cells and even indenting some cells, like pushing a finger into a soft balloon, until they terminate within a few μm of mitochondria that actually use the oxygen. Tracheoles indented into a cell are said to be intracellular, although in reality they are not really within the interior of the cell, for the tracheal epithelium cell

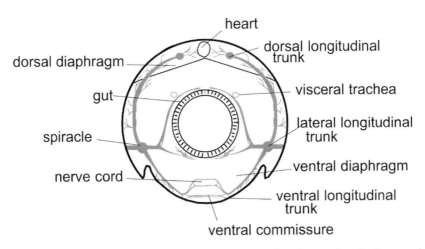

Fig. 1063 Cross section of the abdomen of an insect, showing the principal tracheae and tracheal connections (adapted from Chapman, 1998. The insects: structure and function).

layer and the cell membrane separate the tracheole from the cytoplasm of the cell. The chitinous lining of tracheae are shed at each molt, but the lining of tracheoles is not molted. Spiracular valves at the body surface often can be closed to reduce water loss from the system. Although all parts of the tracheal system allow oxygen to diffuse out to the tissues and carbon dioxide to enter, most of the gas exchange occurs across the smaller tracheae and tracheoles because they are the parts of the system in intimate contact with cells. The extensive ramification of tracheae and tracheoles and their relationship to all cells are similar to the pattern of the vertebrate circulatory system, with the exception that tracheoles end blindly and vertebrate capillaries do not. The tracheal system is remarkably efficient for insects. Insects do not accumulate an oxygen debt during vigorous activity, such as flight. In some very small insects (smaller than a *Drosophila* fly), simple diffusion of gases through the tracheae may suffice, but most insects,

large or small, actively ventilate the system by rhythmic compression of the abdomen. Muscular movements during flight or other vigorous muscular and body movements compress the tracheal system and act like a pump to ventilate the system.

Tracheae develop from embryonic ectodermal tissue, and all parts, including the smallest tracheoles, have an epicuticular lining or intima that is continuous with the external cuticle. In some larger insects one can see larger tracheal trunks that are pulled out and still attached to the old exoskeleton at a molt. Larger tracheae may contain a thicker endocuticle layer that gives more strength to the tubular structure. The internal surface of tracheae contains a hydrophobic substance that helps prevent water from entering the tracheae, and reduces evaporative water loss from the internal tracheal surfaces. Thickened, tight spirals of the cuticular intima, the taenidia, strengthen tracheae, provide elasticity, and help the tubes resist compression and collapse. Even tracheoles and air

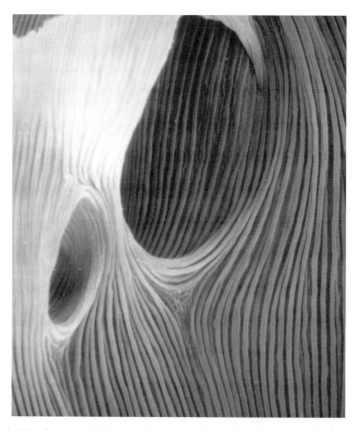

Fig. 1064 A view from inside a large trachea in a mole cricket. This scanning electron micrograph shows the branching of two smaller tracheae and the taenidial windings that strengthen all tracheae and keep them from collapsing.

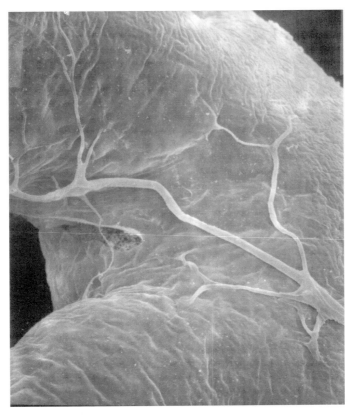

Fig. 1065 A scanning electron micrograph view of a trachea (about 2 μm diameter) that branches several times into smaller tracheae and tracheoles that penetrate the salivary gland tissue of a male Caribbean fruit fly, *Anastrepha suspensa*.

sacs show some evidence of taenidial reinforcements, but thickenings usually are widely spaced apart so that these parts of the system are more flexible.

Tracheae are not simply tubes and tubules; every part of the system contains living cells. Tracheal epithelial cells are thin and flattened, like many small pieces of ribbon glued together and wrapped around the tracheae. A tracheal epithelial cell encloses a tracheole. New tracheae and tracheoles develop as the insect grows. A new tracheole develops as a tubule or cavity within a tracheal epithelial cell that grows out, often in multiple finger-like shapes, from the surface of a larger trachea. The tracheole unites with the larger tracheal branch to which the parent tracheal cell is still attached.

Air sacs, dilated portions of tracheae, occur in many insects. They are variable in size, but frequently are large in flying insects such as honeybees, cicadas, many adult Diptera, and some scarab and buprestid beetles. The rhythmical squeezing action of working flight muscles pumps air sacs like a bellows and increases the flow of air through the system. Air sacs may collapse as growing tissues fill the body space, and by collapsing they make room for the new tissue, with little change in the general body shape. Air sacs serve a hydrostatic function in some aquatic insects, and allow more freedom in vibration of the tympanic membrane in some sound-producing insects. By taking up space and restricting hemolymph volume, air sacs increase solute concentration in the hemolymph without increasing total solute.

Hemocytes, the circulating blood cells, are the only cells in the body that do not have direct tracheal connections. Recently, it was shown that hemocytes tend to accumulate on very thin-walled tufts of tracheae near the last pair of abdominal spiracles and in the compartment at the tip of the abdomen called the tokus. It seems as if the highly branched tracheal system of the last segment and the tokus are sites where the hemocytes can become oxygenated; they probably pass their carbon dioxide directly into the hemolymph.

Discontinuous breathing

Some insects regulate spiracle closing and opening, and keep the spiracles tightly closed and/or apparently closed for a high percentage of the time. Gas exchange occurs in three periods named the open, flutter, and closed periods (often designated as O, F, and C, respectively) because of the action of spiracles over the duration of a cycle. This functional pattern, variously known as discontinuous release of CO_2, passive suction ventilation, discontinuous ventilation cycle, and as the discontinuous gas exchange cycle (DGC) has been known in some insects for more than half a century. The earliest studies were in diapausing pupae of Lepidoptera, and initially DGC was thought to be limited to quiescent insects in a depressed state of metabolism. DGC patterns, however, have been observed in a number of different insects in various states of activity, including ants, the cockroach *Periplaneta americana*, a number of adult tenebrionid beetles, the locust *Schistocerca gregaria*, the lubber grasshopper *Romalea guttata*, and adults of additional species. DGC behavior also occurs in some arthropods other than insects. During DGC, accumulated CO_2 is discharged periodically in bursts during brief intervals when spiracles are open. After a burst, the spiracles are closed for some period of time that varies from species to species. During the closed interval, tissues use oxygen and intratracheal oxygen tension falls. Most insects that exhibit discontinuous gas exchange allow the spiracular valve to flutter with imperceptible amplitude to the unaided eye during a portion of a cycle. The fluttering (F) phase allows small amounts of O_2 to be sucked into the tracheal system by the slight negative pressure arising from O_2 consumption by the tissues. The F phase usually is considered to involve convective transfer of O_2. This has given rise to the name 'passive suction ventilation' that sometimes is applied to the process. The slight internal vacuum retards the outward loss of water vapor and CO_2 during the fluttering phase, and the low influx of O_2 lengthens the time to full opening or 'burst' of spiracles. Tissues produce CO_2 even when the spiracles are closed. The high solubility of CO_2 in aqueous solutions enables insects to accumulate bicarbonate ion, HCO_3^-, in the hemolymph. Buffering capacity of hemolymph aids in solubilizing CO_2 as bicarbonate, and this keeps gaseous CO_2 from building up rapidly in the tracheal system. At some point, probably different for different insects, the relationship between gaseous O_2 and CO_2, and HCO_3^- in solution reaches an equilibrium at which tracheal tension of CO_2 and O_2, and/or pH change in the hemolymph, trigger spiracle opening and release of CO_2 from the hemolymph as a gas. It often has been assumed, but not proven, that the functional benefit of discontinuous ventilation is conservation of water. In diapausing pupae that must pass a long winter under the soil, leaf litter, or other pupation site, water conservation seems quite necessary and a reasonable driving mechanism for the evolution of discontinuous ventilation, since the pupa is a closed (to food/water intake) system. Most adult insects that exhibit discontinuous ventilation do so only intermittently, and the rest of the time they ventilate the system continuously. Surprisingly, some insects do not show DGC respiration under conditions that might be expected (based upon assumptions) to promote the behavior. For example, the lubber grasshopper *Romalea guttata*, which discontinuously ventilates at times, tends to ventilate continuously when dehydrated, a physiological condition in when it presumably has a great need to conserve water. No good explanation has been provided for such behavior. The ant *Camponotus vicinus*, which exhibits discontinuous gas exchange, actually loses more water than CO_2 during the period when the spiracles are open. The harvester ant, *Pogonomyrmex rugosus*, and similar desert ants have a relatively high percentage (up to 13%) of body water loss through the tracheal system, even though they exhibit discontinuous ventilation at times. Whether the higher rate of water loss from the tracheal system in these ants is a significant stress for them may depend upon how much of the time the ants exhibit discontinuous gas exchange, how much access they have to food with high water content, and duration of exposure to environmental extremes. Although it seems somewhat intuitive that water loss might be high from the tracheal system, actual measurements in some insects indicate that more than 90% of total body water loss occurs through the cuticle, with only 2 to 5% typically lost from the tracheal system. In summary, a definitive selection mechanism for evolution of discontinuous gas exchange cycling is not evident, but the mechanism is fairly widespread in many insects including larvae, pupae, and large and small adults. Water conservation, ecological niche occupied by insects, and interactions of sensory components may be important factors.

Gas exchange in aquatic insects

The tracheal system of most aquatic insects is structurally the same as that of terrestrial insects, i.e., with open spiracles and an extensive network of tracheae and tracheoles. These aquatic insects breathe air by coming to the surface. Water is prevented from entering the system by the hydrophobic surface of the tracheae, and in some cases, by spiracles that can be closed. Many aquatic larvae have a tracheal system with only one functional pair of spiracles near the tip of the abdomen. When the insect comes to the surface, it does so with the posterior end uppermost and only the tip of the abdomen bearing the spiracles is held above the surface. A large number of aquatic insects submerge with a bubble or film of air around a spiracle. These gas bubbles or films of air are called gas gills, and they may be temporary or permanent. Temporary ones must be replenished periodically as the oxygen is used. Permanent gas gills are called plastrons. They do not collapse and oxygen can be extracted from water into the gill (if the water is well aerated), allowing the insect to live underwater. A plastron consists of an extensive physical meshwork, either of fine hairs or setae, or a meshwork of small pores and channels that can hold a volume of air and can present a large water-air interface. Plastrons are common in aquatic insects and can take many physical forms. Some aquatic insects are able to capture and utilize gas bubbles of oxygen released by aquatic plants. Certain species of Diptera, Coleoptera, and Lepidoptera independently evolved modifications for piercing aquatic plants for air.

Some aquatic insects have a closed tracheal system without functional spiracles. The lack of functional spiracles eliminates any chance that water will enter the system, but oxygen must diffuse into the tracheal system through the cuticle, a breathing mechanism called cutaneous respiration. Some aquatic insects have tracheal gills, thin flaps of cuticle with many tracheoles just under the cuticle. Larvae of Trichoptera, Plecoptera, Odonata, and some Lepidoptera utilize cutaneous respiration, facilitated by extensive elaboration of thin hair-like or flap-like tracheal gills from the body surface. Internally these insects still have an extensive tracheal system, and gas transport through the tracheae and tracheoles to internal body tissues is the same as in terrestrial insects. Movement of water over the gill and body surface of aquatic insects is important in maintaining a fresh supply of oxygenated water in contact with the body, and most use undulations of the body and/or movements of the gills themselves to create ventilatory currents of water. Larvae of some dragonflies (Anisoptera) draw water into the rectum by elastic expansion of the body as dorso-ventral compressor muscles relax. Typically there are six main gill folds of the cuticular intima in the anterior part of the rectum with extensive tracheoles just beneath the intima that extract O_2 from the water. The water is pumped out by dorso-ventral compression of the abdomen. The rate of ventilation varies with several factors including the O_2 content of the water. About 85% of the water in the rectum is renewed during each pumping cycle, and 25 to 50 cycles/min have been recorded. Larvae also will come to the surface and ventilate the rectum with air when oxygen content of the water is very low.

Many hymenopterans and dipterans are parasitic on other insects, and have been little studied with respect to respiration, perhaps for the obvious reasons that they are usually small and hidden in the body of the host. Cutaneous respiration is probably very important. Many chalcid wasps and tachinid flies hatch from eggs laid on the surface of the host insect and they eat their way into the host. They orient the posterior pair of spiracles at the body surface of the host so that they breathe air directly. Bot fly larvae, *Hypoderma* spp., migrate to the skin of the vertebrate host where they bore a tiny opening to the surface through which gas exchange occurs. Respiration is cutaneous in earlier instars that are migrating through the body of the host.

Respiratory pigments occur in only a few insects. *Chironomus* spp. larvae have a small hemoglobin molecule composed of two chains with a MW of 31,400. The hemoglobin is not in the cells, but occurs as a circulating hemolymph protein. The molecule has extraordinarily high affinity for oxygen and is 50% saturated at a pO_2 of 0.6 mm Hg at 17°C. This means that it will not give up its oxygen to tissues except under extremely low oxygen tension. Its principal function may be to aid recovery from anaerobic conditions and to provide limited O_2 to some critical tissues, such as the nervous system. Hemoglobin occurs in certain cells of *Gastrophilus* spp. (horse bots) and in larvae and adults of some beetles (Family Notonectidae, certain species of the genera *Anisops* and *Buenoa*). In contrast to the situation in *Chironomus* larvae, the hemoglobin of the beetle *Anisops pellucens* is only 50% saturation at a pO_2 of 28 mm Hg at 24°C (i.e., it will give up half of its oxygen at 28 mm Hg, a pressure that might occur in actively metabolizing tissues), giving it much more functional potential. The beetle may get up to 75% of the O_2 consumed during a normal dive from its hemoglobin. Increased temperature also causes the hemoglobin to release more oxygen, which could be important to actively working tissues such as muscles.

The embryo developing inside the egg must obtain sufficient oxygen for development. The majority of aquatic and semi-aquatic insects lay eggs with no special respiratory structures incorporated into the shell, while eggs of a majority of terrestrial insects contain special structures for respiration, including an extensive, inner chorionic meshwork that can function as a plastron when the egg is submerged in well aerated water. Gas exchange in eggs with no special respiratory structure occurs by simple diffusion through the egg shell.

Non-respiratory functions of the tracheal system

Tracheae serve some important functions other than gaseous transport. Tracheae act as the connective tissue of insects, helping to tie cells and tissues together. Tracheae are important as a structural base for at least two important endocrine tissues: (1) the prothoracic glands whose cells are attached to the tracheae near the prothoracic spiracle in larvae of Lepidoptera, and (2) cells of the epitracheal glands attached to the surface of the major ventrolateral tracheal trunk to each spiracle in lepidopterous larvae. Air sacs lie behind sound producing organs in insects and are important to sound modulation. A large air sac backs up to the tymbal located on each side of the first abdominal segment of male cicadas. The size of the air sac varies with different species, and its size and tuning are partly responsible for the species-specific quality of the sound produced by male cicadas. The interconnected prothoracic tracheae extending across the prothorax and into the prothoracic legs of crickets act as a resonator of sounds. They aid the cricket in discriminating direction of sounds reaching the tympanal membrane located just below the joint of the femur with the tibia (i.e., the knee joint) on each prothoracic leg. Hissing cockroaches produce a hissing sound by expelling air forcefully from certain spiracles when they are disturbed. Lubber grasshoppers, *R. guttata*, release quinones from the

tracheal system on the side of the attack when an ant attacks, or when stimulated by probing.

James L. Nation
University of Florida
Gainesville, Florida, USA

References
Chapman, R. F. 1998. *The insects: structure and function.* Cambridge University Press, Cambridge, United Kingdom. 770 pp.
Klowden, M. J. 2002. *Physiological systems in insects.* Academic Press, New York, New York. 415 pp.
Nation, J. L. 2002. *Insect physiology and biochemistry.* CRC Press, Boca Raton, Florida. 485 pp.

TRACHEOLE. A very narrow trachea (one μm or less in diameter) that serves as the actual site of gas exchange between the tracheal system and the body tissues because they penetrate organs and lie so close to the cells.

See also, TRACHEAL SYSTEM AND RESPIRATORY GAS EXCHANGE.

TRACHYPACHIDAE. A family of beetles (order Coleoptera). They commonly are known as false ground beetles. See also, BEETLES.

TRADE NAME. The name given a product that is marketed by a company. The same product may be marketed by more than one company using different names.

TRAIL PHEROMONE. A pheromone that is deposited along the ground or other substrate that allows insects to maintain the group or to communicate to others in their group (usually nest- or tent-mates) directionality, often to a feeding location.

See also, PHEROMONES, SOCIALITY OF INSECTS.

TRANSCRIPT. A mRNA copy of a gene.

TRANSCRIPTION. The process of producing a mRNA copy of a gene.

TRANSFORMATION. The process of changing the genetic makeup of an organism by introducing foreign DNA. Transformation may be transient or stable (transferred to succeeding generations.)

TRANSGENE. The DNA that is inserted into the genome of a cell or organism by recombinant DNA methods.

TRANSGENIC ARTHROPODS FOR PEST MANAGEMENT PROGRAMS. Transgenic organisms contain genes that have been introduced into the genome using biotechnology (recombinant DNA) methods. Transgenic plants (including corn, rice, cotton and squash) have been developed using recombinant DNA methods and currently are grown commercially. Development of the tools to develop transgenic insects and mites has lagged behind that of plants and microorganisms, but genetic modification using recombinant DNA methods have been used to transform a number of pest and beneficial insects and mites. The successful transformation of the fruit fly *Drosophila melanogaster* by a genetically modified transposable element called the *P* element in 1982 stimulated subsequent research to genetically modify other insects (and mites) using similar methods. [See the entry on Transformation of *Drosophila melanogaster* by *P* elements for details of this method.]

Domesticated and semidomesticated insects have been modified by traditional breeding methods for hundreds of years. Genetic manipulation has improved disease resistance and silk production in silk moths, and disease resistance and pollination attributes in honey bees. More recently, natural enemies of insects and mites used in biological control programs in agriculture and forestry have been modified by traditional breeding methods and by hybridization of different strains to achieve hybrid vigor. A pesticide-resistant predatory mite (*Metaseiulus occidentalis*), developed with traditional breeding methods, was incorporated into a integrated mite management program in almonds in California. These predators provided effective control of spider mites, reduced the need for costly pesticides, reduced production costs and saved almond growers approximately $22 million per year, most of which was due to fewer applications of pesticides to control the spider mites. This project demonstrated that genetic improvement of natural enemies could result in improved pest management programs. Genetic improvement of natural enemies for biological control of pest insects and mites by traditional genetic methods has involved selecting for resistance to pesticides, lack of diapause, and increased tolerance to temperature extremes, although modification of other traits, such as sex ratio, theoretically could result in improved biological control.

During the past 40 years, a number of pest insects have been sterilized by irradiation or chemicals for use in genetic control programs. This approach to pest management has been called the sterile insect release method (SIRM) or the sterile insect technique (SIT). Male insects are mass reared and sterilized, usually by irradiation, and released in to control a number of serious pests, including the Mediterranean and Caribbean fruit flies, mosquitoes and the New World screwworm *Cochliomyia hominivorax*. The SIRM has been used to eradicate pests or to reduce pest populations.

Sex and the sorted insects: a case study

Genetic control of pest insects represents an attractive alternative to chemical control of some pests in terms of safety, specificity and the limited negative effect this control tactic has upon the environment. The screwworm (*Cochliomyia hominivorax*) eradication campaign demonstrates what can be achieved with mass releases of males sterilized by irradiation.

The genetic control method used to eradicate the screwworm was called the 'sterile insect release method' (SIRM) or 'sterile insect technique' (SIT), and involves mass rearing, sterilization of males by chemicals or irradiation, and their subsequent release to mate with wild females. Because females of the screwworm mate only once, any wild female mating with a sterile male fails to contribute progeny to the next generation. By releasing an excess of sterile males (compared to the number of wild males), populations decline in a predictable manner, ultimately becoming extinct. Because absolute population densities were often low in the USA, the number of sterile males that had to be released could be produced in 'fly factories'.

The screwworm eradication program in the USA was initiated in Florida with small scale trials on Sanibel Island in 1951. The results were promising and the project was geared up to cover the state of Florida and then the southeastern USA. The screwworm was declared eradicated from the southeastern USA in 1959, one year ahead of schedule. Eradication was achieved in a surprisingly short time due to the combined effects of a severe winter in Florida during 1957–1958, which greatly reduced the over-wintering screwworm population, and a 17-month eradication program beginning in July 1958 that cost approximately $7 million and involved the release of almost 9 billion sterile screwworm flies over an area of approximately 56,000 square miles.

Since 1959, the livestock industry of Florida and adjacent states have saved at least $20 million each year because the screwworm is no longer present; actual benefits are even greater in today's dollars. Furthermore, the elimination of losses due to the deaths of livestock and labor and control costs are only part of the benefits; loss of wildlife to screw-worm attack also was eliminated.

The success of the SIRM program in the south-eastern USA led the cattle growers of Texas to mount, in collaboration with the state and the USDA, a similar but much more ambitious program in the southwestern USA in the 1960s. This program required more time and effort because the area in which the screwworm was to be eradicated bordered on a front 2400 kilometers long, stretching from the Gulf of Mexico to the Pacific Ocean. Despite this challenge, and some setbacks with quality control and reinvasion of flies from Mexico, both Texas and New Mexico were declared 'screwworm free' in 1964.

The SIRM program was moved into Arizona and California in 1965, and in 1966 the entire USA was declared free of screwworms. To reduce the likeli-hood that the screwworm would reinvade the USA from Mexico, the program was expanded into Mexico in 1972, with the goal of eradicating the screwworm all the way south to the Isthmus of Tehuantepec.

After successfully eliminating the pest in Mexico, the SIRM program was expanded to cover all of Cen-tral America. Screwworms were eliminated from Guatemala between 1988 and 1994, from Belize between 1988 and 1994, from El Salvador between 1991 and 1995, from Honduras between 1991 and 1996, from Nicaragua between 1992 and 1998, from Costa Rica between 1995 and 1999, and from Panama between 1997 and 2000. These eradication programs were carried out so that a barrier zone could be set up at the Isthmus of Panama, which is only 190 km wide as compared to the 2,400 km border that the USA and Mexico share. This barrier zone has been maintained by a combination of quarantines and mass releases of sterile screwworms.

Benefits of this massive, and expensive, screw-worm eradication program are large. In 1996, the annual producer benefits in the USA, Mexico and Central America were estimated to be $796 million, $292 million and $77.9 million, respectively. These benefits were due to decreases in deaths of livestock, reduced veterinary services, medicines, insecticides, inspections, handling costs, and increases in meat and milk production. The estimated benefit to cost ratios for the eradication programs average 12.2:1 for Central America to 18:1 for the USA and Mexico.

The principle of sterile insect releases has been applied to other pest insect species, including the Mediterranean fruit fly (*Ceratitis capitata*), tsetse flies (*Glossina palpalis* and *G. morsitans*), mosqui-toes (*Anopheles albimanus*), codling moth (*Cydia pomonella*) and ticks.

Sterile insect release programs usually require only males, but both sexes must be reared. Not only is it expensive to rear large numbers of 'useless' females, but, in the case of species that vector disease or annoy or bite humans or domestic animals, it is undesirable to release any females, sterile or not! As a result, various genetic methods have been used to develop 'genetic sexing strains'. Slight differences in size or color of pupae have been used to sort out undesirable females during mass rearing. Most genetic sexing strains are based on maintaining the desired genes (such as *white pupa* or a temperature sensitive lethal) within translocations. However, because translocations can undergo recombination in the region between the translocation breakpoint and the sexing gene, the strains are not completely stable. As a result, if no practical means exist to remove the recombinants, an increasing number of the undesirable females will be reared and released.

Ideally, a genetic sexing method using recombi-nant DNA methods will become available that could produce only males of high quality and vigor to com-pete with wild males for female mates. Since an all-male colony will be difficult to maintain (!), this character ideally would be a conditional trait, perhaps dependent upon temperature or some other environ-mental cue. Because the released males would be sterile they could not persist in the environment, which could reduce potential risks. Other potential improvements using recombinant DNA methods could result in insects containing a marker gene, such as green fluorescent protein, that would allow the released sterile males to be identified readily in the traps used to monitor progress in the program; because such males would be sterile, risks associated with their release would also be low. Finally, recombinant DNA methods might be developed that would allow the insects to be sterilized without radiation; this could reduce the damage produced by irradiation to the entire body and allow fewer insects to be reared and released, resulting in considerable cost savings.

Why use transgenic methods?

Traditional genetic methods have limitations, and recombinant DNA methods offer new opportunities for improving pest management programs. For example, significant benefits could accrue if recombinant DNA methods allowed sterile insects to be produced without incurring the negative effects of irradiation. During the sterilization process, the insect's whole body is irradiated, which damages all tissues. As a result, the SIRM requires rearing very large numbers of males for release because the irradiated males are damaged by the sterilization process. Commonly, pest populations are first reduced by pesticide applications or through natural seasonal (winter) mortality to reduce the number of insects that have to be released. The number of males released is usually a multiple of the estimated density of wild males, with a 100:1 ratio of sterile to wild males commonly used. Rearing such huge numbers of insects is costly and difficult.

Recombinant DNA methods also could allow unique molecular markers to be inserted into the sterile insects, which would allow SIRM program managers to more easily discriminate between released sterile males and wild fertile males caught in the traps used in monitoring the progress of the program. Current marking methods using fluorescent dusts are unsatisfactory because they can reduce fitness of the insects and the dyes do not always adhere, which could falsely indicate that the program is not working well. Other significant benefits could be obtained if recombinant DNA methods make it possible to control the sex of insects being reared in SIRM programs, to introduce lethal genes or genetic loads into pest populations, or to produce vectors of human and animal diseases that are unable to transmit diseases such as malaria, dengue, yellow fever and sleeping sickness.

Recombinant DNA techniques could make genetic improvement of beneficial insects, such as silkworms (*Bombyx mori*), honey bees (*Apis mellifera*) or biolo-

gical control agents, more efficient and less expensive. Once a useful gene has been cloned, it could be inserted into a number of beneficial species in a relatively short time. Furthermore, recombinant DNA methods broaden the number and type of genes potentially available for use; no longer is a project dependent upon the intrinsic genetic variability of the species under study.

Many have speculated about the role that recombinant DNA methods could play in the genetic control of insects that serve as the vectors of human and animal diseases or pests of agricultural crops. Some consider transgenic technology to be a new and vitally important pest management tool for the control of serious pests that cannot be controlled by any other means. Others have expressed reservations about the goals and methods suggested.

There are limitations to transgenic methods at present. For example, traits primarily determined by single major genes are most appropriate for manipulating insects by recombinant DNA techniques. Methods for manipulating and stabilizing traits that are determined by complex genetic mechanisms are not yet feasible with insects, although such methods could be developed using procedures similar to those developed by plant molecular geneticists.

Components of a genetic manipulation project

Genetic manipulation with recombinant DNA methods requires methods for efficient and stable insertion of foreign genes into the genome of the target insect or mite, and the availability of useful genes and appropriate promoters and other regulatory elements to obtain effective expression of the inserted gene in the appropriate tissue at the appropriate time. Inserting cloned DNA into insects can be accomplished using several different techniques. If the inserted DNA is incorporated into the chromosomes in the cells that give rise to the ovaries and testes, the foreign gene(s) or transgenes could be transmitted faithfully and indefinitely to successive generations (stable germline transformation).

Initial research on stable transformation methods was accomplished with *Drosophila melanogaster* when it was discovered that the *P* element could be genetically manipulated to serve as a vector to carry foreign genes into the chromosomes of germ line cells. In molecular genetics, vectors are self-replicating DNA molecules that transfer a DNA segment between host cells. *P*-element vectors were

Pest management goals that might be achieved with transgenic methods.

Project type	Objective(s)
Biological control of insect and mite pests	
• Improve survival of natural enemies in environment	Enhance ability to control pests by introducing resistance genes; modify diapause or temperature tolerance traits
• Improve effectiveness of natural enemies	Alter traits such as sex ratio (more females), host/prey specificity; restrict ability to fly
Disease control method	
• Develop insects that introduce a vaccine into their hosts when taking a blood meal Improve domesticated and semi-domesticated insects	Provide low cost vaccination against widespread, serious diseases such as malaria
• Improve silk production in silkworms (*Bombyx mori*, *Philosamia*, *Anthaerea*)	Improve quantity, quality or type of silk, introduce disease resistance genes, introduce silk genes from spiders or other silk producing arthropods into moths to produce special types of silk
• Improve honey bees, *Apis mellifera*	Insert genes for resistance to bacterial, viral and fungal pest resistance; introduce resistance to mite (*Varroa*) pests; modify pollination behaviors, modify aggressive behaviors
Population control methods	
• Population replacement	Eliminate traits that make a pest, such as make a pest unable to transmit diseases such as malaria, dengue, sleeping sickness or yellow fever, by altering ability of the pathogen to pass gut or salivary gland barriers; eliminate the need for a blood meal by vector mosquitoes; alter behavior so the vector feeds on only one host. Release altered strain to *replace* the pest population. Replacement will require some type of drive mechanism or a way to select for the released population in the field.
• Insert useful/deleterious genes	Release genetically modified individuals to mate with a 'driver' such as wild individuals in correct ratio to insert genes transposable elements or *Wolbachia* into populations. Proposals include releasing (introgression model) insects with active transposable elements, lethal genes or genes that cause sterility, with the goal of causing so much genetic damage that the pest population crashes (as laboratory populations can do when a new transposable element invades)
Sterile insect release method (SIRM)	
• Sterilize males by recombinant DNA methods to reduce damage from radiation/chemosterilization	Introduce genes and regulatory sequences to cause sterility that can be stimulated by light, diet or other environmental cues during mass rearing, as needed, allowing colony to be reared normally until the transgene(s) is activated
• Mark released males with molecular such as green fluorescent protein, that can be detected in trapped dead males so that the effectiveness of the SIRM program can be assessed	Insert benign marker gene that is expressed or visible in dead insects
• Develop genetic sexing method so that females do not have to be reared, reducing program costs	Introduce a conditional lethal gene so that females can be killed during the mass rearing program when a particular cue is provided

(Modified from Insect Molecular Genetics, 2nd Edition, 2003. by M. A. Hoy, Academic Press).

investigated as possible gene vectors for insects other than *Drosophila*, but failed to function in other insects. Other transposable element vectors such as *mariner*, *piggyBac*, *Hermes* and *hobo* have been isolated from insects and genetically modified for use as vectors to transform insects other than *D. melanogaster*. Another approach to genetic modification involves the genetic engineering of insect gut symbionts. For example, a bacterial symbiont of the Chagas disease vector *Rhodnius prolixus* lives in the insect's gut lumen and is transmitted from adult to progeny by contamination of egg shells or of food with infected feces. The symbionts have been genetically engineered and transmitted to hosts lacking any symbionts. The goal is to reduce the ability of *Rhodnius* to transmit pathogens when it takes a blood meal, perhaps by infecting the bugs with gut bacteria that produce antibiotics that kill the Chagas disease agent. Likewise, gut symbionts of tsetse flies *(Glossina* species), which are vectors of both animal and human African sleeping sickness, also have been transformed. Proposals have been made to release tsetse flies carrying transgenic symbionts so the released flies could replace or out-compete native populations, but fail to transmit the disease.

The ability to introduce cloned genes into the germ line at a predictable chromosomal site is especially desirable, as it reduces the likelihood of 'position effects' on gene expression. Genes introduced by transposable element and viral vectors insert more or less randomly into the chromosomes, making it difficult to predict how well the transgene will be expressed. One method for accomplishing precise insertion is based on a system found in the yeast *Saccharomyces cerevisiae*. A gene for yeast recombinase, called FLP recombinase, and two inverted recombination target sites (FRTs) that are specifically recognized by the FLP enzyme have been cloned. The FLP-FRT system has been modified to insert DNA into a specific site in a *Drosophila* chromosome. If the FRT sites can be inserted into other insects, the system could reduce concerns about unstable transformation that may be elicited by the use of transposable element vectors.

A few experiments have delivered linear or circular plasmid DNA into the genome of insects without using a specific vector. This approach has the advantage of eliminating potential risks of introducing vector sequences into the insect genome, which could result in increased stability of the inserted genes in the genome. This approach assumes that the inserted gene is no more likely than any other gene to be moved by 'wild' transposable elements or viruses.

A variety of methods have been evaluated for delivering genes into insects in order to achieve transformation. These include microinjection of transposable element vectors and other vectors into dechorionated insect eggs, microinjection of plasmids directly into the abdomen of female mites or insects (maternal microinjection), soaking eggs in DNA, using sperm to carry foreign DNA into eggs of the honey bee, using microprojectiles (gene gun technology) to insert DNA into insect eggs, electroporation of DNA into insect eggs, transplanting nuclei and cells, and transformation of an insect microbial gut symbiont (which has been called paratransgenesis because the insect genome has not been modified).

What genes are available to insert?

Cloned DNA can be isolated from the same or other species. It is technically feasible to insert genes from microorganisms into arthropods and have the DNA transcribed and translated, although coding sequences isolated from microorganisms must be attached to promoters (controlling elements) and other regulatory sequences derived from a higher organism so that the gene can be expressed in insects. The regulatory sequences determine when a gene will be transcribed, at what level, in what tissues, and how long the messenger RNA can be used for translation. Considerable research is under way to identify regulatory sequences that regulate genes in specific insect tissues, such as the salivary glands and gut. It also may be possible to isolate a gene from the species being manipulated, alter it, and reinsert it into the germ line, although this approach has not yet been attempted in insects other than *Drosophila* to date.

Project goals will dictate what type of regulatory sequences might be most useful to regulate transgene expression. In some cases, a low level constitutive production of transgenic proteins will be useful, while in other cases high levels of protein production will be required after inducement by a specific cue. Researchers may have to evaluate the trade-offs between high levels of protein production and the subsequent negative effects these could have on fitness of the transgenic arthropod strain based on the specific goals of each program.

After inserting the desired genes and regulatory elements, the next issue is how to detect whether the gene has in fact been incorporated into the germ line. Because transformation methods are relatively inefficient, a screening method is needed to identify transformed individuals. This process is relatively simple in *Drosophila*, where there is a wealth of genetic information and visible markers can be used to identify transgenic individuals. Most pest or beneficial arthropods lack such extensive genetic information or markers. Identifying transformed individuals could be achieved using a pesticide resistance gene as the selectable marker. However, the release of pesticide-resistant pests into the environment would create concerns about risk. Another option is to use antibiotic resistance genes as selectable markers to identify transgenic insects. However, horizontal movement of resistance genes into microbes in the environment could result in increased levels of antibiotic resistance; the likelihood of this potential risk has not been quantified. Antibiotic resistance gene markers are no longer considered safe for release into the environment in transgenic crops, and methods have been developed for their removal. Another potential marker is the ß-galactosidase gene (*lacZ*) isolated from *E. coli*, which can be detected by an assay that produces a blue color in the transformed insects and mites. This construct has been present in a number of released organisms and it has been concluded that risks associated with their release are low. Eye color and green fluorescent protein (GFP) genes also are considered to be safe selectable markers if transgenic insects are to released into the environment. Unfortunately, transgenic insects with mutant eye colors may exhibit abnormal behavior, which could reduce their effectiveness in the field. The effects of GFP on vision could be important when the GFP gene is expressed in the eyes of insects. Normal behavior often is crucial to the function of released insects in pest management programs. It is probably desirable to eliminate unneeded marker genes from transgenic insect strains prior to their release into the environment.

Once transgenic strains have been produced, they should be contained in the laboratory using effective procedures and containment facilities until permits have been obtained from the appropriate governmental agencies that would allow the transgenic arthropods to be released into the environment in a pest management program. Transgenic strains must be evaluated for fitness and the expression of the desired traits should be stable in the laboratory. However, efficacy of the transgenic strain for the intended purpose will have to be evaluated eventually under field conditions. At present, transgenic strains are first evaluated in greenhouses, field cages, or some other contained environment to be sure that they perform as expected and do not exhibit unintended traits. Purposeful permanent releases of transgenic insects or mites have, as of winter 2004, not been requested.

Risk assessments of transgenic arthropods

Risk equals the potential for damage and the likelihood of its occurrence. Risk estimates may be different for pest versus beneficial insects and may depend on whether the insect is expected to persist in the environment or is unable to reproduce and cannot persist. Risks also will vary with the specific transgene(s) inserted. At present, it is easier to identify potential types of harm than to quantify the likelihood of its occurrence.

Relative risks. The least risky transgenic insects could be the domesticated silkworm (*B. mori*), which is unable to survive on its own in the wild. Transgenic *B. mori* are unlikely to have a negative effect on the environment because they should not be able to persist even if they were accidentally released. Also, transgenic pest or beneficial insects that are sterile and unable to reproduce should pose lower risk than insects that are able to reproduce and persist in the environment. Transgenic pest or beneficial insects that are unable to persist because the environment is unsuitable during a portion of the year also are likely to pose a lower risk. Honey bees, *Apis mellifera*, are only semi-domesticated and thus can escape human management to survive in the wild. Transgenic honey bees could pose a greater environmental risk than the domesticated silkworm for this reason.

Evaluating the risks associated with releasing insects and mites that have been manipulated with recombinant DNA techniques will likely change as we learn more about risk assessment procedures and gain experience. Current concerns can be summarized as:

- Is the transgenic population stable?
- Has its host or prey range has been altered?
- Does it have the potential to persist in the environment?
- Will the transgenic strain will have unintended effects on other species or environmental processes?

Another question we may need to ask is how far and how quickly can the transgenic strain disperse from the experimental release site? Less is known about dispersal behavior of many insects than might be needed. The first three questions are relatively easy to answer with a variety of laboratory experiments. The last issue is much more difficult to answer. For the near future, releases of transgenic arthropods in the USA will be evaluated by regulatory agencies on a case-by-case basis. Initial permits for releases will be for short term releases in controlled situations so that unexpected outcomes might be mitigated.

Horizontal gene transfer. One risk issue that is unusually difficult to quantify is the risk of horizontal transfer of transgenes or transposable elements, or of drive elements such as *Wolbachia*, to other organisms. Our knowledge of horizontal transfer and transposable elements only began in the 1950s when Barbara McClintock discovered transposable elements in maize. Horizontal gene transfer results in the movement of genes or transposable elements or drive elements from one insect population to another of the same species, from one insect species to another, or to other organisms in the environment. It is difficult to quantify this risk because we lack fundamental information on the frequencies and mechanisms of horizontal gene transfer; furthermore, the specific genes transferred will be important in determining potential harm as will the site into which the transgene or transposable element inserts. If the insertion is into noncoding DNA in the chromosome, no harm may occur to the recipient. If the insertion is into functional and essential gene(s), then the recipient may be harmed or killed. If the insertion is into somatic cells the effect, if any, may be minimal but if the insertion is into germ line cells then the unintended genetic modification could be transmitted to future generations. The whole topic of horizontal gene transfer in insects has received limited scientific attention until relatively recently.

We do know that horizontal transfer of genes may occur between insect species by movement of naturally occurring transposable elements. Horizontal transfer is thought to be rare, yet more than one such transfer has been observed within historical times in *D. melanogaster* and may have been missed in other species because no one was looking. The *P* element appears to have invaded *D. melanogaster* populations within the last 50 years, perhaps from a species in the *D. willistoni* group. Controversy exists as to whether *P* elements were transferred between *Drosophila* species by the semiparasitic mite *Proctolaelaps regalis*. Another transposable element, *hobo*, appears to have invaded natural populations of *D. melanogaster* around the 1960s, representing the second invasion of this well-studied insect in the past 40 to 50 years.

Transfer of transposable element vectors from insects to other organisms, including humans, is potentially feasible, although these transfers would be expected to occur very rarely. Recall that risk is determined by frequency of occurrence and the damage that might occur. In this case, the frequency is expected to be very low if natural invasions represent a realistic estimate of frequency in the case of purposeful releases of active transposable elements as drive mechanisms or of conversion of inactive transposable element vectors into active ones.

We are still discovering new aspects of the biology and ecology of transposable elements and this lack of knowledge makes it difficult to predict what would happen if insects were released that contained either active or inactive transposable element vectors. The safest course might be to remove any introduced transposable element vector sequences from a transgenic insect strain prior to its permanent release into the environment to reduce the probability that the transgene will move, either within the strain or horizontally between different populations or species.

Some questions to answer when developing a genetic manipulation project if it is to be deployed successfully

PHASE I. Defining the problem and planning the project

- What genetic trait(s) limit effectiveness of beneficial species or might reduce damage caused by the pest?
  - Do we know enough about the biology, behavior, genetics and ecology of the target species to answer this question?
  - Is the potential trait determined by single or multiple genes?
- Can alternative control tactics be made to work more effectively and inexpensively than genetic manipulation projects, and are they more environmentally friendly?

- The costs of genetic manipulation projects are high and the time to develop a functional program can be quite long.
- Transgenic technology may not be appropriate if traditional genetic or other control methods can be used because issues surrounding risk assessment of releasing transgenic arthropods into the environment for permanent establishment have not been resolved.

• How will the genetically-manipulated strain be deployed?
  - Will releases be inoculative and some type of selection or drive system used to replace the wild strain?
  - Will the desired genes be introgressed (introduced) into the wild population?
  - What selection mechanism will be used?
  - Will augmentative releases of very large numbers be required?
  - Will multiple releases be required over many years?

• What risk issues, especially of transgenic strains, should be considered in planning?
  - If pesticide resistance genes are used as a selectable marker for beneficial species is there a possibility of the resistance gene moving to a pest?
  - What is known about the potential for horizontal gene transfer?
  - If transposable element or viral vectors are used in the transformation process, what risks might they pose if the transgenic strain is released into the environment?
  - What health or other hazards might be imposed on human subjects if the transgenic strain were released?

• What advice do the relevant regulatory authorities give regarding your plans to develop a transgenic strain?
  - Which agencies are relevent to consult for your project?

PHASE II. Developing the genetically-manipulated strain and evaluating it in the laboratory

• Where will you get your gene(s)?
  - Should the transgene(s) sequence be modified to optimize expression in the target species if it is from a species with a different codon bias?

• Is it important to obtain a high level of expression in particular tissues or life stages?
  - Where can you get the appropriate regulatory sequences?

• How can you maintain or restore genetic variability in your transgenic strain?
  - Because both artificial selection and transgenic methods typically involve substantial inbreeding to obtain pure lines, how will you outcross the manipulated strain with a field population to improve its adaptation to the field or otherwise increase genetic variability.

• What methods can you use to evaluate 'fitness' in artificial laboratory conditions that will best predict effectiveness in the field?
  - Have life table analyses and laboratory studies of the stability of the trait under no selection been correlated with efficacy in the field?
  - Is it possible to carry out competitive population cage studies?

• Do you have adequate containment methods to prevent premature release of the transgenic strain into the environment?
  - Have these containment methods been reviewed by appropriate regulatory authorities?

• Do you have adequate rearing methods developed for carrying out field tests?
  - Are artificial diets available to reduce rearing costs?
  - Are quality control methods available to maintain quality during mass rearing?

• What release rate will be required to obtain the goals you have set?
  - Do you have an estimate of the absolute population density of the target species in your field test?
  - What release model are you applying: inundative, inoculative, introgressioin, complete population replacement?

• Have you tested for mating biases, partial reproductive incompatibilities or other population genetic problems?

• If the strain is transgenic, have you obtained approval from the appropriate regulatory

authorities to release the strain into the green-house or small plot?
  - Can you contain it in the release site?
  - Can you retrieve it from the release site at the end of the experiment?
  - Can you mitigate if unexpected problems arise?
- How will you measure effectiveness of the modified strain in the field trials?

PHASE III. Field evaluation and eventual deployment in practical pest management project

- If the small-scale field trial results were promising, what questions remain to be asked prior to the deployment of the manipulated strain?
  - Are mass rearing methods adequate?
  - Is the quality control program in place?
  - Is the release model feasible?
  - Were there unexpected reproductive incompatibilities between the released and wild populations?
- If permanent releases are planned, have relevant risk issues been evaluated?
- How will the program be evaluated for effectiveness?
- Will the program be implemented by the public or private sector?
- What did the program cost and what are the benefits?
- What inputs will be required to maintain the effectiveness of the program over time?

(Modified from Hoy 2000.)

Steps in developing a transgenic arthropod

The above description indicates that a number of steps are involved in a program designed to control pest insects through transgenic methods. The target species probably should be identified as a significant pest for which conventional control tactics are ineffective because genetic manipulation is usually more expensive and difficult than other pest management approaches. Furthermore, genetic manipulation with recombinant DNA techniques may generate concerns about risk, requiring additional time and resources.

How best might our knowledge about the pest species' physiology, ecology, or behavior be used

against it? How will the transgenic strain be deployed in a pest management program?

Once a target trait has been identified, it must be genetically altered using appropriate genes and genetic regulatory sequences to ensure that the new trait is expressed at the appropriate time and in appropriate tissues. After a modified strain has been developed, it must be evaluated in the laboratory for fitness and stability. If ultimate deployment requires mass rearing of very large numbers of high quality insects, mass rearing and release models will need to be developed. Eventually the manipulated strain must be released into greenhouses or small field plots in the field for evaluation.

Permission to release a transgenic insect will have to be obtained from (several) regulatory agencies. Short term releases initially will be made into small plots, perhaps in cages. Initial releases of transgenic insects into the environment in the USA are intended to be short term experiments, and current regulation of such releases by the US Department of Agriculture require the researcher to retrieve all transgenic insects from the environment at the end of the experiment.

If the transgenic insect strain(s) perform well and risk assessments are completed satisfactorily, permanent releases into the environment may be allowed, but the guidelines for such releases are lacking as of spring 2002. Many pest management programs, especially those involving replacement of pest populations by the transgenic population, will require permanent establishment in the environment. The use of several 'drive mechanisms' have been proposed for replacement. Analysis of the potential risks of such drive mechanisms has not been carried out.

Could 'gene silencing' reduce program effectiveness?

There is always the risk that a transgenic insect population could be released into the field and fail to function as expected due to a phenomenon called 'gene silencing'. Transgenic plants and mammals have been shown to be able to inactivate (silence) transgenes that overexpress proteins or are otherwise novel. Gene silencing is thought to be due to genetic systems that evolved as a means to prevent high levels of expression of transposable elements or viruses that can cause genetic damage when they invade new hosts. In fungi and plants, gene silencing is associated with several mechanisms, including methylation of the DNA or posttranscriptional or transcriptional processes. Multiple mechanisms of transgene silencing also occurs in *Drosophila melanogaster*. Thus, methods may have to be developed to eliminate transgene silencing in insects or this phenomenon could reduce the effectiveness of a pest management program. The use of genetic sequences called insulators or boundary elements may limit gene silencing.

Gene silencing might be turned into a positive attribute if specific genes in insects could be turned off. Gene silencing has purposefully induced in *D. melanogaster* by introducing a sequence that codes for an extended hairpin-loop RNA by *P*-mediated transformation. Perhaps endogenous gene expression and developmental processes could be modified in other insects by a similar genetic process.

The ultimate utility of transgenic insects and mites for pest management programs remains to be resolved in the coming years. Research on improved transformation methods, isolation of additional useful genes and regulatory elements and development of improved risk assessment methods based on an international consensus must be achieved before transgenic arthropods are widely used in pest management programs.

See also, GENETIC MODIFICATION OF *DROSOPHILA MELANOGASTER* BY P ELEMENTS, STERILE INSECT TECHNIQUE.

Marjorie A. Hoy
University of Florida
Gainesville, Florida, USA

References

Atkinson, P. W., A. C. Pinkerton, and D. A. O'Brochta. 2001. Genetic transformation systems in insects. *Annual Revue of Entomology* 46: 317–346.

Handler, A. M., and A. A. James (eds.). 2000. *Insect transgenesis. methods and applications.* CRC Press, Boca Raton, Florida.

Hoy, M. A. 2000. Transgenic arthropods for pest management programs: risks and realities. *Experimental and Applied Acarology* 24: 463–495.

Letourneau, D. K., and B. E. Burrows (eds.). 2002. *Genetically engineered organisms. assessing environmental and human health effects.* CRC Press, Boca Raton, Florida.

U.S. Department of Agriculture, Animal and Plant Health Inspection Service. 2001. The regulation of transgenic arthropods. http://www.aphis.usda.gov:80/bbep/bp/arthropod/#tgenadoc

TRANSGENIC ORGANISM. An organism whose genome contains genetic material originally derived from an organism (not its parents) or from a different species. The transgene(s) can be transmitted to subsequent generations (stable transformation) or can be lost subsequently (unstable transformation).

TRANSLOCATION. A type of mutation in which a section of a chromosome breaks off and moves to a new position in that or a different chromosome.

TRANSMISSION OF PLANT DISEASES BY INSECTS. Plant diseases appear as necrotic areas, usually spots of various shapes and sizes on leaves, shoots, and fruit; as cankers on stems; as blights, wilts, and necrosis of shoots, branches and entire plants; as discolorations, malformations, galls, and root rots, etc. Regardless of their appearance, plant diseases interfere with one or more of the physiological functions of the plant (absorption and translocation of water and nutrients from the soil, photosynthesis, etc.), and thereby reduce the ability of the plant to grow and produce the product for which it is cultivated. Plant diseases are generally caused by microscopic organisms such as fungi, bacteria, nematodes, protozoa, and parasitic green algae, that penetrate, infect, and feed off one or more types of host plants; submicroscopic organisms such as viruses and viroids that enter, infect, spread systemically and affect the growth of their host plants; parasitic higher plants which range from about an inch to several feet in size and penetrate and feed off their host plants. Plant diseases are also caused by abiotic, environmental factors such as nutrient deficiencies, extremes in temperature and soil moisture, etc. that affect the normal growth and survival of affected plants.

Of the aforementioned causes of disease, many of the microscopic organisms and of the viruses are transmitted by insects either accidentally (several fungi and bacteria) or by a specific insect vector on which the pathogenic organism (some fungi, some bacteria, some nematodes, all protozoa causing disease in plants, and many viruses) depends on for transmission from one plant to another, and on which some pathogens depend on for survival.

TYPES OF PLANT DISEASES

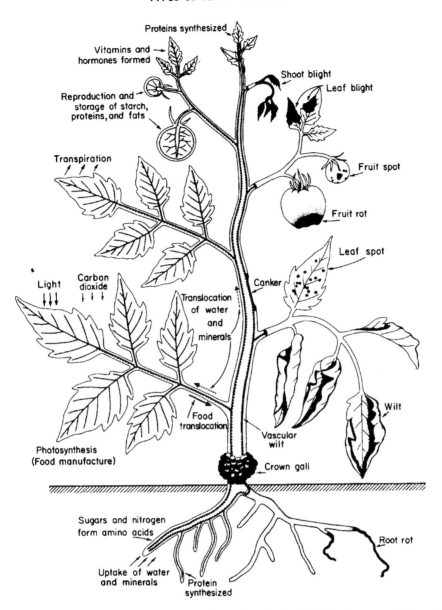

Fig. 1066 Schematic representation of the basic functions in a plant (left)and the interference with these functions (right) caused by some common types of plant diseases. (From Agrios, G.N. 1997. Plant pathology (4th ed.). Academic Press, San Diego, California.)

The importance of insect transmission of plant diseases has generally been overlooked and greatly underestimated. Many plant diseases in the field or in harvested plant produce become much more serious and damaging in the presence of specific or non-specific insect vectors that spread the pathogen to new hosts. Many insects facilitate the entry of a pathogen into its host through the wounds the insects make on aboveground or belowground plant organs. In some cases, insects help the survival of the pathogen by allowing it to overseason in the body of the insect. Finally, in many cases, insects make possible the existence of a plant disease by obtaining, carrying, and delivering into host plants pathogens that, in the absence of the insect, would have been unable to spread, and thereby unable to cause disease. It is offered as a guess that 30–40% of the damage and losses caused by plant diseases is due to the direct or indirect effects of transmission and facilitation of pathogens by insects.

Insects and related organisms, such as mites, are frequently involved in the transmission of plant pathogens from one plant organ, or one plant, to another on which then the pathogens cause disease. Equally important is that insects can and do transmit pathogens among plants from one field to another, in many cases even when the fields are several to many miles apart. Almost all types of pathogens, that is, fungi, bacteria, viruses, nematodes, and protozoa, can be transmitted by insects. Insects transmit

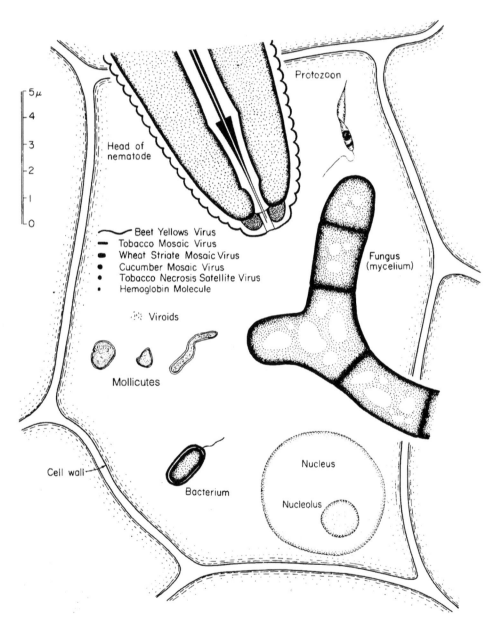

Fig. 1067 Schematic diagram of the shapes and sizes of certain plant pathogens in relation to a plant cell. (From Agrios, G.N. 1997. Plant pathology (4th ed.). Academic Press, San Diego, California.)

pathogens, such as many fungi and bacteria, mostly externally on their legs, mouthparts, and bodies. Almost all plant pathogenic viruses, all phytoplasmas, xylem- and phloem-inhabiting fungi and bacteria, some protozoa, and some nematodes are also transmitted by insects, and they are usually carried by the insect internally. The insects that transmit fungi and bacteria externally on their bodies and legs belong to many orders of insects. On the contrary, the insects that transmit the other pathogens listed above internally are very specialized and specific for the pathogen they transmit and belong to a certain species or genus of insects.

Insects transmit pathogens in three main ways. (1) Many insects transmit bacteria and fungal spores passively by feeding in or walking through an infected plant area that has on its surface plant pathogenic bacteria or fungal spores as a result of the infection. The bacteria and spores are often sticky, cling to the insect as it moves about, and are carried by it to other plants or parts of the same plant where they may start a new infection. (2) Some insects transmit certain bacteria, fungi, and viruses by feeding on infected plant tissues and carrying the pathogen on their mouthparts as they visit and feed on other plants or plant parts. (3) Several insects transmit specific viruses, phytoplasmas, protozoa, nematodes, and xylem- and phloem-inhabiting bacteria by ingesting (sucking) the pathogen with the plant sap they eat. Subsequently, the pathogen circulates through the body of the insect until, with or without further multiplication in the insect, the pathogen reaches the salivary glands and the mouthparts of the insect through which it is injected into the next plant on which the insect feeds.

Role of insects in bacterial diseases of plants

In most plant diseases caused by plant pathogenic bacteria (especially in those that cause spots, cankers, blights, galls, or soft rots, bacteria), which are produced within or between plant cells, escape to the surface of their host plants as droplets or masses of sticky exudates (ooze). The bacteria exudates are released through cracks or wounds in the infected area, or through natural openings such as stomata, nectarthodes, hydathodes, and sometimes through lenticells, present in the infected area. Such bacteria are then likely to stick on the legs and bodies of all

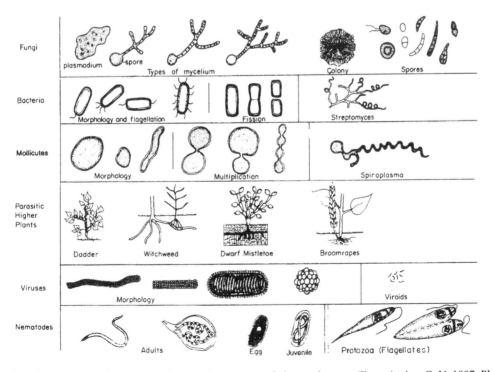

Fig. 1068 Morphology and multiplication of some of the groups of plant pathogens. (From Agrios, G. N. 1997. Plant pathology (4th ed.). Academic Press, San Diego, California.)

sorts of insects, such as flies, aphids, ants, beetles, whiteflies, etc., that land on the plant and come in contact with the bacterial exudates. Many of these insects are actually attracted by the sugars contained in the bacterial exudate and feed on it, thereby further smearing their body and mouthparts with the bacteria-containing exudate. When such bacteria-smeared insects move to other parts of the plant or to other susceptible host plants, they carry on their body numerous bacteria. If the insects happen to land on a fresh wound or on an open natural opening, and there is enough moisture on the plant surface, the bacteria may multiply, move into the plant, and begin a new infection. The same happens if the insects happen to create a fresh wound on the plant.

The type of insect transmission of bacteria is probably quite common and widespread among bacterial diseases of plants, but it is passive and haphazard, depending a great deal on the availability of wounds or moisture on the plant surface. In any case, there are few data on how frequently such transmission occurs, and many conclusions about it are the result of conjecture. A further point that has been made is that insects which, whether above or below ground, wound the host plant organs (roots, shoots, fruit, etc.) by feeding or by ovipositing in them, increase the probability of transmission of plant pathogenic bacteria. This occurs because such insects place the bacteria, with their mouthparts or the ovipositor, in or around wounded plant cells, where they are surrounded by a suspension of nutrients (plant cell sap) in the absence of active host defenses and where they can multiply rapidly and subsequently infect adjacent healthy tissues.

Numerous plant diseases could be listed among those in which bacteria are spread by insects passively as described above, for example, the bacterial bean blights, fire blight of apple and pear, citrus canker, cotton boll rot, crown gal, bacterial spot and canker of stone fruits, etc. In several bacterial diseases, however, the causal bacterium has developed a special symbiotic relationship with one or a few specific types of insects and depends a great deal on these insects for its spread from infected to healthy host plants. Some of the better known bacterium - insect associations are described briefly below.

Bacterial soft rots

Bacterial soft rots cause tremendous losses worldwide, particularly in the warmer climates and the tropics. They are caused primarily by the bacterium *Erwinia carotovora pv. carotovora*, to some extent by *Pseudomonas fluorescens* and *Ps. chrysanthemi*, and, occasionally, by species of *Bacillus* and *Clostridium*. The last two genera of bacteria cause rotting of potatoes and of cut fleshy leaves in storage while *Pseudomonas fluorescens* and *Ps. chrysanthemi* cause soft rots of many fleshy fruits and fleshy vegetables. The species *E. c. pv. carotovora* causes the vast majority of soft rots on fleshy plant organs of any type (leave, blossoms, fruit, stems, or roots), especially in storage and under cover or in plastic bags. Affected fleshy fruits are, for example, strawberries and other berries, cantaloupes, peaches, pears, etc.; vegetables, for example, tomatoes, potatoes, spinach, celery, onions, cabbage, etc.; and ornamentals, for example, cyclamen, iris, lily, etc. Nearly all fleshy vegetables are subject to bacterial soft rots. The soft rot bacteria enter the plant organ through a wound, sometimes in the field but more commonly during storage, and there they multiply rapidly, secrete enzymes that separate the cells from each other and macerate the plant cell walls, which causes the tissues to become soft and to rot. In many cases, these bacteria are accompanied in the rotting tissues by other saprophytic bacteria that further degrade the softened plant tissue and cause it to give off a foul odor. In all cases, rotting tissues become soft and watery, and slimy masses of bacteria ooze out from cracks in the tissues.

The soft rotting bacteria survive in infected fleshy organs in storage and in the field, in plant debris, in infected roots and other plant parts of their hosts, in ponds and streams from where irrigation water is obtained, and to some extent in the soil and in the pupae of several insects. The seedcorn maggot, *Delia platura* (Meigen) (Diptera: Anthomyiidae), was shown to play an important role in the dissemination and development of bacterial soft rot in potatoes both in storage and in the field. The soft rot bacteria are usually introduced into a potato field on infected or contaminated seed pieces but they can also live in all stages of the insect, including the pupae, and there they may survive cold or dry weather conditions. The insect larvae become contaminated with the bacteria as they feed in, or crawl about on, infected seed pieces; they also carry the bacteria to healthy plants and there they deposit them into wounds they create. Even when the plants or storage organs are resistant to soft rot bacteria and can normally stop the advance of the bacteria by developing a barrier of cork layers, the maggots destroy the cork layers as fast as they are formed and the soft rot continues to spread. Some other related flies, for example, the bean seed maggot *Delia florilega* (Zetterstedt), *Drosophila busckii* Coquillett (Diptera: Drosophilidae), and probably others, seem to have analogous relationship to the soft rot of potato and other fleshy organs. It has also been shown that several other flies have similar relationships with soft rot bacteria and the host plants on which they prefer to feed. Such relationships, for example, exist between the cabbage maggot, *Delia radicium* (Linnaeus) and soft rot in the Brassicaceae; the onion maggot, *Delia antiqua* (Meigen), the onion black fly, *Tritoxa flexa* (Weidman) (Diptera: Otitidae), the seedcorn maggot, and the onion bulb fly, *Eumerus strigatus* (Fallen) (Diptera: Syrphidae) and the soft rot of onion; and the iris borer, *Macronoctua onusta* (Grote) (Lepidoptera: Noctuidae) and soft rot of iris.

The exact relationship between soft rot in each host and each specific insect found to possibly be involved in the transmission of soft rot bacteria from one organ or plant to another is not clear. There is little doubt, however, that insect transmission of soft rot bacteria does occur, that insects help introduce the bacteria into wounds they open, and that the presence of insects in soft-rotting tissues inhibits the defense reaction of the plants against the bacteria. The insects also, by carrying the soft rot bacteria internally in their bodies, help the bacteria survive adverse environmental conditions. On the other hand, the bacteria seem to help their insect vectors by preparing for them a more nutritive substrate through partial maceration of the host plant tissues.

Bacterial wilts of plants

In several bacterial diseases of plants, the bacteria enter the xylem conductive system of the plant and there they move, multiply, and clog up the vessels. The clogging of the xylem vessels is further increased by substances released from cell walls by bacterial enzymes and interferes with the translocation of water through the stems to the shoots of the plant. As a result of insufficient water, the leaves and shoots loose turgor, wilt, and eventually turn brown and die. In some bacterial wilts, the bacteria destroy and dissolve parts of the xylem walls and move into the adjacent tissues where they form pockets full of bacteria from which the bacteria ooze out onto the plant surface through cracks or natural openings. In other bacterial wilts, the bacteria remain confined in the xylem and do not reach the plant surface until the plant is killed by the disease.

The wilt-causing bacteria overwinter in plant debris in the soil, in the seed, in vegetative propagative material, and in some cases, in their insect vector. They enter plants through wounds, and they spread from plant to plant through the soil, through tools and direct handling of plants, or through insect vectors. The most important bacterial wilts in which insects play a significant role in the transmission of the bacteria from plant to plant are described briefly below.

Bacterial wilt of cucurbits. Bacterial wilt of cucurbits has been reported from most developed countries but it probably occurs throughout the world. It affects many species of cucurbits, including cucumber, muskmelon, squash, and pumpkin. Watermelon is resistant or immune to bacterial wilt. Diseased plants develop a sudden wilting of their foliage and vines and eventually die. Diseased squash fruit develops a slimy rot in storage. Losses from bacterial wilt vary from an occasional wilted plant to destruction of 75 to 95% of the crop.

Bacterial wilt of cucurbits is caused by the bacterium *Erwinia tracheiphila*. The bacterium survives in infected plant debris for a few weeks but it survives over winter in the intestines of its two insect vectors, the striped cucumber beetle (*Acalymma vittatum* [Fabricius]) and the spotted cucumber beetle (*Diabrotica undecimpunctata* Mannerheim [Coleoptera: Chrysomelidae]). The bacterium depends on these two vectors for its transmission to and inoculation of new plants. In the spring, striped cucumber beetles and, to a lesser extent, spotted cucumber beetles, that carry bacteria, feed and cause wounds on the leaves of cucurbit plants. The insects deposit bacteria in the wounds through their feces and the bacteria enter the wounded xylem vessels in which they multiply rapidly and through which they move to all parts of the plant. In the xylem, the bacteria excrete polysaccharides, secrete enzymes that break down some of the cell wall substances, and induce xylem parenchyma cells to produce tyloses in the xylem. All of them together form gels or gums that clog the vessels, especially at their end walls, thereby reducing the upward flow of water in the xylem by up to 80% and causing the leaves and vines to wilt. Beetles feeding on infected cucurbit plants pick up bacteria on their mouthparts and when they feed onto healthy

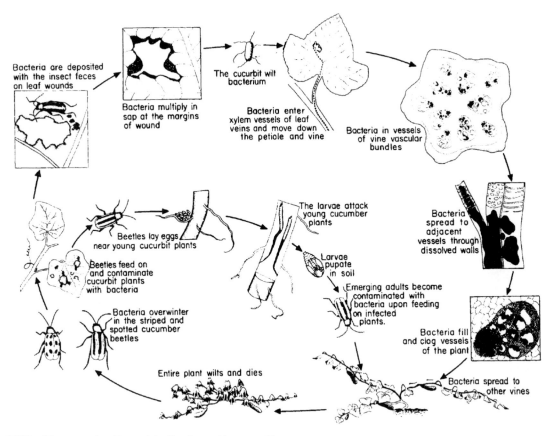

Fig. 1069 Disease cycle of bacterial wilt of cucurbits caused by *Erwinia tracheiphila* and transmitted by the striped cucumber beetle (*Acalymma vittatum*). (From Agrios, G. N. 1997. Plant pathology (4th ed.). Academic Press, San Diego, California.)

plants they deposit the bacteria in the new wounds they have made. Thus, the bacteria start a new infection. Each contaminated beetle can infect several healthy plants after one feeding on an infected plant. It appears that a relatively small percentage of beetles become carriers of the bacteria through the winter. Spotted cucumber beetles transmit the wilt bacteria rather late in the season, therefore they are considered less important vectors of this disease than the striped cucumber beetles.

Bacterial wilt of corn. This disease is also known as Stewart's wilt of corn. It is caused by the bacterium *Pantoea* (formerly *Erwinia*) *stewartii*. It occurs in North and Central America and also in Europe and China. It is more severe in the northern states. The bacterium invades the vascular tissues but it also spreads into other tissues. When sweet corn plants are affected at the seedling stage they may wilt rapidly and may die, or they develop pale green wavy streaks

on the leaves, become stunted, wilt, and may also die. If infected plants survive, they often tassel prematurely, the tassels become bleached and may die, and produce deformed ears. Bacteria also enter the stalk pith, which they macerate in places near the soil line and form cavities. From there the bacteria invade all vascular tissues and spread throughout the plant. Field corn is more resistant to early infection but becomes more severely infected later in the season. Some hybrids are susceptible and their symptoms parallel those of sweet corn. Later infections, after tasseling, produce irregular streaks on the leaves that originate at feeding points of the corn flea beetle, *Chaetocnema pulicaria* Melsheimer. The corn wilt bacteria are also transmitted by the toothed flea beetle (*Chaetocnema denticulata* Illiger), the spotted cucumber beetle (*Diabrotica undecimpunctata howarti* Barber), and by the larvae of the seed corn maggot (*Delia platura* Meigen), wheat wireworm (*Agriotes mancus* Say), and the May beetle

(*Phyllophaga* sp.). It appears, however, that overwintering and spread of the bacteria in the field is carried out primarily by the corn flea beetle.

These beetles cause direct damage to corn leaves and seedlings but their main damage comes from harboring and transmitting the bacteria from plant to plant. The beetles pick up the bacteria when they feed on infected corn plants. The bacteria survive in the digestive tract of the insect as long as the latter lives. The insects are also the main place where the bacteria overwinter. The corn flea beetles overwinter as adults in the upper 2–3 cm of soil in grass sod. They are rather sensitive to low temperatures, however. In mild winters, when the sums of mean temperatures for December, January, and February are above 3 to 4°C, large numbers of beetles survive. When the soil warms up to about 17 to 20°C, they begin to feed on corn seedlings, which they infect with bacteria. Following mild winters, bacterial wilt of corn is spread rapidly by corn flea beetles, and corn losses can be quite severe. During cold winters that average temperatures below 0°C, many of the beetles do not survive and the incidence and spread of bacterial wilt of corn the following spring and summer are quite limited.

Southern bacterial wilt of solanaceous and other crops. This vascular wilt is caused by the bacterium *Ralstonia solanacearum*. It occurs in the warmer regions around the world and is particularly severe in the tropics. It is known by different names in different hosts, for example, southern wilt or brown rot in potato and tomato, Granville wilt in tobacco, and Moko disease in banana. Insects, primarily bees (*Trigona corvine* Cockerell, Hymenoptera: Apidae), wasps (*Polybia* spp., Hymenoptera: Vespidae), and flies (*Drosophila* spp., Diptera: Drosophilidae) have been implicated as vectors. Because these and other insects visit infected stem wounds and natural abscission sites oozing out bacteria, they are considered as playing a role in the transmission of the bacteria to natural infection courts and in providing wounds for bacterial entry, but their importance as vectors has not been documented.

Fire blight of pears, apples and other rosaceous plants. The disease is caused by the bacterium *Erwinia amylovora*. Fire blight occurs in North America, Europe and countries surrounding the Mediterranean Sea, and in New Zealand. It continues, however, to spread into new countries. Fire blight is the most devastating diseases affecting rosaceous plants. The symptoms consist of infected blossoms and young shoots becoming discolored and water-soaked, then being killed rapidly and appearing brown to black as though scorched by fire. The disease spreads rapidly into larger twigs and branches, which it also kills, and parts of or entire trees may be killed. At the base of twig or branch infections, cankers develop at the margins of which the bacteria overwinter. Fruit also become infected and ooze droplets of bacteria. The bacteria kill and macerate the contents of primarily parenchyma cells on flowers and in the bark of young shoots and twigs, but as they destroy these cells they move on mass in the bark. The bacteria also enter the phloem and xylem vessels through which they may move over relatively short distances.

The fire blight bacteria overwinter at the margins of cankers of twigs and branches. In the spring, the bacteria around cankers multiply and their byproducts absorb water and build up internal pressure. This results in droplets of liquid containing masses of fire blight bacteria oozing out of the cankers. The bacteria in the ooze are disseminated by splashing rain and also by flying and crawling insects, several of which are attracted to the bacterial ooze, and their legs, bodies, and mouthparts become smeared with bacteria. More than 200 species belonging to many insect groups, including aphids, leafhoppers, psyllids, beetles, flies, and ants, have been shown to visit oozing cankers and healthy blossoms, although bees and wasps seem not to visit oozing cankers routinely. Insects smeared with bacteria oozing out at cankers carry the bacteria to young shoots where they deposit them in existing wounds or in fresh wounds they make upon feeding, or in the nectar of the flowers. Once the fire blight bacteria are transmitted to blossoms by rain or insects, they enter the flower tissues through nectarthodes or wounds, multiply rapidly in them, and ooze out of them and commingle with the nectar in the flower. The same kinds of insects apparently can transmit fire blight bacteria from infected to healthy flowers but flower to flower transmission of fire blight bacteria is carried out so much more efficiently by pollinating insects, namely bees, that the contribution of other insects to that type of transmission seems to be relatively insignificant. As honeybees, wild bees, bumblebees, wasps, and other insects visit pear, apple, and other flowers infected with fire blight bacteria, their mouthparts, legs, and other body parts become smeared with the bacteria in the nectar. The insects then carry the bacteria and deposit them in the nectar of healthy flowers they visit and there the bacteria start new infections. The bacteria, however, do not survive on or in the insects for more than a few days and do not appear to overwinter in association with the insects.

Olive knot. Olive knot is caused by the bacterium *Pseudomonas savastanoi*. It occurs in the Mediterranean region, in California, and probably the other parts of the world where olive trees grow. The disease occurs as rough galls of varying sizes developing on leaves, branches, roots, on leaf and fruit petioles, and on wounds in tree branches and trunks. Sometimes the galls are so numerous on twigs that the twigs decline and may die back. The galls are the result of growth regulators being produced by the bacteria, which grow and multiply in the intercellular spaces of the outer cells of the galls. In California, the bacteria are spread by running and splashing rain water that carries the bacteria to existing wounds, pruning wounds, and leaf scars. In other parts of the world, however, such as the Mediterranean region, the olive knot bacteria are also spread by the olive fly or olive fruit fly, *Bactrocera* (formerly *Dacus*) *oleae* (Gmelin) (Diptera: Tephritidae), which is the most destructive pest of olive in its own right.

The bacterium and the olive fly have developed a close symbiotic relationship that contributes to the transmission of the olive knot bacteria from tree to tree. The bacteria are carried by all stages (larvae, pupae, and adults) of the olive fly. The adult olive flies, and related fly species, have specialized structures along their digestive tract that are filled with bacteria. There is even a connection of the digestive tract with the oviduct that insures contamination of the eggs before oviposition. Transmission of the bacteria by the insect takes place during feeding and oviposition into olive tissues. The bacteria actually penetrate the egg through the micropyle, thereby ensuring that when the larvae hatch they are contaminated with the bacteria. It appears that while the olive fly plays a significant role in the transmission of the olive knot bacteria, the bacteria contribute to the insect by hydrolyzing proteins and making available to the insect certain amino acids needed by the insect for survival of the larvae and for development of adults.

Insect transmission of xylem-inhabiting bacteria

Quite a few important bacterial diseases of plants, primarily trees, are caused by the fastidious bacterium *Xylella fastidiosa*. These bacteria inhabit the xylem of their host plants and are rather difficult to isolate and to grow on the usual culture media. The diseases they cause differ from the vascular wilts caused by conventional bacteria in that instead of wilt they cause infected plants to decline, some of their twigs to die back, and in some cases the whole plant to die. The xylem-inhabiting fastidious bacteria are transmitted in nature only by xylem-feeding insects, such as sharpshooter leafhoppers (Cicadellidae: Cicadellinae) and spittlebugs (Cercopidae). *Xylella* bacteria seem to be distributed in tropical and semi-tropical areas worldwide. Among the most important diseases caused by *Xylella* are Pierce's disease of grape and citrus variegated chlorosis. Other diseases caused by xylem-inhabiting bacteria include phony peach, plum leaf scald, almond leaf scorch, bacterial leaf scorch of coffee, oak leaf scorch, and leaf scorch diseases of oleander, pear, maple, mulberry, elm, sycamore, and miscellaneous ornamentals, as well as the alfalfa dwarf disease.

Pierce's disease of grape. Pierce's disease is a devastating disease of European-type grapevines (*Vitis vinifera*). It is caused by the xylem-inhabiting bacterium *Xylella fastidiosa*. It occurs in the Southern United States and in California, in Central America, and parts of northwestern South America. The presence of Pierce's disease in an area precludes the production of European-type grapes in that area but some muscadine grapes and hybrids of European grapes with American wild grapes are tolerant or resistant to Pierce's disease. The Pierce's disease bacterium moves and multiplies in the water-conducting (xylem) vessels of shoots and leaves, some of which become filled with bacteria and reduce the flow of water through them. Leaves beyond such blocked vessels become stressed from lack of sufficient water and develop yellowing and then drying or scorching along their margins. During the summer, the scorching continues to expand towards the center of the leaf, while some or the entire grape clusters begin to wilt and dry up. Scorched leaves fall off leaving their petioles still attached to the vine, while the vines mature unevenly and have patches of brown (mature) and green bark. In the following season(s), affected grapevines show delayed growth and stunting. The

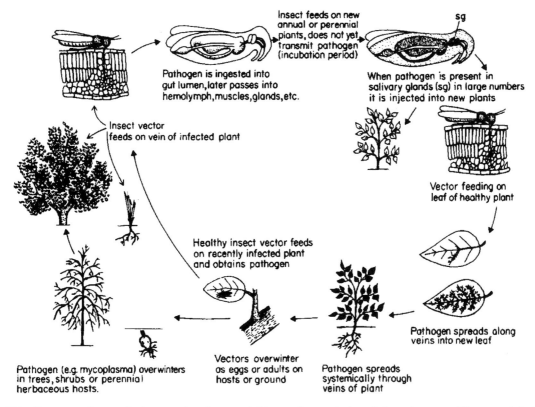

Fig. 1070 Sequence of events in the overwintering, acquisition, and transmission of plant viruses, mollicutes, and fastidious bacteria by leafhoppers. (From Agrios, G. N. 1997. Plant pathology (4th ed.). Academic Press, San Diego, California.)

leaves and vines repeat the symptoms of the first year and both, the top of the plant as well as the root system, decline and die back.

The bacterium that causes Pierce's disease of grape is *Xylella fastidiosa*. The bacterium apparently consists of various host specific strains. The strain that causes Pierce's disease of grape also causes alfalfa dwarf disease and almond leaf scorch. Apparently related but different strains of the bacterium cause citrus variegation chlorosis, the other related leaf scorch diseases of fruit and forest trees and of ornamental trees and shrubs. The identity and taxonomy, as well as the host range and vector preference of the possible strains of *Xylella fastidiosa*, are unknown. In all cases, the bacterium is transmitted from plant to plant through vegetative propagation, such as cuttings, budding, and grafting, and by one or more of several closely related insects. The known vectors of *Xylella fastidiosa* are sharpshooter leafhoppers (family Cicadellidae, subfamily

Cicadellinae) or spittlebugs (family Cercopidae). It is possible that other or all sucking insects that feed on xylem sap, for example, the cicadas (family Cicadidae), are also vectors of *Xylella fastidiosa*. In California, there are at least four important sharpshooter leafhopper vectors of *Xylella*: The blue-green (*Graphocephala atropunctata*), the green (*Draeculacephala minerva*), the red-headed (*Carneocephala fulgida*), and the glassy-winged (*Homalodisca coagulata*) sharpshooters. The vectors may be different in other parts of the world. All vector insects acquire the bacteria when they feed on infected plants. Ingested bacteria seem to adhere to the walls of the foregut of the insect and when the insect moves to and feeds on the next healthy plant, the insect transmits the bacteria into the xylem vessels of that plant where they multiply and cause a new infection. Once a vector acquires bacteria from a diseased plant, it remains infective indefinitely. When, however, infective insects shed their external skeleton by molting, they loose the bacteria and must feed on a diseased plant again before they can transmit the bacteria to healthy plants.

Insect transmission of phloem-inhabiting bacteria

At least four plant diseases are caused by bacteria that inhabit only the phloem of their host plants. These diseases include the destructive citrus greening disease, the severe papaya bunchy top disease, the cucurbit yellow vine disease, and the infrequent clover club leaf disease. The bacteria causing these diseases have not yet been grown on nutrient media and so far many of their properties remain unknown. All of them, however, are transmitted from plant to plant only by specific insect vectors. The citrus greening bacterium is transmitted by a psyllid, while the papaya bunchy top disease bacterium and the clover club leaf bacterium are transmitted by leafhoppers, and the cucurbit yellow vine disease bacterium is transmitted by the squash bug. In the psyllid and leafhopper vectors, the bacterium also multiplies in and is passed from the mother insect to its offspring through the eggs (transovarial transmission). It is not known what happens to the bacteria transmitted by the squash bug.

Citrus greening disease. Citrus greening is a very destructive disease of all types of citrus. It occurs in most citrus producing areas of Asia, including the Arabian Peninsula, and in Africa. The disease is caused by the bacterium *Liberobacter asiaticum* in Asia, and *L. africanum* in Africa. Both bacteria are limited to the phloem of the host plants, and have yet to be cultured. The disease first appears as a chlorosis and leaf mottling on one shoot or branch, which it has given it the name "huanglongbing", or "yellow shoot", in Chinese. Later on, entire trees become chlorotic as though they are suffering from zinc deficiency, their twigs die back, and the trees decline rapidly and become non-productive. Fruit on diseased trees is small, lopsided, and does not color uniformly as it ripens but large parts of it remain green even when mature, thereby the "greening" name of the disease. Diseased fruit is also quite bitter.

Citrus greening is spread by vegetative propagation with buds and grafts, and by at least two citrus psyllids: *Diaphorina citri* Kuwayama, which is the principal vector of the more severe and more destructive Asian form of the citrus greening bacterium that occurs at higher temperatures (30 to 35°C), commonly found at lower elevations; and *Trioza erytreae* Del Guercio, which is the principal vector of the milder, less severe, lower temperature (27°C) African form of the bacterium, which is normally found at higher elevations. Both vectors, however, can transmit both forms of the bacterium. Asian psyllids acquire the bacterium within 30 minutes of feeding while African psyllids require 24 hours. The bacterium apparently multiplies in the vector and can be transmitted within 8 to 12 days from acquisition.

Infected plants and vectors have been introduced into several citrus- producing countries but in most cases it was eradicated before it could become established. The vector of the greening bacterium *Diaphorina citri* was introduced in Brazil in the early 1980s and in Florida in 1998 but, so far, the causal bacteria apparently have not been introduced and no trees have been found in either place to be infected with citrus greening.

Bunchy top of papaya disease. Bunchy top is a devastating disease of papaya. It occurs in most or all islands of the Caribbean Basin and, probably, also in Central America and in the northern part of South America. Young leaves of infected plants show mottling, then chlorosis and marginal necrosis, and become rigid. Internodes become progressively shorter, further apical growth stops, and the plant develops a "bunchy" top. Older leaves may fall off, any fruits that are set are bitter, and the entire plant may die.

Bunchy top of papaya is caused by a rickettsia-like phloem-limited bacterium that moves and multiplies in the phloem elements of the plant. The bunchy top bacteria are transmitted from diseased to healthy papaya plants by the leafhoppers *Empoasca papayae* Oman and *E. stevensi* Young. Symptoms appear 30–45 days after inoculation.

Cucurbit yellow vine disease. Yellow vine disease affects watermelon, melon, squash, and pumpkin. It was first reported in the Texas-Oklahoma area and has since been found in Massachusetts, New York, and Tennessee. Affected plants show vines with yellow leaves, the phloem of leaves and vines becomes discolored, and the leaves and vines collapse and die. The disease has been severe in the Texas and Oklahoma areas where it annually destroys thousands of acres of cucurbits costing millions of dollars.

Cucurbit yellow vine disease is caused by a phloem-limited bacterium that has been placed in the species *Serratia marcescens* and its properties are still being characterized. The bacterium is most probably transmitted by insect vectors. The squash bug, *Anasa tristis*, is considered to be a vector of this bacterium, but its involvement in transmitting this bacterium has been questioned.

Insect transmission of plant diseases caused by mollicutes

Mollicutes are prokaryotes (bacteria) that lack cell walls. In nature, plant pathogenic mollicutes are limited to the phloem of their host plants. Plant pathogenic mollicutes are generally classified as belonging to the genus *Phytoplasma*. Most phytoplasmas have an irregular spherical to elongated shape and have been obtained and maintained on complex nutrient media, although they do not readily grow or multiply on them. A few plant pathogenic mollicutes typically have spiral morphology and belong to the genus *Spiroplasma*. Spiroplasmas grow and multiply readily on specialized nutrient media. Plant diseases caused by mollicutes appear as yellowing of leaves, proliferation of shoots (witches' brooms) and of roots, stunting of shoots and whole plants, greening of flowers, abortion of flowers and fruit, dieback of twigs, and decline and death of trees. Numerous important diseases of annual crops are caused by mollicutes, mostly phytoplasmas, for example, aster yellows of vegetables and ornamentals, tomato big bud (stolbur), corn stunt, etc. Phytoplasmas cause even more, and more severe, diseases on trees, including X-disease of peach, peach yellows, apple proliferation, elm yellows, pear decline, and lethal yellowing of coconut palms. Spiroplasmas also cause severe diseases, for example, corn stunt, and citrus stubborn disease.

All mollicutes, that is, phytoplasmas and spiroplasmas, are spread from plant to plant through vegetative propagation and, in nature, these pathogens depend for their transmission on phloem-feeding, sap-sucking insects, mainly leafhoppers, planthoppers, and psyllids. These insects can acquire the pathogen after feeding on appropriate infected plants for several hours or days, or if they are artificially injected with extracts from infected plants or insects. More insects become vectors when they feed on young leaves and stems of infected plants than on older ones. The insect vector cannot transmit the pathogen immediately after feeding on the infected plant but it begins to transmit it after an incubation period of 10 to 45 days, depending on the temperature. The quickest transmission (10 days) occurs at about $30°C$, while the slowest (45 days) takes place at about $10°C$.

The reason for the incubation period is that the acquired phytoplasmas or spiroplasmas must first multiply in the intestinal cells of the insect vector and then move through the insect by passing into the hemolymph, then infect internal organs and the brain, and finally reach and multiply in the salivary glands. When the concentration of the pathogen in the salivary glands reaches a certain level, the insect begins to transmit the pathogen to new plants and continues to transmit it with more or less the same efficiency for the rest of its life. Insect vectors are not generally affected adversely by the phytoplasmas or spiroplasmas multiplying in their cells, but in some cases they show severe pathological symptoms. Phytoplasmas and spiroplasmas can be acquired as readily or better by nymphs than by adult leafhoppers, etc., and they survive through subsequent molts of the insect. The pathogens, however, are not passed from the adults to the eggs and to the next generation. For this reason, young insects of any stage must feed on infected plants in order to become infective vectors.

Some of the most important plant diseases caused by mollicutes and their insect vectors are described briefly below.

Aster yellows. Aster yellows is caused by phytoplasmas and occurs worldwide. It affects numerous annual crops, mostly vegetables and ornamentals, for example, tomato, lettuce, carrot, onion, potato, chrysanthemum, aster, and many others, on which it causes severe symptoms and serious losses, in some crops amounting to 10–25% of the crop and occasionally up to 80-90% of the crop. Plants infected with aster yellows develop general chlorosis (yellowing) and dwarfing of the whole plant, abnormal production of shoots and, sometimes, roots, sterility of flowers, malformation of organs, and a general reduction in the quantity and quality of yield. The aster yellows phytoplasma is transmitted by several leafhoppers, one of which is the aster leafhopper *Macrosteles fascifrons*. The various leafhopper vectors have a wide host range, as does the aster yellows phytoplasma. The phytoplasmas survive in perennial ornamental, vegetable, and weed plants. The vector leafhopper acquires the phytoplasmas while feeding by inserting its stylet into the phloem of infected plants and withdrawing the phytoplasmas with the plant sap. After an incubation period, when the insect feeds on healthy plants it injects the phytoplasmas through the stylet into the phloem of the healthy plants where they establish a new infection and multiply. The phytoplasmas move out of the leaf and spread throughout the plant causing the symptoms characteristic of the host plant.

Tomato big bud. The disease occurs in many parts of the world but except for a few areas, it is of little economic importance. It affects most Solanaceous vegetables and lettuce. The symptoms include small, distorted, yellowish green leaves and production of numerous thickened, stiff, and erect apical stems that have short internodes. The flower buds are excessively big, green, and abnormal looking, and fail to set fruit. Fruit present when infection takes place becomes deformed.

Tomato big bud is caused by a phytoplasma that is transmitted by several leafhoppers, the main one of which is the common brown leafhopper *Orosius argentatus*. The insect feeds and breeds on infected weed hosts and when they become undesirable the insects move into tomato or other crops bringing with them the big bud phytoplasmas.

Apple proliferation. It is the most important insect transmitted disease of apple in most of Europe. It may also occur in South Asia and South Africa. Depending on prevalence in an orchard, apple proliferation may cause economic losses of 10–80% due to reduction in fruit size, total yield, and vigor of trees. The most conspicuous symptoms of apple proliferation are the production of witches' brooms or of leaf rosettes, and of enlarged stipules at the base of leaves. Affected trees leaf out earlier in the spring but flowering is delayed. The leaves, fruit, and entire trees are smaller, and fruit color and taste are also poor. Proliferating shoots are often infected with powdery mildew.

Apple proliferation is caused by a phytoplasma that also infects other wild and ornamental apple species, and possibly pear and apricot. The phytoplasma is spread in nature by several leafhoppers, including *Philaenus spumarius*, *Aphrophora alni*, *Lepyronia coleoptrata*, *Artianus interstitialis*, and *Fieberiella florii*. The leafhopper vectors acquire the phytoplas-

mas when they feed on the phloem elements of young leaves and shoots of infected apple trees and, after an incubation period, transmit the phytoplasmas into healthy apple trees. The time between inoculation and appearance of symptoms varies with the size of the inoculated tree. Young nursery trees may develop symptoms within a year while large established trees may do so two or three years after inoculation.

Pear decline. It is a serious disease of pear resulting in significant crop losses and also in stunting and death of affected pear trees grown on certain rootstocks. The disease, which is caused by a phytoplasma, occurs in North America and in Europe, and probably in many other parts of the world where pears are grown. The symptoms of pear decline may develop as a quick decline, that is, sudden wilt and death of a tree within a few days or weeks, with or without first showing reddening of leaves, or a slow decline. Quick decline usually develops on trees propagated on certain hypersensitive rootstocks in which a brown necrotic line develops at the graft union of the tree. Slow decline also occurs on trees grafted on the same or other rootstocks, and appears as a progressive weakening of the tree of varying severity. Slow declining trees have reduced or no terminal growth, have few, small, leathery, light green leaves whose margins are slightly rolled up and may be yellow or red in the autumn. Such trees may or may not die a few years after infection. Some infected pear trees, however, show primarily a reddening of the leaves in late summer or early autumn, and mild reduction in vigor.

The pear decline phytoplasma is transmitted from tree to tree by grafting and by the pear psylla (*Psylla pyricola* Forster) and in Europe, probably by *P. pyri* and *P. pyrisuga*. Pear psylla acquires the phytoplasma after feeding on infected trees for a few hours and remains infective for several weeks. Young trees inoculated with phytoplasma by the insect show symptoms the same or the next year, while older trees may take longer. The phytoplasma is sensitive to low temperatures and dies out in the above-ground parts of the tree but survives in the tree roots. In the spring, the phytoplasma recolonizes the stem, branches and shoots and from the latter it can be acquired and transmitted again by the insect vectors.

Lethal yellowing of coconut palms. Lethal yellowing is a blight that kills coconut and some other palm trees within 3 to 6 months from the time the trees show the first symptoms. It occurs in Florida, Texas, Mexico, most Caribbean islands, in West Africa, and elsewhere. The disease appeared for the first time in Florida in 1971 and killed 15,000 coconut palm trees the first two years, 40,000 by the third year, and by the fourth year (1975), 75% of the coconut palms in the Miami area were dead or dying from lethal yellowing. The disease appears as a premature drop of coconuts followed by blackening and death of the male flowers. Subsequently, first the lower and then the other leaves turn yellow and then brown and die, as does the vegetative bud, and the entire top of the tree falls off leaving the palm trunk looking like a telephone pole.

Lethal yellowing is caused by a phytoplasma that lives and multiplies in the phloem elements of the plant. The main means of spread of lethal yellowing from tree to tree is through the planthopper *Myndus crudus*, although other vectors are also possible. As with the other phytoplasma diseases, the vector acquires the phytoplasma while sucking juice off the phloem of palm leaf veins, the phytoplasma multiplies in the vector during an incubation period, and the insect then transmits the phytoplasma when it visits and feeds on leaf veins of healthy palm trees.

Corn stunt. Corn stunt causes severe losses where it occurs although losses vary from year to year. It occurs in the southern United States, Central America, and northern South America. Symptoms consist of yellow streaks in young leaves followed by yellowing and reddening of leaves, shortening of internodes, stunting of the whole plant, and sterile tassels and ears.

Corn stunt is caused by the spiroplasma *Spiroplasma kunkelii*. The spiroplasma invades phloem cells from where it is acquired by its leafhopper vectors *Dalbulus maidis*, *Dalbulus eliminatus*, and other leafhoppers after feeding on infected plants for several days. The vectors transmit the spiroplasma after an incubation period of 2 to 3 weeks, during which the spiroplasma moves and multiplies in the insect.

Citrus stubborn disease. It is present and severe in hot and dry areas such as the Mediterranean countries, the southwestern United States, Brazil, Australia, and elsewhere. It is one of the most serious diseases of sweet orange and grapefruit. It is hard to diagnose but reduces yield, quality, and marketability of fruit dramatically. Infected trees grow upright but are stunted. There is less fruit and it

is smaller, lopsided, green, and sour, bitter, and unpleasant.

Citrus stubborn disease is caused by the spiroplasma *Spiroplasma citri*, which is found in the phloem of affected orange trees. It is transmitted by budding and grafting and, in nature, by several leafhoppers such as *Circulifer tenellus*, *Scaphytopius nitrides*, and *Neoaliturus haemoceps*. Role of insects on fungal diseases of plants

As with bacteria, many insects are involved in the transmission of numerous plant pathogenic fungi from diseased to healthy plants. Insects are also involved in plant diseases by breaking the epidermis and other protective tissues of plants with their mouthparts or with their ovipositor and thereby allowing the fungus to enter. Most of the insect transmissions of fungi are accidental, that is, they occur because the insects happen to become externally or internally contaminated with the fungus or its spores when they visit infected plants and then carry the spores with them to the plants or plant parts they visit next. In some cases, insect transmission of a fungus occurs as the insect visits blossoms during pollination, in others it occurs while wounding plants during oviposition, and in other and most frequent cases, transmission occurs while wounding the plant during feeding. In relatively few cases, the insect and the fungus it transmits develop a symbiotic relationship in which each benefit from its association with the other.

Root-infecting fungi. It should be pointed out that there are innumerable cases for which there is circumstantial evidence that insects are apparently involved in the transmission of many plant pathogenic fungi and in the development of disease by them, but this has not been proven experimentally. In this category belong, for example, root infections by fungi such as *Pythium*, *Fusarium*, and *Sclerotium*, being facilitated by billbugs such as *Calendra parvula* and *Anacentrus deplanatus*, by the Hessian fly *Phytophaga destructor*, and by the southern and northern corn root worms *Diabrotica undecimpunctata howardii* and *Diabrotica longicornis*, respectively. In the black stain root disease of pines, hemlock, and Douglas fir is caused by the fungus *Leptographium wageneri*, the teliomorph of which is *Ophiostoma wageneri*, and is transmitted by the root-feeding bark beetle *Hilastes nigrinus* and two root and crown weevils, *Steremnius carinatus* and *Pissodes fasciatus*.

Stalk or stem-infecting fungi. Many fungi infecting stalks or stems, for example *Gibberella*, *Fusarium*, and *Diplodia* in corn, are apparently aided by various insects, for example, the widespread European corn borer, *Pyrausta nubilalis*.

Trunk and branch canker-causing fungi. Many fungi, such as species of *Neofabrea*, *Nectria*, *Leucostoma* (*Cytospora*), *Ceratocystis*, and *Lepto-sphaeria*, causing tree cankers, are apparently also often associated with and assisted by insects in the initiation and development of the cankers. The insects involved vary with the particular host and fungus. For example, the fungus *Neofabrea perennans* (*Gloeosporium perennans*), the cause of the perennial canker of apple, is transmitted by the wooly aphid *Eriosoma lanigerum*. The woolly aphids feed on the bark at the base of the trunk where they cause the formation of galls within which they multiply. In early spring the galls burst, the aphids come out and the fungus attacks the injured tissue and from it advances into healthy tissue and produces a canker. In the summer, the apple tree produces callus tissue and seals off the fungus and the spread of the canker stops. The aphids, however, grow into the callus tissue and form a new gall, and the process is repeated.

The spittlebug *Aphrophora saratogensis* seems to be involved in the *Nectria* canker of pines, the nitidulid beetle *Carpophilus freemani* and the drosophilid fly *Chymomyza procnemoides* in the *Ceratocystis* cankers of stone fruit trees, while the tree cricket *Oecanthus niveus* and the raspberry midge *Thomasianna theobaldi* are involved in the tree cricket canker of apple and the midge canker of raspberry, respectively. Many more such insect-pathogen associations could be mentioned.

In the beech bark canker, caused by the fungus *Nectria coccinea* var. *faginata*, the fungus is transmitted to some extent by the scale insect *Cryptococcus fagi* but the main effect of the insect is in weakening the tree and reducing its defenses to the fungus. Thus, after beech trees have been heavily infested by scale insects for about three to five years, the fungus invades and kills the bark and the tree forms a canker that may girdle the tree partially or completely and may kill it.

In the birch constriction disease, the lower parts of shoots become constricted at the point where the apical birch woodwasp (*Pseudoxiphydria betulae* Ensl.) feeds on the shoots. The leaves above the constriction wither and die but cling to the twigs past the autumn. Almost all (92%) of the constrictions are also infected with the anthracnose fungus *Melanconium bicolor*.

A similar case in which twig canker initiation and development are facilitated by insects is the cacao dieback disease in which the fungi *Calonectria* (*Fusarium*) *rigidiuscula* and/or *Botryodiplodia theobromae* enter the twigs through wounds created by the feeding of the capsid insects *Sahlbergella singularis* and *Distantiella theobroma*. In isolated infections the tree defenses take over, isolate the fungus, and its further spread stops. In trees massively infested with the insect, however, the fungus develops unchecked in the insect-infested tissues and causes a chronic dieback of twigs. Control of the insects also halts the invasion by the fungus and the tree recovers.

In mango malformation disease, presumably caused by the fungus *Fusarium moniliforme*, the fungus is transmitted by the eriophyid mite *Aceria mangifera*, while other fungi seem to be carried in the digestive tract of certain termites.

Sooty molds. These are black-colored fungi that grow on the surfaces of mostly leaves of plants, especially in the tropics and subtropics. Sooty mold fungi do not penetrate and infect plants but cause disease by blocking the light from reaching the leaves. Sooty mold fungi do not parasitize plants but feed off the honeydew excreted by insects such as whiteflies, scales, mealybugs, aphids, etc. The sooty mold fungi are disseminated through their spores being blown about by wind. However, they are also spread by the honeydew-producing insects and, also, by several other types of insects such as flies, wasps, bees, and ladybug beetles, all of which seek honeydew as a source of food and in the process become smeared with fungus spores which they carry about.

Wood rots. Rotting of wood is carried out primarily by wood-rotting basidiomycete fungi. The shelf or conk-shaped fruiting bodies of many of these fungi are visited routinely by many types of insects and it is believed that many of these insects act as vectors of the wood-rotting fungi. Insects and mites have also been implicated in the spread of some pine rust diseases, while at least three common scolytid beetles have been shown to be involved in the transmission of the scleroderis canker of pine and spruce.

Wood-stain diseases. Wood stain or wood discoloration diseases occur in conifer trees and felled timber. They are caused by the so-called blue-stain fungi, of which the most common are species of *Ceratocystis* and *Ophiostoma*. The blue-stain fungi are associated with several species of bark beetles, such as *Dendroctonus ponderosae*, *Ips pini*, etc., which serve as vectors of the fungi and provide them with wounds for penetration. On the other hand, the fungi reduce the water content of the tree and otherwise improve the microenvironment for the developing brood of insects. Such a fungus-insect relationship is described as true mutualistic symbiosis. In other blue-stain diseases, like the ones caused by the fungi *Trachosporium tingens* and *T. t.* var. *macrosporum*, the fungi are constantly associated with their bark-beetle vectors *Myelophilus* (*Blastophagus*) *minor* and *Ips acuminatus*, respectively, and are found regularly in the breeding places of the insects in pine stems. Such fungal-insect associations are known as symbiotic ambrosia cultures.

In the Southern United States, attacks of short-leaf pines by beetles like *Dendroctonus frontalis* are quickly followed by heavy fungus infection soon after the beetles tunnel through the bark and outer wood. Several fungi, including *Ceratocystis pini*, *Saccharomyces pini*, *Dacryomyces* sp. and *Monilia* sp. can be isolated from the infected wood and are carried by the same insects both externally and internally. A similarly complex association seems to occur in spruce attacked by *Dendroctonus engelmani*, followed by the fungi *Leptographium* sp., *Endoconidiophora* sp., or *Ophiostoma* sp. infecting the wood and causing a gray stain in the sapwood of the infected trees.

Vascular wilts

Several vascular wilts affect trees and some of them cause extensive death of trees, because the fungus responsible for the disease is transmitted from diseased to healthy trees by specific insect vectors. The spores produced by the causal fungi are sticky and are produced primarily inside the tree, therefore, they can be spread by no other means but only by certain insects closely associated with the disease. These vascular wilts include: (1) persimmon wilt, a devastating disease caused by the fungus *Cephalosporium diospyri* which enters through all kinds of wounds but is also transmitted by the powder-post beetle. *Xylobiops basilaris* and the twig girdler beetle

Oncider cingulatus, and (2) mango wilts, one caused by the fungus *Diplodia recifensis* and transmitted by the beetle *Xyleborus affinis*, and another caused by the fungus *Ceratocystis fimbriata* and transmitted by the scolytid beetle *Hypocryphalus mangiferae*. The other two vascular wilts are oak wilt and the Dutch elm disease and will be discussed in some detail below.

Oak wilt. It is caused by the fungus *Ceratocystis fimbriata* and is one of the most important diseases of forest trees. The fungus enters the xylem vessels of trees through fresh wounds to which it is carried by air or insects, and through natural root grafts. Tree parts beyond the point of infection wilt, turn brown and die while newly infected wood shows dark streaks. The fungus is spread to healthy trees by nitidulid beetles such as *Carpophilus lugubris*, *Colopterus niger*, *Cryptarcha ample*, and several species of *Glischrochilus*. These fungi breed in the mycelial mats of the fungus between the bark and wood and carry the fungus both externally on their bodies and internally through their digestive tract. In addition to the nitidulid beetles, several scolytid beetles, such as *Monarthrum fasciatum* and *Pseudopityophthorus minutissimus*, the brentid beetle *Arrhenodes minuta*, the buprestid *Agrillus bilineatus*, the flat-headed borer *Chrysobothrys femorata*, and others, have been shown to carry the spores of the fungus, both externally and internally, when they emerge from the tunnels in diseased trees in which they breed and overwinter and to carry them to susceptible trees in the spring. Transmission of the oak wilt fungus by insects not only spreads the fungus and the disease to new trees and into new areas, it also increases the ability of the fungus to produce new variants and new races more virulent than the existing ones. This is accomplished by the insects bringing together in the same tree the compatible self-sterile mating types which results in the production of perithecia containing the sexual spores ascospores. The latter express any new characteristics brought together during the formation of the spores, some of the characteristics possibly being increased virulence.

Dutch elm disease. It is caused by the fungus *Ophiostoma ulmi* and the most recent variant *Ophiostoma novo-ulmi*, which is replacing the earlier species. The disease was first described in the Netherlands in 1921, found in Ohio and New York in the 1930s, and has since spread throughout the

United States and much of the rest of the world. It kills elm tree twigs, branches and whole trees by clogging their xylem vessels and blocking movement of water from the roots to these parts. Dutch elm disease has been particularly devastating in the United States where the native elm tree *Ulmus americana* is extremely susceptible to the pathogen. The disease has killed almost all trees in its path, especially elm trees planted along streets and parks. Elm trees in forests have also been killed but many of them have escaped infection so far.

The fungus causing Dutch elm disease is spread from diseased to healthy trees by the European elm bark beetle *Scolytus multistriatus* and the native elm bark beetle *Hylurgopinus rufipes*, and by natural root grafts. The fungus overwinters in the bark of dying or dead elm trees and logs as mycelium and spores. The elm bark beetles lay their eggs in galleries they make in the intersurface between bark and wood of weakened or dead elm trees. If the tree is already infected with the Dutch elm disease or if the insects carry with them spores of the fungus, the fungus grows and produces new spores in the tunnels. After the eggs hatch, the larvae make tunnels perpendicular to that made by the adult female and pupate. The adults emerge, carrying thousands of fungal spores on their body. The emerging adults prefer to feed on young twigs of trees and the crotch of small branches. As the beetles burrow into the bark and wood for sap, the spores they carry on their body

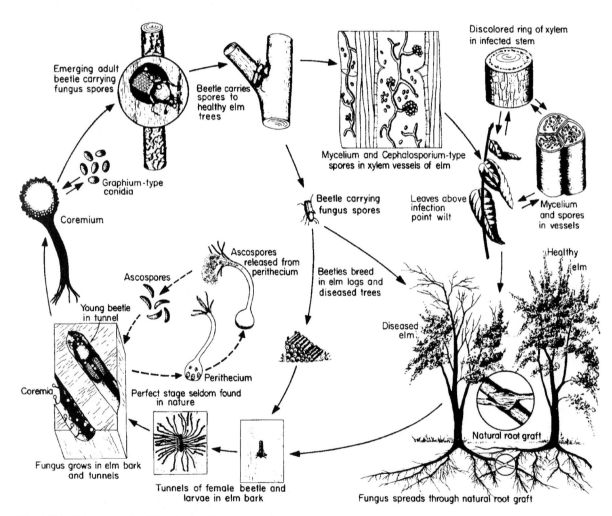

Fig. 1071 Disease cycle of Dutch elm disease caused by the fungus *Ophiostoma ulmi* and transmitted by the European and the American elm bark beetles. (From Agrios, G. N. 1997. Plant pathology (4th ed.). Academic Press, San Diego, California.)

are deposited in the wounded moist tissues of the tree. There the spores germinate and grow into the injured bark and wood and the fungus reaches the xylem vessels of the tree in which it grows producing mycelium and spores. The latter are carried upward by the sap stream where they can start new infections. Shoots beyond the infected areas turn brown, wilt, and die, and as their number increases the tree shows more browning and more wilted branches. Eventually large parts of or entire trees wilt and die, while the fungus continues to grow and spread in the dead tree. Such trees are then visited by adult female beetles that lay their eggs in them and the cycle is repeated.

In Dutch elm disease there is a clear dependency of each organism, the fungus and the insect, on the other. Probably more than 99% of the elm tree infections are caused by the fungus being carried to the elm trees by the elm bark beetles. On the other hand, the elm bark beetles depend on the fungus for causing many elm trees to weaken and die, thereby becoming available as breeding grounds for the two species of elm bark beetles that transmit the fungus. The interdependence of the two organisms has provided the most effective means of managing the Dutch elm disease by burning or debarking dead elm trees and logs, thereby denying the insects the breeding ground they need and, through the reduced number of insects produced, reducing the number of elm trees to which they spread the disease.

Foliar diseases

Many foliar diseases are probably spread by various insects visiting and moving about on leaf surfaces that are exhibiting infections by spore-producing fungi. Spores or the spore containers of many such fungi are sticky or have appendages that cling to the legs or other body parts of such insects and are carried by them to other plants or plant parts they visit next. A few examples of foliar diseases in which insect transmission of the fungal pathogen has been shown to occur are described briefly below.

Powdery mildews. These are diseases that affect most annual and perennial plants. They are characterized by white superficial mycelial growth and sporulation by a small group of fungi that cause symptoms on leaves, shoots, blossoms and fruit of their host plants. Powdery mildews serve as food for many mycophagous fungi and produce large numbers of loosely attached spores. Such spores become attached to, and are disseminated by, insects with which they come in contact. Examples include the feeding of thrips on and the transmission of spores of the fungi *Sphaerotheca panosa* and *Uncinula necator* that cause powdery mildew on rose and grape, respectively. Although these fungi are disseminated readily by wind, it is likely that transmission is aided by insects.

Rust diseases. Most rust diseases produce several types of superficial spores on their host plants that, like those of the powdery mildews, are easily disseminated by air currents but are also visited, eaten, and transported by a wide variety of insects. Furthermore, many rust fungi produce spermatia and receptive hyphae in the same spermagonium but they are self-sterile. Many insects, when visiting such spermagonia become smeared with sticky spermatia. When the insects visit successive spermagonia, they transfer to the receptive hyphae spermatia from the opposite, compatible type. These spermatia can fertilize the receptive hypha which then produces dikaryotic mycelium and spores that contain two nuclei. These dikaryotic spores have entirely different properties. For example, they can infect an entirely different host plant from the plant on which they were produced. The involvement of insects in rust diseases is, therefore, important in both the dissemination of the spores to new hosts and, more importantly, in the fertilization of the fungus and, thereby, in increasing the potential of the fungus to produce more new and possibly more virulent races.

Some other examples of foliar diseases in which insects have been shown to play a role in their transmission include: red pine needle blight caused by the fungus *Pullularia pullulans* and transmitted and aided in penetration of pine needles by a cecidomyiid midge; cucurbit anthracnose, caused by the fungus *Colletotrichum lagenarium* and transmitted and aided in penetration by the spotted cucumber beetle *Diabrotica undecimpunctata*; oil palm leaf spot, caused by the fungus *Pestalotiopsis palmarum* and transmitted and aided in penetration by the oviposition punctures of a tinged of the genus *Gargaphia*.

Diseases of buds and blossoms

Insects often overwinter in buds of plants and many also visit blossoms to feed on the nectar they produce. Buds also often contain mycelium and spores of plant pathogenic fungi, and blossoms are often the first plant organ such fungi attack in the spring. Examples of bud infections in which insects have been implicated to play a role include: bud-blast disease of rhododendron, caused by the fungus *Pycnostyamus azalea* and aided by the leafhopper *Graphocephala coccinea*; bud rot disease of carnations caused by the fungus *Fusarium poae* and aided by the mite *Siteroptes graminum*.

Several diseases of blossoms have been associated with insect vectors. In most cases, transmission of the fungus by the insect is related to the activities of the insect during pollination. Some of the better-known examples include:

Anther smut of carnations. This is caused by the fungus *Ustilago violacea*. In this disease, the pollen is replaced by the teliospores of the fungus but the petals remain unaffected and continue to attract insects. The visiting insects become smeared with the smut spores, which they transfer to previously healthy flowers.

Blossom blight of red clover. This is caused by the fungus *Botrytis anthophila* and is transmitted primarily by pollinating bees.

Ergot of cereals and grasses. This is caused by the fungus *Claviceps purpurea*, which develops in the flowers and produces spores that are contained in a sweet and sticky substance. That substance is attractive to many insects, particularly flies and beetles. The insects feed on the spores and also become smeared with them externally and carry them, externally and through their feces, to healthy flowers. Although primary infections by the ergot fungus are primarily from ascospores produced by sclerotia overwintering on the ground and carried via air currents, insect transmission of conidia is important for secondary transmission of the disease and for transmission over long distances. Some beetles, however, feed on ergot sclerotia on the ground and may carry mycelium and ascospores on their bodies to healthy plants and through them may cause primary infections.

Anthracnose disease of Musa balsamiana. This is caused by the fungus *Gloeosporium musarum* and is transmitted by the hymenopterans *Polybia occidentalis*, *Synoeca surinama*, and *Trigona* sp. The fungus infects the floral parts of the plant but these are dropped while still producing conidia and sweet exudate. The insects

are attracted to the exudate on the fallen flowers and there they become smeared wit conidia, which they subsequently carry to healthy flowers, which the spores infect.

Flower spot of azalea. This is caused by the fungus *Ovulina azaleae* and is transmitted by several species of bees, thrips, and ants. The insects carry spores on their bodies and drop them off on healthy flowers they visit which, in addition, may injure directly and facilitate penetration and infection by the fungus.

Diseases of fruit and seeds in the field

Fruits and seeds are the source of food and the breeding grounds of many insects. Insects puncture fruit and seeds to obtain food and to lay their eggs in them. Although insects often cause direct damage to fruits and seeds that make them unsalable, the damage increases manifold when the insects also carry to the fruits and seeds fungi that infect and cause these organs to rot or to develop other symptoms. Numerous examples of fruit-insect-pathogen interactions could be cited, although in many cases no hard data of such interactions exist. Some of the better-studied cases are described briefly below.

Rots of fleshy fruits

Fig rots. Several kinds of fungi attack figs and cause rotting of fruits in the field. In most such cases, certain types of insects play a more or less important role in the transmission and introduction of the fungus into the fig.

Endosepsis of figs. This is caused by the fungus *Fusarium moniliforme*, and results in the entire fruit content turning into a pulp. The fungus is transmitted from fruit to fruit by the fig wasp *Blastophaga psenes*, which also plays a crucial role in the pollination of figs. Fig trees, being dioecious, have male trees that produce staminate flowers around the opening and gall flowers in the cavity, and female trees that produce only pistilate flowers. The fig wasp lays its eggs in the ovules of the gall flowers of male plants, which are thereby stimulated to grow. The eggs hatch and the larvae parasitize the galls until they pupate. The adults emerge from the pupae and the females are fertilized while still in the male fig. When they come out of the fig, the females brush against the staminate flowers that surround the opening and become smeared with pollen. The female wasps carry the pollen to male and female flowers they subsequently visit for oviposition. In female flowers, however, because of the length of styles, oviposition fails but pollination is nevertheless successful and the fruit develop into edible figs. If, however, as the female wasp visits some infected figs it becomes smeared with spores of the fungus, it transmits the spores to male and female figs it visits, and the fungus then causes endosepsis of the female figs.

Souring of figs. This is caused by yeast fungi that cause fermentation, and appears as discoloration and wateriness of the fig contents which then exude from the fig opening. Such figs shrivel, dry, and cling to the tree. The fermenting yeasts are transmitted to figs externally and internally on the bodies of the two most common visitors of figs still in the tree, the sap beetle *Carpophilus hemipterus* and the fruit fly *Drosophila melanogaster*.

Fig smuts and molds. These are caused by the black mycelium and spores of *Aspergillus niger* and the variously colored growths of other fungi. These fungi are carried into green figs on the bodies of predatory mites and, to a lesser extent, by thrips.

Brown rot of stone and pome fruits. This is caused by the three related fungi *Monilinia fructicola*, *M. fructigena*, and *M. laxa* and affects all the stone fruits and, to a lesser extent, the pome fruits. The fungi are aided in their penetration of the fruit by the feeding and oviposition wounds made by the insects plum curculio (*Conotrachelus nenuphar*) and the oriental fruit moth (*Grapholitha molesta*), and the feeding wounds of the dried fruit beetle (*Carpophilus hemipterus*) and two nitidulid beetles (*Carpophilus mutilatus* and *Haptonchus luteolus*). The insects also become smeared with spores of the brown rot fungi which they carry on their bodies and deposit at the wounds they make on the fruits they visit. In pome fruits, the fungus is facilitated in penetrating the fruit by the feeding holes made by the earwig *Forficula auricularia* at the beginning of ripening of the fruit, at which time they are susceptible to brown rot.

Gray mold of grapes. This is caused by the fungus *Botrytis cinerea*. The fungus spores are generally spread by *air* currents. Penetration of the grapes and shoots, however, seem to be increased by the wounds made on them by the larvae of the lepidopterans *Argyrotaenia pulchellana* and *Lobesia botrana*.

Black pod of cacao. This is caused by the fungus *Phytophthora palmivora* and results in devastating losses of yield. Insects of several different families play a role in the transmission of this disease. At least ten species of ants, especially *Crematogaster striatula* and to a lesser extent *Camponotus acvapimensis* and *Pheidole megacephala*, appear to spread the fungus vertically within the tree, especially during the rainy season, when they carry spore-containing soil particles up the cacao tree for nesting purposes. Certain coleoptera, such as the nitidulid beetle *Brachypeplus pilosellus*, and certain dipterans, such as the fly *Chaetonarius latifemur*, colonize black pods in the field and may carry the fungus internally or externally on their bodies to healthy pods. Because of their large numbers on cocoa trees, their habit of visiting wounded pods, and their proven efficiency to transmit the fungus, these insects are considered the main vectors of the fungus locally and over long distance.

Boll rots of cotton. These are caused by several fungi including *Fusarium moniliforme*, *Alternaria tenuis*, *Aspergillus flavus*, and *Rhizopus nigricans*. Various insects are apparently involved in the transmission of these fungi and they seem to use different mechanisms of transmission. Thus, in boll rot due to *Fusarium* and *Alternaria*, the fungi penetrate cotton bolls through feeding and oviposition wounds made by the boll weevil (*Anthonomus grandis*), the cotton bollworm (*Heliothis zeae*), and the tarnished plant bug, *Lygus lineolaris*, or they are brought to and penetrate through the nectarines by nectar feeding flies such as *Drosophila* and cabbage looper, *Trichoplusia ni*. In boll rots caused by *Aspergillus flavus* and other aflatoxin-producing species, the fungus is primarily wind disseminated but is also carried internally and externally by insects, such as the lygus bug *Lygus hesperus* and the stink bug *Chlorochroa sayi*, that frequently visit cotton bolls. The latter fungus, however, seems to depend for entrance on the presence of large wounds like the large exit holes made by the mature larvae of the pink bollworm, *Pectinophora gossypiella*. On the other hand, boll rots by *Rhizopus stolonifer* occur when wounds made by the bollworm *Earias insulana* and by the pink bollworm are available. In the lint rot of cotton, caused by the fungus *Nigrospora oryzae*, the fungus is transmitted very efficiently by the mite *Siteroptes reniformis*. In Stigmatomycosis or internal boll disease, caused by the fungus *Nematospora gossypol*, the cotton fibers are stained in the absence of external symptoms. This disease is associated with the feeding of several species of plant bugs primarily of the genus *Dysdercus*, often referred to as cotton stainers. The insects carry the fungus spores externally on the mouthparts and internally in their deep stylet pouches and introduce it via their proboscis through the wall of young cotton bolls.

Coffee bean rot. This is caused by the related fungi *Nematospora corylii and N. gossypii*, which cause berries to turn black and subsequently to rot. The fungi are introduced into the berries through the feeding wounds made by the insects *Antestia lineaticolis and A. faceta*. The insects feed on small and large berries and if they carry the fungus the latter causes infection of the bean. The number of infected berries is proportional to the number of insects, approximately 300 insects per tree resulting in infection of all the berries on the tree.

Molds and decays of grains and legumes

Numerous decays and molds affect the various grains and legumes while still in the field and their frequency and severity increase as the number of insects infesting the crops, and feeding on the seeds, increases. In corn, for example, seed rots can be caused by species of the fungi *Fusarium, Gibberella, Diplodia, Cephalosporium, Nigrospora, Physalospora, Cladosporium, Penicillium, Aspergillus, Rhizopus, Trichoderma,* and others. The insects most commonly involved in transmitting and facilitating infection of corn kernels by these fungi are the corn earworm, *Heliothis zea*, and the European corn borer, *Pyrausta nubilalis*, but other borers and other insects also play important roles as vectors and, most importantly, as facilitators of infection by these fungi by creating wounds that allow the fungus to enter the seed. In seed infections by *Aspergillus* and by *Fusarium* there is the added adverse effect of production of debilitating mycotoxins. Similar, although less studied situations have been reported for rice infections by fungi, e.g., *Nematospora corylii*, transmitted and facilitated by wounds made by the rice stinkbug *Oebalus pugnax*; wheat and corn infections by *Nigrospora* sp. and *Fusarium poae,* transmitted by large numbers of *Peliculopsis* mites feeding on and transporting spores of the fungus in their abdominal sacs; and in various legume infections by the fungi *Nematospora, Cladosporium, Aureobasidium,* etc.

transmitted and facilitated in their penetration and infection of the seeds by the stinkbugs *Acrosternum hilare* and *Thyanta custator*, the lygaeid *Spilostethus pandurus*, by thrips, aphids, and other insects.

Molds and decays of harvested fruits and seeds

Generally little is known definitively about the roles of specific insects on the transmission and facilitation of rots of specific fruits and vegetables, and of molds and decays of seeds of specific grains, legumes, or nuts by specific fungi. It is generally accepted, however, that postharvest infections of plant products are greatly increased in numbers and in severity if insects are also present in the same or adjacent containers. There is agreement that insects moving about among stored fruits, seeds, etc., transport externally and internally on their bodies spores of fungi infecting such fruits and seeds and deposit such spores on the next fruit or seed they feed on. There is also agreement that by creating feeding or oviposition wounds on harvested fruit and seeds, the insects create openings through which the fungi can penetrate and release sap and additional nutrients. The fungi then can grow and build momentum to eventually infect and rot the entire fruit or seed.

The fungi that cause most rots of fleshy fruits and vegetable after harvest include *Penicillium, Fusarium, Botrytis, Rhizopus, Alternaria, Sclerotinia, Monilinia,* and *Colletotrichum*, while the molds and decays of grains and legumes involve primarily *Aspergillus, Fusarium,* and *Penicillium*. The insects involved in transmission and facilitation of infection of fleshy organs after harvest include larvae and adults of various Lepidoptera such as the oriental fruit moth, *Grapholitha molesta*, Diptera such as the apple maggot, *Rhagoletis pomonella*, the Mediterranean fruit fly, *Ceratitis capitata*, the house fly, and others. The insects involved in the transmission and facilitation of infection by fungi causing molds and decays of grains and legumes are the larvae and adults of various Coleoptera such as the rice weevil *Sitophilus oryzae*, the granary weevil *Sitophilus granarius*, and the confused grain beetle, *Tribolium confusum*, and also Lepidoptera such as the Angoumois grain moth, *Sitotroga cerealella*, the European corn borer, *Pyrausta nubilalis*, the ear cornworm, *Heliothis zea*, and other insects.

Insect transmission of plant pathogenic nematodes

Two very serious plant diseases caused by nematodes of the genus *Bursaphelenchus* are transmitted by insects. In both diseases there is a symbiotic relationship between the fungal pathogen and the insect vector.

Pine wilt. This is a lethal disease of many species of pines and other conifers. It is caused by the nematode *Bursaphelenchus xylophilus*, known as the pinewood nematode. The nematode is about 800 μm long by 22 μm in diameter and it develops and multiplies rapidly, each female laying about 80 eggs and completing a life cycle in as short as 4 days. The nematode produces the four juvenile stages and the adults. The juvenile stages develop in the resin canals of infected pine trees, feeding at first on plant cells and later on fungi that invade the dying or dead tree. Later, the nematode produces special fourth-stage dispersal juveniles that are adapted to survive in the respiratory system of the cerambycid beetles *Monochamus carolinensis* and *M. alternatus* by which they are transmitted to healthy pine trees.

The pinewood nematode overwinters in the wood of infected dead trees, which also contain larvae of the beetle vectors of the nematode. Early in the spring, the larvae dig small cavities in the wood in which they pupate. As the adult beetles emerge from the pupae later in the spring, large numbers of fourth-stage juvenile nematodes enter the beetles and almost fill the tracheae of the respiratory system of each insect with about 15,000 to 20,000 juveniles. These nematode-carrying adult beetles emerge and fly to young branch tips of healthy pine trees where they feed for several weeks. As the beetles strip the bark and reach the cambium, the nematode juveniles emerge from the insect and enter the pine tree through the wound. The juveniles in the tree then undergo the final molt and produce adult nematodes. The latter migrate to the resin canals, feed on their cells and cause their death, and then they move in the xylem and in the cortex where thy reproduce quickly and build enormous populations of nematodes and kill twigs, branches and entire trees.

After the adult *Monochamus* beetles, the vector of the pine wilt nematode, have fed on young twigs for about a month, they are ready to breed and look for stressed and dead pine trees, including trees showing symptoms or dying from infection by the pinewood nematode. The female beetles deposit their eggs under the bark of such trees where the first two instars develop and feed. The third instar penetrates the wood where it undergoes the next molt and produces the fourth instar, which overwinters there. In early spring, the fourth instar digs a cavity in the wood where it pupates and to which numerous third-stage nematode juveniles are attracted and congregate. The juvenile nematodes undergo the next molt and produce the fourth-stage dispersal juveniles, which by the thousands infect the tracheae of the adult insects as soon as they emerge from the pupae and are carried by them to healthy pine trees, thus completing the cycle.

Red ring of coconut palms. This disease kills coconut palm trees from Mexico to Brazil and in the Caribbean islands. It is caused by the nematode *Bursaphelenchus cocophilus*, which is transmitted from palm to palm by the American palm weevil, *Rhynchophorus palmarum*, the sugarcane weevil, *Metamasius* sp., and probably other weevils. The nematodes infect, discolor, and kill the palm tissues in a ring 3 to 5 cm wide about 5 cm inside the stem periphery over the length of the stem.

The nematode pathogen lays its eggs and produces all its juvenile stages and the adults inside infected palm trees, completing a life cycle in about 9 to 10 days. Female weevil vectors are attracted to red ring-diseased trees but they also lay eggs on healthy or wounded palm trees. If the female carries red ring nematodes, it deposits them in its feeding wounds at bases of leaves or at internodes. The nematodes then enter the palm tissues and undergo repeated life cycles and spread intercellularly in the parenchyma cells of the petioles, stem, and roots, where the cells break down and form a flaky, orange to red discolored tissue with cavities. Red ring nematodes do not invade xylem and phloem tissues but cause tyloses to develop in xylem vessels within the red ring that block the upward movement of water and nutrients. In the meantime, the weevil larvae of the insect vector feed on the red ring tissue and swallow several hundred thousand nematode third-stage juveniles. Of these, however, only a few hundred of the nematodes survive and pass through the molt, internally or externally, to the next stage weevil larvae and to the adult weevil. As weevil females emerge from rotted palms, a small percentage of them carry with them third-stage juveniles of the nematodes. Nematode populations increase rapidly at first but later they decline and about 3 to 5 months after infection there are hardly any red ring nematodes or their eggs left in decomposed stem tissue of infected, dead palm trees. The nematodes, however, survive in newly infected palm trees and, briefly, in their insect vector.

Insect transmission of plant pathogenic protozoa

Three plant diseases: phloem necrosis of coffee, heartrot of coconut palms, and sudden wilt of oil palms, are caused by flagellate protozoa of the genus *Phytomonas*. In all three diseases, protozoa invade the phloem elements of infected plants and multiply in them, reaching populations of varying densities. Some of the sieve tubes become plugged by protozoa. Generally, the more severe the symptoms of infected plants, the higher the populations of protozoa in their phloem. The pathogen is transmitted from infected to healthy plants occasionally through natural root grafts, and primarily by stink bugs (Pentatomidae) such as the genera *Lincus* and *Oclenus*, and possibly others.

Insect transmission of plant pathogenic viruses

Plant viruses cause many and severe diseases of plants, their number and importance being second only to fungal diseases of plants. Most viruses infect their host plants systemically, that is, the virus multiplies internally throughout the plant. Almost all viruses enter and multiply in phloem and in parenchyma cells. Viruses do not produce spores, nor do they come to the surface of the plant. All plant viruses are transmitted to new plants that are propagated from infected plants vegetatively (that is, by grafting or budding, by cuttings, by bulbs, corms, roots, tubers, etc.), and many can be transmitted artificially by mechanical inoculation, that is, by rubbing sap from infected plants onto leaves of healthy plants. Some plant viruses can be transmitted from diseased to healthy plants by pollen or seed produced by infected plants, some by the parasitic higher plant dodder when it is infecting both virus-infected and healthy plants, and some plant viruses are transmitted from plant to plant by certain plant pathogenic fungi, nematodes, or certain mites. More than half of the plant viruses, numbering more than 400, are transmitted from diseased to healthy plants by insects.

The number of insect groups that are vectors of plant viruses is relatively small. The most important vector groups, with the number of vector species and viruses transmitted, are listed below. Hemiptera, which includes the aphids (Aphididae, 192 species, 275 viruses), leafhoppers (Cicadellidae, 49 species, 31 viruses), the planthoppers (Fulgoroidea, 28 species, 24 viruses), the whiteflies (Aleurodidae, 3 species, 43 viruses), the mealybugs (Pseudococcidae, 19 species, 10 viruses), and some treehoppers (Membracidae, 1 species, 1 virus), contain by far the largest number and the most important insect vectors of plant viruses, but the true bugs (Hemiptera, 4 species), the thrips (Thysanoptera, 10 species, 11 viruses) and the beetles (Coleoptera, 60 species, 42 viruses) also are implicated. Grasshoppers (Orthoptera, 27 species) seem to occasionally carry and transmit a few viruses. Unquestionably, the most important virus vectors are the aphids, leafhoppers, whiteflies, and thrips. These and the other groups of Hemiptera have piercing and sucking mouthparts, although several thrips have rasping, sucking ones. Beetles and grasshoppers have chewing mouthparts, but many beetles are quite effective vectors of certain viruses. Generally, viruses transmitted by one type of vector are not transmitted by any other type of vector.

Aphids and aphid-transmitted viruses

Aphids have evolved as the most successful exploiters of plants as a food source, particularly in the temperate regions. Many species of aphids alternate between a primary and a secondary host, although there are many variations of aphid life cycles depending on the aphid species and on climate. Some aphids overwinter as parthenogenetic viviparous forms while others go through their life cycle on one host species or on several related species. On the other hand, there are several aphid species, such as *Myzus persicae*, that have as many as 50 primary and alternate species of host plants.

Aphids have mouthparts that consist of two pairs of flexible stylets held within a groove of the labium. During feeding, the stylets are extended from the labium and, through a drop of gelling saliva, the stylets rapidly penetrate the epidermis. Penetration may stop at the epidermis or it may continue into the middle layers of leaf cells with a sheath of saliva forming around the stylets. The stylets move between the cells until they reach and enter a phloem sieve tube from which the aphids obtain their food. Individual aphids vary in their ability to transmit the virus to individual plants. Infection of a plant with a virus often makes the plant more attractive for aphids to grow on and to reproduce. Both acquisition and transmission of virus by aphids are affected by temperature, humidity and light.

Virus-vector relationships

Insect vectors that have sucking mouthparts carry plant viruses on their stylets, and such viruses are known as stylet-borne, externally borne, or non-circulative, because they do not pass to the vector's interior. The remaining viruses are taken up internally within the vector and are called internally borne persistent circulative or persistent propagative viruses.

Stylet-borne non-persistent transmission. Most externally borne viruses can be transmitted in the typical stylet-borne non-persistent manner. In such a transmission the virus is assisted in its transmission by a specific configuration of its coat protein or by a non-structural virus-encoded protein. The insect acquires the virus from the plant by feeding on it for only seconds or, at most, minutes. The insect can transmit the virus immediately after the acquisition feeding, that is, without any incubation period required for transmission. The insect retains the virus and is usually able to transmit it for only a few minutes after it acquired it. Most of the nearly 300 known aphid-borne plant viruses are stylet-borne non-persistent. Some of the most important groups of plant viruses, such as those in the genera *Potyvirus*, *Cucumovirus*, *Alfamovirus*, and the *Caulimovirus* transmitted by *Myzus persicae* are stylet-borne non-persistent viruses. In the few seconds in which aphids acquire the virus, the aphid stylet usually penetrates only the epidermal cell. Actually, deeper penetration of the stylet into leaf tissues reduces the ability of aphids to transmit the virus. Aphids vary greatly in their ability to transmit viruses, each particular virus being transmitted by one or a few species of aphids. Sometimes, certain virus strains are transmitted by distinct aphid species. Also, even individual aphids in a population vary in their ability to transmit the same virus, some of them being incapable of transmitting the virus.

All non-persistently transmitted viruses have simple structures of elongated or isometric particles with the nucleic acid encapsidated by one or more kinds of coat proteins. In some viruses, the coat protein interacts directly with the binding site of virus retention in the aphid. In other viruses, the virus encodes a non-structural protein which interacts with the aphid-virus retention binding site and forms a bridge between the virus and the aphid stylet. However viruses are bound to the aphid stylet, there must also be a mechanism for release of the virus when the aphid feeds on the next plant. It appears that saliva alone may carry out this function.

Semi-persistent viruses. Some externally borne non-persistent viruses are known as semi-persistent because they reach but do not seem to go past the foregut of the vector; the vector must feed on an infected plant (acquisition period) for several minutes or hours before it can transmit the virus; and the vector can then retain (retention time) and transmit the virus to healthy plants for several hours. Semi-persistent viruses are also assisted in their transmission by a transmission helper protein or coat protein configuration. The best known semi-persistent viruses are caulimoviruses, which occur in most cell types, and the closteroviruses beet yellows virus and curly top virus, which are found primarily in phloem cells. In several of the semi-persistent viruses, a helper component seems to be involved in their transmission. In cauliflower mosaic virus, the helper component consists of two non-capsid proteins, one of which is associated with the virus particles and the other has two binding domains that interact strongly with microtubules. In some cases, certain viruses can be transmitted only in the presence of a second virus which acts as the helper virus.

Persistent viruses. Internally borne viruses are either persistent circulative or persistent propagative. Persistent circulative viruses are acquired from the plant by the vector after an acquisition feeding period of several hours to several days, and then they are retained by the insect vector and can be transmitted by it for several days or weeks. Persistent circulative viruses require a latent period of several hours to several days beyond the acquisition time before they can be transmitted by the insect vector, they reach the hemolymph of the vector, and pass through the various stages of the insect, but not through the ovaries to the egg. Persistent propagative viruses are acquired by the insect after a feeding period of several hours to several days, are retained by the vector for several weeks to several months, they multiply in the vector, they have a latent period of a few to several weeks, and can pass through the various stages of the insect, including transovarial passage to the egg. Persistent viruses are generally transmitted by one or a few species of aphids and cause symptoms characterized by leaf yellowing and leaf rolling.

Persistent circulative viruses. These include primarily the luteoviruses, such as barley yellow dwarf virus, and the nanoviruses, such as banana bunchy top virus. The luteoviruses are acquired after a feeding period as short as 5 minutes but it usually takes 12 hours. After an incubation period of an additional 12 hours, the vector can transmit the virus within a 10 to 30 minute inoculation feeding and can continue transmitting it for several days or a few weeks. In the vector, the virus particles seem to associate only with the hind gut of the aphid, entering its cells by endocytosis into coated pits and vesicles and accumulating in tubular vesicles and lysosomes. Virus particles are then released into the hemolymph by fusion of the vesicles with the plasmodesmata and enter the salivary glands of the aphid via invaginations with two plasma membranes on the hemocoel side of the salivary gland accessory cells. It appears that persistent circulative viruses do not require a non-capsid protein for helper component but they require a protein produced via a read-through of the coat protein stop code if they are to advance beyond the hemocoel. Some persistent circulative viruses also require a helper virus to be present for them to be transmitted by their aphid vector.

Persistent propagative viruses. Propagative viruses are transmitted primarily by leafhoppers and planthoppers but several members of the Rhabdoviridae multiply in and are transmitted by their aphid vector. These bacilliform viruses replicate in the nucleus and the cytoplasm of cells in the brain, salivary glands, ovaries, and muscle of the insect vector. The virus goes through the egg to about 1% of the nymphs. Infection of aphids with rhabdoviruses results in increased mortality of the aphids.

Leafhoppers and planthoppers, and transmission of plant viruses

Leafhoppers lay eggs that hatch to nymphs which pass through several molts before becoming adults. Some of them overwinter as eggs, some as adults, and some as immature forms. They all feed by sucking sap from phloem elements of plants. Their feeding behavior is similar to that of aphids in that the mouthparts, surrounded by the salivary sheath, penetrate the phloem of host plants.

Virus-vector relationships

All hopper-transmitted viruses are persistent circulative or persistent propagative, and are transmitted by only one or by a few closely related species of the hopper vectors. Only two of the 60 sub-families of leafhoppers (Cicadellidae) contain species that are vectors of viruses: the Agalliinae feed on herbaceous dicotyledonous hosts and the Deltocephalinae that feed on monocots. Of the 20 planthopper families (Fulgoroidea), only one, the Delphacidae, have species that are vectors of viruses all of which infect monocotyledonous plants and many of them cause severe diseases on cereal crops such as rice, wheat, and corn.

Semi-persistent transmission. Two viruses, maize chlorotic dwarf virus (MCDV) and rice tungro spherical virus (RTSV), are acquired by their vectors (*Graminella nigrifrons* and *Nephotettix virescens*, respectively) from their hosts within about 15 minutes and are retained by their vectors for one to a few days. MCDV particles have been seen in the foregut and a few other tissues but not beyond. Hoppers egest material from the foregut once in a while during feeding and it is thought that transmission occurs during this ingestion-egestion process.

Persistent transmission. This involves the internal movement of the virus obtained from the plant to the salivary glands of the insect vector. Some of these viruses are circulative while others are propagative.

Circulative viruses. Only two genera of geminiviruses (*Mastrevirus* and *Curtovirus*) are transmitted by leafhoppers in the persistent circular manner. The viruses are acquired by the vector after feeding for a few seconds to an hour. There is a latent period of about a day, presumably for the virus to reach the salivary glands. The internal movement of these viruses is determined by the viral coat protein and by receptor-mediated endocytosis.

Propagative viruses. There are four families and genera of plant viruses that replicate within the cells of their insect vectors as well as the cells of their host plants. Two of these families, Rhabdoviridae and Reoviridae, contain viruses that infect animals, and their virus members that infect plants have been considered as animal viruses that infect plants. The propagative viruses have a latent period of about two weeks. During this period the virus replicates and invades most tissues of the insect vector. When the virus reaches the salivary glands of the vector, the latter can transmit the virus to new plants and can continue to transmit it for the rest of their life. Only a small percentage of the hoppers feeding on infected plants become vectors and of these only about 1% pass the virus through their eggs to the next generation. Various capsid proteins seem to be necessary for passage of viruses through the organs of the vector and are required for transmission.

The two genera that have propagative viruses are *Tenuivirus*, members of which are transmitted by delphacid planthoppers, and *Marafivirus*, which is vectored by the leafhopper *Dalbulus maydis*. These viruses have an acquisition feeding period of 15 minutes to 4 hours, a latent period of 4–31 days, inoculation periods as short as 30 seconds, and can transmit the virus for as long as they live. Almost all of these viruses are transmitted transovarially to the egg.

Whitefly transmission of plant viruses

Whiteflies transmit the viruses in the genus *Begomovirus* of the family Geminiviridae, and all the viruses in the genus *Crinivirus* and some in the genus *Closterovirus* of the family Closteroviridae. Whitefly adults are winged but only the first instar among the larvae is mobile. Whiteflies produce many generations in a year and reach high populations. Only a few species of whiteflies transmit viruses, mostly in the tropics and subtropics, but the viruses they transmit cause very severe diseases. Begomoviruses are transmitted by *Bemisia tabaci* whiteflies, while the criniviruses and the whitefly-transmitted closteroviruses are vectored by the whiteflies *Trialeuroides vaporariorum*, *T. abutilonea*, *B. tabaci*, and the type B of *B. tabaci* (also referred to as *B. argentifolii*). Whitefly mouthparts and feeding behavior resemble those of aphids.

Begomoviruses are bipartite geminiviruses and are transmitted by whiteflies in the persistent circulative manner. A helper factor coded by the virus seems to be involved in the transmission. The whitefly-transmitted monopartite closteroviruses and the bipartite criniviruses reach only the foregut of the vector and are transmitted in the semi-persistent manner. These viruses are retained in the vector for about 3–9 days. Two capsid proteins help the virus in its transmission by the vector.

Thrips transmission of plant viruses

About 10 species of thrips of the family Thripidae are the vectors of about a dozen viruses belonging to four genera (*Carmovirus*, *Ilarvirus*, *Sobemovirus*, and *Tospovirus*) of four families. Thrips are polyphagous insects that have many hosts. Some of the vector species reproduce mainly parthenogenetically. The larvae are rather inactive but the adults have wings and are very active. Thrips adults feed by sucking the contents of subepidermal cells. Adults live up to 3 weeks and there may be as many as 20 generations per year. The tospoviruses are transmitted in the persistent propagative manner, while the viruses of the other genera are transmitted in the pollen carried by the thrips vectors and by mechanical damage during feeding of the vector.

By far the most important thrips-transmitted viruses are the tospoviruses, which include the widespread and severe tomato spotted wilt virus and the impatience necrotic spot virus. In tospoviruses, only the larvae but not the adults can acquire the virus, and their ability to acquire it decreases with age. Larvae sometimes acquire the virus after feeding on a diseased plant for as little as 5 minutes, but usually they must feed for more than an hour both in acquiring and in inoculating the virus. There is a latent period of 3–4 days before the larvae can transmit the virus. The virus is passed from the larvae to the adults which can transmit it, although erratically, for as long as they live. These viruses appear to multiply in the vector but are not passed through the egg. Several structural proteins of the virus seam to be associated with the acquisition, passage through, and inoculation of the virus by its larval and adult insect vector.

Mealybug and other bug transmission of plant viruses

Mealybugs are important as virus vectors primarily on some perennial plants in the tropics and subtropics. They move slowly on plants and therefore are not as efficient virus vectors as those discussed previously. They move from plant to plant, mostly as crawling nymphs, through leaves of adjacent plants being in contact with each other; by ants tending the mealybugs and moving them from one plant to the other; and occasionally by wind.

Mealybugs feed on the phloem and they are vectors of the badnaviruses, such as the cacao swollen shoot virus (CSSV), several closteroviruses, such as grapevine leafroll associated viruses and the pineapple mealybug wilt associated virus, and the tri-choviruses, such as grape viruses A and B. Mealy-bugs acquire the viruses after feeding on diseased plants for only a few, about 20, minutes and retain the virus for a few, up to 24 hours, so the transmission resembles the non-persistent or semi-persistent mechanism of transmission by aphids.

Other bugs that transmit plant viruses include the mirid bugs, which transmit some sobemoviruses in manners that have characteristics of non-persistent, semi-persistent, and beetle-like transmission, and the piesmatid bugs, which transmit beet leaf curl virus in a persistent propagative manner.

Virus transmission by insects that have biting/chewing mouthparts

Although there are a few vectors in the orders Orthoptera and Dermaptera, there are more than 60 vector species in the order Coleoptera (beetles), 30 of them in the family Chrysomelidae. Most beetle vectors tend to eat plant cells between the leaf veins and regurgitate during feeding, thereby bathing their mouthparts with sap and virus. Virus transmission by beetles, however, is specific between each virus and its vector. Beetle-transmitted viruses belong to the genera *Tymovirus*, *Comovirus*, *Bromovirus*, and *Sobemovirus*. Most of these viruses are small (25 to 30 nm in diameter), stable, reach high concentrations, and are easily transmitted by sap. These viruses can also be translocated through the xylem of the plant. Beetles can acquire and can transmit the virus after feeding for a few seconds and they can retain the virus from 1 to 10 days.

Virus transmission by mites

Several members of the mite family Eriophyidae transmit viruses of the genus *Rymovirus* which cause many serious diseases in grain crops. Two mite species of the family Tetranychidae transmit two plant viruses, one of them transmitting the peach mosaic virus. All mites in these families feed by piercing plant cells and sucking their contents.

Eriophyid mites are small (0.2 mm long), move little by themselves and, instead, they are spread by wind. They have two nymphal instars followed by a resting pseudopupa. They complete a life cycle within two weeks. Mites can acquire virus from infected host plants within 15 minutes from the start of feeding and can transmit it to healthy plant within a similar duration. Mites acquire the virus as nymphs

but not as adults. They carry the virus through molts and remain infective for 6 to 9 days.

Tetranychid mites are larger (0.8 mm long). Pre-adult mites readily acquire the virus and they, as well as the adults, transmit the virus efficiently.

Virus transmission by pollinating insects

Honey bees and other pollinating insects seem to play a role in distributing virus-infected pollen from infected plants to healthy ones. It appears, however, that no special mechanisms or involvement of the insect are present in such virus transmission.

Summary

Insects play various roles in the transmission of plant pathogens, and in the initiation and development of disease in plants. In some diseases, the insects incidentally carry pathogens on their bodies or in their feces and deposit them on healthy plants where they cause disease, without developing any special relationships with the pathogens. In several cases, the insects weaken the plants on which they feed and make them much more susceptible to attack by pathogens. In other cases, the pathogens depend on the insects to carry them to healthy plants and to deposit them on fresh wounds through which they penetrate and infect the plants. While pathogens seem to be the beneficiaries of these actions, insects also derive advantages by the pathogen making the diseased plant more attractive to the insect for feeding or breeding purposes, and in some cases, by the insect feeding on the pathogen growing in the cavities made by its insect vector. Also, while in most cases the pathogen does not affect its insect vector directly, there are several plant viruses and mollicutes that multiply in the insect vector as well as in their plant host, and such vector insects often show histopathological symptoms, reduced reproduction, and shorter life span. Most of the insect/pathogen associations are highly specific and involve sophisticated molecular mechanisms that regulate the uptake, retention, and transmission of the pathogen by its insect vector.

See also, PLANT VIRUSES AND INSECTS, MANAGEMENT OF INSECT-VECTORED PATHOGENS OF PLANTS, TRANSMISSION OF *XYLELLA FASTIDIOSA* BATERIA BY XYLEM-FEEDING INSECTS, VECTORS OF PHYTOPLASMAS.

George N. Agrios
University of Florida
Gainesville, Florida, USA

References

Agrios, G. N. 1997. *Plant pathology* (4th ed.). Academic Press, San Diego, California.

Anonymous. Compendium of Diseases of...A series of books on diseases of individual crops published periodically by APS Press, St. Paul, Minnesota.

Capinera, J. L. 2001. *Handbook of vegetable pests.* Academic Press, San Diego, California.

Hull, R. 2001. *Matthews' plant virology (4th ed.).* Academic Press, San Diego, California.

Harris, K. F., and K. Maramorosch. 1980. *Vectors of plant pathogens.* Academic Press, San Diego, California.

Hiruki, C. (ed.). 1988. *Tree mycoplasmas and mycoplasma diseases.* University of Alberta Press, Edmonton, Alberta, Canada.

Nault, L. R. 1997. Arthropod transmission of plant viruses: a new synthesis. *Annals of the Entomological Society of America* 90: 521–541.

Schowalter, T. D., and G. M. Filip (eds.). 1993. *Beetle-pathogen interactions in conifer forests.* Academic Press, San Diego, California.

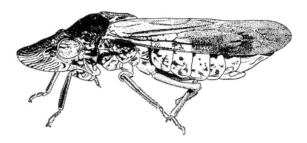

Fig. 1072 The glassy-winged sharpshooter (*Homalodisca coagulata*) a vector of the xylem-limited bacterium *Xylella fastidiosa*. Drawing by and with permission of Rosser Garrison.

TRANSMISSION OF *Xylella fastidiosa* BACTERIA BY XYLEM-FEEDING INSECTS.

Insects that feed predominantly on the sap of plants' water-conducting systems (xylem) can transmit bacteria that are specific parasites of the plant xylem. Most xylem sap-feeders are spittlebugs (families Cercopidae and Machaerotidae) or sharpshooters (subfamily Cicadellinae) belonging to the leafhopper family (Cicadellidae) in the order Hemiptera. Xylem sap is extremely low in nutrition, and xylem sap-feeders have the highest feeding rates of any terrestrial animals, consuming up to 1,000 times their body weight per day.

The best known xylem-limited bacterium that causes plant diseases is *Xylella fastidiosa*. Various strains of *X. fastidiosa* cause disease in grape, peach, almond, orange, alfalfa, and numerous tree species. The bacteria and associated gums plug the xylem system, leading to discoloration and killing or stunting of leaves and fruits. Pierce's disease of grape, phony disease of peach, and citrus variegated chlorosis are among the best known and most economically damaging of these diseases. For example, Pierce's disease is so severe and widespread in the southeastern United States that it has prevented commercial viticulture there using susceptible European grapes. Less

appreciated, but economically and ecologically important, are the numerous and more widespread leaf scorch diseases caused by *X. fastidiosa* in numerous forest trees such as oak, sycamore, maple, and elm. This bacterial species has an enormously wide range of plant species – about 75 to 90% of all plant species tested – in which it can multiply, but it causes symptoms in relatively few. Except for a leaf scorch disease of pear in Taiwan, these diseases appear to be limited to the Americas. However, in 1996, Pierce's disease was confirmed to occur in southeastern Europe. Numerous other bacterial pathogens invade the xylem system, but not exclusively. The widespread *Leifsonia* (*Clavibacter*) *xyli*, a xylem-limited bacterium that causes ratoon stunt disease of sugar cane, is not transmitted by insects.

The great majority of insect species that are xylem sap-feeding specialists that have been tested as vectors (transmitters) of *X. fastidiosa* have been shown to be able to transmit this bacterium to plants. Only a small fraction of the species that are xylem sap-feeding specialists have been tested as vectors, so several thousand species are probably vectors of *X. fastidiosa*. This broad degree of vector specificity, as well as other characteristics of vector transmission point to the method by which insects transmit the bacterium. There is no time delay (latent period) between a vector's acquiring the bacterium by feeding on infected plants and its introducing the bacterium to other plants. Once infective, a vector can transmit for the rest of its life, but immature vectors lose the ability to transmit when they shed their skin. All of these characteristics can be best explained if the bacteria are transmitted from the foregut of the vector. The exact location in the foregut from which the bacteria are transmitted is not yet known. The

Examples of plant diseases caused by strains of the bacterium *Xylella fastidiosa*. See Purcell (1997) for a more complete list

Crop	Disease	Where found
grape (*Vitis vinifera* and other *Vitis* species)	Pierce's disease	southern North America, Central America (& northern South America?)
citrus	variegated chlorosis	South America
peach	phony disease (dwarfing)	southeastern United States
oak, sycamore, elm, others	leaf scorch diseases	southeastern United States
pear	leaf scorch	Taiwan
alfalfa	dwarf	California, probably other locations in North and Central America
coffee	leaf scorch	South America, possibly Central America

lack of a latent period and the persistence of infectivity in vectors makes control of diseases caused by *X. fastidiosa* especially difficult. Because a few infective insects can quickly transmit the bacterium to susceptible plants, insecticides to kill vectors within vineyards has not proven to be effective.

Alexander H. Purcell
University of California
Berkeley, California, USA

References

Goodwin, P., and A. H. Purcell. 1992. Pierce's disease. pp. 76–84 in *Grape pest management* (2nd ed.). University of California, Division of Agriculture and Natural Resources, Oakland, California.

Purcell, A. H. 1989. Homopteran transmission of xylem-inhabiting bacteria. pp. 243–266 in K. F. Harris (ed.), *Advances in disease vector research*, Vol. 6. Springer-Verlag, New York, New York.

Purcell, A. H. 1997. *Xylella fastidiosa*, a regional problem or global threat? *Journal of Plant Pathology* 79: 99–105.

Purcell, A. H., A. H. Finlay, and D. L. McClean. 1979. Pierce's disease bacterium: mechanism of transmission by leafhopper vectors. *Science* 206: 839–841.

Purcell, A. H., and D. L. Hopkins. 1996. Fastidious xylem-limited bacterial plant pathogens. *Annual Review of Phytopathology* 34: 131–151.

Raven, J. A. 1984. Phytophages of xylem and phloem: a comparison of animal and plant sap-feeders. *Advances in Ecological Research* 13: 135–234.

TRANSMISSION THRESHOLD. The level of abundance of a parasite, or the abundance of its vector, that is necessary for disease to spread.

TRANSOVARIAL TRANSMISSION. Passage of a material through the ovariole and within the egg.

TRANSOVUM TRANSMISSION. Passage of a material on the surface of the egg.

TRANSPIRATION. The evaporation of water from the surface of plant foliage.

TRANSPOSABLE ELEMENT. An element that can move from one site to another in the genome. Transposable elements have been divided into two classes, those that transpose with an RNA intermediate, and those that transpose as DNA.

TRANSPOSON. A transposable element carrying several genes including at least one coding for a transposase enzyme. Many elements are flanked by inverted repeats. *Drosophila melanogaster* contains multiple copies of 50–100 different kinds of transposons.

TRANSSTADIAL TRANSMISSION. The transmission of pathogens or parasites through successive stages of the host's life cycle, as from the egg to the larva, pupa and adult.

TRAP CROP. A crop or portion of a crop that is intended to lure insects away from the main crop.

TRAPS FOR CAPTURING INSECTS. Traps developed for capturing insects are as varied as the purpose for the trapping, the insects targeted and the habitats in which they are used. Traps are used for general survey of insect diversity, and these usually are simple interception devices that capture insects moving through an area. Traps also are used for detection of new invasions of insect pests in time and/or space, for delimitation of area of infestation, and for monitoring population levels of established pests. This information is used to make decisions on the initiation of control measures or to measure effectiveness of a pest management program. Traps may be used as direct control measures, for example, by mass trapping (use of a high density of traps throughout the infested area) or by perimeter trapping (use of traps as a barrier around a pest-free area to intercept insects moving into the area) to remove a large number of individuals, with the goal of preventing or suppressing population buildup. Traps can be used as direct control measures if they are highly effective, if they capture a high percentage of females (especially if they capture them before they have a chance to lay eggs), and if they are integrated with other pest management approaches. Factors such as cost per trap, the need to service the traps frequently and the high reproductive rate of individuals that escape capture, however, prevent widespread use of traps as a stand-alone pest control measure.

Traps for specific insect species or pest groups use combinations of cues to lure the target insect and exploit aspects of the insect's behavior to facilitate movement of insects into the trap. Several factors influence the effectiveness of a specific trap. The ability of the trap to mimic and present those cues to the insect, the strength of those cues in influencing the insect's behavior, and the proper placement of the trap in the habitat all are important.

Following is an overview of the basic trap types, and variations of those basic trap types for specific uses. Traps may be used with or without attractant cues, and may use a combination of cues, including visual (color, light, shape), chemical (food/host, pheromone, parapheromone, oviposition) and acoustic stimuli to make them more specific and/or more effective. Automated monitoring systems exist that will transmit information on trap capture to an off-site station. Insect traps are an important part of insect pest management programs. Although a number of trapping systems are discussed, the intent is to provide a general framework of types of traps that are used with some representative examples. It is not intended to be a complete listing of all traps that have

been developed or are in use. Representative literature is presented at the end of this section that will provide additional information on insect traps and specific uses.

Trap types

There are a few basic designs that describe almost all insect traps. They may be a surface that is presented flat or is formed into baffles or cylinders; containers with holes on either the sides, top or bottom; and funnels leading up or placed over a container to hold captured insects. However, different types of traps are used for insects moving through air (that is, flying or wind-borne insects), ground-dwelling or walking insects, subterranean insects, or aquatic insects. A number of these are described below.

Traps for flying insects and wind-blown insects

Interception traps. Interception traps are commonly used for faunal surveys in ecological studies, although they also can have pest management applications. In its simplest form, it is a suspended net with an invagination along the top that leads to a collecting tube. The Malaise trap is an example of an interception trap. Fixed interception traps have been used to study insect migration, with separate collecting tubes for north-bound versus south-bound insects. Or, the trap may be used with a wind vane attachment so that the flat surface of the net swivels to face into the wind. Large, funnel-shaped nets have been mounted on moving trucks for sampling flying insects such as biting midges (Ceratopogonidae) or attached to a suction device to sample Russian wheat aphids, *Diuraphis noxia* (Mordvilko).

Sticky traps. Panels, cylinders or spheres covered with sticky material are probably the most commonly used traps for faunal surveys in agricultural studies. In the simplest form, they may be clear panels that are coated with a material that will retain insects that are blown onto the panel or fly into it. Panels also may be used with a color and/or a shape and a chemical attractant. Very small insects will be retained by a thin coating of motor oil, however larger insects may escape this substance. To capture both small and large insects, sticky material such as Tangle-trap is applied to the surface. The traps can be serviced by using a small tool to scoop insects of interest off of the trap and onto a card or the entire trap can be

Fig. 1073 Some insect traps: top left, Malaise trap, an interception trap for capture of flying insects; top right, Steiner trap, a lure-based trap used for capturing fruit flies; second row left, PC floor trap, a lure-based trap for stored product insects; second row right, pitfall trap for capture of insects walking on the soil surface; third row left, solar bait station for early detection of wireworms; third row right, suspended black ball coated with adhesive for capture of tabanid flies; bottom left, red sphere on yellow panel coated with adhesive for detection of apple maggot flies; bottom right, boll weevil trap.

replaced. For transport, the panel with sticky material can be covered with clear plastic wrap or the panel can be placed in a box with spacers to keep panels from touching other surfaces. To reuse the trap, a paint scraper or thinner can be used to remove sticky material from the surface so that new material can be applied. The advantage of this trap is that it is inexpensive and will capture a variety of insects that are moving through an area. The surface area and trap orientation can be increased by using two panels that are crossed into baffles. The primary disadvantage is that the trap can be very messy, can become coated with dirt or debris and will no longer capture insects, and the sticky material is difficult to remove from the captured insects. These traps, as well as other traps for flying insects described below, are usually placed above the ground at the height of the vegetative growth. When used in row crops, the traps usually are attached to a wooden or metal stake, and the traps are moved higher as the plants grow over the season. When used in trees, traps normally are hung within the canopy. Trap placement and orientation within the canopy, as well as amount of vegetation near the trap, will vary among the trap types and target species.

Three-dimensional triangular traps, diamond traps and wing traps. This group is another set of fairly inexpensive traps that are used with an attractant. These are basically sticky traps with the sticky surface protected on the interior of the trap and are either disposable or used with a sticky component that is replaced at servicing. Delta traps are tent-shaped triangular traps that are small and light weight, which makes them easy to hang in trees and to transport. The Jackson trap is a delta trap that is used for a number of tropical tephritid fruit flies, and has a removable base coated with sticky material which can be replaced at time of sampling. Diamond traps are, as the name states, diamond in shape and are used to monitor indoor pests in public areas, such as stores and supermarkets. Wing traps are composed of a roof and a floor and typically are used for pest Lepidoptera. These are larger and more cumbersome than delta traps, but the larger surface is more suitable for the larger pest moths. Because the sticky surface is contained within the trap, they are less susceptible to dirt and dust and capture fewer non-target insects than

sticky traps, but the surface still may be coated with dust or debris that gets blown into the trap.

Water pan traps. A simple collection method for aphids and other small flying insects is a water pan trap. Insects flying over the traps are attracted to the reflective surface of the water and are captured. The traps are made from rectangular baking pans, storage containers or dish-washing pans partially filled with aqueous solutions of soap or car antifreeze. Care should be used in selecting soap to be used so that it does not contain odors that may be repellent to target insects or that the antifreeze is environmentally safe. (Most antifreeze solutions are poisonous, and traps containing antifreeze should be placed so that animals are unable to gain access to them.) These traps are open containers, so there may be problems with movement in wind, evaporation during dry periods or overflowing during periods of heavy rain.

Bucket traps. Another simple and inexpensive trap is a bucket trap. The bucket trap may be used without a lid, or as a closed container with holes on either the top or sides for insects to enter. Small drainage holes may be placed near the bottom of the trap to allow trapped water to drain out. The Nadel trap and Mission trap are examples of bucket traps that are translucent closed containers with entrance holes around the periphery. The corn rootworm trap has a large funnel-shaped top and a gap between the top and the container. Size of the container and placement, diameter and number of entry holes is dictated by the target insect. These usually are used with a chemical lure of some type. These may be used without a killing agent so that live insects can be obtained, however, captured insects may escape through the entrance openings. A variety of retention devices can be used with bucket-type traps. A pesticide such as dimethyl 2,2-dichlorovinyl phospate (DDVP, mothballs) can be used inside, but if the concentration is too high it may prevent insects from entering. Aqueous solutions of soap, car antifreeze or surfactants such as triton may also be used. Again, care should be used in selecting soap so that it does not contain odors that may be repellent to target insects and environmentally safe antifreeze should be used. Sticky material also may be used on either the interior or exterior surface. When used on the exterior, this may increase effectiveness of lure-baited traps because usually only a proportion of the insects

attracted to the trap will proceed to enter the trap. However, the same problems outlined above for sticky panels will apply to sticky-coated bucket traps. Open bottom cylindrical traps are essentially bucket traps used upside down. The Phase 4 trap is a green cylinder that uses a yellow panel as a sticky insert to retain captured insects. Clear versions of the trap also have been used. In addition to the opening on the bottom of the trap, entrance openings also are located around the periphery of the cylinder.

Bucket traps with funnels. Probably the most common traps for agricultural use are combinations of bucket traps with funnels. They are more costly than simple bucket traps and they often are used with some type of odor attractant. The funnel essentially provides an enlarged hole that directs movement of the attracted insect into the bucket. There are differences among this group of traps in orientation of the funnel and the size of the funnel in relation to the size of the bucket. The Steiner trap is a clear plastic horizontally oriented cylinder with small funnels centered on the flat sides of the cylinders. For insects that tend to approach from the underside of the trap and move upward, the funnel is used with the large opening facing downward and the funnel leading up into a bucket or some type of container. Because it is difficult for the insect to find the top of the funnel for exiting and these insects tend to move upward, they can be used with or without a pesticide or liquid to retain the insects. One example is the McPhail trap, which is a bell-shaped invaginated glass trap that is used for tephritid and drosophilid fruit flies. The original trap was made from glass and there are several plastic versions of McPhail-type traps including dome traps, International Pheromones McPhail traps and Multilure traps available that typically have a clear top and a yellow base. Another example is a wasp trap, such as the Victor yellow jacket trap. This trap has holes around the sides of the base for wasps to enter, and then a funnel that leads to a separate upper container that retains the wasp. The boll weevil trap is another version of a bucket trap with funnel,

but the funnel is mounted on a cylinder. Insects land on the cylinder and move up the sides of the trap, enter the funnel and move into the small collection bucket above the small funnel top.

Other insects approach from the top and either move or fall downward. For these insects, the bucket traps with funnels that have the large opening facing upward and the small opening leading down into the bucket are appropriate. The universal moth trap (unitrap) and Multipher trap are examples of this trap and they commonly are used for Lepidoptera adults. This trap has a lid held above the funnel opening to shelter the contents from rain and is available in a variety of colors. A trap developed for the Japanese beetle uses a funnel with the large opening facing upward and the small opening over a container, but the funnel is topped by two panels crossed to form a baffle. Insects are intercepted by the baffles and fall into the funnel and are captured in the container. The Lindgren funnel trap consists of four, eight or twelve plastic funnels stacked vertically over the container. It is used for ambrosia beetles and bark beetles. Fast-flying beetles hit a funnel and are deflected down into the collection container.

Cone traps. Cone traps are essentially bucket traps with funnels, but the funnel is very large in relation to the size of the bucket or collecting tube. These are used with the large opening of the funnel oriented towards the ground and the top of the funnel leading up to a container. Examples of these traps include *Heliothis* traps and butterfly bait traps. These traps are made from a light cloth or mesh material and the design takes advantage of the moth or butterfly's tendency to move upward. The butterfly bait traps have a large container for the butterfly to move into so that the insect is undamaged.

Traps for walking arthropods and soil-dwelling insects

Pitfall-type traps. Pitfall traps are useful for collecting insects and other arthropods that are walking across a surface. This surface is usually a soil surface,

Fig. 1074 Additional insect traps: top left, grain probe traps for detection of stored grain insects; top right, blacklight trap for detection of nocturnal flying insects; second row left, wing trap, a lure-based trap with a sticky interior for sampling flying insects; second row right, wasp trap, a food lure-based trap; third row left, unitrap, a popular pheromone-based trap for capture of moths; third row right, a bucket trap baited with pineapple and pheromone for capture of palm-infesting weevils; bottom left, a cylindrical yellow panel coated with adhesive for detection of flying insects; bottom second from left, New Jersey light trap for capturing mosquitoes; bottom third from left, mosquito emergence trap; bottom right, McPhail trap, a food lure-based trap for capture of fruit flies.

for capturing beetles, spiders and other ground-dwelling organisms, although they also may catch flying insects that are walking across the soil surface. Pitfall traps placed into the top of stored grain also have been used to sample stored product insects. Pitfall traps consist of a container that is buried in the substrate, and into which insects and arthropods fall and are captured. The traps used in soil consist of an upper funnel, a collecting container, and perhaps a liner that makes servicing easier. Upper funnels can be made from disposable plastic funnels with the bottoms removed to enlarge the hole. Collecting containers can be made from plastic cups, and can be designed for dry catches with screens in the bottom to permit rain water to flow through, or can be made to hold a glycol (antifreeze)/water/detergent solution. Dry containers need to be serviced several times a week to minimize destruction of the sampled insects from other insects entering the trap. They have the advantage of providing live specimens for further studies. Wet containers can be serviced at longer intervals, but have the disadvantage of filling with rain water. Traps should be designed to produce minimal impact on the nearby soil, because insects may be repelled or attracted to the disturbance. To do this, holes initially can be cut with golf-hole cutters and lined with 4'' diameter polyvinylchloride (PVC) pipe. The top of the pipe should be carefully leveled with the soil, and the traps serviced without disturbing the soil. Covers can be installed over the top to prevent rainfall from entering. Enhancement-fences or guides can be installed to guide insects to the trap. A cone with a gradual slope and smooth edge is necessary because insects may back away from the void of a direct hole. Captured insects should be removed at least weekly (wet) or twice weekly (dry).

Grain probe traps. Grain probe traps are a special modification of a pitfall trap for use in stored grain. Grain probe traps consist of an elongated cylinder with holes drilled into the sides that are above a funnel and insect receptacle. An early version of the trap was machined from solid brass and included a hollow cylinder made from 14 gauge brass sheet. Subsequently, probe traps have been made from clear polycarbonate (Lexan) plastic and from a perforated section of tubular polyethylene. The receptacle is coated with liquid Teflon (polytetrafluoroethylene) to prevent captured insects from escaping, however,

insects remain alive for a while and may damage previously captured insects. These traps may be used at any depth within the grain mass, with long rods used to push the traps into place. A rope connected to the trap should be affixed to the roof of the grain bin to allow removal of the grain probe trap and to prevent loss of the trap during grain bin filling and emptying operations. Traps should be inspected at one to two week intervals to remove the trapped insects. Another type of grain probe trap is the PC trap. It is a cone-shaped trap that usually is used on the grain mass surface, although it too can be pushed into the grain mass as long as it remains in an upright position. The top is covered by a convex lid that is covered with concentric circles of small holes to allow insects to enter the trap. When used at the grain surface, the traps are easy to remove for servicing, but webbing produced by larvae of lepidopteran grain pests, such as the Indianmeal moth, may block the openings into the trap and render them ineffective.

Shelter traps. Shelter traps are used for insects that prefer a dark harborage and are useful in areas where it is preferred that the trap be inconspicuous. These usually are used to intercept insects as they walk over a surface and have small openings that encourage insects to enter the trap. Roach motels are the most well-known examples of such traps. They usually have a sticky material inside to trap insects that have entered or contain an oil or other substance to retain attracted insects. Shelter traps are used for stored product insects, and designs include dome traps, stealth traps and corner traps. The PC floor trap has been modified from the PC trap for use as a shelter trap by replacing the cone-shaped bottom with a flat container, which allows the trap to be placed on the ground or hung on a wall. Swarm traps are shelter traps that are used to capture unwanted swarms of domestic honey bees and to detect invading swarms of Africanized honey bees. These traps are fairly large bucket-shaped traps made of molded fiber material, have a single entrance hole, and are hung at least one meter above the ground. They provide a nesting site for bees, and the captured bees can be kept alive and moved to standard bee hives for honey production or pest bees can be identified and destroyed.

Emergence traps. Emergence traps are a type of cone trap used for capturing adults that are in the soil for their larval and/or pupal stages. These traps are made from aluminum screening shaped into a funnel, with the top opening leading to a collecting tube or vial. The trap is placed flush with the ground and soil is pushed up around the edges to seal the trap to the ground. Adults emerging from below-ground stages move up into the trap and are captured in the collecting tube. The number of insects per unit area can be determined and the source of infestations can be identified. However, these traps interfere with ground maintenance activities that prevent use in certain situations. Circle traps are another type of emergence trap. They are wire cone traps that are attached to the trunk or branch of a tree. Insects moving up a tree are captured, and because the traps are off of the ground, they do not interfere with ground maintenance activities.

Solar bait stations. The above traps can be used to capture the above-ground stages of subterranean insects, however the below-ground stages of wireworms (Elateridae), and false wireworms (Tenebrionidae) can be captured with solar bait stations. These are used to estimate wireworm populations and to make decisions on seed treatment/nontreatment. Larvae of these insect groups can be concentrated by creating microenvironments that have favorable moisture, temperatures, and food. A handful of grain (wheat, etc.) is buried a few inches below the surface of the soil. The bait is covered with a mound of soil, about 18″ high, and is covered with clear plastic. Edges of the plastic should be covered with soil to prevent them from blowing away. Stations should be constructed in the fall before soil freezes, and they can be examined in the spring before planting time. Surveyors' flags can be used to mark sites for easy location.

Traps for aquatic insects

Interception traps. Interception traps capture aquatic insects moving through the water and are generally similar to interception traps used for windborne insects. These traps tend to be tapered nets with either round or rectangular openings which either can be fixed to sample insects in moving water or pulled through the water manually. The size of the mesh governs the size of insects retained, but use of too fine a mesh may impede movement of water through the trap and retain too much debris.

Emergence traps. Floating emergence traps are used to capture insects with aquatic larval stages as they emerge as adults. Construction and use is similar to that for emergence traps used for soil-dwelling insects.

Attractant cues

Attractant cues are signals that are used by insects to locate resources for feeding, members of the opposite sex for mating and oviposition sites for egg laying. Cues are added to traps to increase insect specificity and effectiveness of the traps. Attractive cues from natural substrates may be perceived by one or several of the senses including sight, sound and smell. Semiochemicals are naturally occurring, message-bearing chemicals that are used by insects (and other organisms) for communication and for perception of their environment. Semiochemicals that have a behavioral effect on insect orientation, that is, chemicals that cause an insect to move toward the source, are used in insect traps. Of specific interest are kairomones, signals emitted and received by members of different species (interspecific) that give an advantage to the receiving species, and pheromones, signals emitted and received by members of the same species (intraspecific). Traps may use single cues to lure insects, but often a combination of cues is used to improve insect capture.

Visual cues

Color. Color can serve as a strong attractant for use in a trap. Insects may use a specific color to locate host fruit or plant material, with both hue and intensity affecting insect response. Contrasts between light and dark also can play a role, with either the trap in contrast with the background color or through the use of lines with insects orienting to an edge between a light and dark area on a trap. Most sticky traps use a color to target certain insects. Yellow is the most commonly used attractant color, and is used to capture hemipterans such as whiteflies and aphids, but also is used for almost every order of flying insect. The Rebell trap is constructed from two yellow panels as baffles, and is used for walnut husk flies, *Rhagoletis completa* Cresson, and cherry fruit flies, *Rhagoletis cingulata* (Loew). A disadvantage of using yellow for sticky traps is that it also attracts beneficial insects such as hymenopteran parasitoids, and the traps may fill up with these and other nontarget insects. Yellow-colored pan traps are used to capture Hymenoptera, especially the parasitoid species. Orange sticky traps are used for carrot rust fly, *Psila rosae* (Fabricius), and blue sticky traps for western flower thrips, *Frankliniella occidentalis* (Pergande). Blue also is highly attractive to tsetse flies (Glossinidae). White sticky traps are used as panels and as trunk wraps for capture of tarnished plant bug, *Lygus lineolaris* (Palisot de Beauvois), and eastern apple sawfly, *Hoplocampa testudinea* (Klug), and red trunk wraps for apple blotch leafminer, *Phyllonorcycter crataegella* (Clemens). Use of contrasts also can increase insect capture, such as the use of insect silhouettes added to the surface of white sticky panels to increase capture of house flies, *Musca domestica* L.

Shape. Shape also can be used as an attractant, although usually it is combined with appropriate color for the target insect. Spheres commonly are used as traps for tephritid fruit flies. Red spheres are used for apple maggot flies, *Rhagoletis pomonella* (Walsh), small green spheres for blueberry maggot flies, *Rhagoletis mendax* Curran, large green spheres for papaya fruit flies, *Toxotrypana curvicauda* Gerstaecker. A Ladd trap is a red sphere that is mounted on a yellow panel and is used for apple maggot flies and cherry fruit flies, *Rhagoletis cingulata* (Loew). Black spheres have been used to capture biting flies in the family Tabanidae, which are attracted to the movement of the black ball hanging in a tree. A tree silhouette is mimicked by a Tedders trap, which is a baffle constructed from two dark isosceles triangles, with the unequal side forming the base. The size of the Tedders trap can be adjusted to optimize capture of tree-dwelling beetles. Root-infesting weevils, such as pecan weevils, *Curculio caryae* (Horn), and several citrus root weevils, as well as wood-boring beetles, may be captured by these traps. Beetles emerging from soil or moving along the soil surface respond to the trap as if it were the trunk of a tree. The insect moves up the baffles and is captured in a small collection container at the top of the trap.

Light. Attraction of insects to light is easily observed by standing next to a porch light or street light during the night. Night-flying insects orient to the light of the moon, and artificial light produced by man-made devices interferes with this response and brings the insect toward the source of the light. A variety of insect traps have been developed and used that are based on a light as the attractant. Lights used include mercury lamps and black lights (UV) for moths, incandescent lights for flies and mosquitoes, green lights for stored product insects and cyalume lightsticks for aquatic light traps. Light traps, also called electric traps, are typically bucket traps with funnels. When used with fluorescent bulbs, the traps have baffles alongside the bulb that knock insects attracted to the light down into the trap. The Pennsylvania light trap is an example of this trap. Rothamsted and Robinson-type traps use incandescent bulbs, and these have a reverse funnel over the light bulb. The New Jersey trap and the CDC trap are light traps developed for mosquito surveillance. They also use incandescent bulbs but with the addition of a fan that creates a suction to draw attracted mosquitoes down into the bucket and a filter to prevent larger insects from entering the trap. Lights also may be used with sticky traps to retain attracted insects. These more commonly are used for control of indoor pests such as house flies, stored product insects and fleas. A LED-CC trap for whiteflies uses a green light-emitting diode (LED) to attract insects into an open-bottom clear cylindrical trap with a yellow ring on the bottom.

Chemical cues

Food/host lures. Volatile chemicals emitted from host plants, animals and other materials are used by insects to locate food sources. These chemicals are referred to as kairomones, that is, chemicals produced by a plant or animal that are advantageous to the receiving individual of a different species. Host material can be used to bait traps for insect capture, and the spectrum of use of food-based baits for insect control and pest management are covered in separate chapters. Examples of use of host material in a trap include use of carrion in pitfall traps or fruit in butterfly bait traps as survey tools, or fruit and meat in traps for yellow jackets (Hymenoptera). Grain or grain products (e.g., wheat germ, wheat germ oil, corn oil, etc.) commonly are used for stored product insect pests, and grain oils used in shelter traps provide both an attractant and a method to kill attracted insects. However, host material may decay rapidly or may release attractive chemicals for only a short time period after initial placement. The insect may be using only a few of the sometimes numerous volatile chemicals emitted by a host and traps baited with synthetic chemical attractants can be used to lure an insect into a trap. The quality and quantity of the chemical can be controlled by the method of formulation, providing a standard release rate for a known time period, which improves trap performance.

Carbon dioxide is used by mosquitoes to locate vertebrate hosts, and addition of dry ice to a trap can be used to survey blood-feeding mosquitoes. Octenol (1-octen-3-ol) is a naturally occurring chemical emitted by oxen and cows, because they ingest large amounts of vegetable matter, and combinations of octenol and carbon dioxide are used to bait traps for mosquitoes, biting midges and no-see-ums (Ceratopogonidae). To make these chemicals more attractive, they are combined with heat and moisture in a trap such as the Mosquito Magnet. Host kairomones are used with blue sticky traps to capture tsetse flies and in rootworm traps for corn rootworm adults. Floral cues are used in traps for Japanese beetles, *Popillia japonica* Newman, and for nectar-feeding female moths (Lepidoptera). These are used in bucket traps with funnels. Fruit lures that provide volatile chemicals emitted from apples and plums are used in traps for apple maggot flies and for plum curculio, *Conotrachelus nenuphar* (Herbst). Fruit lures for apple maggot flies are hung near red sphere traps, fruit lures for plum curculio are used in a variety of traps, including Tedders, circle and boll weevil traps. McPhail-type traps can be used with a variety of liquid baits. They have been used with fermenting sugar solutions to capture small fruit flies (Drosophilidae) and moths (Lepidoptera), and with aqueous protein solutions for fruit-infesting fruit flies (Tephritidae). Ammonia has been found to be the primary chemical responsible for attraction of tephritid fruit flies to protein solutions, and ammonia alone or in combination with other synthetic volatile chemicals emitted from protein baits are used in sticky panels, sticky spheres or McPhail traps to catch these fruit flies.

Pheromone lures. There are two main types of pheromones that are used with insect traps. Sex pheromones are produced by one sex to attract the opposite sex. The sex pheromones most commonly used in traps are ones that are produced by females. These are used in either wing traps or bucket with funnel traps, and there are lures available for numerous species of pest Lepidoptera and for sweetpotato weevils, *Cylas formicarius elegantulus* (Summers). The moths tend to be active at night, so visual cue is less important, however, the contrast of a white or light colored trap versus the dark background often increases capture. The advantage of traps baited with these sex pheromones is that they are highly specific and very

effective, however, they capture only males and so no information on or samples of females are obtained. Some male tephritid fruit flies produce a sex pheromone that is attractive to female flies, and the papaya fruit fly pheromone increases capture of both male and female flies on green sticky sphere traps.

Aggregation pheromones also are used with insect traps. These are attractive to both sexes and are used to bring both sexes to a common location for both feeding and mating. Thus, both males and females can be captured in traps baited with aggregation pheromones, however, they tend to be less effective than sex pheromones or they need to be presented with other cues. Exceptions are the pheromone produced by boll weevils (Grandlure), *Anthonomus grandis grandis* Boheman, and the lesser and larger grain borers, *Rhyzopertha dominica* (Fabricius) and *Prostephanus truncatus* (Horn). Sticky, shelter and wing traps baited with these pheromone lures capture both sexes of these borers. Aggregation pheromones often are produced by males in conjunction with feeding, thus since both insect pheromone and host kairomones are emitted as signals, both are needed to elicit attraction. Combinations of synthetic aggregation pheromone and host material as the source of kairomones are used in shelter traps for stored product beetles and cockroaches, in Lindegren or panel traps for bark beetles, and in bucket traps for palm weevils. A combination of honey bee pheromones are used in swarm traps to increase bee capture. These include honey bee queen mandibular pheromone (BeeBoost) and an orientation pheromone produced by worker bees (Nasonov).

Parapheromone lures. Parapheromones are a special group of lures that are used for some species of tephritid fruit flies. They act as sex pheromones because they are highly attractive to male fruit flies, similar to female-produced sex pheromones. However, these are not insect-produced compounds and do not appear to play a dominant role in the biology of the responding species. They may be synthetic kairomones, as they are similar in structure to some plant compounds, and access to the synthetic or natural versions of these compounds have been shown to increase sexual competitiveness of males. These compounds include trimedlure for Mediterranean fruit flies, *Ceratitis capitata* (Weidemann), cuelure for melon flies, *Bactrocera cucurbitae* (Coquillett), and methyl eugenol for oriental fruit flies, *Bactrocera dorsalis* (Hendel). These lures are deployed most commonly in white Jackson traps (triangle traps) or on yellow sticky panels.

Oviposition lures. Oviposition lures are chemicals that attract egg-laying (gravid) females. There is a gravid mosquito trap that uses a baited water solution to attract and capture the adult females. An alternative approach is a trap and bait for navel orangeworms, *Amelyois transitella* (Walker) on which the female moth lays eggs.

Acoustic cues

Just as many insect species produce volatile chemicals to attract members of the opposite sex, some species produce sound alone or in combination with chemicals for this purpose. Such cues have been incorporated successfully into panel or bucket traps of several different types. The different sounds broadcast as attraction cues have included songs recorded from conspecifics of the targeted insect and synthetic mimics of the songs. A variety of different speaker systems have been used, including standard loudspeakers, piezoelectric boards with extensive surface area, and piezoelectric cylinders.

Sound traps have been developed that produce highly amplified synthetic or recorded calls of male mole crickets (Gryllotalpidae). A bucket or bucket with funnel trap is placed under the sound emitter to capture responding crickets. These traps also capture tachinid flies that parasitize adult crickets and that locate them by responding to the call. Sounds that mimic buzzing from feeding mosquitoes also have been incorporated into mosquito traps along with chemical cues, heat, moisture and light for backyard use. Sticky traps broadcasting the recorded male lesser wax moth, *Achroia grisella* (Fabricius), calling song have been used to attract virgin females.

Another use of acoustic cues is as an attractant for males of some mosquito and midge species that swarm to attract females. When females fly into the swarm, the males are attracted by their wingbeats, which are of distinctly lower frequencies than those of the males in the swarm. Consequently, males can be attracted in great numbers by placing a black cloth or other swarm marker on the ground and broadcasting recorded or synthetically generated female wingbeats at high intensity from a speaker inside or at the edge of a sticky panel, cylinder, or cup hung about 1 meter above the swarm marker. Acoustic traps can greatly reduce male populations of sedentary mos-

quito species, and also have been used to chemosterilize and re-release males rather than killing them. However, they are not yet in common use in isolated field environments because they require electricity and some technical skill to operate, and the sound that must be broadcast at high amplification for optimal trap catch can be a nuisance.

Automated monitoring systems

Advances in information technology are adding to the field of precision agriculture; that is, use of computers is aiding in management decisions. The ability to rapidly move data from traps into computer databases or spreadsheets is an integral component of precision agriculture, and will facilitate making pest management decisions in a timely manner. Bar codes can be added to traps and a bar code scanner can be taken to the field to expedite data entry on trap type, trap location, etc., so that only insect counts need to be entered manually. The next step in automation is directly recording data from traps into a computer. This can be done by storing data on a local computer or handheld device for later downloading, or data can be transmitted to an off-site computer. One approach to off-site monitoring is the use of gravimetric analysis of flight trap captures of red flour beetles, *Tribolium castaneum* (Herbst). Beetles responding to cone-shaped flight traps fall into a small container coated with liquid Teflon which rests on the weighing pan of a digital pan balance. Signals from the balance are sent over a cable to a personal computer, and weight, which is recorded at sequential intervals, is converted to the number of insects captured over time. Insect movements that interrupt an infrared light beam can be used for computerized data acquisition. Infrared beams can be used to monitor insect movement in activity chambers as a type of actigraph. Infrared beams also are used in an electronic grain probe insect counter system. The beams are located below the bottom of the funnel in a cylindrical grain probe trap. Insects falling through the funnel are counted electronically and time-stamped data is transmitted to off-site computers. Use of counts from electronic traps, along with information from automated temperature and relative humidity probes, can be collected together and used for management decisions.

Trap uses in integrated pest management

The goals of trapping are highly variable. Traps may be used for general survey of biota or for detection of exotic invasive insect pests that have entered a previously uninfested area. Trapping systems for insects are important components in integrated pest management programs. Trapping data are used to make decisions on the initiation or termination of control measures, as well as to assess efficacy of control approaches that have been implemented. Detection trapping is used to alert personnel to the presence of a new insect pest in a previously pest-free area so that control measures can be implemented in a timely manner. Early detection will facilitate pest management strategies such as biological control approaches that are more effective against low pest densities. With the availability of sufficiently effective traps that capture both female and male pest insects, trapping systems may be used as control measures and, thus, could be added to the growing list of biologically based technologies for insect control. Traps can be used as toxicant delivery systems, with insects that visit the trap taking a slow-acting poison back to the rest of the population. This approach is used with social or gregarious insects such as termites, ants and cockroaches. Mass-trapping is the use of large numbers of traps in an effort to suppress the population. Sticky traps usually are used in this approach as a high percentage of responding insects are captured. This approach has been used to suppress populations of a pest, such as apple maggot flies, that spend part of their life cycle away from the host and can be intercepted by traps placed around the periphery of the orchard. However, these traps require frequent servicing to maintain activity. An alternate approach is the development of attract and kill systems, sometimes called attracticides. In this approach, insects responding to traps consume or contact a toxicant, but then exit the trap and die away from the trap. Examples include addition of insecticide to artificial cows for control of tsetse flies, methyl eugenol mixed with insecticide for control of oriental fruit flies, addition of insecticide to pheromone to control codling moth in apples, and addition of insecticide to feeding stimulant in corn rootworm traps. All of these control approaches should be combined with other pest management strategies to be fully successful.

Nancy D. Epsky
USDA/ARS, Subtropical Horticulture Research Station
Miami, Florida, USA
and
Wendell L. Morrill
Montana State University
Bozeman, Montana, USA
and
Richard Mankin
USDA, ARS, Center for Medical, Agricultural and
Veterinary Entomology
Gainesville, Florida, USA

References

Hienton, T. E. 1974. *Summary of investigations of electric insect traps*. Technical Bulletin 1498. Agricultural Research Service, U.S. Department of Agriculture. 136 pp.

Mayer, M. S., and J. R. McLaughlin. 1991. *Handbook of insect pheromones and sex attractants*. CRC Press, Boca Raton, Florida. 1083 pp.

Muirhead-Thomson, R. C. 1991. *Trap responses of flying insects*. Academic Press Inc., San Diego, California. 287 pp.

Pedigo, L. P., and G. D. Buntin. 1993. *Handbook of sampling methods for arthropods in agriculture*. CRC Press, Boca Raton, Florida. 714 pp.

Southwood, T. R. E. 1966. *Ecological methods*. Chapman and Hall, New York, New York. 524 pp.

TREE CRICKETS. A subfamily of crickets (Oecanthinae) in the order Orthoptera: Gryllidae. See also, GRASSHOPPERS, KATYDIDS AND CRICKETS.

TREHALOSE. A polysaccharide found in insects that is one of the two most common carbohydrate stored reserves (the other is glycogen) for insect flight. It occurs principally in the hemolymph, fat body, and gut tissues. Trehalose is usually the first metabolite used when energy is needed. Each molecule of trehalose is hydrolyzed into two molecules of glucose. Trehalose also is rapidly synthesized from glucose as it is absorbed in the midgut. Glucose is not usually present in high concentrations in the hemolymph because of this synthesis, so glucose absorption is easily accomplished.

TREHERNE, JOHN E. John Treherne was born on May 15, 1929, near Swindon, England. He was educated at Bristol University. After military service he was invited by Vincent Wigglesworth the join the Unit of Insect Physiology at Cambridge. Treherne did so, and also served as lecturer and reader at the

University. Upon Wiggleworth's retirement, Treherne headed up a new Unit of Invertebrate Chemistry and Physiology at Cambridge. Treherne's reputation was based on research of insect neurobiology, the blood-brain barrier of insects, gut physiology, circadian rhythms, hormones, cuticle permeability, osmoregulation, and other physiological subjects. His subjects were not just insects, as he worked on molluscs and annelids. Indeed, he was a physiologist first and foremost, and never pretended to be an entomologist in the classic sense. Nevertheless, he enjoyed field entomology, especially marine and salt marsh insects, and made insightful behavioral and evolutionary contributions in this area. Treherne served as editor of "The Journal of Experimental Biology," and edited "University Reviews of Biology," "Advances in Insect Physiology," and others. He served as president of Downing College from 1985 to 1988. Treherne also wrote popular novels, some of which included entomological elements. He died on September 23, 1989.

Reference

Foster, W. 1990. John E. Treherne (1929–1989). *Antenna* 14: 6–9.

TREEHOPPERS (HEMIPTERA: MEMBRACIDAE).

Most treehoppers are readily distinguished from their close relatives, the leafhoppers, by their enlarged pronotum, which extends posteriorly over the remaining thoracic segments and often bears horns, spines, bulbs, or other projections. A few treehoppers have a pronotum that is more modest in size; these differ from leafhoppers in that the scutellum is enlarged with either a median longitudinal keel, or a distinct median longitudinal groove or notch.

Order: Hemiptera
   Infraorder: Cicadomorpha
      Superfamily: Membracoidea
         Family: Membracidae

Membracidae, a large and diverse family of plant-feeding insects, comprises approximately 3,100 described species in 400 genera. Their closest living relatives are Melizoderidae (a small group of treehopper-like insects endemic to chile) and Aetalionidae (a small, mostly neotropical family with one endemic southeast asian genus). Aetalionidae, meli-

Zoderidae and Membracidae together constitute a specialized lineage apparently derived from leafhoppers (Cicadellidae). Treehoppers are mainly a tropical group, but a sizeable fauna of mostly oak-feeding species occurs in temperate North America. Most of the subfamilies of Membracidae are endemic to the new world. Centrotinae, the only treehopper subfamily that occurs worldwide, is well represented in tropical and subtropical regions of Africa, Asia and Australia, but the palearctic fauna is extremely depauperate. The oldest fossil treehoppers are known from Tertiary-age Dominican and Mexican amber.

External morphology

Adult treehoppers range from 3 to 30 millimeters in length. in addition to differences in coloration and the shape of the pronotum, genera and species of treehoppers differ in the shape and proportions of the head, the wing venation, the arrangement of setae on the legs (particularly the hind tibia), the shape of the female ovipositor and the male genitalia. Sexual and seasonal dimorphism in pronotal shape and color is common in many species. Treehopper nymphs are often even more bizarre than the adults. some species have large spines on the head, thorax and abdomen; others are well adapted for crypsis, being strongly depressed with the flattened tibiae fitting like puzzle pieces into notches in the sides of the thorax.

Life history and habits

Details of the life cycle vary from species to species. The female deposits several eggs on the bark or into the living tissue of a woody host plant and may cover the eggs with a frothy substance that hardens when dry. The eggs either remain dormant for a period ranging from a month to over a year, or they develop and hatch within a few weeks. The young, known as nymphs, feed on plant sap by inserting their mouthparts into the phloem vessels of the host plant and go through a series of five molts reaching the adult stage after a period of several weeks. The adult males and females seek each other out for mating, locating each other through vibrational signals made by sound producing organs at the base of the abdomen called tymbals. These signals are transmitted through the substrate and are usually too faint to be heard by human ears. Some treehoppers exhibit anointing behavior similar to that of their relatives, the leafhoppers, but, unlike leafhoppers, treehoppers

Fig. 1075 Treehoppers (Membracidae). (a), *Heteronotus* sp. adult; (b), *Smilia fasciata* Amyot & Serville adult; (c), *Enchenopa binotata* (Say) adult; (d), *Microcentrus caryae* (Fitch) adult; (e), *Microcentrus caryae* nymph; (f), *Stictocephala taurina* (Fitch) nymph. (Photos by C. H. Dietrich.)

do not produce brochosomes (a hydrophobic, granulated coating).

As implied by their name, most treehoppers are strong jumpers as adults. However, treehopper nymphs cannot jump and avoid predation through crypsis, ant-mutualism, or parental care. Although many species are solitary as nymphs and adults, numerous treehopper species are gregarious and exhibit various degrees of parental care (or presocial) behavior. In the most advanced form of parental care, females guard their eggs and remain with the nymphs throughout their development, repelling invertebrate predators by kicking with the legs or buzzing the wings.

Predators and parasites

Treehoppers are a food source for vertebrate predators such as birds and lizards, as well as invertebrate predators such as spiders, assassin bugs, wasps and robber flies. Treehoppers also are attacked by various parasitoids such as dryinid and mymarid wasps, epipyropid moths, pipunculid flies and strepsipterans. They also are attacked frequently by entomopathogenic fungi.

Economic importance

Several treehopper species are minor pests, particularly of tropical fruit trees such as papaya, cacao and palm. A few species are known to transmit plant pathogens. The buffalo treehopper (*Stictocephala bisonia* Kopp & Yonke), a North American species that injures apple and related fruit trees, was introduced accidentally into Europe and now is well established in the temperate zone of the palearctic region as far east as central Asia.

Control

Treehoppers rarely inflict economic damage on crops, but when they do, control usually involves the use of conventional contact insecticides.

See also, BUGS.

Chris H. Dietrich
Illinois Natural History Survey
Champaign, Illinois, USA

References

Deitz, L. L. 1975. Classification of the higher categories of the new world treehoppers (homoptera: membracidae). *North Carolina Agricultural Experiment Station Technical Bulletin* 225: 1–177.

Deitz, L. L., and C. H. Dietrich. 1993. Superfamily membracoidea (homoptera: auchenorrhyncha). I. Introduction and revised classification with new family-group taxa. Systematic Entomology 18: 287–296.

Dietrich, C. H., and L. L. Deitz. 1993. Superfamily Membracoidea (Homoptera: Auchenorrhyncha). II. Cladistic analysis and conclusions. *Systematic entomology* 18: 297–312.

Dietrich, C. H., S. H. Mckamey, and L. L. Deitz. 2001. Morphology-based phylogeny of the treehopper family Membracidae (Hemiptera: Cicadomorpha: Membracoidea). *Systematic Entomology* 26: 213–239.

Mckamey, S. H. 1998. Taxonomic catalogue of the Membracoidea (exclusive of leafhoppers). Second supplement to fascicle I – Membracidae of the general catalogue of the Hemiptera. *Memoirs of the American Entomological Institute* 60: 1–37.

TRIASSIC PERIOD. A geological period at the beginning of the Mesozoic era, extending from about 250 to 213 million years ago.

TRICHOCERIDAE. A family of flies (order Diptera). They commonly are known as winter crane flies. See also, FLIES.

TRICHODECTIDAE. A family of chewing lice (order Mallophaga). They sometimes are called mammal chewing lice. See also, CHEWING LICE.

TRICHOGRAMMATIDAE. A family of wasps (order Hymenoptera). See also, WASPS, ANTS, BEES, AND SAWFLIES.

TRICHOMES. Hairs or small spines on the surface of a plant, and an important morphological defense against attack by insects. See also, TRICHOMES AND INSECTS.

TRICHOMES AND INSECTS. Trichomes, also called plant hairs, are found on vegetative and reproductive structures in all higher plant families. They have evolved independently in a number of plant families. In general, monocotyledons are less pubescent (hairy) than dicotyledons, but trichome production can be induced in many glabrous (hairless or smooth) plant species. Several factors, including various light and temperature regimes, moisture availability and soil conditions affect the development and expression of trichomes. Trichomes are the first structure an insect encounters when landing on a plant, and provide the initial arena for complex and varied plant-insect interactions.

Types and functions of trichomes

The structure and function of trichomes are highly variably within and among plant species. There are two general types of trichomes: glandular, which produce, secrete or contain chemicals, and non-glandular, which are simple hairs and do not produce or contain chemicals. Trichomes have also been classified morphologically. Various morpho-types have been observed including: simple unicellular, multicellular uniseriate, multicellular multiseriate, 2 to 5 branched, stellate, dendritic or arboriform, and peltate. Several different classifications have been proposed and there is much variation and overlap in the description of trichome types. Developing a generalized classification system based on trichome morphology has not been possible because of their highly polymorphic nature.

Trichomes may have either or both physiological and defensive functions for the plant, though may be utilized for a different purpose by an insect. For example, a dense covering of hairs on a leaf may have a physiological role for the plant but may provide a preferred ovipositional surface for an insect. The following physiological functions have been identified for trichomes: altering optical properties of the leaf surface, deflecting solar radiation and thus reducing leaf temperature, maintaining water balance, acquiring nutrients and water from the atmosphere, and excreting excess salts. The focus of this discussion is insect/trichome interactions; therefore, the physiological functions of trichomes will not be considered further. Trichomes can affect both primary (herbivores) and secondary (predators and parasitoids) consumers.

The impact of trichome density, length, and orientation on insect behavior and performance has been well documented. Trichome density is perhaps more important than length or orientation but all three characteristics have been shown to impact insect behavior and/or performance. Wild relatives are often good sources of pubescence and have been utilized to develop insect resistant crop cultivars.

Trichomes as an ovipositional substrate

Oviposition has been positively correlated with increasing trichome density in many herbivorous species. For example, more eggs have been observed on trichome-dense cotton cultivars compared to smooth cotton cultivars for several insects including *Lygus hesperus*, *Helicoverpa* spp., *Earias fabia* and *E. vitella*. A similar situation has been noted for *Laspeyresia glycinivorella* and *Ophiomyia phaseoli* on soybean pods and leaves, respectively. Hairy surfaces can also positively alter egg and larval development and survival by ameliorating environmental factors such as increasing humidity levels. Conversely, trichomes inhibit oviposition in some insect species. At least one species of bruchid, *Callosobruchus chinensis*, prefers smooth as opposed to hairy pigeonpea (a tropical legume) pods for laying eggs. The cereal leaf beetle, *Oulema melanopus*, lays fewer eggs on wheat genotypes with highly pubescent leaves compared to genotypes with glabrous leaves. Both egg laying and larval survival are lower on cultivars with relatively longer and/or denser trichomes.

Impact of trichomes on herbivores

Trichomes interact with insect herbivores in many ways. The role of trichomes in providing defense against insect herbivores has been well documented and has been the topic of considerable investigation. Trichomes provide protection to the plant by acting as a physical barrier limiting an insect herbivore's contact with the plant, or as a chemical barrier by producing toxic compounds that poison the insect or by producing gummy, sticky or polymerizing chemical exudates which impede the insect. The effectiveness of the first mechanism, as a simple physical barrier to reaching the plant surface, is

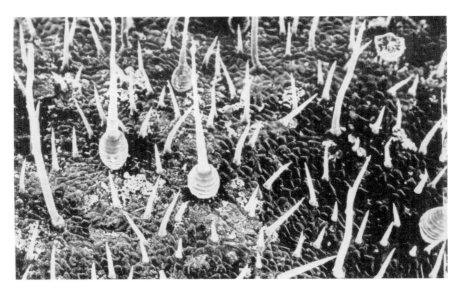

Fig. 1076 Trichomes on the surface of a pigeonpea (*Cajanus cajan*) pod. Three types of trichomes are visible: short and long non-glandular (simple hairs), and a multicellular glandular trichome with a bulb-like base. Photo credit: International Crops Research Institute for the Semi-Arid Tropics, Patancheru, A.P., India.

dependant on the length, density, and orientation of the trichomes, and on the insect's size, mode of locomotion, and type of mouthparts. In general, longer, denser and/or more erect hairs provide a better barrier to insect herbivores than shorter, sparser or recumbent hairs. As a purely physical phenomena, trichomes may also interfere with the insect digestion.

Wild relatives of cultivated plant species often have higher densities of trichomes than the cultivated relatives. A wild relative of pigeonpea possesses twice the density of small, non-glandular trichomes on pods as cultivated pigeonpea pods. When newly emerged pod borer (*Helicoverpa armigera*) larvae are placed on pods of both species, mortality is increased by 85% relative to the cultivated pigeonpea. The high density of non-glandular trichomes prevent the larvae from reaching the pod surface and they starve or desiccate before feeding.

Trichomes have been used to develop soybean cultivars with resistance to several insect pests. Soybean leaves possess simple non-glandular trichomes and cultivars vary in their degree of pubescence. Feeding and survival of at least four different caterpillars and two species of beetles were lower on pubescent leaves than on smooth leaves. Trichome length and shape are also important, especially for small-bodied insects such as the potato leafhopper. Potato leafhopper populations decreased with increasing trichome length, regardless of trichome density. The irregular shape produced by the highly pubescent accessions prevents potato leafhopper from attaching normally for feeding and oviposition.

Reduced movement of 1st instar larvae of pink bollworm on cotton has been observed on highly pubescent versus glabrous leaves. Larvae pause regularly to swing their heads and sample the substrate while moving. This resulted in higher larval mortality due to starvation or desiccation. The larvae of *Heliothis virescens* are also more likely to die from biotic (predators and parasites) and abiotic (high temperature and insecticides) factors because of slower movement on pubescent genotypes.

In contrast to non-glandular trichomes, glandular trichomes can act as both physical and chemical barriers to insect herbivores. The fluidity and volume of the exudates, whether toxic or sticky, may vary with weather, time of day and age of the plant. There is also tremendous diversity in the structure of glandular trichomes and the composition of glandular exudates.

Glandular trichomes confer resistance to several phytophagous insects and, because of the toxic and deterrent properties of their exudates, are often more effective than non-glandular trichomes in providing resistance to insect pests. Glandular trichomes secrete and/or accumulate a variety of compounds that generally act either as insect repellents or immobilize insects by entrapment. In addition to protecting plants by deterring herbivores, glandular trichomes may also attract pollinators; these structures contribute to the flavor and aroma of many plants.

The commercial potato, *Solanum tuberosum*, and its wild relatives, *Solanum* spp., possess two types of glandular trichomes. Type A is short, globular, and releases exudates after rupturing; Type B is longer, more hair-like, and continuously exudes a viscous fluid. After landing, an insect first encounters Type B exudates. In struggling to escape from the sticky coating, the insect disturbs and ruptures Type A trichomes. The exudates in Type A trichomes combine with the Type B exudates to harden the sticky exudate. Thus, small bodied insects such as aphids and leafhoppers are trapped and starve to death. The Type B exudates also increase the mortality rate and inhibit settling and probing behavior of some aphids. These trichomes provide greater resistance to wild *Solanum* spp. than in the cultivated *S. tuberosum*. In general, wild species are more pubescent and have more glandular trichomes, and are hence more resistant to insect pests than the cultivated relatives.

Another type of glandular trichome are those found on chickpea. They secrete highly acidic (pH = 1) exudates, containing primarily malic and oxalic acids. The quantity of malic acid in the exudate reportedly confers resistance to the leafminer, *Liriomyza cicerina*, and the pod borer, *H. armigera*, though others have reported an antibiotic effect of oxalic acid, but not malic acid.

Impact of trichomes on parasitoids and predators

Trichomes can have a direct negative impact on the predators and parasitoids that attack insect herbivores. The efficacy of these natural enemies can be impaired by: (1) increasing search time, (2) decreasing residence time in the host/prey habitat, (3) chemical and/or physical entrapment, and (4) chemical repellents. As with herbivores, small bodied natural enemies are generally more affected by trichomes than larger ones.

The time parasitoids and predators spend searching for hosts/prey is directly related to their walking speed; more hosts/prey are encountered when parasitoids and predators walk faster. Trichomes, even simple non-glandular hairs, interfere with movement and reduce walking speed. Faster walking speeds of several egg parasitoids (*Trichogramma* spp.) have been measured on smooth versus hairy cotton leaves and pigeonpea pods. In addition, parasitoids which walk on the plant surface can be slowed by exudates secreted by glandular trichomes. Predators are similarly hindered by glandular and non-glandular trichomes. The walking speed of coccinellid larvae are inversely related to the density of trichomes on potato and tomato.

On pigeonpea, eggs of an important pest (*Helicoverpa armigera*) are readily parasitized (greater than 55%) by an egg parasitoid (*Trichogramma chilonis*) when placed on leaves but are rarely attacked (less than 1%) when placed on pigeonpea pods. Parasitoids walked significantly faster on leaves than on pods, where their movement was inhibited by long trichomes. The higher density of glandular trichomes on pods compared to leaves resulted in the parasitoids being trapped by glandular exudates. The same pest (*H. armigera*) feeds on chickpea, but is never attacked by the egg parasitoid, *T. chilonis*. The highly acidic trichome exudates secreted by vegetative and reproductive parts of chickpea deter, and may entrap, the egg parasitoid. Other examples of *Trichogramma* spp. being trapped by sticky trichome exudates have been reported from tomato and potato.

The negative effects of glandular trichomes on natural enemies may not be as strong under field conditions as often observed in greenhouse or laboratory experiments. Glandular trichome exudates may be dried by the sun or wind, washed off by rain, or even be rendered ineffective by dust under field conditions.

Conclusions

Trichomes are found on vegetative and reproductive structures in many plant species. There are two types, non-glandular and glandular; both are found in a variety of shapes and forms. Trichomes and insects interact in numerous ways. Trichomes on a number of plants have evolved a defensive function and may protect plants from insect herbivores. Plants are protected from insect feeding when trichomes form a chemical or physical barrier, preventing herbi-

vores from reaching the surface. Trichomes may also provide a preferred oviposition site and/or may interfere with an insect's movement on the plant surface. Both positive and negative interactions, from the insect's point of view, have been documented. Generalizing about the role and function of trichomes and how these structures interact with insects is difficult because trichomes vary greatly across plant species and because the interactions may be specific or unique to the plant-insect association.

See also, PLANT RESISTANCE TO INSECTS.

Thomas G. Shanower
USDA-Agricultural Research Service
Sidney, Montana, USA

References
Gutschick, V. P. 1999. Biotic and abiotic consequences of differences in leaf structure. *New Phytologist* 143: 3–18.
Juniper, B. E., and R. Southwood. (eds.). 1986. *Insects and the plant surface*. Edward Arnold Publishers Ltd., London, United Kingdom. 360 pp.
Levin, D. A. 1973. The role of trichomes in plant defense. *Quarterly Review of Biology* 48: 3–15.
Peter, A. J., and T. G. Shanower. 2001. Role of plant surface in resistance to insect herbivores. pp. 107–132 in T. N. Ananthakrishnan (ed.) *Insects and plant defense dynamics*. Science Publishers, Inc., Enfield, New Hampshire.
Romeis, J., T. G. Shanower, and A. J. Peter. 1999. Trichomes on pigeonpea (*Cajanus cajan*) and two wild *Cajanus* spp. *Crop Science* 39: 564–569.

TRICHOMYCETES. The orders Amoebidiales, Eccrinales, Asellariales, and Harpellales, are generally non-pathogenic, entomogenous fungi found among the class Trichomycetes (Zygomycota). As in the Laboulbeniamycetes, the Trichomycetes are associated with their arthropod hosts as obligate commensals, i.e., the host provides a protected environment as well as nutrients for the fungus, but does not itself benefit (or suffer) from the relationship. In most cases these fungi are associated with aquatic insects. With few exceptions, these fungi attach to their hosts on the cuticle lining the digestive tract. Attachment of thalli is usually by a holdfast secreted by the fungus.

Amoebidium parasiticum (Amoebidiales) binds to the external surface of its freshwater hosts (e.g., mosquito larvae) rather than to the gut. Although not pathogenic, the life cycle of this fungus is closely coordinated with that of the host insect. During the intermolt period of the host, the thallus produces rigid asexual spores that may attach to other areas of the cuticle. Prior to molting or at death, production of amoeboid sporangiospores rather than rigid sporangiospores begins. The amoebae, which lack pseudopodia and do not engulf particulate material, swarm from the thallus, encyst, and produce cytospores. These then form new thalli upon contact with naive insects. It is significant that the amoeboid stage, which is host-independent and produces a resistant (cyst)-type structure, is initiated at a time when the host can no longer provide an ideal environment for the fungus.

Some members of the Harpellales have been described as potential pathogens. In this order, infection was thought to be restricted to digestive tracts of freshwater insect larvae and nymphs. Ingested zygospores or trichospores (asexual monosporous sporangia) germinate, attach to gut cuticle, grow, and sporulate in the digestive tract. Some workers have, however, reported the presence of Harpellales in pupae and adults and outside the host (Dipteran) gut in other tissues. Therefore, the thalli initially localized in the digestive tract have the capacity to invade other tissues, and a pathological relationship may exist.

TRICHOPSOCIDAE. A family of psocids (order Psocoptera). See also, BARK-LICE, BOOK-LICE, OR PSOCIDS.

TRICHOPTERA. An order of insects. They commonly are known as caddisflies. See also, CADDIS-FLIES.

Tricorythidae A family of mayflies (order Ephemeroptera). See also, MAYFLIES.

TRIDACTYLIDAE. A family of grasshoppers (order Orthoptera). They commonly are known as pygmy mole crickets. See also, GRASSHOPPERS, KATYDIDS AND CRICKETS.

TRIGONALIDAE. A family of wasps (order Hymenoptera). See also, WASPS, ANTS, BEES, AND SAWFLIES.

TRIMENOPONIDAE. A family of chewing lice (order Mallophaga). See also, CHEWING LICE.

TRINOTONIDAE. A family of whewing lice (order Mallophaga). See also, CHEWING LICE.

TRIOZIDAE. A family of bugs (order Hemiptera, superfamily Psylloidea). See also, BUGS.

TRITOCEREBRUM. The portion of the brain that innervates the labrum and stomatogastric nervous system, and the most posterior portion of the brain. See also, NERVOUS SYSTEM.

TRITROPHIC INTERACTIONS. Two central issues in ecology are: (i) how do organisms interact with their environment, and (ii) what interactions determine the composition and dynamics of communities. Ecologists have long debated the relative importance of bottom-up and top-down effects on communities. An example of a bottom-up effect is that nutrient availability to plants can affect the composition of insect communities and this can subsequently influence top-down forces. Top-down effects comprise, for instance, the effects of carnivores on herbivore populations with consequences for plant population biology. However, it is becoming more and more clear that this is not an either/or issue, but rather one of the degree to which bottom-up and top-down forces are integrated. Bottom-up forces, such as nutrient availability to plants, can affect the composition of animal communities and this can subsequently influence top-down forces. Furthermore, plants have many characteristics that influence the effectiveness of carnivores in reducing herbivore numbers, i.e., bottom-up and top-down forces can be linked.

Although it is long-known that food webs are composed of more than two trophic levels, it was Price and coworkers who, in 1980, pointed to the important effects of plants on interactions between herbivores and carnivores, and of carnivores on herbivore-plant interactions. They introduced the term 'three-trophic level interactions' or 'tritrophic interactions.' With this term, they indicated that apart from food web considerations where one trophic level feeds on another (lower) level, there are interactions between alternate trophic levels that can be of decisive importance to the outcome of interactions between two subsequent trophic levels. Thus, within

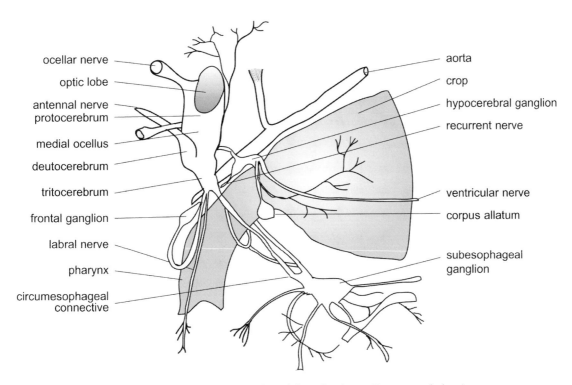

ocellar nerve

optic lobe

antennal nerve
protocerebrum

medial ocellus

deutocerebrum

tritocerebrum

frontal ganglion

labral nerve

pharynx

circumesophageal
connective

aorta

crop

hypocerebral ganglion

recurrent nerve

ventricular nerve

corpus allatum

subesophageal
ganglion

Fig. 1077   Diagram of the insect brain, lateral view (adapted from Snodgrass, Insect morphology).

a tritrophic context, indirect interactions in a community are considered in addition to direct interactions.

To obtain a sound ecological understanding of each bitrophic interaction (one between plants and herbivores and the other between herbivores and carnivores), it is essential to consider at least three trophic levels and all their interrelationships. For instance, plant defense may be aimed directly against herbivores, but, in addition, carnivores may be a component of the plant's battery of defense. Thus, there are two distinct kinds of plant defense: (1) direct defense where plant characteristics have a negative effect on herbivores, and (2) indirect defense that affects herbivores by promoting the effectiveness of their carnivorous enemies. Direct defense may consist of such elements as repellents, toxins or digestibility reducers, while indirect defense may consist of the provision of shelter, alternative food, or chemical information that promotes the abundance and/or activity of carnivores.

The majority of research has been done on plants and arthropods and, therefore, this discussion concentrates on these organisms. The arthropods involved are herbivorous arthropods and carnivorous arthro-

pods, i.e., predators and parasitoids (parasitic wasps). Tritrophic interactions between plants, herbivores and carnivores will be addressed. However, tritrophic interactions are not restricted to these three trophic levels; they may also encompass interactions between, for example, herbivores, first order carnivores and second order carnivores.

Plant characteristics that affect herbivores and indirectly affect natural enemies of the herbivores

Any plant condition that reduces the growth rate of an herbivore makes the herbivore available to natural enemies for a longer period of time and, thus, raises the probability of mortality. This is especially true for mortality from parasitoid attack, because parasitoids are usually stage-specific and, thus, prolongation of the vulnerable growth stage may be essential for the parasitoid to find that stage. For instance, digestibility reducers have growth-retarding effects. This theory was tested using several soybean varieties, the Mexican bean beetle (*Epilachna varivestis* Mulsant) as the herbivore and some predatory pentatomid bugs. It was found that the herbivores that grew more slowly were more effectively

Direct *versus* Indirect Defense of Plants

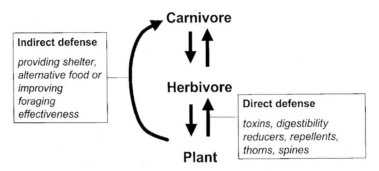

Fig. 1078 Direct and indirect defenses of plants in a tritrophic context.

regulated by the predator than those on the soybean variety that was more favorable for herbivore growth. In addition, it was found that the reduced growth rate of the herbivore also affects the functional response of the predator (the relationship between predation rate and prey density) because more small prey are needed for satiation than large prey, and small prey are usually easier to handle by the predator.

Digestibility reducers produced by plants may affect herbivore-natural enemy interactions in several ways:

(a) prolonged developmental time of herbivores and, thus, prolonged risks of attack by carnivores,

(b) reduced resistance of herbivores to pathogens and natural enemies, and

(c) reduced herbivore body size and, thus, fecundity, resulting in a lower rate of population increase. This may determine the ability of natural enemies to exterminate the herbivore population.

The influence on enemy effectiveness through a direct influence on herbivores is not limited to digestibility reducers. Any plant characteristic that can affect growth, resistance to disease, or fecundity can have an influence. Variation in plant nutrient composition is probably important in this regard. While such variation probably seldom involves complete presence or absence of essential nutrients, changes in their proportions appear to be commonplace. Effects that are sublethal should, therefore, also be commonplace. Negative effects of imbalanced diets on insect survivorship, growth, and

development, not surprisingly, have been demonstrated in controlled artificial diets for a number of insects and nutrients.

On the other hand, retardation of herbivore growth may hamper some predators or parasitoids. For instance, some predators do not take small prey because of a low energetic (nutritional) reward in combination with high capture costs. For example, an insectivorous bird, the European Great Tit, rejects small insect prey when large prey are available.

Another effect of growth rate and size results from the minimum size of an herbivore that is acceptable to its natural enemies; however, this critical size is different for different natural enemy species. Luck and colleagues at the University of California showed that host size is a very important factor in the coexistence of *Aphytis* parasitoid species attacking Californian red scale, *Aonidiella aurantii* (Maskell). *Aphytis lignanensis* needs larger hosts for female progeny than *Aphytis melinus* DeBach. Thus, *A. melinus* can attack hosts before they become available to *Aphytis lignanensis* Compere. In addition, host size depends on plant part. Host size is smallest on wood, largest on fruits and intermediate on leaves. Thus, many more scales are available to *A. melinus* than to *A. lignanensis*, and the degree of this difference is dependent on the location in the tree. As a consequence, *A. melinus* can effectively replace *A. lignanensis* in many citrus areas where red scale is a problem. It is simple to imagine that the three plant parts (wood, leaves and fruits), actually represent different 'cultivars' of a species with different levels of defense, and their different impacts on host-parasitoid interactions.

Retardation of growth can also result from induced resistance in plants. In many cases, plant defense is not active prior to herbivore infestation. The damage inflicted by herbivores may induce the defensive actions of plants. Thus, the production of phytoalexins in response to infestation by pathogens is well known. Such induced responses also may occur after an herbivore attack. For instance, tomato plants that are damaged, either by herbivores or by mechanical means, produce proteinase inhibitors that reduce the digestibility of the plant material. These proteinase inhibitors are produced systemically throughout the damaged plant. Induced resistance (or induced direct defense) appears to be rather nonspecific. Infestation with one species of herbivore often affects the performance of other herbivore species as well. Induced resistance lasts for some time after the herbivores are gone. It may last for hours or days, but in some cases (e.g., for perennial plants such as trees), induced resistance has been reported to last for one to several years. Thus, previous and current herbivory may affect the growth rate of other herbivores and, thus, their vulnerability to parasitoids and predators. In addition, mechanical characteristics of plants can be subject to induction also (e.g., trichomes of nettles).

Plant characteristics that directly affect the effectiveness of natural enemies of herbivores

Mutualisms between plants and animals have been known for a long time but, until the 1980s, the focus had been almost exclusively on mutualisms between plants and herbivores, e.g., involving pollination or seed dispersal. The best and longest (more than 100 years) known mutualism between plants and members of the third trophic level is that between *Acacia* plants and protective ants. The *Acacia* plant provides sugar-containing secretions from extrafloral nectaries and other macronutrients from food bodies on their stems, leaves, or buds. Some *Acacia* species even provide hollow stems, thorns, or bulbs that the ants penetrate to use as shelter for their colonies. The ants guard the source of nutrients and simultaneously defend the plant against herbivorous insects or even encroaching vegetation. Recently, it was shown that ants may even reduce the feeding by large herbivores such as giraffes. Also, less conspicuous structures, such as pits and pouches at vein cross-sections, can function as shelter for carnivorous enemies of herbivores. Such structures are called 'domatia' (Greek for 'little rooms') and are usually inhabited by predatory or fungivorous mites, whose activities can be important in reducing the abundance of plant attackers.

The protection of plants by ants is far more general than the well-known example of *Acacia* suggests. Many plant species have well-developed nectaries on stems, stipules, or leaves, or they secrete sugar-rich liquid from unopened flower or leaf buds. Such nectaries attract virtually any sugar-loving ants, most of which also eat insects. In temperate climates, seedlings and developing buds often secrete nectar, and are defended by ants. For example, seedlings of cherry trees (*Prunus* sp.) secrete nectar that encourages their defense by a variety of opportunistic ants.

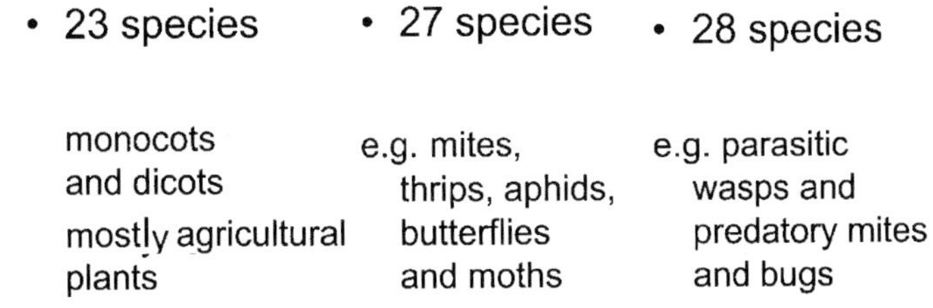

Herbivore-induced carnivore attractants: a *common* phenomenon

Plants	Herbivores	Carnivores
• 8 families	• 13 families	• 8 families
• 23 species	• 27 species	• 28 species
monocots and dicots	e.g. mites, thrips, aphids,	e.g. parasitic wasps and
mostly agricultural plants	butterflies and moths	predatory mites and bugs

Fig. 1079   The induction of carnivore attractants by herbivory has been reported for a wide variety of tritrophic systems.

Ants are generalist protectors that can defend a plant against a wide variety of herbivores. However, recent evidence shows that mutualisms between plants and members of the third trophic level also involve much more specialized predators and parasitoids. When a plant is attacked by an herbivore, the plant starts to produce and emit volatiles that attract natural enemies of the herbivore. This appears to be a common phenomenon among plants and many different herbivore species have been shown to induce carnivore attractants in plants. Herbivory is not always necessary. Hilker and colleagues of the Free University of Berlin showed that even the deposition of herbivore eggs can induce a plant to emit induced volatiles that attract egg parasitoids. It has been shown for several tritrophic systems that the natural enemy discriminates between different plant species infested by the same herbivore species and between different herbivore species on individuals of one plant species. Chemical analyses have shown this specificity on the molecular level. The composition of the blend is determined to the greatest extent by the plant species/cultivar. Not only natural enemies of the herbivore seem to use the herbivore-induced synomone, herbivores may respond also; spider mites avoid locations with high concentrations of induced plant volatiles that are related to conspecific damage. In addition, for a tritrophic system consisting of lima bean or cotton plants, spider mites and predatory mites, it was recorded that downwind, uninfested neighbors of infested plants are more attractive to predatory mites than upwind neighbors or other control plants.

Apart from mutualistic interactions, plants also may hamper the performance of natural enemies. For instance, a thick bark hinders parasitoids that attack herbivores that forage under the bark. The same holds true for the effect of gall thickness on parasitoids of gall-forming herbivores. Hairs on stems and leaves may affect the locomotion and the survival of natural enemies of herbivores (in addition to adverse effects on herbivore performance). For instance, cucumber leaves have many long, non-flexible hairs that interfere with parasitoid movement. In addition, herbivore products, such as honeydew, may adhere to these hairs, which enhances the adverse effects on parasitoid performance. The capability of the parasitoid *Encarsia formosa* Gahan to reduce populations of the greenhouse whitefly (*Trialeurodes vaporariorum*) on cucumbers depends on the density of these hairs. Also, trichomes that hamper herbivore survival may have strong adverse effects on the survival of natural enemies. Hairs, usually seen as a component of direct defense against herbivores thus may be incompatible with natural enemies that are a component of indirect defense.

Effect of dietary specialization

An animal's degree of dietary specialization is an important characteristic frequently used in comparative ecological and evolutionary studies. This concept also has been applied to herbivores and their natural enemies, although knowledge on herbivore range used by natural enemies is generally sparse. Dietary specialization of herbivores is assigned to the plant level, while dietary specialization of natural enemies can be assigned to both the herbivore and the plant level.

Specialist herbivores are usually better adapted to plant defenses, such as toxins, than are generalized herbivores. Specialist herbivores can usually tolerate higher levels of toxins and can sequester the defensive plant chemicals. As a result, they usually are much better protected against their natural enemies such as pathogens, predators and parasitoids. For example, consider the interactions between the nicotine in tobacco, the specialist herbivore *Manduca sexta* L. (tobacco hornworm), the generalist herbivore *Trichoplusia ni* (Hübner) (cabbage looper), and a pathogenic bacterium, *Bacillus thuringiensis*. The generalist herbivore is adversely affected by the nicotine in the plant, while the nicotine has only minor effects on the fitness components of the specialist herbivore. Higher nicotine concentrations in the plant are actually beneficial to the specialist herbivore because infection by the pathogen *B. thuringiensis* is reduced. Among predators and parasitoids, the specialists also are better adapted to the defenses of their prey/host. Thus, specialist herbivores are better protected against generalist natural enemies than against specialist natural enemies. This example shows that tritrophic interactions also may extend to pathogens of herbivores. The following example relates to the dietary specialization of parasitoids. Nicotine may be sequestered by *Spodoptera frugiperda* (J. E. Smith) (the fall armyworm), and adversely affects the development and survival of the polyphagous wasp parasitoid, *Hyposoter annulipes* (Cresson). In contrast, the specialist wasp parasitoid, *Cotesia congregata*

(Say), is much less affected by the sequestered nicotine.

The first step in the chain of events leading to interactions between an herbivore and its predators or parasitoids is the location of an herbivore-containing habitat from a distance by its natural enemies. Predators and parasitoids are faced with a great variety of stimuli they may use to locate their victim, even when considering only one sensory modality, i.e., chemoreception. Both plants and herbivores produce odors and, thus, potential information. The appropriateness and usability of information ultimately depends on two factors: (1) its reliability in indicating herbivore presence, accessibility and suitability, and (2) the degree to which stimuli can be detected. It is assumed that the use of information that is both reliable and easy-to-detect enhances searching efficiency and consequently, Darwinian fitness.

Stimuli derived from the herbivore itself are generally the most reliable source of information. Ideally, the infochemicals should reliably tell natural enemies whether a herbivore is present, to which species it belongs, whether there is more than one individual and, if so, how many, whether it is suitable to be parasitized or eaten, and whether it is readily accessible or hidden. For parasitoids, whose host continues to develop, and for predators that search for an oviposition site, there is the additional need to know whether the food plant is suitable for the development of their offspring. In fact, the more intimate the herbivore-carnivore interaction, the more specific the information that the carnivore needs.

However, the use of herbivore-derived stimuli often is limited by low detectability, especially at longer distances. Stimuli from plants, on the other hand, are usually more readily available because of the plants' relatively larger biomass, but are less reliable predictors of herbivore presence. Reliability of plant cues depends on the predictability of plant infestation over space and time. So, natural enemies are challenged to combine the advantageous characteristics of information from both trophic levels. It has been hypothesized that detectability of stimuli from the second trophic level puts a major constraint on the evolution of herbivore location by their natural enemies. Natural enemies may approach the reliability-detectability problem in three ways: (1) by resorting to the use of infochemicals from herbivore stages different from the one under attack, that are more conspicuous in their infochemical release (infochemical detour), (2) by focusing their responses to herbivore-induced plant volatiles (HIPV) that originate from specific interactions of the herbivore and its food, and (3) by learning to link easy-to-detect stimuli (plant cues) to reliable but hard-to-detect stimuli (herbivore cues). Which of these options is/are employed is dependent on the ecological context in which the parasitoid has evolved, and the degree of dietary specialization is an essential aspect of this.

By using the comparative approach, one can speculate whether and, if so, how the natural enemy species' use of herbivore-induced synomones correlates with their dietary specialization. The more natural enemies are specialized at a certain trophic level, the more they innately use infochemicals from that trophic level. In contrast, polyphagy at a trophic level is expected to result in more plastic responses to infochemicals from that trophic level. Thus, it is hypothesized that: (a) natural enemies that are specialists on specialist herbivores have strong fixed responses to kairomones from their victims, to uninduced plant volatiles and to HIPV, (b) natural enemies that attack several herbivore species on one host plant species have fixed responses to general kairomone components shared by the different herbivores and strong fixed responses to uninduced plant volatiles and HIPV, (c) natural enemies that are specialists on a polyphagous herbivore have strong fixed responses to kairomones and have to learn to respond to plant volatiles and HIPV, and (d) natural enemies that are extremely polyphagous on polyphagous herbivores are not expected to use infochemicals in finding their prey or hosts. These four categories are extremes in a continuum. The ecological settings of many natural enemies are intermediates along the continuum. The way these species use infochemicals will depend on the relative importance of the diet breadth at each trophic level.

Relative importance of plant and natural enemies to herbivore fitness

Herbivorous arthropods are affected by both their food plants and by predators and parasitoids. They have to deal with plant defenses such as toxins and digestibility reducers, but also with predator attacks. In general, it seems that there is a wide variation in effects of plants and natural enemies on the fitness of herbivores. This has led to speculation on which effects are the most important: 'top-down effects' (effects of natural enemies) or 'bottom-up effects' (effects of plants). The answer to this question is not likely to receive a general answer. It seems that the relative importance of the two effects is dependent on system-specific characteristics and the two effects are integrated, rather than alternatives.

The discussion about top-down and bottom-up effects is remarkable with respect to herbivore specialization. Two main selective forces have been mentioned: plant secondary chemicals have long been assumed to be the selective force determining the high degree of specialization among herbivores. More recently, natural enemies of herbivores have been mentioned as another important selective.

The role of toxic plant secondary chemicals has centered on the so-called 'arms race' of plants and herbivores. Less than 10% of all herbivorous insects are polyphagous species that feed on plants in more than three different plant families. The high degree of specialization of insect herbivores usually is explained by the existence of secondary plant chemicals. However, this has received criticism by L. Bernays and colleagues at the University of Arizona who argue that generalist natural enemies, especially predators of the herbivores, may be an important factor in the evolution of narrow host range. Bernays and colleagues argue that oviposition choice and/or host utilization ability can be significantly altered within only 10 to 16 generations, which implies that current patterns of host use should probably be seen as ecologically dynamic, and not as an end result of co-evolutionary processes over millennia. Moreover, host shifts by insects with a restricted host range often are to unrelated plant species, suggesting that behavioral barriers do not simply reflect a need to avoid toxins. Plant secondary chemicals tend to provide the proximate, mechanistic bases for narrow host range, but do not necessarily provide the ultimate, functional basis for the patterns presently seen. An important aspect in their arguments is the concept of 'enemy-free space,' which indicates that, on some plants, an herbivore is not, or is less, affected by natural enemies compared to other plant species. Within a given place and time, preference for a plant less likely to be visited by natural enemies could evolve rapidly. Specializations for continued avoidance of predators also may then be rapidly selected and established. For example, it is self-evident that the most effective insect crypsis is developed on specific substrates, and the most

celebrated instances of sequestration of compounds toxic to vertebrates from plants are reported for insects with specialized feeding habits. Both of these specializations are predator related. Because many parasitoids are relatively restricted in host range, or restricted in the ranges of plants searched, they might be more important in causing host switches or broadening herbivore host range. This could leave the predator element as the most important factor in pushing insect herbivores toward narrow host range and the specializations that can then follow.

The plant's perspective: a paradox in its defense against herbivores and the answer in a tritrophic context

The previous section considered the herbivore's viewpoint of being in between plant defense and natural enemy attack. When taking the plant's perspective of defense against herbivores, a bitrophic view may yield a paradox. If plants affect herbivore fitness by reducing the food quality through digestibility reducers or nutrient composition (e.g., low nitrogen content), the herbivores will respond by inflicting more damage to compensate (e.g., by prolonged feeding, or more intense feeding). Natural enemy attack may be the essential factor needed to turn digestibility reducers into a positive trait. Digestibility reducers result in longer duration of herbivore stages. This may affect attacks by parasitoids that are usually rather restrictive as to the herbivore instar they attack, or attacks by (insect) predators that consume more small prey than large prey. However, experimental evidence for this often-mentioned scenario is still limited.

Plants also may face another dilemma. If the secondary chemicals the plants produce are sequestered by specialist herbivores, that means the herbivores exploit the plant's defense for their own protection against carnivores, and the plant faces a net ecological cost. Many plants only start the (increased) production of these secondary metabolites in response to herbivory. A recent example shows that the induction can be dependent on the herbivore that feeds on the plant. Wild tobacco plants initiate nicotine production in response to wounding by such factors as rabbit feeding or mechanical wounding. However, Kahl and colleagues at the Max Planck Institute for Chemical Ecology showed that when caterpillars of a specialist herbivore, i.e., tobacco hornworm (*M. sexta*), feed on the plant, the plant does not initiate a high nicotine biosynthesis, but rather, initiates the production of volatiles that attract carnivorous enemies of the specialist herbivore.

Hidden players

Currently, tritrophic interactions are mostly considered for above-ground interactions of plants, and herbivorous and carnivorous arthropods. However, it is increasingly apparent that microorganisms play an important role in communities in general, and in plant-insect communities in particular. Microorganisms may compete with insects for food, and microorganisms may be symbionts or pathogens of insects. Therefore, all interactions in tritrophic systems can be affected by microorganisms. Research in this area is rapidly increasing.

Moreover, most knowledge on tritrophic interactions relates to above-ground systems, especially because they are more easily investigated than below-ground interactions. However, above-ground and below-ground interactions cannot be considered individually because they are intimately linked. Intensified research on below-ground interactions, as well as their link to above-ground interactions, is foreseen for the near future.

Theories on plant defense

Classical plant defense theory describes the evolution of secondary plant substances in direct defense against herbivores as the result of an arms race between plants and herbivores: plants evolve secondary metabolites in response to attacks by insects, while insects meet the challenge by evolving new detoxification systems. In the 1970s, P. Feeny of Cornell University, and D. Rhoades and R. Cates at the University of Washington, developed a theory that centers on plant 'apparency,' which emphasizes the herbivore's perspective. A third plant defense theory is the 'resource availability theory,' developed by P. Coley and colleagues at the University of Utah. This theory emphasizes the plant's perspective and centers around the possibilities of defense as affected by the availability of nutrients and energy.

The relatively young field of three-trophic-level interactions has not been the subject of many theoretical considerations. Yet, Price at Northern Arizona University has integrated the plant apparency theory with the actions of natural enemies (indirect defense). He argues that many of the broad patterns relating to natural enemies discovered to date relate to gradients

from herbs to shrubs to trees, from unstable to stable environments, and from herbivores with external to concealed feeding habits. When using plant succession as a starting point, and concentrating on parasitoids as natural enemies, Price notes five patterns:

(1) With plant succession, the number of parasitoid species per host species increases.

(2) With plant succession, the prevalent parasitoid species becomes more generalistic.

(3) With an increase in the number of parasitoid species per host species, host mortality increases; this has not been well-studied.

(4) With an increase of the number of parasitoid species per host species, the probability increases that herbivore populations are regulated.

(5) Host-specific parasitoids (that are more often present in early succession stages) have innate and strong responses to kairomones and plant volatiles, whereas generalistic species (more frequently present in later successional stages) have more plastic responses that are subject to learning.

Price then theorizes on the differences in plant defenses in early and late succession. In early succession, plants are unapparent to herbivores and defend themselves through specific toxins. Thus, host location by parasitoids is constrained and the parasitoids need specialized enzymes to cope with the sequestered toxins in the herbivores. Thus, selection will favor specificity of the parasitoids. It is exactly host-specific parasitoids that are predicted to use specific synomones and kairomones in host location in a genetically fixed pattern. In late succession, plants are apparent and defend themselves in similar ways (convergence) through digestibility reducers. This increases the chance that an herbivore will include a new plant species in its diet; herbivores are less plant-specific. Parasitoid species may be more generalistic than in early successional stages. For these parasitoids, the recognition of, and innate response to, a large number of specific synomones and kairomones is less likely (physiological constraint) and, thus, learning will be more important.

Applied aspects

The importance of considering interactions between insects and their food and enemies in a multitrophic context has been realized since the beginning of the 1980s. Thus, the field of tritrophic interactions is a relatively young one. However, developments in this field are quick and the knowledge gained is not only important from a scientific point of view, but also from an applied point of view. For instance, in environmentally benign pest control, herbivores can be controlled through plant breeding for resistance and through biological control. The combination of these two methods may be explicitly exploited, or may occur without special intervention, because naturally occurring, carnivorous arthropods often play an important, but unnoticed, role in the reduction of pest populations. Therefore, it is important that these two methods are synergistic rather than antagonistic. However, synergistic interaction between host plant resistance and biological control is by no means self-evident. If certain plant characteristics are selected because they have a negative influence on herbivores in the absence of biological control agents, this does not automatically mean that the plant contributes to pest control in the presence of biological control agents. A remarkable example was recorded for cabbage plants. Two cabbage varieties differed in resistance against the cabbage aphid, *Brevicoryne brassicae* (L.), which resulted in reduced aphid populations when parasitoids had been excluded. However, the susceptible cultivar emitted larger quantities of a parasitoid-attracting volatile and, therefore, under conditions where biological control was applied, aphid densities were lower on the susceptible variety than on the resistant variety.

Along similar lines, current developments should be considered in the field of transgenic insect-resistant plants. So far, the main emphasis is on the effects of the transgenic plants on pest insects, without much emphasis on tritrophic effects. However, an important question is: Do plants that have been engineered to express certain toxins affect non-target organisms, including non-target herbivores and carnivorous arthropods that attack target and/or non-target herbivores?

Agricultural plants have to produce under conditions where they are members of a community. Agroecosystems harbor a diverse community – even when they are usually less diverse than the community in a natural ecosystem – and if important carnivores are eliminated or hampered by introducing crop plants that have different characteristics, the result may be that pest control is hampered rather than improved. This is a lesson that is older than the research on tritrophic interactions. The dusting of crop plants with broad-spectrum pesticides in the 1960s and 1970s can be seen as an artificial alteration of plant characteristics that had severe impacts on pest control, where even novel pests were induced as a result of the elimination of carnivorous arthropods.

Conclusion

Interactions in communities comprise direct and indirect interactions. Characteristics of community members can have effects on interactions with their direct enemies or resources, but also on non-producer/consumer interactions. Indirect interactions are by no means less important than direct interactions between producers and consumers. Indirect interactions can decisively influence producer-consumer interactions. Tritrophic interactions show that the debate on the role of bottom-up versus top-down forces in shaping communities is not an either/or debate, but rather one of the degree of integration of the two forces. Considering tritrophic interactions is not only interesting from a basic point of view, it is also important from an applied point of view (for example, in developing novel methods of environmentally benign pest control). Finally, the study of tritrophic interactions is not an end-point but only a beginning. After initiating studies on interactions among plants, herbivores and carnivores (not exclusively insects), it is now clear that microorganisms can play important roles in these interactions. Moreover, second-order carnivores and interactions with organisms that cannot be exclusively linked to a

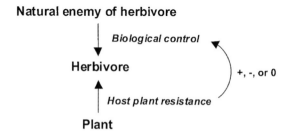

Fig. 1080 Durable and environmentally benign pest management can incorporate host plant resistance and biological control. Host plant characteristics may affect biological control positively, negatively, or not at all. Which of these interactions between host plant resistance and biological control occurs can be important for the success of environmentally benign pest control.

single trophic level, such as omnivores, should be included as well. Therefore, the study of tritrophic interactions is evolving into the research area of multitrophic interactions.

Marcel Dicke
Wageningen University
Wageningen, The Netherlands

References

Barbosa, P. 1988. Natural enemies and herbivore-plant interactions: influence of plant allelochemicals and host specificity. pp. 201–229 in P. Barbosa and D. K. Letourneau (eds.), *Novel aspects of insect-plant interactions.* Wiley and Sons, New York, New York.

Bernays, E., and M. Graham. 1988. On the evolution of host specificity in phytophagous arthropods. *Ecology* 69: 888–892.

Chadwick, D. J., and J. A. Goode (eds.). 1999. Insect-plant interactions and induced plant defense. Novartis Foundation Symposium 223. Wiley, Chichester, United Kingdom.

Dicke, M. 1999. Direct and indirect effects of plants on performance of beneficial organisms. pp. 105–153 in J. R. Ruberson (ed.), *Handbook of pest management.* Marcel Dekker, New York, New York.

Dicke, M. 1999. Evolution of induced indirect defense of plants. pp. 62–88 in R. Tollrian and C. D. Harvell (eds.), *The ecology and evolution of inducible defenses.* Princeton University Press, Princeton, New Jersey.

Groot, A. T., and M. Dicke. 2002. *Insect-resistant transgenic plants in a multitrophic context.* Plant Journal 31: 387–406.

Hilker, M., and T. Meiners. 2002. Induction of plant responses towards oviposition and feeding of herbivorous arthropods: a comparison. *Entomologia Experimentalis et Applicata* 104: 181–192.

Karban, R., and I. T. Baldwin. 1997. *Induced responses to herbivory.* Chicago University Press, Chicago, Illinois.

Price, P. W., C. E. Bouton, P. Gross, B. A. McPheron, J. N. Thompson, and A. E. Weis. 1980. Interactions among three trophic levels: influence of plant on interactions between insect herbivores and natural enemies. *Annual Review of Ecology and Systematics* 11: 41–65.

Price, P. W. 1991. Evolutionary theory of host and parasitoid interactions. *Biological Control* 1: 83–93.

Turlings, T. C. J., J. H. Loughrin, P. J. McCall, U. S. R. Rose, W. J. Lewis, and J. H. Tumlinson. 1995. How caterpillar-damaged plants protect themselves by attracting parasitic wasps.. *Proceedings of the National Academy of Sciences of the United States of America* 92: 4169–4174.

Van der Putten, W. H., L. E. M. Vet, J. A. Harvey, and F. L. Wackers. 2001. Linking above- and belowground multitrophic interactions of plants, herbivores, pathogens, and their antagonists.. *Trends in Ecology & Evolution* 16: 547–554.

van Emden, H. F. 1986. The interaction of plant resistance and natural enemies: effects on populations of sucking insects. pp. 138–150 in D. J. Boethel and R. D. Eikenbary (eds.), *Interactions of plant resistance and parasitoids and predators of insects.* Ellis Horwood, Chichester, United Kingdom.

Vet, L. E. M., and M. Dicke. 1992. Ecology of infochemical use by natural enemies in a tritrophic context. *Annual Review of Entomology* 37: 141–172.

TRIUNGULIN. The active first instar of stylopids (Strepsiptera) and certain beetles (Coleoptera) that undergo hypermetamorphosis, becoming less active after they find a host and commence feeding. (contrast with planidium)

TRIUNGULOID LARVA. A larval body form with minute, active, spiny first instars, and found in certain predaceous beetles and stylopids.
See also, TRIUNGULIN.

TROCHANTER. A small leg segment found between the coxa and femur. The second segment of the leg.
See also, LEGS OF HEXAPODS.

TROCHANTIN. The basal part of the trochanter when it is subdivided. In Coleoptera and some other groups, it refers to a structure present on the outer side of the coxa.

TROCHILIPHAGIDAE. A family of chewing lice (order Mallophaga). See also, CHEWING LICE.

TROCTOPSOCIDAE. A family of psocids (order Psocoptera). See also, BARK-LICE, BOOK-LICE, OR PSOCIDS.

TROGIDAE. A family of beetles (order Coleoptera). They commonly are known as skin beetles. See also, BEETLES.

TROGIIDAE. A family of psocids (order Psocoptera). See also, BARK-LICE, BOOK-LICE, OR PSOCIDS.

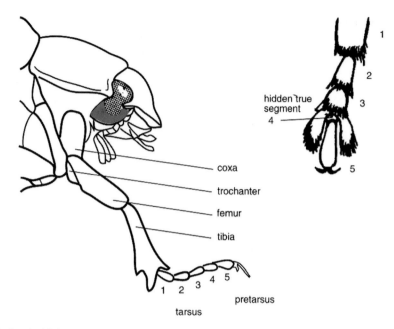

Fig. 1081 Leg of a beetle (Coleoptera: Scarabaeidae) leg showing its component parts, and a close-up of one type of beetle tarsus (foot).

TROPHALLAXIS. A behavior found in social insects involving the exchange of food between members of the colony, or among colony members and guests. The food may originate from the mouth (stomodeal trophallaxis) or from the anus (proctodeal trophallaxis).

TROPHIC EGG. A degenerate egg that is inviable, and used to feed other members of the colony.

TROPHIC LEVEL. The classification of organisms in a community according to their feeding relationships. The first trophic level usually is green plants. The second tropic level includes herbivores, and insects are a principal component. However, insects also are carnivores, parasites, and detritivores, so they appear in various tropic levels.

TROPHIC STRUCTURE. Organization of a community in terms of energy flow through the constituent trophic levels.

TROPHOGENIC POLYMORPHISM. Polymorphism in social insects due to differential feeding of the larvae, or different sized eggs.

TROPICAL BURNET MOTHS (LEPIDOPTERA: LACTURIDAE). Tropical burnet moths, family Lacturidae, total 138 species, mostly Indo-Australian but with a few in the southern United States; actual world fauna probably exceeds 250 species. The family is in the superfamily Sesioidea in the section Tineina, subsection Sesiina, of the division Ditrysia. Adults small to medium size (11 to 65 mm wingspan), with head scaling average; haustellum naked; labial palpi upcurved; maxillary palpi small, 1 to 2-segmented; antennae filiform. Wings elongated. Maculation usually colorful, especially with orange, red and yellow colors, and with various bands or spots, and with darker hindwings. Adults nocturnal but some may be crepuscular. Larvae are leaf skeletonizers and are colorful, but most are not known biologically; somewhat slug-like, with concealed head. Host plants include families Celastraceae, Moraceae, and Sapotaceae.

Trophic relationships of some insect orders

Primary consumers(herbivores)	Secondary consumers(insectivores and carnivores)	Decomposers
Coleoptera (some)	Coleoptera (some)	Coleoptera (some)
Diptera (some)	Diptera (some)	Diptera (some)
Embiidina	Ephemeroptera (some)	Ephemeroptera (some)
Grylloblattodea	Hemiptera (some)	Isoptera
Hemiptera (some)	Hymenoptera (parasitic and provisioning wasps)	Plecoptera (some)
Hymenoptera (sawflies, horntails, some bees)	Mallophaga	
Mecoptera	Mantodea	
Lepidoptera	Megaloptera	
Orthoptera	Neuroptera	
Phasmatodea	Odonata	
Thysanoptera (most)	Plecoptera (some)	
	Raphidioptera	
	Siphunculata	
	Thysanoptera (some)	
	Trichoptera	

John B. Heppner
Florida State Collection of Arthropods
Gainesville, Florida, USA

References

Busck, A. 1913. Notes on the genus *Mieza* Walker, with descriptions of three new species from Costa Rica. *Insecutor Inscitiae Menstruus* 1: 70–73.

Common, I. F. B. 1990. Family Zygaenidae (burnets, foresters) [part]. pp. 294–299. In *Moths of Australia*. Melbourne University Press, Carlton.

Heppner, J. B. 1995. Lacturidae, new family (Lepidoptera: Zygaenoidea). *Tropical Lepidoptera* 6: 146–148.

TROPICAL CARPENTERWORM MOTHS (LEPIDOPTERA: METARBELIDAE). Tropical

carpenterworm moths, family Metarbelidae, include 103 species, mainly Afrotropical and Oriental, with one species in the Palearctic region; actual world fauna likely exceeds 150 species. The family is in the superfamily Cossoidea (series Cossiformes) in the section Cossina, subsection Cossina, of the division Ditrysia. Some recent views place the group as a subfamily of Cossidae. Adults small to medium size (21 to 48 mm), with head small and rough scaled; labial palpi upcurved; maxillary palpi minute; antennae short and bipectinate; legs short. Body robust. Wings elongated and rounded (frenulum-retinaculum usually absent). Maculation mostly shades of brown and gray, even nearly white background, with dark spotting (often also with a preapical darker forewing

discal spot); hindwings darker. Adults may be crepuscular. Larvae nocturnal borers on tree bark or in tree trunks, but most species remain unknown biologically. Host plants include various trees in Anacardiaceae, Guttiferae, Lauraceae, Leguminosae, Myrtaceae, Rhamnaceae, Rutaceae, Sapindaceae, and Sterculiaceae. A few have minor economic status.

John B. Heppner
Florida State Collection of Arthropods
Gainesville, Florida, USA

References

Dalla Torre, K. W. von, and E. Strand. 1923. Lepidarbelidae. In *Lepidopterorum Catalogus*, 28:1–14 [part]. W. Junk, Berlin.

Holloway, J. D. 1986. Family Metarbelidae. In *Moths of Borneo*, 1:42–46, pl. 4. Malayan Nature Society: Kuala Lumpur (Malayan Nature Journal, 40: 42–46, pl. 4).

Janse, A. J. T. 1925. A revision of the South African Metarbelidae. *South African Journal of Natural History* 5: 61–100, pl. 4–8.

Seitz, A. 1929. Familie: Indarbelidae. In *Die Gross-Schmetterlinge der Erde*, 10: 803–806, pl. 93 (1933); 14: 501-513, pl. 78 (1929). A. Kernen, Stuttgart.

Srivastava, A. S. 1962. A preliminary study of the life-history and control of *Indarbela quadrinotata* Wlk. (Metarbelidae: Lepidoptera). *Proceedings of the National Academy of Sciences of India* 328: 265–270.

TROPICAL ERMINE MOTHS (LEPIDOPTERA: ATTEVIDAE). Tropical ermine moths, family Attevidae, include 48 species, mostly tropical

Fig. 1082 Example of tropical ermine moths (Attevidae), *Atteva niveigutta* Walker from India.

and in the genus *Atteva* (the single partially non-tropical species known occurs across the southern United States and into the Caribbean and Mexico); actual fauna probably at least 60 species. The family is part of the superfamily Yponomeutoidea in the section Tineina, subsection Tineina, of the division Ditrysia. Adults small to medium size (20 to 34 mm wingspan), with head smooth-scaled; haustellum naked; labial palpi upcurved; maxillary palpi 1 to 2-segmented. Wings elongated with termens rather rounded. Maculation usually colorful (many with red or orange forewings), with various spots on the forewings. Adults diurnal or crepuscular. Larvae are leaf webbers and leaf skeletonizers on Araliaceae and Simaroubaceae. Minor economic species occur on *Ailanthus* trees in India and the United States.

John B. Heppner
Florida State Collection of Arthropods
Gainesville, Florida, USA

References

Clarke, J. F. G. 1965. *Atteva* Walker. In J. F. G. Clarke, Hyponomeutidae [sic], in *Catalogue of the type specimens of Microlepidoptera in the British Museum (Natural History) described by Edward Meyrick,* 5: 292–293. British Museum (Natural History), London.

Mathur, R. N. 1960. Setal arrangement of *Atteva fabriciella* Swederus (Yponomeutidae, Lepidoptera). *Indian Journal of Entomology* 21: 1–5.

Powell, J. A., J. A. Comstock, and C. F. Harbison. 1973. Biology, geographical distribution, and status of *Atteva exquisita* (Lepidoptera: Yponomeutidae). *Transactions of the San Diego Society of Natural History* 17: 175–186.

Taylor, O. R. 1967. Relationship of multiple mating in *Atteva punctella* (Lepidoptera: Yponomeutidae). *Annals of the Entomological Society of America* 60: 583–590.

TROPICAL FRUIT PESTS AND THEIR MANAGEMENT. Tropical fruits are important in several production areas of south and southeast Asia, Australia, Africa, the Mediterranean, the Americas and the Caribbean region. Tropical fruits are regularly grown in different climates from latitude 23°27′ north to 23°27′ south of the equator, while some are grown to approximately 37° north in Spain. Proximity to the sea, sea currents, altitude, direction of prevailing winds, rainfall and air humidity all have modifying effects on these crops. Most tropical fruits are perennial plants which persist for several years without abrupt, major changes other than seasonal leaf formation, flowering, and fruit development. However, pineapple, papaya, and passion fruit are grown for shorter periods of time and their arthropod management is influenced by their persistence in the field.

In general, tropical fruit crops provide a relatively stable environment over many years, offering continuing habitats for both pests and natural enemies and providing opportunities for biological control and effective pest management programs. Avocado, mango, pineapple, banana, passion fruit, litchi, guava, *Annona* spp., durian, mangosteen, rambutan, acerola and carambola, and their most common pests are included here.

Annona fruits (Annonacea family)

The genus *Annona* embraces several valuable fruit trees and is centered in the Neotropics, with about 110 species. The most common species are *A. muricata* L., *A. montana* McFadden, *A. glabra* L., *A. cherimola* P. Miller, *A. squamosa* L., *A. reticulata* L., and *A. longiflora* Watts. Fruits of *Annona* are fleshy aggregates that arise from coalesced carpels and contain large seeds with reticulate endosperm.

Most species of *Annona* have specific climatic requirements for growth, flowering, and fruit maturation. The origins of the tropical species, such as the sugar apple (*A. squamosa*), are the warm lowland regions of Brazil, Guyana, Venezuela, Mexico, and the West Indies. A distinctly subtropical species is the cherimoya (*A. cherimola*), which originates from the cool Andean valleys of Peru and Ecuador at elevations of around 2,000 m. The hybrid atemoya (*A. cherimola* x *A. squamosa*), which appeared spontaneously when parent trees were cultivated side by side and also through manual pollination, exhibits intermediate climatic requirements for growth and fruiting.

Within the family Annonaceae, fruits of the cherimoya, sugar apple, atemoya, and soursop (*A. muricata*) have the greatest potential for utilization and export in the American, Caribbean, Asian and Australian countries. Much of the production of commercially grown species has spread from their indigenous areas to tropical and subtropical parts of Australia, New Zealand, Asia and around the Mediterranean. Thus, production methods vary from more sophisticated systems to less intensive, small farm or backyard type production. The key pests are the annona seed borers, *Bephratelloides* spp., *B. cubensis* (Ashmead), *B. pomorum*, *B. paraguayensis* (Crawford), and *B. petiolatus* Grissell and Schauff (Hymenoptera: Eurytomidae), destroying seeds and pulp; fruit borers, *Cerconota anonella* Sepp (Lepidoptera: Oecophoridae), *Talponia batesi* Heinrich (Lepidoptera: Tortricidae) and *Thecla ortyginus* (Lepidoptera: Lycaenidae); fruit flies, the Queensland fruit fly, *Bactrocera tryoni* (Froggatt), *Anastrepha* spp., and *Ceratitis capitata* (Diptera: Tephritidae); and recently, the pink hibiscus mealybug, *Maconellicoccus hirsutus* (Green), causing defoliation.

Avocado, *Persea americana* Mill. (Lauraceae family)

The avocado, *Persea americana* Mill. (Lauraceae), is of Central American origin and well known to the native inhabitants of Mexico, Central America and northern South America. Today, avocado is grown commercially not only in North America and throughout tropical America and the larger islands of the Caribbean, but also in Polynesia, the Philippines, Australia, New Zealand, Madagascar, Mauritius, Madeira, the Canary Islands, Algeria, tropical Africa, South Africa, southern Spain, southern France, Sicily, Crete, Israel and Egypt. The most important pests include the tea red mite, *Oligonychus coffeae* (Nietner), avocado brown mite, *Oligonychus punicae* (Hirst), the persea mite, *Oligonychus perseae* (Tuttle, Baker and Abbatiello), and the avocado red mite, *Oligonychus yothersi* (McGregor) (Acari: Tetranychidae), as well as the eriophyid mite *Tegolophus perseaeflorae* (Keifer) (Acari: Eriophyidae). Severe damage to the leaves can be produced by the psyllids *Trioza anceps* Tuthil and *Trioza perseae* Tuthill (Hemiptera: Psyllidae), and by the large blacktip wilter, *Anoplocnemis curvipes* F., injuring new flush,

and by the leaffooted bugs, *Amblypelta nitida* Stål and *A. lutescens lutescens* (Distant) (Hemiptera: Coreidae), attacking fruits. Other important species are the avocado lace bug, *Pseudacysta perseae* (Heidemann) (Hemiptera: Tingidae), causing leaf chlorosis and necrosis, the mirids, *Dagbertus fasciatus* (Reuter), *D. olivaceous* (Reuter), and *Rhinacloa* sp. (Heteroptera: Miridae), affecting flowers and fruitlets, the thrips, *Scirtothrips aguacatae* Johansen and Mojica and *Scirtothrips kupae* Johansen and Mojica, *Scirtothrips perseae* Nakahara, the red-banded thrips, *Selenothrips rubrocinctus* (Giard) (Thysanoptera: Thripidae), affecting leaves and sometimes fruits. Among lepidopterans, several species affect production, including the avocado seed moth, *Stenoma catenifer* Walsingham (Lepidoptera: Oecophoridae), the western avocado leaf roller, *Amorbia cuneana* Walsingham (Lepidoptera: Tortricidae), and the avocado loopers *Anacamptodes defectaria* (Guenée), *Epimeces detexta* (Walker), *Epimeces matronaria* (Guenée), *Oxydia vesulia transponens* (Walker), and *Sabulodes aegrotata* Guenée (Lepidoptera: Geometridae), which consume foliage and eventually chewing on small fruit.

Other important pests are the weevils, *Conotrachelus perseae* Barber, *Conotrachelus aguacatae* Barber, *Conotrachelus serpentinas* (Klug) and *Heilipus lauri* Bohemann (Coleoptera: Curculionidae), boring into fruits and seeds.

Banana and plantain, *Musa* spp. (Musaceae family)

Bananas and plantains (*Musa* spp.) are among the most important crops in tropical and subtropical climates. The genus *Musa* evolved in Southeast Asia where numerous undomesticated *Musa* species still grow as opportunistic weeds. Edible bananas (*Musa* spp., Eumusa series) originated within this region from two wild progenitors, *Musa acuminata* and *M. balbisiana*, producing a series of diploids, triploids and tetraploids through natural hybridization. Additionally, man has selected for parthenocarpy (development of fruit without pollination or seeds). Hybridized bananas may be divided into six genome groups (i.e., AA, AAA, AAB, AB, ABB, ABBB) based on the relative contributions of *M. acuminata* and *M. balbisiana*. Domesticated bananas include a wide range of dessert, cooking and brewing cultivars. The most extensively grown bananas are triploids. In terms of gross value of production, bananas and

Fig. 1083   Some tropical fruit pests: upper left, banana weevil larva (photo Rita Duncan); upper right, papaya fruit fly (photo R. Swanson); second row left, papaya scale (photo M. Shepard); second row right, papaya leafhopper (photo Rita Duncan); third row left, Annona seed borer (photo H. Nadel); third row right, avocado lacebug (photo J. Peña); bottom left, Caribbean fruit fly (photo Rita Duncan); bottom right, Acerola weevil larva (photo Rita Duncan).

plantains are the fourth most important global food crop. Bananas grown for export are almost exclusively of one variety, 'Cavendish'; this cultivar accounts for slightly more than 10% of global production. Bananas are rhizomatous herbaceous plants ranging in height from 0.8 to 15 m. A mat (= banana stool) consists of an underground corm (rhizome) from which one or more plants (shoots) emerge. Adventitious roots spread extensively 4 to 5 m from the parent and downward 75 cm or more; however, most roots are near the soil surface. Plants represent a single shoot (pseudostem, stem, leaves, flower and bunch). Yield is normally expressed in kg/area/year reflecting both number and size of bunches harvested.

Pests of the corm and pseudostem include the banana weevil, *Cosmopolites sordidus* (Germar), banana pseudostem borer, *Odoiporous longicollis* (Olivier) and the West Indian sugarcane borer, *Metamasius hemipterus sericeus* (Olivier) (Coleoptera: Curculionidae). The following are pests of flowers and fruits: banana thrips *Hercinothrips bicinctus* (Bagnall), *Caliothrips bicinctus* Bagnall, *Chaetanaphothrips orchidii* (Moulton), *C. signipennis* (Bagnall), *Thrips hawaiiensis* (Morgan) and *Tryphactothrips lineatus* Hood (Thysanoptera: Thripidae), while the banana fruit scarring beetles, *Colaspis hypochlora* Lefebre (Coleoptera: Chrysomelidae) and the 'Irapua' bee, *Trigona spinipes* Fabricius (Hymenoptera: Apidae), are considered troublesome in some areas in the Neotropics. The banana moths, *Opogona sacchari* Bojer, *Opogona glyciphaga* (Meyrick) (Lepidoptera: Tineidae), the banana scab moth, *Nacoleia* (*Notarcha, Lamprosema*) *octasema* (Lepidoptera: Pyralidae) damage flowers and sometimes bunches. Important pests of foliage include the bagworm, *Oiketikus kirbyi* Guilding (Lepidoptera: Psychidae), *Caligo* spp., *Opsiphanes* spp. (Lepidoptera: Nymphalidae). The banana aphid, *Pentalonia nigronervosa* Coquerel. (Heteroptera: Aphidae) is important as the sole vector of the bunchy-top disease.

Barbados cherry or acerola, *Malpighia glabra* (L.) (Malpighiaceae family)

Barbados cherry is a tropical fruit native to the West Indies, Central America, and South America. Recently, the Barbados cherry has received attention because its fruits are exceptionally high natural source of ascorbic acid (vitamin C). Its cultivation has extended throughout the subtropics and tropics

and some of the largest plantings are in Brazil. Estimated commercial acreage in the Caribbean region is more than 400 acres with a potential crop value of several million US dollars. Flowering and fruit set occurs almost continuously from April through November in Florida, and fruits mature in approximately 30 days. Acerola's most important pests are the acerola weevil, *Anthonomus macromalus* Gyllenhal (= *A. flavus*, = *A. bidentatus*, = *A. malpighia*) (Coleoptera: Curculionidae), injuring fruits and flowers; fruit flies, *Anastrepha* spp., and the Mediterranean fruit fly, *Ceratitis capitata* Wiedemann (Diptera: Tephritidae), and the coreids, *Leptoglossus* spp., *Crinocerus* spp. (Hemiptera: Coreidae), and the stink bug, *Nezara viridula* L (Hemiptera: Pentatomidae), causing fruit deformation.

Carambola, *Averrhoa carambola* L. (Oxalidacea family)

Known as carambola, star apple, or five corner, this fruit tree is a 5 to 12 m high evergreen tree native to Southern Asia. Leaves imparipinnate, flowers in axillary or cauliflorous panicles, pentamerous. The fruit is a large berry, ovoid to ellipsoid in outline, with five pronounced ribs, stelate in cross section. The tree grows well in tropical or subtropical lowland conditions and flowers abundantly. Most of the information on pests of carambola comes from Southeast Asia and Australia. Important arthropods include the carambola fruit fly, *Bactrocera carambolae* Drew and Hancock (Diptera: Tephritidae), whose larvae feed on the fruit and make tunnels in the fruit; fruit piercing moths, *Gonodonta* spp. and *Eudocima* spp. (Lepidoptera: Noctuidae), pierce the skin of the ripe or ripening fruit with their strong proboscis, the damage resulting in crop loss or unmarketable fruit; the armored scale, *Morganella longispina* (Morgan) (Hemiptera: Diaspididae), feeds on buds and bark causing severe die back; the stink bug, *Nezara viridula* (L.), extracts fruit fluids leaving a small puncture and eventual fruit deformation; and the false spider mite, *Brevivalpus phoenicis* (Geijskes) (Acarina: Tenuipalpidae), causes bronzing of the fruit.

Durian, *Durio zibethinus* Murray (Bombacaceae family)

Durian, a common fruit in Southeast Asia, originated in Borneo and Sumatra, and is held in high esteem from Sri Lanka and southern India to New

Guinea. The ripe fruits, or rather the arils which form the edible part, are generally eaten fresh. Its strong and unmistakable smell is present in every market and street stall during the fruiting season. The fruit also can be preserved, deep frozen or consumed in ice creams, cakes and cookies. The fruit is a globose, ovoid or ellipsoid capsule, up to 25 cm long and 20 cm diameter, green to brownish, covered with numerous broadly pyramidal, sharp, up to 1-cm-long spines; usually with five thick, fibrous valves. Seeds are up to 4 cm long and completely covered by a white or yellowish, soft, very sweet aril. The tree is propagated by seeds, or superior cultivars by budding on seeding stock. Thailand is the largest producer with 444,500 tons in 1987, followed by Indonesia.

The key pests of durian are *Hypomeces squamosus* F. (Coleoptera: Curculionidae), a polyphagous weevil in which the adult stage feeds on leaves, causing extensive defoliation of young plants; the durian psyllid, *Allocaridara malayensis* (Crawford) (Hemiptera: Psyllidae), inserts eggs into young leaves, and both nymphs and adults feed on the underside of the leaf; the durian borer, *Conogethes punctiferalis* (Guenée) (Lepidoptera: Pyralidae), damages both young and mature fruits by boring into the pulp. The newly hatched larva of the durian seed borer, *Mudaria luteileprosa* Holloway (Lepidoptera: Noctuidae), feeds initially on the skin of the fruit and later bores into the husk and then into the seeds. The mite *Eutetranychus africanus* (Tucker) (Acari: Tetranychidae) feeds on the upper side of the leaf, especially along the mid-vein, causing withering spots that later spread over the whole leaf.

Guava, *Psidium guajava* L. (Myrtaceae family)

The guava, *Psidium guajava* L., occurs naturally from southern Mexico to South America and the Caribbean. The guava is a small tree that is grown worldwide in the tropics and warm subtropics for its edible fruit. Guavas can flower and bear fruit continuously in the tropics, however, there are normally two crops a year. In Florida and Puerto Rico, there is a large crop in early to mid-summer (June and July), and a smaller crop in late winter (February). In Hawaii, the small harvest peak occurs in April to May and the heavy peak in September and November. In India and Malaysia, the main crop is in mid-winter and the lesser crop during the rainy season (July–September). Fruits mature 90 to 150 days after flowering depending on the variety or clone of the fruit and cultural

and weather conditions. The large white flowers attract a large variety of pollinating insects including many species of bees. In many tropical regions the honey bee, *Apis mellifera* L., is the most important pollinator species.

Guavas are either eaten fresh, or used in juice, ice cream, jellies, pastes or preserves. Key pests involve fruit flies of the genera *Anastrepha* and *Bactrocera*, and several species of fruit boring weevils, such as *Conotrachelus psidii* Marshall. Fruit flies injure the fruit by ovipositing eggs and the developing larvae consume the pulp, leaving the fruit unmarketable. A secondary pest is *Heliopeltis theobromae* Miller (Miridae), which feeds on many parts of the plant, including the fruit. Damage to the fruit results in necrotic lesions which render the fruit unmarketable. Many species of mealybugs, including *Ferrisia virgata* (Cockerell), *Planococcus citri* (Risso), *Planococcus pacificus* Cox, *Pseudococcus citriculus* Cox, *Pseudococcus nipae*, *Planococcus minor* (Maskell), and *Pseudococcus lilacinus* Cockerell affect guava. Normally these insects do not damage the host severely, but large populations cause the fruit to become misshapen and deformed. For instance, honeydew produced by mealybugs often results in sooty mold, which lowers the fruit's market value. The redbanded thrips, *Selenothrips rubrocinctus* (Giard) and *Scirtothrips dorsalis* Hood, also attack guavas by feeding on the leaves and fruit and cause russeting or bronzing of the plant surface.

Mites can cause serious damage, and they include *Oligonychus yothersi* (McGregor), *Panonychus* sp., *Eotetranychus* sp., and *Oligonychus biharensis* (Hirst). These mites are reported to feed on leaves, causing leaves to become dull green, and then bronzed. The mites, *Tegolophus guavae* and *Brevipalpus* spp., cause damage to fruits and tender leaves.

Litchi and longan (Sapindaceae family)

The litchi (*Litchi chinensis* Sonn.) and longan (*Dimocarpus longan* Lour.) are closely related species belonging to the family Sapindaceae. Of Southeast Asian origin, they thrive in subtropical areas with cool dry winters and warm wet summers, but may be grown in tropical areas at high elevation. Around the world, litchis have been successfully grown commercially in latitudes from 15 to 35°. The litchi is a traditional fruit in China, and it occupies a special place in Chinese culture. Indo-China is the center of origin for this species and many old specimen trees have been identified. Litchi and longan fruits are best when eaten fresh, but they also can be processed by canning, juicing or drying into litchi nuts. In Asia they are usually harvested as whole panicles in order to maintain freshness, and are sold in street markets or at roadside stalls within a few days of harvest.

Apart from China and India, large litchi industries have been developed in Taiwan, Thailand, Vietnam, Madagascar, South Africa and Réunion. Smaller but expanding industries exist in Australia, Bangladesh, Mauritius, Mexico, the Seychelles, Spain and the USA (California, Florida and Hawaii).

In general, longans can be grown successfully wherever litchis are cultivated. Although both species flower at about the same time, longans take longer than litchis to mature. Major world producers are China (400,000 tons), Thailand (150,000 tons) and Vietnam (20,000 tons).

Litchi and longan are adapted to the warm subtropics and produce the best crops when winters are short, dry and cool, but frost-free. Such climatic conditions initiate the development of flower panicles. The inflorescences emerge during late winter and the flowers open in early spring. Most cultivars set fruit far in excess of what an individual tree can carry through to maturity and will shed the excess at various times during fruit development. Longan flowers about two to three weeks later than litchi, and matures four to eight weeks later. Harvesting fruit as whole panicles has the effect of pruning and stimulates new leaf growth after harvest. It also reduces tree size. Ideally, trees should produce one or two vegetative flushes after harvest. The aim is to have the second or third flush commence in winter. If conditions are cool during the early part of the flush, development of the new growth will be floral. However, if warm weather is encountered, the new growth will produce leaves. Litchi and longan often are cultivated in the same geographical areas, and many pests are common to both crops.

Pests include the fruit borers *Conopomorpha sinensis* Bradley, *Conopomorpha litchiella* Bradley (Lepidoptera: Gracillariidae), *Cryptophlebia peltastica* Meyr., *Cryptophlebia leucotreta* Meyr., *Cryptophlebia bactrachopa* Meyr. (Lepidoptera: Tortricidae), and in the New World, *Crocidosema* n. sp. (Lepidoptera: Tortricidae). These insects lay new eggs on the fruit anytime after fruit set, as well

as on new leaves and shoots. Larvae can bore into leaf buds and fruits. Fruit-piercing moths (Lepidoptera: Noctuidae) attack a range of fruits throughout Southeast Asia, the South Pacific and Australia. The larvae of fruit-piercing moths develop on a variety of host plants. In the Pacific Islands, *Eudocima* (*Othreis*) *fullonia* (Clerck), *Eudocima salaminia* (Cramer) and *Eudocima jordani* (Holland) are primary feeders on litchi fruit on the wet tropical coast of Queensland in Australia. Unlike most lepidopterous pests, of which the larva is the damaging stage, in this case it is the adult that causes the damage through its feeding on the fruit. The loopers *Oxyodes scrobiculata* F. and *Oxyodes tricolor* Guen. (Lepidoptera: Noctuidae) occupy similar niches in Thailand and Australia, respectively. The caterpillars feed on the foliage of litchi trees and can cause severe defoliation. Although they will eat leaves of any age, they prefer the younger ones. The litchi longicorn beetle, *Aristobia testudo* (Voet), is a serious pest of both litchi and longan in Guangdong Province in China. In Taiwan, the white spotted longicorn beetle, *Anoplophora maculata* (Thomson), has a one year life cycle. These beetles girdle branches by chewing 10 mm strips of bark. The larvae bore into the xylem and create tunnels up to 60 cm long in the wood. Several bugs belonging to the family Tessaritomidae attack litchis and longans throughout China, Southeast Asia and Australia. *Tessaritoma papillosa* Drury occurs in southern China, Vietnam, Thailand, Burma, the Philippines and India, although there are reports that *Tessaritoma javanica* Thunberg and *Tessaritoma quadrata* Distant are the species found on litchi in India. In litchis and longans, adults and nymphs feed on terminals, which may be killed, and also on flowers and fruit, causing these to fall.

The litchi erinose mite, *Aceria litchii* (Keiffer), also known in China as litchi hairy mite, hairy spider, or dog ear mite, occurs throughout China and Taiwan, India, Pakistan, Hawaii and Australia. The litchi erinose mite attacks new growth, causing a felt-like erineum to be produced on the leaflets. This may form as several blisters, but if the infestation is severe, it may eventually cover the entire leaflet, causing it to curl. Whole terminals may be deformed. The erineum is at first silver-white, changing as it ages to light brown and then dark reddish brown. The longan erinose mite, *Aceria longana* Boczek and Knihinicki, is a sporadic but major pest of longan in Thailand. *A. longana* is specific to longan, severely affecting the terminals and flowers. The longan gall mite, *Aceria dimocarpi* (Kuang), is associated with longans in China where it is recorded as causing erineum on leaves and also witches' broom symptoms. The litchi leaf midge, *Dasyneura* sp., is regarded as one of the major pests of litchi in China, and *Litchiomyia chinensis* Yang and Luo has been reared from galls collected on litchi leaves in Guangdong.

Mango, *Mangifera indica* L. (Anacardiacea family)

The mango most likely originated in Southeast Asia, particularly in the Malay Archipelago. However, it was probably first grown as a food crop in India, where its cultivation is thought to have commenced at least 4,000 years ago. Mangoes are now grown as an important crop all over the world, in both tropical and subtropical areas. Mangoes may be eaten ripe or green, or the fruit is juiced or processed into preserves, chutney, frozen puree, or dried. World production of mangoes in 2000 was estimated to be about 24.5 million tons. Worldwide, mango is affected by many pests. For instance, mango flowers are attacked by a variety of insects such as midges, leafhoppers, caterpillars and thrips. These include leafhoppers (Hemiptera: Cicadellidae) infesting mango flowers throughout Asia, such as *Idioscopus clypealis* Lethierry, *Idioscopus niveosparsus* Lethierry, *Idioscopus magpurensis* Pruthi and *Amritodius atkinsoni* Lethierry. The mango blister midge or mango gall midge (Diptera: Cecidomyiidae), *Erosomyia mangiferae* Felt, infests mangoes in the West Indies. Five cecidomyiid species, including *Erosomyia indica* Grover and Prasad, are reported to attack mango flowers and to cause severe damage in India, while *Dasyneura mangiferae* (Felt) does the same in Hawaii.

Numerous species of lepidoptera have been found to infest mango flower panicles in all production areas, but in few cases have the species involved been identified. In Florida, *Pococera atramentalis* Lederer (Pyralidae) and *Platynota rostrana* (Walker) (Tortricidae) are the two most damaging species of a complex that also includes *Pleuroprucha insulsaria* (Guenée) (Geometridae), *Tallula* spp. (Pyralidae) and *Racheospilla gerularia* (Hübner). In the Philippines, the tip borer, *Chlumetia transversa* Walker (Noctuidae), is second only to mango hoppers as a pest of flowers. The geometrids, *Eupithecia* sp., *Chloropteryx glauciptera* Hampson and *Oxydia*

vesulia (Cramer), as well as *Penicillaria jocosatrix* Guenée (Noctuidae), also can be found feeding on mango panicles. Other blossom pests are *Frankliniella bispinosa* (Morgan) and *Frankliniella kelliae* (Sakimura) in Florida, the western flower thrips, *Frankliniella occidentalis* (Pergande), in Israel, and the chili thrips, *Scirtothrips dorsalis* Hood, in Thailand.

Because of the high price paid for unblemished fruit, fruits need to be protected from a range of insects that may cause physical damage or loss, or merely affect its outward appearance. Flies of the Tephritidae are pests of mango in many parts of the world. In the Neotropics, *Anastrepha* spp., *A. obliqua* (Macquart), the Mexican fruit fly, *A. ludens* (Loew), *A. serpentina* (Wiedemann), *A. striata* Schiner, *A. suspensa* (Loew), *A. ocresia* Walker, *A. distincta* Greene, *A. fraterculus* Wiedemann, and *A. chiclayae* Greene are considered important pests. *Bactrocera tryoni* (Frogatt), *B. neohumeralis* (Hardy), *B. jarvisi* (Tryon), *B. zonata* (Saunders), *B. frauenfeldi* Schiner and *B. dorsalis* (Hendel) are all reported to attack mango. The Mediterranean fruit fly, *Ceratitis capitata*, is reported to attack mango throughout the world, whereas the marula fruit fly, *Ceratitis cosyra* (Walker), and the Natal fruit fly, *Ceratitis rosa* Karsch, are important in Africa.

The mango seed weevils, *Sternochetus mangiferae* (F.) (Coleoptera: Curculionidae), are widely distributed in Africa, Asia, Australia, the Pacific Islands and in some Caribbean islands, whereas *Sternochetus gravis* (F.) and *Sternochetus frigidus* (F.) occur in India and Bangladesh. Seed weevils generally lay eggs in small green fruit, and the larvae tunnel to the seed, where they feed and develop. *Deanolis sublimbalis* Snellen (Lepidoptera: Pyralidae), the mango seed borer, is also an important pest of mango fruits in the Philippines, Vietnam, China, Thailand, Indonesia and Papua New Guinea. The distinctive red-banded larvae feed on and bore through the pulp to the seed.

Fruitspotting coreid bugs, such as the yellowish-green coreid bugs *Amblypelta lutescens lutescens* (Distant) and *Amblypelta nitida* Stål, and the tip wilter, *Anoplocnemis curvipes* (Fabricius), can be serious pests of young mango trees in South Africa. *Amblypelta lutescens* feeds on the young fruit, causing black lesions to develop and the fruit to fall.

Pest of leaves and buds include gall midges (Diptera: Cecidomyiidae) such as *Protocontarinia matteiana* Kieffer and Cecconi, *Erosomyia* spp., *Procontarinia schreineri* Harris, *Amradiplosis echinogalliperada* Mabi and *P. matteiana*. The galls dry up and fall out, leaving a typical 'shot-hole' effect. Red-banded thrips, *Selenothrips rubrocinctus* (Giard), and the Mediterranean mango thrips, *Scirtothrips mangiferae* Priesner, feed on the leaf, especially adjacent to the midrib, where they cause a silvering that develops into necrosis, eventually leading to leaf drop. The mango bud mite, *Aceria mangiferae* Sayed is reported to attack the buds of terminals, and spider mites belonging to the genus *Oligonychus* (*O. mangiferae* Rahman and Sapra, *O. punicae* (Hirst), and *O. yothersi* McGregor) feed on the upper surface of mango leaves.

The mango scale, *Aulacaspis tubercularis* (Newstead), is regarded as a key pest mostly because it infests the fruit. Among the mealybugs, *Rastrococcus invadens* Williams, and the margarodid *Drosicha stebbingii* (Green) (Margarodidae) remain as important pests of leaves in different mango growing areas.

Mangosteen, *Garcinia mangostana* L. (Guttiferae family)

Mangosteen is known only as a cultivated species, although there have been occasional observations of wild specimens in Malaysia. The scarcity of mangosteen orchards, limited fruit supply and the fruit's short shelf life are major marketing problems. The long juvenile period of 10 to 15 years, which discourages commercial production, contributes to much of the limited fruit supply. The mangosteen fruit is a globose and smooth berry, turning dark purple at ripening. It is a crop of the humid tropics, often found in association with durian. The fruit has a diameter of 4 to 7 cm. The edible part is the sweet white aril that envelops the seeds within the pericarp.

Floral initiation to anthesis takes about 25 days and the fruit ripens 100 to 120 days later. The most serious pests are *Hyposidra talaca* (Walker) (Lepidoptera: Geometridae), a highly polyphagous insect. *Hyposidra talaca* is a typical looper, causing defoliation in tropical lowlands and highlands. The citrus leafminer, *Phyllocnistis citrella* (Lepidoptera: Phyllocnistidae), mines leaves of mangosteen, causing leaf deformation and often leading to early fall. *Sictoptera cucullioides* Guenée (Lepidoptera: Noctuidae) is reported to feed voraciously on young flushes of mangosteen.

Papaya, *Carica papaya* L. (Caricaceae family)

Papaya, *Carica papaya* L., is a major tropical fruit cultivated through the tropical and Neotropical regions of the world between 32° north and south. Although papaya was probably cultivated by early civilizations in the New World, no botanical records are available prior to the arrival of Columbus to America. This herbaceous plant is also known as papaw, paw paw, kapaya, kepaya, lapaya, tapaya, papayao, papaya, papaia, papita, lechosa, fruta bomba, mamon, mamona, mamao, and tree melon. Papaya is cultivated mainly for its edible fruit, but medicinal and industrial uses also have been documented.

Fruit flies (Diptera: Tephritidae) are the only group of insects that actually penetrate the pulp or seeds. Twenty-six species from seven genera, *Anastrepha* (two species), *Bactrocera* (17 species), *Ceratitis* (three species), *Dacus*, *Euphranta*, *Myoleja*, and *Toxotrypana* (one species each) attack papaya fruits. *Toxotrypana curvicauda* Gerstaecker is the most important fruit fly species attacking papaya in the Americas and Caribbean Basin, whereas *Bactrocera papayae* (Drew and Hancock) is one of the most threatening pests to papaya in Australia. *Anastrepha* spp., *C. capitata* (Wiedemann), *C. catoirii* Guerin-Meneville, and *C. rosa* Karsch also are known to attack papaya.

Arthropods affecting the foliage and trunk of papaya include the white peach scale, *Pseudaulacaspis pentagona* (Targioni-Tozetti), the papaya scale, *Philephedra tuberculosa* Nakahara and Gill, and the papaya mealybug, *Paracoccus marginatus* Williams and Granara de Willink. Several species of mites feed on papaya. They include the carmine mite, *Tetranychus cinnabarinus* (a key pest), the red and black flat mite, *Brevipalpus phoenicis* (an occasional pest), and the papaya leaf edge roller mite, *Calacarus brionese*. Damage by the broad mite, *Polyphagotarsonemus latus* (Banks) (Tarsonemidae), is confused with virus-like symptoms. Nine cicadellid species from three genera (*Empoasca*, *Poeciloscarta* and *Sanctanus*) can affect papaya, such as *Empoasca papayae* Oman, which transmits bunchy-top, *E. stevensi* Young, and *E. insularis* Oman. Aphids do not colonize papaya plants and are considered minor pests, but several species, *Aphis coreopsidis* (Thomas), *Aphis nerii* Boyer de Fonscolombe, *Aphis gossypii* Glover, *Aphis spiraecola* Patch, *Myzus persicae* (Sulzer), and *Toxoptera aurantii* (Boyer de Fonscolombe), can be found on papaya plants. Several aphid species [*M. persicae*, *M. euphorbiae*, *A. spiraecola* (= *A. citricola*), *A. gossypii*, *A. craccivora*, *A. nerii*, *R. maidis*, and *T. auranti* (Boyer de Fosc)] are capable of transmitting papaya ringspot virus. Among the defoliators, hornworms (Lepidoptera: Sphingidae), *Eryinnis alope* (Drury), *E. ello*, and *E. lassuxi merianae* Grote, are considered important.

Passion fruit (Passifloraceae family)

Passion fruits belong to the genus *Passiflora* (family Passifloraceae), which has a wide genetic base. While some species are undomesticated, others are cultivated as ornamental plants, for nourishment and for medicinal purposes. The majority of *Passiflora* are indigenous to the tropical and subtropical regions of South America. Of the 400 known species of *Passiflora*, about 50 or 60 bear edible fruits. A few species are economically important, e.g., *Passiflora edulis* Sims., botanical form *flavicarpa* Deneger, the yellow passion fruit, whose juice and pulp are used extensively as ingredients of beverages, salads, fruit cocktails and desserts. The major producers of passion fruits are found in South America, mainly Brazil, Colombia, Peru and Ecuador. Commercial plantations of passion fruits are found also in Australia, Hawaii, India, New Guinea, Kenya, South Africa, Sri Lanka and Costa Rica. *Passiflora edulis* f. *flavicarpa*, *P. edulis* (purple passion fruit), and *P. alata* (sweet passion fruit) are the main species cultivated in the world. *Passiflora ligularis* (granadilla) and *P. quadrangularis* (badea) are cultivated in the Andean region of South America and in Central America.

Although the passion fruit crop has great economic potential, its establishment and expansion have been hindered by various problems. For example, passion fruit is attacked by a wide host range of diseases, insects and mites. Some pest species cause significant losses, reaching the status of key pests or secondary pests. They include lepidopterous defoliators such as three heliconiine species, *Dione juno juno* Cramer, *Agraulis vanillae vanillae* Linnaeus, and *Eueides isabella huebneri* Ménétries (Nymphalidae). Many species of bugs attack passion fruit; the majority belong to the Coreidae (leaf-footed bugs), such as *Diactor bilineatus* Fabricius, *Leptoglossus* spp., and *Holhymenia* spp. *Diactor bilineatus* is the most common species in Brazil and Venezuela, and is known as the passion fruit bug because it feeds only on fruit of *Passiflora* spp. Among the *Holhymenia*, *Holhymenia clavigera* (Herbst.) and *H. histrio* (Fabricius) are reported attacking passion fruit. *Leptoglossus gonagra* Fabricius and *L. australis* Fabricius are reported from several passion fruit producing regions.

Flies of the genera *Anastrepha* Schiner (Tephritidae) and *Lonchaea* Fallén (Lonchaeidae) cause economic losses. *Anastrepha consobrina* (Loew), *A. ethalea* (Walker), *A. grandis* (Macquart), *A. kuhlmenni* Lima, *A. lutzi* Lima and *A. pseudoparallela* (Loew) are the most frequent species followed by *Anastrepha pallidipennis* Guerne on yellow passion fruit. The oriental fruit fly, *Bactrocera dorsalis* (Hendel), melon fly, *Dacus cucurbitae* Coquillett, and the Mediterranean fruit fly, *Ceratitis capitata* Wiedemann, are known to attack passion fruit in Hawaii. The Queensland fruit fly, *Dacus tryoni* (Froggatt), is the most important insect pest of passion fruit in Australia.

Neosilba pendula (Bezzi) and *Dasiops* sp. (Lonchaeidae) attack flowers and buds of passion fruit. The species of *Dasiops* include *D. curubae* Steykal, *D. inedulis* Steykal and *D. passifloris* McAlpine. Other flies that may also feed upon flower and buds are *Lonchaea cristula* McAlpine (Lonchaeidae) and *Zapriothrica salebrosa* Wheeler (Drosophilidae).

Several species of mites have been reported from passion fruit including *Brevipalpus phoenicis* (Geijskes) (Tenuipalpidae) and the red spider mites, *Tetranychus mexicanus* (McGregor) and *T. desertorum* Banks (Tetranychidae), causing general discoloration of the leaves and necrosis, culminating in leaf drop. The broad mite, *Polyphagotarsonemus latus* (Banks), induces malformations in developing leaves, which later dry and drop. It may attack flowering buds, causing a reduction in the number of flowers.

Pineapple, *Ananas comosus* (L.) Merr. (Bromeliaceae family)

The pineapple, *Ananas comosus* (L.) Merr., is a self-sterile herbaceous, perennial, monocotyledonous plant originating in the New World. The commercial cultivars are placed in five distinct groups, i.e., 'Cayenne', 'Spanish', 'Queen', 'Pernambuco' and 'Perolera'. It is currently grown in many countries with tropical and subtropical climates. Most pineapple pests are endemic to the new countries and adapted to utilize the pineapple to support their life-cycles.

More than 100 species of plant-parasitic nematodes have been recorded in association with pineapple roots, and between three and five species are usually found in most pineapple fields. The key nematode pests include the root-knot nematode (*Meloidogyne javanica* and *M. incognita*), the reniform nematode (*Rotylenchulus reniformis*) and the lesion nematode (*Pratylenchus brachyurus*). The key nematode pests of pineapple are all endoparasites. They either establish a permanent feeding site within the root and become sedentary, e.g., root-knot and reniform nematodes, or they live inside roots but remain migratory, e.g., lesion nematodes. Key arthropod pests are the pineapple leathery pocket mite/pineapple fruit mite, *Steneotarsonemus ananas* (Tryon), *Dolichotetranychus floridanus* (Banks), and the symphylids *Hanseniella* (Symphyla: Scutigerellidae), e.g., *H. unguiculata*, *H. ivorensis*, and *Scutigerella sakimurai* (Symphyla: Scolopendrellidae). *Steneotarsonemus ananas* is important because of its association with the fruit diseases, interfruitlet corking, leathery pocket and fruitlet core rot/black spot. The pineapple flat mite, *D. floridanus*, attacks plants of all ages. Damage is primarily to the whitish leaf bases, which develop brown/black necrotic lesions and appear progressively dehydrated, with feeding mites around lesion peripheries, or as orange patches on leaf bases.

The symphilids *H. unguiculata*, *H. ivorensis* and *S. sakimurai* feed on young meristematic tissues causing newly established plants to stop root development and causing severe witches' broom symptoms. Other important pests are the onion thrips or yellow spot thrips, *Thrips tabaci* Lindeman, responsible for yellow spot disease of pineapples, the pink pineapple mealybug, *Dysmicoccus brevipes* (Cockerell), associated with the devastating disease, pineapple mealybug wilt, the pineapple scale, *Diaspis bromeliae* (Kerner), infesting the lower pineapple leaves, the pineapple caterpillar, *Thecla basilides* (Geyer), causing larval galleries and rendering the fruit unmarketable, and the white grubs, *Lepidiota grata*, *Rhopaea magnicornis*, *Adoretus ictericus*, and the melolonthids, *Macrophylla ciliata* and *Asthenopholis subfasciata*, destroying the root system.

Rambutan, *Nephelium lappaceum* Blume (Sapindaceae family)

Rambutan is a tropical relative of litchi, with a distribution that ranges from southern China through the Indo-Chinese region, Malaysia, and Indonesia to the Philippines. As the name suggests (the Malay word 'rambut' meaning hair), the fruit is glabrous, resembling a burr. Rambutan, while highly prized in its region of origin, remains a minor fruit internationally. The fruits are consumed fresh or canned. The rambutan fruit is an ellipsoid to subglobular schizocarp, up to 7 cm × 5 cm, usually consisting of one nutlet. The skin color varies from yellow to purplish red. The seed usually is covered by a thick, sweet, juicy, white to yellow, translucent sarcotesta, which is the edible part.

The main flowering period occurs during the dry season. Fruits ripen about 110 days after bloom. The most important pests are the durian borer, *Conopomorpha* (= *Acrocercops*) *cramerella* (Snellen) (Lepidoptera: Gracillariidae). This insect tends to damage only the part of the fruit next to the fruit stalk. The cockchafer beetle, *Adoretus* (= *Lepadoretus*) *compressus* (Weber) (Coleoptera: Scarabaeidae), feeds on the interveinal areas close to the center of the leaf. Adults of *Apogonia cribricollis* Burmeister (Coleoptera: Scarabaeidae) are active at night, feeding on foliage from the leaf margin inwards. *Chalcocelis albiguttatus* (Snellen) and *Parasa lepida* (Cramer) (Lepidoptera: Limacodidae) cause severe defoliation of rambutan trees.

Integrated pest management (IPM) for tropical fruit

Tropical fruit crops provide a relatively stable environment over many years, offering continuing habitats for both pests and natural enemies, and providing opportunities for biological control and effective pest management programs. Any attempt to develop integrated control programs in fruit crops must take into account the following: (1) knowledge of native or resident arthropod fauna; (2) arthropod fauna affecting the tree crop in its area of origin or domestication; and (3) presence of natural enemies. The basis for integrated pest management includes the pest's biology and ecology, sampling and monitoring techniques, economic thresholds and the application of management tactics, i.e., chemical, biological, autocidal, plant resistance, etc.

Sampling and monitoring

Sampling and monitoring methods for tropical fruits are well established for some direct pests such as fruit flies and fruit borers, as well as for some indirect pests (i.e., banana weevil) and defoliators. However, the inability to relate results of sampling to infestation of the fruit at the time of harvest is still a problem. Attempts have been made to develop sampling techniques for several pests of mango, avocado and banana; however for some pests and crops, adequate sampling techniques are not available. This, of course, may be a reflection of the place where the crop is grown, the purpose of tree cultivation (export vs. internal consumption) and grower economic solvency (Australia, USA, South Africa, Israel) vs. growers from other regions of the world.

Economic thresholds

As export commodities, tropical fruit crops have a consistently high value, with highest prices for undamaged fruit of high quality; thus, the lack of knowledge of economic thresholds, or the low existent economic thresholds, requires control programs to focus on preventing damage by these pests.

Chemical control

Pesticides have been significant for protecting tropical fruit from insect attack and increasing tropical fruit productivity, therefore the control of fruit and foliar insect pests tends to be heavily dependent on chemical insecticides/acaricides. Information regarding proper timing, spray volumes, and knowledge of the pest complex is available for some crops such as pineapple, and scarce for others such as papaya. Regular, heavy sprays for controlling fruit flies and leafhoppers in papaya can cause heavy outbreaks of mites and other pests. Widespread use of nonselective pesticides continues to be the rule, but currently there is a trend toward evaluating a new generation of pesticides, adoption of selective spraying, timing of spray applications, and determining the effect of pesticides on predators and parasitoids. In contrast, there is minimal to nonexistent information on the effect of pesticides on pollinators. For instance, the deleterious effects on pollinators can be reduced by timing spray applications in passion fruit, according to the cultivar. Purple passion, whose flowers open during the morning hours, should be sprayed during late afternoon, while the yellow cultivar, whose flowers open in the afternoon, should be sprayed in the morning.

Attractants (pheromones)

Use of sexual attractants (pheromones) is mostly limited to fruit flies, while little information is

available for other insect groups, such as curculionids and Lepidoptera.

Crop sanitation

Crop sanitation is an important factor for maintaining low banana weevil densities in banana farms, and is used as a tentative effort to control important pests of avocados (i.e., weevils), annona (annona seed borer) and mango (mango seed weevil).

Biological control

Biological control has great potential as a major tactic for regulating pest populations in fruit orchards. The ability to apply biological control effectively has increased in recent decades because of greater knowledge of the arthropod fauna of some tropical fruits (citrus, avocado, pineapple and mango). While biocontrol agents are recorded for most pests of other tropical fruit crops, concentrated efforts to use bio-regulators are rarely observed. For instance, efforts toward developing systems for biocontrol of pink mealybug, *Maconellicoccus hirsutus*, and carambola fruit fly, *Bactrocera carambolae*, in the Caribbean were not initiated until these pests were major threats to crops such as citrus. If a stenophagous insect (i.e., avocado weevils, papaya fruit fly, avocado thrips and annona fruit moths) is the major constraint to a single commodity, efforts for biological control are halfhearted. The exception to this rule is the current strong effort toward biocontrol of banana weevil and the search for natural enemies of avocado thrips. With time, it is likely that more extensive use will be made of biological agents to control pests of tropical fruit crops.

Host plant resistance

Host plant resistance offers considerable promise as a tactic in pest management. Even though it appears that most efforts are concentrated toward plant pathogens, use of this tactic merits attention for some crops and some pest species. Resistance to arthropods in tropical fruit germplasm collections should be a high priority. Most efforts are directed at insect vectors (papaya and pineapple) and pests of avocado, mango and guava. Several fruits, i.e., durian and acerola, continue to be largely unnaturalized, and improved selections are rare.

Cultural practices

Untreated backyard trees and neglected plantings are considered to be major sources of pests, such as fruit flies. For instance, in Australia, hygiene and attention to alternative host plants that can increase pest pressure on the custard apple orchard are important. Mature fruit infested with yellow peach moth or with fruit fly should be collected and destroyed. Preferred fruit fly hosts like guava and loquat should not be planted in or near the orchard. Cultural practices generally do not offer a direct means for controlling pests, but used properly they can enhance natural enemy activity or retard pest population growth to a degree that is important in integrated control programs.

Jorge E. Peña
University of Florida
Homestead, Florida, USA

References

Aguiar-Menezes, E., E. B. Menezes, P. C. R. Cassino, and M. Soares. 2002. Passion fruit. pp. 361–390 in J. E. Peña, J. L. Sharp, and M. Wysoki (eds.), *Tropical fruit pests and pollinators*. CABI Publishing, Wallingford, United Kingdom.

Gold, C. S., B. Pinese, and J. E. Peña. 2002. Pests of banana. pp. 13–56 in J. E. Peña, J. L. Sharp, and M. Wysoki (eds.), *Tropical fruit pests and pollinators*. CABI Publishing, Wallingford, United Kingdom.

Gould, W. P., and A. Raga. 2002. Pests of guava. 2002. pp. 295–314 in J. E. Peña, J. L. Sharp, and M. Wysoki (eds.), *Tropical fruit pests and pollinators*. CABI Publishing, Wallingford, United Kingdom.

Ooi, P. A. C., A. Winotai, and J. E. Peña. 2002. Pests of minor tropical fruit. pp. 315–330 in J. E. Peña, J. L. Sharp, and M. Wysoki (eds.), *Tropical fruit pests and pollinators*. CABI Publishing, Wallingford, United Kingdom.

Pantoja, A., P. A. Follett, and J. Villanueva. 2002. Pests of papaya. pp. 131–156 in J. E. Peña, J. L. Sharp, and M. Wysoki (eds.), *Tropical fruit pests and pollinators*. CABI Publishing, Wallingford, United Kingdom.

Peña, J. E., H. Nadel, M. Barbosa-Pereira, and D. Smith. 2002. Pollinators and pests of *Annona*. pp. 197–222 in J. E. Peña, J. L. Sharp, and M. Wysoki (eds.), *Tropical fruit pests and pollinators*. CABI Publishing, Wallingford, United Kingdom.

Petty, G., G. Stirling, and D. P. Bartholomew. 2002. Pests of pineapple. pp. 157–196 in J. E. Peña, J. L. Sharp, and M. Wysoki (eds.), *Tropical fruit pests and pollinators*. CABI Publishing, Wallingford, United Kingdom.

Waite, G. K. 2002. Pests and pollinators of mango. pp. 103–130 in J. E. Peña, J. L. Sharp, and M. Wysoki (eds.), *Tropical fruit pests and pollinators*. CABI Publishing, Wallingford, United Kingdom.

Waite, G., and J. S. Hwang. 2002. Pests of litchi and longan. pp. 331–360 in J. E. Peña, J. L. Sharp, and M. Wysoki (eds.), *Tropical fruit pests and pollinators*. CABI Publishing, Wallingford, United Kingdom.

Wysoki, M., M. A. van den Berg, G. Ish-Am, S. Gazit, J. E. Peña, and G. Waite. 2002. Pests and pollinators of avocado. pp. 223–294 in J. E. Peña, J. L. Sharp, and M. Wysoki (eds.), *Tropical fruit pests and pollinators*. CABI Publishing, Wallingford, United Kingdom.

TROPICAL FRUITWORM MOTHS (LEPIDOPTERA: COPROMORPHIDAE).

Tropical fruitworm moths, family Copormorphidae, are a small family of 58 species, mostly tropical; actual fauna probably exceeds 100 species. The family is part of the superfamily Copromorphoidea in the section Tineina, subsection Tineina, of the division Ditrysia. Adults small to medium size (12–37 mm wingspan), with head somewhat smooth-scaled; haustellum naked; labial palpi prominent and porrect; maxilary palpi 3 to 4-segmented (rarely 1-segmented). Wing venation with forewing veins long and typically equidistant near the termen; wings typically rounded along forewing termen. Maculation mostly dull gray with dark markings, and many have forewing scale tufts. Adults nocturnal. Larvae are leaf feeders using a leaf web, or are borers (one feeds beneath bark), but few biologies are known. Host plants include Berberidaceae, Ericaceae, Moraceae, Podocarpaceae, and Rubiaceae.

John B. Heppner
Florida State Collection of Arthropods
Gainesville, Florida, USA

References

Clarke, J. F. G. 1955. Copromorphidae. In J.F.G. Clarke, *Catalogue of the type specimens of Microlepidoptera in the British Museum (Natural History) described by Edward Meyrick*, 2: 509–531. British Museum (Natural History), London.

DeBenedictis, J. A. 1984. On the taxonomic position of *Ellabella* Busck, with descriptions of the larva and pupa of *E. bayensis* (Lepidoptera: Copromorphidae). *Journal of Research on the Lepidoptera* 23: 74–82.

DeBenedictis, J. A. 1985. The pupa of *Lotisma trigonana* and some characteristics of the Copromorphidae (Lepidoptera). *Journal of Research on the Lepidoptera* 24: 132–135.

Heppner, J. B. 1984. Revision of the Oriental and Nearctic genus *Ellabella* (Lepidoptera: Copromorphidae). *Journal of Research on the Lepidoptera* 23: 50–73.

Heppner, J. B. 1986. Revision of the New World genus *Lotisma* (Lepidoptera: Copromorphidae). *Pan-Pacific Entomologist* 62: 273–288.

TROPICAL LATTICE MOTHS (LEPIDOPTERA: ARRHENOPHANIDAE).

Tropical lattice moths, family Arrhenophanidae, total 30 species, mostly Neotropical but recently with some Southeast Asian additions; actual fauna probably exceeds 50 species. The family is part of the superfamily Tineoidea, in the section Tineina, subsection Tineina, of the division Ditrysia. Adults small to medium size (12 to 69 mm wingspan), with robust bodies and roughened head scaling; haustellum reduced or absent; maxillary palpi absent; antennae serrate to bipectinate. Maculation varies from light colored (often yellow) with translucent and colored wing marks, to dark and somber colored. Adult activity nocturnal but some are diurnal (e.g., species in Taiwan). Biologies are unknown except for one Neotropical species with casebearing larvae that feed on fungi.

Fig. 1084 Example of tropical longhorned moths (Lecithoceridae), *Lysipatha diaxantha* Meyrick from Taiwan.

John B. Heppner
Florida State Collection of Arthropods
Gainesville, Florida, USA

References

Bradley, J. D. 1951. Notes on the family Arrhenophanidae (Lepidoptera: Heteroneura), with special reference to the morphology of the genitalia, and descriptions of one new genus and two new species. *Entomologist* 84: 178–185.
Davis, D. R. 1984. Arrhenophanidae. In J. B. Heppner (ed.), *Atlas of Neotropical Lepidoptera. 2. Checklist: Part 1 (Micropterigoidea-Immoidea)* 6. W. Junk, The Hague.
Davis, D. R. 1991. First Old World record of Arrhenophanidae. *Tropical Lepidoptera* 2: 41–42.

TROPICAL LONGHORNED MOTHS (LEPIDOPTERA: LECITHOCERIDAE).

Tropical longhorned moths, family Lecithoceridae, total about 1,038 described species, mostly tropical Oriental, but also with one group in the Palearctic; actual fauna probably exceeds 1,500 species. There are 4 subfamilies: Ceuthomadarinae, Oditinae, Lecithocerinae, and Torodorinae. The family (previously known as Timyridae) is part of the superfamily Gelechioidea in the section Tineina, subsection Tineina, of the division Ditrysia. Most species are in Lecithocerinae (475 sp.). The most colorful and largest species are in the tropical subfamily Torodorinae. Adults small (5 to 30 mm wingspan), with head mostly smooth-scaled, and antennae mostly long; haustellum scaled; labial palpi long; maxillary palpi 4-segmented and folded over haustellum base (rarely reduced to one segment). Maculation varies from dull to brightly colored and variously marked. Adults are mostly diurnal and many have the habit of holding the antennae together to the front when at rest. Larvae may mostly be leaf litter feeders or leaf tiers, but few species are known biologically. A few varied host plants are recorded, such as Fagaceae, Myrtaceae, Rosaceae, and Rubiaceae.

John B. Heppner
Florida State Collection of Arthropods
Gainesville, Florida, USA

References

Clarke, J. F. G. 1965. Timyridae. In J. F. G. Clarke, *Catalogue of the type specimens of Microlepidoptera in the British Museum (Natural History) described by Edward Meyrick*, 5: 1–255. London: Br. Mus. (Nat. Hist.).
Gozmány, L. A. 1978. Lecithoceridae. In H. G. Amsel, F. Gregor and H. Reisser (eds.), *Microlepidoptera Palaearctica*, 5 : 1–306, 93 pl. G. Fromme: Vienna. [in German]
Wu, C.-S. 1997. Lepidoptera. Lecithoceridae. In *Fauna Sinica. Insecta*. Vol. 7. Science Press, Beijing. 302 pp.
Wu, C.-S., and K.-T. Park. 1999. A taxonomic review of the family Lecithoceridae (Lepidoptera) in Sri Lanka. *Tinea* 16: 61–72; *Korean Journal of Systematic Zoology* 15: 1–9, 205–220; *Insecta Koreana* 16: 1–14, 131–142.

TROPICAL PLUME MOTHS (LEPIDOPTERA: OXYCHIROTIDAE).

Tropical plume moths, family Oxychitoridae, include only six species, all Indo-Australian and South Pacific. Some specialists include this family as part of the Tineodidae. The family is in the superfamily Pterophoroidea in the section Tineina, subsection Tineina, of the division Ditrysia. Adults small (10 to 12 mm wingspan), with head scaling average; haustellum naked; labial palpi porrect; maxillary palpi 4-segmented. Wings extremely linear, with long fringes, or wider and with both wings split into two plumes each. Maculation shades of brown or gray, with coordinated spotting or banding in the split-wing species. Adults may be crepuscular. Larva of one species feeds on seeds of white mangrove (Avicenniaceae); the remainder are unknown biologically.

John B. Heppner
Florida State Collection of Arthropods
Gainesville, Florida, USA

References

Clarke, J. F. G. 1986. Family Oxychirotidae. In Pyralidae and Microlepidoptera of the Marquesas Archipelago. *Smithsonian Contributions to Zoology* 416: 14–16.

Common, I. F. B. 1970. Oxychirotidae. In *Insects of Australia*, 835. Melbourne: Melbourne University Press.

Common, I. F. B. 1990. Family Tineodidae. In *Moths of Australia*, 322–325 [part]. Melbourne University Press, Melbourne.

Heppner, J. B. 1997. Oxychirotidae. In *Lepidopterorum Catalogus*, (n.s.). Fasc. 62. Association for Tropical Lepidoptera, Gainesville. 8 pp.

TROPICAL SLUG CATERPILLAR MOTHS (LEPIDOPTERA: DALCERIDAE).

Tropical slug caterpillar moths, family Dalceridae, include 84 Neotropical species (only one sp. occurs north of Mexico, in southern Arizona). Two subfamilies are known: Acraginae and Dalcerinae. The family is in the superfamily Cossoidea (series Limacodiformes) in the section Cossina, subsection Cossina, of the division Ditrysia. Adults small to medium size (11 to 50 mm wingspan), with head small and scaling roughened; labial palpi 2-segmented and very short; maxillary palpi vestigial; antennae short and bipectinate. Body robust. Wings quadratic and rounded; hindwings notably ovoid; frenulum sometimes absent. Maculation mostly brown hues with few markings but can be colorful, with yellow, orange, or pink, or lustrous white and silvery. Adult activity uncertain; possibly only nocturnal or crepuscular. Larvae slug-like, often with translucent gelatinous wart-like surface; feeding as leaf feeders (early instars as leaf skeletonizers), but few are known biologically. Various host plants are used and some

Fig. 1085 Example of tropical slug caterpillar moths (Dalceridae), *Acraga coa* (Schaus) from Mexico.

larvae are polyphagous. Few have any economic status.

John B. Heppner
Florida State Collection of Arthropods
Gainesville, Florida, USA

References

Epstein, M. E. 1996. Dalceridae Dyar, 1898. pp. 79–81 In Revision and phylogeny of the limacodid-group families, with evolutionary studies on slug caterpillars (Lepidoptera: Zygaenoidea). *Smithsonian Contributions to Zoology* 582: 1–102.

Hopp, W. 1928. Beitrag zur Kenntnis der Dalceriden. *Deutsche Entomologische Zeitschrift Iris* 42: 283–287.

Miller, S. E. 1994. Systematics of the Neotropical moth family Dalceridae (Lepidoptera). *Bulletin of the Museum of Comparative Zoology* 153: 301–495.

Orfila, R. N. 1961. Las Dalceridae (Lep. Zygaenoidea) argentinas. *Revista Investigaciones Agricultura* 15: 249–264, 1 pl.

Seitz, A. (ed.) 1938–39. Familie: Dalceridae. In *Die Gross-Schmetterlinge der Erde*, 6: 1303–1304 (1938); 1305–1312, pl. 168, 185 (1939). A. Kernen, Stuttgart.

TROPICAL THEILERIOSIS. See also, PIRO-PLASMOSIS.

TROPIDUCHIDAE. A family of insects in the superfamily Fulgoroidea (order Hemiptera). They sometimes are called planthoppers. See also, BUGS.

TROUT STREAM BEETLES. Members of the family Amphizoidae (order Coleoptera). See also, BEETLES.

TRUE KATYDIDS. A subfamily (Pseudophyllinae) of katydids in the order Orthoptera: Tettigoniidae. See also, GRASSHOPPERS, KATYDIDS AND CRICKETS.

TRUMPET. The respiratory horn or tube of the mosquito pupa.

TRUMPET LEAFMINER MOTHS (LEPIDOPTERA: TISCHERIIDAE). Trumpet leafminer moths, family Tischeriidae, total 81 known species from all regions except Australia, but the Nearctic has most of the known species (48 sp.). The family forms a monobasic superfamily, Tischerioidea, in the section Nepticulina, of the division Monotrysia, infraorder Heteroneura. Adults small (6 to 11 mm wingspan), with head rough-scaled, with very large head tuft (no eye-caps on the antennal bases); haustellum is short, scaled basally; labial palpi 3-segmented, short and porrect; maxillary palpi minute and 1-segmented. Wing venation is very reduced, with frenular bristles as the wing coupling. Maculation is generally somber but often with iridescences. Adults are diurnally active. The few larvae known are leafminers, usually trumpet-shaped mines or blotch mines, on a variety of host plants.

John B. Heppner
Florida State Collection of Arthropods
Gainesville, Florida, USA

References

Braun, A. F. 1972. Tischeriidae of America north of Mexico (Microlepidoptera). *Memoirs of the American Entomological Society* 28: 1–148.
Diskus, A. 1998. Review of the Tischeriidae (Lepidoptera) of Central Asia. *Acta Zoologica Lituanica* 8(3): 23–33.
Grandi, G. 1929. Contributo alla conoscenza della *Tischeria gaunacella* Dup. ed appunti sulla *Tischeria complanella* Hbn. (Lepidoptera – Tischeriidae). *Bolletino del Laboratorio di Entomologia Bologna* 2: 192–243, 5 pl.
Hering, O. 1926. Die Blattminierer-Gattung *Tischeria* in ihren palaearktischen Arten. *Entomologische Jahrbücher* 35: 99–106, 1 pl.
Sato, H. 1993. *Tischeria* leafminers (Lepidoptera, Tischeriidae) on deciduous oaks from Japan. *Japanese Journal of Entomology* 61: 547–556.

TRUMPET-NET CADDISFLIES. Members of the families Polycentropodidae and Psychomyiidae (order Trichoptera). See also, CADDISFLIES.

TRUNCATE. Cut off squarely.

TRUNK AND CANKER-CAUSING FUNGI. See also, TRANSMISSION OF PLANT DISEASES BY INSECTS.

TRYPANOSOMES. Trypanosomes are microscopic unicellular protozoa that are ubiquitous parasites of plants, invertebrates, and vertebrates. These parasites have existed for more than 300 million years and have evolved with their natural hosts. Most of the trypanosomes cause no harm to their hosts and are found in locations throughout the world. Some, however, cause serious diseases in their hosts and are of major medical and veterinary significance. Only a few species of trypanosomes are pathogenic to man. These trypanosomes are not only found in sub-Saharan Africa, where they infect both man and livestock, but also occur in Canada, Latin America, and extend to the southern borders of the United States.

The pathogenic trypanosomes of mammals are called hemoflagellates. These trypanosomes require the blood of their host (human, livestock, etc.) in order to undergo changes pertaining to their life cycles. In addition, the parasite may be transmitted to its host by various techniques, depending on the species of the parasite. For example, *Trypanosoma equiperdum*, a trypanosome known to infect horses, is transmitted via venereal contact. However, the vast majority of the different species of trypanosomes affecting mammals are transmitted via an insect vector.

The most important insect vectors associated with trypanosomiasis (disease caused by protozoa belonging to the genus *Trypanosoma*) in vertebrates are the tsetse flies belonging to the genus *Glossina*. There are 33 known and recognized species of this genus. These vectors can be further divided ecologically into three major groups: the riverine flies of the *palpalis* group; the savanna flies of the *morsitans* group; and the forest flies of the *fusca* group.

Tsetse flies can be distinguished from other flies by the presence of a needle-like forward projecting proboscis, a hatchet-like cell in the middle of the wings and a lateral plumose like arista on the antennae. They are closely related to the muscoid (i.e., house flies, stable flies) flies but may be given their own family name, Glossinidae.

Unlike most flies that are oviparous, the female tsetse fly retains the larva in the uterus and gives birth to one 3rd instar larva at a time. During gestation, the larva is fed from so-called uterine milk-like glands found in the body. When the female larviposits, the larva burrows quickly in the soil and transforms quickly into a pupa. Adult emergence is dependent on the soil temperature and moisture conditions. Adult tsetse flies mate only once and a female produces about 10 offspring during her life span. Both sexes are active bloodsuckers.

Rarely are tsetse flies found in open country because all need some tree/vegetation cover. They are mostly confined to a specific type of habitat and the physical conditions prevailing therein. The flies feed on those animals most readily available in their specific habitat, and therefore carry trypanosomes harbored by their preferred hosts. As a result, infection rates and kinds of trypanosomes vary from one fly habitat to another.

Another important insect vector associated with trypanosomiasis in vertebrates is the assassin bug, cone-nose bug, or kissing-bugs belonging to the family Reduviidae of the order Hemiptera (true bugs). They are characterized by having 2 pairs of wings; the forewings have leathery basal portions and a membranous distal portion. The second pair of wings is membranous. Hemiptera belonging to this family have piercing sucking mouthparts adapted for feeding on other arthropods or blood. The most important genera of reduviid bugs with some medical importance are *Triatoma*, *Panstronglyus*, and *Rhodnius* which serve as vectors of *Trypanosoma cruzi*, the causative agent for Chagas' disease.

General life stages of trypanosomes

Trypanosomes appear in four basic forms or stages as they undergo changes in the life cycle. These forms or stages differ from species to species, in that some species may exhibit only 2 of the four forms and others may exhibit all four forms. Each form has specific structures that enable it to adapt to its environment. One form is the amastigote or leishmanial stage. This stage has a kinetoplast, which is a structure that is always located at the place of entry into the cell of the flagellum and is opposed to the basal body of the later. The kinetoplast is further composed of a blepharoplast and parabasal body. This structure is analogically similar to the structure of the mitochondria, which is the powerhouse of most

types of cells. In addition to a kinetoplast, the amastigote stage has a nucleus. Although the orientation of the kinetoplast can be detected by using the flagellum as a landmark, the leishmanial stage does not have a flagellum. A second stage is the promastigote or leptomonad stage. This stage is found within the tissues of the infected host. In this stage, a flagellum is present. From this stage, the parasite can and does change back and forth between the amastigote and the leptomonad stage in a phenomenon known as transition or morphogenesis. A third stage is the epimastigote or the crithidial stage. The unique characteristic or defining feature of this stage in the history of hemoflagellates is that its undulating membrane or flagellum originates anteriorly to the nucleus. Finally, the last stage is the trypomastigote or trypanosomal stage. The unique or defining feature of this stage is that its undulating membrane originates posterior to the nucleus.

In addition to the morphological differences, these stages can also show metabolic differences. For example, another interesting aspect to the life cycle of the trypanosomes involves the cytochrome oxidase system. The cytochrome oxidase system is one that requires the presence of the metal iron (Fe) in order to work. The only stage that uses this system (i.e., the cytochrome oxidase system) is the epimastigote stage or form. Once the epimastigote stage is placed in a culture that contains other sources of nutrients apart from iron, there is an immediate change or transformation that occurs. The epimastigote form of the parasite immediately changes to the trypomastigote stage, which does not use the cytochrome oxidase system. The trypomastigote stage then use the other forms of nutrients in the culture in order to survive.

Pathogenic trypanosomes in humans

The three major species of this parasite that are extremely pathogenic to humans are *Trypanosoma rhodesiense*, *Trypanosoma gambiense*, and *Trypanosoma cruzi*. These parasites were named based on, or according to, either the region they were first found or by the person who first found them. All are hemoflagellates, and thus require the blood of their host in order to undergo their life cycles. Furthermore, the three species also have one or two of the same life cycle stages or forms present. Notwithstanding, different vectors transmit them and all three parasites run a totally different course of infection in man.

T. rhodesiense and *T. gambiense* are found primarily in Africa while *T. cruzi* is located throughout Latin America and extends to the southern borders of the United States of America.

Trypanosoma rhodesiense

Trypanosoma rhodesiense is the causative agent for East African sleeping sickness or Rhodesian trypanosomiasis and is found primarily in the eastern region of Africa. It produces a virulent form of sleeping sickness and often results in death within a matter of weeks if not treated.

The principal insect vectors (tsetse flies) responsible for the transmissions of *T. rhodesiense* are *Glossina morsitans*, *Glossina pallidipes* and *Glossina swynnertoni*. When feeding on an infected host, the flies obtain the trypanosomal stage of the protozoa, which undergoes a period of development in the midgut region and eventually transforms into the epimastigote stage. The epimastigote stage then migrates to the insect's salivary glands and then changes to the elongated infective (metacyclic) trypanosomal stage. This stage can now be passed to another host during a blood meal and can initiate infection or disease. The transformations usually are completed in about three weeks.

Once in the human blood, the parasite multiplies rapidly. In most infected individuals, there is an enlargement of the lymph nodes (Winterbottom's sign) in the post-cervical region. Irregular fever, headache, joint and muscle pains, and a rash characterize the disease. *T. rhodesiense* may be found in is the blood, lymph nodes and tissues of humans.

Trypanosoma gambiense

Trypanosoma gambiense is the causative agent of the disease known as West African sleeping sickness and is found in the western part of Africa. *T. gambiense* is essentially transmitted between humans and tsetse flies sharing riverine vegetation habitats. In the case of *T. gambiense*, two tsetse flies (*Glossina palpalis* and *Glossina tachinoides*) serve as the primary vectors and are capable of transmitting the trypanosome.

West African sleeping sickness runs a more chronic course of infection in humans. In the first or early phase of the disease, infected individuals show similar clinical and pathological symptoms

when infected with *T. rhodesiense*. However, the second or late phase of the disease is unique to *T. gambiense*. This phase involves severe damage to the central nervous system, and may be accompanied by other physical and mental impairment (mental retardation, speech impediments, swelling of the brains, etc.). At this time, the terminal sleeping stage develops and gradually the patient becomes more and more lethargic.

A person can be infected for months or even years without expressing obvious symptoms of the disease. When symptoms do appear, the disease is already at an advanced stage and, without treatment can lead to death. During the early or febrile stage of the sleeping sickness caused by *T. gambiense*, the organism occurs in the blood and lymph node of the infected individual. In the late stages where there is a development of cerebral symptoms, and the organisms can be found in the cerebrospinal fluid.

In infections with *T. rhodesiense* or *T. gambiense*, the presence of trypanosomes in chancre fluid, lymph node aspirates, blood, bone marrow, or, in the late stages of infection, cerebrospinal fluid (in the case of *T. gambiense*), using microscopic examination and other assays (such as testing for specific antibodies in the blood) are helpful in diagnosing trypanosomiasis.

In trypanosomiasis with *T. rhodesiense* or *T. gambiense*, treatment should be started as soon as possible and is based on the infected person's symptoms and laboratory results. The drug regimen depends on the infecting species and the stage of infection. Pentamidine isethionate and suramin (under an investigational New Drug Protocol from the CDC Drug Service) are the drugs of choice to treat the hemolymphatic stage of West and East African Trypanosomiasis, respectively. Melarsoprol is the drug of choice for late disease of *T. gambiense* with central nervous system involvement.

Tsetse flies do not normally fly far from their breeding sites (i.e., streams, lakeshores, lowland forests with dense growth of shrubs and trees, etc.) and are attracted to moving objects during the day. Control methods for the tsetse flies have included the removal of breeding sources, the use of a variety of insecticides, and well as traps.

Trypanosoma cruzi

Trypanosoma cruzi was first discovered in 1909 in Brazil by the scientist Carlos Chagas, in the midgut of its insect vector the kissing bug (*Triatoma infestans*). It is known to infect man and a wide range of domestic and wild species, including dogs, cats and rodents.

Chagas disease (named after Dr. Chagas) or American Trypanosomiasis is caused by *Trypanosoma cruzi* and is mostly prevalent in Brazil, Argentina, Uruguay, Chile, Venezuela, Central America and the Caribbean and afflicts approximately 15–20 million people. It is a disease primarily affecting low-income people living in rural areas. Chagas disease is also the leading cause of cardiac failure for men age 20-40 in Brazil due to population and workforce movements into area infested with the insect vector. Houses built of adobe, mud, or thatch with cracks in the walls provides a suitable habitat and breeding habitat for the insect vectors.

Transmission of the trypanosomes depends on the presence of the parasite, vectors, reservoirs, and the host being present in the same location. Important reservoir animals in the United States are raccoons, opossums and armadillos. Also, in addition to animal reservoirs, insect vectors (*Triatoma infestans*, *Panstrongylus megistus*, *Rhodnius prolixus*) belonging to the family Reduviidae may also transmit the parasite. These bugs tend to inhabit cool, dark and damp areas, which makes underdeveloped communities with dilapidated, inadequate or simply unsanitary housing especially at risk for the infection.

Transmission of the parasite occurs when reduviids or kissing bugs that live in the cracks and crevices of substandard housing ingest the parasite in the blood from infected humans or animals. The protozoan *Trypanosoma cruzi* multiplies in the digestive tract of the insect and is eliminated in the feces on a person's skin, usually while the individual is sleeping at night. The individual often rubs the contaminated feces into a bite wound, an open cut, the conjunctiva or other mucous membranes. In addition, transmission through blood transfusion and the placenta have been reported. Animals may also become infected with the protozoa via similar methods, or by actually eating infected bugs. Transmission does not occur through the bite of the insect vector.

Some individuals may be infected and not show any symptoms of the disease until many years after infection (chronic form). The most recognized symptom of acute Chagas disease (especially in children) is the Romana's sign, or the swelling of the eye on one side of the face. Other symptoms may include fever, fatigue, enlarged liver and spleen and swollen lymph nodes, followed by convulsions and cardiac involvement. In infants and very young children with acute Chagas disease, cardiac arrest and death may result.

Beside the Romana's sign, and because many individuals do not show symptoms of the disease, xeno-diagnosis (process whereby a small sample is injected into the body of an animal and after a period of time, the animal is tested for the amastigote form) may be able to determine the presence of the protozoan. Another form of laboratory diagnosis involves the culture of tissue aspirates especially from the lymph nodes on culture plates (N.N.N culture plates).

No vaccination or drug is both safe and effective in the prevention and treatment of Chagas disease. Some drugs (anti-inflammatory doses of glucocorticoids and other supportive measures) may help to alleviate the symptoms associated with Chagas disease.

The usual method of insect control is the application of residual insecticides to the interior surfaces and roofs of the houses. In rural areas where the bugs may live and breed in substandard housing, plastering the walls to cover up cracks and crevices, or replacing homes with bricks and cement blocks can significantly contribute to the elimination of the insects inside of the home. The use of bed nets while sleeping has also shown to be effective in preventing bites from the bug. Outside, removal of the breeding sites of animals where the insects are found may also help to reduce the population of the insect vector, thereby reducing possible contact with the human host.

Trypanosomes in animals

Some trypanosome species have been reported from wild and domesticated animals. These trypanosomes may be transmitted biologically or mechanically by various biting insects. *Glossina* or tsetse fly have been shown to transmit *Trypanosoma rhodesiense*, *T. gambiense*, *T. brucei*, *T. congolense*, *T. vivax*, *T. evansi* and *T. suis*. While insects such as tabanids have been shown to mechanically transmit *T. evansi* in animals, *T. equiperdum* in horses and camels is transmitted by direct blood contact during copulation. Some other trypanosomes, such as *T. thelieri* and *T. cervi*, are found in animals in the United States, but these species of the parasite are not pathogenic.

Of the various diseases associated with trypanosomes in which wild and domesticated animals are involved, *T. brucei* produces a disease called nagana which infects a wide range of animals, especially cattle. This trypanosome normally produces chronic infections in animals and may also produce different clinical signs. A severe acute case may be seen in horses, donkeys, dogs, goats and camels. In cattle, a chronic but sometimes fatal form of infection exists when the animals are infected by *T. brucei*.

When livestock are infected with *T. brucei*, they are mass-treated with drugs such as ethidium, isomethamidium or berenil. These drugs are effective, both for treatment and for prophylaxis, but may also be mutagenic. Slaughtered cattle fed these drugs may only be used for human consumption several months after drug treatment in order to avoid any residual effect of the drugs.

Recent findings and issues associated with the trypanosomes

Despite some progress with the trypanosomes, resurgence of the parasite remains a concern. During the 19th and 20th centuries, trypanosomiasis was among the vector-borne diseases that prevented the development of large areas in the tropics, especially Africa. Fortunately, by 1910, it was shown that trypanosomes require blood sucking insect vectors, and over the next 50 years prevention and control programs were applied to curb the vectors. These programs placed emphasis on the elimination of the vector breeding sites, environmental hygiene and the limited use of chemical insecticides. However, the success of curbing the vectors was short lived, because by 1970 a reemergence of the vectors was seen and later intensified, over the next 20 years. Trypanosomes now infect over two million people each year.

Several factors have been implicated for the increase and reemergence of trypanosome related diseases. These factors include the following:

Civil and political unrest. Many of the affected African countries are currently experiencing wars, leading to the disruption of the available forms of controlling both the spread of the vector, and the parasite. War has also aided in the disruption of the ecosystem, and the displacement of the vectors from their native breeding grounds. In addition, many of the citizens of the affected countries have also been displaced, and are affected by the ongoing poverty.

Competing national health priorities. Currently, most of the available local resources are being put into research involving the prevention and cure of AIDS and malaria. Due to the efforts to solve the problems related to these diseases, funding for trypanosomiasis is limited.

Lack of funding support to aid in the availability of new drugs, vector control and diagnostic tests.
There has been very limited funding to support new drugs, vector control, and diagnostic tests. Due to the high level of poverty in most of the affected countries, most individuals cannot afford to buy the drugs to treat their illnesses. It is estimated that less than 10% of the infected individual are treated. Furthermore, there are no vaccines to protect the uninfected population from the parasite, because the parasite is able to mutate in order to evade the immune system.

Recently, several organizations including the World Health organization, the World Bank and some pharmaceutical companies have decided to look into the problem of the reemergence of trypanosomiasis. They have decided to employ the following methods to control the disease: increase research on vaccines, provide environmentally safe insecticides, educate the masses on the parasite and how to deal with it, and increase the availability of drugs to the infected areas so that people can afford them. Only by these concerted efforts will we be able to have a significant affect on trypanosomiasis.

See also, TSETSE FLIES, *Glossinia* SPP. (DIPTERA: GLOSSINIDAE).

Clarence M. Lee
Howard University
Washington, DC, USA
and
Earlene Armstrong
University of Maryland College Park
Maryland, USA

References
Barrett, M. P. 1999. The fall and rise of sleeping sickness. *Lancet* 353: 1113–1114.
Beaty, B., and W. C. Marquardt. 1966. *The biology of disease vectors.* University Press of Colorado, Boulder, Colorado. 632 pp.

Bogitsh, B. J., and T. C. Cheng. 1990. *Human parasitology*. Saunders College Publishing, Philadelphia, Pennsylvania. 435 pp.

Lambrecht, F. L. 1985. *Trypanosomes and hominid evolution*. Bioscience 35.

Mulligan, H. W. 1970. *The African trypanosomiasis*. Wiley-Interscience (John Wiley and Sons), New York, New York. 950 pp.

Welburn, S. C., E. Fevre, and P. Coleman. 1999. Sleeping sickness rediscovered. *Parasitology Today* 15: 303–305.

TRYPANOSOMIASIS. A disease of vertebrate animals caused by trypanosomes.

See also TRYPANOSOMES.

TRYPSIN MODULATING OOSTATIC FACTOR. (TMOF) In female mosquitoes, the decapeptide hormone TMOF is synthesized by the ovarian follicle after a blood meal, affects the midgut cells, and regulates digestion.

TSETSE FLIES, *Glossina* SPP. (DIPTERA: GLOSSINIDAE). Tsetse flies (*Glossina* spp.), pronounced "set-see" or "tet-see," are found only in Africa where they range discontinuously from coast to coast, limited primarily by environmental and ecological factors. They infest 37 countries and about 10 million km^2 of sub-Saharan Africa. Their negative impact on the potential for economic development is immense, and they often have been blamed for the widespread poverty that exists in tropical and subtropical Africa.

Tsetse flies have a significant impact on human activities because they are obligatory blood feeders and they transmit blood parasites of the genus *Trypanosoma*, which cause sleeping sickness. When untreated or treated too late, sleeping sickness is a fatal disease. Because many tsetse species feed on humans and domestic animals, as well as on the wild animals that serve as immune reservoirs of the parasites, the potential for transmission can be very high. The incidence in fly populations of trypanosomes that cause animal trypanosomosis (trypanosomiasis), for example, often exceeds 20% and sometimes is as high as 90%.

Animal sleeping sickness

The major impact of animal sleeping sickness, nagana, is to preclude maintenance of domestic animals. Thus, this disease has restricted cattle production in the tsetse-infested areas to less than 15% of the carrying capacity of the land. Without draught animals, inhabitants have been limited to subsistence level farming because it is necessary to till the land by hand. Without cattle and other domestic animals, severe protein deficiency is widespread. The estimated annual cost and losses in potential animal and crop production is U.S. $4 billion.

Preventive or curative drugs for control of nagana in livestock are only partially effective and often uneconomical due to the development of resistance. Trypanotolerant livestock, such as the N'Dama cattle in western Africa, have provided partial relief, but under heavy fly challenge these small cattle also succumb to the parasite. Immunization against trypanosomes has not as yet been successful due to the antigenic plasticity of the parasites. Thus, the primary means of preventing transmission is by elimination of contact between vectors and hosts by control of the vectors. Tsetse flies are the only significant vectors of trypanosomosis, although some limited mechanical transmission results from the feeding activities of other biting flies (*Stomoxys*, *Tabanus*, etc.).

Human sleeping sickness

Human sleeping sickness, transmitted by tsetse flies from man to man in western and central Africa (*Trypanosoma gambiense*) and from animal

Fig. 1086 Adult tsetse fly, *Glossina morsitans*, engorged with blood. (Photo by D.F. Lovemore.)

reservoirs to man in eastern and southern Africa (*Trypanosoma rhodesiense*), has a lower incidence than animal trypanosomiasis. This is because of the relatively fewer encounters between flies and humans than between flies and livestock, and also because of the much lower rate of *T. gambiense* and *T. rhodesiense* infections in fly populations. In 1999, over 45,000 new human cases were reported, but due to incomplete reporting it is estimated that the actual human incidence was probably closer to 300,000 to 500,000. Historically, millions have been killed by trypanosomosis. Curative drugs are harsh, but can be effective with early diagnosis. Neurological disorders are common following recovery from infection. Due to civil unrest and disruption, combined with reduced control effort in many infested regions, the 60 million individuals estimated to be at risk of infection are subject to increased likelihood of severe disease outbreaks.

Tsetse biology

Vector control is complicated by such a wide distribution of the flies, which usually are found in broad belts that are more or less continuous and often cross international boundaries. They can be difficult to detect in areas where the fly populations are advancing or at low density. Furthermore, there are over 20 distinct vector species, although not all are important vectors and often several species co-exist in the same areas.

Members of the *morsitans* group, which inhabits savannah grasslands, and the *palpalis* group, which is generally found in lacustrine and riverine situations, are the primary vectors of both human and animal trypanosomosis. Because of ecological requirements, the *palpalis* group is generally restricted to areas near the dense vegetative cover found in aquatic situations, whereas species of the *morsitans* group have a wider range of ecological niches and are more widely dispersed.

All species are larviparous, not laying eggs but producing living young that have been nurtured in the uterus of the female fly for several days. They deposit only one offspring at a time and, even though the adults occasionally survive for as long as 90 to 100 days in nature, their biotic potential is quite low compared to that of other dipteran species. The larvae usually burrow into the soil or other suitable substrate, where they pupate and develop for 3 to 4 weeks or more before emerging as adults. Thus, at any given time a significant portion of the fly population is underground, a fact that plays an important role in control strategy.

The adults move about in a diurnal pattern, governed to a great extent by temperature. For example, in the summer in southern Africa, *G. m. morsitans* feeds in the mornings when the temperature is moderate, seeks shelter during midday when the temperature is high and the insect exhibits negative phototaxis, and resumes feeding as the temperature declines in the latter part of the day. At night the flies rest. Nutritional requirements cause the flies to seek blood meals about every third day, and the males also actively visit host animals in search of

Habitat relationships of *Glossina* spp.: savannah (*morsitans*), riverine (*palpalis*), and forest (*fusca*) group tsetse flies

Morsitans	Palpali	Fusca
austeni	calignea	brevipalpis
longipalpis	fuscipes fuscipes	fusca congolensis
morsitans centralis	fuscipes martini	fusca fusca
morsitans morsitans	fuscipes quanzensis	fuscipleuris
morsitans submorsitans	pallicera pallicera	haningtoni
pallidipes	pallicera newsteadi	longipennis
swynnertoni	palpalis gambiensis	medicorum
	palpalis palpalis	nashi
	tachinoides	nigrofusca hopkinsi
		nigrofusca nigrofusca
		schwetzi
		severini
		tabaniformis
		vanhoofi

mates feeding there. In this process, the sexually appetitive males move about several times daily, often resting on host animals grazing or traversing the bush. In the course of these activities, the flies alight on numerous resting sites in their preferred habitat. The preferred resting sites vary somewhat for the different species.

Tsetse behavior and ecology in relation to control

The association of tsetse flies with particular vegetation types has played a major role in determining control practices. Riverine species of the *palpalis* group are generally found within a short distance of their breeding sites, in or close to the riverine vegetation canopy that provides protection from sunlight, high temperature, and lower humidity of the surrounding, more open vegetation. This habitat affords the fly an environment replete with a variety of hosts, and incidentally provides direct contact with humans and domestic animals when they come to the water. Partial or complete removal of the riverine vegetation results in the reduction or elimination of these tsetse flies. However, in more humid areas, species of the *palpalis* group may be found well away from surface water. *Glossina tachinoides*, for example, has been observed to breed in peridomestic habitats, where bush encroachments around villages have allowed flies to find suitable resting and larviposition sites well away from the typical riverine situation. Similar observations with species of the *morsitans* group have revealed their preferences for selected resting sites and their dependence upon certain

Fig. 1087 Adult female tsetse fly depositing fully developed larva.

ENCYCLOPEDIA OF ENTOMOLOGY

characteristics of the habitat for survival. Ruthless removal or discriminative clearing of selected portions of the habitat results in major reductions of fly density because the flies cannot then find suitable protective cover when environmental stresses are maximal.

Similarly, feeding behavior characteristics of the tsetse fly have provided a means for reducing population density with a corresponding reduction in the risk of trypanosomosis. Most tsetse species feed on a variety of animals but prefer certain hosts, sometimes riverine. For example, in Zimbabwe, *G. m. morsitans* feeds on numerous hosts, but prefers bushpig, warthog, bushbuck, and kudu. Thus, by selective reduction of these four species, the fly density could be reduced along with the incidence of trypanosomosis.

However, for economic, environmental and aesthetic reasons, game reduction and wide-scale bush clearing are no longer acceptable control measures.

Awareness of specific tsetse resting niches spurred the development of selective application methods for the organic insecticides that first became readily available in the 1940s. Because most species have well defined resting site preferences, such as the lower boles of trees of certain diameters, application of persistent insecticides to these sites provided excellent control without resort to broadspread applications. Furthermore, since the flies have a long life span, during which the majority probably move less than 1,000 meters from their origin, not all resting sites needed to be sprayed to achieve a high level of control. Flies below ground in the pupal stage eventually emerge and sooner or later come to rest on a treated surface. Treatment of 10% to 20% of the preferred resting sites in some tsetse habitats was sufficient to eliminate or substantially reduce the fly population when the toxicant persisted for several months.

Approaches to fly management

Glossina spp. are extremely susceptible to insecticides. Aerial applications of very low rates of nonpersistent aerosols repeated at 2 to 3 week intervals have been used to initially control the ambient fly population and then the flies that have recently (0 to 15 days) emerged, thereby precluding reproduction and eliminating the fly population after 6 or 7 spray cycles. Alternatively, selectively placed helicopter applications of persistent insecticides have been used with dramatic results. These applications were discriminative in that they included only those habitats that offer refuge for the fly in the dry season, when tsetse fly distribution typically is more restricted. Most of the application was deposited on the leaves of the upper canopy, where tsetse rest at night. These residual types of application were particularly effective in areas with discrete wet and dry seasons.

But the broadspread application of pesticides has given way to the use of attractants and trapping to expose the flies to spot treatments of pesticides. Both visual and olfactory components are involved in tsetse host-seeking behavior. In the 1970s, animal emanations attractive to tsetse flies were found to be highly effective for trapping several species of tsetse, especially when combined with suitable visual attractants. By utilizing a persistent insecticide in conjunction with attractant devices, significant population reductions can be achieved in a matter of months. Such trapping provides a relatively inexpensive method of control, and because the pesticide is incorporated into the attracting device the probability of environmental contamination is low.

However, the attractant approach does not guarantee elimination of the vector, and thus the disease, even though fly population density in most situations can be reduced to well below the threshold level necessary to maintain regular transmission. Extensive study and operational level trials of the sterile insect technique have demonstrated the feasibility of this approach for area-wide elimination of tsetse where geographically isolated fly populations already have been reduced to low density. Used to eliminate *Glossina austeni* from the main island in Zanzibar, the inhibition of natural reproduction with sterile fly releases after reducing fly density by trapping is an environmentally friendly method of control that has potential application for most of Africa.

See also, TRYPANOSOMES.

David A. Dame
University of Florida
Gainesville, Florida, USA
and
Anthony M. Jordan
University of Bristol
Bristol, England

References
Buxton, P. A. 1955. *The natural history of tsetse flies.* H. K. Lewis & Co., Ltd., London, United Kingdom. 816 pp. + 47 plates.

Food and Agriculture Organization. 1992. *Programs for the control of African animal trypanosomiasis and related development.* FAO Animal Production and Health paper 100. Proceedings of Symposium, Harare, Zimbabwe. www.fao.org/docrep/004/to559e/to599e00.htm

Leak, S. G. A. 1998. *Tsetse biology and ecology: their role in the epidemiology and control of trypanosomosis.* CABI Publishing, Wallingford, Oxon, United Kingdom. 592 pp.

Mulligan, H. W. (ed.) 1970. *The African trypanosomiases.* George Allen and Unwin, Ltd., London, United Kingdom. 950 pp.

World Health Organization. 1998. Control and Surveillance of African trypanosomiasis. Technical Report Series No. 881. vi + 113 pp. www.who.int/health-topics/

TUBERCLE. A small raised area or extension of the integument. In caterpillars, a hair often originates from these raised areas.

TUBE MOTHS (LEPIDOPTERA: ACROLOPHI-DAE). Tube moths, family Acrolophidae, total 270 species in the New world, mostly in the large genus *Acrolophus*; actual fauna likely exceeds 350 species. The family is part of the superfamily Tineoidea, in the section Tineina, subsection Tineina, of the division Ditrysia. Two subfamilies are used: Amydriinae and Acrolophinae (formerly included in Tineidae). Adults small to medium size (9 to 60 mm wingspan), with rather robust bodies and roughened head scaling and usually large recurved labial palpi; haustellum naked (unscaled); maxillary palpi minute, 2-segmented.

Fig. 1088 Example of tube moths (Acrolophidae), *Acrolophus plumifrontellus* (Clemens) from Florida, USA.

Maculation is mostly somber hues of brown or black, sometimes with some spotting and other markings. Adults are mostly nocturnal, but some may be crepuscular. Larvae are root feeders, mostly of grasses, and construct long silken tubes to feed on hostplant roots. A few are economic, mainly as turf grass pests.

John B. Heppner
Florida State Collection of Arthropods
Gainesville, Florida, USA

References

Becker, V. O., and G. S. Robinson. 1981. Neotropical taxa referable to *Acrolophus* (Lepidoptera: Tineidae). *Systematic Entomology* 6: 143–148.

Davis, D. R., and E. G. Milstrey. 1988. Description and biology of *Acrolophus pholeter*, (Lepidoptera: Tineidae), a new moth commensal from gopher tortoise burrows in Florida. *Proceedings of the Entomological Society of Washington* 90: 164–178.

Hasbrouck, F. F. 1964. Moths of the family Acrolophidae in America north of Mexico (Microlepidoptera). *Proceedings of the United States National Museum* 114:487–706.

Hinton, H. E. 1955. On the taxonomic position of the Acrolophinae, with a description of the larva of *Acrolophus rupestris* Walsingham (Lepidoptera: Tineidae). *Transactions of the Royal Entomological Society of London* 107: 227–231.

TURGOR. The distension of living tissue, usually plant tissue, due to internal pressures (hydration).

TSETSE FLIES. Members of the family Glossinidae (order Diptera). See also, FLIES, HISTORY OF INSECTS.

TUBE-MAKING CADDISFLIES. Members of the families Polycentropodidae AND Psychomyiidae (order Trichoptera). See also, CADDISFLIES.

TULAREMIA. See also, TICKS.

TUMBLING FLOWER BEETLES. Members of the family Mordellidae (order Coleoptera). See also, BEETLES.

TUNGIDAE. A family of fleas (order Siphonaptera). See also, FLEAS.

TURFGRASS INSECTS OF THE UNITED STATES: BIOLOGY AND MANAGEMENT.

Turfgrasses are grown throughout the United States, covering over 30 million acres (12.2 million hectares). As urban environments have grown, the use of turfgrass in the U.S. has increased over the past several decades. Subsequently, large areas of land have been developed with dense, dark green, uniform turf providing numerous practical, recreational, and ornamental uses. Uses for turfgrass include dense turf for soil stabilization and erosion prevention; runways for rural airports; cemeteries; parks and recreational facilities for relaxation, picnics and general family activities; sports facilities and complexes including baseball, cricket, football, lawn bowling, polo, soccer, tennis and, especially, golf courses. Turfgrasses often are considered the most intensively managed plantings in urban landscapes. It is estimated that $45 billion is spent per year in the U.S. for turfgrass culture in its many forms.

Turfgrass can be designated as either cool-season or warm-season grasses depending on their climatic adaptations. These two categorical designations can be further divided into four distinct turfgrass adaptation zones in the U.S. including (1) cool, humid; (2) cool, arid or semi-arid; (3) warm, humid; and (4) warm, arid or semiarid. Despite these partitioned zones, there is considerable overlap of the species of turfgrass grown in various regions of the U.S. As a result, it is rather difficult to effectively provide an accurate description for the boundaries of specific insect pests and nearly impossible to give a comprehensive classification for all turfgrass pests including disease pathogens, insects and weeds.

Some insect pests are host specific, feeding only on certain turfgrass species or types, while others are non-discriminate, infesting a diversity of grass and broadleaf plant materials. An insect pest such as the European chafer typically only infests cool-season areas, whereas mole crickets and fire ants infest warm-season turfgrasses. Preference of an insect pest can be either for a host (e.g., turf-type or species) or it may be a reflection of the geographic range of the pest. Destructive turfgrass insects can be divided into three primary groups according to the habitat in which the destructive life stage spends its life in the turfgrass ecosystem. They include (1) leaf and stem, (2) thatch, and (3) root zone/soil. This method of grouping is highly valuable because specific control tactics or strategies implemented for each

group have a direct bearing on the effectiveness or satisfaction of control for the respective insect pest. Some insects are capable of occupying more than one habitat during its development, and may even occupy all three habitats during its various life stages. Thus, a modified designation of turf habitats has been suggested including (1) foliage and stem, (2) crown and thatch, and (3) soil.

Effective control of any pest (plant pathogen, insect or weed) can be accomplished successfully only with a comprehensive understanding of both the host and the pest. Factors such as growth habits and cultural requirements of the host (turfgrass) must be understood. Knowledge of the biology, behavior, ecology, life history for the host and pest are critical for making decisions. Symptomology or type of damage caused by the pest(s), accurate identification of the pest, information regarding the time of year, growth stage of the host and the pest, and environmental conditions under which pest damage is most likely to occur are all important factors that also should be understood. This information is the foundation for the Integrated Pest Management (IPM) philosophy. IPM is a commonsense approach to pest management that is effective and environmentally responsible. IPM relies on a combination of preventative and corrective measures to keep pest densities below levels that would cause unacceptable damage. IPM includes sampling and monitoring, accurate pest identification, decision-making, appropriate intervention, follow-up, and detailed record keeping. In terms of the appropriate intervention tactic, several control options are available including biological, chemical, cultural, and plant resistance. IPM is a decision- making process that does not preclude the use of pesticides, however, it does not rely upon chemical control as its first line of defense. The ultimate goal of IPM is to manage pests effectively, economically, and with minimal risks to people and the environment. For this reason, turfgrass managers must constantly be aware of potential pest problems that they may experience. The listing that follows provides vital information for the major insect and mite pests affecting turfgrass in the U.S.

Foliar and stem inhabitants: surface chewing and sucking insects

Armyworms (Lepidoptera: Noctuidae). Armyworms get their name because of their gregarious nature of crawling in large numbers from one field or

Fig. 1089 Some important turfgrass pests: top left, green June beetle; top right, Japanese beetle; second row left, fall armyworm; second row right, bronzed cutworm; third row left, a June beetle larva (white grub); third row right, black turfgrass ataenius adults and larva; bottom row left, glassy cutworm; bottom row right, black cutworm.

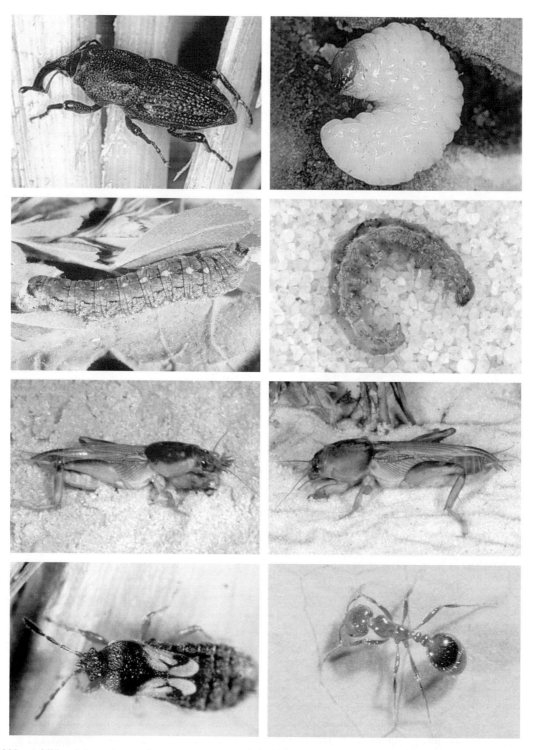

Fig. 1090 Additional important turfgrass pests: top row left, billbug adult; top row right, billbug larva; second row left, variegated cutworm; second row right, tropical sod webworm; third row left, tawny mole cricket adult; third row right, southern mole cricket adult; bottom row left, immature chinch bug; bottom row right, red imported fire ant.

feeding site to another after they have exhausted their food supply. Although these pests are often sporadic, they do have the potential for outbreaks. Most armyworm larvae are thick-bodied, hairless, striped caterpillars that chew on the foliage of turfgrass. There is one armyworm species that frequently attacks turfgrasses in both the northern and southern United States, the 'common' or 'true' armyworm.

Armyworm, *Pseudaletia unipuncta* (Hawthorn). Frequently referred to as the 'common' or 'true' armyworm in order to differentiate it from other armyworm species, the armyworm is a native species that is distributed throughout the United States and southern Canada east of the Rockies.

Armyworm adults are evenly pale-brown to grayish-brown moths with a wingspan of approximately 38 to 40 mm. The distinct white spot in the center of their front wings readily identifies them. The hind wing is dirty white. The female adult moths lay pearly white eggs, 0.5 mm diameter, in masses containing 20 to several hundreds of eggs. Young larvae, more than 2 mm long, are pale green. Mature larvae range from 35 to 50 mm in length, grayish to greenish-brown, with two pale-orange stripes along each side of the body and another pale-colored, broken stripe down the middle of the back. The head capsule is honeycombed with dark lines. Pupae are reddish-brown, about 16 to 19 mm long and shaped like a football.

Armyworm caterpillars feed on a wide range of grasses. They especially like barley, corn, millet, oats, rice, rye and wheat. When such food supplies are depleted, they will attack nearby turfgrass. Young larvae initially feed on tender foliage resulting in a skeletonized appearance. Third instar and older larvae are primarily nocturnal and consume all or part of the leaf.

Armyworms overwinter as partially grown larvae or pupae in the southern half of the U.S. and possibly as partially grown larvae in the northern half. Infestation in the North may also be the result of spring migration flights of adults. In the temperate United States, there are typically two generations (broods) per year. The number of annual broods depends primarily on latitude. The first generation is a result of annual moth migration in the spring (i.e., late April through May). Once the eggs are laid, they typically hatch in as few as three days. Larvae usually go through six instars over a period ranging from 20 to 48 days. Pupation occurs in the soil and

the duration of the pupal stage averages 15 days. Adult armyworm moths are nocturnal, with most flight activity occurring within two hours after sunset. A second generation occurs during June and July. In the southern regions, four to five generations may occur. Infestations in turfgrass are often most severe following drought conditions. Populations have a tendency to fluctuate widely from generation to generation and year to year.

Fall armyworm, *Spodoptera frugiperda* (J. E. Smith). The fall armyworm is found throughout much of the South and is a frequent problem in turfgrass, especially in the Southeast westward into Texas, New Mexico, Arizona and southern California. It does not overwinter outside of Florida and the immediate Gulf Coast area, and must disperse to reinfest areas each year. Populations usually do not reach the more northern areas until very late summer or fall.

Fall armyworm moths have a wingspan of about 38 mm and the forewings are generally a dark gray and mottled in appearance. There is a white spot near the tip of the forewing and a light diagonal mark in the middle of each wing. The back wings are white. The eggs are laid in clusters of 50 to 100 and are initially greenish white, but soon turn dark. The egg clusters often look gray and fuzzy from the scales of the female's wings. The caterpillars are green to brown, to almost black. The head is marked by a typical inverted "Y" on the "face" or front. There is a longitudinal dark stripe that runs the length of the body and a fainter, pale mid-dorsal stripe. Each abdominal segment has four small black dots. The fully grown larvae are 35 to 50 mm long. The pupa is 13 mm long and is found in the soil. It is reddish-brown, somewhat football shaped, and becomes black prior to emergence of the moth.

Fall armyworms feed on a wide range of plants, but prefer grasses of all types. However, this pest is most commonly associated with bermudagrasses in the southern U.S. They can and do feed on both cool and warm season turfgrasses and are often most likely to attack lush, green, dense areas of turf. Fall armyworms are very common problems in areas that have recently been sodded or sprigged. The small larvae skeletonize the leaf blades while larger larvae may consume all above-ground plant parts. The turf also may turn brown and look dead from drought stress. The larvae often move from heavily damaged areas into new food sources. Warm season turf usually recovers from such feeding, but if it occurs late in the season, the turfgrass may be stressed going into the winter.

Fall armyworm moths are active at night and may be attracted to lights. They lay eggs on a wide range of objects, usually light-colored, that are adjacent to turf areas. These sites may include fence posts, metal buildings, flagpoles, the underside of leaves of various landscape plants, gutters, etc. Eggs hatch quickly in warm temperatures and the small larvae spin down on silken thread to the turf surface and begin feeding. Small caterpillars feed at any time of the day, but larger ones feed more at night to avoid predation by birds. The fall armyworms will feed for 2 to 3 weeks and then burrow into the soil to pupate. New moths emerge two weeks later and the cycle starts again. Multiple infestations can occur in the same location and population as high as 100 per square meter have been observed. The number of generations per year depends upon location and migration of the moths. Migration varies each year depending upon weather conditions, which are responsible for enhancing the migration of the moths. In the extreme South, four or more generations may occur and the damage from the larvae may begin in the spring. In locations further north, only one generation will occur and it may not be seen until August or September. Wet springs sometimes increase the likelihood of fall armyworm outbreaks.

Chinch bugs (Hemiptera: Lygaeidae). Chinch bugs damage turfgrass by sucking plant juices from stems (i.e., leaf sheaths) and crowns, causing gradual yellowing and eventual dead patches of turf. Damage occurs predominantly during hot, dry periods of time in mid- to late summer. Sunny areas are often most heavily infested. There are four chinch bug species that are important turfgrass pests in the United States: (1) the hairy chinch bug, (2) the common chinch bug, (3) the 'buffalograss' chinch bug, and (4) the southern chinch bug.

Hairy chinch bug, *Blissus leucopterus hirtus* Montandon. The hairy chinch bug occurs across the eastern Canadian provinces and from the northeastern United States to the Middle Atlantic states south to Virginia and west to Minnesota.

Chinch bugs have gradual metamorphosis, thus the immature (nymph) life stage is similar in physical appearance and feeding habits to the adult, however,

there is variation in size, color and wing development. There are five nymphal stages; the first instar nymphs are tiny, approximately 2.5 mm long, bright red with a white band across the abdomen. As the nymphs develop and mature, their color changes from bright red to orangish-brown and ultimately black once they reach the fifth instar. Hairy chinch bug adults are approximately 6.0 to 6.4 mm long and 1.0 mm wide, and females typically are slightly larger than males. Adults range in color from grayish-black to black and white and often are covered with fine hairs and the legs are typically a dark, burnt-orange tint. There is a triangular-shaped marking in the middle of the outer edge of each wing. Populations of hairy chinch bugs may consist of mostly long-winged (macropterous) or short-winged (brachypterous) individuals, or contain both wing forms. The long-winged chinch bugs have wings that extend to the tip of the abdomen, and short-winged chinch bugs have wings that extend only halfway to the tip. The adults lay eggs that are tiny, elongate, and bean-shaped; initially they are white, but become orange-red within a few days prior to hatching.

The hairy chinch bug is one of the most important insect pests of northern turfgrasses, especially home lawns. It prefers cool-season turfgrasses including creeping bentgrass, fine fescues, Kentucky bluegrass, and perennial ryegrass. However, hairy chinch bugs will feed also on zoysiagrass, a warm-season turfgrass species that occasionally occurs where cool-season turfgrasses dominate. Both the adults and nymphs cause damage to turfgrass plants by inserting their piercing-sucking mouthparts into stems and crowns, extracting plant juices while pumping toxic salivary fluids into the plant. Subsequently, turfgrass plants are damaged by the loss of plant fluids as well as the clogging of the conducting tissues within the stems. Generally, hairy chinch bug occurs sporadically in scattered aggregations rather than uniformly distributed across the turf. Damage typically occurs during hot, dry periods in mid- to late summer when turf is commonly experiencing drought stress symptomology. Additionally, hairy chinch bug prefers open, sunny areas of turf that have heavy thatch accumulation as well as high percentages of perennial ryegrass and fine-leaf fescue. Initial damage often appears as irregular patches of wilted, yellowish-brown turf that frequently is mistaken for drought stress. As populations grow, often reaching 200 to 300 bugs per 0.1 square meter, feeding damage intensifies and damaged patches of turf begin to coalesce into large areas of dead or dying turf that does not recover regardless of irrigation or rainfall.

Typically there are two generations per year in the temperate United States; however, in the most northern portion, as well as Canada, there is only one generation per year. The adults overwinter in the turfgrass thatch, leaf litter, and similar sites. They become active in the early spring as temperatures reach 10°C. Thereafter, the adults feed and mate for approximately two weeks before females begin laying eggs in mid-April to May. In areas where there is only one generation per year, egg-laying may be delayed by several weeks and occurs over an extended period. Females lay as many as 20 eggs per day for a period of about 2 to 3 weeks. Eggs typically are laid in leaf sheaths or in the thatch. Because insects are cold-blooded animals, they are dependent on temperature, thus the developmental rate of eggs and nymphs varies within species and geographic range. Eggs that are laid during April may require as long as a month to hatch, and eggs laid in midsummer may hatch in as few as 7 to 10 days. Where two generations occur, the first generation typically matures in 4 to 6 weeks, usually by mid-July. The second-generation adults then begin laying eggs from mid-July through late August. These eggs hatch soon thereafter, and the nymphs complete development by September or October. As cooler temperatures prevail, the adults seek out protected sites to overwinter. The most extensive damage to turf commonly occurs during periods of heat and drought stress, typically in late July and August when the first generation adults are actively feeding and laying eggs.

Common chinch bug, *Blissus leucopterus leucopterus* (Say). The common chinch bug is an occasional pest of turf, especially in the Great Plains region. It feeds primarily on small grains and other field crops, however, sometimes it damages turfgrass, especially when located in close proximity to maturing small grain fields. Several turfgrass species that are likely to be attacked by the common chinch bug include creeping bentgrass, fescues, Kentucky bluegrass, perennial ryegrass, and zoysiagrass. This species closely resembles the hairy and the buffalograss chinch bugs both in appearance and damage.

Buffalograss chinch bug, *Blissus occiduus*. The buffalograss chinch bug was first identified in Nebraska feeding and causing damage on buffalograss. Upon its discovery, it was understood that the buffalograss chinch bug was limited to buffalograss as a host. However, it has recently been reported damaging zoysiagrass, and may have the potential also to feed on and damage several other turfgrasses and small grain crops. This species closely resembles other chinch bugs species in appearance and damage.

Southern chinch bug, *Blissus insularis* Barber. The southern chinch bug is a major pest throughout the range of its primary host, St. Augustinegrass. Distribution is generally from the Carolinas down through all of Florida and west to Texas. They also can be found in California, Mexico and throughout the Caribbean. Like other chinch bugs, they damage the turfgrass by sucking out plant juices.

Southern chinch bugs are similar to other chinch bugs, but slightly smaller than the hairy chinch bug. The adults are about 1 mm wide and 3 to 3.5 mm long and are black with shiny white wings. Most characteristics are similar to other chinch bugs including the presence of long- and short-winged adults.

The southern chinch bug is one of the most important insect pests of home lawns in the deep South. Virtually all St. Augustinegrass will be infested with southern chinch bugs, but not necessarily at damaging population levels. Southern chinch bugs will feed also on bermudagrass, centipedegrass, bahiagrass, and zoysiagrass. As with the other species of chinch bugs, their feeding on stems, crowns, and stolons removes plant juices, but also injects a toxin. This causes the grass to turn yellow and often die. Populations often are aggregated and damage appears in patches. Drought or heat-stressed areas of lawns are areas that usually are attacked first and most severely.

There are probably two to three generations in the most northern areas of the southern chinch bug range. Further south in northern Florida and the Gulf Coast area, three to four generations are common. In south Florida, seven or more generations can occur. Overwintering in the most northern areas is primarily as adults while all stages are present in the more southern areas.

Cutworms (Lepidoptera: Noctuidae). Cutworms are plump, smooth, dull-colored caterpillars that hide in the turf profile in burrows or aerification holes during the day, emerging at night to feed on the foliage and leaf sheaths of turfgrass plants.

Black cutworm, *Agrotis ipsilon* (Hufnagel). The black cutworm is considered the most destructive cutworm that attacks turfgrasses. It is a major pest of creeping bentgrass putting greens, tees, and occasionally fairways throughout the United States and worldwide. Black cutworm larval feeding damage typically results in small, irregular sunken patches or pockmarks that often are mistaken for golf ball marks. Subsequently, black cutworm damage reduces the smoothness and uniformity of the putting green surface. Because aesthetics and playability are high priorities for golf course superintendents, multiple insecticide applications are done each growing season to control this important turfgrass insect pest.

Adult black cutworm are robust, hairy (scaly) moths with a wingspan of approximately 35 to 45 mm. The forewings are dull-brown to grayish-black, and slightly lighter in color or pale towards the tip of the wings. A distinctive, black, dagger-shaped marking is located in the center of each forewing, approximately 6 mm from the tip. The hindwings are considerably lighter, uniformly cream to dirty white, with darker veins. Black cutworm adult moths typically hold their wings flat over the back in a triangular position when resting. Males can be differentiated easily from females by their comb or feather-like antennae, whereby female moth antennae are filiform or slender and thread-like.

Black cutworm females lay eggs predominantly singly near the tip of grass blades. Eggs are approximately 0.5 mm in diameter. They are cream colored initially, later becoming darker as the embryo develops. Larvae are hairless with the exception of a few scattered bristles. The dorsal side, above the spiracles, ranges in color from gray to black, and the ventral side is typically lighter gray. Spiracles are typically black and extend from the prothorax to the terminal abdominal segment. As a whole, the body of a black cutworm larva is without distinct stripes or marking except for an indistinct, pale stripe down the middle of the dorsal side. Under magnification, the cuticle (skin) has a pebble-like surface, and is generally greasy in appearance. Like many other caterpillars, there are three pairs of true legs on the thorax and five pairs of fleshy prolegs on the abdomen. Neonate larvae, less than 24 hours old, are approximately 3.5 mm long whereas mature larvae range from 30 to 45 mm long and are approximately 7 mm wide. Black cutworm larvae typically have six instars, however, they sometimes have seven. Upon maturity, black cutworm larvae pupate in the turf profile predominantly in the soil, and occasionally in the thatch associated with the turf. Pupae are reddish-brown to dark brown in color and are about 19 mm long.

The black cutworm is distributed throughout most of North America, as well as in Europe, Asia, Africa and elsewhere. Depending on geographic location, latitude and temperature, there can be two to six generations per year. In the southern U.S., in states such as Louisiana and Alabama, there are five to six overlapping generations, whereas in the northern U.S., in states such as Minnesota and Wisconsin, there may be only one or two generations per year. The black cutworm has difficulty surviving subfreezing soil temperatures, thus, it may be unable to overwinter in some years north of the transition zone (a narrow range between the cool-season temperate region and warm-season zone, extending from southern Illinois, Missouri and Ohio to Tennessee and North Carolina). As a result, spring infestations in northern areas begin with the arrival of migratory adults from southern states. Black cutworm moths can be carried several hundred miles in a few days by strong southerly winds. In Wisconsin, the first spring migrants typically arrive in late April to early May; damage from their offspring begins appearing in early to mid-June. Damage from the second generation (brood) starts appearing late July through early August. Occasionally a third generation will appear in late summer or early fall if temperatures are conducive, otherwise it is understood that third generation adults will migrate south as temperatures are less favorable. Multiple larval sizes (instars) may be present at a given time because black cutworm populations tend to spread out their emergence and egg-laying as the season progresses. In addition, the first generation appearance of black cutworm in states located in the northernmost temperate U.S. may have a later arrival of migrating adults than states in the southernmost location of the temperate U.S. In the southernmost states (e.g., Alabama, Louisiana, Mississippi and Georgia), first generation larvae may be present as early as February and March and cutworm problems may persist into November and December. In North Carolina, adults have been captured in pheromone traps every month of the year except January. Turfgrass in Florida and Gulf Coast areas may experience problems with cutworms all year. Areas with warm season turfgrass that are closer to the transition zone may experience cutworms as early as March and April through late October.

Both black cutworm moths and larvae are nocturnal. Adults do not cause any turf damage, however, they do feed by sucking nectar from flowers. Soon after emergence, adult females call (attract) males by releasing a sex pheromone (chemical sex attractant). Once fertilized, female moths begin laying eggs. Individual females can lay as many as 1,200 to 1,600 eggs over a 5 to 10 day period. Eggs are predominantly attached to the tips of grass blades singly. Eggs typically hatch in 3 to 6 days depending on temperature. Once hatched, the young larvae (first and second instars) begin feeding on leaf blades, both day and night. Young larvae feed on both the top and bottom sides of the of leaf blades in a skeletonizing manner. As larvae develop and mature, older larvae (third to 6th instars), they become nocturnal (feeding only during the scotophase) as well as developing a subterranean habit, forming silk-lined burrows in the turf thatch or soil. On golf course turf, the larvae will commonly occupy aerification holes, spike marks, or golf ball injury sites. Black cutworm larvae hide in the aforementioned burrows during the daylight, and venture out to feed at night. The larvae typically pass through six molts, maturing in 20 to 40 days. Young larvae do not cause measurable damage, however, large larvae are highly destructive, consuming relatively large amounts of turfgrass foliage in a single night. Once black cutworm larvae reach maturity, pupation occurs within the larval burrow or the turfgrass profile. After approximately 14 to 21 days, the adult moth emerges.

Most black cutworm larvae feeding occurs from around midnight until just before dawn. On golf course putting greens, wandering larvae can move approximately 20 m or more in a single night. This ability to move helps explain why golf course putting greens and tees are sometimes reinfested soon after being treated with a short-residual insecticide. It is likely that black cutworm larvae are moving onto putting greens from the surrounding area.

Bronzed cutworm, *Nephelodes mimians* Guenée. The bronzed cutworm is only an occasional pest of home lawns and golf course turf, and unlike the black cutworm, they rarely damage golf course putting greens and tees. Bronzed cutworm is distributed across the northern half of the U.S. east of the Rocky Mountains, as far south as Tennessee.

Adult bronzed cutworms are similar in size to black cutworm, however, they lack the black dagger-shaped mark on the wings. The forewings are highly variable in color ranging from brown to purplish-gray to maroon, and there is a wide, darker-brown band across the middle of each forewing. The hindwings are buff-colored similar to the black cutworm. Bronzed cutworm eggs resemble black cutworm eggs. Larvae are fat-bodied, light to dark brown dorsally and lighter ventrally, and they have a distinctive bronze sheen. Larvae also have a light-yellow stripe that extends longitudinally the entire length of the body on the center of the dorsal side, with another pale longitudinal stripe on each side of the body. Spiracles are black and are located on each thoracic and abdominal segment. Fully developed, mature larvae are 35 to 45 mm long. Bronzed cutworm and black cutworm pupae look similar in color and size.

Although the bronzed cutworm has been well studied in field crops, little is understood about its biology in turf. Unlike most turf-infesting cutworm species, the bronzed cutworm has only one generation per year. Adult moths emerge, mate, and lay eggs in late summer or early fall. The eggs, however, do not hatch until the following spring. After eggs hatch in April, larvae begin feeding and developing until fully developed in mid- to late May, when most damage occurs. As bronzed cutworm larvae reach maturity, they burrow into the soil and form pupal cells where they remain until pupation in mid-August. Thereafter, new adult moths emerge in approximately 30 days.

Variegated cutworm, *Peridroma saucia* (Hübner). The variegated cutworm is an occasional pest in lawns and golf course roughs, especially in rural areas where turfgrass areas are bordered by field crops where infestations may originate. It occurs throughout both North and South America.

Variegated cutworm moths have yellowish to brownish forewings that frequently have a row of small black dashes on the leading edge of the wing. Similar to the bronzed cutworm, they lack the black dagger-shaped marking present on black cutworm moths. Variegated cutworm moth hindwings are whitish, with dark-shaded veins. Variegated cutworm eggs are similar in size and color to black cutworm eggs. However, variegated cutworm eggs are laid in groups of several hundred eggs in a single layer on foliage. Mature larvae are approximately 35 to 46 mm long and vary in color from pale gray to dark brownish-gray. A row of pale yellow dots and dashes extend longitudinally on the dorsal side of the body, and a distinct brownish W-shaped marking is apparent on the eighth abdominal segment, as well as a yellowish or orange area near the terminal abdominal segment. The larvae also have a rather indistinct, black, yellowish or orange marking that extends longitudinally on both sides. Pupae are similar to both bronzed and black cutworm.

The variegated cutworm has from 3 to 6 generations per year depending on geographic location. Adult moths emerge from overwintered pupae in March and April. Larvae from eggs deposited by the first generation moths are typically fully developed by late May. The subsequent generations occur from June to November, with the last generation producing overwintering pupae. A complete generation, egg to adult, typically requires 8 to 9 weeks. Compared to other cutworm species, the variegated cutworm is less subterranean and nocturnal, thus is sometimes feeds fully exposed during daylight, especially on cloudy days.

Sod webworms. There are more than 20 species of sod webworms that feed on turfgrasses in North America, however, only about three or four species are pests in specific geographical regions. Among the more common sod webworm species that attack turfgrasses in the northern U.S. are the bluegrass webworm, *Parapediasia teterrella* (Zincken); the larger sod webworm, *Pediasia trisecta* (Walker); and the western lawn moth, *Tehama bonifatella* (Hulst). All of the aforementioned sod webworms species at one time were grouped together in the genus *Crambus*. Because their biology is similar, they can be discussed together. The tropical sod webworm, *Hertopetogramma phaeopteralis* (Guen(e), has a wide distribution only in the southeastern U.S. and will be discussed separately.

In general, sod webworms are relatively small larvae that live in silk-lined tunnels in the turf canopy, specifically in the thatch and soil, hence the name 'webworms.'

Adult sod webworms are small, buff-colored moths that have wingspans that range in size from 19 to 25 mm. The forewings are primarily whitish, dull gray to tan, with longitudinal stripes as well as other indistinct markings of brown, black, gold,

silver, and even yellow. Such markings are used to identify individual species. Hindwings are typically lighter, whitish to light-gray, with delicate fringes on the outer margins of the wing. When resting, the wings are usually folded over the body. Possibly the most distinguishing characteristic of sod webworm moths is the presence of two small, snout-like projections that extend forward from the front of the head. These snout-like projections are merely mouthparts, but are the reason why sod webworm moths are frequently referred to as 'snout moths.' Moreover, the combination of folded wings and the snout enable sod webworm adults to be distinguished from other turfgrass moths. Sod webworm eggs are extremely tiny, oval to barrel-shaped, with fine, longitudinal ribbing on the surface. Ribbing is distinctive for each species. Sod webworm larvae range in color from beige to gray to brown to greenish, depending on species. Nearly all sod webworm species have characteristic dark, circular spots and coarse hairs scattered randomly over the body. Fully developed larvae are typically 16 to 25 mm long. Pupae are 'torpedo-shaped,' approximately 10 to 13 mm long, and vary in color from tan to dark brown.

Several species of sod webworms are distributed throughout the temperate U.S., however, damage appears to be greatest in the Midwest and the eastern U.S. Sod webworms attack and damage several turfgrass species including creeping bentgrass, Kentucky bluegrass, perennial ryegrass, as well as fine-leaf and tall fescues, however, they will also feed and cause damage on other grasses that are considered weeds. Certain turfgrasses, such as perennial ryegrass and fescues that contain endophytes, are relatively resistant to sod webworms. Sod webworm larvae hide in silk-lined burrows in the turf canopy, emerging at night to feed on turfgrass foliage. Feeding damage results in leaves and stems being chewed off just above the crown; thereafter, the plant material is pulled into the burrow where it is consumed. Initial signs of sod webworm feeding damage typically appears as general thinning of the turf, followed by small irregular patches of brown, closely cropped grass. Close inspection of the damaged area reveals silk-lined tunnels and clumps of green insect frass (fecal pellets) near or around the burrow. When infestations are high and damage escalates, the small irregular patches begin coalescing into large irregular patches of brown, closely cropped grass. Early symptoms of sod webworm damage are typically masked, especially when the turf is dormant from drought stress, and as a result the damage rapidly becomes apparent after the turf recovers from the environmental stress and the sod webworm damaged area fails to recover. Moreover, sod webworm damage frequently is mistaken for several fungal diseases, golf ball damage, golf shoe (spikes) damage, as well as black cutworm damage.

Sod webworms in the northern U.S. overwinter as partially grown larvae in silk-lined chambers in the turf canopy. In the early spring when soil temperatures become conducive, sod webworm larvae resume feeding and growing, eventually pupating. Thereafter, adults typically emerge in 10 to 20 days, depending on species and temperature. Immediately after adult emergence, mating occurs, and females typically begin laying eggs by the following night. The adults are nocturnal, active at dusk or after dark. Adult females are rather unique in their oviposition behavior. They fly rather close to the turf surface (30 to 60 cm above), erratically for short distances, fluttering or hovering over the turf dropping approximately 60 eggs like bombs from a military aircraft. Each female lives for about two weeks or less, laying several hundred eggs during her lifetime. The dropped eggs do not contain any sticky substrate for adhesion, therefore, they tend to settle into the turf canopy where they are rarely seen. Eggs typically hatch in about seven days, and there are usually six to eight instars. Newly hatched and early instar larvae feed by scraping surface tissues from leaf blades. Soon thereafter, larvae drop to the ground to form the silk-lined burrows or tunnels. As larvae develop and mature, they feed predominately at night. A complete life cycle, egg to adult, requires 6 to 10 weeks, and depending on geographic location, most temperate region sod webworm species have two to three generations per year.

Adult tropical sod webworms do not roll their wings around their bodies when they are at rest. Instead they hold them in more of a delta-winged fighter plane configuration. The dingy brown moths have a wingspread of about 20 mm. The larvae is a dingy cream color, but will take on a green appearance when the larvae are feeding. They develop through 7 or 8 instars, reaching a maximum length of 19 mm. The head is yellowish brown. The larvae feed only at night and require about 25 to 45 days to complete development. Tropical sod webworms attack and damage all species of warm season turf-grass. Highly maintained bermudagrass is attacked most often and damaged most severely. The larvae give the turf a ragged, notch appearance from their feeding and eventually the turf will take on a close-cropped appearance and turn yellow. Damage is often first noticed near flower beds and shrubs. Damage is common in southern Florida in the spring and throughout the rest of its range in the Southeast by late summer. Generations may continue well into the fall and damage often is mistaken for fall armyworm feeding.

Crown and thatch inhabitants: burrowing insects

The larval stage of numerous types of billbugs and weevils damage turfgrasses by burrowing into the stems (leaf sheath) or by damaging the crowns (apical meristems). The apical meristem is the most vulnerable part of the plant because it is the source that contains the growing points whereby roots, shoots, and leaves originate. Subsequently, when burrowing insects attack turfgrasses, plant death often results.

Annual bluegrass weevil, *Listronotus maculicollis* (Dietz). The annual bluegrass weevil, formerly called the hyperodes weevil, is a serious pest of closely cut annual bluegrass (*Poa annua*) on golf courses, bowling greens, and tennis courts in the northeastern U.S. Although both the adults and larvae cause damage to turfgrasses, adult damage is minor compared to larval damage.

Annual bluegrass weevil adults are small, 3.5 to 4.0 mm long, generally black or dark charcoal-gray beetles. The body is covered with fine, yellowish hairs and scales that typically wear off as the beetle ages. Thus, older adults often appear shiny black. Newly emerged adults are light reddish-brown, and often do not darken for several days. The thorax is approximately one-third as long as the abdomen, and the head is prolonged into a blunt (broad and short) snout. As a result, the annual bluegrass weevil is regularly confused with other turf-infesting billbugs that are similar in morphological characteristics. However, annual bluegrass weevil can be accurately differentiated by its shorter and broader snout; billbugs have a longer, narrower snout. Annual bluegrass weevil antennae are attached at the tip of the snout and can be folded back along the side of the snout in a compact groove. Annual bluegrass weevil eggs are rice-shaped, approximately 1 mm long, rounded at both ends, yellow initially but eventually becoming smoky gray to black before hatching. Eggs are laid between leaf sheaths of annual bluegrass. Females typically lay two to three eggs end to end within the leaf sheath. Larvae are creamy white, legless, with a distinct sclerotized brown head capsule. Larvae are approximately 1 mm long when newly hatched, and about 5 mm long when fully grown. There are five instars, and all are similar in appearance, differing only in size. Pupae are approximately 3.5 mm long, whitish at first, but later become reddish brown before the adults emerge. The snout, legs and wing pads are visible on the pupa, but are folded close to the body.

The only host that annual bluegrass weevil attacks is closely cut annual bluegrass. Although damage is only minor, the adults chew notches or holes in grass blades. The larva is the primary damaging life stage. Young larvae feed and tunnel within the plant stems. As larvae develop and become too large to feed within plant stems, they burrow out and feed externally on the crown of the plant. One larva can kill several plants during its lifetime. Damage typically begins along the edges of golf course fairways, especially those bordering wooded areas or other overwintering sites, as well as around the edges of putting greens and tees. Damage first appears as yellowish-brown, wilting or scattered dead patches of turf that eventually coalesce into larger dead areas as larvae develop and mature. The tunneled stems easily break off near the crown of the plant. Large infestations have the potential to cause severe damage to golf course putting greens, tees and fairways, especially where annual bluegrass is prevalent.

The annual bluegrass weevil has one to two generations per year depending on geographical locations; in more northern areas of its range it has only one generation per year. Adults overwinter in refuges such as leaf litter under trees, tufts of tall fescue, or other sheltered sites including golf course roughs where the turf is typically higher. Adults become active in the spring (mid-April), when they begin to crawl or fly to closely cut annual bluegrass hosts to begin feeding. Adult annual bluegrass weevil typically hide in the foliage during daylight, and later climb up turfgrass plants to feed at night. In early May, adults begin laying eggs. The eggs typically hatch in 4 to 5 days, depending on temperature. Immediately thereafter, the newly hatched larvae begin burrowing within the grass stems until they reach the third instar. Older larvae, third to fifth

instars, burrow out of the plant to feed along the exterior, primarily on the plant crown. Upon maturation, annual bluegrass weevil larvae pupate sometime in mid- to late June in earthen cells just under the soil surface. The second generation adults emerge in late June or early July to feed, mate, and lay eggs. In areas where there are two generations per year, most populations pupate by late August and adults emerge sometime in September, and migrate back to overwintering sites.

Billbugs (Coleoptera: Curculionidae). Billbugs are one of the most misdiagnosed pests of turfgrass. Turfgrass managers often confuse billbug damage with symptoms of drought stress, disease or other insect damage such as chinch bugs and white grubs. There are four species of billbugs that are considered major insect pests within their range in the U.S. These species include the bluegrass billbug, Denver billbug, Phoenician billbug, and the hunting billbug. Billbug damage is similar to that of weevil damage. Both the adults and larvae feed and cause damage to turf, however, adult damage is minor compared to larval damage.

Bluegrass billbug, *Sphenophorus parvulus* Gyllenhal. The bluegrass billbug is among the most serious pests of Kentucky bluegrass and perennial ryegrass in the temperate U.S. It is distributed throughout most of the U.S. and southern Canada where cool-season turfgrasses are grown. Damage frequently results in areas of brown, dead turf.

Adult bluegrass billbugs are characteristic billbugs that have a long, slender, beak-like snout. They are approximately 7 to 8 mm long, excluding the length of the snout. Adults are hard-bodied, sclerotized, usually slate gray to black in color, and may sometimes appear brownish from dried soil adhering to their body. Newly emerged adults are initially reddish-brown, but eventually become dark after a few days. Bluegrass billbug eggs are elongate, bean-shaped, translucent white eggs approximately 1.6 mm long. Larvae are plump, legless, creamy white grubs with a sclerotized brown head capsule. All larval instars are similar except for size; early instars (first and second) range in size from 1.3 to 2.4 mm long and mature larvae (fifth instar) are approximately 6 to 10 mm long. Pupae are approximately 8.5 mm long, creamy colored, gradually changing to reddish brown prior to adult emergence. The pupa has several morphological characteristics of the adult, the snout and legs are tucked under the thorax, and the wings are folded along the side of the abdomen.

Kentucky bluegrass is the preferred host of the bluegrass billbug, but it also will feed on and cause damage to perennial ryegrass as well as occasionally fine-leaf and tall fescues, especially when they are near heavily infested Kentucky bluegrass sites. Damage is usually worst from late June to early August, especially when turf is undergoing heat and drought stress. Although damage is only minor, the adults chew notches or holes in grass blades. The larva is the primary damaging life stage. Young larvae feed and tunnel within the plant stems. As larvae develop and become too large to feed within plant stems, they burrow out and feed externally on the crown of the plant. One larva can kill several plants during its lifetime. Damage first appears as yellowish-brown, wilting or scattered dead patches of turf that eventually coalesce into larger dead areas as larvae develop and mature. The tunneled stems easily break off near the crown of the plant. Lower populations of bluegrass billbugs typically produce scattered brown patches of turf, whereas heavier infestations can completely destroy areas of turf.

Adult bluegrass billbugs overwinter practically everywhere, in areas including thatch, soil crevices, under bark mulch or leaf litter, as well as other sheltered locations. However, it has been reported that crevices between the sidewalks and lawns is a preferred winter refuge. Adults become active in late April to mid-May when soil temperatures reach approximately 18°C. Once active, they are frequently seen crawling over sidewalks, curbs and driveways on warm spring days as they are seeking out suitable turfgrass in which to feed and lay eggs. After mating, adult female bluegrass billbugs begin laying eggs into small crevices created by adult feeding, chewed in grass stems just above plant crowns. Eggs typically are laid singly, and occasionally in groups of two or three. Each female can lay up to 2 to 5 eggs per day, and as many as 200 in her lifetime. Nearly all eggs are laid by early July, however, some females may continue laying eggs as late as August. Eggs usually hatch in approximately six days, and young larvae begin feeding within the grass stems, later burrowing down to feed on the plant crown as they mature. Infested turfgrass plants are hollowed out and packed with a powdery frass. As the bluegrass billbug larvae become too large (second or third

instar) to feed within the turfgrass stems, they burrow out and move to the soil to feed externally on both the crowns and roots. An accurate indicator of bluegrass billbug activity is the presence of fine, whitish, sawdust-like frass near the feeding site. Larvae are most abundant in the soil from early June to early August, and they typically require 35 to 55 days to mature, depending on temperature. Thereafter, they pupate in small earthen cells in the soil. The adults emerge in approximately 8 to 10 days. Thus, adults are abundant in late summer and fall, briefly feeding before seeking out overwintering sites as cooler temperatures prevail. Occasionally, some early emerging adults may begin to lay eggs for a second generation, however, resulting larvae do not develop fast enough to mature before the onset of winter.

Denver billbug, *Sphenophorus cicatristriatus* Fahraeus. Compared to other billbug species, little is known about the biology and life history of the Denver billbug. However, damage from this insect pest has been reported in Colorado, Kansas, and central and western Nebraska. It is considered the most serious pests of Kentucky bluegrass in Colorado. Damage frequently results in areas of brown, dead turf.

Adult Denver billbugs are considerably larger than either the bluegrass or hunting billbug, reaching 8 to 12.5 mm long. The Denver billbug adult is differentiated from the other two species by its larger size and the presence of distinctive, double-lobed markings on the wing covers. The Denver billbug may overwinter as an adult, but is more likely to spend the winter as a mid-to-late stage larva. After a spring feeding period by the larvae, eggs are laid by adults, and subsequent larvae develop over the summer. Larvae are plump, legless, creamy white grubs with a sclerotized brown head capsule. All larval instars are similar except for size, early instars (first and second) range in size from 1.3 to 2.4 mm long and mature larvae (fifth instar) are approximately 6 to 10 mm long. Pupae are approximately 9.0 mm long, creamy colored, gradually changing to reddish brown prior to adult emergence. The pupa has several morphological characteristics of the adult, the snout and legs are tucked under the thorax, and the wings are folded along the side of the abdomen.

Kentucky bluegrass is the preferred host of the Denver billbug, but it will also feed on and cause damage to perennial ryegrass. Damage typically occurs in the fall and early spring. Damage symptoms of the Denver billbug are comparable to that of the bluegrass and hunting billbug. The larva is the primary damaging life stage. Young larvae feed and tunnel within the plant stems. As larvae develop and become too large to feed within plant stems, they burrow out and feed externally on the crown of the plant. One larva can kill several plants during its lifetime. Damage first appears as yellowish-brown, wilting or scattered dead patches of turf that eventually coalesce into larger dead areas as larvae develop and mature. The tunneled stems easily break off near the crown of the plant. Lower populations of bluegrass billbugs typically produce scattered brown patches of turf, whereas heavier infestations can completely destroy areas of turf.

Phoenician billbug, *Sphenophorus phoeniciensis* Chittenden. The Phoenician (or Phoenix) billbug occurs primarily in southern California and Arizona. The adults and larvae have a typical billbug appearance. The adult Phoenician billbug has an M-shaped raised area on the pronotum. It is primarily a pest of bermudagrass and zoysiagrass, particularly in those areas that are stressed or poorly maintained.

The injury from Phoenician billbug feeding is greatest during the summer months, at times of maximum temperatures and drought stress. The larval stage is the principal stage that causes damage. The small larvae usually feed inside the plant stems. The larvae continue to feed and once they become larger, they feed outside the stem around the crown and stolons of the plant. Larvae can kill a plant (or several plants) but usually the bermudagrass is growing quite rapidly during the summer and damage goes unnoticed until fall. Damage usually appears as stressed turf that turns yellowish-brown and wilts, which is easily mistaken for drought stress. Scattered dead patches may occur later in the season or be observed the following spring as the bermudagrass fails to grow. The damaged stems break off easily at the crown of the plant. Populations often are highest and damage most severe in sunny, drought-prone locations.

Hunting billbug, *Sphenophorus venatus vestitus* Chittenden. The hunting billbug closely resembles the bluegrass billbug but is slightly larger and has parenthesis-like markings on the back of the thorax. This billbug species prefers warm-season turfgrass species including but not limited to bermudagrass and zoysiagrass. On occasion, hunting billbug will damage Kentucky bluegrass. The hunting billbug is primarily a pest of the southeastern U.S., however, it is also found in the mid-Atlantic states as well as further west and north into Missouri, Kansas and southeast Nebraska.

Adult hunting billbugs are characteristic billbugs that have a long, slender, beak-like snout. They are approximately 6 to 11 mm long, excluding the length of the snout. Adults are hard-bodied, sclerotized, usually slate gray to black in color, and may sometimes appear brownish from dried soil adhering to their body. Hunting billbug eggs are elongate, bean-shaped, translucent white eggs approximately 1.6 mm long. Larvae are plump, legless, creamy white grubs with a sclerotized brown head capsule. All instars are similar except for size; early instars (first and second) range in size from 1.3 to 2.4 mm long and mature larvae (fifth instar) are approximately 7 to 10 mm long. Pupae are approximately 8.5 mm long, creamy colored, gradually changing to reddish brown prior to adult emergence. The pupa has several morphological characteristics of the adult; the snout and legs are tucked under the thorax, and the wings are folded along the side of the abdomen.

Damage is usually greatest from mid-June through early August during the period of maximum heat and drought stress. The larva is the principal damaging life stage. Young larvae feed and tunnel within the plant stems. As larvae develop and become too large to feed within plant stems, they burrow out and feed externally on the crown of the plant. One larva can kill several plants during its lifetime. Damage first appears as yellowish-brown, wilting or scattered dead patches of turf that eventually coalesce into larger dead areas as larvae develop and mature. The tunneled stems easily break off near the crown of the plant. Lower populations of hunting billbugs typically produce scattered brown patches of turf, whereas heavier infestations can completely destroy areas of turf.

As is the case with the Denver billbug, little is known about the biology of this pest. The hunting billbug has been reported to overwinter as dormant adults in the soil. Adults become active in late April to mid-May when soil temperatures reach approximately 18°C. Once active, they are seen frequently crawling over sidewalks, curbs, and driveways on warm spring day as they are seeking out suitable turfgrass in which to feed and lay eggs. After mating, adult female bluegrass billbugs begin laying eggs into small crevices created by adult feeding, chewed in grass stems just above plant crowns. Eggs typically are laid singly, and occasionally in groups of two or three. Each female can lay up to 2 to 5 eggs per day, and as many as 200 in her lifetime. Nearly all eggs are laid by early July, however, some females may continue laying eggs as late as August. Eggs usually hatch in approximately 3 to 10 days, and young larvae begin feeding within the grass stems, later burrowing down to feed on the plant crown as they mature. Infested turfgrass plants are hollowed out and packed with a powdery frass. As the bluegrass billbug larvae become too large (second or third instar) to feed within the turfgrass stems, they burrow out and move to the soil to feed externally on both the crowns and roots. An accurate indicator of bluegrass billbug activity is the presence of fine, whitish, sawdust-like frass near the feeding site. Larvae are most abundant in the soil from early June to early August, and they typically require 35 to 55 days to mature, depending on temperature. Thereafter, they pupate in small earthen cells in the soil. The adults emerge in approximately 8 to 10 days, thus adults are abundant in late summer and fall, briefly feeding before seeking out overwintering sites as cooler temperatures prevail.

Soil inhabitants: root-infesting insects

White grubs (Coleoptera: Scarabaeidae). White grubs are the most widespread, and considered the most destructive, insect pests of turfgrasses in the continental U.S. Most species of white grubs damage turfgrass by chewing off the roots near the soil surface. Most often, white grub damage occurs during periods of hot and dry weather. Subsequently, turf loss can be relatively abrupt and severe. To compound this problem, vertebrate predators such as armadillos, badgers, birds, moles, raccoons, skunks and various other animals may dig up infested areas of turf in search of grubs. In many cases these animals cause more damage than the grubs themselves.

Because white grubs feed on the roots below ground, they often go undetected until measurable loss to the root system has occurred. Moreover, grubs can be relatively difficult to control because soil insecticides must penetrate the turf canopy and thatch layer in order to effectively make contact with the grubs located in the soil. To achieve maximum control, turfgrass managers, through appropriate applica-

tion equipment, adjuvants or surfactants, gravity, and irrigation or natural rainfall, must effectively place respective insecticides into the target zone where the grubs are located.

The larval stage (grub) typically is the damaging life stage to turfgrass, however, some species of beetles cause damage to ornamental plant materials as adults. White grubs are stout-bodied beetle larvae. About 15 species are pests of turfgrasses in North America, with about 10 species considered to be important: *Aphodius* grubs, Asiatic garden beetle, black turfgrass ataenius, European chafer, green June beetle, Japanese beetle, May beetles, northern and southern masked chafers, and Oriental beetle. These species of white grubs are generally similar in appearance, habits, and the damage they cause. An overview of the biology of individual species follows.

Aphodius grubs, *Aphodius granaries (L.)* and *Aphodius paradalis* Le Conte. Two species that belong to the genus *Aphodius* are occasional pests of turfgrass, especially golf course fairways. *Aphodius* grubs are relatively small grubs compared to most other white grubs. They closely resemble black turfgrass ataenius grubs, and because they are frequently found in association with black turfgrass ataenius, often they are confused or misidentified. Although they seem to be less frequently associated with turf than black turfgrass ataenius, they are capable of causing serious damage to turf.

Adults are typically black with a reddish tinge, reddish-brown legs, and pale antennae. The adults have two triangular projections on the outer edge of the tibia and hind leg, whereas black turfgrass ataenius lack these projections. *Aphodius* eggs are tiny (less than 0.7 mm when fully hydrated), and pearly white. Larvae are typically white grubs, however, they are considerably smaller than most other common turf-infesting species. Newly hatched first instars are approximately 2.4 mm long, and are difficult to see. Second instar larvae are about 5 mm long and third instars are approximately 7.0 mm long. Fully mature third instars often are mistaken for young grubs of other species such as European chafers, Japanese beetles, and masked chafers. A distinguishing morphological characteristic that differentiates the two species of *Aphodius* larvae, as well as black turfgrass ataenius and other white grubs, is the raster pattern. The raster pattern is a distinctive pattern of hairs, spines, and bare spaces on the raster located on the ventral side of the last abdominal segment, anterior of the anus. The raster of *Aphodius* grubs has two rows of short spines forming a distinctive V-shaped pattern, whereas black turfgrass ataenius grubs have a random arrangement of spines as well as two distinctive, pad-like structures at the tip of the abdomen, anterior of the anal slit.

Both the adults and larvae of *Aphodius* feed primarily on decaying organic matter, especially animal manure. However, occasionally they will feed on living roots of cool-season turfgrasses such as annual bluegrass, creeping bentgrass, and Kentucky bluegrass. *Aphodius* grubs cause sporadic, severe damage predominantly to golf courses in the temperate U.S. Damage to home lawns is uncommon. The initial symptoms of *Aphodius* larval damage are patches of thin or wilted turfgrass that resemble drought stress, however, the turf does not respond or recover with the application of irrigation. As turfgrass root loss continues, the turf typically dies in irregular patches that eventually coalesce into larger dead areas. Heavily infested areas of turf can be rolled up similar to a loose piece of carpet. Upon close inspection or sampling, the grubs, pupae, and adults can be found under the dead patches of turf in the soil. Irrigated turfgrass with a high proportion of annual bluegrass are especially susceptible. Populations of as many as 200 to 300 per square meter are not uncommon. In addition, vertebrate predators frequently forage for grubs, resulting in additional turf damage.

The life history of *Aphodius* is poorly understood. Observations from Ohio, Michigan and Ontario, Canada suggest that there is only one generation per year with adults becoming active during the first warm days of spring; egg-laying apparently begins 2 to 3 weeks earlier than that of black turfgrass ataenius. However, other published reports indicate the possibility that two annual generations may be possible, especially in the more southern parts of the species range. Nonetheless, because *Aphodius* species are frequently associated with black turfgrass ataenius, similar biological attributes may be extrapolated to effectively manage *Aphodius*.

Asiatic garden beetle, *Maladera castanea* (Arrow). The Asiatic garden beetle is a relatively minor pest of turfgrass, however, it can be locally abundant and damaging, especially in the temperate northeastern U.S.

Asiatic garden beetle adults are dull chestnut-brown, with a velvety appearance and a slight iridescent sheen. The beetles range in size from 8 to 11 mm long and 5 to 6.4 mm wide. The elytra (wing covers) do not quite reach the tip of the abdomen, leaving the terminal two segments of the abdomen exposed. The ventral side of the Asiatic garden beetle adult is partially covered with yellow hairs, and each visible segment of the abdomen has a row of backward-pointing yellow hairs that extend across the width of the body. The dorsal side of the elytra is bald with the exception of a row of fine hairs on the outer margins. Another distinctive morphological characteristic is the presence of scattered, small hairs on the dorsal (top) side of the head. Asiatic garden beetle eggs are pearly white, oval and are approximately 1 mm in diameter. After becoming hydrated from soil moisture, they become almost spherical. The eggs are typically laid in clusters of 3 to 19 eggs that are loosely held together by a gelatinous secretion. Newly hatched first instars are approximately 1.4 mm long and reach 19 mm long when fully mature. They are like most other white grubs, having a C-shaped body, a brown head capsule, and six jointed legs. However, the body color of Asiatic garden beetle larvae typically remains somewhat lighter than in other white grub species. The important distinguishing feature is the single, transverse curved row of spines on the raster, together with a Y-shaped anal opening. The hind legs have tufts of hairs, and extremely small claws relative to those on the pro- and meso- legs (front and middle legs, respectively). The presence of a whitish, enlarged, bulbous morphological structure (the stipes) on each maxilla beside the jaws is another distinctive feature of Asiatic garden beetle grubs. The pupa is approximately 7 to 10 mm long, and appears white initially but turns tan as it matures. Initially the pupa is enclosed in the final instar skin, which soon splits and is pushed back over the terminal portion of the abdomen; thus, it lies exposed in an earthen cell created by the larva. This characteristic is common also in the European chafer and May beetles.

Asiatic garden beetle grubs feed on the roots of all cool-season turfgrasses as well as weeds, woody ornamentals, herbaceous perennials, and vegetables. Adult Asiatic garden beetles prefer box elder, butterfly bush, cherry, Devil's walkingstick, Japanese barberry, oriental cherry, peach, rose, strawberry, sumac, and viburnum. Asiatic garden beetle larvae feed on the roots of various turfgrass species, causing typical white grub injury resulting in thinning, wilting, irregular dead patches of turf. Well or highly maintained turf that is irrigated regularly and is located near weedy areas containing adult Asiatic garden beetles' preferred food sources is more likely to be infested. Compared to other white grub species such as Japanese beetles, an equal number of Asiatic garden beetle grubs is typically less destructive. It is likely that this is a result of the fact that Asiatic garden beetle grubs typically feed deeper in the soil profile, thus leaving more of the turfgrass root system intact. Where Asiatic garden beetle infestations are heavy, it is not uncommon to observe more than 100 grubs per square meter, resulting in severe damage. Not only do Asiatic garden beetle grubs cause damage, but the adults feed on more than 100 species of woody and herbaceous ornamental plant material. When beetles are abundant, preferred plants can be stripped of foliage and flowers.

Asiatic garden beetle has a univoltine (one life cycle per year) life cycle similar to the Japanese beetle. Adults are most abundant from mid-July to mid-August. Adult females burrow into the turf canopy to lay eggs in the soil at about 2.5 to 5 cm in depth. Each female is capable of laying up to 60 eggs in her approximately 30 day lifetime. Once laid, eggs typically hatch in about 10 days; immediately thereafter the young larvae commence feeding on the tender, succulent roots of turfgrass and decaying organic matter until sometime in the fall when the first measurable frost occurs. The majority of Asiatic garden beetle grubs will have attained the final instar (third) by the first frost, however, approximately 25% will overwinter as second instars. As the soil temperatures continue to decline, the grubs will continue to burrow deeper into the soil profile, ultimately overwintering approximately 20 to 43 cm below the turf surface in a semi-dormant, non-feeding state. Sometime in mid- to late April, the grubs will move slowly back up to the root zone where they will begin feeding until mid-June. Thereafter, mature grubs will begin preparing earthen cells or cavities 4 to 10 cm below the turf surface where they will pupate. Pupation occurs mainly from mid-June through mid-July, and typically lasts for approximately 10 days. New adults remain in the pupal cell for a few days until they become fully sclerotized (hardened) and their color changes from whitish to chestnut-brown. Soon after, the adults burrow upward and emerge from the soil to feed, mate, and lay eggs in susceptible areas of turf.

Black turfgrass ataenius, *Ataenius spretulus* Haldeman. Although sporadic in occurrence, the black turfgrass ataenius causes severe damage to golf courses in the temperate U.S., especially where cool-season turfgrasses are grown. They are also occasional pests where bentgrass greens are utilized in warm season turfgrass areas of the U.S. Black turfgrass ataenius has many of the same biological characteristics as several other white grub species, however, its season life cycle differs in that there are typically two generations per year throughout much of its range. In the more southern areas of the U.S., its life cycle is poorly understood, but more than two generations may occur in some areas.

Adult black turfgrass ataenius are relatively small, shiny black beetles ranging in size from 3.6 to 5.5 mm long and approximately half as wide. There are distinct longitudinal grooves on the elytra. Newly emerged adults are reddish-brown initially, and darken in a few days. Eggs are minute, less than 0.7 mm in diameter after absorbing water from the soil, and pearly white. Larvae are typical white grubs in appearance, however, they are considerably smaller than other turfgrass-infesting species. Newly hatched larvae are quite small, approximately 2.4 mm long, thus they are relatively difficult to see. Second and third instars are about 5 mm and 7.0 mm long, respectively. Fully mature black turfgrass ataenius grubs are mistaken frequently for young grubs (first instars) of several other white grub species such as European chafer, Japanese beetle, and both northern and southern masked chafers. However, black turfgrass ataenius larvae can be distinguished easily by the two distinctive, pad-like structures located at the terminal end on the ventral side of the abdomen anterior to the anal slit. Upon maturation, black turfgrass ataenius grubs pupate into small, approximately 4.2 to 5.7 mm long, pupae that are initially cream colored, but eventually become reddish-brown before the adult beetle emerges. A distinctive characteristic of black turfgrass ataenius pupae is that the wings and legs are folded close to the body.

Black turfgrass ataenius larvae typically attack and cause damage to the roots of annual bluegrass, creeping bentgrass, and Kentucky bluegrass on golf courses. They also feed on decaying organic matter, and the adults feed on manure and decaying organic matter. Because black turfgrass ataenius grubs feed on the roots of turfgrasses, the first symptoms appear as patches of thin or wilted turf that resembles drought stress, however, the turf does not recover with the application of irrigation or rainfall. As damage and subsequent root loss increases, the turf typically dies in irregular patches that eventually coalesce into larger dead areas. Heavily infested areas can be pulled or rolled up similar to loose carpet. Upon close inspection of the turf, large numbers of larvae, pupae and adults can be observed in the soil under the turf. Black turfgrass ataenius seems to prefer, and are most common in, areas of turf that are close-cut, with a moist, compacted layer of thatch. Additionally, golf course fairways that have a high proportion of annual bluegrass composition are especially susceptible. Because black turfgrass ataenius grubs are relatively small, they typically are found in higher densities than most other white grub species, with populations of 215 to 325 per square meter not uncommon. Accenting the damage associated with black turfgrass ataenius grubs, vertebrate predators frequently forage for grubs, resulting in additional turf damage.

Depending on geographic location, black turfgrass ataenius has one to two generations per year. There is typically only one generation per year in the Great Lakes states, New York, northern New England, Ontario, and other parts of its range. Adults reportedly overwinter along the edges of wooded areas along the perimeter of golf courses. The adult beetles seek refuge in leaves, pine needles, grass clippings, and other debris in the upper 2.5 to 5 mm of the soil. Most of the overwintering black turfgrass ataenius adults have mated. Subsequently, in the early spring, from late March to early April, the adults become active and can be observed in swarms flying over golf course putting greens and fairways on warm afternoons as well as around lights at night. Soon thereafter, black turfgrass ataenius adults begin laying eggs in early May continuing until mid-June. Black turfgrass ataenius eggs are laid in clusters of approximately 11 to 12 eggs within cavities formed by the female near the soil-thatch interface. Eggs typically hatch within 7 days, and the larvae immediately begin feeding on fine, succulent roots and organic matter. This first generation of grubs typically is present from late May until early July, with damage commonly appearing in late June. Larvae require

approximately four weeks to mature; thereafter, they burrow down into the soil profile to pupate and emerge as adults sometime in late June through early July. Upon emergence, these beetles mate and begin laying eggs in July through early August, producing a second generation of grubs which typically causes damage in August and early September. These grubs develop, mature, and pupate by late August or early September, producing adults that mate and emigrate to overwintering sites in the late fall. Generations appear to overlap and more than one life stage is common at a given time since the adults lay eggs over a period of several weeks. In geographic areas where there is not enough time to complete development of a second generation, these beetles mate but will not lay eggs. In areas of southern California, there is great variation from year to year in the occurrence of black turfgrass ataenius. Damaging populations of grubs do not occur until late in the summer in most years, but occasionally early summer damage may occur. Frequent insecticide applications for cutworms as well as high temperatures in the summer may reduce the likelihood of black turfgrass ataenius infestation on putting greens in the southern areas of the U.S.

European chafer, *Rhizotrogus majalis* (Razoumowsky). The European chafer is a native of Europe, where it is a serious turfgrass pest. In the U.S. as well, it is a turfgrass insect pest, primarily in the northeast. Where European chafer occurs, often it is considered the most damaging grub species. European chafer grubs are considerably larger than most other white grub species, especially Japanese beetle larvae. Thus, they commonly are more destructive than equal numbers of other species. In addition, they typically feed later into the fall and resume feeding earlier in the spring than other grub species.

European chafer adults are medium-sized, light-reddish beetles measuring approximately 13 to 15 mm long, with a slightly darker head and pronotum. The posterior edge of the dorsal side of the pronotum has a narrow band of light-yellow hairs, and the ventral side of the thorax is covered with pale yellow hairs. The terminal portion of the abdomen protrudes beyond the elytra, and they have distinct longitudinal grooves and minute punctuations. Both European chafers and May beetles resemble one another, however, European chafers are slightly smaller and they lack a tooth on the tarsal claws of the mesothoracic legs. Additionally, the European chafer has more distinct grooves on the elytra. Newly laid eggs are oval, shiny, milky white initially, but become dull gray after a few days. As they absorb water from the soil, the eggs become spherical, swelling to approximately 2.3 to 2.7 mm in diameter prior to hatching. European chafer larvae are typical C-shaped white grubs, with a yellowish-brown head and six distinct, jointed legs. The larvae can be differentiated easily from other white grub species by the raster pattern. It has two distinct, nearly parallel rows of small spines that diverge outward at the tip of the abdomen, similar to a slightly open zipper. The raster pattern, in combination with a Y-shaped anal slit, readily distinguishes European chafer larvae from other turfgrass-infesting white grub species in the temperate U.S. First instar European chafer grubs are initially translucent white, though the terminal portion of the abdomen becomes dark after feeding. Fully mature larvae are approximately 23 mm long. Both the prepupa and pupa resemble most other white grub species. They are approximately 16 mm long, smaller than May beetle pupae, yet larger than Japanese beetle pupae. European chafer pupae shed the larval exoskeleton, whereas Japanese beetle and Oriental beetle pupae lie within the shed larval skin.

European chafer larvae feed on the roots of all cool-season turfgrasses, as well as numerous grassy and broadleaf weed species in pastures and nurseries. The grubs cause thinning, wilting, and subsequent irregular patches of turf that can be readily pulled back or rolled up from the soil like a roll of carpet. Population densities of 18 to 28 grubs per square meter are not uncommon in lawns and golf course turf. Damage by European chafer grubs typically appears in September, especially when the turf is already stressed by both heat and drought. Moreover, damage also can be observed in the spring when the grubs resume feeding on the previously weakened turf. Although European chafer adults occasionally will sample the margins of tree leaves, they rarely cause any measurable damage.

The European chafer has a life cycle that closely resembles those of other white grub species such as Japanese beetle and May beetle. In northern portions of Michigan and New York, adult European chafers begin emerging in mid-June and are most abundant from late June until mid-July, and terminate by late July. In areas further south, such as New Jersey, Ohio and Pennsylvania, adult activity typically occurs about two weeks earlier. Adults are most active on warm, clear nights when the temperature is above 19°C, and on favorable nights large numbers of adults begin emerging from the turf canopy near sunset. The adults fly to nearby trees, swarming about by the thousands. After a courtship period of approximately one hour, mating pairs begin to make their way to the turf canopy, ultimately returning into the soil before sunrise. Female European chafer adults continue to return to trees several times to re-mate during their 1 to 2 week lifetime. Each female lays 20 to 40 eggs in her life. Eggs are typically laid singly at 5 to 10 cm depth in the moist soil. Eggs hatch in approximately two weeks, and most hatch by late July. There are three instars of European chafer grubs; first instars are predominant until mid-August, second instars are present in early September, and third instars continue to feed and develop into November. Similar to other white grubs species, European chafers move down into the soil profile just below the frost line to overwinter. In the spring, when soil temperatures become conducive, the grubs return to the upper 2.5 to 5 cm near the soil-thatch interface to vigorously feed until late May. In early June the European chafer larvae begin moving down into the soil profile, to a 5 to 25 cm depth, to form earthen cells to pupate. The prepupal life stage requires approximately 2 to 4 days, while the pupal stages typically takes about two weeks. Thereafter, adults begin emerging in mid-June.

Green June beetle, *Cotinus nitidia* L. The green June beetle is a native of the eastern U.S., and is widely distributed east of the Mississippi river, north to St. Louis, Missouri, and south to Texas. They are most common in the southernmost edge of the temperate region where cool-season turfgrasses are grown. Because of their relatively large size, both the green June beetle adults and grubs regularly attract attention where they are abundant. Adult green June beetles are rather prolific fliers that are mistaken often for wasps as they swarm in a 'buzzing' manner over the turf in mid-summer. Green June beetle grubs are relatively large compared to other turfgrass-infesting white grub species.

Adult green June beetles are considerably larger than Japanese beetles. They are approximately 19 to 25 mm long and 12.5 mm wide. They vary in color on the dorsal side of the body and elytra from dull brown with longitudinal stripes of green, to uniform, velvety forest green. The outer margins of the elytra range in color from tan to orange-yellow. The ventral side of the body is shiny metallic green or gold. Newly laid eggs are dull white, approximately 1.5 mm in diameter. As the eggs absorb water from the soil, they become larger, 3 mm in diameter, and more spherical in shape. First instar grubs are approximately 6 to 7 mm long, second instars are 15 to 17 mm long, and fully mature larvae are relatively large (45 to 48 mm long), robust, and more parallel-sided compared to other white grub species. Green June beetle grubs have short, stubby legs and mouthparts in relation to their overall body size. Upon maturation, the green June beetle forms a cocoon-like cell composed of soil particles held together by a sticky secretion, within which pupation occurs. Green June beetle pupae are large, approximately 25 mm long and 12.5 mm wide. The pupae are initially whitish, but gradually darken over time and eventually exhibit tints of the adult coloration before the adults emerge. They can be identified readily by their distinct locomotion; essentially, they scurry along on their backs, upside-down.

Green June beetle grubs primarily feed on decomposing organic matter, including compost, thatch, and grass clippings. There are indications that poorly managed turfgrass with excessive thatch may be at greater risk, and that the use of organic fertilizers just prior to the adult flights may make a turfgrass area more attractive for egg-laying. Green June beetle grubs do not feed on the roots like many other scarabs, however, they do cause considerable damage by their burrowing and tunneling behavior. The grubs make distinct, open vertical burrows with a surface hole approximately the size of a human's thumb. The burrowing and tunneling action of green June beetle larvae causes loose soil to be excavated out at the mouth of the burrow at night, forming a small mound approximately 50 to 75 mm across. The aforementioned mounds are frequently mistaken for ant mounds except that the soil particles are considerably coarser. This action disrupts the roots system, dislodging the turfgrass, and loosens the surface topsoil, permitting the turf to desiccate. As a result, the turf typically wilts and thins, allowing the invasion of broadleaf and grassy weeds. Adult green June beetles occasionally damage tree leaves, though they mostly feed on ripening fruits, tree sap and other sugary food sources. Similar to the grubs, the adults create little piles of soil as they burrow into and out of the turf canopy for egg-laying and resting.

Like many other scarabs, the green June beetle has a one-year life cycle. Adults typically emerge in late June with peak activity occurring over a 2 to 3 week period, usually in mid-July in Kentucky; this event may be 2 to 3 weeks earlier in the south and 1 to 2 weeks later in areas further north. Green June beetle adults are active during the daytime, and rest on vegetation as well as under the thatch during the night. After mating, green June beetle female adults seek out turf areas with moist soils containing high levels of organic matter. The females burrow down into the turf approximately 5 to 13 cm, excavating a small cavity where she lays a cluster of 10 to 30 eggs. The cluster of eggs is enclosed in a small-sized sphere of soil held together by a sticky secretion. Each green June beetle female makes numerous egg chambers, laying as many as 60 to 75 eggs over a two-week period of time. Eggs hatch in approximately two weeks. Subsequently, the young larvae begin feeding at the soil/thatch interface by early August. After continuing to feed and grow for several weeks, green June beetle larvae attain the third instar in late October or early November in the northern regions of their distribution. Third instars can be seen as early as the end of August in some Mid-Atlantic states, but also will remain active until November. As the soil temperatures continue to decline, the grubs will continue to burrow deeper into the soil profile, ultimately overwintering approximately 20 to 76 cm below the turf surface in a semi-dormant, non-feeding state. In some of the southern states, where soil temperatures do not drop significantly, green June beetle grubs can remain active throughout the winter on warm nights. Sometime in mid- to late April, the grubs will move slowly back up to the soil/thatch interface to feed until late May. Thereafter, mature grubs will begin preparing earthen cells or cavities, 20 to 40 cm below the turf surface, where they will pupate. Pupation occurs mainly in early June, and typically lasts for approximately 10 to 20 days. After about three weeks, adult green June beetles burrow upward and emerge from the soil to feed, mate, and lay eggs in susceptible areas of turf.

Japanese beetle, *Popillia japonica* Newman. The Japanese beetle is a native to the main islands of Japan, where it is not considered a pest. However, in the eastern U.S. it is considered by many to be the worst insect pest of turfgrasses and ornamental landscapes. As a result, countless dollars and resources are aimed at controlling Japanese beetle adults and larvae. Japanese beetle larvae are white grubs that cause typical root-feeding damage. As for the adults, unlike most scarab adults, Japanese beetles attack and cause serious damage to over 300 ornamental plant species.

Japanese beetle adults are attractive, broadly oval insects that are approximately 8 to 11 mm long and 5 to 7 mm wide. The head, thorax, and abdomen are shiny, metallic green. With the exception of the head, much of the dorsal side of the body is covered with hard, coppery-brown elytra that do not fully extend to the tip of the abdomen. Another distinguishing character is the presence of five patches of white hairs that are located on each side of the abdomen, as well as another pair of white tufts on the dorsal side of the terminal abdominal segment posterior to the elytra. These tufts of hairs differentiate Japanese beetles from all other beetles that they closely resemble. Like many other insect species, the females are typically slightly larger than their male counterparts. Newly laid eggs are pearly white and oblong, approximately 1.5 mm long. Upon absorption of water from soil moisture, the eggs become spherical, almost doubling in size within a few days. The larvae are typical white grubs with three pairs of distinct, jointed legs, and a yellowish-brown head capsule; they too assume the C-shaped position in the soil and when handled. Newly hatched grubs are translucent white, becoming darker once they have fed. Like all white grub species, there are three larval instars. First instars are approximately 1.5 mm long, and fully mature third instars are 25 to 30 mm long. When compared to European chafers, masked chafers, and May beetles, they are considerably smaller. Japanese beetle larvae can be identified easily by their distinctive raster pattern that exhibits two rows of short spines that are arranged in the shape of a truncated V pattern. Pupae are typically scarab-like, and are cream colored initially, but gradually become tan to light brown. They measure about 14 mm long and 7 mm wide. Within a few days of adult emergence, the pupa appears metallic green.

Japanese beetle adults feed on over 300 species of herbaceous and woody ornamental plant species, including numerous popular shade trees. Initial feeding damage usually occurs in the upper canopy of preferred trees, on the upper leaf surface, leaving only a lace-like skeleton of the leaf veins. The grubs feed on the roots of equally as many plants, including all cool-season turfgrasses, weed species and ornamental plants. Damage to turfgrass by grubs results in thinning, wilting, and subsequent irregular patches of turf that can be readily pulled back or rolled up from the soil like a roll of carpet. Feeding damage by larvae to ornamental plant material frequently results in poor plant health and in some cases eventual death of the plant material.

Much like many other scarab species, the Japanese beetle has a one-year life cycle throughout most of its range in the U.S. Depending on geographic location, adult emergence is most common in mid- to late June with peak activity occurring in early July. Emergence may occur about two weeks earlier in the Carolinas and two to three weeks later in the more northern states. Within days of emergence, mating and egg-laying occurs. Virgin female adults call males with a chemical sex attractant (pheromone). As many as 20 to 100 males may aggregate around a single female attempting to mate with her. After mating, the female will seek out suitable turfgrass sites in which to lay eggs. Such sites include areas with moist, loamy soil covered with well-maintained, lush, closely cut turf. Throughout this mating and egg-laying process, both male and female adults continue to feed gregariously on ornamental plants, usually beginning in the upper canopy of the tree. Females re-mate after each egg-laying episode, laying eggs in clutches of 1 to 4 eggs in the upper 7.5 cm of soil. This process is repeated every few days during the normal lifespan of a female, 30 to 45 days, during which she may lay as many as 40 to 60 eggs. After about a two week incubation period, the first instar grubs begin feeding on fine, succulent roots and organic matter, usually sometime in late July or early August. In approximately 2 to 3 weeks, the grubs will molt, becoming second instars, and eventually reaching mature grubs (third instars) after another 3 to 4 weeks. Throughout this time, as periods of dry soil moisture or drought exist, the grubs may burrow deeper in the soil profile. As cooler temperatures prevail in the late fall, and the first measurable frost occurs, the grubs will begin to burrow deeper into the soil to depths of about 5 to 20 cm below the soil surface where they will overwinter in a semi-dormant, non-feeding state. As soil temperatures warm to above

10(C in the spring (late March or early April), the grubs will begin to move back up into the root zone to resume vigorously feeding for approximately 4 to 6 weeks before going back down into the soil profile to form an earthen cell in which to pupate, normally in late May or early June. Pupation typically requires approximately 2 to 3 weeks; thereafter, adult Japanese beetles emerge. At a latitude of Kentucky, Virginia and Maryland, adults and eggs will be present in July, first and second instars present by mid-August, and third instars from late September, and overwintering until May. Prepupae and pupae are seen at this latitude in May and early June. Timing of these developmental stages will be 2 to 3 weeks later in northern regions and about 1 to 2 weeks earlier in southern regions.

May beetles, *Phyllophaga* spp. There are approximately 25 species of May beetles that may infest and feed on the roots of turfgrasses in North America. They are rather destructive as a result of their large size, and because they destroy the roots relatively close to the soil surface. Compared to most other white grub species, May beetles typically occur at considerably lower densities than other turf-infesting grub species.

May beetle adults are brownish, reddish-brown to almost black, medium- to large-sized, heavy-bodied beetles ranging in size from 11 to 24 cm long. Depending on species, body pubescence varies greatly; some species are almost hairless, while other are quite fuzzy. It is very difficulty to differentiate among the various species. Egg are pearly white and oval when first laid, but become more spherical after absorbing water from the soil. When fully mature and hydrated, eggs are nearly 2.5 to 3 mm wide. In most species, fully mature or third instar grubs are approximately 25 to 38 mm long. May beetle grubs have a V- or Y-shaped anal slit with the stem of the Y shorter than the arms. The most distinguishing characteristic is the raster pattern that exhibits two parallel rows of short spines, and resembles a zipper. The pupa is about 7 to 10 mm long, and appears white initially but turns tan as it matures. The pupa is initially enclosed in the final larval instar skin, which soon splits and is pushed back over the terminal portion of the abdomen. Thus, it lies exposed in an earthen cell created by the larva. This characteristic is common with the Asiatic garden beetle and European chafer.

May beetle grubs attack the roots of nearly all common turfgrasses. They cause extensive damage to turfgrass much like many of the other turf-infesting scarab species. The adults feed on tree leaves, chewing on the tissue between the veins, similar to Japanese beetles. Preferred hosts include birch, elm, hickory, oak, persimmon, poplar and walnut. In extreme situations, whole trees are sometimes defoliated in the spring.

Depending on the species, May beetles have 1- to 4-year life cycles. Nearly all the May beetle species that occur in the temperate U.S. have a 3-year life cycle, however, certain species have a 2-year life cycle in the transition zone and father south where warm-season turfgrasses are grown. There are a select few species that have only 1-year cycles. All *Phyllophaga* species have a life cycle that includes the developmental stages of an egg, three larval instars, a pupa, and an adult. The adults of most species are active from April to June depending on geographic location. In some southern regions, though, flight activity may be seen during any month of the year. May beetle adults emerge and are active just after sunset, when they fly to the tops of trees to feed and mate before returning to the turf before sunrise. Subsequently, they are highly attracted to lights, and often are observed bouncing off screen doors. After mating, females fly to the turf to burrow down approximately 8 to 10 cm to lay eggs singly in moist soil. An individual female typically lays approximately 20 to 50 eggs in her lifespan. After an incubation period of about 3 to 4 weeks, the eggs hatch and first instar grubs begin feeding on the succulent roots of turfgrass and organic matter. The larvae then quickly develop, growing and molting into fully developed second instars by early fall (September). Thereafter, they continue feeding into the late fall before the first measurable frost occurs, when they migrate downward to lower depths for overwintering. In the early spring, the following April, the second instars migrate upward to resume feeding on the roots of turfgrass, eventually molting into third instars by mid-June. The third instars continue to feed and grow, causing most of their damage in July and August. For May beetle species with a 2-year life cycle, they will reach their full size and pupate by the end of the second summer. Most of these 2-year species reach full maturity as adults by late fall of the second year, but remain in the soil throughout the winter, emerging the following spring. For species with a 3-year life cycle, as cooler temperatures prevail during the fall of the second year, the nearly fully mature larvae slowly migrate downward again in late September and early October to overwinter. Finally, in the third year of the life cycle, in the early spring during late March and early April, the third instars again migrate upward to feed throughout April and May before they complete their larval development. In June, the larvae migrate slightly downward to a depth of approximately 15 to 25 cm where they form an earthen cell in which to pupate. By late June, the third instars become prepupae, transform into pupae in July and August, becoming adults sometime in late August or early September. The young adults remain in the aforementioned earthen cell until the following spring, when they emerge in May or June. Regardless of whether the species has a two or three year life cycle, most of the damage is occurs during the second year.

Northern and southern masked chafers, *Cyclocephala borealis* (Arrow) and *Cyclocephala lurida* (Bland). Both northern and southern masked chafers are native to North America, and are widely distributed from the Atlantic seaboard westward to the Rocky Mountains and into the southwestern U.S., and from southern New York to as far south as South Carolina. Northern and southern masked chafers are among the most destructive insect pests of turfgrass in the Midwest and Central U.S., causing typical white grub damage. In addition, there is often significant indirect injury due to the feeding of vertebrate predators on these grubs. Distribution of both species overlaps throughout much of the Midwest. However, southern masked chafers are more common in the southern extent of their range.

Northern and southern masked chafers are quite similar in appearance and size in all life stages. Northern masked chafer adults are dull yellow-brown beetles, approximately 11 to 12 mm long and 6 to 7 mm wide. The only distinction from the southern masked chafer is that southern masked chafer is more shiny, and reddish-brown. Males and females of both species have distinguishable darker, chocolate brown color across the head and eyes that enables one to differentiate them from most other scarab species that are similar in size and color. Other important morphological characteristics that help to identify northern masked chafer beetles are the dense hair on the ventral side of the thorax, and the scattered arrangement of erect hairs on the wing covers. The adults of southern masked chafer lack such hairs. Newly laid eggs of both the northern and southern masked chafers are pearly white, oval, approximately 1.3 to 1.7 long, that like other white grub species, become considerably larger and more spherical after absorbing water from the soil. The larvae of northern and southern masked chafers are typical white grubs with a C-shaped body and six jointed legs. It is practically impossible to differentiate the two species. Newly hatched grubs are translucent white, becoming grayish after feeding, and they are about 4 to 5 mm long. Fully mature northern and southern masked chafer grubs are considerably larger and more robust than Japanese beetle larvae, ranging in size from 23 to 25 mm in length. They also have a reddish-brown head capsule compared to a more yellowish-brown head capsule in Japanese beetles. Their raster pattern exhibits an evenly spaced, non-uniform arrangement of approximately 20 to 30 stout hairs. Pupae are approximately 17 mm long and are creamy initially, later becoming reddish-brown as they mature.

Although northern and southern masked chafer grubs eat the roots of all turfgrasses, including endophytic cool-season species, they also feed on decaying organic matter and frequently are found in mulched plant beds, heavier organic soils and compost heaps. When damage does occur to turf, typical white grub feeding damage is common. The adults do not feed or cause damage to turf or other plant material.

The northern and southern masked chafers both have a 1-year life cycle like many other scarab species. The adults are active in June and July, with peak activity occurring in late June through early July in areas such as southern Ohio. They may be observed up to a month earlier in more southern locations. There is a distinct difference in adult behavior between northern and southern masked chafers. Southern masked chafer is predominantly active from just after sunset until around 11:00 p.m., whereas northern masked chafer is mostly active after midnight. Otherwise, their behavior is quite similar in that they swarm over the turf surface on warm, humid nights, especially after a heavy rain, in search of mates. As soon as mating occurs, the female will burrow into the turf canopy to lay her eggs in the soil. The eggs are laid singly or in small clusters in the upper 2.5 to 5 cm of the soil. The incubation period is approximately 14 to 18 days depending on temperature, and most eggs hatch by late July to mid-August. When soil moisture is adequate, the larvae will grow and develop relatively quickly, molting into second instars in approximately 3 weeks. By early September, the grubs develop into third instars, continuing to feed and grow until about mid- to late October when cooler soil temperatures force them to go deeper into the soil profile to overwinter. Similar to other white grub species, they overwinter in a earthen cell just below the frost line. In the spring, when temperatures become conducive, they migrate back up into the root zone to resume feeding until fully mature in mid- to late May. Thereafter, the fully mature larvae begin moving down into the soil profile to form earthen cells to pupate. The prepupal life stage proceeds, requiring approximately 4 to 6 days, while the pupal stage typically takes about 14 to 20 days to complete. Thereafter, adult beetles begin emerging in mid- to late June.

Oriental beetle, *Exomala orientalis* (Waterhouse). The Oriental beetle is native to the Philippines, but was accidentally introduced into the U.S. in Connecticut sometime before 1920. It was initially called the 'Asiatic beetle' when it was first discovered in the U.S. The Oriental beetle is primarily a pest of regional importance in the northeastern temperate U.S. However, in recent years it has caused extensive damage to turfgrass (both warm- and cool-season turfgrass) as far south as northern Georgia. It is present throughout the Appalachian mountains and even the foothills in North Carolina. Like most turf-infesting scarabs, it causes typical grub damage.

Adult Oriental beetles are broadly rounded beetles approximately 9 to 10 mm long with relatively spiny legs. The adult beetles range in color from predominantly straw-colored to nearly entirely brownish-black. Their head is typically solid dark brown and the elytra have distinct longitudinal grooves. The eggs are somewhat milky-white, oblong, approximately 1.5 mm long, becoming slightly larger and more spherical after absorption of water. Oriental beetle larvae closely resemble Japanese beetle grubs. Both species have a transverse anal slit, though they can be differentiated easily by their raster patterns. Oriental beetle grubs have two parallel rows of 10 to 16 short, stout inward-pointing spines. Fully mature (third instar) Oriental beetle grubs are approximately 20 to 25 cm long. The prepupa resembles that of the Japanese beetle. Pupae are approximately 10 cm long, cream colored at first, but eventually become light brown as they age. The pupa also has as distinct thick fringe of hairs at the terminal segment of the abdomen.

Oriental larvae feed and destroy the roots of all cool-season turfgrasses as well as nursery stock and strawberry. The resulting damage is typical of most turf-infesting white grub species. The adults cause very little damage to plant material.

Like most scarab species that infest turf, the Oriental beetle has a 1-year life cycle. Its life cycle is similar to that of the Asiatic garden beetle and the Japanese beetle, with adults emerging in late June to early July. In many areas, the adults emerge a week or two earlier than the Japanese beetles. Because they are relatively weak fliers as compared to the Japanese beetle, they often are found crawling around, especially on flowers. Adult females lay their eggs singly 2.5 to 23 cm deep in moist soils. Each female lays about 25 eggs in her lifetime, throughout July and into early August. The eggs typically hatch after an incubation period of about three weeks. The larvae immediately begin feeding on the succulent roots of turfgrass, growing and developing into second instars by early September, and third instars by early October. Grub damage is most apparent in early September, especially when the turf is experiencing heat and drought stress. As cooler soil temperatures prevail, the nearly mature grubs burrow down into the soil to hibernate in earthen cells. In the spring, when soil temperatures are conducive in April and early May, they migrate back to the root zone to feed until early June. Thereafter, the grubs again burrow down slightly into the soil profile to pupate in earthen cells. The prepupal and pupal periods last for approximately 1 to 2 weeks, respectively. Nearly all of the grubs pupate by mid- to late June, and the adult beetles subsequently begin emerging in late June. Timing of these developmental changes may be approximately one to two weeks week earlier in the southern region of its distribution (North Carolina) than in the more northern states.

Mole crickets *(Scapteriscus spp.)* (Orthoptera: Gryllotalpidae). Mole crickets are undeniably some of the most destructive insect pests of turfgrass in the southeastern United States. Four species of mole crickets can be found in the U.S., although only two of the four cause considerable damage to golf courses, lawns and sod farms. The tawny mole cricket, *Scapteriscus vicinus* Scudder, and the southern mole cricket, *Scapteriscus borellii* Giglio-Tos (formerly *S. acletus* Rehn and Hebard) are believed to have been introduced through ports in the early 1900s from South America, and are the two most prominent pest species. Both tawny and southern mole crickets can be found in the coastal region of the Southeast, from the southern portion of North Carolina over to eastern Texas. Two other species of mole crickets, the short-winged mole cricket (*S. abbreviatus* Scudder) and the northern mole cricket (*Neocurtilla hexadactyla* Perty) can be found in the U.S. as well. The northern mole cricket is a native species that rarely, if ever, reaches pest status. It can be found from southern New England south to Florida and west to the Central Plains. The short-winged mole cricket is incapable of dispersing through flight due to its shortened wings, and thus, is typically found in coastal areas near ports of entry. When present in abundant numbers, the short-winged mole cricket can cause significant damage to turfgrass. At this time, its distribution is limited to Georgia, Florida, and some Caribbean islands, including Haiti, Nassau, Puerto Rico, Cuba and the Virgin Islands.

Tawny mole crickets are a more significant pest than southern mole crickets due to the fact that their diet consists mostly of plant material, including both the above-ground and below-ground parts of the turfgrass, while the southern mole crickets are primarily predaceous. The tunneling nature of both species makes them very damaging to turfgrass. As the crickets tunnel in their underground burrows, mechanical damage of the root system occurs, resulting in desiccation and susceptibility to other types of damage from foot traffic or golf carts.

Mole crickets are difficult to mistake for other turfgrass pests due to their unique appearance. Adults are relatively large, ranging from 3.2 to 3.5 cm long, and have characteristic shovel-like front legs. The pronotum of mole crickets is heavily sclerotized and their bodies are covered with a dense coat of fine hairs. The nymphs resemble the adults, but are smaller in size, and lack fully grown wings. At maturation, the front wings of the southern mole crickets and tawny mole crickets are folded back and almost reach the tip of the abdomen. To differentiate between the two species, one can look at the coloration or the tibial dactyls. Tawny mole crickets are typically golden brown with a mottled coloration on the pronotum and more robust, while southern mole

crickets are grayish with four pale-colored dots on the pronotum. The most reliable distinguishing characteristic between the two pest species is the arrangement of the tibial dactyls. Both species have two dactyls, with the separation between the two for the tawny mole crickets being V-shaped and narrower than the width of one dactyl, while the southern mole crickets have a larger U-shaped space that is about the size of the width of one dactyl.

There are three distinct developmental stages of mole crickets: eggs, nymphs and adults. Both species produce one generation in most areas of their distribution, although southern mole crickets may have two generations per year in south Florida. Adults of both species have significant dispersal flights in the spring and minor flights in the fall. Tawny mole crickets typically fly from February to May in Florida, while southern mole crickets fly from March to June, and it is not uncommon to have flights of both species occurring on the same night. Flights for both species generally occur up to a month later in their northern range in the Carolinas. Tawny mole crickets will complete most flight activity before oviposition, while southern mole crickets will fly between clutches. The tendency of southern mole crickets to fly between egg-layings may have contributed to their quick spread after introduction into the U.S. Flights begin soon after sunset and continue for approximately 60 to 90 minutes in response to the calls of the males. Although the majority of flying mole crickets appear to be females, it is not uncommon for males to comprise a percentage of the flying crickets. Some of the flights are performed in an attempt to find a suitable mate, but many of the females that fly have already mated and use the intensity of the males' calls to indicate good oviposition sites. The moisture levels present in the soil influence the intensity of the calls, and because desiccation is a major factor in egg mortality, it is necessary for the females to find adequate soil moisture for successful oviposition.

Once the female finds a suitable site, she will spend about 2 weeks feeding and tunneling before laying a clutch of 25 to 60 eggs. Once the eggs have been laid, the female seals the tunnel and leaves the eggs to develop and hatch in 3 to 4 weeks, depending on the temperature. Females of both species can lay as many as 10 clutches of eggs. Oviposition begins as early as March for tawny mole crickets in northern Florida, and may continue well into April and June.

Some southern mole crickets begin laying eggs as early as April in northern Florida, with peak egg-laying occurring in May and June. Timing for egg-laying and hatch is usually delayed by a month in the northern regions. Typically all tawny eggs have been laid by the beginning of July, while it is not unusual to see newly hatched southern mole cricket nymphs into September. After hatching, new nymphs will tunnel closer to the surface and begin feeding. As the nymphs continue to grow and go through 7 to 10 instars, damage will become more evident. By December, approximately 85% of tawny mole crickets have molted to adults, while only 25% of the slower-developing southern mole crickets reach maturity before winter. Those nymphs that do not reach adulthood by winter will complete development in the spring. Even though there are adults of both species present by fall, no mating or egg-laying appears to occur after fall flights. Crickets of both species overwinter as large nymphs or adults and do not become active again until the spring when temperatures rise. Mole cricket activity is most significant at night following heavy rains or irrigation when the crickets come to the surface to forage for food.

Effective management of mole crickets involves scouting, mapping, sampling and treatment. In the spring when adults are present, mating and flying, considerable damage may appear. Due to the large size of the crickets, unpredictable weather and large dispersal flights, it is difficult to effectively control crickets in the spring. At this time, spot treatment is advised for sensitive or critical areas. It is also very important to map those areas that appear to have significant adult activity for treatment of nymphs later. The time period after egg hatch when nymphs are small and nearer the surface is the best time to treat, usually in mid-June or early July. It is best to concentrate on those areas where adult activity was seen in the spring. Soapy water flushes can be used to confirm the presence of small nymphs in areas where adult activity was seen earlier in the year. An important aspect of control is to get the insecticide treatment to come into contact with the crickets. Therefore, areas of mole cricket activity should be allowed to dry for several days and then pre-irrigated the night before treatment, thus bringing the nymphs closer to the surface and making control easier. As crickets continue to grow and increase in feeding, damage may reappear in the fall, but effective control

of small nymphs is the best way to minimize their impact.

Ground pearls *(Margorodes* spp.) (Hemiptera: Coccidae). Ground pearls are a common name for several species of unique scale insects that spend most of their life in the soil and feed on the roots of various species of warm-season turfgrass. The two most common species in turf are *Margarodes meridionalis* Morrill and *Eumargaraoides laingi* Jakubski. These pests are present almost everywhere in the U.S. that warm season turfgrass is most common, from North Carolina across to California.

Ground pearls are subterranean, and the female spends most of its life in the soil. The pinkish-white eggs are laid in clusters in the soil and covered with a white waxy substance. The first instar nymphs that hatch are only 0.2 mm long and this crawler stage attaches itself to a root. Once attached it covers itself with a waxy substance and begins formation of the 'pearl.' As the pearl grows it may appear to have a purple to yellowish globular covering. The pearls range in size from 0.5 mm to 2 mm in size. Adult females may appear in the late spring and throughout the summer. They are pinkish sac-like creatures with well-developed forelegs and are about 1.6 mm in length. The male is rarely seen and has a gnat-like appearance.

Ground pearls attack a range of warm-season turfgrasses, but appear to be most common on bermudagrass, centipede, zoysiagrass, and St. Augustinegrass. Centipede often appears to be the most severely damaged. The distribution is sporadic, but where infestation does occur, the damage can be quite serious. The nymphs (pearls) feed on the roots by extracting plant juices (and possibly injecting toxic substances), which results in an initial appearance of unhealthy, yellow turfgrass. This appearance usually worsens during hot, dry summers and the plant often will turn brown and then die. This rapid death would indicate toxin may be involved. The damaged areas vary from a few centimeters to several meters in diameter. The turfgrass usually does not regrow in these areas even the next summer. Weeds often grow in these irregular, but somewhat circular areas. Through time, the damaged areas may eventually coalesce and a significant portion of the turf may die.

The life history of this pest is not well understood across its range, but one generation per year probably occurs throughout its range. Under unfavorable conditions, the ground pearls may take longer than a year to complete their life cycle. The nymphs overwinter in the soil, and some nymphs will mature in the spring and summer, emerge from the cyst and move toward the soil surface as adults. The adults may be seen moving on the soil surface in late spring and all summer. After moving for a short period of time, they reenter the soil and move down only about 5 to 8 mm to secrete the waxy coat in which the eggs will be laid a few days later. The females lay eggs without mating, as they reproduce parthenogenically. They lay about 100 eggs over a two week period and the eggs hatch in about 10 days to two weeks. The new crawlers move to the roots, attach and begin to encyst once feeding is initiated. Ground pearls have been found as deep as 25 to 30 cm in the soil. A fungal disease of ground pearls has been reported, and ants probably play a significant role as predators.

Red imported fire ant *(Solenopsis invicta* Buren) (Hymenoptera: Formicidae). The red imported fire ant has become a pest of great significance throughout the warm season turfgrass areas of the South, particularly the Southeast west to Texas. This introduced species of ant was accidentally introduced from South America into the Mobile, Alabama, area around 1930. This ant has spread rapidly throughout the South in all areas with mild winters and adequate moisture. States as far north as Virginia and Tennessee now battle this problem.

The mounds from fire ants are typically conical in shape and vary considerably in size, but large mounds are usually 25 to 40 cm in diameter and may be 20 to 30 cm in height. The mounds may penetrate more than a meter into the soil. In very sandy soils the mounds may be less well developed. They usually are found in open sunny areas and often appear following a disturbance of the soil (e.g., construction). Mounds commonly appear at the base of trees, structures, near rotting logs and stumps, power boxes, and along curbing and sidewalks. A mound contains three forms of adult ants as well as the brood (immature ants). Adult ants can be black reproductive males, queens or worker ants. The large egg-laying queens have no wings while the unmated queens and males are winged. The worker ants (which are the ants most commonly seen foraging near the mound) are wingless, sterile females and they vary in size from 1.5 to 4 mm in length. These workers have a shiny black abdomen and reddish-brown head and thorax.

Red imported fire ants forage and feed on a variety of materials, often on insects and related organisms, seeds, carrion and discarded food items. These ants can both sting and bite, but the venom from their sting is the biggest problem. They can sting repeatedly and are quite aggressive against anything that disturbs their mound. A relatively small percentage of people are hypersensitive to the venom in the imported fire ant sting and can become quite ill from even one sting. Fire ants also cause problems by creating unsightly mounds that may smother the turfgrass, create an uneven surface, and damage mowing equipment. They do not feed directly on the turfgrass, but they can be found infesting any turfgrass species that grows in a climate appropriate for fire ant survival.

Mating flights can occur any time of year, but are most common in the spring and early summer. Mating flights often occur after a rainy period. Reproductive males and unmated females (both winged) fly in the spring and after completion of the mating flight, the newly mated queen lays eggs in a small chamber in the soil. The males die soon after the flight, while the queen sheds her wings and begin the task of starting a new colony. Approximately 10 to 15 eggs are laid and once they hatch in a little over a week, the queen takes care of them until they are adults in about three weeks. These adults then begin taking care of the queen and egg production increases up to 200 eggs per day. These new fire ants mounds are inconspicuous, and often are not noticed until fall. During the summer, the colony increases in size and by late in the year, the above-ground portion of the mound may be quite prominent. Mature colonies may contain anywhere from 100,000 to more than 500,000 ants. The workers typically live two months, but a queen can survive for five years. Some colonies may contain more than one queen. Colonies with single queens are more territorial and areas with such colonies usually will have 40 to 150 mounds per acre. Multiple queen colonies are less protective and workers roam freely from one mound to another. There may be more than 200 mounds per acre in these situations.

Management of turfgrass insect pests

Effective and environmentally responsible insect pest management begins with a planned, well-designed control strategy that considers a multitude of tactics including but not limited to biological, chemical, cultural, and plant resistance. This pest management strategy is referred to frequently as Integrated Pest Management (IPM). IPM relies on a combination of preventative and curative (i.e., corrective) measures to keep pest densities below levels that would cause unacceptable damage. The goal of IPM is to manage pests effectively, economically, and with nominal risk to people, animals and the environment. IPM is not a rigid pest management program that precludes the use of pesticides, nor is it a biological or organic pest control program. IPM is merely a decision making process that considers all control tactics that will provide acceptable control of a respective pest. When successful, IPM typically reduces dependence on pesticides.

The process of IPM involves the following seven principal steps:

Sampling and monitoring. Regularly inspecting areas of turf throughout the growing season enables a turfgrass manager to detect pests early, before they reach damaging levels. Monitoring can aid in assessing the need for action, provide an evaluation of the success of previously implemented control tactic(s), and develop site history information that will provide invaluable insight for potential future problems.

Pest identification. Accurate identification is essential. The biology of a pest cannot be understood if the pest is not correctly identified.

Decision-making. Management decisions are guided by action thresholds (density of pests that will cause unacceptable damage). As decisions are made, factors such as likelihood of success, treatment cost and environmental consequences are considered.

Appropriate intervention. All appropriate management tactics, including biological, chemical, cultural control and plant resistance, must be considered. Turfgrass managers first must attempt to determine the cause of the pest outbreak, and make the necessary cultural management practice adjustments to reduce the risk of future problems. Occasionally, insecticide treatments may be warranted, however, always consider less toxic products when practical.

Follow-up. Through consistent and disciplined sampling and monitoring, turfgrass managers can effectively evaluate the success of a control tactic as well as determine the need for further action.

Record-keeping. Accurate and detailed record-keeping enables a turfgrass manager to recall when and where specific pest problems have occurred and to plan to deal with them in the future. Such information can help managers to evaluate previous management practices and may provide insight in developing and/or modifying future pest management programs. Record-keeping also provides managers with a document that may circumvent potential litigation.

Employee and client information. Another crucial component of a successful IPM program is communication. This involves educating employees, clients and the public. Education of employees can be accomplished by developing ongoing training programs that focus on pest recognition, and biology and agronomic factors that affect pest management decisions. Other educational opportunities, such as field days, workshops and short-courses, also provide valuable training that helps induce employees to work more safely and effectively as well as helping to increase professionalism and environmental stewardship. As for effective client communication, information vehicles such as newsletters, fact sheets, bulletins, fliers, personal communication, electronic mailings, etc., all provide excellent opportunities to convey valuable information. Finally, communication to the public sector can be accomplished through newspaper articles, bulletins, open houses, town meetings, television promotions, etc.

IPM considers all appropriate management options, including biological, chemical, cultural and plant resistance.

Biological control

Biological control is the use of predators, parasitoids, and disease-causing microbes or pathogens to suppress pest populations. Compared to chemical control research, to date there has been very little research on the impact and use of predators and parasitoids in the turfgrass system. Most biological control research efforts have been directed at microbial control (i.e., bacterial, fungal, viral, and entomopathogenic nematodes). Moreover, of this work, much has focused on entomopathogenic nematodes (i.e., insect parasitic nematodes). Entomopathogenic nematodes are microscopic roundworms that attack and kill insect larvae, and reproduce within the dead

Beneficial invertebrates: predators and parasitoids.

Organism	Taxonomy	Type	Host(s)
Ant	Hymenoptera: Formicidae	Predator	Eggs and larvae of numerous insects
Big-eyed bug	Hemiptera: Geocoridae	Predator	Chinch bug, greenbug, mites, eggs and larvae of small arthropods
Entomopathogenic nematodes	Secernentea: Rhabditida	Predator/ Pathogen	Numerous insect larvae
Green lacewings	Neuroptera: Chrysopidae	Predator	Aphids, mealybugs, and other soft-bodied arthropods
Ground beetles	Coleoptera: Carabidae	Predator	Eggs and larvae of numerous arthropods
Tiger beetles	Coleoptera: Carabidae	Predator	Eggs and larvae of numerous arthropods
Lady beetles	Coleoptera: Coccinellidae	Predator	Aphids and mealybugs
Rove beetles	Coleoptera: Staphylinidae	Predator	Eggs and larvae of arthropods
Scoliid wasps	Hymenoptera: Scoliidae	Parasitoid	Green June beetle larvae and other white grubs
Spiders	Arachnida: Araneida	Predator	Most insects
Tiphiid wasps	Hymenoptera: Tiphiidae	Parasitoid	Most white grub species

host. They are beneficial organisms that naturally occur in most soils, and they are practically harmless to humans, plants and animals. There are numerous species of insect parasitic nematodes; different species and strains of nematodes vary in their activity against different species of insects. As with many biological control agents, specific environmental conditions must be met in order for nematodes to be effective: they are sensitive to direct sunlight and high temperatures, and they require thin films of water that enable them to move over soil particles.

Despite the potential that biological control agents possess, only a limited number of companies currently market biological control agents. As a result, product availability and cost often discourage turfgrass managers from integrating them into their integrated pest management program. To further confound the reluctance of the use of biological control agents, they are frequently directly compared to conventional insecticides, and because they do not consistently perform at the same level, they are often discounted or not considered as a viable control option. Nonetheless, more consideration should be given to biological control agents.

Chemical control

When used judiciously, selectively and responsibly, insecticides are tremendously important tools in an overall IPM program. To date, chemical control continues to be the most common control strategy. This is likely due to numerous factors such as convenience, economics, effectiveness, familiarity and/or

confidence of turfgrass managers with this strategy. Nonetheless, chemical control should not be the primary line of defense in managing turfgrass insect pests. When effective, alternative control strategies are available, always consider tactic(s) that have the least risk to people and the environment. Conversely, in some instances alternative, non-chemical control strategies may not be available. Thus, insecticide treatments may be the only control option. Subsequently, when selecting an insecticide, always consider products that are least hazardous to people and the environment, and that will provide the most effective control in an economical manner.

As previously mentioned, turfgrass insects can be categorized into three primary groups: (1) leaf and stem, (2) thatch, and (3) root zone/soil. Subsequently, chemical control of insect pests in respective groups require specific considerations and application procedures.

Leaf and stem insect pests. The turf area should be mowed and the clippings removed prior to the insecticide application to enhance the insecticide penetration into the turf canopy. A thorough irrigation before an insecticide application will help move insects out of the turf canopy (i.e., thatch and soil) and bring them to the surface. For night feeding insects such as black cutworm or sod webworm larvae, insecticide applications should be made in the late afternoon or early evening to maximize potential for exposure to the target pest and minimize the likelihood of photodegradation (i.e., sunlight decomposition) and volatilization of the insecticide, as well as reduce pesticide exposure to humans and animals. When liquid (i.e., sprayable) insecticides are used, avoid irrigation for at least 12 hours to maximize contact and subsequent control. Regardless of insecticide formulation, avoid mowing turf after application for at least 24 hours. When granular products are used, a moderate application (i.e., one-quarter inch or 6 mm) of irrigation is recommended to help dissolve and activate insecticide granules. Liquid and granular applications have their respective advantages and disadvantages. Liquids typically provide greater mortality from initial contact and they frequently leave a residue that results in some residual control. In contrast, granules have little initial contact activity, and because they have a tendency to bounce off the foliage, they are not effective against foliage dwelling insects such as mites and greenbugs. However, they provide somewhat longer residual control of thatch dwelling insect pests.

Thatch inhabiting insect pests. Turfgrass insects that inhabit thatch can be difficult pests to control depending on the quantity of thatch accumulation, as well as the active ingredient and formulation of the insecticide. When thatch accumulation is one-half inch (12 mm) or greater, the effectiveness of insecticide applications is greatly impeded. Consequently, reducing the thatch layer by de-thatching or aerification prior to an insecticide application will increase insecticide efficacy. Another important tactic for effective management of thatch inhabiting insect pests is to apply a moderate (i.e., one-quarter inch or 6 mm) amount of irrigation prior to the insecticide application to maximize the likelihood of the treatment reaching the target pest. Overall, granular insecticides tend to perform better than liquid formulations, however, the active ingredient of the product is the overriding factor that influences the effectiveness.

Root zone/soil inhabiting insect pests. Soil dwelling or root feeding insect pests are often the most difficult to control. Pests such as white grubs and mole crickets can be controlled with either liquid or granular insecticides. The two primary strategies for controlling soil inhabiting or root feeding insects are: (1) preventative and (2) curative (i.e., corrective) control.

Preventative control is a strategy in which an insecticide is applied before a possible insect problem and subsequent damage develops. This approach can be compared to an insurance policy; potential damage is avoided or minimized. This approach to insect pest management can be quite attractive to turfgrass managers since it is relatively easy to implement and it requires dramatically less time and effort in monitoring, sampling and decision-making. Preventative control requires the use of insecticides that have a relatively long residual activity (i.e., more than 100 days). Nonetheless, turfgrass managers must still understand the biology of the pest in order to accurately time the treatment to ensure maximum control of the target pest.

Curative control is the complete opposite of preventative control; this approach is essentially a reactionary or corrective strategy. This approach heavily relies on monitoring and sampling of turf areas. When populations of a turfgrass insect pest are determined to be above an unacceptable level (i.e., threshold), an insecticide application is made to control the pest. Curative control applications typically are applied after damage is evident, though ideally before severe damage results. Regardless of the formulation or active ingredient of the insecticide, the application must be watered in thoroughly with a minimum of one-half to three-quarters of an inch (12 to 18 mm) of post treatment (i.e., after the application) irrigation to move the residue through the thatch to the target zone where the pest is located. If an irrigation system is not available, simply rely on anticipated rainfall. Another effective strategy for optimal control of soil inhabiting and root feeding insects is the application of pretreatment irrigation. The application of one-half inch (12 mm) of water 24 to 48 hours before insecticide treatment will encourage the target insects to move closer to the surface and to decrease the absorbency of the thatch. This approach is especially important if conditions have been hot and dry and insects are deeper in the soil.

When using any chemical control agent or product, always read and follow the label directions. Be conscientious of the potential impact on humans, animals and the environment. Consider using products that are the least toxic and hazardous. Ultimately, remember that there are no 'silver bullets' that will eliminate all pests.

Cultural control

Cultural control is described as suppressing pest populations or reducing their damage by normal or slightly modified management practices. These measures must be implemented before the insect reaches pest status or damage occurs. Essentially, cultural control is a proactive approach to pest management. Such management practices include, but are not limited to: (1) mowing, (2) irrigation and drainage, (3) fertility, and (4) thatch management. Ultimately, healthy turfgrass is more tolerant to potentially damaging insect populations.

Mowing. Generally, proper mowing practices such as sharp mower blades and removing no more that one-third of the grass blade at each mowing will make the turf healthier and subsequently more tolerant to insect damage. Excessively close mowing and/or scalping greatly reduces plant vigor, thus decreasing tolerance to environmental stresses. As a result, close-cut turf tends to have greater susceptibility to damage from certain turfgrass insect pests. Close and consistent mowing may remove the eggs of some insect pests such as the black cutworm.

Irrigation and drainage. Irrigation and drainage can have both a positive and negative impact on the potential for insect damage. Certain insect pests prefer hot and dry conditions while other prefer irrigated turf where soils have adequate moisture. For example, Japanese beetle adults will seek out moist areas of turf to lay their eggs. This is likely due to the fact that moist soils are required for egg and young larval survival. Thus, withholding irrigation during adult egg-laying may reduce the potential for survival of eggs and subsequent grub damage. Unfortunately, this cultural management strategy is not practical for a golf course manager since adult activity occurs in late June through August. Conversely, turfgrass insect pests such as chinch bugs prefer hot and dry conditions for survival and optimal reproduction. Therefore, irrigating or watering when chinch bugs are present will help to minimize damage and subsequent survival and reproduction. Overall, sound irrigation contributes to healthy, vigorous turf, thus, the turf is able to withstand higher pest populations and recover more rapidly from insect damage.

Fertility. Similar to the effects of irrigation and drainage, fertility also can have both a positive and negative impact on the potential for insect damage. For example, high levels of nitrogen fertility often result in rapid, succulent growth, subsequently increasing the likelihood of insect damage. Excessive nitrogen fertilization coupled with impro-

per irrigation and mowing also can result in a thatch accumulation which provides a hospitable environment for insects such as billbugs, cutworms, chinch bugs, and sod webworms. Additionally, heavy fertilization of cool-season turfgrasses in the spring stimulates rapid shoot growth at the expense of a deep, healthy root system. These weak, shallow-rooted turfgrass plants are less likely to recover from insect damage. A proper, well balanced fertility program can result in health, vigorous turf that has a greater potential to recover from insect damage. Moreover, in some instances, fertilization coupled with appropriate irrigation can aid in the ability of turf to recuperate from insect injury such as white grub damage. For specific fertility and irrigation recommendations, consult your local cooperative extension agent or state specialist.

Thatch management. Thatch is described as a tightly intermingled layer of living and dead roots, crowns, rhizomes, stolons and organic debris that accumulates between the zone of green vegetation and the soil surface. Certain levels of thatch (i.e., less than one-half inch or 12 mm) accumulation are beneficial, however, this layer in the turf canopy also provides a prime habitat for insect pests. In addition, thatch also acts as a barrier to the penetration of soil insecticides targeted at soil dwelling and root feeding insects. As a result, turfgrass managers frequently spend valuable time and resources attempting to reduce thatch to an acceptable level. Thatch accumulation can be managed effectively through proper irrigation, fertility, aerification and verticutting (i.e., vertical mowing or de-thatching).

Plant resistance

Establishing insect-resistant turfgrasses is another valuable IPM tool. Resistance to insects has been discovered in numerous plants, though the degree of resistance may vary considerably from one plant species to another. Even within a species, one cultivar or variety can have varying levels of resistance to particular insect pests. When it comes to plant genetics, there are often trade-offs associated with plant resistance. While a plant may be developed for a characteristic such as color, other beneficial traits may be sacrificed or lost to attain this trait (e.g., insect resistance). Unfortunately, plant resistance to insects is a relatively low priority of both managers and plant breeders. There is higher priority or value in developing a plant material that exhibits improved color, texture, growth, cold-hardiness, disease resistance, etc. Subsequently, only a few insect resistant turfgrasses exist. Nonetheless, of the insect resistant turfgrasses that are available, most are quite effective in managing pests. One mechanism by which plant resistance is expressed is via endophyte-infected turfgrasses. Endophytes are fungi that live within healthy grass plants, but have no adverse effect on the turfgrass plants. To date, they are known to occur only in perennial ryegrass, and tall and fine-leaf fescues. Endophytes produce alkaloids that are toxins that do not harm the plant, but are either deterrents or toxicants to the insects that feed on the above-ground plant parts, including stems, leaf sheaths and leaves. Only nominal levels of the alkaloids are present in root tissues, thus, endophyte-infected turfgrasses do not provide meaningful control of root feeding insects. In addition to the insect resistance attributes that endophytes have, they also may improve stress tolerance and enhance resistance of turfgrass to some diseases. Unfortunately, there are no endophytic cultivars of Kentucky bluegrass or creeping bentgrass, nor do they occur in warm-season turfgrasses.

R. Chris Williamson
University of Wisconsin
Madison, Wisconsin, USA
and
Rick Brandenburg and Sarah Thompson
North Carolina State University
Raleigh, North Carolina, USA

References

Brandenburg, R. L., and M. G. Villani. 1995. *Handbook of turfgrass insect pests.* Entomological Society of America, Lanham, Maryland.

Niemczyk, H. D., and D. J. Shetlar. 2000. *Destructive turf insects.* H.D.N. Books, Wooster, Ohio.

Potter, D. A. 1998. *Destructive turfgrass insects.* Ann Arbor Press, Chelsea, Michigan.

Vittum, P. J., M. G. Villani, and H. Tashiro. 1999. *Turfgrass insects of the United States and Canada.* Cornell University Press, Ithaca, New York.

Watschke, T. L., P. H. Dernoeden, and D. J. Shetlar. 1995. *Managing turfgrass pests.* Lewis Publishers, Ann Arbor, Michigan. 361 pp.

TURTLE BEETLES. Members of the family Chelonariidae (order Coleoptera). See also, BEETLES.

Fig. 1091 Example of tussock moths (Lymantriidae), *Lymantria dispar* (Linnaeus) from Germany.

TUSSOCK MOTHS (LEPIDOPTERA: LYMANTRIIDAE). Tussock moths, family Lymantriidae, total 2,490 species worldwide; actual fauna likely exceeds 3,000 species. Most of the fauna is Old World tropical (ca. 2,090 sp.). Two subfamilies are used, Orgyiinae and Lymantriinae, but this classification is uncertain. The family is in the superfamily Noctuoidea, in the section Cossina, subsection Bombycina, of the division Ditrysia. Adults small to very large (16 to 135 mm wingspan); some with brachypterous females (some are apterous). Maculation mostly somber browns and grays, but some mostly yellow or white, or more colorful; a few species even with hyaline wings. Adults mostly nocturnal but some are diurnal or crepuscular. Larvae are leaf feeders, sometimes gregariously. Host plants include many different plant families. Many species are serious defoliators of forest trees.

John B. Heppner
Florida State Collection of Arthropods
Gainesville, Florida, USA

References

Bryk, F. 1934. Lymantriidae. *In Lepidopterorum catalogus* 62: 1–441. W. Junk, The Hague.

Bryner, R. 2000. Lymantriidae – Trägspinner. In *Schmetterlinge und ihre Lebens räume: Arten – Gefährdung – Schutz. Schweiz und angrenzenden Gebiete*, 3: 529–580, pl. 24-25. Pro Natura-Schweizerische Bund fuer Naturschutz, Basel.

Ferguson, D. C. 1978. Noctuoidea. Lymantriidae. In R. B. Dominick et al. (eds.), *The moths of America north of Mexico including Greenland.* Fasc. 22.2. E. W. Classey, London. 110 pp, 9 pl.

Griveaud, P. 1977. Insectes Lépidoptéres. Lymantriidae. In *Faune de Madagascar*, 43: 1–588.

Holloway, J. D. 1999. Family Lymantriidae. *In The moths of Borneo,* 5: 1–188, 63 + 12 pl. Malayan Nature Society: Kuala Lumpur (Malayan Nature Journal, 53).

Seitz, A. (ed.). 1910–34. Familie: Lymantriidae. In *Die Gross-Schmetterlinge der Erde,* 2: 109–141, pl. 19–22 (1910); 2(suppl.): 95–106, 283–284, pl. 8 (1932–34); 6: 535–564, pl. 72–74 (1927); 10: 291–387, pl. 38–47 (1915–25); 14: 127–205, pl. 20–28 (1926). A. Kernen, Stuttgart.

TWIG BORER.

An insect that enters the shoot tip of a growing woody plant, causing the twig to wilt and die.

TWIRLER MOTHS (LEPIDOPTERA: GELE-CHIIDAE).

Twirler moths, family Gelechiidae, are a very large family, with over 4,830 species described, however, possibly with a fauna exceeding 10,000 species worldwide. Subfamily arrangements have varied but now include four subfamilies, plus many tribes: Physoptilinae, Gelechiinae, Pexicopiinae, and Dichomeridinae. The family is part of the superfamily Gelechioidea in the section Tineina, subsection Tineina, of the division Ditrysia. Adults small (4 to 35 mm wingspan), with head smooth-scaled and labial palpi recurved; haustellum scaled; maxillary palpi 4-segmented. Hindwings usually with distinctive falcate apical point and with long fringes. Maculation varied, but mostly somber colors, but some can be colorful and with metallic-iridescent markings. Adults mostly nocturnal but some are diurnal or crepuscular. Many adults tend to twirl in circles on leaves when disturbed. Larvae have a range of feeding habits but most are leaf skeletonizers, using a leaf fold or leaf

tie as protection. A large variety of plants are used as hosts. Some species are economically important.

John B. Heppner
Florida State Collection of Arthropods
Gainesville, Florida, USA

References

Clarke, J. F. G. 1969. Gelechiidae. In J. F. G. Clarke, *Catalogue of the type specimens of Microlepidoptera in the British Museum (Natural History)* described by Edward Meyrick, 6: 219–537; 7: 1–531. British Museum (Natural History), London.

Elsner, G., P. Huemer, and Z. Tokár 1999. *Die Palpenmotten (Lepidoptera, Gelechiidae) Mitteleuropas: Bestimmung – Verbreitung – Flugstandort – Lebensweise der Raupen.* F. Slamka, Bratislava. 208 pp (85 + 28 pl.).

Gaede, M. 1937. Gelechiidae. In *Lepidopterorum Catalogus,* 79:1–630. W. Junk, The Hague.

Hodges, R. W. 1986. Gelechioidea. Gelechiidae (part). In *The moths of America north of Mexico including Greenland.* Fasc. 7.1: 1–195, 4 pl. (1986); 7.6: 1–339, 5 pl. (1999). Wedge Entomological Foundation, Washington.

Huemer, P., and O. Karsholt. 1999. *Microlepidoptera of Europe. Volume 3. Gelechiidae I (Gelechiinae: Teleiodini, Gelechiini).* Apollo Books, Stenstrup. 356 pp, 14 pl.

Janse, A. J. T. 1949–54. Gelechiadae [sic]. In A. J. T. Janse, *The moths of South Africa,* 5: 1–464, 202 pl. Pretoria.

Li, H. -H. 2002. *The Gelechiidae of China (I) (Lepidoptera: Gelechioidea).* Nankai University Press: Tianjin, 538 pp (32 pl.).

Piskunov, V. I. 1981. Gelechiidae. In *Identification keys to the insects of European Russia. 4. (Lepidoptera),* 2: 659–748. Academie Nauk. [in Russian], St Petersburg.

TWISTED-WING PARASITES.

Members of the order Strepsiptera. See also, STYLOPIDS.

TWOSPOTTED SPIDER MITE.

See also, CITRUS PESTS AND THEIR MANAGEMENT, SMALL FRUIT PESTS AND THEIR MANAGEMENT, POTATO PESTS AND THEIR MANAGEMENT.

TWO-SPOTTED STINK BUG, *Perillus bioculatus* (FABRICIUS) (HEMIPTERA: PENTATOMIDAE, ASOPINAE).

The two-spotted stink bug or double-eyed soldier bug, *Perillus bioculatus,* is a predatory bug belonging to the pentatomid subfamily Asopinae. It is believed to have originated somewhere from the southeastern Rocky

Fig. 1092 Example of twirler moths (Gelechiidae), *Filatima albilorella* (Zeller) from Florida, USA.

Mountains to the Plains region and to have followed the eastward migration of the Colorado potato beetle, which appears to be its primary prey, at least in agricultural settings. The insect is presently found from Mexico into Canada.

Like those of other asopines, adults of *Perillus bioculatus* are broadly oval and more or less shield-shaped. Adult males are 8.5 to 10 mm long and the females are 10.5 to 11.5 mm long. Basically, there are three color forms: in the white form, the background is white with black and brown markings; in the yellow form, there are less pale areas, which are yellow rather than white, and all markings are black; in the red form, the background is red, and markings are black. The adults may gradually change from white or yellow to red, but not vice versa. The adults live for 1 to 2 months and are capable of laying 100 to 300 eggs. The bug has 2 to 3 generations per year and overwinters as an adult from September to October to April to May; adults usually hibernate in litter, but have also been found to enter buildings.

The eggs are blackish and barrel-shaped to somewhat elongate; the operculum is encircled by a row of about 15 micropylar processes, which are much smaller than in the spined soldier bug, *Podisus maculiven-*

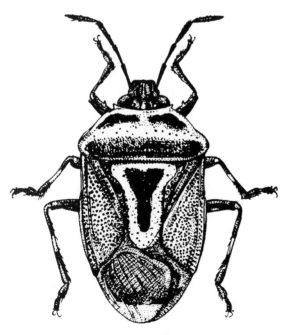

Fig. 1093 Adult of two-spotted stink bug, *Perillus bioculatus.*

tris. The egg is on average 1.2 mm long and is 0.9 mm at its widest diameter. The eggs are laid in clutches consisting of 10 to 25 eggs, often arranged in a double row. There is no embryonic development below 15°C. At temperatures between 20 and 25°C, eggs hatch within 5 to 8 days.

The nymphs are round to oval in shape. There are five nymphal instars. The body length of the five instars averages 1.5, 3, 4 to 5, 6 to 7 and 7 to 9 mm, respectively. First and second instars have a black head and thorax and a tomato red abdomen with dark patches. In third and fourth instars, the red background color may vary from tomato red to orange red or reddish yellow; the black color of the thorax often shows a greenish luster. In the fifth instar, there are two distinct forms: the dark form is greenish black and red, whereas in the pale form, the thorax and abdomen are mainly white to yellowish with dark markings. In the fifth instar, distinct wing pads are developed. During the last three instars, the red color may gradually change to yellow and white; on the other hand, nymphs with a yellowish background color may also become red. The carotinoid pigment responsible for the red coloration in the nymphs and the adults is believed to originate from their food. At high temperatures, the red pigment of the nymphs is quickly oxidized, yielding paler individuals.

The first instars hardly move about and form close groups. The gregariousness of the nymphs is more pronounced than in *Podisus maculiventris*, and even in later instars, the nymphs tend to form clusters, especially when they are preparing for the next molt and when temperatures are low. The first instars usually do not feed on prey and only require moisture, mainly in the form of plant juices. The nymphs start to attack prey from the second instar on. The nymphs and adults have a thickened rostrum or beak which they use to kill their prey and to suck up the liquefied prey tissues. Smaller nymphs in particular are often observed to attack prey collectively; on the other hand, gregariousness can also enhance the development of smaller nymphs by their opportunistic feeding on prey killed by larger, conspecific nymphs. Like other asopines, all predatory stages regularly feed on plant sap. Development of the nymphal stage requires about 3 weeks at 20 to 25°C, partly depending on food availability and quality.

Perillus bioculatus appears to be more of a specialist pentatomid predator compared to the spined soldier bug, *Podisus maculiventris*. In the field, the predator is usually found in association with the eggs and larvae of coleopterans, mainly from the Chrysomelidae family. It has been reported to feed on other insects, including the larvae of several lepidopterans. The bug appears to be, however, a rather timid predator with a dislike for highly mobile or aggressive prey. The two-spotted stink bug is best known for its predation on the Colorado potato beetle, *Leptinotarsa decemlineata*. The predator attacks all life stages of the beetle, but some scientists found that it has a preference for the eggs. Interestingly, the bugs usually suck out every egg in a mass, unlike some of the pest's predators with chewing mouthparts. It has been estimated that *Perillus bioculatus* can consume over 300 eggs during its nymphal period. This suggests a great potential for eliminating Colorado potato beetle populations provided that the pentatomid is present early enough in the season. Because natural populations of the predator in potato fields are usually too low early in the season, augmentative releases of laboratory-reared individuals may help in suppressing outbreaks of the pest in early spring. However, some workers have questioned the predator's dispersal and predation capacities, particularly under cool climates. Nonetheless, in small scale field plots, significant reductions of high density populations of Colorado potato beetle larvae have been achieved by releasing 1 to 3 second- or third-instars per plant. Like the spined soldier bug, *Perillus bioculatus* has been introduced in different parts of Europe for biological control of the Colorado potato beetle, but none of these introductions were successful.

High numbers of predators needed for augmentative field releases necessitate an economically viable mass production. The two-spotted stink bug can be reared on larvae of several noctuids (including *Trichoplusia* and *Spodoptera* spp.), a method that is cheaper than rearing it on the Colorado potato beetle, because the noctuids are easily mass-reared on artificial diets. The availability of an artificial diet for the predator itself may further reduce costs of mass propagation. Artificial diets have been developed that can support consecutive generations of the predator, but development and survival rates are reduced and fecundity is only one tenth of that raised on live prey.

See also, STINK BUGS (HEMIPTERA: PENTATOMIDAE) EMPHASIZING ECONOMIC IMPORTANCE, STINK BUGS, PREDATORY (HEMIPTERA: TENTATOMIDAE, ASOPINAE), BUGS (HEMIPTERA).

Patrick De Clercq
Ghent University
Ghent, Belgium

References

De Clercq, P. 2000. Predaceous stinkbugs (Pentatomidae: Asopinae). pp. 737-789 in C. W. Schaefer and A. R. Panizzi (eds.), *Heteroptera of economic importance*. CRC Press, Boca Raton, Florida.

Hough-Goldstein, J. A. 1998. Use of predatory pentatomids in integrated management of the Colorado potato beetle (Coleoptera: Chrysomelidae). pp. 209–223 in M. Coll and J. R. Ruberson (eds.), *Predatory Heteroptera: their ecology and use in biological control*. Proceedings, Thomas Say Publications in Entomology, Entomological Society of America, Lanham, Maryland.

Knight, H. H. 1923. Studies on the life history and biology of *Perillus bioculatus* Fabricius, including observations on the nature of the color pattern. *19th Report, State Entomologist of Minnesota*: 50–96.

Rojas, M. G., J. A. Morales-Ramos, and E. G. King 2000. Two meridic diets for *Perillus bioculatus* (Heteroptera: Pentatomidae), a predator of *Leptinotarsa decemlineata* (Coleoptera: Chrysomelidae). *Biological Control* 17: 92–99.

Tamaki, G., and B. A. Butt. 1978. Impact of *Perillus bioculatus* on the Colorado potato beetle and plant damage. *U.S. Department of Agriculture Technical Bulletin 1581*.

TYMPANUM. A membrane-covered cavity on the thorax, abdomen, leg or other part of the body that functions like an ear to perceive sound. These chordotonal organs are optimized to perceive high frequency sound.

TYPHUS. There are several types of typhus, all of which are associated with arthropods. The most common are epidemic and endemic typhus Epidemic typhus (also known as Brill disease) is associated with poor hygiene, and often is associated with cold temperatures. It is spread by lice. Historically, it is associated with periods of warfare and human disaster. Endemic typhus (also known as murine typhus) is associated with rat fleas and rat feces. It is most common during the warm weather when rats and fleas are most abundant. These forms of typhus are caused by rickettsia, either *Rickettsia*

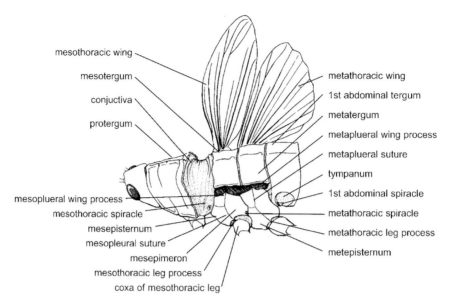

Fig. 1094 Head and thorax of a grasshopper (Orthoptera).

prowazekii (epidemic tphus) or *Rickettsia typhi* (endemic typhus).

Symptoms include high fever, severe headache, nausea and vomiting, abdominal pain and a transient rash. It is a severe illness, and may result in death if it is not treated, particularly in the case of epidemic typhus. Symptoms occur within 14 days of exposure, and the disease is treated with antibiotics. Sanitation and avoidance of flea-infested rats and rat feces is recommended.

Other forms of typhus include tick typhus and scrub typhus. Tick typhus, transmitted by ticks, is a form of spotted fever and is found in Africa and the Indian subcontinent. Scrub typhus is transmitted by mites, and is found in Southeast Asia and the Pacific islands. Avoidance of wild animals and animal habitats is recommended for avoidance of these diseases.

See also, MITES, TICKS, HISTORY OF INSECTS.

TYPE SPECIMEN. The original specimen ('type') from which a new species description is created. The holotype.

U

UENOIDAE. A family of caddisflies (order Trichoptera). See also, CADDISFLIES.

UHLER, PHILIP REESE. Philip Uhler was born at Baltimore, Maryland, USA, on June 3, 1835. He is known as America's greatest hemipterist. Uhler was born into a prominent merchant family but was not interested in his father's business, preferring natural history. In 1864 he was placed in charge of the insects at the Museum of Comparative Zoology at Harvard University, and the library at Cambridge, Massachusetts. He resigned in 1867 to return to Baltimore where he became assistant librarian at the Peabody Institute. In his spare time he studied entomology, economic entomology at first but then on Hemiptera. A book-buying trip to Europe in 1888 allowed him to examine many of Europe's insect collections, and he soon became a leading authority in this field. However, failing eyesight caused him to give up these pursuits in 1890. Uhler named many of Hemiptera collected on the early explorations of western North America. Notable publications include "List of Hemiptera of the region west of the Mississippi River, including those collected during the Hayden explorations of 1873," and "Hemiptera, Standard Natural History, Vol. 2." Uhler died at Baltimore, Maryland, on October 21, 1913.

References
Essig, E. O. 1931. *A history of entomology*. The Macmillan Company, New York. 1029 pp.

Mallis, A. 1971. *American entomologists*. Rutgers University Press, New Brunswick, New Jersey. 549 pp.

ULIDIIDAE. A family of flies (order Diptera). See also, FLIES.

ULTRA LOW VOLUME (ULV). Delivery of a liquid formulation of undiluted pesticide in very small droplets.

ULTRASTRUCTURE OF INSECT SENSILLA. The cuticle of insects is largely responsible for the success of these terrestrial arthropods. It is rigid and hard in areas requiring support and protection and flexible in regions associated with locomotion and the detection of mechanical stimuli. It is highly impermeable and restricts water loss from the body surface. Cuticle does not expand, thereby limiting growth, making it necessary for growing insects to periodically shed the existing cuticle (a process called molting) and to replace it with another, larger one. The insensitivity of insect integument also limits the reception of stimuli. This insensitivity is remedied by the presence of transcuticular sensory mechanisms, each specifically designed to detect one of a wide range of stimulus types.

Sensory perception for taste, smell, touch, sound, vision, proprioception, and geo-, thermo-, and hygroreception involves a three-stage process consisting of coupling, tranduction, and encoding. Coupling connects the stimulus to the sensory neuron.

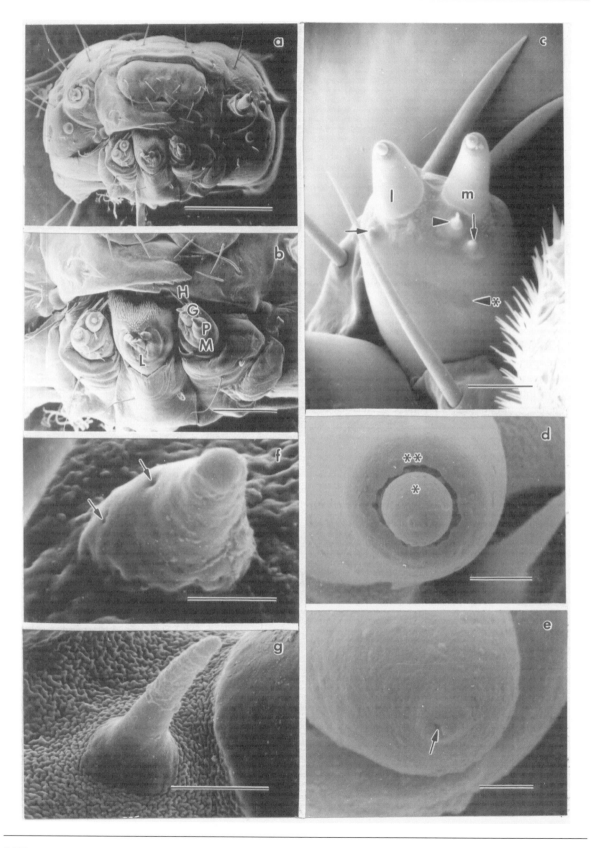

Transduction converts the stimulus energy to electrical energy (receptor current). Encoding is the process that generates action potentials from the receptor current.

All type I sensilla (see below, Classification of sensilla), with the exception of certain scolopidial sensilla, consist of a specialized sensory cuticle innervated by the dendrites of one or more sensory neurons and usually three or four accessory cells that enwrap the neurons and associated sinuses. The exoskeleton and small size of insects impose certain restrictions on the nervous system, which are reflected in their behavior. Small size and short transmission fibers reduce the distance for impulse conduction and decrease reaction time. This allows them to respond very rapidly to stimuli. Size limitation also implies a decrease in neuronal number and a consequent reduction in informational capacity. The insect nervous system is organized in such a way that stimulation of a single sense cell may trigger a series of responses. Insects possess primary sense cells, which not only receive the stimulus, but initiate and transmit information by means of a direct connection with the central nervous system.

Cuticular parts

The components of an insect sensillum, that are considered to be cuticular, are those that are shed with the exuvium at a molt. These include the sensory cuticle and intracuticular structures (pore canals and pore tubules or filaments), the socket or insertion in the surrounding insensitive cuticle, and the sleeves or sheaths that extend inward around the distal part of the dendrite.

Lipoidal pore canals are found in the body cuticle, as well as in the walls of the sensory cuticle, but are modified to varying extents. The canals usually contain pore filaments or tubules possessing an osmiophilic, lipoidal wall and a non-osmiophilic, non-lipoidal core. In most sensilla, pore canals open into the sensillar sinus liquor of the lumen within the sensory cuticle and in some sensilla, pore canals extend the length of the cuticular wall and open into the liquor of the sensillar sinus, under the sensory cuticle.

The dendritic sheath is cuticular in origin and is shed at a molt. Examination by transmission electron microscopy (TEM) reveals that the dendritic sheath is an invagination of the surface layer of epicuticle and resembles the superficial cuticulin layer in composition and density. The dendritic sheath is thought to possibly play a significant role in the stimulus transfer mechanism between the outer sensory cuticle and the enclosed dendrite in sensilla that are mechanosensitive (see below, Mechanosensilla).

Classification of sensilla

The sensory neurons of insect sensilla are classified into two types: type I and type II. Type I neurons are bipolar, contain a ciliary structure in the distal dendrite, and innervate the sensory cuticle. The distal dendrite can be simple, branched, or lamellated in different types of sense organs. These neurons originate *de novo* within the epidermis. The neuronal perikaryon (cell body) remains peripheral in location, while the axon grows inward toward the central nervous system (CNS). A Type I sensillum is derived from an epidermal mother cell, which divides to form the nerve and sheath cells.

Type II neurons are bipolar or multipolar, lack a ciliary structure, and are never associated with cuticular processes. These neurons innervate the body wall, alimentary canal, muscles, epidermis, connective tissues, and sheath cells. They may have originated within the central nervous system and then migrated outwards to innervate mesodermal and epidermal tissues.

Fig. 1095   SEM micrographs. Figs. a, b. *Spodoptera frugiperda*. Figs. c–e. *Mamestra configurata*. Figs. f, g. *Choristoneura fumiferana*. Fig. a. Frontal view of whole head. Scale bar = 1 mm. Fig. b. Frontal view of ventral mouthparts. G, galea; H, dorsal guard hair on the galea; L, labium; M, maxilla; P, maxillary palp. Scale bar = 400 μm. Fig. c. Fronto-ventral view of a right galea showing lateral (l) and medial (m) uniporous styloconic sensilla and three aporous basiconic sensilla: spire-shaped (arrowhead), lateral (horizontal arrow) and medial (vertical arrow). An aporous campaniform sensillum (arrowhead with asterisk) is located approximately midventrally on the galeal wall. Scale bar = 25 μm. Fig. d. Higher magnification of the tip of a styloconic sensillum. Each sensillum consists of a small terminal cone or peg (asterisk) inserted into a socket (double asterisk) on a large style. Scale bar = 4 μm. Fig. e. Apical view of the cone of a styloconic sensillum with a terminal pore (arrow). Scale bar = 1 μm. Fig. f. Aporous basiconic sensillum. Beneath the pores (arrows), are longitudinal pore canals filled with pore tubules that extend to the surface of the cuticle. Scale bar = 2 μm. Fig. g. Aporous spire-shaped basiconic sensillum. Scale bar = 5 μm. Both types of sensilla in Figs. f and g are inserted into inflexible sockets.

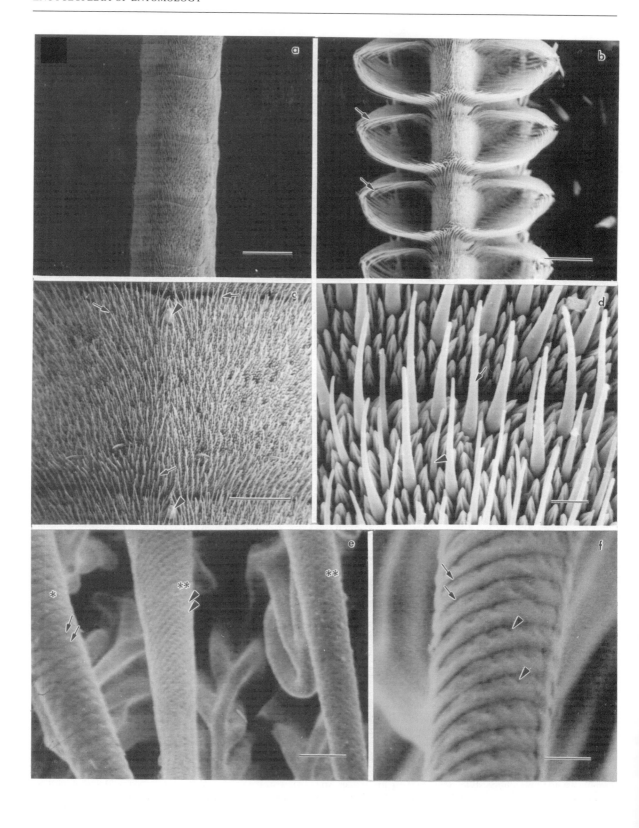

A type I sensillum can be defined as a sense organ that has one or more Type I bipolar neurons whose dendrites are enveloped by at least two sheath cells. All type I insect sensilla presumably evolved from an epidermal hair (seta) and all their component parts are believed to be homologous.

Traditionally, sensillar types have been classified on the basis of the morphology of their cuticular parts and their position on, within, or under the cuticle.

Sensilla trichodea. These are hairs innervated by one to several neurons. They can vary greatly in length and are freely moveable on a basal membrane. They can be solely mechanosensitive, dually mechano- and contact chemosensitive, olfactory, or thermosensitive.

Sensilla chaetica. These are bristles or spines innervated by one or more neurons. They are similar to the sensilla trichodea, but have thicker walls. They are typically set in a socket and can be mechano- or contact chemosensitive.

Sensilla basiconica. These are pegs, cones, or papillae innervated by one to several neurons. They are similar in function to trichoid sensilla, as well as being hygrosensitive.

Sensilla coeloconica. These are basiconic pegs or cones set in shallow pits innervated by two to several neurons. They are most often chemo-, thermo-, or hygrosensitive.

Sensilla ampullacea. These are basiconic pegs set in deep pits innervated by two to several neurons. The surface opening is often much narrower (flask-shaped) than that of coeloconic sensilla. They have sensory characteristics similar to the sensilla coeloconica.

Sensilla squamiformia. These are sensory scales innervated by one or more neurons and may be mechano- or chemosensitive. They have not been identified in immature insects.

Sensilla campaniformia. These are dome, bell, or cupola-shaped structures positioned at or below the surface of the cuticle. They are characteristically innervated by one neuron and are solely mechanosensitive.

Sensilla placodea. These are plate-like sensilla innervated by several to many neurons. They are positioned level with, slightly raised above, or depressed below the surface cuticle, and are olfactory.

Sensilla scolopophora. Also called chordotonal organs, these are subcuticular and are not associated with an external modification of the cuticle, though they maintain an attachment to it. Each sensilla unit is innervated by one to three neurons. There are two types of scolopidial sensilla: amphinematic and mononematic. Amphinematic scolopidia are drawn into a distal thread or tube and may be either integumental (direct attachment with the cuticle) or subintegumental. Mononematic scolopidia lack such a thread or tube and are subintegumental (without any direct attacment to the cuticle). Some chordotonal sensilla are propriceptive in function and respond to stretch

Fig. 1096 SEM micrographs. Figs. e, f. FESEM micrographs. Figs. a–f. *Manduca sexta*. Fig. a. Adult female antennal flagellum showing numerous hair-like trichoid sensilla. Scale bar = 250 μm. Fig. b. Adult male antennal flagellum showing long multiporous pheromonal trichoid sensilla (arrows). Scale bar = 250 μm. Fig. c. Higher magnification of an annulus from Fig. a showing many long, multiporous trichoid sensilla (arrows). The majority of these sensilla have ultrastructural characteristics typical of olfactory sensilla. Each arrowhead indicates an aporous styliform complex sensillum. Scale bar = 100 μm. Fig. d. Higher magnification of a region from Fig. 2c, showing long multiporous trichoid (arrow) and shorter basiconic (arrowhead) olfactory sensilla. Scale bar = 10 μm. Fig. e. Higher magnification of a multiporous trichoid (asterisk) and two multiporous basiconic sensilla (double asterisks) sensilla. The shaft of the trichoid sensillum bears circumferential cuticular ridges (arrows), more apparent at a higher magnification in Fig. f. The cuticular shafts of both trichoid and basiconic sensilla bear numerous pores, more apparent at a higher magnification in Figs. f and Figs. a–c. The pores of the basiconic sensillum are distributed in oblique linear rows (arrowheads) along the long axis of the sensillum. Scale bar = 1 μm. Fig. f. Higher magnification of the middle portion of the cuticular shaft of a trichoid multiporous sensillum showing that the shaft of the sensillum bears circumferential cuticular ridges (arrows). These ridges form a helical pattern over the basal quarter of the length of the sensillum and a more circular pattern over the remaining length. The cuticular shaft is also perforated by pores (arrowheads) in cuticular depressions that are arranged in a single row along the midline of the ridges. Scale bar = 1 μm.

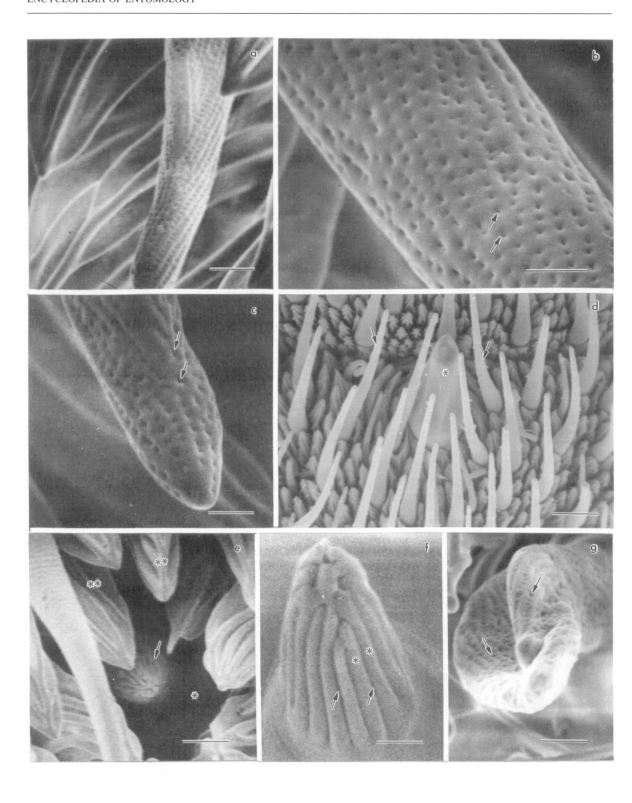

during movements of body parts, while others respond to air-borne vibration.

Sensilla styloconica. These are cones, pegs, or squat hairs, inserted at the tip of a conical or cylindrical projection (style) of insensitive cuticle. They are innervated by one or a few neurons and are mechano- and/or chemosensitive.

Other uniquely-shaped sensory sturctures have been described in insects as large conical sensory appendices, domes, knobs, partitioned plates, knobbed rods, flower-shaped domes, ear-like sensilla auricularis, clubbed hairs, and coniform or styliform complexes.

There are three main sensillar categories: AP (aporous) or NP (no-pore) sensilla, which are either mechanosensitive or hygro- and thermosensitive; UP (uniporous) or TP (terminal-pore) sensilla, containing gustatory neurons alone, or in combination with a mechanosensitive cell, and MP (multiporous) or WP (wall pore) sensilla, occurring as 2 main types: single-walled (SW) sensilla, with pore tubules, and double-walled (DW) sensilla, with spoke canals and the sensillar liquor as stimulus conducting systems. SW sensilla contain olfactory neurons, whereas DW sensilla contain olfactory and/or thermohygro-sensitive neurons.

Aporous (AP) sensilla

Mechanosensilla. AP sensilla lack a permeable pore in their sensory cuticle, but may have a typical cuticular pore canal system with pore tubules extending to the surface cuticle. During a molt, the exuvial dendritic sheath is pulled through the new cuticle at its point of cuticular insertion. When the exuvium is cast, this sheath is broken off, often resulting in a molting scar or pore. This pore can easily be misinterpreted as one that is permeable and belonging to UP sensilla.

AP sensilla are mainly mechanosensitive and are innervated by a single, unbranched neuron that is enclosed by a relatively thick dendritic sheath. The characteristic feature of these mechanosensilla is an accumulation of microtubules, called a 'tubular body,' usually located in the distal region of the distal dendrite. The tubular body generally consists of 50 to 100 tubules lying parallel to one another in an electron-dense material, although considerable variation in complexity and structure has been noted in numerous insect species. It is thought to be the site of sensory transduction.

Thermo-hygro sensilla. AP sensilla may also serve a thermo-hygrosensitive function. Essential features regarded as adaptations to hygro- and thermoreception are the lack of pores; an inflexible socket; the occurrence of dendritic outer segments which fill the lumen of the peg (type 1, moist/dry receptors), or the occurrence of one or two dendritic outer segments which terminate below the peg (type 2, cold receptors or type 3, with unknown function). Typically, AP thermo-hygrosensilla are innervated by three bipolar neurons referred to as a 'triad,' consisting of two type-1 and one type-2 neurons.

Uniporous (UP) sensilla

UP chemosensilla are usually termed contact chemosensitive, gustatory, or taste sensilla. Their chemosensitivity is generally predicated on contact with chemicals in solution. They are typically innervated by three to five or six unbranched distal dendrites and have one permeable apical or subapical pore, through which chemical communication between the dendrites and the external environment occurs. This pore also serves as the molting pore remaining open after the exuvial dendritic sheath is shed.

Pores of UP sensilla may have either a simple pit pore (UPP) or a sculptured porous point (UPS). The latter may vary from simple grooves leading to the

Fig. 1097 FESEM micrographs. Fig. 3d. SEM micrograph. All figures are of *Manduca sexta* sensilla. Fig. 3a. Multiporous basiconic sensillum. Scale bar = 1 µm. Fig. b. Higher magnification of a cuticular shaft taken from the middle of a multiporous basiconic sensillum. Note the pores (arrows). Scale bar = 1 µm. Fig. c. Higher magnification of the distal end, near the tip, of a multiporous sensillum. The pores (arrows) extend along the entire length of the sensillum. Scale bar = 0.5 µm. Fig. d. Aporous styliform complex sensillum (asterisk) surrounded by multiporous trichoid sensilla (arrows). Scale bar = 10 µm. Fig. e. Aporous coeloconic sensillum (arrow) lying in a deep pit (asterisk) and surrounded by microtrichia (double asterisks). Scale bar = 1.5 µm. Fig. f. Higher magnification of coeloconic sensillum displaying numerous longitudinal ridges (asterisks). There are grooves (arrows) between the ridges. Scale bar = 0.5 µm. Fig. g. Multiporous auriculate (spoon-shaped) sensillum endowed with numerous pores (arrows). The upper surface is deeply concavely indented. Scale bar = 1 µm.

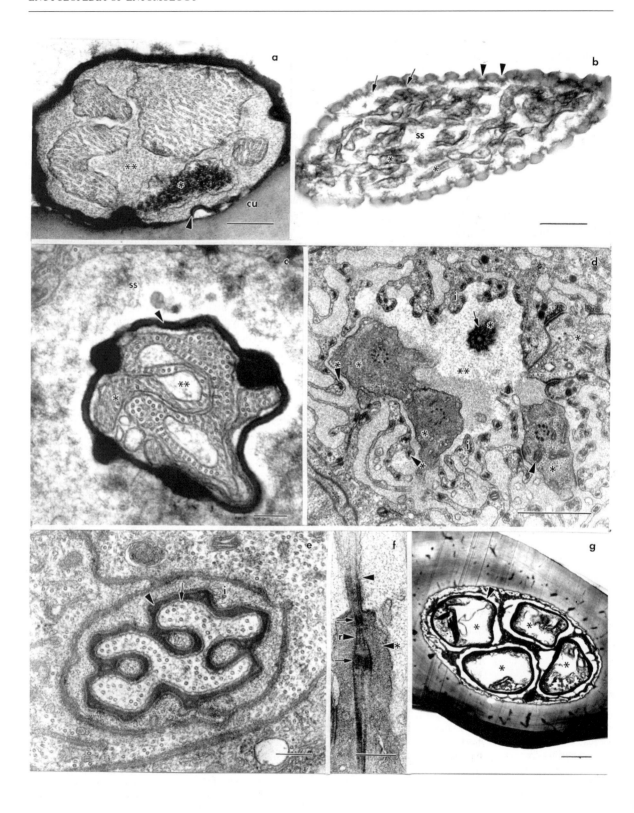

pore to finger-like cuticular projections surrounding the opening. Pores vary in shape and size, ranging in diameter from 10 nm to 200 nm. They may contain a viscous mucoid extrusion from the dendritic channel that can cover the external opening. It is thought that this extrusion continuously fills the pore and may serve to conduct chemical stimulants within the dendritic channel to the dendrites below. Pores contain pore tubules or plugs of fenestrated fibrils of cuticular or other origin, in addition to the dendritic channel liquor. Such plugs may function by conferring selectivity to the conduction mechanism and a specificity of response to the sensillum.

Differences in the permeability of the galeal lateral and medial styloconic pegs (cones) to heavy metal ions, such as cobalt, mercury, and lead, have been shown to occur in several larval lepidopteran species, including *Mamestra configurata*, *Trichoplusia ni*, *Spodoptera frugiperda*, *Choristoneura fumiferana*, *Lymantria dispar*, and *Malacosoma lutescens*. The differences in permeability to these ion markers are of significance in interpreting the varying electrophysiological response to certain stimulants or deterrents applied to the terminal pore.

Some or all of the dendrites from UP sensory neurons extend through the dendritic channel towards the tip. Many UP sensilla contain a mechanosensitive neuron, whose dendrite terminates in a tubular body at the base of the sensory cuticle. Such sensilla are dually contact chemo- and mechanosensitive.

The dendritic sheath in UP sensilla is fused to the inner wall of sensory cuticle at the base of the pore and the liquor in the terminal dendritic channel is continuous with the liquor in the ciliary sinus. The dendritic sheath physically separates the dendritic chamber or channel from the sensillar sinus. The sheath is thought to be permeable to some small ions, but impervious to larger molecules. A permeable dendritic sheath would enable the sensillar sinus to act as a reservoir for ions, required by dendrites in producing the receptor potential and maintaining a resting potential. The dendritic channel is often appressed to the sensory cuticle along one side of the sensillum along much of its length.

Fig. 1098 TEM micrographs. Figs. a, c–f. *Mamestra configurata*. Figs. b, 4g. *Manduca sexta*. Fig. a. Cross section near the base of the peg of a uniporous styloconic sensillum showing four chemosensory distal dendrites and the tubular body of the mechanosensory dendrite (asterisk). These dendrites all contain longitudinally oriented microtubules. The tubular body of the mechanosensory dendrite is composed of an accumulation of microtubules lying parallel to one another within an electron-dense matrix. The mechanosensory dendrite is closely apposed to the cuticular wall (cu) of the cone. At this level, the dendritic sheath (arrowhead) is thin on the side adjacent to the cone cuticle, prior to becoming thicker, proximally. The dendrites are surrounded by the dendritic channel (double asterisk). Scale bar = 0.1 μm. Fig. b. Multiporous auriculate sensillum. Note the numerous distal dendritic branches (asterisks) and extremely thin cuticular wall bearing numerous pores (arrowheads) and underlying pore tubules (arrows). The pore tubules, associated with each pore, extend from the base of a small circular pore kettle toward the dendrites through the surrounding sensillar sinus (ss). Some of the pore tubules appear to contact a few of the dendritic branches. Scale bar = 0.4 μm. Fig. c. Cross section cut near the base of an aporous spire-shaped sensillum. At this level, two distal dendrites, one of which is lamellated (asterisk), are enclosed by a thick dendritic sheath (arrowhead) and surrounded by the dendritic channel (double asterisks). ss, sensillar sinus. Scale bar = 0.2 μm. Fig. d. Cross section showing five proximal dendrites (asterisks) from a uniporous styloconic sensillum. One of the dendrites (arrow) is cut at the apex of the distal basal body of the proximal dendrite. The basal body bears a 9 × 3 + 0 microtubular configuration and nine radiating alar spokes of electon-dense material. One alar spoke arises from each of the nine triplet microtubules. At this level, the inner sheath cell (i) projects longitudinal folds with discrete complexes of microfibrils and microtubules (arrowheads with asterisks) around the proximal dendrites and encloses a ciliary sinus (double asterisks). The cytoskeletal elements are located along the inner border of the highly infolded inner sheath cell and adjacent to the ciliary sinus. Arrowhead, mitochondrion. Scale bar = 0.1 μm. Fig. e. Cross section of a distal dendrite enclosed by a highly convoluted dendritic sheath (arrowhead) from an aporous basiconic sensillum. The microtubules within the dendrite are arranged in a single row along the periphery of the dendritic membrane (arrow). This membrane is in close apposition to the dendritic sheath. The inner sheath cell (i) enwraps the sensory neuron from its origin, near the proximal end of the dendritic sheath, to the level of the neuronal cell bodies. Scale bar = 0.2 μm. Fig. f. Longitudinal section showing a distal dendritic segment (arrowhead) inserting into a proximal dendritic segment (arrowhead with asterisk) from a uniporous sensillum. Distal (upper arrow) and proximal (lower arrow) basal bodies are located in the distal end of the proximal dendrite. From the distal basal body, nine ciliary rootlets (r with arrowhead) extend proximally in a ring around the proximal basal body to which they attach. Scale bar = 0.5 μm. Fig. g. Cross section showing four similar contiguous sensilla (asterisks) housed in an aporous styliform complex sensillum. Each sensillum unit is innervated by three bipolar sensory cells. Two of the distal dendrites are cylindrical and one is lamellate in shape. At this level, each of the distal dendritic bundles is individually enclosed by a thick dendritic sheath (arrowhead). Scale bar = 1 μm.

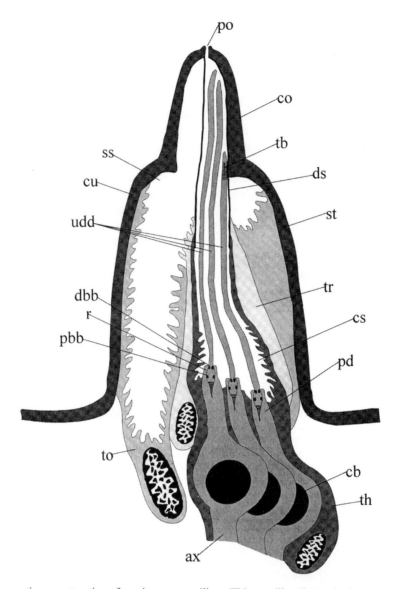

Fig. 1099 Diagrammatic reconstruction of a uniporous sensillum. This sensillum has a single pore positioned at the apex of the cone. Only three sensory cells are shown. ax, axon; cb, cell body; co, cone; cs, ciliary sinus; cu, cuticle; dbb, distal basal body; ds, dendritic sheath; pbb, proximal basal body; pd, proximal dendrite; po, pore; r, rootlets; ss, sensillar sinus; st, style; th, thecogen cell; to, tormogen cell; tb, tubular body of mechanosensitive neuron; tr, trichogen cell; udd, unbranched distal dendrites.

In UP sensilla, pore tubules, comprising the cuticular sidewall lipoidal pore canal system, extend from the sensillar sinus through the sensory cuticle to the surface of the sensillum and selectively transport materials. The sidewall pore canal system around the terminal pore and distal portion of the dendritic sheath has been shown to be be permable to cobalt and mercury ions in some lepidopterous larvae, but not in others. The functional significance of the permeability of the sidewall pore canal system into the sensillar sinus is not known.

Multiporous (MP) sensilla

MP sensilla all possess an abundance of pores in their sensory cuticle. A mechanosensitive dendrite is not typically associated with these sensilla. There

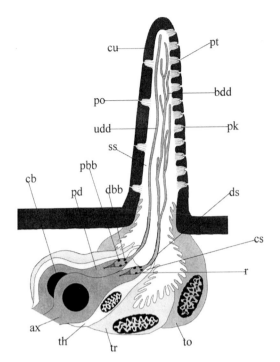

Fig. 1100 Diagrammatic reconstruction of a multiporous sensillum showing the features of both a thick- and thin-walled sensillum in the same drawing for comparative purposes. Only two sensory cells are shown. The unbranched distal dendrite, thick cuticular wall, and pores with underlying pore tubules (left half of drawing) show features typically found in a thick-walled trichoid sensillum. The branched distal dendrite (bdd), thin cuticular wall, pore kettles, and pores with underlying pore tubules (right half of drawing) show features typically found in a thin-walled basiconic sensillum. Basiconic sensilla are typically shorter in height, have a cuticular wall that is generally thinner and pierced by a higher density of pores, have a greater number of pore tubules associated with each pore; and have many dendritic branches that fill the lumen of the sensillum, in comparison to trichoid sensilla. ax, axon; bdd, branched distal dendrite; cb, cell body; cs, ciliary sinus; cu, cuticle; dbb, distal basal body; ds, dendritic sheath; pbb, proximal basal body; pd, proximal dendrite; po, pore; pk, pore kettle; pt, pore tubule; r, rootlets; ss, sensillar sinus; th, thecogen cell; to, tormogen cell; tr, trichogen cell; udd, unbranched distal dendrite (modified from Shields and Hidebrand 1999a with permission from publisher)

are two basic types of MP chemosensilla: single-walled sensilla with pore tubules and double-walled sensilla with spoke canals. MP sensilla are commonly referred to in the literature as MPP and MPG, since examination by scanning electron microscopy and TEM revealed a pitted and grooved surface, respectively.

Multiporous pitted (MPP) sensilla

MPP sensilla can be subdivided into two types based on thickness of the sensory cuticle and the conduction mechanism beneath the effective pore. Typically, thin- and thick-walled MPP sensilla have cuticles approximately 100 to 300 nm and 200 to 1000 nm, respectively. Pore density and diameter varies in thin and thick-walled sensilla, from 15 to 60/μm^2 to 2 to 20/μm^2, and 20 to 25 nm to 6 to 15 nm, respectively. The pores in both thin- and thick-walled types are funnel-shaped and flare outward from a narrowed opening just below the cuticular surface. In thin-walled MPP, each pore usually opens into a small circular chamber (pore kettle), 100 to 200 μm^2 in diameter, which abuts the sensillar sinus. Generally, more than 15 (sometimes as high as 50 to 60) pore tubules (averaging 15 nm in diameter) per pore kettle extend from its base toward the dendrites through the sensillar sinus.

Pore tubules are considered to be component parts of the sensory cuticle. The sensory pore tubule system of sensory cuticle is likely homologous with that of the pore canal and pore tubule system of the body integument. It is thought that sensory pore tubules are maintained either through the sensillar sinus or via the dendritic terminations. The cuticular pore canal system may function to transport lipoidal material from the epidermis to the surface for cuticular maintenance, in addition to the fomation of a superficial lipoidal layer. This layer may act as a barrier against moisture loss through the cuticular pores and may contain proteinaceous components. It may also trap specific stimulant molecules that diffuse to the nearest pore along the pore tubules directly to a dendrite or, indirectly through the sensillar liquor.

In MPP sensilla, the dendrites emerge from the open end of the dendritic sheath, which terminates at or near the base of the sensory cuticle, into the sensillar channel. This channel is continuous with the sensillar sinus. A molting scar may be visible on the cuticular surface near the point of dendritic sheath termination.

Thin-walled MPP sensilla are typically innervated by 2 or 3 dendrites. The dendrites emerge beyond the open, distal end of the dendritic sheath and branch (about 7 to 150 branches). Generally, 10 to 12 pore tubules extend from the base of large pore kettles, which lie in close proximity to the sensillar sinus. These sensilla have a wider range of chemosensitivity than thick-walled MPP sensilla and are referred to as 'generalist' sensilla.

In thick-walled MPP, the pore-kettles are reduced and cylindrical pockets often extend from the sensillar sinus toward the pore. Generally, 3 to 8 pore tubules extend from the base of the pore funnel into the sinus either through the cuticular wall or pocket, where present. Thick-walled MPP sensilla are typically innervated by 2 to 5 unbranched neurons. These sensilla generally have a more selective range of chemosensitivity than thin-walled MPP sensilla and are referred to as 'specialist' sensilla.

MPP sensilla are considered to be primarily olfactory. The transduction process is thought to involve six 'steps': (1) adsorption of the molecule into a superficial lipoidal and/or proteinaceous layer at the outer surface of the sensillar wall to the nearest pore; (2) diffusion along the pore tubules directly to the dendritic membrane or into the sensillar fluid; (3) binding to a receptor molecule on the dendritic membrane; (4) possible configurational change of the receptor protein; (5) resulting increase in conductance generating a receptor potential, and (6) rapid inactivation of the odorant molecule.

Multiporous grooved (MPG) sensilla

MPG sensilla possess double concentric walls; the inner wall is smooth and surrounds the distal dendrites, and the outer wall is scalloped or fluted with longitudinal groves. The inner wall extends from the tip of the sensillum to near its base. The two walls are connected by cuticular spokes. Each spoke contains a radial spoke channel that is continuous with the sensillar sinus liquor. These channels, usually 20 to 30 nm in diameter, extend inward towards the distal dendrites. The sensillar sinus liquor bathes the dendrites. Small pores (10 to 20 nm in diameter) are found at the bottom of each longitudinal groove and are connected with the inside of the sensillum via the spoke channels. Material from the sensillar sinus appears to flow along the spoke channels over the grooved surface of the sensillum, presumably representing the trapping and conduction mechanism for chemical stimulant molecules.

MPG sensilla are typically innervated by 2 to 5 neurons with branched distal dendrites. In some species, the dendritic sheath may extend from near the base of the sensillum and fuse radially with the cuticle. In others, it may continue to near the ciliary region of the dendrites. The dendritic sheath

is molted through an apical pore and pore tubules are absent. Wall thickness in MPG sensilla ranges from 300 to 400 nm and pore density is estimated at just over $200/\mu m^2$. These sensilla are considered to be primarily olfactory.

Sensory neurons

Insect sensory neurons are typically bipolar, with an enlarged perikaryon that is usually located in or below the epidermis. The dendrite extends distally from the distal pole of the cell body toward the dendritic sheath and sensory cuticle. The axon extends proximally from the perikaryon towards the central nervous system. The dendrite is abruptly constricted about midway along its length at the ciliary region (see also, below). This point demarcates the thin, distal dendritic segment from the thicker, proximal dendritic segment. The distal segment is also referred to as the ciliary segment or cilium. Typically, there is only one ciliary process per neuron.

Distal dendritic segments contain evenly dispersed longitudinal microtubles as the sole organelle. Vesicular inclusions and multivesicular bodies may also be present between the microtubules or just above the ciliary region. The distal dendritic segments are enclosed in a dendritic sheath from near their base to a variable point distally. Invaginations of the dendritic sheath produce longitudinal folds that partially or completely separate the dendrites along a portion of their length. In UP sensilla, the tip of the mechanosensitive dendrite often becomes completely separated from the other dendrites. Tight or gap junctions may also occur between the dendrites within the dendritic sheath.

The ciliary region is usually short and may be distinctly funnel-shaped, assuming a $9 \times 2 + 0$ microtubular arrangement near its point of origin. This microtubular arrangement is maintained distally in most scolopophorous and some aporous thermohygrosensilla. The ciliary region dilates into the distal dendritic segment, at its point of entry into the dendritic sheath, and the doublet microtubules separate and increase in number. Vesicular inclusions and multivesicular bodies may be present within the ciliary region.

The narrow distal dendritic segment inserts centrally into the wider tip of the thicker proximal segment. The distal and proximal, centriole-like basal bodies lie directly below the ciliary segment and are arranged in tandem. Each basal body has a $9 \times 3 + 0$

microtubular configuration. The distal basal body is located at the apex of the proximal dendritic segment, whereas the proximal basal body lies in tandem and slightly below it. The nine tubule doublets extend distally into the ciliary segment, presumably originating from the distal basal body. A circle of doublet tubules is embedded in a thin, dense 'collar' or 'necklace' of electron-dense material, just above the distal basal body. Rootlets with periodic cross-striations extend proximally from within the distal basal body. They pass around and attach to the proximal basal body and proceed toward the perikaryon. In scolopophorous sensilla, the rootlets are well developed and usually fuse to form a solid striated rod or ciliary root that extends toward the perikaryon. Vesicles, multivesicular bodies, and mitochondria typically surround the distal and proximal basal bodies. The vesicular contents are presumably used in the development and maintenance of the sensillar cuticular components. The cytoplasm of the proximal dendrite and perikaryon contains an abundance of rough and smooth endoplasmic reticulum, longitudinal microtubules, vesicles, Golgi bodies, free ribosomes, and mitochondria. The nuclei of the neuronal cells bodies are typically large, rounded, centrally located, and contain finely dispersed chromatin, indicative of high metabolic activity.

Axons from the sensory cells project to the CNS. The axons merge along common tracts to form nerves. Typical organelles found in the axon are mitochondria and evenly distributed longitudinal microtubules. Several wraps of glial cells individually enwrap the basal portions of each of the neuronal cell bodies and axons, typically inserting under the wrap of the inner sheath cell. In larger nerves, collections of axons are enclosed by a perineurial cell layer, which is further enclosed by a basement membrane in the hemocoel.

Sheath cells

Accessory or sheath cells are specialized epidermal cells that envelop the neurons of insect chemosensilla. Typically, there are four sheath cells: a thecogen (inner), trichogen (intermediate), tormogen (outer), and a basal or glial cell. The inner, intermediate, and outer sheath cells enwrap the dendrites and the distal part of the nerve cell body. A basal glial cell enwraps the basal part of the cell bodies and axons. The number of sheath cells varies from 2 to 4 depending on the sensillar type. Each sheath cell forms at least one complete wrap around the neuron along most of its length and often overlaps.

The inner sheath cell typically wraps once around the sensory neuron(s), from the base of the dendritic sheath to some part of the cell body. The proximal dendrite and perikaryon of each sensory neuron are each individually wrapped by the inner sheath cell. This sheath cell usually terminates to one side of the neuronal perikaryon in an expansion containing its nucleus. The dendritic sheath is secreted by the inner sheath cell. The inner sheath cell may also be involved in the formation of the sensory cuticle if only two sheath cells are present in the sensillum.

In the ciliary region, the inner sheath cell encloses a small, ciliary sinus and also extends small microvilli into it. Microvillation indicates a possible secretory activity. The dendritic sheath and inner sheath cell form a continuous tube enclosing the contents of the ciliary sinus. A sinus liquor (fluid secreted into the ciliary sinus by the inner sheath cell) bathes the proximal and distal ends of the distal and proximal dendritic segments, respectively. This liquor is usually finely granular and electron-lucent. It is thought to contain nutrients secreted by the inner sheath cell.

The inner sheath cell also forms well-developed longitudinal desmosomal junctions with the distal termination of the proximal dendritic segments. Bundles of longitudinally-oriented microtubules and microfibrils, located along the inner border of the inner sheath cell, are associated with these junctions. The microfibrils are about 6 nm in diameter and resemble actin fibers. The microtubule and actin cytoskeletal elements may provide cellular rigidity and serve as mechanically stabilizing elements, respectively. The dendrites presumably grow through the inner sheath cell during development.

Both intermediate and outer sheath cells are very similar to each other in structure and function. The intermediate sheath cell enwraps the inner sheath cell along most of its length). The intermediate sheath cell also lines part of the sensillar sinus and extends microvilli, from its outer surface, into this sinus. This cell may secrete the sensory cuticle. The outer sheath cell envelops the intermediate sheath cell, from near the neuronal cell body to the sensory cuticle. This cell is thought to secrete the base or socket of the seta. The outer sheath cell encloses the sensillar sinus and similarly, extends microvilli from its outer surface into this sinus. Both intermediate and outer sheath cells presumably sequester nutrients from the underlying hemolymph and actively transport and secrete them into the sensillar sinus.

Since the dendritic sheath is thought to be selectively permeable to ions, the sensillar sinus, located adjacent to the dendritic channel, may function as an ion reservoir to replenish ions lost during dendritic activity. The intermediate and outer sheath cells are also thought to maintain a chemical environment appropriate for the stimulus conduction and tranduction processes of the dendrites. These cells may be involved in cuticular maintenance and the provision of nutrients for the inner sheath cell. The sensillar sinus is usually larger than the ciliary sinus, however, its size is variable depending upon the type of sensillum. Similar to the inner sheath cell, the intermediate and outer sheath cells successively draw to one side near the level of the neuronal cell bodies and each terminates in an expansion containing the nucleus. The proximal extremities of the intermediate and outer sheath cells usually terminate in close proximity to the hemocoel, from which they are separated by a basement membrane. The inner, intermediate, and outer sheath cells are all rich in mitochondria, vesicles, mutivesicular bodies, rough endoplasmic reticulum, free ribosomes, longitudinal microtubules, and Golgi bodies.

The glial sheath cell individually enwraps the basal portions of each of the neuronal cell bodies, always inserting under the wrap of the inner sheath cell. The glial cell also individually enwraps the axons. The cell body of the glial cell is small, as is its nucleus. Within its nucleus, the chromatin is condensed into large granules. The cytoplasm contains abundant rough endoplasmic reticulum and few mitochondria. A cytoplasmic layer of either the inner or glial sheath cell, proximally, separates all parts of every neuron from one another, from the base of the cilium. The axons within the nerve are individually enwrapped by the glial cells. The axon presumably grows through the glial cell in development. See also, TASTE AND CONTACT CHEMORECEPTION.

Vonnie D.C. Shields
Towson UniversityTowson,
Maryland, USA

References

Altner, H. 1977. Insect sensillum specificity and structure: an approach to a new typology. Pp. 295–303 in J. Magnen and P. MacLeod (eds.), *Olfaction and taste IV* (Paris). Information Retrieval, London, United Kingdom.

Altner, H., and L. Prillinger. 1980. Ultrastructure of invertebrate chemo-, thermo-, and hygroreceptors and its functional significance. *International Review of Cytology* 67: 69–139.

McIver, S. B. 1985. *Mechanoreception.* Pp. 71–132 in G. A. Kerkut and L. I. Gilbert (eds.), *Comprehensive insect physiology, biochemistry and pharmacology*, Vol. 6. Pergamon Press, Oxford, United Kingdom.

Schneider, D. 1964. Insect antennae. *Annual Review of Entomology* 9: 103–122.

Shields, V. D. C. 1994. Ultrastructure of the uniporous sensilla on the galea of larval *Mamestra configurata* (Walker) (Lepidoptera: Noctuidae). *Canadian Journal of Zoology* 72: 2016–2031.

Shields, V. D. C. 1994. Ultrastructure of the aporous sensilla on the galea of larval *Mamestra configurata* (Walker) (Lepidoptera: Noctuidae). *Canadian Journal of Zoology* 72: 2032–2054.

Shields, V. D. C. 1996. Comparative external ultrastructure and diffusion pathways in styloconic sensilla on the maxillary galea of larval *Mamestra configurata* (Walker) (Lepidoptera: Noctuidae) and five other species. *Journal of Morphology* 228: 89–105.

Shields, V. D. C., and J. G. Hildebrand. 1999. Fine structure of antennal sensilla of the female sphinx moth *Manduca sexta* (Lepidoptera: Sphingidae). I. Trichoid and basiconic sensilla. *Canadian Journal of Zoology* 77: 290–301.

Shields, V. D. C., and J. G. Hildebrand. 1999. Fine structure of antennal sensilla of the female sphinx moth *Manduca sexta* (Lepidoptera: Sphingidae). II. Auriculate, coeloconic, and styliform complex sensilla. *Canadian Journal of Zoology* 77: 302–313.

Slifer, E. H. 1970. The structure of arthropod chemoreceptors. *Annual Review of Entomology* 15: 121–142.

Zacharuk, R. Y. 1985. Antennae and sensilla. Pp. 1–69 in G. A. Kerkut and L. I. Gilbert (eds.), *Comprehensive insect physiology, biochemistry and pharmacology*, Vol. 6. Pergamon Press, Oxford, United Kingdom.

Zacharuk, R. Y., and V. D. Shields. 1991. Sensilla of immature insects. *Annual Review of Entomology* 36: 331–354.

UNAPPARENT RESOURCES. Food resources (either insect or plant) that are "hard to locate" or unapparent to potential predators or herbivores. Unapparent resources often are protected against consumption by generalist predators or herbivores by possessing a toxin that deters consumption. (contrast with apparent resources)

UNDERWING MOTHS – THE GENUS *Catocala* (LEPIDOPTERA: NOCTUIDAE). The genus *Catocala* Schrank is a speciose group of colorful and large moths in the family Noctuidae. The adults range from 4 to nearly 14 centimeters in wingspan. *Catocala* are commonly known as "underwings," in reference to the contrast between their drab, bark-colored forewings and their racily patterned hindwings, which are usually jet black with prominent yellow, orange, or red bands, and white fringes.

Distribution

On a worldwide basis, *Catocala* is the second most speciose genus in the Noctuidae, exceeded in number of species only by cutworm moths in the genus *Euxoa* Hübner. *Catocala* occur throughout the northern hemisphere, though most frequently in deciduous forests, and the slightly more than 200 described species are split nearly evenly between the Nearctic and Palearctic regions. The *Catocala* faunas of the eastern and southern United States are among the most diverse, with 40–50 species occurring sympatrically in many areas. Interestingly, in contrast to most other Holarctic noctuid genera, there is no single species of *Catocala* whose geographic range includes both the Nearctic and Palearctic.

Fig. 1101 Details of adult and larval *Catocala* species, from Plates VI and XIII, in the monograph of North American species in the genus published by Barnes & McDunnough (1918).

Phylogeny

Catocala contains several monophyletic species groups, comprising species with similar larval food-plant use. However, the relationships among these groups and of *Catocala* to other catocaline noctuids remain unclear, mirroring the incomplete understanding of phyletic relationships within the expansive subfamily as a whole. The species-level taxonomy of the Nearctic taxa has been examined in detail, but the corresponding Palearctic literature is less comprehensive, and no monographic treatment of the world fauna has been published. Most *Catocala* species were described during the 19th century, and the early workers in the genus, who were male, followed a tradition of coining specific epithets reflecting themes of romance, sorrow, and/or notable women. Thus, for example, *Catocala amatrix* Hübner (sweetheart), *C. lacrymosa* Guenee (tearful), *C. insolabilis* Guenée (inconsolable), *C. titania* Dodge and *C. cordelia* Hy. Edwards (from Shakespeare).

Biology

All species of *Catocala* whose life histories are known are single brooded. The adults are nocturnal and on the wing in mid to late summer in most localities. When at rest during the daytime on tree trunks and branches, adult *Catocala* conceal their brightly colored hindwings under the forewings, and are highly cryptic. However, adult *Catocala* are also wary when at rest, and will fly off suddenly and rapidly if disturbed. They usually settle again instantly upon reaching another perch, often vanishing from plain view. *Catocala* species have characteristic escape flights and resting habits: for example, some species circle the trunk and land again on the same tree, whereas others invariably fly to another tree. In studies with avian predators, the hindwings of *Catocala* have been shown to serve both as an initial startle mechanism and as a subsequent deflective device that focuses attacks away from the body. Many patterns of lepidopteran wing damage have been categorized quantitatively using *Catocala* as test subjects, and particular damage patterns have been correlated with behavioral responses on the part of predators. Field-collected *Catocala* regularly bear such wing damage patterns that can be traced to attacks by birds, bats and lizards.

Winter diapause in *Catocala* takes place in the egg stage. The eggs are laid singly or in rows or clumps on the larval foodplants. The eggs of most species are strongly dorsolaterally flattened, and the female moths have elongate, modified ovipositors which they use to secrete the eggs under exfoliating bark, or in crevices on branches and tree trunks. The larvae of most species hatch in spring and commence feeding on young foliage in April and May. Considering the genus as a whole, hatching from individual egg clutches varies from synchronous (2 to 3 days) to staggered (7 to 8 weeks), but each *Catocala* species exhibits a characteristic spread in its egg hatch that mirrors the foliating schedule of its larval foodplants.

Catocala larvae are active, semi-loopers that feed nocturnally. Most have cryptic bark-like coloration and morphologies, especially in the later instars. Mature larvae often have variously developed projections on the dorsal surface of the fifth abdominal segment that closely match leaf scars, buds, or broken branch tips; as well as a dense row of fine, sublateral filaments that are pressed against the substrate. Both of these characteristics break up the outline of the resting larva and enhance its crypsis. The young larvae rest preferentially on the midribs of leaves and on petioles, and shift to resting on branches and tree trunks as they mature. Most *Catocala* species pass through five larval instars, with some species having six or seven instars. Larval development is usually completed within five to seven weeks of egg hatch. Pupation occurs in a loose silken cocoon of leaf litter and debris, with adults hatching in three to four weeks.

Unlike most other noctuid genera, the high species diversity in *Catocala* is coupled to a quite limited larval foodplant diet. In total, the genus is known to use only nine dicotyledonous families as larval hosts. However, even that liberally estimates foodplant breadth, because the overwhelming majority of *Catocala* species feed on plants in the families Fagacaeae, Salicaceae, Rosaceae, and Juglandaceae. No *Catocala* species is known to feed on plants from more than one of these plant families, and the foodplants of a given *Catocala* species are normally limited to one or more closely related plant genera. *Catocala* that use Fagaceae, Salicaceae, and Rosaceae are found throughout the northern hemisphere, but Juglandaceae-feeding is essentially limited to North America, to a complex of two dozen closely related species apparently representing a single phyletic radiation. Although the larvae of many *Catocala* species can be collected commonly, the genus is not known to cause significant economic effect (at least in North America; possible exceptions include *C. maestosa* Hulst and *C. agrippina* Strecker, which sometimes occur in high density on pecan trees).

Collection

Because of their size and beauty, *Catocala* moths have been favorites with collectors for centuries. The adults are readily captured at both mercury vapor and ultraviolet light traps, and many species can be attracted to artificial bait sources. The term "sugaring" for moths refers to collecting using sweetened artificial bait, and *Catocala* have always figured prominently in the sugaring lore (the technique also goes by many other synonyms, such as "wine roping"). A functional artificial bait mixture requires a source of sugar as well as alcohol, but the lepidopterological literature is richly punctuated with collectors' idiosyncratic additions to lure *Catocala* and other prizes, like treacle, pureed banana, and expensive Barbados rum. Sugaring is a notoriously fickle and as yet poorly quantified collecting technique, producing divergent results in certain geographic regions, at some seasons, and under particular climatic conditions; sultry and unsettled weather immediately preceding electrical storms generally seems to be the most productive. Another time-honored, and sporting, collecting method that focuses specifically on *Catocala* is "tree tapping," in which one attempts to locate and creep up on adults without startling them as they rest during the day on tree trunks and branches.

Lawrence F. Gall
Yale University
New Haven, Connecticut, USA

References

Barnes, W., and J. McDunnough. 1918. Illustrations of the North American species of the genus *Catocala*. *Memoirs of the American Museum of Natural History* 3: 1–47.

Gall, L. F. 1991. Evolutionary ecology of sympatric *Catocala* moths (Lepidoptera: Noctuidae). I. Experiments on larval foodplant specificity. *Journal of Research on the Lepidoptera* 29: 173–194.

Gall, L. F., and D. C. Hawks. 2002. Systematics of moths in the genus *Catocala* (Noctuidae). III. The types of William H. Edwards, Augustus R. Grote, and Achille Guenée. *Journal of the Lepidopterists' Society* 56: 234–264.

Grote, A. R. 1872. On the North American species of *Catocala*. *Transactions of the American Entomological Society* 4: 1–20.

Mitter, C., and E. Silverfine. 1988. On the systematic position of *Catocala* Schrank (Lepidoptera: Noctuidae). *Systematic Entomology* 13: 67–84.

Sargent, T. D. 1976. *Legion of night: the underwing moths.* University of Massachusetts Press, Amherst, Massachusetts. 222 pp.

UNGULATE LICE. Members of the family Haematopinidae (order Siphunculata). See also, SUCKING LICE.

UNIQUE-HEADED BUGS. Members of the family Enicocephalidae (order Hemiptera). See also, BUGS.

UNIT OF HABITAT SAMPLING. An absolute sampling technique wherein the sampling unit is the habitat inhabited by the insect (e.g., apple fruit for apple maggot, ear of corn for corn earworm).

UNIVOLTINE. Having a single generation (life cycle) per year.

URANIIDAE. A family of moths (order Lepidoptera). They commonly are known as swallowtail moths. See also, SWALLOWTAIL MOTHS, BUTTERFLIES AND MOTHS.

URIC ACID. The principal form of waste product elimination by insects. Uric acid, a purine, is synthesized by fat bodies from protein nitrogen and nucleic acid nitrogen. It is excreted from the Malpighian tubules. Though uric acid is energetically costly to produce, it is very conservative of water, and so aids greatly in water conservation.

URODIDAE. A family of moths (order Lepidoptera). They commonly are known as false burnet moths. See also, FALSE BURNET MOTHS, BUTTERFLIES AND MOTHS.

UROGOMPHUS. (pl., urogomphi) Paired processes located dorsally on the 9th abdominal segment.

UROSTYLIDAE. A family of bugs (order Hemiptera, suborder Pentamorpha). See also, BUGS.

URTICARIA. Itchy, and often elevated or discolored spots on the skin; wheals or hives

URTICATING HAIRS. Hairs that cause a stinging or burning sensation in humans or other animals upon contact because they contain poison glands, or simply because the barbed hairs are irritating.

USINGER, ROBERT LESLIE. Robert Usinger was born at Ft. Bragg, California, USA, on October 24, 1912. He attended the University of California-Berkeley where he received both his B.S. in 1935 and his Ph.D. in 1939. At that time he accepted a faculty position at University of California-Davis, where he worked until World War II. Following military service, he transferred to Berkeley, where he taught and conducted research. Usinger was a systematist, with expertise in Hemiptera. He traveled the world in pursuit of insects. He published several important books and monographs on systematics and Hemiptera, including ''Methods and principles of systematic zoology'' (with Mayr and Linsley, 1953), ''Aquatic insects of California'' (1956), ''General zoology'' (with Storer, 1965), and ''Monograph of the Cimicidae'' (1966). He was an active member of the International Commission on Zoological Nomenclature, and the Entomological Society of America, including service as president of the latter organization. Usinger died prematurely on October 1, 1968, in San Francisco.

References

Linsley, E. G., and R. F. Smith. 1969. Robert Leslie Usinger 1912–1968. *Annals of the Entomological Society of America* 62: 1218–1219.

UVAROV, (SIR) BORIS PETROVICH. Boris Uvarov was born on November 5, 1888, in Ural'sk

Province, Russia. He was interested in insects from an early age, and attended Petersburg (Leningrad) University from 1906 to 1910. He studied Orthoptera, and was already publishing on the importance of habitats as early as 1910 (though this was not his first paper, having already published three botanical papers). After graduating, Uvarov immediately obtained an appointment as an entomologist in southern Russia in a cotton-growing district, where he had the opportunity to work on biology and control of migratory locust and Moroccan locust. In 1915 he transferred to another office that had responsibility for Georgia and Armenia. He also lectured at Tbilisi University, and published numerous papers on local grasshoppers/locusts. This was an important period in Uvarov's formation of phase theory, in that he gained an appreciation of the importance of local conditions in insect biology. With the collapse of the Russian Empire, Uvarov relocated to London, where he worked for the Imperial Bureau of Entomology. Here he published "Locusts and grasshoppers" in English (1928), as well as taxonomic work and important reviews on nutrition and climate. Here he also began to publish profound insight into locust phases and polymorphism, and the importance of solitary versus gregarious mode of life. This was an entirely unprecedented insight into insect biology, a variation that is matched only by sociality in terms of behavior, and equivalent to climate and parasitism in terms of importance in population regulation. In England, Uvarov implemented data collection that produced important information on several locusts, including range of several species and the important concept of locust management by preventing the locusts from reaching the gregarious stage. Starting in 1945 Uvarov and a small team of acridologists received official designation as the "Anti-locust Research Centre" and conducted important studies on African and Asian locusts until his official retirement in 1959. This did not stop his entomological contributions, however, and he then published "Grasshoppers and locusts" Volume 1 in 1966, and Volume II in 1977. Uvarov's contributions totaled over 400 publications; about half were taxonomic and about half were population biology and suppression. His insight into phase biology serves as a unique and long-lasting tribute to his creativity and analytical abilities. His influence has been profound, perhaps unprecedented among grasshopper and locust workers, and for this reason he has been called the 'father of acridology.' He was knighted in 1961. He died in London on March 18, 1970.

References

Bey-Biyenko, G. Y. 1970. Sir Boris Uvarov (1888–1970) and his contribution to science and practice. *Entomological Review* 49: 559–562.

Waloff, N., and G. B. Popov. Sir Boris Uvarov (1888–1970): The father of acridology. *Annual Review of Entomology* 35: 1–24.

UZELOTHRIPIDAE. A family of thrips (order Thysanoptera). See also, THRIPS.

V

VAGINA. The rectum, a common chamber into which the anus and gonopore open; the cloaca.

See also, REPRODUCTION.

VALDIVIAN ARCHAIC MOTHS (LEPIDOP-TERA: HETEROBATHMIIDAE). Valdivian archaic moths, family Heterobathmiidae, include about nine species, with three named thus far. The family is the single member of the superfamily Heterobathmioidea and the sole member of the suborder Heterobathmiina. Adults small (10 to 11 mm wingspan), with head rough-scaled, with small mandibles; labial palpi are 3-segmented and short; maxillary palpi are 5-segmented and folded. Maculation is usually iridescent purple with fine spots. All are diurnally active as adults, feeding at *Nothofagus* (Fagaceae) flowers. Larvae are leafminers of *Nothofagus* trees in southern South America.

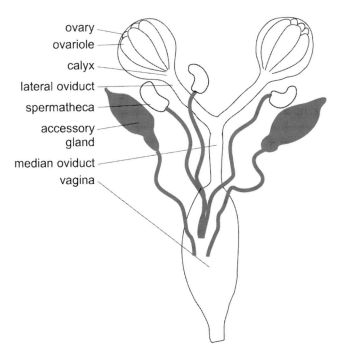

Fig. 1102 Diagram of the female reproductive system, as found in *Rhagoletis* (Diptera) (adapted from Chapman, The insects: structure and function).

John B. Heppner
Florida State Collection of Arthropods
Gainesville, Florida USA

References

Kristensen, N. P., and E. S. Nielsen. 1983. The *Heterobathmia* life history elucidated: immature stages contradict assignment to suborder Zeugloptera (Insecta, Lepidoptera). *Zeitschrift fur Zoologie, Systematik und Evolution* 21: 101–124.

Kristensen, N. P., and E. S. Nielsen. 1998. *Heterobathmia valvifer* n. sp.: a moth with long apparent 'ovipositor valves' (Lepidoptera: Heterobathmiidae). *Steenstrupia* 24: 141–156.

VALDIVIAN FOREST MOTHS LEPIDOPTERA: ANDESIANIDAE).

Valdivian forest moths, family Andesianidae, total only 3 known species from southern Andean forest zones of Argentina and Chile. The family is the most primitive of the Monotrysian moths, in its own monobasic superfamily Andesianoidea, and in the section Nepticulina. Adults are robust and of medium size (27 to 61 mm wingspan), with head scaling roughened; haustellum very short (with vestigial mandibles present); labial palpi elongated (with elongated 2nd segment); maxillary palpi 5-segmented (unfolded); and antennae bipectinate in males (filiform in females). Wing maculation is lustrous gray with numerous darker vertical striations, plus some dark spotting. Wing venation is heteroneurous but somewhat primitive with added veins. Adults apparently are nocturnal (possibly crepuscular), mainly in *Nothofagus* forests. Biologies are unknown, but larvae are likely to be stem borers.

John B. Heppner
Florida State Collection of Arthropods
Gainesville, Florida USA

References

Davis, D. R. 2003. Andesianidae, a new family of monotrysian moths (Lepidoptera: Andesianoidea) from austral South America. *Invertebrate Systematics* 17: 15–26.

Gentili, P. 1989. Revision sistemática de los Cossidae (Lep.) de la Patagonia Andina. *Revista de la Sociedad Entomológica Argentina* 45: 3–76 (part).

VAN DEN BOSCH, ROBERT.

Robert van den Bosch was born on March 31, 1922, in Martinez, California, USA. He was educated at the University of California, Berkeley, where he received a degree in physical education before shifting his emphasis to entomology. He received his Ph.D. in 1950. He worked at the University of Hawaii from 1949 to 1951, the University of California-Riverside from 1951–1963, and the University of California-Berkeley from 1963 to 1978. van den Bosch had great influence in the fields of integrated pest management and biological control. Along with colleagues V.M Stern, R.F. Smith and others at the University of California, van den Bosch espoused an integrated approach to pest management that used chemical insecticides as only one element of pest suppression, relying on biological and cultural control for much mortality among pest populations. More specifically, chemical pesticides were viewed as a necessary but undesirable tactic that would be used only when absolutely necessary. The deleterious effects of pesticides including pesticide resistance among the target pests, inadvertent mortality to nontarget organisms such as beneficial insects and wildlife, and hazards to pesticide applicators were noted as reasons to avoid pesticide use. Though many have now come to accept this view, this was a radical viewpoint in the 1950s and 1960s when insecticides seemed to many to be miracle products. van den Bosch published more than 150 technical articles and several landmark books including "Biological control" (with P.S. Messenger, 1973), "Introduction to integrated pest management" (with M.L. Flint, 1981), and "The pesticide conspiracy" (1978). The latter was particularly provocative and served to sensitize both the research community and public to the lack of sound policy with regard to pesticide use in the United States. van den Bosch also served as chair of the Division of Biological Control at the University of California-Berkeley, from 1969 until his premature death at age 56, on November 19, 1978.

Reference

Hoy, M. A. 1993. Biological control in U.S. agriculture: back to the future. *American Entomologist* 39: 140–150.

VAN DUZEE, EDWARD PAYSON.

Edward Van Duzee was born at New York, New York, USA, on April 6, 1861. The son of a scientist, Edward was encouraged to work in the sciences, and he did so quite successfully, eventually becoming known as the greatest hemipterist in the United States. He became assistant librarian of the Grosvenor Library, a research library, in Buffalo, New York, in 1885, and then head librarian 10 years later. This environment allowed him to devote his attentions to the Hemiptera, and the bibliography of Hemiptera. Van Duzee collected throughout the United States. About 1900 he began a great monograph of the Hemiptera, a project that took 10 years to complete. It proved to be an outstanding compilation and serves as a model of completeness. Van Duzee left the Grosvenor Library after 27 years of employment, and before the catalogue was completed, due to ill health. He accepted employment as an instructor at the University of California, Berkeley, and the catalogue ''Hemiptera of America north of Mexico, excepting the Aphididae, Coccidae, and Aleyrodidae'' was printed by the University. Van Duzee did not enjoy teaching, and in 1916 accepted a position as assistant librarian and curator of entomology at the California Academy of Sciences. He spent the next 24 years building the insect collection, and also served as editor of the journal ''The Pan-Pacific Entomologist'' for the first 14 years of its existence. In all, Van Duzee published more than 250 articles over the course of his career. He died on June 2, 1940.

References

Essig, E. O., and R. L. Usinger. 1940. The life and works of Edward Payson Van Duzee. *The Pan-Pacific Entomologist* 16: 144–177.

Mallis, A. 1971. *American entomologists.* Rutgers University Press, New Brunswick, New Jersey. 549 pp.

VANHORNIIDAE. A family of wasps (order Hymenoptera). See also, WASPS, ANTS, BEES, AND SAWFLIES.

VANNUS. A region of the wing marked by a large fan-like expansion of the posterior region; the vannal region.

VARIABLE INTENSITY SAMPLING PLAN. A sampling plan that shares the characteristics of fixed-sample size and sequential sampling plans in which prior sampling information is used to evaluate the number and allocation of subsequent sample units, but ensures that sample units are collected throughout the sample universe.

See also, SAMPLING ARTHROPODS.

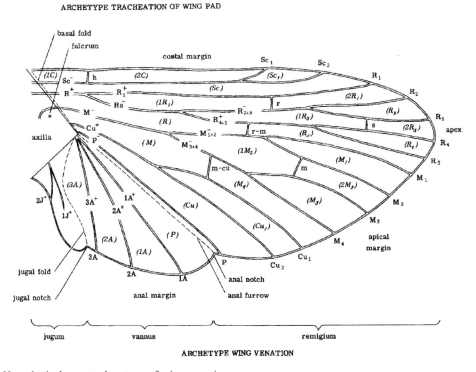

Fig. 1103 Hypothetical ancestral pattern of wing venation.

VARIEGATED CUTWORM. See also, TURF-GRASS INSECTS AND THEIR MANAGEMENT.

VARIEGATED MUD-LOVING BEETLES.
Members of the family Heteroceridae (order Coleoptera). See also, BEETLES.

VARLEY, GEORGE C. George Varley was born in 1910 and educated at Cambridge University, Cambridge, England. As a student, he became interested in mathematical models relating to population dynamics, a subject that would fascinate him for his entire career. He completed his Ph.D. in 1935 and then held a series of positions at University of California at Riverside, Cambridge University, King's College at Newcastle, and finally he was appointed Hope Professor at Oxford. He served in this latter capacity for 30 years, retiring in 1978. At Oxford, Varley worked on winter moth for 20 years, producing an unusually complete understanding of its biology, and developing key factor analysis as a method of analyzing life tables. It was a technique that would become a basic ecological method for analyzing animal populations. Varley's work proved to be very influential in population ecology. He died in 1983.

Reference

Hassell, M. P. 1983. George C. Varley, 1910–1983. *Antenna* 7: 121–122.

VASCULAR SYSTEM. The system of plant tissues that conducts water, minerals, and products of photosynthesis within the plant.

VASCULAR WILTS. See also, TRANSMISSION OF PLANT DISEASES BY INSECTS.

VAS DEFERENS. (pl., vasa deferentia). A tube that collects the sperm from the testicular follicles and transfers them to the seminal vesicle for storage and ejaculation.
See also, REPRODUCTION.

VECTOR. An agent, normally an animal, carrying a microorganism pathogenic for members of another

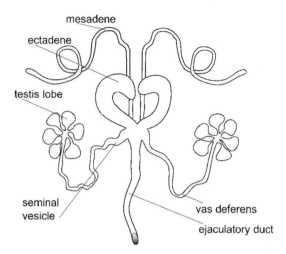

Fig. 1104 Diagram of a male reproductive system as found in *Tenebrio* (Coleoptera) (adapted from Chapman, The insects: structure and function).

species. The vector may or may not be essential for the completion of the life cycle of the pathogenic microorganism.

VECTOR CAPABILITY OF BLOOD-SUCKING ARTHROPODS: A FORECASTING MATRIX.
The ability to predict the potential for hematophagous arthropods to become vectors of pathogens is important in assessing their potential hazard to human or livestock health. The question of why one or another pathogen can be transmitted by a particular blood-feeding arthropod (hematophage), but not vectored by others, was first posed by Russian scientist Evgeny Pavlovsky, who formulated a theory of natural focality of infections and parasitic diseases.

Vladimir Beklemishev, a specialist in invertebrate evolution, comparative anatomy and systematic, was apparently the first who attempted to answer this question. He proposed a matrix based on the taxonomy of the vectors and the pathogens. According to this matrix, the ability to transmit a certain pathogen is limited to certain taxa. For example, this matrix forecasted that *Ixodes* ticks can transmit spirochetes, rickettsiae, and rickettsia-like microorganisms. However, the forecasting methodology was disregarded by the scientific community so Berlemishev's paper was never translated into English. As a result, the discovery of the vectors of Lyme disease and ehrlichiosis agents was made as late as the 1980s. Had the matrix received greater attention, the search for vectors might have progressed more rapidly.

A new, updated, prognostic matrix is presented (see Forecasting Matrix) that takes into account food consumption of vectors. Most of the vector groups are insects, and more than half are Diptera.

According to the type of pre-adult feeding, all hematophagous arthropods can be divided into two large groups. In one group (Forecasting Matrix, Nos. 9–15), only blood (or lymph, as in bush mites) serves as a source of food, or at least part of food components must contain blood (as in flea larvae). Preimaginal (pre-adult) stages using many different sources of food, but not blood, compose the second group (Forecasting Matrix, Nos. 1–8). Only representatives of the first group (blood consumers on all stages) appear to serve as relevant or potential vectors of rickettsiae or rickettsia-like (*Ehrlichia*, *Cowdria*) microorganisms, with a single exception. The exception is triatomid bugs known as vectors of only one pathogen, *Trypanosoma cruzi*, which is the Chagas disease agent in South and Central America. This representative of the insect order Heteroptera shows a very specific type of quasi-peritrophic, chitin-like membrane in its midgut, allowing for no other parasites but pure-blood-consumers to develop. The infective form of *T. cruzi* is excreted with bug feces. These vectors might be called specific contaminators if the pathogen would not develop in the bug gut to achieve the infectious stage.

Rickettsiae develop either in digestive tract epithelium of lice and fleas (Forecasting Matrix – *dte*) or both there and in other cells, including salivary glands and ovarial cells in infected ticks and mites. This peculiar capability of rickettsiae for development and "conservation" in the gut epithelium of hematophagous arthropods allows them to survive during the vector's molting periods (Forecasting Matrix, Nos. 9–10; 11–14) and thus to be transferred to the next stage of arthropod development, and even into the next generation. The phenomenon of transstadial transmission leads to understand how tsutsugamushi (or river, or bush) fever agent survives inside the free-living predaceous (entomophagous) nymphs and adults of trombiculid mites. Why rickettsiae are able to develop only in the vectors whose larval stage is hematophagous is not yet clear. Worm-like flea larvae are mainly detrivorous and only get the necessary components of blood from the blood-containing excreta of adults. Nonetheless, clearly this kind of diet is necessary to make the adult flea gut epithelium an available medium for pathogenic rickettsial penetration and proliferation.

Arboviruses are represented by numerous families with very different core and envelope structures. Arboviruses can reproduce only intracellularly. The ability of an arthropod to serve as a virus vector seems to be based less on a potential vector's properties and more on arbovirus specificity. Ticks and gamasid mites, however, are known not only as vectors of many arboviruses but also reservoirs, being capable of transmitting them both transstadially and transovarially.

The majority of true, or potential, vectors are polyphagous, polyxenous ectoparasites, which are able to obtain blood from various vertebrate species. There is only one exception that confirms the rule, i.e., the louse *Pediculus humanus corporis* (Anoplura), an obligate hematophage at all stages and, according to the matrix, a monoxenous human parasite that can be a vector of rickettsiae. Indeed, these lice provide a very good environment for the accumulation and multiplication of the specific human typhus agent, *Rickettsia provazeki*. Being obligate hematophages, lice serve as a very good reservoir for still another specific human pathogen belonging to the spirochete (*Borrelia*) group. Both these pathogens are specific human disease agents and are transmitted from man to man. Rickettsiae in lice only develop in the gut epithelium and the pathogen's transmission by lice as well as fleas is a result of an infectious insect being crushed and rickettsia-infected feces penetrating into skin scratches. Yet *Borrelia*, the above obligate blood-consuming pathogen, is only capable of development in the arthropods whose ontogenetic stages are strictly hematophagous, i.e., lice, ticks, and gamasid mites. Trombiculid mites as lymph consumers must be excluded from this group as well as fleas, whose larvae feed not only on blood containing "parental excreta" but also detritus.

Borreliae develop in the gut content of lice, ticks and mites. In lice, however, they are capable of neither invading the louse body cavity nor being transmitted specifically through the bite of an infected insect. In contrast, borreliae in infected ticks are able to migrate through the body cavity into salivary glands. Infected ticks transmit this pathogen specifically (e.g., Lyme disease pathogen by nymphs or adults of *Ixodes* ticks attached to the host skin).

Potential vector capacity forecasting matrix. Variants of transmission ability: ⊕=proved for man; +=proved for animal; −=absent; P= potential.

No. of group	Taxon	Common name	Pre-adult: Type of feeding	Pre-adult: Source of food	Adult: Source of food	Adult: Speed of digestion	Viruses: Arboviruses	Bacteria: Rickettsiae & rickettsia-like	Bacteria: Spirochetae (Borrelia)	Bacteria: Bacillae (Yersenia)	Pyroplasmids	Protozoa: Trypanosomatids	Protozoa: Haemosporids	Helminths: Filariidae
1	Glossinidae (Insects)	Tsetse flies	adenotrophia	"mother milk"	Bl	hours	−	−	−	−	−	⊕dtl	−	−
2	Hyppoboscidae	Blood-sucking flies	adenotrophia	"mother milk"	Bl	days	−	−	−	−	−	+dtl	−	−
3	Muscidae		detritophagia	D, M	Bl, Su	days	P	−	−	−	−	+dtl	−	−
4	Tabanidae	Horse flies	detritophagia, predatoriness	D, F, M, Flo	Bl, Su	days	P	−	−	−	−	+dtl	+ bcs	⊕bc
5	Culicidae	Mosquitoes	detritophagia, planktonophagia	D, M, Zpl, Phpl	Bl, Su	days	⊕bcs	−	−	−	−	+dtl	⊕bcs	⊕bc
6	Ceratopogonidae	Gnats	detritophagia	D, F, M	Bl, Su	Days	+ bcs	−	−	−	−	+dtl	bcs	+ bc
7	Phlebotomidae	Sand flies	detritophagia	D, F, M	Bl, Su	Days	⊕bcs	−	−	−	−	⊕dtl	+ bcs	P
8	Simuliidae	Black flies	Planktonophagia	M, Zpl, Phpl	Bl,	Days	P	−	−	−	−	+dtl	−	⊕bc
9	Siphonaptera	Fleas	detritophagia, hematophagia	D, F, M, Bl (adult's excreta)	Bl	Days	−	⊕dte	−	⊕dtl	−	+dtl	P	+ bc
10	Anoplura	Lice	hematophagia	Bl	Bl	hours	−	⊕dte	⊕dtl	−	−	−	−	−
11	Triatomidae	Kissing bugs	hematophagia	Bl	Bl	Days	−	−	−	−	−	⊕dtl	−	−
12	Ixodidae (Ticks and mites)	Hard ticks	hematophagia	Bl	Bl	weeks, months	⊕bcs	⊕bcs	⊕dtl,	bcs P	⊕bcs	+dtl	−	+ bc
13	Argasidae	Soft ticks	hematophagia	Bl	Bl	months	⊕bcs	+ bc	⊕dtl,	bcs P	⊕bcs	P	−	+ bc
14	Gamasina	Mites	hematophagia	Bl	Bl	days	+ bcs	+ bc	+	−	−	+dtl	−	−
15	Trombiculidae	Bush mites	lymphophagia	L	Hl	days	−	⊕bcs	−	−	−	−	−	−

Part of arthropod organism where pathogen develops (multiplies): bc = body cavity; bcs = body cavity and salivary glands; dtl = digestive tract lumen; dte = digestive tract epithelium. Sources of food: Bl = blood; D = detritus; F = fungi; Flo = free living organisms; Hl = haemolymph of insects; L = lymph of vertebrate hosts; M = microorganisms; Phpl = phytoplankton; Su = sugar; Zpl = zooplankton.

The peculiarities of *Borrelia* biology suggest that other blood-sucking insects are unable to transmit these pathogens.

Bacilli, e.g., the plague agent *Yersenia pestis* as a natural focal disease of man, develop successfully in the flea gut and are transmitted through flea regurgitation during attempts to suck blood from their host. This pathogen can develop and even be transmitted by ticks as well, but these acari fail to play any part in plague transmission. All other arthropods must be excluded as potential vectors of bacilli.

Protozoan agents have a much wider range of potential and real vectors. Only pyroplasmids from the genera *Theileria* and *Babesia* that show a specific mode of sporogonia in the salivary gland cells of ticks are vectored by hard and soft ticks. As no other blood-sucking arthropod has the necessary type of cells in their salivary glands, this means that it is useless to search for vectors of these specific pathogens other than ticks.

Trypanosomatids are capable of development in all blood-sucking arthropods, irrespective of preimaginal food consumption, exclusive of lymphophagous trombiculid mites and obligate human ectoparasites like body lice. Only insects, not ticks or mites, can serve as hemosporid vectors. Most of them transmit these pathogens to birds. Of special note are the *Anopheles* (Culicidae) mosquitoes, vectors of human malaria agents (i.e., *Plasmodium falciparum*, *Plasmodium vivax*, *Plasmodium malariae*, and *Plasmodium ovale*). Interestingly, inclusion of the *Pseudomonas* group of microorganisms in the larval diet is essential for the successful development of these haemosporids in the potential vector organism.

Helminths, predominantly *Filarioidea* nematodes, are transmitted by a very wide range of vectors, mainly insects. Among these helminths, some are very dangerous parasites (e.g., loaosis by Tabanid flies and *Onchocerca* species by black flies) which cause blindness in humans. Mosquitoes also are vectors of dangerous filarial diseases such as brugiosis and elephantiasis.

The forecasting matrix, though gaining little attention in the past, contains important patterns that help predict likely disease-vector relationships. For example, new species of rickettsial, spirochetal and pyroplasmid pathogens might be discovered among ticks and gamasid mites. New arboviruses might be revealed among tabanids and black flies, but not likely from lice or kissing bugs. This matrix virtually precludes a successful search for hemosporid protozoans among lice and other purely hematophagous arthropods at all ontogenetic stages.

Andrey N. Alekseev
Russian Parasitological Society
St. Petersburg, Russia

References

Alekseev, A. N. 1985. The theory of connections of feeding types and digestion of blood-sucking arthropods with their ability to be specific vectors of transmissive diseases agents. *Parazitologiya* 19: 3–7. [In Russian]

Beklemishev, V. N. 1948. Interface between taxonomical position of pathogens and vectors of transmissive diseases of animals and man. *Meditsinkaya parazitologiya i parazitarnye bolezni* 17: 385–400. [In Russian]

Goddard, J. 2000. *Infectious diseases and arthropods*. Humana Press, Totowa, New Jersey. 231 pp.

Jadin. J., I. H. Vincke, A. Dunjic, J. P. Delville, M. Wery, J. Bafort, and M. Scheepers-Biva. 1966. Rôle des Pseudomonas dans la sporogonie de l'hématozoaire du paludisme chez le moustique. *Bulletin de la Société de Pathologie Exotique* 59: 514–525.

Pavlovsky, Y. N. (ed.). 1963. *Human diseases with natural foci*. Foreign Languages Publishing House, Moscow, Russia. 346 pp.

VECTORS OF PHYTOPLASMAS. Phytoplasmas are phytopathogenic Mollicutes. These minute wall-less bacteria are found exclusively in phloem and primarily on sieve elements and cannot be cultured artificially. They were first associated with 'yellows' diseases in 1926, although at the time they were thought to be viruses. Since then, phytoplasmas have been found in hundreds of plant species. Over 300 plant diseases are known worldwide, including economically important flowers, vegetables and orchards.

Phytoplasmas are vectored from plant to plant by a select group of Hemipterans feeding in the phloem. Of these insects, the leafhoppers and planthoppers have been shown to vector phytoplasmas, and more recently, a psylla (*Bactericera trigonica*) has been shown to vector phytoplasmas in carrots.

Once an insect has fed in the phloem of a phytoplasma-infected plant, a number of events must take place for it to be a successful vector. There are a number of anatomical and physiological barriers that the phytoplasma must overcome: it must successfully survive salivary and digestive enzymes; penetrate and replicate in the midgut cells; penetrate out of the midgut cells into the hemolymph; survive any phago-cytes or encapsulation by hemocytes while in the hemolymph; travel to, penetrate and replicate in the salivary glands; and finally be released into the phloem of a plant as the insect feeds. The period of time from when the insect first acquires the phyto-plasma until the time it can vector it to a new plant is the latent period. While a minimum amount of time is required for the physiological processes, it is also affected by temperature. Furthermore, vectors can acquire more than one type of phytoplasma.

There are a number of techniques available to detect and identify specific phytoplasmas in insects; however, all molecular (i.e., DNA probes, PCR/RFLP, and chromosome sequencing) and immunolo-gical (i.e., ELISA, dot hybridization and southern hybridization) tests cannot distinguish a vector spe-cies from an insect that has simply fed on an infected plant and thus has phytoplasma in its alimentary canal. To determine if an insect is a vector, it must be allowed to feed and transmit the pathogen. A rela-tively new technique combines molecular techniques with insect feeding. The potential vector, contained in a small tube, is allowed to feed through Parafilm on a sucrose solution. If the insect is an infective vector, PCR analysis of the sucrose will yield positive results. This technique allows rapid determination of the type of phytoplasma and the percent of infec-tive vectors in any population.

To date, there are no efficient methods to control the spread of phytoplasma diseases, only to reduce its impact. Efforts to develop such control measures focus either on the plant or the insect vector. Phyto-plasmas are not seed-transmitted but can be trans-mitted by grafting. Planting resistant varieties of plants or grafting to resistant rootstock is the best option to reduce spread of the disease, but resistant rootstock varieties do not exist for all plants. Tetracy-cline solutions have been variously applied to control infections in orchards. Such methods are usually effective only for the duration of the treatment; once stopped, symptoms return. However, currently there is research on genetically engineering plants to express antibodies, which may be a solution in the future. In cases where there are no resistant source plants, clean plants must be protected from insects with physical (nets or screening) or chemical (insec-ticides) means. Thorough knowledge of the biology of the vector and epidemiology of the disease will eliminate unnecessary insecticide treatments when the vector is either not present or is in a non-infective phase. If the vector is monophagous, using clean source plants and eliminating wild reservoir plants can achieve control of the disease. If the vector is polyphagous, border rows of a plant more 'desirable' to the insect could be planted and heavily treated to protect the crop.

Phytoplasmas offer a great challenge to the researcher because they cannot be cultured and because there is no effective control. New molecular biological techniques are aiding in the quest for more understanding about the pathogen, pathogen-vector interaction, and pathogen-plant interaction.

Phyllis G. Weintraub
Agricultural Research Organization
Gilat Research Station Negev, Israel

References

Abad, P., I. Font, E. L. Dally, A. I. Espino, C. Jorda, and R. E. Davis. 2000. Phytoplasmas associated with carrot yellows in two Spanish regions. *Journal of Plant Pathology* 82 (1). www.agr.unipi.it/sipav/jpp/journals/abs0300.htm

Chen, Y. D., and T. A. Chen. 1998. Expression of engineered antibodies in plants: a possible tool for spiroplasma and phytoplasma disease control. *Phytopathology* 88: 1367–1371.

Kunkel, L. O. 1926. Studies on aster yellows. *American Journal of Botany* 23: 646–705.

Lee, I.-M., R. E. Davis, and DE. Gundersen-Rindal. 2000. Phytoplasma: phytopathogenic mollicutes. *Annual Review of Microbiology* 54: 221–255.

Zhang, J., S. Miller, C. Hoy, X. Zhou, and L. Nault. 1998. A rapid method for detection and differentiation of aster yellows phytoplasma-infected and -inoculative leaf-hoppers. http://esa.cos.com/cgi-bin/itinerary/index/author/ zhang:::jiangua/

VEGETABLE LEAFMINER, *Liriomyza Sativae* BLANCHARD (DIPTERA: AGROMYZI-DAE).

Vegetable leafminer is found commonly in southern North America, and in most of Central and South America. Occasionally it is reported in more northern areas because it is transported with plant material. It cannot survive cold areas except in greenhouses.

Host plants

Vegetable leafminer attacks a large number of plants, but seems to favor those in the plant families Cucurbitaceae, Leguminosae, and Solanaceae. Nearly 40 hosts from 10 plant families are known in Florida.

Among the numerous weeds infested, the nightshade, *Solanum americanum*; and Spanishneedles, *Bidens alba*; are especially suitable hosts in Florida. Vegetable crops known as hosts include bean, celery, eggplant, onions, pepper, potato, squash, tomato, watermelon, cucumber, beet, pea, lettuce and many other composites. Vegetable leafminer was formerly considered to be the most important agromyzid pest in North America, but this distinction is now held by American leafminer, *L. trifolii*.

Natural enemies

Vegetable leafminer is attacked by a number of parasitoids, with the relative importance of species varying geographically and temporally. In Hawaii, *Chrysonotomyia punctiventris* (Crawford) (Hymenoptera: Eulophidae), *Halicoptera circulis* (Walker) (Hymenoptera: Pteromalidae), and *Ganaspidium hunteri* (Crawford) (Hymenoptera: Eucoilidae) are considered important in watermelon. In California and Florida, the same genera or species were found attacking vegetable leafminer on tomato or bean, but *Opius dimidiatus* (Ashmead) (Hymenoptera: Braconidae) also occurred commonly in Florida. Levels of parasitism are often reported to be proportional to leafminer density, but parasitoid effectiveness can be disrupted when insecticides are applied. Steinernematid nematodes can infect *L. trifolii* larvae when the nematodes are applied in aqueous suspension and the plants are held under high humidity conditions.

Life cycle and description

The developmental thresholds for eggs, larvae, and pupae are estimated at 9 to 12°C. The combined development time required by the egg and larval stages is about 7 to 9 days at warm temperatures (25 to 30°C). Another 7 to 9 days is required for pupal development at these temperatures. Both egg-larval and pupal development times lengthen to about 25 days at 15°C. At optimal temperatures (30°C), vegetable leaf miner completes development from the egg to adult stage in about 15 days.

Egg. The white, elliptical eggs measure about 0.23 mm in length and 0.13 mm in width. Eggs are inserted into plant tissue just beneath the leaf surface and hatch in about three days. Flies feed on the plant secretions caused by oviposition, and also on natural exudates. Females often make feeding punctures, particularly along the margins or tips of leaves, without depositing eggs. Females can produce 600 to 700 eggs over their life span, although some estimates of egg production suggest that 200 to 300 is more typical. Initially, females may deposit eggs at a rate of 30 to 40 per day, but egg deposition decreases as flies grow older.

Larva. There are three active instars, and larvae attain a length of about 2.25 mm. Initially the larvae are nearly colorless, becoming greenish and then yellowish as they mature. Black mouthparts are apparent in all instars, and can be used to differentiate the larvae. The average length and range of the mouthparts (cephalopharyngeal skeleton) in the three larval feeding instars is 0.09 (0.6–0.11), 0.15 (0.12–0.17), and 0.23 (0.19–0.25) mm, respectively. The mature larva cuts a semicircular slit in the mined leaf just prior to formation of the puparium. The larva usually emerges from the mine, drops from the leaf, and burrows into the soil to a depth of only a few cm to form a puparium. A fourth larval instar occurs between puparium formation and pupation, but this is generally ignored by authors.

Pupa. The reddish brown puparium measures about 1.5 mm in length and 0.75 mm in width. After about 9 days the adult emerges from the puparium. Mating initially occurs the day following adult emergence.

Adult. The adults are principally yellow and black in color. The shiny black mesonotum of *L. sativae* is used to distinguish this fly from the closely related American serpentine leafminer, *Liriomyza trifolii* which has a grayish black mesonotum. Females are larger and more robust than males, and have an elongated abdomen. The wing length of this species is 1.25 to 1.7 mm, with the males averaging about 1.3 mm and the females about 1.5 mm. The small size of these flies serves to distinguish them from pea leafminer, *Liriomyza huidobrensis* (Blanchard), which has a wing length of 1.7–2.25 mm. The yellow femora of vegetable leafminer also help to distinguish these species, as the femora of pea leafminer are dark. Flies normally live only about a month. Flies are uncommon during the cool months of the year, but often attain high, damaging levels by mid-summer. In warm climates they may breed continuously, with many overlapping generations per year.

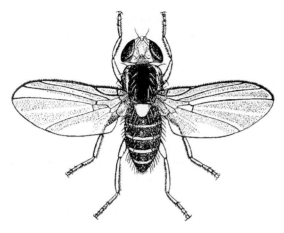

Fig. 1105 Vegetable leafminer adult, *Liriomyza sativae*.

Damage

Foliage punctures caused by females during the acts of oviposition or feeding may cause a stippled appearance on foliage, but this damage is slight compared to the leaf mining activity of larvae. The irregular mine increases in width from about 0.25 mm to about 1.5 mm as the larva matures, and is virtually identical in appearance and impact with the mines of *L. trifolii*. Larvae are often easily visible within the mine where they remove the mesophyll between the surfaces of the leaf. Their fecal deposits are also evident in the mines. Researchers have found that 30-60% yield increases are possible when effective insecticides were applied.

Management

Sampling. Several methods for population assessment have been studied, and collecting puparia in trays placed beneath plants and placement of yellow sticky traps to capture adults are most popular. Yellow sticky traps, however, have the advantage of being able to quickly detect invasion of a field by adults from surrounding areas.

Insecticides. Foliar application of insecticides is often frequent in susceptible crops. Insecticide susceptibility varies greatly both spatially and temporally. Many organophosphate and carbamate insecticides are no longer effective. Insecticides are disruptive to naturally occurring biological control agents, and leafminer outbreaks are sometimes reported to follow chemical insecticide treatment for other insects.

Cultural practices. Some crops vary in susceptibility to leaf mining. This has been noted, for example, in cultivars of tomato, cucumber, cantaloupe, and beans. However, the differences tend to be moderate, and not adequate for reliable protection. Placement of row covers over cantaloupe has been reported to prevent damage by leafminer. Sometimes crops are invaded when adjacent crops are especially suitable. Weeds are a source of flies, but also a source of parasitoids.

See also, VEGETABLE PESTS AND THEIR MANAGEMENT.

References

Capinera, J. L. 2001. *Handbook of vegetable pests.* Academic Press, San Diego. 729 pp.

Johnson, M. W. 1987. Parasitization of *Liriomyza* spp. (Diptera: Agromyzidae) infesting commercial watermelon plantings in Hawaii. *Journal of Economic Entomology* 80: 56–61.

Palumbo, J. C. 1995. Developmental rate of *Liriomyza sativae* (Diptera: Agromyzidae) on lettuce as a function of temperature. *Southwestern Entomologist* 20: 461–465.

Parrella, M. P. 1987. Biology of *Liriomyza. Annual Review of Entomology* 32: 201–224.

Petitt, F. L. and D. O. Wietlisbach. 1994. Laboratory rearing and life history of *Liriomyza sativae* (Diptera: Agromyzidae) on lima bean. *Environmental Entomology* 23: 1416–1421.

Schuster, D. J., J. P. Gilreath, R. A. Wharton, and P. R. Seymour. 1991. Agromyzidae (Diptera) leafminers and their parasitoids in weeds associated with tomato in Florida. *Environmental Entomology* 20: 720

VEGETABLE PESTS AND THEIR MANAGEMENT. Throughout the world, vegetables are an important element of the human diet, providing a vital source of carbohydrates, vitamins, and minerals. Some vegetables remain unique to specific cultures, whereas others have gained wide acceptance and have been transported to several continents where they are grown extensively. In many cases, insect pests have been transported to new continents as well. In some cases, the insects apparently were transported along with the initial plant material. In other cases, the insects were introduced as ''hitch-hikers'' on other products, but once gaining entrance to their new home found it quite suitable because their host plants had preceded them. Another important source of pests is host adaptation or host switching, wherein insects adapted to feeding on a particular plant (often a weed or other non-crop plant) begin feeding on introduced crops. Such insects usually adapt to crop plants in the same plant family as their original host, or inherently have a broad host range.

The major groups of vegetable crops and some key pests follow. A more complete list of pests, their hosts and geographic ranges, and the plant tissues damaged are listed in the accompanying table.

Artichoke (family Compositae)

Artichoke, more correctly known as globe artichoke, is a thistle-like plant grown for the edible blossom bud. It is a perennial, with new growth arising annually from the roots. Like some other perennials, it can be grown with some success as an annual crop by planting roots, but this is not a common practice. Although artichoke can be grown over a broad geographic area, it is not cold-hardy. Commercial production is generally limited to Mediterranean climates, where the cool, moist environment favors its growth. It is not a popular vegetable, and is considered by many to be a ''luxury'' vegetable. The origin of artichoke appears to be the western Mediterranean region of Europe. A similar plant grown for the leaf stalks is cardoon, but this plant is not grown extensively. The key pest of artichoke is the artichoke plume moth, *Platyptilia carduidactyla* (Riley) (Lepidoptera: Pterophoridae), though such aphids as the artichoke aphid, *Capitophorus elaeagni* (del Guercio), and the bean aphid, *Aphis fabae* Scopoli (both Hemiptera: Aphididae), can be quite damaging at times.

Asparagus (family Liliaceae)

Asparagus is a hardy perennial plant, and once established, remains productive for 15 to 20 years. It is grown in many temperate climates, and is winter-hardy in most climates. It is usually grown from crowns (roots), but seeds may be used. Seeds are spread freely by birds, and it is common to find asparagus growing wild along roadsides, fences and irrigation ditches. Asparagus is popular in home gardens grown in northern areas, probably because it is one of the first crops available for harvest in the spring. It has been a popular vegetable since ancient times, and originated in the Mediterranean region.

The spears or stems are harvested as they first push up from the soil; once they begin to branch, they become tough and inedible. The most serious pests are the asparagus aphid, *Brachycorynella asparagi* (Mordvilko) (Hemiptera: Aphididae), and the asparagus beetles, *Crioceris* spp. (Coleoptera: Chrysomelidae).

Bean and related crops (family Leguminosae) (bush bean, chickpea, cowpea, dry bean, English pea, faba bean, lentil, lima bean, mung bean, pigeon pea, snap bean, etc.)

Legumes are known for their ability to harbor nitrifying bacteria; nitrogen enhances soil productivity. The cultivated legume vegetable crops are not particularly efficient as a source of nitrogen for plant growth, however, so fertilization is still required. They are, however, a relatively good source of vegetable protein, so they are an important dietary component in areas where animal protein is limited. Most of the leguminous vegetable crops are warm-weather crops and are killed by light frosts. English pea is a notable exception, thriving under early season and cool weather conditions, but killed by heavy frost. The legume vegetables are annual crops. The *Phaseolus* bean crops, such as snap bean and lima bean, are native to Central America. English pea and cowpea likely originated in Asia. The legumes are cultivated for their seeds or seed pods, and are eaten fresh or dried.

There are many important pests of legumes, and they vary among crops and geographically. The seed-attacking maggots, *Delia* spp. (Diptera: Anthomyiidae), can be important pests under cool weather conditions. The Mexican bean beetle, *Epilachna varivestris* Mulsant (Coleoptera: Coccinellidae) is important in North America. The Old World bollworm, *Helicoverpa armigera* (Hübner) (Lepidoptera: Noctuidae), affects pigeon pea crops in Africa and Asia. English pea is quite susceptible to infestation by the pea aphid, *Acyrthosiphon pisum* Shinji (Hemiptera: Aphididae), the pea leaf weevil, *Sitona lineatus* (L.) (Coleoptera: Curculionidae), and the pea midge, *Contarinia pisi* (Winnertz) (Diptera: Cecidomyiidae); they are most damaging in northern areas. Cowpea is plagued by cowpea weevils, *Callosobruchus* spp. (Coleoptera: Bruchidae). Locally, a number of other pests can be important, particularly thrips, leaf miners, leafhoppers, and flea beetles.

Beet and related crops (family Chenopodiaceae) (beet, chard, spinach, Swiss chard)

Beet apparently originated in the Mediterranean region, and spinach originated in Iran. The beet and its relatives are biennial crops, requiring more than one year, but less than two, to complete their natural life cycle. They are grown as annuals when cultivated as vegetables. Though originally grown entirely for its foliage, cultivars of beets with edible below-ground portions (the edible "root" is mostly thickened stem material) became popular in Europe beginning about 1800. Beet and chard are not important crops except in cool-weather climates. Spinach is more popular as a commercial crop relative to other chenopods, but is relatively unpopular compared to many warm-weather vegetables. In home gardens, these crops thrive nearly everywhere. The principal pests are the green peach aphid, *Myzus persicae* (Sulzer) (Hemiptera: Aphididae), and the beet and spinach leafminer, *Pegomya* spp. (Diptera: Anthomyiidae). In western North America and in Europe, the beet leafhoppers, *Circulifer tenellus* (Baker), and *C. opacipenis* (Leth.), (both Hemiptera: Cicadellidae), can be very damaging.

Cabbage and related crops (family Cruciferae) (broccoli, Brussels sprouts, cabbage, cauliflower, Chinese cabbage, collards, kale, kohlrabi, mustard, radish, rutabaga, turnip, etc.)

Cruciferous vegetables, often called "cole" crops, are grown for their leaves. These are cool-season crops, and tolerate light freezes and even brief heavy freezes, but prolonged deep freezes are fatal. Though naturally biennials, they are grown as annuals. Cabbage and its many forms originated along the shores of Europe; mustard and radish are from Asia. Some are popular foods throughout the world, others are of regional significance. Perhaps the most interesting vegetable in this group is broccoli, which has become popular only since the 1950s. Cauliflower is a moderately important crop. Cabbage and turnip have declined in importance, though cabbage remains a significant crop. Rutabaga is an important dietary element in northern climates, and collards are important in southern areas. The principal pests of cabbage and its closest relatives include the root maggots, *Delia* spp. (Diptera Anthomyiidae); the cabbage aphid, *Brevicoryne brassicae* (Linnaeus), and the turnip aphid, *Lipaphis erysimi* (Kaltenbach), (both Hemiptera: Aphididae); the diamondback moth, *Plutella xylostella* (Linnaeus), (Lepidoptera: Pyralidae); the cabbage looper, *Trichoplusia ni* (Hübner) and the cabbage moth, *Mamestra brassicae* (L.), (both Lepidoptera: Noctuidae); and the imported cabbageworm, *Pieris rapae* (Linnaeus), (Lepidoptera: Pieridae). Mustard and radish tend to be plagued more by the green peach aphid, *Myzus persicae* (Sulzer), (Hemiptera: Apididae).

Carrot and related crops (family Umbelliferae) (carrot, celery, celeriac, chervil, coriander, fennel, parsley, parsnip)

The umbelliferous vegetables are biennial, but are grown as annuals. They require cool weather to develop properly. Most survive heavy frost, but are killed by prolonged freezing weather. The major crops of this group, carrot and celery, originated in the Mediterranean region. Carrot and parsnip are grown for their roots; celery, celeriac and fennel for the swollen stem bases; and parsley and coriander (cilantro) for their foliage. Carrot and celery, and to a lesser degree parsley and coriander, are popular vegetables. Parsnip is an important crop in northern climates because it stores well during the winter months, and can even be left in the soil during freezing weather. Fennel is growing in popularity, but as yet, is a minor crop awaiting discovery by some cultures. Chervil is grown in Europe, and different types are grown for their roots and foliage. In some areas, the carrot weevil, *Listronotus oregonensis* (LeConte) (Coleoptera: Curculionidae), the willow-carrot aphid, *Cavariella aegopodii* (Scopoli) (Hemiptera: Aphididae), and the carrot rust fly, *Psila rosae* (Fabricius) (Diptera: Psilidae), are important pests. Leafhoppers (Hemiptera: Cicadellidae), sometimes transmit aster yellows disease. The American serpentine leafminer, *Liriomyza trifolii* (Burgess), and the serpentine leaf miner, *L. sativae* Blanchard (both Diptera: Agromyzidae), are often a serious threat to commercially-grown celery.

Lettuce and related crops (family Compositae) (celtuce, chicory, endive, escarole, lettuce, radicchio)

Lettuce is a popular salad vegetable. Lettuce apparently originated in Europe or Asia, and has been grown for over 2,000 years. Lettuce and most related crops are grown for their leaves, though, in the case of celtuce, the stem is eaten. Lettuce and related crops are cool-season annuals. Although killed by heavy frost, these crops are also susceptible to disruption by excessive heat. Hot weather causes lettuce to flower and become bitter tasting. Several insects are important pests of lettuce. The aster leafhopper, *Macrosteles quadrilineatus* Forbes, (Hemiptera: Cicadellidae), is an important vector of aster yellows in some production areas. Several species of aphids (Hemiptera: Aphididae), including the lettuce aphid, *Nasonovia ribisnigri* (Mosley), and the lettuce root aphid, *Pemphigus bursarius* (L.), may be damaging, though the green peach aphid, *Myzus persicae* (Sulzer), generally is the most important pest. Numerous caterpillars such as the corn earworm, *Helicoverpa zea* (Boddie), and the cabbage looper, *Trichoplusia ni* (Hübner) (both Lepidoptera: Noctuidae), also threaten lettuce crops.

Okra (family Malvaceae)

Okra is thought to be native to Africa and is an important crop in tropical countries. It is an annual plant and is killed by light frost. Okra is grown for the seed pods which, like snap beans, are harvested before they mature. Okra is unusually tall for a vegetable crop, often attaining a height of 2 m. The pods are subject to attack by several pests. Some of the most damaging are: the red imported fire ant, *Solenopsis invicta* Buren (Hymenoptera: Formicidae); the cotton aphid, *Aphis gossypii* Glover (Hemiptera: Aphididae); the Old World bollworm, *Helicoverpa armigera* (Hübner) (Lepidoptera: Noctudae); the southern green stink bug, *Nezara viridula* (Linnaeus) (Hemiptera: Pentatomidae); cotton stainers, *Dysdercus* spp. (Hemiptera: Pyrrhocoridae); and leaffooted bugs, *Leptoglossus* spp. (Hemiptera: Coreidae).

Onion and related plants (family Alliaceae) (chive, garlic, leek, onion, shallot)

The onion and its relatives are biennial or perennial plants, but cultivated as annuals. These have long been important crops, with the use of onions documented for nearly 5,000 years. Their origin is thought to be Asia. Onion is grown principally for the below-ground leaf bases, which form a bulb, but the tops are also edible. They are tolerant of cool weather, but also thrive under hot conditions. Onion and garlic are cultivated widely. The key pests of onion are onion thrips, *Thrips tabaci* Lindeman (Thysanoptera: Thripidae), and the onion maggot, *Delia antiqua* (Meigen) (Diptera: Anthomyiidae).

Rhubarb (family Polygonaceae)

Rhubarb is cultivated as a perennial. Its origin is northern Asia, and its use can be traced back about 5,000 years. Rhubarb thrives where summers are cool. The stalks, or leaf petioles, are used as food, though this vegetable is infrequently consumed. Most commercial production occurs in northern regions, but it remains principally a home garden crop. There are few important pests of rhubarb, with the bean aphid, *Aphis fabae* Scopoli (Hemiptera; Aphididae), and the rhubarb curculio, *Lixus concavus* Say (Coleoptera: Curculionidae), perhaps the most serious.

Squash and related crops (family Cucurbitaceae) (cucumber, melons, pumpkin, squash, watermelon, etc.)

The cucurbit crops are important vegetables, though they vary in economic importance. They are annual plants, and are warmth-loving crops. Light frost will kill cucurbits, and even cool weather will permanently disrupt growth. All are cultivated for their fruit. Squash and pumpkin originated in Central and South America; cucumber, melons and watermelon are from Africa, or perhaps, Asia. Cucurbit crops have some serious insect pests, including the squash vine borer, *Melittia cucurbitae* (Harris) (Lepidoptera: Sesiidae); the squash bug, *Anasa tristis* (De Geer) (Hemiptera: Coreidae); the melonworm, *Diaphania hyalinata* (L.), the pumpkin caterpillar, *Diaphania indica* (Saunders), and the pickleworm, *Diaphania nitidalis* (Stoll) (all Lepidoptera: Pyralidae); the red pumpkin beetle, *Raphidopalpa foveicollis* (Lucas), *Diabrotica* spp., *Aulacophora* spp., and cucumber beetles, *Acalymma* sp. (all Coleoptera: Chrysomelidae). Whiteflies (Hemiptera: Aleyrodidae) and aphids (Hemiptera: Aphididae) are plant virus vectors.

Sweet corn (family Graminae)

Corn, which is usually known as maize outside of North America, apparently was domesticated in Mexico, and perhaps is descended from a similar grain, teosinte. Corn originally was cultured because it was productive, a good source of carbohydrates and other nutrients, and the grain stored well. Sweet corn is a recent innovation that was first developed in the mid 1700s, and lacks the storage characteristics of the older types, or grain corn. Corn is grown for the seeds, which are clustered in a structure called the ''ear.'' Sweet corn is a popular vegetable, and the development of fresh corn that does not quickly lose its sweetness (supersweet cultivars) has increased demand for whole-ear corn. Many pests feed on corn, including the corn earworm, *Helicoverpa zea* (Boddie), the fall armyworm, *Spodoptera frugiperda* (J.E. Smith), and several cutworms (all Lepidoptera: Noctuidae); the *Diabrotica* rootworms (Coleoptera: Chrysomelidae); the European corn borer, *Ostrinia nubilalis* (Hübner) (Lepidoptera: Pyralidae); the Japanese beetle, *Popillia japonica* Newman (Coleoptera: Scarabaeidae); and the corn leaf aphid, *Rhopalosiphum maidis* (Fitch) (Hemiptera: Aphididae).

Sweet potato (family Convolvulaceae)

Sweet potato is an immensely important crop in some parts of the world, principally in tropical areas. The sweet potato probably originated in Mexico and is well adapted to tropical growing conditions. Some moist-fleshed "sweet potatoes" are called yams, but yams are a separate species normally found in Polynesia. A perennial crop, sweet potato is normally grown as an annual. It is cultivated for its tuber. The sweet potato cannot tolerate prolonged cool weather and perishes if exposed to light frost. The sweetpotato weevil, *Cylas formicarius* (Fabricius) (Coleoptera: Curculionidae), is the most damaging pest of this crop, but the sweetpotato vine borer, *Omphisa anastomosalis* (Guenee) (Lepidoptera: Pyralidae), wireworms (Coleoptera: Elateridae), whiteflies (Hemiptera: Alyrodidae), and the flea beetle and the cucumber beetle larvae (both Coleoptera: Chrysomelidae) are economically threatening.

Tomato and related plants (family Solanaceae) (eggplant, pepper, potato, tomatillo, tomato)

The solanaceous crops are among the most important vegetable crops. Both the crops and the pests are quite diverse. Potato ranks as the most valuable, but tomato and pepper also rank among the most highly valued crops. The potato is grown for its tuber, and the other crops for their fruit. The origins of the solanaceous crops are diverse; eggplant originated in India, potato and tomato in Peru, and tomatillo and pepper in Mexico or Guatemala. Tomato, tomatillo, pepper and eggplant are warm-season perennials that are cultivated as annuals. The potato is a cool-season perennial cultured as an annual. Potato crops are at risk from the Colorado potato beetle, *Leptinotarsa*

decemlineata (Say) (Coleoptera: Chrysomelidae), but several aphids (Hemiptera: Aphididae) also are commonly damaging, particularly the green peach aphid, *Myzus persicae* (Sulzer) and the potato aphid, *Macrosiphum euphorbiae* (Thomas) (both Hemiptera: Aphididae). Beet leafhoppers (Hemiptera: Cicadellidae), *Circulifer tenellus* (Baker) and *Circulifer opacipennis* (Leth.), can be damaging to potatoes and tomatoes. Tomato is affected by the silverleaf whitefly, *Bemisia argentifolii* Bellows and Perring (Hemiptera: Aleyrodidae); the corn earworm, *Helicoverpa zea* (Boddie) (Lepidoptera: Noctuidae); and many other pests. The pepper weevil, *Anthonomus eugenii* Cano (Coleoptera: Curculionidae), is the key pest of pepper in warm-weather areas of North America.

The characteristics of pests

Pests vary from place to place, and from time to time. In some cases, pests are consistently abundant or damaging (so-called key pests), but in other cases, many years may pass before a particular species attains pest status. Most vegetable producers plan for the regular, or key, pests (many of them are mentioned above or are listed in the table) and take preventative actions to keep the pests from inflicting injury. It also is wise to monitor crops and to be alert for the less regular pests that come along periodically and may also require some corrective action.

The pest complex

Irrespective of the crop, there usually is a diversity of pests attacking, often including one or more aphids, flea beetles, leaf beetles, weevils, and cutworms or leaf-feeding caterpillars. The pest complex is often quite similar within related crops (members of the same plant family), and so it is not surprising to see the potato aphid (Hemiptera: Aphididae) on tomato, or the Colorado potato beetle (Coleoptera: Chrysomelidae) on tomato and eggplant, because they are all members of the family Solanceae and have similar plant chemistry. Some pest species distinguish among related plants, however, and so specialists, such as the pepper weevil (Coleoptera: Curculionidae), are limited to a single host, in this case, pepper.

In contrast, the pest complex often is quite different among crops from different plant families. Thus, the squash bug (Hemiptera: Coreidae) and the pickleworm (Lepidoptera: Pyralidae), are limited to Cucurbitaceae; the tuber flea beetle and the Colorado

Fig. 1106 Some vegetable pests of international significance. All have crossed national boundaries and most have spread to new continents. Top row left, asparagus beetle, *Crioceris asparagi*; top row right, Colorado potato beetle, *Leptinotarsa decemlineata*; second row left, Mexican bean beetle, *Epilachna varivestris*; second row right, whitefringed beetle, *Naupactus* sp.; third row left, yellowmargined leaf beetle, *Microtheca ochroloma*; third row right, cabbage aphid, *Brevicoryne brassicae*; bottom row left, corn leaf aphid, *Rhopalosiphum maidis*; bottom row right, asparagus aphid, *Brachycorynella asparagi*.

Fig. 1107 Additional vegetable pests of international significance. Top row left, harlequin bug, *Murgantia histrionica*; top row right, southern green stink bug, *Nezara viridula*; second row left, diamondback moth, *Plutella xylostella*; second row right, European corn borer, *Ostrinia nubilalis*; third row left, imported cabbageworm, *Pieris rapae*; third row right, corn earworm, *Helicoverpa zea*; bottom row left, pickleworm, *Diaphania nitidalis*; bottom row right, beet armyworm, *Spodoptera exigua*.

potato beetle (both Coleoptera: Chrysomelidae), are limited to Solanaceae; and the asparagus beetle (Coleoptera: Chrysomelidae), and the asparagus aphid (Hemiptera: Aphididae), are limited to asparagus.

Polyphagous species, insects with a very wide host range, display different behavior. Insects, such as the melon aphid, the green peach aphid (both Hemiptera: Aphididae), the silverleaf whitefly (Hemiptera: Aleyrodidae),and the Japanese beetle (Coleoptera: Scarabaeidae), seem to feed on almost everything. This is not true, of course, as they have preferred, or more suitable, hosts, but their host range includes hundreds of species. Thus, they can be expected to appear on, and perhaps be damaging to, many crops.

Crops are not the only factor determining the presence of pests. Weather and climate are quite important because some pests are adapted for warm climates and others for cool climates. Thus, their tolerance of weather and climate determines the extent of their geographic range as surely as an ocean or a mountain range. For instance, silverleaf whitefly (Hemiptera: Aleyrodidae), melon thrips (Thysanoptera), or broad mite (Acarina) are not seen in cool climates; nor are pea leaf weevil (Coloptera: Curculionidae), Japanese beetle (Coleoptera: Scarabaeidae), or European crane fly (Diptera: Tipulidae) found in subtropical and tropical environments. They are adapted to certain environments and cannot live outside their pre-adapted geographic range. Over time, they may evolve a broader tolerance and expand their geographic range, but this is a slow process. Interestingly, plants tend to be more tolerant of weather conditions than do their insect parasites. Therefore, the extremes of the host plant's geographic range often lack the ''normal'' complement of insect herbivores.

Weather also influences the dispersal of vegetable insects. Some species overwinter only in warm climates, dispersing northward annually. Their ability to disperse varies from year to year depending on weather patterns, though, in many areas, it is not a question of whether or not they will disperse, but only when the dispersal will occur. Examples of annual long-range dispersal are common in North America, where species such as the black cutworm, the fall armyworm (both Lepidoptera: Noctuidae), the pickleworm (Lepidoptera: Pyralidae), the aster leafhopper, and the beet leafhopper (both Hemiptera: Cicadelidae) disperse northward annually.

Types of damage

Insects usually affect vegetable crops by feeding directly on the harvested part of the plant (called direct pests), or on a part of the plant that is important to its productivity, such as the roots or the leaves. This latter group is known as indirect pests because they cause damage indirectly, by weakening the plant. Thus, caterpillars act as direct pests when they feed on the actual tomato fruit, but as indirect pests when they consume the tomato plant's foliage. Typically, species that function as direct pests are more damaging. Leaf feeding can be direct injury, however, when the foliage is harvested, as in lettuce or spinach. Foliage and fruit can be damaged aboveground, or plant tissues (roots, tubers, basal stem tissue) may be damaged below-ground. Damage can be manifested in leaf rolling or webbing, speckling, or growth deformities in addition to tissue removal.

A third important type of damage is plant disease transmission. Plant disease, particularly plant viruses, can be transmitted through the activities of insects. Some insects, such as aphids, may transmit diseases even to plants they do not normally feed upon. Disease transmission can occur while an aphid is sampling its environment, flying from plant to plant seeking an appropriate host. Even if the aphid alights on a plant that is unsuitable for its growth and reproduction, a short probe by the aphid with its mouthparts while it is testing the plant for suitability can be enough to transmit some viruses. Some insects also cause disease-like injury while feeding, even if they do not transmit a pathogen. This occurs when a component of the insect's saliva is toxic. Such insects are said to be toxicogenic and produce ailments such as hopperburn and psyllid yellows.

Approaches to pest management

Vegetable crops, like fruit crops, have high cosmetic standards. Thus, there is considerable economic pressure on vegetable producers to maintain their crops free of pests and pest-related injury. Vegetable crops, especially when marketed as fresh crops, also command a high price. The high cosmetic standards and prices cause producers to spare no effort to protect their crops from pest damage. This usually results in frequent insecticide applications. Even with the rapidly escalating costs of pesticides, insecticide costs usually are a minor component of the total costs of crop production. Thus, many growers do not hesitate to apply insecticides. It is not always necessary to

Some important vegetable crop pests, the continents where they are found, and the crops and parts of the crop plants affected by these pests.

Taxa and name	Continents affected[a]	Crops damaged[b]	Parts of crops damaged
Coleoptera: Chrysomelidae			
Acalymma vittatus (Fabricius), striped cucumber beetle	NA, SA	cucurbits, legumes	blossom, fruit, foliage, root
Aulacophora abdominalis (F.), pumpkin beetle	AU, AS	cucurbits	foliage
Ceratoma ruficornis (Oliver), redhorned leaf beetle	NA, SA	legumes	foliage
Crioceris asparagi (L.), asparagus beetle	EU, NA	asparagus	foliage, stem
Diabrotica balteata LeConte; banded cucumber beetle	NA, SA	numerous crops, but especially crucifers, cucurbits, and solanaceous crops	foliage, flower, fruit, root
Diabrotica undecimpunctata Mannerheim, spotted cucumber beetle	NA	numerous but primarily legume, solanaceous and cucurbit crops	blossom, fruit, foliage, root
Diabrotica virgifera LeConte, western corn rootworm	NA	cucurbits, sweet corn	blossom, fruit, foliage, root
Epitrix cucumeris (Harris), potato flea beetle	NA	solanaceous crops	foliage, root
Epitrix tuberis Gentner, tuber flea beetle	NA	solanaceous crops	foliage, root
Leptinotarsa decemlineata (Say), Colorado potato beetle	EU, NA	solanaceous crops	foliage
Phyllotreta cruciferae (Goeze), crucifer flea beetle	AF, EU, NA	crucifers	foliage, root
Raphidopalpa foveicollis Lucas, red pumpkin beetle	AF, AS	cucurbits	blossoms, foliage
Coleoptera: Bruchidae			
Acanthoscelides obtectus (Say), bean weevil	AF, EU, NA, SA	legumes	fruit
Coleoptera: Curculionidae			
Anthonomus eugenii Cano, pepper weevil	NA	pepper	blossom, fruit
Apion spp., weevils	AF, AS, AU, SA	numerous crops	blossom, fruit, foliage, stem
Ceutorhynchus assimilis Paykull, cabbage seedpod weevil	EU, NA	crucifers	fruit
Cylas formicarius (Fabricius), sweetpotato weevil	AF, AS, AU, NA, SA	sweet potato	root, tuber
Cylas puncticolis Boh., African sweetpotato weevil	AF	sweet potato	foliage, stem, tuber
Listronotus oregonensis (LeConte), carrot weevil	NA	carrot	root, stem
Naupactus spp., whitefringed beetles	AF, AU, NA, SA	numerous crops	foliage, root, tuber
Sitona lineatus (L.), pea leaf weevil	AS, EU, NA	legumes	foliage, root

(Continued)

Taxa and name	Continents affected[a]	Crops damaged[b]	Parts of crops damaged
Coleoptera: Coccinellidae			
Epilachna chrysomelina (F.), 12-spotted melon beetle	AF, EU	cucurbits	foliage
Epilachna varivestris Mulsant, Mexican bean beetle	NA	bean	foliage
Coleoptera: Scarabaeidae			
Popillia japonica Newman, Japanese beetle	AS, NA	numerous crops	foliage, fruit
Diptera: Agromyzidae			
Liriomyza bryoniae (Kaltenbach), tomato leaf miner	AF, AS, EU	numerous crops	foliage
Liriomyza huidobrensis (Blanchard), pea leafminer	AS, NA, SA	numerous crops	foliage
Liriomyza sativae Blanchard, vegetable leafminer	EU, NA	numerous crops	foliage
Liriomyza trifolii (Burgess), American serpentine leafminer	EU, NA, SA	numerous crops	foliage
Melanagromyza sojae (Zehn.), bean fly	AF, AS	legumes	stem
Melanagromyza obtusa Mall., bean pod fly	AS	legumes	seed
Ophiomyia phaseoli (Tyron), bean fly	AF, AS, AU	legumes	stem
Phytomyza horticola Goureau, pea leaf miner	AF, AS, EU	numerous crops	foliage
Diptera: Anthomyiidae			
Delia antiqua (Meigen), onion maggot	AS, EU, NA	onion and related crops	root, stem
Delia platura (Meigen), seedcorn maggot	AS, AF, AU, EU, NA	numerous crops	roots, stem
Delia radicum (L.), cabbage maggot	AS, EU, NA	crucifers	root, stem
Pegomya betae Curtis, beet leafminer	AF, AS, EU, NA	chenopods	foliage
Pegomya hyoscyami (Panzer), spinach leafminer	AF, AS, EU, NA	chenopods	foliage
Diptera: Cecidomyiidae			
Contarinia pisi (Winnertz), pea midge	EU	legumes	blossom
Diptera: Psilidae			
Psila rosae (Fabricius), carrot rust fly	AS, EU, NA	carrot	root
Diptera: Tephritidae			
Bactrocera cucumis (French), cucumber fly	AU	cucurbits and tomato	fruit
Bactrocera cucurbitae (Coquillett), melon fly	AF, AS	cucurbit, legume, and solanaceous crops	blossom, fruit, vine
Bactrocera dorsalis (Hendel), oriental fruit fly	AF, AS	cucurbit and solanaceous crops	fruit
Daucus ciliatus Loew, lesser pumpkin fly	AF, AS	cucurbits	fruit

(Continued)

Taxa and name	Continents affected[a]	Crops damaged[b]	Parts of crops damaged
Daucus frontalis Becker, African melon fly	AF	cucurbits	fruit
Eulia heraclei (L.), celery fly	EU	umbellifers	foliage
Myopardalinis pardalina Bigot, cantaloupe fruit fly	AS	cucurbits	fruit
Diptera: Tipulidae			
Tipula paludosa Meigen, European crane fly	AS, EU, NA	numerous crops	roots
Hemiptera: Coreidae			
Anasa tristis (De Geer), squash bug	NA	cucurbits	foliage, fruit
Leptoglossus spp., leaffooted bugs	AF, AS, AU, NA	numerous crops	foliage, fruit
Hemiptera: Miridae			
Lygus lineolaris (Palisot de Beauvois), tarnished plant bug	NA	numerous crops	blossom, fruit, stem, foliage
Hemiptera: Pentatomidae			
Nezara viridula (L.), southern green stink bug	AF, AS, AU, EU, NA, SA	numerous crops but principally legume, crucifer, and solanceous crops	blossom, fruit
Hemiptera: Pyrrhocoridae			
Dysdercus spp., cotton stainers	AF, AS, AU, NA, SA	okra	fruit
Hemiptera: Aleyrodidae			
Aleyrodes proletella (L.), cabbage whitefly	AU, EU	crucifers	foliage
Bemisia argentifolii Bellows and Perring, silverleaf whitefly	E, NA	numerous crops but especially solanaceous and cucurbit crops	foliage
Bemesia tabaci (Gennadius), cotton whitefly	AF, AS	numerous crops	foliage
Trialeurodes vaporariorum (Westwood), greenhouse whitefly	E, NA	numerous crops but especially solanaceous and cucurbit crops	foliage
Hemiptera: Aphididae			
Acyrthosiphon pisum (Harris), pea aphid	AF, AS, AU, EU, NA, SA	legumes	foliage, stem
Aphis fabae Scopoli, bean aphid	AF, AS, EU, NA, SA	chenopods, legumes	foliage, stem
Aphis gossypii Glover, melon aphid	AF, AS, AU, EU, NA, SA	numerous crops	foliage, stem
Brachycorynella asparagi (Mordvilko), asparagus aphid	AS, EU, NA	asparagus	foliage
Brevicoryne brassicae (L.), cabbage aphid	AF, AS, AU, EU, NA, SA	crucifers	foliage
Cavariella aegopodii Scopoli, willow-carrot aphid	EU, NA	umbellifers	foliage

(Continued)

Taxa and name	Continents affected[a]	Crops damaged[b]	Parts of crops damaged
Lipaphis erysimi (Kaltenbach), turnip aphid	AF, AS, AU, EU, NA, SA	crucifers	foliage
Macrosiphum euphorbiae (Thomas), potato aphid	AF, AS, AU, EU, NA, SA	numerous crops	foliage, stem
Myzus persicae (Sulzer), green peach aphid	AF, AS, AU, EU, NA, SA	numerous crops	foliage, stem
Nasonovia ribisnigri (Mosley), lettuce aphid	EU, NA, SA	lettuce and related crops	foliage
Pemphigus bursarius (L.), lettuce root aphid	AF, AS, AU, EU, NA	lettuce and related crops	root
Hemiptera: Cicadellidae			
Circulifer tenellus (Baker), beet leafhopper	AF, EU, NA	chenopod, cucurbit, and solanaceous crops	foliage
Circulifer opacipennis (Leth.), beet leafhopper	AF, EU, AS	chenopod, cucurbit and solanaceous crops	foliage
Dalbulus maidus (DeLong and Wolcott)	NA, SA	sweet corn	
Empoasca fabae (Harris), potato leafhopper	NA	numerous crops	foliage
Macrosteles quadrilineatus Forbes, aster leafhopper	NA	several crops, especially lettuce	foliage
Hemiptera: Delphacidae			
Peregrinus maidis (Ashmead), corn delphacid	NA, SA	sweet corn	foliage
Lepidoptera: Gelechiidae			
Keiferia lycopersicella (Walsingham), tomato pinworm	NA, SA	solanaceous crops	foliage
Phthorimaea operculella (Zeller), potato tuberworm	AF, AS, AU, EU, NA, SA	solanaceous crops	foliage, tuber
Lepidoptera: Noctuidae			
Agrotis ipsilon (Hufnagel), black cutworm	AF, AS, AU, EU, NA, SA	numerous crops	foliage, fruit, stem
Agrotis segetum (Denis & Schiffermuller), turnip moth	AF, AS, EU	numerous crops	stem, root
Autographa gama (L.), silver Y moth	AF, AS, EU	numerous crops	foliage
Helicoverpa armigera (Hübner), old world bollworm	AF, AS, AU, EU	numerous crops	foliage, fruit
Helicoverpa zea (Boddie), corn earworm	NA, SA	numerous crops	foliage, fruit
Hydraecia micacea (Esper), potato stem borer	AS, EU, NA	numerous	stem
Mamestra brassicae (L.), cabbage moth	AS, EU	crucifers	foliage
Noctua pronuba (L.), large yellow underwing	AF, AS, EU	numerous	foliage, root, stem, tuber

(Continued)

Taxa and name	Continents affected[a]	Crops damaged[b]	Parts of crops damaged
Peridroma saucia (Hübner), variegated cutworm	AF, AS, EU, NA, SA	numerous crops	foliage, fruit, stem
Spodoptera eridania (Cramer), southern armyworm	NA, SA	numerous crops but especially cucurbit and solanaceous crops	foliage, fruit
Spodoptera exigua (Hübner), beet armyworm	AS, NA	numerous crops	foliage
Spodoptera frugiperda (J.E. Smith), fall armyworm	NA, SA	numerous crops but especially sweet corn	foliage
Spodoptera littoralis (Boisduval), African cotton worm	AF, EU	artichoke, crucifers, and solanaceous crops	foliage
Spodoptera litura (Fabricius), rice cutworm	AU, AS	legumes, sweetpotato, and solanaceous crops	foliage
Trichoplusia ni (Hübner), cabbage looper	AF, AS, EU, NA, SA	numerous crops	foliage
Xestia spp. (L.), spotted cutworm	AF, AS, EU, NA	numerous crops	foliage, stem
Lepidoptera: Nymphalidae			
Acraea acerata Hew.	AF	sweet potato	foliage
Lepidoptera: Pieridae			
Pieris brassicae (L.), large cabbage white butterfly	AF, AS, EU	crucifers	foliage
Pieris canidia (L.), small white butterfly	AS	crucifers	foliage
Pieris rapae (L.), imported cabbageworm	AF, AS, AU, EU, NA	crucifers	foliage
Lepidoptera: Pterophoridae			
Platyptilia carduidactyla (Riley), artichoke plume moth	NA	artichoke	stem
Lepidoptera: Pyralidae			
Diaphania hyalinata L., melonworm	NA, SA	cucurbits	foliage
Diaphania indica (Saunders) pumpkin caterpillar	AF, AS, AU, NA, SA	cucurbits	foliage
Diaphania nitidalis (Stoll), pickleworm	NA, SA	cucurbits	blossom, fruit
Etiella zinckenella (Treitschke), limabean pod borer	AF, AS, AU, EU, NA, SA	legumes	fruit
Evergestis forficalis (L.), garden pebble moth	AS, EU	crucifers	foliage
Hedylepta indicata (Fabricius), bean webworm	AF, AS, SA	legumes	foliage
Leucinodes orbonalis Guenée, eggplant fruit borer	AF, AS	numerous crops	foliage, fruit

(Continued)

Taxa and name	Continents affected[a]	Crops damaged[b]	Parts of crops damaged
Maruca testulalis (Geyer), bean pod borer	AF, AS, AU, NA, SA	legumes	fruit
Ostrinia nubilalis (Hübner), European corn borer	AF,AS, EU, NA	sweet corn and solanaceous crops	fruit, stem
Omphisa anastomosalis Guenée, sweetpotato vine borer	AS	sweet potato	stem
Plutella xylostella (L.), diamondback moth	AF, AU, AS, EU, NA, SA	crucifers	foliage
Lepidoptera: Sesiidae *Melittia cucurbitae* (Harris), squash vine borer	NA, SA	cucurbits	stem
Lepidoptera: Sphingidae *Agrius convulvuli* (L.), sweetpotato hornworm	AF, AS, AU, EU	sweetpotato, legumes	foliage
Lepidoptera: Tortricidae *Cydia nigricana* (Fabricius), pea moth	EU, NA	legumes	fruit
Lepidoptera: Yponomeutidae *Acrolepiopsis assectella* (Zeller), leek moth	AS, EU	onion and related plants	foliage
Hymenoptera: Tenthredinidae *Athalia* spp., cabbage sawfly	AS, AF, EU	crucifers	foliage
Thysanoptera *Frankliniella occidentalis* (Pergande), western flower thrips	AS, EU, NA, SA,	primarily cucurbit and solanaceous crops	blossom
Frankliniella schulzei (Trybom)	AF	legumes, sweet potato, and solanaceous crops	foliage
Thrips palmi Karny, melon thrips	AU, AS, NA, SA	cucurbit and solaceous crops	blossom, fruit, foliage
Thrips tabaci Lindeman, onion thrips	AF, AU, AS, EU, NA, SA	onion and related crops	foliage, fruit
Acari *Aculops lycopersici* (Massee), tomato russet mite	AF, AS, AU, EU, NA, SA	solanaceous crops	foliage, stem
Polyphagotarsonemus latus (Banks), broad mite	AF, AS, AU, EU, NA, SA	numerous cops	foliage
Tetranychus urticae Koch, twospotted spider mite	AF, AU, AS, EU, NA, SA	numerous but principally cucurbit, legume, and solanaceous crops	foliage

[a]Regions are AF, Africa; AU, Australia/New Zealand; AS, Asia; EU, Europe; NA, North America; SA, South America.
[b]Crops are artichoke, family Compositae; asparagus, family Liliaceae; chenopods, family Chenopodiaceae; crucifers, family Cruciferae; cucurbits, family Cucurbitaceae; legumes, family Leguminosae; lettuce, family Compositae; okra, family Malvaceae; onion, family Amaryllidaceae; pepper, family Solanaceae; solanaceous, family Solanaceae; sweet corn, family Graminae; sweet potato, family Convolvulaceae; umbellifers, family Umbelliferae.

apply insecticides, however, because there are times and places when insects are not sufficiently abundant to cause injury, or there are other alternative actions that can be taken to limit injury. There is adequate justification for careful monitoring of vegetable crops, and application of the economic injury level concept, which would eliminate unnecessary insecticide application.

Cultural practices, such as modifying the location or time of planting, tillage, weed control, level and type of irrigation and fertilization, and physical barriers, such as floating row covers, are techniques that can be used to reduce the ability of insects to infest and reproduce within crops. Such practices ideally will maximize environmental resistance, and eliminate the threat of pest damage. Sometimes these practices are completely adequate, and other times they are an important component of pest management that requires additional, complimentary intervention. In southern regions of the United States, for example, early-planted cucurbit crops usually escape infestation by the pickleworm (Lepidoptera: Pyralidae). Late-planted or autumn crops are routinely infested by this pest, necessitating the application of insecticides to prevent crop injury.

Host plant resistance involves the use of vegetable plant cultivars that inherently are less supportive of insect growth and development, or that are less injured by insect feeding. For example, the glossy varieties of crucifers that lack the waxy bloom and are, therefore, green rather than the normal grayish green color, are somewhat resistant to the diamondback moth (Lepidoptera: Pyralidae). Unfortunately, the genetic diversity in horticulturally acceptable vegetables (growth, shipping, appearance and taste characteristics affect ''acceptability'') is often quite limited, and cultivars with insect resistance often are unavailable. Also, plant breeders have not concentrated on producing pest-resistant vegetable crops to the degree that they have worked on field and forage crops. This lack of attention is due to the ready availability of insecticides and the high value of vegetables, which allows economic use of insecticides.

Biological control involves the use of biotic agents to suppress pest populations. In the case of invading pest species, this can be done by introducing natural enemies that have been ''left behind'' in the native land of the pest. Assuming successful introduction and establishment of an effective natural enemy, this approach sometimes results in adequate and continuous suppression, allowing cultivation of the crop without additional release of the natural enemy, and without insecticidal intervention. On the other hand, some pest problems require regular intervention. For example, many crops infested with caterpillars are treated several times during each cropping cycle with the bacterium *Bacillus thuringiensis*.

The most commonly used approach to pest suppression in vegetable crops is the use of chemical insecticides. Insecticides are used because there is almost always an effective product available; this is not true for cultural practices, host plant resistance, and biological control. Insecticides can be applied to the soil to prevent seed, seedling and young plant injury; to the foliage and fruit to protect against direct and indirect pests; and to baits or traps to reduce the number of pests available to injure crops. Insecticides can be very selective, or have a wide range of activity, depending on the need. Insecticides can be formulated to be systemic in action, to function as fumigants, and to have short or long residual activity. Thus, despite concern about the adverse effects of insecticides (primarily threats to humans and wildlife), they remain a commonly used tool for vegetable production.

John L. Capinera
University of Florida
Gainesville, Florida, USA

References

Capinera, J. L. 2001. *Handbook of vegetable pests*. Academic Press, San Diego, California.

Hill, D. S. 1983. *Agricultural insect pests of the tropics and their control (2nd ed.)*. Cambridge University Press, Cambridge, United Kingdom.

Hill, D. S. 1987. *Agricultural insect pests of temperate regions and their control*. Cambridge University Press, Cambridge, United Kingdom.

Howard, R. J., J. A. Garland, and W. L. Seaman (eds.). 1994. *Diseases and pests of vegetable crops in Canada*. The Canadian Phytopathological Society and the Entomological Society of Canada, Ottawa, Ontario, Canada.

McKinlay, R. G. (ed.) 1992. *Vegetable crop pests*. CRC Press, Inc., Boca Raton, Florida.

VEIN. A tube running through the wings of insect, through which blood (hemolymph) is pumped.

See also, WINGS OF INSECTS, HEMOLYMPH.

ARCHETYPE TRACHEATION OF WING PAD

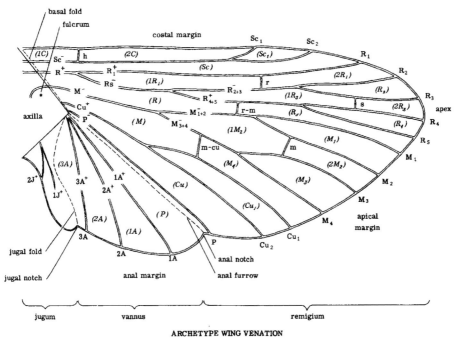

Fig. 1108 Hypothetical ancestral pattern of wing venation.

VELIIDAE. A family of bugs (order Hemiptera). They sometimes are called broad-shouldered water striders. See also, BUGS.

VELOCIPEDIDAE. A family of bugs (order Hemiptera). They sometimes are called fast-footed bugs. See also, BUGS.

VELVET ANTS (HYMENOPTERA: MUTILLIDAE). Because of their generally bright colors and conspicuous appearance, velvet ants have long been favorite targets of insect collectors. Thus, it is with some degree of surprise that so little is known about them and their biology. Female mutillids, which are much more frequently observed than males, generally look very much like ants. They are wingless and can be seen scurrying about on the ground. They are, however, actually wasps. Their ant-like appearance, coupled with the fact that they are covered by a dense pubescence (hairs), has led to the common name of velvet ants.

Linnaeus included eight species of mutillids in his tenth edition of *Systema Naturae* in 1758. By the lat-

ter part of the nineteenth century, an estimated 3,500 to 4,000 species had been described, nearly all of which were included in the genus *Mutilla* Linnaeus. It was not until near the beginning of the 20th century that many new genera were established.

The family Mutillidae is found worldwide but is predominantly tropical. As of 1975, there were approximately 8,000 described species. Brothers (1975) investigated the phylogeny of Mutillidae and recognized seven monophyletic subfamilies. That phylogeny is generally accepted. The family is now known worldwide from approximately 230 genera and subgenera, included in those seven subfamilies.

The higher-level classification is as follows:

Order: Hymenoptera
  Superfamily: Vespoidea
    Subfamily: Myrmosinae
    Subfamily: Pseudophotopsidinae
    Subfamily: Ticoplinae
    Subfamily: Rhopalomutillinae
    Subfamily: Sphaeropthalminae
    Subfamily: Myrmillinae
    Subfamily: Mutillinae

General characteristics

Velvet ants can be divided into nocturnal and diurnal. The nocturnal mutillids are known almost exclusively from the males. They tend to be more wasp-like in appearance, brown or reddish-brown in color, and from 1 to 2 cm in length. Males almost always have two pairs of membranous wings.

Females of the diurnal velvet ants tend to be more ant-like in appearance. So far as known, all are wingless. Body color is most often black, brown, or reddish-brown. However, the body color is often not visible due to the dense pubescence. The hairs (more properly known as setae) are often brightly colored (red, orange, or yellow), or they may be a combination of one of the bright colors and black. Occasionally, the hairs may be entirely black or entirely white. The pattern of pubescence is usually species-specific. Males of even the diurnal velvet ants tend to be more wasp-like in appearance. Body color is most often black. Although some males may be brightly colored like the females, they are more frequently covered with black hairs over most of their body.

Sexual dimorphism

As noted above, the males and females of velvet ants can look quite different. Males generally have wings; females are wingless. Males tend to be more wasp-like in appearance; females tend to be more ant-like in appearance. Color of the hairs may be similar or quite different. Size may be the same or quite different. In some groups, males carry the females in flight during mating. Where this occurs, males are much larger. In other groups, females are generally much larger.

Fig. 1109 A velvet ant female, *Dasymutilla occidentalis* (photo, L.J. Buss).

Fig. 1110 A velvet ant male, *Dasymutilla* sp. (photo, L.J. Buss).

Taxonomy

The extreme sexual dimorphism in velvet ants has led to problems in taxonomy. Some genera are known from males only; others are known from females only. The same is frequently true for species within a genus. The genus *Dasymutilla* Ashmead is one of the most common genera in North America and includes all of the large, fuzzy velvet ants that most people readily recognize. The genus includes approximately 150 described species. Of those, approximately one third are known from males only, about one third from females only, and about one third are known from both sexes. In many cases, those species that exhibit a lesser degree of sexual dimorphism are known from both sexes, such as the large, red *Dasymutilla occidentalis* Linnaeus of the eastern United States commonly referred to as a 'cowkiller.' In that species, males and females look almost identical except for the presence of wings in the males.

Biology

Relatively little is known about mutillid biology. It is known that mutillids are parasites (more correctly parasitoids) of other insects. Worldwide, the hosts are probably known for less than 10% of all described species. Of those that are known, most seem to be other ground-nesting bees and wasps, although there are some exceptions. A few species have been reported to parasitize tsetse flies in Africa.

Adults are generally active for only a few months out of the year, although specimens have been collected in every month of the year. Most observations have been of females, which are more limited in their mobility. Only a few observations have been recorded of mating behavior in this group. These have been almost entirely of diurnal species, as might be expected. There has been some consistency in those observations that have been made. Males usually fly a few centimeters above the ground into a slight breeze or wind, suggesting that a pheromone (or sexual attractant) is involved. Courtship and mating are usually very brief with the entire process not lasting more than a couple of minutes.

Female mutillids apparently attack the larval stage of their host after the latter has spun its cocoon. Females enter the nests by digging through the soil or by breaking the walls of mud nests. Once they have gained entry, they chew a hole in the cocoon, turn around, and insert a single egg. They then plug the hole using salivary fluids and particles of soil or mud. After oviposition, about three days are required for the egg to hatch. The mutillid larva then proceeds to eat the host larva until the latter has been entirely consumed. This usually takes only a few days. The full-grown larva then spins a cocoon in the chamber prepared for the host, and pupates. Pupation is thought to take about twenty days.

In tropical and subtropical regions, there may be two to several generations a year. In the United States, mutillids apparently have only one generation a year. In colder regions, overwintering occurs in the pupal stage for most species. However, at least one species is known to overwinter in the adult stage.

Defense

One of the most interesting things about velvet ants is their vast array of defense mechanisms. First and foremost, they are wasps, and females are able to deliver a very painful sting. Interestingly, although males are not capable of stinging, they, too, have a very elaborate 'stinging behavior.' They do a very good acting job that will very frequently lead a potential predator to believe that they are being stung, thus affecting release of the male mutillid. Coupled with the sting, the often bright warning coloration will discourage many potential predators from attacking. Other defense mechanisms include a very tough exoskeleton, stridulation (the production of a squeaking sound that is used in mating as well as defense), quick zig-zagging movements by females as they run on the ground, flight by males, and a good set of mandibles. It has been said that "an animal that

discovers that the mutillid is hard-bodied and a powerful stinger is likely to remember its brilliant and unusual color pattern and to avoid it in the future" (Evans and Eberhard 1970:221).

See also, WASPS, ANTS, BEES, AND SAWFLIES.

Donald G. Manley
Clemson University
Florence, South Carolina, USA

References

Brothers, D. J. 1975. Phylogeny and classification of the aculeate Hymenoptera, with special reference to Mutillidae. *University of Kansas Science Bulletin* 50: 483–648.

Brothers, D. J. 1993. Family Mutillidae. pp. 188–203 in H. Goulet and J. T. Huber (eds.), *Hymenoptera of the world: an identification guide to the families*. Centre for Land and Biological Resources Research, Ottawa, Ontario, Canada.

Evans, H. E., and M. J. West Eberhard. 1970. *The wasps.* University of Michigan Press, Ann Arbor, Michigan.

Manley, D. G. 2000. Defense adaptations in velvet ants (Hymenoptera: Mutillidae) and their possible selective pressures. pp. 285–289 in A. D. Austin and M. Dowton (eds.), *Hymenoptera: evolution, biodiversity and biological control.* CSIRO Publishing, Collingwood, Australia.

Mickel, C. E. 1928. Biological and taxonomic investigations of the Mutillid wasps. *United States National Museum Bulletin* 143: 1–351.

VELVET WATER BUGS. Members of the family Hebridae (order Hemiptera). See also, BUGS.

VELVETY SHORE BUGS. Members of the family Ochteridae (order Hemiptera) See also, BUGS.

VENOMS OF ECTOPARASITIC WASPS. The complex interactions that occur between ectoparasitic wasps and their hosts ultimately lead to altered host development, physiology, and behavior. Precisely how these host changes come about is poorly understood, but in most cases involving ectoparasitic wasps, a venom is injected into the host. Venoms of ectoparasitic non-aculeate Hymenoptera ('Parasitica') fall into two categories based on the impact on the host: those that evoke paralysis and those that are non-paralytic. The vast majority of ectoparasitic species studied produce paralyzing venoms. Such species are considered idiobionts and the venom confers an adaptive advantage in the relationship with the host: a mobile host can be permanently or temporarily paralyzed during egg laying by the female wasp. A still-mobile host is also a potential threat to developing parasitoids, consequently these venoms are more often permanent paralyzing agents. Paralyzing venoms are typically produced by the adult female wasp, however, for some ectoparasitic species (e.g., Chalcidoids), the wasp larvae produce salivary venoms/secretions that apparently can halt host development in a manner consistent with paralysis.

In contrast, non-paralyzing venoms do not render the host immobile. In fact, several species utilize non-mobile stages of the host (e.g., pre-pupae and pupae). These venoms seem to function by altering host physiology for the benefit of the developing parasites. In most cases, venom-induced host changes center around nutritional considerations of the immatures; female wasps must ensure that their progeny maximally use the host since for idiobiont species, the host does not feed following parasitism, and is thus a finite pool of nutrients. Venoms from ectoparasitic species belonging to the Pteromalidae and Eulophidae also are known to inhibit host molting, suppress immune responses, elevate lipid metabolism, slow or completely inhibit respiratory physiology, and alter protein expression in a variety of host tissues. Such host effects are consistent with parasitoids referred to as koinobionts, a condition more typical of endoparasitic species that depend on polydnaviruses.

Active venom proteins from ectoparasitoids, whether paralyzing agents or non-paralytic toxins, are much larger in molecular mass than those of the aculeate Hymenoptera, and perhaps this is related to the different usage of venoms among hymenopterans: aculeates use venom primarily in defense whereas venom is associated with reproductive strategies of non-aculeates. For example, low molecular weight proteins and peptides are typical of the venoms of social Hymenoptera (ants, bees, and wasps). The venoms of bees and social wasps are also abundant in neutral and basic amino acids. Comparatively few studies have examined the composition of ectoparasitic wasp venoms, but for those examined, the venom proteins are considerably larger than the aculeates. *Habrobracon hebetor* (Braconidae) produces a potent venom with at least two highly paralytic protein toxins (Brh-I and Brh-V, estimated molecular masses 71–73 kDa) and one smaller, less

insectidical toxin (about 20–40 kDa). All three toxins appear to be glycosylated, a feature characteristic of mid- to high-molecular weight venom proteins of hymenopterans, but not of low molecular mass toxins. A 66-kDa venom protein has been isolated from *Euplectrus comstockii* (Eulophidae) that does not induce paralysis but does arrest host development by inhibiting larval-larval ecdysis. This arrestment protein is thought to be comprised of at least two subunits, suggesting that the native protein is possibly a dimer composed of two 33 kDa proteins. *Nasonia vitripennis* (Pteromalidae) produces a complex venom with over 9 major venom proteins visualized by denaturing electrophoresis (SDS-PAGE). A 69-kDa venom protein has been isolated from crude venom that retards host development and appears ultimately responsible for triggering death of the host.

The constituents of the venom are synthesized by venom glands, which are sometimes referred to as poison glands or acid glands. The venom glands of ectoparasitic wasps may connect to a large reservoir, or may attach directly to the distal portion of the vagina or common oviduct. Both the venom gland and reservoir are of ectodermal origin and are lined with chitin. The chitinous lining is derived from squamous epithelial cells and is an apparent necessity since many species store the venom in an active form as opposed to a precursor. Secretory cells are found throughout the venom apparatus (gland and reservoir), but localization of specific venom protein synthesis in any parasitic species has yet to occur. In at least *N. vitripennis*, the bulk of the venom is synthesized in the columnar epithelial cells lining the acid gland and is discharged directly into the lumen of the venom reservoir apparently packaged into small vesicles.

The precise tissues targeted by ectoparasitic wasp venoms have not been clearly identified, nor have the pathways activated/suppressed by these venoms been deciphered. Neuromuscular junctions most likely are targeted by paralytic venoms. For example, venom from *H. hebetor* is thought to evoke paralysis by binding to a receptor associated with the excitatory glutamatergic system on pre-synaptic membranes. Venom causes an accumulation of pre-synaptic neurotransmitter vesicles, thus blocking neurotransmitter release by inhibiting vesicle fusion with post-synaptic membranes. No other paralytic venoms from non-aculeate hymenopterans have been sufficiently studied for comparison with *H. hebetor* to determine if this is the norm or the exception among the parasitic Hymenoptera. Paralysis is, however, the norm for venoms of parasitic aculeate hymenopterans, but the mode of action of these venoms does differ from *H. hebetor*. The aculeate venoms trigger paralysis by blocking cation channels in pre- and post-synaptic membranes of host nervous and muscle tissues.

Non-paralytic venoms from ectoparasitic wasps seem to target a variety of tissues, most likely in a receptor-mediated fashion. Inhibition of larval and pupal ecdysis by *E. comstockii*, *E. plathypenae*, and *N. vitripennis*, respectively, suggests that these venoms disrupt tissue responsiveness to ecdysteroids, a condition that is not corrected by application of exogenous 20-hydroxyecdysone. This effect implies that some aspect of the receptor-hormone complex is disrupted or that the complex is not able to bind to and/or activate response elements within the cell nucleus. Venom from *N. vitripennis* also elevates *de novo* synthesis of host lipids in fat body and hemolymph, implicating an involvement of adipokinetic hormone activation and/or synthesis in response to envenomation. *N. vitripennis* venom also induces fly hosts to enter a developmental arrest that is characterized by a suppression of respiratory metabolism and altered protein expression, both of which show tremendous similarity to the diapause physiology of the host. The latter observation has led to speculation that the venom targets the host brain to redirect the cell-cycle control system. Consistent with this prediction is the observation that heat shock protein synthesis (hsp 23 and hsp 70) in host brains are highly upregulated following venom injection. The host response differs from a typical stress response in that all hsps are not uniformly upregulated due to envenomation by *N. vitripennis*, and the response persists for an extended duration.

Venom from at least two species, *Eulophus pennicornis* (Eulophidae) and *N. vitripennis*, alter hemocyte behavior in the host. In the case of *E. pennicornis*, venom is not thought to be directly associated with alterations in the numbers of circulating hemocytes, changes in phagocytic abilities, or loss of hemocyte membrane integrity, although it seems that the altered host immune response is only evident when both venom injection and oviposition occur. In contrast, venom from *N. vitripennis* specifically targets plasmatocytes and granulocytes independent of oviposition. Host plasmatocytes are destroyed within one hour of envenomation by a

mechanism consistent with apoptosis. Granulocytes remain viable throughout parasitism, but lose the capacity to adhere to foreign objects and spread when cultured *in vitro*.

All ectoparasitic wasp venoms have the ability to ultimately kill the host. The most thoroughly investigated pathway eliciting host death involves the venom from *N. vitripennis*. Oncosis is the primary mechanism of host cell death. The venom apparently binds to a G-protein-sensitive receptor, activating one or more signal transduction pathways. The currently accepted mode of action for the venom involves the activation of phospholipase C, thereby elevating cAMP levels and eventually triggering release of free intracellular calcium from either mitochondria or smooth endoplasmic reticulum. Elevated intracellular calcium levels are thought to have a multiplicity of effects on target cells, including activation of cytosolic proteases that break down cytoskeletal proteins, and stimulation of endonuclease activity resulting in degradation of genomic DNA. Any of these effects can destroy the cell from within and contribute to the ultimate fate of the host, death. This mode of operation parallels the action of at least one toxin (mastoparan) isolated from several social wasps.

David B. Rivers
Loyola College
Baltimore, Maryland, USA

References
Beard, R. L. 1963. Insect toxins and venoms. *Annual Review of Entomology* 8: 1–18.
Piek, T. 1986. *Venoms of the Hymenoptera*. Academic Press, London, United Kingdom. 570 pp.
Quicke, D. L. J. 1997. *Parasitic wasps*. Chapman and Hall, London, United Kingdom. 470 pp.
Rivers, D. B., L. Ruggerio, and J. A. Yoder. 1999. Venom from *Nasonia vitripennis*: a model for understanding the roles of venom during parasitism by ectoparasitoids. *Trends in Entomology* 2: 1–17.
Schmidt, J. O. 1982. Biochemistry of insect venoms. *Annual Review of Entomology* 27: 339–368.

VENTRAL DIAPHRAGM. A ventral layer of thin cells and muscle that assists in circulation of hemolymph around the nerve cord.

VENTRAL NERVE CORD. The series of interconnected ganglia lying along the ventral surface of the insect, and providing nervous coordination.

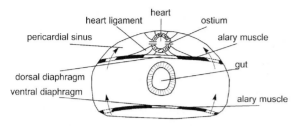

Fig. 1111 Cross section of an insect abdomen, showing components of the insect circulatory system and direction of hemolymph flow (adapted from Evans, Insect biology).

VENTRAL. The lower surface.

VERMIFORM. A body form that is worm-like, lacks a distinct head, and is legless. This term often is used to describe fly (Diptera) larvae, but also applies to newly hatched grasshoppers. Such newly hatched grasshoppers, called vermiform larvae, are enclosed in a membrane, so the legs are not functional. Vermiform larvae wiggle from the below-ground egg pod to the soil surface, where they free themselves from the membrane and possess functional appendages. Only after escaping from the membrane is the young hopper considered to be a first instar.

VERMILEONIDAE. A family of flies (order Diptera). They commonly are known as worm lions. See also, FLIES.

VERMILEONIDAE. A family of flies (order Diptera). See also, FLIES.

VERMIPSYLLIDAE. A family of fleas (order Siphonaptera). They sometimes are known as carnivore fleas. See also, FLEAS.

VERTEX. The top of the head between the compound eyes.
  See also, HEAD OF HEXAPODS.

VERTICILLIUM LECANII. *Verticillium* (Deuteromycotina: Hyphomycetes) is a heterogenous genus that includes a number of devastating plant-pathogenic fungi, including the well-characterized soil fungi *V. dalhliae* and *V. albo-atrum*. The best-known entomopathogenic species, *V. lecanii*, was first described on coffee scales from Java in the late 1800s. Since that time, the fungus has been reported to infect several other insects, most notably whiteflies, thrips, aphids and grasshoppers. In addition, *V. lecanii* can infect other fungal phytopathogens such as rusts, powdery mildews and pest nematodes. Although the fungus has been identified worldwide, it is most common in tropical and subtropical regions due to its water requirement.

There are distinctly different morphotypes within the species; conidial size varies depending upon the isolate, and the colonies vary from white to yellow, with a cottony to compact or mealy texture. That *V. lecanii* is a diverse species also is evidenced by experimental attempts at recombination by parasexual-type methods (i.e., hyphal anastomosis and protoplast fusion), which often results in complete incompatibility between many isolates. *Verticillium lecanii* appears to be an anamorph of the ascomycete *Torrubiella*.

Verticillium is characterized by the presence of verticillate conidiophores that bear loose whorls of phialides. The phialides are usually awl-shaped, and conidia are smooth-walled, hyaline, and one-celled. Conidia can be produced in slimy masses or sometimes in chains.

Verticillium lecanii has been used in a number of studies to test its effectiveness in greenhouses. Some results have been promising and have led to the commercial production of strains specific for whiteflies and aphids. The whitefly strain has not proven to be as efficient a control agent as the aphid strain, perhaps because whiteflies are sessile, so the fungus is not dispersed as easily from one insect to another as in the case of mobile aphids. *Aschersonia aleyrodis*, another Deuteromycete, has been suggested to be a better control agent of greenhouse pests than *V. lecanii*. For example, *A. aleyrodis* appears to be more virulent to whiteflies, and the spores survive on host plant leaves longer than those of *V. lecanii*. *Verticillum lecanii* conidia are relatively unstable and must be stored frozen or at 4°C. Multiple passaging of the fungus on mycological media does not seem to cause attenuation, but can alter colonial morphology and growth rates, which indirectly affect the use of the fungus in biocontrol programs.

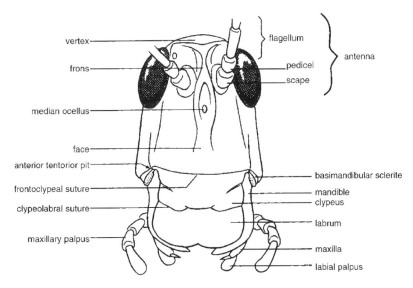

Fig. 1112 Front view of the head of an adult grasshopper, showing some major elements.

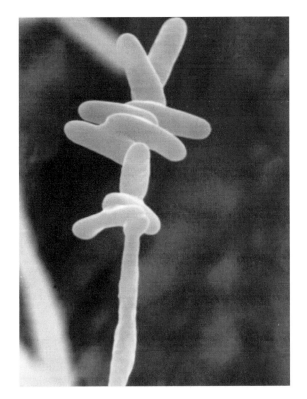

Fig. 1113 Scanning electron micrograph of *Verticillium lecanii* conidia. Note how the conidia are stuck together in clusters.

References

Hall, R. A. 1981. The fungus *Verticillium lecanii* as a microbial insecticide against aphids and scales. pp. 483–498 in H. D. Burges (ed.), *Microbial control of pests and plant diseases.* Academic Press, London, United Kingdom.

Heale, J. B. 1988. The potential impact of fungal genetics and molecular biology on biological control, with particular reference to entomopathogens. pp. 211–234 in M. N. Burge (ed.), *Fungi in biological control systems.* Manchester University Press, Manchester, United Kingdom.

Meyer, S. L. F., and R. J. Meyer. 1996. Greenhouse studies comparing strains of the fungus *Verticillium lecanii* for activity against the nematode *Heterodera glycines. Fundamental and Applied Nematology* 19: 305–308.

Verhaar, M. A., T. Hijwegen, and J. C. Zadoks. 1996. Glasshouse experiments on biocontrol of cucumber powdery mildew (*Sphaerotheca fuligines*) by the mycoparasites *Verticillium lecanii* and *Sporothrix rugulosa. Biological Control* 6: 353–360.

VESICATING SPECIES. Insect species that cause production of blisters on an animal that contacts the insect. This condition is sometimes caused vesicular dermatitis, and is most often induced by blister beetles containing cantharidin.

See also, CANTHARIDIN.

VESPIDAE. A family of wasps (order Hymenoptera). They commonly are known as paper wasps, yellow jackets, hornets, mason wasps and potter wasps. See also, WASPS, ANTS, BEES, AND SAWFLIES.

VESTIGIAL. The remnants of a formerly functional organ, usually reduced in size.

VETERINARY PESTS AND THEIR MANAGEMENT. Pests of veterinary importance are unique in their associations with animals hosts. Unlike the many pests that utilize plants or plant materials for their survival, pests of veterinary importance feed on hosts that can for the most part move from place to place. Thus, the hosts can live in a variety of habitats, move from one habitat to another during a daily cycle, and persist through a variety of climatic conditions during an annual cycle. The host range can be relatively small and associated with a central nest or burrow, or it can be practically endless in the case of herds of range cattle or migratory antelope. Likewise, the range of the pest can be limited geographically to certain locations, climates, or altitudes. Conversely, the pest can be cosmopolitan and affect similar hosts on almost every continent of the planet. Specifically, highly specialized pests can be limited to one or two host species, such as the sheep ked, *Melophagus ovinus*, a widely distributed parasite of sheep and goats.

Host response

Pests of veterinary importance confront a host that has the ability to resist their presence through their blood chemistry and they must resist the hosts' attempts to wall them off and otherwise limit their feeding activities. Pests of veterinary importance also confront a host that has the ability to physically resist their presence, and most pests face the possibility of bodily injury or death each time they visit a host to feed. A number of recognized host responses may be elicited by the presence of a pest. These include muscle twitching, scratching, chewing, rolling on the ground, tail swishing, head tossing, bunching, foot stomping, and running away. The hematophagous pests are particularly vulnerable because they have no choice. There is no alternative food source. Without taking a blood meal, most cannot live or reproduce.

Economic losses – physical. Besides the direct effects of the pests' feeding on hosts, the host response can add to, or be solely responsible for, economic loses from direct or indirect contact with the pest. The feeding activities of some pests (e.g., lice and mites) cause itching, and host animals scratch or rub themselves incessantly, causing additional damage to affected areas as they try to gain relief. Fly worry can distract animals from eating or resting adequately, thus causing decreases in weight gains.

Physical loss factors caused by selected pests to livestock and poultry.

Physical loss factors	Pests	Affected animals
Fly worry	Flies: *Musca domestica, Stomoxys calcitrans, Haemeatobia irritans, Hippelates* sp. *Musca autumnalis*	Cattle, horses, other livestock
Exsanguination	Flies: *Stomoxys calcitrans, Haematobia irritans,* Tabanidae, Simuliidae. Mosquitoes: various species Ticks: *Boophilous* spp., various species	Cattle, horses, other livestock
Suffocation	Flies: Simuliidae, Chironomidae, Mosquitoes: various species	Cattle, horses, other livestock
Disfiguration	Ticks: *Boophilous* spp., various species	Cattle, horses, other livestock
Itching/Skin Destruction	Mites: *Sarcoptes scabiei, Demodex* spp., *Psoroptes* spp., *Chorioptes* spp., *Ornithonyssus syilviarum*	Cattle, horses, poultry, other livestock
	Lice: *Haematopinus* spp., *Bovicola* spp., *Menacanthus stramineus, Menapon gallinae*	Cattle, horses, poultry, other livestock
	Flies: *Melophagous ovinus*	Sheep and goats
Gadding	Flies: *Hypoderma* spp.	Cattle

Economic losses – disease. Pests can also transmit diseases to the host, resulting in sickness or death. Pests such as mosquitoes, phlebotomine sand flies and tsetse flies are the cause of incalculable amounts of suffering by humans and animals alike because of the disease organisms they transmit in so many parts of the world. Ticks constitute a major threat to livestock industries in many countries in the western hemisphere and in Australia and Africa. New threats arise constantly as little known diseases suddenly increase to epidemic levels (e.g., Lyme disease) and diseases prevalent in one part of the globe appear unexpectedly in another (e.g., West Nile Fever). Vector species reside in locations where associated disease organisms do not exist and the threat of infestation of the vector by the disease organism is always a possibility. Chances for dispersal of diseases and vectors have been increased greatly by international air travel.

Economic losses – mechanical. Host species can be killed by suffocation when certain insects, notably mosquitoes, black flies, and chironomid midges, eclose in large numbers. If insect numbers are not large enough to cause hosts to suffocate, the hematophagous pests such as mosquitoes and black flies can debilitate or kill hosts by exsanguination and/or by the amount of toxic saliva that is injected during feeding attempts. Ticks in large numbers can cause economic losses through disfiguration, blindness, and death, in addition to the possibility of transmitting diseases. Millions of dollars are spent for tick eradication programs in the US and other countries to protect the livestock industries.

Diseases transmitted by selected vectors and the hosts that are affected.

Diseases	Vectors	Hosts
West Nile Fever	Mosquito: *Culex* spp., other species	Birds, Poultry, Horses, Humans
St. Louis Encephalitis	Mosquito: *Culex* spp.,	Birds, Humans
Equine Encephalitis	Mosquito: *Culiseta melanura*, *Aedes* spp., *Culex* spp.	Birds, Horses
Heartworm Disease (Dirofilariasis)	Mosquito: numerous culicine and anopheline species	Dogs, Cats
Bluetongue	Biting midges: *Culicoides variipennis*, *C. pallidipennis*	Sheep, Cattle
Sand Fly Fever (Leishmaniasis)	Sand fly: *Phlebotomus* spp. and *Lutzomyia* spp.	Dogs, Humans
Mastitis	Flies: *Haematobia irritans*	Cattle
Keratoconjunctivitis	Flies: *Musca autumnalis*, *M. domestica*, *Hippelates* spp.	Cattle
Nagana (Trypanosomiasis)	Tsetse fly: *Glossina morsitans* group	Domestic animals other than poultry
Plague	Flea: *Xenopsylla cheopis*, various other flea genera	Rodents, e.g., *Rattus rattus*, dogs, cats, humans
Babesiosis	Ticks: *Boophilus* spp., *Haemaphysalis punctata*, *Rhipicephalus* spp.	Cattle, Sheep, Goats
	Demacentor spp., *Anocentor nitens*	Horses
Anaplasmosis	Ticks: *Boophilus* spp., *Rhipicephalus* spp. *Ixodes* spp, *Demacentor* spp., Tabanidae (mechanical transmission by several species)	Cattle
Heartwater Fever	Ticks: *Amblyomma lepidum*	Sheep, Goats, Camels
Sarcoptic Mange	Mites: *Sarcoptes scabiei*	Humans
	S. scabiei varieties	Swine, Horses, Mules, Dogs, Sheep
Demodectic Mange	Mites: *Demodex folliculorum*, *D. brevis*,	Humans Humans
	D. spp.	Dogs, Cattle, Swine, Goats, Horses

Management of pests of veterinary importance.

Management technique	Targeted pests	Situation
Pesticide		
space sprays	flies	barns
surface applications	flies, mites	animal housing
animal applications		
sprays	flies, ticks, lice,	cattle, horses
	cattle grubs,	pastured cattle
	sheep keds,	sheep, goats
	poultry mites,	caged laying hens
	other mites	cattle, sheep
pour-ons	lice, flies, cattle grubs	cattle
wipe-ons	mosquitoes	horses
back rubbers	*Haematobia irritans*	pastured cattle
dusts	*H. irritans*	pastured cattle
ear tags	*H. irritans*	pastured cattle
larvicide	*Musca domestica,*	caged layer houses
	other nuisance Diptera	confined poultry
dips	ticks	pastured cattle
baits	*M. domestica*	animal housing
Biological control		
Parasitic wasps		
Muscidifurax raptor	*M. domestica*	caged laying hens, indoor confinement dairy
Spalangia endius	*M. domestica*	caged laying hens
Spalangia spp.	*M. domestica,*	caged laying hens
	Stomoxys calcitrans	cattle feedlots
Predaceous flies		
Hydrotaea (Ophyra)		
aenescens	*M. domestica,*	caged laying hens
	other nuisance Diptera	confined poultry
Predaceous beetles		
Carcinops pumilio	*M. domestica*	caged laying hens
Predaceous mites		
Macrocheles		
muscadomesticae	*M. domestica*	caged laying hens
Bacteria		
Bacillus sphaericus	Mosquitoes	larval habitats
B. thurigiensis israelensis	Mosquitoes	larval habitats
Fungi		
Beauveria bassiana	*M. domestica,*	larval habitats
Entomophthora muscae	*M. domestica*	animal housing
Mechanical control		
trapping	*M. domestica,*	confined livestock
	S. calcitrans,	and poultry
	H. irritans	dairy farms
sanitation	*M. domestica,*	dairy, poultry and swine farms
	S. calcitrans	dairy farms and cattle feedlots
manure management	*M. domestica*	dairy, poultry and swine farms, and cattle feedlots.
habitat management	ticks	range cattle

Fig. 1114 Some arthropods of veterinary importance: top left, saltmarsh mosquito; top right, horse fly; second row left, deer fly; second row right, house fly; third row left, stable fly; third row right; biting midge; bottom left, flea, bottom right, tick (photos by J.L. Castner; biting midge by Dick Axtell).

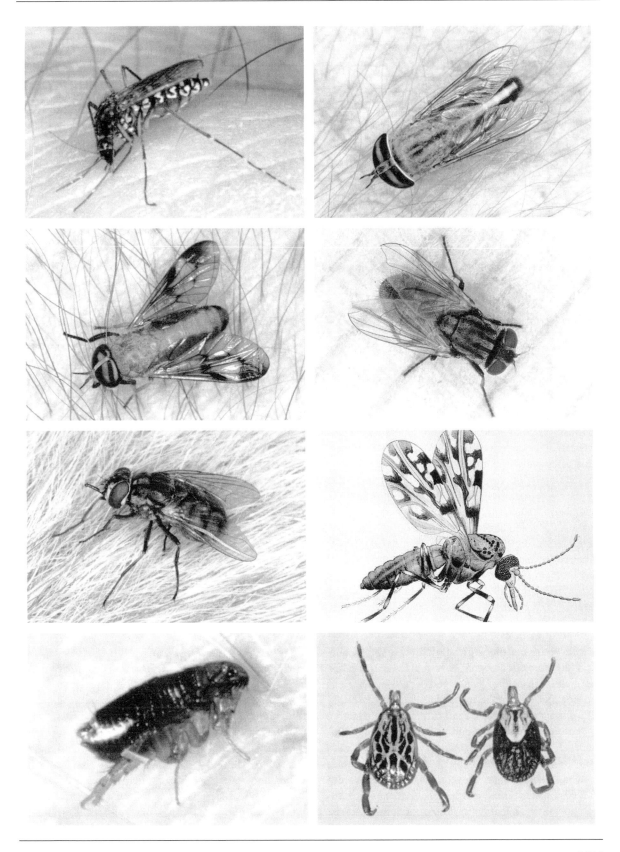

Relationship between pest and host

The relationship between the pest and the host can vary. Some pests remain closely associated with the host and rarely if ever leave the host. The horn fly, *Haematobia irritans*, is termed a continuous pest because after finding a host, it remains almost always on the host. The relationship with *H. irritans* is so close that adults will die if separated from the host for just a few hours. Other pests associate with the host only when feeding. The stable fly, *Stomoxys calcitrans*, is termed an intermittent pest because it takes a blood meal from the host then leaves the host until another blood meal is required. Stable fly mates, oviposits and performs daily activities other than feeding without any association with a host.

The face fly, *Musca autumnalis*, and the horn fly are so closely related with their host animal (i.e., cattle) that the immature stages of these flies can develop only in cattle manure. Horn flies are sometimes observed residing and feeding on hosts other than cattle, namely horses. However, these flies cannot lay eggs in horse manure and will therefore not produce more horn flies unless cattle are nearby.

Feeding sites

Most pests of veterinary importance have preferred feeding sites on a host. With this approach, pests effectively divide a host and minimize feeding interference. On cattle, ticks such as *Amblyoma americanum* localize in the ears and around the escutcheon beneath the tail. Horn flies feed mainly on the back or the undersides of the body. Stable flies feed mainly on the lower legs. Face flies, *Musca autumnalis*, feed around the head and eyes. Mosquitoes feed on the sides of the body. Lice are found on the neck and tail. Cattle grubs are found beneath the skin on the back.

Method of attack

The method of attack or infestation can be unique, as is the corresponding host response. Some pests, like ticks, are able to crawl undetected up the legs and body of an animal and attach themselves in the ears. Other pests, like the adult cattle grub, are unable to cause the host direct bodily harm, yet their presence as they attempt to oviposit on the host may send the host galloping away in a panic. Mosquitoes inject an anesthetic as they probe the host. Painful effects of their feeding are thus reduced, but the subsequent allergic reaction at the bite site may persist for days. Stable flies inject no anesthetic when feeding and their bites are extremely painful. However, the pain usually ceases immediately when the fly stops feeding. Pests like the horse flies (family Tabanidae) inject not only an anesthetic, but also ample amounts of anticoagulant to keep the blood flowing. This often results in a host with dripping wounds long after the horse flies have finished feeding.

Myiasis

The infestation of living or dead tissues by fly larvae is called myiasis, and this type of development is found not only in the bots and grubs, but in other flies as well. The primary screwworm fly, *Cochliomyia hominivorax*, oviposits exclusively in open wounds on living animals. Larvae emerge to feed on the living tissues and cause debilitating to life-threatening effects on the host. By contrast, the secondary screwworm fly, *Cochliomyia macellaria*, generally oviposits on hosts that are dead. Some flies oviposit on necrotic tissues in living hosts and feed only on the dead tissues. These flies secrete an antibiotic, allantoin, and were used to clean battlefield wounds before the advent of pharmaceutical antibiotics in the 1940s.

Although most pests of veterinary importance are considered to be external parasites, several fall into the category of internal parasites. These include the bot flies of the genus *Gasterophilus*. Larvae are ingested orally or nasally and migrate to the intestinal tract, where they spend most of their lives. The larvae of cattle grubs, *Hypoderma* spp., migrate through the tissues of cattle and eventually lodge in the skin of the back with the posterior spiracles exposed to the air. Besides being painful to the host, this pest ruins the value of the hide. The human bot fly, *Dermatobia hominis*, actually captures hematophagous flies on their way to feed on a host, oviposits on them, and they in turn passively and inadvertently infest the host with *Dermatobia* larvae.

Management of pests of veterinary importance

Management of the number and variety of pests has at times been difficult and it is still practically impossible to keep many of the economic pests at tolerable thresholds. During the latter part of the 20th century, the most common form of management involved the use of pesticides. In the late 1940s and into the 1950s, DDT was relied upon for control of veterinary pests. DDT was followed by a succession of other pesticides in various pesticide groups as pests became resistant to one chemical after another. Chemical pesticides are still the basis of control for many pests (for example, ticks on cattle), but this approach becomes more difficult as resistance continues to intensify. The development of new pesticides has been stifled by the high cost of registration and the potential for a short effective period in the field because of resistance and cross resistance. The most common methods and devices for application of pesticides include sprays, dusts, dips, pesticide-impregnated cattle ear tags, back rubbers, wipe-ons, pour-ons, and baits.

Biological control, particularly for flies, has been investigated by many scientists and used most successfully in confinement poultry and swine operations. Many studies have been done with confined cattle, but results are difficult to evaluate because of the extreme variations in field conditions. The agents most commonly used are small parasitic wasps, such as *Muscidifurax raptor* and *Spalangia endius*, on poultry, dairy, and confined beef operations, and the facultative predator, *Hydrotaea aenescens*, on poultry and swine farms. The bacteria *Bacillus thurengiensis* var *israelensis* (Bti) and *B. sphaericus* have been used successfully for mosquito management and the fungi *Entomophthora muscae* and *Beauveria bassiana* have been tested against house flies and a variety of pests, respectively, with varying degrees of success.

Growth regulators have also gained wide acceptance and are classified as biological control by some workers in the field. Growth regulators are used mainly for control of mosquitoes and flies in the immature stages, and their activity is based on interference with the chemicals involved with the molting precess. Methoprene is widely used for mosquito control, and cyromazine is added to the feed of poultry and used successfully for controlling house flies and other nuisance Diptera. Diflubenzuron has also been tested for a number of uses in fly control.

Mechanical control, consisting mostly of sanitation and trapping, has been used, particularly for flies. Source reduction (i.e., removal of hay, manures, and other breeding sites) can be very successful for fly control on farms. Trapping programs have been advocated, but have not been widely accepted because of related costs.

Integrated pest management (IPM) has been discussed and used for many years and successful programs have been developed. Because of the variety in farm design, many IPM programs must be custom designed for each farm. The exceptions may be confined swine and poultry where the farms are quite similar because of standardized housing designs. IPM is usually based on mechanical and biological control methods, and the focused use of selected pesticides as needed.

Management can be very difficult because of the wide-spread movement of pests, such as flies. Also, the use by ticks, for example, of alternate hosts such as deer and other feral animals complicates management. Thus, the success of pest management on individual farms is not easy to assess. Area-wide treatment of animals has been tried with limited success because the ability of the pests to disperse has been under-estimated. However, this may be the method of the future if wide-spread pest outbreaks are to be contained.

See also, MYIASIS, BLUETONGUE DISEASE, AFRICAN HORSE SICKNESS, CAT FLEA, MITES, TICKS, DIROFILARIASIS, MOSQUITOES, TICK PARALYSIS, TSETSE FLIES, HORSE FLIES, DEER FLIES.

Jerome A. Hogsette
U.S. Department of Agriculture, ARS
Gainesville, Florida, USA

References

Farkas, R., and J. A. Hogsette. 2000. Control possibilities of filth-breeding flies in livestock and poultry production. pp. 889–904 in L. Papp and B. Darvas (eds.), *Manual of Palearctic Diptera*, Vol. 1. General and applied dipterology. Science Herald, Budapest, Hungary.

Greenberg, B. 1971. *Flies and disease*, Vol. I. Princeton University Press, Princeton, New Jersey.

Greenberg, B. 1973. *Flies and disease*, Vol. II. Princeton University Press, Princeton, New Jersey.

Harwood, R. F., and M. T. James. 1979. *Entomology in human and animal health* (7th ed.). Macmillan Publishing Co., Inc., New York, New York.

Hogsette, J. A., and R. Farkas. 2000. Secretophagous and haematophagous higher Diptera. pp. 769–792 in L. Papp and B. Darvas (eds.), Manual of Palearctic Diptera, Vol. 1. General and applied dipterology. Science Herald, Budapest, Hungary.

Kettle, D. S. 1995. *Medical and veterinary entomology (2nd ed.)*. CAB International, Wallingford, United Kingdom.

Wall, R., and D. Shearer. 1997. *Veterinary entomology*. Chapman and Hall, London, United Kingdom.

VIBRATIONAL COMMUNICATION. Insects are small life forms in large macrocosms, and therefore require specific methods of communication to overcome the relatively large distances separating individuals from mates, food resources and optimal habitats. This small size poses the inherent problem of separation of the sexes by relatively huge distances. A voluminous literature exists on how various groups have evolved tactics and systems, largely acoustical, chemical or visual, to insure effective mate finding. The large diversity of insects suggests that other exciting and dynamic systems of communication may await discovery.

Intersexual communication using low-frequency, substrate-borne vibrations is a mode of communication that has long been recognized, but little explored in arthropods such as scorpions, spiders and insects. Only in the past few decades has there been much effort to differentiate this mode from air-borne sound communication, determine how widespread it is in insects and to explore its quantitative aspects and evolution in particular groups. Termites, ants and several families of land bugs use vibrations to signal alarm; hornet workers use them to stimulate oviposition by the queen; most of the other groups use vibrational signals for locating mates or in conjunction with copulation. Some bush and tree crickets and land bugs use a combination of vibrational signals and stridulation-produced acoustical songs for intersexual communication.

Substrate-borne vibrations are produced by insects in various ways including percussion (drumming, tokking), tremulation (body pushups or jerking), tymbal clicking, stridulation or combinations of these methods. Percussion involves striking some part of the body, usually the abdomen, directly against the substrate; this has traditionally been termed drumming or tokking (Afrikaans word meaning "to knock"). The contact may be quick, producing a single, low-frequency vibration wave in the substrate,

Insect orders known to use vibrational communication. A more detailed analysis of taxa, sexes or castes producing signals, signal functions and key references for each order were presented by Stewart (1997).

Order	Primary method(s) of vibration production	Major function(s)
Orthoptera; Longhorned Grasshoppers, Katydids, Crickets	Percussion, Stridulation	mate finding
Blattodea; Cockroaches	Percussion, Stridulation	mate finding, mating inducement
Isoptera; termites	Percussion	colony alarm
Plecoptera; Stoneflies (representatives of all nine Arctoperlarian families)	Percussion, Stridulation, Tremulation	mate finding
Psocoptera; Booklice	Percussion	mate finding
Hemiptera/Heteroptera; True Bugs (representatives of 16 families)	Stridulation, Tremulation Tymbal Clicking	mate finding
Hemiptera; Leafhoppers, Planthoppers (representatives of 3 families)	Tymbal Clicking	mate finding
Neuroptera; Green Lacewings, Alderflies, Spongillaflies (representatives of 3 families)	Tremulation	mate finding, mating inducement
Coleoptera; Death Watch Beetles, Darkling Beetles	Percussion	mate finding
Diptera; Chloropid Flies	Tremulation	mate finding
Trichoptera; Caddisflies (representatives of 7 families)	Percussion, Tremulation	mate finding
Hymenoptera; Ants, Wasps	Percussion	colony alarm, queen Stimulation

or prolonged as a scrape or rub, producing a series of waves whose frequency and time intervals depend on the texture of the body part and substrate making contact. Such a rub is actually a body-substrate stridulation. Tremulation consists of some form of body movement (jerking, pushups or wagging) not involving direct contact with the substrate. Movement energy is transferred through the planted tarsi and converted to low-frequency, transverse waves in the substrate. Some genera of longhorned grasshoppers, katydids, crickets, stoneflies, female planthoppers, alderflies, and chloropid flies utilize tremulation in intersexual communication. An advantage of this over percussion is that it does not produce associated sounds in resonant substrates that might be audible and attractive to potential predators. Vibrations produced by stridulation and tymbal activity in some Hemiptera are of great importance for intersexual signaling, and their use by singing Orthoptera (crickets, grasshoppers, katydids) is widespread.

Vibrational communication is a viable, alternative evolutionary strategy for small insects that are limited in body size to house elaborate song-producing structures and the frequencies that can be effectively air-transmitted by them. The biophysical aspects of vibrations in solid substrates are complex, but the vibrations offer a good directional communication medium enabling location of a signal emitter by a receiver, and for recognition of various parameters of the emitters signal. In ways little researched and understood, one individual may be able to measure the vibrational output of another through some selectively arrived-at neuronal capacity encoded to respond only to a potential conspecific mate or sibling. Scorpions, and presumably insects, can determine direction of a signal emitter by integrating time delays as small as 0.2 milliseconds received by sense organs of the different legs. The planted tarsi represent positions like points on a compass, and the source of a directional vibrational signal is read from the time delays of the signal in differentially reaching these positions. Distance to an emitter may be determined by intensity of wave signals received above some threshold.

A disadvantage for insects that have adopted vibrational communication is the complication of wave transmission in the heterogeneous substrates present in their habitats. The variety of waveforms produced (longitudinal, transverse, bending, surface) are distorted and dampened by natural substrates,

restricting the range of communication, but this apparently has been overcome because transmission through plants, dead plant items and connected leaf mats has been demonstrated for distances up to 8 meters, that may translate to 800–1,000 times the length of the insects' bodies. Such substrates are, therefore, effective transmission mediums for vibrational signals.

Little is known of the energy cost of vibrational signaling, as an important component of reproductive effort, but in general it is considered less than for producing high-frequency air-borne sounds. The cost of percussion-assisted mate-finding in some beetles is 10% or less than that of random searching over the communicable area. Such cost-saving must represent an advantage for vibrational communication and probably plays a major role in the early stages of behavioral evolution. The vibrational signals of most insects consist of simple volleys of evenly spaced signals. Out-group comparisons have indicated that these types of signals produced by percussion are ancestral, and that derived signaling to achieve species-specificity, and to possibly enable some degree of sexual selection for fitness, has involved evolution toward increased signal complexity. Complex signals have been arrived at through: (1) more sophisticated signaling methods, (2) changes in the rhythm of signaling, or (3) possibly increased selection of the type of natural substrates used for signal transmission. Modern species have either retained the relatively simple ancestral pattern, with slight time interval modification for species specificity, or have developed various complex systems using combinations of these three derived behaviors. The high degree of species-specificity of vibrational signals in insects make them valuable behavioral lines of evidence for delineating morphologically similar species and resolving phylogenetic relationships (see also DRUMMING COMMUNICATION OF STONEFLIES).

Kenneth W. Stewart
University of North Texas
Denton, Texas, USA

References

Gogala, M. 1985. Vibrational communication in insects (biophysical and behavioral aspects). pp. 117–126 in K. Kalmring and N. Elsner (eds.), *Acoustical and vibrational communication in insects*. Proceedings, XVII International Congress of Entomology. Verlag Paul Parey, Berlin, Germany.

Heady, S. E., L. R. Nault, G. F. Shambaugh, and L. Fairchild. 1986. Acoustic and mating behavior of *Dalbulus* leafhoppers (Homoptera: Cicadelidae). *Annals of the Entomological Society of America* 79: 727–736.

Henry, C. S. 1980. The importance of low-frequency, substrate-borne sounds in lacewing communication (Neuroptera: Chrysopidae). *Annals of the Entomological Society of America* 73: 617–621.

Lighton, J. R. B. 1987. Cost of tokking: the energetics of substrate communications in the toktok beetle, *Psammododes striatus*. *Journal of Comparative Physiology* B 157: 11–20.

Maketon, M., and K.W. Stewart. 1988. Patterns and evolution of drumming behavior in the stonefly families Perlidae and Peltoperlidae. *Aquatic Insects* 10: 77–98.

Stewart, K.W., and M. Maketon. 1991. Structures used by Nearctic stoneflies (Plecoptera) for drumming and their relationship to behavioral pattern diversity. *Aquatic Insects* 13: 33–53.

Stewart K.W. 1997. Vibrational communication in insects, epitome in the language of stoneflies? *American Entomologist* 43: 81–91.

VIBURNUM LEAF BEETLE, *PYRRHALTA VIBURNI* (PAYKULL) (COLEOPTERA: CHRYSOMELIDAE).

Pyrrhalta viburni (Paykull), commonly known as the viburnum leaf beetle, is a landscape pest of Eurasian origin that is quickly spreading through the northeastern United States. Breeding populations actually became established in North America (near Ottawa, Ontario, Canada) in the late 1970s, but did not reach the U.S. until 1994, when it was discovered in Maine. The pest was found in New York State in 1966, apparently the result of a separate immigration event, because the two founding locations are very widely separated. As of 2001, the beetle could be found throughout the southern half of Maine, in many of the northern and western counties of New York, as well as the northwestern tip of Pennsylvania and the northwestern corner of Vermont. Given its distribution across continental Europe, it seems likely that this pest will become commonplace throughout much of the range of *Viburnum*.

The only known hosts for *P. viburni* are in the plant genus *Viburnum*. Within the genus, however, species differ greatly in susceptibility. Some species, most notably arrowwood (*V. recognitum*) and cranberrybush (*V. trilobum* and *V. opulus*) viburnums, are devoured by the pest and are killed after only a few years of repeated infestation, while others,

especially leatherleaf (*V. rhytidiophylloides*) and Koreanspice (*V. carlesii*) viburnums and their relatives, are virtually immune. Arrowwood is particularly important in the spread of *P. viburni* because it is widely distributed in native habitats, especially along large bodies of fresh water. The spread of the pest has been most rapid in such habitats. Spread of the pest may also be hastened by transport of infested nursery stock since *P. viburni* is not subject to quarantine restrictions.

The insect is univoltine, and overwinters in the egg stage. Eggs hatch in early May in upstate New York, around the time of bud break of the host plant. Larvae crawl to the tips of shoots, and begin feeding on the underside of newly expanding leaves. The larvae, which are at first pale yellow, go through three instars, changing to solid black between instars. As the second and third instars grow, the black sclerites form spots and stripes as the intersegmental membranes expand. The larvae complete their development in 4 to 5 weeks, then crawl to the soil and burrow down several centimeters to pupate. Adults emerge several weeks later.

Adults, which resemble smaller, more drab versions of elm leaf beetle, feed on the leaves of the same host species as the larvae. Feeding damage by adults is distinctly different from that of larvae, however. Whereas larvae feed on leaf tissue between veins, leaving skeletonized leaves when populations are heavy, adults create oblong cutouts in the leaf, measuring about 2 mm wide by several mm long. Adults are active until late September or early October, all the while feeding and laying eggs on host plants. Oviposition sites of the beetle are quite distinctive; eggs are laid in masses on the underside of young shoots, generally in a linear array of several egg clusters per shoot. The female first chews a small

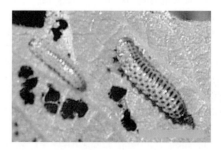

Fig. 1115 Larvae of *Pyrrhalta virburni*. Second instar is shown on left, third instar on right. (Photo by Paul A. Weston.)

Fig. 1116 Adult of *Pyrrhalta virburni* on twig. The adults range in size from 4.5 mm (male) to 6.5 mm (female), and are medium brown in color. (Photo by Paul A. Weston.)

hole into the pith of the stem, and then deposits about half a dozen eggs in the hole. She then caps the eggs with a mixture of feces and chewed foliage or bark. The caps protrude noticeably from the surface of the twig, and probably help to retain moisture and protect the eggs from natural enemies. The eggs remain dormant until the following spring.

Several species of parasitoids are known to use *P. viburni* as a host in Europe, but none of these natural enemies occurs in the U.S. Generalist predators, such as lady beetles and lacewing larvae, also feed on larvae, and at least one species of

Fig. 1117 Feeding damage by adult *Pyrrhalta virburni*. Note that feeding occurs in discrete cutouts with rounded ends, randomly distributed across the leaf surface. (Photo by Paul A. Weston.)

nematode (*Heterorhabditis bacteriophora*) effectively utilizes *P. viburni* as a host.

Paul A. Weston
Cornell University
Ithaca, New York, USA

References
Balachowsky, A. S. 1963. *Entomologie appliquée à l'agriculture. Traité. Tome I, Coléoptères* (second volume). Masson, Paris, France. 1391 pp.
Becker, E. C. 1979. *Pyrrhalta viburni* (Coleoptera: Chrysomelidae), a Eurasian pest of viburnum recently established in Canada. *Canadian Entomologist* 111: 417–419.
Weston, P. A., B. C. Eshenaur, and R. E. McNiel. 2000. Viburnum resistance. *American Nurseryman* 192: 51–53.

VIERECK, HENRY LORENZ. Henry Viereck was born at Philadelphia, Pennsylvania, USA, on March 28, 1881. In 1900 he began studying Hymenoptera at the Academy of Natural Sciences in Philadelphia. After a brief attempt at studying medicine he returned to entomology and worked at the Connecticut Agricultural Experiment Station, the Pennsylvania Department of Agriculture, Bureau of Entomology, the California State Horticultural Commission, United States Department of Agriculture Biological Survey, and the Canada Department of Agriculture. He published 92 papers on Hymenoptera, principally on Ichneumonidae and Andrenidae. He died on October 8, 1931, when he was struck by a hit-and-run motorist while collecting along a road near Loudenville, Ohio.

Reference
Mallis, A. 1971. *American entomologists*. Rutgers University Press, New Brunswick, New Jersey. 549 pp.

VINEGAR FLIES. Members of the family Drosophilidae (order Diptera). See also, FLIES.

VINE MEALYBUG, *Planococcus ficus* SIGNORET (HEMIPTERA: PSEUDOCOCCIDAE).
Vine mealybug, *Planococcus ficus* Signoret, is a recently introduced pest in California vineyards. Vine mealybug was first recorded from the Crimea region in the 1890s; currently, it is found in the Middle East, North African and Mediterranean countries, Pakistan,

Argentina and California in the United States. While vine mealybug infestations are most damaging in vineyards, the mealybug is also an occasional pest of apple, avocado, banana, date palm, fig, mango and citrus. In the laboratory, vine mealybug can be reared on a number of hosts, including butter-nut squash (*Cucurbita moschata*) and potatoes (*Solanum tuberosum*). The vine mealybug is one of several mealybugs species that attack vineyards, e.g., *Pseudococcus maritimus* Erhorn (grape mealybug), *P. viburni* (Signoret) (obscure mealybug), *P. longispinus* (Targioni-Tozzeti) (long-tailed mealybug), and occasionally by *Planococcus citri* Risso (citrus mealybug) and *Maconellicoccus hirsutus* Green (pink hibiscus mealybug). Of these mealybugs, the vine mealybug is most closely related to the citrus mealybug. While they look very similar and can coexist in the same regions, citrus and vine mealybug populations can be separated by their sex pheromones.

Mealybugs are related to aphids and whiteflies and feed in a similar manner with long piercing-and-sucking mouthparts to remove plant juice. The vine mealybug can feed on the vine roots, trunk, canes, leaves and fruit. An adult female vine mealybug is 3 to 4 mm long, wingless, oval-shaped body, and covered by white wax. The secreted wax forms long filaments protruding along the body margin, which can be used to identify some mealybug species. The adult female lays from 300 to 500 eggs in a cotton-like (formed by secreted wax) ovisac. Under a controlled environment at 28°C, eggs hatch in 7 to 9 days. The emerging crawlers are bright-orange to yellow-orange colored. These crawlers, or first instars, are very mobile and move around the vine until they find a suitable feeding site. Once the crawler settles and begins feeding, the bright-orange body is covered by the waxy secretions and the mealybug takes on a more subdued pale white color. Females have four instars and, in the field, they can be distinguished only by their relative sizes. Each stage lasts 4 to 7 days, depending on temperature. Adult male mealybugs are very different in appearance. Males are about 2 mm long and have a pair of functional wings, and long caudal filaments. Males look similar to females in the egg and crawler stages, whereas the second and third instars are bit more narrow and oblong than females. Males pupate after the third instar, whereas females become reproductive in the apparent fourth instar. In California's San Joaquin Valley, there are 6 generations of vine -mealybugs per year.

Within-vine distribution

Vine mealybugs are found on all parts of the vine, including the roots. In spring, as soon as the leaf buds form and break, vine mealybug crawlers start to move up towards crown and the canes. By mid-summer, all stages of the vine mealybug can be found throughout the vine including the leaves. Starting in mid-summer and continuing through the harvest time, mealybugs begin to infest the grape bunches. (Infestation of the grape bunch begins earlier in the season as you move further south and into the desert agricultural regions in the Coachella Valley.) In heavy infestations, i.e., when the mealybug density is high, the pest can foul the grape bunches with honeydew, ovisacs, and live and dead bugs, making the fruits unmarketable. Excessive honeydew secretion also promotes the growth of sooty mold, further reducing marketability. After harvest and leaf drop, the mealybugs are found either on the trunk, typically deep under the bark or in feeding holes made by moth or beetle wood borers, or on the roots; there they remain at low densities until the following spring when the cycle is repeated.

Biological controls

There are many species of ladybird beetles that feed on mealybugs, the most well-know is the mealybug destroyer, *Cryptolaemus montrousieri* Mulsant (Coleoptera: Coccinellidae). Both larvae and adults of the mealybug destroyer can feed on all stages of the vine mealybug; however, the adult beetle prefers to feed on eggs in the mealybug ovisac to increase reproduction and egg deposition. The mealybug destroyer was first introduced into California from Australia in 1892. The beetles larvae also produce waxy filaments that cover the body and disguise the beetle as a mealybug destroyer (that can help the beetle evade ants which protect the mealybugs). The beetle larva can be distinguished by its distinct body segments and non-uniform length waxy filaments. Because the beetle prefers cool and relatively humid conditions, its numbers never increase to high levels in the interior parts of California where there is dry and warm climate. Lacewings also feed on mealybugs. There are brown and green lacewings resident in California vineyards. The green lacewings (*Chrysoperla* and *Chrysopa* species) are most common. Like the beetles, the lacewings are more effective when they attack the smaller mealybug stages. Both the mealybug destroyer and some green lacewing species can be purchased from commercial insectaries and released to suppress mealybugs.

Worldwide, there are several species of small (less than 2 mm) parasitic wasps (Hymenoptera: Encyrtidae) that have been found attacking the vine mealybug. These include *Anagyrus pseudococci* (Girault), *Allotropa* sp., *Leptomastidea abnormis* (Girault), *Leptomastix flavus* Mercet, *L. dactylopii* Howard, and *Coccidoxenoides peregrinus* (Timberlake). In the mid 1990's *A. pseudococci* and *L. abnormis* were released in California's Coachella Valley to suppress vine mealybug population growth and spread. *A. pseudococci* and *L. abnormis* have been recovered; however, parasitism levels rarely exceed 25%. Further north, in the San Joaquin Valley, we found that, by late-summer, resident *A. pseudococci* parasitized 80–95% of mealybugs that were exposed on leaves, bunches, canes, and trunk sections – although we have not seen any parasitized mealybugs on the roots or under refugia deep under the bark. *Allotropa* sp. contributed 1% to the total parasitism levels. *Anagyrus pseudococci* is a small wasp about 1.5 mm long. The female is brown in color and has white antennae with black at the base; the male is black, 0.7 to 0.9 mm long, and has hirsute, black antennae. *Anagrus pseudococci* can oviposit in 2nd instar to adult stage mealybugs; with male offspring typically emerging from the smaller mealybugs (2nd–3rd instar) and female parasitoids emerging from the larger mealybugs. Parasitoid development time is about 18 days during the summer. *Leptomastidea abnormis* as another biological control is under investigation. It completes its life cycle in 20 to 22 days under laboratory conditions. *Leptomastidea abnormis* is about half the size of *A. pseudococci*, with two black bands on clear wings.

Both *A. pseudococci* and *L. abnormis* are commercially available. Researchers from University of California Berkeley and Riverside are currently investigating augmentation programs in the Coachella Valley and San Joaquin Valley. Augmentation may be needed because, although the percentage vine mealybug parasitism from naturally occurring parasitoids can be high, the parasitoids appear in the field, late in the season to effectively control the mealybug before the population moves up the vine and infests the grape bunches. Research is currently underway to investigate area-wide release programs early in the season (spring) and targeting the over-wintered mealybug population.

Insecticide controls

To control vine mealybugs, delayed-dormant insecticides (e.g., chloropyrifos [Lorsban®]) can be used to kill over-wintered mealybugs; however, control is incomplete as some over-wintered mealybugs may remain in protected regions of the vine. During the season, a number of foliar materials are available (e.g., methomyl [Lannate], imidacloprid [Provado®]) but control is still dependent on complete coverage. For this reason, one of the best insecticidal controls is a systemic application of imidacloprid [Admire®]. Effectiveness of this systemic material is dependent on good delivery to the entire root zone, which can be influenced by the irrigation system used or soil type and condition.

Cultural controls

Sanitation is also a tool to prevent spread of the mealybugs. Weed control during the winter and early spring time may help prevent a build up of the mealybug population. Vineyard equipment, such as a grape harvester, used in an infested field should be washed thoroughly with water and bleach before moving into another field. Similarly, it is good idea for field crews to clean and change their clothing and shoes after visiting an infested field to remove any mealybugs that may cling onto their skin or clothes.

Sampling methods

It is easy to detect the vine mealybugs during harvest in a heavily infested vineyard; the presence of shiny, honeydew and dark sooty mold on the vine and the leaves close to the trunk is an indicator of the presence of vine mealybugs on that vine in that year. For quick monitoring purposes, it is advisable to look for the sooty mold on the leaves just before the leaf fall and rain (October–November), and make plans for treatments accordingly. Currently, researchers are developing a vine mealybug sex pheromone that may greatly improve sampling for low-density mealybug populations. This is critically important because chemical treatment is recommended if only a few vines are infested in order to catch new infestations.

Other mealybug pest species

The other mealybugs that are commonly found in the vineyards are grape mealybug (*Pseudococcus maritimus*), long-tailed mealybug (*P. longispinus*) and obscure mealybug (*P. viburni*). Long-tailed and obscure mealybugs are found only in the coastal grape growing regions of California. The grape mealybug is commonly found both inland and the Pacific coast of California as well as other grape growing regions like New York, Oregon and Washington states. Pears and apples are other host plants of grape mealybugs. These species can be distinguished by their two caudal filaments that are longer than the rest of the filaments and form a V-shaped tail. The grape and obscure mealybugs look very similar; however, they can be differentiated by the color of the ostiolar fluids, which are secretions from lateral ostiolar pores. The long-tailed mealybugs, as their name suggest, they have longer caudal filaments as compared with the other mealybug species.

The grape and obscure mealybugs generally have two to three generations per year, depending on temperature. The eggs and crawlers from 2nd summer generation (August/September) overwinter. None of these mealybug species produce as much honeydew (or economic damage) as the vine mealybugs. Both obscure and long-tailed mealybugs produce very little honeydew compared to both vine and grape mealybugs.

Raksha Malakar-Kuenen and Kent M. Daane
University of California
Berkeley, California, USA

References

Cox, J. 1989. The mealybug genus *Planococcus* (Homoptera: Pseudococcidae). *Bulletin of British Museum of Natural History* 58: 1–78.

Malakar-Kuenen, R., K. Daane, W. Bentley, G. Yokota, L. Martin, K. Godfrey, and J. Ball. 2001. Population dynamics of vine mealybug and its natural enemies in the Coachella and San Joaquin Valleys. University of California Plant Protection Quarterly 11: 1–4. (online: http://www.uckac.edu/ppq).

McKenzie, H. L. 1967. *Mealybugs of California*. University of California Press. Berkeley and Los Angeles, California.

VIRAL FLACHERIE. An infectious flacherie of silkworm larvae caused by a small nonoccluded virus.

See also, FLACHERIE.

VIRGINOPARA. In aphids, the viviparous female occurring during the summer months (after the fundatrix generation). See also, APHIDS.

VIREMIA. The presence of virus in the hemolymph or blood.

VIRION. The mature virus, the ultimate phase of viral development. The virion is either a naked or an enveloped nucleocapsid. The term 'virus' embraces all phases of the viral development, and it includes the virion.

VIRULENCE. The quality or property of being virulent, or the ability of a pathogen to infect a host and cause disease; the quality of being poisonous; the disease-producing power of a microorganism.

VISUAL ATTRACTANTS AND REPELLENTS IN IPM. In essence, Integrated Pest Management (IPM) is a decision-based process involving the coordinated use of multiple tactics for optimizing the control of all classes of pests in an ecologically and economically sound manner. Tactics available for use in the management of pest insects include insecticidal control and behavioral control. For insecticidal control, monitoring population levels of pest insects and their natural enemies is a key component in deciding whether or not insecticidal treatment is necessary. Visual traps that are attractive to pests or beneficial insects can be useful monitoring tools. For behavioral control, visually attractive traps, or visual repellents may be used in direct suppression of insect pests.

Development and effective deployment of visual attractants and repellents in IPM call for an understanding of several variables that influence attractancy and repellency. These include properties of insect vision, visual properties of resources sought by insects and the role of visual stimuli in the resource selection process.

Compound eyes contain the principal photoreceptors of insects. Photoreceptors of many insects are capable of perceiving a bandwidth of energy extending (in human terms) from ultraviolet (approximately 300 to 400 nanometers) through blue (400 to 500 nm), green (500 to 560 nm), yellow (560 to 590 nm) and orange (590 to 630 nm). Few kinds of insects are able to perceive red (630 nm and above). Insects with large compound eyes may possess a greater degree of visual acuity (and, therefore, are more capable of perceiving the visual properties of hosts against the background) than insects with small compound eyes. For most insects, visual acuity is highest under high intensity natural light (for example, full sunlight at midday) and lowest under low light intensity (for example, at dusk, nighttime and dawn).

There are three properties of resource items for insects that may serve as visual cues to foragers: spectral quality, pattern and dimensions. Resource items may absorb, transmit or reflect incoming light. It is the reflected light that provides the most useful information to foragers. The color of reflected light includes elements of brightness (intensity), hue (dominant wavelength) and saturation (spectral purity). The foliage of most plants reflects light at comparatively low intensity, with maximum reflectance at 550 nm and nearly no ultraviolet reflection. Some flowers may reflect moderate or high levels of ultraviolet light.

Insects may use a variety of stimuli as cues for finding resources. These include olfactory, visual and auditory stimuli. The distance at which each type of stimulus is perceived varies widely among different kinds of insects. Generally, however, olfactory and auditory stimuli are perceived at greater distances than visual stimuli, with the latter often being the principal close-range cue used by the forager. Thus, visual attractants and repellents in IPM usually are not operative at distances exceeding a meter or so (in exceptional cases, a few meters). The greater the degree of hue or intensity contrast between a visual stimulus and the background, the greater the distance at which the insect may be able to detect the stimulus.

Visually attractive traps developed to monitor pest or beneficial insects in IPM usually incorporate characteristics that mimic the color or pattern (shape) of the visual cues used in finding resources. Pests including aphids, whiteflies, leafhoppers, anthomyiid flies and some beetles, as well as beneficials including coccinellids and hymenopterous parasitoids, often are monitored by yellow traps that are believed to mimic the plant foliage used by these insects as a resource cue. Yellow may constitute a ''super-normal'' foliage type stimulus, because yellow emits peak energy at about the same bandwidth of the insect-visible spectrum as green foliage emits peak energy, but at a greater intensity.

Several kinds of pest thrips and some kinds of pest sawflies are monitored by white or blue traps whose reflectance mimics the color of the flowers from which the adults obtain nectar or pollen. Similarly, several kinds of biting flies are monitored by white or blue traps that appear to mimic the reflectance properties of either resting sites or host animals. In most cases, the absence of ultraviolet reflectance from white or blue traps enhances the capture of pest insects, but the converse is true in the case of some anthomyiid flies and biting flies.

The shape of a trap can be an important attribute in the monitoring of certain pests. For example, several kinds of scolytid beetles are monitored using vertical, black, cylindrical traps that mimic the trunks of coniferous host trees. Similarly, some kinds of frugivorous tephritid flies are monitored using fruit-sized spheres that mimic the shape of the host fruit.

When baited with attractive odor, visually attractive traps can be used to achieve direct behavioral control of certain pest insects. The tsetse fly (*Glossina pallidipes*), for example, can be effectively

Fig. 1118 Black cylindrical funnel traps mimic the silhouette of pine trees and are used, in conjunction with odor baits, to capture bark beetles (photograph by J. Foltz).

suppressed in parts of southern Africa by deploying odor-baited, insecticide-impregnated targets of blue cloth at densities of 3 to 5 targets per km². Several species of tabanid flies can be suppressed by deploying large, odor-baited, dark-colored spheres, or other objects that roughly resemble the color and shape of potential hosts. The apple maggot fly, *Rhagoletis pomonella*, can be controlled effectively by ringing apple orchards with red spheres that are odor-baited and are either sticky or are treated with insecticide.

Visual stimuli have proven very effective in repelling aphids, whiteflies and thrips that vector viral pathogens of row plants. Rows of young plants are protected from viral inseminations when they are underlain with ultraviolet-reflecting mulch, such as aluminum foil, or materials sprayed with aluminum paint. Reflected ultraviolet light resembles sunlight and signals immigrating adults to keep on flying rather than settle on foliage. The use of ultraviolet-reflective mulch in protecting row crops against viral infection is expected to increase substantially when effective, biodegradable mulches replace the current non-degradable types.

Ronald J. Prokopy
University of Massachusetts
Amherst, Massachusetts, USA

References

Allan, S. A., J. F. Day, and J. D. Edman. 1987. Visual ecology of biting flies. *Annual Review of Entomology* 32: 297–316.

Kring, J. B., and D. J. Schuster. 1992. Management of insects on pepper and tomato with UV-reflective mulches. *Florida Entomologist* 75: 119–129.

Lindgren, B. S. 1983. A multiple funnel trap for scolytid beetles. *Canadian Entomologist* 115: 299–302.

Muzari, M. O. 1999. Odour-baited targets as invasion barriers for tsetse flies: a field trial in Zimbabwe. *Bulletin of Entomological Research* 89: 73–77.

Prokopy, R. J., and E. D. Owens. 1983. Visual detection of plants by herbivorous insects. *Annual Review of Entomology* 28: 337–364.

Prokopy, R. J., S. C. Wright, J. L. Black, X. P. Hu, and M.R. McGuire. 2000. Attracticidal spheres for controlling apple maggot flies: commercial-orchard trials. *Entomologia Experimentalis et Applicata* 97: 293–299.

Fig. 1119 Aluminum-coated plastic mulch underneath young tomato seedlings repels alighting aphids and whiteflies (photograph by D.J. Schuster).

VISUAL MATING SIGNALS. Evolution has sculpted and colored the surfaces of the arthropod body. Natural selection favored appearances that mislead predators through disguise and camouflage, i.e., 'cryptic coloration,' and those that warn enemies to stay away through bright advertisements of distastefulness or dangerous weapons, i.e., 'aposematic coloration.' In addition, body surfaces may help modify the effects of the physical environment, e.g., to absorb or reflect heat. Certain colors, shapes and movements also influence the behavior of the conspecifics in contexts including social interactions, e.g., honeybee dances. Other signals evolved in the context of mating.

Sexual selection and the evolution of displays

These sex-related signals, or 'displays,' are directed either toward rivals (usually males attempting to repel other males) or potential sexual partners (usually males attempting to attract mate-choosing females). Display surfaces, and the behaviors associated with their exhibition, have evolved through sexual selection, the process in which genes are favored because they give greater access to the opposite sex. Competition among members of one sex for access to the other results in 'intrasexual selection,' and mate choice and the displays directed to the choosing sex evolve through 'intersexual selection.'

Sexually selected displays presumably resolve competitions and/or facilitate mate choices because receivers evaluate and act on information contained in the display's colors, size and motions. The experimental evidence that arthropods perceive visual displays and then modify their reproductive activities is phylogenetically widespread. Some examples include the sexual response of fireflies (Lampyridae) to patterned electric lights, and female wolf spiders' (Lycosidae) reactions to videotapes of the stylized foreleg movements the males perform during courtship.

The basis of sexually selected signals

Ultimately, both inter–and intrasexual selection occur because of differences in 'parental investment' between the sexes. Parental investment is anything, material or behavioral, that a parent invests in an offspring that increases the offspring's chance of surviving at the cost of the parent being able to invest in other offspring. There is a fundamental sexual dimorphism in parental investment due to the female's initial investment in large, resource rich eggs, and the male's in small, inexpensive sperm. All other things being equal, a female's capacity to reproduce is limited by the resources she can obtain

Fig. 1120 There are dangers as a well as opportunities inherent in visual signals. The eye(s) of this spider may have first been caught by the firefly's bioluminescent sexual display (photo by Steven Wing).

and use to produce her eggs. Multiple mates do not necessarily result in more offspring, although she may be able to choose a mate whose genetic or other qualities, sometimes illuminated through displays, result in her bearing more, or more fit, offspring. On the other hand, ejaculates are cheap and males invest little in the insemination of a female. The more females a male copulates with, the more offspring he will father. As a result, selection favors males that successfully compete with one another for access to mates, often through displays directed both to females and to other males.

There are exceptions to this "grand generality." For example, males of some species of katydids (Tettigonidae) and balloon flies (Empididae) provide females with nutritious 'nuptial gifts,' such as huge protein-rich ejaculates and insect cadavers. As a result, a female's reproduction is influenced by her ability to gain access to gift-bearing males. In these instances there may be sex-role reversals and males choose mates from among competing, displaying females.

Sexual dimorphism and sexually selected displays

In some species, members of one sex differ in appearance from members of the opposite sex. Sexual dimorphism in color is suggestive of a display in one of the sexes. Charles Darwin, for example, was struck by the brilliancy of male butterfly coloration and the relative drabness of females. He, and subsequently others, proposed that sexual selection had generated colors of greater saturation and more precise demarcation in males because they were used in sexual communications while females maintained the ancestral, more cryptic, coloration. Some exceptions to the 'rule' that females bear the more ancestral coloration tend to support the idea that sexual selection has been more important in the evolution of male than female coloration. There are instances in butterflies (Papilionidae), e.g., the African swallowtail, *Papilio dardanus*, where relatively palatable females are 'Batesian mimics' of one or more aposematic-unpalatable species, but males maintain a pattern more typical of closely related species. It seems likely that male-signalers have conserved a coloration that is important in a visual display, and that females, either because they are the display-receiving sex or because they are not involved in intrasexual competitions, were free to abandon their ancestral coloration for one that offered more protection from predators.

However, not all sexual dimorphisms in color are the direct result of sexual selection. For example, male and female stick insects (Phasmatodea), such as *Diapheromera covilleae*, sometimes bear special resemblance to different types of foliage on their host plants, and while the two sexes look quite dissimilar, there is little in their mating behavior to suggest a visual display is the cause of the difference. On the other hand, surfaces and colors that are not sexually dimorphic may be used in displays. Movements of strikingly patterned wings are typical of sexual encounters in fruit flies (Tephritidae), but sexual dimorphism in wing coloration and marking is rare.

The advantages and disadvantages of light in communications

In comparison to chemicals and sounds, reflected light perceived as color and pattern has both advantages and disadvantages as a means of communication in insects. Among its advantages are its capacity to carry large amounts of reliable information. A signal is able to transmit a message through modulation, predictable and specific variations, e.g., Morse code, or a particular color pattern on the surface of a wing. Visual cues may be more reliable in this sense than chemicals whose combinations and densities can be blended and obscured by even slight motions of the air. It is difficult to imagine a message consisting of patterned puffs of odor making sense to a distant receiver on a breezy day.

Some modes of communication work better than others in a given habitat and context. Among the potential disadvantages of visual displays are that they can be blocked by opaque objects that chemicals

Fig. 1121 The head of a male antlered fruit fly (*Phytalmia cervicornis*) from New Guinea. The deer-like antlers projecting from the cheeks are as much visual propaganda as weapons (drawing by Kevina Vulinec from Sivinski 1997).

or sounds could go around, and are unlikely to reach as far as pheromones and acoustic signals. Visual signals also lack persistence in the absence of the signaler, e.g., ant recruitment trails. They can even be dangerous to the signaler, as they may be more easily tracked by predators than pheromones or sounds, and even provide an unusual opportunity for predators to lure mate-seekers. For example, fireflies (Lampyridae) can be tricked by 'aggressive mimics' that produce the luminous sexual display-response that attracts males of other firefly species to their final rendezvous.

Arthropod visual signals are potentially detailed and informative, but only at short distances. The effective range of visual signals is generally limited because only a small body-surface area is available for a display and because of the poor resolution of the insect compound eye. Insect eyes can resolve details of little objects such as other insects, but only at extremely close range (''[compound eyes]...would give a picture about as good as if executed in rather coarse wool-work and viewed at a distance of a foot'' [Henry Mallock]). One possible exception to the generally short range of arthropod visual signals is bioluminescence, e.g., the flashes of fireflies. An extreme example is the mass display of male *Pteroptyx* fireflies, which gather by the thousands in certain trees and flash synchronously.

Idiosyncratic details of a particular arthropod's vision also influence the forms of its visual displays. For example, jumping spiders (Salticidae) see colors and are able to detect motion while moving, but wolf spiders (Lycosidae) are color-blind and must be still to perceive movement. Jumping spiders use bright colors and 'dances' in their displays, while wolf spider courtships consist largely of raising and waving darkly pigmented or hairy legs.

UV markings

Human and arthropod vision differ in the range of light wavelengths perceived. Insects are generally insensitive to red, but can see ultraviolet light (UV). Insects that appear drab to us may, in fact, be brightly reflective in the ultraviolet range and able to communicate among themselves in color patterns that are invisible to us and perhaps invisible to some of their vertebrate predators (although diurnal birds typically see UV, as do some mammals). For example, males of certain butterflies (e.g., *Colias* spp. [Pieridae]) bear brilliant UV markings, but the extensive sexual dimorphism in these species was only discovered through specialized photography. As a result, use caution in interpreting a particular arthropod's colors until the entire spectrum is described.

The content of intrasexual displays

As Charles Darwin noted, ''The season of love is that of battle...,'' and conflicts among sexual competitors, usually males, are sometimes resolved by threats communicated through visual displays. The intensity of aggression ('agonistic behaviors'), and the investment in the colors, structures and behaviors that convey the sender's ability to fight, follows a pattern. The greater the control a male can exercise over females or the resources females require, such as oviposition or feeding sites, the more profitable it is for males to invest in weapons and advertisements of prowess. Typically, controllable resources are relatively small and rare and, thus, easily defended. As resources become common, large and spread out, they become less and less defendable and less and less likely to be associated with aggressive displays.

Antlered flies (Diptera) are classic examples of displaying males guarding a discrete resource. Projections from the head, either eye-stalks or horns sprouting from the cheeks, occur in species from nine families of flies. Perhaps the most elaborate are found in the large New Guinean fruit flies (Tephritidae) of the genus *Phytalmia*, one species of which the great 19th century naturalist Alfred Wallace described in the following manner: ''The horns of (*P. megalotis*) are about one third the length of the insect, broad, flat, and of an elongated triangular form. They are of a beautiful pink color, edged with black, and with a pale central stripe. The front of the head is also pink, and the eyes violet pink, with a green stripe across them, giving the insect a very elegant and singular appearance.''

Female *Phytalmia* oviposit in pinhole sized punctures in certain species of freshly fallen trees. These very small and rare resources attract females and are easily defended, and the owner's initial defense is accomplished through visual display. Resident (defending) and intruder males of *P. mouldsi* clash by rising up on their legs and pushing hard against each other's remarkable heads, although the antlers themselves do not play a major role in the battle. However, males whose horns are experimentally

lengthened or shortened are respectively more and less likely to win fights. In addition, those whose antlers are removed altogether are treated by their rivals as females. Thus, antler-weapons serve to a great degree as displays whose size opponents appear to correlate with fighting ability.

Rival males pay attention to each other's visual displays when deciding how they will go about the all-important activities of holding resources and inseminating females. The reason is either the display is an accurate indicator of another's ability to defend its resource, or the opponent accepts that it is. 'Honest advertisement' is an important principle in both intra- and intersexual signals. Over evolutionary time, a signal that exaggerates the capabilities of its sender will eventually be ignored by increasingly discriminating receivers. However, a signal that is expensive to build and display, or risky to use, is likely to honestly reflect the capacity of the sender for violence. Paradoxically, displays that 'handicap' the emitter are the most useful in deciding whether it would be prudent to escalate a conflict or withdraw.

Stalk-eyes in flies are a case of what may have been a dishonest display evolving into a handicap and an honest advertisement. Smaller flies typically retreat from face-to-face confrontations with larger opponents and fights are normally between similar sized individuals. Suppose that the size of a rival is assessed by the breadth of the head, as gauged by the degree of overlap between the two sets of eyes. If so, males can appear large and conquer psychologically simply by broadening their faces. As deceitfully widened heads become more common, even further exaggeration is required, and the resulting 'arms race' pulls the eyes farther and farther apart until they are at the ends of extraordinary stalks (in one 8 mm long species from Borneo, the combined length of the stalks is 20 mm). In the process, the stalks have become honest advertisements, genuine burdens that only the most robust males can produce and maneuver. Not only do males use stalk eyes to decide which opponents can be expelled and whether to fight or flee, but females also consider their size when choosing a mate.

The content of intersexual displays
Location cues: long-distance signals. Because males usually benefit more from multiple matings than do females, they typically spend more energy and take greater risks than females to locate mates.

Fig. 1122 The eyes of this male fly (*Achias* sp.) are at the ends of stalks whose combined width exceeds the length of its body. This condition, although usually less extreme, is found in a number of fly families and may be the result of trying to appear as large as possible in the opposing eyes of sexual rivals (drawing by Kevina Vulinec in Sivinski 1997).

When females are scattered through the environment, this often means males travel considerable distances searching for cues to find females. Because it is also in the female's interest to be found by one or more males and proceed with oviposition more quickly than other females, she may emit a relatively inexpensive signal that improves her chances of being located. Sexual pheromones are common female location cues that are typically produced in small, economical amounts and sent out with little risk as few predators or parasitoids specialize in tracking down the emitters of these chemically dilute and sporadic signals.

Modifications of female form or coloration that serve as signals to orient searching males may be less common than pheromones because the distance at which a visual cue can be perceived will generally be less than that of volatile chemicals. In addition, natural enemies without any special sensory adaptations might recognize visual signals as well as, or even better than, conspecific males. However, female location cues do seem to occur. For instance, in certain luminescent beetles like the phengodid *Phengodes nigromaculata*, larviform adult females retain the elaborate system of larval light organs that prob-

ably served as an aposematic advertisement of unpalatability. But the adult's lights are much brighter than the larva's and could act as a beacon to searching males whose large, complex male antennae suggest a female pheromone is produced as a longer distance signal as well. An alternative explanation for the brighter lights of adult females may be that chemical signaling requires more time spent exposed on the surface of the ground, and because of their greater vulnerability, females invest more in their aposematic display.

Competition among females to attract males from a distance typically will be less intense than competition among males to attract females. When males produce long-distance signals, whether chemical, acoustic, or visual, it is assumed that the signals are now more expensive than the travels females must undergo to reach the signaler, that the sender may be in greater danger from predators and parasitoids, and that females will be more likely to judge males on the basis of their signals than vice versa and respond selectively to the signals they deem most attractive.

While there are many familiar male long-distance chemical and acoustic signals (e.g., Mediterranean fruit fly, *Ceratitis capitata*, and crickets [Gryllidae]), there seem to be fewer obvious visual examples. Part of the reason may again be the relatively limited range of vision in insects which is better suited to close-up examinations. Even so, there are a number of interesting male visual-location cues. The lights of perched male fireflies (Lampyridae) are visual equivalents of a katydid's (Tettigoniidae) song, and diurnal 'beacons' that appear to advertise male positions occur in certain long-legged flies (Dolichopodidea). Male *Chrysotus pallipes* have much enlarged

Fig. 1123 A larviform female phengodid beetle (*Phengodes nigromaculata*) (a), and the same insect photographed by its own bioluminescence (b). The relatively bright lights of adult glow-worms may be beacons that help attract mate searching males (photos by J. Sivinski).

and shiny labial palps that emit silver flashes as they signal from the surface of leaves, and the reflected light from the tiny insect is surprisingly bright.

Courtship displays. Courtship implies that the sender of a signal is attempting to overcome resistance in the receiver, and that the display will somehow influence (or manipulate) the receiver to copulate. This element of 'seduction' can occur in long-distance signals as well, particularly those produced by males, and this blurs the distinction between long-distance beacons and courtship performances. However, greater amounts of detailed information, particularly visual information, would be available to insects in close proximity that are aware of each other's presence. Certain sexual behaviors will be unique to such close-up situations and so deserve separate consideration.

The meaning of communications with the opposite sex often is ambiguous, and there are several competing theories that attempt to explain the evolution of courtship displays. These hypotheses can be divided into the following categories: 1) displays promise a material benefit to the receiver, such as a nuptial gift of food or high female fecundity, 2) displays advertise the genetic qualities of the performer through self imposed handicaps, and 3) displays provide neither direct material nor genetic information, but are due to either genetic feed-back ('Fisherian run away selection') or manipulation of the receiver's perceptions ('sensory bias'). It may be that all of these are involved in the evolution of different displays.

Material benefits. The exchange of 'goods' from males to females, i.e., nuptial gifts, have been well documented in several insect groups. Examples include scorpionfly (Mecoptera) males that are required by females to produce captured prey of sufficient size, or else successful sperm transfer is not allowed. Some male Lepidoptera provide large, nutritious spermatophores during mating. The same is true of certain orthopteran males, while other orthopteran and coleopteran males provide nutritious secretions from the mouth or glands.

Such nuptial feeding invites deceit through false appearance. Things can be promised and then withheld to tempt another mate, or be promised but be completely absent. On the other hand, a female attempting to acquire a male's investment might exaggerate her fecundity, and appear to be a bigger mother and better recipient than she really is. When

either sex has materials the other requires, be it protein or access to eggs, advertisements, including visual advertisements, are likely to be elaborated through exaggeration. As in stalk-eyes, the expense of elaboration may eventually become a handicap and forge an honest display.

Empidid or dance flies (Diptera) provide examples of both male and female visual displays evolving in the context of sexual 'commerce.' Predaceous male dance flies often provide mates with a prey item, but sometimes this is wrapped in a silk balloon that might have originally magnified the appearance of the gift and which, in some cases, may no longer hold anything and is perhaps only an empty promise. Even more peculiar are the odd outgrowths of the midlegs of male *Rhamphomyia scurissima*. This mass of swellings suggests the male has a small dead insect in its grasp, and, with this deception, he may be able to lure a gift-seeking female into a sexual encounter.

On the other side of sexual bartering are females whose apparent fecundity influences whether or not they are offered nuptial gifts. In some species, there are sex-role reversals where females swarm and males carrying insect prey choose the largest females to both feed and inseminate. Females may call attention to their abdomens, and the numbers of eggs they contain, with garish, bright silver markings. In several empidid genera, females inflate their abdomens with air until they are nearly spherical, and so may falsely appear to carry extraordinary numbers of eggs.

Genetic benefits. *Drosophila* (Diptera) females that are free to pick their own mates produce more mature offspring over their lifetimes than do females that are provided with randomly chosen males. The implication is that certain males are better sires, and that females are able to recognize their genetic qualities. But what sort of qualities are they choosing and how are they recognized? It may be that phenotypes that have survived to old age or acquired the resources to attain unusually great size are reflections of an underlying genetic capacity to survive and forage effectively. For example, in certain tephritid fruit flies females prefer larger males, although the role of vision in judging size is unknown.

In general, the problem with signals that advertise such characteristics is that they can be exaggerated. Increasingly discriminating females will demand honest advertisements from prospective mates, and

a.

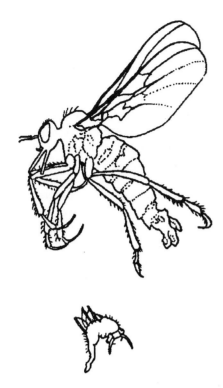

b.

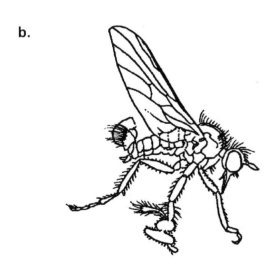

Fig. 1124 Males of many empidid species provide potential mates with an insect cadaver to eat (a). The peculiar outgrowths on the midlegs of the dance fly *Rhamphomyia scaurissima* (b) originally may have mimicked such a nuptial gift and attracted females through dishonest advertisement (drawing by Kevina Vulinec in Sivinski 1997).

signals that are expensive, or dangerous, or otherwise impossible to counterfeit should become preferred. For example, it is hard for a male to appear to be more symmetrical than it actually is. The complexity of embryonic development often results in less than perfect bilateral symmetry, and the greater the number of deleterious recessive alleles in the genotype the more unlikely it becomes. As a result, 'fluctuating asymmetries' in males are windows through which the value of their genes can be perceived and judged. Male *Calopteryx splendens* (Odonata: Calopterygidae) with more homogeneous wing spots perform more displays and obtain more mates, and male cerambycid beetles with more symmetrical antennae are more sexually successful. However, it is not always clear that female insects can visually discriminate fine degrees of asymmetry, and it is possible that they are attracted to male characteristics that co-vary with symmetry. For instance, a male that is more symmetrical may also have the physical capacity to signal for longer periods of time, and it is the signaling duration and not the symmetry that females find attractive.

Expense also insures honesty. A display that requires substantial material to produce, considerable energy to perform, and that places the emitter in danger is likely to be the production of a genuinely fit individual, one that had access to resources and was able to escape predators. Thus, displays are expected to be handicaps that are a burden to the insect outside of its sexual life.

Concentration on sexual matters can make males more vulnerable to predators, and it may be that the riveted attention that males focus on potential mates during courtship is in itself a handicap whose message is that "I possess such strength and agility that I can afford to give you my undivided attention and survive." Other physical trials might reveal genetic quality, e.g., females of the wolf spider *Hygrolycosa rubrofasciata* choose males on the basis of their leg-drumming rates which, in turn, are predictors of their viability.

In addition to risky, arduous or complicated behavior, physical structures may prove both costly to the signaler and attractive to the opposite sex. There are certainly any number of peculiar sexual dimorphisms in insects that could be both displays and handicaps. Many courting insects energetically use their wings, which are also important organs of predator-escape, in displays that might jeopardize their safety. For instance, potentially compromising wing motions are ubiquitous in the courtships of the tephritoid Diptera; e.g., *Callopistromyia annulipes* (Ulidiidae) males 'strut' with wings upraised like a peacock's tail, and rapid wing movements are nearly universal in the sexual interactions of other large taxa like the parasitic Hymenoptera (such motions may also waft pheromones and/or produce acoustic signals).

Colors can attract both the opposite sex and the attention of predators, and in addition, some colors may be difficult or even dangerous to produce. In birds for instance, red pigments are derived from carotenoids that the animal cannot manufacture and must be obtained from food sources. Further, carotenoids are important components of the immune system and their removal from a vital health-function in order to be 'painted' on the body surface may send a powerful message about both the foraging capacities and disease resistance of the signaler. Males often suffer more injuries and greater mortality than females, and this may be due to either more conspicuous colors or more risky mate-searching behaviors that expose them to predators. An illuminating case of dangerous bright coloration, as opposed to dangerous behavior, occurs in *Ischnura* damselflies (Coenagrionidae). Females may be either drably colored or brightly colored like the males ('andromorphs'). Brightly colored females are less likely to be recognized as potential mates and so avoid pestering by males. Because of their male-like appearance, they are able to lay their eggs more efficiently, but, unfortunately, they are also more likely to attract the attention of predators and, in some populations, they have only one-third the life expectancy of their plainer sisters.

Finally, color may indicate the ability of the signaler to avoid or survive debilitating diseases and larger parasites. Females may prefer such males either because they are likely to sire healthy offspring or because healthy males are less likely to transmit a disease during copulation. It has been suggested that the colors and forms of certain dung-feeding male scarab beetles might highlight the mite population they carry and allow females to choose males with lower mite loads.

Fisherian run away selection. If the genes for the expression of a male trait (e.g., an element of a courtship display) and the female preference for the trait are linked, then females that choose the most extraor-

dinary examples of the male characteristic will tend to have daughters with a still more extraordinary 'taste' for the display. The extravagance of the display and the female demand for the extravagant will continue to increase hand-in-hand ("run away" in the sense of positive feedback) until the spiraling expense and danger of the signal are so great that natural selection puts an end to further elaboration. In this process, the display has no 'meaning,' females gain no insights into male genetic or material qualities from the display and choose mates only for their capacity to produce a more or less arbitrarily attractive set of colors, structures and behaviors.

Recently, the popularity of this theory of the origin of displays has suffered somewhat for two reasons. First, phylogenetic studies reveal that male displays have been lost in a number of lineages, much more frequently than previously had been imagined, and this loss is inconsistent with linked traits running away with one another. Second, displays increasingly are found to be reliable indicators of male quality and not simply matters of haphazard fashion. Particularly telling cases of informative signals occur when the same displays that males use to persuade females are also employed in competitive interactions with other males. A sexual rival has no reason to be intimidated by a structure or behavior that is only fashionable and not an advertisement of the ability to acquire a resource or to defend associated females. For example, in the stalk-eyed fly (Diopsidae), *Cyrtodiopsis whitei*, female harems 'roost' in the company of individual males, and females prefer associations with long-stalked males because stalk-eye length is correlated to offspring quality. Eye stalks are also used in antagonistic interactions.

Sensory bias - . 'Sensory bias' theories of the origin of displays suppose that male sexual displays take advantage of preferences that have evolved in females for other reasons, and so manipulate females into responding. This is another type of male signal that tells females nothing useful about male genetic or material qualities, at least in the early stages of its evolution. For example, females of the water mite *Neumania papillator* (Acari: Parasitengona) use vibrations to locate their copepod prey, and males make similar vibrations with their forelegs. Females move toward and clutch vibrating males, and are more likely to do so when they are hungry. It seems that males have exploited the female's feeding behaviors to increase their rate of sexual contact.

Manipulative visual displays occur in the wolf spider (Lycosidae) genus *Schizocosa*. Males perform leg-waving courtships, but the various species have different degrees of foreleg ornamentation. Female spiders will orient toward video tapes of male courtships, and prefer the images of males with tufts of hair on their legs, regardless of whether or not their conspecific males have such hair-tufts. Thus, females in both tufted and untufted species appear to have a pre-existing preference for hairy legs that males of some species take advantage of and that males of other species do not.

This taste for leg-tufts might be a byproduct of hunting behaviors where female spiders must orient toward and approach small moving objects in their environment. Because such a response is necessary for the display-receiver to obtain food, it is difficult for her to completely ignore the signaler, but females might become more discriminating over time and sensory bias signals could evolve to carry information about male quality. For example, female *Schizocosa* spp. are more responsive to males with symmetrical forelegs, and such symmetry may be the result of developmental stability or a successful fighting career. Foreleg displays that now are showcases for symmetry may be elaborations of what began as lures.

Copulatory and postcopulatory displays

A female insect may choose the sire of her offspring at several points prior to oviposition. Her responses to precopulatory courtships determine whether or not she will mate. But even while coupled, she may choose to direct sperm to different portions of her reproductive tract where they will either be used for fertilizations, or be shunted away into places where they will languish or perish. Following mating she may simply eject sperm from a particular male, or copulate again with a more suitable male whose new ejaculate will block the old sperm from reaching the ova.

Because of these opportunities for 'cryptic female choice,' males may guard mates or continue courting mates with displays during and even after insemination. There are a number of male copulatory behaviors that might qualify as visual appeals to females making cryptic choices. *Sabathes cyaneus* is an unusually brightly colored mosquito whose mid legs end in a feathery 'paddle' of elongate iridescent blue and gold scales. Males fly toward resting females with their paddles extended, land nearby, suspend themselves by their forelegs, and then swing and wave their ornaments. 'Waggling,' during which the midlegs continue to rise and fall, continues through the copulation. Mounted males of the long-legged fly *Scapius platypterus* (Dolichopodidae) insure their visual signals will be observed by resting their front legs over the female's head while the midlegs are held to the side near her eyes and waved back and forth. Males of the micropezid fly *Cardiacephala myrmex* close the gap between their foreleg motions and their mates by alternatively scratching and regurgitating on her eyes. Not only flies apply their ornaments to female eyes during copulation. Male wasps in the genus *Crabo* (Hymenoptera: Sphecidae) have the front tarsi dilated into horny plates punctured by membranous dots giving them a sieve-like appearance. During mating, these are placed over the female eyes, perhaps resulting in a specialized visual signal similar to what would be obtained by shining a light through an antique computer 'punch card.'

Mating systems and visual displays

Arthropods seek mates in a number of ways and in a variety of places, some suited to the production of visual displays and others not. Swarms, where multiple males (or rarely females) fly in place, sometimes by the thousands, over a spot or 'marker' are both common and seemingly poor places to advertise to potential mates. With numerous insects swooping about each other, signals, particularly pheromones, would be difficult to track back to their emitter. And, if the emitter cannot be identified, then cheaters who do not invest in a signal but fly about in another's 'perfume-cloud,' have an energetic advantage that would eventually lead to their increase and then to the collapse of the entire display system. Visual displays are somewhat more likely, but still suffer from a poor capacity for individual recognition. For example, swarming mosquitoes are typically drab, but non-swarming species are often colorful. However, if flight is slow enough and swarmers separate enough, then visual signals can and do evolve. Small species of Mayflies (Ephemeroptera) sometimes have glassy wings and their swarms resemble falling snowflakes in the sunshine. Some of the most

North American genus *Calotarsa* have enlarged hind-legs that bear a curious collection of projections and glittering aluminum-colored flags. They fly in a slow and dignified manner while allowing their ''...hind feet to hang heavily downward and look as if they were carrying some heavy burden.''

Aerial displays by non-swarming insects are easier to evolve and maintain because the complications of confusing neighbors are diminished. The brighter colors, and occasionally unique markings, of male butterflies are displayed in intrasexual encounters and some of these occur during aerial conflicts. In many species of *Photinus* fireflies (Lampyridae), males make nocturnal flights, advertising their availability with flash patterns. These patterns of flashes elicit response flashes from perched females, and through flash dialogues males locate females, land near them, and mating occurs. Some species of *Photuris*, another genus of fireflies, are aerial predators of signaling *Photinus* males. The flashing signal is more difficult for the predator to track in flight than a continuous glow. Flashing may have evolved from an ancestral continuous glow in response to this predation pressure.

Once courtship occurs on the ground or leaf surfaces or some other platform, there is finally a stable stage on which males can easily perform elaborate behaviors with complexly patterned ornaments and be relatively easily recognized as individuals with their own particular messages to send. Ground-based mating systems can consist of single males searching or occupying a signaling site (often a resource) or groups of males either aggregated at a resource or gathering in non-resource based 'leks.' Visual displays occur in these different mating systems. Many of the highly decorated male dolichopodid flies appear to be solitary, or at least not tightly aggregated. Striking sexual dimorphisms are also encountered in lekking species in other families. An unusual example of the latter is the scuttle fly, *Megaselia aurea* (Phoridae). Like most phorids, males are nondescript flies, cream and grey in color, but females are yellowish orange with an iridescent orange patch on their dorsal surface. In addition to their unusual coloration, females have the peculiar habit (for their sex) of gathering on leaf surfaces in groups of up to a dozen or so. Males visit these sex-role reversed leks, pair with a female, and fly off to copulate on nearby vegetation. The bright orange of the lekking females suggests a visual display directed either to female lek-mates or male visitors.

Fig. 1125 Mosquitoes are rarely ornamented. An exception are males of the genus *Sabathes* whose midlegs are decorated with iridescent 'feathers.' This feather is waved before females both prior to and during mating (drawing by Kevina Vulinec in Sivinski 1997).

remarkable insect ornaments occur in certain swarming, flat-footed flies (Platypezidae). Species of the

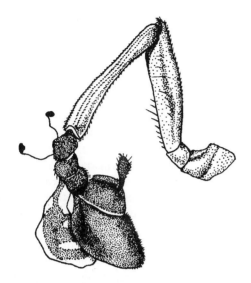

Fig. 1126 Swarming insects are seldom in a position to emit visual signals. The intertwining flights of numerous insects are simply too confusing. However, there are exceptions and the hindlegs of the swarming flatfooted fly *Calotarsa insignis* are a spectacular example of insect ornamentation (drawing by Kevina Vulinec in Sivinski 1997).

The locations of ornaments and colors

The structures and colors used in visual displays occur all along the arthropod body, but there are certain points of concentration. Some display locations emphasize a particular body part. Female dance flies that call attention to, or exaggerate, the size of their ovaries enlarge and color their abdomens, and ornaments used in male-male interactions are often on the head, perhaps because the head is used in the pushing style of combat typical of Diptera. Elaborations such as eyes elongated into cow-like horns (*Zygotricha dispar* [Drosophilidae]) and antlers and eye-stalks may be embellishments of actual weapons that now also advertise sexual competitiveness.

Movement can be an important component of a signal either because receivers are likely to notice the motion and so be more likely to see the ornament, or because the motion itself is the message and the movement is made more obvious by the ornament. In either case, ornaments or motions could serve as 'amplifiers' that attract the attentions of the opposite sex to the display. It has been argued that a simple preference for the first male a receptive female might see could select for such amplifiers, and that competitions among amplified males to be the first noticed might act as 'mate-filters.' On average, females would be most likely to couple with active, easily seen males whose apparency is a correlate of a good genetic constitution.

This premium on movement has served to concentrate signaling onto moveable body parts such as legs, wings, labial palps, antennae, and even the external genitalia. In dolichopodid flies, the male genital region is sometimes much enlarged (the hypopygium) and this may be raised and lowered during the male's courtship advance (e.g., *Dolichopus omnivagus*). In one unidentified species, males rise up on their long legs, beat their wings, lower the hypopygium until it hangs perpendicular to the body, and then slowly twirl their genitals. Dipteran legs are adorned with signaling devices that range from simple flags of enlarged setae to bizarre banners of hairy projections and strange devices. In yet another dolichopodid, *Campsicnemus magius*, the front legs are so swollen, pendanted and hairy that critics suggested the original specimens were deformed by fungi. Likewise, the fore tibia of the flower fly (Syrphidae), *Tityusia regulas*, are "enormously thickened, grooved, twisted and distorted" with an "extremely long, extremely matted" dark pile of fringe, and tarsi are extravagantly flattened with the lateral edges of the second, third, and fourth segments prolonged into narrow, down-curving lobes. Other examples occur outside of the Diptera, e.g., some male leaf-footed bugs (Coreidae) may use colorful expansions of both hind leg and antennae in defense of territories such as flower heads.

As noted earlier, while color can appeal to the opposite sex or deter rivals, it may also attract predators, and displays that can be produced only when the appropriate receiver is in a position to appreciate the signal may be safer than those that are on all the time. Ornaments on some moveable structures have the advantage of concealing and then revealing a signal in all its splendor. Consider any number of butterflies (e.g., the white M hairstreak, *Parrhasius m-album* [Lycaenidae] or the Florida purplewing, *Eunica tatila* [Nymphalidae]) whose outer wing coloration is dull and cryptic, but whose inner surface is dazzlingly bright. Such an insect at rest is difficult to detect, but with its wings opened while on its perch or in flight, it is an iridescent beacon (although one must consider the habitat, as periodic flashes of iridescence in the dappled light of a forest may be misleading or disguising).

The phyletic distribution of visual sexual displays

Different arthropods have different capacities and opportunities for the production of visual sexual displays. Diurnal butterflies have a greater chance to employ visual signals than do nocturnal moths. As noted earlier, jumping spider (Salticidae) vision lends itself to the reception of color and complex movement, while the eyes of the related wolf spiders (Lycosidae) do not. The frenetic mass flights of small swarming flies are incompatible with visual signals, while the largely surface-bound foraging habits of dolichopodid flies may give them a ready stage on which to perform elaborate displays and so be the reason they are among the most ornamented Diptera.

As a result of these and other differences, there appears to be a clumped distribution of visual displays. There is hardly enough research to quantify this distribution, but it seems that the more derived flies (Brachycera) have a disproportionate number of apparent visual displays. Butterflies, luminous beetles and Odonata are other obvious hotspots of visual communication.

Conclusion

The diversity of arthropod visual signals is immense, owing to the evolutionary plasticity of the arthropod body and the behaviors used in combination with visual signals to modify and enhance their effects. Visual signals are adapted to a wide variety of habitats and contexts. As noted above, not all arthropod signaling evolved by sexual selection, as not all colorful arthropods are signaling to sexual partners or rivals. In fact, given the limitations of insect eyes, colors and patterns more often than not may be directed to visually acute predators such as birds and serve to startle, warn of unpalatability, or to mislead.

John Sivinski
USDA-ARS
Gainesville, Florida, USA
and
Steven R. Wing
University of Florida
Gainesville, Florida, USA

References

Choe, J. C., and B. Crespi. 1997. *The evolution of mating systems in insects and arachnids.* Cambridge University Press, Cambridge, United Kingdom.

Espmark, Y., T. Amundsen, and G. Rosenqvist. 2000. *Animal signals.* Tapir Academic Press, Trondheim, Norway.

Hauser, M. D., and M. Kinishi. 2000. *The design of animal communication.* MIT Press, Cambridge, Massachusetts.

Sivinski, J. 1997. Ornaments in the Diptera. *Florida Entomologist* 80: 142–164.

Thornhill, R., and J. Alcock. 1983. *The evolution of insect mating systems.* Harvard University Press, Cambridge, Massachusetts. 547 pp.

Vain-Wright, R. I., and P. R. Ackery. 1984. *The biology of butterflies.* Academic Press, London, United Kingdom.

VITAMINS. Vitamins are nutrients needed for normal growth of insects. Those required are thiamine, riboflavin, pyridoxine, niacinamide, pathothenic acid, biotin, folic acid and choline. Some insects also require carnitine, ascorbic acid, carotene, vitamin B-12, vitamin E, and others.

VITELLINE MEMBRANE. A membrane surrounding the egg, which in turn is surrounded and protected by the chorion.

VITELLOGENESIS. The formation of yolk in the developing egg.

See also, OOGENESIS, ENDOCRINE REGULATION OF REPRODUCTION, REPRODUCTION.

VITELLOGENIN. The major yolk proteins are called vitellogenins while being transported to the hemolymph, and consist of large glycolipoproteins which are produced by the fat body and secreted for uptake by maturing oocytes. Not all yolk proteins are vitellogenins, however, and in higher Diptera the yolk proteins are small polypeptides. See also, ENDOCRINE REGULATION OF REPRODUCTION

VIVIPAROUS. Organisms that bear living young, as opposed to eggs.

VOLATILITY. Ability of a substance to evaporate or vaporize.

VOLUNTEER PLANTS. The unexpected and undesired emergence of plants, usually self-seeded by the previous plants.

VULVA. The opening of the vagina.

W

WAGGLE DANCE. A dance performed by honey bees in which workers communicate with other workers the location of new food and nest sites.

WALKER, FRANCIS. Francis Walker was born at Southgate, England, on July 31, 1809. He is known principally for his collecting and writing skills. His entomological publishing commenced in 1832 when he contributed a chapter on chalcids to the "Entomological Magazine." Shortly thereafter, he undertook the editorial management of the magazine, though he remained in this capacity for much more than a year. He collected a great amount of material for the British Museum, produced several catalogues such as "List of the specimens of the dipterous insects in the collection of the British Museum," and described numerous species from both Europe and North America. Between 1851 and 1856 he published "Insecta Saundersiana" in four parts. He prepared over 20,000 descriptions in several orders, though many have been found to be faulty. Indeed, his faulty descriptions caused something of a scandal at the British Museum. He is not identified with a particular group, having described neuropterans, orthopterans, termites, hemipterans, flies, wasps, moths and beetles, but he likely had more effect on the study of Coleoptera than any other group. Walker died at Wanstead, England, on October 5, 1874.

References

Essig, E. O. 1931. *A history of entomology*. The Macmillan Company, New York. 1029 pp.

Papavero, N. 1971, 1973. Essays on the history of Neotropical Dipterology with special reference to collectors (1750–1905). Museu de Zoologia, Universidade de São Paulo.

WALKINGSTICK DEFENSIVE BEHAVIOR AND REGENERATION OF APPENDAGES. Some 2,500 species, mostly from warmer parts of the world, comprise the order Phasmatodea, the so-called stick insects. Often up to 8 cm long and frequently apterous (wingless), stick insects are slow and clumsy and, for that reason alone, are vulnerable. Many, however, are cryptically colored, armed with spines, or protected by their resemblance to twigs and tree leaves. Others that comprise only a few known species, are protected by defensive glands. In Taiwan, the walking stick insect *Megacrania tsudai* Shiraki, causes considerable damage to screw pine along the southern coast. It has three distinctive characterizations. First it can secrete a monoterpene alkaloid actinidine as a defensive chemical. Secondly, it can regenerate lost appendages, and lastly, it can reproduce individuals parthenogenetically for long periods of time.

Chemicals such as irrdodials, nepetalactone, benzaldehyde, benzothiazole, limonene and quinoline have been reported as defensive chemicals produced by other phasmid insects from different parts of the world.

Megacrania tsudai has two large bilateral thoracic glands and, when molested, it will spray a white discharge toward the disturbance. But unlike the other stick insects, the main chemical secreted was identified as a nitrogen containing alkaloid actinidine, with other related minor compounds such as boschniakine

Fig. 1127   Regeneration of appendages (top row) in the larva of stick insect *Megacrania tsudai* Shiraki. (Left) First instar of the insect in which the right metathoracic leg was amputated. (Right) After ecdysis, the right metathoracic leg of the second instar appeared as a miniature leg. Meanwhile, the right prothoracic leg was amputated. (Center) The third instar 3 days after ecdysis. The right metathoracic leg looked like its normal left leg, but is short in length. Note that the regenerated right prothoracic leg was regenerated as a miniature leg. Female *Megacrania tsudai* Shiraki showing white discharge being sprayed (below) when the insect was tapped on the abdomen.

and two stereoisomers of 1-acetyl-3-methylcyclo-pentane. Gas chromatography-mass spectrometry was used in these identifications.

Another feature of the insect is its ability to regenerate an amputated leg as an immature. An appendage can be regenerated during subsequent instars, indicating that at least some of the cells surrounding an appendage are sufficiently undifferentiated to retain the ability to reform that appendage. Since regeneration of a leg requires a molt, it might be related to the combination of molting hormone, juvenile hormone and signal transduction.

Megacrania tsudai can produce offspring by parthenogenesis. In an enduring study, female insects have been cultivated in a laboratory for more than 10 years, with no males present in the laboratory culture or in the field.

Y. S. Chow
National Museum of Natural Science
Taiwan, Republic of China

References

Bouchard, P., C. C. Hsiung, and V. A. Yaylayan. 1997. Chemical analysis of defense secretions of *Sipyloidea sipylus* and their potential use as repellents against rats. *Journal of Chemical Ecology* 23: 2049–2057.

Brockes, J. P. 1997. Amphibian limb regeneration: rebuilding a complex structure. *Science* 276: 81–87.

Chow, Y. S., and Y. M. Lin. 1986. Actinidine, a defensive secretion of stick insect, *Megacrania alpheus* Westwood (Orthoptera: Phasmatidae). *Journal of Entomological Science* 21: 97–101.

Eisner, T., R. C. Morgan, A. B. Attygalle, S. R. Smedley, K. B. Herath, and J. Meinwald. 1997. Defensive production of quinoline by a phasmid insect *Oreophoetes peruana*. *Journal of Experimental Biology* 200: 2493–2500.

Gu, S. H., Y. S. Chow, and C. M. Yin. 1997. Involvement of juvenile hormone in regulation of prothoracicotropic hormone transduction during the early last larval instar of *Bombyx mori*. *Molecular and Cellular Endocrinology* 127: 109–116.

Ho, H. Y., and Y. S. Chow. 1993. Chemical identification of defensive secretion of stick insect, *Megacrania tsudai* Shiraki. *Journal of Chemical Ecology* 19: 39–46.

WALKINGSTICKS. Members of an order of insects (order Phasmatodea). See also, STICK AND LEAF INSECTS.

WALOFF, NADEJDA. 'Nadia' Waloff was born at St. Petersburg, Russia, on September 2, 1909. Her family fled the Bolsheviks in 1919 and became refugees in Britain. In 1926 her mother died, and afterwards her father and oldest brother emigrated to Romania. Nadia and three siblings were taken in by Sir Boris Uvarov, director of the Anti-Locust Research Center in London. It is likely that this experience encouraged Nadia and her sister Zena to pursue careers in entomology. Waloff received a bachelor's degree from Imperial College and taught school for several years. In 1946 she began work on insect pests at Imperial College and the Pest Infestation Laboratory in Slough. Thus she embarked on a distinguished career at Imperial College and also attained a Ph.D. and Sc.D. before retiring in 1978. Both before and after her retirement, Waloff was known as an outstanding instructor. However, she conducted some very thorough and landmark research, alone and in collaboration, on flour moth diapause, population dynamics of grasshoppers, yellow meadow ant, and broom-infesting insects. She died on June 5, 2001.

Reference

Jones, T., and R. Southwood. 2002. Dr. Nadejda Waloff FRES (Hon) (1909–2001). *Antenna* 26: 6–9.

WALSH, BENJAMIN DANN. Benjamin Walsh was born at Frome, Somerset, England, on September 21, 1808. He graduated from Cambridge University, Cambridge. His parents wished him to enter the ministry, but he disavowed the hypocrisies of the church. He emigrated to the United States in 1838 and settled into a farming life near Cambridge, Illinois, for about 12 years. In 1850 he moved to Rock Island, Illinois, and opened a lumber yard. He retired from this in 1858 and entered the real estate business and, to a lesser degree, politics. It is at this time in his life that Benjamin Walsh became interested in insects, and by 1860 he was reported to have given an extemporaneous 2-hour lecture to the Illinois State Horticultural Society. Soon he began publishing in farm newspapers and scientific journals. He did such a fine job at publicizing the dangers from insects that the Illinois State Legislature appointed him state entomologist. Walsh and C. V. Riley started the "American Entomologist" from 1868 to 1880. It consisted of three volumes and was considered to be an outstanding popular periodical on insects. In November 1869 Walsh was absentmindedly walking along the railroad tracks when he was struck by a train, and perished a few days later, on November 18, 1969.

Reference

Mallis, A. 1971. *American entomologists.* Rutgers University Press, New Brunswick, New Jersey. 549 pp.

WALSINGHAM, (LORD) THOMAS DE GRAY. Thomas Walsingham was born at London, England, on July 29, 1843. He was educated at Cambridge University and served as a member of the British House of Commons. He succeeded to the title and estates of his father in 1870, and was appointed a trustee of the British Museum in 1876. His wealth allowed him to travel and to purchase specimens at will, and he developed considerable expertise in the Microlepidoptera. His collecting trips included most of southern Europe, northern Africa, and the western United States. Walsingham served as president of the Entomological Society of London, and was a member of numerous other scientific societies. Among his noteworthy publications are "Illustrations of the Lepidoptera Heterocera in the British Museum" (1879), and "North American Tortricidae" (1884). He died in London on December 3, 1919.

Reference

Essig, E. O. 1931. *A history of entomology.* The Macmillan Company, New York. 1029 pp.

WARBLE FLIES. Members of the family Oestridae (order Diptera). See also, FLIES.

WASMANN, ERICH. Erich Wasmann was born on May 29, 1859, at Meran, Austria. He studied for the priesthood and entered the Jesuit order in 1875. From 1890 to 1892 he studied zoology at the University of Prague. Wasmann suffered from a lung disorder and was advised to spend as much time as possible out-of-doors. This, perhaps, led to his interest in ants, and he eventually became an authority on myrmecophiles and termitophiles. He also wrote on instinct and intelligence of ants, and the relationship of ants to other animals. He authored 433 publications with over 280 on myrmecophiles and termitophiles. Noteworthy publications included "Kritisches Verzeichniss der myrmecophilen und termitophilen Arthropoden" (1894), "Die moderne Biologie und die Entwicklungstheorie" (1910), and "The Berlin discussion of the problem of Evolution" (1909). The University of Freiburg in Germany awarded him an honorary doctoral degree in 1921, and the Jesuits established a biological journal honoring him, the "Wasman Journal of Biology." He died on February 27, 1931, in Valkenburg, Holland.

References

Anonymous. 1931. Obituary, Father Erich Wasmann, S. *J. Entomological News* 42: 240.
Herman, L. H. 2001. Wasmann, Erich. *Bulletin of the American Museum of Natural History* 265: 154–155.

WASSMANNIAN MIMICRY. The mimicry of ants by staphylinid beetles that live within the nests of ants. See also, MIMICRY.

WASP MOTHS. Some members of the family Arctiidae (order Lepidoptera) also known as tiger moths. See also, TIGER MOTHS, BUTTERFLIES AND MOTHS.

WASPS. Certain members of an order of insects (order Hymenoptera). See also, WASPS, ANTS, BEES, AND SAWFLIES.

WASPS, ANTS, BEES AND SAWFLIES (HYMENOPTERA). The order Hymenoptera includes insects commonly known as ants, bees and wasps. Most authors believe that the name has been derived from the Greek *hymen*, meaning 'membrane' and referring to the parchment-like transparent wings, without any scales or hair, and which often may be clouded. However, it also has been suggested that the name is derived from *hymeno* = god of marriage (referring to the union of fore and hind wings) + *ptera* = wings.

The Hymenoptera is one of the largest orders of insects. It is estimated that about 300,000 species occur in the world, of which not more than 30% have

been described. Most hymenopterans are useful and beneficial to mankind. The honey bees give us honey. They, along with other bees, are important pollinators of flowering plants that include commercial crops, particularly the fruit trees. There has been a process of co-evolution between the flowers and the hymenopterans, resulting in better pollination and fertilization, and the development of colorful and intricate flowers. Many hymenopterans are predators of insect pests, and a group of hymenopterans, commonly referred to as 'parasitic Hymenoptera' are utilized in biological control projects for the control of insect pests. This method is environmentally friendly and minimizes the use of insecticides. On the other hand, phytophagous hymenopterans, such as the sawflies, have resulted in the destruction of many coniferous forests in North America. Some wasps and ants are dreaded because of their venomous stings, which cause serious reactions in some people, even resulting in death.

The order Hymenoptera is also biologically rather important. Members of the order exhibit great diversity in their way of life and behavior, and include the highly evolved social insects. The social organization of bees and wasps has been a subject of many books and comparisons with the social life of mammals, including man, and represents a peak of evolutionary behavior.

The Hymenoptera alone exhibit the haploid-diploid mechanism of sex determination. The males are haploid, having an unpaired set of chromosomes. The females are diploid, possessing a set of paired chromosomes. The females have a normal gametogenesis so that the eggs are haploid, while in the males the first meiotic division generally is abortive so that haploid sperms are produced. Normally, the unfertilized eggs develop into males, while the fertilized eggs develop into females.

The social Hymenoptera exhibit polymorphism, and sexual dimorphism is evident. They also exhibit a 'caste system' where the majority of the individuals are 'workers' which are undeveloped females lacking the capacity to reproduce. They are responsible for all the chores of the colony, including food gathering, maintenance of the colony, defense, etc. The 'queen' is the normal female which is capable of reproducing and, depending upon the type of social organization, one to many queens may be in one colony. The haploid male is only responsible for the fertilization of the queen, and may develop and be tolerated in the colony for this function alone. The social hymenopterans also have developed the capacity to store food for the progeny in the colony. This behavior of the bees has been exploited to harvest honey from the bee colonies.

The solitary bees also provision their nests with food for their progeny. The progressive evolution of social behavior in Hymenoptera, and the mechanisms of communication amongst the members of a colony, have been studied by many entomologists.

Morphological characters

Members of the order Hymenoptera have mouthparts adapted for biting and chewing, biting and lapping, or biting and sucking. The antennae usually are long, with 10 or more segments (some exceptions), and often geniculate. There are two pairs of membranous wings; the hind wings are smaller than the fore wing and both wings are interlocked by means of hooklets called hamuli. The bases of fore wings are covered by a crescent-shaped sclerite called the tegula. The fore wing usually has a sclerotized thickened area midway along its anterior margin, called the stigma. The wings may be reduced or absent. Apterous forms occur in several families. Ants exhibit dimorphism with winged forms occurring during the swarming phase of their life cycle. The wing venation is greatly modified as many veins take a transverse course. There is often a fusion of principal veins, and in higher forms the venation is reduced. The leg is five segmented. In several families (commonly grouped as 'parasitica') the base of the femur is cut off, giving the impression that there are 2 trochanters. This structure is called trochantellus but in most taxonomic publications the trochanters are said to be 1 or 2-segmented. The tarsus is typically 5-segmented, but in some chalcid wasps the number may be reduced. The basitarsus is modified into a pollen basket in bees. The abdomen is basally constricted, and either broadly or narrowly attached to the thorax.

One characteristic feature of the hymenopterans is that the first abdominal segment is fused with the metathorax to form the propodeum. This is quite evident in higher Hymenoptera. There is a flexible joint between the first and the second apparent abdominal segments. In the Symphyta (sawflies, etc.) the abdomen is broadly attached to the propodeum, while in Apocrita (ants, bees, wasps, etc.) the abdomen is narrowly attached to the propodeum by a short or long petiole.

The females have an ovipositor which is built on a lepismatid form from the ancestral appendages of abdominal segments 8 and 9. It is modified in different groups for sawing, boring, piercing, stinging, or it may be reduced.

There are two main types of larvae. The Symphyta larvae are caterpillar-like, with three pairs of thoracic legs and at least 5 pairs of prolegs on the 2nd to 6th abdominal segments; these prolegs are devoid of crochets (found in lepidopterous larvae). They are seen crawling and feeding on foliage. The Apocrita larvae are legless, grub-like, living in nests, or in or on the body of other insects and spiders. The metamorphosis is complete (holometabolous). Pupation occurs in cocoons.

Classification

The classification adopted in *Hymenoptera of the world: an identification guide to families*, edited by Goulet and Huber (1993) is given below. They group many aculeate superfamilies and recognize only three: Apoidea, Chrysidoidea and Vespoidea. Also, they do not recognize the two convenient but informal groupings Aculeata and Parasitica. A beginner and a general reader is perhaps better served by recognizing the traditional classification of Hymenoptera into many superfamilies. The classification given in Borror, Triplehorn and Johnson (1989) is still a good general summary of the classification of the Hymenoptera.

An outline of the classification of the Hymenoptera

Suborder SYMPHYTA (= CHALASTOGASTRA) – sawflies, horntails
  Superfamily: CEPHOIDEA
    Family: Cephidae – stem sawflies
  Superfamily: MEGALODONTOIDEA
    Family: Megalodontidae
    Family: Pamphiliidae – leaf-rolling and web-spinning sawflies
  Superfamily: ORUSSOIDEA
    Family: Orussidae – parasitic wood wasps
  Superfamily: SIRICOIDEA
    Family: Siricidae – horntails
  Superfamily: TENTHREDINOIDEA
    Family: Argidae
    Family: Blasticotomidae
    Family: Cimbicidae

Family: Diprionidae
Family: Pergidae
Family: Tenthredinidae – common sawflies
Superfamily: XYELOIDEA
Family: Xyelidae
Unplaced
Family: Anaxyelidae
Family: Xiphydriidae
Suborder: APOCRITA
Group PARASITICA (not recognized in Goulet and Huber)
Superfamily: CERAPHRONOIDEA
Family: Ceraphronidae
Family: Megaspilidae
Superfamily: CHALCIDOIDEA
Family: Agaonidae
Family: Aphelinidae
Family: Chalcididae – chalcid wasps
Family: Elasmidae
Family: Encyrtidae
Family: Eucharitidae
Family: Eulophidae
Family: Eupelmidae
Family: Eurytomidae
Family: Leucospidae
Family: Mymaridae – fairy flies
Family: Ormyridae
Family: Perilampidae
Family: Pteromalidae
Family: Rotoitidae
Family: Signiphoridae
Family: Tanaostigmatidae
Family: Tetracampidae
Family: Torymidae
Family: Trichogrammatidae
Superfamily: CYNIPOIDEA
Family: Charipidae
Family: Cynipidae – gall wasps
Family: Eucoilidae
Family: Figitidae
Family: Ibaliidae
Family: Liopteridae
Superfamily EVANIOIDEA
Family: Aulacidae
Family: Evaniidae – ensign wasps
Family: Gasteruptiidae
Superfamily: ICHNEUMONOIDEA
Family: Braconidae – braconid wasps
Family: Ichneumonidae – ichneumon wasps

Superfamily: MEGALYROIDEA
Family: Megalyridae
Superfamily: MYMAROMMATOIDEA
Family: Mymarommatidae
Superfamily: PLATYGASTROIDEA
Family: Platygastridae
Family: Scelionidae
Superfamily: PROCTOTRUPOIDEA
Family: Austroniidae
Family: Diapriidae
Family: Heloridae
Family: Monomachidae
Family: Pelicinidae
Family: Peradeniidae
Family: Proctotrupidae
Family: Roproniidae
Family: Vanhorniidae
Superfamily: STEPHANOIDEA
Family: Stephanidae
Superfamily: TRIGONALYOIDEA
Family: Trigonalyidae
Suborder APOCRITA (= CLISTOGASTRA)
Group ACULEATA (not recognized in Goulet and Huber)
Superfamily APOIDEA – bees
Series: APICIFORMES (formerly APOIDEA)
Family: Andrenidae – andrenid bees
Family: Anthophoridae – digger bees, cuckoo bees, carpenter bees
Family: Apidae – honey bees, bumble bees
Family: Colletidae – yellow faced bees; plaster bees
Family: Ctenoplectidae
Family: Fideliidae
Family: Halictidae – halictid bees
Family: Megachilidae – leaf-cutting bees
Family: Melittidae – melittid bees
Family: Oxaeidae
Family: Stenostritidae
Series SPECIFORMES (formerly SPHECOIDEA)
Family: Ampulicidae – ampulicid wasps
Family: Astatidae
Family: Crabronidae
Family: Heterogynaidae
Family: Mellinidae
Family: Nyssonidae
Family: Pemphredonidae
Family: Philanthidae
Family: Sphecidae – mud daubers, digger wasps

Superfamily CHRYSIDOIDEA (= Bethyloidea)
    Family: Bethylidae – bethylid wasps
    Family: Chrysididae – cuckoo wasps
    Family: Dryinidae – dryinid wasps
    Family: Embolemidae
    Family: Plumaridae
    Family: Sclerogibbidae
    Family: Scolebythidae
Superfamily: VESPOIDEA
    Family: Bradynobaenidae
    Family: Formicidae – ants (formerly
        Formicoidea)
    Family: Mutillidae – velvet ants, mutillids
    Family: Pompilidae – spider wasps (formerly
        Pompiloidea)
    Family: Rhopalosomatidae – rhopalosomatid wasps
    Family: Sapygidae
    Family: Scoliidae – scoliid wasps (formerly
        Scolioidea)
    Family: Sierolomorphidae
    Family: Tiphiidae – tiphiid wasps
    Family: Vespidae – wasps, yellow jackets,
        hornets, etc.

Key to the superfamilies

This is a simplified key to the superfamilies ignoring taxa showing exceptions to the general character states. Also, the traditional superfamilies are retained. It is based on the winged adults.

1. Abdomen joined broadly to the thorax (no waist or petiole). Hind wing with 3 or 4 closed cells (cells not open to wing margin). Fore wing with one or more distinct anal veins. Trochanters 2-segmented. Sawflies, horntails. SUBORDER SYMPHYTA 2
Abdomen joined narrowly to thorax (a narrow waist or petiole present, abdomen flexible). Hind wing with 0–2 cells at base. Fore wing without anal veins except rarely. SUBORDER APOCRITA 5
2. Fore tibia with 2 apical spurs. Ovipositor blade-like and usually short. 3
Fore tibia with 1 apical spur. Ovipositor usually long and thin 4
3. Hind tibia with 1 preapical spur or none. Pronotum in dorsal view with posterior margin strongly concave (narrowed medially behind). . . . . TENTHREDINOIDEA
Hind tibia with 2–3 preapical spurs. Pronotum in dorsal view with posterior margin weakly

concave (not strongly narrowed medially behind). . . . . MEGALODONTOIDEA
4. Last tergum of female and last sternum of male each with apical, median, cylindrical projection with concave tip. Pronotum somewhat rectangular with its hind margin concave. Costal cell present and distinct . . . . SIRICOIDEA
Last tergum and last sternum of both sexes apically thin and without cylindrical projection. Pronotum somewhat triangular with its hind margin weakly concave. Costal cell absent or narrow . . . . CEPHOIDEA
5. Legs, at least the hind leg, with 2 trochanters (trochanter + trochantellus) 6
Legs with 1 trochanter (Proctotrupoidea runs through either half of the couplet) 14
6. Antenna usually with more than 16 segments, not elbowed. Hind wing with usually 2 closed cells. 7
Antenna with 15 or fewer segments, often elbowed. Hind wing with 1 closed cell or none. [One family of Cynipoidea – Sclerogibbidae with 17–40 segments, then costal vein absent.] . . . 9
7. Costal cell of fore wing absent; costal and subcostal veins touching each other . . . . ICHNEUMONOIDEA
Costal cell present, may be narrowed in some rare families 8
8. Head with a crown of tubercles around median ocellus. Ovipositor long. Abdomen elongate. Costal cell narrow . . . . STEPHANOIDEA
Head without tubercles. Costal cell broad. Ovipositor short. Stouter insects . . . . TRIGONALYOIDEA
9. Abdominal petiole with 1 or 2 scale-like or node-like projections. Females with a sting . . . . FORMICOIDEA
Abdominal petiole without scales. Females with a sting or an ovipositor 10
10. Abdomen inserted high on propodeum much above the level of hind coxae EVANIOIDEA
Abdomen attached normally, near hind coxae 11
11. Hind corner of pronotum not reaching tegula, separated from tegula by an extra sclerite, "prepectus". Middle tibia with 1 or 0 spur. Antennae usually elbowed. Coloration often metallic greenish. . . . . CHALCIDOIDEA
Hind corner of pronotum reaching tegula,

prepectus absent. Middle tibia often with 2 spurs. Antennae elbowed or not. Coloration never metallic greenish 12

12. Costal vein absent. Radial cell large and deep, surrounded by strong veins except often the costal margin. Abdomen compressed laterally, often covered by a single large tergite . . . . .CYNIPOIDEA
Costal vein present. Radial cell present or absent; if present, not as above 13

13. Fore tibia with 1 apical spur. Antennae inserted on face usually above clypeus. Abdomen rounded or margined laterally . . .PROCTOTRUPOIDEA
Fore tibia with 2 apical spurs. Antennae inserted on face next to clypeus and abdomen rounded laterally . . . . CERAPHRONOIDEA

14. Hind wing with 1 closed cell or none 15
Hind wing with 2 closed cells. Aculeata . . . 17

15. Hind margin of hind wing nearly always with a sharp notch near base cutting off a basal lobe (anal lobe) . . . . CHRYSIDOIDEA
Hind margin of hind wing without a sharp notch, not cutting off an anal lobe 16

16. First hind tarsal segment 0.25 as long as second. Segments 2–4 of female abdomen separate, each very long. Male abdomen club-shaped . . . . . PROCTOTRUPOIDEA: PELECINIDAE
First hind tarsal segment longer than second. Abdominal tergites 2–4 fused together. . . . . .PROCTOTRUPOIDEA

17. Hind corner of pronotum reaching tegula, the hind edge of pronotum without a semicircular projecting lobe just below the level of tegula . . . 18
Hind corner of pronotum not reaching tegula, the hind edge of pronotum with a projecting lobe just below the base of tegula 23

18. Tergite 1 with 1 or 2 erect scales or nodes. Second recurrent vein absent. Hind wing without an anal lobe. ANTS . . . . FORMICOIDEA
[Now included under VESPOIDEA]
Tergite 1 without scales or nodes. Second recurrent vein usually present. Hind wing usually with an anal lobe 19

19. First discoidal cell (1M or discal cell) of fore wing very long, much longer than submedian cell (1M + Cu1). Fore wing usually folded lengthwise. Wasps [except Masarinae] . . .VESPOIDEA
First discoidal cell (1M) of fore wing shorter than submedian cell (1M + Cu1). Fore wing not folded lengthwise 20

20. Mesopleuron with a straight horizontal groove across its mid-height . . . . POMPILOIDEA
[Now under VESPOIDEA]
Mesopleuron without a straight horizontal groove across its mid-height 21

21. Apex of antennal segments 3–8 or more with a pair of stout bristles on upper and inner sides. Tarsal segments 2–4 of female expanded . . . . VESPOIDEA: RHOPALOSOMATIDAE
Apex of antennal segments 3–8 without a pair of stout bristles. Tarsal segments 2–4 of female not expanded 22

22. Mesosternum and metasternum together forming a plate divided by a transverse suture, and overlapping bases of middle and hind coxae. Apical half of wing membrane with fine longitudinal wrinkles (imbrications). . . . . SCOLIOIDEA
[Now under VESPOIDEA]
Mesosternum and metasternum not forming such a plate. Mesosternum with 2 lobe-like extensions projecting between and partly covering middle coxae. Hind wing with a jugal lobe, not imbricated . . . . VESPOIDEA: TIPHIIDAE

23. First segment of hind tarsus enlarged and more or less flattened. Head and body with branched or plumose hairs. Labium usually produced as a long tongue. Bees . . . . .APOIDEA
First segment of hind tarsus not noticeably enlarged. Head and body with simple hairs (hairs unbranched or plumose). Labium usually not produced as a long tongue . . . . . . SPHECOIDEA

Suborder: Symphyta (Chalastogastra)

The abdomen is broadly attached to the thorax. There is no constriction between them. The first abdominal segment is only partly incorporated in the thorax. The venation is more complete with numerous cross-veins, forming many wing cells. The fore wing has at least one closed anal cell and most species have a closed anal cell in the hind wing. The metanotum has a pair of raised structures known as cenchri that engage with a scaly area on the underside of the fore wings and keep them in place. These are absent, however, in the superfamily Cephoidea. The ovipositor is adapted for sawing and boring. The syphytans are phytophagous except members of the family Orussidae. Members of this suborder commonly are called sawflies, horntails, etc.

Suborder: Apocrita (Clistogastra)

The abdomen is narrowly attached to the thorax by a short or a long petiole. The first abdominal segment (usually the tergite) is intimately fused with the metathorax to form the propodeum and the basal portion of the second morphological abdomen is constricted to form a short or a long petiole by which it joins the thorax (propodeum). The thorax thus has four segments and the first apparent abdominal segment is morphologically the second segment. Because of this confusion in counting the abdominal segments, many authors refer to the three hymenopteran body parts as prosoma (head), mesosoma (thorax) and metasoma or gaster (abdomen). The wing venation and wing cells are reduced. Sometimes there are no veins and cells.

Among the Apocrita two different groups may be distinguished. Because of exceptions, these groups, called Parasitica and Aculeata, have not been accorded formal status, but still they are useful groupings.

The Parasitica includes hymenopterans the larvae of which are parasitic in or on the immature states of other insects and arachnids and develop at the expense of their hosts. The adults are free living. These insects are referred to commonly as parasitoids. They have an ovipositor to lay eggs, usually have a trochanter and trochantellus (commonly referred to, though incorrectly, as having a 2-segmented trochanter), have long filiform antennae, and have reduced venation when compared to the Symphyta. They exert natural control of unwanted insects in forests and fields and often are used for the biological control of insect pests.

The Aculeata, on the other hand, usually are either social or solitary hymenopterans, construct hives or nests, provision their nests with food for the developing larvae, have a sting (modified ovipositor), and usually have a fixed number of antennal segments and lack a trochantellus. The venation is more complete when compared with the Parasitica. There are exceptions, however, as among the superfamily Chrysidoidea, some members are parasitoids but show characteristics of the aculeate hymenopterans. Aculeata includes ants, bees and wasps exhibiting social life, polymorphism, etc.

Treatment of representative superfamilies and families

Suborder: Symphyta

Superfamily: Cephoidea, Family Cephidae – Stem sawflies

The Cephidae is a small family occurring in the Holarctic Region, Old World tropics and Madagascar. The larvae are stem-borers and damage the growing tip of the stem. Several species attack cereal crops.

Superfamily: Megalodontoidea

Members of this superfamily are characterized by having a very large, almost prognathous head that is widest near the clypeus. Two families are recognized, the Megalodontidae and Pamphilidae. The Megalodontinae occurs in the temperate regions of Eurasia. Their antennae are either saw-like or comb-like, and tergites 2 to 5 are not folded above spiracles. The larvae feed on herbaceous plants. Members of the family Pamphilidae are characterized by having thread-like or filiform antennae and tergites 2 to 5 having a longitudinal fold above spiracles. They are Holarctic in distribution. The larvae roll leaves or spin silk to form webs in which they feed. The larvae are gregarious and attack conifers and fruit trees.

Superfamily: Orussoidea, Family Orussidae – Parasitic wood wasps

Members of this group are quite different from other sawflies. They are parasitic in habits. The adults are cylindrical in cross-section and have several spines around the median ocellus. The antennae are inserted below the ventral margin of the eye. The venation is reduced. The ovipositor is very long and thin, and coiled within the abdomen when the insect is resting. The Orussidae is a very small, relict, and widely distributed family. The members are seen with an ant-like gait over tree trunks. They are parasitoids of wood boring Coleoptera and Hymenoptera larvae.

Superfamily: Siricoidea, Family Siricidae – Horntails, wood wasps

This group contains large wood boring sawflies occurring in the Northern Hemisphere. They have been introduced accidentally into Australia and New Zealand and are serious pests of conifers. The tip of the abdomen is produced into a long horn-like

Fig. 1128 Drawing of horntail in the family Siricidae (Source: Hymenoptera of the world: an identification guide to the families. Agriculture and Agri-Food Canada, 1993. Reproduced with the permission of the minister of Public Works and Government Services, 2003).

projection, giving the name horntails to these insects. The adults are large (up to 50 mm long), brightly colored, and they also are referred to as wood wasps. The pronotum is transversely folded and is collar shaped. They have a strong, drill-like ovipositor by which they drill holes in hard wood or conifer trees and insert their eggs into dead or decaying trees. They also deposit spores of a symbiotic fungus along with the eggs.

Superfamily: Tenthredinoidea

Members of the superfamily Tenthredinoidea are distinguished from other Symphyta by having a short pronotum that is concave posteriorly, the fore tibia having two apical spurs, and hind tibia with one pre-apical spur. The antennae are thread-like, usually with 9 segments (varying from 9 to 15 in some cases). The scutellum has a transverse furrow cutting off triangular post-tergite posteriorly. The ovipositor generally hardly projecting beyond the apex of abdomen. This is the largest superfamily of Symphyta, and they are distributed worldwide.

Family: Tenthredinidae. Common sawflies. The adults are brightly colored, wasp-like insects, common during spring and early summer but some species occur also during late summer and early fall.

The larvae feed gregariously on foliage orienting themselves in a characteristic manner. Some species are serious pests of coniferous trees and cause heavy defoliation of forests. A few species are gall makers and leaf miners. Many species exhibit sexual dimorphism.

Suborder: Apocrita

Group: Parasitica

Superfamily: Chalcidoidea

The superfamily Chalcidoidea is probably the largest in the order, with about 2,500 known genera and 20,000 known species. It also contains the smallest known insects. The majority of the species are parasitoids or hyperparasitoids of other insects and of great significance in biological control. Non-parasitic, plant feeding chalcidoids are found in the families Agaonidae, Torymidae, Perilampidae and Eurytomidae. The agaonids are found in figs; some others infest seeds and some produce galls. In the families Chalcididae and Leucospidae the hind femora are enlarged. In Mymaridae the wings have elongate fringe of hairs and these insects often are called 'fairy flies.'

The Chalcidoidea may be recognized by having an elongate antennal scape, so that the antennae appear elbowed. The flagellum usually is differentiated into a proximal funicle and an apical club. The pronotum

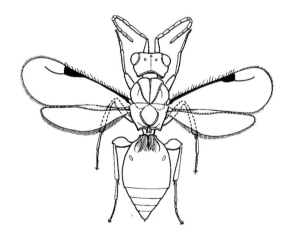

Fig. 1129 Drawing of wasp in the family Megaspilidae (Source: Hymenoptera of the world: an identification guide to the families. Agriculture and Agri-Food Canada, 1993. Reproduced with the permission of the minister of Public Works and Government Services, 2003).

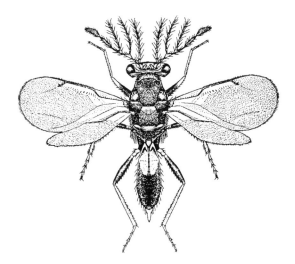

Fig. 1130 Drawing of wasp in the family Eulophidae (U.S. Department of Agriculture).

does not extend back to the tegula and a triangular sclerite, prepectus, is present between the hind margin of pronotum and the tegula. The wing venation is highly reduced so that generally only one vein is present in the fore wing.

Superfamily: Cynipoidea

The cynipoids are generally small insects (except *Ibalia*), brown or black in color. The antennae are filiform, the pronotum extends back to the tegula, wing venation is reduced, with the radial cell triangular and pterostigma absent except in some rare families. The abdomen in laterally compressed.

The majority of cynipoids are primary endoparasitoids developing in the larvae of holometabolous insects, or in the nymphs of Psylloidea. Most species of the family Cynipidae are gall producers and are called gall wasps. These galls commonly are seen on oak and other trees. In the family Ibalidae, the radial cell is very long and thin and the abdomen is strongly compressed and knife-like. Species of *Ibalia* may be up to 30 mm long.

Superfamily: Evanioidea

In Evanioidea the abdomen is attached to the thorax high above the hind coxae, antennae are with 13 to 14 segments, fore wing is with a costal cell and the wing venation is fairly complete. Members of the family Evaniidae are called ensign wasps. These are black or black and red insects with the abdomen small and oval, carried like a flag high above the hind

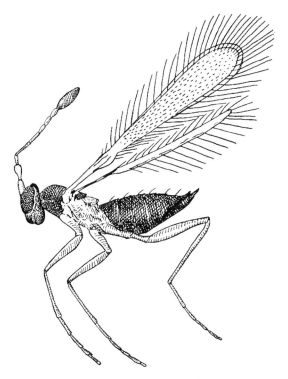

Fig. 1131 Drawing of wasp in the family Mymaridae (U.S. Department of Agriculture).

coxae. They are parasitoids of cockroach oothecae. Members of the family Aulacidae have a long, well exserted ovipositor and are endoparasitoids of wood boring coleopterous and hymenopterous larvae. In Gasteruptiidae the hind tibia is clavate, abdomen

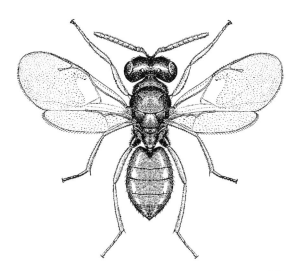

Fig. 1132 Drawing of wasp in the family Pteromalidae (U.S. Department of Agriculture).

Fig. 1133 Drawing of wasp in the family Trichogrammatidae (U.S. Department of Agriculture).

long, clavate, and ovipositor well exserted. They are ectoparasitoids of larvae of Sphecidae, Vespidae and Apidae.

Superfamily: Ichneumonoidea

This superfamily includes two families, the Ichneumonidae and Braconidae. These are small to very large insects, mostly usually fully winged, with antennae more than 13 segments, not geniculate or apically clavate. The pronotum extends back to the tegula. Wing venation is well developed and characteristic.

The Ichneumonidae is perhaps the largest family of insects, with over 60,000 known species. Most species are ecto- or endoparasitoids of the larvae and pupae of holometabolous insects, particularly of Lepidoptera and Symphyta. Some others parasitize spider egg masses or attack bee larvae in hives. They are distinguished from the family Braconidae by the presence of second recurrent vein (2m-cu) and by the absence of the cubitus vein (1/Rs + M) in the fore wing. Abdominal tergites 2 and 3 are not fused. It includes Agriotypidae and Paxylommatidae, often considered as separate families.

The family Braconidae appears to be the second largest family of insects with over 40,000 known species. It differs from the Ichneumonidae in the absence of the second recurrent vein (2m-cu) (except in Apozyginae), and by the presence of the cubitus vein (1/Rs + M) in the fore wing. Tergites 2 and 3 are fused (weakly so in Aphidiinae). It includes Aphidiidae, which are aphid parasitoids.

Superfamily: Proctotrupoidea

Members of this superfamily are parasitoids in the immature stages of other insects. They are small-sized insects with much reduced venation, but may be distinguished from the chalcids by the nature of the thorax and the abdomen. The pronotum in

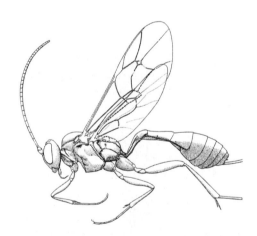

Fig. 1134 Drawing of wasp in the family Ichneumonidae (Source: Hymenoptera of the world: an identification guide to the families. Agriculture and Agri-Food Canada, 1993. Reproduced with the permission of the minister of Public Works and Government Services, 2003).

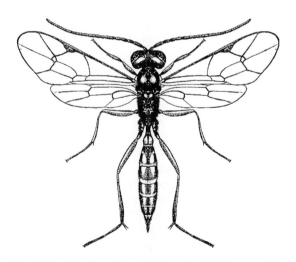

Fig. 1135 Drawing of wasp in the family Ichneumonidae (U.S. Department of Agriculture).

proctotrupids appears triangular in shape in lateral view and extends back to the tegula. The ovipositor issues from the tip of the abdomen.

One member of the family Pelicinidae, the North American *Pelicinus polytrurator* (Drury) is a large (the female about 50 to 60 mm long), shiny black insect. The abdomen is very long and thin. The males are rare and only about 25 mm long, with the tip of abdomen swollen. They are parasitic in the larvae of June beetles, and the adults emerge in summer. Members of the family Proctotrupidae are small, about 3 to 10 mm long insects, with a large pterostigma and a very narrow marginal cell. They are solitary or gregarious parasites in the larvae of Coleoptera and Diptera.

Superfamily: Platygasteroidea

These are small to minute, usually black and shiny insects, and usually are classified under Proctotrupoidea. Two families, Diapridae and Platygasteridae, are placed under it. Species of Diapridae have a shelf-like protuberance on the face. They are parasitoids of the immature stages of Diptera. Members of the Scelionidae attack eggs of spiders and several groups of insects. They have highly reduced wing venation and a flattened abdomen.

Superfamily: Stephanoidea

Only one family, Stephanidae, is included here, which was classified earlier under Ichneumonoidea. The head is spherical and with a crown of about 5 teeth around the median ocellus. There is a slender long neck. The hind coxae are elongate and the hind femora are swollen and toothed ventrally. They are common in tropics and are parasitoids of wood boring coleopterous larvae.

Suborder: Apocrita

Group: Aculeata

Superfamily: Apoidea

The superfamily Apoidea includes insects commonly referred to as the bees. They exhibit sexual dimorphism in the number of antennal segments — 13 in male and 12 in female. Tergite VII of the female is deeply cleft medially or separated into 2 lateral lobes. The female gonocoxite IX is constricted but not divided into articulating proximal and distal plates. The pronotum is not extending back to tegula, usually forming a lobe covering the mesothoracic spiracle. The body hairs are branched, plumose, and the hind tarsi are usually widened. They are anthophilous, solitary or social insects.

Two main groups of bees are recognized: the long tongued bees and the short tongued bees.

Family: Apidae.   Long tongued bees: the honey bees and the bumble bees. The first two segments of the labial palp are elongate and flattened, glossa is long and slender, and the maxillary palp is vestigial. In addition, the hind leg is modified to have a 'pollen basket' composed of the enlarged tibia and the first tarsal segment, and the hind tibia has an apical margin of short, stiff setae. The jugal lobe in hind wing is shorter, and there are three submarginal cells.

The bees are important pollinators of fruit and other important crops. Their role as pollinators far exceeds their importance as providers of honey and bees wax. There are three main subfamilies. The subfamily Apinae includes the honey bees. Three species are known in the world: *Apis mellifera*, the European honey bee, widely domesticated worldwide; *A. dorsata*, the wild honey bee that is ferocious and makes a single large hive high on trees, buildings,

Fig. 1136   Photographs of some Apidae: left, bumblebee, *Bombus* sp.; right, honeybee, *Apis* sp. (photos, J. L. Castner).

etc., in the tropics; and a gentle, timid honey bee, *Apis florea* that makes a single hive in bushes in the tropics. The subfamily Bombinae includes the bumble bees. They are large, robust, black with yellow hairs on thorax and abdomen. They nest in hollows in the ground, have annual colonies and are important pollinators of certain kinds of clover. Members of the subfamily Euglossinae are called 'orchid bees.' They are brightly colored metallic bees distributed in the tropical parts of the world.

Family: Anthophoridae. These insects are also long-tongued bees that are robust and hairy, about 10 to 20 mm long, and usually brownish in color. The hind tibiae have apical spurs, genae are short and narrow, and the maxillary palps are well developed. They nest in the ground.

The family Anthophoridae is divided into several subfamilies. Members of the subfamily Nomadinae are called 'cuckoo bees.' They are parasitic in the nests of other bees, are wasp-like in appearance, and have fewer hairs on the body. Members of the common genus, *Nomada*, are reddish in color and are small to medium in size. The subfamily Anthophorinae contain bees known as 'digger bees.' They are robust and hairy. Most of them are solitary but some species nests in burrows in the ground ,and the cells are lined with a thin wax-like substance. The subfamily Xylocopinae includes 'carpenter bees.' These bees nest in wood or stems of plants. The common genera are *Xylocopa* and *Ceratina*. *Xylocopa* species are large, black, shiny, resembling the bumble bees but without tufts of hair on the abdomen. The *Ceratina* species are small, about 6 mm long, dark bluish-green in color.

Family: Megachilidae. These commonly are called the 'leaf cutting bees.' They are medium-sized, moderately robust bees, differing from other bees in having two submarginal cells in the fore wing that are about equal in size. The pollen carrying species have a tuft of hairs on the ventral side of the abdomen and not on the legs. They usually make nests in natural cavities in wood or ground and line their cells with neatly cut pieces of leaves. Most species are solitary. A few species are parasitic.

Family: Colletidae. They are 'short-tongued bees.' Segments 1 and 2 of the labial palp are short and not flat, similar to segments 3 and 4, the mesopleurum usually is without epistomal groove, and

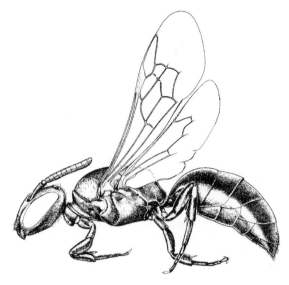

Fig. 1137 Drawing of bee in the family Colletidae (Source: Hymenoptera of the world: an identification guide to the families. Agriculture and Agri-Food Canada, 1993. Reproduced with the permission of the minister of Public Works and Government Services, 2003).

the volsella of the male genitalia is usually well developed. This family includes the yellow-faced bees (Hylaeinae) and the plaster bees (Colletinae). The glossa is bilobed or truncate. They nest in the ground or in various natural cavities.

Family: Andrenidae. These are also short-tongued bees. The glossa is pointed and there are two subantennal sutures below each antennal socket. They also nest in burrows in the ground, often in colonies and usually in areas with sparse vegetation.

Family: Halictidae. Members of this family resemble other short-tongued bees but have only one subantennal suture below each antennal socket. The basal vein is rather strongly arched. They also nest in the ground and some species, the 'sweat bees,' are attracted to perspiration.

Superfamily Sphecoidea – Sphecoid wasps

Members of this superfamily resemble Apoidea in having sexual dimorphism in the number of antennal segments and in the nature of the genitalia. However, they have simple body hairs that are not branched, the pronotum is short and collar-like, and hind tarsi are not widened. They are solitary in habits, mostly fossorial, but some construct mud nests. They are predacious and provision their nests with lepidopterous larvae, Hemiptera, Orthoptera, Arachnida, etc. The prey is stung and the eggs laid upon them in the cell. The adults visit flowers. They commonly are known as digger wasps and mud daubers. The classification of the superfamily differs considerably: some classify it into several families, while others accord subfamily status to most families.

Family: Sphecidae. Mud daubers; digger wasps. The posterior margin of the pronotum, in dorsal view, is straight. It appears like a collar. The hind margin of pronotum, in lateral view, ends in a rounded lobe that does not reach the tegula. The episternal sulcus on the mesopleurum is more vertical than horizontal. The inner margin of the eyes is not notched in most species.

Members of the Sphecidae are solitary wasps, although some species tend to nest in a small area. A small number of species in the tropics show a eusocial behavior. Most species nest in hollows in the ground and in plant stems, wood cavities, etc. Some construct mud nests. The females hunt for other insects and spiders that serve as food for the developing larvae. The prey is stung and paralyzed before placing it in the nest, and the eggs are laid upon them. The subfamily Sphecinae includes genera like *Sceliphron* and *Chalybion* that commonly are called mud daubers. They construct nests of mud and provide them with spiders. They have a very long and slender petiole. Another common species is *Sphex ichneumoneus* (L.), which is reddish brown in color with the abdominal tip black. *Trypoxylon* belongs to the subfamily Larrinae, and commonly is referred to as organ-pipe mud daubers because they construct long tubes, attaining the length of 25 mm or more. The Crabroninae contains small to medium sized species that have black and yellow markings.

Members of the family Nyssonidae includes the cicada killer, *Sphecius speciosus* (Drury). It is a large wasp, about 40 mm long, black or rusty brown with yellow abdominal bands. It provisions the nest with cicadas.

Superfamily: Chrysidoidea

In the superfamily Chrysidoidea, the number of antennal segments is the same in the male and the female, usually 10 or 11 (more than 19 in Sclerogibbidae); tergite VII of female is complete, not cleft medially; female gonocoxite IX consists of proximal

and distal portions which articulate at their point of contact; the hind wing has an anal lobe, or if the anal lobe is unclear, then the body is metallic; the hind wing is usually without closed cells; the pronotum extends back to tegula except in some Chrysididae, and the hind femur is without a trochantellus. Some species are metallic in coloration. The cuckoo wasps (Chrysididae) are external parasitoids of full grown wasp or bee larvae. Others are parasitoids of larvae of Lepidoptera and Coleoptera (Bethylidae), or parasitoids of auchenorrhynchous Hemiptera (Dryinidae).

Family: Chrysididae. The insects belonging to the family Chrysididae commonly are called the cuckoo wasps. They are small, about 8 to 12 mm in length, metallic blue or green in color, and the body bears coarse punctures. The adults fly in bright sunlight, visiting flowers in search for nectar. They females are seen flying in search of holes and crevices in search of their hymenopterous host eggs and larvae (usually aculeate Hymenoptera) upon which they lay their eggs. Upon hatching, the chrysid larva feeds on the host eggs or the larvae, as well as the stored food for the host's offspring. This phenomenon is called cleptoparasitism.

Family: Bethylidae. Members of the family Bethylidae are small to medium sized, usually black in color. The females of many species are apterous and ant-like but in other species both males and females are fully winged. These wasps are parasitoids of the larvae of Lepidoptera and Coleoptera.

Family: Dryinidae. The dryinids usually have 10-segmented antennae and large heads, and broad toothed mandibles. They are parasitoids of auchenorrynchous Hemiptera. They exhibit sexual dimorphism: the males are always winged but the females may be winged to apterous. The fore tarsi of females are developed into pincer-like chelae used to grasp and hold the planthoppers as hosts. Their life history is complex and interesting.

Superfamily: Vespoidea – Wasps and ants

The antennae bear 12 segments in the female and 13 in the male. The pronotum extends back to the tegula, its posterodorsal margin shallowly to rather deeply concave, without any lobe covering the spiracle. The metapostnotum is short and transverse, fused with the propodeum. The legs are without trochantellus (second trochanter). The fore wing has

9 to 10 closed cells. The hind wing has 2 closed cells (sometimes fewer), and has a jugal (anal) lobe. Sternites 1 and 2 are separated by a constriction. The ovipositor is modified into a short sting, issuing from the tip of the abdomen and concealed at rest. The body is devoid of plumose hairs.

The current view is that the Vespoidea includes the ants, mutillids, vespids, pompilids, scoliids and the tiphiids, all which were earlier assigned their own superfamilies. The important families are treated below.

Family: Vespidae. Yellow jackets, hornets, paper wasps, potter wasps. The fore wing has a long discal (1M) cell, much longer than the submedian (1M + Cu) cell, usually about half as long as wing, and with three submarginal cells. Wings usually folded lengthwise at rest. Posterior margin of pronotum U-shaped.

This is a relatively large group including the well-known wasps. Most species are black and yellow colored. Some species are eusocial, living in colonies. The females and the workers have a well developed sting. The social species construct a nest made out of papery material consisting of chewed up wood and plant material. Some make nests of mud.

The subfamily Vespinae includes the social wasps known as the yellow jackets and hornets belonging to the genera *Vespa*, *Vespula* and *Dolichovespula*. They construct papery nests consisting of many tiers of hexagonal cells, all enclosed in a paper envelope. These can become very large. They may be constructed in open branches, hollows in the ground or

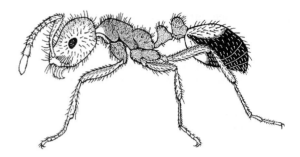

Fig. 1138 Drawing of an ant (family Formicidae) (U.S. Department of Agriculture).

tree trunks, or at any other hollow space or projecting surface around houses. The subfamily Eumeninae includes solitary wasps that construct mud nests resembling pots (potter wasps, mason wasps), commonly seen attached to houses. Most species have an elongate petiole. The subfamily Polistinae are called the paper wasps. The nests consist of a single, horizontal comb of paper cells, hanging by a slender stalk. The common genus is *Polistes*.

Family: Formicidae. Ants. One of the most distinctive features of ants is the presence of one or two node-like protuberances on the petiole. The antennae usually are elbowed, with a long first segment, and the males may have filiform antennae. The ants are eusocial insects, living in colonies, where there are three distinctive castes: the queen, males and the workers. Winged forms, consisting of males and females, develop during the swarming season and after mating the females start a colony. The female sheds the wings, locates a nesting site, makes an

Fig. 1139 Photographs of some Vespidae: left, paper wasp, *Polistes* sp. (photo J. L. Castner); right, yellow jacket, *Vespula* sp. (photo L. Buss).

excavation, and produces her first brood, which consists of workers. The workers are sterile females.

Fire ants are notorious for their burning sting. Two species, *Solenopsis invicta* Buren and *S. richerti* Forel, are common in the southeastern United States. These ants are rather aggressive and are quick to attack people and animals when disturbed. In some persons they may cause serious allergic reactions and may even lead to the death of the individual.

Family: Pompilidae. Spider wasps. The pompilids are slender wasps with spiny legs. They have a transverse suture across the mesopleurum. They prey on spiders, sting to paralyze them, and store them in their underground cells as food for the developing larvae.

Family: Mutillidae. Velvet ants. The females are wingless, ant-like, and are covered with dense hairs. The males are winged, usually larger than the females, and also have dense hairs. A usual distinguishing feature is a structure called the 'felt-like' along the side of the second abdominal segment. Most mutillids are external parasitoids of the larvae and pupae of wasps and bees.

See also, ANTS, BEES, FAIRY FLIES, HARVESTER ANTS, HONEY BEE, SAWFLIES, PARASITICA, VELVET ANTS, TIPHIID WASPS.

Virendra K. Gupta
University of Florida
Gainesville, Florida, USA

References

Arnett, R. H., Jr. 1993. *American insects. A handbook of the insects of America north of Mexico.* The Sandhill Crane Press, Inc., Gainesville, Florida. 850 pages.

Borror, D. J., C. A. Triplehorn, and N. F. Johnson. 1989. *An introduction to the study of insects.* Saunders College Publishing, Philadelphia, Pennsylvania. 872 pages.

Gauld, I., and B. Bolton (eds.) 1988. *The Hymenoptera.* British Museum (Natural History) and Oxford University Press, Oxford, United Kingdom. 332 pages.

Goulet, H., and J. T. Huber (eds.) 1993. *Hymenoptera of the world: an identification guide to families.* Agriculture Canada, Canada Communication Group Publishing, Ottawa, Canada. 668 pages.

Richards, O. W., and R. G. Davies. 1957. *A general textbook of entomology.* Methuen & Co. Ltd., London, United Kingdom. 886 pages.

WATER BUGS. Members of the family Corixidae (order Hemiptera). See also, BUGS.

WATER MEASURERS. Members of the family Hydrometridae (order Hemiptera). See also, BUGS.

WATER PENNY BEETLES (COLEOPTERA: PSEPHENIDAE).

Anyone who picks up rocks out of streams or from lakeshores is probably familiar with water pennies, the tiny beetles that are found on the undersides of rocks in the water. The common water penny, larva of *Psephenus*, may be difficult to see because it resembles the surface of a rock and can cling tightly to the substrate. During daylight hours, water pennies are normally on the underside of stones, but at night they creep to the upper surface to dine upon the algae that thrive there in the sunlight. Over-wintering occurs during the larval stage. When mature, the larva crawls out of the water and settles upon a protected solid substrate such as the side or the underside of a rock near the water. Here, the larva becomes attached to the substrate all around the ciliated perimeter. The pupa then develops beneath the protective carapace of the larval skin. After 10 to 12 days, the adult water penny beetle emerges from the pupal skin. A median dorsal longitudinal crack in the tent-like larval carapace enables the adult to escape. Weather permitting, it flies to a rock projecting above the surface of the stream in a nearby riffle. If it is male, it joins the other male water pennies hurrying around in search of females. The females do not seek mates. They wait for the eager males to find them. After mating, the female water penny crawls down into the water and spends the rest of its life (a few days at most) seeking suitable oviposition sites and laying eggs on the underside of rocks in the riffles. The eggs, usually yellow, form a layer one egg deep. Clusters on small rocks may contain only a few dozen eggs, but clusters on large rocks usually contain hundreds of eggs, and may be deposited by several females.

After 10 to 15 days, tiny, almost invisible water pennies hatch. As they grow larger, they must shed their outgrown skins about six times before they leave the water to pupate. Their entire life span is about 21 to 24 months in Ohio and Michigan, but probably less in warmer regions. All significant feeding is done during the larval stage. Only one species of *Psephenus*, *P. herricki*, occurs in eastern and Midwestern United States and Canada, but several other species occur in western states, Mexico, and some Central and South American countries. In Neotropical

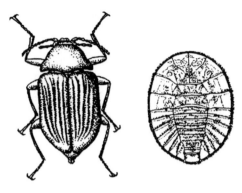

Fig. 1140 *Psephenus texanus* (Subfamily Psepheninae) left, adult male, 4 to 5.5 mm long, brown to black; found chiefly on rocks projecting above the water in stream riffles in southwestern Texas and north central Mexico, flying readily from rock to rock; right, larva, common water penny, usually found clinging tightly to the undersides or sides of relatively smooth stones under water, in riffles.

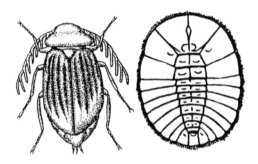

Fig. 1141 *Eubrianax edwardsii* (Subfamily Eubrianacinae) left, adult male, 3 to 4.5 mm long, yellowish to black; along rocky streams in California and Oregon; right, larva, another type of water penny, usually found clinging to submerged stones in stream riffles.

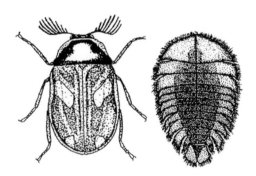

Fig. 1142 *Acneus quadrimaculatus* (Subfamily Eubriinae or Family Eubriidae) left, adult male, 3.5 to 4.5 mm long, dark brown or black with yellow or orange spots on elytra; found along small, rapid, low-elevation shady streams in California and Oregon, usually near small waterfalls; right, larva, false water penny, found in such streams or in spray zone of falls.

regions and Asia, there are also other genera of Psephenidae with larvae like those of *Psephenus*.

The larvae of members of the subfamily Eubrianacinae are enough like those of *Psephenus* to share the common name, "water pennies." In both subfamilies, the larvae have paired, feather- like, tracheal gills on the ventral surfaces of four or more abdominal segments. *Eubrianax* occurs in all zoogeographic realms but Australia. Members of the subfamily *Eubriinae* (or to many coleopterists, the separate Family Eubriidae) are called "false water pennies." Larvae of some Eubriine genera (e.g., *Dicranopselaphus*) look so different that no one would be likely to call them "water pennies" of any sort, and they are too uncommon to be given a common name. Seen from above, the larvae of *Afroeubria* (residents of Africa) greatly resemble the larvae of *Psephenus*, but the underside of all Eubriine larvae gives them away. Instead of paired feathery gills on several abdomimal segments, as in *Psephenus* and *Eubrianax* larvae, Eubriines have a single, retractable, tripartite gill located medially on the underside of the last (ninth) abdominal segment. Adult Eubriines are less conspicuously associated with water than are adult Psephenines.

None of the water penny beetles are of known economic importance to man, but they are of interest to most people who are familiar with them.

See also, BEETLES.

Harley P. Brown
University of Oklahoma
Norman, Oklahoma, USA

References

Brown, H. P. 1983. A catalog of the Coleoptera of America north of Mexico, Family: Psephenidae. U.S. Department of Agriculture, Agriculture Handbook Number 529–41, prepared by Agricultural Research. 8 pp.
Brown, H. P. 1987. Biology of riffle beetles. *Annual Review of Entomology* 32: 253–273.
Brown, H. P., and C. M. Murvosh. 1974. A revision of the genus *Psephenus* (water penny beetles) of the United States and Canada (Coleoptera, Dryopoidea, Psephenidae). *Transactions of the American Entomological Society* 100: 289–340.
McCafferty, W. P. 1981. *Aquatic entomology (subtitled The fishermen's and ecologists' illustrated guide to insects and their relatives)*. Science Books International, Boston, Massachusetts. 488 pp.
Murvosh, C. M. 1971. Ecology of the water penny beetle *Psephenus herricki* De Kay. *Ecological Monographs* 41: 79–96.

WATER POLLUTION AND INSECTS. The impact of human activities on water resources became clear with the advent of the Industrial Revolution. Prior to this, it was mainly a localized phenomenon. For example, in the 18th century, the River Thames, which flows through London, England, produced such a terrible stench from raw sewage that sheets soaked in vinegar were hung in the Parliament to offset the nauseating smell. However, the English government only started to control the sewage pollution that produced these odors in the face of typhoid fever and cholera epidemics. The protection of human health and the management of ecosystems to provide essential ecological services (e.g., drinking water, waste elimination, forestry products) remain the current basis of pollution control regulation and technology in the developed world.

Pollution has been defined in various ways, but essentially it is the wrong substance in the wrong concentration, at the wrong place, at the wrong time. Water pollution can have many sources, but the most common inputs are human sewage and outputs from industrial and agricultural practices. Typically, pollution is defined as the result of anthropogenic (or human-induced) activities.

Pollution can cause changes in the structure (e.g., biodiversity) and function (e.g., primary and secondary productivity) of ecosystems. Insects are one of the major, if not the major group of organisms, that has been studied in terms of pollution. Because they are at or near the bottom of food webs, changes in insect populations will affect the remainder of the food web. In addition, insects can mark the entry point of noxious chemicals into the food web, and the economic consequences can be profound. For example, the destruction or diminishment of insect populations in fresh waters will affect the vertebrate predators such as fish and waterfowl that feed on them.

The effects of pollution on insects have to be studied at a variety of spatial and temporal scales. Effects can occur at the molecular level in fractions of seconds (biochemical effects), at the ecosystem level over several decades, and at various scales in between these extremes.

Because some insect species are sensitive to various forms of pollution while others are tolerant of these pollutants, the responses of insects have been used to detect, measure, and judge environmental disturbance. Insects are useful in biomonitoring because they are ubiquitous and have many pertinent, readily measurable biological characteristics. Food is the principal source of many pollutants for terrestrial insects, whereas aquatic insects acquire residues from both food and the abiotic environment. In this entry, we will consider both the effects of water pollution on insects and the use of insects in the biological monitoring of water quality.

Detection of pollution effects in insects

A pollutant may cause deleterious effects if it influences the physiology or behavior of an organism, alters its capacity for growth, reproduction, or mortality, or changes its capacity for dispersal. Successful identification of pollution effects in insects depends on a number of conditions surrounding most studies. First is the need for adequate baseline information. Long-term monitoring of undisturbed populations is a prerequisite for detection of disturbance because natural successional changes, catastrophes, and population fluctuations may be mistakenly interpreted as pollution-caused trends in short-term studies. Historical data or paleoenvironmental approaches may provide the needed information in the absence of active, long-term, field-monitoring programs.

Second, causation can only be proven using an experimental approach. Experiments can also pinpoint early stage responses to pollution through an understanding of initial changes at different spatial and temporal scales. There is a need to link laboratory results and field surveys through field experimentation, but the experimental design must be suitable to the questions being asked to provide unequivocal answers.

Third, the species is the basic biological unit, so it should be the level at which responses to pollution are measured. There are many examples of species in the same genus responding differently to the same pollutant and dose; the lumping of taxa is a disservice in studying pollutant effects. Better autoecological knowledge of species is a prerequisite to understanding of pollution effects. Having expressed the ideal situation, practical realities such as taxonomic problems and cost often mean that higher taxonomic levels are used in biomonitoring programs.

Many abiotic and biotic factors directly or indirectly affect an organism's response to a pollutant, so it is difficult to predict the effects of pollutants and their different doses in nature. Some of the abiotic factors include: type of pollutant, concentration, duration and degree of exposure (chronic acute, lethal versus sublethal), time of year (temperature), quality of the water and sediment in which the organism lives, and interactions with other pollutants. Some of the biotic factors include: life-cycle stage (instar, age, rate of growth and development), health of the organism (parasitized or not), genetic composition (resistance), behavior with regard to exposure (cryptic or directly exposed), interactions with other species, and host alterations (important for phytophagous insects). Some of these factors are discussed more fully below.

Acute toxicity may have locally or temporarily significant effects on mortality, but sublethal effects may be more important in terms frequency of occurrence and length of exposure. Sublethal effects include disturbed enzyme functions, reduced reproduction, and altered behavior, all of which act to reduce an organism's fitness and affect its survival.

Indirect effects of pollutants on ecosystems, as mediated by insects, are often more profound than direct effects. Pollutants can influence the functioning of ecosystems by reducing species diversity, modifying food chains, and changing patterns of energy flow and nutrient cycling. For example, numbers of insects in groups naturally suppressed by fish predation tend to increase if fish are reduced by pollution such as acidification. Increased numbers of insect predators can then consume substantial amounts of small crustacean zooplankton, changing the natural relationships among trophic levels and, perhaps, affecting secondary productivity.

Life-cycle stages vary in their sensitivity to pollutants. For example, eggs are generally resistant stages, except when the pollutant is heat. Early instars seem to be more susceptible than later instars, perhaps because early instars have higher surface to volume ratios, they are more active, and they have thinner cuticles. Diapausing and pupal insects are also usually resistant stages.

It is difficult to generalize about the effects of pollution for all but the best-studied insect taxa. Most typically, some species within an insect taxon will respond positively, some will respond negatively, and some will be unaffected by a pollutant. In aquatic habitats, midges (Chironomidae), especially members of the subfamily Orthocladinae, seem to be tolerant of metal pollution. Some species of midges thrive in habitats heavily contaminated by organic matter (sewage, nutrients). Some species of mayflies (Ephemeroptera), stoneflies (Plecoptera), and caddisflies (Trichoptera) are sensitive to a number of pollutants, and have been used as indicator taxa in biomonitoring activities. In addition, most pollutant-induced changes in predators are negative, whereas changes in herbivores are often positive and are mediated through the host plant.

Relationships among exposure, dose, and effects on individual organisms (the effects of pollutants on insect populations are mediated through individuals) are complex. Different species, sexes, and life stages may react in different ways to the same exposure for both genetic and environmental reasons. Biotic interactions between individuals in a population and species within a community are also complicating factors. Therefore, generalizations for most insect groups will only be possible with the acquisition of much more taxon-specific data.

Responses of aquatic insect groups to pollution

Aquatic insects comprise all or part of 11 orders of the class Insecta and characteristically show different responses to pollution inputs, both among orders and within orders. A study done in the 1970s even demonstrated that when water quality tolerances were available for more than a single species within the same genus, the majority of genera had species listed as tolerant, facultative, or intolerant to pollution!

Typically, at the order level, three groups are considered as particularly intolerant to pollution effects: the Ephemeroptera, the Plecoptera, and the Trichoptera. The 'EPT' orders (initials taken from the first letter of each order name) are also among the best known taxonomically; consequently, differential responses have been the best studied. For example, mayflies are known to be sensitive to acidification, and show reduced densities and/or richness at low-pH sites, which often result from acid deposition through precipitation (i.e., acid rain). Stoneflies are intolerant of oxygen loss (and they tend to be only found in highly oxygenated, cool-water habitats). Caddisflies are sensitive to both organic and inorganic contaminants.

However, not all EPT species are sensitive to all pollutants. Some mayflies are among the most tolerant of all aquatic insects to acidification and metal pollution. Some caddisflies occur in heavily polluted sections of European rivers. If the EPT orders are considered the most sensitive of the insects to pollution, the Diptera represent the other end of the spectrum. Increases in the percentage composition of the insect community (or even the benthic macroinvertebrate community, which includes insects, mollusks, Crustacea and other invertebrates) typically occur below releases of poorly functioning sewage treatment plants, industrial plants, and other anthropogenic inputs. Typically, these increases are in individuals of the family Chironomidae, the bloodworms (based on the hemoglobin pigment that gives a bright red color to some of the larvae). However, as mentioned for EPT, there is a broad range of tolerances not only among the species in this order, but even in the family Chironomidae.

Among the other aquatic insect orders, pollution effects have been noted as well. Species of Odonata (dragonflies and damselflies) have had their distributional range reduced because of increased eutrophication caused by the discharge of nutrients from agriculture and sewage. Hemiptera (bugs) have surface-dwelling species that are adversely affected by inputs of soap and other surfactants, and many of the water-column dwelling taxa are known to be sensitive to pesticide applications (either direct or as runoff).

Although it is evident that pollution response is best explained at the species level, tolerance values for water pollution have been developed for the different families of aquatic insects.

Tolerance values for aquatic insects. Average values for all taxa of aquatic insects reflect water quality, with 0–3.7 Excellent; 3.8–4.2 Very good; 4.3–5.0 Good; 5.1–5.7 Fair; 5.8–6.5 Fairly poor; 6.6–7.2 Poor; 7.3–10.0 Very poor.

Odonata
    Aeshnidae 3
    Calopterygidae 5
    Coenagrionidae 9
    Cordulegastridae 3
    Corduliidae 5
    Gomphidae 1
    Lestidae 9
    Libellulidae 9
    Macromiidae 3
Ephemeroptera
    Baetidae 4
    Baetiscidae 3

    Caenidae 7
    Ephemerellidae 1
    Ephemeridae 4
    Heptageniidae 4
    Leptophlebiidae 2
    Metretopodidae 2
    Oligoneuridae 2
    Polymitarcyidae 2
    Potomanthidae 4
    Siphlonuridae 7
    Trichorythidae 4
Plecoptera
    Capniidae 1
    Chloroperlidae 1
    Leuctridae 0
    Nemouridae 2
    Perlidae 1
    Perlodidae 2
    Pteronarcyidae 0
    Taeniopterygidae 2
Trichoptera
    Brachycentridae 1
    Calamoceratidae 3
    Glossosomatidae 0
    Helicopsychidae 4
    Hydropsychidae 4
    Hydroptilidae 4
    Lepidostomatidae 1
    Leptoceridae 4
    Limnephilidae 4
    Molannidae 6
    Odontoceridae 0
    Philpotamidae 3
    Phryganeidae 4
    Polycentropodidae 6
    Psychomyiidae 2
    Rhyacophilidae 0
    Sericostomatidae 3
    Uenoidae 3
Megaloptera
    Corydalidae 0
    Sialidae 4
Lepidoptera
    Pyralidae 5
Coleoptera
    Dryopidae 5
    Elmidae 4
    Psephenidae 4
Diptera
    Athericidae 2

Blephariceridae. 0
Ceratopogonidae. 6
Blood-red
    Chironomidae 8
    Other Chironomidae 6
    Dolichopodidae 4
    Empididae. 6
    Ephydridae. 6
    Psychodidae. 10
    Simuliidae 6
    Muscidae. 6
    Syrphidae 10
    Tabanidae 6
    Tipulidae. 3

Responses of aquatic insects to pollution

The continued presence of aquatic insects in a given habitat depends on their ability to survive the water quality and habitat conditions of the aquatic environment present there. Contaminants in the water can range from deficiencies (e.g., oxygen) to excess (e.g., copper) of certain components, to contamination with substances that are increasingly deleterious as concentrations increase (e.g., pesticides). However, while most species of aquatic insects react negatively to most contaminants (e.g., oil, sedimentation) there are some species that thrive. As a result, the characteristic response to contamination is that intolerant species decline, species tolerant to that specific type of contamination will appear over time or if present in low numbers increase over time, and the evenness of the community (or the distribution of individuals among the taxa present) and the composition change.

It is important to remember that, in addition to water quality, the continued presence of aquatic insects depends on habitat conditions as well. For example, if water quality conditions are not changed (i.e., no contaminants are added) but habitat declines (e.g., the riparian vegetation is removed), aquatic insect composition will change just as if water quality impairment occurred. Of course, habitat and water quality are related; for example, a change in bank stability will result in increased siltation, and consequent water quality changes caused by suspended sediments.

Below, we describe various pollution inputs that affect aquatic insects. The list is not all-inclusive

and the individual effects where contaminants are co-occurring may not just be additive but may be synergistic. In each case described below, the response of insects depends on the severity of the contamination (which can vary depending on past history at a site), habitat conditions, and a variety of other factors. It is important to remember that in all these examples the response is similar: species that are intolerant of that contaminant will be eliminated, those that are tolerant will survive, those that are tolerant but not represented in the habitat before contamination may appear if they are present in the local species pool, and the composition of the aquatic insects will change.

The disposal of domestic and industrial wastes has been a major challenge of civilization. Anthropogenic inputs can be divided into three types: organic, which includes domestic sewage, industrial wastes from food processing plants, and fertilizers; chemical, which includes industrial byproducts and agricultural chemicals such as insecticides; and physical, which includes habitat disruption and its consequences (e.g., erosion and siltation). Each input differs in its effects on aquatic insects, depending on concentration, duration of exposure, and various other factors.

Organic wastes. With inputs of organic material, the general effect is a decrease in dissolved oxygen. This effect has led to the development of the classic pollution-recovery pattern in streams that has been well known for over 100 years. In this pattern, dissolved oxygen concentrations, water conditions, and the fish, insect, and algae assemblages are described over zones that represent areas above the pollution source, within the source, below the source, and far below the source. In the upstream-most area, oxygen concentration is high, the water is clear, and fish populations are characterized by a mixture of predaceous, algivorous, and detritivorous fish; insect groups are dominated by species of Ephemeroptera, Plecoptera, and Trichoptera, and the periphyton includes *Navicula* diatoms and *Oedegonium* and *Dinobryon* algae. Where pollution inputs enter, oxygen decreases, the water becomes darker and turbid, and only tolerant fish occur – carp, buffalo, gar, and catfish. The insect assemblage becomes dominated by chironomid midge larvae and simuliid (black fly) larvae. *Paramecium, Beggiatoa* and *Stentor* dominate planktonic assemblages. As pollution worsens, oxygen drops to extremely low levels, the water is septic with noxious odors and floating sludge. No fish occur and only mosquito larvae (e.g., *Culex*), very tolerant dipterans such as *Eristalis*, the rat-tailed maggot, and oligochaetes such as *Tubifex* thrive. Pollution-tolerant plankton such as sewage fungus (the colonial bacteria *Sphaerotilus*), the blue-green algae *Oscillatoria* and the diatom *Melosira* predominate.

Of course, the solution to pollution has always been dilution. Recovery occurs downstream as oxygen concentration increases. If no new organic pollution sources are added, stream water quality returns through the second and then to the first pollution-recovery zone.

Probably the most widespread anthropogenic change to aquatic environments worldwide is the input of organic matter and nutrients such as nitrogen and phosphorus. Usually referred to as eutrophication, the typical causes of this type of pollution are sewage inputs containing human or animal wastes, agricultural fertilizers, or even industrial detergents. In North America and Western Europe, direct inputs of untreated sewage have been greatly reduced but inputs of mineral nutrients, which result in increases in algae, remain. Although algae produce oxygen by photosynthesis, their respiration requires oxygen and night-time reductions in oxygen concentrations are the result of eutrophication.

Chemical wastes. In contrast to organic wastes, chemical inputs usually affect aquatic organisms through their direct toxic action, although indirect effects such as changes in pH or osmotic pressure may affect aquatic insects. Our understanding of the effects of chemical wastes on aquatic insects largely comes from toxicity tests conducted with single or multiple species of insects for either short or long durations. Toxicity testing is also a tool used in biomonitoring, and is described later.

Not unexpectedly, given that insects are their target organisms, most insecticides used to control agricultural pests exhibit toxicity to at least the aquatic insects that are related to the target pests. Even the widely used *Bti* biological insecticide (*Bacillus thuringiensis israelensis*) used to control black flies and mosquitoes shows some toxic effects on other, closely related Diptera.

A type of pollution that has caused major effects on aquatic insects is acidification. The main cause of this type of pollution is considered to be fossil fuel combustion and the long-distance transport of SO_2 and NO_x pollutants with their eventual deposition during precipitation. Acidification affects aquatic insects through changes in physiology, increases in trace metal concentrations that are toxic to some organisms, and indirectly through changing food availability by altering either photosynthetic or decomposition pathways.

Mayflies are the best studied of all invertebrates in terms of acid stress, and responses to low pH range from increases in drift rates (i.e., in which mayflies leave the substrate and 'drift' downstream), reduced growth, higher respiration rates, and avoiding oviposition in acidic streams. While most Ephemeroptera species decline as pH is lowered, other more tolerant species may increase but, in general, there is an overall decrease in insect richness and productivity.

The differential responses of aquatic insects to changes in oxygen concentration and pH are evident from the results of numerous studies. Only Diptera predominate in the reduced oxygen areas of streams (the poly- and meosaprobic zones) and many mayflies are limited in their distribution to waters with pH > 5.0.

Physical wastes. The third category of pollution, physical wastes, generally affects aquatic insects by being abrasive or injurious to their gills, interfering with respiration, reducing food supply by reducing light penetration, or filling the interstitial spaces of substrates in which aquatic insects dwell. Habitat alterations that increase erosion are the major sources of these inputs.

Pollution sources range from municipal wastes and other consequences of urbanization, to livestock and other agricultural activities, to industrial inputs of toxic metals. To some, the changes induced by damming streams fall into the area of pollution, as does the issue of introduced species and their impact on native fauna. The literature on pollution is vast, and the individual pollutants are almost infinite in number. While specifics vary, the same pattern emerges: intolerant species are eliminated, tolerant species thrive, and the composition of aquatic insect communities under the stress is altered.

Use of insects in biomonitoring

Biological monitoring is the systematic use of living organisms, or their responses, to determine

Number of species in the different orders of aquatic insects showing different tolerance to organic pollution and acidification. 'Poly' refers to the polysaprobic zone, which is a zone grossly polluted by organic matter; 'α-meso' and 'β-meso' refer to zones of increasing oxygen, and 'oligo' refers to a zone of high oxygen content. (from Rosenberg and Resh 1993)

Order	Poly	α-meso	β-meso	Oligo	pH > 5.5	5.5 < 5.0	5.0 < 4.7	< 4.7
Odonata	0	2	17	5	0	0	5	2
Ephemeroptera	0	6	20	23	21	20	9	13
Plecoptera	0	0	4	24	0	8	0	14
Hemiptera	0	4	2	0	0	0	2	0
Trichoptera	0	4	18	31	1	10	7	33
Megaloptera	0	2	3	2	0	0	1	2
Diptera	4	44	36	62	9	3	35	41
Coleoptera	0	8	19	18	0	0	1	0

the quality of the environment. Biological monitoring has been a major component of water quality assessments for the past two decades. Although prior to this, assessments primarily involved chemical and physical measurements, the inclusion of a biological component has greatly improved the process. First, because water pollution is essentially a biological problem, it only makes sense to evaluate it by biological means. Second, and perhaps best explained by analogy, water quality monitoring that only involves chemical and physical measurements is akin to a photograph, an instantaneous documentation of conditions at the time a sample was taken. In contrast, biomonitoring is akin to a videotape; organisms are present through the range of environmental conditions present, so they add a temporal component of past conditions to the monitoring program.

Of all groups that have been recommended for use in water quality assessments (such as fish, algae, and protozoans), macroinvertebrates are the group most frequently used and recommended. Why is this so? First, they are ubiquitous in occurrence and consequently are affected by perturbations in many different kinds of aquatic habitats. Second, there is a large number of species (perhaps 25,000 species of aquatic insects in the US and Canada alone) that exhibit a range of responses to environmental stress. Third, their sedimentary nature (or the sudden departure from it) tends to illustrate the spatial extent of a problem. Finally, they have relatively long life cycles, relative to many other groups of organisms, which allow the examination over time of features such as age structure and abundance. To be fair, the use of macroinvertebrates in biomonitoring also has disadvantages, but most of these can be accounted for in the design of biomonitoring programs.

The tradition of biological monitoring is over a century old and initially was based on the idea of indicator organisms, in which the presence of certain types of organisms in themselves were indicative of pollution status. In part, this approach was an outgrowth of the lake classification system developed by German limnologists (especially August Thienemann) that described lake categories in terms of the dominant species of Chironomidae present. Today we think of the indicator concept as rather simplistic; the assemblage of organisms present is now viewed as the best measure indicating pollution.

In the past 30 years that aquatic insects have been used in biomonitoring, there has been a shift in the type of study that has been conducted in North America. In the 1970s, emphasis was on qualitative assessments with relative (rather than absolute, i.e., number of individuals/area) samples being compared from control and test areas. However, by the next decade emphasis had switched to studies involving replicated absolute (fixed-area) samples and using inferential statistical tests. By the 1990s, emphasis was again on non-quantitative sampling and analysis involving comparisons among sites in terms of specific measures (usually referred to as metrics) or multivariate statistics in which each taxon is a variable in the analysis. Part of the key to this shift was the development and use in the 1990s of Rapid Biomonitoring Protocols, an approach encouraged by the US Environmental Protection Agency, described in detail below.

The 'reference-condition approach' (RCA) has provided a powerful approach for large-scale biomonitoring programs. The 'reference condition' represents a group of minimally disturbed sites organized by selected physical, chemical, and biological characteristics. The RCA is an alternative to more traditional comparisons of control vs. impacted sites in bioassessment. Sites suspected of being impacted are compared to suitable reference sites using multivariate statistics. The underlying models produce output that shows the degree of impairment of impacted sites compared to reference sites. The RCA is currently being used in national biomonitoring programs using macroinvertebrates in countries like the United Kingdom and Australia. A slightly different RCA, using 'multimetrics', has been developed for water quality monitoring in the U.S.A.

The practice of biomonitoring

Biomonitoring approaches are available that cover hierarchical scales ranging from processes that occur at the molecular level to those that occur at the ecosystem level. In practice, however, most biomonitoring with aquatic insects is conducted at the species assemblage and community levels, and these levels are what we will describe in detail.

Species assemblages and community level biomonitoring involves studies that include replicated sampling and statistical analysis. The distinction between qualitative and quantitative methods is often blurred in that many rapid bioassessment programs also involve replicate sampling.

Species richness. No matter which approach is used, similar measures (or metrics) are commonly used to distinguish between impaired and unimpaired aquatic insect assemblages and communities. The most commonly used measure is that of species richness (i.e., the number of species present), whether used for the whole community or perceived intolerant groups such as the EPT taxa. The use of species richness is based on the premise that the number of taxa decreases as water quality decreases because of the elimination of the taxa that are intolerant of a certain type of pollution. An older idea was that the number of individuals of these taxa also decreased but current evidence indicates that elimination occurs. A difficulty with using species richness is that identification, or at least the ability to distinguish taxa to count them, is required.

Abundance. Enumerations of numbers of individuals (either of the whole communities or of insect assemblages such as EPT) are often used but, in practice, are often difficult to interpret because of the high natural variability in aquatic systems. Spatial variation in substrate, flow conditions, and a myriad of

Biomonitoring approaches used at different spatial scales. EPT = Ephemeroptera, Plecoptera, and Trichoptera.

Hierarchical level	Examples	Advantages/disadvantages
Molecular	Changes in enzyme activity of respiratory metabolism	Subcellular levels should be sensitive, early warning indicators of stress; limited by a basic knowledge of biochemical and physiological processes in most aquatic insects
Individual	Morphological deformities; deviation from normal behavioral patterns; survival, growth, and reproduction success	Methods well developed, but indicators of stress are often qualitative
Population	Numbers of individuals; size spectra of population	Taxonomic difficulty; separation of natural variability from that caused by perturbation
Species Assemblages	Proportion of individuals in perceived pollution-intolerant groups (e.g., EPT)	Most groups viewed as intolerant are actually a mixture of intolerant and moderately tolerant taxa
Community	Taxa richness; diversity and similarity indices; biotic indices	Taxonomic difficulties; sampling problems to adequately represent proportional occurrences
Ecosystem	Structure of food webs; chemical cycling; primary and secondary production	Provides holistic view of problem; time consuming and costly

other factors produce high- and low-density patches of organisms.

A classic pattern in pollution response is the elimination of certain intolerant taxa and drastic increases in tolerant taxa. This response commonly occurs with certain species of Oligochaeta and even some chironomid midge larvae in response to severe pollution conditions, especially where high amounts of organic matter result in drastic decreases in oxygen concentrations. However, this type of pollution is not the only one that induces density increases. The chironomid midge larva, *Cricotopus bicinctus*, which tends to commonly occur in low densities, often increases by orders of magnitude when the habitats it occurs in are exposed to contaminants ranging from oil to chromium.

Evenness and diversity. Shifts in abundance of insect populations are the basis for evaluating the evenness or diversity (as opposed to richness). Evenness refers to the distribution of individuals among the taxa present. For example, two communities of stream insects may each contain 10 species and 100 individuals; however, their evenness will differ if the first community has 10 individuals of each species and the second has one species with 91 individuals and the rest with one individual each.

Evenness is often calculated along with diversity indices, an approach that was widely used in the 1970s but is considered as a less reliable tool for biomonitoring by many researchers today. Depending on the formula used for calculating the diversity of a community (or species assemblage), the index value can be weighted more towards the richness or evenness component.

Biotic indices. Biotic indices are the evaluation tool that is both widespread in use and unique to water quality monitoring programs using aquatic insects. In calculating a biotic index score, organisms collected are assigned values according to their tolerance or intolerance (ideally) to the pollutant being evaluated, but in actuality most values are based on responses of the insect to organic pollution. Examples of tolerance values are often used at the family level for many rapid bioassessment procedures in North America but in the rest of the world they are usually used at genus or species level, an approach that is more commonsensical given that there is high variability in pollution tolerance and intolerance within

aquatic insect families. The actual calculation of a biotic index score involves multiplying the number of individuals of each taxon by their tolerance values, summing them, and dividing this sum by the number of individuals collected.

Functional feeding groups. A recent addition to biomonitoring procedures has been the incorporation of functional feeding group (FFG) measures into programs. FFG measures differ from those described above in that the former emphasizes characterization of the structural aspects of the community, while FFGs emphasize the functional aspects. The use of FFGs in biomonitoring is based on the premise that organisms evolved certain morphological-behavioral food-gathering mechanisms and can be placed in particular groups. Consequently, each of these groups is expected to occur in proportionally higher abundance in accumulations of particular food sources (or habitat types). Most biomonitoring programs characterize the species encountered as belonging to one of several functional groups: scrapers (consume attached algae); shredders (consume decomposing vascular plant tissue); collectors (consume decomposing fine particulate organic matter); and predators (consume living animal tissue). Leaf packs, for example, should have a high proportion of shredders; algae-covered rocks should have more scrapers. Departures from expected proportions could indicate contamination or disturbance (e.g., a decrease in shredder abundance could indicate deleterious changes in the riparian zone that provides the food staples for the shredders to use).

A characteristic of community based approaches used in rapid bioassessment procedures involves calculation of combination indices of many of the measures described above. While this spreads the risk of incorrect assessments using single measures, it is also perceived as giving a balance to the various measures used. The inclusion of combination indices of the various measures has given rise to the name 'multimetric approaches', which (incorrectly) has been considered as the same as rapid bioassessment procedures.

Other techniques. Two other approaches must be mentioned in the use of aquatic insects in biomonitoring: toxicity studies and paleolimnological reconstruction. Toxicity testing has the longest history of any technique in biomonitoring, and is supposedly traceable to Aristotle placing a freshwater fish in seawater to observe its reaction! Today, toxicity tests form a continuum of complexity from single-species, laboratory-based, short-term exposures, to long-term single- and multiple-species tests, to field-based toxicity testing in outdoor experimental systems. Many species of aquatic insects are used in this testing, and results of studies are used to regulate discharges, compare toxicants, and predict environmental effects of their use.

Paleolimnological approaches to biomonitoring involve using sediment cores to reconstruct the history of aquatic systems that have been exposed to impacts such as eutrophication, acidification, or climate changes. This approach is confined to lakes and large rivers, and is not widely used by regulatory agencies.

Conclusion

Water pollution is a biological problem. Consequently, the effects of pollution on insects, the most speciose group usually present in fresh waters, is expected to be a significant outcome of anthropogenic impacts. Likewise, because they show varied responses to pollution, insects are the logical basis for monitoring effects. This fact has been recognized by the existence of thousands of government-sponsored and volunteer programs all over the world that use aquatic insects to monitor water quality.

Vincent H. Resh
University of California
Berkeley, California, USA
and
David M. Rosenberg
Freshwater Institute
Winnipeg, Manitoba, Canada

References

Fjellheim, A., and G. G. Raddum. 1990. Acid precipitation: biological monitoring of streams and lakes. *Science of the Total Environment* 96: 57–66.

Hall, R. J., and F. P. Ide. 1987. Evidence of acidification effects on stream insect communities in central Ontario between 1937 and 1985. *Canadian Journal of Fisheries and Aquatic Sciences* 44: 1652–1657.

Heliövaara, K., and R. Väisänen. 1993. *Insects and pollution*. CRC Press, Boca Raton, Florida.

Hellawell, J. M. 1986. *Biological indicators of freshwater pollution and environmental management*. Elsevier Applied Science Publishers, London, United Kingdom.

Hilsenhoff, W. L. 1988. Rapid field assessment of organic pollution with a family-level biotic index. *Journal of the North American Benthological Society* 7: 65–68.

Kerr, M., E. Ely, V. Lee, and A. Desbonnet. 1994. *National Directory of Volunteer Environmental Monitoring Programs*. 4th edition. EPA 841-B-94-001. US Environmental Protection Agency and Rhode Island Sea Grant, University of Rhode Island, Narragansett, Rhode Island.

Lenat, D. R., and V. H. Resh. 2001. Taxonomy and stream ecology – The benefits of genus- and species-level identifications. *Journal of the North American Benthological Society* 20: 287–298.

Resh, V. H., M. J. Myers, and M. J. Hannaford. 1996. Macroinvertebrates as biotic indicators of environmental quality. pp. 647–667 in F. R. Hauer and G. A. Lamberti (eds.), *Methods in stream ecology*. Academic Press, San Diego, California.

Reynoldson, T. B., R. H. Norris, V. H. Resh, K. E. Day, and D. M. Rosenberg. 1997. The reference condition: a comparison of multimetric and multivariate approaches to assess water-quality impairment using benthic macroinvertebrates. *Journal of the North American Benthological Society* 16: 833–852.

Rosenberg, D. M., and V. H. Resh (eds.) 1993. *Freshwater biomonitoring and benthic macroinvertebrates*. Chapman and Hall, New York, New York.

Saether, O. A. 1979. Chironomid communities as water quality indicators. *Holarctic Ecology* 2: 65–74.

WATER SCAVENGER BEETLES. Members of the family Hydrophilidae (order Coleoptera). See also, BEETLES.

WATERSCORPIONS. Members of the family Nepidae (order Hemiptera). See also, BUGS.

WATER SPRINGTAILS. A family of springtails (Poduridae) in the order Collembola. See also, SPRINGTAILS.

WATER STRIDERS. Members of the family Geridae (order Hemiptera). See also, BUGS.

WATER TREADERS. Members of the family Mesoveliidae (order Hemiptera). See also, BUGS.

WAX. A substance secreted by insects from glands located on various parts of the body. Wax serves a protective function for most insects, but bees use wax to build cells in their hives. The term "wax," as generally applied to the lipids on an insect's cuticle, is a complex mixture of lipids, most of which do not fit the chemical definition of "wax."

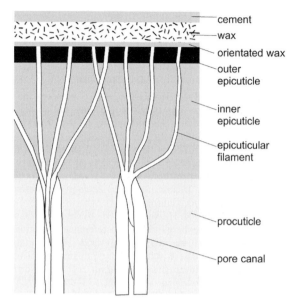

Fig. 1143 Cross section of the insect epicuticle (adapted from Chapman, The insects: structure and function).

WAX CHANNELS. Narrow-diameter channels in the epicuticular layer of the insect integument. They deliver lipids to the surface.

WAX SCALES. Some members of the family Coccidae, superfamily Coccoidae (order Hemiptera). See also, BUGS.

WEB-LOVERS. Members of the family Plokiophilidae (order Hemiptera). See also, BUGS.

WEBSPINNERS (EMBIIDINA). Embiidina is the oldest name for a peculiar order of semi-social, largely tropical insects later named Embiodea, Embioptera, or Embiaria. They also are known as footspinners and embiids. With less than 500 named species, it is regarded as a small order, but worldwide collecting indicates that it is moderately large, potentially with about 2,000 species. Because embiids tend to remain in their self-produced silk galleries, they rarely are collected by usual methods except for males attracted to lights. One must search for the often-obscure colonies and rear series, especially males, in laboratory cultures.

Characteristics

These are small insects, generally measuring 4 to 10 mm in length, though a few attain nearly one centimeter. Webspinners are soft-bodied and elongate, and pale to dark in color. They bear compound eyes, but lack ocelli. The antennae are filiform, and consist of 15 to 32 segments. They have chewing mouthparts. The prothorax is narrower than the head. The fore legs and hind legs are stouter than the middle legs. The tarsi have three segments, and the basal segment is markedly swollen on the forelegs. Female webspinners are wingless. Males are winged or wingless, and if winged they bear two pairs of smoky-colored membranous wings, with both pairs about equal in size and shape. The abdomen bears 10 segments, and a pair of 2-segmented cerci is present at the apex. The immatures undergo four nymphal instars, and in the case of winged males, the last two bear wing pads. Webspinners display incomplete metamorphosis. All life-stages of all species have the basal segment of the fore tarsi greatly enlarged by scores of tightly-appressed, globular, multicellular silk glands. By means of a thread-like duct, the secretion of each gland is conveyed to one of a multitude of hollow, dermal, seta-like ejectors scattered on the ventral surface of the tarsus. With rapid, criss-cross strokes in many directions, even dorsally, the fore-tarsi produce dense-walled silk galleries ramifying in or on the food supply, usually the outer bark of trees, leaf-litter, and lichens on rocks. In arid regions, galleries are shielded beneath stones and extend down soil cracks.

The galleries are just narrow enough to insure constant body hair contact with the walls and are increased in diameter as the embiids grow. Most of the order's peculiarities augment predation-escape by rapid reverse movement. These include a slender supple body with short legs, a forward-projected head, enlarged depressor muscles of hind tibiae to power reverse movement, highly sensitive cerci to guide such movement, and adaptations reducing friction of protruding structures, such as wings and ovipositor, against gallery walls.

Because females must live long enough to produce eggs and guard young, they gained maximum speed in reverse movement by becoming universally apterous by neotenic retardation of development of adult structures. In arid, open environments where flight would expose them to adverse conditions, especially predators, males of many species likewise are

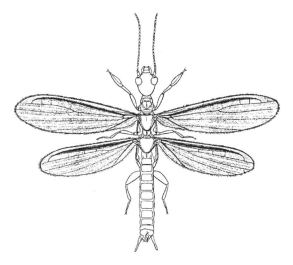

Fig. 1144 Typical alate male embiid, *Oligotoma saundersii* (Westwood) (Oligotomidae), body length 10 mm.

apterous, or subapterous, through degrees of neoteny. Fortunately, in all but one species, the male's genitalia fully develop and serve as the most useful characters in classification. Females are difficult to identify without associated adult males.

Evolution

Neotenic apterism appears to be a relatively recent adaptation. Undoubtedly, during early evolution of the order, perhaps in the Permian period, both sexes must have possessed wings. The first stage of reducing the 'barb effect' of wings during reverse movement was evolution of flexibility, a weakening of

Fig. 1145 Characteristic silk galleries under stone in a semi-arid environment, California.

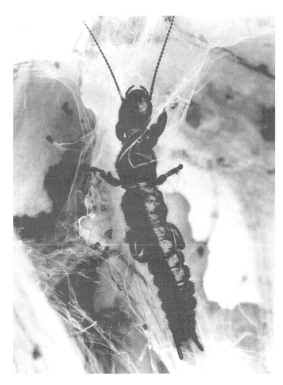

Fig. 1146 Typical adult female, *Pararhagadochir trachelia* (Navás), body length 12 mm, from Argentina. Shows neotenic apterism of all females of the order.

most longitudinal veins. This enabled the wings to fold forward, even crumple, over the back and thereby reduce friction. However, because stiff-winged flight must have remained useful, wings evolved an ability to temporarily stiffen when extended in flight. The means was blood pressure in the full length of certain veins, particularly in the anterior radius (RA), which became a terminally closed sinus, termed radial blood sinus (RBS). Such a conspicuous vein characterizes the wings of all wing-bearing males of the order. Presence or absence of wings in males occurs throughout the order by random convergence, usually associated with climate.

Embiids primarily occur in subtropical and tropical environments from sea level to almost 4,000 meters. A few species range into temperate regions, but many suitable regions haven't been occupied due to female apterism and hazards of movement outside of galleries. Lack of dispersal of females by flight restricts the order to continental lands and explains the absence of the order on oceanic islands except for a few species, particularly of the genus *Oligotoma*, which have been widely spread in ancient and modern human commerce.

Although the order probably had an early origin, the fossil record is weak. The only possible pre-tertiary fossil, *Burmitembia venosa*, was found in Burmese amber (questionably Cretaceous). All Tertiary fossils, mostly in amber, are modern types.

Classification

There are various taxonomic arrangements possible, but the following eight families seem to represent a consensus currently:

Class: Insecta
  Order: Embiidina
    Family: Clothodidae
    Family: Embiidae
    Family: Notoligotomidae
    Family: Embonychidae
    Family: Anisembiidae
    Family: Australembiidae
    Family: Teratembiidae
    Family: Oligotomidae

Because so many taxa await description, it isn't possible to develop an adequate classification at this time. Species of Clothodidae, confined to South America, are the most plesiomorphic. The large family Embiidae is highly diverse and requires division into subfamilies and, perhaps, families. The large family Anisembiidae is confined to the Americas. Tropical Asia has a number of families, particularly Oligotomidae, and some very distinct families, such as Embonychidae. Australia also has many oligotomids, a few Notoligotomidae and an abundance of Australembiidae, a family confined to eastern Australia. The large family Teratembiidae is best represented in the Americas and Africa, with only a few Asian species. A number of additional families await description. Strangely, Madagascar doesn't appear to have endemic embiids. Apparently there are no Southern Hemisphere origins. No species is of economic importance.

Edward S. Ross
California Academy of Sciences
San Francisco, California, USA

References

Ross, E. S. 1984. A synopsis of the Embiidina of the United States. *Proceedings of the Entomological Society of Washington* 86: 82–93.

Ross, E. S. 2000. EMBIA Contributions to the biosystematics of the insect order Embiidina. Part 1, Origin relationships and integumental anatomy of the insect order Embiidina. *Occasional Papers of the California Academy of Sciences* 149: 1–53.

Ross, E. S. 2000. EMBIA Contributions to the biosystematics of the insect order Embiidina. Part 2, A review of the biology of Embiidina. *Occasional Papers of the California Academy of Sciences* 149: 1–36.

WEB-SPINNING SAWFLIES. Some members of the family Pamphiliidae (order Hymenoptera, suborder Symphyta). See also, WASPS, ANTS, BEES, AND SAWFLIES.

WEDGE-SHAPED BEETLES. Members of the family Rhipiphoridae (order Coleoptera). See also, BEETLES.

WEEDS IN CROP SYSTEMS FOR PEST SUPPRESSION. Weeds are generally considered to be deleterious in crop systems because of their competition with crop plants for limited water, nutrients, light, and space, and occasionally because of allelopathy. Weeds are also known to be important reservoirs for some crop pests, including bacterial, fungal, and viral pathogens, nematodes, and arthropods. In addition, weeds can obstruct harvest operations. However, weeds support beneficial arthropod populations and vegetative diversity is associated with faunal diversity. Species diversity has been related to ecosystem stability and a lower frequency of pest outbreaks, although some pests [e.g., tsetse, (*Glossina* spp.), and rhinoceros beetles, (*Oryctes* spp.)], maintain low, relatively stable densities that are still economically injurious.

Vegetational diversification in polycultures to suppress pest arthropods has at times produced positive results. For example, corn/bean (*Zea mays* L./ *Phaseolus vulgaris* L.) dicultures sustained less injury by banded cucumber beetles, *Diabrotica balteata* LeConte; fall armyworm, *Spodoptera frugiperda* Smith; and leafhoppers, *Empoasca kraemeri* Ross., than corn monocultures. Maize/ legume/squash (*Cucurbita* spp.) polycultures in tropical Mexico produced higher yields and harbored 38% fewer melonworms, *Diaphania hyalinata* (L.), larvae than squash monocultures. Cover crops, such as beans, *Phaseolus* sp., in apple, *Malus* sp., orchards, have been associated with reduced codling moth, *Cydia pomonella* L.; rosy apple aphid, *Anuraphis roseus* Baker; and leafhopper infestations. Modifications of the polycultural concept include 'strip cropping' to accumulate natural enemy populations or to attract pest populations into concentrated areas where they can be sprayed with insecticides. Crop rotation is another mode of vegetational diversification, conducted on a temporal rather than on a concurrent basis. A modification of crop rotation was demonstrated when canola, *Brassica napus* L., and wheat, *Triticum aestivum* L., in the winter 'relayed' aphid predators to sorghum in the spring, and from sorghum to cotton in the summer. Theories that attempt to explain the lower pest infestations in many vegetatively diversified crops include: (1) spatial dilution of the pest's primary food or shelter resource, (2) chemical or structural interference with host location and use by herbivores, and (3) enhanced natural enemy populations resulting from increased herbivore prey abundance and diversity.

Native weed species and communities have been linked with the suppression of insect pests in cropping systems, including fall armyworm in corn; imported cabbageworm, *Pieris rapae* L., cabbage aphid, *Brevicoryne brassicae* L., and cabbage whitefly, *Aleyrodes brassicae* (Wlk.) in Brussels sprouts, *Brassica oleracea* L. cv. 'Brussels sprouts'. Weed growth was associated with reduced infestations of the potato leafhopper, *Empoasca fabae* (Harris), in

alfalfa, *Medicago sativa* L.; velvetbean caterpillar, *Anticarsia gemmatalis* Hübner, green stink bug, *Nezara viridula* L., and corn earworm, *Heliothis zea* Boddie, in soybeans, *Glycine max* (L.) E. Merrill; and green stink bug in macadamia, *Macadamia* sp., orchards. Two cropping systems, Louisiana sugarcane, *Saccharum officinarum* L., and south Texas cotton, *Gossypium hirsutum* L., are presented as case studies because they have been more extensively examined regarding weed interactions with arthropod populations than other cropping systems, and because they demonstrate different outcomes and principles.

Louisiana sugarcane

The intrusion of weeds, especially the perennial johnsongrass, *Sorghum halepense* (L.) C. Persoon, can cause sugarcane yield losses of up to 84%. Most weeds in sugarcane are less competitive annuals that decline after crop canopy closure blocks sunlight, but even annual weed infestations, particularly hairy crabgrass, *Digitaria sanguinalis* (L.) J. Scopali, can reduce sugarcane density, biomass, and commercial sugar yield by 24, 19, and 15%, respectively.

The importance of natural enemies in sugarcane in Louisiana was emphasized when applications of nonselective and persistent insecticides were associated with increased populations of the sugarcane borer, *Diatraea saccharalis* (F.), the key pest. Sugarcane fields in Louisiana that were infested with monocot, dicot, or monocot/dicot mixtures were found to be

Fig. 1147 Weedy fields such as this sugarcane field in southern Louisiana produce greater yields than weed-free crops because weeds harbor populations of beneficial insects that suppress key pests.

significantly more diverse in terms of soil surface- and foliage-associated arthropods, including predators, than weed-free sugarcane habitats. Herbivorous prey arthropods, which support populations of natural enemies, were significantly more numerous on the ground, sugarcane stalks, and sugarcane leaves in weedy habitats, and the abundance of arthropod prey was significantly greater on weed foliage than on the soil surface, or sugarcane in weedy habitats.

The red imported fire ant, *Solenopsis invicta* Buren, has been identified as the most important predator of the sugarcane borer, and was found to establish significantly more colonies per unit area (up to 85%) in weedy than in weed-free sugarcane habitats. Imported fire ant workers were significantly more abundant on the soil surface, sugarcane stalks, and weed foliage in weedy habitats. A study using a rare earth element (samarium) as a stable-activable tracer ingested by selected colonies in weedy sugarcane had 28% smaller foraging area (because of significantly greater prey availability) than weed-free sugarcane. The reduction in territorial area and the 25% increase in foraging activity per unit area permitted more dense colonization by fire ant in weedy habitats.

At least 18 other formicid species have been identified as being natural enemies of the sugarcane borer in Louisiana, and it has been suggested that a multiple predatory ant complex is more effective than one dominated by fire ant. One study determined that fire ant foraging activity among weedy and weed-free sugarcane habitats was negatively correlated ($r = -0.069$, $P \leq 0.05$) with the abundance of the formicids *Paratrechina vividula* Nylander, but failed to influence *Pheidole moerens* Wheeler. The fire ant has displaced other indigenous formicid species from many habitats throughout its distribution in the United States.

Spiders (Araneae) are the chief egg predators of the sugarcane borer, and second only to the fire ant among overall natural enemies of the sugarcane borer. Louisiana sugarcane fields are known to support at least 84 spider species in 18 families. Vegetational complexity involving weeds is associated with increased spider diversity and abundance on the soil surface and on sugarcane foliage. Predators of the sugarcane borer such as carabids, elaterid larvae, staphylinids, dermapterans, and neuropterans are found in Louisiana sugarcane, but weed cover has not been associated with increased populations of these insect groups.

Although grass species, including *Paspalum* spp., and johnsongrass can serve as hosts to the sugarcane borer, increased numbers of predatory arthropods in weedy sugarcane appears to have overridden any deleterious influence of weeds as secondary hosts of the pest. Sugarcane borer injury (numbers of bored internodes) was reduced by 28 to 42% in weedy sugarcane. Economic analysis and sugarcane biomass yields in weedy versus weed-free habitats were not consistently different in plots where the sugarcane borer was controlled with fenvalerate, a pyrethroid insecticide.

Also, soil incorporated aldicarb, used for nematode control, in weedy habitats resulted in low predator populations, particularly fire ants, that were similar to levels encountered in weed-free sugarcane habitats. Sugarcane borer injury was 19 to 32% greater in aldicarb treated weedy areas. Therefore, the hypothesis that plant chemical or structural complexity in weedy systems interferes with host location by herbivores does not adequately explain sugarcane borer infestations trends. Predation appears to be more important.

A study in southern Louisiana revealed that weedy sugarcane field plots, without chemical control of the sugarcane borer, can yield about 12% ($P \leq 0.05$) less commercial sugar per hectare than weed-free plots mainly because of weed-induced sugarcane stand reduction. Net return (US$/ha), however, was at least 14.6% higher than in the weed-free habitats. The same study suggested that research is needed to study the effect of limited areas of natural weed growth in sugarcane fields for enhancing natural enemy populations. Similarly, in sugarcane grown in Hawaii, the association of the dipteran sugarcane borer parasite *Lixophaga sphenopheri* (Villen.) with weeds resulted in reduction of New Guinea sugarcane weevil, *Rhabdoscelus obscurus* Boisd. It was suggested that weeds be permitted to grow in limited areas or 'islands' within sugarcane fields to reduce pest populations.

The greenbug, *Schizaphus graminum* (Rondani), which can obtain sugarcane mosaic virus (SCMV) from infected weeds and transmit the disease to sugarcane, was more abundant in the weedy sugarcane habitats in Louisiana, and, possibly as a result, SCMV infection of sugarcane was 14% higher than in weed-free habitats. Weeds have also been reported to be associated with higher populations of the phytophagous nematodes *Criconemella* spp., *Meloidogyne* spp., and *Tylenchorhynchus* sp. collected from sugarcane rhizospheres in southern Louisiana. In the case of *T. annulatus*, populations appeared to be influenced by SCMV infection of the host sugarcane plant.

South Texas cotton

In south Texas, the key pest of cotton production is the boll weevil, *Anthonomus grandis grandis* Boheman, which can commonly cause greater than 50% yield losses if populations are unchecked. In nature, boll weevils have been shown to reproduce on seven *Gossypium* spp., (including *G. hirsutum* and *G. barbadense* L.), three *Cienfugosia* spp., two *Hampea* spp., and *Thespesia populnea* (L.) D. Solander *ex* J. I. Corrêa da Serra in Texas, 486 boll weevils were examined and found to have ingested 8,900 pollen grains from 58 plant families of which the majority were in the families Salicaceae (28%), Fabaceae (13%), and Poaceae (8%).

Weedy cotton field plots in the Lower Rio Grande Valley were associated with significantly higher populations of nine of eleven prey arthropod groups counted, including cicadellids, herbivorous hemipterans, lepidopteran larvae, and dipterans; and nine of the thirteen natural enemy arthropod groups counted, including *Georcoris* spp., *Orius* spp., *Nabis* spp., and neuropterans. For the most part, differences among arthropod populations in weedy and weed-free cotton plots became more substantial as the season progressed and weed biomass increased as compared to biomass early in the season. Diversity among the arthropod groups that were counted did not differ for ground associated arthropods, but was significantly higher for samples collected from cotton plant canopies in weedy habitats than on cotton plant canopies in weed-free habitats. Numbers of ant colonies in the weedy and weed-free cotton habitats were low and did not differ, and were mainly comprised of leafcutter ants, *Atta* spp., and tropical fire ants, *Solenposis geminata* (F.). The most effective predator of boll weevils in Texas cotton is the fire ant. It is reported to cause an average of 84% boll weevil mortality as compared to 0.14%, 6.9%, and 7% caused by parasitism, desiccation, and egg infertility, respectively, and some researchers claim that the fire ant, originally from South America, prefers boll weevils to other food sources in the cotton agroecosystem. In weedy and weed-free cotton habitats in the Lower Rio Grande Valley, fire ants were not found in any samples. Examination of fallen boll weevil infested squares revealed that, in weedy and weed-free habitats, 50% of the boll weevils were killed by heat and desiccation as compared to 8% and 2% killed by ants and disease, respectively (40% survived and emerged).

Silverleaf whitefly, *Bemisia argentifolii* Bellows and Perring, numbers were significantly higher on weed-free cotton plant foliage than on weedy cotton foliage and weed foliage in the later part of the growing season. This probably reflects the whitefly's preference for cotton plants over weeds. It is possible that dense weed growth might impede or deter some whiteflies from settling on cotton plants in weedy areas. It is also possible that parasitism of silverleaf whiteflies, reported to be higher on weeds (especially hirsute species) than on cotton, contributed to the lower whitefly populations on the weeds. Higher cotton aphid, *Aphis gossypii* Glover, densities in weedy cotton habitats might have resulted from the same conditions as for whiteflies. Ants and coccinellid beetles, both of which consume aphids, were more abundant in the weedy cotton habitats. Whether more abundant in weedy or weed-free habitats, heightened prey availability was associated with twelve of the fourteen predator groups counted.

Despite a general trend for higher populations of natural enemies in the weedy habitats, boll weevil oviposition injury to squares was unaffected. As a possible explanation, predatory arthropod populations built to significantly greater levels in weedy habitats during late May and June when weed biomass was highest and when most squares had become bolls that are less vulnerable to boll weevil oviposition. Also, most natural enemies indigenous to the United States are not considered to be major causes of mortality to boll weevils. The numbers of squares counted on cotton plants in the study were not significantly different between the weedy and weed-free treatments because boll weevil damage was unaffected by the treatments and because weed growth was too light in the early season to have caused a weed related decline in square production. The significantly fewer bolls counted in the weedy habitats later in the season resulted from reduced cotton plant densities caused by weed competition and lower plant heights that were likely induced by thigmomorphogenesis (shortening of internodes resulting from physical contact with objects or wind) and shading caused by taller weeds. The significantly lower

($\geq$50%) cotton yields reflected the lower numbers of bolls.

The results found in the weed-diversified field cotton in the Lower Rio Grande Valley of Texas agree with polyculture research in Asia where cotton yields were either unaffected or reduced when interplanted with groundnuts, *Arachis hypogaea* L.; soybeans, *Glycine max* (L.) E. Merrill; greengram, *Phaseolus aureus* W. Roxburgh; and blackgram, *P. mungo* (L.) Hepper. Further, a study in Central America found that there were fewer lacewing (Chrysopidae) eggs on cotton plants intercropped with corn and weeds than in cotton monocultures. A study in Africa showed that yields of intercropped cotton and groundnuts grown on alternating ridges did not improve in low rainfall areas, but yields were better from intercropped plants in wetter areas.

Although boll weevil injury to cotton was not suppressed in the Lower Rio Grande Valley of Texas, other insect pests of cotton might be more vulnerable to natural enemies. The differences in weed biomass and the cost to lint production suggest that the quantity of natural weed growth needed to build significantly larger numbers of natural enemy populations in weedy habitats is detrimentally competitive with the cotton crop. Although the fire ant is a key predator of boll weevils elsewhere, the hot, dry habitat in the Lower Rio Grande Valley cotton fields appears to have limited fire ant colonization to the extent that it is not a factor in boll weevil mortality.

Weed communities in cropping systems can affect arthropod populations in a number of ways, as illustrated by the cases of sugarcane in Louisiana and cotton in the Lower Rio Grande Valley of Texas. Both cases, however, illustrate that uncontrolled weed growth is deleterious to crop yield, but in the case of sugarcane, this loss can be offset by the advantage conferred primarily by the substantially higher populations of the fire ant and associated reduction in sugarcane borer injury, and lower field maintenance dollar inputs in weedy habitats. In some situations, vegetative complexity might be associated with ecological stability. In others, various parameters have interceded, such as the introduction of an exotic pest without effective natural enemies in its new habitat, to render void the benefits of ecosystem richness. For weeds to become widely integrated into IPM strategies, development of techniques for maintaining appropriate densities and diversities of innocuous weed populations in or around cropping systems will likely be a key concern.

Allan T. Showler
U.S. Department of Agriculture, ARS SARC
Weslaco, Texas, USA

References

Altieri, M. A., and W. H. Whitcomb. 1979. The potential use of weeds in the manipulation of beneficial insects. *Hortscience* 14: 12–18.

Altieri, M. A., and W. H. Whitcomb. 1980. Weed manipulation for insect pest management in corn. *Environmental Entomology* 4: 483–489.

Jones, G. D., and J. R. Coppedge. 1999. Foraging resources of boll weevils (Coleoptera: Curculionidae). *Journal of Economic Entomology* 92: 860–869.

Showler, A. T., and S. M. Greenberg. 2003. Effects of weeds on selected arthropod herbivores and natural enemy populations, and cotton, *Gossypium hirsutum* L., growth and yield. *Environmental Entomology* 32: 39–50.

Showler, A. T., and T. E. Reagan. 1991. Effects of sugarcane borer, weed, and nematode control strategies in Louisiana sugarcane. *Environmental Entomology* 20: 359–370.

Showler, A. T., T. E. Reagan, and R. M. Knaus. 1990. Sugarcane weed community interactions with arthropods and pathogens. *Insect Science and Its Application* 11: 1–11.

Zandstra, B. H., and P. S. Motooka. 1978. Beneficial effects of weeds in pest management – a review. *Pest Articles News Summaries* 24: 333–338.

WESMAEL, CONSTANTIN.

Professor Wesmael was born at Brussels, Belgium, in 1798. He was a noted hymenopterist, concentrating principally on the parasitic forms. However, he also studied the Neuroptera, Coleoptera, and other groups. He died at Josse-ten-Noode, Belgium, on October 25, 1872.

Reference

Anonymous. 1872. Obituary, Prof. Wesmael. *Entomologist's Monthly Magazine* 9: 167.

WESTERN BALSAM BARK BEETLE, *Dryocoetes confusus* SWAIN (COLEOPTERA: CURCULIONIDAE, SCOLYTINAE).

The western balsam bark beetle is distributed throughout the range of its host, subalpine fir, *Abies lasiocarpa*, in western North America from British Columbia and Alberta to New Mexico and Arizona. Apparently, it is the only species of the genus that is capable of causing tree mortality of healthy trees. Populations may increase after windstorms provide suitable downed host material for the insects to exploit or during prolonged periods of drought. To a lesser extent it also attacks other species of *Abies* and Engelmann spruce, *Picea engelmanni*. Akin to other tree-killing bark beetles, this insect causes tree mortality by feeding and developing in the phloem of the tree. Utilization of the phloem for habitat and nourishment results in girdling of the tree so that translocation of nutrients is impeded. In addition, upon invading its host, the insect inoculates the tree with a pathogenic fungus, *Ceratocystis dryocoetidis*. It is most likely the combined action of the insect and the fungus that cause the tree to succumb. Extensive mortality of subalpine fir, which is typically not a long-lived species, has occurred throughout its range over the last five decades to a number of disturbance agents including the western balsam bark beetle and its associated fungus.

Initial symptoms of attack include; an accumulation of a mixture of light brown boring dust and frass at the base of the tree and resin flow from the bole of the tree. Copious resin flow may also indicate that the insects have been prevented from successful entry into the tree, commonly referred to as a pitch out. Western balsam bark beetles do not produce the characteristic pitch tubes, an accumulation of resin and frass at the point of entry into the tree, that many bark beetles produce. Attacks are often seen above 2 meters on the bole. Identification of successfully attack trees is often difficult. As the tree dies, the foliage changes in color to a light yellow and then a brick colored red before falling off the tree. In some areas, foliage fading can take 2 to 3 years to become noticeable.

Adult beetles are reddish brown to black and 3 to 4 mm in length. The females have a prominent, erect, dense, circular brush of red-brown to yellow setae in the frons. In British Columbia, and most likely throughout most of its range, the insect has a two-year life cycle. However, under certain climatic conditions a one-year life cycle may occur. The primary dispersal flight occurs during the end of June and July. A second, smaller peak in flight activity also occurs in the fall. Males initiate tree attack, excavate a nuptial chamber, and mate with several females. Invasion and infestation of a new tree involves the use of pheromones and host chemicals. Each female then constructs an egg gallery; with females radiating from the nuptial chamber commonly producing 4 to 9 galleries. Soon after initiation of gallery construction the females begin ovipositing on the sides of the gallery. This process continues into the fall. Eggs begin hatching in the summer and this brood spends the first winter as a young larvae. The following year the larvae resume development, pupate by late summer, and become adults by the fall but remain in the tree until the following spring. After the first year of attack, egg-laying mated females overwinter in the galleries and continue egg gallery extension and egg laying well into the second year. These females may re-emerge from the tree and attack a new tree or the same tree to deposit a third compliment of eggs.

In a British Columbia study, the average diameter of subalpine fir killed by the western balsam bark beetle ranged from 31 to 60 cm; and mortality was concentrated around the largest trees in the stands. In a Wyoming study from the Bighorn Mountains, average subalpine fir diameter of infested areas was 20 cm; but mortality did not appear to be concentrated around the largest trees. The Wyoming study also indicated that increased mortality levels were observed in dense stands with a higher host type component.

Jose F. Negron
USDA Forest Service
Ft. Collins, Colorado, USA

References

Bright, D. E., Jr. 1963. Bark beetles of the Genus *Dryocoetes* (Coleoptera: Scoloytidae) in North America. *Annals of the Entomological Society of America* 56: 103–115.

Garbutt, R. 1992. *Western balsam bark beetle.* Forest Pest Leaflet 64. Natural Resources Canada, Canadian Forest Service, Pacific Forestry Centre, Victoria, British Columbia, Canada.

Mathers, W. G. 1931. The biology of Canadian barkbeetles. The seasonal history of *Dryocoetes confusus* Sw. *The Canadian Entomologist* 63: 247–248.

Molnar, A. C. 1965. Pathogenic fungi associated with a bark beetle on alpine fir. *Canadian Journal of Botany* 43: 563–570.

McMillin, J. D., K. K. Allen, D. F. Long, J. L. Harris, and J. F. Negron. 2003. Effects of western balsam bark beetle on spruce-fir forests of North-central Wyoming. *Western Journal of Applied Forestry.* 18: 259–266.

Stock, A. J. 1991. The western balsam bark beetle, *Dryocoetes confusus* Swaine: impact and semiochemical-based management. Ph.D. Thesis. Simon Fraser University, Burnaby, British Columbia, Canada.

WESTERN BLOTS. Proteins are separated electrophoretically and a specific protein is identified with a radioactively labeled antibody raised against the protein in question.

WESTERN CORN ROOTWORM, *Diabrotica virgifera virgifera* **LECONTE (COLEOPTERA: CHRYSOMELIDAE).** The western corn rootworm, *Diabrotica virgifera virgifera* LeConte is one of the most economically important pests of corn in the U.S. Corn Belt and southern Canada. Larvae can cause substantial injury by feeding on corn roots. Larval tunneling interrupts the integrity of root systems and may destroy individual roots or root nodes. Root feeding by larvae can adversely affect plant growth and development and reduce plant stability and grain yield. If extensive root injury coincides with heavy rains and strong winds, plants may lodge (lean over), reducing light interception by plants and resulting in mechanical harvest losses. Adults are less likely to cause economic loss in field corn, but if high beetle densities are present when silks emerge, excessive silk feeding (i.e., silk clipping) can occur which may lead to reduced pollination.

D. v. virgifera is widely distributed from northwest Mexico into the southwest U.S., across much of the north central to northeastern U.S., and is present in southern Canada. This chrysomelid beetle also has become established as an invasive species in central Europe. A subspecies, the Mexican corn rootworm, *Diabrotica virgifera zeae* Krysan and Smith occurs from Central America to Oklahoma in the U.S. The subspecies are closely related; they are morphologically very similar with no pheromonal or structural barriers to mating, and there is no ecological or temporal isolation between them. However, when male *D. v. virgifera* from a South Dakota (U.S.) population have been mated with female *D. virgifera zeae* from Texas or central Mexico many eggs are laid but most do not hatch. The reciprocal cross always produces fertile eggs. A *Wolbachia* bacteria strain that is present in most U.S. populations of *D. v. virgifera* is the cause of the unidirectional reproductive incompatibility between subspecies. *Wolbachia* could be functioning as an isolating mechanism between the subspecies in hybrid zones.

The western corn rootworm is a univoltine insect species. In the north central U.S., eggs are laid in soil (up to 30 cm deep) from late July to mid-

Fig. 1148 Western corn rootworm females; examples of the variability in the black elytral pattern that can occur in a population. (Photo by L. Campbell and J. Kalisch.)

September. Eggs are white to light yellow (size: 0.3×0.5 mm), remain in diapause over the winter, and hatch over approximately a 5 week period beginning in late May to early June. Newly hatched larvae are about 3 mm long and move short distances through air filled pores in the soil to locate and establish on suitable plant roots. Larval development progresses through 3 instars. Larvae move off of roots for brief periods to molt between instars. Third instar larvae (approximately 12 mm long) are cream colored, with a sclerotized head capsule and anal plate. When larval development is complete, larvae pupate in earthen cells. White exarate pupae remain in earthen cells for 6 to 13 days before adults emerge. Rate of development of immature stages is temperature and density dependent. Adults can be present from late June to September. Adults have an elongate body shape and range from 4.2 to 6.8 mm long. The elytra are typically straw to yellow with black markings. The black elytral pattern can be reduced to distinct vittae on the inner (sutural) and outer (humeral) margins or cover most of the elytra except for the outer edge and a patch at the apex.

The western corn rootworm has a narrow larval host range that is restricted to certain grass species but appears to be ecologically monophagous on corn. This is primarily because the western corn rootworm exhibits a strong affinity for corn fields as feeding and ovipositional sites. The generally accepted hypothesis is that the western corn rootworm evolved with corn and corn relatives in Mexico and/or Central America and only moved into the U.S. after corn was introduced. Adults are highly mobile, with dispersal occurring on two scales. Adults exhibit local or trivial movement within and among adjacent fields during the entire time adults are present. A small portion of most populations also exhibits migratory behavior which leads to dispersal over longer distances. Migrating females often are mated, but have not yet developed mature eggs. Many migration events occur during or just after peak emergence, but long-range movement has been observed throughout the adult activity period. The mobility of adults enables rootworm populations to quickly colonize first-year cornfields that do not have a resident rootworm population. A recent dramatic eastern range expansion of this species occurred from western Kansas and Nebraska (eastern range limit in 1940s) to the U.S. eastern seaboard in less than 50 years. Corn had been grown in the eastern U.S. for many years prior to the range extension by *D. v. virgifera*.

Many factors can interact to determine adult *D. v. virgifera* population densities in a cornfield at different times during the season. The number of beetles emerging within a cornfield is often a primary contributor to the total population level in the field. Planting date significantly affects when beetle emergence begins. In late planted fields, as compared to early planted fields, initial emergence is often delayed and total emergence reduced. The relative attractiveness of food sources can greatly affect immigration and emigration rates. Adult *D. v. virgifera* feed preferentially on succulent reproductive tissues of corn (i.e., pollen, silks, kernels) but will feed on corn leaves and other pollen sources when preferred corn tissues are unavailable, or present in a less preferred phenological stage. Pollinating corn fields are highly attractive to adult *D. v. virgifera*. If corn in a field pollinates later than corn in surrounding fields, beetles often move into the pollinating field from the fields that have finished pollination. Contrasts in plant phenology within and among fields can significantly affect *D. v. virgifera* distribution patterns.

The female preovipositional period is at least 10 to 14 days when optimal food is available (e.g., corn silks, pollen). Maximum egg production per female under optimal conditions ranges from 800 to 1,000 eggs. In most of its range, the western corn rootworm

will oviposit in cornfields. In the eastern Corn Belt (i.e., area centered around east central Illinois, northwest Indiana) a population of western corn rootworm has adapted to the long-term practice of rotating corn and soybean (larval host to non-host plant) by annually depositing a significant number of eggs outside of corn. Beetles in this area also exhibit much movement back and forth between corn and soybean which is not seen in other geographic areas where *D. v. virgifera* occurs.

Biotic and abiotic factors can influence immature and adult corn rootworm survival. Although various vertebrate and arthropod predators, parasitoids, and pathogens have been reported as natural enemies of *Diabrotica* species, none appear to regulate population densities in commercial cornfields. Weather patterns can greatly influence *D. v. virgifera* survival. In the northern parts of the *D. v. virgifera* range, prolonged cold temperatures during winter can kill a high percentage of eggs. Very wet conditions (e.g., water-logged soils) during the larval/pupal period can reduce larval establishment on plants or kill larger larvae and pupae.

Various western corn rootworm management options are available to producers. Rotating corn with a crop that is not a suitable larval host will effectively prevent larval injury in most parts of the species range (exception is in the U.S. eastern Corn Belt where crop rotation has become ineffective as a management tool, see previous discussion). If corn is planted in a field for two or more successive years (continuous corn), a corn rootworm beetle scouting program and available economic thresholds can be used to determine if a chemical control tactic is needed. When necessary, chemical control tactics can be directed toward larval or adult stages. A commonly used strategy is to apply a soil insecticide at planting or first cultivation to reduce larval feeding damage and protect the primary root structure of the corn plant. Newer technologies that also target the larval stage include insecticide-treated seed and corn rootworm protected transgenic plants (i.e., corn tissues express a toxin that is lethal to rootworm larvae). An alternative strategy targets the adult stage. Insecticide applications are used to suppress beetle populations and reduce egg-laying so that larval populations the following year will not cause economic loss. To date, natural enemies (i.e., predators, parasitoids) have not been useful as biological control agents for *D. v. virgifera*.

Biological agents often have provided insufficient or inconsistent control and have been too cost prohibitive to include in commercial management programs.

Lance J. Meinke
University of Nebraska
Lincoln, Nebraska, USA

References

Darnell, S. J., L. J. Meinke, and L. J. Young. 2000. Influence of corn phenology on adult western corn rootworm (Coleoptera: Chrysomelidae) distribution. *Environmental Entomology* 29: 587–595.

Giordano, R., J. J. Jackson, and H. M. Robertson. 1997. The role of *Wolbachia* bacteria in reproductive incompatibilities and hybrid zones of *Diabrotica* beetles and *Gryllus* crickets. *Proceedings of the National Academy of Sciences of the USA* 94: 11439–11444.

Krysan, J. L., and R. F. Smith. 1987. Systematics of the *virgifera* species group of *Diabrotica* (Coleoptera: Chrysomelidae: Galerucinae). *Entomography* 5: 375–484.

Krysan, J. L. 1999. Selected topics in the biology of *Diabrotica*. pp. 479–513 in M. L. Cox (ed.), *Advances in Chrysomelid biology*. Backhuys Publishers, Leiden, The Netherlands.

Levine, E., J. L. Spencer, S. A Isard, D. W. Onstad, and M. E. Gray. 2002. Adaptation of the western corn rootworm to crop rotation: evolution of a new strain in response to a management practice. *American Entomologist* 48: 94–107.

Meinke, L. J. 1995. Adult corn rootworm management. University of Nebraska Agricultural Research Division Miscellaneous Publication 63-C.

WESTERN GRAPELEAF SKELETONIZER *Harrisina brillians* BARNES AND McDUNNOUGH (LEPIDOPTERA: ZYGAENIDAE).

The western grapeleaf skeletonizer is one of only several species in the genus *Harrisina* in North America. Western grapeleaf skeletonizer is native to the southwestern United States and northern Mexico. Its original distribution is believed to include regions of Arizona, New Mexico, Colorado, Nevada and the Mexican states of Sonora, Chihuahua, Coahuila and San Luis Potos.

The other common species, *Harrisina americana* (Guérin), occurs over the eastern half of North America and is an occasional vineyard pest. Its biology is very similar to the western grapeleaf skeletonizer. In fact, it is attracted to the (R)-(-) enantiomer form of the primary sex pheromone component attractive to the western grapeleaf skeletonizer.

Grape production in California was first impacted by the western grapeleaf skeletonizer soon after its introduction into southern California in 1941. Although severe damage in San Diego County vineyards was reported for several years, western grapeleaf skeletonizer densities decreased by the mid-1950s because of several biological control agents.

In 1961, the western grapeleaf skeletonizer infestation extended into the San Joaquin Valley of central California. Despite attempts at eradicating this pest, first in the 1940s in southern California and later in the San Joaquin Valley, the western grapeleaf skeletonizer became firmly established in both locations.

Both in southern and central California, western grapeleaf skeletonizer populations expanded slowly from their initial points of introduction. The slow increase in their range is mainly attributed to the tendency of moths to remain near where they developed as immatures if adequate vegetation is present. The inability to eradicate this pest in the San Joaquin Valley may be due in part to several features of its biology; for example, a small percentage of the population enters diapause during early seasonal generations, and some pupae may remain in diapause for over 18 months.

Life stages

Western grapeleaf skeletonizer eggs are laid in clusters commonly containing 50 to 150 eggs. Eggs are cream to light yellow in color and shaped like capsules. They are laid on their sides and do not touch one another. Moths oviposit almost exclusively on the bottom surface of leaves located most commonly at the basal portion of each shoot in the spring, and on new, fully expanded leaves as the season progresses. Healthy females lay nearly 300 eggs during their lifetime.

Although there are typically five larval instars, a few individuals have only four instars. The first instar has a cream-colored head capsule and pronotal shield that is difficult to see with the unaided eye. Their body is light cream and no color pattern is present. Second instar head capsules are dark, and the body lacks dark pigmentation until larvae are close to molting, at which time a faint brown mottling is visible. Furthermore, second instar larvae have pronounced body hairs. Third instars are identified by their distinctive medium brown bands that become more pronounced through the course of this life stage.

Fig. 1149 Western grapeleaf skeletonizer female moth adjacent to recently laid egg cluster.

Fig. 1151 Late stage larvae demonstrating atypical feeding pattern signifying a viral infection.

Bands become blue as third instars develop. Head capsules in third and later instars are retractile and relatively difficult to see, but the pronotal shield is dark and prominent. Fourth and fifth instars have distinct black and blue bands upon a bright yellow background.

Although early instars have body hair, later instars (third to fifth) have urticating hairs that are highly irritating to human skin, and of particular importance to farm workers when harvesting grapes. Urticating hairs are characteristic of the family Zygaenidae. The irritating factor is believed to be hydrogen cyanide, resulting from the breakdown of cyanoglucosides.

Fifth instar western grapeleaf skeletonizer larvae pupate under the bark of grapevine trunks or at the base of trunks in dried leaf litter or most anything that provides cover. The process of moving from a feeding site to a pupation site and initiation of spinning a cocoon occurs within 36 hours. The cocoons are a cream color and have an anterior flap-like portion used as an exit by emerging moths. This 'flap' can be teased open with forceps for removing pupae that can be relatively easily sexed.

Adult western grapeleaf skeletonizer moths are bluish-black in color. Males have a slender abdomen terminating in a v-shape as a result of soft spines. The female abdomen is considerably broader throughout with a distinctively square shaped terminus. Although both male and female moths have antennae with lateral bristles, male bristles are longer and have fine lateral hairs.

Life cycle

The western grapeleaf skeletonizer has a very restricted host plant range. It seldom infests plant species other than several in the family Vitaceae, including commercial and wild grape (*Vitis* spp.), Virginia creeper (*Parthenocissus quinquefolia* (L.)) and Boston ivy (*P. tricuspidata* (Plauchon)).

In Arizona, western grapeleaf skeletonizer has three generations each year. Moths first occur in May, and the first brood of larvae is present during June and early July.

In Fresno County, California, moth emergence corresponding to the first seasonal flight is initiated

Fig. 1150 Grape leaf with early instar larvae.

approximately 71% of the time from mid-March to early April. Moth emergence then increases gradually and continues into early May. The second flight of moths in California's Central Valley commonly begins in early June. A one-to-two-week period occurs between the end of the first and beginning of the second moth flights, during which time virtually no moths are caught in traps. The third moth flight period commonly begins during the last week in July. There is a small degree of overlap between second and third moth flights. Pheromone trap data suggest that a small fourth generation occurs during some years.

In San Diego County coastal areas and cooler highland areas, western grapeleaf skeletonizer spring moth flights occur approximately one month later than in the Central Valley. As a result, only 2 to 3 generations occur each year in the cooler regions.

Western grapeleaf skeletonizer egg, larval and pupal development share similar, lower and upper temperature thresholds (combined life stage estimate: 9.0 and 28.2°C, respectively). Moth longevity is about seven days. Although longevity has been noted to vary in relation to mean daily temperature, longevity and oviposition schedules are only loosely linked to daily temperatures under climatic conditions in California's San Joaquin Valley. Male flight activity occurs almost exclusively from sunrise to late morning, whereas female flight activity gradually increases from late morning to early afternoon. It has been suggested that daytime flight behavior and black coloration enables this species to thermoregulate, thereby having a similar level of activity over a broad range of moderate to high temperatures under clear skies. This is likely to differ in coastal climates subject to frequent cloudiness. Western grapeleaf skeletonizer larvae feed extensively, if not exclusively, during daytime hours.

Pheromone systems

The primary component of the western grapeleaf skeletonizer sex pheromone is [sec-butyl-1-(z)-7-tetradecenoate (racemic)] .The pheromone is very stable when contained in a rubber septum or sheet-like polymer laminate, and effective for four weeks during high ambient temperatures.

Wing, delta and bucket traps (unitrap) work well for catching western grapeleaf skeletonizer moths. The bucket trap is very convenient and is capable of catching far more moths. Traps are particularly reliable in identifying when early season moth emergence first occurs, and work well when located upwind, several rows into a vineyard and approximately five vines down the row.

Biological control

In southern California, two parasitoids have played an important role in western grapeleaf skeletonizer biological control. A braconid wasp, *Glyptapanteles harrisinae* (Muesebeck), and a tachinid fly, *Ametadoria misella* (Wulp) [= *Sturmia harrisinae* (Coquillett)], were introduced in the early 1950s. The wasp was collected in southeast Arizona, while the fly was obtained from Arizona and Mexico. *Ametadoria misella* established rapidly in the San Diego area, and parasitism of 30 to 70% was common within a year of release. *Glyptapanteles harrisinae* also established rapidly, however, parasitism levels dropped several years following initial establishment.

Parasitoids were first released in the San Joaquin Valley in the mid-1970s, with an intensive release effort occurring in the late 1970s. Parasitism occurred at low levels(less than 1.0%) during the first several years following initial releases. By the early 1990s, parasitism levels of 30% by *A. misella* were common.

A granulosis virus of western grapeleaf skeletonizer (HbGV) has played a key role in controlling western grapeleaf skeletonizer in the southern and interior valley regions of California. It was introduced inadvertently as a contaminant associated with parasitoids imported from Arizona and Mexico. Viral disease symptoms are very evident in most life stages of infected individuals. During oviposition, infected females frequently lay eggs in unorganized groups, with the eggs clumped into small groups several layers high. In addition, infected young larvae feed as unorganized groups, within which dead individuals are common. The appearance of the leaf area fed upon by young and old infected western grapeleaf skeletonizer larvae is considerably different from that resulting from feeding by healthy larvae. Feeding by young infected larvae on grape leaves occurs in small, more or less individual patches that do not result in well formed feeding 'windows' (translucent areas consisting of leaf epidermis) typical of early feeding by healthy uninfected larvae. Fourth and fifth instar feeding by infected individuals is also patchy. This contrasts greatly with feeding by healthy larvae of similar age, in which all leaf tissues are consumed except primary and secondary leaf veins.

Within field distribution of western grapeleaf skeletonizer

Larvae are commonly observed in greatest abundance along vineyard perimeters. This is particularly apparent where light infestations result in damage that is almost exclusively on the edge of vineyards. In contrast to the well defined edge effect in cases of low to moderate abundance, populations are readily observed throughout the vineyard when population densities are high.

For research purposes, western grapeleaf skeletonizer pupae can be monitored easily using corrugated cardboard with cell diameters equal to the diameter of fifth instars (approximately 5 mm) .The corrugated cardboard can be cut into 4 to 8 cm wide strips and placed as bands around canes or trunks. When seeking a pupation site, the larvae will readily (if not preferentially) enter the cell-like entrance holes along the edge of a band and pupate inside.

Pheromone traps as monitoring tools

Pheromone traps can be used for several purposes, including the determination of first emergence of western grapeleaf skeletonizer moths in spring, identifying the duration of primary activity during a flight period, and providing a rough approximation of relative population density. For these data to be of greatest value, they should be compared to previous years of data from the same site. To successfully identify the first date of emergence, traps should be placed in the field by mid-February in central California. Early trap placement and checking traps twice each week are necessary to accurately determine first moth emergence. For most uses, checking traps at six or seven day intervals and estimating emergence as having occurred midway between intervals is adequate.

Simulation phenology model

A computer simulation model was developed for predicting western grapeleaf skeletonizer seasonal phenology. The model incorporates a logistic spring emergence function, a non-varying estimate of moth longevity (i.e., seven days), and a schedule of percent daily oviposition by an average moth over a seven day life span. In addition, egg, larval and pupal development is based on degree-day sums required to complete each life stage (145, 385, 278 degree-days, respectively). Degree-days per day are calculated from minimum/maximum temperatures utilizing a lower threshold of 9.0°C and a high temperature, horizontal threshold of 28.2°C.

The western grapeleaf skeletonizer model is initialized with the date of the first pheromone trap catch (i.e., the BIOFIX point) signifying initiation of the first annual moth flight.

W. J. Roltsch and B. Villegas
California Department of Food and Agriculture
Biological Control Program
Sacramento, California, USA

References

Carr, W. C., W. J. Roltsch, and M. A. Mayse. 1992. Diurnal and generational flight activity of the western grapeleaf skeletonizer (Lepidoptera: Zygaenidae): comparison of monitoring methods. *Environmental Entomology* 21: 112–116.

Curtis, C. E., P. J. Landolt, R. R. Heath, and R. Murphy. 1989. Attraction of western grapeleaf skeletonizer males (Lepidoptera: Zygaenidae) to S-(+)-2-Butyl-(Z)-7-Tetradecenoate. *Journal of Economic Entomology* 82: 454–457.

Roltsch, W. J., and M. A. Mayse. 1993. Simulation phenology model for the western grapeleaf skeletonizer (Lepidoptera: Zygaenidae): development and adult population validation. *Environmental Entomology* 22: 577–586.

Stark, D. M., A. H. Purcell, and N. J. Mills. 1999. Natural occurrence of *Ametadoria misella* (Diptera: Tachinidae) and the granulovirus of *Harrisina brillians* (Lepidoptera: Zygaenidae) in California. *Environmental Entomology* 28: 868–875.

Stern, V. M., and B. A. Federici. 1990. Biological control of western grapeleaf skeletonizer, *Harrisina brillians* Barnes and McDunnough (Lepidoptera: Zygaenidae), with a granulosis virus in California. pp. 167–176 in N. J. Bostanian, L. T. Wilson, and T. J. Dennehy (eds.), *Monitoring and integrated management of arthropod pests of small fruit crops*. Intercept Limited, Andover, United Kingdom.

WESTERN HARVESTER ANT, *Pogonomyrmex occidentalis* (CRESSON) (HYMENOPTERA: FORMICIDAE).

The western harvester ant is common in semi-arid shrublands and grasslands throughout much of the western interior of North America. This species, like others in the genus *Pogonomyrmex*, is called a harvester ant because the workers collect seeds for food. The western harvester ant also removes plants near nests by clipping leaves and stems. Nests of the western harvester ant are large, conical soil and gravel mounds, surrounded by a circular area cleared of vegetation. In some western rangelands where soils are favorable, the population densities of this harvester ant may exceed 50 colonies per ha and the denuded area due to nest clearings may affect 10% of the total ground surface area. Ants may also harvest a considerable portion of the seeds produced by plants. The western harvester ant is therefore both ecologically and economically important, and is sometimes considered a rangeland pest because of the reduction in the plant forage available for grazing by livestock. These ants can also inflict a painful sting when their nest is disturbed.

The significance of the mound-building and plant-clearing habits of the western harvester ant has been the subject of considerable speculation, but is likely related to temperature regulation within their nests. The conical nest mound is noticeably flattened on the side with the nest entrance, which usually faces southeast to maximize exposure to the morning sun. Removal of surrounding plants also reduces shading of the nest, especially when the sun is at a low angle. The adult workers move larvae up or down within the nest mound daily in a manner that reflects a optimal range of temperature for brood development.

The foraging activity of the workers is also influenced by the soil temperature. Most seed harvesting by ants occurs when the surface temperature is between $30°$ and $40°C$, but some foraging may be observed over a more extreme range of temperatures ($20°$ to $50°C$). When soil temperatures are high ($40°$ to $50°C$), ants frequently collect arthropods that apparently die of heat stress on the surface. Thus, although the western harvester ant is primarily seed eating, it also scavenges considerable amounts of arthropod forage, plant litter and feces.

Ants harvest a wide variety of seeds available in their foraging areas, but workers consistently prefer, or avoid, the seeds of some plant species. This selective seed harvesting, along with soil modification of the nest site, often results in a different admixture of plant species near ant nests compared to surrounding areas. Ants attract nestmates to dense patches of preferred seeds by laying a recruitment pheromone along the soil surface. Workers are recruited to seeds by detecting differences in the concentration of the pheromone with their antennae as they move sinuously along the chemical trail.

The reproductive behavior, life cycle and colony longevity of the western harvester ant are similar to those in most other *Pogonomyrmex* species. During the summer, ant colonies produce numerous winged males and females (''alates'') that are reproductively viable and morphologically distinct from the sterile workers. In contrast to the iron-red color of the workers, male alates are black and have disproportionately smaller heads. Female alates resemble workers in color and proportion, but are 2 to 3 times larger in size. In most areas, alates leave the nest between late July and early August for their nuptial flights. Alate flight activity is synchronous among colonies, usually occurring in the late morning on two or three days following a significant rainfall. The alates aggregate

Fig. 1152 Nest mound and clearing of the western harvester ant, *Pogonomyrmex occidentalis* (Cresson), in the sagebrush steppe of southwestern Wyoming.

on hills or cliffs, and in the uppermost portions of shrubs, trees, or structures such as fenceposts, windmills, or buildings. Males "lek" together in groups (form mating aggregations) of 5 to 20 individuals, and females are attracted to male aggregations by a sex pheromone. Females typically mate with several males, a behavior that later results in genetically variable workers after the queen begins reproducing in colony establishment.

After locating a suitable site, the foundress queen sheds her wings and digs a simple, vertical tunnel 10 to 15 cm deep. During the first few weeks, the queen must rear the first brood as well as forage for food. Most foundress queens die before the first offspring develop into adult workers. Colony mortality continues to be high during the first year when there are few workers, especially if a young colony is in close proximity to older, larger ones with well-established territories. Over time, this selective mortality near existing colonies tends to produce a regular spacing pattern between neighboring ant colonies, although territories often overlap and foraging trails may interdigitate. If a colony survives for two years, then it generally has a low mortality and may persist for 10 to 50 years. A thriving colony typically has 2,000 to 3,000 workers, but larger ones may harbor 10,000 workers. Colonies are monogynous (one queen) and a single queen (except for the foundress queen) lives for the entire lifespan of the colony.

Because colonies may occupy a site for several years, the digging activity of ants gradually alters the chemical and physical characteristics of soils. Levels of nutrients essential for plant growth are usually higher in nest soils than in surrounding areas. The numerous tunnels and chambers in nests also increase water infiltration from the surface, and higher levels of organic matter facilitate the retention of soil water and nutrients. As a result, the size and species of plants adjacent to nest clearings or on abandoned nest sites may differ from surrounding areas. Ants, therefore, have beneficial effects on soil productivity and plant species diversity.

The western harvester ant is usually found in deep, well-drained soils with a sandy or gravelly texture because ants excavate some nest tunnels and chambers 2 to 3 m deep in soil. Colony densities are highest where these soils occur, usually on upland plains or sloping terraces. In contrast, ants are often absent from shallow soils, or fine-textured clays where workers are precluded from digging deep nest tunnels, or where soil expansion and contraction makes the tunnels unstable. Repeated soil disturbance, such as heavy erosion or trampling by livestock, is also unfavorable to ants because workers spend more effort on nest reconstruction and maintenance at the expense of food collection. Under these conditions, ants are present in low densities due to high rates of colony mortality or nest abandonment. From the standpoints of soil productivity and integrity, therefore, high densities of the western harvester ant indicate a healthy range condition rather than a degraded one.

The effects of harvester ants on total plant production are usually insufficient to justify broadscale chemical controls of ants on western rangelands. Control measures using chemically treated baits are economically feasible only on intensively managed ranches, or in rural settlements where humans commonly encounter ant colonies. Given its potential roles in soil productivity, plant species diversity, and as an indicator of soil disturbance, the absence of the western harvester ant actually might be less desirable than its presence on western rangelands.

Thomas O. Crist
Miami University
Oxford, Ohio, USA

References

Cole, B. J. 1994. Nest architecture in the western harvester ant, *Pogonomyrmex occidentalis* (Cresson). *Insectes Sociaux* 41: 401–410.

Crist, T. O., and J. A. MacMahon. 1991. Foraging patterns of *Pogonomyrmex occidentalis* in a shrub-steppe ecosystem: the roles of temperature, trunk trails, and seed resources. *Environmental Entomology* 20: 265–275.

MacMahon, J. A., J. F. Mull, and T. O. Crist. 2000. Harvester ants (*Pogonomyrmex* spp.): their community and ecosystem influences. *Annual Review of Ecology and Systematics* 31: 265–291.

Rogers, L. E. 1987. Ecology and management of harvester ants in the shortgrass plains. pp. 261–270 in J. L. Capinera (ed.). *Integrated pest management on rangeland*. Westview Press, Boulder, Colorado.

Wiernasz, D. C., and B. J. Cole. 1995. Spatial distribution of *Pogonomyrmex occidentalis*: recruitment, mortality, and overdispersion. *Journal of Animal Ecology* 64: 519–527.

WESTERN THATCHING ANT, *Formica obscuripes* **(FOREL) (HYMENOPTERA: FORMICIDAE).** The western thatching ant, *Formica obscuripes*, is a relatively large mound-building ant. Its distribution reportedly extends from northern Indiana and Michigan westward across the northern United States and southern Canada to Oregon and British Columbia. This ant also is found in an area extending southward including Utah, Colorado, northern New Mexico and California. The host range of the western thatching ant includes various vegetation types, such as forested areas, grasslands, and sagebrush. The dome-shaped nests may vary considerably depending on age of the colony and the habitat; however, they are typically 0.5 meter in height and 1.0 to 1.5 meters in diameter. The main brood chambers typically extend to a depth of one meter or more below the soil surface, and the thatch to a depth of one-third of a meter or less. The thatch nests are constructed from dry plant materials found in the area, such as pine needles and twigs in a pine forest or sagebrush twigs and grass in a semi-arid region. The thatch may frequently be surrounded by a sandy or bare soil ring. Another characteristic of the ant is that it constructs slightly excavated, and even partially covered, trails on the ground leading in nearly straight lines from the nest to feeding sites in trees or other plants. The ants feed upon honeydew collected from homopterous insects such as aphids, and also consume other arthropods. Ants are one of the groups of social insects that practice a division of labor, each working for the good of the whole.

The life cycle of the ant is one of complete metamorphosis, with egg, larva, pupa, and adult stages. The eggs are about 0.6 mm in length and 0.3 mm in diameter. Larvae are about 6 mm in length at maturity. Worker pupae range from 3.5 to 7 mm in length, and sexual male and female pupae about 9 mm. The larvae are small legless white worms. The adult ant workers appear to have at least two, and maybe three, different morphological forms. The major (maxima) workers are larger than the media workers, which in turn are larger than the minor workers (minima). The major workers are approximately 6 mm or more in length and have an orange-red head and thorax, and a black abdomen. The smaller media and minor workers (tenders) typically collect honeydew from homopterous insects (such as aphids, treehoppers, and scales on trees and low growing vegetation) and also carry brood in the nest. It has been suggested

that the small black minor workers are largely restricted to working in the primary nests. They have been observed, however, along with media workers, to transport honeydew from homopterous insects on plants to small secondary ant nests in soil surrounding the basal area of plants. These secondary nests may then be utilized as transit sites for subsequent honeydew transport to larger primary nests. The small minor workers may also perform other duties in the primary nest, especially as nurse-maids for the brood and queens. The major orange-red and black workers are known to then collect honeydew from smaller workers and transport it along ant trails leading to primary nests. These workers have larger crops (stomachs) and can therefore accommodate transporting larger amounts of honeydew. They also collect other arthropods (mostly insects) to use as food for the colony, and are extensively involved in nest repair. It is thought all three morphological forms can perform the same duties when necessary. During cold months of the year adult ants hibernate in the nest below ground.

It has been estimated that the number of adult workers per colony may reach as high as 35,000 to 40,000. There is apparently a minimum spacing requirement between colonies that will be tolerated by the ants. If that space is violated, colonies tend to disperse. When disturbed, this ant species can release a stream of formic acid from the distal end of the abdomen. This appears to be a defensive mechanism, but could also serve some other function as well. It has been noted these ants will chew off the bark from the base of some plants, especially when in close proximity to the nest, and then eject formic acid directly onto the cambium layer. This seems to

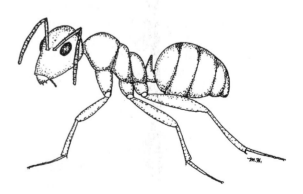

Fig. 1153 Western thatching ant, *Formica obscuripes* (Forel).

facilitate eventual death of the plant. As the plant dries, branches and stems break off, which may subsequently be used to further enlarge the thatch on the nest. In addition to workers, other members of the ant colony include queens and males. The number of queens per colony may vary; however, it is thought there are typically two or more in each colony. Alate (winged) sexual forms appear on top of the mounds at specific times of year. This may vary from May to August depending on the region and climatic conditions. Once the fertilized alate queen returns to earth from her nuptial flight, she breaks off her wings and then commences to seek an appropriate place to establish a new colony. This is a natural process of propagating new colonies within given species. It has been noted, however, that nuptial flights of *Formica obscuripes* may not occur in all regions of its range. The queens may live for a number of years; however, once they die, the colony will generally die within a few months. The male ants die shortly after the mating flights have been completed. Eggs that have been fertilized develop into female ants and unfertilized ones into males. The type of nutrition larvae receive is thought to have some effect on whether fertilized eggs become queens or workers. Pheromones are used very effectively as means of communication between members within the colony.

The western thatching ant should be considered a beneficial insect. It tunnels into the soil at its nesting site, thereby facilitating mixing and aerating of the soil. These ants feed upon a number of other insect species, including some pest species such as the western spruce budworm. It has been noted that nectaries of North American black cherry are most active during several weeks following budbreak. The western thatching ant is attracted to the nectar as a food source. Young eastern tent caterpillars, which may also be present at that time, may be captured and fed upon by these ants. In addition to feeding on other insects, including some pest species, these ants provide a measure of protection for the nectar-secreting homopterous insects they tend. This relationship may be viewed as a mutualistic one between the host plant, the aphids (for example), and the ants. Some defoliating insect pests are fed upon by the ants, thereby benefiting the ants and the plant. The ants protect aphids from their natural enemies and receive honeydew in return; and finally, the plant provides food for the aphids and thatch material for the ant nest.

Mark Headings
The Ohio State University Agricultural Technical Institute
Wooster, Ohio, USA

References

Bradley, G. A., and J. D. Hinks. 1968. Ants, aphids, and jack pine in Manitoba. *The Canadian Entomologist* 100: 40–50.

Cole, A. C. 1932. The thatching ant, *Formica obscuripes* (Forel). *Psyche* 39: 30–33.

Headings, M. E. 1971. The tree foraging activity of *Formica obscuripes* (Forel) and its influence on arthropoda of jack pine. M. S. Thesis. Michigan State University, East Lansing, Michigan. 117 pp. (unpublished).

Holldobler, B., and E. Wilson. 1990. *The ants*. The Belknap Press of Harvard University Press, Cambridge, Massachusetts. 732 pp.

Weber, N. A. 1935. The biology of the thatching ant, *Formica rufa obscuripes* (Forel) in North Dakota. *Ecological Monographs* 5: 165–206.

Wilson, E. O. 1971. *The insect societies*. The Belknap Press of Harvard University Press, Cambridge, Massachusetts. 548 pp.

WEST INDIAN SWEETPOTATO WEEVIL.
See also, SWEETPOTATO WEEVILS AND THEIR ERADICATION IN JAPAN.

WEST NILE FEVER.
West Nile Fever is an arthropod-borne human illness characterized by fever, headache, rash, muscle pains, swollen lymph nodes, and rarely, meningoencephalitis. The disease is caused by the West Nile virus. This was one of

Fig. 1154 Mound of western thatching ant, *Formica obscuripes* (Forel) (approximately 0.6 m high and 1.5 m in diameter at base).

the earliest arboviral infections to be documented, the virus being isolated from the blood of an infected woman in the West Nile province of Uganda – hence the name of both the virus and the disease.

West Nile virus belongs to a group of viruses commonly known as arboviruses because they are carried (transmitted) by arthropods (jointed-legged invertebrates such as insects, ticks, etc.). It is scientifically classified under the *Flavivirus* genus within the family Flaviviridae. It is related to other encephalitis-causing viruses like Japanese encephalitis virus, Murray Valley Encephalitis virus, Saint Louis encephalitis virus and eight other viruses that form the West Nile subgroup (complex) of viruses.

West Nile virus has been isolated from vertebrates and/or arthropods in some 25 countries within the Ethiopian, Palearctic and Oriental regions of the world, extending from Portugal in the west, to South Africa to the south, and Russia and India to the east. Very recently, in 1999, it spread to North America. Apart from humans, infected vertebrate animals include cattle, sheep, horses, goats, dogs, rabbits, hares, rodents, game animals, chickens, pigeons, sparrows, and wild birds (especially wetland and migratory birds). The generally accepted view of virus transmission is that it cycles between mosquitoes and birds. Most birds are unaffected by the virus, but can amplify the virus in their blood and thus pass it on when mosquitoes bite. Periodic transmission occurs to other vertebrates and humans when mosquitoes bite them accidentally or as a result of changing their feeding behavior.

Most of the virus detections from arthropods have been in mosquitoes and ticks. Mosquitoes are recognized as the major vectors of the virus. *Culex univittatus* has been implicated as the major mosquito vector with *Culex pipiens* and *Culex neavei* as secondary vectors in the Ethiopian zoogeographical region. *Culex modestus* and *Culex pipiens* have been implicated as vectors in the Palearctic region, and *Culex tritaeniorhynchus* and *Culex quinquefasciatus* as vectors in the Oriental region. Many other species of *Culex*, *Aedes* and *Anopheles* mosquitoes are experimentally capable of transmitting West Nile virus to mammal or bird hosts, but their importance in natural transmission is unknown. West Nile virus also has been isolated from wild caught ixodid and argasid ticks in the Palearctic region, but the role of ticks in natural transmission has not been clearly defined.

The prevalence of West Nile virus infections in humans varies with geographical region, but generally ranges from 20 to 80% in endemic countries. Infection is most often without symptoms, but sporadic cases, or clusters of cases or outbreaks of West Nile fever do occur. When disease occurs, the incubation period is 1 to 6 days. The onset of illness is abrupt, with fever, headache, and swollen lymph nodes being the most common symptoms. Muscle, eye and abdominal pains, sore throat and vomiting may occur. A rash appears on the trunk and upper extremities of the body during or toward the end of the period of fever. Recovery usually occurs after about 5 to 6 days. The clinical course of the disease can be more severe or even fatal in the elderly. Meningoencephalitis is a well known complication of West Nile fever. This is associated with neck stiffness, vomiting, confusion, abnormal reflexes, disturbed consciousness, tremor of extremities, convulsions and coma.

There has been a recent upsurge of a more severe form of West Nile fever in Europe. An outbreak in 1996/97 in Bucharest, Romania, involved more than 500 cases and a case fatality rate of almost 10%. There are reports of an outbreak in Israel involving more than 150 cases and 12 deaths in September 2000. The virus has now crossed the Atlantic ocean: an outbreak of West Nile fever occurred for the first time ever in New York City, USA, in September 1999, involving 56 cases and 7 deaths. Dozens of crows in the area also died (crows are especially sensitive to the virus), and West Nile virus was isolated from several birds, including crows and exotic birds from a local zoo. To date, some 18 species of birds including crows, robins, blue jays, and bald eagles have been infected throughout the eastern United States. Clearly, the virus is there to stay. This underscores an important current issue in public health – the rapid and frequent spread of microorganisms into new areas as a result of modern air travel and international trade. The danger is that microbes invading new territory often find hosts with no immunity, with catastrophic consequences to the hosts. West Nile fever is a relatively mild disease but its recent spread is a warning of trouble to come when more dangerous pathogens find their way into new areas.

Felix P. Amerasinghe
International Water Management Institute
Colombo, Sri Lanka

References

Hurlburt, H. S. 1956. West Nile infection in arthropods. *American Journal of Tropical Medicine and Hygiene* 5: 76–85.

Karabatsos, N. (ed.) 1985. International catalog of arboviruses, including certain other viruses of vertebrates (3rd edition and supplements 1986–98). *American Society of Tropical Medicine and Hygiene*, San Antonio, Texas.

Peiris, J. S. M., and F. P. Amerasinghe. 1994. West Nile fever. pp.139–148 in G. W. Beran and J. H. Steele (eds.), *Handbook of zoonoses*. Section B: Viral (2nd ed.). CRC Press, Inc., Boca Raton, Florida.

Hubalek, Z., and J. Halouzka. 1999. West Nile fever – a reemerging mosquito-borne viral disease in Europe. *Emerging Infectious Diseases [serial online]* 1999 5: [15 screens]. Available at http://www.cdc.gov/ncidod/EID/eid.htm

Howe, L. M. 1999. Kunjin West Nile fever update. Available at http://www.earthfiles.com/earth092.htm

West Nile Fever in the United States (update on St. Louis encephalitis). Available at http://www.who.int/disease-outbreak-news/n1999/oct/o6oct1000.html

WESTWOOD, JOHN OBADIAH.

John Obadiah Westwood was born at Sheffield, England, on December 22, 1805. He studied law, was admitted to the bar, and became a partner in a law firm. However, his principal interests were archaeology and entomology. Westwood was a noted dipterist, working on such groups as Mydidae, Bombyliidae, and Acroceridae, but possessing immense and broad knowledge on many aspects of Diptera. His artistic ability was a particular strength, and his illustrations were noted for accuracy. Of particular note is his publication ''An introduction to the modern classification of insects'' (1840), in which he proposed a systematic arrangement of the British insect genera. Other noteworthy publications include ''Arcana entomologica,'' ''Oriental entomology,'' and ''Exotic insects.'' In 1858, a wealthy patron of Westwood's (Hope, see above) established a professorship in invertebrate zoology at the University of Oxford. Westwood was named to the professorship, where he continued his taxonomic work for 35 years. He died at Oxford on January 2, 1893.

References

Anonymous. 1893. The late professor Westwood. *The Canadian Entomologist* 25: 261–262.

Papavero, N. 1971, 1973. *Essays on the history of Neotropical Dipterology with special reference to collectors (1750–1905)*. Museu de Zoologia, Universidade de São Paulo.

WETAS AND KING CRICKETS. A family of crickets (Anostostomatidae) in the order Orthoptera. See also, GRASSHOPPERS, KATYDIDS AND CRICKETS.

WETTABLE POWDER. Finely divided, dry pesticide material that forms a stable suspension when mixed with water.

WHEAT BULB FLY. See also, WHEAT PESTS AND THEIR MANAGEMENT.

WHEAT CURL MITE. See also, WHEAT PESTS AND THEIR MANAGEMENT.

WHEAT GROUND BEETLE. See also, WHEAT PESTS AND THEIR MANAGEMENT.

WHEAT HEAD ARMYWORM. See also, WHEAT PESTS AND THEIR MANAGEMENT.

WHEAT JOINTWORM. See also, WHEAT PESTS AND THEIR MANAGEMENT.

WHEAT PESTS AND THEIR MANAGEMENT.

Bread wheat, *Triticum aestivum* L., is one of the major cereal crops, along with rice and corn. One of the first domesticated crops, it is widely grown in temperate climates, and accounts for approximately 1/5 of the acreage devoted to cereal production. Wheat is often referred to as the 'staff of life,' and accounts for about 20% of the calories in the human diet.

There are several other types of wheat, e.g., durum wheat and club wheat, and there are several other cereal crops with similar growth habits and production practices. Arthropod pests of wheat commonly attack these related crops as well. Barley, *Hordeum vulgare* L., is a widely adapted temperate cereal. It is used primarily for feed and in the production of malt. Oat, *Avena sativa* L., is similar to barley in adaptation, and is used primarily as a feed grain. Rye, *Secale cereale* L., is notable for its adaptation to harsh environments. However, its production is declining due to concerns about grain quality, weediness, and poor agronomic traits. Triticale, X *Triticosecale* Wittmack, is a new crop derived by crossing wheat and rye in an attempt to combine the grain quality of wheat with the hardiness of rye. It has become a useful feed and forage crop in certain areas.

Wheat may be sown either in the fall (winter wheat) or spring (spring wheat). Winter wheats are preferred over spring wheats because of greater yield potential, with the latter predominating where climatic conditions preclude the use of winter types. Seeding rates are quite flexible relative to other crops, because of the wheat plant's ability to compensate for changes in competition from its neighbors. Wheat generally responds to nitrogen fertilization, while phosphorus and potassium requirements vary with local soils. While wheat has few equals as a dryland crop, irrigated production systems are important in some regions. Most systems are designed to optimize grain production; however, wheat is also a high quality forage and is sometimes grown in dual purpose systems intended to optimize both animal and grain production.

Growth and development of wheat and related crops begins with germination and emergence of the seedling. After the appearance of several leaves, the plant enters the tillering phase, in which additional stems (tillers, culms) are formed. The spike (head, ear) starts to form near the end of the tillering phase. A rapid stem elongation phase follows. The end of stem elongation is marked by the appearance of the flag leaf (highest leaf on the stem) and the emergence of the spike. The final phase before maturity and harvest is kernel (seed) growth.

Yield of these crops is determined by the relative magnitude of three components. The priority of pest management should be to protect these components to the point that is economically feasible.

1. Spikes per unit area is determined by the number of plants established during emergence and by the number of stems formed during the tillering phase. Pests that might affect this process include wheat bulb fly, *Delia coarctata*, and wireworms.

2. Number of kernels per spike is determined during stem elongation. Pests that might affect this process include Hessian fly, *Mayetiola destructor*, and Russian wheat aphid, *Diuraphis noxia*.

3. Kernel size is determined during the period from spike emergence to plant maturity. Pests that attack the flag leaf or that feed on the developing kernels, e.g., Sunn pest, can reduce kernel size.

Various pests are important to the production of wheat and related cereal crops because of their ability to affect one or more of these yield components. The following table lists some of the more important wheat pests. Some are currently major pests and other have been in the past or have the potential to cause important yield reductions.

Wheat production systems vary greatly in the amounts of external inputs, e.g., fertilizers and pesticides, used as well as in the degree of mechanization. Limited resource growers in the Middle East, for example, produce wheat with few inputs or mechanization. Production in western North America is highly mechanized because of the large acreages involved, but other inputs are limited because of low production potential. European production is very input intensive because land is scarce and production potential is high. These differences will determine, in large part, the best pest management approaches to employ in a given production system.

Pest management strategies can be categorized as either preventive, or reactive (therapeutic). Low input wheat production systems tend to achieve pest management objectives through low cost, preventive strategies such as plant resistance, biological control or modified cultural practices. More input-intensive systems tend to emphasize the judicious use of chemical controls.

Plant resistance can be defined as the management of crop pests through the use of cultivars that are relatively resistant to a given pest. This is a relatively inexpensive but effective method employed to prevent losses to such key pests as Hessian fly and Russian wheat aphid. Limitations to this approach include the lack of resistance genes to a given pest, as well as the time required to develop an agronomically acceptable resistant cultivar. Recent advances in molecular biology may help overcome some of these obstacles.

Biological control can be defined as the use of one organism to control another. In the case of crop pests, the controlling organism typically is another arthropod or a pathogen such as a bacterium, fungus or virus. The most common strategy in wheat and

Fig. 1155   Some wheat insects: top left, wireworms; top right, Say stink bug, *Chlorochoa sayi*; second row left, corn leaf aphid, *Rhopalosiphum maidus*; second row right, greenbug, *Schizaphis graminum*; third row left, Russian wheat aphid, *Diuraphis noxia*; third row right, army cutworm, *Euxoa auxiliaris*; bottom left, armyworm, *Pseudaletia unipuncta*; bottom right, pale western cutworm, *Agrotis orthogonia*.

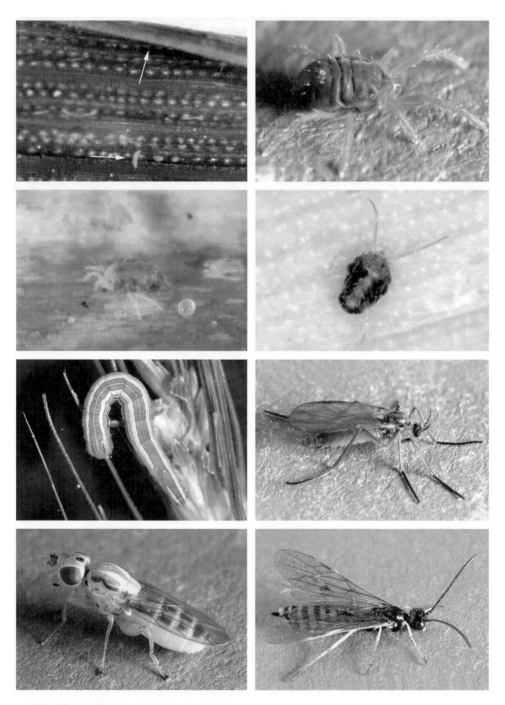

Fig. 1156 Additional wheat pests: top left, wheat curl mite, *Aceria tosichella*. Note curling of leaf above. Mites usually found in interveinal spaces; top right, winter grain mite, *Penthaleus major* (Ken Gray Image Collection, Oregon State University, modified); second row left, Banks grass mite, *Oligonychus pratensis*, egg and adult female; second row right, brown wheat mite, *Petrobia latens*, adult female; third row left, wheat head armyworm, *Faronta diffusa*; third row right, Hessian fly adult, *Mayetiola destructor* (Ken Gray Image Collection, Oregon State University, modified); bottom left, wheat stem maggot adult, *Meromyza americana* (Ken Gray Image Collection, Oregon State University, modified); bottom right, wheat stem sawfly adult, *Cephus cinctus* (Ken Gray Image Collection, Oregon State University, modified).

Insect and mite pests of wheat and related crops.

Common name[1]	Scientific name	Order	Family
Wheat curl mite	*Aceria tosichella* (Keifer)	Acari	Eriophyidae
Winter grain mite	*Penthaleus major* (Dugès)	Acari	Eupodidae
Banks grass mite	*Oligonychus pratensis* (Banks)	Acari	Tetranychidae
Brown wheat mite	*Petrobia latens* (Müller)	Acari	Tetranychidae
Grasshoppers	*Melanoplus* spp. (usually)	Orthoptera	Acrididae
Grain thrips	*Limothrips cerealium* (Haliday)	Thysanoptera	Thripidae
Say stink bug	*Chlorochroa sayi* (Stål)	Hemiptera	Pentatomidae
Sunn pest (and other common names)	*Eurygaster integriceps* (Puton), other *Eurygaster* and *Aelia* spp.	Hemiptera	Scutelleridae
Bird cherry-oat aphid	*Rhopalosiphum padi* (L.)	Hemiptera	Aphididae
Corn leaf aphid	*Rhopalosiphum maidis* (Fitch)	Hemiptera	Aphididae
English grain aphid	*Sitobion avenae* Fabricius	Hemiptera	Aphididae
Greenbug	*Schizaphis graminum* (Rondani)	Hemiptera	Aphididae
Rose-grain aphid	*Metopolophium dihrodum* (Walker)	Hemiptera	Aphididae
Russian wheat aphid	*Diuraphis noxia* (Mordvilko)	Hemiptera	Aphididae
Western wheat aphid	*Holcaphis tritici* (Gillette)	Hemiptera	Aphididae
Army cutworm	*Euxoa auxiliaris* (Grote)	Lepidoptera	Noctuidae
Armyworm	*Pseudaletia unipuncta* (Haworth)	Lepidoptera	Noctuidae
Pale western cutworm	*Agrotis orthogonia* (Morrison)	Lepidoptera	Noctuidae
Wheat head armyworm	*Faronta diffusa* (Walker)	Lepidoptera	Noctuidae
Wheat ground beetle	*Zabrus tenebroides* (Goeze)	Coleoptera	Carabidae
Cereal leaf beetle	*Oulema melanopus* (L.)	Coleoptera	Chrysomelidae
Wheat wireworm	*Agriotes mancus* (Say)	Coleoptera	Elateridae
Plains false wireworm	*Eleodes opacus* (Say)	Coleoptera	Tenebrionidae
Wheat bulb fly	*Delia coarctata* (Fallén)	Diptera	Anthomyiidae
Hessian fly	*Mayetiola destructor* (Say)	Diptera	Cecidomyiidae
Lemon wheat midge	*Contarinia tritici* (Kirby)	Diptera	Cecidomyiidae
Wheat midge	*Sitodiplosis mosellana* (Géhin)	Diptera	Cecidomyiidae
Frit fly	*Oscinella frit* (L.)	Diptera	Chloropidae
Wheat stem maggot	*Meromyza americana* (Fitch)	Diptera	Chloropidae
Black grain stem sawfly	*Trechelus tabidus*	Hymenoptera	Cephidae
European wheat stem sawfly	*Cephus pygmaeus* (L.)	Hymenoptera	Cephidae
Wheat stem sawfly	*Cephus cinctus* (Norton)	Hymenoptera	Cephidae
Wheat jointworm	*Tetramesa tritici* (Fitch)	Hymenoptera	Eurytomidae
Wheat strawworm	*Tetramesa grandis* (Riley)	Hymenoptera	Eurytomidae

[1]Common names approved by the Entomological Society of America are used where possible.

similar crops is to conserve naturally occurring biological control agents. Less common are releases of biological control agents into a pest's environment. Releases may be inoculative (i.e., sufficient agents to establish a population), augmentative (i.e., sufficient agents to raise an existing population to an effective level), or inundative (i.e., sufficient agents to control the pest infestation). Releases into wheat tend to be inoculative or augmentative in nature.

Modified cultural practices can be defined as changes in normal crop production operations that are intended to either make the crop environment less conducive to a given pest or more favorable for biological control agents. Examples in wheat production systems might include delayed sowing dates to avoid pest infestation, or increased crop diversification to provide additional food and shelter for biological control agents.

Most reactive or therapeutic measures for pest management in wheat involve chemical control, the application of insecticides or miticides to control existing pest infestations. Chemical controls are unique in that they are capable of eliminating existing infestations over large areas in a relatively short time span. However, insecticides and miticides are poisons, and their use has associated health and environmental risks. Also, chemical controls are expensive, and are cost effective only if they are applied when the economic return from the treatment is expected to be greater than the treatment cost.

There are many examples of pests adapting to a given type of control, particularly resistant cultivars and insecticides, rendering them ineffective. In addition, many preventive controls are not sufficiently effective to completely control pest infestations. For these and other reasons, a combination of controls is generally recommended for the management of wheat pests. This multi-tactic approach is often referred to as integrated pest management or IPM.

Wheat curl mite

The wheat curl mite, *Aceria tosichella* (Kiefer), is found throughout Europe and North America. It is most important as a vector of Wheat Streak Mosaic virus disease. However, it can damage wheat directly by curling leaves. Leaf curling is a tight roll of the leaf margin in contrast to the looser roll of the entire leaf caused by Russian wheat aphid. The emerging spike may become trapped in the flag leaf if the latter is tightly curled.

Wheat curl mite is member of a group of microscopic, wormlike plant-feeding mites known as eriophyids. They have two pairs of legs and a few setae. Eriophyid mites are very difficult to identify because of their size and simplified external morphology, and there are many unresolved questions regarding the identities of the eriophyids associated with a given plant species.

Eggs, immature stages, and adult wheat curl mites are found in the winter on wheat and other nearby perennial grasses. As temperatures rise in the spring, mite populations develop under leaf sheaths, inside newly emerged leaves, and eventually within the wheat head glumes. Eggs are placed in rows along leaf veins. An average complete generation requires 10 days. Most mites are found on the terminal leaves and move to each new leaf as it emerges.

As the wheat plant dries down, the wheat curl mites congregate on the upper parts of the plants where they are picked up by wind currents and are carried to their oversummering grass hosts. As summer hosts start to dry down, the reverse process occurs and mites are carried by winds to newly emerged winter wheat.

The wheat curl mite attacks a wide variety of grasses, mostly in the *Agropyron*, *Elymus*, *Hilaria*, *Hordeum*, *Lolium*, *Muhlenbergia*, *Triticum*, and *Zea* genera. Corn, volunteer wheat and wheat grasses

are potential important oversummering hosts. This species is also associated with garlic, onions, and other bulb crops.

Destruction of volunteer wheat and the maintenance of a volunteer-free period prior to planting winter wheat in the fall is the most effective management practice for this mite and the diseases that it vectors. Some effective varietal resistance (due to resistance to the mite) is available. Chemical control of wheat curl mites has not been shown to be an effective or economical practice.

Winter grain mite

The winter grain mite (also known as blue oat mite or pea mite), *Penthaleus major* (Dugès), is found in temperate areas throughout the world. It has higher moisture requirements than other mite pests of winter wheat and thus is found in more humid production regions. Crops heavily infested with winter grain mite take on a grayish or silvery cast. Plants may be stunted. Leaf tips appear scorched and entire leaves or plants can be killed.

Winter grain mite is a large dark brown to black mite, with reddish legs. A tan or orange spot is often visible on the back. This mite has two generations per year. In North America, the first begins in the fall as the oversummering eggs hatch after rains provide adequate moisture. Feeding peaks in December or January. Second generation feeding peaks in March or April. Populations decrease and oversummering eggs are produced as temperatures become unfavorably warm.

Newly-hatched mites feed on leaf sheaths and tender shoots near the ground. Older immatures and adults feed higher up on the plants at night and on cloudy or cool days. They move to the soil surface and into the soil to seek moisture and to avoid warm temperatures. Winter grain mite activity is greatest between 5°C and 21°C. On hot, dry days it may be necessary to dig 10 to 12 cm into the soil to find the mites. The mites are not harmed by high humidity, rainfall, short periods of sleet or ice cover, or by ground frozen to a depth of several inches.

Favored hosts include small grains and grasses. Winter grain mite also has been reported to damage legumes, vegetables, ornamental flowers, cotton, peanut, and various weeds.

Little is known about winter grain mite management. Crop rotation is recommended. Infestations probably should exceed several mites per plant before chemical treatment is considered.

Banks grass mite

The Banks grass mite, *Oligonychus pratensis* (Banks), is a spider mite that feeds on grasses in many parts of the Americas and Africa. It attacks a wide variety of grasses, and is considered to be a serious pest of corn and an occasional pest of grass hay, sorghum, turf and wheat in irrigated crops grown in semi-arid regions.

Damaged wheat leaves are chlorotic and small plants may be killed. Damaged areas on larger plants may have a bright yellow appearance. Infested plants in the fall are often near drying corn. The undersides of infested leaves may be covered with noticeable amounts of webbing.

In North America, fertilized female Banks grass mites move into winter wheat in the fall as their summer hosts, primarily field corn, begin to dry down. Overwintering mites are bright orange. With the onset of winter conditions the mites move to the crowns of wheat plants where they feed until spring. Small pearly white eggs then are laid that mature into pale to bright green male and female adults. Mites then breed continuously on wheat and summer hosts until their return to winter wheat in the following fall. Banks grass mites produce heavy webbing to protect colonies consisting of eggs, immatures and adults. Colonies usually are found on the undersides of leaves. The time required to complete a generation varies with temperature and is usually 10–20 days.

Damaged leaves first become yellow, then brown and necrotic. Heavy populations can kill small plants and reduce kernel size in larger plants.

Crop rotation patterns that place corn adjacent to wheat should be avoided. Biological control is quite useful when Banks grass mite is on its summer host. Chemical control is commonly used against it in higher value crops, e.g., irrigated field corn.

Brown wheat mite

The brown wheat mite, *Petrobia latens* (Müller), is a pest of wheat and similar crops in most small grain-producing areas of the world. The economic importance of this species has been difficult to determine because it is closely associated with drought conditions. Brown wheat mite has also been reported from many other cultivated plants, including turf grasses, sorghum, onions, fruit trees, carrots, cotton, lettuce, iris, alfalfa, and clover.

Wheat leaves damaged by brown wheat mite have a finely mottled appearance, and may have chlorotic tips. Heavily infested crops appear drought stressed, and often have an overall yellowish to bronzed discoloration.

In North America, brown wheat mite spends the summer in the soil as a white egg resistant to hot, dry conditions. In the fall, as cooler, wetter conditions return, these eggs start to develop and hatch after a 10-day incubation. Adult brown wheat mites can be distinguished from other mites found on wheat by their extremely long front legs. Female mites mature after feeding on wheat for about two weeks and then lay round, red eggs which give rise to further fall (one or two) and spring (two or three) generations. Males have not been observed in this species. As summer conditions return, white oversummering eggs again are produced. Both red and white eggs are placed on soil particles adjacent to wheat plants.

Brown wheat mites feed on plant sap during the day and spend the night in the soil or under surface debris. Their activity peaks at about mid-afternoon on warm, calm days. This mite is not affected by cold temperatures, but populations are quickly reduced by driving rains of about 1 cm or more.

Management of volunteer wheat is an important preventive measure for brown wheat mite, as it is with many winter wheat pests. Any management practices that serve to minimize drought stress also are important. Once an outbreak occurs, however, chemical control is the only effective management practice. It is often difficult to justify a chemical treatment because brown wheat mite infestations are associated with drought stress. Also, if white eggs are present and red eggs are mostly hatched, the population is in natural decline and treatment is not economically justifiable.

Thrips

Several thrips attack wheat and other small grains throughout the world, including the barley thrips, *Limothrips denticornis* (Haliday), grain thrips, *Limothrips cerealium* (Haliday) and wheat thrips, *Haplothrips tritici* (Kurdjumov). Thrips feeding results in discolored or distorted leaf tissue. More importantly these thrips feed on developing grains, resulting in sterility or poorly formed or discolored kernels.

These thrips overwinter as adults. Females are light brown to black in color, elongate, and have fringed wings. Males are smaller, and, in the case of *Limothrips*, wingless. Fertilized females move to cereals and other grasses in the spring and begin to feed on the youngest leaves. They insert their eggs in leaf tissue in sheaths or axils, and in the spike when it becomes available. Wheat thrips oviposition tends to be near or on the spike. The pale yellow (*Limothrips*) or red (*Haplothrips*) nymphs are similar in shape to the adults, and feed in leaf sheaths or in the spike. Pupation occurs in the soil. Females mate and then disperse to overwintering sites after the cereal host matures. There are one to two generations per year.

The only available management strategy for these thrips is the use of well-timed insecticide applications. Treatments must be made before flowering, and should be based on scouting and an economic threshold, if available.

Say stink bug

The Say stink bug, *Chlorochroa sayi* (Stål), is likely a complex of species which also includes *C. granulosus* and *C. uhleri*. Much of the research on this pest group was conducted before recent taxonomic studies revealed that three species were involved. The name Say stink bug will be used for all members of this group. It is found in the western United States and Canada, and attacks several crops including wheat, barley, alfalfa, and cotton. Russian thistle appears to be an important non-cultivated source of food and shelter throughout the year.

Damage to wheat and barley is caused by feeding on developing seeds. This results in reduced numbers of grains and shrunken, discolored and shriveled grain, and therefore reduced yield and quality. The type and amount of damage is closely related to the stage of plant and grain development, with greatest vulnerability between spike emergence and the grain 'dough' stage.

Stink bugs are so called because they release a foul-smelling fluid when disturbed. The Say stink bug is large, reaching about 20 mm in length as adults. They are shield-shaped and bright green in color, and they are marked on the back with raised white dots on back as well as three white or orange spots. Say stink bug eggs are barrel-shaped and marked with white circles, when viewed from above. The eggs are usually attached to the leaf in two-row clusters.

Adult Say stink bugs overwinter in sheltered areas such as underneath tree bark or at the base of native grasses. They become active in the spring and move to various host plants, including Russian thistle, where they feed, mate and lay eggs. As their original hosts are depleted or mature, they move to cultivated crops, particularly small grains. Large groups of adults may move from field to field, remaining only as long as grains are in an acceptable developmental stage. There may be several generations during the summer before the adults move to overwintering sites in the fall.

Because of the sporadic and transitory nature of Say stink bug infestations in small grains, management strategies are generally limited to chemical control. Treatment thresholds are not well defined, but they are quite low because of the great damage potential per individual stink bug.

Sunn pest

The Sunn pest, *Eurygaster integriceps* Puton, is the most important member of a group of shield bugs that are major pests of small grains in eastern Europe and the Near and Middle East. The information given here applies to *E. integriceps*, but the other species in the complex are quite similar in appearance and biology. These insects feed on all aboveground parts of wheat and other cereals, but the most important damage is due to direct feeding on developing grains. This results in loss of kernels and shriveled or discolored kernels. Damaged kernels have very poor baking quality, and can affect the baking quality of undamaged kernels when milled together.

Sunn pests are large insects, about 1.5 times longer than they are wide. They get their family name (shield bugs) from the fact that the scutellum (part of the thorax) extends back to cover most of the abdomen. Sunn pests are quite variable in color, ranging from yellow-brown to black.

Sunn pest has one generation per year. Eggs are laid in small clusters on leaves of cereals and many other grass hosts. Nymphs feed on leaf tissues until maturation, which usually coincides with kernel development. After two weeks or so of feeding, the adults fly to higher elevations to spend the summer and following winter. In more northern latitudes this period is spent in deciduous forests. In the fall they tend to move to protected overwintering sites, where large numbers can be found clustered in relatively small areas. In the spring, adults return to cereal fields and other host locations to mate and reproduce.

Chemical control is the predominant management strategy. Biological control with egg parasitoids and cultural controls, primarily the use of short season cultivars and modified harvest practices, have also been recommended.

Bird cherry-oat aphid

The bird cherry-oat aphis, *Rhopalosiphum padi* (L.), is an important pest of cereals in most production areas of the world. It is important for the damage that it causes as well as for vectoring Barley Yellow Dwarf virus (BYDV). Yield loss relationships are complex, being affected by crop type, crop growth stage, crop condition, presence of other aphid species, and presence of BYDV. Grain yield losses as high as 50% have been observed in winter wheat infested at the two leaf stage.

The wingless form of the bird cherry-oat aphid is broadly oval in shape, is yellowish-green to dark green in color often with a reddish patch at the rear, and long, black-tipped cornicles. The life cycle is complex where severe winter conditions and suitable primary hosts are present. Reproduction occurs on the primary host, several species of *Prunus* trees, in the fall, and the winter is passed in the egg stage. Colonies are formed on new growth in the spring, which give rise to winged females that disperse to grasses for the summer. Their progeny are wingless, parthenogenetic females, and this form predominates on various grass hosts throughout the spring and summer. Winged males and winged, parthenogenetic females are produced in the fall for the return flight to the primary host. If winter conditions are sufficiently mild and *Prunus* plants are absent, then a simplified life cycle of parthenogenesis on grass hosts is followed throughout the year.

Management strategies include modified cultural practices such as delayed planting to avoid virus transmission and promotion of biological controls. In more intensive production systems, chemical controls based on economic thresholds are often used.

Corn leaf aphid

The corn leaf aphid, *Rhopalosiphum maidis* (Fitch), is an important pest of cereals in many production areas of the world, particularly in warmer climates. It is more important as a vector of viruses such as Barley Yellow Dwarf virus (BYDV) than for the plant damage it causes. Corn leaf aphids prefer to feed on the young leaves of many types of grasses, including small grains and maize. Yield losses in wheat or other cereals have not been well documented.

The wingless form of the corn leaf aphid is elongate and varies in color from yellowish green to bluish green with black legs, antennae and cornicles. The antennae and cornicles are relatively short. The cornicles are long and black in color. A simplified life cycle of parthenogenesis on grass hosts is followed in most parts of the world, although males are observed occasionally.

Management strategies include cultural practices such as modified planting dates to avoid virus transmission. Biological controls are often effective. Chemical controls are rarely recommended because of this aphid's limited damage potential.

English grain aphid

The English grain aphid, *Sitobion avenae* (Fabricius), is an important pest of cereals in many production areas of the world. It is important for the damage that it causes to the plant, as well as for vectoring of Barley Yellow Dwarf virus (BYDV). English grain aphids prefer to feed on upper leaves and the spikes, once they have emerged. Late season infestations can reduce grain yield and quality. Yield loss relationships are complex, being affected by crop type, crop growth stage, crop condition, presence of other aphid species, and presence of BYDV. Grain yield losses greater than 30% have been observed in wheat.

The wingless form of the English grain aphid is moderate in size and spindle-shaped. It is generally green to brown in color. The cornicles are long and black in color. The life cycle occurs entirely on grasses. Sexual reproduction and overwintering eggs occur where severe winter conditions are present. If winter conditions are sufficiently mild then a simplified life cycle of parthenogenesis on grass hosts may be followed throughout the year.

Management strategies include cultural practices such as modified planting dates. Biological controls can be effective in some situations. Plant resistance has been identified but not used commercially. In more intensive production systems, chemical controls based on economic thresholds are used.

Greenbug

The greenbug, *Schizaphis graminum* (Rondani), is an important pest of cereals in many production areas of the world. It is most important for the damage that it causes to the plant, but is also a vector of Barley Yellow Dwarf virus (BYDV). Leaf tissue affected by greenbug is often discolored yellow or red due to the phytotoxic nature of its saliva. Yield loss relationships are complex, being affected by crop type, crop growth stage, crop condition, presence of other aphid species, and presence of BYDV. Grain yield losses greater than 50% have been observed in several crops.

The wingless form of the greenbug is relatively small and more elongate than oval in shape. It is generally light green in color, with a darker green dorsal stripe. The cornicles are mostly colorless with darker tips. The life cycle occurs entirely on grasses. Sexual reproduction and overwintering eggs occur where severe winter conditions are present. If winter conditions are sufficiently mild then a simplified life cycle of parthenogenesis on grass hosts may be followed throughout the year.

Management strategies include modified cultural practices such as delayed planting to avoid virus transmission and reduced tillage. Biological controls can be effective in some situations. Greenbug resistant varieties have been used widely in sorghum, but the development of biotypes has limited the sustainability of this approach. In more intensive production systems, chemical controls based on economic thresholds are often used; however, greenbug resistance to carbamate and organophosphate insecticides has been documented.

Rose-grain aphid

The rose-grain aphid, *Metopolophium dirhodum* (Walker), is an important pest of cereals in many non-tropical areas of the world. It is important as a direct plant pest as well as a vector of Barley Yellow Dwarf virus (BYDV). Rose-grain aphids prefer to feed on the upper leaves of wheat and other grasses, but rarely on the spike. Yield loss relationships are complex, being affected by crop type, crop growth stage, crop condition, presence of other aphid species, and presence of BYDV. Grain yield losses of 10 to 20% have been observed in wheat.

The wingless form of the rose-grain aphid is spindle shaped, and is yellowish green to green with a dorsal stripe on the abdomen. The antennae, legs

and cornicles are generally pale. The life cycle is usually complex, although a simpler life cycle of parthenogenesis on grasses is common in western Europe. Reproduction occurs on the primary host, several species of *Rosa,* in the fall, and the winter is passed in the egg stage. Colonies are formed on new growth in the spring, which give rise to winged females that disperse to grasses for the summer. Their progeny are wingless, parthenogenetic females, and this form predominates on various grass hosts throughout the spring and summer. Winged males and winged, parthenogenetic females are produced in the fall for the return flight to the primary host.

Management strategies include modified cultural practices such as delayed planting to avoid virus transmission and promotion of biological controls. In more intensive production systems, chemical controls based on economic thresholds are often used.

Russian wheat aphid

The Russian wheat aphid, *Diuraphis noxia* (Mordvilko), is an important pest of wheat and barley in western North America and South Africa. It also damages these crops sporadically in North Africa, southern Europe, the Middle East, Central Asia, Chile and Argentina. It is important for the direct damage that it causes to its hosts, but is not considered an important vector of plant viruses. Damage symptoms include tightly rolled leaves, white or purple leaf streaking and stunted plants. Yield loss relationships are complex, being affected by crop type, crop growth stage, crop condition, and the presence of other aphid species. Grain yield losses as high as 80% have been observed in malt barley.

The wingless form of the Russian wheat aphid spindle-shaped, light green in color, and may have a somewhat waxy appearance. The antennae are short and the cornicles are greatly reduced. The supracaudal process is unique among the cereal aphids. In the New World and South Africa a life cycle of parthenogenesis on a variety of cool season grass hosts is followed. Sexual reproduction and an overwintering egg stage are commonly observed in Old World populations. The closely related western wheat aphid, *Diuraphis tritici* (Gillette), is found in western North America, has a similar life cycle and causes similar damage.

Management strategies include modified cultural practices and promotion of biological controls. Resistant varieties have been developed in wheat and are grown widely in areas with consistent Russian wheat aphid infestations. In more intensive production systems, chemical controls based on economic thresholds may be used.

Army cutworm

The army cutworm, *Euxoa auxiliaris* (Grote), occurs in the cereal production areas of much of western North America. This is a climbing cutworm found feeding at night and on cloudy days on the aerial portions of small grains and many other plants. Yield loss relationships are poorly understood, with the degree of loss being particularly influenced by crop growth stage, plant population and drought stress. Losses in winter wheat in excess of 30% have been observed. More losses observed when there is relatively little foliage, which increases the likelihood of damage to the crown.

Army cutworm larvae are nearly 5 cm long when fully grown, generally colored light gray with lighter markings and a pale stripe running down the back. Adults are small moths, with a wingspan of about 3 to 4 cm. They are nuisance invaders of homes and other structures during their spring migration to higher elevations.

The army cutworm has one generation per year. Eggs hatch in the fall following a rainfall. The winter is passed as partially grown caterpillars, which feed on warmer days. They feed more frequently in the spring and develop more rapidly. As daytime temperatures rise, the army cutworm is found under soil clods and other debris during the day. This is a climbing cutworm that always feeds above ground during the night and on cloudy days. Pupation occurs in a small chamber constructed several inches below the soil surface. Moths emerge in May and June and migrate to higher elevations to escape high summertime temperatures. In late summer and early fall, the moths return to the plains to lay their eggs in wheat fields and other cultivated areas.

Because of the sporadic nature of outbreaks, management strategies are mostly limited to insecticide treatments based on economic thresholds. Pyrethroid insecticides, as a group, have been quite effective.

Armyworm

The armyworm, *Pseudaletia unipuncta* (Haworth), is a pest of small grains and other grass crops in North America east of the Rocky Mountains. Armyworm outbreaks occur only occasionally, more commonly during cool, wet springs, because natural population regulation usually prevents the development of economically significant infestations. The common name comes from this species' habit of moving in mass once they deplete local food sources. Damage commonly occurs when armyworm larvae defoliate the lower portions of various crops, although complete defoliation may occur during severe infestations. Defoliation is often complete, with only the leaf midribs left uneaten. Armyworm feeding in small grains near crop maturation is even more serious because it can result in spikes being clipped off and lost when larvae chew through the last remaining green tissue just below.

Mature larvae are about 4 cm in length, smooth-bodied, and dark grey to greenish-black. They have five stripes, three on the back and two on the sides, running the length of the body. While the stripes on the back are variable in color, the stripes on the sides are pale orange with a white outline. The head capsule is remarkable for its 'honeycomb' of black markings.

Armyworms pass the winter as partially grown larvae in the southern part of their range. Armyworm moths migrate to the northern part of their range in early summer. The moths have grayish-brown forewings, each with a white spot near the center, and grayish-white hind wings. The wingspan is approximately 4 cm. Armyworm moths lay their eggs in rows or clusters on the lower leaves of various grass crops. They prefer dense, lush vegetation for egglaying. Newly hatched larvae move with a looping (inchworm) action. Larvae feed at night and on cloudy days, and hide under crop debris during sunny periods. One or more generations may occur per year.

Pale western cutworm

The pale western cutworm, *Agrotis orthogonia* Morrison, occurs in North America from the western High Plains to the Pacific and from Canada to New Mexico. It is a more consistent problem in the northern plains states and southern Canada. It is potentially one of the most devastating insect pests of wheat in the region. Pale western cutworm is a subterranean cutworm, and feeds on stems below ground just above the moisture line. In drier soils larvae tend to feed further down on the stem and consume little tissue before moving to the next stem. This results in many severed stems and entire fields may be lost in a matter of days. In moister soils larvae feed further up on the stem and consume most of a single stem, which results in less overall damage to the crop.

Pale western cutworm larvae have greyish white, unmarked abdomens. They are distinguished from other similarly unmarked cutworms by the presence of two distinct vertical brown bars on the front of the head capsule. Eggs are deposited in loose soil. Some eggs may hatch within a few weeks, but most wait until the following spring when air temperatures at the soil surface reach 21 °C. Hatch may be delayed for up to several months if moisture and temperature conditions are unfavorable. Larvae prefer loose, sandy or dusty soil and are found most easily in the driest parts of the field, such as hilltops. Pupation occurs in chambers constructed several inches below the soil surface. Emergence of the gray to brownish white adult moths occurs in late summer and fall. They have an approximately 3 to 4 cm wingspan and are distinguished from other cutworm moths by the white undersurface of the wings.

Outbreaks are associated with dry conditions in the previous spring, because rainfall events of more than 6 to 7 mm tend to drive the cutworms to the soil surface and expose them to increased predation and parasitism. If the preceding May and June had fewer than 10 days with 6 to 7 mm or more inches of rainfall, then pale western cutworm population densities can be expected to increase. If the preceding May and June had more than 15 such days then pale western cutworm abundance will be greatly reduced.

Because of the sporadic nature of outbreaks, management strategies are mostly limited to insecticide treatments based on economic thresholds. Pyrethroid insecticides, as a group, have been quite effective.

Wheat head armyworm

The wheat head armyworm, *Faronta diffusa* (Walker), is found in most cereal production areas of North America north of Mexico and east of Arizona. Wheat head armyworm feeds on various grasses and small grains crops, with the first observed outbreaks occurring on timothy. While wheat head armyworm is considered to be a minor pest, its habit of feeding on developing spikes and grains gives it significant damage potential. Damaged kernels have the attachment end gouged out, as if they had been damaged in storage by *Sitophilus* spp. weevils.

Little is known about the life history of this insect. It spends the winter as a pupa in the soil. The small brown moths emerge to lay eggs in rows within leaf sheaths in the spring, and larvae can be found in wheat in June. First generation larvae feed on the heads of wheat at night and hide near the base of the plant during the day. Fully grown larvae are variable in color, ranging from light green to brown. They are tapered towards the rear, and have several stripes that vary in width and color.

No chemical control data or economic threshold studies are available for this insect. Infestations often are limited to field margins. If an outbreak occurs, any registered contact insecticide should be effective against this insect.

Wheat ground beetle

The wheat ground beetle, *Zabrus tenebroides* (Goeze), is an important pest of seedling winter wheat in many parts of Europe. Larvae feed on the leaves of young plants, causing a frayed or skeletonized appearance, or tunnel in the crown. Loss of plant stand often results. Adults feed on developing grains, but this damage is not considered to be as important as that caused by the larvae. Yield loss relationships can be significant under conditions unfavorable for compensatory plant growth.

The adults are small (about 1 cm) reddish brown ground beetles. Eggs are laid in soil when moisture content is adequate, usually from July through September. The larval stage is found from late September through the following spring, when the adults appear. The larvae, up to 3 cm in length, have a whitish body and brown head, thorax and legs. They overwinter in chambers in the soil, and feed actively on host plants during periods of active crop growth.

Management strategies emphasize cultural practices including adequate fertilization to encourage crop compensation in thinned stands, and crop rotation. Insecticide treatments targeting the larvae also are recommended.

Cereal leaf beetle

The cereal leaf beetle, *Oulema melanopus* (L.), has been known in Europe as an occasional pest of small grains for more than a century. It was introduced into North America in 1962, and has spread to most cereal production in the northern half of the United States. Adults and larvae defoliate small grains, especially those sown in the spring. Losses as high as 75 percent in barley and oats have been reported. Reduction in malt barley quality is of special concern.

Adult cereal leaf beetles prefer to feed on young, actively growing leaves. They chew completely through the leaf, between the veins, resulting in linear streaking of the leaf. This damage is easily confused with that of flea beetles. Larvae feed only on the upper surface of the leaves. They feed down to the leaf cuticle, staying between the veins, resulting in distinctive linear 'window-pane' damage. Larval feeding differs from adult damage in that it is wider and limited to the upper surface of the leaf. Tips of damaged leaves may turn white, and heavily infested fields may have a frosted appearance.

Adult cereal leaf beetles are about 5 to 7 mm long, with a metallic blue head and wing covers, a red pronotum (neck) and orange-yellow legs. The eggs are less than 2 mm long, yellow when first laid, but darkening to yellow-brown when about to hatch. Larvae appear dark and slug-like. Their skin is yellowish brown, and is covered by a mass of slimy, dark fecal material.

Cereal leaf beetles spend the winter in protected sites like grass crowns, grain stubble, wooded areas, or under house siding. They become active when air temperatures are above 10 °C. They feed on wild grasses and move to small grains when available. Females mate in the spring, and begin to lay eggs about two weeks later. A single female may lay up to 300 eggs over a six week period. The eggs hatch in four to 23 days, depending on temperature. Larvae feed for 10 to 21 days, depending on temperature, before crawling down the plant to pupate in the upper two inches of soil. The larval feeding period can last up to two months, due to continuous overwintering adult emergence and egg laying. The pupation period lasts two to three weeks.

Newly emerged adult beetles feed for two to three weeks on the new succulent growth of a wide variety of grasses, after which they disperse to overwintering sites where they enter dormancy. Winter mortality, due extreme temperatures and natural enemies, ranges to from 40 to 70 percent. There is one generation of cereal leaf beetle per year.

Effective biological controls are available, and are used preventively and successfully. The larval parasite *Tetrastichus julis* (Walker) has been an effective biological control agent in the western United States. The egg parasite *Anaphes flavipes* (Forester) has worked well in combination with larval parasites in the eastern United States. Insecticide treatments are used to control existing infestations based on the potential for economic losses greater than the cost of treatment.

Wireworms and false wireworms

Wireworms and false wireworms attack wheat in most production regions of the world. Important North American wireworm species include the wheat wireworm, *Agriotes mancus* (Say), the prairie grain wireworm, *Ctenicera aeripennis destructor* (Brown), and the Great Basin wireworm, *C. pruinina* (Horn). Important false wireworms include several *Eleodes* species, such as the prairie false wireworm, *E. opacus* (Say), are found in drier production regions of North America. Wireworms and false wireworms damage wheat by feeding on seeds, seedlings and young plants, resulting in lost stand. Yield losses occur if plant population losses are large enough to overcome the compensatory ability of the crop.

Adult wireworms are known as click beetles, which get their name from the sound they make as they jump. They are elongate brown or grey beetles, with parallel sides and rounded at the head and rear. The damaging larval stage is cylindrical, slender, hard-bodied, usually a shiny brown or pale yellow in color. The flightless adult false wireworms are known as darkling beetles. These are large black or reddish brown beetles, which can be recognized by the odd angle that the body is held at when they run. The larvae are similar in appearance to wireworm larvae.

Wireworm adults lay their eggs in soil and most of the life cycle is spent in the larval stage. Larvae will be found at varying depths in the soil, depending on temperature and moisture. Life cycles are variable, lasting from one to nine years. The false wireworms have a similar life cycle, although most are completed within three years.

Management strategies emphasize cultural practices that promote rapid germination and seedling growth to shorten the period that the plant is most vulnerable to attack. Insecticidal seed treatments also are commonly used.

Wheat bulb fly

The wheat bulb fly, *Delia coarctata* (Fallén), is an important European pest of wheat and other small grains. It is also found in North America, but generally is not considered to be an important pest there. Larvae tunnel in the base of the main stem or tillers, resulting in the loss of plants or spikes.

The adults are small, slender gray flies with relatively large wings and a bristly appearance. They lay their eggs in the soil in mid summer. The elongate white eggs hatch during the winter and the larvae move to grass hosts and bore into the base (bulb) of the main stem or one of the tillers. After several weeks, when about 3/4 grown, the maggots move to another stem, either on the same plant or a new plant. Pupation occurs in late spring, and within a few weeks adults emerge from the soil to mate and lay the eggs of the next generation.

Management strategies include early planting of winter crops to obtain larger more tolerant plants at the time of larval hatch, and delayed planting of spring crops to avoid the hatch and infestation period. Chemical controls include treatment of the seed, seedbed and plants. Expert systems and other decision support tools are available to help improve the efficacy and cost effectiveness of insecticide treatments.

Hessian fly

The Hessian fly, *Mayetiola destructor* (Say), has spread from its original home in the Caucasus region of Eurasia to most major wheat producing regions of the world. It has been a pest of wheat in the United States since the 18th century, and is considered to be one of the most serious pests of wheat in the world. Damage to the plant is caused by the larval stage, which sucks juices from the plant. Feeding results in stunted and dead tillers and plants, and broken stems resulting in unharvestable grains. Hessian fly prefers wheat but also attacks barley and rye.

The adult is a small grey or black midge. The eggs are cylindrical and glossy red in color. The wormlike larvae vary in color, but are greenish white when fully grown. Pupation occurs within a glossy brown protective case (puparium) referred as a 'flaxseed.'

In the fall, adults emerge from stubble and volunteer winter wheat to lay several hundred eggs on the leaves of young plants. The adults die within a few days of emergence. At about the same time the eggs hatch and the larvae move to the crown of the plant to feed. The larvae usually finish their development before the onset of cold weather. Winter is passed in the 'flaxseed' stage. Pupation and a second flight of adults occurs in the spring at about the time of jointing. Larvae of the spring generation feed in leaf sheaths and then form the puparia in which they spend the summer.

The use of resistant varieties is the most effective method of managing Hessian fly in areas where it is a key pest. However, Hessian fly biotypes often develop in response to the deployment of resistant varieties. Delaying planting to avoid the short egg-laying period is also an effective management practice. The 'fly-free' date is a term for the time after which wheat can be planted to avoid Hessian fly infestations. Destruction of volunteer wheat also helps to reduce Hessian fly infestations.

Wheat midge

The wheat midge, *Sitodiplosis mosellana* (Géhin), is found in most wheat production areas in the Northern Hemisphere, and can be a key pest of wheat under favorable conditions. The lemon wheat blossom midge is a related species, *Contarinia tritici* (Kirby), that is similar in appearance, life cycle and damage. However, its distribution is limited to Europe and Asia. Wheat midge larvae feed on developing kernels of wheat and other grasses, causing them to shrivel, crack and become deformed. These losses in yield and quality can be great enough and the midge difficult enough to control in certain areas that ceasing wheat production is the only economic solution.

The adult wheat midge is a small, fragile orange fly with prominent black eyes, long legs, and oval fringed wings. They emerge from the soil in late June or early July. Females are active during the evening hours, laying eggs on freshly emerged spikes. The orange maggots hatch within one week and move to the interior of the spike to feed on the surface of a developing kernel. After two to three weeks, fully grown larvae drop to the soil to overwinter. However, under dry conditions they may enter a resting stage that may last even to the time of harvest. Overwintering larvae remain in the soil for one to several winters, usually in the top several inches, until conditions are favorable for pupation. Larvae then become active and move to the soil surface to pupate.

Management of wheat midge has relied on insecticide treatments to control adults and prevent egg laying, and to kill larvae before they enter the spikes. The most effective treatments are applied between 70% spike emergence and flowering. Larval surveys may be used to assess wheat midge risk and the advisability of sowing an alternate crop. Some sources of resistance in wheat to wheat midge have been identified.

Frit fly

The frit fly, *Oscinella frit* (L.), is found in cereal production areas throughout the Northern Hemisphere, but is considered to be an important pest only in Europe. The nature of the damage caused by frit fly depends on the growth stage of the crop. Infestation of small plants results in dead plants or stems, while damage to the spike by later generations results sterility.

The adult is a small, robust black fly. Elongate, white eggs are deposited on cereals and other grass hosts. The pale yellow larvae mine leaves when newly hatched and later bore in stems or feed on undeveloped florets. Pupation occurs on the plant. There typically are three generations per year, with adults active in April–May, June–July and August—September.

Management recommendations include control of volunteer plants and other alternate hosts, as well as early sowing of spring grains to promote larger, more tolerant plants at the time of first generation infestation. In Europe, insecticides may be used to protect seedlings.

Wheat stem maggot

The wheat stem maggot, *Meromyza americana* Fitch, is found throughout North American cereal production areas. Damage to small plants results in dead tillers or plants, and damage to larger plants results in white, sterile spikes. Yield losses of 10–15% have been observed in heavily infested fields.

Wheat stem maggot passes the winter in the larval stage, in the lower parts of the stems of wheat and other hosts. They pupate in the spring and the greenish yellow, striped adults emerge in June. The white, cylindrical eggs are laid on the leaves and stems of wheat and other hosts. The newly hatched pale green maggots of this generation enter the leaf sheaths and tunnel into the tender tissues of the stem. Tunnels in wheat are usually 5 to 8 cm in length, and result in white, sterile spikes. Another generation of flies emerges in midsummer to lay eggs on volunteer cereal plants and other grasses. The fall generation emerges in late August to early September and lays eggs in the new winter wheat crop.

The use of delayed planting, following the dates recommended to escape Hessian fly infestations, is an effective management practice. Destruction of volunteer plants is also recommended. The effectiveness of chemical control is unknown, and currently not recommended.

Sawflies

Three sawflies attack cereals in the Northern Hemisphere: The black grain stem sawfly, *Trechelus tabidus* (Fabricius); the European wheat stem sawfly, *Cephus pygmaeus* (L.); and the wheat stem sawfly, *Cephus cinctus* Norton. The wheat stem sawfly is prevalent in western North America, particularly in the northern Great Plains, while the other two species are found in eastern North America and Europe. All three have similar field biology and cause similar damage. The following information is specific to wheat stem sawfly but should apply in general to the other species as well.

When they are mature, wheat stem sawfly larvae move to the base of the stem and gnaw a ring around the inside of the stem. The weakened stems break easily, making grain unavailable for harvest. Complete crop losses have been observed.

The wheat stem sawfly overwinters in the larval stage within wheat stems in a ground-level tunnel. Pupation occurs in May and the slender black and yellow adult sawflies are present from mid June to mid July. Females place single white, crescent-shaped eggs within stems of cereals and other host grasses. These hatch into pale yellow, legless worms similar in appearance to a caterpillar. The sawfly larvae tunnel up and down the stem. Mature larvae move to the base of the stem to form their overwintering chamber. The inside of the stem is chewed away, and the stem usually breaks at the point of this damage. The larvae then plug the area just below the break, forming the chambers in which they will pass the winter and pupate the following spring.

The use of solid stem wheat varieties is the most common management recommendation for wheat stem sawfly. Solid stems are, however, associated within lower yield potential. Several cultural controls also have been suggested, including destruction of stubble in the late fall or early spring. This is effective in reducing sawfly overwintering, but may conflict with soil and water conservation goals. Delayed planting of spring wheat and the use of trap crops are additional recommended cultural practices. Biological controls are considered to be of limited effectiveness. There are no effective chemical controls.

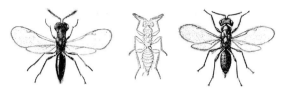

Fig. 1157 Wheat jointworm adult, *Tetramesa tritici* (left), and wheat strawworm, *Tetramesa grandis* summer generation adult (center) and spring adult (right) (modified from Webster, F. M. 1903. USDA Entomology Bulletin 42.)

Wheat jointworm

The wheat jointworm, *Tetramesa tritici* (Fitch), is most common in eastern North America and the Pacific Northwest. Once considered one of the major wheat pests, on a par with Hessian fly, it is now thought to be of only minor importance. Larval feeding causes the formation of small stem galls, resulting in malformed tillers prone to breakage.

The wheat jointworm overwinters in the pupal stage within stem galls in wheat stubble. The small black adult wasps gnaw their way out of the pupal chambers in April and May and spend the following month in wheat laying their eggs in wheat stems just above one of the nodes. The small white larvae form cells in the stem wall where they feed on plant sap. The swelling formed by the presence of the cell is referred to as a stem gall. Larvae cease feeding at about the time of wheat ripening.

The only available management recommendations are late fall or early spring stubble destruction and growing the current year's crop at some distance from the previous year's crop. Effective insecticide treatments are not available.

Wheat strawworm

The wheat strawworm, *Tetramesa grandis* (Riley), is found in most North American wheat production areas. Once considered one of the major wheat pests, on a par with Hessian fly, it is now thought to be of only minor importance. Spring generation larvae feed on and destroy the developing spike. Summer generation larvae feed within the stem, just above one of the nodes, reducing yield and quality to an undetermined degree.

The wheat strawworm overwinters in the pupal stage within a feeding cell in the wheat stem, usually located near a node. In the spring, the small, black wingless spring adults emerge and lay the small, white, bulb-like eggs near the growing point of a til-ler. The robust, pale yellow larvae hatch and destroy the growing point and form a feeding cell at the base of the tiller. Pupation occurs within the cell, and within two weeks the larger, winged summer adults (females only) emerge. The summer adults lay their eggs near one of the upper joints where the larvae feed and mature within the stem. Summer-form larvae are longer and more slender than those formed in the spring generation. A cell is formed near one of the joints where the larvae pass the summer and fall until pupation.

Recommended cultural controls include late fall or early spring stubble destruction, growing the current year's crop at some distance from the previous year's crop, and control of volunteer wheat plants. The use of early maturing varieties allows escape from the summer generation. No chemicals controls are known to be effective.

Frank B. Peairs
Colorado State University
Ft. Collins, Colorado, USA

References

Blackman, R. L., and V. F. Eastop. 2001. *Aphids on the world's crops: an identification and information guide* (2nd ed.). John Wiley and Sons, New York, New York.

Burton, R. L., K. J. Starks, and D. C. Peters. *The army cutworm.* Oklahoma State University Agricultural Experiment Station Bulletin B-749. Stillwater, Oklahoma.

Caffrey, D. J., and G. W. Barber. 1919. *The grain bug.* USDA Bulletin 779, Washington, DC.

Cook, R. J., and R. J. Veseth 1991. Wheat health management. APS Press, St. Paul, Minnesota.

Critchley, B. R. 1998. Literature review of sunn pest *Eurygaster integriceps* Put. (Hemiptera, Scutelleridae).. *Crop Protection* 17: 271–287.

Foster, J. E., P. L. Taylor, and J. E. Araya. 1986. The Hessian fly. Purdue Agricultural Experiment Station Bulletin 502. Purdue University, West Lafayette, Indiana.

Hatchett, J. H., K. J. Starks, and J. A. Webster. 1987. Insect and mites pests of wheat. pp. 625–675 in E. G. Heyne (ed.), *Wheat and wheat improvement* (2nd ed.). Agronomy Monograph 13. American Society of Agronomy, Madison, Wisconsin.

Haynes, D. L., and S. H. Gage. 1981. The cereal leaf beetle in North America. *Annual Review of Entomology* 26: 259–287.

Hein, G. L., J. B. Campbell, S. D. Danielson, and J. A. Kalisch. 1993. *Management of the army cutworm and pale western cutworm.* University of Nebraska NebGuide C-35, Lincoln, Nebraska.

Jeppson, L. R., H. H. Keifer, and E. W. Baker. 1975. *Mites injurious to economic plants.* University of California Press, Berkeley, California.

Jones, T. H., J. E. B. Young, G. A. Norton, and J. D. Mumford 1990. An expert system for management of *Delia coarctata* (Diptera: Anthomyiidae) in the United Kingdom. *Journal of Economic Entomology* 83: 2065–2072.

Kryazheva, L. P., Y. Ovsyannikova, V. N. Pisarenko, M. V. Sorokina, G. V. Shokh, and N. N. Dubrovin 1990. Main features of the damage caused by the grain beetle *Zabrus tenebrioides* Gz (Coleoptera, Carabidae) in the southern USSR. *Entomological Review.* 69: 157–167.

Lewis, T. 1997. *Thrips as crop pests*. CAB International, Wallingford, Oxon, United Kingdom.

McPherson, J. E., and R. M. McPherson. 2000. *Stink bugs of economic importance in America north of Mexico*. CRC Press, Boca Raton, Florida.

Painter, R. H., H. R. Bryson, and D. A. Wilbur. 1954. *Insects and mites that attack wheat in Kansas*. Kansas Agricultural Experiment Station Bulletin 367. Manhattan, Kansas.

Peairs, F. B. 1986. *Management of spider mites in corn*. Colorado State University Cooperative Extension Service in Action 5.555. Ft. Collins, Colorado. 4 pp.

Peairs, F. B. 1996. *Armyworms*. Colorado State University Cooperative Extension Fact Sheet 5.503. (http://www.ext.colostate.edu/pubs/insect/05503.html). Fort Collins, Colorado.

Peairs, F. B. 1999. Mites in wheat. Colorado State University Cooperative Extension Service in Action 5.578. Ft. Collins, Colorado. 4 pp., online only (http://www. ?ext.colostate.edu/pubs/insect/05578.pdf).

Peairs, F. B., and T. L. Archer 1999. Spider mites. pp. 104–106 in K. L. Steffey, M. E. Rice, J. All, D. A. Andow, M. E. Gray, and J. W. Van Duyn (eds.), *Handbook of corn insects*. Entomological Society of America, Lanham, Maryland.

Phillips, W. J., and F. W. Poos. 1940. *The wheat jointworm and its control*. USDA Farmers Bulletin 1006. U.S. Government Printing Office, Washington, DC.

Phillips, W. J., and F. W. Poos. 1953. *The wheat strawworm and its control*. USDA Farmers Bulletin 1323. U.S. Government Printing Office, Washington, DC.

Quisenberry, S. S., and F. B. Peairs (eds.) 1998. *A response model for an introduced pest – the Russian wheat aphid*. Thomas Say Publications in Entomology, Entomological Society of America, Lanham, Maryland.

Webster, R. L. 1911. *The wheat-head army-worm as a timothy pest*. Iowa State College Agricultural Experiment Station Bulletin 122. Ames, Iowa.

Weiss, M. J., and W. L. Morrill. 1992. Wheat stem sawfly revisited. *American Entomologist* 38: 241–245.

WHEAT STEM MAGGOT. See also, WHEAT PESTS AND THEIR MANAGEMENT, FLIES.

WHEAT STEM SAWFLIES: *Cephus cinctus* **NORTON,** *Cephus pygmaeus* **(L.) AND** *Trachelus tabidus* **(F.) (HYMENOPTERA: CEPHIDAE).**

Several species of sawfly (Hymenoptera: Cephidae) are pests of wheat and other cereal crops worldwide. The Cephini, the tribe to which these agriculturally important species belong, includes 50 to 60 grass-feeding species. This tribe evolved in the Old World, and only four species are found in North America: *Calameuta clavata* (Norton), a non-pest species, *Cephus cinctus* Norton, the wheat stem sawfly, *C. pygmaeus* (L.), the European wheat stem sawfly, and *Trachelus tabidus* (F.), the black grain stem sawfly.

Cephus cinctus is widely distributed across North America, from California to the Mississippi River and British Columbia to Manitoba. Many authorities consider it a native North American insect which adapted to wheat as European settlers began large-scale cultivation of cereal crops. Alternatively, there is some evidence that *C. cinctus* may have been inadvertently introduced into North America from northeastern Asia. A sawfly species from eastern Siberia, *C. hyalinatus* Konow, has recently been synonymized with *C. cinctus* using morphological features in preserved specimens. This has greatly expanded the distribution of *C. cinctus* to include northeastern Asia. Several other species, including *C. camtschatcalis* Enslin and *C. zahaikevitschi* Ermolenko, may also be conspecific with *C. cinctus*. The co-occurrence of *C. cinctus* in Asia and North America, and

Fig. 1158 A female wheat stem sawfly (*Cephus cinctus*) ovipositing into a wheat stem.

preponderance of *Cephus* species in Asia, suggests the possibility that *C. cinctus* may have been introduced from Asia to North America.

Cephus pygmaeus and *Trachelus tabidus* are found in Europe, Asia and North Africa. *Cephus pygmaeus* has a wider distribution than *T. tabidus*, but the latter species appears to be more common in the Mediterranean region. Both species were introduced into North America in the late 1800s, and became pests of small grain crops. In North America, *C. pygmaeus* has been found from northern Virginia to Ontario, but not west of Pennsylvania. *Trachelus tabidus* spread east and south from a probable point of entry somewhere between New York City and Richmond, Virginia, and is now found in Ohio, Pennsylvania, Maryland, Delaware, Virginia and North Carolina, but apparently not in Canada. Following a successful biological control effort against *C. pygmaeus* and *T. tabidus* in the 1930s, these two species were reduced to non-economic status.

Host plants

Wheat, barley, and rye are the main cereal crops attacked by sawflies, though infestation in other small grains such as triticale, spelt, and others have been observed. Cultivated oats do not support sawflies. Until recently, in western North America, *C. cinctus* attacked primarily spring wheat, while winter wheat suffered less damage because of its earlier sowing date and more advanced growth stage when sawflies emerge. Winter wheat now sustains extensive damage in the northern Great Plains. The choice of host plant appears to be related as much to stem diameter and phenological growth stage as to particular grass species. *Cephus cinctus* feeds on a number of grasses including species of *Agropyron*, *Bromus*, *Elymus*, and *Elytrigia*, in addition to cereal crops. The known host range for *Cephus pygmaeus* is wider than reported for *C. cinctus* and includes *Agropyrum*, *Avena*, *Bromus*, *Hordeum*, *Phleum*, *Secale*, and *Triticum*. The range of non-cultivated host plants for *T. tabidus* is unknown.

Biology

The biology of the agriculturally important sawflies is very similar and, at least for the three pest species found in North America, has been well studied. Females deposit eggs into the elongating stems of host plants in early summer and the developing larvae

feed by moving up and down the length of the stem. Though several eggs may be laid within a stem, only a single larva survives to maturity. As the plant matures, and usually prior to harvest, the larva moves down to the base of the stem and chews a V-shaped notch around the inside of the stem. The stem lumen is plugged with frass and sawdust below the notch, forming a chamber or hibernaculum. The notch weakens the stem, which usually breaks, producing a 'stub' that remains anchored in the ground. The sawfly larva undergoes an obligatory diapause within this chamber during the winter or the dry season. The stub is often covered with debris or soil, and is therefore well protected from excessive cold or dry conditions. The diapausing larva is contained within a thin membranous cocoon. Diapause is broken as temperatures, moisture levels and/or photoperiods increase in the spring. Post-diapause development includes pre-pupal and pupal stages lasting approximately 20 to 30 days depending on ambient temperatures. Adults emerge from the cocoon and use their mandibles to exit the stub through the plug. The agriculturally important species have a single generation per year, with a short-lived (7 to 10 d) and apparently non-feeding adult stage.

Damage and yield loss

Feeding by sawfly larvae reduces vascular efficiency resulting in fewer kernels per head and lower kernel weight, and can also reduce the protein content of the grain. Additional grain loss is the result of lodging after the inside of the stem is girdled. *Cephus cinctus* was first reported as a pest in Canada's Prairie Provinces in the late 1890s, and continues to be a key constraint to cereal production in the northern Great Plains. Damage is greatest in Montana, North Dakota, Alberta, Saskatchewan, and Manitoba. Infestation levels of more than 70% have been recorded with the weight of grain per head reduced by more than 20%. Losses of approximately 80% have been reported from central Montana. Annual losses to wheat, barley, rye, triticale and other cereal crops may exceed US $100 million in the region.

Prior to the successful biological control campaign against *C. pygmaeus* and *T. tabidus*, infestation levels as high as 45% for *C. pygmaeus* and 65% for *T. tabidus* were reported in eastern North America. *Trachelus tabidus* may have caused greater damage to wheat than *C. pygmaeus*, though the latter species is more widely distributed. Neither species is currently considered a serious pest of wheat in North America. Pest surveys from the 1980s detected less than 5% infestation in wheat from *C. pygmaeus*.

Cephus pygmaeus is also a pest in Europe and the North Africa/Mediterranean region, with reported infestation levels of up to 40% in Morocco, and reductions of 6 to 17% in kernel weight and 30 to 45% in kernel number. Damage and grain weight reductions of more than 45% have been reported from Switzerland, France, Germany, Belgium, Poland and Romania.

Several other sawfly species are pests of small grains in other parts of the world. Wheat yield losses in China of 6% have been reported for *C. fumipennis*, with some cultivar losses exceeding 30%. The lack of published literature on many of these other species may result from being less well studied or may be due to incorrect field identification.

Natural enemies

The most important sawfly natural enemies are parasitic Hymenoptera, though 5 to 10% of overwintering sawfly larvae may be killed by pathogenic fungi. There are also unpublished reports of at least one predator attacking sawflies but predators and pathogens have relatively little impact on sawfly population dynamics compared to parasitoids. The list of parasitoids attacking cephids appears extensive, though little is known about most species. Many species attack other hosts found in grasses or are obligate or facultative hyperparasitoids. Primary parasitoids of gramivorous Cephidae are found in the following families: Braconidae, Eulophidae, Ichneumonidae, and Pteromalidae. The most important, and the most well-studied are the braconids and ichneumonids. Most natural enemy complexes include one or more members of each family. The exception is the natural enemy complex attacking *C. cinctus*. Thus far, no ichneumonids have been recovered from *C. cinctus*. This is interesting because ichneumonids dominate the natural enemy complexes of other *Cephus* species. In terms of biology, the braconids are larval ectoparasitoids, while the other species are larval or egg-larval endoparasitoids.

Management

Management of wheat stem sawfly has focused on the development of resistant wheat cultivars, various tillage operations to destroy larvae and/or pupae in the stubble, and biological control using exotic parasitoids. Pesticides are generally too costly, given the relatively low value of wheat, and are ineffective because the larvae are concealed within the stem. Efforts to develop resistant wheat cultivars with solid-stems began in the 1930s following earlier reports that this trait provided resistance to *C. pygmaeus* in Europe. The first solid stemmed cultivar ('Rescue') was released in 1946, and this has been followed by a number of other solid stemmed cultivars. The performance of these cultivars has been variable. Sawfly larvae apparently suffer higher mortality and/or cause less damage in solid stemmed versus hollow stemmed cultivars, though resistant cultivars generally have lower yield and protein content than alternative hollow stemmed cultivars. Recent studies suggest that there may be a negative association between protein content and stem solidness but that yield *per se* should not be affected by the solid stemmed character. In addition, the degree of stem solidness in a given cultivar may be reduced by light, temperature, moisture supply, and plant spacing. The variable field performance resulting from interactions among environmental variables and resistant cultivars has reduced farmers' confidence in these cultivars.

Tillage, either in the spring or fall, has also produced equivocal results with significant limitations. Several reports indicate that tillage can reduce populations of wheat stem sawfly. A number of tillage implements have been evaluated, but the critical factor seems to be separating soil from the base of the wheat stem. Soil attached to the base of the stem insulates sawfly larvae from low temperatures and low humidity. Unfortunately, though tillage operations may be somewhat effective in controlling wheat stem sawfly, there are several important disadvantages associated with this strategy. Tillage requires an additional field operation, increasing wheat production costs and reducing profitability. Second, tillage operations also increase soil erosion rates, and run counter to no-till recommendations.

Biological control efforts have been directed against the three agriculturally important sawfly species in North America: *Cephus cinctus*, *C. pygmaeus*, and *T. tabidus*. Exotic parasitoids were established against the latter two species, and continue to maintain host populations below economic levels. Some authors have questioned whether the introduced natural enemies are the key regulating factor, or whether other factors may be involved, but without further study this question cannot be answered. All attempts, including three different projects covering more than 15 years, to find and establish exotic natural enemies against *C. cinctus* have ended in failure. There has been a renewed interest in North America to manage wheat stem sawfly using exotic natural enemies, particularly from eastern Asia.

See also, WASPS, ANTS, BEES, AND SAW-FLIES, (HYMENOPTERA).

Thomas G. Shanower
U.S. Department of Agriculture,
Agricultural Research Service
Sidney, Montana, USA

References

Holmes, N. D. 1977. The effect of wheat stem sawfly, *Cephus cinctus* (Hymenoptera: Cephidae), on the yield and quality of wheat. *Canadian Entomologist* 109: 1591–1598.

Ivie, M. A. 2001. On the geographic origin of the wheat stem sawfly (Hymenoptera: Cephidae): a new hypothesis of introduction from northeastern Asia. *American Entomologist* 47: 84–97.

Morrill, W. L., G. D. Kushnak, P. L. Bruckner, and J. W. Gabor. 1994. Wheat stem sawfly (Hymenoptera: Cephidae) damage, rates of parasitism, and overwinter survival in resistant wheat lines. *Journal of Economic Entomology* 87: 1373–1376.

Shanower, T. G., and K. A. Hoelmer. 2002. Biological control of wheat stem sawflies: past and future. *Journal of Agricultural and Urban Entomology*. (in press)

Wallace, L. E., and F. H. McNeal. 1966. *Stem sawflies of economic importance in grain crops in the United States.* U.S. Department of Agriculture Technical Bulletin 1350. 50 pp.

Weiss, M. J., and W. L. Morrill. 1992. Wheat stem sawfly (Hymenoptera: Cephidae) revisited. *American Entomologist* 38: 241–245.

WHEAT STRAWWORM. See also, WHEAT PESTS AND THEIR MANAGEMENT.

WHEAT THRIPS. See also, WHEAT PESTS AND THEIR MANAGEMENT, THRIPS.

WHEELER, WILLIAM MORTON. William Morton Wheeler was born on March 19, 1865, at Milwaukee, Wisconsin, USA. He had a diverse career, but established himself as one of the foremost authorities on ants and other social insects. Wheeler's early schooling in Milwaukee was marked by behavioral problems, so his father enrolled him in a strict German school with a reputation for extreme severity of discipline. He stayed on to attend the German-American College, which led to a position in 1884 at Ward's Natural Science Establishment in New York, a vendor of natural history supplies to schools and museums. Wheeler returned to Milwaukee in 1885, taught high school for two years and became custodian of the new Milwaukee Public Museum. There he developed proficiency in morphology, embryology and cytology. He left in 1890 to accept a fellowship at Clark University. He began to publish on insects and received his Ph.D. from Clark in 1892. He followed his professor, C.O. Whitman, to the University of Chicago, where he became an instructor. Following a study trip to Europe, Wheeler was made an assistant professor of embryology, though his entomological interested began to expand to include more than embryology. In 1899 he was made professor of zoology at the University of Texas. In 1903 he became curator of invertebrate zoology at the American Museum of Natural History, which allowed him to devote most of his time to ants. In 1908 Wheeler returned to teaching, and became professor of economic entomology at Harvard University, and where he became dean of the Bussey Institution. Wheeler traveled over much of the world, and published prodigiously, authoring almost 500 publications. Among his noteworthy publications are "Ants: their structure, development and behavior" (1910), "Social life among the insects" (1923), and "A study of insect behavior" (1930). Wheeler died On April 19, 1937, at Cambridge, Massachusetts.

References

Essig, E. O. 1931. *A history of entomology.* The Macmillan Company, New York. 1029 pp.

Mallis, A. 1971. *American entomologists.* Rutgers University Press, New Brunswick, New Jersey. 549 pp.

WHIRLIGIG BEETLES. Members of the family Gyrinidae (order Coleoptera). See also, BEETLES.

WHITEFLIES (HEMIPTERA: ALEYRODIDAE). This group consists of only about 1200 species, though it occurs mostly in the tropics and is not well studied, so this is likely an underestimate of its actual size. There are 126 genera described, but only seven have more than 50 species, and many genera contain only a single species. Because adults are so similar, classification is based on the structure of the fourth instar, the 'pupal case.' Though there are advantages to this system, one of the serious disadvantages is that the form of the pupal case is affected by the form of the host plant cuticle, resulting in variable appearance. The classification is as follows:

Order: Hemiptera
    Suborder: Sternorrhyncha
        Superfamily: Aleyrodioidea
            Family: Aleyrodidae
                Subfamily: Udamoselinae?
                Subfamily: Aleurodicinae
                Subfamily: Aleyrodinae

The Udamoselinae is based on a single specimen (which has subsequently disappeared), and has never again been collected; thus, its existence is questionable. The Aleurodicinae are found principally in South America, are relatively large for whiteflies, and are considered to be primitive. The wing venation is not as reduced as in Aleyrodinae, and the pupal cases are more complex.

Appearance and biology

Whiteflies typically are only one to three mm in length. The wings are transparent, or clouded. Wing venation is always reduced. The adult is often covered with a powdery white wax, the basis for the family name. However, not all species are white, and some are black (e.g., citrus blackfly, *Aleurocanthus woglumi*). Adults have two pairs of wings. The antennae usually are 7-segmented. The tarsi bear two tarsomeres, with the apical tarsomere bearing two claws and possessing an empodium between the claws. The immatures resemble scale insects; they are flattened, oval, and secrete a waxy material. A conspicuous fringe is sometimes present. The anal opening (vasiform orifice) of the immature opens dorsally on the last abdominal segment, and has diagnostic value.

Parthenogenesis occurs in some species. Females deposit their small, oval or elongate eggs on a short stalk that is inserted into the leaf tissue of the host. Sometimes the female deposits waxy material near the eggs, or deposits the eggs in a circle on the surface of a plant leaf. The first instars are very small, but have long legs and antennae, and crawl actively. They are transparent to translucent. Thereafter the appendages are reduced and the two nymphal stages are sessile. This is followed by the so-called pupal stage, during which some feeding occurs. Pupae often have large dorsal setae, though a few are present in nymphs. The adult typically emerges from a T-shaped slit on the dorsal surface of the pupal case. The mouthparts of the piercing–sucking type; both sexes feed.

Damage

Whiteflies can be very destructive pests, particularly in the tropics and subtropics. They injure plants directly through their feeding (removal of plant sap) and indirectly through transmission of plant viruses. To a much lesser degree, the production of honeydew by these insects (and subsequent growth of sooty mold) also affects the value of crops because it coats the foliage and other plant parts.

Whiteflies are the tropical equivalent of the aphids found in more temperate regions. They are vagile and opportunistic, building to high densities when weather and host plant condition are suitable. They feed almost entirely on angiosperms, avoiding gymnosperms. Among angiosperms they feed mostly on dicotyledonous plants, though the Graminae, Palmae and Smilacaceae among the monocots support whiteflies. The whiteflies do not display a great degree of host specificity.

Some of the more common, and important species, are treated below:

Greenhouse whitefly, *Trialeurodes vaporariorum* (Westwood)

Greenhouse whitefly is found widely around the world, including most of the temperate and subtropical regions North America, South America, Europe, Central Asia and India, northern and eastern Africa, New Zealand and southern Australia. It does not thrive in most tropical locations, and occurs in colder regions only by virtue of its ability to survive winter in greenhouses. In colder areas, it overwinters only in such protected locations, but in mild-winter it survives outdoors throughout the year. The origin of this species is not certain, but is thought to be Mexico or the southwestern United States.

Host plants. This species has a very wide host range, with over 300 species recorded as hosts. However, some hosts are more suitable. Vegetable plants often serving as good hosts are bean, cantaloupe, cucumber, lettuce, squash, tomato, eggplant, and occasionally cabbage, sweet potato, pepper, and potato. Among greenhouse-grown vegetables, the most common hosts are tomato, eggplant, and cucumber. When adults land on a favored host plants such as eggplant, they almost always remain to feed and oviposit; on a less preferred host such as pepper they usually take flight after tasting the plant. Many ornamental plants serve as good hosts, including ageratum, aster, chrysanthemum, coleus, gardenia, gerbera, lantana, poinsettia, salvia, verbena, zinnia and many others.

Natural enemies

Natural enemies of greenhouse whitefly are numerous, but few are consistently effective, especially under greenhouse conditions. Greenhouse whitefly is attacked by the common predators of small insects, including minute pirate bugs (Hemiptera: Anthocoridae), some plant bugs (Hemiptera: Miridae), green lacewings (Neuroptera: Chrysopidae), brown lacewings (Neuroptera: Hemerobiidae), and ladybirds (Coleoptera: Coccinellidae). Parasitic wasps attacking greenhouse whitefly are largely confined to the family Aphelinidae, but many species are involved and they vary regionally. Those known from North America, including Hawaii, are *Encarsia formosa* Gahan, *Aleurodophilus pergandiella* (Howard), *Eretmocerus haldemani* Howard, *Prospaltella transvena* Timberlake, and *Aphidencyrtus aphidivorus* (Mayr). Although these agents exercise considerable control on whitefly populations in weedy areas or on crops where insecticide use is minimal or absent, they do not survive well in the presence of most insecticides. *Encarsia formosa* has been used successfully under greenhouse conditions, and to a lesser extent field conditions, to affect biological suppression; for more information, see the section on biological control.

The pathogens of greenhouse whitefly are principally fungi, particularly *Aschersonia aleyrodis, Paecilomyces fumosoroseus,* and *Verticillium lecanii.* All occur naturally and can cause epizootics in greenhouses and fields, and also have been promoted for use in greenhouses as bioinsecticides. *Aschersonia* is specific to whiteflies, *Verticillium* has a moderately wide host range, and *Paecilomyces* has a broad host range. For optimal development of disease, high humidity is required. *Aschersonia* is spread principally by rainfall, so often fares poorly in greenhouse environments.

Life cycle and description. The development period from egg to adult requires about 25 to 30 days at 21 degrees C. Thus, because the preoviposition period of adults also is short, less than 2 days above 20 degrees C, a complete life cycle is possible within a month. Greenhouse whitefly can live for months, and oviposition time can exceed the development time of immatures; this results in overlapping generations. Optimal relative humidity is 75 to 80%. The developmental threshold for all stages is about 8.5 degrees C.

Eggs are oval in shape, and suspended from the leaf by a short, narrow stalk. The eggs initially are green in color and dusted with white powdery wax, but turn brown or black as they mature. The eggs measure about 0.24 mm in length and 0.07 mm in width. Eggs are deposited on the youngest plant tissue, usually on the underside of leaves in an incomplete circular pattern. Up to 15 eggs may be deposited in a circle which measures about 1.5 mm in diameter. This pattern results from the female moving in a circle while she remains with her mouthparts inserted into the plant. This pattern is less likely on plants with a high density of trichomes because plant hairs interfere with the oviposition behavior. Duration of the eggs stage is often 10–12 days, but eggs may persist for over 100 days under cool conditions. When cultured at 18, 22.5, and 27 degrees C, egg development requires an average of 15, 9.8, and 7.6 days, respectively. Maximum fecundity varies according to temperature; optimal temperature is 20–25 degrees C regardless of host plant. When feeding on eggplant, greenhouse whitefly produces over 500 eggs, on cucumber and tomato about 175 to 200 eggs (Drost et al. 1998).

The newly hatched whitefly nymph is flattened, oval in outline, and bears functional legs and antennae. The perimeter is equipped with waxy filaments. The first instar measures about 0.3 mm in length. It is translucent, usually appearing to be pale green in color but with red eyes. After crawling one cm or so from the egg it settles to feed and molt. Development if the first instar requires 6.5, 4.2, and 2.9 days, respectively, when cultured at 18, 22.5, and 27 degrees C. The second and third nymphal stages are similar in form and larger in size, though the legs and antennae become reduced and nonfunctional. They measure about 0.38 and 0.52 mm in length, respectively. Duration of the second instar requires about 4.3, 3.2, and 1.9 days whereas third instars require 4.5, 3.2, and 2.5 days, respectively when cultured at 18, 22.5, and 27 degrees C. The fourth nymphal stage, which is usually called the 'pupa,' differs in appearance from the preceding stages. The fourth instar measures about 0.75 mm in length, is thicker and more opaque in appearance, and is equipped with long waxy filaments. The pupal stage actually consists of the fourth nymphal instar period, which is a period of feeding, plus the period of pupation, which is a time of transformation to the adult stage. Thus, pupation occurs within the cuticle of the fourth instar. Duration of the fourth instar period and pupal period are 8.7 and 5.9, 5.9 and 4.0, and 4.5 and 2.8 days, respectively, at 18, 22.5, and 27 degrees C.

Individuals of greenhouse whitefly which develop on lightly or moderately pubescent leaves tend to be relatively large and to have 4 pairs of well developed dorsal waxy filaments. In contrast, whiteflies developing on densely pubescent leaves tend to be smaller, and to bear more that 4 pairs of dorsal filaments. These morphological variations are not entirely consistent, and have led to considerable taxonomic confusion.

Adults are small, measuring 1.0 to 2.0 mm in length. They are white in color, with the color derived from the presence of white waxy or mealy material, and have reddish eyes. They bear 4 wings, with the hind wings nearly as long as the forewings. The antennae are evident. In general form, viewed from above, this insect is triangular in shape because the distal portions of the wings are wider than the basal sections. The wings are held horizontally when at rest; this characteristic is useful for distinguishing this species from the similar-appearing silverleaf whitefly and sweetpotato whitefly, which hold their wings angled or roof-like when at rest. Mating may occur repeatedly, though females can also produce eggs without mating.

Damage

Adult and nymphal whiteflies use their piercing-sucking mouthparts to feed on the phloem of host plants. This results in direct damage, resulting in localized spotting, yellowing, or leaf drop. Under heavy feeding pressure wilting and severe growth reduction may occur. Whiteflies also secrete large amounts of sugary honeydew, which coats the plants with sticky material, and must be removed from fruit before it is marketed. The honeydew also provides a substrate for growth of sooty mold, a black fungus that interferes with the photosynthesis and transpiration of plants.

Greenhouse whitefly is, as the common name suggests, primarily a pest in greenhouses, and is a serious limitation to the production of plants grown in such structures. However, it can also be a field pest, often in warmer climates but also in cool climates when seedlings contaminated with whiteflies are transplanted into the field. For example, in Hawaii field-grown tomatoes suffered a 5% reduction in fruit weight with as few as 0.7 whiteflies per square cm of leaf tissue, and a 5% reduction in grade-A fruit due to contamination with honeydew at densities of about 8.3 whiteflies per square cm of leaf.

Greenhouse whitefly is capable of transmitting viruses to plants, but is not considered to be a serious vector, particularly relative to the *Bemisia* spp. However, greenhouse whitefly transmits beet pseudo-yellow virus to cucumber in greenhouse culture.

Management. Although whitefly nymphs and adults can be detected readily by visual examination of foliage, most monitoring systems take advantage of the attraction of adults to yellow, and use yellow sticky traps to capture flying insects. Sticky cards or ribbons are suspended at about the height of the crop for optimal monitoring. Traps must be placed close to plants or close to the ground or population densities will be underestimated.

Applications of insecticides are often made to minimize the effects of whitefly feeding on crops in greenhouses. Greenhouse whitefly feeds on the lower surface of foliage and is sessile throughout most of its life, habits that minimize contact with insecticides, and resulting in frequent applications and effectiveness mostly against the adult stage. In greenhouse culture, application intervals of only 4 to 5 days are common, and sytemic insecticides are often used to increase the likelihood of insect contact with toxins. Thus, whitefly resistance to nearly all classes of insecticides is known, and rotation of insecticide classes is encouraged. Mixtures of insecticides are often used, which is indicative of high levels of resistance among whiteflies to insecticides. Field populations of greenhouse whitefly invariably are derived from greenhouse populations, and possess similar resistance to many insecticides. Applications of petroleum oils and biological control agents help to avoid difficulties with insecticide resistance.

Few cultural practices are available, but disruption of the whitefly population with host-free periods is important. Continuous culture of plants allows whiteflies to move from older to younger plants. Similarly, weeds may allow whiteflies to bridge crop-free periods, and should be eliminated.

Seasonal inoculative release of the parasitoid *Encarsia formosa* Gahan into crops infested with greenhouse whitefly has been used extensively for suppression of whiteflies on greenhouse-grown vegetable crops. Excellent suppression of whiteflies is attainable, but on host plants such as cucumber and eggplant, which are very favorable for whitefly reproduction and have hairy leaves that interfere with parasitoid searching, frequent releases must be made. Alternatively, cucumber varieties with reduced trichome density have been developed which favor parasitism. Another critical factor is temperature, because low greenhouse temperatures are more suitable for whitefly activity than parasitoid activity. Daytime temperatures of about 24 degrees C seem to be optimal; temperatures of 18 degrees C or less suppress parasitoid searching. A cold-tolerant *Encarsia* strain that is active at 13 to 17 degrees C has also been used to overcome this temperature problem. Interference from pesticides can markedly affect parasitoid survival, so other pests such as mites must be managed biologically also. Lastly, release rates are important because if too many parasitoids are released the host whiteflies are driven nearly to extinction, leading to disappearance of the parasitoids; this is most likely to occur in small greenhouses. Alternatively, parasitoid releases can be made throughout the season, irrespective of whitefly presence. Although the protocols and technologies for whitefly management using *E. formosa* have been perfected for use in greenhouses, management under outdoor conditions awaits further research.

Silverleaf whitefly, *Bemisia argentifolii* Bellows and Perringand sweetpotato (or cotton) whitefly, *Bemisia tabaci* (Gennadius)

Silverleaf whitefly and sweetpotato whitefly are closely related whitefly species which cannot be distinguished easily by appearance, although there are some biological differences. Much literature from tropical environments around the world referring to sweetpotato whitefly probably pertains to silverleaf whitefly and other closely related species. Until the identities of these species are accurately determined, their distribution will remain unknown and some aspects of their biologies will remain confused. Cooler climates do not experience major problems with silverleaf whitefly except when overwintering in greenhouses occurs or vegetable transplants originate from whitefly-infested areas.

Host plants. Silverleaf whitefly has a very wide host range. There may be 500 worldwide. Sweet potato, cucumber, cantaloupe, watermelon, squash, eggplant, pepper, tomato, lettuce, broccoli and many other crops are hosts, but suitability varies. For example, sweet potato, cucumber, and squash are much more favorable for whitefly than are broccoli and carrot. Various weeds and field crops may favor survival of whiteflies during vegetable-free periods. Wild lettuce, *Lactuca serriola*, and sowthistle, *Sonchus* spp., are examples of suitable weed hosts. Cotton, soybean, and to a lesser degree alfalfa and peanut are field crop hosts. Sweetpotato whitefly has a narrower host range than silverleaf whitefly.

Weather. Silverleaf whitefly thrives under hot, dry conditions. Rainfall seems to decrease populations, although the mechanism is not known. Silverleaf whitefly is not a strong flier, normally moving only short distances in search of young tissue. However, under the proper weather conditions dramatic, long-distance flights involving millions or billions of insects are observed. Such flights normally occur in the morning as the sun heats the ground, and the insects invariably move downwind.

Natural enemies. Numerous predators, parasitoids, and fungal diseases of silverleaf whitefly are known. General such as minute pirate bugs (Hemiptera: Anthocoridae), green lacewings (Neuroptera: Chrysopidae), and ladybirds (Coleoptera: Coccinellidae) are important, as are many parasitic wasps, particularly in the genera *Encarsia* and *Eretmoceris* (both Hymenoptera: Aphelinidae).While these agents exercise considerable control on whitefly populations in weedy areas or on crops where insecticide use is minimal or absent, they do not survive well in the presence of most insecticides.

Life cycle and description

Silverleaf whitefly can complete a generation in about 20–30 days under favorable weather conditions. In tropical countries up to 15 generations per year have been reported, but in the United States there are considerably fewer.

The egg is about 0.2 mm long, elongate, and tapers distally; it is attached to the plant by a short stalk. The whitish eggs turn brown before hatching, which occurs in 4 to 7 days. The female deposits 90 to 95% of her eggs on the lower surfaces of young leaves.

All instars are greenish, and somewhat shiny. The flattened first instar is mobile, and is commonly called the 'crawler' stage. It measures about 0.27 mm long and 0.15 mm wide. Movement is usually limited to the first few hours after hatch, and to a distance of 1–2 mm. Duration of the first instar is usually 2 to 4 days. The feeding site is normally the lower surface of a leaf, but sometimes greater than 50% of the nymphs are found on the upper surface, and feeding location seems not to affect survival. The second and third instars are similarly flattened, but their leg segmentation is reduced and they do not move. Duration of these instars is about 2 to 3 days for each. Body length and width are 0.36 and 0.22 mm, and 0.49 and 0.29 mm, for the second and third instar, respectively. The sessile fourth instar is usually called the 'pupa' although this is not technically correct because some feeding occurs during this instar. The appearance of the fourth instar is variable, depending on the food plant; this stage tends to be spiny when developing on a hairy leaf but has fewer filaments or spines when feeding on smooth leaves. The fourth instar measures about 0.7 mm in length and 0.4 mm in width. Duration of the fourth instar is about 4 to 7 days. Total pre-adult development time averages 15 to 18 days in the temperature range of 25–32 degrees C, but increases markedly at lower temperatures. The lower and upper developmental thresholds are considered to be about 10 and 32 degrees C.

The adult is white in color and measures 1.0 to 1.3 mm in length. The antennae are pronounced and the eyes red. Oviposition begins 2 to 5 days after emergence of the adult, often at a rate of about 5 eggs per day. Adults typically live 10 to 20 days and may produce 50 to 150 eggs, although there are records of over 300 eggs per female. Females may reproduce without fertilization but males are common, sometimes outnumbering females, so most females are probably fertilized.

Damage. Adult and nymphal whiteflies use their piercing-sucking mouthparts to feed on the phloem of host plants. This results in direct damage, which is manifested in localized spotting, yellowing, or leaf drop. Under heavy feeding pressure wilting and severe growth reduction may occur. Systemic effects also are common, with uninfested leaves and other tissue being severely damaged as long as feeding whiteflies are present on the plant. A translocated toxicogenic secretion by nymphs, but not by adults, is implicated. Usually it is the young, developing tissue that is damaged by whiteflies feeding on older tissue. Once the whiteflies are removed, new plant growth is normal if a disease has not been transmitted. Damaged foliar tissue, however, does not recover once injured. Among leafy vegetables and crucifers, white streaking or discoloration, especially of veins, is common. Studies in Texas and Arizona demonstrated similar losses, and indicated that yields could be optimized if plants were treated with insecticide at whitefly densities of 3 adults per leaf or 0.5 large nymphs per 7.6 sq cm of leaf area.

A disorder called irregular ripening affects tomato fruit when whiteflies feed on tomato foliage. Although the tomato foliage is not damaged, the internal portions of the fruit do not ripen properly and the surface is blotched or streaked with yellow.

Squash silverleaf, a disorder responsible for the common name of *B. argentifolli*, has been known from Israel since 1983. Silverleaf symptomology includes blanching of the veins and petioles, and eventually the interveinal areas of the leaf. The fruit of both yellow and green fruited varieties also may be blanched.

In addition to direct damage, silverleaf whitefly also causes damage indirectly by transmitting plant viruses. Over 60 plant viruses, most belonging to a group called geminiviruses, are known to be transmitted to crops by silverleaf whitefly. Some viruses, such as tomato yellow leaf curl virus, cause more damage than the insect feeding alone, so the effects are devastating. Unfortunately, unlike the case with the phyotoxemia caused by the whitefly salivary secretions, once viruses are inoculated into the plant there is no recovery by the host even if the whiteflies are eliminated.

Lastly, whiteflies cause injury by excreting excess water and sugar in the form of honeydew. This sticky substance accumulates on the upper surface of leaves and fruit, and provides a substrate for growth of a fungus called sooty mold. The dark mold inhibits photosynthetic activity of the foliage, and may also render the fruit unmarketable unless it can be washed thoroughly and the residues removed.

Management. Eggs tend to be concentrated on young foliage and mature larvae on older foliage. Large nymphs are a good stage for population assessment because they do not move and are large enough to see without magnification. Adults tend to be concentrated close to the soil. Such distributions must be considered in population assessment prior to initiating management practices. Visual observation of the lower leaf surface and vacuum sampling are less time consuming, and in some cases more precise, than yellow sticky traps.

In southern states, where silverleaf whitefly can be the most important insect problem on some vegetable crops, frequent applications of insecticides are often made to minimize the direct and indirect effects of whitefly feeding. Whitefly resistance to nearly all classes of insecticides is known, and rotation of insecticide classes is encouraged. Mixtures of insecticides are often used, which is indicative of high levels of resistance. Most agriculturalists suggest that whitefly numbers be maintained at low levels because once they become abundant they are difficult to suppress; this, of course, exacerbates development of insecticide resistance. The phytotoxemia and disease transmission potential of this insect exaggerates its damage potential, further justifying frequent application of insecticides.

Silverleaf whitefly feeds on the lower surface of foliage and is sessile throughout most of its life — habits that minimize contact with insecticides. Frequent insecticide application also disrupts naturally occurring biological control agents. In an attempt to minimize the cost and disruptive effects of insecticides, and to reduce the evolution of insecticide resistance, soaps and oils have been extensively studied for whitefly control. The mechanism of control by surfactants such as soaps and oils is not clearly understood, but disruption of the insect cuticle, physical damage, and repellency are postulated. In any event, mineral and vegetable oils alone, or in combination with soaps and detergents, can provide some suppression of whiteflies. Combination of insecticide and oil often enhances whitefly control.

Although a great number of predators, parasitoids, and fungal diseases are known to attack silverleaf and sweetpotato whitefly, no biotic agents are known to provide adequate suppression alone. Under greenhouse conditions, parasitoids can be released at high enough densities to provide some suppression, especially when insecticidal soap and other management techniques are also used. Under natural field conditions, parasitism does not usually build to high levels until late in the growing season. Insecticides often interfere with parasitoids, of course, and effective use of biological control agents will probably be limited to cropping systems where insecticide use is minimized and other management techniques are used which favor action of predators, parasitoids, and disease agents. *Verticillium, Paecilomyces*, and other fungi similarly show some promise under greenhouse conditions, but are limited by low humidity under field conditions.

Cultural controls can be vitally important in managing silverleaf whitefly. Incorrect crop management, in particular, can create or exacerbate whitefly problems. Whiteflies can move from crop to crop, and area-wide crop-free periods help diminish populations. Thus, prompt tillage of land and destruction of crop residues after crop maturity is recommended. Similarly, weeds can harbor whiteflies, whitefly-transmitted diseases, and whitefly parasitoids, so weed management is a consideration. Trap crops such as cucumbers can be used to provide temporary protection to less preferred crops such as corn.

Row covers and other physical barriers can reduce infestation of crops, and infection with disease. Screen hole sizes of about 0.19 square mm or smaller are required to successfully exclude silverleaf whitefly. Colored and aluminum mulches provide only temporary reduction in whitefly abundance and disease transmission.

Disease transmission. Growers generally rely on whitefly suppression to manage disease incidence. This is not entirely satisfactory, however, and removal of virus-infected plants is often suggested to minimize within-field spread of viruses. Since whiteflies may transmit disease from one crop to another, or from weeds to crops, vegetation management is important. As noted above, reflective mulches have not produced economic benefits consistently. Mineral and vegetable oils inhibit virus transmission. Row covers or other physical barriers can substantially prevent disease transmission, but are often not practical.

References

Capinera, J. L. 2001. *Handbook of vegetable pests.* Academic Press, San Diego. 729 pp.

Cock, W. J. W. 1993. *Bemisia tabaci* – an update 1986–1992 on the cotton whitefly with an annotated bibliography. CAB International, United Kingdom. 78 pp.

Gerling, D. 1990. *Whiteflies: their bionomics, pest status and management.* 348 pp.

Mound, L. A. and S. H. Halsey. 1978. *Whitefly of the world.* British Museum, Chichester, United Kingdom. 340 pp.

van Lenteren, J. C., H. J. W. van Roermund and S. Sutterlin. 1996. Biological control of greenhouse whitefly (*Trialeurodes varorariorum*) with the parasitoid *Encarsa formosa:* how does it work? *Biological Control* 6: 1–10.

WHITE GRUBS.
Larvae of scarab beetles. See also, TURFGRASS INSECTS AND THEIR MANAGEMENT.

WHITE MUSCARDINE.
A mycosis of various larval, pupal, and adult insects, caused by the fungus *Beauveria bassiana.*

See also, MUSCARDINE.

WHITES.
Some members of the family Pieridae (order Lepidoptera). See also, YELLOW-WHITE BUTTERFLIES, BUTTERFLIES AND MOTHS.

WHORL.
The arrangement of leaves in a circle around the stem.

WIEDEMANN, CHRISTIAN RUDOLPH WILHELM.
Christian Wiedemann was born in Braunschweig, Germany, in 1770. He obtained a medical degree in Jena, Germany, in 1792, and beginning in 1805 obtained a series of positions at the University of Kiel. His medical publications included books on anatomy, resuscitation of asphyxiated and drowned persons, and midwifery. In addition to his medical work, Wiedemann became very interested in Diptera. From 1800 to 1806 he edited the "Archiv für Zoologie und Zootomie," and from 1817 to 1825 the "Zoologisches Magazin." About this time he began a thorough study of exotic Diptera in several collections, including a redescription of Fabricius' flies, which was published in both Latin and German. Wiedemann also published an "Analecta Entomologica," with new descriptions, and then the "Aussereuropäische zweiflügelige Inseckten." He had access to rich collections from throughout South America and the West Indies, and named many new species. He died in 1849 at Kiel, Germany.

Reference

Papavero, N. 1971, 1973. *Essays on the history of Neotropical Dipterology with special reference to collectors (1750–1905).* Museu de Zoologia, Universidade de São Paulo.

WIGGLESWORTH, (SIR) VINCENT BRIAN.
Vincent Wigglesworth was born at Kirkham, England, and educated at Cambridge University. He was employed by the London School of Hygiene and Tropical Medicine, the University of London, and Cambridge University. He also directed the unit of Insect Physiology at Cambridge from 1943–1967 and was named Quick professor of biology. He was knighted for his profound contributions to insect physiology. Wigglesworth was one of the foremost insect physiologists of the 20th century, and is best known for his work with *Rhodnius prolixus*, a blood-sucking bug. His research included work on the hormonal basis for growth and molting, the role of brain secretions in regulating growth, function of sense organs, respiration in eggs, the role of the cuticle in preventing water loss, diel periodicity, and others. Probably his most important contribution, however, was the early work documenting the presence and importance of hormones regulating insect growth and development. His most important publications were "The physiology of insect metamorphosis" (1954), "The control of growth and form" (1959), and "The principles of insect physiology" (6 editions, 1939 to 1965). Wigglesworth's insect physiology was the standard insect physiology text for decades. He died in 1994.

Reference

Locke, M. 1994. Professor Sir Vincent B. Wigglesworth, C.B.E., M.D. F.R.S. (1899–1994). *Journal of Insect Physiology* 40: 823–826.

WILLIAMS, CARROLL MILTON. Carroll Williams was born on December 2, 1916, in Richmond, Virginia, USA. He received a bachelor's degree from the University of Richmond in 1937, and a Ph.D. from Harvard University in 1941. He also earned a M.D. from Harvard. Appointed to the faculty of biology at Harvard in 1946, he was named Bussey Professor of Biology in 1965. Williams chaired the Biology Department from 1959 to 1962 and the Cellular and Developmental Biology Department from 1972 to 1973. Williams was a pioneer in insect endocrinology. He was the first to extract and characterize juvenile hormone. He also discovered brain hormone, and with the collaboration of others, partially characterized ecdysone. Williams also popularized the idea of 'third generation pesticides,' using synthetic hormones or hormone analogs as a means of specifically regulating insect populations without harming other forms of life. Although 'third generation' pesticides have not supplanted the second, there have been significant inroads made toward the use of insect hormones for pest management. Williams received numerous awards and honors over the course of his career, including the George Leslie Award, the Boylston medal, the AAAS-Newcomb Cleveland prize, the Howard Taylor Ricketts Award, and the Entomological Society of America's Founders Memorial Award. He also was a member of the prestigious National Academy of Sciences. Williams died on October 1, 1991.

Reference

Branton, D., F. C. Dafatos, and E. O. Wilson. 1996. Memorial minutes on Carroll Williams (FAS) and Aaron Gissen (HMS) and Louis Zetzel. The Harvard University Gazette. Seen at www.news.Harvard.edu/gazette/1996/03.21/MemorialMinutes.htm (August 2002).

WILLISTON, SAMUEL WENDELL. Samuel Williston was born at Boston, Massachusetts, USA, on July 10, 1852. He is known as an authority on Diptera and paleontology. He was raised in Kansas, however, and attended Kansas State University and left before completing his degree to seek his fortune. After contracting malaria, he returned to collect and complete his degree in 1872. He then commenced the study of medicine, and became interested in fossils, attaining an M.S. degree from University of Kansas in 1875. He became an assistant paleontologist at Yale University, and then received his M.D. in 1880 and his Ph.D. in 1885. At Yale University, he specialized in anatomy, and remained there until 1890. Williston then moved back to the University of Kansas, where he held the rank of professor and dean of the Medical School. In 1902 he moved to the University of Chicago, where he remained until his death. Although Williston's career did not include entomology, it was his recreational pastime – not an uncommon situation with 'entomologists' of the era. And even as a part-time entomologist he attained recognition as a dipterist, and a renowned authority of Syrphidae. In fact, he turned down an offer from C. V. Riley to become a professional entomologist. He was a prolific writer, authoring 283 publications, most in paleontology but 97 in entomology. Among his important publications were ''Synopsis of North American Syrphidae'' (1886), and ''Manual of the families and genera of North American Diptera'' (1896). Williston died at Chicago on August 30, 1918.

Reference

Essig, E. O. 1931. *A history of entomology*. The Macmillan Company, New York. 1029 pp.

WILT DISEASE. Infection of lepidopteran larvae by many nuclear polyhedrosis viruses; the expression of infection is a drooping or wilting of larvae, followed by death.

WINDOW FLIES. Members of the family Scenopinidae (order Diptera). See also, FLIES.

WING. The paired membranous structures used for flight in insects. One pair (called the primaries) is normally attached to the mesothorax, the other pair (called the secondaries) to the metathorax. Either or both pairs may be absent. Some wings are modified for purposes other than flight.

WING COVERS. The forewings of adult insects when they are thicker than the hind wings and cover the hind wings when the insect is at rest.

See also, WINGS OF INSECTS.

WINGED WALKINGSTICKS. A family of walkingsticks (Phasmatidae) in the order Phasmatodea. See also, WALKINGSTICKS AND LEAF INSECTS.

WING PAD. The incompletely developed wings of nymphs.

WINGS OF INSECTS. The ability to fly is one of the factors responsible for the biological and evolutionary success of insects. Typically, adult insects bear two pairs of wings that articulate with the thorax, though some have but one pair and others are wingless. Wings may be present in immature hemimetabolous insects, but they are incompletely developed (and called wing pads) until the adult stage is attained. Mayflies (Ephemeroptera) are exceptional in having a winged stage (the subimago) prior to reaching the adult stage, which also is winged.

The wings are flattened regions of the integument, arising dorsolaterally between the nota and pleura of the meso- and metathoracic segments. They receive rigidity, in part, from the veins contained within the wings. The veins are thickened regions of the integument that remain separated, allowing blood (hemolymph) to be pumped through the veins. This is critical during expansion of the wings following emergence of the adults. The veins also contain trachea and nerves. Folds also are present in the wings, and take two forms: flexion lines, where bending occurs during flight, and folding lines, where the wing folds during periods of rest. The wings are normally transparent and membranous, though they may be iridescent, pigmented, or even thickened and opaque. A thickened, pigmented spot is found on the anterior edge of some insect wings, and is called the pterostigma or stigma. This region of greater mass reduces wing flutter during gliding, and enhances control of wing movement during wing beats. When at rest, the wings are held over the back, and in many insects involves folding. Longitudinal folding is most common, but transverse folding also occurs. The wings of Coleoptera and Dermaptera fold transversely so they can fit beneath the elytra.

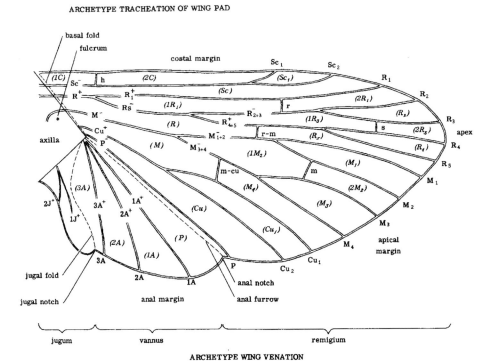

Fig. 1159 Generalized wing venation showing wing regions and major veins.

Areas of the wing

Most of the wing veins are concentrated in the anterior region of the wings; this gives maximum efficiency and support during flight. The anterior edge is called the costal margin. The posterior region often bears separate lobes. The posterior region at the base of the front wing in Lepidoptera and Trichoptera is called the jugal lobe or jugum. Flies (Diptera) often have three separate lobes, the thoracic squama, alar squama, and the alula. The posterior wing edge is called the anal margin; the distal edge is the apical margin. The angle between the costal and apical margins is called the apical angle. The angle at the base of the anterior portion of the wing is the humeral angle. The anal area of the wing os the posterior region including the anal veins. The costal area is the anterior region of the wing immediately behind the anterior edge. A fold in the wing often occurs in the cubito-anal area, and is called the anal furrow. A notch may also be present along the posterior edge, canned the anal notch.

Wing form

Wing form varies among the orders of insects. The front wings and hind wings are very similar in shape in the Odonata, Isoptera, Embiidina, and Mecoptera. In contrast, the hind wing is wider in Plecoptera, Mantodea, Blattodea, and Orthoptera. In some orders the hind wings are considerably smaller than the front wings, such as in Ephemeroptera and Hymenoptera. In Diptera, the hind wings (called halteres) are so reduced as to not aid directly in flight; in male Strepsiptera it is the front wings that are reduced to haltere-like structures. Wings also may be branched, a common condition in some Lepidoptera, or fringed,

as in Thysanoptera, some Lepidoptera (e.g., Tinaeoidea), and some Hymenoptera (e.g., Mymaridae).

Wings may be shortened; when both pairs are reduced they are said to be brachypterous or micropterous. Wingless insects are called apterous. The size or occurrence of wings may vary seasonally or geographically within the same species; such polymorphism is common in Hemiptera, and to a lesser degree in Orthoptera.

The front wings may be modified in form, and often are more sclerotized than the hind wings. The front wings of Coleoptera and Dermaptera are heavily sclerotized and called elytra; the wing venation is lost in these structures. In some beetles, particularly the Curculionidae, the front wings are fused together and cannot open. In Orthoptera, Mantodea, and Blattodea the front wings are thickened, but veins are evident; they are called tegmina. In the Hemiptera/Heteroptera, the basal portion of the front wings are thickened, but the apical region is membranous; these are called hemelytra.

Wing coupling

The front and hind wings of insects are affected primarily by the distortions of the thorax, and do not move independently. Although not physically linked in Odonata and Orthoptera, many orders have anatomical coupling mechanisms that assist in synchronization of wing beat. Apparently the two-winged condition is more efficient than the four-winged condition, so they move together. The coupling takes the form of a jugal lobe at the base of the front wings, the humeral lobe at the base of the hind wings, and setae called frenular bristles. There are many variations among the different insect groups.

Wing venation

In some primitive insects the pattern of insect veins is irregular. In most insects, however, the pattern of venation is dominated by a few longitudinal veins that radiate from the wing base to the broader wing tip. The principal longitudinal veins (and their standard abbreviation), ordered from leading to trailing edge, are: the costa (C), subcosta (Sc), radius (R), media (M) but often divided into anterior media (MA) and posterior media (MP), cubitus (similarly divided into CuA and CuB or Cu1 and Cu2), and anal veins (numbered 1A, 2A, 3A, etc.). These major

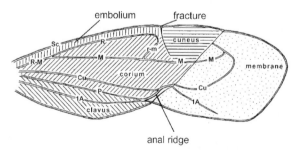

Fig. 1160 A typical hemelytron, showing the thickened basal portion and the membranous distal (apical) portion of the front wing of a bug (Hemiptera).

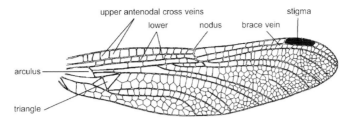

Fig. 1161 The front wing of a dragonfly, showing net-winged venation and other characters found in Odonata.

veins may branch, with the given numerical sub-scripts. Cross veins connect the major veins transversely, and normally are named after the longitudinal veins they connect. The area enclosed by veins is called a cell, and wing cells are named after the anterior vein.

See also, LOCOMOTION.

References

Chapman, R. F. 1982. *The insects, structure and function* (3rd edition). Harvard University Press, Cambridge, Massachusetts.

Comstock, J. H. 1918. *The wings of insects*. Comstock Publishing Company, New York.
Dudley, R. 2000. *The biomechanics of insect flight*. Princeton University Press, Princeton, New Jersey.
Snodgrass, R. E. 1935. *Principles of insect morphology*. McGraw-Hill, New York.
Wootton, R. J. 1979. Function, homology, and terminology in insect wings. *Systematic Entomology* 4: 81–93.

WING VENATION. The pattern of tubular vessels (veins) lying between the upper and lower surfaces of the wing in adult winged insects.

See also, WINGS OF INSECTS.

WINTER CRANE FLIES. Members of the family Trichoceridae (order Diptera). See also, FLIES.

WINTER GRAIN MITE. See also, WHEAT PESTS AND THEIR MANAGEMENT.

WINTER STONEFLIES. Members of the stone-fly family Taeniopterygidae (order Plecoptera). See also, STONEFLIES.

WIREWORM. The larva of click beetles (order Coleoptera, family Elateridae).

WIREWORMS OF POTATO. See also, POTATO PESTS AND THEIR MANAGEMENT.

WIREWORMS OF WHEAT. See also, WHEAT PESTS AND THEIR MANAGEMENT.

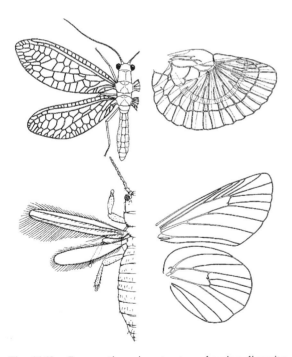

Fig. 1162 Comparative wing structure showing diversity among insect wings: top left, lacewing wings (Neuroptera); top right, hind wing of an earwig (Dermaptera); bottom left, thrips wings (Thysanoptera); bottom right, skipper butterfly wings (Lepidoptera).

WIRTH, WILLIS WAGNER. Willis Wirth was born on October 17, 1916, near Dunbar, Nebraska. He studied biology at Peru (Nebraska) State Teachers College and received a B.S. degree from Iowa Sate University in 1940, a M.S. degree from Louisiana State University in 1947, and a Ph.D. from the University of California at Berkeley in 1950. He served as an officer in the U.S. Public Health Service during World War II, working on malaria control and quarantine. From 1949 to 1983, Wirth was a research entomologist with USDA, ARS at the National Museum of Natural History, Smithsonian Institution, in Washington, DC. He served on many special assignments over the years, working in Australia, Florida, Texas, New York, Panama, and several locations in Europe. Wirth became a renowned authority on biting midges, studying the taxonomy and public health implications of Ephydridae, Chironomidae, Canaceidae, Dolichopodidae, and Ceratopogonidae. Wirth provided committee service for the Entomological Society of America, and was an adjunct professor at the University of Maryland and the University of Florida. After retiring in 1984, Wirth moved to Gainesville, Florida, to take an active role in the Florida State Collection of Arthropods. He published over 400 papers, mostly on biting midges, and particularly on Ceratopogonidae. He died September 3, 1994, in Gainesville, Florida.

Reference

Mount, G. A., D. L. Kline, D. V. Hagan, and W. Grogan. 1995. Willis Wagner Wirth. *American Entomologist* 41: 63–64.

WITCHES BROOM. An abnormal growth form in plants characterized by dense clustering of branches and often caused by the feeding secretions of piercing-sucking insects.

WOLCOTT, GEORGE N. George Norton Wolcott was born in Utica, New York on July 12, 1889, to David Clinton Wolcott and his wife Marion Delia Benedict. He attended Utica Free Academy in Utica, New York, and upon graduation in 1905, attended New York State College of Agriculture (now Cornell University). He received the B.S. in either 1907 or 1909, and the M.S. in Agriculture in 1915. In 1925 Wolcott was awarded the Ph.D. in

Entomology from Cornell University. He married Magdalen Hall in 1919 and the couple had three children: Ann, David, and Oliver. Although he initially (in 1910) worked as a Federal Agent in the Bureau of Entomology of the Department of Agriculture in Texas, his most significant contributions to agricultural entomology were made in Puerto Rico. There, he worked with the Sugar Producers Association from 1910 to 1912 and as Director of Entomology of the Insular Experiment Station, Rio Piedras, from 1914–1916 and again from 1932–1956, when he retired and returned to the mainland. Between 1919 and 1929 he held various positions as Entomologist, initially in Puerto Rico, then in the Dominican Republic, Haiti and Peru, returning each time to Puerto Rico, the place he seemed to love. One of his most notable undertakings was to control the lesser sugarcane borer, *Diatraea saccharalis* (F.) (Lepidoptera: Pyralidae) in Puerto Rico. He initiated the release of the egg parasitoid *Trichogramma minutum* Riley (Hymenoptera: Trichogrammatidae) for *D. saccharalis* egg population reduction and eliminated the burning of sugarcane fields to prevent the destruction of parasitoid populations. He was equally successful in control of *D. saccharalis* in Haiti. Dr. Wolcott also conducted extensive studies on several coleopteran pests of sugarcane, including the sugarcane rhinoceros beetle, *Strategus barbigerus* Chapin, the white grubs *Lachnosterna* sp. and *Phyllophaga* sp. (Coleoptera: Scarabaeidae), and the sugarcane root weevil (= sugarcane rootstock borer), *Diaprepes abbreviatus* L. (Coleoptera: Curculionidae). Wolcott promoted the use of beneficials, including parasitoids, in pest control programs and to that end, traveled extensively through the Caribbean and Latin America, collecting various parasitoids such as *Cryptomeigenia aurifacies* Walton and *Eutrixoides jonesii* Walton (Diptera: Tachinidae) and *Tiphia parallela* Smith (Hymenoptera: Scoliidae) for white grub control. One of Dr. Wolcott's contributions to entomology in Puerto Rico has also had a significant and longlasting beneficial impact on biocontrol in the southeastern United States. He introduced the wasp *Larra americana* Saussure (later identified as *L. bicolor* F.) (Hymenoptera: Sphecidae) into Puerto Rico from Brazil for the control of the mole cricket *Scapteriscus didactylus* (Latreille) (Orthoptera: Gryllotalpidae) ('La changa'), a pervasive agricultural pest in Puerto Rico. This single classical biological control effort by Dr. Wolcott laid the foundation for current biological control successes against mole cricket pests of turfgrass throughout the southeastern United States. More than 200 of Dr. Wolcott's publications and manuscripts continue to serve as important references for scientists who study pest problems in the Caribbean region. In particular, his ''Insectae Borinquenses'', a revised annotated check-list of Puerto Rican insects, is a classic. At one time some of his books had been adopted as texts by several agricultural colleges and institutes in the Caribbean. Dr. George N. Wolcott was not only a prolific writer and meticulous researcher, but he trained many in entomological techniques. He also revived the once defunct Entomological Society of Puerto Rico, founded by Dr. Van Dine, to further promote the exchange of ideas among entomologists and other agricultural scientists throughout the Caribbean Basin.

Pauline O. Lawrence
University of Florida
Gainesville, Florida, USA

References

Lawrence, P. O. 2000. The pioneering work of George N. Wolcott: implications for U.S.-Caribbean entomology in the 21st century. *Florida Entomologist* 83: 388–399.
Wolcott, G. N. 1936. Insectae Borinquenses. A revised annotated check-list of the insects of Puerto Rico. *Journal of Agriculture of the University of Puerto Rico* 20: 1–600.
Wolcott, G. N. 1941. The establishment in Puerto Rico of *Larra americana* Saussure. *Journal of Economic Entomology* 34: 53–56.

WOLLASTON, THOMAS VERNON.

Thomas Wollaston was born at Scotter, Lincolnshire, England, on March 9, 1822. He obtained a B.A. and M.A. from Jesus College, Cambridge. Suffering from ill health, he wintered at Madeira for several years and visited Cape Verde and St. Helena, setting the stage for his studies of island fauna. Among his important publications were: ''Insecta Maderiensia'' (1854), a treatise on the variation in species (1856), a museum bulletin of the Coleoptera of the Canary Islands (1864), ''Coleoptera Atlantidum'' (1865), ''Coleoptera Hesperidum'' (1867), and '' Coleoptera Sanctae-Helenae'' (1877). On these islands he elucidated the endemic fauna, discovering numerous new species of beetles. He died January 4, 1878, at Teignmouth, England.

References

Anonymous. 1878. Obituary, Thomas Vernon Wollaston. *Entomologist's Monthly Magazine* 14: 213–215.
Anonymous. 1878. Obituaries, Mr. T.V. Wollaston, M.A., F.L.S. *The Canadian Entomologist* 10: 34–35.

WOOD-ATTACKING INSECTS.

The relentless advance of western civilization has involved, among other things, deforestation concomitant with construction of centrally heated homes, businesses, and other facilities in place of the once-extensive woodlands. The termites, carpenter ants, beetles, and other

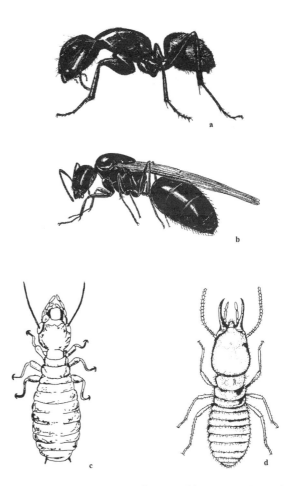

Fig. 1163 Wood-feeding insects: (a) a carpenter ant worker, (b) a carpenter ant queen, (c) a termite worker, (d) a termite soldier, specifically *Prorhinotermes simplex* (Hagen) (a is after USDA, b is after U.S. Bureau of Entomology, c is modified and redrawn from Duncan & Pickwell, d is modified and redrawn from U.S. National Museum).

insects that formerly dwelled within woodlands, of course, proceeded to invade the structural lumber used in today's construction. This lumber is obtained both from increasingly expensive hardwoods such as oak, hickory, ash, and maple, and from less desirable softwoods such as pine, hemlock, fir, redwood, and other evergreen trees. Both have a central, dark-colored, non-living heartwood and an outer light-colored, living sapwood, and both are susceptible to attack by wood-destroying insects, except for the heartwood of black locust, cypress, cedar, redwood, and certain other trees which is relatively resistant to attack.

Carpenter ants, termites, powderpost beetles, and the old house borer are among the many insects with destructive habits found in and around wooden buildings. They share little phylogenetic relationship other than joint inclusion in the overall class Insecta and a predilection to burrow in and chew wood. Carpenter ants, unlike termites, are true ants. Termites are insects whose ant-like appearance is responsible for the common name ''white ant'' often applied to them. Reproductive forms of ants and termites have membranous wings, lacking in non-reproductive forms, and both exhibit a highly developed social behavior. This is as opposed to powderpost beetles and the old house borer which are true beetles and not in the least ant-like in appearance. These beetles have thick forewings (elytra) covering membranous hindwings and can be distinguished from one another by, among other attributes, their antennae, which are relatively short in the powderpost beetles and relatively long in the borer.

Carpenter ants

Among the commonly encountered carpenter ants are the black carpenter ant, *Camponotus p. pennsylvanicus* (De Geer), and the reddish brown carpenter ant, *C. ferrugineus* (F.), of the eastern United States; the red carpenter ant, *C. novaboracensis* (Fitch), of the northern states; *C. herculeanus* (L.) and the west coast carpenter ant, *C. modoc* (Wheeler), of the western states; the Florida carpenter ant, *C. abdominalis floridanus* (Buckley); and the Hawaiian carpenter ant, *C. variegatus* (F. Smith).

In their natural woodland habitat, carpenter ants play an important role as predators of defoliating caterpillars, aphids, and other insects. They also break down vulnerable, dying trees and logs and stumps resulting therefrom. Unfortunately, they have

turned to colonization of structural wood as mankind has increasingly replaced woodland with heated buildings. Carpenter ants require wood with a relatively high moisture content, for which reason the ''Minimum Property Standards and Building Codes'' of HUD (U.S. Department of Housing and Urban Development) specify that only wood with a moisture content of less than 20% is appropriate for construction. One can predict with some certainty that a carpenter ant infestation in or around a specific structure stems from a preexisting or current moisture problem at that site.

Carpenter ants have an evenly rounded thorax separated from the apparent abdomen by a pedicel, or 'waist'-like constriction that is lacking in termites. The pedicel in carpenter ants bears a single peg-like node. This node, together with the elbowed antennae, the circle of 'hair' at the abdominal apex, and the large front and the small hind pair of strongly veined wings, helps identify them.

As the name carpenter suggests, these insects chew and tunnel within wood, but they do not actually eat it. At the outset they tunnel within moist, weathered wood, and then they may extend their activities into adjacent sound wood. They make nests in which to rear their young, whether in an infested house or in a nearby tree, post, or utility pole. They capture caterpillars, aphids, and other insects to feed their larvae which, like helpless fledglings, await the arrival of food.

Carpenter ant colonies consist of three kinds of individuals: numerous wingless, sterile female workers, a few winged reproductive males called drones, and usually only a single winged reproductive female called the queen. Workers resemble reproductive individuals except for their lack of wings. They either make chambers in wood in which tend the eggs and larvae of the next generation, or they forage outside the nest. They may travel over 90 m in search of prey insects, animal remains, plant juices, and honeydew from aphids or scale insects to feed the queen and young back in the nest. They are prone to invade homes in early spring in absence of insect prey, foraging for sugars, meats, fats, and other food scraps found on kitchen counters, within dishwashers, in other food preparation and storage areas, and even search potted indoor plants for any infesting aphids or scale insects. These invaders may become so troublesome that they prompt a rash of calls to local pest control operators for help in eradicating them.

The relatively longer hours of daylight and the spring rains that follow are cues for adult carpenter ants to emerge from overwintering pupae. The winged reproductive female and male ants, like many other insects, are attracted to artificial light. They fly into brightened areas from tree stumps, logs, or other infested areas and gain access to the structural wood of buildings through the eaves, attic, sliding doors, or windows. Once established at these points, which are usually moist due to exposure to direct rainfall or condensation, they then may colonize adjacent areas of sound wood. They may also gain entry through sturdy tree roots penetrating a basement or garage wall, through tree branches in contact with the roof, or through firewood carelessly stacked against outside walls.

Numbers of winged swarmers appear and take flight in late spring, again causing alarm on the part of homeowners who fear that these troublesome insects are termites rather than ants. This triggers a second rash of calls to professional pest control operators. The winged ants in question consist of drones and of females destined to become queens. They mate, and the drones die soon afterward. Each inseminated queen breaks off her wings, burrows into wood with a favorable moisture content (usually 20% or more), settles within a hollow destined to become the new nest, and rears her first young. New workers emerge some weeks later from the initially laid eggs and soon take over all nest building, tending of young, and foraging, leaving the new queen to devote herself exclusively to egg laying. It takes more than 3 to 4 years to establish a productive colony, a single one of which may eventually have as many as 3,000 individual members.

Carpenter ants overwinter in the pupal stage either in the nest, in soil, or at times under flower pots and other places of concealment. Their total life cycle takes about 50–70 days for completion.

As noted, carpenter ants do not eat wood, but they may severely damage it through their chewing. Workers make slit-like openings in the surface of the infested wood. They use these holes to eject loosened debris, leaving the chambers within with a clean, almost 'sandpapered' appearance. This is opposed to the 'dirty' appearance of termite tunnels. The ejected wood may be in the form of shavings or a fine sawdust depending on the nature and condition of the wood. Occasional cone-shaped mounds of sawdust piled up to 30 cm high on the floor beneath infested wood are evidence of carpenter ant activity. The ants damage windows, doors, flooring, attic joists, sill plates, fascia boards, eaves, beams, baseboards, decks, indoor swimming-pool boardwalks, wooden columns supporting porches, entrances, roofs, and fences. They also bite with their powerful mandibles, though this behavior is of negligible concern.

Suggested remedial methods for use against carpenter ants include:

- Prevention of condensation on the building's wood components
- Keeping the roof intact and leakproof, especially at the chimney flashing, around vents, and where roof levels change
- Checking gutters and down spouts to see that they properly collect roof runoff and spill it onto ground-level splash blocks directed away from the house
- Installing window wells that slope away from the house and, if necessary, covering them with a plastic bubble
- Checking for leaks in plumbing, heating pipes, hot water heater, and clothes washer
- Ventilating and/or de-humidifying if dampness is a problem
- Using yellow exterior light installations where practical
- Trimming off tree branches touching the building
- Eliminating nearby tree stumps or logs with potential for carpenter ant activity
- Finally, locating the likely site of an infesting colony and its foraging runways and applying registered, EPA-approved insecticide formulations at that site after reading and carefully following the product label directions. Assuming the preceding is beyond the resources of the householder, as it often is, the control effort should be entrusted to a state-licensed, certified pest control operator.

Termites

Termites are small, reclusive, true wood-eaters belonging to insect order Isoptera. They lack the 'wasp waist' of ants, from which they are further distinguished by their straight, bead-like rather than elbowed antennae and their usually colorless, dirty

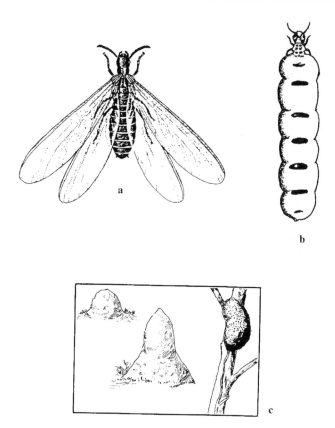

Fig. 1164 Wood-feeding insects: (a) a termite queen before having shed her wings, (b) a tropical termite queen visibly swollen with eggs, her abdominal tergites appearing as dark, segmentally repeated suffusions, (c) the above-ground termitaria or nests of three different species of tropical termite, with the far-right tree nest showing a pencil-like, Y-shaped tunnel leading to the ground (a is modified and redrawn from Elzinga, b is modified and redrawn from Skaife, c is modified and redrawn from Snodgrass).

white appearance. Termites convert dead and decaying wood and plant derivatives into useful humus in nature, reducing and recycling the debris. Moreover, they serve as a much sought-after, wild food delicacy in some countries. They are of greatest concern, however, because of the extensive damage they cause to crops, trees, and buildings. They are a far more serious problem than are either carpenter ants or powderpost beetles.

Termites occur worldwide. About 2,200 species are known, mostly from tropical or subtropical regions, where some construct massive nests called termitaria, some up to 6 m high, composed of earth, saliva, and fecal materials. There are about 45 species in the United States, primarily in warmer regions. They are not nearly as well established in the colder, boreal parts of the country, but the advent of centrally heated buildings has provided an opportunity for them to occupy some of those areas too. Most of

our local species either live in galleries within moist decaying logs or build nests in the ground.

Subterranean termites including *Reticulitermes* spp., damp wood termites including *Zootermopsis* spp., dry wood termites such as *Incisitermes* spp. and *Cryptotermes* spp., and the desert dry wood termite, *Marginitermes hubbardi* (Banks), are among the 30 native species known for their damage to plants and buildings. Notable for their destructiveness are *Reticulitermes flavipes* (Kollar) of the eastern, midwestern, and southern states, *Reticulitermes hesperus* Banks of the western states, and the Formosan subterranean termite, *Coptotermes formosanus* Shiraki, that has spread from the Orient into Honolulu and from there into many southern states.

Termites have three distinct castes: reproductives, workers, and soldiers. The juveniles or nymphs have the potential of developing, as needed, into functional reproductives, workers, or soldiers. Caste determination is determined not at the time of hatching, but is mediated through a pheromone exchange dependent on the specific requirements of the colony at a given time. This trophallaxis or mutual grooming, like that of ants, is the means by which wood particles and pheromonal secretions are spread throughout the colony, and the process is ultimately responsible for the colony's social organization. The fully winged original founders of the colony, the king and queen, are the primary reproductives. However, short-winged or wingless supplementary reproductives may be produced if the king or queen dies or if groups of termites become isolated from the parental colony. These supplementary reproductives may then take over colony formation.

A mature colony of termites produces large numbers of winged kings and queens, called swarmers or alates, in spring with the onset of rain showers, usually in May–June but in heated buildings as early as March–April. Incidentally, this timing coincides with that of the winged carpenter ants in and around homes. The alate termites fly about weakly before landing and shedding their wings. The female emits a pheromone which attracts the male, and they pair. Together they excavate a chamber within the soil beneath a rotting log or wood, and they remain underground in this 'nuptial chamber' for life. The king mates with the queen, she lays eggs, and the royal pair then takes initial care of the hatched young, feeding them predigested nutrients. Later, when able to feed on wood, the nymphs and new workers assume all nest responsibilities except reproduction. The king and queen continue mating from time to time, and the queen's abdomen may become visibly distended with eggs. The delicate-bodied, colorless, wingless, blind, moisture- and darkness-requiring workers carry out all foraging, feeding, nest building, and other non-reproductive duties. They excavate passageways through the soil and reach structural wood directly if it is in contact with the soil, or they gain access to it indirectly through cracks in concrete slabs and foundations or by constructing earthen shelter tubes over concrete and masonry not in contact with soil. These tubes, which are reliable indicators of termite infestation, offer protection from desiccation and help the termites reach structural wood over foundations and other obstructions. They are constructed of a grayish, plaster-like sealant of saliva, excrement, soil particles, and bits of wood. They include: (1) narrow, branching, fragile exploratory tubes which are abandoned after contact with the above-surface wood or other cellulose material, (2) lighter colored drop tubes suspended from floor joists or crawl spaces from wood to ground and composed exclusively of wood particles, and (3) more substantially built working tubes used to facilitate the workers' movements from soil to wood.

Though termites eat wood, a dietetic feat that few other animals can perform, they lack the digestive enzymes required to break down and assimilate the nutrients of this cellulose-laden substance. How do they eat a food they cannot digest? The answer involves minute single-celled microorganisms housed within the termites' specialized alimentary canal that provide the required enzymes. The relationship between the two organisms is a mutualistic one in which the protozoans receive a dark, moist sheltering home and an abundance of masticated food both for their use and for that of the host termite that provides the food. Neither can live without the other. Newly hatched termites lack these intestinal protozoans and die if unable to become infected. Newly molted termites not only shed their skin but also cast the entire lining of the hind gut, including the protozoans lodged within. However, like the newly hatched termites, they quickly infect themselves through oral and anal food exchanges with their nest mates and become functional wood eaters themselves.

Termite foragers are delicate-bodied, colorless, wingless workers that require darkness and moisture. They construct pencil-like runways to bridge any masonry, sub floors, support piers, or metal supportive joists that they must traverse from their nest in the ground, where they find needed moisture, to the house above, whose wood they eat. These tunnels, like their above-ground galleries, are lined with a grayish, plaster-like sealant of saliva, excrement, and soil and so are reliable indicators of infestation except, of course, when concealed (as they sometimes are) within the foundation's hollow cinder blocks. Thus, one cannot rely upon discovery of a warning sawdust trail such as that left behind by carpenter ants or powderpost beetles. Damage may be far advanced, and the wood close to structural failure, before discovery of termite infestation.

Termite soldiers are characterized by an enlarged yellowish-brown, hard head with powerful jaws. The function of this wingless, blind caste is to defend the colony from enemies, particularly ants. Termite soldiers, kings, queens, and young nymphs are unable to feed themselves, so they must be fed, mouth-to-mouth, by other nymphs and workers.

Mark-release-recapture methods indicate that often it is not a single colony that is active in a given area, as previously thought, but sometimes several decentralized entities interconnected by a network of underground tunnels with different nesting and foraging sites. The foraging populations of six *Reticulitermes flavipes* entities have been estimated as ranging from 200,000 to 5 million termites. Once foragers initially locate food, they secrete a pheromone resulting in an increase both in numbers of individuals and in increased activity. The foraging territory may encompass up to one-third acre, and the foraging distance may exceed 70 m.

How can one determine whether an infestation is a carpenter ant or a termite one? Termite tunnels are obvious indicators of infestation. In absence of them, however, large numbers of accumulated, similar-sized, delicately veined wings in and around lighted places are indicative of termites. If, on the other hand, one finds dead ants with attached, strongly veined, dissimilar-sized wings, this is indicative of carpenter ant infestation.

Termites have been called hidden invaders because they can eat away the wood of a door or window without leaving the slightest hint of their presence. They leave a thin papery covering of wood over such wooden elements as they infest, which consequently may crumble under slight pressure. Usually 3 to 8 years of untreated termite infestation is sufficient to cause extensive damage to a house. The annual cost of termite treatment in the United States is estimated at about 2 billion dollars.

Periodical inspection is indicated if one is to avoid the consequences of extensive termite infestation. This inspection should include general areas such as entrance doors and frames, walls, ceilings, sagging or buckling floors, slightly raised wallpaper, baseboards, closets, areas around showers, tubs, and window frames. In basements, foundation walls, sill plates, window sills, wooden stairwell bases, and plumbing should be inspected, along with floor joists and plumbing running through the floor; in attics, areas around chimneys, vent pipes, and roof rafters should be examined; and at the exterior of the house, firewood, form boards, or wood debris stacked against the house, planters, railroad ties, landscape timbers, fence posts, trellises, crawl space access doors, and windows and openings near pipes should

be inspected. One should also tap the wood elements of the house with the handle of a screwdriver. A hollow sound indicates a hidden problem. HUD's ''Minimum Property Standards and Building Codes'' approves use only of pressure-treated, dry (moisture content less than 20%) lumber for construction. The above procedures and standards notwithstanding, no wood is ever entirely safe from the ravages of termites. These insidious creatures gain entry into buildings whenever they find favorable conditions.

Remedial methods for termites involve correction of moisture problems as soon as detected and scrupulously avoiding the soil-wood contact of which termites avail themselves to enter a building. Specifically:

- Grading the building lot so as to drain water away from the foundation
- Maintaining the gutters, keeping them free of leaks, clear of debris, and with the down spouts discharging away from the foundation
- Embedding deck, fence, and other posts in cement
- Assuring proper attic ventilation
- Covering three-quarters of the crawl space by polyethylene sheet (leaving one-fourth open) and providing proper crawl space ventilation
- Applying immediate control treatment if termites are seen indoors or immediately outside the building
- Establishing an effective termite barrier between buildings (including garages) and the surrounding lot. This step may involve employment of termite-certified, licensed pest control operators who apply registered, EPA-approved termiticides per the label directions.

Real estate agencies in some states are required by law to have buildings inspected by licensed pest control operators for the incidence of termites and other wood-destroying organisms. A formal report of corrective measures, if indicated, is required before finalization of the real estate transaction. In some states, builders are required by law to treat lots against termite infestation before undertaking construction, and they must provide appropriate warranties to that effect.

Powderpost beetles
The related taxa of Coleoptera, the Anobiidae and Bostrichidae (some consider Lyctinae to be a separate family also), include species collectively termed

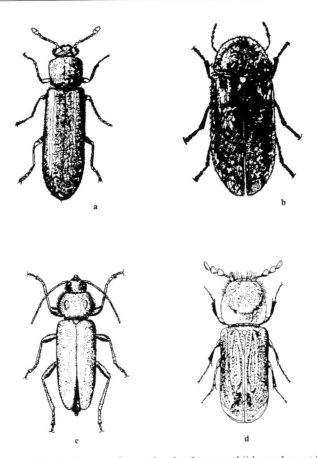

Fig. 1165 Wood-feeding insects: (a) a lyctine powderpost beetle, (b) an anobiid powderpost beetle, specifically the death-watch beetle *Xestobium rufovillosum* (De Geer), (c) the old house borer *Hylotrupes bajulus* (L.), (d) a bostrichid powderpost beetle, specifically the lead-cable borer *Scobicia declivis* Le Conte (a is modified and redrawn from USDA, b is modified and redrawn from Ohio Biological Survey, c is modified and redrawn from Bennett, Owens, & Corrigan, d is modified and redrawn from Mallis).

powderpost beetles. All of them damage timber, reducing its inner tissues to powder and perforating it with 'pinholes' or 'shot holes,' hence the common name. However, the nature of their frass (which is a mixture of sawdust and excrement) and the 'shot holes' that they produce vary with the species. Powderpost beetles rank next to termites with respect to the overall damage done to buildings and to the dry seasoned timber that they infest and re-infest.

Ten of the 66 known species of Lyctinae, or 'true powderpost beetles,' occur in the United States, of which the brown lyctus beetle *Lyctus brunneus* Stephens and the European lyctus beetle *L. linearis* Goeze are ones of major concern. They infest seasoned hardwood with at least a 3% starch content. They attack the seasoned sapwood of such trees as ash, hickory, oak, elm, pecan, maple, poplar, cherry,

sycamore, and walnut, as well as lumber, wood paneling, window and door frames, hardwood floors, furniture, tool handles, pallets, imported hardwood products, and even bamboo. They rarely infest lumber older than five years, and they never attack painted or varnished wood. They produce a powdery frass that, unlike that of the other powderpost beetles, is so fine as to resemble talcum powder. Development from egg to adult takes from 2 to 4 years, except in the south of the U. S. where two generations per year may occur. As the adult lyctines emerge from their holes they eject their powdery frass which often trickles out like raindrops. The adult is a small (usually under 5 to 6 mm long), flattened, reddish-brown or black beetle with a prominent head constricted behind the eyes. Each antenna has a long basal segment and a 2-segmented club.

Bostrichids other than Lyctinae, or 'false powderpost beetles,' are of less consequence than powderpost beetles because the hardwoods that they infest are not commonly used for interior floors, woodwork, and trims, and they do not re-infest seasoned wood. Most false powderpost beetles are larger (some up to 52 mm long) than other powderpost beetles. They are usually black, elongate, and subcylindrical, with the head usually deflexed and covered by the hood-like thorax. Each antenna has a 3 to 4 segmented club. Their frass is coarse, meal-like, and tightly packed. Among important bostrichids are the black polycaon, *Polycaon stouti* (LeConte), cylindrical beetles up to 25 mm long whose prominent head, unlike that of the other bostrichids, is visible from above, not being concealed beneath the thorax, and the lead cable borer, *Scobicia declivis* LeConte, a reddish-brown beetle, 5 to 6 mm long, that normally bores in oak, maple, and other trees but may bore through the lead covering of aerial telephone cables.

There are nearly 260 species of anobiids in the United States, of which the furniture beetle, *Anobium punctatum* (DeGeer), and the deathwatch beetles *Xestobium rufovillosum* (De Geer), *Hemicoelus* spp., and the commonly encountered *Euvrilletta peltata* (Harris) are the most important. The furniture beetle is a reddish-brown cylindrical beetle 4 to 6 mm long. It lays its eggs either on the wood surface or inside emergence holes in finished wood. The eggs hatch in 6 to 10 days, and the larvae tunnel into wood. The larval stage typically lasts 2 to 5 years in processed wood, but in nature in dead tree branches the life cycle is sometimes completed within a single year, probably because fungal and/or bacterial attack on the tree branch may make more nutrients available. The deathwatch beetle *Xestobium rufovillosum* is an insect that derives its name from its habit of tapping the head or jaws against wood. This ticking sound, actually a mating signal, becomes audible during periods of prolonged quiet, as during a wake or in the night, hence the common name. The emerging adults are slender, 6 mm long, dark beetles with lengthened, but not clubbed, antennae. They attack building timbers in poorly ventilated areas often previously damaged by moisture or fungi. *Euvrilletta peltata* is a reddish or brownish powderpost beetle covered by yellow hairs. It attacks both seasoned and unseasoned wood and is a serious pest of crawl space timbers in southeastern U.S.

Old house borer

Over 1,200 species of the long-horned beetle family Cerambycidae are known from the United States, yet of that total only the old house borer, *Hylotrupes bajulus* (L.), of the Atlantic seaboard states causes significant damage to structural wood. Other long-horned species occasionally found in structural lumber emerge from the timbers and infest outside trees, but the old house borer attacks both seasoned and unseasoned lumber and re-infests previously attacked wood. This unfortunate habit may be responsible for its 'old house' appellation, though it infests new homes too, particularly when infested lumber is used in construction. Its creamy white, peg-shaped larvae hollow out extensive galleries in the seasoned softwoods of the sills, floor joists, studs, door jams, flooring, siding, and rafters of old buildings. It attacks conifers, pines, spruce, fir, and hemlock but not hardwoods. The adult is a slightly flattened, dark, gray-black to brown-black beetle up to 25 mm long with yellowish gray hair on its body. Two raised, black glossy knobs on the prothorax help to distinguish it.

Remedial measures suitable both for wood-damaging powderpost beetles and for the old house borer include:

- Selecting quality, kiln-dried, pressure-treated wood for construction
- Making periodic inspections for infestation
- Applying varnishes or paints to lumber

Notwithstanding these precautions, these insects may gain entry into the household and then usually can be eradicated only by fumigation carried out by trained professional help.

Other wood-damaging insects

Some other wood-destroying insects, though economically insignificant compared to carpenter ants, termites, powderpost beetles, and the old house borer, deserve mention. Among the wood-damaging Coleoptera of minor importance are the golden buprestid, *Buprestis aurulenta* L., which attacks fire, lightning, or disease-damaged wood in the western states; the wharf borer, *Nacerdes melanura* (L.), whose larvae bore into the heartwood and sapwood of the seasoned softwoods used in fresh-water and marine wharf pilings; the parthenogenetic and paedogenetic telephone pole beetle, *Micromalthus debilis*

LeConte, whose larvae attack oak and chestnut logs, wooden panels, as well as telephone poles; the curculionid *Hexarthrum ulkei* (Horn) whose larvae occasionally infest wooden buildings in eastern U.S., excavating small, irregular tunnels filled with a fine frass composed of distinct ovoid fragments; the spider beetle *Sphaericus gibboides* Boieldieu, which damages stored animal and vegetable products and tunnels in the wood of kitchen cabinets made of Douglas fir to pupate; the ambrosia beetle *Gnathotrichus sulcatus* (Le Conte) that bores complex tunnels into freshly cut wood and feeds on a fungus growing on the tunnel walls; and the wide-headed beetle *Platypus wilsoni* Swaine that infests fire, lightning, and disease-damaged hardwood and coniferous trees. Among the wood-damaging Hymenoptera of minor concern are the large carpenter bee *Xylocopa v. virginica* (L.), known for its characteristic round entry holes of about one-half inch in diameter in doors, porches, window sills, wooden beams, and fences from which pour a fine stream of sawdust; the horntail *Tremex columba* (L.), sometimes accidentally transported in firewood and poor grades of lumber used for construction; and the leaf-cutting bee *Megachile relativa* Cresson, which does not directly damage the sidings of homes but attracts woodpeckers that, in the act of predation on the bees, themselves cause serious destruction.

See also, TERMITES, ANTS, POWDERPOST BEETLES.

S. K. Gangwere
Wayne State University
Detroit, Michigan, USA
and
S. Sastry
U.S. Department of Health and Human Services
Washington, DC, USA

References

Christensen, C. 1983. *Technician's handbook to the identification and control of insect pests.* Franzak & Foster Co., Cleveland, Ohio. 208 pp.

Garry, B. W., J. M. Owens, and R. M. Corrigan. 1982. *Pest control operations* (4th ed.). Purdue University/Edgell Communications Project, Randolph, New Jersey. 495 pp.

Hickin, N. E., and R. Edwards. 1975. *The insect factor in wood decay.* St. Martin's Press, New York, New York. 383 pp.

Mallis, A. 1997. *Handbook of pest control - the behavior, life history, and control of household pests.* Mallis Handbook & Technical Training Co., Cleveland, Ohio. 1456 pp.

Moore, H. B. 1979. *An introduction to wood-destroying insects (their identification, biology, prevention, and control).* Book redistributed by Pest Control Magazine, Cleveland, Ohio. 133 pp.

WOOD GNATS. Members of the family Anisopodidae (order Diptera). See also, FLIES.

WOOD NYMPHS. Some members of the family Nymphalidae, subfamily Satyrinae (order Lepidoptera). See also, BUSH-FOOTED BUTTERFLIES, BUTTERFLIES AND MOTHS.

WOOD ROT. See also, TRANSMISSION OF PLANT DISEASES BY INSECTS.

WOODY VEGETATION. Plants with rigid, woody, stems that do not die back annually. (contrast with herbaceous vegetation)

WOOD WASPS. Members of the family Xiphydriidae (order Hymenoptera: suborder Symphyta). See also, WASPS, ANTS, BEES, AND SAWFLIES.

WOOD-STAIN DISEASES. See also, TRANSMISSION OF PLANT DISEASES BY INSECTS.

WOOLLY APHIDS. Members of the family Eriosomatidae (order Hemiptera). See also, APHIDS, BUGS.

WOOLLY WHITEFLY. See also, CITRUS PESTS AND THEIR MANAGEMENT, WHITEFLIES.

WORKER. A member of the nonreproductive caste in social insects that contributes to the welfare of the colony through nest building, foraging for food, and brood and queen maintenance. Workers often are armed with stings. See also, ANTS, BEES, and WASPS, ANTS, BEES, AND SAWFLIES (HYMENOPTERA).

WORM LIONS. Members of the family Vermileonidae (order Diptera). See also, FLIES.

WRINKLED BARK BEETLES. Members of the family Rhysodidae (order Coleoptera). See also, BEETLES.

X

X-CHROMOSOME. A sex chromosome that usually is present in two copies in females (XX) and in one copy (unpaired) in males (XO or XY).

XENASTEIIDAE. A family of flies (order Diptera). See also, FLIES.

XENIC CULTURE. Culture of insects when an unknown number of species (usually of microorganisms) are present.

XEROPHYTIC. Living in dry places.

XIPHOCENTONIDAE. A family of caddisflies (order Trichoptera). See also, CADDISFLIES.

XIPHYDRIIDAE. A family of wasps (order Hymenoptera, suborder Symphyta). They commonly are known as wood wasps. See also, WASPS, ANTS, BEES, AND SAWFLIES.

XYELIDAE. A family of sawflies (order Hymenoptera, suborder Symphyta). See also, WASPS, ANTS, BEES, AND SAWFLIES.

XYLOMYID FLIES. Members of the family Xylomyidae (order Diptera). See also, FLIES.

XYLOMYIDAE. A family of flies (order Diptera). They commonly are known as xylomyid flies. See also, FLIES.

XYLOPHAGID FLIES. Members of the family Xylophagidae (order Diptera). See also, FLIES.

XYLOPHAGIDAE. A family of flies (order Diptera). They commonly are known as xylophagid flies. See also, FLIES.

XYLOPHAGOUS. Feeding on or in woody plant tissue.

XYPHIOPSYLLIDAE. A family of fleas (order Siphonaptera). See also, FLEAS.

Y

Y-CHROMOSOME. A sex chromosome that is characteristic of males in species in which the male typically has two dissimilar sex chromosomes (XY).

YELLOW DOG TICK. See also, TICKS.

YELLOW-FACED BEES. Members of the family Colletidae (order Hymenoptera, superfamily Apoidae). See also, BEES, and WASPS, ANTS, BEES, AND SAWFLIES.

YELLOW FEVER. The mosquito borne yellow fever virus causes severe disease in humans. Yellow fever was first described as a disease in the early part of the 17th century. The disease was one of the most devastating and important diseases in Africa and the Americas during the next 300 years. There were yellow fever outbreaks in many port cities in the Americas that involved thousands of human cases. The specter of yellow fever was much feared throughout Africa and the Americas due to the havoc caused by the disease and the accompanying economic disruption. The last major yellow fever epidemic in the United States occurred in New Orleans in 1905. There were about 4,000 human cases of yellow fever and nearly 500 people died from yellow fever during this epidemic.

The history of yellow fever epidemics provides numerous examples why this disease inspired dread and fear in Africa, the U.S., Central America, the Caribbean and South America. Examples of the numbers of deaths during outbreaks are startling: 6,000 dead in Barbados in 1647; 3,500 deaths in Philadelphia; 1,500 in New York City in 1798; 29,000 deaths in Haiti in 1802; 20,000 deaths in over 100 American towns in 1878. Yellow fever has re-emerged with an upsurge of human cases in the latter half of the 20th century. There were 100,000 cases and 30,000 deaths in Ethiopia in 1960–62; 17,500 cases with 1,700 deaths in Upper Volta in 1983; Cameroon 20,000 cases with 1,000 deaths in 1990. The World Health Organization officially reported 18,735 yellow fever cases with 4,522 deaths for the period 1987 to 1991.

Insect vectors

Several different mosquito species can be involved with the transmission of yellow fever depending on the geographic region and habitat. The most important mosquito species involved worldwide in the transmission of yellow fever to humans is *Aedes aegypti*, also known as 'the yellow fever mosquito.' The association between yellow fever transmission to humans by *Aedes aegypti* was a major breakthrough in understanding this dread disease. Major Walter Reed, U.S. Army, was the leader of the scientists who were able to show the role of *Aedes aegypti* in their work while stationed in Cuba in 1901. The subspecies *Aedes aegypti aegypti* is widely distributed throughout the tropics and subtropics of the world. This form is able to breed in a variety of artificial containers, i.e., flower pots, tires, water jars, many commonly found around human habitats. *Aedes aegypti aegypti* also has a distinct preference for humans as a source of blood that makes this species particularly efficient in circulating a pathogen from infected humans to uninfected humans. As a result, this highly urbanized mosquito species has the capability of sustaining the urban epidemics of yellow fever that have occurred during the last 300 years. The control of *Aedes aegypti aegypti* populations is considered of primary importance in reducing the risk of urban areas to yellow fever.

There are other mosquitoes that can be involved in yellow fever virus transmission. In tropical regions of the Western Hemisphere, mosquitoes in the genus Hemagogus transmit yellow fever virus to monkeys in the forest canopy. This is called the jungle yellow fever cycle in the Americas. The danger occurs when these canopy mosquitoes infect humans when trees are felled for clearing, and the infected human returns to an urban environment where *Aedes aegypti aegypti* can become infected, resulting in an urban epidemic. In tropical Africa, the mosquito *Aedes africanus* transmits yellow fever virus between forest dwelling canopy monkeys; *Aedes bromeliae*, *Aedes vittatus* and *Aedes furcifer-taylori* transmit the virus to monkeys in the savanna and gallery forest regions of Central Africa. The epidemic danger to humans is when infected humans bring the virus to urban centers inhabited by *Aedes aegypti*.

The virus

Yellow fever virus is a member of the group of viruses called Flavivirus. The virus has been found in the tropical regions of the Americas and Africa, and there have been historical yellow fever incursions for brief outbreaks in parts of Europe. The yellow fever virus has never been demonstrated in Asia, Australia or the Pacific despite the presence of *Aedes aegypti* in these regions. The reason for this is unknown and the subject of much speculation. Although Walter Reed showed *Aedes aegypti* transmits the agent of yellow fever in 1901, and that the agent was smaller than bacteria, it was not until 1927 that yellow fever was shown to be a virus.

The disease

Yellow fever symptoms may range from clinically inapparent to fatal. Studies have shown that in some regions of Latin America as much as 90% of the population have been infected with yellow fever virus and have no clinical symptoms of disease. Once an infected mosquito bites a human, the incubation period is generally 3 to 6 days. Mild cases appear similar to many other common illnesses that produce a fever. However, when clinical disease occurs, the onset of the disease is very sudden. There is high fever (102° to 104°F), headache, malaise, back pain, chills, prostration, nausea, slow pulse and vomiting. Yellow fever virus can be found in the blood of the patient for about 4 days after infection, and as a result the patient is capable of infecting mosquitoes during this time. Some individuals show a rapid recovery at this point and the symptoms stop. This phase can last from 3 to 4 days. More severe cases also have their symptoms subside for a brief remission but symptoms return in a day or so with fever, vomiting, abdominal pain, prostration, dehydration, jaundice due to liver involvement, internal bleeding, bleeding of the nose, mouth, gums or in their urine, kidney or liver failure. The internal bleeding results in blood in the vomit, called 'black vomit' due to the color, and dark stools. No virus is in the blood at this point so the patient is not infectious to mosquitoes. There is no cure; treatment is supportive to try to reduce the severity of these symptoms. This is the diphasic part of yellow fever. Twenty to fifty percent of people who enter the second phase die from yellow fever. Death usually occurs between the 7th and 10th day of the illness. Some very severe atypical cases of yellow fever may die as early as 3 days after the onset of symptoms. Mortality from yellow fever is about 10% of clinical cases but has reached as high as 50% of those people developing symptoms.

Dr. Max Theiler developed a vaccine against yellow fever, called the 17D vaccine, in the 1930s. In 1951, Dr. Theiler received a Nobel Prize for his work. This vaccine is still widely used and provides excellent protection against yellow fever for 10 years post-vaccination. There are studies showing that protection may be as long lived as 30 to 35 years after being vaccinated in some people.

Monkeys also can be infected with yellow fever virus. In most tropical regions of the world monkeys serve as the wild host, maintaining the virus in the absence of human involvement. Many monkey species experience yellow fever disease and suffer deaths. One of the early signs of yellow fever in a region may be the discovery of dead monkeys.

Impact and problems

Yellow fever remains a serious and dread disease in many parts of the world despite advances in understanding mosquito transmission, advances in mosquito control, advances in understanding human risk and the development of a very effective and safe vaccine. The continued appearance of yellow fever epidemics in Africa, and the appearance of yellow fever cases in the Americas are all cause for concern that this dread disease remains very dangerous. In addition, the historical absence of yellow fever in Asia despite the presence of *Aedes aegypti* is also cause for concern since Asian populations would be extremely susceptible due to lack of any native immunity to the yellow fever virus. Asian populations may be extremely vulnerable to yellow fever with potentially catastrophic consequences.

The ability to interrupt a yellow fever outbreak will depend on being able to bring to bear a diverse array of tools including an efficient and effective mosquito control program and a massive vaccination program. Both are extremely difficult to accomplish in many regions of the world where the risk of a yellow fever outbreak may be greatest. Successful mosquito control against *Aedes aegypti* has reduced the number of yellow fever cases in many cities. However, mosquito control resources may be non-existent and delivery of vaccine insufficient. In the 1990s, the worldwide annual production of yellow fever vaccine was about 15 million doses with demands on vaccine extremely unpredictable. A vaccination program that is geared to regions in advance of an expected epidemic is cost effective, but it is unlikely to be successful because of the time delay in identifying the epidemic and because it takes 5 to 7 days for the vaccine to provide any protection after inoculation. On the other hand, a campaign to vaccinate the entire population in the absence of yellow fever would be extremely costly and require a long-term commitment to vaccinate anyone entering the population through birth or immigration. The challenges of yellow fever remain formidable.

See also, MOSQUITOES, HISTORY AND INSECTS.

Walter J. Tabachnick
University of Florida
Vero Beach, Florida, USA

References

Monath, T. P. 1989. Yellow fever. pp. 139–231 in T. P. Monath (ed.), *The arboviruses: epidemiology and ecology*. CRC Press, Boca Raton, Florida.

Strode, G. K. (ed.). 1951. *Yellow fever*. McGraw-Hill Book Co., Inc., New York, New York. 710 pp.

Tabachnick, W. J. 1991. Evolutionary genetics and insect borne disease. The yellow fever mosquito, *Aedes aegypti*. American Entomologist 37: 14–24.

YELLOW FLIES. Some members of the family Tabanidae (order Diptera). See also, FLIES, HORSE FLIES AND DEER FLIES (DIPTERA: TABANIDAE).

YELLOW JACKETS. Members of the family Vespidae (order Hymenoptera). See also, WASPS, ANTS, BEES, AND SAWFLIES.

YELLOW MEALWORMS. See also, STORED GRAIN AND FLOUR INSECTS.

YELLOWS. Some members of the family Pieridae (order Lepidoptera). See also, YELLOW-WHITE BUTTERFLIES, BUTTERFLIES AND MOTHS.

YELLOW-WHITE BUTTERFLIES (LEPIDOPTERA: PIERIDAE). Yellow-white butterflies, family Pieridae (including jezebels, orangetips, sulphurs, whites, and alfalfa and cabbage butterflies), total about 1,275 species worldwide, most being Indo-Australian (ca. 515 sp.). Four subfamilies are recognized: Pseudopontiinae (a single African species), Dismorphiinae, Pierinae, and Coliadinae. The family is in the superfamily Papilionoidea (series Papilioniformes), in the section Cossina, subsection Bombycina, of the division Ditrysia. Adults small to large (23 to 100 mm wingspan); antennae often with weak clubs. Wings mostly triangular or rounded; hindwings usually small and rounded, but sometimes larger than forewings and somewhat pointed (Dismorphiinae); body usually slender but sometimes robust. Maculation usually variously yellow or white, with darker spots or patches of various shape (rarely without markings); rarely hyaline; sometimes very colorful, and many with special UV coloration ve different from what is evident under white light. Pierinae are mostly white species and Coliadinae have more of the colorful and yellow species; many

Fig. 1166 Example of yellow-white butterflies (Pieridae), *Anthocharis sara* Lucas from California, USA.

pierid colorations are from pterin or flavone deposits. Adults diurnal. Larvae leaf feeding (one species feeds gregariously in Mexico). Various plants are utilized, including Capparidaceae, Leguminosae, Loranthaceae, and Santalaceae, among others, but especially Cruciferae. Some economic species are known, particularly on cabbages and other crucifers.

John B. Heppner
Florida State Collection of Arthropods
Gainesville, Florida USA

References

Courtney, S. P. 1986. The ecology of pierid butterflies: dynamics and interactions. *Advances in Ecological Research* 15: 51–131.

Eitschberger, U. 1984. Systematische Untersuchungen am *Pieris napi-bryoniae-* Komplex (s.l.) (Lepidoptera, Pieridae). In *Herbipoliana. Marktleuthen.* 2 volumes (1104 pp., 101 pl.).

Feltwell, J. 1982. *Large white butterfly. The biology, biochemistry and physiology of Pieris brassicae (Linnaeus).* W. Junk, The Hague. 538 pp. (In Series Entomologica, 18).

Krzywicki, M. 1962. Pieridae. In *Klucze do Oznaczania Owadów Polski. 27. Motyle – Lepidoptera,* 65: 1–45, 1 pl. [part]. Polskie Towardzystwo Entomologiczne [in Polish], Warsaw.

Pleisch, E., and P. Sonderegger (eds.). 1987. Pieridae – Weisslinge. In *Schmetterl inge und ihre Lebensräume: Arten – Gefährdung – Schutz. Schweiz und angrenzenden Gebiete,* 1: 136–162, pl. 2–4. Pro Natura-Schweizerische Bund fuer Naturschutz, Basel.

Seitz, A. (ed.). 1906–31. Familie: Pieridae. In *Die Gross-Schmetterlinge der Erde,* 1: 39–74, pl. 17–27 (1906–07); 1(suppl.): 93–125, 332–335, pl. 7 (1930–31); 5: 53–111, 1014–1026, pl. 18–30 (1908–24); 9: 119–190, pl. 50–73 (1909–10); 13: 29–69, pl. 10–22 (1910). A. Kernen, Stuttgart.

Talbot, G. 1932–35. Pieridae. *In Lepidopterorum Catalogus,* 53, 60: 1–697. W. Junk, Berlin.

YOLK. The nutritive matter of the egg, from which the developing embryo derives sustenance.

See also, EMBRYOGENESIS, ENDOCRINE REGULATION OF INSECT REPRODUCTION, OOGENESIS.

YPONOMEUTIDAE. A family of moths (order Lepidoptera). They commonly are known as ermine moths. See also, ERMINE MOTHS, BUTTERFLIES AND MOTHS.

YOUNG, JR., DAVID A. David Young was born May 26, 1915, at Wilkinsburg, Pennsylvania, USA. He obtained a B.A. degree from the University of Louisville in 1939 and then taught science in the public school system from 1939 to 1941. In 1942 he was awarded a M.S. from Cornell University. Serving in the military during World War II, he returned to serve as an instructor at the University of Louisville from 1946 to 1948. He also studied at the University of Kansas, and was awarded a Ph.D. in 1950. Young then joined the United States National Museum and was employed by the Insect Identification and Foreign Parasite Introduction Section of USDA from 1950 to 1957. Young joined the faculty of North Carolina State University in 1957, where he retired in 1980 but continued his writing until 1986. Young was an internationally acclaimed leafhopper specialist. He described 807 new species, 207 new genera, and a new tribe. Among other publications, he authored the monumental "Taxonomic study of the Cicadellinae (Hemiptera: Cicadellidae)," which was published in three parts and treated 292 genera from throughout the world. He died on June 8, 1991.

Reference

Deitz, L. L. 1991. David A. Young, Jr. *American Entomologist* 37: 251.

YUCCA MOTHS (LEPIDOPTERA: PRODOXI-DAE). Yucca moths, family Prodoxidae, total 65 species, mostly western Nearctic, but now also including some genera (Lamproniinae) from other regions that previously were placed in Incurvariidae. The family is in the superfamily Incurvarioidea, in the section Incurvariina, of division Monotrysia, infraorder Heteroneura. There are two subfamilies: Prodoxinae and Lamproniinae. Adults small (5 to 33 mm wingspan), with rough head scaling; haustellum reduced, scaled; labial palpi porrect; maxillary palpi folded, 5-segmented (rarely 4-segmented). Maculation often is white with various darker markings, or more colorful and iridescent. Adults are diurnal. Larvae are seed, flower stalk, or stem borers; rarely gall makers. Host plants are various yucca plants (Agavaceae) for the well-known yucca pollinators of North America, while other species (Lamproniinae) are on hardwood trees and bushes.

John B. Heppner
Florida State Collection of Arthropods
Gainesville, Florida USA

References

Davis, D. R. 1967. A revision of the moths of the subfamily Prodoxinae (Lepidoptera: Incurvariidae). *Bulletin of the United States National Museum* 255: 1–170.

Nielsen, E. S., and D. R. Davis. 1985. First Southern Hemisphere prodoxid and the phylogeny of the Incurvarioidea (Lepidoptera). *Systematic Entomology* 10: 307–322.

Pellmyr, O. 1999. Systematic revision of the yucca moths in the *Tegeticula yuccasella* complex (Lepidoptera: Prodoxidae) north of Mexico. *Systematic Entomology* 24: 243–271.

Powell, J. A., and R. A. Mackie. 1966. Biological interrelationships of moths and *Yucca whipplei* (Lepidoptera: Gelechiidae, Blastobasidae, Prodoxidae). *University of California Publications in Entomology* 42: 1–59.

Z

ZELLER, PHILIPP CHRISTOPH. Philipp Zeller was born at Steinheim-on-the-Mur, Würtemberg, Germany, on April 8, 1808. He began collecting insects as a boy, and though he received a degree from the University of Berlin, he obtained no formal instruction in entomology. He began working at the gymnasium at Frankfort on the Oder in 1830, and devoted his leisure time to study of insects, especially the Lepidoptera, eventually becoming known as an authority on Microlepidoptera, including Pyralidae and Tineidae. He began publishing in 1833, and was named Professor by the King of Prussia in 1852. Among his important publications were "North American Micro-Lepidoptera" (1872, 1873, 1875) and "Natural history of the Tineina" (with H.T. Stainton) (1855). He named many of the economically important Microlepidoptera from around the world. Zeller died at Grunhof, Germany, on March 27, 1883.

References

Anonymous. 1883. Obituary notices, Prof. P.C. Zeller. *Canadian Entomologist* 15: 176–177.
Essig, E. O. 1931. *A history of entomology.* The Macmillan Company, New York. 1029 pp.
Stainton, H.T. 1883. Philipp Christoph Zeller. *Entomologist's Monthly Magazine* 20: 1–8.

ZETTERSTEDT, JOHANN WILHELM. Johann Zetterstedt was born on May 20, 1785, near Mjölby, Sweden. Interested in biology from his youngest days, Zetterstedt entered Lund University in 1805, and received a doctoral degree in 1808. He was named docent in botany in 1812, but there was no salary attached to this position, so he supported himself with private lessons. In 1822, he was awarded a position at the University, and worked there until his retirement in 1853. In 1869 he received an honorary degree. Zetterstedt's list of publications is not very long, but includes some important works, including "Fauna insectorum Lapponica" (1828), "Insecta Lapponica descripta" (six issues between 1838 and 1840), and the monumental "Diptera Scandinaviae disposita et descripta" (14 volumes between 1842 and 1860). He died December 23, 1874, at Lund, Sweden.

Reference

Herman, L. H. 2001. Zetterstedt, Johann Wilhelm. *Bulletin of the American Museum of Natural History* 265: 157–158.

ZOOGEOGRAPHIC REALMS. Dissimilar distributions of existing animals (normally illustrated by vertebrates), usually isolated geographically and defined by continents, but sometimes separated by mountain ranges or other physiographic features. The principal realms are:

Australian realm. Australia and nearby islands, with a preponderance of marsupials, large flightless birds, and parrots, as well as an absence of mammals. This is sometimes called the Austalasian realm.

Oriental realm. India and Southeast Asia south through Indonesia, with tree shrews, orangutan, and gibbon.

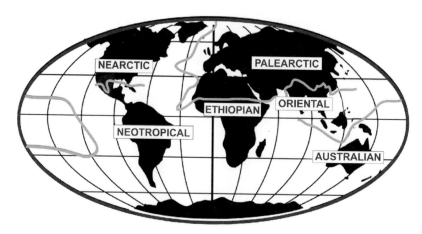

Fig. 1167 The traditional zoogeographic realms.

Ethiopian realm. Africa, though northernmost Africa is more similar to Europe (Palearctic realm), with antelopes, giraffes, elephants, rhinoceros, gorillas, dogs, and cats. This is sometimes called the African or Afrotropical realm.

Neotropical realm. South and Central America including the Caribbean Islands, with sloths, armadillos, anteaters, tapirs, toucans, and hummingbirds.

Holarctic realm. (Nearctic and Palearctic realms) — the Holarctic realm includes most of the northern hemisphere, but is often subdivided into the Nearctic realm (North America south to central Mexico) and the Palearctic realm (Europe and Asia except for Southeast Asia). The faunas of the Nearctic and Palearctic are really quite similar, with such animals as vireos, wood warblers, deer, bison and wolves.

Some authors, however, subdivide the realms further, treating separately the African island of Madagascar (Malagasy realm) and the contact area of the Australian and Oriental realms, particularly the Indonesia-Papua New Guinea area (Indo-Australian realm). The zoogeographic realms are sometimes called the biogeographic realms but this is not particularly desirable because plant distribution does not entirely conform to the regions formed by animal distribution.

See also, FLORISTIC KINGDOMS.

ZOOGEOGRAPHY. The distribution of animal groups in space.

See also, ZOOGEOGRAPHICAL REALMS.

ZOONOSIS. (pl., zoonoses) Diseases of animals that may be transmitted to humans, often by arthropods.

ZOOPHAGOUS. Feeding on animals or animal products.

ZOPHERIDAE. A family of beetles (order Coleoptera). They commonly are known as ironclad beetles. See also, BEETLES.

ZORAPTERA. An order of insects. They commonly are known as angel insects or zorapterans. See also, ANGEL INSECTS.

ZORAPTERANS. Members of the insect order Zoraptera. See also, ANGEL INSECTS.

ZOROTYPIDAE. A family of angel insects (order Zoraptera). See also, ANGEL INSECTS.

ZYGAENIDAE. A family of moths (order Lepidoptera). They commonly are known as smoky moths and burnet moths. See also, BURNET MOTHS, BUTTERFLIES AND MOTHS.

ZYGOTE. A fertilized egg formed as the result of the union of the male and female gametes.